Plant Metabolism

Dedication

This book is dedicated to the memory of Tom ap Rees who died in Cambridge on October 3rd 1996 while riding his bicycle. Tom contributed a chapter to the first volume but his duties as Head of Plant Sciences in Cambridge left him no time to contribute to this second edition. However, throughout this volume are numerous references to his work on plant metabolism. Many eminent plant biologists started their careers in Tom's laboratory and instilled into them was Tom's insistence for first rate results. Those who never had the pleasure of working with him were still affected by his insistence on high quality information and he was often heard in seminars asking for better controls or more efficient recoveries.

For many of us, having had Tom as a friend will be something we will always cherish. His enthusiasm and energy were infectious. He delighted in his garden and the village of Little Eversden where he lived. He was an avid supporter of the local horticultural society and was as much at ease with farmers and villagers as he was with scientists.

Tom will be greatly missed both as a scientist and as a friend.

David Dennis

Plant Metabolism

edited by
DAVID T. DENNIS,
DAVID B. LAYZELL,
DANIEL D. LEFEBVRE
and
DAVID H. TURPIN

Addison Wesley Longman

Addison Wesley Longman Limited,
Edinburgh Gate, Harlow,
Essex CM20 2JE, England
and Associated Companies throughout the world

First published 1990
Second edition published 1997

British Library Cataloguing in Publication Data
A catalogue entry for this title is available from the British
Library

ISBN 0-582-25906-1

Library of Congress Cataloging-in-Publication Data
A catalog entry for this title is available from the Library of
Congress

Set by 32 in 9/11½ pt Times
Produced by Longman Singapore Publishers (Pte) Ltd.
Printed in Singapore

Contents

Preface

In the preface to the first edition, we commented on the rapid expansion in our knowledge of plant metabolism during the previous decade. We could not have anticipated, at that time, the even more rapid expansion that would occur in the years subsequent to its publication. It became apparent that if the textbook was to remain valuable to advanced undergraduates, postgraduate students and researchers in general, it would have to be updated. In addition, as was pointed out by reviewers and readers, there were some topics that should have been covered in a text on plant metabolism. We have attempted to remedy these deficiencies.

The title of the book has been changed to more closely describe the contents. The previous title, *Plant Physiology, Biochemistry and Molecular Biology*, was chosen to emphasize our intention to integrate these disciplines. However, the major theme of the book was metabolism. It is still the objective of the volume to show the relevance of molecular biology and physiology to biochemistry and metabolism. For many of us, molecular biology is an essential tool to be used to gain an understanding of metabolism and it is this aspect of the discipline that is included here. Similarly, plant physiology is a subject that requires a whole book of its own and cannot be covered adequately in this volume. We hope sufficient attention is paid to this area to reveal the importance of a knowledge of metabolism as a component of the complexity of physiological processes in plants.

All chapters of the present edition have been extensively revised and updated. In addition, obvious omissions from the first volume have now, we trust, been adequately covered. A major component of plant metabolism is the cell wall which accounts for a considerable utilization of the carbon in plant cells. In addition, cell walls are not just an inert structural framework but a complex and dynamic component of the cell. To address this, a chapter on the structure and biosynthesis of cell walls by Nick Carpita has now been included. In addition, an extensive metabolic activity that is present in most plant cells and often called secondary metabolism, was missing from the first edition. The importance of this type of metabolism to the well-being of plants is now recognized and some aspects of this vast subject have now been included in a chapter entitled 'Metabolism of defense and communication' by Brian Ellis. The remarkable advances that have taken place in the area of plant development clearly merited inclusion in this edition and this has been achieved by a chapter written by Peter McCourt and Kallie Keith. A new chapter, 'The manipulation of resource allocation in plants' by Steve Blakeley has been added to Section X. The prospect of manipulating plant carbon metabolism by genetic manipulation is becoming a reality and it seemed appropriate to include this development in a textbook entitled *Plant Metabolism*. Finally, to remedy a deficit in Section I we have a new chapter on protein synthesis by Kenton Ko.

The book is divided into basically the same sections as in the previous edition, with some reorganization to make them more rational. Section I deals in general terms with basic ideas about control of metabolism through the control of gene expression, enzyme regulation and compartmentation. This section now includes the new chapter on protein synthesis by Kenton Ko. Finally, to complement this new chapter, the chapter on protein degradation has been moved from Section X in the previous edition to this section.

Section II deals with cytosolic carbon metabolism and has been expanded to include the chapters on cell walls and communication and defense. The other chapters in this section cover basic carbon metabolism and, although the section is on cytosolic metabolism, it is impossible also not to include plastid metabolism. Sections III and IV on mitochondrial metabolism and mitochondrion-cytosol interactions are basically the same as before with some changes in organization and authors. Similarly, the chapters in Sections V, VI and VII have the same format as in the first edition. Sections VIII and IX on nitrogen metabolism and assimilate partitioning and storage are the result of a major rearrangement of Section VIII in the first edition. This has eliminated overlap and makes this whole area more cohesive. The final section, X, has a new chapter on carbon allocation and completely revised chapters on gene transfer and plant improvement through genetic engineering.

As in the previous edition, our hope is that this book will provide students of plant metabolism, whether they be undergraduates or postgraduates, with an awareness of plant metabolism and an appreciation of its integration and control. In addition, we hope that the essential linkages between the disciplines of molecular biology, biochemistry and physiology have been made evident. As before, we trust that this text will be a reference book for everyone in the field.

We would like to thank all the authors of the previous edition who extensively revised and updated their chapters. We also thank and welcome all the new authors who have made major contributions to this edition. We hope they will all be available for the third edition of *Plant Metabolism* when this edition becomes out-of-date, as it surely must in this field.

David Dennis

Contributors

J. Andrews
5402 Retana Drive, Madison, WI 53714, USA

C. A. Atkins
Department of Botany, University of Western
Australia, Nedlands 6009, Western Australia,
Australia

S. Aubert
Laboratoire de Physiologie Cellulaire Vegétale, Unité
Associeé au CNRS No. 521, Department Recherche
Fondamentale, Centre d'Etudes Nucleaires,
Université Joseph Fournier, 85X 38041 Grenoble,
Cedex, France

A. Baker
Department of Biochemistry, University of
Cambridge, Cambridge CB2 1QW, UK

J. D. Bewley
Department of Botany, University of Guelph,
Guelph, Ontario, N1G 2W1, Canada

S. D. Blakeley
Department of Biology, Queen's University,
Kingston, Ontario, K7L 3N6, Canada

L. Bonen
Department of Biology, University of Ottawa,
Ottawa, Ontario, K1N 6N5, Canada

B. B. Buchanan
Department of Plant Biology, University of
California, 111 Koshland Hall, Berkeley, CA 94720-
0001, USA

D. T. Canvin
Department of Biology, Queen's University,
Kingston, Ontario, K7L 3N6, Canada

N. Carpita
Botany and Plant Pathology, Purdue University,
West Lafayette, IN 47907, USA

R. Casey
Department of Botany, University of Guelph,
Guelph, Ontario, N1G 2W1, Canada

D. T. Dennis
Department of Biology, Queen's University,
Kingston, Ontario, K7L 3N6, Canada

R. Douce
Laboratoire de Physiologie Cellulaire Vegétale, Unité
Associeé au CNRS No. 521, Department Recherche
Fondamentale, Centre d'Etudes Nucleaires,
Université Joseph Fournier, 85X 38041 Grenoble,
Cedex, France

B. E. Ellis
Department of Plant Science, University of British
Columbia, 2357 Main Mall, Vancouver, British
Columbia, V6T 1Z4, Canada

M. J. Emes
Department of Cell and Structural Biology,
University of Manchester, School of Biological
Sciences, Williamson Building, Oxford Road,
Manchester, M13 9PL, UK

K. B. Freeman
Department of Biochemistry, McMaster University,
Hamilton, Ontario, L8N 3Z5, Canada

K. S. Gellatly
NRC Biotechnology Institute, 110 Gymnasium Place,
Saskatoon, SK, S7N 0W9, Canada

R. M. Gifford
CSIRO, Division of Plant Industry, GPO 1600,
Canberra ACT 2601, Australia

M. W. Gray
Department of Biochemistry, Dalhousie University, Halifax, Nova Scotia, B3H 4H7, Canada

J. S. Greenwood
Department of Botany, University of Guelph, Guelph, Ontario, N1G 2W1, Canada

C. Halpin
Zeneca Seeds, Plant Biotechnology Section, Jealott's Hill Research Station, Braknell, RG12 6EY, UK

R. Hartlen
Department of Biochemistry, McMaster University, Hamilton, Ontario, L8N 3Z5, Canada

P. M. Hatfield
Department of Horticulture, University of Wisconsin-Madison, 1575 Linden Drive, Madison, WI 53706-1590, USA

W. D. Hitz
DuPont Experimental Station, A. Box 80402, Wilmington, DE 19880-0402, USA

Y. Huang
Department of Biology, Queen's University, Kingston, Ontario, K7L 3N6, Canada

H. C. Huppe
Department of Biology, Queen's University, Kingston, Ontario, K7L 3N6, Canada

D. L. Inglis
Department of Biochemistry, McMaster University, Hamilton, Ontario, L8N 3Z5, Canada

R. Ireland
Biology Department, Mount Allison University, Sackville, New Brunswick, E0A 3C0, Canada

V. N. Iyer
Department of Biology, Carleton University, Ottawa, Ontario, K1S 5B6, Canada

K. Keegstra
MSU-DOE Plant Research Lab., Michigan State University, East Lansing, MI 48824-1312, USA

K. Keith
Department of Botany, University of Toronto, 25 Wilcocks Street, Toronto, Ontario, M5R 3B2, Canada

K. Ko
Department of Biology, Queen's University, Kingston, Ontario, K7L 3N4, Canada

N. J. Kruger
Department of Plant Sciences, University of Oxford, Oxford, OX1 3RB, UK

H. Lambers
Department of Plant Ecology and Evolutionary Biology, Utrecht University, P.O. Box 800.84, 3508 TB, Utrecht, The Netherlands

D. B. Layzell
Department of Biology, Queen's University, Kingston, Ontario, K7L 3N6, Canada

R. C. Leegood
Department of Animal and Plant Science, University of Sheffield, P. O. Box 601, Sheffield, S10 2UQ, UK

D. D. Lefebvre
Department of Biology, Queen's University, Kingston, Ontario, K7L 3N4, Canada

M. Lord
Department of Biological Sciences, University of Warwick, Coventry, CV4 7AL, UK

P. McCourt
Department of Botany, University of Toronto, 25 Wilcocks Street, Toronto, Ontario, M5R 3B2, Canada

F. D. Macdonald
Department of Plant Biology, University of California, 111 Koshland Hall, Berkeley, CA 94720-0001, USA

B. L. A. Miki
Plant Research Centre, Genetic Engineering Section, Agriculture Canada, Ottawa, Ontario, K1A 0C6, Canada

K. A. Mott
Department of Biology, Utah State University, Logan, UT 84322-5335, USA

J. E. Mullet
Department of Biochemistry/Biophysics, Texas A&M University, College Station, TX 77843, USA

F. B. Negm
Department of Biology, Queen's University, Kingston, Ontario, K7L 3N6, Canada

N. Nelson
Bermuda Biological Station for Research, Inc., Ferry Reach, St. George's, Bermuda, GE 01, USA

M. Neuberger
Laboratoire de Physiologie Cellulaire Vegétale, Unité Associeé au CNRS No. 521, Department Recherche Fondamentale, Centre d'Etudes Nucleaires, Université Joseph Fournier, 85X 38041 Grenoble, Cedex, France

W. Newcomb
Department of Biology, Queen's University, Kingston, Ontario, K7L 3N6, Canada

J. B. Ohlrogge
Department of Botany, Michigan State University, East Lansing, MI 48824-1312, USA

M. Peoples
CSIRO, Division of Plant Industry, GPO 1600, Canberra, ACT 2601, Australia

J. W. Pierce
DuPont Experimental Station, A. Box 80402, Willmington, DE 19880-0402, USA

W. C. Plaxton
Departments of Biology and Biochemistry, Queen's University, Kingston, Ontario, K7L 3N6, Canada

B. B. Prézelin
Department of Ecology, Evolution and Marine Biology, University of California, Santa Barbara, CA 93106, USA

C. Salon
Department of Biology, Queen's University, Kingston, Ontario, K7L 3N6, Canada

K. Schmid
Department of Biological Sciences, Butler University, 4600 Sunset Avenue, Indianapolis, IN 46208-4385, USA

M. Stitt
Botanisches Institut, University of Heidelberg, In Neuecheimerfeld 360, Heidelberg 6900, Germany

D. H. Turpin
Office of the Vice-Principal Academic, Room 239, Richardson Hall, Queen's University, Kingston, Ontario, K7L 3N6, Canada

C. P. Vance
United States Department of Agriculture, Agricultural Research Service and Department of Agronomy and Plant Genetics, University of Minnesota, St. Paul, MN 55108, USA

R. D. Vierstra
Department of Horticulture, University of Wisconsin-Madison, 1575 Linden Drive, Madison, WI 53706-1590, USA

H. G. Weger
Department of Biology, University of Regina, Regina, SK, S4S 0A2, Canada

C. A. West
Department of Chemistry and Biochemistry, University of California, 405 Hilgard Avenue, Los Angeles, CA 90024-1569, USA

I

The Control
of Metabolism

1 Fundamentals of gene structure and control

Daniel D. Lefebvre and Kevin S. Gellatly

Introduction

This chapter presents the basic concepts of gene structure and regulation. It does not deal with recombinant DNA methodology, but is an outline of current knowledge so that chapters on plant genes and their expression can be more easily understood. In addition, since the examples employed in this chapter include those from organisms other than plants, the reader will be presented with a perspective from which to judge the present state of understanding of gene regulation in plants.

The field of molecular genetics is enormous and extremely dynamic. Areas with the greatest relevance to plant molecular genetics will therefore be emphasized. These topics include the composition of prokaryotic and eukaryotic genes, and the production and characterization of messenger RNA. Some details will be given about the organization of genes with a particular focus on the identification of consensus sequences of regulatory regions especially those involving plant genes.

The gene is the basic hereditary unit in all living organisms. It is defined as a segment of deoxyribonucleic acid (DNA) involved in the production of a polypeptide chain. Each gene is composed of a coding region and regions preceding and following it that are involved in regulating its expression. Genes may also contain intervening, non-coding, sequences which are situated within the coding region. The coding region provides the information for the amino acid sequence of a protein and is also referred to as the open reading frame (ORF). Regulatory sequences, which usually precede the ORF, generally consist of specific DNA sequences which act as recognition sites for various proteins involved in ribonucleic acid (RNA) synthesis. In this way, regulatory sequences control the rate and amount of RNA that is synthesized.

Transcription is the process by which RNA is synthesized using a single strand of DNA as the template. The DNA sequence (genetic code) is comprised of triplet base coding units called codons. Each codon directs the incorporation of a particular amino acid into the nascent protein primary sequence by the translational apparatus. Translation of mRNA (messenger RNA) into their corresponding proteins occurs on ribosomes situated either in the cytoplasm or within an organelle; the latter occurs when the gene is within the mitochondrial or chloroplast genome.

Transcription and translation are the two major genetic events where the control of gene expression can be asserted. Translational control is discussed in Chapter 2. Post-translational control also occurs and this aspect of metabolic regulation is dealt with in Chapters 3 and 30.

Prokaryotic genes

The operon

The structure and regulation of the genes of prokaryotic organisms were the first to be studied. In bacteria, genes for proteins catalyzing related activities are often arranged in tandem within the genome so that they are transcribed on a single polycistronic mRNA in which each cistronic component is the messenger template for a single protein. Such a collection of genes is referred to as an operon (Miller and Reznikoff, 1980).

Bacteria must be able to adapt to rapid environmental changes in a similar manner as plants. Therefore, bacterial genes are constructed to enable highly variable rates of expression. When required, mRNA molecules can be produced at high levels when their genes receive the appropriate signals. The compounds responsible for these signals are called inducers.

The classical study of Jacob and Monod in 1961 on the lactose (*lac*) operon in the bacterium *Escherichia coli*, paved the way to an understanding of the regulatory components of prokaryotic genes (Fig. 1.1). The *lac* operon contains structural genes for three proteins: β-galactosidase, which hydrolyzes lactose into glucose and galactose; lactose permease, which is the transporter for the uptake of lactose into the cell; and thiogalactosidase transacetylase, an enzyme involved in the detoxification of non-metabolizable β-galactosides which can then be secreted from the cell.

Under normal conditions, *E. coli* utilizes glucose as a readily available carbon source. Lactose is an inferior carbon source and is not usually available to the bacterium. Hence, in the absence of lactose or the presence of glucose, the *lac* operon is dormant or is in a repressed state. A specific regulatory gene is involved in this repression by facilitating production of a repressor protein. There are two domains on this repressor protein: one has an affinity for the *lac* operator which is the DNA sequence of the operon situated just before (upstream of) the open reading frames and the other domain binds lactose, the inducer molecule. Lactose-free repressor binds to the operator and prevents transcription by RNA polymerase. When lactose forms a complex with the repressor it can no longer bind to the operator and the *lac* operon mRNA is transcribed.

The promoter

The promoter is the site where RNA polymerase physically binds to genes. In the *lac* operon, the promoter is only slightly upstream of the operator site. Binding of the repressor appears to sterically interfere with RNA polymerase binding, thereby preventing transcription.

The DNA sequence of a promoter determines the maximum possible rate of gene transcription. Between different genes this rate can vary by greater than three orders of magnitude. In *E. coli* promoters, there are two separate sets of nucleotide sequences which are highly conserved and therefore essential for transcription. If the transcriptional start site is designated base position +1, at −35 there is the RNA polymerase recognition and binding site. At −10 the double-stranded DNA is thought to be open for transcription of the mRNA using the DNA complemenary minus strand as the template. Transcription is in the 5′ to 3′ direction based on the sugar-phosphate bonds of the DNA coding (or plus) strand, that which contains the protein coding sequence.

Consensus sequences

The consensus sequence of a particular functional component of a gene is a DNA sequence that appears to be conserved between different genes with only minor changes. Consensus sequences tend to occur in a group of closely related genes; however, some are ubiquitous. Hundreds of promoter regions have been sequenced in *E. coli* and only the −35 and −10 regions exhibit consensus sequences. These RNA polymerase binding sites have the consensus sequence TTGACA at the −35 box and TATAAT at the −10 box (or Pribnow box). Conservation of the base at each position varies from 45 to 100% and alterations can have variable (mild to severe) affects on the rate of transcription. The distance between these two consensus sequences is 16–18 base pairs (bp) in 90% of promoters, with limits appearing to be as low as 15 or as high as 20 bases. This distance is critical because of the geometry of the interaction between the DNA strand and RNA polymerase.

In a few bacterial promoters, one of the consensus sequences is missing and ancillary proteins are required for RNA polymerase to initiate transcription. However, these are rare exceptions and typically the −35 and −10 boxes are present.

The site for initiation of RNA polymerase activity – the start base for transcription – is usually a purine (A or G). Commonly it is the centre of the sequence CAT, although this is certainly not an essential signal.

Although consensus sequences are useful and attractive means of identifying essential DNA regions, they tend to over-simplify the signal concept. Promoter efficiency cannot be predicted entirely from the degree of homology with the consensus. Almost all bacterial promoters vary from the consensus and, in addition, adjacent base partners within the consensus can have dramatic positive or negative effects on transcription rates.

The turning off of the *lac* operon caused by the repressor binding to the operator is a prime example of negative genetic control. Bacteria also employ positive regulation of gene transcription whereby a transcriptional activator is required. The best understood system is that of cyclic AMP (cAMP). In the absence of glucose the concentration of cAMP builds up and binds to CAP, a DNA-binding protein. This complex in turn binds to promoters which activate operons involved in the utilization of alternative energy sources to glucose.

Cis- and trans-acting factors

In both prokaryotes and eukaryotes, the control of transcription involves DNA sequences and factors which can act in two ways: (1) directly, by being physically associated with the gene undergoing regulation (*cis*-regulating) or (2) indirectly, by producing a diffusible protein product that recognizes a target gene or genes (*trans*-regulating). *Cis*-regulating elements include promoters and closely associated DNA regions such as operators which exert their effects as a consequence of their DNA sequences being associated with the DNA of the gene. The genes for *trans*-acting factors need not be in close proximity with the target genes. The concept of *cis*- and *trans*-acting factors applies equally well to eukaryotic genes. In eukaryotes the genes for *trans*-acting factors are often located on a different chromosome from the target gene. These genes specify proteins which can migrate throughout the nucleus and act on one or a number of separate target sites simultaneously.

Mutations which result in an inability to transcribe a gene can be *cis*- or *trans*-acting. For example, a *cis* defect in *lac P* of the *lac* operon (Fig. 1.1) prevents the binding of RNA polymerase to this DNA region,

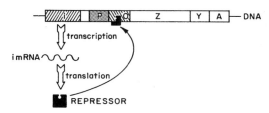

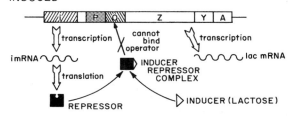

Fig. 1.1 Lactose operon. In the absence of lactose inducer the repressor binds to the operator (O) preventing transcription of the structural genes. When lactose is present it binds to form an inducer–repressor complex which cannot bind the operator and transcription proceeds to produce a single messenger encoding β-galactosidase (Z), β-galactoside permease (Y) and β-galactoside transacetylase (A).

effectively preventing transcription of the gene. This is genotypically distinct from a *trans*-acting *lac I* mutation which alters this gene's product, the repressor, in such a way that it is no longer able to bind lactose and hence remains attached to the gene preventing transcription. These mutant cells, however, are phenotypically identical.

Transcriptional attenuation

The process of transcription is controlled not only at initiation as described above but also by the process of transcriptional attenuation. Attenuation of the *trp* operon, which is responsible for synthesizing the amino acid tryptophan, can be used as an example (Fig. 1.2).

The structural genes of the *trp* operon are preceded by a leader peptide which codes for a sequence 14 residues in length. Under normal nutritional conditions, transcription of this leader sequence produces an mRNA transcript which forms a hairpin structure that causes the adjacent RNA polymerase

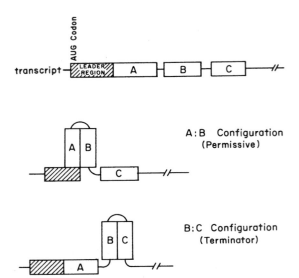

Fig. 1.2 Transcription attenuator in operons of amino acid biosynthesis. Translation of the messenger RNA begins shortly after the mRNA starts to be transcribed from the DNA. When the amino acid of the operon's biosynthetic pathway is at a deficient level, translation stops on the leader sequence of the protein because this region is rich in this amino acid. Stalling allows the A:B secondary structure to form and prevents the B:C configuration. This A:B secondary structure enhances the production of the messenger. If translation of the leader region proceeds without stalling because of sufficient levels of the amino acid, the B:C terminator forms instead and this prevents further transcription.

to release the DNA template and terminate transcription. The energetically expensive process of tryptophan synthesis is thereby averted.

However, when tryptophan levels are low, the *trp* operon must be activated so that the bacterial cell can produce its own tryptophan. In this instance, attenuation control is overridden through the exploitation of the timing of transcription versus translation. As the leader peptide sequence is transcribed by the RNA polymerase, ribosomes bind to the transcript and initiate translation. Within the leader peptide are two adjacent sequences calling for the addition of tryptophan to the amino acid chain. When tryptophan levels are low in the bacterial cell, the ribosome must pause during translation and wait for tryptophan to become available. The site on the

mRNA transcript upon which the ribosome waits is immediately adjacent to the region which forms the hairpin structure. The presence of the ribosome prevents hairpin loop formation, hence preventing the termination of transcription and allowing the transcription of the *trp* operon to proceed.

Attenuation control therefore provides another level of transcriptional control of prokaryotic gene expression in addition to repressor–operator control.

Translational control

Control of gene expression can also occur during translation. Initiation of translation requires a purine-rich nucleotide region of six to eight bases just upstream from the AUG start codon. Shine and Dalgarno identified this as a ribosome-binding site in 1974. They found that there was homology between this RNA region and a section of the 16S ribosomal RNA. Base pairing between the mRNA and the 16S RNA probably implements initiation. mRNA is translated most efficiently if this sequence is exactly eight bases upstream from the start codon.

In certain circumstances proteins bind to specific mRNA Shine–Dalgarno sequences, thereby preventing the initiation of translation. For example, by this means ribosomal (r) protein synthesis is kept strictly in parallel with the production of ribosomal RNA. R-proteins form a strong association with ribosomal RNA and a weaker association with their own mRNA at the Shine–Dalgarno sequences. Thus when r-protein production is higher than that of rRNA, the excess r-proteins bind to their own mRNAs and physically prevent translocation of the mRNA strand along the ribosome during translation.

Secondary structure loops in mRNA can also modulate translation. These override the presence of an otherwise effective translation initiation region by physically preventing the binding of the Shine–Dalgarno region to the 16S rRNA.

Translation can also be affected by the frequency of codon usage. The levels of the different transfer RNA (tRNA) species for each amino acid can vary and result in different translation rates depending on whether the messenger requires tRNAs with prevalent or rare anticodons (codon-binding domain of tRNA).

Structure of cyanobacterial genes

Although the available knowledge of cyanobacterial gene organization is not as extensive as that of non-photosynthetic prokaryotes, there is reason to believe that the genomes of cyanobacteria and bacteria are very similar and are probably regulated identically (Haselkorn, 1986).

In the cyanobacterium *Anabaena*, the genes of the nitrogenase complex are encoded in the *nifHDK* operon which contains three open reading frames; each coding for one polypeptide component of the complex. Nitrogenase activity occurs in specialized nitrogen-fixing structures called heterocysts, which differentiate from normal vegetative cells. During this differentiation an interesting gene rearrangement occurs which results in a novel means of gene regulation (Fig. 1.3). In vegetative cells the *nifD* ORF is interrupted by 11 kilobase pairs (kbp) of DNA which effectively prevents transcription. Removal of the insertion creates the mature functional operon in the heterocyst. The gene which controls excision, *xisA*, is located on the 11 kbp sequence which, following its excision, circularizes into a stable plasmid during differentiation.

Anabaena genes sometimes possess more than one promoter. This is the case with *glnA*, a gene involved in ammonium assimilation (Turner *et al.*, 1983). When ammonia is plentiful, promoter sequences

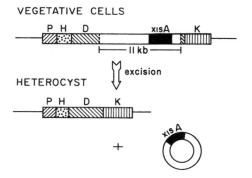

Fig. 1.3 Gene rearrangement in the *nif* operon of *Anabaena*. In vegetative cells which do not fix nitrogen, *nif D* is interrupted by an 11 kbp insert, which is excised during cellular differentiation to heterocysts. Excision is performed by the gene product of *xisA*, itself located within the insert. Excision results in a functional operon. P, promoter; H, D, and K are open reading frames.

similar to those of *E. coli* function as the regulators of *glnA* expression. However, under conditions where nitrogen fixation occurs (by the action of *nif* gene products), a promoter region similar to those of the *nif* genes controls *glnA* transcription. These two different promoters allow the synthesis of *glnA* and related enzymes to be maintained in the same proportion under the different conditions.

Eukaryotic genes

Eukaryotic genes differ in a number of ways from those of lower organisms. For example, they are not arranged in operons and, therefore, each coding region is associated with its own regulatory sequence. The coding regions are often interrupted by introns (non-coding intervening sequences) requiring post-transcriptional splicing in order to form mature mRNAs that are functional. Transcription of mRNA is performed by RNA polymerase II, one of three eukaryotic RNA polymerases. RNA polymerase I and III are responsible for the synthesis of ribosomal and transfer RNA, respectively. In prokaryotes, all of these functions are performed by a single RNA polymerase.

The messenger

Eukaryotic transcription initially results in the formation of heterogeneous nuclear RNA (hnRNA) in the nucleoplasm which is the precursor of mRNA. The average hnRNA molecule is four to five times longer than the average mRNA which is typically 1800–2000 bases (1.8–2.0 kb). During transcription hnRNA is rapidly 5′ capped with 7-methylguanosine residues linked to the RNA by a triphosphate bridge (m^7Gppp). Poly(A)polymerase uses ATP as a substrate to add a poly(A)-tail to the 3′ end of certain mRNAs. The 5′ caps and 3′ poly(A)-tails remain in the resultant mRNA species, although a few bases may be removed from the 3′ end during transport from the nucleus to the cytoplasm.

Processing of hnRNA into mature mRNA involves the removal of the non-translated regions, introns or intervening sequences, situated within the coding regions. Exons are RNA segments found both in the precursor hnRNA and in the mature RNA. The presence of introns may have resulted from genetic recombination bringing together domains from originally separate genes. Possibly, introns, which can be very long, also have the function of ensuring that coding sequences are kept intact during genetic crossing-over events (recombination). Long introns would increase the probability of genetic recombination events without damage to the exons which in turn would have given rise to novel alleles.

Specific base sequences delineate exon–intron boundaries, but RNA sequence complementation between the two boundaries is not involved in the bringing together of the appropriate sites for splicing. Instead, adapter RNAs which are attached to the splicing enzymes have been implicated in this process. The two consensus sequences in all higher eukaryotes for the left and right intron junctions, the donor and acceptor sites respectively, are

Left Junction
↓
$5'$ . . . exon . . . C/AAG   GUA/GAGU . . . intron . . .

Right Junction
↓
$Py_{10}NAG$ G . . . exon . . . $3'$

where Py is a pyrimidine (C or T). Splicing requires ATP and occurs in two stages (Fig. 1.4). Initially the left junction is cut and the free $5'$ end of the intron forms a $5'–2'$ bond with an A which is always present at position 6 in a consensus sequence called the TACTAAC box situated 18–40 nucleotides upstream of the right junction. This circular DNA, a lariat, is released and linearized when the right junction is cut and spliced to the left junction in stage two of the process. Except for the invariant A, the TACTAAC box sequence is not particularly well conserved between genes and this is essentially only a convenient label.

Most eukaryotic genes contain introns. Two notable plant examples which are intron free are the genes for zein storage proteins and the chlorophyll a/b binding protein of wheat.

Almost all eukaryotic mRNAs are monocistronic, but each messenger contains many more nucleotides than are necessary to encode the protein. Plant $5'$ leader sequences, which are highly AU-rich, range in length from 9–193 bases with a typical length of 40–

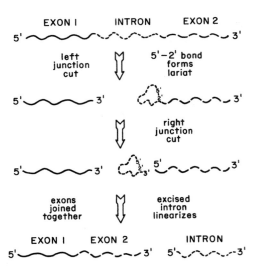

Fig. 1.4 Removal of introns produces mature messenger RNA. *In vitro* studies have shown that the left exon–intron junction is cleaved and the intron forms a lariat with a $5'$ to $2'$ bond. The right junction is then cut and the exons are covalently linked together. The intron assumes a linear form which is probably rapidly degraded *in vivo*.

80 bases (Joshi, 1987). The $5'$ methylated cap is involved in ribosome binding. Following addition of the cap, initiation factors eIF-4F (cap recognition protein), eIF-4A (ATP-dependent RNA-helicase) and eIF-4B (eIF stimulator) remove secondary structure from the mRNA leader region ensuring that the appropriate sites are exposed in single-stranded form (Jaramillo *et al.*, 1990). The cap is necessary to complex the initiation codon to the ribosome even in the rare cases where the cap and the initiation codon are as much as 1000 bases apart.

The 40S ribosomal subunit recognizes the $5'$ cap and travels along the mRNA until the AUG initiation codon is encountered. It appears that the AUG triplet must be in the right environment which is probably determined by nearby upstream and downstream bases. As a result of these requirements not all possible start codons are recognized for translation initiation, although in the great majority of instances, the AUG triplet closest to the $5'$ end is the actual start site (Joshi, 1987). In plants and animals there is a requirement for a purine at position −3 (A:75%, G:20%), and a G at position +4 for optimal initiation of translation (Kozak, 1991). The optimal sequence for initiation codon recognition for animals (Grünert and Jackson, 1994) and plants (Joshi, 1987) respectively are

gccgccAcc<u>AUG</u>GAu and uaaaca<u>AUG</u>Gcu

where the upper case residues have been found to exert the strongest positive influence. It has been proposed that the animal sequence may function by being complementary to the 18S rRNA sequence 3′ GGUGG, but this clearly cannot be the case with plant transcripts. Subsequently, the 60S large subunit associates with the mRNA–40S to form the initiation complex and begin translation of the mRNA.

The AU-rich, 3′ non-translated trailer can be very long – for example, up to 2000 bases – and contains many distinct *cis* elements necessary to direct termination of transcription and post-transcriptional polyadenylation. Although mammalian genes often have a single polyadenylation site, multiple sites are commonly found in plant genes (Mentoliu *et al.*, 1990). In most cases plant genes will consistently use a single polyadenylation site, however, differential use of sites in response to developmental cues has been observed (Williamson *et al.*, 1989; Sanfaçon and Wieczorek, 1992). Incorrect functioning of animal poly(A) signals in transgenic plant tissues (Hunt *et al.*, 1987) and of plant signals in transgenic yeast (Irniger *et al.*, 1992) suggests plant signals are distinct from those in the other eukaryotic kingdoms.

Polyadenylation signals have the consensus sequence AAUAAA in plants and animals. Plant gene polyadenylation signals maintain their function after much greater sequence deviation than do those of mammalian genes such that most plant transcripts do not possess an intact AAUAAA motif.

Most plant and animal mRNAs are polyadenylated. One-third of animal non-poly(A) message is histone mRNA, which needs to be rapidly turned-over during the cell cycle. In contrast, higher plant histone mRNA is polyadenylated (Wu *et al.*, 1989). Polyadenylation of plant transcripts is often necessary for the stability and translational efficiency of the message (Gallie, 1991). The level of stability imparted by the poly(A)-tail increases in proportion to its length. In addition, translation efficiency of polyadenylation increases with the presence of a 5′ cap, implying some communication between cap and poly(A)-tail is occurring (Gallie, 1991). A recent study of the mRNA transcript of the mammalian protooncogene *c-myc* has found evidence that a 39-base U-rich domain in the 3′ trailer is a signal responsible for mediating its rapid degradation (Alberta *et al.*, 1994).

The presence of poly(A) can be exploited experimentally for the separation of mRNA from rRNA since oligo(dT) sepharose can be used to bind poly(A)-tails. Also, small oligo(dT) fragments can be used to prime reverse transcriptase on the poly(A)-tail for first strand synthesis of complementary DNA. Although this is a useful characteristic of eukaryotic messengers, researchers must bear in mind that not all mRNAs are polyadenylated.

Regulation by antisense RNA

Antisense RNA is a transcript that is the complement of the sense or coding mRNA transcript. *In vivo* expressed antisense RNA interacts with sense RNA in a way that results in a reduction or elimination of target gene expression. Two mechanisms have been proposed to explain antisense RNA inhibition. An annealed antisense RNA and sense mRNA complex can be unstable and preferentially targeted for rapid degradation or antisense RNA may sterically block transcription, transport or translation of the sense transcript.

The prokaryotic transposable element Tn10 provides an illustration of antisense RNA translational regulation. Translation of Tn10 is regulated by a small antisense RNA transcript (pOUT) which contains a 35 base stretch exactly complementary to a region of the transposable gene (pIN). This region includes the Shine–Dalgarno sequence and the translation–initiation codon. In most cases, regulation is effected by complementation interference.

In manipulating gene expression by antisense RNA, a molar excess of antisense transcript must be introduced or produced transgenically to see an appreciable reduction in mRNA translation. Complete inhibition is not usually achievable, however, it is often sufficient to produce a mutant phenotype. Inhibition by antisense RNA may be due to several mechanisms. Assuming antisense RNA has no effect on the half-life of a mRNA species, the physical interference of the antisense transcript either at the ribosome or in transport across the nuclear envelope may delay translation such that few functional gene products are produced. Alternatively, reduced mRNA expression may be the result of decreased stability of the mRNA and antisense RNA hybrids.

In theory, the mechanism of antisense RNA regulation should function equally well in prokaryotic and eukaryotic cells and in fact, this is the case. Ecker and Davis (1986) introduced into carrot cells synthetic gene constructs composed of promoters associated with a reporter gene (chloramphenicol acetyltransferase) in correct orientation with or without the reverse orientation construct (Fig. 1.5). Protein synthesis was effectively inhibited when the reverse orientation was present. Antisense RNA inhibition is currently a standard technique for observing the effects of transient inhibition of a target protein and for producing stable mutants. For example, an ethylene mutant of the plant *Arabidopsis thaliana* was produced by antisense RNA expression of a key ethylene biosynthetic enzyme (Oeller *et al.*, 1991).

Codon usage in higher plants

All organisms exhibit a unique codon bias. Monocots, dicots, chloroplasts, green algae and cyanobacteria all preferentially utilize different subsets of the 64 available codons that code either for one of 20 amino acids or for termination (Campbell and Gowrie, 1990). Dicot nuclear genes predominantly use a set of 44 codons that tend to end in A or U. In contrast,

Fig. 1.5 Gene constructs employed in the study of plant antisense RNA by Ecker and Davis. The promoters employed in this construct were from nopaline synthase, cauliflower mosaic virus 35S RNA, and phenylalanine ammonia lyase. The reporter gene was chloramphenicol acetyltransferase (CAT). The nopaline synthase polyadenylation signal was used in all constructs. Experiments were performed by the electroporation of sense and antisense genes in different proportions to one another into carrot protoplasts followed by the measurement of transient CAT expression.

monocot nuclear genes prefer a more restrictive 38 codons which usually end in C or G. Termination codons also differ between monocot and dicot plants (Angenon *et al.*, 1990). With the capability of researchers to design and construct genes and degenerate (deduced from amino acid sequence data) oligonucleotides, an understanding of the host-specific codon bias may maximize transgenic expression and oligonucleotide hybridization.

Transcriptional regulation

Cis-*regulatory elements*

Promoters Eukaryotic promoters consist of three important regions: (1) the transcriptional start (designated position +1), (2) a sequence 20–30 bp upstream of this start point and (3) a region further upstream at about −75 (Fig. 1.6).

The TATA box is located approximately at position −25 and has the consensus sequence

$$
\begin{array}{c}
\text{A A} \\
\text{TATA A} \\
\text{T T}
\end{array}
$$

It is surrounded by GC-rich sequences which may play a role in its function. Except for its position at −25 it could pass for the bacterial −10 (Pribnow) box located at 10 bp upstream of the prokaryotic transcriptional start. In some genes such as those for zein (a maize storage protein) two TATA boxes may be found as close as 10 bp apart.

The function of the TATA sequence is to direct RNA polymerase II to the correct transcriptional start site. Mutant genes that lack a TATA box will initiate transcription from many sites. However, a TATA box is insufficient to control transcription alone; at least one other important upstream element is necessary for transcription initiation.

Centered around −75 are one or more elements responsible for regulating the frequency of initiation. These upstream elements often show homologies between genes. Two of the most common of these elements are the GC and CAAT boxes which have the respective animal consensus sequences

$$
\text{GGGCGG} \quad \text{and} \quad
\begin{array}{c}
\text{C} \\
\text{GG CAATCT} \\
\text{T}
\end{array}
$$

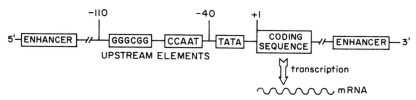

Fig. 1.6 *Cis*-regulatory DNA elements. The promoter region includes the TATA box and upstream elements. Enhancer elements may be situated further upstream or downstream of the protein coding region.

These elements may be present in the same gene in either orientation and in single or numerous copies.

CAAT boxes are often absent from plant genes. Cereal genes have a different consensus sequence

CCATCTCNACC

at position −90 called the CATC box which may serve as a substitute for CAAT (Kreis *et al.*, 1986). Certain ribulose bisphosphate carboxylase small subunit genes (*rbcS*) may have no requirements for CAAT or functionally equivalent sequences (Morelli *et al.*, 1985; Kuhlemeier *et al.*, 1987). The regions between the TATA box and the upstream elements, when present, are not involved in promoter function and the distance between them can be artificially lengthened to some degree without impairing promoter function.

Other transcription factors may be found within promoter regions. In response to wounding or infection, transcriptional expression of the chalcone synthase genes leads to the production of flavonoid anti-microbial compounds. Two *cis*-acting regions found upstream of the TATA box are critically dependent for the expression of a bean chalcone synthase gene (Arias *et al.*, 1993). The sequences of the G-box and H-box are

CACGTG   and   CCTACC

respectively. A G-box conserved among the ribulose bisphosphate-1,5-carboxylase gene family has the consensus sequence

C
CACGTGGCA
A

A G-box from the higher plant *Arabidopsis thaliana* was capable of activating expression of a reporter gene in yeast (Donald *et al.*, 1990). In conjunction with the identification of DNA-binding transcription

factors (see below), an ever-increasing number of *cis*-acting elements are being characterized that may otherwise not be evident from sequence conservation. This is true for promoter elements and the other transcriptional regulatory *cis*-acting elements.

Enhancers and silencers   Many eukaryotic genes contain sequences that have a positive (or negative) effect on transcription but do not have the sequence conservation and strict positional requirements of promoters. These sequences have a strong tendency to be unique or shared by a small group of related genes. Therefore, most enhancer sequences are identified by studying the effects of numerous deletion mutants of a particular gene or gene family sharing a common transcriptional regulatory profile. Enhancers are typically 100–200 bp in length and can act in either orientation and at a considerable distance – literally thousands of base pairs – from the target gene's promoter. They are linked to the target genes (on the same chromosome) and, as such, are *cis*-acting DNA sequences. They may be 5′, or 3′ to the transcriptional start sequence and usually act by stimulating the nearest promoter(s). These elements commonly contain repeated sequences which individually possess a small stimulatory ability although no DNA consensus sequences are common to all enhancers. One sequence which occurs upstream of several animal genes has the consensus

AA
NTGTGG
TT

In plants, many *rbcS* genes from dicots contain a closely related sequence to the above animal enhancer,

TT A T
GTGTGG A TA G
CC T A

at −140 from the transcriptional start (Kuhlemeier *et al.*, 1987). A similar sequence is also present at −215 indicating that redundancy of information is an advantage for at least some plant enhancers.

A number of other enhancer regions have been identified in specific plant genes. To find an enhancer region, expression vectors are constructed with sequential deletions or rearrangements of upstream elements placed in front of a sensitive reporter gene such as GUS (β-glucuronidase synthase). Reporter gene activity in the transgenic plants correlates to the transcriptional regulation imposed on the gene of interest (for methods see Chapter 35).

Histone genes (H1, H2A, H2B, H3 and H4) exhibit cell cycle-dependent (S-phase) expression at the transcriptional level. Hexamer and octamer *cis*-acting elements, with the respective consensus sequences

ACGTCA   and   CGCGGATC,

are essential for transcription of wheat histone genes during S-phase of the cell cycle (Nakayama *et al.*, 1992). Plant histone genes fall into one of two groups based on organization of these two elements. Type I genes have a hexamer two base pairs upstream of a reverse-oriented octamer with the resultant sequence

ccACGTCAncGATCCGCG

where the hexamer and octamer are shown in upper case respectively. Type II genes possess only an octamer element.

In the α-subunit of the β-conglycinin gene there are five 6 bp repeats with the consensus

A
AGCCCA.
C

One repeat is located at position −149 relative to the site of transcriptional initiation, while the other four repeats occur at −255, −204, −182, and −171. The pair of repeats at −204 and −182 and their intervening sequence form a larger imperfect direct repeat of 28 nucleotides with the −171 and −149 repeats and their intervening sequence. An additional characteristic of this 5′ region is that the −149 and −171 repeats as well as the −182 and −204 repeats are 16 bp apart. This separation of 16 nucleotides places them on the same face of the DNA helix. This

could be significant to gene regulation since both sites would be accessible in concert to *trans*-acting factors (see next section). Petunia plants which were transformed with a truncated form of this gene containing only the 6 bp repeat at −149 had only 5% of the expression level found in plants transformed with the complete gene. Similar repeats are present in the genes for the β-subunit of β-conglycinin and phaseolin (Chen *et al.*, 1986).

The Ti-plasmid of *Agrobacterium tumefaciens* has evolved the ability to transfer functional genes into plants (T-DNA). The promoters of these genes are eukaryotic in nature and one that has been studied extensively is responsible for the synthesis of the amino acid octopine. Between positions −193 to −178 an element was found to possess perfect dyad symmetry

5′ . . ACGTAAGC   GCTTACGT . . 3′
3′ . . TGCATTCG   CGAATGCA . . 5′

where the palindromic sequences to each side of a central axis are the same. This 16 bp palindrome possesses all the attributes of a conventional enhancer sequence (Ellis *et al.*, 1987). Its enhancing effect diminishes with increasing distance from the promoter and it has been shown to be functional when placed at the 3′ end of genes. As would be expected from a palindrome, reversing its orientation has no effect. Another plant *cis*-acting element which forms a perfect palindrome is the G-box motif

GCCACGTGGC

conserved in several *rbcS* genes.

Plant genes may require 3′ sequences for expression. Potato proteinase inhibitor II does not function in transgenic tobacco when a 200 bp downstream region is not included in chimaeric gene constructs (Sanchez-Serrano *et al.*, 1987).

Trans-*acting factors*

Transcriptional regulation of gene expression is a complex process involving the interaction of *trans*-acting proteins with *cis*-acting DNA sequences. The DNA-binding proteins generally possess one of three characteristic structural motifs: the helix–loop–helix (or helix–turn–helix), the zinc finger, or the leucine zipper.

The helix–loop–helix motif (HLH), characterized by two alpha helices separated by a beta turn, was first encountered in various bacterial and bacteriophage DNA-binding proteins (including *E. coli* CAP) upon resolution of their secondary structures. While one helix (the recognition helix) binds to a target DNA sequence within the major groove, the other alpha helix locks the recognition helix in position (Hochschild *et al.*, 1986). The helix–loop–helix motif has since been found in several eukaryotic proteins involved in developmental regulation. Examples include the yeast mating type gene products MATa1 and MATα2, and the extremely conserved homeodomain gene products (Laughton and Scott, 1984).

Analysis of the structure of transcription factor IIIA (TFIIIA), which is responsible for binding to the promoters of 5S rRNA genes, led to the discovery of the zinc finger-motif. Within the amino acid sequence of TFIIIA there are nine occurrences of the sequence

$$Cys\text{-}N_2\text{-}Cys\text{-}N_{12}\text{-}His\text{-}N_3\text{-}His$$

where N is any amino acid. The paired cysteine and histidine residues form a tetrahedral bond with a zinc ion while the 12 intervening residues loop out to form a DNA-binding structure (Rhodes and King, 1986).

The basic/leucine zipper (bZIP) motif was discovered when DNA-binding proteins from yeast (GCN4) and rat (C/EBP) showed homology to the proto-oncogene products MYC, FOS and JUN, that are involved in *trans*-activation of cell proliferation. Highly stable alpha helical regions of two polypeptides dimerize by the interdigitation of leucine residues projecting from each helix (Landschulz *et al.*, 1988). Once this zipper structure has formed, a basic helix from each polypeptide is in position to interact with the target DNA sequence. DNA-binding activity of bZIP proteins can either be observed when two identical polypeptides homodimerize (e.g. C/EBP), or when two distinct polypeptides form a heterodimer (e.g. FOS and JUN).

These three structural motifs may not be as distinct as previously thought. A plant regulatory protein which combines a homeodomain (helix–turn–helix motif) with a bZIP motif may represent a new DNA-binding protein class (Ruberti *et al.*, 1991).

There are three general strategies used to identify plant transcription factors (Katagiri and Chua, 1992).

Isolation of genes from regulatory or developmental mutants can be achieved either by chromosome walking from known genetic markers, or by screening for a gene disrupted by T-DNA insertion (using a T-DNA-specific probe). Another approach involves using *cis*-acting elements to bind *trans*-acting factors *in vitro*. Any bound proteins can be recovered with the *cis*-acting elements and characterized. Finally, transcription factor genes homologous to previously characterized *trans*-acting genes can be isolated either by heterologous screening for close relatives or polymerase chain reaction amplification between two primers corresponding with known consensus sequences.

An ever-increasing number of *cis*- and *trans*-acting factors are being identified in animals (Mitchell and Tjian, 1989) and plants (Katagiri and Chua, 1992). Determining their *in vivo* involvement in transcriptional regulation is a formidable task. Nevertheless, progress has been made in resolving some regulatory systems. The different experimental approaches and structural motifs discussed above are apparent in the four examples of plant regulatory systems mentioned below.

Wheat histone gene regulation is one of the best characterized plant regulatory systems. Two HBP (histone-promoter-binding protein) protein subfamilies (HBP-1a and HBP-1b) are *trans*-acting factors that bind to either of two hexamer motifs of type I histone genes (Mikami *et al.*, 1994). HBP-1a and HBP-1b proteins bind the hexamer sequences

Hex-a	CCACGT	and
Hex-b	ACGTCA	

respectively using bZIP (basic/leucine zipper) domains. Additional HBP-like *trans*-acting factors have been isolated from a variety of plant species such as maize, tobacco, parsley and wheat. These *trans*-acting factors, involved in a diverse variety of gene regulation events, may be members of a bZIP superfamily.

The *trans*-acting factor 3AF1 is associated with the regulation of rubisco (ribulose bisphosphate carboxylase), one of the most abundant proteins on earth. In response to light, the zinc-finger motif of 3AF1 binds to an AT-rich sequence of the promoter for the *rbcS* gene (rubisco small subunit) (Lam *et al.*, 1990). In addition, other transcriptional activators likely bind to this same *cis*-acting sequence.

Another light-responsive gene, chalcone synthase (CHS), is induced by interactions with three *trans*-acting proteins (Weisshaar *et al.*, 1991). The *chs* promoter has four *cis*-acting elements (boxes I, II, III, and IV) involved with CHS expression (Schulze-Lefert *et al.*, 1989). CPRF (common plant regulatory factor)-1, -2 and -3 are bZIP proteins with homology to HBP-1a. These proteins bind to the heptameric *cis*-acting sequence

ACGTGGC

which is a portion of the conserved G-box *cis*-acting element.

Maize pigment varies according to the expression of anthocyanin pigments. The anthocyanin biosynthetic pathway genes are regulated by a family of helix–loop–helix *trans*-acting factors called *R* (Ludwig and Wessler, 1990). Multiple allelic variants of *R* regulatory genes have been identified by Mendelian genetics due to the easily selectable and non-destructive nature of the pigment mutants. The so-called *R* gene is comprised of two tightly linked loci, S and P, that control seed and plant pigmentation. All members of the *R* gene family have a high degree of homology to the mammalian *myc* gene product and other HLH regulatory proteins.

In addition to *R*, the *trans*-acting factors *C1* and *Pl* are necessary for expression of the maize anthocyanin biosynthetic gene *Bz1* (UDP glucose:flavonoid 3-O-glucosyltransferase) (Roth *et al.*, 1991). *C1* and *Pl* are highly homologous to the mammalian DNA-binding protein MYB. Moreover, the promoter region of the *Bz1* gene has *cis*-acting sites homologous to consensus sequences of the mammalian *myc* and *myb* genes. These *Bz1 cis*-acting sites

myb-like	TAACTG	and
myc-like	GGCAGGTGC	

are the *C1* and *R* binding sites, respectively.

Negative regulation of transcription

The *cis*-acting factors involved in negative control are referred to as blockers or silencers. Here the mammalian SV40 virus provides the best characterized example. A protein isolated from cells infected with SV40 and encoded on the viral genome has been termed the T antigen. It regulates its own

production and that of other virally encoded proteins by reducing the synthesis of their mRNAs. T antigen achieves this by binding to sites near the viral replication origin.

The gene for pea chlorophyll *a/b* binding protein and *rbcS-3A* have upstream sequences which inhibit transcription. Normally, constitutive photosynthetic genes have their expression greatly reduced in the dark and in non-green tissue as a result of the activity of these elements (Simpson *et al.*, 1986; Nagy *et al.*, 1988). Interestingly, both these silencer elements appear to have enhancer properties in the light.

It is generally felt that *trans*-acting factors act on virtually all eukaryotic genes. Stereospecific interactions by the resultant protein-DNA complexes either facilitate or hinder initiation of transcription by RNA polymerase II. Interactions with other DNA-associated proteins must also be implicated in these processes.

Organelle genes

Mitochondrial and plastid genomes code for a small number of proteins found in these organelles. They produce their own ribosomal and transfer RNAs, and a small number of mRNAs (i.e. around ninety in chloroplasts and eight or more in mitochondria). The nuclear genome codes for most of the organellar proteins which are then imported from the cytoplasm. In some cases separate protein subunits are encoded in the different genomes (e.g. nuclear *rbcS* and chloroplast *rbcL* genes of ribulose bisphosphate carboxylase), which poses interesting questions regarding regulatory interactions.

Organelle genes are essentially prokaryotic in nature. They possess promoter sequences which have been shown to be recognized by *E. coli* RNA polymerase. They are usually polycistronic, although they do occasionally contain introns, a characteristic very rarely seen in prokaryotic genes. In the case of ribosomal protein S12 of tobacco chloroplasts, mature mRNA is formed by the *trans*-splicing of two pre-mRNAs. In this case, exon 1 is encoded by a separate gene than exons 2 and 3 (Hildebrand *et al.*, 1988). The lack of a membrane separating the genome from the ribosomes suggests that there is some process other than physical compartmentation involved in determining which RNAs are translated into protein: recall that the nuclear membrane separates the eukaryotic hnRNA from the mature mRNA found in the cytoplasm.

The regulation of gene expression in organelles and the intimate relationships of organellar and nuclear genes are the subjects of other chapters in this volume.

Concluding remarks

The regulation of gene expression both during and after transcription involves numerous and diverse mechanisms, some of which are becoming well understood. Although plants, animals and fungi regulate and perform transcription somewhat differently, expression of mammalian genes in plants is highly feasible (Lefebvre et al., 1987). The identification and characterization of cis- and trans-acting factors in plants is proceeding at a considerable pace. In most cases, however, the regulatory roles of these factors still need to be identified. In addition, the cascade of receptors, G-proteins, secondary messengers, protein kinases and phosphatases which regulate the transcriptional regulatory factors, are quickly coming to light. The characterization of regulatory pathways is leading to an understanding of how primary signals elicit differential gene expression.

References

Alberta, J. A., Rundell, K. and Stiles, C. D. (1994). Identification of an activity that interacts with the 3'-untranslated region of c-myc mRNA and the role of its target sequence in mediating rapid mRNA degradation. J. Biol. Chem. **6**, 4532–8.

Angenon, G., Van Montagu, M. and Depicker, A. (1990). Analysis of the stop codon context in plant nuclear genes. FEBS Lett. **271**, 144–6.

Arias, J. A., Dixon, R. A. and Lamb, C. J. (1993). Dissection of the functional architecture of a plant defence gene promoter using a homologous in vitro transcription initiation system. Plant Cell **5**, 485–96.

Campbell, W. H. and Gowrie, G. (1990). Codon usage in higher plants, green algae, and cyanobacteria. Plant Physiol. **92**, 1–11.

Chen, Z.-L., Schuler, M. A. and Beachy, R. N. (1986). Functional analysis of regulatory elements in a plant embryo-specific gene. Proc. Natl. Acad. Sci. USA **83**, 8560–4.

Donald, R. G. K., Schindler, U., Batschauer, A. and Cashmore, A. R. (1990). The plant G box promoter sequence activates transcription in Saccharomyces cerevisiae and is bound in vitro by a yeast activity similar to GBF, the plant G box binding factor. EMBO J. **9**, 1727–35.

Ecker, J. R. and Davis, R. W. (1986). Inhibition of gene expression in plant cells by expression of antisense RNA. Proc. Natl. Acad. Sci. USA **83**, 5372–6.

Ellis, J. G., Llewellyn, D. J., Walker, J. C., Dennis, E. S. and Peacock, W. J. (1987). The ocs element: a 16 base pair palindrome essential for activity of the octopine synthase enhancer. EMBO J. **6**, 3203–8.

Gallie, D. R. (1991). The cap and poly(A) tail function synergistically to regulate mRNA translational efficiency. Genes and Devel. **5**, 2108–16.

Grünert, S. and Jackson, R. J. (1994). The immediate downstream codon strongly influences the efficiency of utilization of eukaryotic translation initiation codons. EMBO J. **13**, 3618–30.

Harland, R. and Weinbraub, H. (1985). Translation of mRNA injected in Xenopus oocytes is specifically inhibited by antisense RNA. J. Cell Biol. **101**, 1094–9.

Haselkorn, R. (1986). Organization of the genes for nitrogen fixation in photosynthetic bacteria and cyanobacteria. Ann. Rev. Microbiol. **40**, 525–47.

Hildebrand, M., Hallick, R. B., Passavant, C. W. and Bourque, D. P. (1988). Trans-splicing in chloroplasts: The rps12 loci of Nicotiana tabacum. Proc. Natl. Acad. Sci. USA **85**, 372–6.

Hochschild, A., Douhan, J. and Ptashne, M. (1986). How λ repressor and λ cro distinguish between O_R1 and O_R3. Cell **47**, 807–16.

Hull, R. and Covey, S. N. (1985). Cauliflower mosaic virus: pathways of infection. BioEssays **3**, 160–3.

Hunt, A. G., Chu, N. M., Odell, J. T., Nagy, F. and Chua, N-H. (1987). Plant cells do not properly recognize animal gene polyadenylation signals. Plant. Mol. Biol. **7**, 23–25.

Irniger, S., Sanfaçon, H., Egli, C. M. and Braus, G. H. (1992). Different sequence elements are required for function of the cauliflower mosaic virus polyadenylation site in Saccharomyces cerevisiae compared with plants. Mol. Cell Biol. **12**, 2322–30.

Jaramillo, M., Browning, K., Dever, T. E., Blum, S. and Traschel, H. (1990). Translational initiation factors that function as RNA helicases from mammals, plants and yeast. Biochim. Biophys. Acta **1050**, 134–9.

Joshi, C. P. (1987). An inspection of the domain between putative TATA box and translational regulation start site in 79 plant genes. Nucleic Acids. Res. **16**, 6643–53.

Katagiri, F. and Chua, N-H. (1992). Plant transcription factors: present knowledge and future challenges. Trends Genet. **8**, 22–7.

Kozak, M. (1991). Structural features in eukaryotic mRNAs that modulate the initiation of translation. *J. Biol. Chem.* **266**, 19867–70.

Kreis, M., Williamson, M. S., Forde, J., Schmutz, D., Clark, J., Buxton, B., Pywell, J., Marris, C., Hendersen, J., Harris, N., Shewry, P. R., Forde, B. G. and Miflin, B. J. (1986). Differential gene expression in the developing barley endosperm. *Phil. Trans. R. Soc. Lond. B.* **314**, 355–65.

Kuhlemeier, C., Green, P. J. and Chua, N.-H. (1987). Regulation of gene expression in higher plants. *Ann. Rev. Plant Physiol.* **38**, 221–57.

Lam, E., Kano-Murakami, Y., Gilmartin, P., Niner, B. and Chua, N-H. (1990). A metal-dependent DNA-binding protein interacts with a constitutive element of a light responsive promoter. *Plant Cell* **2**, 857–66.

Landschulz, W. H., Johnson, P. F. and McKnight, S. L. (1988). The leucine zipper: A hypothetical structure common to a new class of DNA binding proteins. *Science* **240**, 1759–64.

Laughton, A. and Scott, M. P., (1984). Sequence of a *Drosophila* segmentation gene: Protein structure homology with DNA-binding proteins. *Nature* **310**, 25–31.

Lefebvre, D. D., Miki, B. L. and Laliberte, J.-F. (1987). Mammalian metallothionein functions in plants. *Bio/Technology* **5**, 1053–6.

Ludwig, S. R. and Wessler, S. R. (1990) Maize *R* gene family: tissue-specific helix–loop–helix proteins. *Cell* **62**, 849–51.

Montoliu, L., Rigau, J. and Puigdomenech, P. (1990). Multiple polyadenylation sites are active in the α1-tubulin gene from *Zea mays*. *FEBS Lett.* **277**, 29–32.

Mikami, K., Sakamoto, A. and Iwabuchi, M. (1994). The HBP-1 family of wheat basic/leucine zipper proteins interacts with overlapping *cis*-acting hexamer motifs of plant histone genes. *J. Biol. Chem.* **269**, 9974–85.

Miller, J. J. and Reznikoff, W. S. (1980). *The Operon*, 2nd edn., Cold Spring Harbor, New York.

Mitchell, P. J. and Tjian, R. (1989). Transcriptional regulation in mammalian cells by sequence-specific DNA binding proteins. *Science* **245**, 371–8.

Morelli, G., Nagy, F., Fraley, R. T., Rogers, S. G. and Chua, N.-H. (1985). A short conserved sequence is involved in the light-inducibility of a gene encoding ribulose bisphosphate carboxylase small subunit of pea. *Nature* **315**, 200–4.

Nagy, F., Kay, S. A. and Chua, N.-H. (1988). Gene regulation by phytochrome. *Trends Genet.* **4**, 37–42.

Nakayama, T., Sakamoto, A., Yang, P., Minami, M., Fujimoto, Y., Ito, T. and Iwabuchi, M. (1992). Highly conserved hexamer, octamer and nonamer motifs are positive *cis*-regulatory elements of the wheat histone H3 gene. *FEBS Lett.* **300**, 167–70.

Oeller, P. W., Min-Wong, L., Taylor, L. P., Pike, D. A. and Theologis, A. (1991). Reversible inhibition of tomato fruit senescence by antisense RNA. *Science* **254**, 437–9.

Rhodes, D. and King, A. (1986). An underlying repeat in some transcriptional control sequences corresponding to half a double helical turn of DNA *Cell* **46**, 123–32.

Roth, B. A., Goff, S. A., Klein, T. M. and Fromm, M. E. (1991). *C1*- and *R*-dependent expression of the maize *Bz1* gene requires sequences with homology to mammalian *myb* and *myc* binding sites. *Plant Cell* **3**, 317–25.

Ruberti, I., Sessa, G., Lucchetti, S. and Morelli, G. (1991). A novel class of plant proteins containing a homeodomain with a closely linked leucine zipper motif. *EMBO J.* **10**, 1787–91.

Sanfaçon, H. and Wieczorek, A. (1992). Analysis of cauliflower mosaic virus RNAs in Brassica species showing a range of susceptibility to infection. *Virology* **190**, 30–9.

Sanchez-Serrano, J., Keil, M., O'Conner, A., Schell, J. and Willmitzer L. (1987). Wound-induced expression of a potato proteinase inhibitor II gene in transgenic tobacco plants. *EMBO J.* **6**, 303–6.

Shine, J. and Dalgarno, L. (1974). The 3'-terminal sequence of *Escherichia coli* 16S ribosomal RNA: Complementarity to nonsense triplets and ribosome binding sites. *Proc. Natl. Acad. Sci. USA* **71**, 1342–6.

Schulze-Lefert, P., Becker-Andre, M., Schulz, W., Hahlbrock, K. and Dangl, J. L. (1989). Functional architecture of the light-responsive chalcone synthase promoter from parsley. *Plant Cell* **1**, 707–14.

Simpson, J., Schell, J., Van Montagu, M. and Herrera-Estrella, L. (1986). Light-inducible and tissue-specific pea *lhcp* gene expression involves an upstream combining enhancer- and silencer-like properties. *Nature* **323**, 551–4.

Tabata, T., Nakayama, T., Mikami, K. and Iwabuchi, M. (1991). HBP-1a and HBP-1b: leucine zipper-type transcription factors of wheat. *EMBO J.* **10**, 1459–67.

Turner, N. E., Robinson, S. J. and Haselkorn, R. (1983). Different promoters for the *Anabaena* glutamine synthetase gene during growth using molecular or fixed nitrogen. *Nature* **306**, 337–42.

Weisshaar, B., Armstron, G. A., Block, A., da Costa e Silva, O. and Hahlbrock, K. (1991). Light-inducible and constitutively expressed DNA-binding proteins recognizing a plant promoter element with functional revelance in light responsiveness. *EMBO J.* **10**, 1777–86.

Williamson, J. D., Hirsch-Wyncott, M. E., Larkins, B. A. and Gelvin, S. B. (1989). Differential accumulation of a transcript driven by the CaMV 35S promoter in transgenic tobacco. *Plant Physiol.* **90**, 1570–6.

Wu, S.-C., Gyorgyey, J. and Dudits, D. (1989). Polyadenylated H3 histone transcripts and H3 histone variants in alfalfa. *Nucleic Acids Res.* **17**, 3057–63.

2 Protein synthesis in plant cells
Kenton Ko

Plant proteins and their usefulness

Proteins occupy a central role in all organisms, and plants are no exception to this rule. These relatively large, diverse macromolecules account for a substantial portion of the dry weight of most living plant cells and are important constituents of plant tissues. A large number, if not most, of the biochemical, physiological and structural processes are governed and regulated by one or more of these proteins. In many instances, proteins involved in such processes have been termed enzymes, regulators or structural components, as evident throughout the various chapters of this book. Proteins are not only important in the synthesis, regulation and production of a large variety of biochemical molecules but also in the storage of nitrogen-containing nutrients in the form of polypeptides. Since proteins are composed of smaller building blocks termed amino acids, storage proteins are therefore rich stores of carbon, hydrogen, oxygen and sulfur as well as nitrogen. A good example of nutritious storage proteins are those found in dry bean seeds. This particular aspect is discussed in detail in Chapter 34. As one can gather by merely scanning over this or any other book in this field of study, proteins are extremely diverse in size, function and characteristics. This remarkable diversity is reflected in the amino acid sequence of the proteins. However, despite compositional differences, whether they are subtle or massive, all proteins are believed to be synthesized by very similar mechanisms. The general scheme for arriving at a proteinaceous macromolecule is well established and has been dealt with largely in the preceding chapter. The information for all proteins ultimately lies in the nucleotide sequence of the gene that encodes the

specific protein. This genetic information is not ready for use until it is converted into an intermediate molecule, the RNA messenger or transcript, through a process known as transcription. The genetic information is now in a form that can be converted to the appropriate amino acid sequence, a process called translation, or in other words, protein synthesis. The chain of amino acids will eventually be completed from one end to the other and give rise to a full length protein sequence. The completed protein molecule will then proceed through various post-translational events to initiate its function in the plant cell.

Although the general process of arriving at a proteinaceous macromolecule appears rather straightforward and simplistic, it is by no means that. The construction of a protein molecule requires the orchestrated effort of ribosomes and a host of factors. A large amount of our present knowledge on the synthesis of proteins is derived from animal systems but many components of the protein synthesis machinery now have plant analogues identified and characterized (Abramson et al., 1988). It appears that many of the components and events in animal protein synthesis also apply to plant protein synthesis; therefore one can draw information from animal systems to assist in our understanding of this complex process in plant cells.

One may ask at this point, why study the synthesis of proteins in plant cells? One answer is of course the old cliché, the pursuit of knowledge and the understanding of the world in which we exist, but there is an even more pressing need for this knowledge in our modern times – economics. The birth of plant biotechnology has developed a need to learn how to engineer plants to produce products effectively and economically. Products such as specialty oils, plastics and enzymes for food processing can be produced in plants and farmed as a crop (Knauf, 1995). As you know, in order to produce such commodities, one needs to synthesize proteins either for direct use or to direct the synthesis of the desired product. To obtain maximum yield per unit of plant material, the enzymes must be correctly synthesized in optimal quantities. Many problems can be encountered in the synthesis of a protein, especially in the engineered environment of a transgenic plant. Therefore it is very important to understand the protein synthesizing process in order to maximize it for genetic engineering purposes.

Protein synthesis in a plant cell is compartmentalized

The majority of the plant cell's proteins are synthesized in the cytoplasmic compartment, as is also the case in animal cells. The proteins synthesized in this compartment stay either in the cytoplasm where they presumably function or are transported to other compartments of the cell where they can proceed through other post-translational events. These events are discussed in some of the other chapters in the book. Protein synthesis in plants is also compartmentalized, perhaps more than in animal systems. In animal cells, the other compartment where independent protein synthesis takes place is the mitochondrion. Plant cells have in addition to a mitochondrial compartment a plastidic compartment where independent protein synthesis occurs. Therefore plant cells have at least three known separate compartments within the cell that can independently perform protein synthesis; the cytoplasm, the mitochondrion and the plastid (Figure 2.1). Independent protein synthesis means that the compartment can be separated from the rest of the cellular content in test tubes and still possess the machinery and capability to translate full length proteins. Even though the resulting protein products synthesized from each of these compartments are indistinguishable as to their origin, in other words, proteins are proteins wherever they are synthesized, the mechanism for producing proteins are slightly

different. The cytoplasmic compartment protein synthesizing machinery is eukaryotic in nature. Prokaryotic systems that resemble more closely the machinery found in bacteria are used in the mitochondrial and plastidic compartments. Therefore there exist three different protein synthesizing systems in a plant cell. The eukaryotic and prokaryotic systems will be reviewed separately in the following two sections.

Eukaryotic protein synthesis in the cytoplasm

Initiation of translation

Upon transcription of the genetic information from the DNA form to the RNA form, the information dictating the sequence of the protein can then be converted one amino acid at a time (see Fig. 2.2 for an overall scheme). Although translation ultimately occurs on 80S ribosomes, the initiation of translation involves a number of *e*ukaryotic *i*nitiation *f*actors designated as eIFs. The freshly transcribed mRNA must first be recognized by the translational machinery. Recognition of the transcript involves cap binding factor eIF-4F and the 5′ cap (m7GpppN) structure. The cap structure is believed to serve two roles in the transcript: translational efficiency and mRNA stability. This is followed by the binding of an ATP-dependent RNA helicase, eIF-4A, and a stimulatory factor eIF-4B. The eIF-4B factor stimulates the activities of both eIF-4A and eIF-4F (Jaramillo *et al.*, 1990; also see review by Gallie, 1993). The main purpose of these three simultaneously involved factors is to dissolve secondary structures present in the 5′ untranslated leader sequence of the mRNA transcript. The mRNA is now ready to associate with the 40S ribosomal subunit. The transcript is then scanned from 5′ to 3′ for the initiation codon, AUG. The 40S ribosomal subunit does not work alone in the scanning process. Scanning also involves a multi-subunit factor eIF-2 and is assisted by the immediate surrounding sequence context of the initiation codon. The eIF-2 factor helps locate the initiation codon (Dasso *et al.*, 1990).

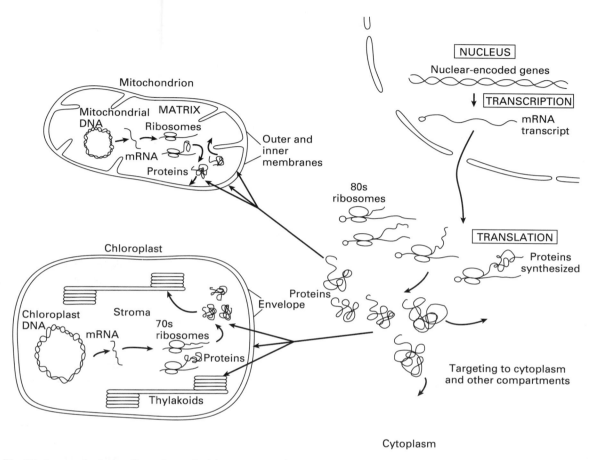

Fig. 2.1 A general scheme of protein synthesizing systems and compartments within a plant cell.

Regulating initiation efficiency

The efficiency of recognizing transcripts, hence the efficiency of translation, can be affected significantly not only by the presence of eIF factors but also by the sequences in the 5′ untranslated leader region and sequences around the initiation codon. Surveys of plant nuclear genes show that the 5′ leader sequence can range from 9 to 193 bases in length (Joshi, 1987). The most common lengths are 40–80 bases long. One obvious feature of these 5′ leader sequences is their richness in AU content. The percentage of AU content can go from 51% to greater than 70%. There is presently evidence that alterations in the 5′ leader sequence, either through length changes or alternate splicing routes, can greatly affect efficiency and function (see review by Gallie, 1993). This raises the possibility that such transcript altering events may have regulatory implications.

The elongation process

Once the appropriate initiation codon is recognized and engaged, the 40S ribosomal subunit-eIFs-mRNA initiation complex enters the next phase of translation, elongation. Although initiation can arguably be the most important determinant of translational efficiency, elongation can encounter problems that may affect the rate of translation as

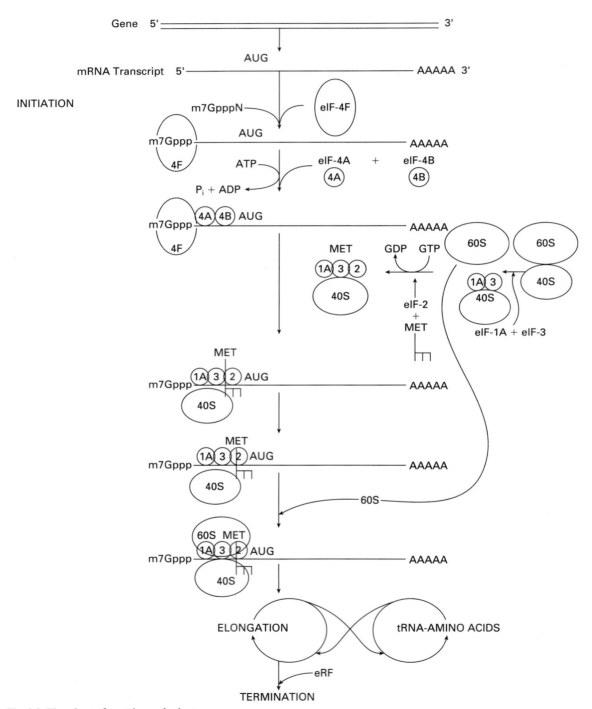

Fig. 2.2 Flowchart of protein synthesis steps.

well. The presence of rare codons (codons that are not used very often in plants) or secondary structures in the downstream sequences of the RNA transcript may cause pauses or stacking of the translation machinery. The pausing and stacking of ribosomes has been observed *in vitro* but has not been reported *in vivo*. Although very little is known about the elongation process in plants, this step is very similar to mammalian systems. To commence the elongation stage, the initiation complex is required to join up to the 60S large ribosomal subunit to form an 80S complex. The elongation phase is primarily a cyclic event, adding one amino acid at a time until the stop codon or end of the protein is reached. In mammalian systems this cyclic process is facilitated by four elongation factors termed eEFs (see review by Hershey, 1991). The four main steps are: (1) binding of aminoacyl-tRNA to the aminoacyl or 'A' site of the ribosome, a process catalyzed by eEF-1α; (2) GTP hydrolysis and ejection of eEF-1α-GTP; (3) formation of the peptide bond between the adjacent amino acid residues, a reaction catalyzed by peptidyl transferase, a component located in the 60S ribosomal subunit; and (4) translocation to the peptidyl or 'P' site with the aid of eEF-2. The A and P sites of the ribosome are named as such to differentiate the entry site of amino acid carrying tRNAs (also known as aminoacyl tRNAs) and the polypeptide carrying tRNA (or also called peptidyl tRNA), respectively. The general scheme of this elongation process is very similar to that in the prokaryotic system; therefore more detail can be found in genetics and biochemistry books. Analogues for eEF-1α have been characterized for a number of plant species and different plant tissues. Therefore it is reasonable to predict that other factors involved in the elongation process will most likely exist in plants and possess similar modes of function. The synthesis of a protein is an energy expensive endeavour requiring at least three GTP molecules per cycle. One high energy containing bond is cleaved at both the binding and translocation steps and two additional GTP molecules are utilized to synthesize the aminoacyl-tRNA that is consumed in each cycle. Although it sounds laborious and cumbersome, elongation is actually quite rapid. An 80S ribosome incorporates up to six amino acids per second and this occurs with high fidelity. The duration of the elongation phase of course depends entirely on the molecular size of the protein being synthesized.

Terminating translation

In comparison to the initiation and elongation stages, the termination phase is relatively straightforward. Termination occurs when the translational machinery encounters a nonsense or stop codon in the A site of the ribosome. A release factor termed eRF in the mammalian system promotes the cleavage of the completed peptidyl-tRNA, thereby releasing the finished protein. The same process most likely functions in plants.

The most frequently encountered stop codon is UGA, occurring in approximately 46% of the 748 plant genes surveyed. The other two stop codons, UAA and UAG, appear in approximately 28% and 26% of the genes, respectively. It is interesting to note that dicotyledonous plants prefer UAA (46%) over UGA (36%) or UAG (18%) (Angenon et al., 1990). The sequences following or downstream of the stop codon may also contribute to the termination process. The 3' downstream untranslated region tends to be AU rich. One of the most noticeable sequences at the 3' end of the mRNA is the poly(A)-tail. The poly(A)-tail is believed to promote transcript stability and may also affect translational efficiency (Gillham et al., 1994). The mechanism for affecting translational efficiency is not known at present.

Post-transcriptional control of gene expression

The synthesis of proteins, as you may already gather from some of the features discussed in the above paragraphs, can also be an opportunity for the plant cell to regulate and control the expression of a gene, by altering how the translational machinery makes a specific polypeptide product. There are many subtle means, some fairly cryptic, in which the control of gene expression can occur at the level of translation (see Fig. 2.3 for a summary of possible controlling mechanisms). Compartmentalization is one very effective physical way of controlling the process. Transcription and translation may occur in different compartments, e.g. the nucleus and the cytoplasm.

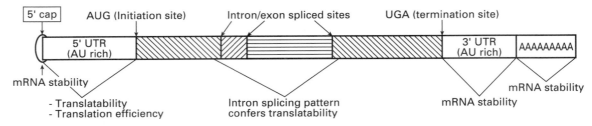

Fig. 2.3 Overview of translational signals that may contribute to the post-transcriptional control of gene expression.

Sequestering an mRNA in one place can physically prevent its translation. The translational process can also be effectively altered by the route in which non-translated intervening sequences (also called introns) in a transcript are manipulated or removed thus changing the resulting product. Whether a protein is synthesized or not may depend on the intron splicing pattern taken by the transcript. Removal of an intron generally confers translatability to the transcript and in some cases a totally different protein may even arise.

Even though there are ingenious ways of controlling translation, mRNA turnover remains the primary and best known regulatory mechanism for controlling gene expression at the protein synthesis level. Features on the transcript itself such as the 5' cap structure, 5' leader sequences and 3' poly(A)-tails, all make a major contribution to this regulatory mechanism by affecting translational efficiency and transcript stability (Yamamoto *et al.*, 1995; Caspar and Quail, 1993). These 'cis-acting' elements are, of course, not restricted to these examples but can be present elsewhere along the RNA transcript. The precise identity of these *cis*-acting elements is dependent very much on the gene/transcript in question.

The regulation of specific RNA transcript turnover is not affected by the *cis*-acting elements alone but must also be facilitated by appropriate *trans*-acting factors that are in turn controlled by the regulatory program in effect (see review by Gillham *et al.*, 1994). Regulation can be under developmental control such as during embryogenesis and germination, light, hormones or aging (see recent review by Gallie, 1993). The turnover of mRNA can also be influenced by temporary regulation such as that encountered during heat shock, hypoxia, wounding, water and nutrient starvation. Examples of mRNA turnover mechanisms in action can be found in a number of seed storage protein genes. The transcription of some seed as well as some non-seed genes in soybean embryos occur at comparable rates but display a 100- to 10,000-fold difference in steady state mRNA levels. In developing barley endosperm or pea cotyledons, some seed protein messages, especially the major ones, are preferentially more stable than other minor forms. In addition to controlling gene expression at the level of translation, the turnover of transcripts may possibly have a physiological importance such as in the case of phosphate limitation (Bariola *et al.*, 1994) and senescence in *Arabiodopsis thaliana*. Specialized RNAses are induced during senescence and phosphate starvation to mobilize inorganic phosphate from RNA. Although this phenomenon appears to lack transcript specificity it is activated under special conditions and other genes activated specifically during these conditions appear to be relatively more stable.

Once the protein is synthesized and released it is theoretically available for structuring, assembly, use, targeting or further modifications. Some of these modifications may also be necessary for activating the protein in the correct place and at the correct time. Many of these aspects are covered to some degree in other chapters.

Prokaryotic protein synthesis in plastids and mitochondria

Although the majority of the proteins in a plant cell are made by the cytoplasmic eukaryotic protein synthesis system, there are two additional compartments that also possess their own protein synthesis machinery. The intracellular compartments that contain their own protein synthesis systems are the plastids and the mitochondria. Both of these organelles are enclosed and delineated from the cytoplasmic contents by selectively permeable double membranes; therefore the three protein synthesizing systems are functionally segregated from one another and operate independently. The plastidic and mitochondrial translational machineries are also present in multiple units, in the average range of 50–100 for plastids and hundreds or perhaps even thousands for mitochondria, so that collectively these compartments occupy a substantial amount of the total cell's protein synthesizing capacity. Although these compartments are totally capable of synthesizing proteins, their role in the cell is to control the production of proteins encoded internally by the plastid chromosome or the mitochondrial genome. The genetics of the organellar genomes are dealt with in Chapters 12 and 18. Since the organellar genomes are small in comparison to the nuclear genome, the number and variety of internally synthesized protein products is limited. Approximately 100 or so products are made from plastid DNA-encoded genes and about 30–50 polypeptides are made from mitochondrial genes (Shinozaki *et al.*, 1986; Newton, 1988). Although the diversity is not like nuclear-encoded proteins, the number and types of internally-made products is significant.

The organellar protein synthesizing machinery

The organelles' ability to independently synthesize proteins, however, does not mean that they are completely autonomous entities. It is perhaps a bit inconsistent that even though internally they synthesize a fair number of products, a large percentage of the components that form the translational machinery is encoded externally by the nuclear genome. These components are made in the cytoplasm and are then targeted to the organelle where they are processed and assembled into their functional sites. Therefore, in this respect, these organellar compartments depend on the cytoplasmic protein synthesizing system in order for the

compartments themselves to have the capability to make their own products. Plastids require import of some translational factors, and some ribosomal proteins for both the 50S and 30S subunits, in order for the compartment to complete its translational machinery (see recent review by Gruissem and Tonkyn, 1993). In addition to what plastids require, mitochondria must also acquire from the nuclear genome a set of at least eight transfer RNAs. The interlinking of all three translation systems has great implications on the way the expression of proteins is controlled and coordinated. The interaction is also very important for communicating the state and needs of each compartment in relation to the state of the plant.

The protein synthesizing process in the two organelles is mechanistically more similar to the prokaryotic system than the eukaryotic system. The bacterial translational process is described in detail in many genetics textbooks and will not be reiterated here. The overall scheme of prokaryotic protein synthesis is similar to that described for the cytoplasm in the preceding section. In fact our knowledge of eukaryotic systems is based on bacterial systems. The plastid is perhaps more similar to the bacterial translational machinery than to mitochondria. Many of the components in plastids, for example, ribosomal proteins or elongation factors, are very similar to bacterial or prokaryotic counterparts. They are so similar that many plastid genome-encoded proteins can be synthesized using bacterial translation systems. Plastidic polypeptides can be synthesized successfully and quite efficiently in *in vitro* bacterial extracts or in living cells. The efficiency of translating plastidic proteins in bacterial extracts is also due in large part to the prokaryotic-like signals in the gene sequence, for example, promoters and Shine–Delgarno sequences. The ability to use bacteria for making proteins has been a very useful feature for researchers to study plastid genes and their protein products (Boyer and Mullet, 1986).

The mitochondrial translational machinery

The prokaryotic-type of translational system in mitochondria is slightly different from the one found in plastids (Douce, 1985). The difference is not so much in how the proteins are made but in the components of the translational machinery. For example, the ribosomal RNAs in the ribosomes are slightly different in size from the prokaryotic system. The 26S and 18S rRNAs in mitochondrial ribosomes are equivalent counterparts of the 23S and 16S rRNAs found in bacterial ribosomes. Another difference is the origin of ribosomal proteins. In mitochondria, none of the ribosomal proteins are encoded internally by the organellar chromosome. Nevertheless the translational process is similar to plastids.

Both organellar translational machineries can also be under the influence of *cis*- and *trans*-elements such as that controlling translation in the eukaryotic system. Organellar systems can also control gene expression through translation much in the same way as in the cytoplasm; therefore this aspect will not be reiterated here. Reviews on translational and post-transcriptional control in plant organelles have been written recently by Gruissem and Tonkyn (1993) and Gillham *et al.* (1994).

Simulating protein synthesis in a test tube

The protein synthesizing activity of the cytoplasm and the organelles can be reconstituted effectively and with relative ease in a test tube. This capability has been a very important tool for researchers, whether they are studying the translational process or are making proteins to answer other biological questions. The cytoplasmic translational process can be carried out simply by using cellular extracts. A commonly used plant tissue for making this cytoplasmic extract is wheat germ (Anderson *et al.*, 1983). The procedure is fairly straightforward and quick. Briefly, fresh wheat germ is finely ground into a simple buffer solution and passed over an ion exchange column to remove small inhibitory products. The eluted extract contains all of the components necessary for synthesizing proteins and for charging the tRNAs with the correct amino acid residue. The extracts are supplemented with energy in the form of ATP and GTP, and with all of the twenty amino acids. The protein being synthesized depends entirely on the transcript added into the reaction. Therefore a researcher can control what polypeptide is being

made using the *in vitro* wheat germ translation system. One can also radiolabel proteins simply by substituting an amino acid with a radiolabeled one.

Protein synthesizing activity can also be carried out *in organello* (Nivison *et al.*, 1986). This is done by isolating the organelle from the plant cells and feeding the intact organelle the twenty amino acids and energy. In order to follow what is happening, one of the amino acids is radiolabeled. This, however, does not allow the researcher to control what is being expressed. In other words the proteins made *in organello* are the ones with transcripts currently inside the organelle. Since many organellar proteins can be made in bacteria, extracts from bacteria or even living bacteria cells can be used to control the synthesis of the desired protein in the same way as with wheat germ extracts (Boyer and Mullet, 1986).

The knowledge of how proteins are made and the technological capability of making proteins, whether it is *in vitro* or *in vivo*, are important steps to understanding the functions of different proteins and the roles they play in the plant cell. The importance of this knowledge is evident in the central role played by proteins in plant metabolism and physiology. This knowledge is also extremely valuable for plant genetic engineers in the design of applications in plant biotechnology.

References and further reading

Abramson, R. D., Browning, K. S., Dever, T. E., Lawson, T. G., Thach, R. E., Ravel, J. M. and Merrick, W. C. (1988). Initiation factors that bind mRNA: a comparison of mammalian factors with wheat germ factors. *J. Biol. Chem.* **263**, 5462–7.

Angenon, G., Van Montagu, M. and Depicker, A. (1990). Analysis of the stop codon context in plant nuclear gene. *FEBS Let.* **271**, 144–6.

Anderson, C. W., Straus, J. W. and Dudock, B. S. (1983). Preparation of a cell-free protein-synthesizing system from wheat germ. In R. Wu, L. Grossman and K. Moldave (eds), *Methods Enzymol*, **101**, 635–44.

Bariola, P. A., Howard, C. J., Taylor, C. B., Verburg, M. T., Jaglan, V. D. and Green, P. J. (1994). The *Arabidopsis* ribonuclease gene *RNS1* is tightly controlled in response to phosphate limitation. *Plant J.* **6**, 673–86.

Boyer, S. K. and Mullet, J. E. (1986). Characterization of *P. sativum* chloroplast *psbA* transcripts produced *in vivo*, *in vitro* and in *E. coli*. *Plant Mol. Biol.* **6**, 229–43.

Caspar T. and Quail, P. H. (1993). Promoter and leader

regions involved in the expression of the *Arabidopsis* ferredoxin A gene. *Plant J.* **3**, 161–74.

Dasso, M. C., Milburn, S. C., Hershey, J. W. B. and Jackson, R. J. (1990). Selection of the 5′-proximal translation initiation site is influenced by mRNA and eIF-2 concentrations. *Eur. J. Biochem.* **187**, 361–71.

Douce, R. (1985). *Mitochondria in Higher Plants*, Academic Press, London.

Gallie, D. R. (1993). Post transcriptional regulation of gene expression in plants. *Ann. Rev. Plant Physiol. Plant Mol. Biol.* **44**, 77–105.

Gillham, N. W., Boynton, J. E. and Hauser, C. R. (1994). Translational regulation of gene expression in chloroplasts and mitochondria. *Ann. Rev. Genetics* **28**, 71–93.

Gruissem, W. and Tonkyn, J. C. (1993). Control mechanisms of plastid gene expression, *Crit. Rev. Plant Sci.* **12**, 19–55.

Hershey, J. W. B. (1991). Translational control in mammalian cells. *Ann. Rev. Biochem.* **60**, 717–55.

Jaramillo, M., Browning, K., Dever, T. E., Blum, S., Trachsel, H., Merrick, W. C., Ravel, J. M. and Sonenberg, N. (1990). Translation initiation factors that function as RNA helicases from mammals, plants and yeast. *Biochim. Biophys. Acta* **1050**, 134–9.

Joshi, C. P. (1987). An inspection of the domain between putative TATA box and translation start site in 79 plant genes. *Nucl. Acids Res.* **16**, 6643–53.

Knauf, V. (1995). Transgenic approaches for obtaining new products from plants. *Curr. Biol.* **6**, 165–70.

Newton, K. J. (1988). Plant mitochondrial genomes: Organization, expression and variation. *Ann. Rev. Plant Physiol.* **38**, 503–32.

Nivison, H. T., Fish, L. E. and Jagendorf, A. T. (1986). Translation by isolated pea chloroplasts. In A. Weissbach, H. Weissbach (eds), *Methods Enzymol.* **118**, 282–95.

Shinozaki, K., Ohme, M., Tanaka, M., Wakasugi, T., Hayashida, N., Matsubayashi, T., Zaita, N., Chunwongse, J., Obokata, J., Yamaguchi-Shinozaki, K., Ohto, C., Torazawa, K., Meng, B. Y., Sugita, M., Deno, H., Kamogasiwa, T., Yamada, K., Kusuda, J., Takaiva, F., Kato, A., Tohdoh, N., Shimada, and Suguira, M. (1986). The complete nucleotide sequence of the tobacco chloroplaast genome: its gene organization and expression. *EMBO J.* **5**, 2043–9.

Yamamoto, Y. Y., Tsuji, H. and Obokata, J. (1995). 5′-leader of a photosystem I gene in *Nicotiana sylvestris*, *psaDb*, contains a translational enhancer. *J. Biol. Chem.* **270**, 12466–70.

3 Protein degradation

Peggy M. Hatfield and Richard D. Vierstra

Introduction

Protein turnover can be defined as the flow of amino acids from existing protein into newly synthesized protein. As such, the extent and rate of protein turnover is controlled by both synthetic and degradative processes. Newly synthesized proteins may be similar or different to those present before; the latter underscores the critical role turnover has in altering cellular constituents during the life of a plant cell. In many cases, these alterations of protein content are essential for normal development and for appropriate responses to environmental stimuli.

Most proteins have a lifetime that is less than that of the cell, and so are degraded and if necessary, resynthesized. In leaves of actively growing duckweed (*Lemna minor*), total soluble protein turns over every 3–7 days (Davies, 1982). This observation implies that each week, plant cells degrade and resynthesize most of their proteins. It also shows that the metabolic flow of nitrogen and carbon within proteins is both cyclic and continuous. Although most proteins, excluding seed and leaf storage proteins, cannot be viewed simply as storage forms of nitrogen, they do constitute an important nitrogen source. In fact, during times of starvation, the rate of protein turnover is accelerated by increasing the rate of degradation relative to synthesis; this switch generates a pool of free amino acids from less essential proteins that can be used to assemble more essential proteins.

The intent of this chapter is to focus on the role of degradation in controlling the protein composition of various cellular compartments in plants. Functions of protein breakdown will be examined, general principles governing proteolysis and its specificity will be introduced, and examples of the degradative

mechanisms utilized by plants presented. As will be seen, protein degradation is a complex process, involving a multitude of pathways to select, mark, and break down target molecules. From analysis of the proteolytic pathways identified to date in plants, it is clear that the level of complexity required to degrade proteins may rival that required to initially synthesize them.

Functions of protein degradation

Superficially, protein degradation appears to be a wasteful event. Why expend enormous amounts of energy synthesizing a protein, only to degrade it 5 minutes later? However, the observation that many intracellular proteins have half-lives less than one hour implies that cells expend this energy for important reasons. One of the most central functions of proteolysis is to help regulate metabolism. In conjunction with protein synthesis, degradation is essential for maintaining correct enzyme levels and for modulating these levels based on internal and external cues. Indeed, rate-limiting enzymes in many metabolic pathways invariably have fast degradation rates (Goldberg and St John, 1976; Rogers *et al.*, 1986; Vierstra, 1993). Examples include the enzymes ACC-synthase, nitrate reductase, ATP sulfurylase, HMG-CoA reductase, and ornithine decarboxylase, which catalyze crucial steps in ethylene biosynthesis, nitrogen and sulfur assimilation, and sterol and spermine biosynthesis, respectively (Table 3.1). For enzymes that assemble into multisubunit complexes, proteolysis helps maintain the correct balance of subunits by removing those that are in excess. For example, chloroplasts efficiently degrade the free large

Table 3.1   Half lives of proteins from plants and animal sources

Protein	Source	Half-Life (h)	Reference
Plant			
Apoplastocyanin ($- Cu^{2+}$)	*Chlamydomonas*	0.3	Vierstra, 1993
ACC synthase	Tomato	1	Kim and Yang, 1992
NADPH-glutamate dehydrogenase ($- NH_3$)	Chlorella	1	Vierstra, 1993
Phytochrome A (Pfr)	Monocots/dicots	1–2	Vierstra, 1994
Nitrate reductase ($+ NH_3$)	Tobacco, corn	1.6–6	Vierstra, 1993
Chloroplast 32-kDa D1 Protein (high light intensities)	Spirodela	~2	Mattoo *et al.*, 1984
NADPH-protochlorophyllide oxidoreductase (+ light)	Barley	~2	Vierstra, 1993
Squalene synthetase (+ elicitor)	Tobacco	2–4	Vierstra, 1993
Phytochrome A (Pr)	Monocots/dicots	>100	Vierstra, 1994
RUBISCO	Corn	~150	Davies, 1982
Animal			
Gα	Yeast	0.1	Madura and Varshavsky, 1994
αMAT2 repressor	Yeast	0.1	Rechsteiner, 1991
Cyclins A and B	Frog, clam, yeast	>0.1	Hershko and Ciechanover, 1992
Ornithine decarboxylase	Mouse	0.2–1	Rechsteiner, 1991
p53 oncoprotein	Human	1	Hulbregtse *et al.*, 1993
c-jun oncoprotein	Mouse	1.5	Treier *et al.*, 1994
Histone H2A, H2B	Rat	>200	Rechsteiner, 1987

subunit of RUBISCO when it is made in excess of that needed to assemble with the small subunit, and vice versa (Schmidt and Miskind, 1983).

Proteolysis can also control the levels of regulatory proteins whose abundance will determine a particular developmental response. Many types of receptors, cell-cycle regulators, homeotic proteins, signal transduction components (protein kinases, G-proteins, etc.), and transcription factors are included in this group (Table 3.1). Two of the best examples are cyclins and phytochrome A. Cyclins are a family of eukaryotic cell-cycle regulators whose concentrations oscillate during the mitotic cycle. They are synthesized constitutively and accumulate prior to mitosis during G1, S, and G2 phases. When a critical threshold level is surpassed, cyclins initiate mitosis by binding to and activating the cdc2 protein kinase. Passing this threshold also triggers the rapid degradation of cyclins within the active complex during the ensuing metaphase (Glotzner *et al.*, 1991). This loss of cyclins then reduces the pool of active cdc2 kinase thereby preventing entry of the daughter cells into another round of mitosis. Once

most cyclins are removed, degradation is repressed and cyclin levels rise again by constitutive synthesis. Increased cyclin levels then begin another round of the cell cycle.

Phytochromes are a family of photoreceptors involved in plant morphogenesis. They exist in two photointerconvertible forms: a red light-absorbing form Pr, and a far-red light-absorbing form Pfr. They are initially synthesized as Pr, which is biologically inactive in most photoresponses, and convert to the biologically active Pfr form only following irradiation with red light. One form of phytochrome, designated phytochrome A, which is abundant in germinating seedlings, is unique among the phytochrome family in that it is stable as Pr (half life > 100 hours) but is rapidly degraded as Pfr ($t/2 \sim 1$ hour) (Vierstra, 1994). Consequently, young seedlings accumulate high levels of phytochrome A as Pr in the dark but lose most of it once sufficient light converts Pr to Pfr. This differential degradation is important for the young plants by allowing the germinating seedling to first be highly sensitive to light when searching for the soil surface and then to be less sensitive as it emerges from the soil where light is no longer limiting.

Apart from its role in regulation, proteolysis also serves important housekeeping functions. Abnormal proteins continually appear in cells by a variety of mechanisms, including mutations, biosynthetic errors, spontaneous denaturation, and free radical-induced damage (Vierstra, 1993). Stressful conditions such as high and low temperatures, dehydration, heavy metals, high light intensities, and starvation can accelerate such damage. These errant proteins could eventually accumulate to toxic levels without mechanisms for their removal. This possibility is especially true for slow-growing plant cells that cannot use cell division to dilute out debris. To deal with defective proteins, plants, like other organisms, have developed specific mechanisms not only for repairing and refolding damaged proteins, but also for degrading those beyond salvage (Hershko and Ciechanover, 1992; Gatenby and Viitanen, 1994). The chloroplast 32-kDa D1 protein, that comprises the central core of the photosystem II reaction complex, is one such example. Under high light stress, chloroplasts continually remove damaged D1 and replace it with functional counterparts (Mattoo et al., 1984). Degradation of abnormal and normal proteins also can provide an important source of amino acids necessary to make new proteins. This is especially evident during seed germination when a specialized group of seed storage proteins, synthesized during seed maturation, are catabolized to provide a pool of amino acids indispensable for growth of the emerging seedling (Fincher, 1989).

Finally, proteolysis is essential for the process of programmed cell death. One of the natural consequences in the development of multicellular organisms is the timed disintegration of specific cells (Elis et al., 1991). In plants, it can either occur on a massive scale (leaf and flower senescence, fruit ripening, xylem and periderm maturation in wood) or it can be confined to individual cells (petiole abscission zones, tapetal and stomial cell layers in developing anthers) (Greenberg and Ausubel, 1993; Hensel et al., 1993; Vierstra, 1993). Cell death process is a highly organized sequence of events with proteolysis serving to recover protein nitrogen and carbon before the cell completely decays. This economy is most evident during leaf and flower senescence where up to 70% of the protein may be retrieved (Davies, 1982; Staswick, 1994).

General features of proteolysis

At first glance, protein breakdown appears to simply involve digestion of a protein by a set of proteases. However, three generalized features of proteolysis indicate that the process is more complex. First, proteolysis is remarkable selective (Goldberg and St John, 1976; Rogers et al., 1986; Rechsteiner, 1991; Hershko and Ciechanover, 1992). Even within the same cellular milieu, specific proteins can have half-lives ranging from minutes to weeks. Therefore, the proteolytic machinery must have mechanisms to help discriminate those proteins that should be degraded from those that should not. Because most purified proteases are not substrate specific, these mechanisms must control or sequester proteases to prevent random digestion of all proteins.

The second feature of protein breakdown is that all intracellular proteolysis requires energy, usually in the form of ATP (Goldberg, 1992; Hershko and Ciechanover, 1992). This requirement was initially surprising given that cleavage of peptide bonds is an exergonic reaction and that the first proteases purified did not require energy (Goldberg and St John, 1976). It is now apparent that this energy dependence is useful for controlling the rate and specificity of proteolysis. The third feature is that the complete disassembly of a protein is fast, so fast that partial breakdown products, expected to appear in vivo during degradation, are rarely observed (Goldberg and St John, 1976; Hershko and Ciechanover, 1992). This rapid catabolism suggests that substrate recognition is the rate-limiting step in proteolysis and that all ensuing cleavages occur shortly thereafter. The failure to identify partial breakdown products has been a major impediment in our abilities to understand how proteins are processively cleaved into amino acids and to identify the proteases that are involved.

Proteolytic pathways within plants

Plants contain a variety of proteases, including some that are commercially available, such as papain (papaya), ficin (fig), and bromelain (pineapple) (Davies, 1982: Matile, 1982; Storey, 1986). Proteases are generally characterized as either endopeptidases, which cleave internal peptide bonds and thus generate a range of fragments, or exopeptidases, which cleave peptide bonds processively from either the C-terminus (carboxypeptidases) or N-terminus (aminopeptidases), thus liberating free amino acids. Proteases have been found within most cellular compartments, including the vacuole, cytoplasm, nucleus, chloroplast, and microbody (Barrett, 1986). Nevertheless, the specific *in vivo* functions of most plant proteases are unknown beyond this simple classification.

Vacuolar proteases

The vacuole contains many of the well characterized proteases (Boller, 1986; Boller and Wiemken, 1986; Mikola and Mikola, 1986; Storey, 1986). Most function optimally *in vitro* under acidic conditions (pH3 to 6) (Ryan, 1973), conditions that exist within the vacuole *in vivo* (Roberts *et al.*, 1980). Because these enzymes account for the bulk of proteolytic activities in crude plant extracts, it was proposed that the vacuole is the 'lytic compartment of the plant cell', much like the animal lysosome. Accordingly, the vacuole was suggested to be responsible for breaking down many cellular constituents, including proteins from the cytoplasm and other organelles (Matile, 1982). However, it is now evident that vacuolar targeting is not a prerequisite for the turnover of proteins from the cytoplasm and other organelles and that most, if not all, cellular compartments have their own proteolytic machinery (Goldberg and St John, 1976; Matile, 1982; Liu and Jagendorf, 1984; 1986; Malek *et al.*, 1984; Boller and Wiemken, 1986; Vierstra, 1993). Thus, the exact functions of vacuolar proteases remain undefined, although a few roles have been postulated. Vacuolar hydrolases (including proteases) may serve a role in the defense against pathogens, parasites, and herbivores by attacking the invader once the plant cell is lysed (Boller, 1986). They may function during times of starvation or stress to generate a supply of free amino acids or to remove abnormal proteins that accumulate. This degradation would necessarily involve the active transport of substrate proteins into the vacuole from other cellular compartments (Boller and Wiemken, 1986; Canut *et al.*, 1986; Staswick, 1994). Finally, vacuolar proteases may act during the final stages of senescence by degrading any remaining cytoplasmic

or organellar substrates after rupture of the vacuolar membrane (Boller, 1986).

Vacuoles may also function in the storage and mobilization of protein reserves during seed germination. During seed maturation, specific storage proteins are synthesized on rough endoplasmic reticulum and subsequently transported (sometimes via the Golgi apparatus) into membrane-bound vesicles called protein bodies (Boller and Wiemken, 1986; Harris, 1986). The origin of the membrane surrounding the protein body is uncertain in monocots, but in dicot seeds it likely arises from fragmentation of the central vacuole (Harris, 1986). Proteins within these bodies are not degraded until germination, at which time specific proteases are synthesized *de novo* and transported into the vesicles to initiate degradation (Davies, 1982). In legume seeds, protein bodies are located in living parenchyma tissues of the cotyledons, from which release of amino acids can be regulated. Conversely, in cereal grains, protein bodies are found primarily in non-living tissues of the subaleurone and outer layers of the endosperm. Proteases are transported into these tissues, whereupon free amino acids generated from digestion of the storage proteins are released without much control (Fincher, 1989).

Cytoplasmic and nuclear proteases

Various plant tissues contain proteolytic activities that display optimal activity *in vitro* under a higher pH than is present in the vacuole, suggesting that these may operate in the more basic environments of the cytoplasm, nucleus, or chloroplast (Ryan, 1973; Matile, 1982). One mechanism resulting in the degradation of many cytoplasmic and nuclear proteins utilizes the small protein ubiquitin (Hershko and Ciechanover, 1992; Vierstra, 1993). This mode of proteolysis is unique in that ubiquitin mediates protein degradation by becoming covalently conjugated to various proteolytic substrates in a way that tags them for destruction.

Ubiquitin is a compact, globular protein of 76 amino acids present in all eukaryotes (Hershko and Ciechanover, 1992; Vierstra, 1993). The structure of ubiquitin is crucial for its function, as suggested by its highly conserved amino acid sequence. The molecule consists of a globular N-terminal domain and a

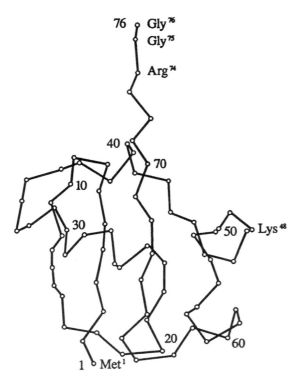

Fig. 3.1 The three-dimensional structure of plant ubiquitin. Amino acids noted are the amino-terminal methionine, the three carboxy-terminal residues, Arg–Gly–Gly, and Lys at position 48 which serves as the binding site of additional ubiquitins during the formation of multi-ubiquitin chains.

flexible, protruding carboxy terminus (Fig. 3.1). Extensive hydrogen bonding within the globular domain makes ubiquitin highly stable to heat, cold, extreme pH conditions, strong denaturants, and most proteases. Through the C-terminal extension, ubiquitin becomes attached to target proteins; removal or alteration of the final two residues (both glycines) completely inactivates the molecule.

The ubiquitin pathway was discovered in rabbits and has since been characterized in other animals, as well as fungi and plants (Fig. 3.2). Ligation of ubiquitin to target proteins requires ATP and is accomplished by the sequential action of two, and sometimes three, enzymes, designated E1, E2, and E3, that each exist as a family of isozymes (Hershko and Ciechanover, 1992; Vierstra, 1993). In the first step of the pathway, E1 (or ubiquitin-activating enzyme)

directs the formation of an high energy, thiol–ester linkage between a cysteine residue of E1 and the carboxyl group of the terminal glycine of ubiquitin, using ATP as the energy source. This activated ubiquitin is donated to a ubiquitin-conjugating enzyme, or E2, again forming an high energy, thiol–ester bond. E2s then promote the conjugation of ubiquitin to the substrate protein, generating an isopeptide bond between the carboxy group of the C-terminal glycine of ubiquitin and the primary amino group on lysine side chains of the target protein. Some E2s can conjugate the target protein alone, but in most cases, the assistance of an E3 (or ubiquitin protein ligase) is required for recognition and ubiquitin transfer (Hershko and Ciechanover, 1992; Scheffner *et al.*, 1995). The different E2 and E3 isozymes present in eukaryotes are responsible, in part, for targeting the vast array of substrates subjected to conjugation (Hershko and Ciechanover, 1992; Vierstra, 1993). A target protein may be modified with one or more ubiquitin molecules; multiple ubiquitins may be attached to different lysine residues within the target, or to other ubiquitin molecules already bound, resulting in a chain of ubiquitins ligated to the protein.

Once a protein is modified with ubiquitin, it is subject to one of three possible fates (Fig. 3.2). Firstly, the protein may be degraded by an ATP-dependent, protease complex called the 26S proteasome. The complex rapidly degrades the conjugated protein into its constituent amino acids and releases ubiquitin intact. In this way, ubiquitin serves as a reusable signal for selective proteolysis. 26S proteasome is a large (1500 kDa), multisubunit complex with numerous protease activities that cleaves peptide bonds on the carboxy side of basic, hydrophobic, or acidic amino acids, yielding small peptides that are eventually hydrolyzed to amino acids by a host of ATP-independent proteases (Hershko and Ciechanover, 1992; Goldberg, 1992; Rechsteiner *et al.*, 1993). Apparently, only those proteins modified with chains of ubiquitins are substrates for degradation. Secondly, ubiquitin conjugates may be disassembled by ubiquitin-protein hydrolases (or isopeptidases) that cleave the peptide bond between ubiquitin and the target protein, liberating both proteins intact (Hershko and Ciechanover, 1992). These enzymes act to release inappropriately targeted substrates, and to liberate ubiquitin from peptides generated during substrate degradation. And thirdly, some proteins modified with a single ubiquitin molecule may be resistant both to degradation by 26S proteasome and to disassembly

by hydrolases. The function of these stable conjugates is not yet known.

Although the specific functions of ubiquitin are still largely unknown in plants, the protein is present in all cell types examined. Enhanced expression of genes

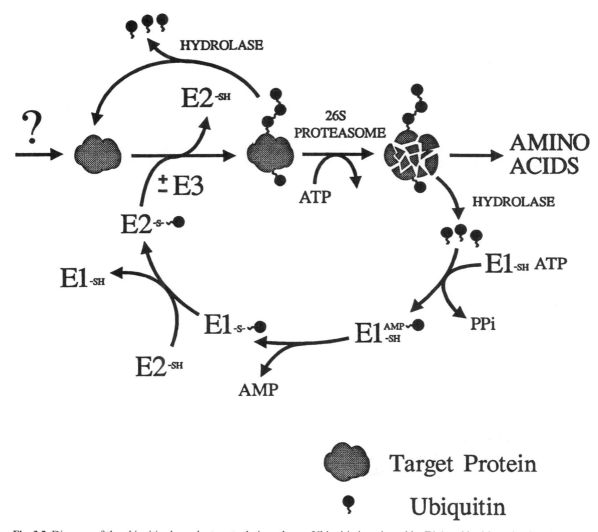

Fig. 3.2 Diagram of the ubiquitin-dependent proteolytic pathway. Ubiquitin is activated by E1 (or ubiquitin activating enzyme); this first involves adenylation of the carboxy-terminal glycine and then binding of the ubiquitin to E1 via a high energy thiol–ester bond between the glycine and a specific cysteine within E1. The bound ubiquitin is transferred to an active-site cysteine in one of many E2s (or ubiquitin conjugating enzymes) via transesterification. Ubiquitin is then attached to the target protein via an isopeptide bond. This transfer can occur with or without an E3 (or ubiquitin protein ligase). The ubiquitin–protein conjugate is either (1) degraded by the ATP-dependent 26S proteasome, (2) disassembled by ubiquitin–protein hydrolases generating free ubiquitin and target protein, or (3) remains stable. During target protein degradation, ubiquitin–protein hydrolases also remove ubiquitin from proteolytic fragments, allowing free ubiquitin monomers to re-enter the pathway.

encoding ubiquitin and E2s in meristematic zones, during xylem maturation, during various stress responses, and during senescence implicate the pathway at several crucial times in development (Vierstra, 1993; Thoma *et al.*, 1996 and references within). Introduction of mutant forms of ubiquitin profoundly affect the development of tobacco, especially vascular tissue (Bachmair *et al.*, 1991). With regard to specific plant substrates, only phytochrome A has been identified as a target (Vierstra, 1994). It has been shown that photoconversion of Pr to Pfr induces the rapid degradation of the receptor concomitant with a transient accumulation of ubiquitin-Pfr conjugates.

Given the high conservation of the ubiquitin pathway among all eukaryotes, additional functions of the pathway in plants can be inferred from studies of other organisms (Hershko and Ciechanover, 1992; Varshavsky, 1992, Vierstra, 1993). Ubiquitin likely mediates the turnover of abnormal proteins and those that are mislocalized or improperly processed; thus it serves an important housekeeping role. Many normal, but short-lived proteins are also likely targets of ubiquitination. As many rate-limiting enzymes and key regulatory proteins have relatively short half-lives, ubiquitin would have a significant role in modulating plant cell metabolism. This is reflected in the growing list of cell regulators from animals and yeast that are degraded by the ubiquitin pathway, including mitotic cyclins, tumor suppressor p53, *mos* and *c-jun* oncogene products, the Gα protein – Gpa1, platelet-derived growth factor receptor, and the yeast αMAT2 mating-type repressor (Rechsteiner, 1991; Vierstra, 1993; Huibregtse *et al.*, 1993; Madura and Varshavsky, 1994; Treier *et al.*, 1994). The activation of ubiquitin and ubiquitin-pathway genes by stress has implicated the pathway in the general stress response of eukaryotes. Other functions controlled, at least in part, by the ubiquitin system include, DNA repair, peroxisomal biogenesis, cell surface recognition, and protein secretion (Hershko and Ciechanover, 1992; Weibel and Kunau, 1992).

Organelle proteases

Organelles in plants such as chloroplasts, mitochondria, microbodies, and endoplasmic reticulum must likewise contain mechanisms for eliminating proteins, but they apparently use proteolytic systems separate from the ubiquitin pathway (Goldberg, 1992). In mitochondria and chloroplasts, the proteases are most similar to those present in bacteria, consistent with the theory of their evolutionary origins. Proteolysis is especially important during the transition of etioplasts to chloroplasts and of microbodies from glyoxysomes to peroxisomes (Vierstra, 1993). During each of these transitions, selective degradation of existing enzymes and import of new enzymes are essential for conversion of the organelle. In most cases, it is likely that proteolysis within the organelle is dependent on the import of nucleus-encoded proteases. With the exception of a few from chloroplasts and one from mitochondria, the identity of these proteases is unknown.

Chloroplasts contain both ATP-independent and ATP-dependent proteolytic activities (Liu and Jagendorf, 1984, 1986; Malek *et al.*, 1984; Greenberg *et al.*, 1987). Whereas the digestion of transit peptides and other small peptide fragments is probably accomplished by ATP-independent proteases, the selective degradation of intact proteins appears to require energy-dependent proteolytic systems. A few of these proteases are encoded by the chloroplast genome but other essential proteins are encoded by the nuclear genome, synthesized in the cytoplasm, and imported into the organelle. Mutations affecting several of these nuclear genes are known to delay leaf senescence by preventing the breakdown of thylakoid membrane components (Stoddart and Thomas, 1982; Dalling and Nettleton, 1986).

Recently, a protease that contains subunits encoded by both the nuclear and chloroplast genomes has been identified. The protease is related to the ATP-dependent protease Clp, first identified from *Escherichia coli* (Gottesman *et al.*, 1990; Maurizi *et al.*, 1990). Clp is composed of two types of subunits, ClpP, which has serine protease activity, and ClpA, which hydrolyses ATP and regulates the activity of ClpP. Protein degradation requires the presence of both subunits and the hydrolysis of ATP to unfold the protein target prior to proteolysis (Goldberg, 1992). In plants, the ClpP subunit is encoded by the chloroplast genome, whereas the ClpA subunit is encoded by the nuclear genome. Thus, even though chloroplasts may contain the ClpP subunit, a functional protease will not be formed until the ClpA subunit is synthesized and transported into the chloroplast. In this way, the nucleus can tightly control chloroplast proteolysis by regulating the levels of Clp A. Additional regulation of ClpP activity may be accomplished by changing the identity of the ClpA subunit; recent evidence indicates that several types of ClpA-like proteins are present, each of which may recognize specific classes of substrates (Gottesman *et al.*, 1993; Kiyosue *et al.*, 1993). Though specific targets of the chloroplast ClpA/P have not been identified, its *E. coli* counterpart rapidly degrades abnormal proteins with aberrant N termini (Varshavksy, 1992; Goldberg, 1992).

Regulation of proteolysis

As mentioned above, individual proteins differ substantially in the rate at which they are degraded *in vivo* (Table 3.1). This poses the question as to how the proteolytic machinery within a single cell compartment is able to selectively recognize the myriad of proteins present, and degrade each with the correct half-life. The mechanisms used to identify various target proteins include: (1) recognition of specific amino acid sequences or inherent physico-chemical characteristics of the target protein; (2) modification of the target to enhance (or repress recognition; (3) *de novo* synthesis or activation of proteases, and/or (4) introduction of either the protease or the target into the cellular compartment containing the other. Several of these mechanisms may be used in combination for a given target.

Substrate characteristics

The characteristics of the substrate play a major role in its degradation. While it was originally thought that more global characteristics (e.g. molecular mass and pI) of the target protein were important, it now appears that less conspicuous properties are involved. This is best exemplified by phytochrome A where a subtle conformational difference between the Pr and Pfr forms is sufficient to alter the degradation of the protein by over 100 fold.

In several cases, even slight changes in amino acid sequence are sufficient to have pronounced effects on half-life. One of simplest determinants involves the amino-terminal residue. As formulated in the N-End Rule, Varshavsky and co-workers discovered that when certain amino acids are at the amino terminus (e.g. Met, Thr, Ser, Gly, Val), proteins are relatively stable, but when others are present (e.g. Lys, Arg, Glu, Asp, Gln, Asn), proteins are rapidly degraded (Varshavsky, 1992). The rule operates in bacteria, fungi, animals and plants, with prokaryotes using the ClpA/P protease and eukaryotes using the ubiquitin pathway for substrate recognition and degradation. In yeast, recognition of the destabilizing amino acids by the ubiquitin pathway is accomplished by a specific E2 working in concert with a specific E3 (Varshavsky, 1992).

Internal amino acids can also be essential for target recognition. In a few cases, small domains called destruction boxes have been identified empirically as degradation signals. For example, the destruction boxes within cyclin B and the *c-jun* oncoprotein consist of only 9 and 27 amino acids, respectively (Glotzner *et al.*, 1991; Treier *et al.*, 1994). Although there appears to be little sequence identity among the destruction boxes identified thus far, several can confer a short half-life on a recipient protein that is normally long lived.

Protein structure also is an important feature, especially for recognizing those proteins with abnormal conformations. The way incorrect folding patterns are detected has not yet been resolved; presumably, it involves general properties of the target such that most abnormal proteins can be identified regardless of their amino acid sequence. The most likely determinant is an increase in hydrophobic surfaces (Gatenby and Viitanen, 1994). In normal proteins, the folding process tends to bury hydrophobic amino acids leaving hydrophilic residues on the surface of the correctly folded proteins. Perturbations in structure or sequence can profoundly alter folding, leaving proteins with substantially more hydrophobic surfaces exposed. These in turn can destabilize the protein further and provide recognition sites for specific proteases.

Regulation of protease activity

Although proteases demonstrate narrow specificity with regard to the cleavage site (e.g. trypsin cleaves after Lys or Arg residues), they often display broad specificity with regard to the protein substrate. Consequently, proteases must be tightly regulated to insure that proteolysis is confined to appropriate substrates. Possible mechanisms include: (1) altering the concentration of the protease by changes in synthesis and/or degradation; (2) changes in protease activity or conversion of the protease from an inactive to an active form (e.g. zymogen activation); (3) changes in the specificity of the protease; (4) changes in the concentration of factors that control the protease (e.g. protease inhibitors); and (5) changes in compartmentation of the protease or substrate.

In many situations, one or more of these methods of regulation are employed. For the ubiquitin pathway, proteolysis is regulated at the level of substrate identification by E2s and E3s. This recognition can be adjusted by changes in the levels or activity of specific E2/E3 isoforms (e.g. oscillating degradation of cyclins) or by structural changes within the substrate (e.g. degradation of phytochrome A after conversion to Pfr) (Hershko and Ciechanover, 1992; Vierstra, 1994). Moreover, modification of substrates by ubiquitination then changes the affinity of the target for the 26S proteasome before catabolism. Changes in the environment of a protease could also affect its activity. For example, the shift in pH and reducing potential that occurs within chloroplasts during photosynthesis could concomitantly activate (or inactivate) chloroplast proteases much the same way these environmental changes alter the activity of enzymes involved in carbon fixation (Buchanan, 1991).

Protein breakdown during seed germination and leaf senescence is associated with increased activitiy of specific proteases. Several proteases are known to be synthesized *de novo* during leaf senescence before the senescence-associated symptoms become evident (Stoddart and Thomas, 1982; Dalling and Nettleton, 1986; Hensel *et al.*, 1993). In germinating mung bean seeds, for example, the major storage protein vicillin is sequestered from most proteases by deposition within protein bodies. Degradation is initiated during seed germination by *de novo* synthesis and import of several proteases into the protein body (Fincher, 1989). One of these proteases, vicillin peptidohydrolase, is very active against vicillin (Mikola and Mikola, 1986). In a few instances, an increase in the proteolytic activity associated with germination results from the activation of a zymogen. In lettuce seed germination, a protease was identified whose activity resulted from proteolytic processing of a larger inactive precursor (Ryan, 1973; Davies, 1982).

Finally, proteinase inhibitors may profoundly affect protein degradation. In tubers and seeds of various plant species, high concentrations of these inhibitors are compartmentalized within the vacuoles. The function of these vacuolar inhibitors is unknown. Whereas, most do not affect endogenous plant proteases, several affect trypsin-like proteases from animal, fungi, and bacteria (Ryan, 1973; Davies, 1982). Although it is possible that these proteinase inhibitors can protect cytoplasmic constituents from inadvertent breakdown, a more likely scenario is that they function in plant defense by inactivating proteases from the invading organism.

Conclusions

Selective proteolysis is an essential process in the homeostasis, growth, and development of plants. Nevertheless, the manner in which plants degrade intracellular proteins is still largely unresolved. Many questions remain to be answered regarding substrate selection and the mechanisms used to degrade them. It is already evident that multiple degradative systems operate within plant cells, with several delineated by compartmental boundaries. Moreover, protein stability can be influenced by a multitude of other factors both extrinsic and intrinsic to the protein, and these factors can change in response to either environmental stimuli or developmental state. A renewed interest in proteolysis in plants will likely result in the realization that protein degradation is a complex and fundamental process.

References

Bachmair, A., Becker, F., Masterson, R. V. and Schell, J. (1991). Perturbations of the ubiquitin system causes leaf curling, vascular tissue alterations, and necrotic lesions in higher plants. *EMBO J.* **9**, 4543–9.

Barrett, A. J. (1986). The classes of proteolytic enzymes. In *Plant Proteolytic Enzymes*, Vol. I, ed. M. J. Dalling, CRC Press, Boca Raton, pp. 1–16.

Boller, T. (1986). Roles of proteolytic enzymes in interactions of plants with other organisms. In *Plant Proteolytic Enzymes*, Vol. I, ed. M. J. Dalling, CRC Press, Boca Raton, pp. 67–96.

Boller, T. and Wiemken, A. (1986). Dynamics of vacuolar compartmentation. *Ann. Rev. Plant Physiol.* 37, 137–64.

Buchanan, B. B. (1991). Regulation of CO_2 assimilation in oxygenic photosynthesis: the ferridoxin/thioredoxin system. Perspective on its discovery, present status, and future development. *Arch. Biochem. Biophys.* 288, 1–9.

Canut, H., Ailbert, G., Carrasco, A. and Boudet, A. M. (1986). Rapid degradation of abnormal proteins in vacuoles from *Acer psuedoplantanus* L. cells. *Plant Physiol.* 81, 460-63.

Dalling, M. J. and Nettleton, A. M. (1986). Chloroplast senescence and proteolytic enzymes. In *Plant Proteolytic Enzymes*, Vol. I, ed. M. J. Dalling, CRC Press, Boca Raton, pp. 125–53.

Davies, D. D. (1982). Physiological aspects of protein turnover. In *Encyclopedia of Plant Physiology*, Vol. 14A, eds D. Boulter and B. Parthier, Springer-Verlag, Berlin, pp. 189–228.

Elis, R. E., Yuan, J. and Horwitz, H. R. (1991). Mechanisms and functions of cell death. *Annu. Rev. Cell Biol.* 7, 663–98.

Fincher, G. B. (1989). Molecular and cellular biology associated with endosperm mobilization in germinating cereal grains. *Annu. Rev. Plant Physiol. Molec. Biol.* 40, 305–46

Gatenby, A. A. and Viitanen, P. V. (1994). Structural and functional aspects of chaperonin-mediated protein folding. *Annu. Rev. Plant Physiol. Plant Molec. Biol.* 45, 469–91.

Glotzner, M., Murray, A. W. and Kirschner, M. W. (1991). Cyclin is degraded by the ubiquitin pathway. *Nature* 349, 132–8.

Goldberg, A. L. and St John, A. C. (1976). Intracellular protein degradation in mammalian and bacterial cells: Part 2. *Annu. Rev. Biochem.* 45, 747–803.

Goldberg, A. (1992). The mechanism and functions of ATP-dependent proteases in bacterial and animal cells. *Eur. J. Biochem.* 203, 9–23.

Gottesman, S., Squires, C., Pichersky, E., Carrington, M., Hobbs, M., Mattick, J. S., Dalrymple, B., Kuramitsu, H., Shiroza, T., Foster, T., Clark, W. P., Ross, B., Squires, C. L. and Maurizi, M. R. (1990). Conservation of the regulatory subunit for the Clp ATP-dependent protease in prokaryotes and eukaryotes. *Proc. Natl. Acad. Sci. USA* 87, 3513–17.

Gottesman, S., Clark, W. P., Crecy-Lagard, V-D. and Maurizi, M. R. (1993). ClpX, an alternative subunit for the ATP-dependent *Clp* protease of *Escherichia coli*: Sequence and *in vivo* activities. *J. Biol. Chem.* 268, 22618–26

Greenberg, B. M., Gaba, C., Mattoo, A. K. and Edelman, M. (1987). Identification of a primary *in vivo* degradation product of the rapidly-turning over 32-kD protein of photosystem II. *EMBO J* 6, 2865–9.

Greenberg, J. T. and Ausubel, F. M. (1993). Arabidopsis mutants compromised for the control of cellular damage during pathogenesis and aging. *Plant J.* 4, 327–41.

Harris, N. (1986). Organization of the endomembrane system. *Annu. Rev. Plant Physiol.* 37, 73–92.

Hensel, L. L., Grbic, V., Baumgarten, D. A. and Bleecker, A. B. (1993). Developmental and age-related processes that influence the longevity and senescence of photosynthetic tissues in *Arabidopsis*. *Plant Cell* 5, 553–64.

Hershko, A. and Ciechanover, A. (1992). The ubiquitin system for protein degradation. *Annu. Rev. Biochem.* 61, 761–807.

Kim, W. T. and Yang, S. F. (1992). Turnover of 1-aminocyclopropane-1-carboxylic acid synthase protein in wounded tomato fruit tissue. *Plant Physiol.* 100, 1126–31.

Kiyosue, T., Yamaguchi-Shinozaki, K. and Shinozaki, K. (1993). Characterization of cDNA for a dehydration-inducible gene that encodes a *CLP* A, B-like protein in *Arabidopsis thaliana* L. *Biochem. Biophys. Res. Comm.* 196, 1214–20.

Liu, X-Q. and Jagendorf, A. T. (1984). ATP-dependent proteolysis in pea chloroplasts. *FEBS Lett* 166, 248–52.

Liu, X-Q. and Jagendorf, A. T. (1986). Neutral peptidases in the stroma of pea chloroplasts. *Plant Physiol.* 81, 603–8.

Madura, K. and Varshavsky, A. (1994). Degradation of Gα by the N-end rule pathway. *Science* 265, 1454–8.

Malek, L., Bogorad, L., Ayers, A. R. and Goldberg, A. L. (1984). Newly synthesized proteins are degraded by an ATP-stimulated proteolytic process in isolated pea chloroplasts. *FEBS Lett* 166, 253–7.

Matile, P. H. (1982). Protein degradation. In *Encyclopedia of Plant Physiology*, Vol. 14A, eds D. Boulter and B. Parthier, Springer-Verlag, Berlin, pp. 169–88.

Mattoo, A. K., Hoffman-Falk, H., Marder, J. B. and Edelman, M. (1984). Regulation of protein metabolism; coupling of photosynthetic electron transport to the in vivo degradation of the rapidly metabolized 32-kilodalton protein in the chloroplast membranes. *Proc. Natl Acad. Sci. USA* 81, 1380–84.

Maurizi, M. R., Clark, W. P., Kim, S-H. and Gottesman, S. (1990). ClpP represents a unique family of serine proteases. *J. Biol. Chem.* 265, 12546–52.

Mikola, L. and Mikola, J. (1986). Occurrence and properties of different types of peptidases in higher plants. In *Plant Proteolytic Enzymes*, Vol. I, ed. M. J. Dalling, CRC Press, Boca Raton, pp. 97–117.

Rechsteiner, M. (1991). Natural substrates of the ubiquitin proteolytic pathway. *Cell* 66, 615–18.

Rechsteiner, M., Hoffman, L. and Dubiel, W. (1993). The multicatalytic and 26S proteases. *J. Biol. Chem.* 268, 6065–8.

Roberts, J. K., Ray, P. M., Wade-Jardetzky, N. and Jardetzky, O. (1980). Estimation of cytoplasmic and vacuolar pH in higher plants by ^{31}P NMR. *Nature* 283, 870–72.

Rogers, S., Wells, R. and Rechsteiner, M. (1986). Amino acid sequences common to rapidly degraded proteins: the PEST hypothesis. *Science* **234**, 364–8.

Ryan, C. A. (1973). Proteolytic enzymes and their inhibitors in plants. *Annu. Rev. Plant Physiol.* **24**, 173–96.

Scheffner, M., Nuber, U. and Hulbregtse, J. M. (1995). Protein ubiquitination involving an E1-E2-E3 enzyme ubiquitin thioester cascade. *Nature* **363**, 81–3.

Schmidt, G. W. and Mishkind, M. L. (1983). Rapid degradation of unassembled ribulose 1,5-bisphosphate carboxylase small subunit in chloroplasts. *Proc. Natl. Acad. Sci. USA* **80**, 2632–6.

Staswick, P. E. (1994). Storage proteins of vegetative plant tissues. *Annu. Rev. Plant Physiol. Plant Molec. Biol.* **45**, 303–32.

Stoddart, J. L. and Thomas, H. (1982). Leaf senescence. In *Encyclopedia of Plant Physiology*, Vol. 14A, eds D. Boulter and B. Parthier, Springer-Verlag, Berlin, pp. 592–636.

Storey, R. D. (1986). Plant endopeptidases. In *Plant Proteolytic Enzymes*, Vol. I, ed. M. J. Dalling, CRC Press, Boca Raton, pp. 119–40.

Thoma, S., Sullivan, M. L. and Vierstra, R. D. (1996). Members of two gene families encoding ubiquitin-conjugating enzymes, *AtUBC1-3 and AtUBC4-6*, from *Arabidopsis* are differentially expressed. *Plant Molec. Biol.* (in press).

Treier, M., Staszewski, L. M. and Bohmann, D. (1994). Ubiquitin-dependent c-jun degradation *in vivo* is mediated by the δ domain. *Cell* **78**, 787–98.

Varshavsky, A. (1992). The N-end rule. *Cell* **69**, 725–35.

Vierstra, R. D. (1993). Protein degradation in plants. *Ann. Rev. Plant Physiol. Plant Molec. Biol.* **44**, 385–410.

Vierstra, R. D. (1994). Phytochrome degradation. In *Photomorphogenesis in Plants*, eds R. E. Kendrick and G. H. M. Kronenberg, Martinus Nijhoff Publishers, Dordrecht, Netherlands. pp. 141–62.

Weibel, F. F. and Kunau, W-H. (1992). The Pas2 protein essential for peroxisome biogenesis is related to ubiquitin-conjugating enzymes. *Nature* **359**, 73–6.

4 Molecular biology of development

Peter McCourt and Kallie Keith

Molecular biology and genetic analysis

Most early attempts to apply the techniques of molecular biology to plant development have centered around the expectation that identification of a sufficient number of genes and characterization of their patterns of expression would yield a mechanistic understanding of development. For example, data demonstrating that root cells have different patterns of gene expression from leaf cells has some use in refining our understanding of the physiological and biochemical details of these two tissues and the differences between them; however, little if anything can be learned about the mechanisms of developmental switching which cause a meristematic cell to become a leaf cell rather than a root cell.

Recently, molecular biological studies have been yielding insights into underlying mechanisms of plant development, mostly because they are being applied in combination with genetic analyses of mutant phenotypes. The strengths of molecular analyses permit us to determine the structure of genes and their products (RNA and protein) and to study the processes that influence gene and gene product expression. These types of studies give us detailed information on the mechanisms by which a process functions. They do not necessarily give us information about all the biological processes a given gene and its product influence, some of which are far from obvious.

Genetic analysis, on the other hand, permits us to probe a process by mutation, thereby identifying genes which in some way influence the process being studied. Each method when used alone can provide a great deal of useful information. However, it is when these methods are used in combination that the full power of the two approaches are fully exploited (Struhl, 1983). Molecular manipulation, which permits us to construct chimaeric genes which when reintroduced into the organism and analyzed genetically enable researchers to test the functions of the modified genes *in vivo*. Once the biological function of a modified gene or its protein product has been tested *in vivo* it is desirable, if possible, to return to *in vitro* methods of analysis in order to clarify the biochemical details of the process being studied. This chapter is devoted to the study of development, events or processes which occur over time and in specific contexts. Emphasis is placed on the combined use of molecular and genetic analyses as a means of testing the function of various genes and their protein products *in vivo*.

The marriage between the disciplines of genetic (functional) analysis and molecular biological (structural) analysis has been extremely productive in other developmental systems such as *Drosophila melanogaster* and *Caenorhabditis elegans*, and at the cellular level in *Saccharomyces cerevisiae*, and has only been applied relatively recently in plant systems. In part, this can be attributed to the technical difficulties of genetically characterizing plant species and the associated ability to clone genes identified by mutant alleles. Although sophisticated genetic analysis has been conducted for decades in a number of crop plants such as barley, pea and tomato, three species, *Arabidopsis thaliana*, *Antirrhinum majus* (snap dragon) and *Zea mays* stand at the forefront of current research in plant development. Like the crop plants mentioned above, Antirrhinum and maize have both been studied genetically for many decades and have the advantages afforded by large collections of well-studied mutations and well-characterized transposable element systems which, in principle, can be used to tag (and then clone) genes solely on the basis of mutant phenotypes (Wienand *et al.*, 1986). Arabidopsis has different advantageous attributes, such as a short life cycle and a much smaller genome than most plants (including Antirrhinum and maize), which make the isolation of new mutations in the confines of a laboratory and the molecular cloning of genes much easier than in most other plants. Furthermore, transformation methods have been developed for Arabidopsis which permit the reintroduction of engineered genetic constructs into various genetic (mutant) backgrounds for *in vivo* characterization of gene and gene-product function.

The volume of information being generated in these three systems alone could easily fill this book so the focus of this chapter has been narrowed to the discussion of a few instructive areas of current active research. As emphasized above, the approaches being used to dissect the developmental problems described here are at this time just as important as the principles of plant development that they reveal. The work to be discussed falls into three general areas of plant development: cell determination in inflorescence and floral meristems, tissue-specific gene expression, and finally, advances in understanding the molecular mechanisms of plant hormone action.

Each genetic system has its own conventions used to differentiate wild-type and mutant alleles of a given gene and the protein products produced by those alleles. In order to remain consistent throughout this chapter the conventions used by Arabidopsis researchers have been applied. In this chapter a gene and its protein product are given the same two or three letter code, sometimes followed by a number. Gene (allele) names are all italicized and protein products are not. The first letter of wild-type allele codes are capitalized and mutant allele codes are written entirely in lower case letters. Wild-type protein product codes are entirely capitalized, and mutant protein product codes are written entirely in lower case letters.

Determination

In higher plants, tissues and organs are mostly produced from groups of rapidly dividing cells called meristems. Different types of meristems are defined by the structures they produce and these change throughout the plant life cycle. For example, after germination the vegetative meristem produces leaf primordia until the plant switches to reproductive growth. At this time the same meristematic region becomes determined to produce inflorescence meristems, which typically produce flower bracts with flower meristems in their axils. The sepals, petals, stamens and carpels are then sequentially produced in whorls from the floral meristems. Determination in this sense is defined as an induced change in the developmental fate of a cell which remains after the induction is removed. Although little is known mechanistically about how meristematic cells change their identity, the identification and characterization of genes that direct the formation and development of new organs once the switch is made is progressing. This is best illustrated in studies of flower development

Aside from the esthetic value of working on flower development, flowers are ideal structures for addressing questions of cell fate. Unlike leaf tissue, for example, the different tissues in a flower can be easily identified. Petal tissue can be distinguished from sepal tissue without a microscope and consequently mutations that alter the fate of a petal can be visually identified in mature flowers. Unfortunately the success of identifying so many mutations and genes in different plants over a short time period has produced a giddy genetic nomenclature which is confusing at best. In the following discussion, specific gene names have been kept to a minimum and the genes that have been studied in Antirrhinum and Arabidopsis have been described collectively using the terms of the model for the control of floral development that has arisen from the study of these genes. For more detailed information reviews on this subject by both Meyerowitz and Coen listed in the references are highly recommended.

The flowers of Antirrhinum and Arabidopsis have very similar organization. Mature flowers are composed of four concentric whorls of organs which are produced one after another sequentially from the floral meristem. The outermost and first whorl is made up of sepals, the second whorl of petals, the third whorl consists of stamens and the fourth and innermost whorl contains the carpels (Fig. 4.1). Intensive screening for mutations which alter whorl patterns in both plant species has resulted in the identification of two major sets of regulatory genes: meristem identity genes and organ identity genes.

Genetic analysis of meristem

Mutations in meristem identity genes generally result in failure to initiate the flowering program, resulting in plants that produce inflorescence meristems, but no floral meristems or flowers. In Antirrhinum, for example, a severe mutation of the *Floricula* (*Flo*) gene causes the plant to continually produce bracts rather than any flower structures (Coen *et al.*, 1990). The Flo gene was cloned by transposon tagging in

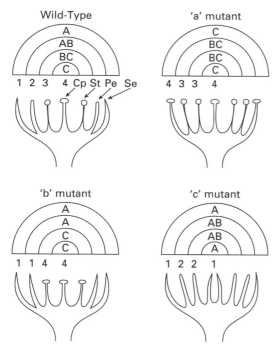

Fig. 4.1 Model illustrating the combinatorial model of flower organ identity. The numbers 1, 2, 3, 4 each represent a whorl region of a flower with the class of genes expressed in each whorl represented by a letter (A, B, C). The side view of a flower is represented. Changes in the whorl pattern due to mutation are represented for a-, b- and c-type mutants. Se, sepal; Pe, petal; St, stamen; Cp, carpel.

Antirrhinum and although it encodes a protein which cannot be identified based on amino acid sequence, it has a number of features that are found in the activation domains of transcription factors. The expression pattern of Flo is intriguing because with the exception of the stamen primordium, which shows no expression, Flo gene activity is only detected transiently in each wild-type floral primordium. This implies that Flo has a role not only in switching the inflorescence meristem to a floral meristem but also in the production of floral organs. However, expression of Flo, or any other cloned gene, in a particular primordium or tissue does not in itself mean that gene has a role in that tissue. As mentioned earlier, there is Flo expression in most of the floral organ primordia including the bracts of wild-type flowers. However, even severe Flo mutants produce normal bracts suggesting the gene does not have a major function in determining bract identity. This principle may turn out to be more general than might be expected in plant development. That is, the expression of a developmental gene (usually a transcription factor) occurs in tissues in which on a gross scale it does not appear to have an effect, and the tissue that is affected is produced immediately following the tissue in which it is expressed.

The isolation of *flo* mutations using transposons provides a number of advantages in addressing developmental questions of signaling between cell layers (Carpenter and Coen, 1990). The transposon that was used in these studies, TAM3, is inherently unstable and excises at a relatively high frequency from genes that it has disrupted. Occasionally this results in the reconstruction of a functional Flo gene in those cells and in all their progeny. On the plant these reversion events are evident from the appearance of the odd normal flower on a phenotypically mutant *flo* plant that otherwise only produces bracts. Many times, however, the seeds from these flowers, when planted, only produce plants with the original flo phenotype suggesting that functional gametes were produced from tissue that was homozygous for the flo mutation. One interpretation of this observation is that the FLO gene product acts in a non-cell autonomous manner in the tissue (and/or perhaps surrounding tissues) that give rise to the gametes. The gametic tissue is derived only from the L2 cell layer in the developing flower and, therefore, it appears that reversion of L1 or L3 tissue back to wild-type Flo gene expression can result in the development of a fertile flower even when the L2 layer is still mutant. Thus, the production of functional FLO protein in one cell or cell layer influences the development of the cells around that cell or cell layer.

Genetic analysis of organ identity

Mutations in the organ identity genes of Arabidopsis and Antirrhinum are phenotypically similar, and can be roughly placed into three phenotypic categories (reviewed by Coen and Meyerowitz, 1991). The first group or type 'a' mutations result in two replacements. The organs in both the first and second whorls are replaced with organs that normally occur in the fourth and third whorls, respectively. In other words, carpels replace sepals and in the same flowers, stamens replace petals. The phenotype of the organ identity pattern is described from whorl 1 to whorl 4, and the organs occur as carpels, stamens, stamens, carpels (designated here 4, 3, 3, 4). The class 'b' mutations replace whorl 2 organs with whorl 1 organs and whorl 3 organs with whorl 4 organs (sepals, sepals, carpels, carpels or 1, 1, 4, 4). Class 'c' mutations express whorl 2 and 1 organs in place of whorl 3 and 4 organs, respectively (sepals petals, petals, sepals or 1, 2, 2, 1). One consistency between all of these phenotypic classes is that a single mutation in any of the classes affects two whorls. The models which have been proposed to account for these results posit three functional domains, each defined by one of the three classes of mutations. The gene products act alone or in different pairwise combinations to produce four different whorls. Correct establishment of whorl 1 organs requires expression of class A genes. Whorl 2 is specified by combined expression of genes in classes A and B. Whorl 3 is specified by combined expression of genes in classes B and C, and the fourth whorl only requires the expression of class C genes (1=A, 2=AB, 3=BC, 4=C).

This combinatorial, or as it is often called, the ABC model of floral organ identity also accounts for the fact that the organs in a given whorl are replaced rather than lost. For example, class b mutants have

the second and third whorl organs replaced with the first and fourth whorl organs, respectively. The loss of B gene function would result in a gene expression pattern from whorl 1 to whorl 4 that can be described as A, A, C, C, respectively. This translates to an organ identity pattern of 1, 1, 4, 4 rather than the wild-type pattern of 1, 2, 3, 4. The class a mutations result in an organ identity pattern of 4, 3, 3, 4, which according to the model should correspond to a gene expression pattern of C, BC, BC, C. If this pattern of gene expression is correct, it suggests that the A gene product has two roles in the first two whorls. One role is a positive or activating function, which permits the production of sepals and petals in whorls 1 and 2 respectively, and the second role is a negative function: the inhibition of C gene expression in those same whorls. Loss of A gene expression permits the C gene to be expressed resulting in the mutant pattern. The restriction of C gene expression by the A gene product in wild-type flowers also applies in reverse, that is, the C gene product appears to inhibit A gene expression, since class c mutants (organ identity pattern 1, 2, 2, 1) produce organs in whorls 3 and 4 that require A gene expression. The mutual incompatibility of A and C gene expression in wild-type flowers implies that these genes are not only required for organ identity but also delimit the tissues in which organ identity genes are to be expressed. Genetic testing of the ABC model by construction of double and triple mutants in Arabidopsis has not only been supportive of the ABC model but was instrumental in its development. Construction of a triple mutant that lacked A, B and C functions result in a flower in which all whorls resemble cauline leaves, suggesting the leaf is the developmental 'ground state' from which the various flower whorls are built (Bowman et al., 1991). The addition of A function which can be made by constructing a 'bc' double mutant results in sepals (whorl 1) in all the whorls. Flowers from an 'ab' double mutant, which therefore only have C function, produce whorls of carpel like structures (whorl 4), but perhaps most interestingly the 'ac' double which only has B function has leaf-like structures in whorls 1 and 4 and structures somewhat between a petal and a sepal in whorls 2 and 3. This indicates that B function is established independently of the A or C genes since only whorls 1 and 4 look like ground state structures.

Molecular studies of floral development

The genetic studies of floral cell fate determination have been very useful in identifying the genes which determine cell fate in flowers and in the construction of models which on a gross scale describe how these gene products may interact to specify a given outcome. However, genetic studies tell us little if anything about how the gene products carry out their functions on a molecular scale. Molecular studies have begun to answer some of these questions and have provided direct tests of the ABC model. With the exception of one gene (Ap2) which is involved in establishing A function in Arabidopsis all the genes cloned so far in both Antirrhinum and Arabidopsis belong to a family of functionally related transcription factors (see review by Weigel and Meyerowitz, 1994). The gene products all share a 56 amino acid domain at the amino terminal which is thought to bind DNA and to be involved in the dimerization of proteins. This domain was first identified in the MCM1 gene of yeast which encodes a DNA-binding transcriptional activator of mating type-specific genes and in serum response factor genes (SRF) which are transcriptional regulators of the c-fos oncogene in humans. When the Deficiens (DEF) gene, a class B gene, in Antirrhinum and the Agamous1 gene (AG1), a class C gene, in Arabidopsis were also shown to contain this motif, the initial letter of the four genes carrying this motif were combined to coin the acronym 'MADS' box gene (MCM1-AGAMOUS-DEFICIENS-SRF) to refer to any gene containing this conserved domain (Schwarz-Sommers et al., 1992). Developmental studies of MADS gene expression suggests the ABC genes are sequentially expressed early after the meristem is induced to flower and the patterns of gene expression fit the genetic predictions of the ABC model reasonably well.

The isolation of genes that establish organ identity has also opened up new approaches for developmental analysis. Ectopic expression of genes in tissues in which they do not normally function allows further testing of the roles of these genes in development. For example, ectopic expression of AG1, a class C gene, causes a phenotype similar to that caused by a class a mutation (Mizukami and Ma, 1992). This result supports the notion suggested by the model, namely that the A and C genes cannot function in the same whorl. Unexpectedly, ectopic expression of Ag1 also causes the ovules inside the carpel to transform into carpels. In a recent genetic screen for mutations causing female sterility a mutation designated *bell* was identified that causes carpeloid ovules (Modrusan et al., 1994). Studies of Ag1 expression in wild-type and *bell* mutants show Ag1 is expressed in mutant but not wild-type ovules, suggesting the wild-type function of the Bell gene is to repress Ag1 expression in ovules.

Although much work remains, these results hint that the genes identified in floral studies may offer insights into other areas of plant development. For example, now that floral specific regulators such as FLO have been identified, will it be possible to identify genes that are responsible for activating Flo gene expression, which may eventually lead back to the identification of the primary signals for floral initiation? Likewise, now that organ identity genes have been defined and mutational analysis has provided a model for how decisions are made in the floral meristem, experiments can be designed to determine whether or not similar mechanisms are used in root and shoot meristems.

Tissue specific gene expression

Once the fate of a cell is determined, specific groups of genes are expressed so that the cell can perform its designated function. Reasonably, but perhaps naively, many researchers have assumed that tissue specific gene expression is governed by the interaction of developmentally controlled *trans*-acting regulators with *cis*-acting DNA sequences in the promoters of particular genes. In this view, the presence or absence of a *cis*- or *trans*-acting element determines the place in the plant where the gene will be expressed. While these principles operate to some extent to control tissue specific gene expression in plants, one of the two examples of tissue specific gene expression presented here, seed storage protein synthesis in maize, demonstrates that tissue specific expression is not always as simple as expected, and in fact, this may turn out to be the rule, rather than the exception.

Seed specific gene expression

Of the many examples of research into what regulates tissue specific gene expression perhaps the best characterized in plants at the molecular level is seed storage protein accumulation (see review by Goldberg *et al.*, 1989). Aside from the agronomic importance of seeds and seed storage products (see Chapter 34), the experimental popularity of this system stems from the ease with which seed storage protein gene expression can be monitored. It is very easy to isolate seed tissue away from all others and reserve proteins accumulate to very high levels relative to other embryo- or seed-specific gene products. Furthermore, reporter gene technology makes it possible to quantitate gene expression by fusing the promoter of any seed storage protein gene to a structural gene, the product of which can be easily assayed. Transgenic plants which express these reporter constructs are then produced, and transcription from the seed storage gene promoter can be indirectly assayed by measuring the enzymatic activity of the reporter protein. The most widely used reporter gene for plant studies has been the UidA gene of *E. coli*, which encodes β-glucuronidase (GUS). The popularity of the GUS system is due to the fact that GUS activity can be easily assayed colorimetrically. It is important to note, however, that the colored product which is produced by GUS does not break down readily in plant tissue; therefore, it is not possible to determine the length of time for which a given promoter is expressed, only that a promoter is active in a specific tissue. Residual GUS product in a tissue may also lead to misinterpretation of promoter expression. Thus, measuring promoter activity via promoter-GUS reporter constructs can give misleading results if the proper control experiments are not done.

By making a collection of fusion genes containing different fragments of various seed storage promoters with GUS structural genes and transforming these reporter constructs into a variety of plant species, a number of groups have determined which *cis*-acting sequences are required for seed specific expression (see Tomas, 1993 for review). Using *in vitro* binding assays, attempts have been made to purify proteins that interact with these sequences in the hope of identifying seed specific trans-acting factors. Unfortunately *in vitro* binding is not always a good measure of *in vivo*

interaction. In maize, for example, mutations in the *Opaque-2* gene (*op2*) cause a 50% reduction in a-zein seed storage protein levels compared to wild-type. The Op2 gene has been cloned by transposon tagging and encodes a protein that belongs to the basic domain leucine zipper (bZIP) class of transcription factors (Schmidt *et al.*, 1987). The basic regions of bZIP proteins have been shown to bind DNA and the leucine zipper repeats are thought to be involved in dimerization of proteins. Transient transformation experiments of maize embryos show the Op2 gene is required for *trans*-activation of zein promoters and *in vitro* binding experiments have defined a region of the zein promoter where OP2 may interact (Schmidt *et al.*, 1992). However, the story is not as simple as it first appeared. In *op2* mutants, expression of another gene (b32) encoding a *Z. mays* seed storage albumin is dramatically reduced (Lohmer *et al.*, 1991). The OP2 protein binds this promoter *in vitro* at five sites that have no sequence identity to the zein Op2 binding sequences. The discrepancy in the target sites may reflect physiological differences in binding between the two promoters *in vivo* or may be a function of conditions used *in vitro*. Whichever the case, these and other *in vitro* binding experiments conducted without corroborating genetic evidence must be interpreted with caution. This type of situation becomes even more complex when *trans*-acting elements which interact in a combinatorial fashion are identified. These certainly add another level of complexity to the design and interpretation of experiments attempting to sort out the interactions of *cis*- and *trans*-acting elements.

Coloration in Antirrhinum

Coloration patterns are extremely useful systems to exploit in understanding mechanisms of tissue specificity because enormous variation in the pattern and intensity of pigmentation can be generated without loss of viability to the organism (reviewed by Coen *et al.*, 1988 and Martin and Gerats, 1993). This is true not only in plants, but in most pigmented multicellular organisms. Unlike analysis of seed storage protein accumulation, subtle genetic changes in coloration in Antirrhinum flowers or maize kernels, for example, can be easily recognized. In Antirrhinum flowers numerous mutations have been catalogued roughly into two categories. Those that are defective in a particular enzyme in pigment biosynthesis and those that affect the expression pattern of the pigment across the flower. These latter mutations are intriguing not only because they may tell us something about tissue specificity but also because they suggest that even within the same tissue there may be internal patterns that must be interpreted by a tissue specific gene. A number of these mutations have now been analyzed at the molecular level and in this section two of these classes will be discussed.

Cis-acting coloration mutations

Early in Antirrhinum flower development anthocyanin pigmentation appears at the ring at the base of the tube and in the lobes of the unexpanded petals. Next, as the tube expands and develops the petals become fully pigmented yielding a mature flower with dark red lobes and a slightly lighter tube (Fig. 4.2). Anthocyanin intensity reflects the mRNA

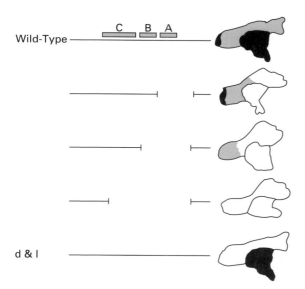

Fig. 4.2 Schematic representation of wild-type and various coloration pattern mutants of Antirrhinum. The line represents the Pallida promoter. Gaps in the lines represent deletions of various parts of the promoter caused by TAM3 excisions. The changes in color pattern of the flower are diagrammed on the right-hand side. A, B and C represent three control regions defined by the various deletion mutants. del, *delila* mutant.

levels of the biosynthetic genes, therefore color patterns reflect specific spatial and temporal patterns of gene expression. Most work in this pathway has centered on mutations affecting two genes: *nivea* (*niv*) which encodes the enzyme chalcone synthase and *pallida* (*pal*) which encodes dihydroflavanol-4-reductase. One mutation in the Pal gene, designated *pal*$^{rec-2}$, has led to a rich source of new mutations that affect the pattern of pallida gene expression (Coen *et al.*, 1986). The original *pal*$^{rec-2}$ mutation is due to an unstable insertion of the TAM3 transposon into the promoter of the Pallida gene. The germinal instability of TAM3 leads to a high frequency of imprecise excision of the transposon, which in turn results in a variety of stably inherited DNA rearrangements in the promoter of the Pallida gene. Thus, comparisons of pigment patterns produced in stable *pal* alleles with sequence analysis of the DNA alterations caused by TAM3 excision has provided a functional assay for determining which DNA regions in the promoter are responsible for pigment patterning. Unlike the promoter-GUS fusion studies used to identify seed specific *cis*-acting regulators, this analysis has the advantage that the phenotype of each mutant determines the molecular analysis of the promoter region. Moreover, whereas transgenic plants contain GUS constructs in random chromosomal positions, all the promoter mutations generated in the *pal* system are in the same natural chromosomal location in the cell.

Comparative analysis of promoter structure with phenotype has defined three regions (A, B and C) which differentially effect the pattern of *Pallida* expression (Fig. 4.2). Mutations in region A decrease in the amount of anthocyanin in the lobes relative to the levels in the tubes. This region contains DNA sequences that are very similar to *cis*-acting regulatory sequences defined in other plant systems termed G-box elements. Mutations in region B, which lies 9–10 bases to the left of region A, completely disrupt Pal expression in the lobes with low expression confined to the base of the tube. This region contains a sequence that is identical to an adenovirus regulatory *cis*-element. Finally, mutations that remove A, B and 79 bases upstream of this region eliminate Pal expression across the flower. This 79 base pair region contains the CAAT box which has been shown to interact with a plethora of *trans*-

acting factors in mammalian systems (Dorn *et al.*, 1987). That regions A, B and C are required for the correct specificity of color pattern in the flower not only shows the diversity of controlling information that is encoded in a promoter but also suggests that all three regions interact with a collection of different *trans*-acting genetic elements.

Trans-acting mutations

In Antirrhinum at least one *trans*-acting factor that regulates Pal gene expression has been identified (Goodrich *et al.*, 1992). The *delila* (*del*) mutation results in a number of changes in floral color pattern in the flower. The most striking is the complete loss of pigment in the tube with little reduction in the lobes (Fig. 4.2). Phenotypic and molecular analysis of Del gene expression in combination with a number of the *pal* mutations mentioned above show that Del does not regulate Pal expression through regions A or B but does require region C (Almeida *et al.*, 1989). Recently Del was cloned by TAM3 transposon tagging and shown to encode a protein with similarity to the myc-related basic helix–loop–helix (HLH) class of DNA transcription factors (Goodrich *et al.*, 1992). The binding sites of many HLH transcription factors as defined *in vitro* identify the consensus sequence in region A. However, as stated earlier, deletions of the A region do not disrupt Del control of the Pal gene and region C, which is essential for Del control of the Pal gene, does not have an HLH consensus sequence. Taken together this suggests Del recognizes a binding site different from those of other HLH proteins and again brings into question the use of 'stare and compare' analysis for understanding promoters.

Interestingly the *DEL* gene product is closely related to the products of the R gene family, which regulate pigment production in maize. The conservation of two *trans*-acting factors that carry out similar functions between two plants that diverged approximately 150 million years ago suggests the evolution of color patterns in plants may simply reflect changes in the pattern of expression of conserved regulators (Goodrich *et al.*, 1992).

The information from both *cis*- and *trans*-acting mutations that affect the color pattern across an *Antirrhinum* flower suggests tissue specificity in general may be controlled in a combinatorial fashion. If this is true, it may be possible to account for complex cellular and tissue specificity by simultaneous binding of different combinations of *trans*-acting factors to the different *cis*-acting elements. Not only would different *cis*-acting elements bring different proteins in close proximity to each other but many of these *trans*-acting factors have dimerization domains which in turn bring other regulators to the promoter. In such a scenario it does not require much imagination to envision how only a few factors differentially expressed in combination with a number of *cis*-elements would be more than enough to specify where and when all the genes in the plant should act. Redundancy is an inherent consequence of a combinatorial system of gene control. In plants, which have highly plastic and environmentally sensitive development, such backup systems are undoubtedly vital in altering developmentally regulated gene expression in response to environmental variation.

Hormone action

As seen from the previous sections changes in the expression of a gene undeniably play a role in controling alternative developmental pathways. However, as we work back through the genetic circuitry at some point the expression of a regulatory gene or the activity of its product must itself be controled by an external stimulus. Thus signals from the environment are intricately involved in patterns of development. In plants, phytohormones influence many diverse processes in plant development ranging from seed germination to root, shoot and flower formation. Research addressing the molecular basis of plant hormone action has been an intense area of molecular and genetic investigation for a number of years and the fundamental principles of the approach are similar; regardless of the hormone or the effects being studied. In this section investigations into the mechanisms of ethylene biosynthesis and ethylene action will be described, but the approaches being used to study the action of abscisic acid auxins, gibberellins and other plant growth regulators are similar.

Ethylene biosynthesis

Ethylene gas regulates many developmental processes in plants ranging from seed germination to senescence of flowers and leaves. Of all the processes affected by ethylene, perhaps the most intensively investigated is its role in fruit ripening. Large losses in fruit and vegetable sales every year are due to ethylene-induced spoilage, resulting in intensive interest in finding inhibitors of ethylene synthesis and action. Biochemical studies have clearly shown that ethylene is produced from methionine via the synthesis of 1-aminocyclopropane-1-carboxylic acid (ACC) (reviewed by Kende, 1989). Although ACC is converted to ethylene by the enzyme ACC oxidase, the rate limiting step in the pathway is the production of ACC which is catalyzed by ACC synthase.

The recent cloning and analysis of ACC synthase expression has shown that induction of ethylene at different times in development and under different environmental conditions actually involves the expression of a small family of ACC synthase genes (Theologis, 1992). In tomato, for example, which has six isozymes, only two genes are expressed during fruit ripening. The redundancy and complexity of ACC synthase regulation via differential gene expression made the matter of determining the role of a given isozyme in a given ethylene regulated processes a difficult task. However, cloning and characterization of ACC synthase at the molecular level permitted direct testing of the roles of ethylene in various plant functions. For example, tomato fruit ripening can now be controlled using antisense RNA to inhibit expression of specific ACC synthase isozymes (Oeller et al., 1991). Expression of antisense RNA for one of the tomato ACC synthase genes reduced enzymatic activity to less than 0.5% of wild-type activity. This results in a severe reduction in ethylene production and in tomato fruits that do not ripen. Addition of exogenous ethylene reverses the antisense induced phenotype, but ethylene must be continually supplied in order for the fruit to ripen fully. This implies that production of threshold levels of ACC synthase are insufficient and that continuous transcription of the genes involved in ethylene biosynthesis are required for ethylene-mediated ripening (Oeller et al., 1991).

Ripening involves the coordinate expression of a number of genes encoding enzymes involved in respiration, chlorophyll degradation, carotenoid synthesis, cell wall degradation and conversion of starch to sugars. So-called 'antisense' fruits have been very useful in assessing which of these functions is under ethylene control and which, if any, are controlled by ethylene-independent mechanisms. Expression of genes that encode polygalacturonase (PG), which is important in softening the cell wall and chlorophylase, a chlorophyll degrading enzyme, were not affected in antisense plants, but respiratory metabolism, biosynthesis of aromatic compounds and synthesis of lycopene (a carotenoid) were inhibited (Oeller et al., 1991). Furthermore, although antisense plants had normal levels of PG RNA, translation of PG message was inhibited in these plants. Therefore, tomato ripening involves both ethylene-dependent and ethylene-independent processes and both transcriptional and translational regulation of these processes mediate ethylene action during fruit ripening. While the cloning of ACC synthase and the manipulation of ethylene concentration in tissues has proven the functional role of ACC synthase in controlling processes such as fruit ripening, it has not revealed extensive information regarding the mechanism of ethylene action. Genetic analysis, on the other hand, has been used to identify genes involved in the perception of ethylene and transduction of that signal to the cell.

Ethylene action

In contrast to our understanding of ethylene biosynthesis, knowledge of the underlying molecular mechanisms of ethylene perception and action have only recently begun to be elucidated. The use of ACC synthase antisense plants clearly demonstrates that ethylene has marked effects on transcription and translation of genes, but it does not tell us how the plant perceives ethylene or how that perception is converted into a cellular response, in this case, gene expression. The isolation and characterization of Arabidopsis mutations which alter the plant's response to ethylene has begun to provide this information.

Seedlings when grown in the dark have dramatically elongated hypocotyls, and exposing such seedlings to exogenous ethylene has long been known to retard hypocotyl elongation. By visually screening mutagenized populations of dark-grown seedlings for plants which are insensitive to ethylene (i.e. had much longer hypocotyls than wild-type seedlings when germinated in the dark and gased with ethylene) mutations in three independent genes designated *etr1*, *ein2* and *ein3* were identified (Bleeker *et al.*, 1991, Keiber and Ecker, 1993). The *etr1* mutants, which have been the most carefully characterized physiologically, not only show altered ethylene responses in dark-grown seedlings but are also insensitive to ethylene at several different stages of development. These include reduced seed germination, retarded ethylene-induced leaf senescence and peroxidase activity and loss of positive feedback regulation of ethylene synthesis.

The Etr1 gene was cloned by chromosome walking and the amino-terminal portion of the putative protein product shows no sequence identity to any known protein (Chang *et al.*, 1993). The carboxy-terminal portion, however, is similar to a family of bacterial sensory genes called two-component regulators, which are involved in signaling the nucleus that an event has occurred on the cell surface. Two component regulators have been implicated in processes as diverse as chemotaxis and osmoregulation (Stock *et al.*, 1989). Two-component regulators coordinate two separate reactions, and usually but not always involve two separate proteins. One component has a kinase activity which is activated after the stimulus is perceived. The kinase component is often referred to as the sensor or modulator and although in some cases these appear to function as classical membrane receptor proteins, in many examples it is quite clear the kinase is not a sensor. After the kinase regulator is stimulated it autophosphorylates a conserved histidine residue. The phosphate group is then transferred to a conserved aspartate residue on the cytoplasmic regulatory component which activates or inhibits regulatory function that influences the response. In many but not all cases the response regulator is a transcriptional activator. Although the kinase and response regulators are the main players in this prokaryotic signal transduction system the term two-component is somewhat misleading since these systems generally contain additional signal

transduction components. Using sequence comparisons the ETR1 protein appears to have both the sensory autokinase domain and the regulatory phosphorylation site, but has no transcriptional activator domain in the second component.

All the *etr1* mutations that cause insensitivity to ethylene alter the novel amino-terminal portion of ETR1, suggesting this domain may be involved directly or indirectly in the recognition of the ethylene signal. Interestingly, all the etr1 mutations presently known result in dominant ethylene insensitive phenotypes. These dominant phenotypes can be accounted for by proposing that perception of ethylene and transmission of the information that ethylene has contacted the cell requires the interaction of a multimeric ETR1 complex, and mutant *etr1* proteins disrupt these associations or interactions and thereby poison the multiprotein complex. In this context, ETR1 must act as a positive regulator in the ethylene signal transduction pathway. A second possibility is that ETR1 normally functions in the absence of ethylene to suppress ethylene responses. In this scenario mutations in the amino-terminal portion of ETR1 somehow causes the protein to be constitutively activated, thereby continuously repressing ethylene responses.

Clues as to how the ETR1 protein might work have come from molecular studies of other Arabidopsis mutations which alter plant responses to ethylene and their interactions with *etr1*. One of these mutations, designated *ctr1*, cause the plant to constitutively behave as if ethylene is present, even when it is absent (Keiber *et al.*, 1993). Cloning of the Ctr1 gene has shown it encodes a protein with similarity to the *Raf* family of serine/threonine protein kinases found in mammals, Drosophila and *C. elegans*. Phenotypes of *ctr1* are recessive, so when CTR1 kinase activity is eliminated the plant behaves as if ethylene is always present, and the ethylene signal transduction pathway is always active. Since loss of CTR1 activity produces a constitutive response, the wild-type CTR1 kinase represses the ethylene signal transduction pathway, or is a negative regulator of this pathway.

Mutations in two genes such as Etr1 and Ctr1 have phenotypes easily distinguishable from wild-type and from each other. Hence with some knowledge of how a pathway operates, it is sometimes possible to order the action of these genes in the process they affect by analyzing the phenotypes of double mutants. In the case of a signal transduction pathway, however, in which the product of one reaction often modifies the activity (either positively or negatively) of one or more subsequent reactions, interpretations of double mutant phenotypes are difficult and must be interpreted with caution. If a double mutant is phenotypically similar to only one of the single mutant parents, the mutation responsible for that phenotype is said to be epistatic to the mutation responsible for the phenotypes that are not apparent in the double mutant. In the simplest sense the mutation that is epistatic more than likely encodes a gene product that acts 'at or downstream' of the other gene product. The *etr1 ctr1* double mutant plants are phenotypically similar to the *ctr1* single mutants and show none of the *etr1* phenotypes (Keiber *et al.*, 1993). Based on the above

interpretation, the CTR1 gene product acts at or downstream of the ETR1 gene product. Although a number of models can be envisioned the dominant nature of *etr1* might mean that in the absence of ethylene, ETR1 constitutively activates CTR1, which in turn negatively prevents the ethylene response pathway from being expressed (Fig. 4.3A). In the wild-type situation, when ETR1 perceives ethylene (or is signaled that the cell has perceived ethylene), CTR1 activity is then down-regulated, shut off or mitigated in some way, so that ethylene-induced responses can occur. The loss of ETR1 response to ethylene means CTR1 is constitutively activated and therefore the cell is constantly in an insensitive state to ethylene (Fig. 4.3B). The loss of CTR1 function causes ethylene responses to be constitutively ON irrespective of the genetic state of *etr1* because CTR1 is required to carry out ETR1 mediated signal transduction (Fig. 4.3C,D). Although their

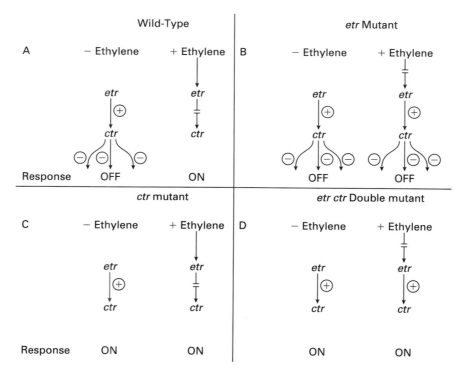

Fig. 4.3 Possible model to explain the role of ETR1 and CTR1 in ethylene signal transduction. −ethylene, the plant was not exposed to ethylene; +ethylene, the plant was exposed to ethylene. OFF, the plant does not show an ethylene response; ON, the plant shows a characteristic ethylene response.

biochemical functions have yet to be shown directly, the fact that both prokaryotic and eukaryotic signal transduction proteins have been identified in a plant hormone signal transduction pathway suggests plants use similar components to transduce signals, which alter gene expression, ultimately resulting in cellular responses.

Conclusions

The complex problem of how a single-celled fertilized zygote becomes a functional mature three-dimensional organism is frequently cited as one of the most exciting frontiers of plant biology. Development is generally believed to be the controller of all other plant processes and therefore to understand it will require fundamentally different paradigms than those that have been defined in other more mature areas of plant research such as metabolism and biochemistry. In this chapter examples of genes have been cited, mostly transcription factors, that appear to control programs in the plant. Mutations in those genes have profound effects on the development of the plant. At first glance then, this chapter on plant development in a book of plant biochemistry and molecular biology may seem out of place. However, as our knowledge of plants increases it is becoming apparent that development is not a separate controller of metabolism but interacts with metabolism as an equal partner in allowing the plant to grow and differentiate. Although it can be argued that the genes mentioned in this chapter do control metabolic pathways (for example, Antirrhinum floral development genes control pigment biosynthesis) the reverse is also true. Genes are now being identified that encode enzymes involved in fundamental metabolism but when mutated have profound effects on the development of the plant. For example, Arabidopsis mutants deficient in the synthesis of the cell wall sugar fucose are dwarf in stature (Reiter *et al.*, 1993). Can these genes and their protein products be considered metabolic regulators of development? It is clear that deep understanding of development will in the future require a thorough understanding of metabolism and vice versa.

References

Almeida, J., Carpenter, R., Robbins, T. P., Martin, C. and Coen, E. S. (1989). Genetic interactions underlying flower color patterns in *Antirrhinum majus. Genes Develop.* **3**, 1758–67.

Bleecker, A. B., Estelle, M. A., Somerville, C. R. and Kende, H. (1991). Insensitivity to ethylene conferred by a dominant mutation in Arabidopsis thaliana. *Science* **241**, 1086–9.

Bowman, J. L., Smyth, D. R. and Meyerowitz, E. M. (1991). Genetic interactions among floral homeotic genes of *Arabidopsis. Develop.* **112**, 1–20.

Carpenter, R. and Coen, E. (1990). Floral mutations produced by transposon mutagenesis in *Antirrhinum majus. Genes Develo*p. **4**, 1483–93.

Chang, C., Kwok, S. F., Bleecker, A. B. and Meyerowitz, E. M. (1993). *Arabidopsis* ethylene-response gene ETR1: Similarity of product to two-component regulators. *Science* **262**, 539–44.

Coen, E. S., Carpenter, R. and Martin, C. (1986). Transposable elements generate novel spatial patterns of gene expression in *Antirrhinum majus. Cell* **47**, 285–96.

Coen, E. S., Almeida, J., Robbins, T. P., Hudson, A. and Carpenter, R. (1988). Molecular analysis of genes determining spacial patterns in *Antirrhinum majus*. In *Temporal and Spacial Regulation of Plant Genes*, eds D. P. S. Verma and R. B. Goldberg, Springer-Verlag, Wein, New York, pp. 63–82.

Coen, E. S., Romero, J. M., Doyle, S., Elliot, R., Murphy, G. and Carpenter, R. (1990). *Floricula*: A homeotic gene required for flower development in *Antirrhinum majus. Cell* **63**, 1311–22.

Coen, E. and Meyerowitz, E. (1991). The war of the whorls: genetic interactions controling flower development. *Nature* **353**, 31–7.

Goldberg, R. B., Barker, S. J. and Perez-Grau, L. (1989). Regulation of gene expression during plant embryogenesis. *Cell* **56**, 149–60.

Goodrich, J., Carpenter, R. and Coen, E. S. (1992). A common gene regulates pigmentation pattern in diverse plant species. *Cell* **68**, 955–64.

Kende, H. (1989). Enzymes of ethylene biosynthesis. *Plant Physiol.* **91**, 1–4.

Kieber, J. J., Rothenberg, M., Roman, G., Feldmann, K. A. and Ecker, J. R. (1993). *CTR1*, a negative regulator of the ethylene response pathway in *Arabidopsis*, encodes a member of the Raf family of protein kinases. *Cell* **72**, 427–41.

Lohmer, S., Maddaloni, M., Motto, M., DiFonzo, N., Hartings, H., Salamini, F. and Thompson, R. D. (1991). The maize regulatory locus *Opaque-2* encodes a DNA binding protein which activates the transcription of the b-32 gene. *EMBO J.* **10**, 617–24.

Martin, C. and Gerats, T. (1993). Control of pigment bisynthesis genes during petal development. *Plant Cell* **5**, 1253–64.

Mizukami, Y. and Ma, H. (1992). Ectopic expression of the floral homeotic gene *AGAMOUS* in transgenic *Arabidopsis* plants alters floral organ identity. *Cell* **71**, 119–67.

Modrussan, Z., Reiser, L., Feldmann, K. A., Fisher, R. L., and Haughn, G. W. (1994). Homeotic transformation of ovules into carpel-like structures inm *Arabidopsis*. *Plant Cell* **6**, 175–86.

Oeller, P. W., Min-wong, L., Taylor, L. P., Pike, D. A. and Theologis, A. (1991). Reversible inhibition of tomato fruit senescence by antisense RNA. *Science* **254**, 437–9.

Reiter, W-D., Chapple, C. C. S. and Somerville, C. R. (1993). Altered growth and cell walls in a fucose-deficient mutant of Arabidopsis. *Science* **261**, 1032–5.

Schmidt, R. J., Ketudat, M., Aukerman, M. J. and Hoschek G. (1992). *Opaque-2* is a transcriptional activator that recognizes a specific target site in 22-kD zein genes. *Plant Cell* **4**, 689–700.

Schmidt, R. J., Burr, F. A., and Burr, B. (1987). Transposon tagging and molecular analysis of the maize regulatory locus *opaque-2*. *Science* **238**, 960–63.

Schwarz-Sommers, Z., Hue, I., Huijser, P., Flor, P. J., Hansen, R., Tetens, F., Lonnig, W.-E., Saedler, H. and Sommers, H. (1992). Characterization of the *Antirrhinum* floral homeotic MADS-box gene deficiens: evidence for a DNA binding and autoregulation of its persistent expression throughout flower development. *EMBO J.* **11**, 251–63.

Stock, J. B., Ninfa, A. J. and Stock, A. M. (1989). Protein phosphorylation and regulation of adaptive responses in bacteria. *Microbiol. Rev.* **53**, 450–90.

Struhl, K. (1983). The new yeast genetics. *Nature* **305**, 391–7.

Theologis, A. (1992). One rotten apple spoils the whole bushel: The role of ethylene in fruit ripening. *Cell* **70**, 181–4.

Tomas, T. L. (1994). Gene expression during plant embryogenesis and germination: An overview. *Plant Cell* **5**, 1401–10.

Weigel, D. and Meyerowitz, E. M. (1994). The ABCs of floral homeotic genes. *Cell* **78**, 203–9.

Wienand, U., Sommer, H., Schwartz, Z., Shephard, N., Saedler, H., Kreuzaler, F., Ragg, H., Fautz, E., Hahlbrock, K., Harrison, B. J. and Peterson, P. A. (1986). A general method to identify plant structural genes among genomic DNA clones using transposable element induced mutations. *Mol. Gen. Genet.* **187**, 195–201.

5 Metabolic regulation
William C. Plaxton

'Regulation is not a late development superimposed on metabolism after catalysis had become well-established . . . Regulation is the most fundamental difference between living and nonliving systems, and it must have coevolved with other properties of life from the beginning' (Atkinson, 1977).

Key concepts

The ability to regulate the rates of metabolic processes in response to changes in the internal and/or external environment is a fundamental feature which is inherent in all organisms. This adaptability is necessary for conserving the stability of the intracellular environment (homeostasis) which is essential for maintaining an efficient functional state in the organism.

Complexity of metabolism and concept of biochemical unity

Even the most primitive of species is metabolically complex. For example, although the unicellular bacterium *Escherichia coli* is approximately 500-fold smaller than a typical eukaryotic cell, each *E. coli* cell contains over 2000 different types of proteins, most of which are enzymes. The metabolic complexity of even the smallest prokaryote is a result of these many separate enzymatic reactions which make up the metabolic pathways which constitute metabolism.

The substantial metabolic complexity of green plants is reflected by their autotrophic nature, extensive biosynthetic capabilities, metabolic redundancy, and the presence of metabolic

compartments not found in bacteria or animals, namely the plastid and cell vacuole. The sessile nature of most plants further complicates their metabolic needs. In order to survive many plants must possess extensive anatomical, physiological, and biochemical adaptations to environmental stresses such as heat, cold, drought, salinity stress, nutrient limitation, and anoxia.

Despite its complexity, a general understanding of metabolism, applicable to all the phyla, has been achieved. This is because common evolutionary 'solutions' to the problem of 'biochemical design' occur in a wide variety of organisms. Thus, the types of substrates, cosubstrates, coenzymes, fuels and metabolic pathways used (as well as the genetic code) are in many cases ubiquitous to all life forms. This is the concept of *biochemical unity*.

In general, biochemical unity still applies to metabolic regulation. Comparative biochemistry has shown that the design of regulation (i.e. the types of regulatory mechanisms found in metabolic pathways) is similar from one organism to the next. However, it is the implementation of these designs – the regulatory details – which not only can differ widely from species to species, but can differ widely for similar metabolic pathways in different cell types of a single organism, or even within different organelles of a single cell.

For example, the enzyme citrate synthase, which catalyzes the following reaction:

acetylCoA + oxaloacetate → citrate + CoA

is regulated in different manners in different organisms. In respiring animal cells, a major function for this enzyme is in the citric acid cycle which is operating in the mitochondria to produce ATP via oxidative phosphorylation. Here, the overall endproduct, ATP feedback inhibits citrate synthase. This is logical since at high levels of ATP, the ATP-generating citric acid cycle will be slowed down, but will speed up again as the ATP levels fall. *E. coli*, which is primarily anaerobic, generates ATP mainly through glycolytic fermentation of glucose. The main role of the citric acid cycle in *E. coli* is in the production of biosynthetic precursors and reducing power (NADH). In this organism, citrate synthase is unaffected by ATP, but is inhibited by the ultimate endproduct, NADH. In contrast, germinating oil seeds contain a glyoxysomal citrate synthase, which is not inhibited by ATP or NADH. Here the enzyme functions are part of the glyoxylate cycle, a key component in the metabolic conversion of storage triglycerides to sucrose.

Thus, the concept of biochemical unity tends to break down when individual metabolic controls are compared. Although the structure and products of a metabolic pathway may be identical in various organisms, the *environment* and *function* of that pathway may not be the same. Nonetheless, all metabolic controls do have a common basis, and certain regulatory strategies are ubiquitous.

The basis of metabolic control

'Pacemaker' enzymes

It is self evident that the flux, or rate of movement, of metabolites through any given pathway must be closely coordinated with the needs of the cell, tissue, or organism for the final endproduct(s) of the pathway. A major way of accomplishing such regulation is through altering the activity of at least one *rate-limiting* enzyme of the pathway (Cohen, 1983; Crabtree and Newsholme, 1985). Therefore, a major focus of metabolic control has been on these key enzyme reactions – the *pacemakers* or *rate-determining steps* – which are believed to be most important in the control of metabolite flow, *in vivo*. Normally, the rate-determining step(s) of a pathway is essentially irreversible (i.e. has a high negative free energy change, *in vivo*), has a low activity overall, and frequently occurs at the first committed step of a pathway, directly after major branch points, and at the last step of a 'multi-input' pathway. Much effort

has been expended in identifying not only the rate-determining step(s) of metabolic pathways, but also the mechanisms which are used to modulate the activity of these key regulatory enzymes.

Control analysis

In pursuing the 'pacemaker' theory, one attempts to formulate a theory of metabolic control by performing accurate and detailed analyses on the *in vitro* kinetic and regulatory properties of purified 'key' enzymes. A potential problem with this reductionist approach is that biological systems may display regulatory properties which are not possessed by their isolated components (i.e. the properties of biological systems are generally greater than the sum of the properties of their isolated parts). Moreover, as discussed by ap Rees and Hill (1994) the demonstration that a particular enzyme exhibits pronounced regulatory properties *in vitro* indicates that it may be important in controling of metabolic flux *in vivo*, but cannot tell us how much of the control of the pathway is due to this one enzyme. Thus, another important approach to examining metabolic control is to analyze the whole system. The control analysis theory developed by Kacser in 1973 attempts to provide a quantifiable mechanism for probing intact biological systems and interprets resultant data without resorting to preconceived notions as to which enzymes in a pathway are 'rate-determining steps' or 'pacemakers'. In fact, an important prediction of the control analysis theory is that metabolic control is shared among many if not all steps in a pathway. Kacser has established the concept of the *flux control coefficient* (C_E^J) whose value specifies the change in metabolic flux (ΔJ) to small changes in the activity of any enzyme (ΔE) in the metabolic system as follows: $C_E^J = (\Delta J / \Delta E)$. If this value is very small (i.e. 0.01), it means that any change in the activity of the chosen enzyme will have little effect on pathway flux, and that this enzyme therefore exercises very little control on overall pathway flux. If the flux control coefficient is determined to be close to 1.0, it shows that the change in pathway flux responds almost proportionally to a small change in the enzyme's activity. Such an enzyme would clearly be important to regulating the overall rate of metabolite movement (flux) through the pathway. In other words, flux through the pathway is peculiarly sensitive to small perturbations in the activity of that particular enzyme. Experimental determination of the magnitude of appropriate flux control coefficients apparently yields an unambiguous evaluation of the existing quantitative allotment of control among the various steps in a pathway, under specified conditions. As the magnitude of any one flux control coefficient is not a property of the enzyme *per se* but depends upon the concurrent activities of all the other enzymes in the system, individual flux coefficients must be determined experimentally from the intact system. Flux control coefficients are therefore determined by measuring how pathway flux changes following alteration of the activity of a specific enzyme *in situ*. Recent advances in molecular biology now allow for direct manipulations of *in vivo* enzyme activities and promise to yield exciting information on the control of metabolism. Readers should refer to reviews (Kacser, 1987; Kacser and Porteous, 1987; ap Rees and Hill, 1994) for further details and insights concerning the 'control analysis' approach to examining metabolic regulation.

Types of metabolic control

Basic mechanisms

The magnitude of metabolite flux through any metabolic pathway will depend upon the activities of the individual enzymes involved. Two basic mechanisms can potentially be used by the cell to vary the reaction velocity of a particular enzyme. These are 'coarse' and 'fine' metabolic control.

Coarse metabolic control

Coarse metabolic control is a long-term (i.e. hours to days), energetically expensive, response which is achieved through changes in the total cellular population of enzyme molecules. Coarse control can be applied to one or all of the enzymes in a particular pathway, and most frequently comes into play during tissue differentiation or long-term environmental (adaptive) changes. The total amount of a given

enzyme is dependent on the rates of its biosynthesis versus degradation. Thus, any alteration in the rates of transcription, translation, mRNA processing or degradation, or proteolysis can be construed as coarse metabolic control. The regulation of gene expression, protein synthesis and protein turnover have been covered in some detail in the first three chapters of this book, and will therefore not be discussed here.

Fine metabolic control

Fine metabolic controls are generally fast (i.e. seconds to minutes), energetically inexpensive, regulatory devices which modulate the activity of the pre-existing enzyme molecule. Operating mainly on a pathway's regulatory, or rate-limiting, enzyme(s), fine controls allow the cell to prevent metabolic chaos. Fine controls can be thought of as 'metabolic transducers' which 'sense' the momentary metabolic requirements of the cell, and modulate the rate of metabolite flux through the various pathways accordingly.

The enormous advances which have been made in our understanding of plant biochemistry over the past 20 years are providing a clear indication that there is a comparable range and sophistication of fine metabolic controls operating in both plant and animal systems. It is important to be aware that the fine controls to be discussed in detail below are not mutually exclusive, but often interact with, or may actually be dependent upon, one another.

Fine control 1. Alteration in substrate or cosubstrate concentration

The rate of an enzyme-catalyzed reaction is dependent on its substrate concentration when that substrate is subsaturating. Substrate concentrations for most enzymes are indeed subsaturating *in vivo*. In other words, enzymes do not normally operate at their V_{max}. Often the *in vivo* substrate concentration is less than or equal to the K_m or $S_{0.5}$ value of the enzyme for that particular substrate (K_m or $S_{0.5}$ = substrate concentration yielding $0.5V_{max}$ for enzymes which show hyperbolic or sigmoidal substrate saturation kinetics, respectively). Hence, an important question arises as to whether or not the rate of pathway flux could be controlled by changes in the substrate concentration for any of the enzymes which comprise a pathway.

Following activation of a metabolic pathway, the concentration of its constituent metabolites may increase by about two-fold. However, the rate of pathway flux may increase up to 10-fold during the same period. In order to obtain a 10-fold change in activity of an enzyme which shows hyperbolic substrate saturation (i.e. Michaelis–Menten) kinetics, the increase in its substrate concentration would have to be about 80-fold (Fig. 5.1A). Thus, variation in substrate concentration cannot be the sole determinant of the activity of enzymes which show hyperbolic substrate saturation kinetics.

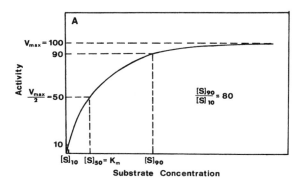

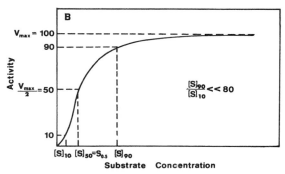

Fig. 5.1 (A) Relationship between substrate concentration and reaction rate for enzymes which show hyperbolic substrate saturation kinetics; (B) relationship between substrate concentration and reaction rate for enzymes which show sigmoidal substrate saturation kinetics.

Not all enzymes show simple Michaelis–Menten substrate kinetics, however. Multisubunit, polymeric enzymes often have more than one substrate binding site which can interact via conformational change to yield a sigmoidal plot of activity versus substrate concentration (i.e. show cooperative binding of substrate) (Fig. 5.1B). Enzymes of this nature have been termed 'allosteric' because they can assume 'other shapes' or conformations by the reversible, non-covalent binding of a specific metabolite. Sigmoidal substrate saturation kinetics has been referred to as *homotropic* allosterism since the allosteric modulator and the substrate are identical. Sigmoidal enzymes can increase their activity 10-fold with as little as a two to five-fold increase in substrate concentration (Fig. 5.1B). The actual increase in substrate concentration which would be required is dependent upon the degree of cooperativity with which the enzyme binds its substrate. With increased cooperativity (i.e. higher values for n_H, the Hill coefficient) smaller-fold increases in substrate concentration are required to effect the same relative change in enzyme activity.

In summary, changes in substrate concentrations which normally occur *in vivo*, could alter the rate of pathway flux, but most significantly for enzymes which show sigmoidal substrate saturation kinetics (i.e. homotropic allosterism). Enzymes of this nature have been found in bacteria, animals and plants. Invariably these enzymes have been identified as regulatory, or rate-limiting. A possible evolutionary 'advantage' of sigmoidal substrate saturation kinetics is that it allows a much more sensitive control of reaction rate by substrate concentration.

Fine control 2. Variation in pH

Most enzymes have a characteristic pH at which their activity is maximal, i.e. the pH optimum. Above or below this pH the activity normally declines, although to varying degrees depending on the particular enzyme. Thus, enzymes can typically show pH activity profiles ranging in shape from very broad to very narrow. The pH optimum of an enzyme is not always the same as the pH of its intracellular surroundings. This suggests that the pH dependence of enzyme activity may be a factor which determines its overall activity in the cell. As cells contain thousands of enzymes, all differentially responsive to pH, the intracellular pH may represent an important element of fine metabolic control.

The light-dependent activation of several of the enzymes of the reductive pentose phosphate pathway (Calvin cycle) provides a well characterized example of how changes in pH can contribute to the overall regulation of plant enzymes. As discussed in Chapter 19 of this book, photosynthetic electron transport has been linked to H^+ uptake into the thylakoid lumen. This establishes a proton gradient between lumen and stroma which is believed to drive photophosphorylation. The transport of H^+ ions into the lumen results in a light-dependent increase in stromal pH from about 7.0 to 8.0. In the absence of light, H^+ ions leak back into the stroma and its pH falls from 8.0 to 7.0. As several of the reductive pentose phosphate pathway enzymes have a relatively sharp alkaline pH optima of between 7.8 and 8.2, the increase in stromal pH which is contingent upon photosynthetic electron transport ensures that these enzymes, and hence the overall cycle, will only be fully operative in the light.

Fine metabolic control of plant enzymes brought about by alterations in intracellular pH is not restricted to the chloroplast. For example, activation of the cytosolic isozyme of the glycolytic enzyme pyruvate kinase during anaerobiosis of germinating castor oil seeds has been attributed, in part, to an anoxia-induced reduction in cytosolic pH (Podestá and Plaxton, 1991).

Fine control 3. Allosteric effectors

Multisubunit regulatory enzymes often have allosteric sites separate from the active or catalytic site, where specific inhibiting or activating metabolites can reversibly bind. By varying the concentration of these non-substrate effector molecules it is possible to markedly alter the velocity of a particular enzyme and, therefore, alter the flux of metabolites through the entire pathway. Allosteric effectors alter enzyme activity by binding to an allosteric site and eliciting a precise change in the enzyme's conformation. This type of conformational change has been termed

heterotropic allosterism, because in contrast to the aforementioned homotropic allosterism, the allosteric modulator is a molecule other than the substrate. The conformational change brought about by the binding of an allosteric effector will thereby promote (in the case of activators) or hinder (in the case of inhibitors) enzyme-substrate interactions such that some or all of the kinetic constants V_{max}, K_m or $S_{0.5}$, and n_H are significantly altered. This regulatory strategy allows for metabolites that are remote from a specific reaction to function as feedforward (activators) or feedback (inhibitors) control signals. The key regulatory significance of this interaction is that, at the low substrate concentrations that usually exist *in vivo*, the rate of reaction, and therefore the pathway flux, can be dramatically increased or decreased by the binding of an effector. This is illustrated graphically in Fig. 5.2 which shows that with the addition of an activator or inhibitor it is possible to vary an enzyme's reaction rate from $0.1V_{max}$ to $0.9V_{max}$ with little or no change in substrate concentration.

Essentially all rate-determining enzymes are at least partially regulated by allosteric effectors. For an effector to have meaningful regulatory significance, the concentration at which it significantly activates or inhibits *in vitro* must be similar to the effectors' actual *in vivo* concentration range. As much of the material discussed in various pages of this book will testify, allosteric regulation is used extensively by plants to regulate pathway flux, *in vivo*.

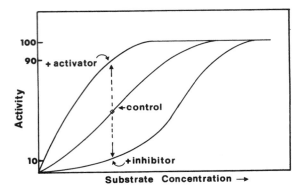

Fig. 5.2 Typical effect of the addition of an activator or inhibitor on substrate saturation kinetics for an allosteric enzyme.

Enzyme activation

Activators interact at an allosteric site and normally increase the reaction rate at subsaturating substrate concentrations by causing a reduction in the K_m or $S_{0.5}$ value (Fig. 5.2). This may or may not be accompanied by a corresponding increase in V_{max}.

One of the first reports of physiologically meaningful allosteric activation of a plant enzyme was provided by Ghosh and Preiss (1966) for spinach leaf ADP-glucose pyrophosphorylase. ADP-glucose pyrophosphorylase, which is found only in plastids, is the rate-limiting enzyme in the pathway of starch biosynthesis from hexose phosphates (Preiss, 1984; also see Chapter 7). This enzyme catalyzes the formation of pyrophosphate and ADP-glucose from ATP and glucose 1-phosphate and shows significant feedforward activation by low concentrations of 3-phosphoglyceric acid (3-PGA), the immediate product of photosynthetic CO_2 fixation by ribulose 1,5-bisphosphate carboxylase. In the case of ADP-glucose pyrophosphorylase the activator 3-PGA causes a very marked enhancement in V_{max} as well as a significant reduction in the K_m values for the substrates ATP and glucose 1-phosphate. The concentration of activator required for half-maximal activation (i.e. $A_{0.5}$) is less than 0.05 mM. This is in the range of 3-PGA concentrations found in the cell and suggests this activating effect is relevant, *in vivo*.

Enzyme inhibition

Competitive inhibition is the commonest form of enzyme inhibition and arises when the inhibitor and substrate compete for a common substrate binding site (i.e. the active site), so that when one binds the other cannot. The effect of a competitive inhibitor is to decrease the apparent affinity of the enzyme for the substrate without any effect on the reactivity of the enzyme-substrate complex once formed (Cornish-Bowden and Wharton, 1988). Thus, competitive inhibition causes an increase in the K_m or $S_{0.5}$ values, does not affect V_{max}, and is reversed by increasing the substrate concentration. Although competitive inhibitors are not true allosteric effectors (because they do not bind to an allosteric site, a site which is distinct from the catalytic site) they nevertheless constitute an important aspect of enzyme regulation, *in vivo*. For

example, potent competitive inhibition of many plant acid phosphatases by their product P_i is in accord with the hypothesis that acid phosphatases are particularly active *in vivo* during nutritional P_i starvation when intracellular P_i levels are greatly depleted (Duff *et al.*, 1994). Conversely, any accumulation of cellular P_i caused by P_i resupply would act as a tight regulatory control to prohibit further hydrolysis of certain phosphorylated compounds by the various intracellular acid phosphatases.

Mixed inhibition is a rarer, but important, form of metabolic regulation in which the inhibitor reversibly interacts with the enzyme, or enzyme-substrate complex, at a true allosteric site. In this instance both V_{max} and the affinity of the enzyme for its substrate(s) are reduced. The inhibition of plant ADP-glucose pyrophosphorylase by P_i is a typical example of mixed type inhibition (Preiss, 1984). Mixed inhibition is frequently misconstrued as noncompetitive inhibition, which theoretically reduces an enzyme's catalytic potential (i.e. V_{max}) without affecting substrate binding (i.e. K_m or $S_{0.5}$). However, according to Cornish-Bowden and Wharton (1988): 'With more information about enzyme mechanisms it has become evident that it is not very likely for an inhibitor to interact with an enzyme and alter its catalytic potential without having any effect on the binding of substrate and, if we exclude pH effects in which the proton can be regarded as an inhibitor, nature has not provided us with any examples.'

Interacting effectors

The activity of an allosteric enzyme, *in vivo*, is dependent on the relative concentrations of its activators versus inhibitors. Often, the presence of an activator can override, or cancel, inhibitory signals. This is shown clearly for the aforementioned ADP-glucose pyrophosphorylase, in which the activator 3-PGA negates the inhibitory effect of P_i (Preiss, 1984). Thus, it is the ratio of plastidic [3-PGA]/[P_i] which is believed to be a major factor in determining the rate ADP-glucose pyrophosphorylase, and thus starch biosynthetic activity, *in vivo*. This hypothesis, formulated in 1966 on the basis of ADP-glucose pyrophosphorylase allosteric properties which were observed *in vitro*, has been subsequently confirmed by various *in vivo* evidence (Preiss, 1984).

The adenine nucleotides as metabolic effectors

Energy transduction and energy storage, involving the adenine nucleotides (ATP, ADP, and AMP) are a fundamental feature of metabolism. Thus, it is of no surprise that an extensive system of allosteric regulators are the adenine nucleotides which operate on a large number of reactions to adjust the rate of formation of ATP to its utilization (Atkinson, 1977; Crabtree and Newsholme, 1985). In many plant and animal tissues, the adenine nucleotides are maintained in equilibrium by the enzyme adenylate kinase which catalyzes the reaction, $ATP + AMP \leftrightarrow 2\ ADP$. Thus, a decrease in the concentration of ATP will occur following an increase in that of AMP, and vice versa. The widespread importance of the adenine nucleotides in metabolic regulation led Atkinson (1977) to introduce the concept of 'energy charge' (energy charge = 'the state of the adenylates'), which he defined as:

$$\text{energy charge} = \frac{[ATP] + \frac{1}{2}[ADP]}{([ATP] + [ADP] + [AMP])}$$

Energy charge can theoretically vary between 0 (i.e. all the adenylate is present as AMP) and 1 (i.e. all of the adenylate is present at ATP). As this parameter does fluctuate to a certain extent *in vivo*, it provides a basis for metabolic control by energy charge. Regulatory enzymes that occur in pathways in which ATP is consumed (i.e. anabolic pathways) respond to changes in energy charge in the general way shown by curve 2 of Fig. 5.3. Regulatory enzymes which participate in pathways in which ATP is regenerated (i.e. catabolic pathways) respond in the general way shown by curve 2, Fig. 5.3. It is clear from Fig. 5.3 that any tendency for the energy charge to fall would be prevented by the consequential increase in metabolic flux through ATP-regenerating pathways (curve 1) and by the decrease in flux through ATP consuming pathways (curve 2). In healthy cells, energy charge is normally controlled at a value of between 0.7 and 0.9. This provides a constant environment for the application of all the other mechanisms of metabolic regulation.

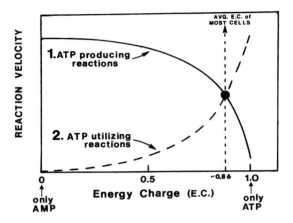

Fig. 5.3 Responses to the cellular energy charge of regulatory enzymes in metabolic pathways in which ATP is produced (curve 1) and in which it is utilized (curve 2) (adapted from Atkinson, 1977).

Fine control 4. Covalent modification

This method of fine control functions cooperatively with allosteric regulation. The general model is that an enzyme is interconverted between a less active and more active form. This interconversion is not due to a relatively freely reversible equilibrium, as encountered with regulation by allosteric effectors, but is governed by two thermodynamically favourable enzyme catalyzed reactions which result in the formation of new stable covalent bonds. Interconversion in either direction can be very fast (i.e. minutes) and very complete (up to 100% conversion). A change in enzyme conformation which is induced by covalent modification leads to an alteration in enzyme-substrate interactions such that kinetic parameters such as V_{max}, K_m and n_H are significantly elevated or reduced. Either interconversion can be used to override or cancel existing allosteric effector signals. This energetically inexpensive control mechanism can quickly provide the cell with an essentially 'new' enzyme form whose kinetic properties are well geared to the cell's momentary metabolic needs.

A key facet of enzyme regulation by reversible covalent modification is that it is the major mechanism in higher eukaryotes whereby extracellular stimuli, such as hormones, light, or environmental stress coordinate the regulation of key enzymes of intracellular pathways (Cohen, 1983, 1985; Ranjeva and Boudet, 1987; Buchanan, 1991, 1992; Poovaiah and Reddy, 1993; Huber *et al.*, 1994). Although some 150 types of post-translational modifications of proteins have been reported *in vivo*, very few appear to be important in enzyme regulation (Cohen, 1983). Dithiol–disulfide interconversions and phosphorylation–dephosphorylation are the most important mechanisms of reversible covalent modification used in higher eukaryote enzyme regulation.

Dithiol-disulfide interconversion

Covalent modification by dithiol–disulfide exchange involves reactions which are chemically similar to those which occur in the formation of disulfide bonds which stabilize the tertiary structure of proteins. The only difference from the latter is that in the native enzyme, disulfides participating in regulation must be accessible to reduction by external thiols since regulation by this mechanism in response to cellular redox state can occur only if the reaction is freely reversible (Ziegler, 1985).

Thioredoxin is a 12 kD heat-stable protein which in its reduced SH form can act as a protein-disulfide reductase (Cohen, 1983). Distributed throughout the phyla, thioredoxin has been found in all cell types in which it has been sought (Holmgren, 1985). This protein appears to play a variety of roles in different cell types. In bacterial and animal systems it may act as a general protein-disulfide reductant, or may participate in the reduction of specific enzymes. In these organisms reduced thioredoxin is regenerated by thioredoxin reductase, at the expense of NADPH (Holmgren, 1985).

Dithiol–disulfide enzyme interconversion appears to have a much greater regulatory significance in photosynthetic organisms. As discussed in further detail in Chapter 21 and elsewhere (Cohen, 1983; Ziegler, 1985; Buchanan, 1991, 1992; Scheibe, 1991; Wolosiuk *et al.*, 1993), this type of reversible covalent modification is extremely important in linking photosynthetic electron flow to light regulation of several key chloroplastic enzymes involved in photosynthetic CO_2 fixation and related processes. Reducing equivalents (electrons) are ultimately generated from H_2O by the photosynthetic electron transport chain. The proteins, ferredoxin, ferredoxin-thioredoxin reductase, and thioredoxin constitute the machinery which shuttles the reducing equivalents from the electron transport chain to selected target enzymes (Fig. 5.4A). Two different thioredoxins, designated *f* and *m*, are part of the ferredoxin/thioredoxin system in chloroplasts. In the reduced state, thioredoxin *f* selectively activates several enzymes of carbohydrate synthesis, whereas thioredoxin *m* preferentially activates and inhibits NADP-malate dehydrogenase and glucose-6-phosphate dehydrogenase, respectively (Buchanan, 1991, 1992). With certain chloroplastic enzymes, the light-dependent alkalization of the stroma and increased concentration of several allosteric effectors appears to enhance the thioredoxin-mediated disulfide–dithiol interconversion process (Buchanan,

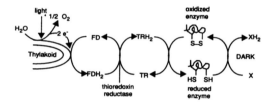

(a)

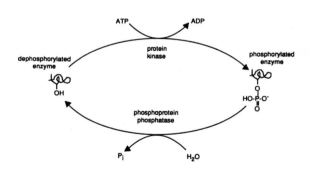

(b)

Fig. 5.4 The regulation of enzyme activity via reversible covalent modification. For details refer to the text. (a) Dithiol–disulfide interconversion. FD and FDH$_2$, oxidized and reduced ferredoxin, respectively; TR and TRH$_2$, oxidized and reduced thioredoxin, respectively; X, some oxidant normally kept reduced in the light; (b) Phosphorylation–dephosphorylation.

1992; Scheibe, 1991). The discovery of a thioredoxin *h* in the plant cytosol and mitochondria (Marcus *et al.*, 1991) will possibly extend this type of regulation to enzymes localized outside the chloroplast.

By mechanisms not yet fully understood, enzymes which are reduced by thioredoxin in the light are oxidized to disulfide forms in the dark. This might be catalyzed by the action of either oxidized thioredoxin or low molecular weight oxidants such as oxidized glutathione, dehydroascorbate, or hydrogen peroxide (Cohen, 1983; Buchanan, 1991, 1992; Wolosiuk *et al.*, 1993).

Phosphorylation-dephosphorylation

Enzyme modification by the reversible covalent incorporation of phosphate is a widespread phenomenon with important consequences for metabolic control *in vivo*. Reversible phosphorylation of proteins on seryl, threonyl, or tyrosyl residues is known to occur in prokaryotes and eukaryotes. As outlined in Fig. 5.4B, the phosphorylation reaction is catalyzed by a protein kinase, usually in an ATP-dependent manner, whereas the reverse reaction is normally catalyzed by a phosphoprotein phosphatase. Both protein kinases and phosphoprotein phosphatases can have wide or narrowly defined substrate specificities. Protein kinases and phosphoprotein phosphatases are always subject to their own fine controls. Each class of protein kinase is allosterically stimulated by a specific 'signal metabolite' (i.e. cyclic AMP or cyclic GMP, diacylglycerol, Ca^{2+}) which makes enzyme phosphorylation responsive to extracellular signals. Protein kinase activation is normally initiated by a much smaller change in effector concentration than with respect to allosteric controls. As the smallest increase in effector concentration yields a maximal effect, a complete on/off control can be obtained. An increase in signal amplification can be made possible by designing an 'amplification cascade' with more enzymes involved (Cohen, 1983, 1985). Animal phosphoprotein phosphatases are frequently controlled by an inhibitor protein which complexes with the enzyme under conditions when the corresponding kinase is activated. Although protein kinases and phosphoprotein phosphatases have been reported to occur in plant tissues, less is known about

their distribution and regulation in plants relative to animals. In both plants and animals, there are significant homologies in amino acid sequences among most of the known protein kinases and protein phosphatases indicating that the interconverting enzymes belong to extended families of proteins that have probably evolved from common ancestral origins (Huber *et al.*, 1994).

It has been definitively shown that in animal systems enzyme phosphorylation serves to coordinate the relative activities of competing metabolic pathways in multifunctional tissues (Cohen, 1983, 1985). At least 50 regulatory enzymes found in all the major metabolic pathways of higher animals are now known to be regulated by reversible protein kinase mediated phosphorylation.

Although our understanding of the role of protein phosphorylation in plant metabolic regulation is still rather limited, there is definitive evidence that plants utilize reversible enzyme phosphorylation to regulate flux through various metabolic pathways. The first plant enzyme which was determined to be regulated by phosphorylation–dephosphorylation control is the pyruvate dehydrogenase complex (PDC) from the mitochondrial matrix of broccoli buds (reviewed by Ranjeva and Boudet, 1987). This enzyme complex can be interconverted between an activated dephosphorylated form and an inactivated phosphorylated form, in a process very similar to that known for animal PDC. This interconversion is believed to play a key role in regulating carbon flow between glycolysis, the citric acid cycle and fatty acid metabolism. The description of this important mechanism for the regulation of higher plant PDC extended our knowledge of this type of metabolic control into the higher plant kingdom and raised the likelihood that other plant enzymes could be controlled in a similar fashion. Not surprisingly, at least eight other plant enzymes have currently been demonstrated, or strongly suggested to be, regulated by reversible phosphorylation. These are: chloroplast stroma pyruvate, orthophosphate dikinase, microsomal hydroxymethylglutaryl-CoA reductase, glyoxysomal malate synthase, plasma membrane H^+-ATPase, as well as the cytosolic enzymes phospho*enol*pyruvate carboxylase, sucrose-phosphate synthase, nitrate reductase, and quinate dehydrogenase (Huber *et al.*, 1994). In addition, several nonenzymatic proteins are known to be reversibly phosphorylated *in vivo*. For example, a 26 kDa protein of the light harvesting chlorophyll–protein complex of photosystem II has been shown to undergo phosphorylation–dephosphorylation and that this probably plays a role in regulating photosynthetic electron flow (Ranjeva and Boudet, 1987; Huber *et al.*, 1994). These findings, along with the existence in plants of several types of protein kinase and phosphoprotein phosphatase (Poovaiah and Reddy, 1993; Duff *et al.*, 1994; Huber *et al.*, 1994), indicates that phosphorylation–dephosphorylation will prove to be a common and important control mechanism in mediating plant cellular responses to external stimuli.

Fine control 5.
Subunit association–disassociation

Usually, multimeric enzymes become less active or inactive when the constituent subunits are dissociated. This property has been exploited to some degree by various organisms to help regulate the activities of certain enzymes. Aggregation or dissociation of subunits is normally induced by the binding of a small molecular weight effector molecule, but may also be instigated by reversible covalent modification.

A plant enzyme that has been proposed to be regulated in this fashion, *in vivo*, is the pyrophosphate-dependent phosphofructokinase (PFP; see Chapters 7 and 8). PFP can apparently be interconverted between an active, approximately 260 kD tetramer, and a less active 130 kD dimer (Black *et al.*, 1987; Dennis and Greyson, 1987). The state of aggregation of this enzyme is dependent on the concentrations of its allosteric activator, fructose 2,6-bisphosphate (Fru-2,6-P_2), and product of the forward reaction, PP_i. The tetrameric form of the enzyme, as well as its activity in the forward, or glycolytic direction, is enhanced by the presence of Fru-2,6-P_2. The dimeric form of PFP, as well as its activity in the reverse, or gluconeogenic direction, is promoted by PP_i (Dennis and Greyson, 1987). Thus, the reversible association of PFP subunits, as determined by the relative cytosolic concentrations of Fru-2,6-P_2, and PP_i, may represent a glycolytic/gluconeogenic regulatory mechanism.

Fine control 6. Reversible associations of metabolically sequential enzymes

Recent studies are providing convincing evidence for subcellular structuring of metabolic pathways which were once thought of as being entirely soluble. For example, various workers have described associations between many of the so-called 'soluble' enzymes of glycolysis and membrane fractions or structural proteins in animal cells (Storey, 1985; Masters *et al.*, 1987; Srere, 1987; Keleti *et al.*, 1989). Of regulatory significance is that the extent of enzyme binding appears to be closely related to the rate of flux through the pathway. During anaerobic mammalian muscle contraction, for example, a marked increase in the binding of key glycolytic enzymes to contractile proteins has been observed. The degree of binding was found to dissipate once the contractile activity, and hence the need for a high rate of glycolytic flux, ceased. Of equal significance are the findings that sequential glycolytic enzymes can also show specific interactions with each other – interactions which have mutual kinetic effects on the enzyme activities. Similar correlations between the degree of enzyme binding and rate of pathway flux have been observed elsewhere in animal systems. Enzymes which show a metabolic dependence in the degree to which they are associated with subcellular particulate structures have been termed *ambiquitous* (Masters *et al.*, 1987).

The micro-compartmentation of enzymes and metabolic pathways which can result from ambiquitous interactions could provide an effective means of metabolic control via (1) direct transfer or 'channeling' of intermediates between consecutive enzymes (Fig. 5.5), and (2) altering enzyme kinetic properties due to conformational changes occurring during binding. It has been suggested, therefore, that metabolic activation of pathways such as glycolysis can occur not only by allosteric regulation and covalent modification of key enzymes, but also by altering the partitioning of enzymes from the soluble to the bound phase (Storey, 1985; Masters *et al.*, 1987; Keleti *et al.*, 1989).

The existence of known stable multienzyme complexes provides favourable evidence for the interaction of soluble enzymes that are metabolically sequential. Due to high intracellular protein concentrations, such interactions are far more likely

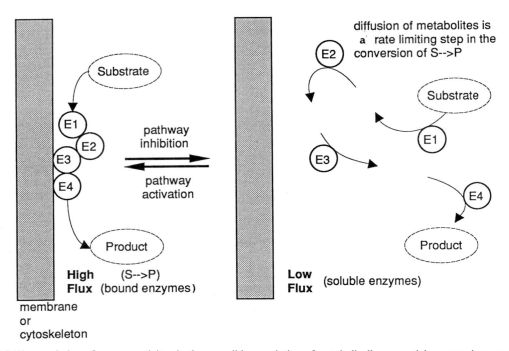

Fig. 5.5 The regulation of enzyme activity via the reversible association of metabolically sequential enzymes into an organized multienzyme complex or '*metabolon*'. E1, E2, E3 and E4 represent the sequential enzymes which catalyze the consecutive reactions of a hypothetical metabolic pathway. 'Substrate' (or S) and 'Product' (or P) depict the pathway's overall starting material(s) and end-product(s), respectively.

in vivo than in dilute *in vitro* enzymological studies (Srere, 1987; Keleti *et al.*, 1989). The limited solvent capacity of the cell also supports interactions between sequential enzymes of a metabolic pathway as the need for a large pool of free intermediates would be eliminated (Atkinson, 1977).

The molecular mechanisms which control the extent of enzyme binding *in vivo* are not yet fully understood. Binding of animal glycolytic enzymes *in vitro* can be influenced by pH, concentrations of substrates, products and allosteric effectors, enzyme phosphorylation, as well as by changes in osmotic and/or ionic strength (Masters *et al.*, 1987; Srere, 1987; Keleti *et al.*, 1989).

Although there is now compelling evidence for complexing and/or binding of so-called 'soluble' enzymes and the direct transfer of substrates between sequential enzymes of various pathways in animal tissues, these possibilities have only recently been addressed in plant systems (Hrazdina and Jensen,

1992). For example, the chloroplastic ferredoxin, ferredoxin–thioredoxin reductase, and thioredoxin system used for disulfide–dithiol enzyme interconversion may exist *in vivo* as an enzyme complex termed 'protein modulase' (Ford *et al.*, 1987). It has been suggested that ionic strength affects the interactions among these proteins and in part could determine the fate of the protein modulase complex *in vitro*. In a subsequent study, Haberlein and coworkers (1992) demonstrated that protein–protein interactions may play a significant role in addition to disulfide–dithiol exchange reactions in the light-dependent activation of several chloroplastic enzymes by thioredoxin. Various studies have also provided compelling evidence that there is direct channeling of substrates between reductive pentose phosphate pathway enzymes in illuminated chloroplasts, and that the enzymes of this pathway exist as large supramolecular protein complexes which may reversibly associate with the thylakoid membrane in the light (Wolosiuk *et al.*, 1993). Similarly, the biosynthesis of flavonoids may take place in a multienzyme complex wholly or partially associated with the cytosolic face of the endoplasmic reticulum membrane (Hrazdina and Jensen, 1992). Furthermore, the well characterized stimulation of respiration which accompanies aging of underground storage organ slices may result, in part, from an association of glycolytic enzymes with a particulate fraction of the cell promoting an elevated glycolytic rate (Moorhead and Plaxton, 1988). Subsequent work indicated that cytosolic aldolase may specifically interact with the metabolically sequential: (1) cytosolic ATP- and PPi-dependent phosphofructokinases in carrot storage roots (Moorhead and Plaxton, 1992), and (2) cytosolic fructose-1,6-bisphosphatase in germinating castor oil seeds (Moorhead *et al.*, 1994). The above studies indicate that before an overall understanding of metabolic control in plants can be achieved we must determine not only how, where and when 'soluble', metabolically sequential plant enzymes might be aggregated into micro-compartments forming an organized multienzyme system (or '*metabolon*'; Srere, 1987), but also how such associations may alter individual enzyme kinetic/regulatory properties. Equally significant will be the evaluation as to what extent the *reversible* formation and dissolution of metabolons contributes to the overall integration and control of plant metabolism.

The central role of calcium in plant metabolic regulation

The free Ca^{2+} ion is now recognized as having an important role in animals and plants as a 'second messenger' or 'signal molecule' which couples various extracellular stimuli such as hormones, light, gravity, or stress with intracellular metabolic events. The coupling of stimulus to response by Ca^{2+} implies that an external stimulus leads to a change in the rate at which Ca^{2+} is transported into or out of a specific subcellular compartment. The resulting alteration of the intracellular free Ca^{2+} concentration thus represents the 'signal' which, through its ability to regulate the activity of specific target enzymes, leads to the appropriate metabolic response.

If Ca^{2+} does act as a second messenger in plants it should be possible to demonstrate that its *in vivo* concentration changes in response to extracellular primary signals such as light or hormones. Through manipulation of external or internal Ca^{2+} concentrations it was initially possible to indirectly demonstrate a role for Ca^{2+} in stimulus-response coupling (Hepler and Wayne, 1985; Kauss, 1987; Poovaiah and Reddy, 1993). More recently, direct measurements of signal-induced alterations in cytosolic Ca^{2+} concentrations which precede a physiological response have been made (Shacklock *et al.*, 1992; Coté and Crain, 1993; Fallon *et al.*, 1993; Poovaiah and Reddy, 1993). Thus, the evidence in support of Ca^{2+} as a universal second messenger in transducing external stimuli in plants is unequivocal.

Regulation of intracellular free calcium concentration

It is generally accepted that animal and plant cytosolic free Ca^{2+} concentrations are normally maintained in the range of 0.1 to 1 μM and that processes are activated by an increase in free Ca^{2+} concentration from 0.5 to 5 μM (Cohen, 1985; Kauss, 1987; Drøbak, 1993; Coté and Crain, 1993; Poovaiah and Reddy, 1993). This very low cytosolic Ca^{2+} concentration is maintained against very high concentrations of Ca^{2+} both outside the cell (i.e. in the cell wall) and inside various subcellular compartments (i.e. cell vacuole, and the lumen of thylakoids and endoplasmic reticulum).

Although there is now direct evidence that many external signals induce changes in cytosolic Ca^{2+}, the mechanisms by which these signals mediate changes in cytosolic Ca^{2+} are still poorly understood. However, recent advances have demonstrated the existence of several types of specific Ca^{2+} channels (for mediating Ca^{2+} influx) and pumps (for mediating Ca^{2+} efflux) in plant membranes that are regulated by plasma-membrane associated and intracellular control mechanisms (Kauss, 1987; Schroeder and Thuleau, 1991; Poovaiah and Reddy, 1993). Thus, all plant cells clearly have a complex 'machinery' in place with which to maintain or alter Ca^{2+} concentration gradients.

The Ca^{2+} channels of higher plant cells belong to

two general classes of ion channels: plasma membrane Ca^{2+} channels, which allow Ca^{2+} influx from the cell wall space into the cytosol, and Ca^{2+} release channels located in the membrane of intracellular organelles, which allow release of sequestered Ca^{2+} (Schroeder and Thuleau, 1991). Likewise, two major categories of Ca^{2+} pumps are thought to exist in higher plants. One is a Ca^{2+}-ATPase which may be partially dependent on calmodulin and is primarily localized in the endoplasmic reticulum and plasma membranes (Kauss, 1987; Poovaiah and Reddy, 1993; Evans, 1994). This system has a very high affinity for Ca^{2+} and may be the 'primary' pump which maintains the low free cytoplasmic Ca^{2+} concentration found in unstimulated cells. The second major class of Ca^{2+} pump is a 'secondary pump' which is localized in the tonoplast of the cell vacuole and utilizes proton gradients generated by the tonoplast H^+-ATPase and/ or H^+-PP_iase as the driving force and has been characterized as a Ca^{2+}/H^+ antiport (Kauss, 1987; Poovaiah and Reddy, 1993). Plastids and mitochondrial membranes are also believed to contain Ca^{2+} transport systems which are used to regulate free Ca^{2+} concentrations within these organelles (Kauss, 1987). Changes in the concentration of free Ca^{2+} probably plays an important role in metabolic regulation in all the subcellular compartments found in plants.

It has become evident that plant and animal Ca^{2+} pumps, while exhibiting areas of marked homology, also differ greatly in properties and regulation. Evans (1994) has summarized evidence that refutes the suggestion (based on work with animal cells) that calmodulin-stimulated Ca^{2+}-ATPases are located exclusively in the plasma membrane of all eukaryotes. As discussed in the preceding paragraph, higher plant endoplasmic reticulum membranes also show this activity, and this has led Evans (1994) to the interesting conclusion that plants have a more complex combination of Ca^{2+} pumps than mammalian cells.

Calcium-modulated proteins

Many lines of evidence accrued since 1970 have demonstrated that a class of Ca^{2+} binding proteins

referred to as 'Ca^{2+}-modulated proteins' are the targets or receptors for Ca^{2+} acting as a signal transducer in eukaryotic cells. These proteins have the ability to bind Ca^{2+} in a reversible manner with dissociation constants ranging from the nanomolar to micromolar range under physiological conditions (Roberts and Harmon, 1992). Stimulation of the cell increases cellular free Ca^{2+} concentrations by up to 5 μM, allowing formation of the Ca^{2+}-modulated protein: Ca^{2+} complex, which is the active species. The Ca^{2+}-modulated protein may be enzymatic, as in the case with Ca^{2+}-dependent protein kinase, or may be a regulatory protein, which when complexed with Ca^{2+} binds to, and thereby regulates the activity of an appropriate enzyme. The enzyme whose activity is altered, in turn, regulates some cellular processes.

Calmodulin

The most important of all known Ca^{2+}-modulated proteins is calmodulin, a highly conserved, heat-stable, small molecular weight protein found in all eukaryotic cells (Roberts and Harmon, 1992; Poovaiah and Reddy, 1993). Containing 4 Ca^{2+} binding sites per molecule, calmodulin is composed of a single subunit and 148 amino acids. The binding of Ca^{2+} induces a conformational change resulting in increased alpha helicity. This Ca^{2+}-induced conformational change allows the protein to interact with and alter the activity of calmodulin-dependent enzymes. The Ca^{2+}-calmodulin complex has a wide range of biochemical activities, including effects on both 'coarse' and 'fine' metabolic regulation (Hepler and Wayne, 1985; Roberts and Harmon, 1992; Poovaiah and Reddy, 1993). Because of these characteristics, calmodulin has become the standard against which all other Ca^{2+}-modulated proteins have been compared.

Ca^{2+} and Ca^{2+}-calmodulin regulated plant enzymes

Plant enzymes which have been reported to be directly regulated by Ca^{2+} or Ca^{2+}-calmodulin include NAD-kinase, β-glucan synthase, Ca^{2+} and H^+-ATPase, quinate dehydrogenase, and protein kinases (Kauss, 1987; Roberts and Harmon, 1992; Poovaiah and Reddy, 1993).

NAD kinase

NAD kinase is found in the plastid, cytosol and mitochondrion of plants and catalyzes the ATP-dependent conversion of NAD^+ to $NADP^+$ The pioneering work of Anderson and Cormier (1978) showed that the activity of NAD kinase was lost during its purification from pea leaves, but could be recovered by adding a Ca^{2+}-dependent, heat-stable, protein activator. This was the first report which demonstrated a probable link between a Ca^{2+}-dependent regulatory protein and a specific enzyme in higher plants. The protein activator was subsequently identified as calmodulin. Figure 5.6 shows a typical calmodulin dose–response curve for NAD kinase. It demonstrates that the activation of NAD kinase occurs at a nanogram levels of calmodulin, which makes this stimulatory effect physiologically relevant. Addition of the Ca^{2+} chelator EGTA abolishes the effect of calmodulin (Fig. 5.6).

The ubiquitous nature of calmodulin-dependent NAD kinase in plants indicates a fundamental role for this enzyme in modulating the pools of NAD versus NADP in various subcellular compartments (Roberts and Harmon, 1993). NAD and NADP concentrations are known to undergo changes *in vivo* in response to a number of environmental stresses and the relative concentrations of these coenzymes could have a profound influence on the metabolic

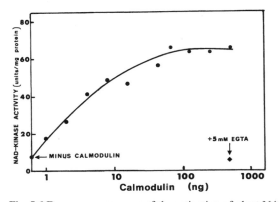

Fig. 5.6 Dose–response curve of the activation of plant NAD kinase by bovine brain calmodulin (figure reproduced with the permission of Prof. D. Marme and Elsevier Science Publishers B.V.).

fluxes of plant cells (particularly that of reductive biosynthetic pathways which require NADPH). However, an integrated picture of the specific metabolic events that are regulated by calmodulin-dependent NAD kinases is still lacking (Roberts and Harmon, 1993).

Protein kinase

Many of the effects of Ca^{2+} on animal metabolism have been found to be mediated by Ca^{2+} acting alone, or by Ca^{2+} acting through calmodulin, to activate protein kinases which results in the phosphorylation of key regulatory enzymes. This has led to the concepts of signal 'amplification' and multienzyme 'cascades' (Cohen, 1983, 1985; Ranjeva and Boudet, 1987). There is now ample evidence that plants also utilize Ca^{2+} to mediate protein phosphorylation, and this is an important mechanism whereby various extracellular stimuli are coupled with intracellular metabolic events (Ranjeva and Boudet, 1987; Roberts and Harmon, 1993; Poovaiah and Reddy, 1993). Relative to animal systems, however, very little is known about plant protein kinase structure or regulation. As numerous protein kinases in animal cells are regulated by calmodulin an initial emphasis was placed on the search for calmodulin-dependent protein kinases in plants. Although a modest effect of exogenous calmodulin has been observed on the protein kinase activity of various impure plant extracts, the isolation and characterization of a homogeneous plant protein kinase that is unequivocally regulated by calmodulin has not yet been achieved (Roberts and Harmon, 1992; Poovaiah and Reddy, 1993). Nevertheless, some of the recent studies coupled with earlier reports strongly suggest the presence of calmodulin-dependent protein kinases in plants. By contrast, a Ca^{2+}-dependent and calmodulin-independent protein kinase, which is unique to plant systems, has been fully purified and highly characterized from soybean (Roberts and Harmon, 1992). The enzyme's primary structure has been deduced from cDNA clones encoding this enzyme; it contains both a protein kinase catalytic domain and a Ca^{2+}-binding regulatory domain similar to calmodulin. This explains the direct activation of the enzyme by Ca^{2+} and has established it as the prototype for a new class of protein kinases. Ca^{2+}-dependent protein kinases appear to be ubiquitous to higher plants and green algae. Moreover, both soluble and membrane bound isoforms of the Ca^{2+}-dependent protein kinase have been demonstrated to exist in a variety of subcellular compartments suggesting that this family of enzymes may be involved in multiple signal transduction pathways in higher plants (Roberts and Harmon, 1992).

Even less is known about the endogenous protein substrates for these kinases, or how phosphorylation affects their activity. Purification and characterization of Ca^{2+} and Ca^{2+}-calmodulin dependent protein kinases and their substrates will be necessary before a comprehensive picture emerges as to the overall role of Ca^{2+} in plant metabolic regulation. Despite its presence in plants, cyclic AMP has no known effect on the phosphorylation of plant proteins (Ranjeva and Boudet, 1987).

Stimulus-response coupling by Ca^{2+}

Light

Photoregulation of plant metabolism can be achieved by effects on cellular energy charge or reducing power, rather than by a classical second messenger. The coupling of photosynthetic electron flow with the disulfide–dithiol interconversion of several chloroplastic enzymes provides a good example of how light can directly regulate plant metabolic processes. However, animal photoreceptor cells have provided plant biochemists with a model whereby Ca^{2+} can act as a second messenger in coupling light with cellular function (Hepler and Wayne, 1985). Indeed, a growing body of evidence is becoming available in plant systems to indicate that light does play an important regulatory role through effects on Ca^{2+} fluxes within the plant cell. Light can stimulate Ca^{2+} uptake by the chloroplast as well as Ca^{2+} efflux from internal stores (Kauss, 1987; Ranjeva and Boudet, 1987; Poovaiah and Reddy, 1993). Of particular interest are studies which suggest that at least some phytochrome-induced responses are mediated by Ca^{2+} and calmodulin (Roux et al., 1986; Shacklock et al., 1992; Fallon et al., 1993).

Phytochrome is an important proteinaceous pigment which regulates many light-mediated growth and developmental responses in plants (Roux et al., 1986). This photoreceptor is found in all green plants and has the ability to interconvert between two spectrally different forms, a red absorbing P_r form and a far-red absorbing P_{fr} form. This interconversion allows phytochrome to initiate responses that are characteristically promoted by red light and inhibited or reversed by far-red illumination. That Ca^{2+} mediates at least some phytochrome initiated responses is suggested by the findings that: (1) P_{fr}, the active form of phytochrome, can affect many membrane properties including promoting the uptake of Ca^{2+} into plant cells; (2) several photomorphogenic effects of red light can be mimicked in darkness by the chemical induction of Ca^{2+} uptake into cells, and these same effects can be blocked by calmodulin antagonists; (3) Ca^{2+}, calmodulin and P_{fr} all play a role in nuclear protein phosphorylation; and (4) red light induces a transient increase in cytosolic Ca^{2+} and activates Ca^{2+} and Ca^{2+}-calmodulin dependent signal transduction pathways (Hepler and Wayne, 1985; Roux et al., 1986; Shacklock et al., 1992; Fallon et al., 1993). These data have supported the hypothesis that phosphorylation of nuclear proteins may be an intermediate step in linking the photoactivation of phytochrome with the photoregulation of gene expression (Roux et al., 1986). Of interest is the discovery that, when prepared from etiolated plant tissues, phytochrome itself is subject to reversible phosphorylation by a protein kinase which copurifies with the photoreceptor (Biermann et al., 1994). The regulatory significance of a phytochrome protein kinase activity, as well as phytochrome phosphorylation, remains to be elucidated.

Plant hormones

Many hormones and neurological stimuli in animals regulate cell metabolism by inducing an alteration in the Ca^{2+} concentration in the cytoplasm. As with animal systems, the role of plant growth hormones (i.e. gibberellins, auxins, cytokinins, and abscisic acid) in manipulating cellular Ca^{2+} levels and/or protein phosphorylation has been firmly established. Consequently, a role for Ca^{2+} as a second messenger important to modulation of hormone-responsive plant systems has been proposed (Hepler and Wayne,

1985; Schroeder and Thuleau, 1991; Trewavas and Gilroy, 1991; Drøbak, 1992; Poovaiah and Reddy, 1993; Coté and Crain, 1993, 1994).

Model of stimulus-response coupling by Ca^{2+}

The current models of how Ca^{2+} mediates transmembrane signaling in plants (Roux et al., 1986; Schroeder and Thuleau, 1991; Trewavas and Gilroy, 1991; Drøbak, 1992; Poovaiah and Reddy, 1993; Coté and Crain, 1993, 1994) are based upon similar models which have been proposed for animal

systems. The initial event in this signaling is the interaction of an extracellular 'signal' molecule with a plasma membrane receptor, or of light with the photoreceptor phytochrome (Fig. 5.7). These interactions are thought to alter Ca^{2+} flux across the plasma membrane or to activate phospholipase C, a membrane-bound enzyme which catalyzes the hydrolysis of a particular class of phospholipids, the phosphoinositides: phosphatidylinositol (PI), phosphatidylinositol 4-phosphate (PIP), and phosphatidylinositol 4,5-bisphosphate (PIP_2). Diacylglycerol (DAG) and inositol 1,4,5-triphosphate (IP_3) are second messenger products of phospholipase C catalyzed PIP_2 hydrolysis that initiate cellular

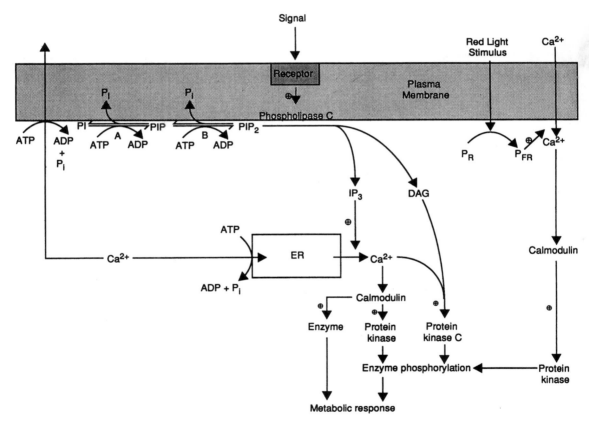

Fig. 5.7 Current model of how Ca^{2+} is believed to mediate transmembrane signaling in plants. For details refer to the text. PI, phosphatidylinositol: PIP, phosphatidylinositol 4-phosphate; PIP_2, phosphatidylinositol 4,5-bisphosphate; A, PI kinase; B, PIP kinase; ER, endoplasmic reticulum; IP_3, inositol 1,4,5-triphosphate; DAG, diacylglycerol; P_R and P_{FR}, red and far-red light absorbing forms of phytochrome, respectively; +, stimulation of Ca^{2+} flux across the plasma and endoplasmic reticulum membranes, or activation of an enzyme.

responses. Both Ca^{2+} and DAG are required to activate protein kinase C which in turn phosphorylates a distinct population of intracellular proteins, some of which may be involved in regulation of the cell cycle. Inositol triphosphate (IP_3), another product of PIP_2 hydrolysis, causes Ca^{2+} release from intracellular Ca^{2+} storage areas (i.e. the cell vacuole). The consequent rise in intracellular Ca^{2+} concentration then activates Ca^{2+} and Ca^{2+}-calmodulin dependent reactions, either by direct activation of an enzyme, or if that enzyme is a protein kinase, then indirectly by phosphorylation of a different subset of cellular proteins. Extracellular signal transduction through PIP_2 hydrolysis is usually very rapid. Although some cellular responses that are mediated by IP_3 and DAG, such as alterations in gene expression, become apparent only after a considerable lag, the initial events triggering these responses are believed to occur within seconds or less. The IP_3 and DAG produced are usually rapidly eliminated by metabolic conversion to other compounds, and low cytosolic Ca^{2+} levels are quickly restored by the action of the Ca^{2+} pumps found in various cellular membranes (see Fig. 5.7).

Coté and Crain (1993, 1994) have reviewed some of the best evidence obtained to date that phosphoinositide hydrolysis mediates transduction of some signals in plants. The evidence is strongest for a role in triggering shedding of flagella by the green alga *Chlamydomonas reinhardtii* under acid stress. Kinetic analysis revealed that PIP_2 hydrolysis occurs within half a second and could trigger the rapid loss of flagella (flagellar loss is an adaptive response that decreases the surface area of the alga, reducing exposure to noxious stimuli). Higher plant responses to pathogenic and osmotic stress, as well as the regulation of turgor changes that underlie stomatal opening and closing and the movements of flowers and leaves, may all be mediated by phosphoinositide hydrolysis.

Although there are still many gaps in applying the model shown in Fig. 5.7 to plants, it is clear that much of the model agrees with the available information – namely the occurrence within plant cells of PIP and PIP_2, PI kinase, PIP kinase, phospholipase C, Ca^{2+} transport mechanisms, calmodulin, Ca^{2+}-dependent enzymes, various types of protein kinases (possibly including protein kinase

C), as well as plant growth hormone-, pathogen-, or light-dependent effects on the transport and intracellular distribution of Ca^{2+}, and protein phosphorylation (Roux *et al.*, 1986; Schroeder and Thuleau, 1991; Trewavas and Gilroy, 1991; Drøbak, 1992; Shacklock *et al.*, 1992; Coté and Crain, 1993, 1994; Fallon *et al.*, 1993; Poovaiah and Reddy, 1993).

Future prospects

It should be apparent from the preceding discussion that remarkable advances have been recently made in the area of plant metabolism and its regulation. Equally apparent, however, is that much has yet to be learned before our understanding is complete. Future workers must continue to identify and purify the key regulatory enzymes of plant metabolic pathways and establish the mechanisms whereby their activities are controlled, *in vivo*. Only once these basic mechanisms are understood can geneticists and molecular biologists fully optimize conditions which will lead to increased crop yields.

Key areas for future research should include the occurrence, mechanism, sites and potential regulatory significance of reversible subcellular structuring, or micro-compartmentation, of so-called 'soluble' plant enzymes and metabolic pathways (i.e. '3D' metabolism). Studies concerning the relationship between enzyme structure and function have necessarily placed much emphasis on enzyme catalytic, allosteric and covalent modification sites. However, as discussed by Srere (1987), if protein:protein interactions are important for normal cell functions, then the evolution and conservation of protein surface binding sites should also be examined.

Continued examination of the mechanisms whereby external environmental stimuli such as light, gravity, and stress are transduced into intracellular metabolic events is another important area for future research. Despite the tremendous progress in Ca^{2+} research in plants over the past 20 years, it is still in its infancy and far more work is required for us to fully understand the precise mechanisms by which Ca^{2+} regulates various physiological processes. In this regard more information on the methods of transmembrane signaling, phosphoinositide metabolism, as well as the concentration of Ca^{2+} in various compartments of living plant cells under changing environmental or physiological conditions is urgently needed. Also needed is the continued identification, purification, and characterization of plant enzymes which are under direct or indirect regulation by Ca^{2+} (i.e. calmodulin regulated enzymes, protein kinases, phosphoprotein phosphatases and their substrates). Of further interest will be the determination of the possible interaction between Ca^{2+}-dependent regulatory pathways and other regulatory pathways. For example, there is a probable 'cross-talk' between calmodulin regulated pathways and those regulated by cyclic AMP in animal systems (Cohen, 1985; Trewavas and Gilroy, 1991). Although efforts by plant physiologists to establish a second messenger role for a cyclic AMP have been unsuccessful (Trewavas and Gilroy, 1991), its prominent involvement in extracellular signal transduction in animal systems, coupled with its reported occurrence in plant cells, indicates that cyclic AMP may have an as yet unknown regulatory function(s) in plants. Undoubtedly, other compounds will be found which fulfill second messenger roles in plant cells. For example, changes in the intracellular concentration of the polyamines spermine and spermidine are known to occur following exposure of plants to environmental stress, changes in light intensity, as well as variations in growth hormones (Flores, 1990). Because polyamines can influence many different physiological and biochemical phenomena, including protein phosphorylation, additional studies should be directed at establishing their potential role as 'signal' metabolites in plants. Resolving these and other problems will contribute greatly to our appreciation as to how the metabolic flexibility of plants permits them to respond appropriately to their immediate environment and energy requirements, thereby providing for efficient maintenance of cellular homeostasis.

Acknowledgment

I am grateful to my mentor and Ph.D. thesis supervisor, Prof. Kenneth B. Storey, who introduced me to the fascinating world of metabolism and its regulation, and whose stimulating lectures provided a foundation for much of this chapter.

References.

Anderson, J. M. and Cormier, M. J. (1978). Calcium-dependent regulator of NAD kinase in higher plants. *Biochem. Biophys. Res. Commun.* **84**, 595–602.

ap Rees, T. and Hill, S. A. (1994). Metabolic control analysis of plant metabolism. *Plant, Cell Environ.* **17**, 587–99.

Atkinson, D. E. (1977). *Cellular Energy Metabolism and its Regulation.* New York, NY, Academic Press.

Biermann, B. J., Pao, L. I. and Feldman, L. J. (1994). Pr-specific phytochrome phosphorylation in vitro by a protein kinase present in anti-phytochrome maize immunoprecipitates. *Plant Physiol.* **105**, 243–51.

Black, C. C., Mustardy, L., Sung, S. S., Kormanik, P. P., Xu, D.-P. and Paz, N. (1987). Regulation and roles for alternative pathways of hexose metabolism in plants. *Physiol. Plant.* **69**, 387–94.

Buchanan, B. B. (1991). Regulation of CO_2 assimilation in oxygenic photosynthesis: the ferredoxin/thioredoxin system. *Arch. Biochem. Biophys.* **288**, 1–9.

Buchanan, B. B. (1992). Carbon dioxide assimilation in oxygenic and anoxygenic photosynthesis. *Photosynthesis Res.* **33**, 147–62.

Cohen, P. (1983). *Control of Enzyme Activity.* London and New York, Chapman and Hall.

Cohen, P. (1985). The role of protein phosphorylation in the hormonal control of enzyme activity. *Eur. J. Biochem.* **151**, 439–48.

Cornish-Bowden, A. and Wharton, C. W. (1988). *Enzyme Kinetics.* Oxford, IRL Press Ltd.

Coté, C. G. and Crain, R. C. (1993). Biochemistry of phosphoinositides. *Ann. Rev. of Plant Physiol. Plant Mol. Biol.* **44**, 333–56.

Coté, C. G. and Crain, R. C. (1994). Why do plants have phosphoinositides? *BioEssays* **16**, 39–46.

Crabtree, B. and Newsholme, E. A. (1985). A quantitative approach to metabolic control, *Curr. Topics Cell. Regulation* **25**, 21–76.

Dennis, D. T. and Greyson, M. F. (1987). Fructose 6-phosphate metabolism in plants. *Physiol. Plant.* **69**, 395–404.

Drøbak, B. K. (1992). The plant phosphoinositide system. *Biochem. J.* **288**, 697–712.

Duff, S. M. G., Sarath, G. and Plaxton, W. C. (1994). The role of acid phosphatases in plant phosphorus metabolism. *Physiol. Plant.* **90**, 791–800.

Evans, D. E. (1994). PM-type calcium pumps are associated with higher plant cell intracellular membranes. *Cell Calcium* **15**, 241–6.

Fallon, K. M., Shacklock, P. S. and Trewavas, A. J. (1993). Detection in vivo of very rapid red light-induced calcium-sensitive protein phosphorylation in etiolated wheat (*Triticum aestivum*) leaf protoplasts. *Plant Physiol.* **101**, 1039–45.

Flores, H. E. (1990). Polyamines and plant stress. In *Stress Responses of Plants: Adaptation and Acclimation Mechanisms*, eds R. G. Alscher and J. R. Cumming, Wiley-Liss, Inc., New York, pp. 217–39.

Ford, D. M., Jablonski, P. P., Mohamed, A. H. and Anderson, L. E. (1987). Protein modulase appears to be a complex of ferredoxin, ferredoxin/thioredoxin reductase, and thioredoxin. *Plant Physiol.* **83**, 628–32.

Ghosh, H. P. and Preiss, J. (1966). Adenosine diphosphate glucose pyrophosphorylase: A regulatory enzyme in the biosynthesis of starch in spinach leaf chloroplasts. *J. Biol. Chem.* **241**, 4491–504.

Häberlein, I., Würfel, M. and Follmann, H. (1992). Non-redox protein interactions in the thioredoxin activation of chloroplast enzymes. *Biochim. Biophys. Acta* **1121**, 293–6.

Hepler, P. K. and Wayne, R. O. (1985). Calcium and plant development. *Annual Review of Plant Physiol.* **36**, 397–439.

Holmgren, A. (1985). Thioredoxin. *Ann. Rev. Biochem.* **54**, 237–71.

Hrazdina, G. and Jensen, R. A. (1992). Spatial organization of enzymes in plant metabolic pathways. *Ann. Rev. Plant Physiol. Plant Mol. Biol.* **43**, 241–67.

Huber, S. C., Huber, J. L. and McMichael, R. W. (1994). Control of plant enzyme activity by reversible protein phosphorylation. *Int. Rev. Cytology* **149**, 47–98.

Kacser, H. (1987). Control of metabolism. In *The Biochemistry of Plants*, Vol. 12, Academic Press, New York, pp. 39–67.

Kacser, H. and Porteous, J. W. (1987). Control of metabolism: what do we have to measure? *Trends Biochem. Sci.* **12**, 5–14.

Kauss, H. (1987). Some aspects of calcium-dependent regulation in plant metabolism. *Ann. Rev. of Plant Physiol.* **38**, 47–72.

Keleti, T., Ovadi, J. and Batke, J. (1989). Kinetic and physico-chemical analysis of enzyme complexes and their possible role in the control of metabolism. *Prog. Biophys. Mol. Biol.* **53**, 105–52.

Marcus, F., Chamberlain, S. H., Chu, C., Masiarz, F. R., Shin, S., Yee, B. C. and Buchanan, B. B. (1991). Plant thioredoxin *h*: an animal-like thioredoxin occurring in multiple cellular compartments. *Arch. Biochem. Biophys.* **287**, 195–8.

Masters, C. J., Reid, S. and Don, M. (1987). Glycolysis – new concepts in an old pathway. *Mol. Cell. Biochem.* **76**, 3–14.

Moorhead, G. B. G. and Plaxton, W. C. (1988). Binding of glycolytic enzymes to a particulate fraction in carrot and sugar beet storage roots: dependence on metabolic state. *Plant Physiol.* **86**, 348–51.

Moorhead, G. B. G. and Plaxton, W. C. (1991). Evidence for an interaction between cytosolic aldolase and the ATP- and pyrophosphate-dependent phosphofructokinases in carrot storage roots. *FEBS Lett.* **313**, 277–80.

Moorhead, G. B. G., Hodgson, R. H. and Plaxton, W. C. (1994). Copurification of cytosolic fructose-1,6-bisphosphatase and cytosolic aldolase from endosperm of germinating castor oil seeds. *Arch. Biochem. Biophys.* **312**, 326–35.

Podestá, F. E. and Plaxton, W. C. (1991). Kinetic and regulatory properties of cytosolic pyruvate kinase from germinating castor oil seeds. *Biochem. J.* **279**, 495–501.

Poovaiah, B. W. and Reddy, A. S. N. (1993). Calcium and signal transduction in plants. *CRC Crit. Rev. Plant Sci.* **12**, 185–211.

Preiss, J. (1984). Starch, sucrose biosynthesis and partition of carbon in plants are regulated by orthophosphate and triose-phosphates. *Trends Biochem. Sci.* **9**, 24–7.

Ranjeva, R. and Boudet, A. M. (1987). Phosphorylation of proteins in plants: regulatory effects and potential involvement in stimulus/response coupling. *Ann. Rev. Plant Physiol.* **38**, 73–93.

Roberts, D. M. and Harmon, A. C. (1992). Calcium-modulated proteins: targets of intracellular calcium signals in higher plants. *Ann. Rev. Plant Physiol. Plant Mol. Biol.* **43**, 375–414.

Roux, S. J., Wayne, R. O. and Datta, N. (1986). Role of calcium ions in phytochrome response: an update. *Physiol. Plant.* **66**, 344–8.

Scheibe, R. (1991). Redox-modulation of chloroplast enzymes. *Plant Physiol.* **96**, 1–3.

Schroeder, J. I. and Thuleau, P. (1991) Ca^{2+} channels in higher plant cells. *Plant Cell* **3**, 555–9.

Shacklock, P. S., Read, N. D. and Trewavas, A. J. (1992). Cytosolic free calcium mediates red light-induced photomorphogenesis. *Nature* **358**, 753–5.

Srere, P. A. (1987). Complexes of sequential metabolic enzymes. *Ann. Rev. Biochem.* **56**, 89–124.

Storey, K. B. (1985). A re-evaluation of the Pasteur effect: new mechanisms in anaerobic metabolism. *Mol. Physiol.* **8**, 439–61.

Trewavas, A. and Gilroy, S. (1991). Signal transduction in plant cells. *Trends Genetics* **7**, 356–61.

Wolosiuk, R. A., Ballicora, M. A. and Hagelin, K. (1993). The reductive pentose phosphate cycle for photosynthetic CO_2 assimilation: enzyme modulation. *FASEB J.* **7**, 622–37.

Ziegler, D. M. (1985). Role of reversible oxidation–reduction of enzyme thiols–disulfides in metabolic regulation, *Ann. Rev. Biochem.* **54**, 305–29.

6 Regulation by compartmentation

M. J. Emes and D. T. Dennis

The sequestration of metabolic pathways

Compartmentation of metabolism and the
sequestration of metabolic pathways into
membrane-bound organelles occurs in all
eukaryotes but is most apparent in the cells of
higher plants. It is a means by which pathways can
be regulated independently from one another and is
a mechanism of controlling the flux of carbon
through the pathways in response to the demands
of cellular metabolism. In addition, the division of
the cell into smaller volumes can serve to
concentrate the intermediates of a pathway and
may prevent futile cycling by opposing reactions. It
may also provide specialized environments more
favourable to certain reactions. Related pathways
that interact through the supply and demand of
substrates may also be kept in close proximity by
being sequestered together.

In some cases, the sequence of reactions that
The major organelles found in plant cells are the
nucleus, mitochondria, Golgi apparatus, microbodies,
peroxisomes, glyoxysomes, vacuoles and plastids. In
addition, it should not be forgotten that the
cytoplasm is also a compartment in its own right. All
these compartments are enclosed by the plasma
membrane. The cell is supported by a rigid cellulose
cell wall, not found in animals, that determines the
shape of the cell. However, the space encompassed by
the cell wall is not inert since it may contain enzymes
(e.g. invertase) essential for metabolism of the cell.

In some cases, the sequence of reactions that
comprises a metabolic pathway may be located in
more than one organelle. An excellent example of this
is photorespiration which requires the chloroplast,
peroxisome, mitochondrion and cytosol (Chapter 22).
Electron microscopy has indicated that these

organelles may be closely associated, facilitating
transfer of common intermediates.

The integration of cell metabolism requires
controlled interaction between the pathways in the
various compartments. This is achieved through a
highly selective permeability of the organellar
membranes so that a limited number of intermediates
can be transported from one compartment to
another. There may also be a close physical
association between the compartments.

Throughout this book, examples of the
compartmentation of plant metabolism are presented.
It is not the intention of this chapter to describe them
all. Examples of a few pathways will be presented as
illustrations of the importance and complexity of
compartmentation.

The unique role of plastids

The most fundamental difference between plants and
other organisms is the presence of an additional
compartment, the plastid, which occurs in all plant
tissues. There is a wide variety of types of plastid,
each with distinct metabolic functions and capacities.
The most commonly studied plastid is the
chloroplast, and a great deal of attention has been
paid to the function of this organelle and its role in
the photochemical generation of reducing power and
ATP that is utilized in the reductive biosynthesis of
carbohydrates from CO_2. Indeed, non-plant
biochemists often regard the reductive pentose
phosphate cycle and the photochemical reactions of
the chloroplast as the only unique aspect of plant
metabolism, since it is generally assumed that plants
are otherwise biochemically more or less the same as
other organisms. However, a major difference

between plant metabolism and that of other organisms is that many activities that are cytosolic in other organisms are localized in plastids in plants which makes a fundamental difference to their organization and control.

The major part of the biosynthetic capacity of a plant cell is localized in plastids, whether they are present as photosynthetically competent organelles or are completely achlorophyllous. This is also true for different tissues such as leaves, roots and seeds. The association of biosynthetic activity with the chloroplast is logical, since this organelle is also the primary source of energy and carbon fixation, both of which are required for biosynthetic activity. In non-green plastids, carbon and energy must be imported for these biosynthetic pathways. The range of other activities found in all plastids include: fatty acid synthesis, the primary assimilation of nitrite into glutamate, the biosynthesis of a number of other amino acids (though not all), starch synthesis, and pigment biosynthesis (Emes and Tobin, 1993). In addition to these anabolic activities, plastids are also able to catabolize hexose phosphates via glycolysis and the oxidative pentose phosphate pathway (see Chapter 8).

In chloroplasts, the primary activity of the organelle is the generation of triose phosphates and hexose phosphates within the reductive pentose phosphate cycle which serve as the precursors of other pathways (see Chapter 21). This is not the case in non-photosynthetic plastids where the predominating metabolic pathways differ depending on the tissue in which the organelle is found. These differences must be a result of differential gene expression occurring both as a function of the plant species and tissue location. For example, in the endosperm of developing castor oil seeds, metabolism is directed predominantly towards fatty acid synthesis which occurs in the plastids. In contrast, the Gramineae store starch in the developing endosperm and this is also synthesized in the plastid. Nevertheless, within the leaves of both these plants, the activity of the plastid must be directed towards production of photosynthate that can be used in the synthesis of such essential intermediates as amino acids for growth and development.

All plastids contain circular molecules of DNA that vary in length from 120–150 kbp and are associated with the inner envelope or with thylakoid membranes. The plastid genome encodes for chloroplast ribosomal RNA genes, transfer RNAs, RNA polymerase subunits, and polypeptides associated with the photosystems as well as the large subunit of ribulosebisphosphate carboxylase. Plastids have a high genome copy number, containing up to 1000 copies of the DNA in some cases, presumably to produce high levels of ribosomal RNA which will be needed for protein synthesis. All plastids within a plant arise from a common progenitor and differentiate in response to the development of the tissue. The rate of transcription of plastid genes varies from one plastid type to another, but the relative rates of transcription are the same between each gene, and mRNA for photosynthetic proteins has been detected in non-photosynthetic plastids (Gruissem and Tonkyn, 1993). It appears, however, that there are differences in RNA stability in different plastids. In addition, the stability of the transcript and the rate of its translation varies during differentiation of one plastid type into another (Mayfield *et al.*, 1995).

Clearly then, there are differences in both the molecular and biochemical properties of different plastids. This difference is important, and one cannot assume that the study of chloroplasts alone will provide an insight into the control of metabolism in all types of plastid. Many of the biosynthetic pathways in different plastids have common features, but the regulation of these pathways and their integration with the rest of cellular metabolism must, inevitably, be different. For example, in chloroplasts, starch is synthesized in the light and degraded in both the light and dark (Preiss, 1982); in root plastids it is synthesized and degraded in the dark but turnover is negligible (Hargreaves and ap Rees, 1988). In contrast, in endosperm amyloplasts starch is synthesized during development but broken down during germination (Duffus, 1984). The only way to obtain information about the regulation of starch metabolism in these different plastids is to study each one separately, since extrapolation from any one case to another is bound to have limited usefulness. The contents of other chapters of this book will deal extensively with metabolism in photosynthetic cells. Hence, the scope of this chapter will be confined to what is known of metabolism in non-photosynthetic plastids.

Metabolism of non-photosynthetic plastids

Transport of metabolites into plastids

The biosynthetic pathways in non-green plastids are dependent on a supply of precursors, ATP and reductant that are generated by the oxidation of carbohydrates. The selective permeability of organelles to metabolites, through the operation of transporter proteins within membranes, offers the means by which metabolism within an organelle may be controlled. It is also a medium through which the metabolic/bioenergetic status of that organelle may be communicated to the rest of the cell.

The best studied of these transporters is the phosphate translocator which has been extensively examined in chloroplasts (see Chapter 25) but to a lesser extent in non-photosynthetic plastids. In chloroplasts, it catalyzes the counterexchange of inorganic phosphate (Pi), triosephosphate and 3-phosphoglycerate, which are competitive with each other at a common binding site. The properties of the phosphate translocator vary among plastid types reflecting the differences in metabolic activities which occur within them (Fig. 6.1). The phosphate translocator of C_4 chloroplasts is able to counterexchange phospho*enol*pyruvate (PEP) and 2-phosphoglycerate in addition to those substrates mentioned above for C_3 chloroplasts (see Chapters 22 and 23). In non-green plastids of roots, developing embryos and seeds it appears that, additionally, this translocator can transport hexose phosphates. In

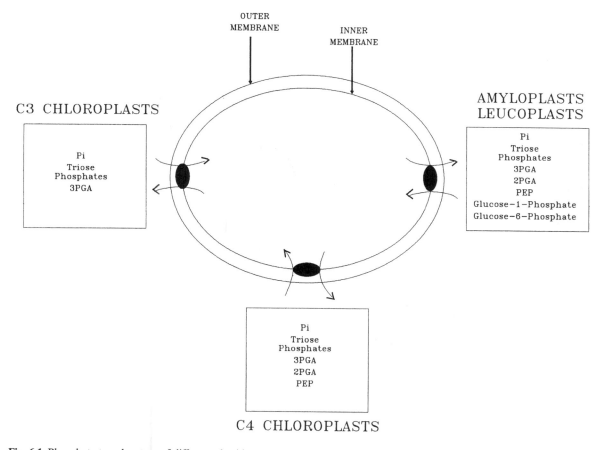

Fig. 6.1 Phosphate translocators of different plastid types.

some cases, this appears to be glucose 6-phosphate, (e.g. in pea roots and developing pea cotyledons) (Bowsher et al., 1989; Hill and Smith, 1991). In others, (e.g. potato) glucose 1-phosphate, (Kosegarten and Mengel, 1994) or both glucose 6-phosphate and glucose 1-phosphate (e.g. wheat and maize endosperms), are transported (Tetlow et al., 1994, Neuhaus et al., 1993).

Recent studies of primary sequences of the phosphate translocator of chloroplasts from related C_3- and C_4-type species have shown remarkable conservation of structure. Molecular modelling implies that perhaps only a very few changes of amino acid residues between the C_3- and C_4-type proteins are needed to alter substrate specificity (Fischer et al., 1994). It is tempting to suggest that the differences between all these different translocators may involve only minor modifications to residues in the binding site.

The observation that hexosephosphates are taken up by non-photosynthetic plastids reflects the fact that, in such organelles, starch is synthesized from imported precursors (unlike chloroplasts, where fixed carbon is generated photosynthetically), and that carbohydrates have to be oxidized to provide the energy for the biosynthesis of fatty acids and amino acids.

The presence of isozymes in the cytosol and plastids

When the same reaction is found in two cell compartments it is generally accepted that it is catalyzed by isozymes present in the compartments (Dennis and Miernyk, 1982) and this has been found to be the case where genes for the isozymes have been isolated (e.g. see Chapter 8). However, there are organellar enzymes where this has not been examined in detail and the two enzymes should more properly be termed isoforms. In the case of plastids, the 80 to 85% of proteins found in the organelle are encoded in the nuclear genome, translated in the cytosol and postranslationally imported into the plastid. The presence of independent genes encoding the plastid and cytosolic forms of the enzyme makes possible the individual control of the expression of the genes allowing the level of enzyme in the organelle to vary

relative to the cytosolic concentration of the isozyme. Since the plastid is the site of much of the biosynthetic activity of the plant cell, the expression of plastid isozymes can modify the biosynthetic capacity of the cell in a tissue specific manner and result in the formation of products that are typical of that particular tissue. The control of the expression of genes for plastid proteins is described in Chapter 18 (see also Mullet, 1988).

It is now generally accepted that plastids are the descendents of endosymbyonts from which, during the course of evolution, most of the genes were transferred to the nucleus. This transfer of genes is a complex process since it would mean the acquisition of eukaryote promoters by the gene and the attachment of a transit peptide to direct the protein into the organelle. A comparison of the plastid enzymes with their cytosolic counterparts has revealed that in some cases (e.g. phosphofructokinase and pyruvate kinase) they have quite distinct properties whereas, in other cases, the isozyme pairs are very similar (e.g. enolase and triose phosphate isomerase). When the DNA or the derived amino acid sequences of cDNA clones for the isozyme pairs are compared a complex pattern emerges (see Chapter 8 for more details). In the case of pyruvate kinase, the plastid isozymes of the enzyme are indeed more prokaryotic in nature as determined by detailed sequence analysis whereas their cytosolic counterparts are more eukaryote in nature (Hattori et al., 1995). However, this simple relationship does not hold for other isozyme pairs.

There appears to have been a considerable rearrangement of the the coding region of the gene for phosphoglycerate kinase so that the cytosolic and plastid isozymes have both prokaryote and eukaryote elements (Longstaffe et al., 1989). In the case of triose phosphate isomerase, the plastid gene appears to have arisen by a gene duplication of the cytosolic isozyme and then to have aquired a transit peptide (Henze et al., 1994). The plastid and cytosolic isozymes of aldolase are the Class I type whereas cyanobacteria have the Class II type of aldolase, a completely different enzyme, showing that during the conversion of the endosymbiont into a plastid the original prokaryotic enzyme was lost (Pelzer-Reith et al., 1994). Hence it must be concluded that several mechanisms were probably in place during evolution for the development of plastid and cytosolic isozymes of the enzymes of the glycolytic pathway.

Protein uptake into different types of plastid

Undoubtedly, the major control of the level of enzymes within plastids is determined by the differential expression of the genes for the plastid isozymes. However, the possibility exists of translational regulation of the uptake of the protein into the organelle. Uptake of proteins into chloroplasts has been extensively studied and is described in Chapter 24. However, there are only a limited number of examples of protein uptake into non-green plastids and for a small number of proteins. The small subunit of rubisco can be transported across the membrane of leucoplasts from developing castor seeds (Boyle *et al.*, 1986). Plastocyanin, a protein only found in chloroplasts, can also be imported into these leucoplasts but the second targeting step which directs the protein to the thylakoids is missing (Halpin *et al.*, 1989). Conversely, the waxy polypeptide that is normally targeted to amyloplasts can be transported into chloroplasts (Klosgen and Weil, 1991). Hence different types of plastid from various tissues have a common uptake mechanism.

The uptake capacity of plastids in both monocots and dicots declines dramatically during the maturation of chloroplasts and etioplasts and this does not result from the senescence of the organelles (Dahlin and Cline, 1991). The import of proteins into etioplasts is reactivated in the presence of light showing that this is a reversible phenomenon. Hence, it appears that the uptake mechanism is correlated with the metabolic needs of the organelle and is an additional control mechanism for determining the protein content of the organelle. Evidence that such a mechanism operates *in vivo* has been found for the compartmentation of starch phosphorylase in potato tubers (Brisson *et al.*, 1989). Using immunogold labelling, it was shown that in young tubers all the phosphorylase is present in amyloplasts, whereas in mature tissues all of it is in the cytosol. Northern blot analysis showed the same transcript to be present in young and old tissues. The protein derived from this transcript has a transit peptide in both tissues but in mature tubers the protein is not transported into the amyloplast. In addition, there is a protease in the cytosol that cleaves the transit peptide resulting in a protein of the same size as that found in the amyloplast. Hence, in this tissue the transport system appears to be controlling the location of the enzyme in the cell.

There are two plastid localized pyruvate kinases in the developing castor seed endosperm (see Chapter 8) which appear to have different uptake mechanisms. One isozyme is transported into chloroplasts and leucoplasts by a mechanism that appears to be identical with that of other plastid proteins (Wan *et al.*, 1995). However, the second isozyme is only transported at higher levels of ATP. This could be a mechanism of correlating the uptake of this isozyme with a high concentration of ATP in the cytosol.

Hence, not only are there different types of metabolite transporter in the various tissues of the plant, there may be changes in the transporters for the proteins that are encoded in the nuclear genome and translated in the cytosol. These data indicate the great variability that exists in plastids from various tissues. This variability determines the specificity and capacity of the metabolic activity in various tissues.

Hexose phosphate metabolism in plastids

Carbohydrate oxidation

In other organisms, the ubiquitous pathways of glycolysis and the oxidative pentose phosphate pathway are not associated with any membrane-bound organelle and it was originally assumed that plants would follow this pattern. With improved methods of cell fractionation, it became increasingly apparent that plant cells are unique in having part or all of these pathways present both in the cytosol and in the plastid. These pathways are shown in Fig. 6.2 where it is illustrated that they share some common intermediates. The regulation of these pathways in the cytosol is dealt with in Chapter 8. From studies of roots, stems, seeds and cell cultures it appears that non-photosynthetic plastids have the enzymic capacity to convert hexose and hexose phosphates at least as far as pyruvate, and in some cases there is evidence for the presence of a pyruvate

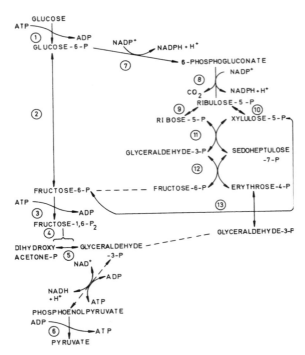

Fig. 6.2 Outline of relationships between the pentose phosphate pathway and glycolysis. Glycolysis: 1, hexokinase; 2, glucose 6-phosphate isomerase; 3, phosphofructokinase; 4, aldolase; 5, triose phosphate isomerase; 6, pyruvate kinase. Pentose phosphate pathway: 7, glucose 6-phosphate dehydrogenase; 8, 6-phosphogluconate dehydrogenase; 9, ribose 5-phosphate isomerase; 10, ribulose 5-phosphate 3-epimerase; 11, transketolase; 12, transaldolase; 13, transketolase.

dehydrogenase complex for its further conversion to acetyl-CoA (Dennis and Miernyk, 1982). However, there are both qualitative and quantitative differences arising between plastids with respects to these pathways. For example, it appears that the first enzyme unique to the oxidative pentose phosphate pathway (OPPP), glucose 6-phosphate dehydrogenase, is absent from some plastids (Frehner et al., 1990).

In the case of glycolysis, it is generally agreed that the reactions for the conversion of glucose 6-phosphate to phosphoglycerate are present in all the non-photosynthetic plastids. However, the level of some of the enzymes for the formation of pyruvate from phosphoglycerate are low in some plastid

preparations. This is particularly the case with phosphoglyceromutase and enolase. There now seems good evidence, from studies of particulate and soluble isozymes, that these two enzymes are present in plastids of developing seeds (Miernyk and Dennis, 1984; Denyer and Smith, 1988; Kang and Rawsthorne, 1994), but whilst they have been reported as latent in plastids from other tissues such as roots, the amount and specific activity associated with the organelle is small and variable (Trimming and Emes, 1993). When the plastid form of enolase was measured in many different tissues from the castor oil plant, it was found to vary from zero to 30% (Miernyk and Dennis, 1992). Chloroplasts from mature leaves had none of the enzyme, whereas plastids from developing seeds had a large amount. Plastids from other tissues, such as young leaves and roots had intermediate amounts, and the level appeared to correlate with the anticipated rate of fatty acid biosynthesis in the tissue. These data indicate that there is a tight control over the level of enzymes found in plastids, probably exerted at the nuclear genome.

The metabolic function of carbohydrate oxidation in plastids

The primary function of plastidic glycolysis and the oxidative pentose phosphate pathway (OPPP) in non-photosynthetic tissues is to provide NAD(P)H and ATP for biosynthetic reactions. Processes within the organelle which depend on these intermediates include inorganic nitrogen assimilation, amino acid synthesis and fatty acid synthesis. Intermediates of glycolysis may also serve as precursors of acetyl units needed for fatty acid synthesis, whereas, as will be seen later, it is likely that precursors for starch biosynthesis are imported directly from the cytoplasm.

Plants take up nitrogen from the soil principally in the form of nitrate (see Chapters 29 and 31) which is first reduced to nitrite in the cytoplasm. All subsequent reactions leading to the synthesis of amino acids then occur within plastids. Two of these, nitrite reductase and glutamate synthase, require reduced ferredoxin for their activity. In chloroplasts, ferredoxin is reduced photochemically by photosystem I in the light. In non-photosynthetic plastids, such as in roots, reductant is generated initially in the form of NADPH by the OPPP (Fig. 6.3). The OPPP releases CO_2 specifically and only from C-1 of glucose 6-phosphate and, when the latter is supplied to purified, intact root plastids as [1-^{14}C]-glucose 6-phosphate, the interaction between carbon and nitrogen metabolism can be observed through the release of $^{14}CO_2$ in the presence of the substrates for nitrite reductase and glutamate synthase (Fig. 6.4). The NADPH generated by the OPPP cannot be used as a substrate by either enzyme but instead is used to reduce the iron-sulfur protein, ferredoxin, in a reaction catalyzed by ferredoxin NADP$^+$-oxidoreductase (FNR). The forms of ferredoxin and FNR found in root plastids are distinct from their chloroplast counterparts (Hase et al., 1991) and are co-induced during nitrate assimilation along with nitrite reductase (Bowsher et al., 1993).

Studies with non-photosynthetic plastids suggest that the carbon units from which fatty acids are derived may be generated through carbohydrate oxidation within the organelle (Miernyk and Dennis, 1983; Kang and Rawsthorne, 1994). All plastids examined to date appear to contain a pyruvate dehydrogenase complex which synthesises acetyl CoA (and generates NADH simultaneously), the precursor for fatty acid synthesis. There are a number of 'entry points' into plastidic glycolysis – hexose, hexosephosphates and triosephosphates – which could lead to the generation of pyruvate. Furthermore, pyruvate could be generated by a NADP$^+$ – malic enzyme reaction which has also been shown to occur in non-photosynthetic plastids (El-Shora and ap Rees, 1991).

Malic enzyme:

$$\text{Malate} + \text{NADP}^+ \xrightarrow{\text{M}^{2+}} \text{Pyruvate} + \text{HCO}_3^- + \text{NADPH} + \text{H}^+$$

Experiments with plastids purified from oil-synthesizing tissues indicate that such plastids may

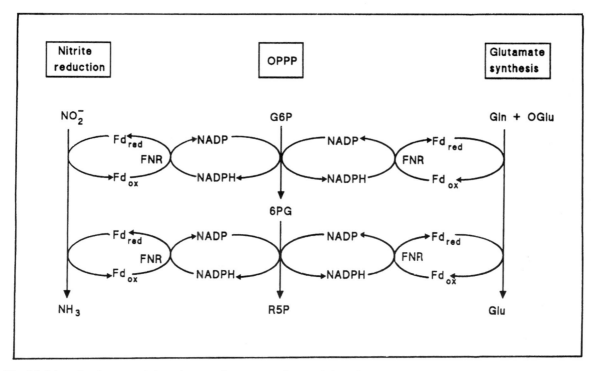

Fig. 6.3 Interaction between nitrite reductase, glutamate synthase and the oxidative pentose phosphate pathway in root plastids. Fd, ferredoxin; OGlu, 2-oxoglutarate; FNR, ferredoxin NADP$^+$-oxidoreductase.

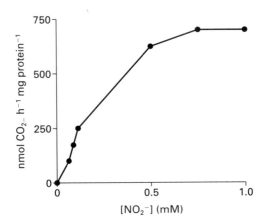

Fig. 6.4 Stimulation of the oxidative pentose phosphate pathway in pea root plastids by nitrite. $^{14}CO_2$ from [1-^{14}C] Glc6P was collected over a 1 h period at different concentrations of nitrite. Note that CO_2 evolution eventually saturates in parallel with the saturation of nitrite reduction by nitrite.

utilize a range of exogenous substrates. In oil seed rape plastids, incorporation of carbon into fatty acids was highest in the order pyruvate > glucose 6-phosphate > triose phosphate > malate > acetate (Kang and Rawsthorne, 1994). By contrast, in plastids from developing castor seeds, malate was a much better substrate for fatty acid synthesis than pyruvate (Smith *et al.*, 1992).

Wood *et al.* (1992) have proposed that acyl groups are transferred from mitochondria to plastids as acylcarnitine derivatives via carnitine/ acylcarnitine translocases. Whilst these have been extensively studied in mammalian systems, there is some dispute as to whether they exist in plants. Masterson *et al.* (1990) demonstrated that carnitine could stimulate fatty acid synthesis from acetyl CoA ninefold when supplied to pea chloroplasts. Similar experiments have yet to be performed with non-green plastids to see how widespread this phenomenon is, though it is worth noting that, in contrast to some non-photosynthetic plastids, chloroplasts may lack a complete glycolytic pathway and the significance of different routes of fatty acid biosynthesis may vary between tissues. *In vivo* it is probable that more than one source will be available and synthesis is likely to result from a multiplicity of substrates.

Plastids also contain an acetyl-CoA synthetase and can directly synthesize acetyl CoA from exogenous acetate. This has been exploited in testing possible sources of ATP for lipid synthesis in pea root plastids (Kleppinger-Sparace *et al.*, 1992). Schunemann *et al.* (1993) have demonstrated an active adenylate translocator in these organelles, capable of counter exchanging ATP and ADP. However, it is possible that ATP could be generated by substrate-level phosphorylation of ADP within plastids. Using similar preparations of plastids to those used by Schunemann *et al.*, Qi *et al.* (1994) found that phospho*enol*pyruvate (PEP) could sustain substantial rates of fatty acid synthesis in the absence of added ATP, implying that pyruvate kinase was able to fulfil this role within the organelle. Glucose and fructose 6-phosphate could also replace ATP to some extent, as could a triosephosphate shuttle utilizing dihydroxyacetone phosphate. Although these experiments were carried out in the presence of NADH and NADPH, it is difficult to be sure whether there is some additional stimulatory effect through reductant provision by carbohydrate oxidation. Curiously, it seems that some preparations of non-photosynthetic plastids are unable to support lipid synthesis without additional NAD(P)H being included in the assay (Kang and Rawsthorne 1994; Qi *et al.*, 1994). This is surprising as they clearly possess the capacity to generate reduced pyridine nucleotides internally and such molecules are not thought to cross plastid membranes.

Starch synthesis

The primary substrate for starch synthesis in non-photosynthetic cells is sucrose imported from the leaf, the initial catabolism of which takes place in the cytoplasm. The synthesis of starch depends upon the transfer of glucosyl units via ADP-glucose to the growing glucan as catalyzed by starch synthase in combination with branching enzyme. The number of isozymes for each step, their expression and function will be touched upon elsewhere (Chapter 7). Since it is generally accepted that these steps are confined to plastids (known as amyloplasts in the case of starch storing organelles), the issues considered here will relate to the source of ADP-glucose and the compartmentation of proposed alternative routes. The premise for what follows is shown in Fig. 6.5, which outlines one model of starch synthesis; in addition, the basis of this model and suggested alternatives to it will be discussed.

By analogy with chloroplasts, it had previously been assumed that starch in the amyloplasts of storage tissues is derived from triosephosphate imported from the cytoplasm. In chloroplasts, triosephosphates generated in the light are converted to glucose 1-phosphate, the immediate precursor of ADP glucose. This gluconeogenic conversion of triosephosphate to hexosephosphate proceeds via the breakdown of fructose 1,6-biphosphate to fructose 6-phosphate and inorganic orthophosphate. Despite earlier claims that the activity of the enzyme responsible for this, fructose 1,6-bisphosphatase, could be detected in non-photosynthetic plastids, Entwistle and ap Rees (1990) showed unequivocally that this enzyme is completely lacking from the cytosol and plastids of starch-storing non-photosynthetic tissues. There is therefore no enzymic route from triosephosphate to hexosephosphate within the amyloplast. There are exceptions to this. For example, in developing pea and oilseed rape embryos, the non-photosynthetic plastids of which have their origins in green cotyledons, the enzymatic

complement is similar to chloroplasts (Denyer and Smith, 1988; Kang and Rawsthorne, 1994). However, although such plastids contain fructose bisphosphatase, as will become apparent, the weight of evidence is still against a major role for triosephosphates in mediating starch synthesis in these organelles.

A significant contribution to this debate has come from studies employing ^{13}C nuclear magnetic resonance to determine the re-arrangement of carbon atoms in starch glucosyl units after feeding labelled glucose to whole tissues (Keeling *et al.*, 1988; Viola *et al.*, 1991). If hexose supplied to whole cells enters the cytoplasm and is converted to starch via triosephosphate metabolism in the amyloplast, then there should be complete randomisation of carbon atoms 1 and 6 as a consequence of the triosephosphate isomerase reaction. Thus the hexose units found in newly synthesized starch should have equal proportions of [1-^{13}C] and [6-^{13}C], irrespective of whether the original hexose supplied was labelled in carbon atom 1 or 6. In fact, Keeling and co-workers observed only a small amount of redistribution, 15–20%, between carbon atoms 1 and 6, which could be accounted for by substrate cycling in the cytoplasm. This argues strongly against the route of sucrose conversion to starch proceeding via triose formation.

The question therefore arises as to the form of carbon entering the amyloplast. Much of the evidence has come from studies of purified amyloplasts. As indicated earlier, a number of groups have now demonstrated that the phosphate translocator of non-photosynthetic plastids is able to transport hexose phosphates, unlike the chloroplast transporter that translocate triosephosphates. Whether glucose 1-phosphate or glucose 6-phosphate is the substrate which enters the plastid seems to depend on the source of the material. In all experiments reported to date, synthesis of starch by amyloplasts supplied with hexosephosphate has been shown to be stimulated by, and usually dependent on, an exogenous supply of ATP. Given the prevailing evidence and what is known about sucrose breakdown in the cytoplasm (Chapter 7) the model presented in Fig. 6.5 is therefore a reasonable interpretation of the experimental evidence.

Recently, however, there have been suggested

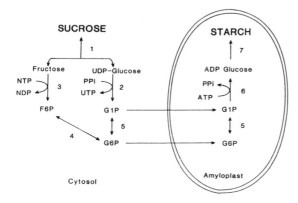

Fig. 6.5 Proposed pathway of starch synthesis in non-photosynthetic plant tissues. (1) sucrose synthase; (2) UDP-glucose pyrophosphorylase; (3) hexokinase; (4) hexosephosphate isomerase; (5) phosphoglucomutase; (6) ADP-glucose pyrophosphorylase; (7) starch synthase. NTP, nucleoside triphosphate (either UTP or ATP).

modifications to even this simplified scheme. Akazawa and co-workers have demonstrated that the adenylate transporter (ATP carrier) is also capable of transporting ADP-glucose, the substrate for starch synthase, and contend that this metabolite is produced in the cytoplasm (see e.g. Pozueta-Romera and Akazawa, 1993). This would require the synthesis of ADP glucose rather than UDP glucose by the sucrose synthase reaction in the cytoplasm. It is difficult to reconcile the view that ADP-glucose is incorporated into starch after direct import from the cytoplasm with the effect of repressing ADP-glucose pyrophosphorylase activity by antisense technology on starch synthesis. (Muller-Rober *et al.*, 1992). In transformed potato, there is a close correlation between activity of this enzyme and starch accumulation, which would not be expected if it had no role in starch synthesis.

There is also a suggestion, though little published direct evidence at this time, that in cells of developing cereal endosperm ADP glucose pyrophosphorylase is located both in the amyloplast and outside in the cytosol (Villand and Kleczkowski, 1994). Since this proposal does not rely on an abnormal sucrose synthase reaction, but would utilize glucose 1-phosphate produced by UDP glucose pyrophosphorylase in the cytoplasm, it may have some merits but awaits the conclusive demonstration of two forms of ADP glucose pyrophosphorylase in different subcellular compartments. How the activities of the two putative forms would be integrated is also a matter for conjecture at the present time.

Conclusions

Compartmention is important in the control of metabolism in all organisms, but, in plants, it has an even greater impact because of the presence of an additional organelle, the plastid. This organelle is the site of much of the biosynthetic activity of the cell in all tissues. In some cases, such as in the permanent and temporary storage of starch and in the synthesis of other storage compounds such as oils and proteins, a major flux of carbon must enter and leave this organelle. As has become increasingly evident, there is a considerable variation in the enzyme complement of different types of plastid to accommodate the

metabolic requirements of the various tissues. In addition, the transport systems for metabolites can be very different in various plastid types and there is some evidence that this might also apply to the uptake of proteins.

All plastids have the same genome and they originate from the same proplastid precursor (Mullet, 1988; and see Chapter 18). The control of the differentiation of the proplastid is not known but must involve a cooperativity of the nuclear and organellar genomes. Since plastids play such a pivotal role in the metabolism of the plant cell, the determination of the mechanism by which this differentiation is controlled should be a primary goal of plant biologists.

References

Boyle, S. A, Hemmingsen, S. M. and Dennis, D. T. (1986). Uptake and processing of the precursor to the small subunit of ribulose 1,5-bisphosphate carboxylase by leucoplasts from the endosperm of developing castor seeds. *Plant Physiol.* **81**, 817–22.

Bowsher, C. G., Hucklesby, D. P. and Emes, M. J. (1989). Nitrite reduction and carbohydrate metabolism in plastids purified from roots of *Pisum sativum* L. *Planta* **177**, 359–66.

Bowsher, C. G., Hucklesby, D. P. and Emes, M. J. (1993). Induction of ferredoxin-NADP$^+$ oxidoreductase and ferredoxin synthesis in pea root plastids during nitrate assimilation. *Plant J.* **3**, 463–7.

Brisson, N., Giroux, H., Zollinger, M., Camirand, A. and Simard, C. (1989). Maturation and subcellular compartmentation of potato starch phosphorylase. *Plant Cell* **1**, 559–66.

Dahlin, C. and Cline, K. (1991). Developmental regulation of the plastid protein import apparatus. *The Plant Cell* **3**, 1131–40.

Dennis, D. T. and Miernyk, J. A. (1982). Compartmentation of nonphotosynthetic carbohydrate metabolism. *Ann. Rev. Pl. Physiol.* **33**, 27–50.

Denyer, K. and Smith, A. M. (1988). The capacity of plastids from developing pea cotyledons to synthesize acetyl CoA. *Planta* **173**, 172–82.

Duffus, C. M. (1984). Metabolism of reserve starch. In *Storage Carbohydrates in Vascular Plants*, ed. D. H. Lewis. Cambridge University Press, London, pp. 231–52.

El-Shora, H. M. and ap Rees, T. (1991). Intracellular location of NADP-linked malic enzyme in C$_3$ plants. *Planta* **185**, 362–7.

Entwistle, G. and ap Rees, T. (1990). Lack of fructose 1,6-bisphosphatase in a range of higher plants that store starch. *Biochem. J.* **271**, 467–72.

Fischer, K., Arbinger, B., Kammerer, B., Busch, C., Briuk, S., Wallmeier, H., Sauer, N., Eckershorn C. and Flugge, U. (1994). Cloning and *in vivo* expression of functional triose phosphate/phosphate translocators from C_3- and C_4- plants: evidence for the putative participation of specific amino acid residues in the recognition of phosphoenolpyruvate. *Plant J.* **5**, 215–26.

Frehner, M., Pozueta-Romero, J. and Akazawa, T. (1990). Enzyme sets of glycolysis, gluconeogenesis and oxidative pentose phosphate pathway are not complete in non green highly purified amyloplasts of sycamore (*Acer pseudoplatanus* L.) cell suspension cultures. *Plant Physiol.* **94**, 538–44.

Gruissem, W. and Tonkyn, J. C. (1993). Control mechanisms of plastid gene expression. *Crit. Rev. Plant Sci.* **12**, 19-55.

Halpin, C., Musgrov, J. E., Lord, J. E. and Robinson, C. (1989). Import and processing of proteins by castor bean leucoplasts. *FEBS Lett.* **258**, 32–4.

Hase, T., Kimata, K., Yonekura, T., Matsumara, T. and Sakakibara, H. (1991). Molecular cloning and differential expression of the maize ferredoxin gene family. *Plant Physiol.* **96**, 77–83.

Hattori, J., Baum, B., Gottlob-McHugh, S. G., Blakeley, S. D., Dennis D. T. and Miki, B. L. Pyruvate kinase isozymes: ancient diversity retained in modern plant cells. *Biochem. Syst. Ecol.* in press.

Henze, K., Schnarrenberger, C., Kellerman, J. and Martin, W. (1994). Chloroplast and cytosolic triose phosphate isomerases from spinach: purification, microsequencing and cDNA cloning of the chloroplast enzyme. *Plant Mol. Biol.* **26**, 1961–73.

Hill, L. M. and Smith, A. M. (1991). Evidence that glucose 6-phosphate is imported as the substrate for starch synthesis by the plastids of developing pea embryos. *Planta* **185**, 91–6.

Kang, F. and Rawsthorne, S. (1994). Starch and fatty acid synthesis in plastids from developing embryos of oilseed rape (*Brassica napus* L.). *Plant J.* **6**, 795–805.

Keeling, P. L., Wood, J. R., Tyson, R. H. and Bridges, I. J. (1988). Starch biosynthesis in developing wheat grain. *Plant Physiol.* **87**, 311–19.

Kleppinger-Sparace, K. F., Stahl, R. J. and Sparace, S. A. (1992). Energy requirements for fatty acid and glycerolipid biosynthesis from acetate by isolated pea root plastids. *Plant Physiol.* **98**, 723–7.

Klosgen, R. B. and Weil, J.-H. (1991). Subcellular location and expression level of a chimeric protein consisting of the maize *waxy* transit peptide and the β-glucuronidase of *Escherichia coli* in transgenic potato plants. *Mol. Gen. Genet.* **225**, 297–304.

Kosegarten, H. and Mengel, K. (1994). Evidence for a glucose 1-phosphate translocator in storage tissue amyloplasts of potato (*Solanum tuberosum*) suspension-cultured cells. *Physiol. Plant.* **91**, 111–20.

Longstaffe, M., Raines, C. A., McMorrow, E. M., Bradbeer, J. W. and Dyer, T. A. (1989). Wheat phosphoglycerate kinase: evidence for recombination between the genes for the chloroplast and cytosolic enzymes. *Nucleic Acids Res.* **17**, 6569–80.

Masterson, C., Wood, C. and Thomas, D. R. (1990). L-acetylcarnitine, a substrate for chloroplast fatty acid synthesis. *Plant Cell Environ.* **13**, 755–65.

Mayfield, S. P., Yohn, C. B., Cohen, A. and Danon, A. (1995). Regulation of chloroplast gene expression. *Ann. Rev. Plant Physiol. Plant Mol. Biol.* **46**, 147–66.

Miernyk, J. A. and Dennis, D. T. (1983). The incorporation of glycolytic intermediates into lipids by plastids isolated from the developing endosperm of castor oil seeds. (*Ricinus communis* L.). *J. Exp. Bot.* **34**, 712–18.

Miernyk, J. A. and Dennis, D. T. (1992). A developmental analysis of the enolase isoenzymes from *Ricinus communis*. *Plant Physiol.* **99**, 748–50.

Muller-Rober, B., Sonnewald, U. and Willmitzer, L. (1992). Inhibition of the ADP-glucose pyrophosphorylase in transgenic potatoes leads to sugar-storing tubers and influences tuber formulation and expression of tuber storage protein genes. *EMBO J.* **11**, 1229–38.

Mullet, J. E. (1988). Chloroplast development and gene expression. *Ann. Rev. Plant Physiol. Plant Mol. Biol.* **29**, 475–503.

Neuhaus, H. E., Henrichs, G. and Scheibe, R. (1993). Purification of highly intact plastids from various heterotrophic plants tissues: analysis of enzymic equipment and precursor dependency for starch biosynthesis. *Biochem. J.* **296**, 395–401.

Pelzer-Reith, B., Penger, A. and Schnarrenberger, C. (1993). Plant aldolase: cDNA and deduced amino-acid sequences of the chloroplast and cytosol enzyme from spinach. *Plant Mol. Biol.* **21**, 331–40.

Pozueta-Romero, J. and Akazawa, T. (1993). Biochemical mechanism of starch biosynthesis in amyloplasts from cultured cells of sycamore (*Acer Pseudoplatanus*). *J. Exp. Bot.* **44**, 297–306.

Preiss, J. (1982). Regulation of the biosynthesis and degradation of starch. *Ann. Rev. Plant Physiol.*, 431–54.

Qi, Q., Kleppinger-Sparace, K. F. and Sparace, S. A. (1994). The role of the triose-phosphate shuttle and glycolytic intermediates in fatty-acid and glycerolipid biosynthesis in pea root plastids. *Planta* **194**, 193–9.

Schunemann, D., Borchert, S., Flugge, U.-I. and Heldt, H. W. (1993). ADP/ATP translocator from pea root plastids. *Plant Physiol.* **103**, 131–7.

Smith, R. G., Gauthier, D. A., Dennis, D. T. and Turpin, D. H. (1992). Malate- and pyruvate-dependent fatty acid synthesis in leucoplasts from developing castor endosperm. *Plant Physiol.* **98**, 1233–8.

Tetlow, I. J., Blissett, K. J. and Emes, M. J. (1994). Starch synthesis and carbohydrate oxidation in amyloplasts from developing wheat endosperm. *Planta* **194**, 454–60.

Trimming, B. A. and Emes, M. J. (1993). Glycolytic enzymes in non-photosynthetic plastids of pea (*Pisum sativum* L.) roots. *Planta* **90**, 439–45.

Tyson, R. H. and ap Rees, T. (1988). Starch synthesis by isolated amyloplasts from wheat endosperm. *Planta* **175**, 33–8.

Villand, P. and Kleczkowski, L. A. (1994). Is there an alternative pathway for starch biosynthesis in cereal seeds. *Z. Naturforsch.* **49c**, 215–19.

Viola, R., David, H. U. and Chudeck, A. R. (1991). Pathways of starch and sucrose biosynthesis in developing tubers of potato (*Solanum tuberosum* L.) and seeds of faba bean. *Planta* **183**, 202–8.

Wan, J., Blakeley, S. D., Dennis, D. T. and Ko, K. (1995). Import characteristics of a leucoplast pyruvate kinase are influenced by a 19-amino-acid domain within the protein. *J. Biol. Chem.* **270**, 16731–9.

Wood, C., Masterson, C. and Thomas, D. R. (1992). The role of carnitine in plant cell metabolism. In A. K. Tobin (ed.), *Plant Organelles. Compartmentation of metabolism in photosynthetic tissues*, Cambridge University Press, Cambridge, pp. 229–63.

Cytosolic Carbon
Metabolism

7 Carbohydrate synthesis and degradation

Nicholas J. Kruger

Introduction

The aim of this chapter is to discuss the pathways of carbohydrate synthesis and breakdown in plants. It is limited to a consideration of metabolism to and from hexose phosphates since the latter form a pool of intermediates in which the reactions of carbohydrate synthesis and degradation converge and interact with other pathways. In particular, the emphasis is on establishing the pathways of metabolism and their interactions rather than detailing our knowledge of the molecular and kinetic properties of the enzymes involved in these processes. This chapter will concentrate primarily, though not exclusively, on the metabolism of sucrose and starch. Together these two compounds provide the major substrates for respiration in most cells and so may be considered to dominate the carbohydrate metabolism of higher plants.

During the oxidation of carbohydrates to provide energy, substantial proportions of many intermediates may be withdrawn for biosynthetic reactions. In this way carbohydrates can contribute to the synthesis of proteins, lipids and organic acids. Although such compounds are often quantitatively important products of carbohydrates metabolism they are not considered here. In general, such compounds are derived from carbohydrates indirectly through intermediates in the pathways of hexose phosphate oxidation. These pathways are discussed in subsequent chapters.

The central role of hexose phosphates

Hexose phosphates are not only intermediates common to the pathways of synthesis and degradation of most carbohydrates, they are the principal site at which these pathways converge. Hexose phosphates are derived either from the breakdown of sugars and polysaccharides or from triose phosphates formed during photosynthesis and gluconeogenesis. They may be used for the synthesis of carbohydrates or for metabolism through the glycolytic and pentose phosphate pathways. The intermediates in this pool are glucose 1-phosphate, glucose 6-phosphate and fructose 6-phosphate, which are interconverted by phosphoglucomutase (EC 5.4.2.2; reaction [7.1]) and glucose-6-phosphate isomerase (EC 5.3.1.9; reaction [7.2]).

$$\text{glucose 1-phosphate} \rightleftharpoons \text{glucose 6-phosphate} \qquad [7.1]$$
$$\Delta G^{\circ\prime} = -7.11 \text{ kJ mol}^{-1}$$

$$\text{glucose 6-phosphate} \rightleftharpoons \text{fructose 6-phosphate} \qquad [7.2]$$
$$\Delta G^{\circ\prime} = 1.67 \text{ kJ mol}^{-1}$$

Hexose phosphates are generally regarded to be close to equilibrium in plant cells, and to form a pool from which intermediates may be either drawn, or added to, as required. Evidence for this is four-fold (ap Rees, 1977; Turner and Turner, 1980). First, the low standard free energy change of the two reactions indicates that both are readily reversible *in vitro*. Secondly, in many tissues the maximum catalytic activities of the two enzymes are high relative to the anticipated flux through these steps. Thirdly, comparison of the apparent equilibrium constants for the reactions with mass-action ratios calculated from measurements of the amount of each intermediate *in vivo* indicate that both enzymes probably operate close to equilibrium. Finally, arguments based on comparative biochemistry and the far more extensive investigations of carbohydrate metabolism in other organisms suggest that the hexose phosphates in plants are near equilibrium.

Much of the above evidence is weakened by the appreciation that carbohydrate metabolism in plants is uniquely organized (ap Rees, 1985). In particular, hexose phosphates are not confined to a single compartment but instead occur in both the cytosol and plastids. This sub-cellular compartmentation may considerably distort our interpretation of measurements of enzymes and metabolites obtained from whole tissues. Moreover, analogy with other systems, in which the organization of hexose phosphate metabolism is considerably simpler, may be misleading.

The crucial feature of the above criticism is the extent to which hexose phosphates may exchange between the pools in the cytosol and plastids. There is strong evidence that in photosynthetic tissues these pools are independent. One indication of this is that the envelope of chloroplasts isolated from a range of plants is largely impermeable to hexose phosphates (see Chapter 25). Further evidence is provided by measurements of the sub-cellular distribution of metabolites following rapid fractionation of plant protoplasts or non-aqueous fractionation of leaf samples (Stitt *et al.*, 1980; Gerhardt *et al.*, 1987). Such studies reveal that the levels of hexose phosphates inside and outside the chloroplasts vary independently. On the other hand, discrete pools of hexose phosphates do not necessarily occur in all tissues. Intact plastids isolated from several non-photosynthetic tissues can import and metabolise glucose 6-phosphate or glucose 1-phosphate (Heldt *et al.*, 1991; Schott *et al.*, 1995). Such exchange probably occurs on a phosphate translocator possessing a broader substrate specificity than that of a conventional mesophyll chloroplast translocator (see Chapter 25). The apparent widespread uptake of glucose 6-phosphate or glucose 1-phosphate by non-photosynthetic plastids implies that, in such tissues, there is a direct link between the hexose phosphate pools in the cytosol and amyloplasts. However, the potential for the phosphate translocator to be kinetically restricted *in vivo* means that it cannot be assumed that the plastidic and extra-plastidic pools of hexose phosphates are necessarily in equilibrium.

The uncertainty about the extent of compartmentation of hexose phosphate in most tissues contributes to the problem of interpreting the evidence that these metabolites are close to equilibrium. This problem can be overcome by considering separately the cytosol and plastids in tissues in which the two pools of hexose phosphates are independent. Data from such systems are extremely limited. In the few photosynthetic tissues that have been studied in sufficient detail, both the cytosol and chloroplasts contain significant activity of phosphoglucomutase and glucose-6-phosphate isomerase (ap Rees, 1985). This suggests that both compartments have the capacity to interconvert hexose phosphates. Stronger evidence is provided by measurements of metabolites following rapid fractionation of spinach protoplasts (Stitt *et al.*, 1980). The mass-action ratios from these measurements are sufficiently similar to the apparent equilibrium constants to suggest that the hexose phosphates are close to equilibrium in both the cytosol and chloroplasts. Nevertheless, some observations indicate that the interconversion of these intermediates is restricted *in vivo*. First, differences between the metabolism of ^{14}C-glucose and ^{14}C-fructose fed to excised plant tissues imply that the reaction catalyzed by glucose 6-phosphate isomerase is unable to equilibrate radiolabel between glucose 6-phosphate and fructose 6-phosphate (ap Rees, 1980). Secondly, measurements of metabolite levels after non-aqueous fractionation suggest that, although the reaction catalyzed by glucose 6-phosphate isomerase is near equilibrium in the cytosol, it is substantially displaced from equilibrium in chloroplasts (Gerhardt *et al.*, 1987). More recently, the differential metabolism of ^{14}C-glucose 6-phosphate and ^{14}C-glucose 1-phosphate to starch and CO_2 by isolated amyloplasts from soybean suspension cultures indicates that the reaction catalyzed by phosphoglucomutase is far from equilibrium in these organelles (Coates and ap Rees, 1994).

Thus, although it is often convenient to consider glucose 1-phosphate, glucose 6-phosphate and fructose 6-phosphate as a readily interconvertible pool of hexose phosphates that is maintained near equilibrium, it must be borne in mind that interconversion of these intermediates, and their exchange between the plastid and cytosol, are often restricted. Under conditions of high metabolic activity such restrictions are appreciable, and can result in both glucose 6-phosphate isomerase and phosphoglucomutase contributing to the control of metabolic flux *in vivo* (Neuhaus and Stitt, 1990).

Sucrose metabolism

The importance of sucrose

Sucrose has three fundamental and interrelated roles in plants. First, it is a principal product of photosynthesis and can account for most of the CO_2 absorbed by a plant during photosynthesis. The kinetics of labelling of sucrose in illuminated leaves exposed to $^{14}CO_2$ indicate that this compound is an important product of photosynthesis rather that an intermediate.

Secondly, sucrose is a major form in which carbon is translocated in plants. This is shown by analysis of the content of sieve tubes. The importance of sucrose in translocation is not simply a reflection of its dominant role as a product of photosynthesis since it is also the principal form in which carbon reserves are exported from non-photosynthetic tissues. During seed germination the mobilization of starch, lipid and, to a lesser extent, even protein involves their conversion to sucrose before translocation from the storage tissues to the developing embryo.

Thirdly, sucrose is the main storage sugar in plants. Chemical analysis reveals that in many plants sucrose is a major component of storage organs, such as the swollen tap root of carrot and sugar beet. However, sucrose is not restricted to such specialized storage tissues and high levels can accumulate in other organs. For example, this sugar may constitute up to 25% of the dry weight of ivy leaves.

The importance of sucrose in plant metabolism is emphasized by the fact that not only does it contribute to the three functions described above, but generally it dominates each of these roles. In certain species sugar alcohols or oligosaccharides are formed during photosynthesis and are transported and stored. However, even where such compounds are of major importance they never entirely replace sucrose.

Although the subsequent sections consider sucrose synthesis and sucrose breakdown separately, there is increasing evidence that these processes interact and that, under normal conditions, there is often

appreciable turnover of sucrose. Direct evidence for concomitant synthesis and degradation of sucrose during periods of net accumulation or utilization has been obtained for a range of tissues. Most notably, in cell suspension cultures of *Chenopodium rubrum*, the rate of sucrose turnover can be up to 30% of the rate of glucose uptake and more than 60% of the rate of respiration (Dancer *et al.*, 1990). Similar studies show that the rate of sucrose turnover is greater than that of sucrose accumulation during the respiratory climacteric in ripening banana fruit (Hill and ap Rees, 1994). Such high rates of turnover are thought to allow the net mobilization or storage of sucrose to respond sensitively to small changes in enzyme or metabolite levels, and emphasize the need to consider both synthesis and breakdown when studying sucrose metabolism.

Sucrose synthesis

Although most apparent in photosynthetic and gluconeogenic tissues, the ability to synthesize sucrose is a widespread, possibly universal, characteristic of higher plant cells. Sucrose is derived from hexose phosphates through the combined activities of UDP-glucose pyrophosphorylase (EC 2.7.7.9; reaction [7.3]), sucrose phosphate synthase (EC 2.4.1.14; reaction [7.4]) and sucrose phosphatase (EC 3.1.3.24; reaction [7.5]). Much of our basic knowledge of the properties of sucrose phosphate synthase and sucrose phosphatase has been reviewed in detail by Avigad (1982) and Hawker (1985), and more recent advances in our understanding of the regulation of sucrose phosphate synthase are summarized by Huber and Huber (1992).

$$\text{glucose 1-phosphate} + \text{UTP} \rightleftharpoons \text{UDP-glucose} + \text{PP}_i \qquad [7.3]$$
$$\Delta G^{\circ\prime} = -2.88 \, \text{kJ mol}^{-1}$$

$$\text{UDP-glucose} + \text{fructose 6-phosphate} \rightarrow$$
$$\text{sucrose } 6^F\text{-phosphate} + \text{UDP} \qquad [7.4]$$
$$\Delta G^{\circ\prime} = -5.70 \, \text{kJ mol}^{-1}$$

$$\text{sucrose } 6^F\text{-phosphate} + \text{H}_2\text{O} \rightarrow \text{sucrose} + \text{P}_i \qquad [7.5]$$
$$\Delta G^{\circ\prime} = -16.5 \, \text{kJ mol}^{-1}$$

The evidence that sucrose is synthesized via this pathway, and not through an alternative route involving sucrose synthase, comes from four sources.

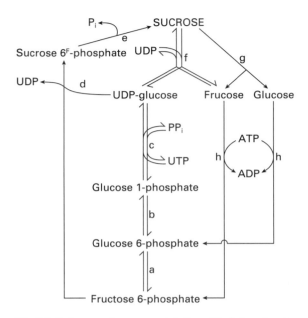

Fig. 7.1 Pathways of sucrose metabolism. The letters denote the following enzymes: a, glucose 6-phosphate isomerase [7.2]; b, phosphoglucomutase [7.1]; c, UDP-glucose pyrophosphorylase [7.3]; d, sucrose phosphate synthase [7.4]; e, sucrose phosphatase [7.5]; f, sucrose synthase [7.6]; g, invertase [7.7]; h, hexose kinase [7.13]. The numbers in brackets refer to the reactions described in the text.

First, the large overall free energy change of the reaction catalyzed by sucrose phosphatase ensures that the production of sucrose is irreversible and allows synthesis to continue in tissues that already contain large amounts of sucrose. Secondly, the kinetics of labelling of pathway intermediates in tissues fed $^{14}CO_2$ or ^{14}C-glucose are consistent with the proposed pathway. During sucrose synthesis hexose phosphates, UDP-glucose and sucrose phosphate are labelled, and both fructose 6-phosphate and the fructosyl moiety of sucrose are labelled before free fructose. The labelling in these latter compounds argues strongly against the involvement of sucrose synthase in sucrose production (ap Rees, 1984). Thirdly, in a range of tissues that make sucrose, the activity of sucrose phosphate synthase greatly exceeds that of sucrose synthase. The significance of this correlation is strengthened by the observation that in some tissues the activity of sucrose phosphate synthase, but not

that of sucrose synthase, is sufficient to account for the rate of sucrose production (ap Rees, 1988). Finally, the vast majority of sucrose synthase activity in leaves is confined to companion cells in zones of phloem loading and unloading (Nolte and Kock, 1993) where it is involved in sucrose catabolism (Lerchl et al., 1995). Thus, the high levels of sucrose synthase commonly reported in whole leaf extracts do not adequately reflect the enzyme activity in mesophyll cells, which are the major site of photosynthetic sucrose production.

There is little doubt that sucrose synthesis from hexose phosphates occurs in the cytosol. The inability of purified isolated chloroplasts to form labelled sucrose from $^{14}CO_2$, and the relative impermeability of the chloroplast inner envelope to sucrose, indicate that sucrose is synthesized outside the chloroplast. This view is confirmed by studies on photosynthesis from $^{14}CO_2$ by isolated protoplasts in which labelled sucrose appears initially in the extra-chloroplastic fraction. Careful fractionation of both photosynthetic and gluconeogenic cells has revealed that all the enzymes necessary for this pathway are present in the cytosol. Furthermore, both sucrose phosphate synthase and UDP-glucose, a metabolic intermediate in the pathway, are restricted to the cytosol (Stitt et al., 1987).

In photosynthetic and gluconeogenic tissues sucrose is predominantly exported from the cells, probably by facilitated diffusion, and subsequently taken up by the phloem complex via a specific sucrose/H^+ cotransport mechanism (Frommer and Sonnewald, 1995). However, sucrose is an important reserve carbohydrate, and can accumulate in the vacuole rather than being exported. The exact mechanism of sucrose uptake by vacuoles in such tissues is unclear. There is modest evidence for the involvement of a sucrose/H^+ antiport system. However, the ability of isolated vacuoles, or tonoplast vesicles, to absorb sucrose is generally poor, suggesting that such a mechanism, if it exists, is extremely labile (Bush, 1993).

Our understanding of the regulation of the pathway of sucrose production is limited almost exclusively to photosynthetic tissues. Comparisons of the apparent equilibrium constants of sucrose phosphate synthase and sucrose phosphatase with the mass-action ratios of reactants in spinach leaves demonstrate that both reactions are far from equilibrium *in vivo* (Krause and Stitt, 1992). The observations that spinach leaves contain very low concentrations of sucrose 6-phosphate, and that these levels do not change significantly during the day as sucrose accumulates in the leaf make it extremely unlikely that inhibition of sucrose phosphatase plays any significant role in the feedback regulation of sucrose production during photosynthesis (Krause and Stitt, 1992). Consequently, of these two enzymes, sucrose phosphate synthase has received by far the greater attention.

Three lines of evidence suggest that sucrose phosphate synthase may have a major role in the control of sucrose production in leaves. First, there is a close correlation between the rate of sucrose synthesis and the extractable activity of sucrose phosphate synthase (Stitt *et al.*, 1987). This correlation is observed between species, between genotypes within a species, and in plants in which the rate of sucrose synthesis is experimentally manipulated by varying factors such as photoperiod, temperature and nutritional status. Particularly significant in this context are the diurnal changes in sucrose phosphate synthase activity which follow closely the rate production during a normal photoperiod. Secondly, 3- to 7-fold over-expression of maize sucrose phosphate synthase in transgenic tomato plants results in a small, but significant, increase in leaf sucrose synthesis (Frommer and Sonnewald, 1995). In contrast, comparable over-expression of the corresponding spinach enzyme in tobacco and potato has no significant influence on the rate of sucrose production. However, the lack of effect in these latter experiments may be attributed to progressive deactivation of excess sucrose phosphate synthase by phosphorylation (Frommer and Sonnewald, 1995). Thirdly, the known regulatory properties of sucrose phosphate synthase are entirely consistent with this enzyme having an important role in the regulation of sucrose synthesis (Huber and Huber, 1982). Although species appear to differ in their responses, in many plants sucrose phosphate synthase can be reversibly (de)phosphorylated producing two kinetically distinct forms. These two forms are distinguished by differences in their substrate affinities, sensitivity to inhibition by phosphate and activation by glucose 6-phosphate. The proportions of the phosphorylated and

dephosphorylated forms of enzyme are determined by the relative levels of a protein kinase and phosphoprotein phosphatase whose activities can be modified by factors related to light, the availability of phosphate and the accumulation of sucrose. Consideration of changes in the proportion of the enzyme in a kinetically active (dephosphorylated) form and variations in the levels of allosteric effectors suggest that both factors contribute to the regulation of sucrose phosphate synthase, and are sufficient to account for the observed variation in the rate of sucrose synthesis throughout a 24-hour period (Huber and Huber, 1992). The ways in which such factors may interact to coordinate sucrose production with the supply of photosynthate and the demand for sucrose by sink tissues are described in detail in Chapter 25.

Sucrose breakdown

Plants contain two types of enzymes capable of cleaving sucrose. One is sucrose synthase (EC 2.4.1.13) which catalyzes the readily reversible reaction shown in reaction [7.6]. Evidence from a wide range of tissues suggests that this enzyme is confined to the cytosol. The other type of enzyme is invertase (EC 3.2.1.26) which catalyzes the essentially irreversible hydrolysis of sucrose to glucose and fructose (reaction [7.7]). Both acid and alkaline invertases are found in plants and are distinguished by having pH optima of about 5 and 7.5, respectively. Acid invertase exists in vacuoles and is associated with plant cell walls, whereas alkaline invertase is probably restricted to the cytosol. The kinetic and physio-chemical properties of these enzymes have been thoroughly reviewed by Avigad (1982) and Hawker (1985).

$$\text{sucrose} + \text{UTP} \rightleftharpoons \text{UDP-glucose} + \text{fructose} \qquad [7.6]$$
$$\Delta G^{\circ\prime} = -3.99\,\text{kJ}\,\text{mol}^{-1}$$

$$\text{sucrose} + \text{H}_2\text{O} \rightarrow \text{glucose} + \text{fructose} \qquad [7.7]$$
$$\Delta G^{\circ\prime} = -29.3\,\text{kJ}\,\text{mol}^{-1}$$

The contribution of these enzymes to the pathway of sucrose breakdown may, in part, be determined by the source of sucrose. Sucrose obtained through translocation can enter a cell via the symplast or the apoplast. Several studies using asymmetrically labelled sucrose have demonstrated that sugar obtained in this way moves primarily through the symplast and is not cleaved into glucose and fructose during transport. It seems likely that many cells receive most of their sucrose via this route (Patrick, 1990). However, in certain tissues sucrose is supplied through the apoplast. This must be the route used by developing seeds in which there are no protoplasmic connections between the maternal and embryonic tissues. Studies on the pathway of sucrose uptake from the apoplast have failed to reveal a consistent pattern. Hydrolysis of sucrose precedes uptake by developing seeds of maize, sorghum and pearl millet; whereas in wheat, rye and barley sucrose is apparently transferred without prior hydrolysis (Thorne, 1985). Somewhat surprisingly, even for tissues in which extensive hydrolysis occurs in the apoplast, such extra-cellular cleavage is not necessarily a prerequisite for sugar uptake. Studies on maize kernels have shown that 1-fluorosucrose, a sucrose analog resistant to hydrolysis by invertase, is absorbed by the developing seeds at rates similar to those of sucrose (Schmalstig and Hitz, 1987b). The purpose of such hydrolysis may be to drive phloem unloading by maintaining the concentration gradient of sucrose between source and sink tissues rather than having a direct role in sugar uptake. Evidence for this possibility is provided by the *miniature-1* mutant of maize, in which the seeds are only one fifth the normal weight due to the absence of apoplastic invertase in the basal endosperm of the developing kernel (Miller and Chourey, 1992).

The second major source of sucrose for metabolism is that stored in the vacuole. At present there is not enough information to exclude the possibility that sucrose itself moves out of the vacuole. However, the more likely route for the mobilization of stored sucrose is through hydrolysis by acid invertase in the vacuole producing hexoses that are released into the cytosol. The main evidence for this proposal is the well-established inverse relationship between intra-cellular acid invertase activity and sucrose content in many plants (ap Rees, 1984). Detailed studies indicate that in beetroot acid invertase and sucrose are both largely confined to the vacuole and that the decline in sucrose during ageing of tissue slices is associated with a corresponding increase in vacuolar invertase activity. These correlations suggest that the level of

invertase can determine the amount of sucrose in vacuoles. Increasing invertase activity in the vacuole may prevent sucrose accumulation and make stored sugar available for metabolism (ap Rees, 1984).

Sucrose entering the cytosol can be metabolized by either alkaline invertase or sucrose synthase. Although the reaction catalyzed by sucrose synthase is readily reversible *in vivo* (Geigenberger and Stitt, 1993) there is good evidence that this enzyme is involved primarily in the breakdown of sucrose. First, the distribution of sucrose synthase in different tissues is that expected for an enzyme concerned in sucrose breakdown rather than synthesis. The activity is generally low in photosynthetic and gluconeogenic cells, and is often high in actively growing tissues that rely on sucrose as their respiratory substrate (ap Rees, 1984). Secondly, in some tissues the activities of the invertases are far lower than that of sucrose synthase, and are insufficient to catalyze the observed rates of sucrose metabolism. A good example of this is developing potato tubers in which the activities of both acid and alkaline invertase are so low that sucrose synthase must be the dominant route of sucrose breakdown (ap Rees, 1988). Thirdly, analysis of the *shrunken-1* mutant of maize reveals that reductions in the level of sucrose synthase in the developing endosperm restrict the ability of this tissue to metabolize sucrose (Preiss, 1982). Similar results are obtained with transgenic potato tubers in which sucrose synthase activity is reduced by antisense inhibition (Zrenner *et al.*, 1995).

At present, the relative roles of alkaline invertase and sucrose synthase in sucrose production are only poorly understood. Excepting potato tubers and a few other tissues, studies based on comparing enzyme activities with the rate of sucrose breakdown have been unable to establish the route of degradation. This is because the capacities of sucrose synthase and alkaline invertase are generally sufficient for both to contribute significantly to sucrose metabolism. In some instances the contribution of these enzymes may be established by genetic manipulation. For example, the dramatic reduction in the extent of starch accumulation, protein content and dry weight of transgenic potato tubers deficient in sucrose synthase confirms that this enzyme is essential for effective sucrose metabolism in the developing tuber (Zrenner *et al.*, 1995). However, even though this approach can yield spectacular results, it has two limitations. One is that it is dependent upon the availability of suitable gene constructs. Since the gene encoding alkaline invertase has not yet been isolated from any plant, the reciprocal experiment involving manipulation of alkaline invertase cannot be performed. The other limitation is that such manipulations can only reveal the extent to which the step that is manipulated is essential for metabolism and not its metabolic role *in vivo*. Failure to affect metabolism by decreasing the activity of an enzyme constitutes only weak negative evidence that the enzyme being manipulated does not contribute to metabolism *in vivo*. The reasons for this are two-fold. One is that antisense inhibition decreases, but rarely eliminates, the target enzyme and the residual activity may be sufficient to catalyze the required flux (e.g. Zrenner *et al.*, 1993). The other reason is that a decrease in flux through a particular pathway resulting from experimental manipulation is often accompanied by an increase in flux through a complementary series of reactions, thus reducing the metabolic consequences of the original manipulation. Such metabolic flexibility appears to be a central feature of plants, and probably accounts for the absence of major metabolic effects of many transgenic manipulations (e.g. Hajirezaei *et al.*, 1994; Burrell *et al.*, 1994).

An alternative method to establish the relative roles of these two enzymes *in vivo* is to compare the rates of metabolism of sucrose and 1-fluorosucrose. The latter compound is an extremely poor substrate for invertase but can be cleaved by sucrose synthase. Such experiments are not simple. Adequate evidence must be provided that the sugars are taken up by the tissue at the same rate. In addition, appropriate allowances must be made for differences in the rates of metabolism of the two substrates and the extent to which the enzymes discriminate between these two compounds. All these points have been addressed in a study on developing soybean leaves which shows that the contribution of the two enzymes to metabolism varies (Schmalstig and Hitz, 1987a). In the youngest leaves sucrose is cleaved almost exclusively by sucrose synthase, whereas invertase accounts for about half of the sucrose metabolism in older leaves. This technique probably offers the best approach to defining the precise roles of sucrose synthase and alkaline invertase. However, we need to establish that

neither fluorosucrose nor its products modify metabolism before we can be sure that such results are an accurate estimate of the two pathways *in vivo*.

The preceding discussion suggests that in most conditions both invertase and sucrose synthase contribute to sucrose breakdown. The free glucose and fructose produced in these reactions are probably phosphorylated by the range of hexose kinases found in plants. These enzymes are discussed in a subsequent section. The fate of UDP-glucose produced by sucrose is less certain. A previous proposal that it may provide the sugar nucleotide donor for starch biosynthesis now seems unlikely. There are two reasons for this. One is that, despite the broad substrate specificity of sucrose synthase *in vitro*, there is good evidence that UDP-glucose is the predominant, if not exclusive, sugar nucleotide produced by this enzyme *in vivo* (ap Rees, 1992). The other reason is that, as will be argued in a subsequent section, ADP-glucose, not UDP-glucose, is the precursor of starch in both photosynthetic and non-photosynthetic cells.

An alternate suggestion is that UDP-glucose is used for the synthesis of cellulose and other structural polysaccharides that are made outside the plastids (Delmer and Amor, 1995). Some UDP-glucose may be metabolized in this way, but the demands of such pathways are likely to be small relative to the amount of UDP-glucose produced by sucrose synthase. Instead the vast majority of UDP-glucose is probably converted to hexose phosphates. Detailed consideration of the fate of sucrose metabolized by potatoes, peas and maize seeds reveals that the activity of invertase is inadequate to support the observed rates of respiration and starch synthesis. This deficit is so great that much of the UDP-glucose produced during sucrose degradation must enter the hexose phosphate pool (ap Rees, 1988).

The most likely route for the conversion of UDP-glucose to hexose phosphate is through the reaction catalyzed by UDP-glucose pyrophosphorylase, generating glucose 1-phosphate (reaction [7.3]). Of the enzymes capable of fulfilling the role, UDP-glucose pyrophosphorylase is the only one known to be present in the cytosol of a wide range of tissues at activities sufficient to support the estimated rates of metabolism (ap Rees, 1988). Although this reaction is often considered to operate in the direction of UDP-glucose synthesis, comparison of the mass-action ratio with the equilibrium constant for UDP-glucose pyrophosphorylase suggests that, at least in spinach leaves and pea embryos, the reaction is close to equilibrium. Consequently, the net flux through this step *in vivo* will depend on the relative levels of the reactants (Weiner *et al.*, 1987; ap Rees, 1988).

A central feature of the scheme outlined above is the need for one mole of pyrophosphate for each mole of sucrose metabolized. However, the source of this pyrophosphate is unknown. Although pyrophosphate is produced in the biosynthesis of protein, isoprenoid lipids and most polysaccharides, it is doubtful that such pathways could provide sufficient pyrophosphate to meet the demands of sucrose metabolism. This opinion is strengthened by the likelihood that much of the pyrophosphate produced in plastids is hydrolyzed by the alkaline pyrophosphatase activity found in these organelles. An alternative is that pyrophosphate is produced by pyrophosphate : fructose 6-phosphate phosphotransferase operating in the direction of fructose 6-phosphate synthesis (ap Rees, 1988; see reaction [7.15]). However, this suggestion now seems unlikely for two reasons. First, transgenic potato plants in which tuber pyrophosphate : fructose 6-phosphate phosphotransferase activity has been reduced to less than 2% of that in normal plants are not significantly restricted in their ability to metabolize sucrose during development (Hajirezaei *et al.*, 1994). This demonstrates that this enzyme is not an essential source of pyrophosphate for sucrose metabolism via sucrose synthase in this tissue. Secondly, comparison of the relative levels of metabolic intermediates with the appropriate mass-action ratio indicate that pyrophosphate : fructose 6-phosphate phosphotransferase operates in a net glycolytic direction in developing tubers, precisely at the time when demand for pyrophosphate by sucrose metabolism is greatest. Thus, far from providing a source of pyrophosphate under these conditions, pyrophosphate : fructose 6-phosphate phosphotransferase is likely to produce an additional demand on the reactions generating this metabolite.

Despite our ignorance about the source of pyrophosphate, elegant confirmation that it is required is provided by transgenic potato plants in which cytosolic pyrophosphate is depleted by ecotopic expression of alkaline pyrophosphatase from *Escherichia coli*. Measurements of metabolic intermediates demonstrate that this manipulation decreases sucrose utilization and its subsequent metabolism to starch by preventing the conversion of UDP-glucose to glucose 1-phosphate (Jelitto *et al.*, 1992).

Starch metabolism

The importance of starch

The pre-eminence of sucrose in plant carbohydrate metabolism is challenged only by the claims of starch. This is by far the dominant storage polysaccharide in plants and is an important metabolic substrate. It is present in all major organs of most higher plants and, in certain tissues, can accumulate to high levels. For example, starch typically accounts for 65–75% of the dry weight of cereal grains and about 80% of that of mature potato tubers. In such tissues starch provided the principal respiratory substrate, and can support high rates of metabolism. A particularly dramatic example of this is the extremely high rate of respiration in the inflorescence of *Arum maculatum* during thermogenesis. This occurs principally, if not exclusively, at the expense of starch stored in the spadix (ap Rees, 1977).

Like sucrose, starch is also a major immediate product of photosynthesis. In the light up to 30% of the $^{14}CO_2$ fixed by leaves is incorporated into starch. However, in general there is an appreciable delay after the onset of photosynthesis before starch begins to accumulate. Moreover this starch is usually mobilized in the ensuing dark period and can be converted to sucrose for export. Such observations have led to the idea that starch accumulation in chloroplasts is primarily a mechanism for storing reduced carbon when the rate of photosynthesis exceeds the capacity of the leaf to export sucrose (Stitt, 1984). Although this view is helpful in explaining some aspects of sucrose and starch metabolism in leaves, we should not consider starch solely as a buffer for sucrose. There is ample evidence that the extent of starch accumulation in leaves is controlled. In tissues exhibiting CAM metabolism the ability to fix CO_2 at night is critically dependent on a continuous supply of phosphoenolpyruvate (see Chapter 23). The latter is often derived from starch accumulated in the preceding light period. Thus, in CAM plants, starch synthesis during the day must be strictly controlled to provide appropriate amounts of phosphoenolpyruvate for CO_2 fixation in the ensuing dark period. Even in plants exhibiting C_3 photosynthesis there is evidence that the level of starch is regulated. In many species, the percentage of photosynthate retained in starch actually rises in plants grown for a few days at lower irradiance or in shorter daylength. Such observations are not consistent with starch being simply an overflow product since under these conditions less, not more, starch should be made. These, and other points, are considered in more detail by Stitt (1984), and argue strongly that starch, in its own right, is an important product of photosynthesis.

Although turnover of starch does occur and could fulfil a similar role to that proposed for sucrose turnover, the available evidence suggests that the rate of cycling of carbon through starch is generally less than that through sucrose (e.g. Dancer et al., 1990; Hill and ap Rees, 1994). However, such estimates may be misleading. Quantifying starch turnover is technically difficult, because starch is a heterogeneous substrate in which different components may not be equally accessible for metabolism, and because the process of starch degradation is ill-defined. Thus, without clear evidence to the contrary, starch metabolism, like that of sucrose, should properly be considered as the balance between the pathways of synthesis and breakdown.

Starch synthesis

Starch is found predominantly, if not exclusively, as semi-crystalline granules that are confined to the plastids. These granules contain a mixture of amylose and amylopectin. Amylose is primarily a linear molecule containing between 600 and 3000 1,4-α-glucosyl residues, although 1,6-α-glucosyl branch points occur about every 1000 residues. Amylopectin is generally much larger and more highly branched, containing between 6000 and 60 000 glucosyl residues with an average of one 1,6-α-glucosyl linkage every 20 to 26 units. Normally the ratio of amylose to

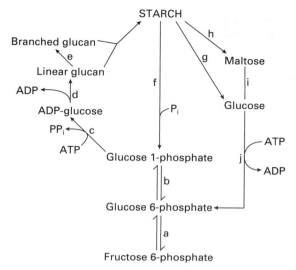

Fig. 7.2 Pathways of starch metabolism. The letters denote the following enzymes: a, glucose 6-phosphate isomerase [7.2]; b, phosphoglucomutase [7.1]; c, ADP-glucose pyrophosphorylase [7.8]; d, starch synthase [9.9]; e, braching enzyme [7.10]; f, starch phosphorylase [7.11]; g, α-amylase; h, β-amylase; i, α-glucosidase; j, hexose kinase [7.13]. The numbers in brackets refer to the reactions described in the text.

amylopectin is about 1 : 3 by weight, although there is appreciable variation in this ratio, in the molecular weight of the constituent polysaccharides, and in the degree and distribution of branching within the amylopectin component. The structure and properties of these constituents of starch and our limited understanding of their arrangement within starch granules are discussed in detail by Smith and Martin (1993).

The pathway of starch synthesis from hexose phosphates involves the following steps. First, ADP-glucose is produced from glucose 1-phosphate by ADP-glucose pyrophosphorylase (EC 2.7.7.27; reaction [7.8]). The glucosyl unit is then transferred from ADP-glucose to the non-reducing end of an α-glucan primer by starch synthase, forming an additional 1,4-α-glucosidic bond (EC 2.4.1.21; reaction [7.9]). Finally, the 1,6-α-glucosyl branch points of amylopectin are introduced by branching enzyme which hydrolyzes a 1,4-α-glucosyl bond and transfers the resulting short oligosaccharide to a primary hydroxyl group in a similar glucan chain (EC 2.4.1.24; reaction [7.10]).

$$\text{glucose 1-phosphate} + \text{ATP} \rightleftharpoons \text{ADP-glucose} + \text{PP}_i \quad [7.8]$$
$$\Delta G^{\circ\prime} = 2.88\,\text{kJ mol}^{-1}$$

$$\text{ADP-glucose} + \alpha\text{-glucan}_{(n)} \rightarrow \alpha\text{-glucan}_{(n+1)} + \text{ADP} \quad [7.9]$$
$$\Delta G^{\circ\prime} = -13.8\,\text{kJ mol}^{-1}$$

$$\text{linear 1,4-}\alpha\text{-glucan} \rightarrow \text{branched 1,6-}\alpha\text{-1,4-}\alpha\text{-glucan} \quad [7.10]$$
$$\Delta G^{\circ\prime} = -9.6\,\text{kJ mol}^{-1}$$

Although we cannot rule out other pathways of starch synthesis, none of the alternatives that have been proposed fit all the available data on starch synthesis (ap Rees, 1992; Okita, 1992). In contrast, evidence for the pathway presented above is overwhelming, and comes from two sources. The first is the study of the activity and location of the relevant enzymes (ap Rees, 1988). The activities of ADP-glucose pyrophosphorylase and starch synthase are sufficient to account for the estimated rates of starch production in tissues that differ widely in their metabolism, and changes in both activities often mirror the rate of starch accumulation during development of storage organs. Moreover, in the relatively few tissues that have been studied carefully, the vast majority of the activity of these two enzymes is confined to plastids, the site of starch accumulation.

The second line of evidence for the pathway of starch synthesis is provided by analysis of mutant plants (Okita, 1992). In maize two endosperm mutants, *shrunken-2* and *brittle-2*, are characterized by having only about 25% of the normal amount of starch. This correlates with a reduction in ADP-glucose pyrophosphorylase activity to less than 10% of that normally found in maize endosperm. Similar starch-deficient mutants have been identified in arabidopsis (*adg-1* and *adg-2*) and pea (*rugosus-b*), and each is associated with reduced ADP-glucose pyrophosphorylase activity. The properties of these mutants are mirrored by the characteristics of transgenic potato plants containing an antisense gene for the large subunit of ADP-glucose pyrophosphorylase (see below). This construct reduces the activity of ADP-glucose pyrophosphorylase in developing tubers to less than 2% of that in untransformed plants, and virtually abolishes starch accumulation. Such plants provide a compelling demonstration that ADP-glucose is required for starch synthesis, and that this metabolite is formed by ADP-glucose pyrophosphorylase.

In photosynthetic tissues the hexose phosphates required for starch synthesis are provided primarily by CO_2 fixation during photosynthesis. This is shown clearly by the ability of isolated intact chloroplasts to incorporate label from $^{14}CO_2$ into starch at rates approaching those of starch accumulation *in vivo*. For non-photosynthetic cells the ultimate source of carbon is almost invariably translocated sucrose. This is metabolized to hexose phosphates in the cytosol and then imported into the amyloplast probably in exchange for P_i on a modified phosphate translocator (Schott *et al.*, 1995). The evidence for this is three-fold. First, many non-photosynthetic tissues contain negligible plastidic fructose 1,6-bisphosphatase activity. Secondly, there is restricted randomization of label between C-1 and C-6 of the glucosyl units of starch when non-photosynthetic tissues are fed specifically labelled glucose. Both observations argue strongly against the metabolism of sucrose to three-carbon intermediates, uptake of triose phosphate or 3-phosphoglycerate, and subsequent resynthesis of hexose phosphate. Such a pathway would require plastidic fructose 1,6-bisphosphatase to convert triose phosphate to hexose monophosphates, and would involve extensive exchange between C-1 and C-6 of hexose monophosphates because of the equilibration of the triose phosphates by triose phosphate isomerase. Thirdly, direct evidence for the uptake of hexose phosphates is provided by the demonstration that plastids from several non-photosynthetic tissues can incorporate label from $[^{14}C]$-hexose phosphates into starch, that such metabolism is dependent on the intactness of the plastid preparation and that it is stimulated by ATP. The precise hexose phosphate used for starch synthesis varies between plastids. For example, plastids from developing pea embryos specifically metabolize glucose 6-phosphate, whereas those from developing wheat endosperm use only glucose 1-phosphate (Smith *et al.*, 1995). At present we do not know whether the ability to import hexose phosphates from the cytoplasm is a general feature of plastids from all non-photosynthetic cells, nor do we understand the significance of the variation between plastids in the specific hexose phosphate taken up and used as a precursor for starch synthesis.

Studies of mutants have revealed that in photosynthetic tissues much of the control of the pathway of starch synthesis from hexose phosphates is vested in ADP-glucose pyrophosphorylase, and only at very high flux do other steps in the pathway appear to contribute to regulation (Neuhaus and Stitt, 1990). ADP-glucose pyrophosphorylase from many photosynthetic tissues has been shown to be inhibited by phosphate, which induces sigmoidal kinetics, and to be activated by 3-phosphoglycerate, which relieves the inhibition by phosphate. These properties form the basis for a hypothesis of the regulation of starch synthesis (Preiss, 1991). The extremely high activity of alkaline pyrophosphatase in plastids (Weiner *et al.*, 1987) implies that the pyrophosphate produced during the synthesis of ADP-glucose is rapidly hydrolyzed. Consequently the reaction catalyzed by ADP-glucose pyrophosphorylase is almost certainly far from equilibrium *in vivo* and therefore flux through this enzyme is sensitive to regulation by allosteric effectors.

In chloroplasts the response of ADP-glucose pyrophosphorylase to 3-phosphoglycerate and phosphate may be the main factor regulating starch synthesis. This view is supported by the observation that in isolated chloroplasts starch synthesis is enhanced when the stromal 3-phosphoglycerate : phosphate ratio is increased. Similar experiments designed to manipulate the relative levels of these metabolites in intact leaves have the predicted effect on the rate of starch synthesis. This ratio probably reflects the availability of fixed carbon for starch synthesis in the light. Thus, the regulation of ADP-glucose pyrophosphorylase by these metabolites allows the rate of starch production to match the supply of substrate provided by photosynthesis. Much of this classic work is summarized by Preiss (1982), and the importance of the stromal 3-phosphoglycerate : phosphate ratio in coordinating carbon metabolism in the cytoplasm and chloroplasts is considered in more detail in Chapter 25.

Less is known about the regulation of starch synthesis in non-photosynthetic cells. ADP-glucose pyrophosphorylase from many such tissues responds to 3-phosphoglycerate and phosphate in the same way as the enzyme from chloroplasts (Preiss, 1991). This observation implies that control of the pathway in non-photosynthetic tissues is similar to that in leaves. However, since carbon for starch synthesis does not enter the amyloplast as three-carbon phosphorylated intermediates, the significance of regulation by the 3-phosphoglycerate : phosphate ratio is uncertain. In fact, much of the evidence suggesting that ADP-glucose pyrophosphorylase has a cardinal role in the regulation of starch synthesis in storage organs is equivocal, and appreciable evidence is accumulating that other reactions in the pathway may contribute to this regulation (Smith *et al.*, 1995). The reduced level of starch found in wrinkled peas containing a mutation at the *rugosus* locus can be attributed to a reduction in the activity of branching enzyme. This suggests that under certain conditions branching enzyme influences starch synthesis. There is similar genetic evidence showing that mutations affecting starch synthase in several species can influence the accumulation of starch. The latter implication is supported by the results of experiments in which the unusual heat-sensitivity of soluble starch synthase has been exploited to selectively alter its activity *in vivo*. In developing wheat ears the decrease in starch synthase activity is accompanied by an almost proportional decrease in starch synthesis, suggesting that this enzyme is very important in regulating starch production. Thus, particularly in storage organs, there is a strong likelihood that control of starch synthesis is shared between the different steps of the pathway.

Consideration of the relative contribution of each step to the regulation of starch synthesis is complicated by the observation that plants contain multiple forms of ADP-glucose pyrophosphorylase, starch synthase and starch branching enzyme which differ in their molecular and kinetic properties (Preiss, 1991). ADP-glucose pyrophosphorylase is a tetrameric enzyme containing two distinct polypeptides, a large subunit and a small subunit, both of which are required for full activity. Both subunits are encoded by multiple genes that are differentially expressed in different tissues. The functional significance of such tissue-dependent expression is not established. However, it is likely that these isozymes differ in their ability to bind allosteric regulators such as 3-phosphoglycerate. If this is so, then different combinations of large and small subunits would show differential sensitivity to allosteric regulation. Such kinetic differences do occur between ADP-glucose pyrophosphorylase isolated from endosperm and leaves of cereals. This variation is likely to have profound effects on the distribution of control between the various steps of starch synthesis in different tissues.

To date all storage tissues that have been studied contain one or more soluble starch synthases that may be partly associated with the starch granules, and another genetically distinct starch synthase that is found exclusively bound to starch granules. The latter enzyme is a 60-kilodalton protein, often called the 'waxy' protein, because it is the product of the *waxy* gene originally identified in maize. Similarly, two classes of starch branching enzyme can be distinguished. Each tissue known to contain two or more isoforms of this enzyme possesses at least one isoform of each class. These forms differ in their substrate specificity, as reflected by differences between forms in the ratio of activities obtained using two different assay methods. Although these branching enzymes are soluble, varying proportions of each form may be associated with starch granules (Martin and Smith, 1995).

Our understanding of the role of the multiple forms of starch synthase and branching enzyme in starch synthesis is limited. However, since the separate forms of starch synthase and branching enzyme differ in their substrate specificity, it is likely that such forms may vary in their contribution to the synthesis of different components of the starch granule (Martin and Smith, 1995). The clearest indication that this is so comes from studies of *waxy* mutants of maize, rice sorghum and amaranthus, and the corresponding *amylose-free* mutant of potato. In these mutants, the almost complete absence of amylose from starch in the storage organ is correlated with a massive reduction in the activity of the major granule-bound starch synthase activity. This clearly implicates the waxy protein in amylose formation, and has led to a proposal to explain how essentially unbranched amylose is synthesized at the same time as highly branched amylopectin (Martin and Smith, 1995). In essence, the suggestion is that soluble starch synthase and starch branching enzyme operate at the periphery of the starch granule to synthesize amylopectin. This amylopectin condenses to form the matrix of the granule, onto which the waxy protein binds. The

latter protein subsequently synthesizes a 1,4-α-glucan which, due to its location within the starch granule, is inaccessible to the soluble starch branching enzymes. Although this is an attractive proposal, explaining several features of the *waxy* phenotype, it is almost certainly too simplistic and will require revision as our understanding increases (Smith *et al.*, 1995).

Similarly, the extent and pattern of branching within amylopectin may be determined by the activities, and relative proportions, of the different forms of starch branching enzyme. In maize the two major isoforms of starch branching enzyme differ considerably in their relative affinities for substrates with different degrees of branching. Consequently, when allowed to react with amylose *in vitro*, branching enzyme II transfers chains that are, on average, shorter than those transferred by branching enzyme I. The biological significance of this is demonstrated by the maize mutant, *amylose extender,* in which a deficiency of branching enzyme II results in amylopectin that is far less branched than that from normal maize (Preiss, 1982). Additional evidence is provided by the correlation between changes in the relative proportions of starch branching isoforms and changes in the average branch length of amylopectin in developing pea embryos (Martin and Smith, 1995). In both examples, the changes in amylopectin structure are precisely those expected from the properties of the different classes of starch branching enzyme. However, further studies suggest that other factors may contribute to the determination of amylopectin structure. Most notable is the *sugary-1* mutant of maize, which produces an abnormal, very highly 1,6-α-branched 1,4-α-glucan, known as phytoglycogen. This mutation appears to alter the activities of several enzymes of starch metabolism, but particularly starch debranching enzyme. This has led to the suggestion that the structure of amylopectin is normally determined by a balance of branching and debranching activities, and that the phytoglycogen seen in the *sugary-1* mutant results from the dramatic reduction in the amount of starch debranching enzyme. At present there is not sufficient evidence to decide whether the branching of amylopectin is determined principally by the properties of the branching enzymes alone, or by a balance between the activities of branching and debranching enzymes (Smith *et al.*, 1995).

The evidence summarized above is consistent with the idea that different forms of starch synthase and starch branching enzyme make major contributions to both the fine structure and relative proportions of amylose and amylopectin within starch. However, we are still a long way from a complete description of the way in which these enzymes interact to determine the range of variation in the proportions of amylose and amylopectin that occurs between starch granules from different species or varieties and, even within plants, between different organs and the same organ at different developmental stages.

Starch breakdown

In contrast to recent advances in our understanding of the regulation of starch biosynthesis, knowledge of starch breakdown in plants is far more rudimentary and is essentially limited to the characterization of starch-degrading enzymes. Plants contain a range of enzymes that may contribute to starch breakdown. In principle, internal 1,4-α-glucosyl bonds can be cleaved hydrolytically by α-amylase (EC 3.2.1.1) or other endoamylases resulting in the production of a mixture of linear and branched oligosaccharides and, ultimately, glucose, maltose, maltotriose and a range of branched α-limit dextrins. Starch can also be hydrolyzed by β-amylase (EC 3.2.1.2) which catalyzes the removal of successive maltose units from the non-reducing end of α-glucan chains. The maltose and other short maltosaccharides produced by these enzymes may be further hydrolyzed to glucose by α-glucosidase (maltase) (EC 3.2.1.20) and subsequently phosphorylated by hexokinase, as described in the next section. Alternatively, 1,4-α-glucosyl bonds can be cleaved phosphorolytically by starch phosphorylase (EC 2.4.1.1; reaction [7.11]). This enzyme produces glucose 1-phosphate from successive glucosyl residues at the end of an α-glucan chain. Although phosphorylase can only attack oligosaccharides larger than maltotetraose, further phosphorolytic cleavage may occur due to the action of a glucosyltransferase such as D-enzyme (EC 2.4.1.25; reaction [7.12]). By increasing the degree of polymerization of short oligosaccharides this enzyme

can make more of the glucan accessible to phosphorylase and hence contribute to starch degradation.

$$\alpha\text{-glucan}_{(n)} + P_i \rightleftharpoons \alpha\text{-glucan}_{(n-1)} + \text{glucose 1-phosphate} \quad [7.11]$$
$$\Delta G^{\circ\prime} = 2.98 \text{ kJ mol}^{-1}$$

$$\alpha\text{-glucan}_{(m)} + \alpha\text{-glucan}_{(n)} \rightleftharpoons \alpha\text{-glucan}_{(m+n-1)} + \text{glucose} \quad [7.12]$$

Branch points in starch are removed exclusively by de-branching enzyme (EC 3.2.1.41) which cleaves 1,6-α-glucosyl bonds hydrolytically releasing linear oligosaccharides for further metabolism. The properties and roles of these enzymes have been reviewed in detail by Preiss (1982) and Beck and Zeigler (1989). As with the enzymes of starch synthesis, plants typically contain multiple isozymes of endoamylase, exoamylase and starch phosphorylase which are differentially expressed (e.g. Sonnewald et al., 1995).

Two aspects of starch breakdown deserve particular attention if we are to elucidate the relative roles of the different enzymes in the mobilization of starch. One is the initial degradation of the insoluble starch granule. It is generally believed that this step is catalyzed only by an endoamylase and that the soluble oligosaccharides released in the reaction provide the substrate for the other enzymes of starch breakdown (Stitt and Steup, 1985). This view is based on several reports which claim that α-amylase is the only enzyme capable of degrading isolated starch grains, and on the fact that starch phosphorylase has a much higher affinity for linear low molecular weight oligosaccharides than for branched polyglucans. This evidence is limited and is considerably weakened by the following considerations. First, in general, the rates of starch breakdown in such studies are low and obtained using α-amylase from bacterial and mammalian sources rather than plants. The characteristics of starch breakdown in such systems may have little relevance to the mechanism of starch breakdown in vivo. Second, we know little about the hydrolysis of starch granules by the Ca^{2+}-dependent heat-labile endoamylase that replaces conventional α-amylase in several tissues. Third, the belief that starch phosphorylase is unable to attack starch granules is questionable. Phosphorylase from pea chloroplasts can release labelled glucose 1-phosphate from ^{14}C-labelled starch granules. Similarly, appreciable amounts of glucose 1-phosphate are produced when spinach leaf starch grains are incubated with extra chloroplastic phosphorylase from the same source (Beck and Zeigler, 1989). Finally, starch phosphorylase has been studied from far too few tissues for us to generalize about its kinetic properties, nor do such properties necessarily indicate the substrate for the enzyme *in vivo*. Together these criticisms emphasize that, at present, there is no compelling evidence to support the idea that the initial degradation of starch is due solely to hydrolysis by endoamylase, although it remains an attractive hypothesis.

The second aspect of starch breakdown that warrants particular attention is the relative importance of hydrolytic and phosphorolytic routes of degradation. This point is difficult to resolve because there is no clear distinction between the two pathways and oligosaccharides released by hydrolysis may, in turn, be metabolized by either amylases or phosphorylase. In addition, the pathway of mobilization may vary depending on the specific metabolic requirements of the tissue. Since starch accumulates in many tissues that have quite different physiological roles it is unlikely that the route of starch degradation will always be the same. For example, during germination of cereal grains starch breakdown coincides with the destruction of the endosperm. There is strong evidence that in such instances starch degradation is extracellular and occurs via a hydrolytic route (Beck and Zeigler, 1989). Glucose produced by the action of α-amylase, de-branching enzyme and α-glucosidase is absorbed by the scutellum, converted to sucrose and then transported to the embryo. In this type of seed the mobilization of starch is controlled primarily by alterations in the levels of the relevant enzymes. During germination the activity of some enzymes increases by *de novo* synthesis whereas for others the increase is due to activation of previously latent enzymes associated with protein bodies. These processes are generally under environmental and hormonal control. They are thought to respond to changes in the level of gibberellic acid (Beck and Zeigler, 1989), although recent evidence implicates sucrose in the regulation of gene expression during this process (Thomas and Rodriguez, 1994).

Other seeds have been studied in which the major storage tissue remains intact throughout germination.

In such examples, starch breakdown occurs intracellularly, although in some the amyloplast membrane is destroyed making the starch grains accessible to cytosolic enzymes. The pathway of starch metabolism in these tissues is variable. The available data suggest that in pea cotyledons starch breakdown is largely phosphorolytic, whereas in soybean and lentil α-amylase is the predominant activity. Although many seeds contain high activity of β-amylase, its role in starch degradation during germination is uncertain. Studies on both rye and soybean have shown that varieties which lack β-amylase germinate normally and mobilize starch at rates similar to those of varieties containing this enzyme. Thus, although β-amylase may contribute to the degradation of oligosaccharides released by α-amylase, it is not essential for starch breakdown (ap Rees, 1988).

In the examples described above, starch degradation is essentially irreversible and continues until all the starch is metabolized. However, this is not so in all tissues. In leaves, starch forms a temporary reserve which accumulates or is mobilized depending on the carbohydrate status of the cell. In such instances cellular integrity must be maintained during starch breakdown and therefore at least the initial stages of starch degradation must occur in the plastids. Studies of the sub-cellular distribution of the enzyme of starch metabolism suggest that, in leaves, starch is mobilized by the combined activities of endoamylase and starch phosphorylase (Stitt, 1984; Stitt and Steup, 1985). Characterization of the products of starch degradation in isolated chloroplasts reveals that glucose 1-phosphate produced by phosphorylase is metabolized within the chloroplast to triose phosphates and 3-phosphoglycerate, whereas the other products, glucose and maltose, are probably exported to the cytoplasm.

In leaves, the relative activities of these two pathways may be important in determining the fate of the carbon released from starch, with only that portion released as neutral sugars being available for sucrose synthesis and subsequent export to the rest of the plant. Comparison of the metabolism of ^{14}C-glucose and ^{14}C-glycerol by leaves indicates that, in the dark, the production of hexose phosphates from triose phosphates is restricted (Trethewey and ap Rees, 1994). This is almost certainly due to inhibition of cytosolic fructose 1,6-bisphosphatase by fructose 2,6-bisphosphate under such conditions (see Chapter 25). This inhibition implies that carbon released from starch as glucose 1-phosphate and subsequently exported to the cytosol as three-carbon phosphorylated intermediates is effectively unavailable for sucrose synthesis. The strongest evidence for this proposal comes from studies on a mutant arabidopsis line, TC265 (now designated *starch excess-1*), in which the chloroplast glucose transporter is disrupted (Trethewey and ap Rees, 1994). Such plants cannot mobilize their leaf starch reserves effectively at night, contain significantly lower amounts of sucrose, and consequently, have lower rates of respiration and lower growth rates than wild-type plants. The latter features are almost certainly due to an inability of the leaves to provide sufficient sucrose to meet the demands of the rest of the plant since addition of exogenous sucrose restores the rate of dark respiration and increases the growth rate. Thus, the pathway of starch degradation appears to be a critical factor in determining the metabolic fate of carbon in leaves.

Unfortunately, we have only a limited understanding of the way in which the pathways of starch degradation are controlled in leaves. Diurnal variation in the level of starch suggests that the mobilization of starch is regulated. In principle this control could be achieved solely by variation in the rate of starch synthesis with the rate of starch breakdown being constant. However, the absence of detectable starch turnover in leaves from both pea and sugar beet in the light argues strongly against this possibility and implies that the pathway of starch degradation is regulated directly (Stitt and Steup, 1985). Although starch breakdown is normally restricted during the day, such regulation is not achieved by light *per se*. This view is supported by the observation that net starch degradation can occur in leaves during continuous illumination, and the demonstration that under certain conditions illuminated spinach chloroplasts catalyze the concomitant synthesis and degradation of starch (Stitt and Steup, 1985).

Several studies have shown that the breakdown of starch at the beginning of the night is often delayed until the level of sucrose in the leaf has declined suggesting that starch mobilization may be controlled by the requirements of the cell for a respiratory substrate. Experiments with isolated chloroplasts from spinach and pea demonstrate that the production of phosphorylated intermediates from starch is stimulated by phosphate. These results raise the possibility that the rate of starch phosphorolysis is determined by the availability of phosphate. Although this may control the production of hexose phosphates in the stroma it cannot, by itself, explain the regulation of starch breakdown. At least in spinach chloroplasts, the reduction in the rate of phosphorolytic breakdown of starch in the absence of phosphate is offset by a corresponding increase in the accumulation of glucose so that the overall rate of starch degradation is unchanged. This interpretation is supported by studies on transgenic tobacco plants in which the chloroplastic (L-form) of starch phosphorylase has been selectively decreased by antisense inhibition. Despite reducing chloroplastic starch phosphorylase to undetectable levels this manipulation has no obvious effect on starch metabolism (Sonnewald *et al.*, 1995). This result implies that, if this activity contributes to starch mobilization under normal conditions, hydrolysis by amylase can compensate for any decrease in phosphorolytic breakdown of starch. Such an interaction suggests that the rate of starch breakdown is largely independent of the pathway of degradation and implies that both hydrolysis and phosphorolysis are controlled. To date, studies of the kinetic properties of the enzymes involved in starch degradation have failed to reveal any obvious regulatory features. Thus, although there is good evidence that starch mobilization in leaves is regulated, we have little idea of how such control is achieved.

The metabolism of free hexose

The cleavage of sucrose and the hydrolysis of starch are the major sources of free hexose in plants. The glucose and fructose produced by these pathways are metabolized principally through conversion to the corresponding hexose 6-phosphate by a hexose kinase (EC 2.7.1.1; reaction [7.13]).

$$\text{hexose} + \text{ATP} \rightarrow \text{hexose 6-phosphate} + \text{ADP} \qquad [7.13]$$

$$\Delta G^{\circ\prime}_{glucose} = -16.7 \, \text{kJ} \, \text{mol}^{-1}$$

$$\Delta G^{\circ\prime}_{fructose} = -14.6 \, \text{kJ} \, \text{mol}^{-1}$$

This reaction is essentially irreversible, but the precise standard free energy change depends on the sugar that is phosphorylated.

The few tissues studied in any detail each contain a variety of hexose kinases (Copeland and Turner, 1987). These are distinguished by differences in their affinity and maximum activity with different sugars. Some are general hexokinases that can phosphorylate a wide range of sugars, such as glucose, fructose and mannose. In contrast, the kinetic characteristics of other forms are such that *in vivo* they are likely to be specific for individual sugars. Although the enzymes may also differ in their ability to use different nucleotide triphosphates as the phosphoryl donor, the comparative nucleotide specificities of the enzymes are such that ATP is likely to be the dominant phosphoryl donor *in vivo*. Differences in the properties of hexose kinases are typified by those presented in Table 7.1 for the six activities identified in developing potato tubers (Renz and Stitt, 1993). At

present we do not know whether any of these forms are artifacts resulting from proteolytic modification during extraction. However, this seems unlikely since similar ranges of hexose kinases possessing comparable molecular and kinetic properties occur in other plants.

The multiplicity of enzymes capable of phosphorylating different hexoses may be due, at least partly, to three features of carbohydrate metabolism in plants. One is the degree to which such metabolism is compartmented. The second is the variety of hexoses that can be metabolized and the range of sources from which these sugars are derived. The third is the consideration that the two dominant sugars in metabolism, glucose and fructose, are produced in variable amounts. These features make it extremely unlikely that a single enzyme would be suitable for the phosphorylation of the range of sugars produced under different conditions. Consistent with this argument, the relative proportions of the six forms of hexose kinase in potato plants vary between organs and between different developmental stages (Renz *et al.*, 1993).

At present we do not know the precise

Table 7.1 Kinetic constants of hexose kinases from developing potato tubers. Fructokinase 1, 2 and 3 display negligible activity with either glucose or mannose. V_{max} values are expressed relative to that obtained using glucose. Phosphorylation coefficients are calculated by expressing the specificity factor (V_{max}/K_m) of the enzyme for a particular hexose relative to that obtained with glucose. Data are from Renz and Stitt (1993).

Enzyme form	Substrate	Kinetic constant		
		K_m (mM)	V_{max} (rel. value)	Phosphorylation coefficient
Hexokinase 1	Glucose	0.041	1.0	1.0
	Fructose	11	1.2	0.0046
	Mannose	0.052	0.83	0.53
Hexokinase 2	Glucose	0.13	1.0	1.0
	Fructose	22	0.59	0.0036
	Mannose	0.29	0.48	0.22
Hexokinase 3	Glucose	0.035	1.0	1.0
	Fructose	8.7	1.27	0.0053
	Mannose	0.038	0.50	0.45
Fructokinase 1	Fructose	0.064	–	–
Fructokinase 2	Fructose	0.090	–	–
Fructokinase 3	Fructose	0.100	–	–

physiological role of each hexose kinase. However, their functions are likely to depend partly on their subcellular location. In most plants that have been investigated, the majority of the fructokinase activity is in the cytosol (ap Rees, 1985; Schnarrenberger, 1990). This enzyme is probably responsible for the metabolism of fructose derived from the cleavage of sucrose by sucrose synthase. The observation that the activity of fructokinase greatly exceeds that of glucokinase in many tissues is consistent with the view that sucrose synthase rather than invertase is the major route of sucrose degradation and that consequently the amount of fructose produced is larger than that of glucose.

In some tissues, a large proportion of the activity that preferentially phosphorylates glucose is associated with mitochondria (ap Rees, 1985). Sub-fractionation studies of mitochondrial preparations, and protection experiments using exogenous proteases, provide strong evidence that this hexokinase is located on the outer mitochondrial membrane. The significance of this association is unclear, although conceivably this arrangement may allow newly synthesized ATP to be used preferentially for the phosphorylation of hexose in the cytosol. In mammals the extent to which a similar hexokinase binds to mitochondria is dependent on the levels of glucose 6-phosphate and ATP, and there is evidence that variation in the amount of binding may contribute to the regulation of glucose phosphorylation. Whether a similar system operates in plants is uncertain. In castor bean endosperm, mitochondrial hexokinase can be released by specific hexose phosphates and nucleotides. However, these metabolites have no effect on the much larger proportion of hexokinase associated with mitochondria in pea leaves and young stems.

The route by which any glucose released during starch hydrolysis is phosphorylated is unknown. This glucose may be exported from the plastid and phosphorylated in the cytosol (Trethewey and ap Rees, 1994). However, isolated chloroplasts from both pea and spinach have the potential to metabolize glucose rapidly (Stitt, 1984). The enzyme responsible for this phosphorylation has not been identified. Although chloroplast preparations contain appreciable hexokinase activity, there is good evidence that, at least in pea leaves, this activity is attached to the outer membrane and probably cannot contribute to the phosphorylation of stromal sugars (Stitt, 1984). Similarly, although one of the forms of hexose kinase in spinach leaves is apparently confined to the plastid stroma, kinetic studies suggest that this form is a specific fructokinase and is unlikely to phosphorylate glucose efficiently *in vivo* (Schnarrenberger, 1990). More detailed studies are needed to see whether plastids contain additional hexokinases.

Synthesis of structural polysaccharides

The biosynthesis of cell walls is one of the most complex areas of plant biochemistry and is a quantitatively important aspect of carbohydrate metabolism, especially during cell growth. Our understanding of the synthesis of the structural polysaccharides has been described in detail elsewhere (Chapter 9) (Delmer and Stone, 1988) and this discussion concentrates on identifying the precursors for the synthesis of these polysaccharides.

Although the precise details of cellulose biosynthesis in higher plants are not established, UDP-glucose is almost certainly the immediate glucosyl donor for this polymer (Delmer and Amor, 1995). Detailed analyses of the levels and labelling of intermediates in developing cotton fibres supplied with radioactive glucose are entirely consistent with UDP-glucose, rather than some other sugar nucleotide, such as GDP-glucose, being the precursor of cellulose (Delmer, 1987). More recently, UDP-glucose has been shown to support 1,4-β-glucan synthesis by digitonin-solubilized enzyme preparations from plasma-membrane-enriched fractions of cotton fibre cells (Kudlicka *et al.*, 1995). However, the rates of cellulose synthesis by such preparations are very low, and are accompanied by the synthesis of large amounts of β-1,3-glucan by a seemingly ubiquitous plasma-membrane associated callose synthase (UDP-glucose: 1,3-β-glucan synthase, EC 2.4.1.34). This report emphasizes just how little we know about cellulose production in higher plants (Delmer *et al.*, 1993). Despite intensive efforts, an enzyme preparation capable of selectively synthesizing cellulose at rates comparable to those observed *in vivo* has not yet been isolated from higher plants.

Most precursors for the other major cell wall polysaccharides are probably derived from UDP-glucose through a series of sugar nucleotide transformations. The principal hexose and pentose residues required for the synthesis of pectins and hemicelluloses can be produced by the following series of reactions: first, oxidation of UDP-glucose to UDP-glucuronate; secondly, decarboxylation of UDP-glucuronate to UDP-xylose; and thirdly, epimerization of these three compounds to form UDP-galactose, UDP-galacturonate and UDP-arabinose, respectively (Fig. 7.3). Evidence that these reactions contribute to polysaccharide synthesis is provided by the demonstration that plants contain the necessary enzymes and that each of these sugar nucleotides is readily incorporated into appropriate polymers in cell-free extracts (Delmer and Stone, 1988).

On the evidence presented above, UDP-glucose may be considered the principal precursor for structural polysaccharide synthesis. This compound is derived from two sources. In cells in which sucrose is metabolized by sucrose synthase, UDP-glucose is produced during the initial cleavage of sucrose and may be used directly for polysaccharide synthesis (Delmer and Amor, 1995). In other tissues UDP-glucose can be formed from the hexose phosphate

pool by UDP-glucose pyrophosphorylase. As described earlier, this reaction is probably close to equilibrium *in vivo*. Therefore the conversion of glucose 1-phosphate to UDP-glucose will be determined largely by the extent to which the latter intermediate is metabolized to form the structural polysaccharides.

Entry of hexose phosphates into glycolysis

In many plant tissues glycolysis is the dominant pathway of carbohydrate oxidation and, under most conditions, it is a major drain on the hexose phosphate pool. The first committed step in this sequence is the conversion of fructose 6-phosphate to fructose 1,6-bisphosphate. Plants contain two enzymes capable of catalyzing this reaction: phosphofructokinase (EC 2.7.1.11; reaction [7.14]) and pyrophosphate : fructose 6-phosphate phosphotransferase (EC 2.7.1.90; reaction [7.15]).

fructose 6-phosphate + ATP $\rightarrow$ fructose 1,6-bisphosphate + ADP
$$\Delta G^{\circ\prime} = -14.2 \, \text{kJ mol}^{-1} \qquad [7.14]$$

fructose 6-phosphate + PP$_i$ $\rightleftharpoons$ fructose 1,6-bisphosphate + P$_i$
$$\Delta G^{\circ\prime} = -2.93 \, \text{kJ mol}^{-1} \qquad [7.15]$$

Pyrophosphate : fructose 6-phosphate phosphotransferase catalyzes a reaction that is near-equilibrium (Weiner *et al.*, 1987). Consequently, assigning a precise role to this enzyme is difficult, and it has been variously implicated in glycolysis, gluconeogenesis, regulation of the pyrophosphate concentration, equilibration of the hexose phosphate and triose phosphate pools, and adaptation to stress (Frommer and Sonnewald, 1995). Much of the evidence for the involvement of the enzyme in glycolysis is circumstantial. In most tissues the activity of pyrophosphate : fructose 6-phosphate phosphotransferase is greater than that of phosphofructokinase, often by 5- to 10-fold. This activity is sufficient to catalyze the observed glycolytic flux. In addition, the enzyme is markedly activated by nanomolar levels of fructose 2,6-bisphosphate, an important regulator of glycolysis in animals and fungi. Based on comparative biochemistry, this metabolite should fulfil a similar role in higher plants. Moreover, the concentration of fructose

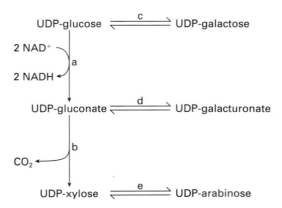

Fig. 7.3 Interconversion of sugar nucleotides. The letters denote the following enzymes: a, UDP-glucose dehydrogenase (EC 1.1.1.22); b, UDP-glucuronate decarboxylase (EC 4.1.1.35); c, UDP-glucose 4-epimerase (EC 5.1.3.2); d, UDP-glucuronate 4-epimerase (EC 5.1.3.6); e, UDP-arabinose 4-epimerase (EC 5.1.3.5).

2,6-bisphosphate often increases when glycolysis is stimulated (Hatzfeld and Stitt, 1991). This correlation is readily explained if pyrophosphate : fructose 6-phosphate phosphotransferase operates in the glycolytic direction. Direct evidence for this proposal is provided by the observation that a large increase in the rate of glycolysis in a suspension culture of *Chenopodium rubrum* is accompanied by an increase in the amount of fructose 2,6-bisphosphate and a decrease in the amount of pyrophosphate (Hatzfeld *et al.*, 1989). The latter change indicates that pyrophosphate : fructose 6-phosphate phosphotransferase has been activated, and contributes to glycolysis when there is a large, rapid increase in the rate of respiration. However, this observation implies neither that pyrophosphate : fructose 6-phosphate phosphotransferase always operates in the direction of fructose 6-phosphate phosphorylation *in vivo* nor that it invariably contributes to net glycolytic flux. Even within *Chenopodium rubrum* suspension cultures the correlation between glycolytic flux and fructose 2,6-bisphosphate level is not absolute, and increases in flux may precede changes in the level of this activator (Hatzfeld and Stitt, 1991). This suggests that the level of fructose 2,6-bisphosphate is changing in response to alteration of glycolytic flux and not *vice versa*. Furthermore, studies on the spadix of *Arum maculatum* indicate that pyrophosphate : fructose 6-phosphate phosphotransferase is not a prerequisite for a large, rapid increase in glycolysis. In this tissue the enormous increase in respiration that occurs during thermogenesis is not accompanied by any detectable change in either fructose 2,6-bisphosphate or pyrophosphate, and the maximum catalytic activity of the enzyme is well below that required to accommodate the observed rate of glycolysis (ap Rees, 1988). Thus, at least in this specialized tissue, pyrophosphate : fructose 6-phosphate phosphotransferase is not essential for glycolysis. Similar conclusions can be drawn from studies on transgenic potato plants in which the activity of pyrophosphate : fructose 6-phosphate phosphotransferase in the tubers is reduced to less than 2% of that in normal potatoes (Hajirezaei *et al.*, 1994). Changes in the relative levels of hexose phosphates and triose phosphates in such tubers

confirm that the enzyme catalyzes a net glycolytic flux in this tissue, at least in the conditions that have been studied. However, removal of the vast majority of the enzyme has no discernible effect on phenotype, growth rate or tuber yield. Although the transgenic plants display modest decreases in starch content compared with wild type, these differences are entirely attributable to indirect effects on metabolism resulting from changes in the relative levels of metabolic intermediates in response to a decrease in pyrophosphate : fructose 6-phosphate phosphotransferase activity. Thus, even in a tissue in which the enzyme operates in the glycolytic direction, pyrophosphate : fructose 6-phosphate phosphotransferase does not appear to have an essential role in the conversion of fructose 6-phosphate to fructose 1,6-bisphosphate. The function of this enzyme remains an enigma, and since it catalyzes a reaction that is near-equilibrium *in vivo* it is entirely possible that it may fulfil different roles in different tissues.

In contrast, there is strong evidence that phosphofructokinase contributes to the entry of hexose phosphates into glycolysis. First, the reaction catalyzed by this enzyme is essentially irreversible *in vitro*, and a comparison of the equilibrium constant for this reaction with the mass-action ratios calculated from the amounts of the relevant metabolites in several tissues indicate that this reaction is probably far from equilibrium *in vivo* (Turner and Turner, 1980). Secondly, there is a good correlation between phosphofructokinase activity and glycolysis in a wide range of plants (ap Rees, 1988). For example, during the development of pea roots, variation in the maximum catalytic activity of phosphofructokinase reflects the changes that occur in the rate of respiration. More detailed investigations of the club of the developing spadix of *Arum maculatum* have shown significant, coordinated increases in the maximum catalytic activities of many glycolytic enzymes, including phosphofructokinase, prior to the massive increase in glycolysis that occurs during thermogenesis. Thirdly, in all tissues in which careful measurements have been made, the activity of phosphofructokinase is sufficient to catalyze the estimated glycolytic flux (ap Rees, 1988). Fourthly, studies of carbohydrate metabolism in isolated chloroplasts and amyloplasts provide evidence for the direct involvement of phosphofructokinase in glycolysis within plastids. Since these organelles lack pyrophosphate : fructose 6-phosphate phosphotransferase, the observed glycolytic flux must be catalyzed by phosphofructokinase (ap Rees, 1985). Finally, our current understanding of the regulation of glycolysis in plants is based, in part, on the known kinetic properties of phosphofructokinase (Turner and Turner, 1980; Dennis and Greyson, 1987).

Despite a wealth of circumstantial evidence implicating phosphofructokinase in the regulation of glycolysis, recent attempts to manipulate the pathways of hexose phosphate oxidation by altering the activity of this enzyme have been unsuccessful. Transformation of potato plants with a chimeric gene encoding phosphofructokinase from *Escherichia coli* under the control of a tuber specific promoter increased the maximum catalytic activity of phosphofructokinase in mature tubers by 14-fold. Despite this change, the rate and pattern of respiration in the transgenic tubers are unaltered (Burrell *et al.*, 1994). These results do not preclude phosphofructokinase from a critical role in catalysing the entry of hexose phosphates into glycolysis. However, they imply that this process is regulated by one, or more, of the subsequent reactions in the pathway. These reactions are considered in detail in the following chapter (Chapter 8).

Concluding remarks

Carbohydrate metabolism in plants differs from that in other organisms in two important aspects. First, plant metabolism is uniquely compartmented. The two principal storage carbohydrates, sucrose and starch, are synthesized in the cytosol and plastids, respectively. Both compartments possess the enzymic capacity to metabolize hexose phosphates, and possess more or less independent pools of these metabolites. Secondly, plant metabolism is redundant, meaning that there is often more than one route by which metabolites may be interconverted. This is particularly so for the pathways of sucrose and starch degradation in which disruption of specific enzymes often has no overall effect on flux. In combination, these unique features considerably hinder our analysis of plant

carbohydrate metabolism. Thus, although we have a reasonable description of the principal pathways of carbohydrate metabolism in plants, our understanding of their regulation is still restricted.

In recent years, molecular techniques have begun to contribute to advances in our knowledge of carbohydrate metabolism. This trend looks set to continue and, in combination with more traditional techniques, the ability to manipulate metabolism by altering the activity of specific enzymes offers a powerful approach to study regulation of these pathways. Such information is valuable in its own right, because of the central role of carbohydrates in the carbon economy of plants. However, a better understanding of the factors regulating metabolism and distribution of the principal storage carbohydrates is attracting increased attention because it also offers the potential for rational improvement of crop plants (Frommer and Sonnewald, 1995; see also Chapter 36).

References.

ap Rees, T. (1977). Conservation of carbohydrate by the non-photosynthetic cells of higher plants. *Symposium of the Society for Experimental Biology* 31, 7–32.

ap Rees, T. (1980). Assessment of the contribution of metabolic pathways to plant respiration. In *The Biochemistry of Plants*, ed. D. D. Davies, Vol. 2, Academic Press, New York, pp. 1–29.

ap Rees, T. (1984). Sucrose metabolism. In *Storage Carbohydrates in Vascular Plants*, ed. D. H. Lewis, Cambridge University Press, Cambridge, pp. 53–73.

ap Rees, T. (1985). The organization of glycolysis and the pentose phosphate pathway in plants. In *Encyclopedia of Plant Physiology*, Vol. 18, eds R. Douce and D. Day, Springer-Verlag, Berlin, pp. 391–417.

ap Rees, T. (1988). Hexose phosphate metabolism by non-photosynthetic tissues of higher plants. In *The Biochemistry of Plants*, Vol. 14, ed. J. Preiss, Academic Press, New York, pp. 1–33.

ap Rees, T. (1992). Synthesis of storage starch. In *Carbon Partitioning: Within and Between Organisms*, eds C. J. Pollock, J. F. Farrar and A. J. Gordon, BIOS Scientific Publishers, Oxford, pp. 115–31.

Avigad, G. (1982). Sucrose and other disaccharides. In *Encyclopedia of Plant Physiology*, eds F. A. Loewus and W. Tanner, Springer-Verlag, Berlin, pp. 217–347.

Beck, E. and Zeigler, P. (1989). Biosynthesis and degradation of starch in higher plants. *Ann. Rev. Plant Physiol. Plant Mol. Biol.* **40**, 95–117.

Burrell, M. M., Mooney, P. J., Blundy, M., Carter, D., Wilson, F., Green, J., Blundy, K. S. and ap Rees, T. (1994). Genetic manipulation of 6-phosphofructokinase in potato tubers. *Planta* **194**, 95–101.

Bush, D. R. (1993). Proton-coupled sugar and amino acid transporters in plants. *Ann. Rev. Plant Physiol. Plant Mol. Biol.* **44**, 513–42.

Coates, S. A. and ap Rees, T. (1994). Metabolism of glucose monophosphates by leucoplasts and amyloplasts from soybean suspension cultures. *Phytochemistry* **35**, 881–3.

Copeland, L. and Turner, J. F. (1987). The regulation of glycolysis and the pentose phosphate pathway. In *The Biochemistry of Plants*, Vol. 11, ed. D. D. Davies, Academic Press, New York, pp. 107–28.

Dancer, J., Hatzfeld, W.-D. and Stitt, M. (1990). Cytosolic cycles regulate the turnover of sucrose in heterotrophic cell-suspension cultures of *Chenopodium rubrum* L. *Planta* **182**, 223–31.

Delmer, D. P. (1987). Cellulose biosynthesis. *Ann. Rev. Plant Physiol.* **38**, 259–90.

Delmer, D. P. and Amor, Y. (1995). Cellulose biosynthesis. *Plant Cell* **7**, 987–1000.

Delmer, D. P. and Stone, B. A. (1988). Biosynthesis of plant cell walls. In *The Biochemistry of Plants*, Vol. 14, ed. J. Preiss, Academic Press, New York, pp. 373–420.

Delmer, D. P., Ohana, P., Gonen L. and Benziman, M. (1993). In vitro synthesis of cellulose in plants: still a long way to go! *Plant Physiol.* **103**, 307–8.

Dennis, D. T. and Greyson M. F. (1987). Fructose 6-phosphate metabolism in plants. *Physiol. Plant.* **69**, 395–404.

Frommer, W. B. and Sonnewald, U. (1995). Molecular analysis of carbon partitioning in solanaceous species. *J. Exp. Bot.* **46**, 587–607.

Geigenberger, P. and Stitt, M. (1993). Sucrose synthase catalyses a readily reversible reaction in vivo in developing potato tubers and other plant tissues. *Planta* **189**, 329–39.

Gerhardt, R., Stitt, M. and Heldt, H. W. (1987). Subcellular metabolite levels in spinach leaves. *Plant Physiol.* **83**, 399–407.

Hajirezaei, M., Sonnewald, U., Viola, R., Carlisle, S., Dennis D. T. and Stitt, M. (1994). Transgenic potato plants with strongly decreased expression of pyrophosphate:fructose 6-phosphate phosphotransferase show no visible phenotype and only minor changes in tuber metabolism. *Planta* **192**, 16–30.

Hatzfeld, W.-D., Dancer, J. and Stitt, M. (1989). Direct evidence that pyrophosphate:fructose 6-phosphate phosphotransferase can act as a glycolytic enzyme in plants. *FEBS Lett.* **254**, 215–18.

Hatzfeld, W.-D. and Stitt, M. (1991). Regulation of glycolysis in heterotrophic cell suspension cultures of *Chenopodium rubrum* in response to proton fluxes at the plasma membrane. *Physiol. Plant.* **81**, 103–10.

Hawker, J. S. (1985). Sucrose. In *Biochemistry of Storage Carbohydrates in Green Plants*, eds P. M. Dey and R. A. Dixon, Academic Press, New York, pp. 1–51.

Heldt, H. W., Flugge, U. I. and Borchert, S. (1991). Diversity of specificity and function of phosphate translocators in various plastids, *Plant Physiol.* **95**, 341–3.

Hill, S. A. and ap Rees, T. (1994). Fluxes of carbohydrate metabolism in ripening bananas. *Planta* **192**, 52–60.

Huber, S. C. and Huber, J. L. (1992). Role of sucrose-phosphate synthase in sucrose metabolism in leaves. *Plant Physiol.* **99**, 1275–8.

Jelitto, T., Sonnewald, U., Willmitzer, L., Hajirezaei, M. and Stitt, M. (1992). Inorganic pyrophosphate content and metabolites in potato and tobacco plants expressing *E. coli* pyrophosphatase in their cytosol. *Planta* **188**, 238–44.

Kudlicka, K., Brown Jr, R. M., Li, L, Lee, J. H., Shen, H. and Kuga, S. (1995). β-Glucan synthesis in cotton fiber. IV. In vitro assembly of cellulose I allomorph. *Plant Physiol.* **107**, 111–23.

Krause, K.-P. and Stitt, M. (1992). Sucrose-6-phosphate levels in spinach leaves and their effects on sucrose-phosphate synthase. *Phytochemistry* **31**, 1143–6.

Lerchl, J., Geigenberger, P., Stitt, M. and Sonnewald, U. (1995). Impaired photoassimilate partitioning caused by phloem-specific removal of pyrophosphate can be complemented by a phloem-specific cytosolic yeast-derived invertase in transgenic plants. *Plant Cell* **7**, 259–70.

Martin, C. and Smith, A. M. (1995). Starch biosynthesis. *Plant Cell* **7**, 971–85.

Miller, M. E. and Chourey, P. S. (1992). The maize invertase-deficient *miniature-1* seed mutation is associated with aberrant pedicel and endosperm development. *Plant Cell* **4**, 297–305.

Neuhaus, H. E. and Stitt, M. (1990). Control analysis of photosynthetic partitioning: impact of reduced ADP-glucose pyrophosphorylase or plastid phosphoglucomutase on the fluxes to starch and sucrose in *Arabidopsis*. *Planta* **182**, 445–54.

Nolte, K. D. and Koch K. E. (1993). Companion-cell specific localization of sucrose synthase in zones of phloem loading and unloading. *Plant Physiol.* **101**, 899–905.

Okita, T. W. (1992). Is there an alternative pathway for starch synthesis? *Plant Physiol.* **100**, 560–64.

Patrick, J. W. (1990). Sieve element unloading: cellular pathway, mechanism and control. *Physiologia Plantarum* **78**, 298–308.

Preiss, J. (1982). Biosynthesis of starch and its regulation. In *Encyclopedia of Plant Physiology*, Vol. 13A, eds F. A. Loewus and W. Tanner, Springer-Verlag, Berlin, pp. 397–417.

Preiss, J. (1991). Biology and molecular biology of starch synthesis and its regulation. In *Oxford Surveys of Plant Molecular and Cell Biology*, Vol. 7, ed. B. J. Miflin, Oxford University Press, Oxford, pp. 59–114.

Renz, A. and Stitt, M. (1993). Substrate specificity and product inhibition of different forms of fructokinases and hexokinases in developing potato tubers. *Planta* **190**, 166–75.

Renz, A., Merlo, L. and Stitt, M. (1993). Partial purification from potato tubers of three fructokinases and three hexokinases which show differing organ and developmental specificity. *Planta* **190**, 156–65.

Schmalstig, J. D. and Hitz, W. D. (1987a). Contribution of sucrose synthase and invertase to the metabolism of sucrose in developing leaves. *Plant Physiol.* **85**, 407–12.

Schmalstig, J. D. and Hitz, W. D. (1987b). Transport and metabolism of a sucrose analog (1′-fluorosucrose) into *Zea mays* L. endosperm without invertase hydrolysis. *Plant Physiol.* **85**, 902–5.

Schnarrenberger, C. (1990). Characterisation and compartmentation, in green leaves, of hexokinases with different specificities for glucose, fructose, and mannose and for nucleotide triphosphates. *Planta* **181**, 249–55.

Schott, K., Borchert, S., Müller-Röber, B. and Heldt, H. W. (1995). Transport of inorganic phosphate and C_3- and C_6-sugar phosphates across the envelope membranes of potato tuber amyloplasts. *Planta* **196**, 647–52.

Smith, A. M. and Martin, C. (1993). Starch biosynthesis and the potential for its manipulation. In *Biosynthesis and Manipulation of Plant Products, Plant Biotechnology Series*, Vol. 3, ed. D. Grierson, Blackie, Glasgow, pp. 1–54.

Smith, A. M., Denyer, K. and Martin, C. R. (1995). What controls the amount and structure of starch in storage organs? *Plant Physiol.* **107**, 673–7.

Sonnewald, U., Basner, A., Greve B. and Steup, M. (1995). A second L-type isozyme of potato glucan phosphorylase: cloning, antisense inhibition and expression analysis. *Plant Mol. Biol.* **27**, 567–76.

Stitt, M. (1984). Degradation of starch in chloroplasts: a buffer to sucrose metabolism. In *Storage Carbohydrates in Vascular Plants*, ed. D. H. Lewis, Cambridge University Press, Cambridge, pp. 205–29.

Stitt, M. and Steup, M. (1985). Starch and sucrose degradation. In *Encyclopedia of Plant Physiology*, Vol. 18, eds R. Douce and D. Day, Springer-Verlag, Berlin, pp. 347–90.

Stitt, M., Wirtz, W. and Heldt, H. W. (1980). Metabolite levels during induction in the chloroplast and extrachloroplast compartments of spinach protoplasts. *Biochim. Biophys. Acta* **593**, 85–102.

Stitt, M., Huber, S. C. and Kerr, P. (1987). Regulation of photosynthetic sucrose synthesis. In *The Biochemistry of Plants*, Vol. 13, eds M. D. Hatch and N. K. Boardman, Academic Press, New York, pp. 327–409.

Thomas, B. R. and Rodriguez, R. L. (1994). Metabolite signals regulate gene expression and source/sink relations in cereal seedlings. *Plant Physiol.* **106**, 1235–9.

Thorne, J. H. (1985). Phloem unloading of C and N assimilates in developing seeds. *Ann. Rev. Plant Physiol.* **36**, 317–43.

Trethewey, R. N. and ap Rees, T. (1994). The role of the hexose transporter in the chloroplasts of *Arabidopsis thaliana* L. *Planta* **195**, 168–74.

Turner, J. F. and Turner, D. H. (1980). The regulation of glycolysis and the pentose phosphate pathway. In *The Biochemistry of Plants*, Vol. 2, ed. D. D. Davies, Academic Press, New York, pp. 279–316.

Weiner, H., Stitt, M. and Heldt, H. W. (1987). Subcellular compartmentation of pyrophosphate and alkaline pyrophosphatase in leaves. *Biochim. Biophys. Acta* **893**, 13–21.

Zrenner, R., Willmitzer, L. and Sonnewald, U. (1993). Analysis of the expression of potato uridinediphosphate-glucose pyrophosphorylase and its inhibition by antisense RNA. *Planta* **190**, 247–52.

Zrenner, R., Salanoubat, M., Willmitzer, L. and Sonnewald, U. (1995). Evidence of the crucial role of sucrose synthase for sink strength using transgenic potato plants (*Solanum tuberosum* L.). *Plant J.* **7**, 97–107.

8 Glycolysis, the pentose phosphate pathway and anaerobic respiration

David T. Dennis, Yafan Huang and Fayek B. Negm

The function of glycolysis in plant cells

The glycolytic pathway is present in almost all organisms and, in plants, the reactions and enzymes of this pathway are fundamentally the same as found elsewhere and are shown in Fig. 8.1 (reviewed by: ap Rees, 1985; ap Rees, 1988; Copeland and Turner, 1987; Plaxton, 1990; Blakeley and Dennis, 1993; Dennis and Blakeley, 1995; Plaxton, 1996). There are, however, very important differences resulting from the nature of plant metabolism. The classical equation for glycolysis is:

$$\text{Glucose} + 2\text{ ADP} + 2\text{ P}_i + 2\text{ NAD}^+ \rightarrow$$
$$2\text{ Pyruvate} + 2\text{ ATP} + 2\text{ NADH} + 2\text{ H}^+$$

This simple equation suggests that glycolysis functions only to produce ATP and reducing power in the form of NADH by the oxidation of glucose to pyruvate. While this is indeed a requirement of this pathway, especially in the dark and under anaerobic conditions, the principal function of the pathway in plants is to provide intermediates for biosynthetic pathways (see Fig. 36.1). The flow of carbon required to satisfy the biosynthetic needs of plants is large because plants are autotrophic and can synthesize all the compounds needed for their growth and development. For example, a major drain on the intermediates of glycolysis is the synthesis of cell wall components such as the polysaccharides and lignin. It has been estimated that lignin alone may account for as much as 20% of the carbon flow through the upper part of the pathway.

Plants are unique among the eukaryotes in containing an additional organelle, the plastid, which is the site of many of the biosynthetic pathways. Hence, not only must there be a flow of carbon through cytosolic glycolysis, but there must also be a supply of carbon for the biosynthetic reactions in the plastid. In photosynthetic plastids (chloroplasts), this carbon can be obtained directly from carbon fixation, but, in young leaves, the biosynthetic demands may exceed the photosynthetic capacity of the chloroplast. In the non-green plastids of roots and seeds, all the carbon for the biosynthetic needs of the organelle has to be imported into the organelle. In seeds, where large amounts of storage compounds are synthesized, the major flux of glycolytic carbon into the plastid may be to supply the needs of these pathways (Dennis et al., 1985).

To supply the demand for carbon for biosynthetic precursors, enzymes of glycolysis are sequestered in the plastid (see Fig. 36.2; also see Chapter 6). In oilseeds, where the demand is very high, all of the enzymes of the pathway have been found in the plastid. In other tissues, the amount of each enzyme of the pathway varies greatly depending upon developmental stage and species (e.g. see Miernyk and Dennis, 1992). Indeed, some enzymes of the pathway may be absent from the organelle.

The principal route for the exchange of carbon between the cytosol and the plastid is the phosphate translocator which exchanges triose phosphate for inorganic phosphate (see Chapter 25). However, more recently, it has been found that related translocators can transport other metabolites such as hexose phosphates and phospho*enol*pyruvate (e.g. Borchert et al., 1993; Trimming and Emes, 1993) so that the interchange between the cytosol and the plastid is complex and may vary between tissues and at different stages of development. These different uptake properties reflect the specific biosynthetic needs of the tissue.

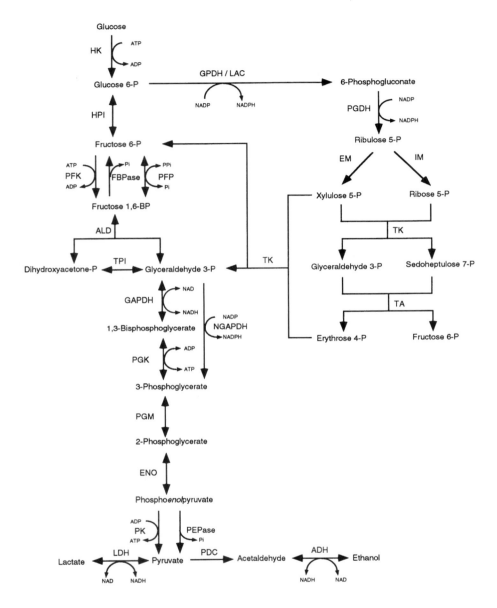

Fig. 8.1 Reactions and interaction between glycolysis and the pentose phosphate pathway. Also shown is the anaerobic reduction of pyruvate to lactate or ethanol. Abbreviations of enzymes used: HK, hexokinase; HPI, hexose phosphate isomerase; PFK, ATP-dependent phosphofructose kinase; PFP, PP$_i$-dependent phosphofructose kinase; FBPase, fructose 1,6-bisphosphatase; ALD, aldolase; TPI, triose phosphate isomerase; GAPDH, glyceraldehyde 3-phosphate dehydrogenase; NGAPDH, NADP-dependent non-phosphorylating glyceraldehyde 3-phosphate dehydrogenase; PGK, phosphoglycerate kinase; PGM, phosphoglyceromutase; ENO, enolase; PK, pyruvate kinase; PEPase, phosphoenolpyruvate phosphatase; LDH, lactate dehydrogenase; PDC, pyruvate decarboxylase; ADH, alcohol dehydrogenase; GPDH, glucose 6-phosphate dehydrogenase; LAC, lactonase; PGDH, 6-phosphogluconate dehydrogenase; EM, phosphopentose epimerase; IM, phosphoriboisomerase; TK, transketolase; TA, transaldolase.

To date, much of the research on plant glycolysis has concentrated on the supply of carbon to satisfy the biosynthetic needs of the tissue. However, it is also important to consider the cofactor requirements for biosynthesis since cofactors are not readily transported across membranes. This is particularly important for pathways such as fatty acid biosynthesis, which only occurs in plastids, and in which the requirement for ATP and reducing power in the form of NADH and NADPH is very great (Dennis, 1989; Dennis and Blakeley, 1993).

In chloroplasts, a supply of cofactors can be derived from photosynthesis, but in non-green plastids and in chloroplasts from young leaves, cofactors are more efficiently supplied by locating pathways that generate them within the organelle. For example, it has been shown that fatty acid biosynthesis in leucoplasts from the developing endosperm of the castor seed is enhanced by the formation of ATP within the organelle. This is generated by plastid localized pyruvate kinase, utilizing phospho*enol*pyruvate imported from the cytosol (Boyle *et al.*, 1990). Hence, the location of glycolysis (and the pentose phosphate pathway) within plastids allows the generation of ATP, NADH, and NADPH within the organelle.

The localization of glycolysis within plastids means that, in plant cells, this pathway is duplicated in the cytosol and plastid compartments. The enzymes catalyzing the reactions within plastids are isozymes of their cytosolic counterparts (Miernyk and Dennis, 1982; Dennis *et al.*, 1991; Emes and Tobin, 1993). In addition, there appear to be multiple forms of translocators to allow communication between the plastid and the cytosol and these translocators vary in different tissues depending upon the individual requirements of each tissue (Emes and Tobin, 1993).

The nature of glycolysis in plants

There is no doubt that the reactions of glycolysis occur in plants, but is there a glycolytic pathway in the classical sense? In most organisms, glycolysis consists of a linear sequence of reactions each of which is essential to the operation of the pathway. Hence, yeast cells are unable to metabolize glucose

if any enzyme of the pathway (e.g. pyruvate kinase) is missing (Maitra and Lobo, 1977). In humans, a fatal condition, hemolytic anemia, occurs if pyruvate kinase is absent from erythrocytes (Miwa, 1990).

In contrast, plants appear to have a much greater flexibility (Black *et al.*, 1987) and to be able to maintain carbon flow through the pathway even in the absence of an 'essential' enzyme such as pyruvate kinase (Gottlob-McHugh *et al.*, 1992). This results from two factors. First, as discussed above, most tissues have at least some of the glycolytic enzymes in both the cytosol and plastid and there are points at which these two pathways communicate through the transporters of the inner plastid envelope. This interaction of the two pathways allows a flexibility that is not found in organisms lacking plastids.

Secondly, there are steps in glycolysis that are catalyzed by multiple enzymes. For example, the conversion of phospho*enol*pyruvate to pyruvate can be catalyzed by cytosolic pyruvate kinase, plastid pyruvate kinase, phospho*enol*pyruvate phosphatase and the successive reactions of phospho*enol*pyruvate carboxylase and malic enzyme (Theodorou and Plaxton, 1995; Plaxton, 1996). In a similar manner, the interconversion of fructose 6-phosphate and fructose 1,6-bisphosphate is catalyzed by three enzymes in plants compared with two in most other organisms.

This duplication of pathways and enzymes provides plants with a flexibility of metabolism not found in other organisms (Black *et al.*, 1987). This is essential to the survival of plants because they have a limited ability to provide cellular homeostasis in which metabolic reactions can occur. Plants are anchored in the soil and each cell is subject to whatever environmental conditions prevail at the time. Hence, in young leaves in the light, triose phosphate for glycolytic reactions will be supplied by photosynthesis whereas, in the dark, sucrose will be the carbon source. Similarly, there may also be rapid changes in water or nutrient status and in temperature. The metabolism of the plant must be flexible to deal with these changes and this flexibility is achieved by multiple enzymes and pathways which are interconnected by elaborate and variable translocators.

The enzymes and isozymes of plant glycolysis

The ATP- and PP$_i$-dependent phosphofructokinases

The two enzymes at the first committed step in the glycolytic pathway have been described in Chapter 7. The ATP-dependent enzyme appears to be the constitutive enzyme for the operation of the pathway. The level of the pyrophosphate-dependent phosphofructokinase is variable although it does appear to have a role in supplying carbon for the pathway (e.g. Hajirezaei et al., 1994). However, transgenic plants in which this enzyme has been almost eliminated grow normally even under conditions of stress so that its function and role in glycolysis are not clear (Hajirezaei et al., 1994; Paul et al., 1995). These enzymes will be described in more detail when the regulation of glycolysis is discussed.

Aldolase

Aldolase catalyzes the aldol cleavage of fructose 1,6-bisphosphate (F1,6BP) to glyceraldehyde 3-phosphate and dihydroxyacetone phosphate. The standard free energy change for this reaction is $+5.5$ Kcal/mol so that, under standard conditions, the equilibrium position strongly favours F1,6BP. However, because there is one substrate and two products, under normal cellular concentrations of metabolites, the reaction is readily reversible. Under some physiological states F1,6BP could accumulate and, since F1,6BP activates the pyrophosphate-dependent phosphofructokinase, this could affect the flow of carbon in glycolysis (Nielsen, 1995).

There are two types of aldolase, termed Class I and Class II, which have very different reaction mechanisms and structural properties. Class I aldolases form a Schiff-base intermediate between the keto group of the sugars and an ε-amino group of a lysine on the enzyme. In contrast, Class II aldolases utilize a divalent cation as an electrophile in the reaction mechanism. Class I aldolases have a broad pH optimum whereas that of the Class II enzymes is very narrow. In addition, Class I aldolases are tetrameric with a subunit M$_r$ of about 40 000 whereas

the majority of Class II enzymes are dimeric with a subunit M$_r$ of about 40 000. There is no sequence homology between Class I and Class II enzymes suggesting that they arose through convergent evolution (Marsh and Lebherz, 1992). All animals and higher plants have Class I type aldolases whereas Class II aldolases are found in yeast and a range of prokaryotic organisms. Some eukaryotes such as *Euglena* have both types of enzyme (Pelzer-Reith et al., 1994).

Higher plants have cytosolic and plastid isozymes of aldolase and both are of the Class I type (Lebherz et al., 1984; Chopra et al., 1990; Razdan et al., 1992). Surprisingly, cyanobacteria have only Class II aldolases. This is in conflict with the concept that plastids are derived from a cyanobacterial endosymbiont. However, the plastid and cytosolic Class I enzymes have only 54% homology at the amino acid level and must have diverged at an early stage of plant evolution (Pelzer-Reith et al., 1993). The cytosolic form of aldolases from different species are more closely related than the plastid and cytosolic isozymes from the same species. In the case of *Cyanophora paradoxa*, an organism that has chloroplasts intermediate between higher plant chloroplasts and those of the cyanobacteria, the cytosolic and plastid forms of aldolase are both type II (Gross et al., 1994). In *Euglena*, which has both types of aldolase, the plastid is a Class I enzyme whereas the cytosol has both Class I and II aldolases (Pelzer-Reith et al., 1994). In contrast, Chlamydomonas has only a single plastid Class I aldolase (Schnarrenberger et al., 1994). Clearly the evolutionary origins of aldolases in plants are complex and different from most other plastid and cytosolic isozymes studied so far.

Triosephosphate isomerase

Triosephosphate isomerase (TPI) catalyzes the interconversion of dihydroxyacetone phosphate and glyceraldehyde 3-phosphate. This enzyme has one of the highest known turnover numbers and the reaction is limited only by diffusion. The standard free energy change for the enzyme is -1.8 kcal/mol in the direction of dihydroxyacetone formation so that at equilibrium, the ratio of dihydroxyacetone phosphate to glyceraldehyde 3-phosphate is about 14:1. It is the favourable equilibrium of the reactions of the lower half of glycolysis that allows glycolysis to proceed in spite of this unfavourable equilibrium. TPI plays a central role in metabolism since it is at the crossroads of glycolysis, gluconeogenesis and the oxidative and reductive pentose phosphate pathways.

TPI from all sources is very similar and is a homodimer with a subunit M_r of about 27 000. The enzyme has been cloned from a number of species and there is a high degree of homology among different plants and between plants and animals suggesting a slow rate of evolution for this gene (Xu et al., 1993; Shih, 1994). There are cytosolic and plastid isozymes of TPI (Henze et al., 1994) although, again, they appear to be very similar and it has been reported that an active heterodimer can be formed between them. Immunological studies have indicated that the plastid enzyme is related to the cytosolic enzyme but is unrelated to the enzyme from prokaryotes suggesting that the plastid isozyme has evolved through a duplication of an ancestral nuclear gene (Henze et al., 1994). In rice, there appears to be only one gene for the cytosolic isozyme (Xu and Hall, 1993). In spinach, there is only one gene for the plastid isozyme (Henze et al., 1994). In contrast, in maize there are multiple copies of the gene for both isozymes (Henze et al., 1994). The gene for cytosolic rice TPI has been sequenced and shown to have eight introns which have identical locations with those in maize (Xu et al., 1993). Interestingly, the first intron of the gene from rice is required for the expression of the gene in monocots but is not required for its expression in dicots, indicating a fundamental difference in the regulation of gene expression in dicots and monocots (Xu et al., 1994).

Glyceraldehyde 3-phosphate dehydrogenase

Glyceraldehyde 3-phosphate dehydrogenase (GAPDH) catalyzes the conversion of glyceraldehyde 3-phosphate to 1,3-bisphosphoglycerate. This is the only oxidation reaction in glycolysis in which the electrons from the substrate are donated to NAD^+ which is reduced to NADH. The free energy change of this reaction is +1.5 kcal/mol so the reaction is readily reversible although the equilibrium favours glyceraldehyde 3-phosphate by about 10:1 over 1,3-bisphosphoglycerate. The equilibrium position of the reaction in the cell is determined by the ratio of NAD^+ to NADH since NADH and NAD^+ combine competitively with the enzyme. In all cases, the enzyme is a homotetramer with a subunit M_r of about 34 000. Each subunit binds one molecule of NAD^+ and the pure enzyme normally contains bound NAD^+. The reaction mechanism has been determined for the enzyme from pea seeds where it was found to have a ping-pong mechanism and involve a bound acyl intermediate (Duggleby and Dennis, 1974). The same reaction mechanism appears to hold for the enzyme from all organisms. The enzyme from plants has been cloned and sequenced and shown to be highly conserved with the enzyme from other organisms.

There are four distinct forms of glyceraldehyde 3-phosphate dehydrogenase in plants. In addition to the cytosolic NAD^+ dependent enzyme which is the enzyme of glycolysis, there is an $NADP^+$ dependent enzyme in the chloroplast that is involved in photosynthesis. The cytosolic NAD^+ dependent enzyme is a homotetramer with four identical subunits, and the chloroplast $NADP^+$ dependent enzyme exists as either homotetramer (A_4), or heterotetramer with two types of subunits (A_2B_2, Brinkmann et al., 1989). This enzyme is quite distinct from the NAD^+ dependent enzyme and is described in Chapter 21. It is noted that a novel NAD^+ dependent enzyme is found in the plastids of the gymnosperm *Pinus sylvestris* L. (Meyer-Gauen et al., 1994). A fourth form of the enzyme is the non-phosphorylating glyceraldehyde 3-phosphate dehydrogenase which forms 3-phosphoglycerate instead of 1,3-bisphosphoglycerate (see below). This step bypasses the formation of ATP by phosphoglycerate kinase and may be important during phosphate starvation when the level of adenylates drops. This enzyme would allow carbon to flow down the pathway for biosynthetic reactions without the need for ADP as a substrate for the second reaction. This enzyme will be described later.

Phosphoglycerate kinase

Phosphoglycerate kinase catalyzes the formation of 3-phosphoglycerate from 1,3-bisphosphoglycerate. The high free energy of hydrolysis of the mixed acid anhydride bond between the phosphate group and the 1-carboxyl of the 1,3-bisphosphoglycerate is used to synthesize ATP from ADP. The standard free energy change for this reaction is −4.5 kcal/mol so that the equilibrium strongly favours 3-phosphoglycerate. However, under conditions of a high ATP/ADP ratio the formation of glyceraldehyde 3-phosphate is possible.

In plants, there are cytosolic and plastid isozymes of this enzyme. The properties of the isozymes are very similar and are also similar to the enzyme from other organisms. A comparison of the nucleic acid and derived amino acid sequences for cytosolic and plastid isozymes of phosphoglycerate kinase revealed a higher than expected level of similarity (Longstaff et al., 1989). In addition, the cytosolic and plastid isozymes appear to have both eukaryote and prokaryote features suggesting a recombination of the plastid and cytosolic genes rather than a direct evolution of the plastid gene from a prokaryote precursor (Longstaff et al., 1989). The chloroplastic phosphoglycerate kinase from Chlamydomonas has similar properties to the enzyme from other plants (Kitayama and Togasaki, 1995) and antiserum raised against the enzyme from the green alga Selenastrum shows a wide cross reactivity with plastid and cytosolic isozymes, consistent with this hypothesis.

Phosphoglyceromutase

Phosphoglyceromutase (PGM) catalyzes the interconversion of 3-phosphoglycerate and 2-phosphoglycerate. The standard free energy change for this reaction is +1.0 kcal/mol so the reaction is readily reversible. There are two forms of PGM. All animals have a cofactor (2,3-bisphosphoglycerate) dependent form whereas higher plants have a cofactor independent type. Among microorganisms, both forms can be found. In addition, the plant enzyme is monomeric with an M_r of 62 000 whereas depending on the sources, the cofactor-dependent form functions as either monomer, dimer or tetramer

with a subunit M_r of approximately 30 000. PGM has been reported to occur in plastids (Simcox et al., 1977; Journet and Douce, 1985; Kang and Rawsthorne, 1994). However, it is not present in all plastids. It is absent from chloroplasts which may prevent the removal of intermediates from the Calvin cycle which shares 3-phosphoglycerate with glycolysis (Stitt and ap Rees, 1979). A study of plastid and cytosolic isozymes of PGM has suggested that they are very difficult to separate and that they are physically and kinetically very similar (Botha and Dennis, 1987).

The cytosolic form of the plant enzyme has been cloned and sequenced from a number of plants (Grana et al., 1992; Huang et al., 1993). There is no sequence homology with the animal enzyme indicating that the two forms of PGM have evolved independently. The animal enzyme has essential histidines at the active site and in vitro mutagenesis and inhibitor studies of the plant enzyme have indicated that histidine residues are also involved in the reaction mechanism of this enzyme (Huang and Dennis, 1995). Hence, although the two types of enzyme are very different, they appear to have undergone convergent evolution with respect to the reaction mechanism. Why one form requires a cofactor and the other is cofactor independent is not clear. The plant enzyme can be overexpressed in an active form in E. coli (Huang and Dennis, 1995). This bacterium has the cofactor dependent PGM. This suggests that the cofactor independent form did not interfere with the metabolism of the bacterium even at very high concentrations, which suggests that there is no metabolic significance to the requirement for the cofactor. The gene for the maize enzyme has been cloned and shown to have nine introns which is different from that of the animal enzyme (Perez de la Ossa et al., 1994).

Enolase

Enolase catalyzes the interconversion of 2-phosphoglycerate and phospho*enol*pyruvate. This reaction is freely reversible since the standard free energy change is +0.4 kcal/mol. There are plastid and cytosolic isozymes of enolase but the level of enzyme varies in different tissues (e.g. Miernyk and Dennis,

1992). In castor plants, the percentage of enolase found in the plastid is highest in the developing seed where it accounts for 30% of the total activity, coincident with the time of maximum oil biosynthesis. In mature leaves of the same plant, enolase is not present in the chloroplasts and only the cytosolic form is found. In general, the level of plastid enolase appears to be correlated with the amount of biosynthetic activity in the tissue, consistent with plastid glycolysis playing a role in the supply of carbon and cofactors to biosynthetic pathways. The enzyme from all sources so far studied is a dimer of identical subunits with M_r values of about 50 000. It requires a divalent metal ion for activity and the physical and kinetic properties of the plant enzyme are similar to the enzyme from other sources.

Enolase has been cloned from both monocots and dicots and shows a remarkable conservation of sequence both between plant species and with other organisms (Van Der Straeten *et al.*, 1991; Lal *et al.*, 1991; Blakeley *et al.*,1994). In *Arabidopsis* only one gene for the enzyme was found and this encoded a cytosolic form of the enzyme. From this it was speculated that a plastid isozyme of the enzyme may not exist, at least in *Arabidopsis* (Van Der Straeten *et al.*, 1991). There is, however, considerable biochemical evidence for presence of enolase in some types of plastid but as yet a clone for a plastid isozyme has not been isolated. In tomatoes, there is only a small increase in the mRNA for enolase under anaerobic conditions and none occurs after heat shock, even though enolase is recognized as a heat shock protein in yeast. In contrast, in maize there was a five fold increase in the mRNA for enolase on anaerobiosis although again no increase on heat or cold shock (Lal *et al.*, 1991).

Pyruvate kinase

Pyruvate kinase (PK) catalyzes the formation of pyruvate from phospho*enol*pyruvate with the concomitant synthesis of ATP from ADP. Hence, it is one of the energy conservation steps in glycolysis. The standard free energy change of this reaction is -7.5 kcal/mol so that it is virtually irreversible. The irreversibility of the reaction stems from the *enol* to *keto* isomerization of pyruvate, the *keto* form being

the most stable in solution whereas the actual reaction product of PK is the *enol* form.

Plastid and cytosolic forms of the enzyme have been isolated (Ireland *et al.*, 1980). The cytosolic enzyme from most organisms is a homotetramer with a subunit M_r of about 55 000 and in some plant tissues, a very similar homotetramer is found (Plaxton, 1996). However, in other cases there are reports that the enzyme may be a heterotetramer (Plaxton, 1996). The homotetrameric enzyme has been purified from developing castor endosperm and does not have pronounced regulatory kinetics as is found for the animal enzyme although it is inhibited by several end products of metabolism (Podesta and Plaxton, 1992). However, from other tissues such as cotyledons (and probably leaves) the enzyme is regulated by the end products of amino acid biosynthesis such as glutamate and glutamine (Podesta and Plaxton, 1994). This would correlate the activity of PK with the supply of carbon in glycolysis with the need for carbon skeletons for amino acid synthesis. These data suggest that there are tissue specific forms of cytosolic PK. A similar regulation has been found for the enzyme from the green alga *Selenastrum* in which the linkage of carbon flow to amino acid synthesis via the regulation of cytosolic PK has been demonstrated (Lin *et al.*, 1989). This regulation bears no resemblance to the regulation of animal PK where the enzyme is activated by F1,6BP. This correlates the activity of the animal PK with the activity of phosphofructokinase so that the entire pathway is activated when phosphofructokinase is turned on in response to the energy needs of the cell. This is a clear indication of the different function of glycolysis in plants and animals.

The cytosolic form of PK has been cloned (Blakeley *et al.*, 1990). Two forms of this enzyme have now been isolated from *Brassica* although it is not known if they correspond to the two proteins found in the heterotetramer (Blakeley and Dennis, pers. commun.). The enzyme has about 38% of the amino acids identical to the enzyme from other organisms and is more related to eukaryotic organisms than prokaryotic. The gene for PK has been sequenced from potato and has three introns one of which is in the 5'-untranslated region (Cole *et al.*, 1992). These introns do not align with any of the 11 introns found in the animal enzyme. There may be at least seven genes for cytosolic potato PK and it is possible that genes are expressed in a tissue-specific manner.

When the transit peptide from the small subunit of rubisco was added to the N-terminus of cytosolic PK and the resulting construct with the 35S CMV promoter was transformed into tobacco, it was found that in some plants cosuppression occurred and the level of cytosolic PK was reduced to zero in leaves (Gottlob-McHugh et al., 1992). These transgenic plants, when grown under normal conditions, showed no phenotype which is remarkable considering the central role this enzyme plays in most organisms. More recently, a detailed analysis of the plants lacking cytosolic pyruvate kinase in their leaves especially under low or intermittent light has shown that growth is inhibited under these conditions and the root-to-shoot ratio is decreased (Knowles et al., 1995; Micallef, Blakeley and Dennis, pers. commun.). This would correlate with a function of the cytosolic PK in the leaves being to supply carbon intermediates for amino acid biosynthesis in the roots.

The plastid form of pyruvate kinase (PK_p) has been purified from developing castor endosperm and other species has been found to be composed of two proteins (Plaxton, 1996). These two proteins have been cloned and shown to be encoded by different genes which have been termed PK_pA and PK_pG (Blakeley et al., 1991). These genes are quite distinct since there is only 35% identity between them and they appear to have diverged early in evolution (Hattori et al., 1995). Both proteins are also quite distinct from cytosolic PK and are more related to the prokaryotic than eukaryotic enzyme. The PK_pG gene from castor has been sequenced and shown to have eleven introns but only one of these aligns with the introns of either the cytosolic gene or the animal gene (Blakeley and Dennis, pers. commun.). The PK_pA gene has been partially sequence and appears to have fewer introns than PK_pG and these are in different locations (Blakeley and Dennis, pers. commun.). So far only one form of PK_pA has been found although it has been cloned from a number of species. In contrast, two forms of PK_pG have been cloned and it is possible that these represent tissue specific forms.

The uptake of the two isozymes of PK_p into plastids is different. The PK_pG protein is imported into both chloroplasts and leucoplasts in a manner that is very similar to other plastid proteins such as the small subunit of rubisco (Wan et al., 1995). However, PK_pA is only imported at much higher levels of ATP. The significance of this difference is not known although it may be a mechanism of correlating the ability of the plastid to synthesize ATP with the energy status of the cytosol. In addition, the relationship between the PK_pA and PK_pG has not been defined. Both proteins appear to occur together in leucoplasts but chloroplasts appear to contain only PK_pG which is probably a different isozyme from the leucoplast PK_pG.

It has been suggested that the plastid form of PK provides ATP for biosynthetic reactions within the organelle. This is supported by the observation that the amount of plastid form of PK increases relative to the cytosolic form during the development of the seed in Brassica (Sangwan et al., 1992). This is an oil seed and a large amount of fatty acid synthesis occurs in the plastid during seed development, a process that requires large amounts of ATP.

Bypasses of enzymes of the pathway

There are three enzymes involved in plant glycolysis that are not found in the majority of other organisms (for reviews see: Theodorou and Plaxton, 1995; Plaxton, 1996). The first step of the pathway, catalyzed by the ATP-dependent phosphofructokinase, is found in all organisms, but, in plants, it is also catalyzed by pyrophosphate-dependent phosphofructokinase. Secondly, there is a non-phosphorylating form of glyceraldehyde 3-phosphate dehydrogenase which converts 3-phosphoglyceraldehyde to 3-phosphoglycerate without the concomitant formation of ATP. Thirdly, there is a phosphatase that appears to specifically dephosphorylate phospho*enol*pyruvate to pyruvate. In addition, the combination of phospho*enol*pyruvate carboxylase, malate dehydrogenase and the NAD^+ malic enzyme could also serve as a bypass to the step catalyzed by pyruvate kinase.

The functions of these bypass enzymes have not been elucidated. It is possible that they allow a rapid re-equilibration of metabolites when the cell is subjected to metabolic changes such as might occur when there is a change from photosynthetically derived carbon in the light to carbon that is imported into the cell in the dark (Dennis and Greyson, 1987). They may also rapidly adjust metabolism during times of stress, which can be a daily occurrence in plants which are directly subjected to environmental changes. This has been suggested in the case of phosphate stress (Theodorou and Plaxton, 1994; Theodorou and Plaxton, 1995). Under this condition, the level of phosphate can be severely reduced in the cytoplasm of the plant cell and this in turn can lead to a serious reduction in the concentration of adenylates. Surprisingly, the level of pyrophosphate appears to be far less affected by phosphate stress and this could lead to the possibility of pyrophosphate being used as an alternative energy source under these conditions. This would allow the flow of carbon down the glycolytic pathway to provide intermediates for essential biosynthetic activities.

The pyrophosphate-dependent phosphofructokinase

The structure and function of the pyrophosphate-dependent phosphofructokinase (PFP) have been described in Chapter 7 and will not be repeated here. PFP is located only in the cytosol of plants. The activity of the enzyme is usually high but can vary greatly in different tissues and at various times of development. In suspension cells of *Brassica nigra*, the activity of the enzyme is very low in cells that are grown in adequate phosphate concentrations and this appeared to be a result of very low levels of the alpha subunit (Theodorou and Plaxton, 1994). The activity of the PFP, however, is greatly enhanced on phosphate starvation as a result of the synthesis of this subunit. Under these conditions the enzyme is stimulated by the activator fructose 2,6-bisphosphate (F2,6BP). This result suggests that the beta subunit of PFP is constitutively expressed whereas the expression gene for the regulatory alpha subunit (Carlisle *et al.*, 1990) is tightly controlled and activated during phosphate stress. This would allow the preferential use of pyrophosphate as an energy source when adenylates are low and allow the continued flux of carbon down glycolysis under phosphate starvation.

However, this is not the sole function for PFP since some species do not show this effect. In addition, transgenic tobacco plants in which the level of PFP has been reduced to very low levels show no reduced tolerance to phosphate stress when compared with wild type plants (Paul *et al.*, 1995). There is no doubt that there is differential expression of the alpha and beta subunits of PFP (Blakeley *et al.*, 1992) and this is difficult to explain when both are needed for maximal activity of the enzyme and to show the response to F2,6BP.

Non-phosphorylating glyceraldehyde 3-phosphate dehydrogenase

The non-phosphorylating glyceraldehyde 3-phosphate dehydrogenase converts 3-phosphoglyceraldehyde directly to 3-phosphoglycerate without the intermediate formation of 1,3-bisphosphoglycerate or the synthesis of ATP by 3-phosphoglycerate kinase (Kelly and Gibbs, 1973). Although the presence of this enzyme has been known for many years in plant tissues, it has not been extensively studied even though it appears to be present in many different plant tissues. It is confined to the cytosol and has very low K_ms for its substrates. Recently, this enzyme has been purified and shown to be different from the regular enzyme of glycolysis (Iglesias *et al.*, 1988; Iglesias and Losada, 1988; Scagliarini *et al.*, 1990).

It is probable that the enzyme acts as a source of NADPH in the cytosol since it has an absolute specificity for $NADP^+$ as cofactor and can act as a shuttle system to transfer reducing equivalents out of the chloroplast. However, more recently it has been found that the non-phosphorylating glyceraldehyde 3-phosphate dehydrogenase is also elevated by about twenty fold under conditions of phosphate stress suggesting that it may act as a bypass at this step in glycolysis when the level of adenylates are low (Duff *et al.*, 1989). However, if this is the case, it is not clear why the enzyme uses $NADP^+$ as the cofactor in contrast to NAD^+ for the normal glycolytic enzyme.

Phospho*enol*pyruvate phosphatase

Phospho*enol*pyruvate phosphatase catalyzes the cleavage of phosphate from phospho*enol*pyruvate without the concomitant formation of ATP from ADP. Hence, the final energy conserving step in glycolysis is bypassed. The enzyme is not absolutely specific for phospho*enol*pyruvate but the K_m for this substrate is low relative to other possible phosphorylated compounds and of the same order as the K_m of the cytosolic pyruvate kinase (Duff *et al.*, 1989). Phospho*enol*pyruvate phosphatase is inhibited by phosphate suggesting that under normal cellular conditions it is inactive. At times of phosphate stress, there is a great increase in the activity of this enzyme and this appears to be the result of a *de novo* synthesis of the protein. Hence, under times of phosphate stress, the final reaction of glycolysis would be bypassed allowing the flux of carbon to continue to pyruvate even though less ATP would be synthesized. Phospho*enol*pyruvate phosphatase appears to be located in the vacuole of the cell which would mean that a transport system across the vacuolar membrane would have to be in place if this bypass was to be effective.

Phospho*enol*pyruvate carboxylase

A final mechanism for the bypass of the reaction catalyzed by pyruvate kinase is the combined reactions of phospho*enol*pyruvate carboxylase, malate dehydrogenase and the mitochondrial NAD^+ malic enzyme. Hence, phospho*enol*pyruvate is converted into mitochondrially-localized pyruvate and in the process NADH is generated in the mitochondrion at the expense of cytosolic NADH. It has been found that the level of phospho*enol*pyruvate carboxylase increases five-fold in suspension culture cells of *Brassica nigra* under conditions of phosphate starvation suggesting this may also act as a bypass reaction when adenylate levels are low.

The regulation of glycolysis

The demand for carbon intermediates to supply the biosynthetic needs of a plant requires that the regulation of the glycolytic pathway be determined by the levels of intermediates that are the precursors of the biosynthetic pathways. This is in contrast with animals where the major function of glycolysis is the provision of intermediates for energy transduction which results in the regulation of glycolysis being determined by the energy needs of the cell through indicators such as energy charge.

The mechanism by which glycolysis is regulated in plants is still not well understood. It is generally accepted that the conversion of fructose 6-phosphate to F1,6BP and phospho*enol*pyruvate to pyruvate are the primary steps for regulation although flux control analysis would suggest that other steps are probably important, at least under some cellular conditions.

In animal cells, the ATP-dependent phosphofructokinase is the enzyme primarily responsible for the control of the flux of carbon through glycolysis. This enzyme is regulated by the energy status of the cell through the level of ATP and ADP or AMP. In turn, the product of the reaction, F1,6BP activates pyruvate kinase. Hence there is a top down regulation of the pathway dependent on the availability of hexose for the cell.

In plant cells, the situation is considerably different. First of all, the supply of carbon in green cells is from photosynthesis which is transported to the cytosol, in most cases, via the phosphate transporter. Hence, carbon can enter glycolysis at the midpoint of the pathway rather than at its beginning. This is also the case during starch breakdown, although it is possible that there is some flow of carbon directly from the plastid to the cytosol at the hexose level. In addition to factors effecting the source of carbon, there are the demands of the biosynthetic pathways that again branch from the glycolytic pathway at the level of triose or phospho*enol*pyruvate. Finally, it is necessary to consider the presence of two glycolytic pathways one in the cytosol and one in the plastid.

The ATP-dependent phosphofructokinase is a major regulatory enzyme in plant glycolysis. However, although there are numerous papers on this enzyme there is little consensus (Dennis and Greyson, 1987). Few studies have described the kinetics of the separated plastid and cytosolic isozymes of the enzyme let alone considering the possibility that there may be tissue specific enzymes. Many different regulatory kinetics have been described from positive to negative cooperativity. However, some regulatory kinetics are clear.

First, the plastid form of phosphofructokinase appears to have more pronounced regulatory kinetics which are fairly similar to the bacterial enzyme (Garland and Dennis, 1980). In contrast, the cytosolic form has less pronounced regulatory kinetics. However, all forms of phosphofructokinase are powerfully inhibited by phospho*enol*pyruvate or metabolites closely associated with phospho*enol*pyruvate, such as glyceraldehyde 3-phosphate (Dennis and Greyson, 1987). The enzyme is activated by phosphate which also reverses the inhibition of phospho*enol*pyruvate. This regulation of phosphofructokinase relates the activity of the enzyme with the activity of the phosphate translocator which exchanges triose phosphate for inorganic phosphate. Phospho*enol*pyruvate is at a major branch point both metabolically and energetically.

A major deficiency in our understanding of the regulation of glycolysis is our knowledge of the manner in which the concentration of phospho*enol*pyruvate is regulated. In animals, pyruvate kinase has pronounced regulatory properties, principally an activation by F1,6BP. In plants, this regulation does not occur and, in most cases, no pronounced regulatory properties have been described. In leaves, cytosolic pyruvate kinase is regulated by amino acids, linking the regulation of the enzyme with the supply of carbon from the phosphate translocator with the demand for amino acids.

In addition to the regulatory properties of the enzymes of the traditional glycolytic pathway, the regulation of the bypass enzymes must be considered. The pyrophosphate-dependent phosphofructokinase is powerfully activated by F2,6BP (Sabularse and Anderson, 1981; Stitt, 1990). In animals, this regulatory metabolite activates the ATP-dependent phosphofructokinase, but it has no effect on the plant ATP-dependent phosphofructokinase. Transgenic plants in which the level of F2,6BP has been artificially modified have altered carbon flux through glycolysis but not to the extent that might be expected (Scott, 1995). In addition, some transgenic plants have been produced with very low levels of the pyrophosphate-dependent phosphofructokinase and these plants show no detectable phenotype with only small effects on carbon flow down the glycolytic pathway (Paul *et al.*, 1995; Hajirezaei *et al.*, 1994).

The mechanism of control of the flow of carbon down the glycolytic pathway is still not understood. In addition to the above regulatory mechanisms,

there have been descriptions of protein–protein interactions, protein aggregation and multiple points of control (see Plaxton, 1996). This is made additionally difficult to interpret by the complexity of the plastid translocators that are now being demonstrated for different tissues and stages of development. This issue will not be resolved until more of the enzymes and transporters have been described and cloned.

The oxidative pentose phosphate pathway

Although the oxidative pentose phosphate pathway is usually depicted as being separate from glycolysis, the two pathways are intimately linked (Fig. 8.1) (reviewed in ap Rees, 1985; Douce, 1985; ap Rees, 1988; Copeland and Turner, 1985). They share the common intermediates glyceraldehyde 3-phosphate, fructose 6-phosphate and glucose 6-phosphate and flow through either of the pathways will be determined by the metabolic needs of the cell. The principal function of the pathway is probably to generate the reduced cofactor NADPH but it must be remembered that plants, in particular, place a major demand on the pathway for intermediates. In all organisms, ribose is required for nucleic acid biosynthesis but in plants there is also a major need for large amounts of erythrose 4-phosphate for aromatic amino acid biosynthesis and the derivatives of these amino acids, the polyphenols and lignin. In some tissues and at certain stages of development, these compounds may account for a major part of the flux of carbon from the hexose phosphate pool. In contrast, the requirement of pentose for cell wall biosynthesis is probably served via UDP derivatives and not through this pathway.

The pathway can be considered as being in two parts. The oxidative reactions of the pathway catalyzed by glucose 6-phosphate dehydrogenase and phosphogluconate dehydrogenase both generate NADPH. These reactions convert glucose 6-phosphate into ribulose 5-phosphate. The reactions catalyzed by glucose 6-phosphate dehydrogenase and phosphogluconate dehydrogenase are essentially irreversible whereas ribulose 5-phosphate epimerase, ribose 5-phosphate isomerase, transketolase, and transaldolase are close to equilibrium. Hence, the cycle can be said to consist of two parts, the oxidative part which produces NADPH and a reversible section that is involved in the regeneration of hexose from ribulose 5-phosphate. However, it must also be remembered that this cycle can also produce the ribose 5-phosphate for nucleoside biosynthesis and erythrose 4-phosphate for aromatic amino acid biosynthesis. These intermediates can be produced from ribulose 5-phosphate but may also be produced by the reverse reactions of the cycle from glyceraldehyde 3-phosphate. Considering the great demand in some cells for aromatic amino acids for polyphenols and lignin, the reversiblity of the non-oxidative part of this pathway may be essential to maintain metabolic balance of these intermediates.

The rate of flux of carbon through the pentose phosphate pathway relative to glycolysis can be estimated by the liberation from the C-1 and C-6 positions of glucose specifically labelled with ^{14}C in these positions, although the methods are in fact very complex if meaningful results are to be obtained (see ap Rees, 1980 for a thorough review of these methods). Estimates of carbon flux through the pentose phosphate pathway suggest that only a relatively small amount of glucose 6-phosphate is oxidized by glucose 6-phosphate dehydrogenase and phosphogluconate dehydrogenase compared with the total amount oxidized by glycolysis and the tricarboxylic acid cycle (ap Rees, 1980).

There is good evidence that this pathway occurs in both the cytosolic and plastid compartments of the cell since the two oxidative enzymes of the pathway are found as cytosolic and plastid isozymes in many different tissues, both green and non-green (see e.g. Dennis and Miernyk, 1982). The two isozymes are distinct and can readily be separated on ion-exchange columns. It has always been assumed that a complete pentose phosphate pathway occurred in both these compartments, but apart from the oxidative enzymes of the pathway, the presence of the rest of the pathway in both compartments is usually just assumed. When the activities of these other enzymes are measured they are often found to be at very low concentrations in the cytoplasm. Recently, the presence of the non-oxidative enzymes of the pathway was carefully examined in spinach leaves (Schnarrenberger et al., 1995). No evidence for the presence of the enzymes of the pentose phosphate

pathway could be found in the cytosol except for glucose 6-phosphate dehydrogenase and phosphogluconate dehydrogenase although large levels of all these activities were found in the chloroplasts. In addition, isozymes of the oxidative enzymes could be found but not of the other enzymes of the pathway. Non-green tissues also appear to have higher levels of the non-oxidative enzymes in the plastid but the cytosolic activities of these enzymes are higher than in green leaves (Simcox et al., 1977). It is possible, therefore that, non-green tissues have a complete pentose phosphate pathway although there are no reports of cytosolic isozymes of these enzymes.

Hence, the role of the pentose phosphate pathway in the cytosol is questionable, at least in green leaves, and, in all tissues, it appears that the pathway is more important to plastid metabolism than it is to that of the cytosol. The shikimate pathway is located principally in the plastid so that the pentose phosphate pathway could supply the erythrose 4-phosphate required by this pathway. Similarly, fatty acid biosynthesis occurs in plastids and the plastid-localized oxidative pentose phosphate pathway could supply the needs for reductant for this pathway. The only problem of a lack of a complete oxidative pentose phosphate pathway in the cytosol is the supply of ribose 5-phosphate for nucleoside biosynthesis and the removal of ribulose 5-phosphate from the oxidative steps catalyzed by glucose 6-phosphate dehydrogenase and phosphogluconate dehydrogenase. These problems could be overcome by the effective transport of pentoses across the inner membrane of the plastid.

There appears to be no complex regulation of the oxidative steps of the pentose phosphate pathway. Glucose 6-phosphate is inhibited by the product of the reaction, NADPH, so that it is the $NADP^+/$ NADPH ratio that controls these steps (Douce, 1985). This would be consistent with the oxidative steps of this pathway being to produce NADPH for biosynthetic reactions. The plastid enzyme is also inactive in the reduced state; the reduction of the enzyme being linked to the photosynthetic electron transport chain via the thioredoxin system. Hence, in the chloroplast, the oxidative steps are not active when the reduction of $NADP^+$ can be accomplished by photosynthesis. However, an interesting variant on this has been found in the green alga *Selenastrum minutum*. This organism has a glucose 6-phosphate dehydrogenase that is also inhibited by reduction using electrons from the ferredoxin/thioredoxin system. It has been found that in the light, if the demand for reducing power is increased by feeding the cells with nitrate, the activity of glucose 6-phosphate dehydrogenase increases to supplement photosynthetic reducing power (Huppe *et al.*, 1992). This shows the interdependence of the supply of reducing power from the photosynthetic electron transport chain and the oxidative reactions of the pentose phosphate pathway in the plastids.

Glucose 6-phosphate dehydrogenase

Glucose 6-phosphate dehydrogenase catalyzes the oxidation of glucose 6-phosphate to the 6-phosphoglucono-lactone with the concomittent reduction of $NADP^+$. There are plastid and cytosolic isozymes of the enzymes which have distinct properties suggesting that they are true isozymes (Schnarrenberger *et al.*, 1973; Fickenscher and Scheibe, 1986). For example, antibodies prepared against the cytosolic form of the enzyme do not cross react with the plastid form (Fickenscher and Scheibe, 1986). As is the case for other organisms, the cytosolic enzyme is a homotetramer with a subunit M_r of about 60 000 whereas the plastid enzyme is slightly smaller. The plastid form of the enzyme is very unstable and little work has been performed on this isozyme.

Recently, the cytosolic isozyme of glucose 6-phosphate dehydrogenase has been purified to homogeneity from potato tubers (Graeve *et al.*, 1994). The enzyme is not inhibited by reducing agents showing that this property is confined to the plastid isozyme of this enzyme. The enzyme shows competitive interaction between $NADP^+$ and NADPH confirming the importance of this ratio in the control of the enzyme. A full-length cDNA clone for the enzyme was isolated and the amino acid sequence determined (Graeve *et al.*, 1994). There was a high homology (52%) between the potato enzyme and the enzyme from yeast and animals but a much lower homology with the bacterial enzyme. The instability of the plastid isozyme has so far prevented a similar study of this isozyme.

Phosphogluconate dehydrogenase

Phosphogluconate dehydrogenase catalyzes the oxidative decarboxylation of 6-phosphogluconate to D-ribulose 5-phosphate and CO_2 with the concomittent reduction of $NADP^+$. There are cytosolic and plastid isozymes of this enzyme in plants that can be easily separated and have been shown by peptide mapping to be distinct isozymes. Even within the cytosol there are distinct isozymes in maize (Bailey-Serres *et al.*, 1992). The enzyme has been purified from several plants and has been shown to be similar to the enzyme from other sources.

Anaerobic Metabolism

The function of anaerobic metabolism

Plants are obligate aerobes and will ultimately die if the energy available from the mitochondrial electron transport chain using oxygen as the terminal acceptor is not available (for reviews see: Ratcliffe, 1995; Perata and Alpi, 1993; Kennedy *et al.*, 1992). Cell death probably occurs through acidification of the cytoplasm, initially through the formation of lactate and subsequently by the release of acid from the vacuole of the cell. However, plants have become adapted to periods of anaerobic stress which can occur at times of flooding when all or part of the plant is submerged. Rice plants are highly tolerant to submersion whereas other plants like peas and maize are very intolerant. In addition, the resistance to anaerobic conditions can be increased by periods of acclimatization under conditions of lowered oxygen tension or hypoxia.

Continued operation of the glycolytic pathway under times of oxygen deprivation requires that the NADH generated by glyceraldehyde 3-phosphate dehydrogenase is re-oxidized to NAD^+. Under aerobic conditions this is achieved by the NADH dehydrogenase complex located on the external surface of the inner mitochondrial membrane. However, this complex is inoperable under anaerobic conditions. Within two minutes of a plant becoming anaerobic, the level of NADH increases greatly showing that the electron transport chain is very susceptible to reduction in oxygen tension.

There are two routes for the anaerobic oxidation of NADH during anaerobiosis. In animals, pyruvate generated by glycolysis is reduced to lactate, by lactate dehydrogenase, which can accumulate and, under aerobic conditions, be converted back to carbohydrates by the liver. No such mechanism appears to have evolved in plants where lactate production appears to play only a transitory role in tolerance to anaerobiosis. In contrast, the decarboxylation of pyruvate to acetaldehyde by pyruvate decarboxylase and the subsequent reduction of acetaldehyde to alcohol by alcohol dehydrogenase are the key features of the plant anaerobic response.

It has been suggested that the pH of the cytosol is responsible for the switching of the flow of carbon from lactate to ethanol production (see Ratcliffe, 1995 and references therein). The initial phase of the anaerobic response stimulates the activity of lactate dehydrogenase which has an alkaline pH optimum. The accumulation of lactate lowers the pH of the cytosol so that the activity switches from lactate to ethanol formation since the pH optimum of pyruvate decarboxylase is acidic. An obligate connection between cell acidity and the switch to alcohol formation involving the initial formation of lactate has been demonstrated in some plants but not in others, suggesting that the effect may be different in various species and may reflect the ability to withstand anaerobic conditions. In some plants, such as rice, the onset of alcohol production appears to occur very early in anaeobiosis and is enhanced by acclimatization to hypoxic conditions. Alcohol itself is toxic to cells and hence there must be a mechanism for its removal. In yeast, alcohol is excreted to the medium and it has been suggested that alcohol is the anaerobic product of choice in plants because it can readily diffuse to the surrounding medium.

Fructose 2,6BP level has been reported to increase in plants during the initial phases of anaerbiosis and since this metabolite activates the pyrophosphate dependent phosphofructokinase, it has been suggested that this enzyme is responsible for increased glycolytic flux during oxygen deprivation (Mertens, 1991). This is supported by the observation that a protist, *Giardia*, an anaerobic gut parasite, only contains the pyrophosphate enzyme

and not the ATP-dependent phosphofructokinase (Rozario *et al.*, 1995). However, there is no direct evidence for this hypothesis and plants in which the level of the enzyme has been reduced to near zero are no less tolerant to anoxia.

In animals, the ATP-dependent phosphofructokinase is activated under anaerobic conditions by the lowering of the energy charge. However, no such regulation has been described for the plant enzyme. However, this enzyme undoubtedly plays a role in the increased flux of carbon through glycolysis. Under anaerobic conditions, the concentration of phosphate is likely to increase because of the inhibition of ATP synthesis in the mitochondrion. In addition, the level of phospho*enol*pyruvate is likely to be reduced due to the flow of carbon to ethanol and the inhibition of glyceraldehyde 3-phosphate dehydrogenase by increased levels of NADH.

Proteins involved in the anaerobic response

Under anaerobiosis, there is an inhibition of the expression of the majority of plant genes. However, the transcription of some genes for the enzymes of glycolytic pathway and those directly involved in the formation of lactate and ethanol are induced under these conditions. These have been termed anaerobic polypeptides (ANPs). Examples of ANPs are sucrose synthase, glucose phosphate isomerase, aldolase, alanine aminotransferase, alcohol dehydrogenase, pyruvate decarboxylase and lactate dehydrogenase (Andrews *et al.*, 1994). The increase in levels of these enzymes may be related to the tolerance to anaerobiosis. For example, the level of sucrose synthase increases greatly in rice, a species tolerant of anoxia, whereas in maize, an intolerant species, there appears to be an increase in the level of mRNA but not of the sucrose synthase protein suggesting different levels of control. The induction of gene expression in response to anaerobic conditions is controlled by an 'anaerobic response element' located upstream of the transcriptional start site. This element has been found in the genes for lactate dehydrogenase, alcohol dehydrogenase, sucrose synthase and aldolase (Perata, 1993).

The most studied ANP is alcohol dehydrogenase (Sachs, 1991). Alcohol dehydrogenase (ADH) catalyzes the reduction of acetaldehyde produced by pyruvate decarboxylase using NADH as the reductant. It is a dimer with a subunit M_r of 40 000. The pH optimum of the enzyme is about 8.5 which is high considering that the enzyme must function under the acidic conditions of anaerobiosis. There are two genes for this enzyme in maize, the products of which can associate to form a dimer so that three isoforms of the enzyme can be found on electrophoresis. The two genes have a high degree of homology with 80% identity at the amino acid level. The plant enzyme has 50% homology with the animal enzyme but lower homology with the yeast enzyme suggesting the plant and yeast enzymes are only distantly related. ADH is constitutively expressed in plants, i.e. even under aerobic conditions. Under these conditions, its activity may be limited by competition for the cofactor NADH or the supply of the substrate acetaldehyde. During anaerobiosis a second isozyme of ADH is expressed which has a lower K_m for NADH. However, the precise role of these isozymes is not clear since even under aerobic conditions, there is sufficient ADH to support the known rates of alcoholic fermentation. However, mutants lacking ADH are very intolerant to flooding indicating that this enzyme is essential for the survival of anaerobiosis although there is no correlation between the levels of ADH and flooding tolerance.

Pyruvate decarboxylase catalyzes the decarboxylation of pyruvate with the formation of CO_2 and acetaldehyde. It is a homotetramer with a subunit M_r of 60 000. It requires both magnesium and thiamine pyrophosphate as cofactors. Cooperative binding of the substrate pyruvate has been reported. Pyruvate decarboxylase is at the principal branch point between aerobic respiration and anaerobic respiration. Pyruvate decarboxylase is produced constitutively at a low level in many species, but the activity of this enzyme is enhanced several fold during the onset of anaerobiosis, as is the level of the mRNA encoding the protein, indicating the effect is through the activation of the gene for this enzyme (Bucher and Kuhlemeier, 1993; Andrews et al., 1994).

Lactate dehydrogenase catalyzes the conversion of pyruvate to lactate using NADH as the reductant. It is a tetrameric enzyme with subunit M_r values of approximately 37-40 000. In most plants there appears to be two genes for the enzyme, the products of which can associate to give five isoforms of the tetrameric enzyme which can be separated by electrophoresis. One of the genes appears to be constitutively expressed whereas the other is activated by anaerobic conditions.

Summary

The pathways of glycolysis, the pentose phosphate pathway and gluconeogenesis are a series of interacting pathways. Complicating the situation further in plants is the fact that all these pathways, at least in part, are located in both the cytosol and plastid compartments which are connected by a series of transporters. In addition, some steps of the pathway are catalyzed by a series of enzymes that may act as bypasses to the regulatory steps of the pathways. At present, little is known about the genes that encode the enzymes of the pathways and whether there are multiple isozymes of these enzymes that show tissue-specific expression. What has been demonstrated is that different tissues have variants of these pathways and clear differences in the transporters that control the flow of metabolites across the membranes of the cellular compartments. These transporters can change depending upon the nutritional status of the tissue.

The most intriguing aspect for the future of this field is the recent demonstration of the regulation of genes through the level of metabolites. This not only links the activation of genes associated with the compartments within a cell but links the metabolism of the cell to other cells in the tissue and cells in other parts of the plant through long distance transport in the xylem and phloem. This linkage is not through a complex system of hormones or secondary messengers as it is in animals but more directly through metabolite levels. Thus plants that are involved with the biosynthesis of all the metabolites required for an autotrophic existence have also evolved integrated regulation of metabolisms to coordinate metabolic activity through all cell compartments and tissues of the organism.

References

Andrews, D. L., MacAlpine, D. M., Johnson, J. R., Kelley, P. M., Cobb, B. G. and Drew, M. C. (1994). Differential induction of mRNAs for the glycolytic and ethanolic fermentative pathways by hypoxia and anoxia in maize seedlings. *Plant Physiol.* **106**, 1575–82.

ap Rees, T. (1980). Assessment of the contributions of metabolic pathways to plant respiration. In *The Biochemistry of Plants*, Vol. 2, eds P. K. Stumpf and E. E. Conn, Academic Press, New York, pp. 1–29.

ap Rees, T. (1985). The organization of glycolysis and the oxidative pentose phosphate pathway in plants. In *Encyclopedia of Plant Physiology*, eds R. Douce and D. A. Day, Vol. 18, Springer, Berlin, pp. 391–417.

ap Rees, T. (1988). Hexose phosphate metabolism by nonphotosynthetic tissues of higher plants. In *The Biochemistry of Plants*, Vol. 14, eds P. K. Stumpf and E. E. Conn, Academic Press, San Diego, pp. 1–30.

Bailey-Serres, J., Tom, J. and Freeling, M. (1992). Expression and distribution of cytosolic 6-phosphogluconate dehydrogenase isozymes in maize. *Biochem. Genet.* **30**, 233–46.

Black, C. C., Mustardy, L., Sung, S. S., Kormanik, P. P., Xu, D.-P. and Paz, N. (1987). Regulation and roles for alternative pathways of hexose metabolism in plants. *Physiol. Plant.* **69**, 387–94.

Blakeley, S. D., Dekroon, C., Cole, K. P., Kraml, M. and Dennis, D. T. (1994). Isolation of a full-length cDNA encoding cytosolic enolase from *Ricinus communis*. *Plant Physiol.* **105**, 455–6.

Blakeley, S. D. and Dennis, D. T. (1993). Molecular approaches to the manipulation of carbon allocation in plants. *Can. J. Bot.* **71**, 765–78.

Blakeley, S. D., Crews, L., Todd, J. F. and Dennis, D. T. (1992). Expression of the genes for the α- and β-subunits of pyrophosphate-dependent phosphofructokinase in germinating and developing seeds from *Ricinus communis*. *Plant Physiol.* **99**, 1245–50.

Blakeley, S. D., Plaxton, W. C. and Dennis, D. T. (1990). Cloning and characterization of a cDNA for the cytosolic isozyme of plant pyruvate kinase: the relationship between the plant and non-plant enzyme. *Plant Mol. Biol.* **15**, 665–9.

Blakeley, S. D., Plaxton, W. C. and Dennis, D. T. (1991). Relationship between the subunits of leucoplast pyruvate kinase from *Ricinus communis* and a comparison with the enzyme from other sources. *Plant Physiol.* **96**, 1283–8.

Borchert, S., Harborth, J., Schunemann, D., Hoferichter, P. and Heldt, H. W. (1993). Studies of the enzymic capacities and transport properties of pea root plastids. *Plant Physiol.* **101**, 303–12.

Botha, F. C. and Dennis, D. T. (1987). Phosphoglyceromutase activity and concentration in the endosperm developing and germinating *Ricinus communis* seeds. *Can. J. Bot.* **65**, 1908–12.

Boyle, S. A., Hemmingsen, S. M. and Dennis, D. T. (1990). Energy requirement for the import of protein into plastids from developing endosperm of *Ricinus communis* L. *Plant Physiol.* **92**, 151–4.

Brinkmann, H., Cerff, R., Salomon, M. and Soll, J. (1989). Cloning and sequence analysis of cDNAs encoding the cytosolic precursors of subunits GapA and GapB of chloroplast glyceraldehyde-3-phosphate dehydrogenase from pea and spinach. *Plant Mol. Biol.* **13**, 81–94.

Bucher, M. and Kuhlemeier, C. (1993). Long-term anoxia tolerance. Multi-level regulation of gene expression in the amphibious plant *Acorus calamus* L. *Plant Physiol.* **103**, 441–8.

Carlisle, S. M., Blakeley, S. D., Hemmingsen, S. M., Trevanion, S. J., Hiyoshi, T., Kruger, N. J. and Dennis, D. T. (1990). Pyrophosphate-dependent phosphofructokinase: conservation of protein sequence between the alpha- and beta-subunits and with the ATP-dependent phosphofructokinase. *J. Biol. Chem.* **265**, 18366–71.

Chopra, S., Dolferus, R. and Jacobs, M. (1990). Cloning and sequencing of the *Arabidopsis* aldolase gene. *Plant Mol. Biol.* **15**, 517–20.

Cole, K. P., Blakeley, S. D. and Dennis, D. T. (1992). Structure of the gene encoding potato cytosolic pyruvate kinase. *Gene* **122**, 255–61.

Copeland, L. and Turner, J. F. (1987). The regulation of glycolysis and the pentose phosphate pathway. In *The Biochemistry of Plants*, Vol. 11, eds P. K. Stumpf and E. E. Conn, Academic Press, San Diego, pp. 107–28.

Dennis, D. T. (1989). Fatty acid biosynthesis in plastids. In *Physiology, Biochemistry and Genetics of Non-green Plastids*, eds. C. D. Boyer, J. C. Shannon, and R. C. Hardison. Current Topics in Plant Physiology: An Amer. Soc. Plant Physiologists Series. Vol. 2, pp. 120–30.

Dennis, D. T. and Blakeley, S. D. (1993). Carbon and cofactor partitioning in oilseeds. In *Seed Storage Compounds: Biosynthesis, Interactions and Manipulation*, eds R. Shewry and K. Stobart. Proc. Phytochem. Soc. Europe: 35, Oxford Univ. Press, Oxford, pp. 262–75.

Dennis, D. T. and Blakeley, S. D. (1995). The Regulation of carbon partitioning in plants. In *Carbon Partitioning and Source Sink Interactions in Plants*, eds M. A. Madore and W. J. Lucas, Current Topics in Plant Physiology: An Amer. Soc. Plant Physiologists Series, Vol 13, pp. 258–67.

Dennis, D. T., Blakeley, S. D. and Carlisle, S. (1991). Isozymes and compartmentation in leucoplasts. In *Compartmentation of plant metabolisim in non-photosynthetic tissues*, Soc. Experimental Biol. Seminar Series: 42, ed. M. J. Emes. Cambridge Univ. Press, Cambridge. pp. 77–94.

Dennis, D. T. and Greyson, M. F. (1987). Fructose 6-phosphate metabolism in plants. *Physiol. Plant.* **69**, 395–404.

Dennis, D. T., Hekman, W. E., Thomson, A., Ireland, R. J., Botha, F. C. and Kruger, N. J. (1985). Compartmentation of glycolytic enzymes in plant cells. In *Regulation of Carbon Partitioning in Photosynthetic Tissue*, eds R. L. Heath and J. Preiss, American Society of Plant Physiologists, Rockville, MD, pp. 127–46.

Dennis, D. T. and Miernyk, J. A. (1982). Compartmentation of nonphotosynthetic carbohydrate metabolism. *Ann. Rev. Plant Physiol.* **33**, 27–50.

Douce, R. (1985). Control of carbohydrate oxidation in plant cells. In *Mitochondria in higher plants: structure, function and biogenesis*, Academic Press, Orlando, pp. 213–31.

Duff, S. M. G., Moorhead, G. B. G., Lefebvre, D. D. and Plaxton, W. C. (1989). Phosphate starvation inducible 'bypasses' of adenylate and phosphate-dependent glycolytic enzymes in *Brassica nigra* suspension cells. *Plant Physiol.* **90**, 1275–8.

Duggleby, R. G. and Dennis, D. T. (1974). Nicotinamide adenine dinucleotide-specific glyceraldehyde 3-phosphate dehydrogenase from *Pisum sativum*: assay and steady state kinetics. *J. Biol. Chem.* **249**, 167–74.

Emes, M. J. and Tobin, A. K. (1993). Control of metabolism and development in higher plant plastids. *Int. Rev. Cytol.* **145**, 149–216.

Fickenscher, K. and Scheibe, R. (1986). Purification and properties of the cytoplasmic glucose-6-phosphate dehydrogenase from pea leaves. *Arch. Biochem. Biophys.* **247**, 393–402.

Garland, W. J. and Dennis, D. T. (1980). Plastid and cytosolic phosphofructokinases from the developing endosperm of *Ricinus communis* I. Separation, Purification, and initial characterization of isozymes. *Arch. Biochem. Biophys.* **204**, 302–9.

Gottlob-McHugh, S. G., Sangwan, R. S., Blakeley, S. D., Vanlerberghe, G., Ko, K., Turpin, D. H., Plaxton, W. C., Miki, B. L. and Dennis, D. T. (1992). Normal growth of transgenic tobacco plants in the absence of cytosolic pyruvate kinase. *Plant Physiol.* **100**, 820–25.

Graeve, K., von Schaewen, A. and Scheibe, R. (1994). Purification, characterization, and cDNA sequence of glucose-6-phosphate dehydrogenase from potato (*Solanum tuberosum* L.). *Plant J.* **5**, 353–61.

Grana, X., de Lecea, L., El-Maghrabi, M. R., Urena, J. M., Caellas, C., Carreras, J., Puigdomenech, P., Pilkis, S. J. and Climent, F. (1992). Cloning and sequencing of a cDNA encoding 2,3-bisphosphoglycerate-independent phosphoglycerate mutase from maize: possible relationship to the alkaline phosphatase family. *J. Biol. Chem.* **267**, 12797–803.

Gross, W., Bayer, M. G., Schnarrenberger, C., Gebhart, U. B., Maier, T. L. and Schenk, H. E. A. (1994). Two distinct aldolases of Class II type in the cyanoplasts and in the cytosol of the Alga *Cyanophora paradoxa*. *Plant Physiol.* **105**, 1393–8.

Hajirezaei, M., Sonnewald, U., Viola, R., Carlisle, S., Dennis, D. T. and Stitt, M. (1994). Transgenic potato plants with strongly decreased expression of pyrophosphate : fructose-6-phosphate phosphotransferase show no visible phenotype and only minor changes in metabolic fluxes in their tubers. *Planta* **192**, 16–30.

Hattori, J., Baum, B., Gottlob-McHugh, S. G., Blakeley, S. D., Dennis, D. T. and Miki, B. L. Pyruvate kinase isozymes: ancient diversity retained in modern plant cells. *Biochem. Syst. Ecol.*, in press.

Henze, K., Schnarrenberger, C., Kellermann, J. and Martin, W. (1994). Chloroplast and cytosolic triosephosphate isomerases from spinach: purification, microsequencing and cDNA cloning of the chloroplast enzyme. *Plant Mol. Biol.* **26**, 1961–73.

Huang, Y. and Dennis, D. T. (1995). Histidine residues 139, 363 and 500 are essential for catalytic activity of cofactor-independent phosphoglyceromutase from developing endosperm of the castor plant. *Eur. J. Biochem.* **229**, 395–402.

Huang, Y., Blakeley, S. D., McAleese, S. N., Fothergill-Gilmore, L. A. and Dennis, D. T. (1993). Higher-plant cofactor-independent phosphoglyceromutase: Purification, molecular characterization and expression. *Plant Mol. Biol.* **23**, 1039–53.

Huppe, H.C., Vanlerberghe, G. C. and Turpin, D. H. (1992). Evidence for activation of the oxidative pentose phosphate pathway during photosynthetic assimilation of NO_3^- but not NH_4^+ by a green alga. *Plant Physiol.* **100**, 2096–9.

Iglesias, A. A. and Losada, M. (1988). Purification and kinetic and structural properties of spinach leaf NADP-dependent nonphosphorylating glyceraldehyde-3-phosphate dehydrogenase. *Arch. Biochem. Biophys.* **260**, 830–40.

Iglesias, A. A., Serrano, A., Guerrero, M. G. and Losada, M. (1987). Purification and properties of NADP-dependent non-phosphorylating glyceraldehyde-3-phosphate dehydrogenase from the green alga *Chlamydomonas reinhardtii*. *Biochim. Biophys. Acta* **925**, 1–10.

Ireland, R. J., DeLuca, V. and Dennis, D. T. (1980). Characterization and kinetics of isozymes of pyruvate kinase from developing castor bean endosperm. *Plant Physiol.* **65**, 1188–93.

Journet, E.-P. and Douce, R. (1985). Enzymic capacities of purified cauliflower bud plastids for lipid biosynthesis and carbohydrate metabolism. *Plant Physiol.* **79**, 458–67.

Kang, R. and Rawsthorne, S. (1994). Starch and fatty acid synthesis in plastids from developing embryos of oilseed rape (*Brassica napus* L.). *Plant J.* **6**, 795–805.

Kelly, G. J. and Gibbs, M. (1973). Nonreversible D-glyceraldehyde 3-phosphate dehydrogenase of plant tissues. *Plant Physiol.* **52**, 111–18.

Kennedy, R. A., Rumpho, M. E. and Fox, T. C. (1992). Anaerobic metabolism in plants. *Plant Physiol.* **100**: 1–6.

Kitayama, M. and Togasaki, R. K. (1995). Purification and cDNA isolation of chloroplastic phosphoglycerate kinase from *Chlamydomonas reinhardtii*. *Plant Physiol.* **107**, 393–400.

Knowles, V. L. *et. al.* (1995). Altered growth under limiting light of transgenic tobacco plants lacking leaf cytosolic pyruvate kinase. *Plant Physiol.* **108**, S-151.

Lal, S. K., Johnson, S., Conway, T. and Kelley, P. M. (1991). Characterization of a maize cDNA that complements an enolase-deficient mutant of *Escherichia coli*. *Plant Mol. Biol.* **16**, 787–95.

Lebherz, H. G., Leadbetter, M. M. and Bradshaw, R. A. (1984). Isolation and characterization of the cytosolic and chloroplast forms of spinach leaf fructose diphosphate aldolase. *J. Biol. Chem.* **259**, 1011–17.

Lin, M., Turpin, D. H. and Plaxton, W. C. (1989). Pyruvate kinase isozymes from the green alga, *Selenastrum minutum*. II. Kinetic and regulatory properties. *Arch. Biochem. Biophys.* **269**, 228–38.

Longstaff, M., Raines, C. A., McMorrow, E. M., Bradbeer, J. W. and Dyer, T. A. (1989). Wheat phosphoglycerate kinase: evidence for recombination between the genes for the chloroplastic and cytosolic enzymes. *Nucleic Acids Res.* **17**, 6569–80.

Maitra, P. K. and Lobo, Z. (1977). Pyruvate kinase mutants of *Saccharomyces cerevisiae*: biochemical and genetic characterization. *Mol. Gen. Genet.* **152**, 193–200.

Marsh, J. J. and Lebherz, H. G. (1992). Fructose-bisphosphate aldolases: an evolutionary history. *Trends Biochem. Sci.* **195**, 110–13.

Mertens, E. (1991). Pyrophosphate-dependent phosphofructokinase, an anaerobic glycolytic enzyme? *FEBS Lett.* **285**, 1–5.

Meyer-Gauen, G., Schnarrenberger, C., Cerff, R. and Martin, W. (1994). Molecular characterization of a novel, nuclear-encoded, NAD^+-dependent glyceraldehyde-3-phosphate dehydrogenase in plastids of the gymnosperm *Pinus sylvestris* L. *Plant Mol. Biol.* **26**, 1155–66.

Miernyk, J. A. and Dennis, D. T. (1982). Isozymes of the glycolytic enzymes in endosperm from developing castor oil seeds. *Plant Physiol.* **69**, 825–8.

Miernyk, J. A. and Dennis, D. T. (1992). A developmental analysis of the enolase isozymes from *Ricinus communis*. *Plant Physiol.* **99**, 748–50.

Miwa, S. (1990). Pyruvate kinase deficiency. *Prog. Clin. Biol. Res.* **344**, 843–52.

Nielsen, T. H. (1995). Fructose-1,6-bisphosphate an allosteric activator of pyrophosphate : fructose-6-phosphate 1-phosphotransferase. *Plant Physiol.* **108**, 69–73.

Paul, M., Sonnewald, U., Hajirezaei, M., Dennis, D. T. and Stitt, M. (1995). Transgenic tobacco plants with strongly decreased expression of pyrophosphate: fructose-6-phosphate 1-phosphotransferase do not differ significantly from wild type in photosynthate partitioning, plant growth or their ability to cope with limiting phosphate, limiting nitrogen and suboptimal temperatures. *Planta* **196**, 277–83.

Pelzer-Reith, B., Penger, A. and Schnarrenberger, C. (1993). Plant aldolase: cDNA and deduced amino-acid sequences of the chloroplast and cytosol enzyme from spinach. *Plant Mol. Biol.* **21**, 331–40.

Pelzer-Reith, B., Wiegand, S. and Schnarrenberger, C. (1994). Plastid class I and cytosol class II aldolase of *Euglena gracilis*. *Plant Physiol.* **106**, 1137–44.

Perata, P. and Alpi, A. (1993). Plant responses to anaerobiosis. *Plant Sci.* **93**, 1–17.

Perez de la Ossa, P., Grana, X., Ruiz-Lozano, P. and Climent, F. (1994). Isolation and characterization of cofactor-independent phosphoglycerate mutase gene from maize. *Biochem. Biophys. Res. Commun.* **203**, 1204–9.

Plaxton, W. C. (1990). Glycolysis. In *Methods in Plant Biochemistry*, ed. P. Lea, Vol. 3, Academic Press, New York, pp. 145–73.

Plaxton, W. C. The organization and regulation of plant glycolysis. *Ann. Rev. Plant Physiol. Mol. Biol.* **47**, in press.

Podesta, F. E. and Plaxton, W. C. (1992). Plant cytosolic pyruvate kinase: a kinetic study. *Biochim. Biophys. Acta* **1160**, 213–20.

Podesta, F. E. and Plaxton, W. C. (1994). Regulation of cytosolic carbon metabolism in germinating *Ricinus communis* cotyledons. II. Properties of phospho*enol*pyruvate carboxylase and cytosolic pyruvate kinase associated with the regulation of glycolysis and nitrogen assimilation. *Planta* **194**, 381–7.

Ratcliffe, R. G. (1995). Metabolic aspects of the anoxic response in plant tissue, In *Environment and plant Metabolism: Flexibility and Acclimation*, ed. N. Smirnoff, BIOS Scientific Publishers, London, pp. 111–27.

Razdan, K., Heinrikson, R. L., Zurcher-Neely, H., Morris, P. W. and Anderson, L. E. (1992). Chloroplast and cytoplasmic enzymes: isolation and sequencing of cDNAs coding for two distinct pea chloroplast aldolases. *Arch. Biochem. Biophys.* **298**, 192–7.

Rozario, C., Smith, M. W. and Muller, M. (1995). Primary sequence of a putative pyrophosphate-linked phosphofructokinase gene of *Giardia lamblia*. *Biochim. Biophys. Acta* **1260**, 218–22.

Sabularse, D. C. and Anderson, R. L. (1981). D-fructose 2,6-bisphosphate: A naturally occurring activator for inorganic pyrophosphate: D-fructose-6-phosphate 1-phosphotransferase. *Biochem. Biophys. Res. Commun.* **103**, 848–55.

Sachs, M. M. (1991). Molecular response to anoxic stress in maize. In *Plant Life Under Oxygen*, eds M. B. Jackson, D. D. Davies and H. Lambers, SPB Academic Publishing, The Hague, The Netherlands, pp. 129–39.

Sangwan, R. S., Gauthier, D. A., Turpin, D. H., Pomeroy,

M. K. and Plaxton, W. C. (1992). Pyruvate-kinase isoenzymes from zygotic and microspore-derived embryos of *Brassica napus*: Developmental profiles and subunit composition. *Planta* **187**, 198–202.

Scagliarini, S., Trost, P., Valenti, V. and Pupillo, P. (1990). Glyceraldehyde 3-phosphate: NADP$^+$ reductase of spinach leaves. Steady state kinetics and effect of inhibitors. *Plant Physiol.* **94**, 1337–44.

Schnarrenberger, C., Flechner, A. and Martin, W. (1995). Enzymatic evidence for a complete oxidative pentose phosphate pathway in chloroplasts and an incomplete pathway in the cytosol of spinach leaves. *Plant Physiol.* **108**, 609–14.

Schnarrenberger, C., Oeser, A. and Tolbert, N. E. (1973). Two isoenzymes each of glucose-6-phosphate dehydrogenase and 6-phosphogluconate dehydrogenase in spinach leaves. *Arch. Biochem. Biophys.* **154**, 438–48.

Schnarrenberger, C., Pelzer-Reith, B., Yatsuki, H., Freund, S., Jacobshagen, S. and Hori, K. (1994). Expression and sequence of the only detectable aldolase in *Chlamydomonas reinhardtii. Arch. Biochem. Biophys.* **313**, 173–8.

Scott, P., Lange, A. J., Pilkis, S. J. and Kruger, N. J. (1995). Carbon metabolism in leaves of transgenic tobacco (*Nicotiana tabacum* L.) containing elevated fructose 2,6-bisphosphate levels. *Plant J.* **7**, 461–9.

Shih, M-C. (1994). Cloning and sequencing of a cDNA clone encoding the cytosolic triose-phosphate isomerase from *Arabidopsis thaliana. Plant Physiol.* **104**, 1103–4.

Simcox, P. D., Reid, E. E., Canvin, D. T. and Dennis, D. T. (1977). Enzymes of the glycolytic and pentose phosphate pathways in proplastids from the developing endosperm of *Ricinus communis* L. *Plant Physiol.* **59**, 1128–32.

Stitt, M. (1990). Fructose 2,6-bisphosphate as a regulatory molecule in plants. *Ann. Rev. Plant Physiol. Mol. Biol.* **41**, 153–85.

Stitt, M. and ap Rees, T. (1979). Capacities of pea chloroplasts to catalyse the oxidative pentose phosphate pathway and glycolysis. *Phytochem.* **18**, 1905–11.

Theodorou, M. E. and Plaxton, W. C. (1994). Induction of PPi-dependent phosphofructokinase by phosphate starvation in seedlings of *Brassica nigra. Plant Cell Environ.* **17**, 287–94.

Theodorou, M. E. and Plaxton, W. C. (1995). Adaptations of plant respiratory metabolism to nutritional phosphate deprivation. In *Environment and plant Metabolism: Flexibility and Acclimation*, ed. N. Smirnoff, BIOS Scientific Publishers, London, pp. 70–109.

Trimming, B. A. and Emes, M. J. (1993). Glycolytic enzymes in non-photosynthetic plastids of pea (*Pisum sativm* L.) roots. *Planta* **190**, 439–45.

Van Der Straeten, D., Rodrigues-Pousada, R., Goodman, H. M. and Van Montagu, M. (1991). Plant enolase: Gene structure, expression, and evolution. *Plant Cell* **3**, 719–35.

Wan, J., Blakeley, S. D., Dennis, D. T. and Ko, K. (1995). Import characteristics of a leucoplast pyruvate kinase are influenced by a 19-amino-acid domain within the protein. *J. Biol. Chem.* **270**, 16731–9.

Xu, Y. and Hall, T. C. (1993). Cytosolic triosephosphate isomerase is a single gene in rice. *Plant Physiol.* **101**, 683–7.

Xu, Y., Harris-Haller, L. W., McCollum, J. C., Hardin, S. H. and Hall, T. C. (1993). Nuclear Gene encoding cytosolic triosephosphate isomerase from rice (*Oryza sativa* L.). *Plant Physiol.* **102**, 697.

Xu, Y., Yu, H. and Hall, T. C. (1994). Rice triosephosphate isomerase gene 5′ sequence directs β-glucuronidase activity in transgenic tobacco but requires an intron for expression in rice. *Plant Physiol.* **106**, 459–67.

9 Structure and biosynthesis of plant cell walls

N. Carpita

Introduction

The plant cell wall is a highly organized network of polysaccharides, proteins, and phenylpropanoid polymers in a slightly acidic solution containing several enzymes and many organic and inorganic substances. It is a firm supporting structure, but cells can enlarge it at different regions of the cell surface to provide many shapes and sizes needed for their different functions. The cell wall is not a static structure, but a dynamic metabolic compartment, and it shares a molecular continuity with the plasma membrane and cytoskeleton. Current models of the structure of primary and secondary walls depict ever clearer features of the cell wall chemistry and organization but give only a generalized view of the walls of the many species of flowering plants (Carpita and Gibeaut, 1993; McCann and Roberts, 1991). These depictions cannot fully capture how the molecular structures and arrangements of the wall polymers differ among species, among tissues of the same species, among single cells, and even among the wall domains within a single cell.

Examples of specialized cell walls are numerous. Guard cell pairs orient cellulose microfibrils in such a way as to generate via turgor the asymmetric curvature needed to make a stomatal pore. Epidermal cells form specialized exterior domains of cutin and suberin to block the loss of water vapor. The endodermal cells suberize their contiguous side walls to force the water to move symplastically into the stele. Specialized epidermal hairs form extraordinarily thick and fluffy secondary walls of almost pure cellulose crystals, whereas sclerenchyma and collenchyma develop very stiff cellulosic but highly lignified walls. These reinforcing secondary walls can be seen in a variety of forms of wall thickening in the water-conducting tracheids of the xylem. Some walls function well after the cells that produce them are dead and dessicated. For example, the orientation of the polymers assembled in the walls of living cells results in mechanical strains upon desiccation, and these strains result in abscission of plant parts and the dehiscence of fruit coats along defined planes. Feathery tuffs of hair cells of dandelion fruits and the stiff 'barbs' of cocklebur seeds form only upon drying, and each in a different way aid in the scattering of seeds.

Not all the specialized features of cell walls are structural. Some cell walls contain surface markers that foretell patterns of development and mark positions within the plant. They contain components for cell–cell and wall–nucleus signaling, and their polymers orient in special ways to provide unique functional shapes (Roberts, 1990). This polarity extends to differentiated cells whose walls maintain firmer connections between the cell wall and plasma membrane resulting from adaptation to stress (Wyatt and Carpita, 1993). Fragments of wall polysaccharides may elicit the secretion of defense molecules and the wall may become impregnated with protein and lignin to armor it from invading fungal and bacterial pathogens.

The plant cell wall is a dynamic compartment, from the cell plate formed during cytokinesis, through cell expansion and elongation when the firm but pliant wall is loosened and expands to many orders of magnitude its original area. Cells so tightly integrate loosening and synthesis that walls extend without change in wall thickness or apparent spacing of their microfibrilar framework.

This chapter encompasses a brief survey of the chemistry of the cell wall molecules and our present state of knowledge on how these molecules are synthesized and assembled to form a functional cell wall. The diversity of structures among different kinds of cells, the dynamic changes in wall composition, the synthesis of cellulose at the membrane surface, and the synthesis and targeting of all other non-cellulosic components via the ER-Golgi apparatus are all parts of this process. New and improved techniques to study cell wall biology are always giving us new perspectives on the structure, biosynthesis, and dynamics of the cell wall, and these methodologies will be discussed with emphasis on the wall's carbohydrate components.

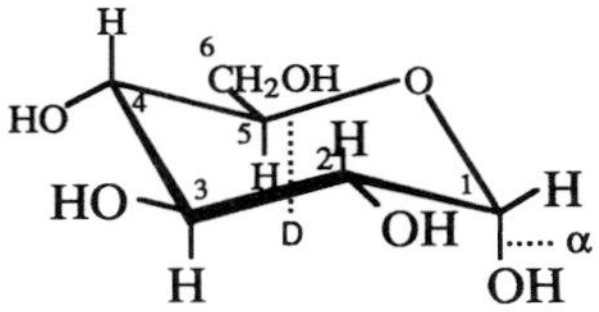

Carbohydrate nomenclature

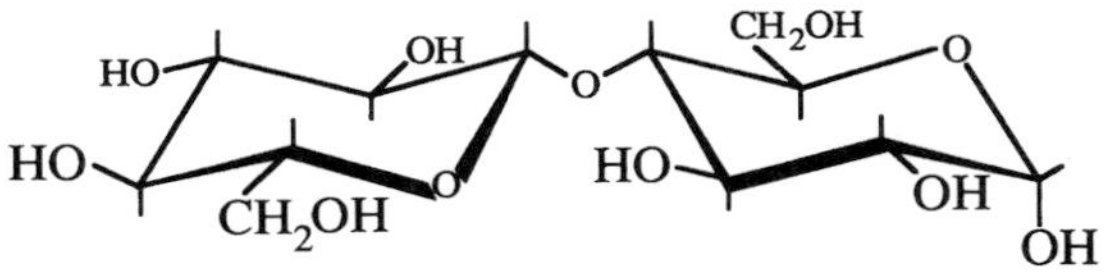

Primary cell walls are mostly carbohydrate

Most of the primary cell wall is composed of chains of sugars. Because sugars are composed of carbon and water, i.e. they have the basic structure $C_n(H_2O)_x$, they are called carbohydrates. Sugars represent a vast spectrum of polyhydroxyl aldehydes and ketones in a straight-chain form or, if they contain four carbons or more, rearrange into heterocyclic rings usually containing four or five carbons and a single oxygen. Sugars in polymers are always locked in these heterocyclic rings. Sugars with five-membered rings, i.e. four carbons and an oxygen, are called furanoses, whereas those in six-membered rings are pyranoses. One must distinguish the ring form from the kind of sugar, because pentoses (five-carbon sugars) and hexoses (six-carbon sugars) can be found in either ring form. D-Glucose is the most stable hexopyranose (Fig. 9.1A). The five tetrahedral carbon atoms forming a pyranose ring project the hydrogen and hydroxyl groups in either equatorial positions, away from the ring, or in axial positions above and below the ring. The anomeric carbon, the only carbon bound to two oxygen atoms, is numbered C-1, and the others are numbered sequentially around the ring. The hydroxyl group of the anomeric carbon can be in the axial (α) or equatorial (β) position. The D or L designation of configuration refers to the position of the hydroxyl group on the asymmetric carbon farthest from the C-1, i.e. the C-5 of hexoses and the C-4 of pentoses. The D-configuration means that the hydroxyl group, when viewed in a Fischer projection, is on the right

Fig. 9.1 The conformational structure of D-glucose and the linkage structure of β-D-glucosyl-($1 \rightarrow 4$)-D-glucose (cellobiose).

side, and L means the hydroxyl group is on the left. The designation does not refer to optical rotation. The designation D or L should always accompany reference to an α or β configuration. In solution, the hydroxyl group of the anomeric carbon will flip-flop between the α and β configurations, but is locked into a specific configuration in the glycosidic linkage. The secondary alcohol groups of the other carbons are locked into either axial or equatorial positions. α-D-Glucose is the most stable of the hexoses because the hydroxyl and the C-6 primary alcohol groups are all in the equatorial position, which is energetically more favorable.

Other sugars can derive from D-glucose. D-Mannose and D-galactose are made by conversion of the C-2 and C-4 hydroxyl groups, respectively, from the equatorial to the axial positions. The C-6 primary alcohols of these sugars can be oxidized to a carboxylic acid group to form D-glucuronic acid, D-mannuronic acid, or D-galacturonic acid. Enzymic removal of the carboxyl group from D-glucuronic acid forms the pentopyranose, D-xylose, in which all of the carbons are now part of the heterocyclic ring. The C-4 epimer of D-xylose is L-arabinose. The D to L conversion occurs because, in this instance, the C4 epimerization happens to be the last asymmetric carbon. The C-6 of some hexopyranoses can also be reduced to form methyl groups. In plants, the two major cell wall deoxysugars are L-rhamnose (6-deoxy-L-mannose) and L-fucose (6-deoxy-L-galactose). All of these interconversions are carried out enzymatically with nucleotide sugars. UDP-Glucose, -glucuronic acid, -galactose, -galacturonic acid, -xylose, and -arabinose are the principal nucleotide sugars used in

synthesis of most polysaccharides (Feingold, 1982). Nucleotide–sugar interconversions beginning with UDP-Glc produce the other UDP-sugars in reactions catalyzed by a UDP-Glc dehydrogenase and C4-epimerase, a UDP-GlcA decarboxyl-lyase, and a UDP-Xyl C4-epimerase (Fig. 9.2). Except for UDP-Glc dehydrogenase, these interconversion enzymes are

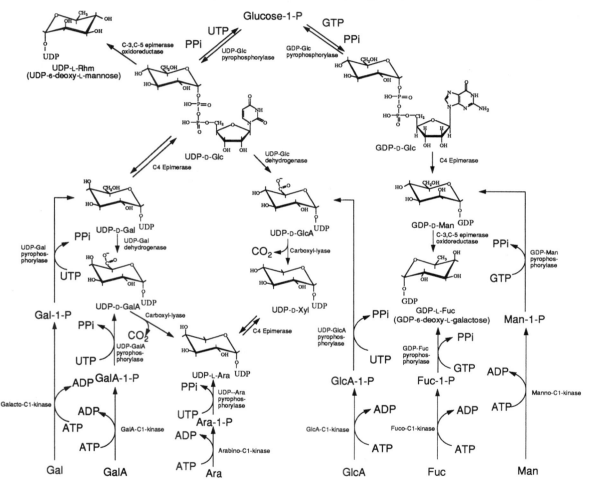

Fig. 9.2 The major nucleotide–sugar interconversion pathways in plants. UDP-Glc and GDP-Glc can be made from Glc-1-P by action of a pyrophosphorylase or directly from sucrose by sucrose synthase. UDP-Gal and GDP-Man are made in equilibrium reactions by a C-4 and C-2 epimerase, respectively. UDP-Xyl and UDP-Ara are interconverted similarly by a C-4 epimerase. There are two other ways UDP-Ara can be synthesized. First, a carboxyl-lyase can convert UDP-Gal to UDP-Ara, and second, arabinose can be phosphorylated by a C-1 kinase, and a pyrophosphorylase can generate UDP-Ara. The latter pathway represents one of several in which C-1 kinases generate sugar-1-phosphates as direct substrates for pyrophosphorylases to regenerate nucleotide sugar end-products. Because the nucleotide-sugar interconversion pathway represents the only pathway for *de novo* synthesis, the C-1 kinase/pyrophosphorylase 'salvage' pathways are testimony to recycling of these sugars after hydrolysis of these monosaccharides from specific polymers.

probably all membrane bound and localized at the ER-Golgi apparatus. Guanosine based nucleotide–sugars can also be substrates for specific polysaccharides, such as GDP-Glc and GDP-Man in the synthesis of glucomannan and GDP-Fuc in the fucosylation of complex glycoproteins, pectins, and some cross-linking glycans.

The anomeric carbon always forms the glycosidic linkage

The anomeric carbon is always linked to the hydroxyl group of another sugar, sugar alcohol, hydroxylamino acid, or phenylpropanoid compound, and this is called a glycosidic linkage. For D-glucose, a sugar can be attached to the O-2, O-3, O-4, or O-6, the O corresponding to the oxygen attached to its respective numbered carbon – only the C-5 position is unavailable as it is the heterocyclic oxygen that constitutes part of the ring structure. A disaccharide can be described with respect to both linkage and anomeric configuration. For example, cellobiose is β-D-glucose $(1 \rightarrow 4)$-D-glucose, and the anomeric carbon of one forms a glycosidic linkage by dehydration to the equatorial hydroxyl at the C-4 position of another D-glucose (Fig. 9.1B). Only one D-glucose is locked in the β-configuration, whereas, in this instance, the other D-glucose is undesignated because its anomeric hydroxyl group is free to flip–flop in either configuration. Note that in this special linkage, the anomeric carbon is linked to the hydroxyl group farthest away from the anomeric carbon of the second glucose. For the β-linkage to occur with this equatorial hydroxyl group, the sugars to be linked must be rotated about 180° relative to each other, and iteration of this linkage produces an almost linear molecule. The large spectrum of sugars, which provide several hydroxyl groups in axial and equatorial positions and two anomeric configurations for attachment, enable cells to construct an enormous repertory of three-dimensional shapes necessary not only as structural materials but also as delicate sensing molecules. By comparison, with nucleic acids, just over 1000 pentamers are possible with the 4 bases (4^5), whereas 3 200 000 pentapeptides can be made from the 20 amino acids (20^5). These are enormous permutations for seemingly small oligomers, but for carbohydrates, this diversity of linkage structures of carbohydrates means that over 5 billion pentasaccharides could be generated using just 10 different hexoses! Glucose alone can be used to form almost 15 000 different pentameric structures. The ability to form the glycosidic linkage at more than one position on a sugar also engenders carbohydrates with branched structures, magnifying the potential diversity of form.

Methods to study cell wall composition and architecture

The linkage structure of pure polymers can be deduced

Fortunately for the carbohydrate chemist, only a small portion of the possible linkages has been selected in Nature. Polymer backbones are formed from one to a few repeated sugars and linkages, and the complexity comes from branching and appendant sugar units. Linkage structure of polysaccharides is deduced by methylation analysis. Polymers are suspended in dry DMSO, and methyl groups are introduced at all the free hydroxyl groups. The polymer is then hydrolyzed to partly methylated monosaccharides, and the C-1 is reduced chemically from an aldehyde to alcohol. The remaining new hydroxyl groups, which formerly were protected in the ring or linked to other sugars, are acetylated. The 'partly methylated alditol acetates' are separated by gas–liquid chromatography, and the structure of each derivative is deduced by electron-impact mass spectrometry (Carpita and Shea, 1989). The analysis can reveal the position at which another sugar was attached, but not which sugar. Polysaccharides are named after the principal sugars that constitute them. Most polysaccharides have a backbone structure, and that backbone is implied by the last sugar in the name. For example, xyloglucan is a glucan backbone with xylosyl units attached to it, glucuronoarabinoxylan is a xylan backbone with glucosyluronic acid and arabinosyl units attached to it, and so forth.

Structural features are revealed by several spectroscopic methods

Nuclear magnetic resonance is used to probe, non-invasively and non-destructively, wall chemical composition, orientation of polymers, relative mobility of polymers in solution, and even how polymers interact with each other. The nuclei of all atoms spin, and for those with an odd number of neutrons and protons, like ^{1}H and ^{13}C, magnetic moments are generated. The frequency of their spins varies depending on the chemical and electrical environment generated by neighboring atoms.

The most straightforward way of deducing anomeric configuration of a sugar is by ^{1}H-NMR. The resonances of all the protons of a sugar are 'coupled', which means that more than one signal appears at frequencies characteristic for each of the hydrogen atoms. With cellobiose, for example, the angle of the axial H-1 of the anomeric carbon and the axial H-2 of the neighboring carbon in the β-linked glucose (cf. Fig. 9.1b) is much larger than the equatorial H-1 of an α-linked glucose in maltose. The larger the angle, the larger the difference in frequency between the proton signals of these coupled hydrogen atoms.

The ^{13}C atoms exhibit a much wider range of chemical shifts than ^{1}H atoms, and the anomeric carbons, the secondary alcohols, and the primary alcohols can be resolved. The natural abundance of ^{13}C is only a little more than 1% of the total carbon atoms, but ^{13}C NMR is sensitive enough to determine the anomeric configurations and linkage-induced chemical shifts of even large and complex polymers. Many ^{13}C-labeled substrates are now available commercially, and the low natural abundance of ^{13}C in one respect can be considered a boon because it makes possible the use of ^{13}C pulse-labeling and detection by NMR *in vivo* as a non-invasive alternative to use of radioactive substances. Further, the metabolic fate of individual carbon atoms can be followed because substrates can be chemically synthesized to place the ^{13}C atom at about any position in a molecule.

Sugars or polymers freely tumbling in solution give strong signals, whereas materials with highly restricted mobility, such as cellulose microfibrils and other polymers attached tightly to them, do not. The relatively immobile substances are revealed by the technique of cross polarization/magic angle spinning.

Fourier transform infrared (FTIR) spectroscopy is a sensitive means to detect specific chemical bonds within the underivatized cell wall (Séné *et al.*, 1995).

In the FT instrument, all frequencies are scanned simultaneously, and informative absorbance spectra can be obtained in minutes. Esters, such as methyl esters of PGA and phenolic esters, amides of peptide bonds, carboxylic acids, and carbohydrates have characteristic absorbances. Unlike ^{13}C-NMR, FTIR is extremely sensitive, and data from a $10 \times 10\,\mu m$ area of isolated cell wall material can be obtained. A microscope accessory allows spectra to be obtained from domains within a single cell wall, and polarized IR is used to determine the orientation of pectins and proteins with respect to cellulose microfibrils. FT-Raman spectroscopy provides complementary spectral information to FTIR, particularly lignins, related phenolic substances, and many substances that do not absorb IR irradiation appreciably (Séné *et al.*, 1995).

Toward sequencing polysaccharides

'Sequence-dependent glycanases' not only cleave specific glycosidic linkages but also their activities either require, or are restricted by, additional structural features of the polysaccharide (Carpita and Gibeaut, 1993). For example, the *Trichoderma viride* cellulase cleaves only unsubstituted $(1 \rightarrow 4)\beta$-D-glucosyl linkages. This feature has made it useful in determining the frequency of contiguous attachment of xylosyl units onto the $(1 \rightarrow 4)\beta$-D-glucan chain of xyloglucan. While the activities of enzymes such as the *Trichoderma* cellulase are blocked by appendant groups, other enzymes actually 'read' structural features around the site of cleavage. A *Bacillus subtilis* endoglucanase will cleave a $(1 \rightarrow 4)\beta$-D-glucosyl linkage but only if preceded by a $(1 \rightarrow 3)\beta$-D-linkage, and a *B. subtilis* xylanase cleaves $(1 \rightarrow 4)\beta$-D-xylosyl linkages only at the sites with appendant glucuronic acid units. Several of these enzymes have been used to yield oligomers characteristic of repeating unit structures that give a reasonable idea of the general sequence and conformation of very large polymers.

Cell wall architecture can be accurately imaged in the electron microscope

The cell wall has little defined structure when

conventional fixation, dehydration and resin-embedding procedures are used, but several reliable methods are used for the visualization of cellulose microfibrils. Generally a purified cell wall fraction is prepared followed by dry cleaving and shadow casting. Material extracted with hydrogen peroxide–acetic acid can be post stained with $KMnO_4$ to reveal the microfibrillar structures.

The fast-freeze, deep-etch, rotary-shadowed replica technique has yielded remarkable preservation of the walls from higher plants (Fig. 9.3A). Because the walls are frozen quickly, there is little ice damage, and since no chemical fixatives or dehydrants are used, the spacing of its architectural components is preserved. All molecules are visualized at high resolution without staining, and the three-dimensional molecular arrangement of the wall is preserved (McCann *et al.*, 1990). After gentle extraction of pectic polysaccharides, the fine thread-like cross-linking glycans can be seen to span the larger microfibrils (Fig. 9.3A). This technique has also allowed more accurate determinations of microfibril diameters and their spacing.

Measurements of length of individual extracted polymers can be made by direct visualization, achieved by spraying the polymers in glycerol onto a

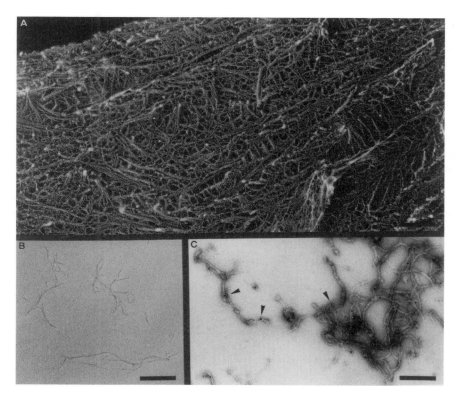

Fig. 9.3 (A) Electron micrograph of cellulose microfibrils in an onion cell wall. Walls were quick-frozen after extraction of pectic polysaccharides, then deep-etched and rotary shadowed to give a three-dimensional appearance to the image. Microfibrils in these expanded but unelongated walls criss-cross in many strata. A lacy appearance results from the appearance of fine thread-like connections from one microfibril to another. It is likely that several xyloglucan chains have coalesced during the etching in order to be able to see these thread-like connections. Bar = 200 nm. (B) Rotary-shadowed replica of aggregates of xyloglucan molecules extracted from onion cell walls. (C) Immunogold negative staining, with a monoclonal antibody JIM 5 that recognizes a relatively unesterified pectic epitope, of rhamnogalacturonans extracted from onion cell walls. Arrows indicate 5 nm colloidal gold particles. Scale bars represent 200 nm. From McCann *et al.*, 1994, with permission from publisher.

freshly cleaved surface of mica, dried *in vacuo* or taken through the fast-freeze, deep-etch procedure, and then imaged by rotary-shadowing (Fig. 9.3B). In contrast to relica plating, these techniques are amenable to labeling with antibodies directed against specific epitopes (Fig. 9.3C).

Specific probes for cell wall polymers reveal tissue, cell, and wall domain specificity

At the light microscopy level, the distribution of various polysaccharides in different cells is detected with traditional histochemical stains such as phloroglucinol for lignin, aniline blue for callose, Calcofluor for cellulose and callose, and ruthenium red for pectins. Gold-labeled disabled hydrolytic enzymes form stable interactions with substrates in etched plastic sections and can be used to localize specific polymers rather than just monosaccharides (Joseleau and Ruel, 1985).

Antibodies against specific polysaccharide epitopes can be used to determine the locations of cross-linking glycans, pectins, AGPs, and extensins in the wall by immuno-electron microscopy. The antibody approach has the advantage that the recognition epitope is often specific for oligomer sequences rather than for single sugars or sugar linkages. For example, gold-labeled antibodies specific to either unesterified or methyl esterified PGAs have been used to localize each of these pectins within the cell walls of several cell types. Monoclonal antibodies which recognize Ca^{2+}-bound PGA but not the unbound polymers have also been characterized. These antibodies are potentially useful to resolve where PGA junction zones are located in the walls of single cells. Antibody probes, coupled to colloidal gold, can be used at the electron microscope level to determine the patterns of polysaccharide distribution in a single cell wall.

Carbohydrate polymers of plant cell walls

The pectin matrix polymers

Pectins are some of the most complex polymers known, and they are thought to perform many

functions: to determine wall porosity, to provide charged surfaces that modulate wall pH and ion balance, and to serve as recognition molecules that signal appropriate developmental responses to symbiotic organisms, pathogens, and insects. An interaction of pectic substances with cross-linking glycans may even control the activity of wall loosening enzymes and proteins and their glycan substrates.

Pectins are defined classically as material extracted from the cell wall by Ca^{2+}-chelators, such as ammonium oxalate, EDTA, EGTA, or CDTA. Two fundamental constituents of pectins are polygalacturonic acid (PGAs), which are homopolymers of $(1 \rightarrow 4)\alpha$-D-galactosyluronic acid (GalA), and rhamnogalacturonan I (RG I), which are rod-like heteropolymers of repeating $(1 \rightarrow 2)\alpha$-L-rhamnosyl-$(1 \rightarrow 4)\alpha$-D-GalA disaccharide units (Fig. 9.4A–C). The PGAs contain up to about 200 GalA units and are about 100 nm long. The best documented RGs are isolated from the cell walls by enzymic digestion with PGAse, but the length of RG I is unknown because there may be stretches of PGA on the ends of the molecule. The rhamnosyl units may also interrupt long runs of PGA. An unusual RG II has the richest diversity of sugars and linkage structures known, and this low-abundance polymer may fulfill a role as a signal molecule.

Other polysaccharides composed of mostly neutral sugars, such as arabinans, galactans, and highly branched arabinogalactans (AGs) of various configurations and sizes, are attached to the O-4 of many of the rhamnosyl residues. In general, about half of the rhamnosyl units of RGs have side chains, but this ratio can vary with cell type and physiological state. Even though the GalA units of RG are not methyl esterified, roughly one-third of the GalA units of RG I can be acetylated at the O-3 position. The arabinans are chains of 5-linked arabinose, but can be branched frequently at the O-2 and O-3 positions to comprise a diverse group of branched arabinans. The galactans and two types of AGs are the typical side chains of RG. Type I AGs are found only in the pectin fractions and are composed of $(1 \rightarrow 4)\beta$-D-galactan chains with mostly *t*-arabinosyl units at the O-3 of the galactosyl units. Type II AGs constitute a broad group of short $(1 \rightarrow 3)$- and $(1 \rightarrow 6)\beta$-D-galactan chains connected to each other by $(1 \rightarrow 3, 1 \rightarrow 6)$-linked branch point

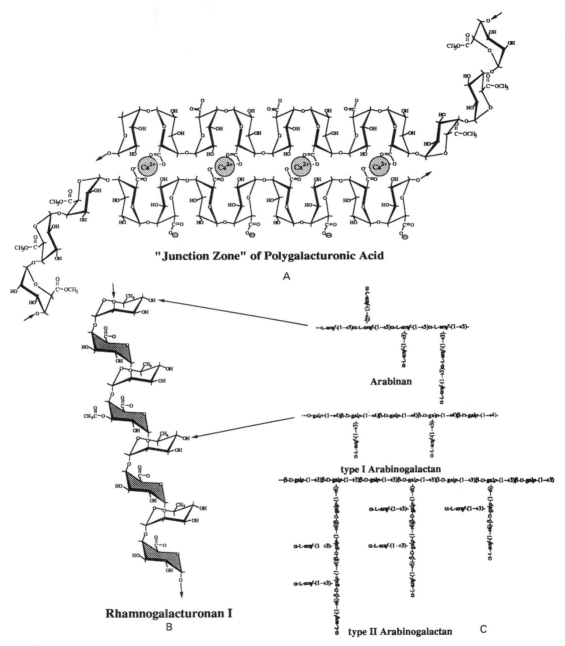

"Junction Zone" of Polygalacturonic Acid

A

Rhamnogalacturonan I

B

Arabinan

type I Arabinogalactan

type II Arabinogalactan   C

Fig. 9.4 Diverse structure of the pectic polysaccharides: (A) An 'egg-box' junction zone: two antiparallel chains of PGA cross-linked at unesterified regions by Ca^{2+}. Esterified portions of the PGA chains block formation of Ca^{2+} cross-links. (B) The contorted rod-like RG I (GalA units are shaded). About one-third of the GalA units are acetylated at secondary alcohols. (C) Three types of side-groups that attach to about half of the rhamnosyl units of RG I: 5-linked arabinans, type I arabinogalactans, and type II arabinogalactans.

residues. Most of the remaining O-3 or O-6 galactosyl positions are filled with *t*-arabinosyl groups.

The type II AGs are also associated with hydroxyproline-rich (non-extensin) proteins (AGPs) that may give them special unknown properties. Some AGPs are localized in the plasma membrane. Other AGPs are soluble in aqueous solutions and constitute over 70% of the soluble polysaccharides in cellular extracts and membrane vesicles from gently macerated tissues. The type II AGPs are a major secretory material found in the medium of cells from all flowering plants grown in liquid culture.

Crosslinking glycans interlock the cellulosic framework

The major cross-linking glycans of primary cell walls of flowering plants are xyloglucans (XyGs), glucuronoarabinoxylans (GAXs), and mixed-linked $(1 \rightarrow 3),(1 \rightarrow 4)\beta$-D-glucans (MGs) (Fig. 9.5A–C). For most plants, XyGs are the major non-cellulosic polysaccharide, but a species- and tissue-dependent diversity of fundamental unit structures has been recognized.

The XyGs are linear chains of $(1 \rightarrow 4)\beta$-D-glucan, but, unlike cellulose, they possess numerous xylosyl units added at regular sites to the O-6 position of the glucosyl units of the chain. Digestion of XyG with *Trichoderma* cellulase, whose action is restricted to unbranched 4-linked glucosyl units, yields hepta- and octasaccharides from seed reserve XyGs of legumes but different oligomers from the primary walls of many species (Fig. 9.5A). In most flowering plants, the basic repeating unit of XyG consists of additions of α-D-xylosyl units upon three contiguous glucosyl units of the backbone linked by a single unbranched glucosyl residue. A common substitution is a β-D-galactosyl unit at the O-2 of one or two of the xylosyl units, and α-L-fucose is added to the O-2 of some of the subtending galactosyl units to produce a trisaccharide side chain attached to alternate heptasaccharide units. A few of the nonasaccharides are substituted further with another α-L-fucosyl-$(1 \rightarrow 2)\beta$-D-galactosyl unit to form an undecasaccharide. The XyGs of tobacco and the Solanaceae lack fucose and galactose but have additions of one or two α-L-arabinofuranose units

directly on the xylose units. Like many other polysaccharides, the XyGs are acetylated after they are synthesized in the Golgi apparatus, and most of the galactosyl units of the trisaccharide side chains bear acetyl groups. Walls of grasses contain small amounts of XyG that bind to cellulose, but the xylosyl units are more randomly added to the glucan backbone.

Glucuronoarabinoxylans (GAXs) are linear chains of $(1 \rightarrow 4)\beta$-D-xylose with single arabinose units and, less frequently, single glucosyluronic acid (GlcA) units at the O-2 of the xylosyl units. The GAXs are the principal cross-linking glycans of the Poaceae, and in these species, the arabinofuranosyl units are attached exclusively on the O-3 positions of the backbone xylose units, whereas the GlcA units are at the O-2 positions. Non-gramineous species have a small amount of GAX but with α-L-arabinosyl units attached mostly at the O-2 and O-3 of the xylosyl units, and sometimes exclusively at the O-2 position.

When grass cells begin to elongate, they accumulate mixed-linkage β-D-glucans (β-D-glucans) in addition to GAX. β-D-Glucans are unbranched homopolymers of glucose containing a mixture of $(1 \rightarrow 4)\beta$-D-glucose linear oligomers connected by single $(1 \rightarrow 3)\beta$-D-glucose units. About 90% of the β-D-glucan macromolecules of about 6×10^5 to 3×10^6 Da is composed of cellotriose and cellotetraose units in a ratio of about $2:1$ and connected via $(1 \rightarrow 3)$-β-D-linkages. During polymerization, the tri- and tetrasaccharides constitute polymers about 50 residues long that are spaced by oligomers of more than four contiguous $(1 \rightarrow 4)$-glucosyl units terminating with $(1 \rightarrow 3)\beta$-D-glucosyl linkages.

Other non-cellulosic polysaccharides, such as gluco- and galactoglucomannans, and galactomannans potentially interlock the microfibrils in some primary walls, but are found in much lower amounts. The (galacto)glucomannans are roughly $1:1$ mixtures of $(1 \rightarrow 4)\beta$-D-mannosyl and $(1 \rightarrow 4)\beta$-D-glucosyl units, with varying amounts of terminal α-D-galactosyl units added to the O-6 position of the mannosyl units. In galactomannans, the backbones are exclusively $(1 \rightarrow 4)\beta$-D-mannan with the α-D-galactosyl units added at the O-6 positions. These polysaccharides accumulate in much greater abundance in differentiating tissue, especially in the thickened walls of the seed endosperm or cotyledon.

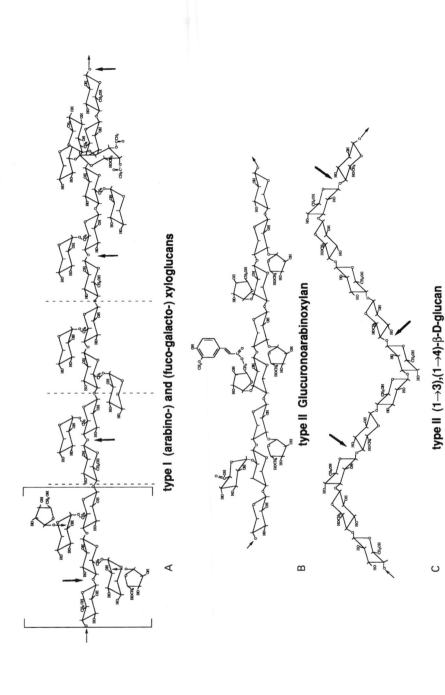

Fig. 9.5 (A) Xyloglucan structures in different Type I cell walls. The basic heptasaccharide repeating unit structure may bear additional substitutions as indicated, the most well known of which is the trisaccharide substitution formed by addition of an α-L-fucose-(1 → 2)β-D-galactosyl unit at the O-2 of the first xylosyl unit of every other heptasaccharide. Other substitutions include the addition of galactosyl, arabinofuranosyl and fucosyl units to xylosyl side groups. Many of the galactosyl units of the trisaccharide side chain are acetylated. *In brackets*: In the Solanaceae and tobacco, the major repeating unit appears to be a pentamer, rather than heptamer, with one or two arabinosyl units added directly to the O-2 position of the xyloses. Unconfirmed is a possibility that the Solanaceae XyG units may be separated by two unbranched glucosyl units rather than one. (B) A unit structure of the highly substituted glucuronoarabinoxylan. In the GAX from Type II walls, the L-arabinofuranosyl units are added strictly to the O-3 position, whereas in Type I walls, the units are attached to O-2 and, to a lesser extent, the O-3 position. The D-glucosyluronic acids are added to the O-2 position in all GAXs and GXs. Feruloyl groups are esterified to a few of the arabinosyl units and subsequently form several phenyl-phenyl and ether linkages to other esterified feruloyl units and to lignin (see Fig. 9.7). (C) Grass β-D-glucans (MG). A *Bacillus subtilis* endo-glucanase cleaves (1 → 4)β-D-glucosyl linkages just in front of (1 → 3)- linkages (arrows) to yield cellobiosyl- and cellotriosyl-(1 → 3)β-D-glucose oligomers in a ratio of about 2 : 1. Polymers of about 15 of these cellotriosyl and -tetraosyl units are linked together by slightly larger cellodextrins to form macromolecules that range from about 250 kDa to several million.

In grass species, MGs not only appear during cell expansion in the seedlings, but also accumulate to substantial amounts in the walls of the endosperm cells.

The cellulose framework

Cellulose is the most abundant plant polysaccharide and provides the framework for all plant cell walls. Cellulose is a paracrystalline assembly of several dozen $(1 \rightarrow 4)\beta$-D-glucan chains that are hydrogen bonded along their length. By electron diffraction, the chains of native cellulose are found to be arranged in parallel (Cellulose I), indicating that the chains are synthesized at one end. Chains solubilized by strong alkali can spontaneously recrystallize into the more stable anti-parallel conformations (Cellulose II). In higher plants, these microfibrils are composed of about 36 to 50 individual chains, but those of algae can form large cables or ribbons of several hundred individual chains. Microfibrils are about 10 nm wide and are spaced about 30 nm apart. Each chain may be just several thousand units long but individual chains begin and end at different places within a microfibril. Callose is a $(1 \rightarrow 3)\beta$-D-glucan chain that can form helical duplexes and triplexes. Callose is made by a few cell types at specific stages of wall development such as that formed in growing pollen tubes and in phragmoplasts of dividing cells. Callose is also made at the surface of the plasma membrane by most cells in response to wounding or the penetration by fungal hyphae.

Other structural elements of cell walls

Structural proteins

Extensin is probably the most well-studied extracellular structural protein of plants (Kieliszewski and Lamport, 1994). It consists of repeating Ser-(Hyp)$_4$ and Tyr-Lys-Tyr sequences that are important for secondary and tertiary structure (Fig. 9.6). The repeating Hyp units predict a 'polyproline II' rod-like

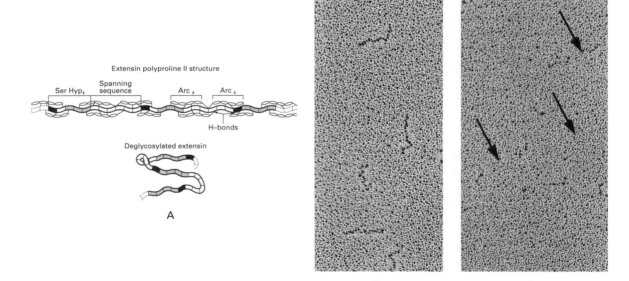

Fig. 9.6 Structural features of major cross-linking proteins of the primary wall. (A) A Ser-(Hyp)$_4$ or related motif found in many flowering plants is glycosylated with mono-, di-, tri-, and tetraarabinosides that associate with the polyproline helix to reinforce a rod-shaped structure. The Tyr–Lys–Tyr motif is the likely position of the intramolecular isodityrosine linkage. (B) Negative staining of isolated extensin precursors reveals their rod-shaped structure. (C) Removal of the arabinosides results in loss of the rod-shaped structure. From Stafstrom and Staehelin, 1986, with permission.

molecule. The flexible regions containing the Tyr-Lys-Tyr can be locked into a reinforced structure by forming an intramolecular isodityrosine unit. The rod-shaped extensins have been imaged by electron microscopy, and putative cross-links to other rods have been proposed. The (Hyp)$_4$ units possess arabinosyl tri- and tetrasaccharides whose (1 → 2) and (1 → 3) linkages cause them to wrap around the protein rods. These arabinans stabilize the structure because removal of these sugars results in loss of the rod-like appearance of extensin and loss of the polyproline II structure in solution (Fig. 9.6). The intramolecular isodityrosines of extensin have been established, but how extensins are cross-linked in the wall is not really known.

Other proteins may be necessary to actually lock the extensins together. One candidate is a 33 kDa protein called the RPRP (or simply PRP), the repetitive proline-rich protein. These proteins are more highly expressed later in cell development and particularly in the same vascular cells as extensin. The structure of PRPs is unknown, but their similarity to extensin suggests they are also rod-shaped proteins. A glycine-rich protein (GRP) is another major cell wall protein in many plants. The proteins, sometimes enriched with over 70% glycine, are predicted to be β-pleated sheets rather than rod-shaped molecules. GRPs are thought to form a plate-like structure at the plasma membrane-cell wall interface, and the cell wall face of the pleated sheet has an arrangement of aromatic amino acids that may serve as initiation sites for lignification. Like extensin, the cell wall GRP is recalcitrant to extraction, but antibodies have been raised against artificial peptides corresponding to hydrophilic regions outside of the pleated sheet. These antibodies localize the GRP precursors to protoxylem elements and other cells that eventually become lignified. Genes for GRPs have now been found in many species, but the proteins constitute a remarkably diverse group that may function as structural elements inside the cell as well as in the wall.

Grasses have a threonine/hydroxyproline-rich protein that appears to be related to extensin. Like extensin, the soluble precursors accumulate early in the cell cycle and become insoluble during cell elongation and differentiation. This polymer is enriched in vascular tissue and in special, reinforced wall structures such as the pericarp.

Lignin and suberin

Lignins are complex networks of aromatic compounds called phenylpropanoids (Chapter 10), more specifically the hydroxycinnamyl alcohols or 'monolignols' p-coumaryl, coniferyl and sinapyl alcohol (Whetton and Sederoff, 1995). The aromatic portions of the monolignols are called p-hydroxyphenyl, guaiacyl, and syringyl moieties, respectively, and the lignin derived mostly from one of these lignol aromatic groups are so named. Grasses and related families also contain substantial amounts of feruloyl species in primary and secondary walls, but in most plants the lignin is co-synthesized with the secondary wall to provide rigidity to supportive and water-conductive cells.

Lignins can be detected in tissue sections by specific strains, such the Mäule reagent (a mixture of $KMnO_4$ and ammonia), the Wiesner reagent (phloroglucinol), and Safranin O. The Mäule reagent is particularly useful because a bright red color is produced specifically by syringyl lignin, whereas the guaiacyl lignin produces a yellow color. Total lignin is quantified by various methods, either chemically, with acetyl bromide or Klason hydrolysis, or by ^{13}C-NMR, UV, FTIR, or FT-Raman spectroscopy. Monomer composition of lignin is normally derived by degradative means, such as acidolysis and thioacidolysis, as well as by permanganate, nitrobenzene and alkali/copper oxide oxidation. These latter oxidation products are p-hydroxybenzaldehyde, vanillin, and syringaldehyde from p-hydroxylphenyl, guaiacyl and syringyl lignin, respectively. The monolignols are linked via ester, ether, or carbon–carbon bonds. A schematic formula giving the major kinds of monolignols and their possible intra-linkages demonstrates the remarkable complexity of structure as well as the formidable task of lignin chemists who try to examine its structure (Fig. 9.7). Because all higher plants contain p-hydroxylphenyl, lignins are classified strictly by their guaiacyl and syringyl content. Gymnosperms are predominately guaiacyl type, whereas the woody angiosperms and grasses exhibit a broad range of ratios of guaiacyl–syringyl residues.

Suberin is a material found in specific tissues and cell types, more notably the root and stem epidermis,

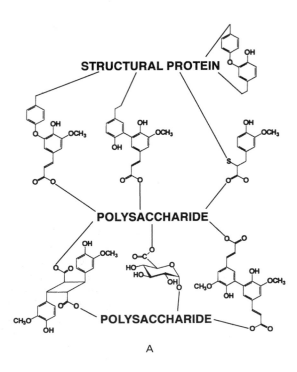

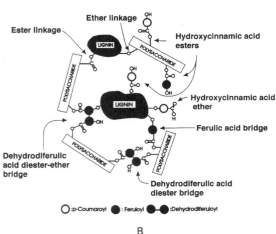

Fig. 9.7 A schematic representation of the many kinds of lignin interactions in plants. (A) Some possible aromatic–carbohydrate and aromatic–protein interactions. (B) A summary of the kinds of aromatic ester and ether cross-links between carbohydrate and lignin. From Iiyama *et al.*, 1994, with permission.

cork cells of the periderm, the surfaces of wounded cells, and in localized domains of the endodermis and bundle sheath cells. Suberins are recognized by lipid specific stains such as Sudan IV owing to the presence of long-chain fatty acids and alcohols, dicarboxylic acids, and hydroxylated fatty acids. The core of suberins are lignin-like, and the attachment of the long-chain hydrocarbons impart a strongly hydrophobic characteristic to the material that prevents water movement in suberized walls.

Cutin and its associated waxes are related to suberin and found on the leaf and stem surfaces, providing a strong barrier to the diffusion of water vapor. Waxes are generally esters of long-chain fatty acids and alcohols, but are more reasonably described as complex mixtures of these hydrocarbon esters, ketones, phenolic esters, terpenes, and sterols.

Assembly of polymers into primary cell walls

Structural domains of the primary wall

In dividing cells, the microfibrils are wound around each cell randomly. When elongation begins, microfibrils are wound more or less transversely around the longitudinal axis. The primary cell wall is made up of two, and sometimes three, structurally independent but interacting domains. One domain, the fundamental cellulose-cross-linking glycan framework is embedded in a second domain of matrix pectic polysaccharides (about 30% of the total mass). The third independent domain consists of the structural proteins or a phenylpropanoid network. The pectin matrix may interact with the cellulose–glycan framework. The various structural proteins could form intermolecular bridges with other proteins without necessarily binding to the polysaccharide components. Grasses, which have very little structural protein compared with dicots and other monocots, can have extensive interconnecting networks of phenylpropanoid compounds. Conceptually, there are three domains where orientation of the components constitute almost independent determinants of strength in elongating cells: the microfibrils arranged in the transverse axis, the cross-linking glycans in the longitudinal axis, with cross-linking networks of structural proteins or phenylpropanoid compounds that tie the two carbohydrate components together and impart radial strength.

Diversity of primary cell walls in flowering plants

The type I wall Most dicots and monocots have cell walls with about equal amounts of XyG and cellulose. Not all of the XyG resides as a monolayer coating the microfibrils, and much of it spans the milieu between microfibrils (Fig. 9.8A). Such spanning regions have been observed by electron microscopy (Fig. 9.3A). Extracted XyGs are from 20 to over 700 nm long with most being about 200 nm, clearly long enough to span the distance between two microfibrils and bind to each of them. XyGs may occupy two distinct regions in the wall: one that binds tightly to the exposed faces of glucan chains in the cellulose microfibrils, and a second that spans the distance to the next microfibril or simply interlocks with other XyGs to space and lock the microfibrils into place. Given the increasing awareness of the diversity of xyloglucan structures among the families of higher plants, sub-classes of type I walls that reflect this evolutionary development may soon be developed.

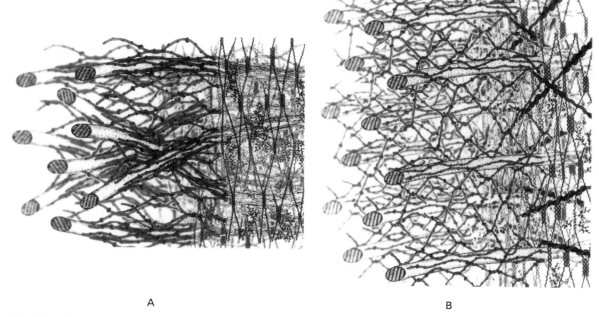

A   B

Fig. 9.8 A depiction of the cross-section of a type I primary cell wall of most flowering plants. (A) The cellulose microfibrils are interlaced with XyG polymers, and this framework is embedded in a matrix of pectic polysaccharides, PGA and RG, the latter substituted with arabinan, galactan, and arabinogalactan. Because XyGs have only a single face which can hydrogen bond to another glucan chain, the XyGs are depicted as woven to interlace the microfibrils. Many other associations are possible, including bridging of two microfibrils by a single XyG. With microfibril diameters of 10 nm and a spacing of about 30 nm, a primary wall 80 nm thick can have only about 3 strata. (B) The expanding type I cell wall: A model of possible alterations in structure that first permit microfibril separation and then lock them into form. Cleavage or dissociation of XyGs by 'growth-relevant' proteins loosens the microfibrils which separate in the long axis of the page. After displacement, extensin molecules, inserted radially, interlock the separated microfibrils to cease further stretching. Additional proteins may also be inserted to cross-link extensin, forming a heteropeptide network. Formation of intramolecular covalent bonds among the individual wall proteins signals the end of elongation. After Carpita and Gibeaut, 1993.

The cellulose–XyG framework is embedded in a pectin matrix. The pore size limiting unrestricted diffusion of molecules through the primary wall is only about 4 nm in diameter, but at least some molecules with Stokes radii larger than this limiting diameter can pass through the wall to the plasma membrane through a few larger pores. The pore size is controlled by the pectin matrix. Location of the de-esterification of PGA, the size and frequency of junction zones, and the size and conformation of the side chains attached to the RG could all influence the porosity of the pectin gel to the extent that movement and activity of enzymes for wall metabolism could be controlled. The helical chains of PGAs can condense by crosslinking with Ca^{2+} to form 'junction zones', linking two anti-parallel chains. Just how many contiguous unesterified GalA residues are needed to form stable junction zones and the extent to which several chains can stack to form the multiple 'eggbox' structures are not known *in vivo* or *in vitro*. PGA is thought to be secreted as highly methyl esterified polymers, and the enzyme pectin methylesterase (PME) located in the cell wall cleaves some of the methyl groups to initiate binding to Ca^{2+}. Regardless, PMEs are expected to have a marked influence on the charge density of the overall pectin matrix. The rhamnosyl units of RG I and their side chains interrupt the Ca^{2+} junctions. Some PGAs and RGs are cross-linked by ester linkages to other polymers held more tightly in the wall matrix.

The type II primary wall   The type II primary wall is composed of cellulose microfibrils similar in structure to those of the type I wall, but instead of XyG, the principal polymers that interlock the microfibrils in dividing cells are GAXs (Fig. 9.9A). The degree of branching greatly affects their ability to bind to each other. The unbranched $(1 \rightarrow 4)$-linked xylan can hydrogen bond to cellulose or to each other. In XyG, the xylosyl units are attached at the O-6, away from the binding plane. This actually stabilizes the linear structure and permits binding to one side of the glucan backbone. In contrast, in GAX the attachment of side groups to the internal O-2 and O-3 secondary alcohols of the xylan backbone constitutes steric hindrance to the formation of hydrogen bonds. Hence, all those arabinosyl and GlcA groups prevent bonding between two unbranched xylan chains or

xylan to cellulose. The type II walls contain small amounts of XyG that bind to cellulose. The type II walls are also notably poor in pectin, but antibodies against either methyl esterified or unesterified PGA detect them in specific cell types.

The type II walls are enriched in ferulate and *p*-coumarate esters formed by attachment of their carboxyl group to the O-5 of arabinosyl units of GAX. Formation of several kinds of diferulate esters from neighboring feruloylated GAX chains can cross-link a portion of the GAX matrix, but a substantial portion of the GAX is held onto the wall by etherified rather than esterified phenolic compounds. The PGAs of maize pectins also contain non-methyl esters, whose formation and disappearance are coincident with the most rapid rate of cell elongation. In general, grasses are pectin-poor, and much of the charge density is provided by the GlcA units on GAX. These units are never esterified.

The constituents of the type II wall change markedly when cells begin to expand. Some arabinans, particularly the 5-linked arabinans, are found in the walls of dividing cells but are no longer made during cell expansion. Instead, MGs are now synthesized along with GAX. The appearance of β-D-glucan during cell expansion, the association of hydrolysis of β-D-glucan *in muro* and acceleration of its hydrolysis rates by growth regulators all implicate direct physical involvement of the polymer in growth.

Just as in the type I walls, the form of the type II walls is fixed during differentiation, but unlike type I walls, the cross-linking polymer is not the Hyp-rich extensin. Much of the cross-linking function probably rests with the esterified and etherified phenolic acids, and formation of these cross-linkages is accelerated at the end of growth.

Cell wall pliancy during growth and wall extension

Tip growth, or the growth and deposition of new wall material strictly at the tip of a cell occurs in some plant cells, notably root hairs and pollen tubes, but for the vast majority of plant cells new material is deposited uniformly along the entire expanding wall. During cell expansion, the wall is a

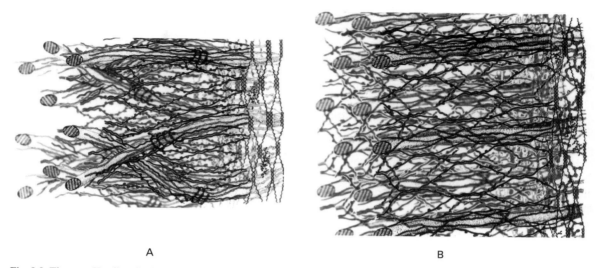

A   B

Fig. 9.9 The type II cell wall of the Poaceae. (A) The microfibrils are interlocked by GAXs instead of XyGs. Unlike XyGs, the xylans are substituted with arabinosyl units which block hydrogen bonding. The xylans are probably synthesized in a highly substituted form, and the units are cleaved in the extracellular space to yield stretches of the xylan that can bind on either face to cellulose or to each other. Porosity of the GAX domain could be determined by the extent of removal of the appendant units. Some highly substituted GAX remains intercalated in the small amount of pectins that also are found in the primary wall. Unlike the type I wall, a substantial portion of the non-cellulosic polymers are 'wired on' the microfibrils by alkali-resistant phenolic linkages. (B) The expanding type II cell wall: A model of possible alterations in structure that first permit microfibril separation and then lock them into form. Absent from the developing wall of dividing cells, the MG appears during cell elongation and is possibly the principal interlocking polymer. Long stretches of $(1 \rightarrow 4)$-linked glucan in MG could bind to cellulose or to other glucan. Cleavage of the long stretches of $(1 \rightarrow 4)$-glucosyl units by endoglucanases would free the cellulose microfibrils. Although some tissues accumulate extensin-like proteins and other structural proteins, the fundamental cross-linking responsibility in the primary wall of growing cells falls on the esterified and etherified phenolic compounds that lock the wall into place and halt further stretching of the microfibrils. After Carpita and Gibeaut, 1993.

metabolically active compartment. The molecules interlacing cellulose are 'loosened' enzymically, and the internal osmotic pressure induces a tangential stress that pulls the fibrillar components apart. New microfibrils and associated polymers are laid down on the innermost surface of the wall, forming a highly stratified and cross-linked fabric. The helical orientation of the cellulose microfibrils constrains the direction of cell expansion, whereas the dynamic interaction of the cellulose and non-cellulosic polysaccharide matrices dictates the rate and extent of cell expansion. An interplay of these factors establishes cell shape.

The biochemical basis for wall expansion is still hotly debated. For many years a consensus among physiologists had been that glycanohydrolases are activated at specific sites within the cell wall to cleave tension-bearing non-cellulosic polysaccharides which interlace the cellulose microfibrils. That consensus is challenged today by two new candidates for wall loosening enzymes. A new enzyme activity was discovered that, under conditions of excess XyG substrate, could carry out a transglycosylation rather than hydrolysis. Hence, one chain of XyG can be cleaved and reattached to the non-reducing terminus of another XyG chain (Fry *et al.*, 1992). In such a mechanism, the microfibrils could undergo a transient slippage but overall tensile strength of the interlocking XyG matrix is not sacrificed. Cosgrove and his colleagues (McQueen-Mason *et al.*, 1992) discovered completely different proteins that catalyze wall extension *in vitro* without detectable hydrolytic or transglycolytic events, something neither purified

XyG hydrolases nor transglycosylases are able to do. Recent demonstration that these proteins are able to induce extension of paper indicates that they probably catalyze breakage of hydrogen bonds, an activity that could disrupt the tethering of cellulose by XyGs in the type I walls and by MGs in the type II walls.

Despite the marked differences in the composition of type I and type II walls, the growth physics of grasses and other flowering plants are similar. All respond to the same classes of growth regulators in the same tissue-specific way, and the walls become more 'extensible' during rapid cell elongation. Hence, the walls respond to the same cues and perform the same physical functions during growth, but they do so with very different kinds of molecules; all flowering plants have a unified mechanism that induce the physical changes in wall structure needed for extension regardless of the kind of molecules that tie the microfibrils together. Because the cereal expansins induce good extension of tissues with type I walls, it is attractive to think that expansins are the enzymes integral to the unified mechanism. How extension occurs enzymically in cells with type II walls is not clear. The role of turnover of the MGs in extension of the type II walls is an additional complicating factor for which a reasonable explanation is lacking to date.

Cell wall biogenesis

The three phases of wall formation

The cell-wall chemical structure and architecture change during the stages of cytokinesis, elongation and maturation. The developing cell plate, or 'phragmoplast', is derived from vesicles of the Golgi apparatus and grows outward until it fuses with the existing primary wall. After the completion of cytokinesis, cellulose microfibrils are woven around the newly formed daughter cells. Formation of independent walls is sometimes marked by a clear transition of the phragmoplast from a wrinkled or wavy structure to one which is firm and flattened (Mineyuki and Gunning, 1990). The exact mechanism for how the new wall partitions the existing parent wall is not known.

The second stage of wall development involves expansion and differentiation of the primary wall. Many cells undergo little change in cell-wall structure after attaining full size, but a third stage of wall development in some cells is the formation of secondary wall thickenings. Secondary walls are well recognized in wood, but even in herbaceous plants, secondary walls are deposited in vascular and sometimes epidermal tissues after cell growth has ceased. The secondary wall is composed mostly of cellulose, but other polysaccharides, such as xylans and glucomannans, are deposited along with the cellulose. Thick secondary walls are also made as reserve materials in cotyledon and endosperm cells of developing seeds. Depending on species, different non-cellulosic polymers are made, such as xyloglucans, galactomannans, arabinogalactans, and $(1 \rightarrow 3),(1 \rightarrow 4)\beta$-D-glucans.

Cellulose is made at the plasma membrane surface

Cellulose and callose are the only polymers known to be made at the plasma membrane surface. Cellulose synthesis is probably catalyzed by multimeric 'terminal complexes' (TCs). The TCs are observed in freeze fracture replicas of plasma membrane specifically at the termini of the impression of growing microfibrils. The TCs of higher plants form 'rosettes', and their appearance in the plasma membrane coincides with the activity of cellulose synthesis (Fig. 9.10). No TCs have been identified in isolated membranes, so the process of membrane isolation may irreversibly disrupt them, an indication of the lability of cellulose synthesis *in vitro*.

Cell-free synthesis of plant cellulose has not been achieved. However, *in vitro* synthesis of extracellular cellulose ribbons by the bacterium *Acetobacter xylinum* has been well characterized (Ross *et al.*, 1990). UDP-Glc is the substrate for this synthesis, and a unique cyclic diguanylic acid (c-di-GMP) strongly activates it. A technique called 'product entrapment' was used to purify the cellulose synthase polypeptides of *Acetobacter*, and subsequently a cellulose synthase operon was proven by complementation. Unfortunately, no higher-plant cellulose synthase genes have been obtained using the *A. xylinum* probes.

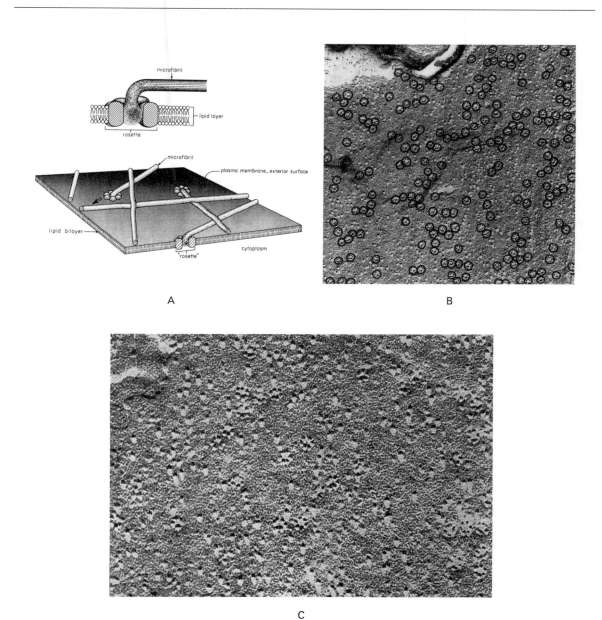

Fig. 9.10 (A) Terminal complexes associated with the formation of microfibrils in flowering plants are 'rosette' structures, six particles each about 8 nm in diameter that form a hexagonal arrangement on the inner leaf of the plasma membrane. A larger isolated particle is associated with the other leaf and fits in a 10 nm cavity within a 25 nm rosette (Giddings *et al.*, 1980). (B) These rosettes are imaged in vessels from cress (*Lepidium sativum* L.) forming the secondary wall thickenings (Herth, 1985). The density of the rosettes increases asymmetrically in the plasma membrane underlying cellulose deposition, and individual microfibrils appear to terminate at these complexes. Each circle represents identification of a rosette, and the density is associated with the spiral thickenings of these developing vessels. (C) In high magnification, the hexagonal shape is clearly seen. From Herth, 1985, with permission.

Cellulose synthase activity from higher plants is lost even under conditions of gentle plasma membrane preparation, suggesting that synthetic capacity depends on the orientation of synthase complexes in the membrane. When this orientation is lost, the complex may fall apart, and the single remaining activity is callose synthase rather than cellulose synthase. While demonstration of cellulose synthesis *in vitro* is still equivocal, UDP-Glc was shown by kinetic studies to be the primary substrate.

Callose synthase is solubilized by detergents and partially purified, but these preparations still contain many polypeptides. Attempts have been made to identify the catalytic subunits by affinity labeling techniques. Currently researchers are purifying putative synthase polypeptides for amino acid sequencing.

There are two eukaryotic systems that make related glucans *in vitro*. The synthesis of the $(1 \rightarrow 3)$- and $(1 \rightarrow 4)\beta$-D-glucans in *Saprolegnia* is stable *in vitro*. Synthesis of the $(1 \rightarrow 3)$-linked glucans is favored by high concentrations of UDP-Glc, whereas the $(1 \rightarrow 4)$-linked glucans is favored by addition of Mg^{2+} and has a lower Km for UDP-Glc. Synthesis of the $(1 \rightarrow 4)$-linked glucans is also stimulated by c-di-GMP. The cellular slime mold *Dictyostelium discoideum* synthesizes cellulose during formation of the sporangium stalk. Rod-like $(1 \rightarrow 4)\beta$-D-glucans are made by a Mg^{2+}-requiring enzyme using UDP-Glc as substrate. This synthase also makes $(1 \rightarrow 3)\beta$-D-glucans *in vitro*, but, unlike higher plants, the $(1 \rightarrow 4)\beta$-D-glucans were a much larger proportion of the product of the *in vitro* reactions (Blanton and Northcote, 1990).

Most non-cellulosic polysaccharides are synthesized at the Golgi apparatus

As in animal cells, the plant Golgi apparatus is a factory for the synthesis, processing, and targeting of glycoproteins, and requires many glycosyl transferases (Staehelin and Moore, 1995; Gibeaut and Carpita, 1994). In addition, all the cross-linking glycans and pectins are polymerized in the Golgi apparatus. Golgi apparatus has been shown by autoradiography as the site of synthesis and export of hexose- and pentose-containing polysaccharides.

Unlike the plasma membrane-localized glucan synthases, several different nucleotide sugars are substrate for multiple synthase reactions in the Golgi apparatus. The location of the synthase enzymes within the organelle has not been established. For branched polysaccharides, the nucleotide–sugar substrate used for synthesis of the backbone may be donated from either the cytosolic or lumenal side of the Golgi, but the branch units are probably added from the lumenal side. Hence, synthesis must be coordinated with transport of the nucleotide sugars into the Golgi apparatus.

The isolated Golgi apparatus can synthesize $(1 \rightarrow 4)\beta$-D-glucans from UDP-Glc. The synthesis of both unbranched $(1 \rightarrow 4)\beta$-D-glucan and XyG is enhanced by mM levels of UDP-Glc and UDP-Xyl, with Mn^{2+} or Mg^{2+} as cofactor. Cleavage of the XyG reaction products showed that substantial amounts of the diagnostic Glc_4Xyl_3 heptasaccharide units were made. The glucosyl- and xylosyltransferases appear tightly coupled, whereas the attachment of additional sugars to the xylosyl units occurs independently. For example, immunocytochemical techniques and gradient separation of the Golgi membranes show that the fucosyl units are added to XyG at later stages of Golgi vesicle formation.

A wall polysaccharide unique to the grasses is MG. The MG differs from other $(1 \rightarrow 4)\beta$-D-glucans and callose because the $(1 \rightarrow 4)$- and $(1 \rightarrow 3)\beta$-D-glucosyl linkages are ordered within the polymer (Fig. 9.5C). UDP-Glc is the substrate for MG synthase, and either Mg^{2+} or Mn^{2+} are required as a cofactor. The macromolecular MG synthesized *in vitro* in the Golgi apparatus is identical to the wall polysaccharide. The Golgi apparatus from grasses synthesizes a considerable proportion of callose, unlike Golgi apparatus from plants with type I walls. This activity was stimulated only 2-fold by $CaCl_2$ as compared with 7-fold by the plasma membrane enzyme, and was somewhat inhibited by Mg^{2+} and Mn^{2+}. The Golgi-specific callose synthase may represent the 'default' synthase, and like that of the plasma membrane, its activity is stable upon solubilization with detergents. Perhaps, like cellulose synthase, MG synthase may revert to callose synthase when disrupted.

There are many reports of the synthesis of other wall polysaccharides suspected to be made in the Golgi apparatus. The problem that most researchers have faced in attempting their *in vitro* synthesis is the lack of convenient methods for verifying the structure of the synthesized products. One of the more intriguing questions yet to be answered is how arabinofuranosyl units are made. L-Arabinose is in the furanose ring conformation in most plant polymers containing this sugar, including GAX, 5-linked arabinans, AGP, and extensin, whereas UDP-Ara is exclusively in the pyranose form. An arabinosyltransferase may be unique from other glycosyltransferases in its ability to permit ring rearrangement before addition of the sugar to the polymer.

Xylans and glucuronoxylans can be made *in vitro*, although the radioactive products have not been fully characterized and the activities recovered are relatively small. The transferases involved in these syntheses are controlled developmentally. Glucomannans can be synthesized *in vitro* with GDP-Glc and GDP-Man as substrates, with either insoluble or solubilized membrane preparations. Galactomannans accumulate in the wall space during seed development in legumes. Membrane preparations from developing seeds catalyze the synthesis of galactomannan. GDP-Man is the substrate for the formation of the $(1 \rightarrow 4)\beta$-D-mannan backbone, whereas the side-groups of Gal from UDP-Gal is dependent upon the simultaneous transfer of mannosyl units.

Incorporation of GalA units from UDP-GalA in products thought to be PGA have been reported, and formation of carboxy methyl esters has been shown to use S-adenosyl methionine as the methyl group donor. Much of the PGA in the Golgi apparatus is methyl esterified before transport to the wall, and this methylation was shown also by immunocytochemistry to occur later in Golgi development.

Structural proteins

As any other secretory protein destined for the cell wall, the structural proteins are translated and inserted into the endoplasmic reticulum and, hence, all cDNAs for cell wall proteins encode signal peptides. Of the four major cell wall structural proteins, all but the GRPs have hydroxyproline as a major constituent (Showalter, 1993). Hydroxylation of the proline residues occurs post-translationally in the endoplasmic reticulum. Glycosylation of the extensins is also thought to occur predominantly in

the ER. The site of glycosylation of the AGPs is unknown, but is likely to occur in the Golgi apparatus because it involves the attachment of large, highly branched galactan chains and subsequent linkage of arabinosyl units. When the polysaccharide contents of the Golgi apparatus and secretory vesicles are characterized, one finds that a majority of the material is AGP.

The complexity of the Golgi apparatus function is manifest in the coordinated secretion of very different materials – polysaccharides, structural proteins, and a broad spectrum of enzymes – which are assembled at the surface and later subjected to enzymic modification. The AGPs may represent a chaperone-like matrix that prevents premature associations and keeps enzymic functions quiescent until the secretory materials are assembled in the wall. The AGPs exhibit a marked turnover after secretion, and salvage pathways are active during growth to efficiently return arabinose, galactose, and certain other sugars to synthetic reactions (Fig. 9.2).

Lignin synthesis

Pathways for the synthesis of the monolignols are fairly well documented in the higher plants (Chapter 10), but how these lignin precursors are assembled in the wall is still unknown. Phenylalanine and tyrosine, produced by the shikimic acid pathway via prephenic, *p*-hydroxyphenylprephenic, and in some plants, arogenic acid, are deaminated by ammonia lyases to cinnamic and *p*-coumaric acid, respectively, the latter being the parent hydroxycinnamic acid from which many aromatic compounds are made. The cinnamic acid made from phenylalanine is converted via a C-4-hydroxylase to *p*-coumaric acid. *p*-Coumaric acid resides at a major branch point in the synthesis of many aromatic compounds. Via a CoA ligase and chalcone synthase, *p*-coumaric acid is directed toward flavonoid compounds, whereas another CoA ligase, a CoA reductase, and a hydroxycinnamoyl alcohol dehydrogenase provide *p*-coumaryl alcohol, a direct lignin precursor. Additional lignol precursors are made by successive hydroxylation of *p*-coumaric acid to caffeic acid and subsequent methylation of the C-3-hydroxyl group to form ferulic acid. This sequential hydroxylation and methylation of the C-5 position yields 5-hydroxyl ferulic acid and sinapic acid. Substrate-specific CoA ligases, CoA reductases, and alcohol dehydrogenases provide the coniferyl and sinapyl alcohol substrates. All of these reactions appear to be in the cytosol.

How these monolignols are used to form lignin is not at all clear. The lignols can be glycosylated in reactions that are associated with the ER and Golgi apparatus, and this glycosylation may be necessary for membrane transport and targeting. Glycosylated lignols also accumulate in the vacuole. The extent to which these monolignols begin to condense and form associations with carbohydrates or other materials during secretion is unclear. Once in the wall, the monolignols and their initial condensation products are polymerized by peroxidases. More recently, data have pointed to laccase, a member of the 'blue copper oxidase' family of enzymes, as another important enzyme in lignin biosynthesis (Dean and Ericksson, 1994). Researchers have few clues to the specificity of aromatic interactions and how polymerizing enzymes achieve the ordered structures of lignin.

Model systems to study cell wall structure and biosynthesis

Cells in liquid culture

Cells in liquid culture have been used for a variety of basic studies of plant metabolism. Most media are devised so that cells divide and elongate but do not differentiate, and are a homogeneous population of cells with only primary walls. Pioneering work with cells of the sycamore-maple provided us with the first clues to the assembly of the pectins and cross-linking glycans of the developing primary wall. Since then, walls of cells of diverse species have been characterized, and the arabinosylated XyGs found in the Solanaceae, the fucose-rich XyGs of other species, and the GAXs of the grasses are essentially identical in the walls of cells in liquid culture and primary walls of the intact plant.

A major difference between cells in intact plants and those in culture is that the latter are bathed in a large amount of extracellular fluid, which the cells fill with several polymers from the walls. They also secrete large amounts of type II AGPs into the medium. Growing evidence reveals that cells in the plant also secrete large amounts of AGPs, but because the extracellular fluid is small the AGPs participate in a turnover cycle that prevents their accumulation during active wall synthesis.

A completely new wall is made during cell division. Using cell synchronization of the tobacco BY-2 cell line, remarkably good cell-free preparations of membranes which form the phragmoplasts have been obtained (Kakimoto and Shiboaka, 1992). Protoplasts also provide a means of looking at the early stages of wall synthesis and regeneration. In carrot protoplasts, highly esterified pectins are secreted within hours after isolation, and de-esterification of the pectins is associated with the formation of a gel matrix that quickly envelops the protoplasts. The cellulose–xyloglucan framework takes several days to form, and a turgor-competent wall, which probably requires the insertion of structural proteins, takes longer still. Hence, synthesis and assembly of each of these domains appears to be independent of the other. The components of each domain can change independently depending on developmental state or in response to stress. For example, walls of tobacco cells adapted to grow under conditions of abnormally high NaCl or a drought-mimicking medium have much less cellulose relative to XyG and have altered pectin constituents. Extensin precursors are synthesized and secreted by the adapted cells but fail to polymerize around the smaller cellulose-xyloglucan framework (Iraki et al., 1989). Remarkably, tomato cells adapt to grow in medium containing dichlorobenzonitrile, an herbicide that specifically inhibits cellulose synthesis. The cells adapt not by circumventing the site of inhibition to make cellulose but by tightly cross-linking the pectin matrix, functionally replacing the missing microfibrillar framework (Shedletzsky et al., 1990).

Cotton fibers cultured *in vitro* provide a biochemical model of wall synthesis

One of the more remarkable examples of cell development *in vitro* is the formation of cotton fibers by unfertilized ovules in liquid culture. The cotton fibers are epidermal hair cells that elongate for about three weeks, achieving a length of almost 3 cm. Toward the end of the elongation phase and primary wall formation, a thick secondary wall of almost pure cellulose is then deposited, increasing the mass over 20-fold and nearly filling the lumen of the fiber cell at maturity, about two weeks later. At maturity the fibers of the plant are more than 95% cellulose. Beasley and Ting (1974) devised a medium containing gibberellin and auxin that maintained cell viability of the ovule without induction of callus but rather the induction of epidermal hair elongation and subsequent cellulose deposition. The fibers cultured *in vitro* form more quickly, achieving a maximum length near that of the plant-grown fibers in a little over two weeks and with cellulose content of over 60%. These fiber cells were used to establish kinetic evidence that UDP-Glc is the substrate of cellulose synthesis, and later as a model to study differential gene expression in cells exhibiting primary and secondary wall formation.

Tracheary elements develop synchronously in *Zinnia* mesophyll cells *in vitro*

More recently, *Zinnia* mesophyll cells have been used to study *in vitro* the development of tracheary elements. Intact cells are obtained aseptically by gently mashing young leaves of *Zinnia* in a mortar and pestle and incubating them in a medium containing moderate amounts of cytokinin and auxin (Fukuda and Komamine, 1980). Culture conditions have been optimized so that up to 60% of the cells undergo xylogenesis. The single cells elongate over about a two to three day period, and a substantial reorganization of the cytoskeletal components ensues. The secondary thickenings, quite reminiscent of the many specific forms of thickenings found in the plant, begin to form about the third day of culture. The development culminates with cell lysis leaving a hollow cylinder complete with a thick secondary wall of carbohydrate, structural protein, and lignin. This system has been used in cytological, biochemical, and molecular studies of aspects of the involvement of the cytoskeleton in wall deposition, to polysaccharide and lignin deposition, and stage-specific gene expression.

Whole plants and tissues as models of wall synthesis and turnover

There have been numerous studies of the carbohydrate and lignin components in a broad range of species with emphasis on seeds, fruits, vegetative and woody tissues of economic importance. All of these can be used to determine developmental changes in wall structures. A few examples will illustrate how these whole-plant models can be useful for study of many aspects of wall biogenesis.

The *Zinnia* system described above offers a manageable system to study xylogenesis. However, carbohydrate and lignin co-formation is quite varied among species, and often the intact plant offers a more than satisfactory system to study wall metabolism. Although pine trees are not convenient genetic models of plant development, gene expression during xylogenesis in loblolly pine has been examined through some ingenious manipulations. During the flush of new xylem growth in the Spring, young trees are felled and their bark is peeled off mechanically. The bark splits at the site of new xylem development. These living cells can be scraped off aseptically and used for studies of their carbohydrate and lignin composition and for extraction of RNA and proteins to examine gene expression related to xylogenesis (O'Malley *et al.*, 1992). This technique has now been used in the study of xylogenesis in several important tree species.

Similarly, tissue sections from developing fruit can be excised and cultured *in vitro* for studies of specific environmental and chemical cues that influence the rate of ripening. John Labavitch and colleagues (Campbell *et al.*, 1990) optimized such a system with the tomato pericarp in order to examine the cell specificity of enzymes thought to be associated with depolymerization of the pectin matrix during fruit ripening.

Genetic models to study wall biogenesis

Arabidopsis has taken center stage as a genetic model to study many aspects of plant development, and the study of the dynamics of cell wall structure and function is no exception. A brute-force approach was used to screen for mutants of *Arabidopsis* that have altered carbohydrate and lignin components of the primary wall. Over three dozen mutants have been classified in which one or several specific sugars are over- or under-represented compared with the distribution in wild type plants (Reiter *et al.*, 1993). Characterization of all of the mutants will be an enormous task. To begin, a mutant lacking fucose in stem and leaf tissue was characterized as a lesion in the nucleotide–sugar transformation pathway. The mutant could be rescued by feeding L-fucose, which entered the nucleotide sugar pool via an alternate pathway (see Fig. 9.2). Phenotypically, the plant is slightly dwarfed but otherwise nearly indistinguishable from wild type. One particularly interesting feature is that the tissues are more brittle and, hence, more easily pulled apart. Although quite low in abundance relative to the other wall sugars, fucose is an important appendant sugar attached to the O-2 of galactose in RG II as well as XyG. Fucose is also a notable appendant sugar in some of the complex carbohydrates of secreted glycoproteins. How loss of this appendant sugar causes the wall brittleness remains a mystery, but these kinds of mutants undoubtedly will prove useful to characterize the function of many specific polysaccharides.

The Somerville group also selected a mutant with the inability to make syringyl lignin, a character detected quite easily with the Mäule reagent (Chapple *et al.*, 1992). Some of the sinapic acid precursor to syringyl is normally shunted to water-soluble products that fluoresce strongly in UV light; this mutant was selected simply by its altered fluorescence behavior. From feeding studies, the mutation was deduced to have an inability to convert ferulic acid to sinapic acid. Total lignin content is seemingly unaltered, a result of an apparent compensatory accumulation of guaiacyl lignin. Lignin–carbohydrate interactions exert a great influence on digestibility of forage crops by animals, and the kind of lignin present rather than the total amount is often a more critical factor. Hence, mutants with altered lignin type may yield a new forage crop that exhibits greater digestibility without sacrifice of the strengthening function of lignin to the water-conducting cells of the plant.

The fiber flax plant could become an equally valuable genetic model specifically for studies of cell-wall development. Flax grows to about a meter in less than three months after planting. The phloem fibers exterior to the vascular bundles develop thick secondary walls that, at maturity, are about 70% cellulose and 20% cross-linking non-cellulosic glycans. The polysaccharide composition of flax cell walls are relatively simple, but distinct stage-specific polysaccharides occur as fibers mature. Flax is similar to *Arabidopsis* as a genetic system. Like *Arabidopsis*, the flax genome is relatively small (1.5 pg DNA/2C nucleus, less than 50% being repetitive DNA) and has a diploid complement of 30 small chromosomes. Generation time is relatively short (60–90 days), and plant size easily manageable under greenhouse conditions. Cultivated flax is self-pollinating, but cross pollination can be easily achieved. Flax is a dicotyledonous plant amenable to transformation by *Agrobacterium fumefaciens* and *A. rhizogenes*, and transgenic plantlets can be successfully propagated.

References

Beasley, C. A. and Ting, I. P. (1974). Effects of plant growth substances on *in vitro* fiber development from unfertilized cotton ovules. *Amer. J. Bot.* **61**, 188–94.

Blanton, R. L. and Northcote, D. H. (1990). A 1,4-β-D-glucan synthase system from *Dictyostelium discoideum*. *Planta* **180**, 324–32.

Campbell, A. D., Huysamer, M., Stotz, H. U., Greve, L. C. and Labavitch, J. M. (1990). Comparison of ripening processes in intact tomato fruit and excised pericarp discs. *Plant Physiol.* **94**, 1582–9.

Carpita, N. C. and Gibeaut, D. M. (1993). Structural models of primary cell walls in flowering plants: consistency of molecular structure with the physical properties of walls during growth. *Plant J.* **3**, 1–30.

Carpita, N. C. and Shea, E. M. (1989). Linkage structure by gas chromatography–mass spectrometry of partially-methylated alditol acetates. In *Analysis of Carbohydrates by GLC and MS*, eds C. J. Biermann and G. D. McGinnis, CRC Press, Boca Raton, FL, pp. 155–216.

Chapple, C. C. S., Vogt, T., Ellis, B. E. and Somerville, C. R. (1992). An *Arabidopsis* mutant defective in the general phenylpropanoid pathway. *Plant Cell* **4**, 1413–24.

Dean, J. F. D. and Ericksson, K-E. L. (1994). Laccase and the deposition of lignin in vascular plants. *Holzforschung* **48**, 21–33.

Delmer, D. P. and Amor, Y. (1995). Cellulose biosynthesis. *Plant Cell* **7**, 987–1000.

Feingold, D. S. (1982). Aldo (and keto) hexoses and uronic acids. In *The Encyclopedia of Plant Physiology, New Series*, Vol. 13A, eds F. A. Loewus and W. Tanner, Springer-Verlag, Berlin, pp. 3–76.

Fry, S. C., Smith, R. C., Renwick, K. F., Martin, D. J., Hodge, S. K. and Matthews, K. J. (1992). Xyloglucan endotransglycosylase, a new wall-loosening enzyme activity from plants. *Biochem. J.* **282**, 821–8.

Fukuda, H. and Komamine, A. (1980). Establishment of an experimental system for the study of tracheary element differentiation from single cells isolated from the mesophyll of *Zinnia* elegans. *Plant Physiol.* **65**, 57–60.

Gibeaut, D. M. and Carpita, N. C. (1994). Biosynthesis of plant cell-wall polysaccharides. *FASEB J.* **8**, 904–15.

Giddings, T. H., Brower, D. L. and Staehelin, L. A. (1980). Visualization of particle complexes in the plasma membrane of *Micrasterias denticulata* associated with the formation of cellulose fibrils in primary and secondary walls. *J. Cell Biol.* **84**, 327–39.

Herth, W. (1985). Plasma-membrane rosettes involved in localized wall thickening during xylem vessel formation of *Lepidium sativum* L. *Planta* **164**, 12–21.

Iiyama, K., Lam, T. B-T. and Stone, B. A. (1994). Covalent crosslinks in the cell wall. *Plant Physiol.* **104**, 315–20.

Iraki, N. M., Bressan, R. A., Hasegawa, P. M. and Carpita, N. C. (1989). Alteration of the physical and chemical structure of the primary cell wall of growth-limited plant cells adapted to osmotic stress. *Plant Physiol.* **91**, 39–47.

Jarvis, M. C. (1984). Structure and properties of pectin gels in plant cell walls. *Plant Cell Environ.* **7**, 153–64.

Joseleau, J-P. and Ruel, K. (1985). A new cytochemical method for ultrastructural localization of polysaccharides. *Biol. Cell.* **53**, 61–66.

Kakimoto, T, and Shiboaka, H. (1992). Synthesis of polysaccharides in phragmoplasts isolated from tobacco BY-2 cells. *Plant Cell Physiol.* **33**, 353–61.

McCann, M. C. and Roberts, K. (1992). Architecture of the primary cell wall. In *The Cytoskeletal Basis of Plant Growth and Form*, ed. C. W. Lloyd, Academic Press, London, pp. 109–29.

McCann, M. C., Wells, B. and Roberts, K. (1990). Direct visualization of cross-links in the primary plant cell wall. *J. Cell Sci.* **96**, 323–34.

McQueen-Mason, S., Durachko, D. M. and Cosgrove, D. J. (1992). Two endogenous proteins that induce cell wall extension in plants. *Plant Cell* **4**, 1425–33.

Mineyuki, Y. and Gunning, B. E. S. (1990). A role for preprophase bands of microtubules in maturation of new cell walls, and a general proposal on the function of preprophase band sites in cell division in higher plants. *J. Cell Sci.* **97**, 527–37.

O'Malley, D. M., Porter, S and Sederoff, R. R. (1992). Purification, characterization, and cloning of cinnamyl alcohol dehydrogenase in loblolly pine (*Pinus taeda* L.). *Plant Physiol.* **98**, 1364–71.

Reiter, W-D., Chapple, C. and Somerville, C. (1993). Altered growth and cell walls in a fucose-deficient mutant of *Arabidopsis*. *Science* **261**, 1032–5.

Roberts, K. (1990). Structures at the plant cell surface. *Curr. Opinion Cell Biol.* **2**, 920–28.

Ross, P., Mayer, R. and Benziman, M. (1991). Cellulose biosynthesis and function in bacteria. *Microbiol. Rev.* **55**, 35–58.

Séné, C. F. B., McCann, M. C., Wilson, R. H. and Grinter, R. (1994). Fourier-transform Raman and Fourier-transform infrared spectroscopy. An investigation of five higher plant cell walls and their components. *Plant Physiol.* **106**, 1623–31.

Shedletzky, E., Shmuel, M., Delmer, D. P. and Lamport, D. T. A. (1990). Adaptation and growth of tomato cells on the herbicide 2,6-dichlorobenzonitrile (DCB) leads to production of unique cell walls virtually lacking a cellulose-xyloglucan network. *Plant Physiol.* **94**, 980–87.

Showalter, A. M. (1993). Structure and function of plant cell wall proteins. *Plant Cell* **5**, 9–23.

Staehelin, L. A. and Moore, I. (1995). The plant Golgi apparatus: structure, functional organization and trafficking mechanisms. *Annu. Rev. Plant Physiol. Plant Mol. Biol.* **46**, 261–88.

Stafstrom, J. P. and Staehelin, L. A. (1986). The role of carbohydrate in maintaining extensin in an extended conformation. *Plant Physiol.* **81**, 242–6.

Whetton, R. and Sederoff, R. (1995). Lignin biosynthesis. *Plant Cell* **7**, 1001–13.

Wyatt, S. E. and Carpita, N. C. (1993). The plant cytoskeleton–cell wall continuum. *Trends Cell Biol.* **3**, 413–17.

10 Metabolism of defence and communication

B. E. Ellis

Introduction

An essential component of higher plant evolution has been the development of an ability to survive numerous challenges from both the environment and other organisms. For example, efficient capture of sunlight is essential for photosynthesis, but this also exposes the plant to damaging UV-B radiation. Similarly, accumulation of photosynthate makes plants metabolically self-sufficient, but also makes their tissues a primary food resource for countless heterotrophic organisms, ranging from biotrophic plant pathogens to mammalian herbivores. Even within the plant community itself, there is constant competition for resources, whether these be sunlight, water, minerals, pollination vectors or symbionts. Plants must deal with these challenges without being able to move from their growing site. Their key to success has been the evolution of chemical versatility.

Plants produce a remarkably wide range of chemicals, and more are being discovered every year. One feature of this biosynthetic virtuosity initially puzzled biologists; few of these chemicals occurred universally throughout the plant kingdom. As a result, this activity was categorized as 'secondary metabolism', i.e. not essential for day-to-day existence of the organism. To more zoocentric minds, the pattern simply reflected poor enzyme specificity in plant cells; perhaps, a metabolic 'rubbish heap' for organisms that had no obvious excretory system. However, as a result of painstaking studies of the specific biological context within which plants must operate, including their constant interaction with other organisms, it is becoming clear that the interface between a plant and its surroundings is essentially defined and controlled by the plant's

chemistry. Signalling between plants and symbionts, deterrence of herbivores, attraction of pollination vectors, protection from pathogens, shielding from UV radiation; all are dynamically mediated by a chemistry that is often species- and context-specific. Thus, the chemical profile displayed by any particular plant at a given time and place will be broadly defined by its genetic inheritance, which has arisen from an on-going refinement of the optimum chemistry for survival in the face of rapidly evolving populations of both microbes and insects. Expression of this genetic potential, however, is conditioned by both the plant's developmental state and its environment. The striking variability in secondary metabolite patterns therefore arises from continuous selection of the most useful combination of defence and communication mechanisms from among a vast repertoire of chemical potential existing in the plant genome.

It is reasonable to ask whether this phenotypic repertoire is supported by a corresponding diversity at the genetic level. The limited evidence available thus far suggests that it is; most enzymes of secondary metabolism display a substrate specificity at least as tight as those of primary metabolism, and they are encoded by unique genes, or gene families. There are, however, indications that particular motifs and/or domains may be 'shuffled' or reorganized at the genetic level to create new catalytic possibilities, and thus open new biosynthetic branchlets. For instance, the enzymes chalcone synthase (CHS) and stilbene synthase (STS) use the same substrates but create quite different products, each of which is the starting point for a large family of metabolites (Fig. 10.1). However, the differences between CHS and STS at the nucleotide and amino acid sequence

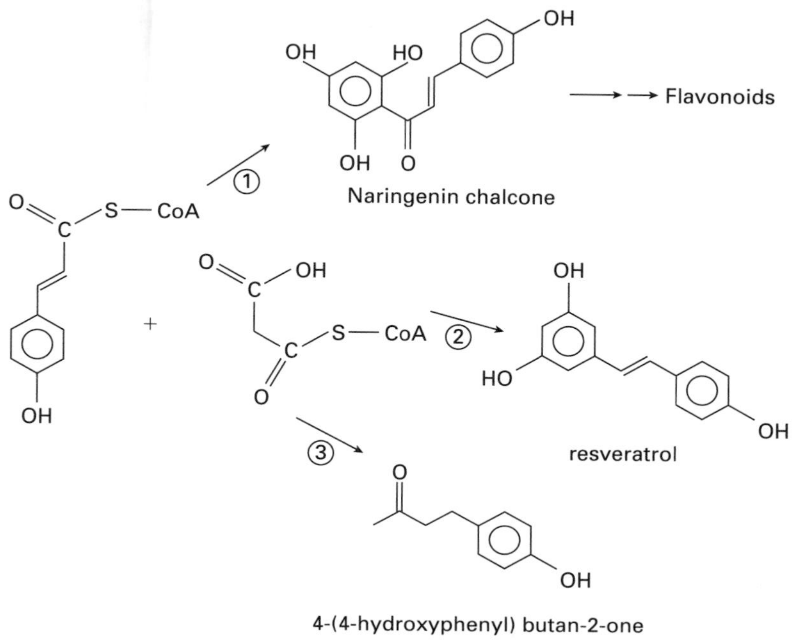

Fig. 10.1 Condensation of 4-coumaryl-CoA and malonyl-CoA can be catalyzed by several different but related enzymes. Catalysis by chalcone synthase (1) produces the precursor for the flavonoid group, while the closely related stilbene synthase (2) creates the basic structure of the stilbene family. p-Hydroxyphenylbut-3-ene-2-one synthase (3) catalyzes an analogous reaction to initiate synthesis of 'raspberry ketone'.

level are relatively modest, and they clearly share common origins (Tropf *et al.*, 1994). As sequence data become available for other enzymes of secondary metabolism, it seems likely that we will see more of this phenomenon of structural parsimony.

Underlying the bewildering array of plant secondary metabolites is a small number of core biosynthetic pathways that generate the precursors for the majority of the known compounds. These core pathways create certain structural themes that are diagnostic for the biogenetic origins of each compound. A carbon skeleton that can be defined largely in terms of isopentane units is probably derived from the **isoprenoid** (terpenoid) pathway, while the presence of one or more aromatic rings suggests an origin in the shikimate/**phenylpropanoid** pathway, and a pattern of alternating oxygenated and reduced carbons along a linear or cyclized skeleton is the hallmark of a **polyketide** chain generated by acetate/malonate condensations (Fig. 10.2). It is important to remember, however, that subsequent

reactions can modify these basic patterns to the point where they are virtually unrecognizable without careful experimental analysis.

The core patterns are also not totally exclusive; for example, aromatic rings are sometimes generated from isoprenoid skeletons or from polyketide cyclizations, and two different core sequences can also contribute separate parts of a given molecule. The flavonoids provide a classic example of such a pattern. Finally, there is a large and diverse group of secondary metabolites that are derived, at least in part, from different amino acids. In the following sections, these four major biosynthetic themes will be developed, together with examples of biological roles that have been defined for products in each group.

The limited number of core biosynthetic sequences that support secondary metabolism implies that there are relatively few interfaces between secondary and primary metabolism in plants. In fact, most of the photosynthetically-fixed carbon directed to secondary metabolite formation flows either through acetyl-CoA

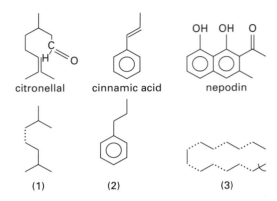

citronellal   cinnamic acid   nepodin

(1)   (2)   (3)

Fig. 10.2 Three of the core biosynthetic patterns that are widely used in plant secondary metabolism. Isoprenoid metabolites (1) are characterized by the head-to-tail linkage of isopentane units (shown in heavy lines), and phenylpropanoid compounds (2) display the basic carbon skeleton of L-phenylalanine, from which most of them originate. Polyketide products (3) arise from sequential condensations of acetyl units, which is reflected in the pattern of alternating oxygenated carbons around the skeleton. One or more terminal carbon atoms is often removed during polyketide biosynthesis, as is the case with nepodin.

(isoprenoid lipids, modified fatty acids, polyketides) or L-phenylalanine (phenylpropanoids). Enzymes such as acetyl-CoA carboxylase, and reaction sequences such as the shikimate pathway, thus play important roles in both primary and secondary metabolism, but whether this duality of function is controlled by different gene products, by post-translationally modified enzyme isoforms, or by some form of subcellular compartmentation remains unknown. Similarly, the high energy demands associated with secondary metabolite synthesis must be somehow integrated with the on-going energy needs for growth and development in the plant cell, but the mechanisms underlying this resource distribution have yet to be determined.

Isoprenoid metabolism

The core reaction sequence that gives rise to isoprenoid compounds is discussed in detail in Chapter 28. Some products of this pathway are found universally in the plant kingdom as

photosynthetic pigments (carotenoids), as growth factors (abscisic acid, gibberellins, brassinosteroids), as hydrophobic side-chains on quinone co-factors (ubiquinone, plastoquinone, tocopherol) or chlorophylls, as carriers of oligosaccharides for glycoprotein synthesis (dolichols), and as the phytosterol component of plant membranes. However, on the biosynthetic platform of the core reaction sequence have been built many structural variations which appear to have narrower distributions, and thus fit the model of classical secondary metabolites. Geranyl pyrophosphate (C_{10}) can, for instance, be used as a substrate by any one of a series of monoterpene cyclases, depending on the species and environment (Fig. 10.3). These cyclization products can, in turn, be acted upon by an array of oxygenases and oxidoreductases to generate a suite of volatile monoterpene metabolites (Chappell, 1995).

Monoterpenes are known to play roles in modulating insect behaviour, acting as repellants or attractants (Rodriguez and Leven, 1976), and they can also influence other plants. An extreme example of this is the ability of *Salvia leucophylla* to inhibit the

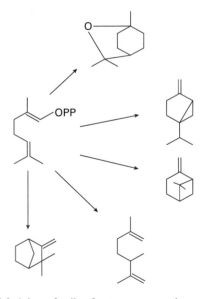

Fig. 10.3 A large family of monoterpene cyclases exists in plants. Some of the structural types generated from geranyl pyrophosphate by different cyclases are shown.

growth of both its own seedlings and those of other species, with the result that thickets of this plant are formed within the chaparral grasslands of coastal southern California. The inhibitory agents is this case appear to be the monoterpenes cineole and camphor.

A different group of cyclases act upon farnesyl pyrophosphate (C_{15}) to yield a series of sesquiterpene skeletons that can then be further modified to generate over 100 distinct end-products (Fig. 10.4). In Solanaceous species (tobacco, tomato, etc.), the activity of this branch of isoprenoid metabolism is rapidly induced by challenge with pathogenic microorganisms, resulting in the synthesis of antimicrobial sequiterpene phytoalexins, such as rishitin. Other members of this class can also affect insect growth and development. For instance, juvabione (from balsam fir) is a potent insect juvenile hormone analogue (Fig. 10.4).

An analogous process results in cyclization of geranylgeranyl pyrophosphate (C_{20}) by diterpene cyclases to produce a host of metabolites as diverse as the phytoalexin, casbene (see Chapter 28), kaurene

(precursor for gibberellins) and tenulin, a potent insect antifeedant in *Helenium amarum*.

The generation of C_{30} and higher isoprenoid metabolites in plants uses a somewhat different mechanism. A tail-to-tail condensation of two molecules of farnesyl pyrophosphate yields a symmetrical C_{30} hydrocarbon (cycloartenol or squalene) that is then activated by epoxidation of one terminal double bond. This epoxide serves as the electrophile for a concerted series of ring-closure electron transfers that are enzymatically controlled, and are triggered by removal of a distal proton. The resulting cholestane ring system is the starting point for biosynthesis of the ubiquitous phytosterols but also for numerous triterpenes, cardenolides and steroidal alkaloids (Fig. 10.5). Cardenolides (also known as cardiac glycosides from the physiological impact of one their members, digitoxin, on human heart function) are central players in a complex interaction between milkweed (*Asclepias* spp.) and Monarch butterflies (*Danaus plexippus*). Larvae of the Monarch feed on milkweed foliage, in the process

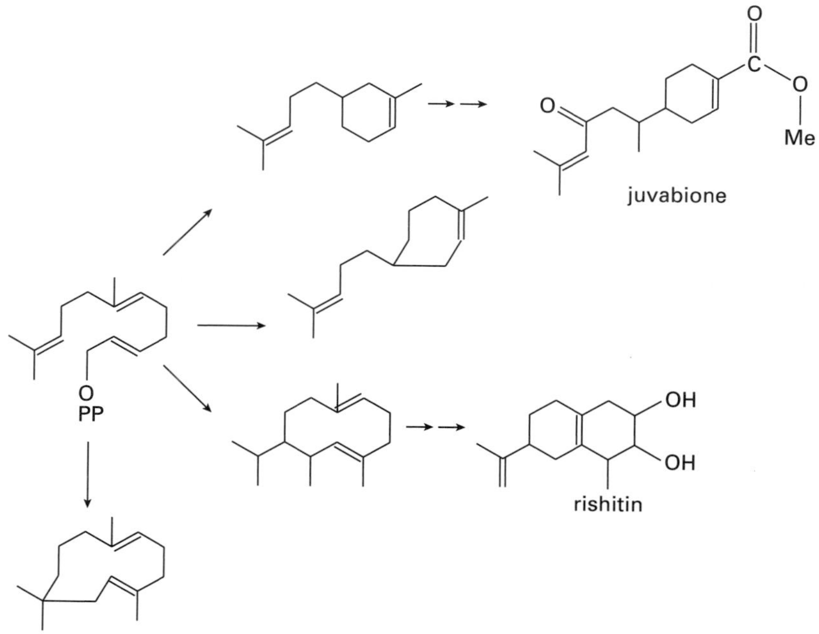

Fig. 10.4 Cyclization of farnesyl pyrophosphate by different sesquiterpene cyclases produces a wide array of structural classes, some of which are shown above.

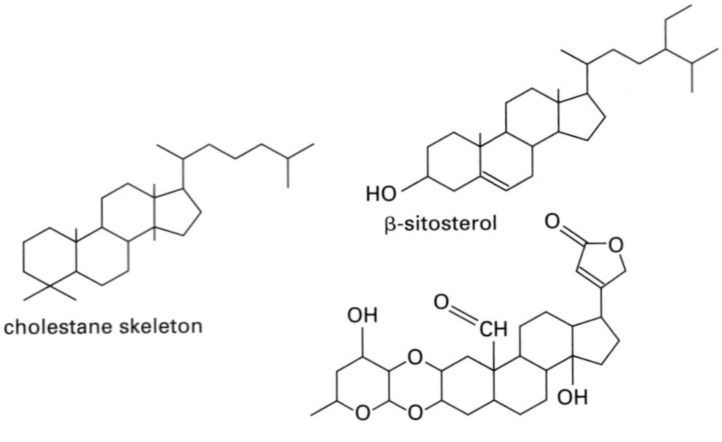

Fig. 10.5 The basic cholestane skeleton arises from condensation of two molecules of farnesyl pyrophosphate. In plants, this structural theme may be only slightly modified to form widespread membrane lipids such as beta-sitosterol, or highly modified to form phylogenetically restricted compounds such as the cardiac glycoside, calotropin.

ingesting a mixture of cardenolides such as calotropin, which are removed from the gut and sequestered. The accumulated cardenolides remain within the body of the insect during metamorphosis and become incorporated into both the abdomen and wings of the resulting adult. Birds that attempt to eat the adult, even a piece of the wing, are poisoned by the cardenolides. They quickly regurgitate the insect parts, and subsequently avoid feeding on Monarch butterflies.

Phenylpropanoid metabolism

When the aquatic ancestors of land plants first began to colonize the shallow littoral zone, and then the adjoining land surface, they faced a bombardment of potentially lethal UV-B radiation. One of the hallmarks of aromatic compounds that possess both an unsaturated propene sidechain and one or more ring-hydroxyls is their efficiency at absorption of UV-B radiation. It has been suggested, therefore, that occupation of these new ecological niches was only possible through co-evolution of the ability to synthesize phenolic compounds (Jorgensen, 1994). The chemical reactivity of phenolics has also proven useful to plants in other ways, and as a result there are few

species of higher plants that do not accumulate substantial levels of phenolics of various kinds.

The key reaction that provides a bridge between primary metabolism and the phenylpropanoid pathways is catalyzed by L-phenylalanine ammonia-lyase (PAL). This unique enzyme acts upon L-phenylalanine to remove both the *alpha*-amino group and the *pro-S* hydrogen from the *beta*-position of the side-chain, yielding *trans*-cinnamic acid. This deamination initiates the 'general' or 'core' phenylpropanoid pathway, a sequence of reactions that involves introduction of a hydroxyl at the 4-position of the ring of cinnamic acid to form 4-coumaric acid, addition of a second hydroxyl at the 3-position to yield caffeic acid, and O-methylation of the 3-hydroxyl to form ferulic acid (Fig. 10.6). The free hydroxycinnamic acids are seldom found in significant amounts in plant tissues, since they are rapidly converted to Coenzyme A esters or glucose esters through the action of the enzymes 4-coumaryl CoA ligase or UDPG : hydroxycinnamic acid glucosyltransferase, respectively. These activated derivatives form a crucial branch-point in phenylpropanoid metabolism, since they can undergo a wide range of reactions. The most prominent product classes are hydroxycinnamyl esters, flavonoids, lignin and hydroxybenzoic acids.

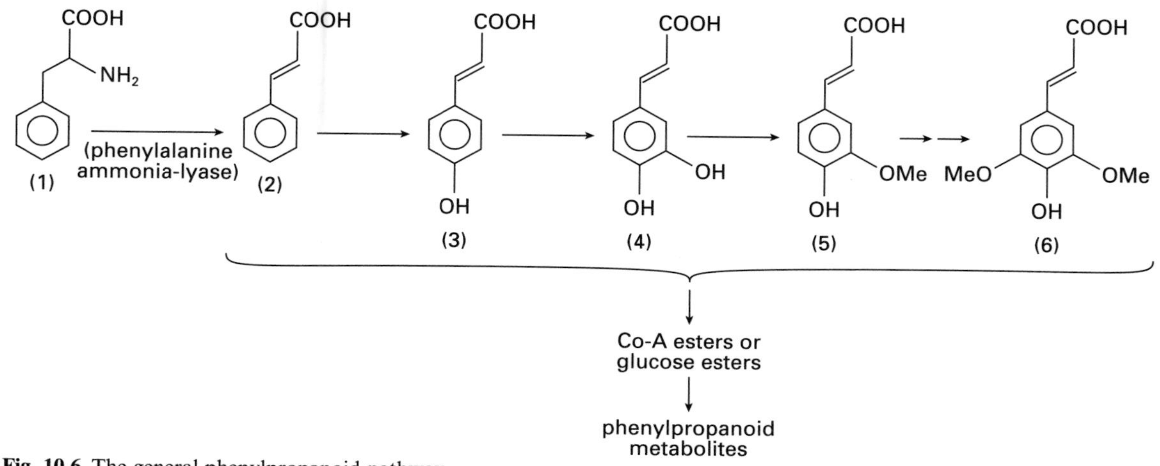

Fig. 10.6 The general phenylpropanoid pathway.

Hydroxycinnamic acid derivatives

Coupling of hydroxycinnamic acids to other metabolites through ester or amide linkages appears to be a ubiquitous phenomenon in higher plants. Both 4-coumaric and ferulic acids are found esterified with cell wall polymers, where they likely serve as nucleation sites for crosslinking with lignin as the latter is deposited into the cell wall matrix. Soluble conjugates are also universally accumulated, but the specific array of conjugates synthesized by a given species, or within a given tissue, shows tremendous diversity. The range of cinnamic acid derivatives involved in these conjugates is relatively limited (most commonly, 4-coumaric, caffeic, ferulic and/or sinapic acids are found), but these are found linked to an extensive series of alcohol or amine substrates in different plants. In particular, caffeic acid esters such as chlorogenic acid, rosmarinic acid and caffeoyl putrescine (Fig. 10.7A) are accumulated to substantial levels in many plant families. The 3,4-dihydroxy nucleus of caffeic acid (and structurally related phenolics such as gallic acid (3,4,5-trihydroxybenzoic acid), makes it chemically reactive under oxidizing conditions, and this property is greatly enhanced by the catalytic activity of the enzyme polyphenol oxidase. For this reason, polyhydroxycinnamic acid derivatives are normally sequestered away from polyphenol oxidase, and held in a mildly acidic environment, within the central vacuole.

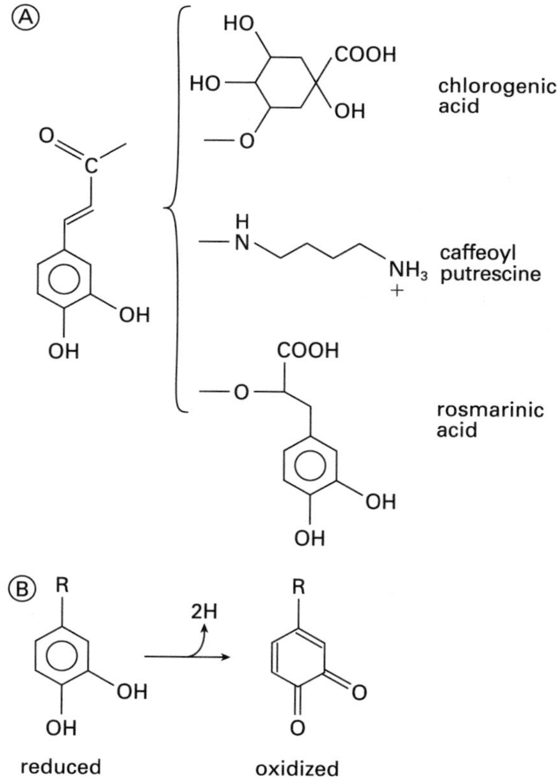

Fig. 10.7 (A) Several of the common caffeic acid esters found in plant tissues. (B) Oxidation of the *ortho*-dihydroxy ring of compounds such as caffeic acid and its derivatives creates highly reactive *ortho*-quinone species.

Loss of cell integrity through wounding, or as part of a hypersensitive response to pathogen invasion, allows polyphenol oxidase to reach its substrate(s) and initiate a rapid oxidation of phenolic esters. Oxidized caffeyl derivatives possess an unstable *ortho*-quinone structure (Fig. 10.7B) that reacts rapidly with other nucleophiles, including other ester molecules, and both thio and amino side-chains on proteins. The inter-ester reaction cycle can generate randomly cross-linked oligomers and polymers, eventually producing the brown/black melanization that is typical of damaged plant tissues. Reaction of oxidized phenolic derivatives with proteins usually causes loss of enzyme function, which is presumed to have defensive value through the disabling of viral proteins, of hydrolytic enzymes secreted by plant pathogens, or of digestive proteins in herbivores.

Flavonoids

Extension and cyclization of the hydroxycinnamyl-CoA side chain occurs universally in higher plants to generate an important class of tri-cyclic phenolics called flavonoids. Plants have developed many variations on the basic flavonoid nucleus. The latter is the product of chain extension of the 4-coumaryl CoA side-chain using three malonyl CoA units in a reaction directly analogous to the extension of fatty acid carbon chains. The enzyme which catalyzes this extension reaction (chalcone synthase) also directs an immediate cyclization of the chain-extended 4-coumaryl derivative, yielding naringenin chalcone (Fig. 10.1). Chalcone synthase appears to be a cytosolic enzyme, but whether it competes with the fatty acid extension reactions for the supply of malonyl CoA (Chapter 22), or operates as a complex with its own acetyl CoA carboxylase, remains to be established. Chalcone synthase seems to be universally distributed in higher plants, but it is accompanied in some plant species by closely related enzymes that catalyze either fewer cycles of chain extension and/or alternative cyclization mechanisms to produce compounds such as anti-microbial stilbenes, or 4-(4-hydroxyphenyl)butan-2-one (Fig. 10.1), a key flavour determinant in raspberries.

The major metabolic fate of naringenin chalcone itself is to undergo enzymatic ring-closure (catalyzed by chalcone isomerase) to form naringenin, which is considered the progenitor of essentially all other flavonoid structures. Depending on the specific reactions carried out on the heterocyclic ring of naringenin and its relatives, an extended resonance structure can be generated over the flavonoid nucleus which allows it to very efficiently absorb UV-B radiation. Flavonoids of this type are typically prominent components of the mixture of phenolic compounds found in plant epidermal tissues, and their concentration rises when the tissue is exposed to increased UV radiation. Conversely, mutational loss of the ability to synthesize flavonoids results in sharply increased UV sensitivity (Lois and Buchanan, 1994). These responses suggest that the capacity to produce flavonoids and the ability of plants to colonize terrestrial niches may have been closely linked in ancestral species. Whatever the original selection pressure that yielded flavonoid biogenesis, plants have since made extensive use of this versatile skeleton.

'Non-hydrolyzable' tannins, another class of potent protein inactivators, form a sub-group of the flavonoids. They consist of carbon–carbon crosslinked oligomers based on the leucoanthocyanidin nucleus, and derive their name from their ability to bind strongly to animal skin proteins through a combination of hydrogen bonds and hydrophobic interactions. The modified proteins are completely resistant to proteolysis. This process formed the basis of leather production for centuries, although the use of 'vegetable tannins' has now been largely replaced by the use of synthetic reagents. The 'hydrolyzable' tannins, which have a completely different structure based on multiple gallic acid residues esterified with glucose, share this high affinity for many proteins. The interaction between tannins and the proteins of the tongue and gut also produces an intense astringency that can be a powerful feeding deterrent to insects and other herbivores.

Most plants are capable of oxidizing and dehydrating the heterocyclic central ring of the flavonoid skeleton to produce the cationic anthocyanidins, molecules whose absorption of electromagnetic radiation extends into the visible region. The result is a range of pigments extending from red–orange to blue, which are especially prominent in ripe fruit, flowers and juvenile leaves.

The basic colour displayed by a particular cell type is determined by the anthocyanidin structural type (Fig. 10.8A), but its hue and intensity are controlled by glycosylation and acylation of the anthocyanidin nucleus, by its association with co-pigment molecules such as other, colourless, flavonoids, by complexation with metal ions in the vacuolar solution, and finally, by the pH of that solution (Fig. 10.8B).

Interestingly, one group of plant families is unable to synthesize anthocyanins, even though all the biosynthetic steps leading to the anthocyanidin precursors appear to be present. This group, which

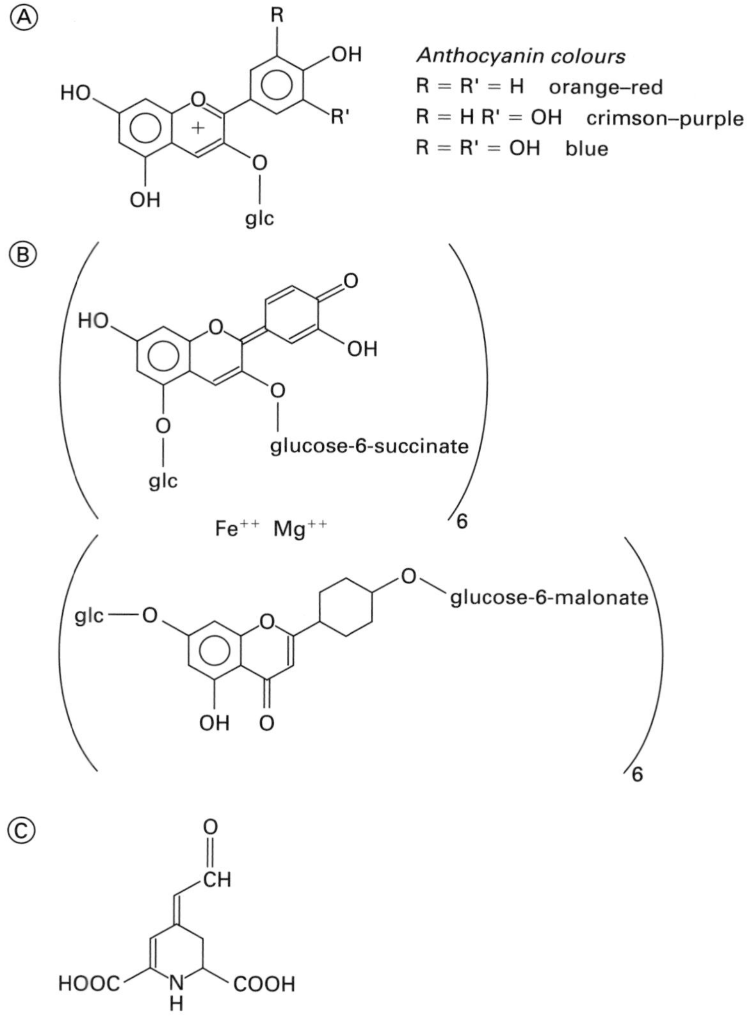

Fig. 10.8 (A) The ring hydroxylation pattern on the anthocyanin molecule has a strong influence on the colour of the molecule in solution. (B) Structure of the blue flower pigment, protocyanin, isolated from the blue cornflower, *Centaurea cyanus*. This supramolecular complex contains six moles of a colourless flavone (co-pigment), six moles of an oxidized anthocyanin (a cyanidin derivative), and one mole each of iron and magnesium ions. It has self-assembles in the correct proportions at pH 5.0 (Kondo *et al.*, 1994). (C) Betalamic acid, the central chromophore of the betalains, originates from tyrosine by cleavage of the aromatic ring and recyclization to a heterocyclic derivative.

includes the cacti, the beet family and ornamental species such as *Amaranthus* and *Bougainvillea*, has evolved to use an unusual ring-cleavage product, betalamic acid, as its central pigment molecule (Fig. 10.8C). Betalamic acid readily reacts with primary and secondary amines to form conjugates (betalains) whose intense colours range from yellow to deep red.

The betalains and anthocyanins are believed to play an important role throughout the plant kingdom in attracting pollen and seed vectors, but non-pigmented flavonoids have also recently been found to be essential for communication and recognition in other plant interactions. Most plants develop symbiotic associations between their root systems and microbial partners, associations that can be vital for gaining access to adequate supplies of scarce nutrients such as nitrogen and phosphorus. The best-known relationships are the nitrogen-fixing nodules formed between legumes and rhizobial bacteria, but far more wide-spread are the intra- and extra-cellular hyphal complexes created when mycorrhizal fungi take up

residence in plant roots. In both cases, it appears that specific flavonoids released into the rhizosphere by the host plant are detected by potential microbial symbionts. Rhizobial associations have been more fully explored, and in these interactions, it has been shown that detection of the correct flavonoids by the bacteria triggers the production and release of specific modified oligosaccharides (Chapter 9). These, in turn, induce dramatic changes in gene expression and morphology in the plant root, and initiate the complex process of bacterial assimilation and nodule formation.

A remarkable analogy to this chemical signalling pattern has also been revealed by recent studies of *Petunia* mutants lacking chalcone synthase. One feature of this mutant phenotype is male sterility, resulting from the failure of the germinating CHS-deficient pollen tube to grow down through the style of CHS-deficient plants. Pollen tube growth resembles hyphal extension in some respects (e.g. tip growth), and chemical signalling between the pollen tube and the stylar tissue is known to be the basis of

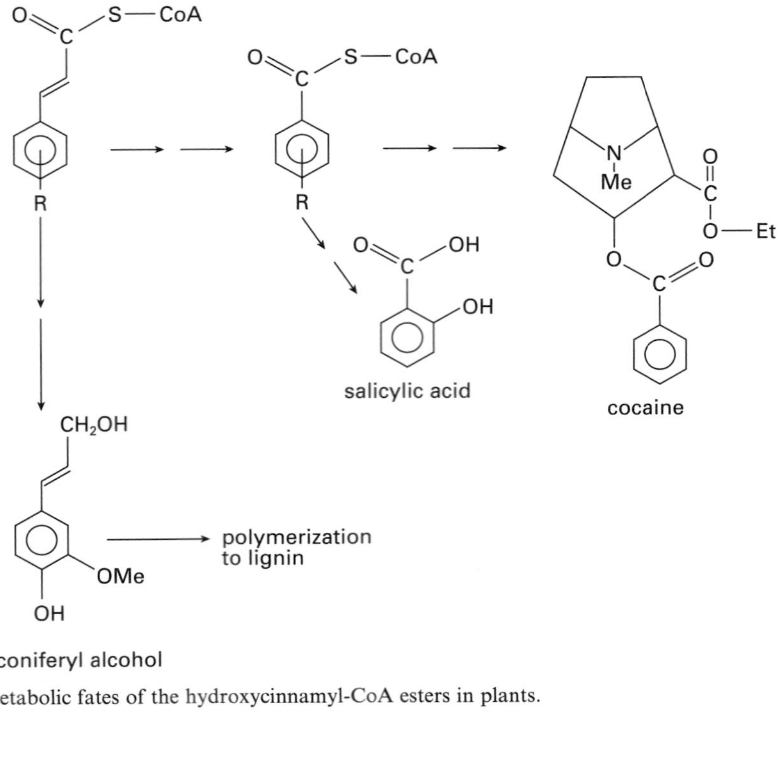

Fig. 10.9 Alternative metabolic fates of the hydroxycinnamyl-CoA esters in plants.

gametophytic self-incompatibility systems in plants. A search for potentially missing signal components in the CHS-deficient genotypes established that supplying specific flavonoids to the mutant pollen restored its ability to fertilize the ovule of both mutant and wild-type flowers (Vogt and Taylor, 1995). It remains to be seen whether this chemical communication system is a universal feature of pollen–pistil interactions in higher plants.

Lignin

The quantitatively most important product of phenylpropanoid metabolism is lignin. It has been estimated that 15–20% of the carbon fixed in the biosphere each year is eventually incorporated into this durable cell wall polymer. Reaction of hydroxycinnamyl-CoA esters with hydroxycinnamyl-CoA oxidoreductase, instead of chalcone synthase, directs the cinnamic acid derivatives to lignin formation. A second reduction of the side-chain completes the conversion of the CoA ester to the alcohol (Fig. 10.9), which can be stored as the 4-O-β-D-glucoside, or oxidized in a free radical reaction that allows several alternative intermolecular bonds to be created, apparently at random. This free radical polymerization takes place within the cell wall, and generates a strong, hydrophobic three-dimensional matrix that surrounds and immobilizes the existing polysaccharide and protein complement of the wall (Yamamoto et al., 1989). The lignin polymer is probably covalently linked to the polysaccharide strands in many places through radical coupling to the aromatic ring of esterified ferulic acid, and to proteins through similar coupling to tyrosine sidechains.

Benzoic acid derivatives

A third possible metabolic fate of the cinnamyl-CoA esters is a cycle of side-chain cleavage that removes the terminal carbons as units of acetyl-CoA and leaves the phenylpropanoid skeleton reduced to a benzoyl-CoA derivative (Fig. 10.9). Benzoic acid and its hydroxylated derivatives occur widely in the plant kingdom, either free or as conjugates. For instance, both cocaine and taxol carry benzoyl acyl groups attached to their central structure. This is not simply

a matter of chemical 'decoration', since removal of the benzoyl groups drastically changes the biological activity of the parent compound.

Perhaps the most intriguing of the benzoic acid derivatives, however, is salicylic acid (2-hydroxybenzoic acid). This compound, originally isolated from the bark of willow (*Salix*), has been valued for hundreds of years for its potent analgesic and anti-inflammatory properties. Many tons of the O-acetyl derivative (ASA, or 'aspirin') are still used medicinally every year, but anecdotal evidence suggested that ASA might also influence plant health. It has recently been confirmed that trauma such as virus infection or wounding of plant tissues can trigger an increase in the intracellular levels of salicylic acid. This metabolic response is essential for establishment of systemic acquired resistance (SAR) within the plant, the phenomenon in which infection of one part of the plant renders other, uninfected, parts of the plant resistant to a subsequent challenge. Salicylic acid does not appear to be the mobile signal that is transmitted to other parts of the plant, but if the increase in salicylic acid levels is blocked, SAR fails to become established (Delaney et al., 1994). The observation that a salicylic acid binding protein has the characteristics of a catalase provides a glimpse of the potential complexity of these signal transduction systems. Catalase efficiently destroys hydrogen peroxide, which is one of the 'activated oxygen species' (AOS) (others include superoxide anion and hydroxyl radical) that rapidly appear in damaged or elicited plant tissues and have been suggested to be part of the signal transduction matrix. Salicylic acid inhibition of catalase activity may potentiate this signal transduction by prolonging the lifetime of the AOS pool in the tissue.

Acetate/malonate metabolism

Sequential condensation of malonyl-CoA with acetyl-CoA is universally employed in both prokaryotic and eukaryotic organisms to generate fatty acids (Chapter 27) and isoprenoids (Chapter 28). Acetyl-CoA is generated in the mitochondrion by the pyruvate dehydrogenase complex that feeds glycolytic carbon to the tricarboxylic acid cycle. However, since the chain extension reactions are primarily located in other cellular compartments, such as the plastids and

ER, a mechanism must exist for acetyl-CoA production in these other compartments.

Plants have adapted the basic fatty acid biosynthetic pathway in many ways to generate specialized products of great importance to their survival. One major challenge for plants is controlling the rate of water loss from their tissues. To deal with this, a water impermeable polymer barrier (cutin) is deposited on the external surface of the epidermal cells in leaves, stems and fruit. Chemically, cutin is a polyester network consisting of cross-linked hydroxylated fatty acids and fatty alcohols. This polyester scaffolding is permeated with wax, which consists primarily of a mixture of long-chain alkanes, also derived biosynthetically from fatty acids. A related polymer, suberin, forms part of the protective surface layer on root tissues. Unlike cutin, however, the suberin polyester network is dominated by C-18 dicarboxylic acid monomers, and includes a substantial component of phenylpropanoid esters. Since suberin is often rapidly deposited in thin layers at, or just beneath, the surface of wound sites in plant tissues, it is thought that its hydrophobic character serves as a barrier to water loss while the phenolic esters may help deter pathogen invasion.

Fatty acid modification occurs widely in the plant kingdom. In addition to the chain hydroxylation and *omega*-oxidation needed for cutin and suberin biosynthesis, many species accumulate unusual fatty acid derivatives in their seeds (e.g. erucic acid in rapeseed oil) or other storage tissues (e.g. polyacetylenic compounds in the laticifer network of *Tagetes*). Specific biological functions for most of these have yet to be defined, although non-standard fatty acids are frequently toxic, and polyacetylenes can act as photosensitizing agents.

Of more general significance in the context of communication and defence is the peroxidation of linolenic acid residues on membrane phospholipids in response to tissue trauma. The resulting hydroperoxy derivative can be enzymatically converted in several steps to a volatile cyclized product, jasmonic acid (Fig. 10.10). This compound is a potent inducer of defence-related genes (Sembdner and Partier 1993), and its even more volatile methyl ester can act as a signal between plants, as well as internally. The jasmonates appear to play an important role in plant signal transduction, but clarification of the interplay

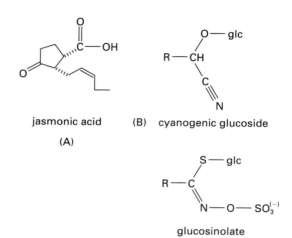

jasmonic acid   (B)   cyanogenic glucoside

(A)

glucosinolate

Fig. 10.10 (A) Jasmonic acid, a chemical messenger that can function both intra- and extracellularly. (B) Two classes of amino acid derivatives that generate toxic products once the glucosidic bond is cleaved.

between jasmonates, activated oxygen species, salicylate and other potential 'second messengers' will require much more research.

The basic chain extension process employed in fatty acid biosynthesis, involving sequential additions of two-carbon units derived from malonyl-CoA followed by reduction, dehydration and saturation, has been adapted to a number of other biosynthetic uses. In these cases, however, the initial condensation product is usually not reduced, leaving an extended carbon chain bearing keto functions on alternating carbons along the backbone. Because of the polarization of the carbon–oxygen double bonds distributed along these 'polyketide' molecules, the intervening methylene groups display a partial carbanion character. As a result, polyketide chains will cyclize with great ease, a tendency that has been exploited in plant metabolism as an alternative route to formation of aromatic rings. The most widespread example of this pattern is the extension of the sidechain of the phenylpropanoid metabolite, 4-coumaryl-CoA, to generate the carbon skeletons of flavonoids, stilbenes and related derivatives. A few plant families also employ the polyketide chain mechanism to build multi-ring aromatic carbon skeletons that are further modified to yield naphtho- or anthraquinones.

Amino acid derivatives

The diversity within this sector of secondary metabolism is particularly marked. Tens of thousands of plant metabolites have been identified that retain an amino nitrogen derived from one of the usual amino acids, especially tyrosine, phenylalanine, tryptophan, lysine, glycine, arginine and methionine. Beyond the retention of the nitrogen function, however, there is little commonality across the biosynthetic origins of these compounds. Many of them are lumped together as *alkaloids*, an ill-defined classification that derives from their *basic* chemical character (as opposed to the many acids found in plant tisues), and the fact that they often display physiological activity in mammalian systems. Cocaine (Fig. 10.9), morphine, nicotine, caffeine and lysergic acid are prominent examples of alkaloids whose effects on human physiology that have been exploited for millenia. By contrast, the biological role(s) of most alkaloids within the plants that synthesize them remains largely unexplored.

Two non-alkaloidal classes of nitrogenous secondary metabolites have attracted special interest because of their unusual structures, and the influence they have on use of certain families of edible crop plants. *Cyanogenic glycosides* are decarboxylated derivatives of amino acids that are stabilized by the attachment of glucose to the cyanohydrin hydroxyl (Fig. 10.10). Normally, the cyanogenic glycosides are sequestered away from glycosidases through tissue compartmentalization, but tissue damage (for example, through herbivory) allows the glycosidases to gain access to their substrates. Hydrolysis of the glucosidic linkage leads to spontaneous decomposition of the cyanohydrin and release of hydrogen cyanide, acute levels of which can be seriously toxic to herbivores. Even chronic lower levels of cyanide release, such as that produced during consumption of some varieties of cassava, can have a long-term impact on the health, and thus the survival, of individuals who make cassava a dietary staple.

Another amino acid-derived class of metabolites is particularly prominent in the cruciferous species. *Glucosinolates* are biosynthesized from decarboxylated amino acids (primarily methionine, tryptophan, phenylalanine and tyrosine) whose skeleton is further modified to incorporate thiol and sulfate groups (Fig. 10.10). The resulting thiohydroxamate derivatives are stabilized as thioglucosides through attachment of a glucose residue to the thiol group. As with the cyanogenic glucosides, the glucosinolates and the corresponding thioglucosidases are stored in separate tissue compartments within the plant, but any tissue trauma that disrupts this compartmentalization will trigger a rapid hydrolysis of the glucosinolate and release of the thiohydroximate derivatives. These spontaneously decompose to form a pungent mixture of volatile thiocyanate, isothiocyanate and nitrile derivatives, which imparts a distinctive flavour and aroma to crucifer plant parts. This characteristic is the basis of the older name for glucosinolates, the 'mustard oil glycosides'. Aside from their olfactory impact, the glucosinolate hydrolysis products are potent protein denaturants and could serve as feeding deterrents for potential herbivores. On the other hand, specialist insects such as the cabbage butterfly (*Artogeia rapae*) have developed the ability to recognize the chemical signature of crucifer species, and use it to locate a suitable host for oviposition.

Future directions

Over the last century, 'natural product' chemists have devoted a tremendous effort to characterizing the chemical diversity of higher plants. While this structure elucidation and cataloguing effort continues, the biological focus has changed to establishing the role of these chemicals at the interface between plants and their environment (Harborne, 1972). That interplay is both dynamic and multi-dimensional, and to date we have barely scratched the surface. We have acquired some information on the way specific plant metabolites impact other plants and other organisms, but our knowledge is crude and fragmentary – a snapshot of a moving target. This challenging task will occupy plant scientists (and entomologists, microbiologists, zoologists, ecologists, etc.) in interdisciplinary studies for many years to come. It is essential, however, that we expand and refine our understanding of *chemical ecology*, if only to keep pace with our ability to manipulate plant genomes, whether through classical breeding or genetic engineering. A major focus of many plant genome alteration efforts is the modification of plant chemistry, but the ultimate consequences of creating such changes can be very unpredictable in the absence of a full understanding of the context of interactions faced by a plant during its lifetime.

Finally, plant biologists are just beginning to uncover the intricate mechanisms by which plant cells perceive enviromental challenges and respond appropriately to them. Detection of pathogens, UV-B, wounding, temperature shock and a multitude of other potential threats to the organism's integrity involves receptor and signal transduction systems, and this field of research is moving ahead at remarkable speed. While there are hints of parallels with better characterized systems in other organisms, it is likely that plant-specific mechanisms will also be unveiled by this research. One of the major challenges in this endeavour is the detection and identification of trace levels of signal metabolites against the rich chemistry background that is so typical of plant cells.

References.

Chappell, J. (1995). Biochemistry and molecular biology of the isoprenoid biosynthetic pathway in plants. *Ann. Rev. Plant Physiol. Plant Molec. Biol.* **46**, 521–48.

Delaney, T. P., Uknes, S., Vernooij, B., Friedrich, L., Weymann, K., Negrotto, D., Gaffney, T., Gut-Rella, M., Kessmann, H., Ward, E. and Ryals, J. (1994). A central role of salicylic acid in plant disease resistance. *Science* **266**, 1247–50.

Harborne, J. B. (1972). *Phytochemical Ecology*, Academic Press, New York.

Jorgensen, R. (1994). The genetic origins of biosynthesis and light-responsive control of the chemical UV screen of land plants. In *Genetic Engineering of Plant Secondary Metabolism. Recent Advances in Phytochemistry*, Vol. 28, eds B. E. Ellis, G. W. Kuroki and H. A. Stafford, Plenum Press, New York, pp. 179–92.

Kondo, T., Ueda, M., Tamura, H., Yoshida, K., Isobe, M. and Goto, T. (1994). Composition of protocyanin, a self-assembled supramolecular pigment from the blue cornflower, *Centaurea cyanus. Angew. Chem. Int. Ed.* **1994**, 978–9.

Lois, R. and Buchanan, B. B. (1994). Severe sensitivity to ultraviolet radiation in an *Arabidopsis* mutant deficient in flavonoid accumulation. II. Mechanisms of UV-resistance in *Arabidopsis. Planta* **194**, 504–9.

Rodriguez, E. and Levin, D. A. (1976). Biochemical parallelisms of repellents and attractants in higher plants and arthropods. In *Biochemical Interaction Between Plants and Insects. Recent Advances in Phytochemistry*, Vol.10, eds J. W. Wallace and R. L. Mansell, Plenum Press, New York, pp. 214–70.

Sembdner, G. and Partier, B. (1993). The biochemistry and the physiological and molecular actions of jasmonates. *Ann. Rev. Plant Physiol. Plant Molec. Biol.* **44**, 569–89.

Tropf, S., Lanz, T., Rensing, S. A., Schroder, J. and Schroder, G. (1994). Evidence that stilbene synthases have developed from chalcone synthases several times in the course of evolution. *J. Molec. Evolution* **38**, 610–18.

Vogt, T. and Taylor, L. (1995). Flavonol 3-O-glycosyltransferases associated with *Petunia* pollen produce gametophyte-specific flavonol diglycosides. *Plant Physiol.* **108**, 903–11.

Yamamoto, E., Bokelman, G. H. and Lewis, N. G. (1989). Phenylpropanoid metabolism in cell walls – an overview. In *Plant Cell Wall Polymers – Biogenesis and Biodegradation*, eds N. G. Lewis and M. G. Paice, American Chemical Society, Washington, pp. 68-88.

Mitochondrial Metabolism

11 Mitochondrial structure
William Newcomb

Introduction

Mitochondria are organelles with a double membrane envelope that occur in eukaryotic cells. Mitochondria are the sites of oxidative phosphorylation, the process in which the oxidation of pyruvate, fatty acids and other compounds releases energy that is coupled to the phosphorylation of ADP to form ATP. Mitochondria can usually be easily distinguished from small plastids because the inner mitochondrial membrane is invaginated, forming long narrow lobes called cristae. In addition, the outer membrane of mitochondria usually stains very faintly in contrast to that of plastids. Two distinct compartments, the intermembrane spaces and matrix, are present in mitochondria. The intermembrane spaces are located between the inner and outer mitochondrial membranes while the matrix is found within the confines of the inner mitochondrial membrane. Some authors can also distinguish the intracristal space which lies within the invaginations of the cristae and thus is an extension of the intermembrane space. In contrast, three distinct compartments are present within chloroplasts.

Shape, distribution and numbers

Mitochondria typically appear as rod-shaped organelles, $0.5 \, \mu m \times 1–2 \, \mu m$, in most transmission electron micrographs of eukaryotic cells. In higher plant cells most mitochondria appear to be rod-shaped with hemispherically-shaped ends, although this is not always the case. Some plant mitochondria appear to be cup- or filament-shaped. Phase-contrast microscopic studies show that

mitochondria can change shape, varying from globular to threadlike or branched. Observations with high voltage transmission electron microscopy of certain animal cells show that the mitochondria can have a long filamentous morphology. Certain algae (*Chorella* and *Chlamydomonas*) have a single mitochondrion in the form of a single branched organelle, the mitochondrial reticulum. Ultrathin sections of these cells usually show small profiles of circular or rod-shaped mitochondria. However, three-dimensional reconstructions using micrographs of serial ultrathin sections of these cells revealed the mitochondrial reticulum. Mitochondrial reticula are

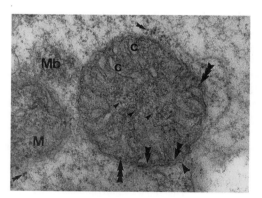

Fig. 11.1 Transmission electron micrograph of a mitochondrion in an infected cell within a root nodule of *Myrica gale*. Shown are the outer mitochondrial membrane (single large arrows), inner mitochondrial membrane (double large arrows), cristae (c), mitochondrial robosomes (single small arrows), cytoplasmic ribosomes (double small arrows) and electron-dense particles (presumably calcium phosphate) (triple large arrows). Also shown are a portion of another mitochondrion (M), a microbody (Mb) and a small vacuole (V). $\times 43\,200$.

probably infrequent in higher plant cells because profiles of branched mitochondria are rarely observed in ultrathin sections of botanical specimens.

Unlike the situation in some animal cells, there does not appear to be a particular spatial arrangement of mitochondria within the cytoplasm of higher plant cells. Transfer cells which contain numerous wall ingrowths believed to facilitate the short distance transport of solutes have numerous mitochondria, presumably for the energy requirements of solute movement across the plasma membrane. While the mitochondria appear to be more numerous in the portion of the cytoplasm near the wall ingrowths, they are not arranged preferentially in close proximity to the plasma membrane bounding the cell wall ingrowths. In some plant cells mitochondria lie near chloroplasts, and in living cells mitochondria can be seen moving inward and away from larger plastids, suggesting that this movement might be associated with an exchange of metabolites. In some plant cells strands of ER surround mitochondria and plastids; again this association may indicate an exchange of solutes between these organelles.

The number of mitochondria per unit volume of cytoplasm appears to remain similar in different developmental stages of the same cell type even though the total number of mitochondria in the cell increases. For example, in young cells of the central zone of the *Zea mays* root cap, about 200 mitochondria are present in each cell; in the mature root cap cells that are greatly enlarged about 2000 to 3000 mitochondria are present. In very active cells such as transfer cells, companion cells of the phloem and secretory cells in nectaries, a large fraction of the cytoplasm (up to *c.* 20%) may be occupied by mitochondria. Some tiny flagellated algae (e.g. *Micromonas*) may have only one mitochondrion per cell. *Chlorella* has one mitochondrial reticulum per cell, as does the yeast *Saccharomyces*. The algae *Chlamydomonas* has 10–15 branched mitochondria per cell.

Outer membrane

The inner and outer membranes are distinct in their appearance. The outer membrane is smooth and occasionally appears to be connected to profiles of smooth ER. In addition, the outer membrane contains less protein and more lipid than the inner membrane. The outer membrane is permeable to all molecules of 10 000 daltons or less because of the presence of a large channel-forming protein, porin.

Inner membrane

The inner membrane, which consists of an unusually high amount of protein (70% protein and 30% lipid by weight), is invaginated forming cristae and contains particles on its inner surface which bounds the mitochondrial matrix. The cristae increase the surface area of the inner membrane. Negative staining has revealed spherically-shaped particles of $\sim$9 nm diameter. These particles are not apparent in freeze-fracture electron microscopic studies, as the fracture plane most likely passes through the middle of the inner membrane rather than through the surface where the particles are located. The particles have been isolated and shown to break down ATP. These particles were originally referred to as the coupling factor because *in vivo* they were thought to couple the energy release during electron transport to the synthesis of ATP. These particles are now known as ATP synthetase (also called $F_0F_1ATPase$) and constitute about 15% of the total inner membrane protein. This protein complex contains a transmembrane proton channel. When protons flow down the channel according to their electrochemical potential, ATP is synthesized. Other integral components of the inner membrane include: succinic dehydrogenase, the only tricarboxylic acid cycle enzyme that is tightly bound into the membrane; the NADH dehydrogenase complex which accepts electrons carried by NADH from tricarboxylic acid cycle enzymes located in the matrix; ubiquinone, which receives hydrogen atoms from succinic dehydrogenase and the NADH dehydrogenase complex; the bc_1 complex that accepts electrons from ubiquinone and transfers them to cytochrome *c*; and the cytochrome oxidase complex which transfers electrons from cytochrome *c* to oxygen.

Mitochondrial matrix

The mitochondrial matrix is enclosed by the inner mitochondrial membrane and consists of a ground substance of fine particles: nucleoids, electron transparent regions containing fine fibrils (mitochondrial DNA) which are digestable by deoxyribonuclease; mitochondrial ribosomes which are smaller than cytoplasmic robosomes; and electron dense particles that consist largely of calcium phosphate. Usually several nucleoids are present in each mitochonrion but there is at least one exception, *Beta vulgaris* (beet) leaf mitochondria, which contain only one nucleoid per mitochondrion. Many of the fine particles within the matrix are likely enzymes associated with the oxidation of pyruvate and fatty acids, the tricarboxylic acid cycle, and the expression of mitochondrial DNA.

Mitochondrial development

Mitochondria are self-replicating organelles and only arise by division from pre-existing mitochondria. The numbers of mitochondria per cell along with mitochondrial fine structure may be indicative of the rates of cellular metabolic activity. For example, root cortical cells have only a few mitochondria, while transfer cells contain many. In addition, the numbers of cristae per mitochondrion may change during development. In the *Arum* flower spadix the number of cristae per mitochondrion increases as the spadix matures and the respiration rate rises.

Further reading

Alberts, B., Bray, D., Lewis, J., Raff, M., Roberts, K. and Watson, J. D. (1994). *Molecular biology of the Cell*, 3rd edn, Garland Publishing Inc., New York.

Gunning, B. E. S. and Steer, M. W. (1975). *Ultrastructure and the Biology of Plant Cells*, Edward Arnold, London.

12 The mitochondrial genome and its expression

Linda Bonen and Michael W. Gray

The genetic function of mitochondrial DNA

In plants, as in all other eukaryotes with mitochondria, the mitochondrial DNA (mtDNA) plays a vital role in mitochondrial biogenesis and respiratory function. It does so by encoding a small number of essential polypeptides of the respiratory chain. Typically, a dozen or more mtDNA-encoded mRNAs, which specify components of the respiratory complexes (Table 12.1), are transcribed from mtDNA and translated by a distinctive mitochondrial protein synthesizing system, the rRNA and most tRNA components of which are also encoded by the mitochondrial genome. The mtDNA, however, carries <10% of the genetic information needed to assemble a mitochondrion; most mitochondrial proteins are encoded in nuclear DNA, synthesized in the cytosol, and imported into the organelle. Nevertheless, the correct expression of genes encoded in mtDNA, and the coordination of mitochondrial and nuclear gene function, are critical for the assembly of an active electron transport chain, which allows the mitochondrion to carry out coupled oxidative phosphorylation (see reviews in Wolstenholme and Jeon, 1992).

Although the basic role of mtDNA is constant, the precise set of genes encoded by it does vary among eukaryotes (Table 12.1). For example, genes that encode subunits of the NADH dehydrogenase of complex I are absent from the mtDNA of *Saccharomyces cerevisiae* and some other yeasts. Genes for subunits of ATP synthase, ribosomal proteins, and cytochrome *c* biogenesis proteins are also variably present. This is most easily explained by the 'missing' mitochondrial genes having been moved to the nucleus in the course of evolution (see below).

A number of animal, fungal and protist mtDNAs have been sequenced in their entirety and their genes identified. Several representatives are shown in Table 12.1. Within the plant kingdom, the mtDNA from the bryophyte, *Marchantia polymorpha* (liverwort), has been completely sequenced (Oda *et al.*, 1992); it contains the largest set of mitochondrial genes reported to date, including novel open reading frames (ORFs) that appear to have no counterparts in the mtDNA of other eukaryotes. In the case of flowering plants, the full set of genes is not yet known, but the number identified so far is relatively large (Table 12.1). Isolated plant mitochondria synthesize approximately 20–30 polypeptides, consistent with plant mtDNA encoding a greater number of proteins than animal mtDNA but fewer than *M. polymorpha* mtDNA. The liverwort mtDNA is slightly larger than the chloroplast genomes of land plants, but it is less tightly packed with genes, having only about two-thirds the number of protein-coding genes.

Mitochondrial genome diversity

One of the most remarkable aspects of mtDNA is the tremendous diversity in its size and physical form in different eukaryotes (reviewed in Gray, 1992). This structural diversity (Table 12.2) contrasts sharply with the basic genetic conservatism of mtDNA, reflected in the 'core' set of mitochondrial genes found in nearly all eukaryotes.

The smallest mtDNAs (14–42 kbp) are found in vertebrate and invertebrate animals. The mtDNA size range is broader and in general larger in fungi (e.g. *S. cerevisiae*, 68–85 kbp) and protists (*Prototheca*

Table 12.1 Identified genes in mitochondrial DNA

	Human	Yeast	Liverwort[a]	Angiosperm
Ribosomal RNAs				
large subunit (*rnl*)	+	+[b]	+[b]	+
small subunit (*rns*)	+	+	+[b]	+
5S (*rrn5*)	−	−	+	+
Transfer RNAs	22	25	27[b]	<20[c]
Respiratory chain:				
Complex I				
nad1	+	−	+	+[b]
nad2	+	−	+[b]	+[b]
nad3	+	−	+[b]	+
nad4	+	−	+[b]	+[b]
nad4L	+	−	+[b]	+
nad5	+	−	+[b]	+[b]
nad6	+	−	+	+
nad7	−	−	(+)[b,d]	+[b]
nad9	−	−	+	+
Complex III				
cob	+	+[b]	+[b]	+
Complex IV				
coxI	+	+[b]	+[b]	+
coxII	+	+	+[b]	+[b,e]
coxIII	+	+	+[b]	+
Complex V				
atp6	+	+	+	+
atp8	+	+	−	?
atp9	−	+	+[b]	+
atpA	−	−	+[b]	+
Ribosomal proteins				
rps1	−	−	+	+
rps2	−	−	+	?
rps3	−	−	+	+[b]
rps4	−	−	+	?
rps7	−	−	+	+[e]
rps8	−	−	+	?
rps10	−	−	+	+[e]
rps11	−	−	+	?
rps12	−	−	+	+[e]
rps13	−	−	+	+[e]
rps14	−	−	+[b]	+[e]
rps19	−	−	+	+
rpl2	−	−	+[b]	+[b]
rpl5	−	−	+	+
rpl6	−	−	+	?
rpl16	−	−	+	+
var	−	+	−	−

Table 12.1 (*cont*)

	Human	Yeast	Liverwort[a]	Angiosperm
Cytochrome *c* biogenesis				
orf509[f]	–	–	+	+
orf322[f]	–	–	+	+
orf277[f]	–	–	+	+
orf228[f]	–	–	+	+
orf183[f] (*orf25*)	–	–	+	+
orf172[f] (*orfB*)	–	–	+	+
orf244[f]	–	–	+	+
other orfs	–	–	25	?
intronic orfs				
group I	–	+	+	?
group II	–	+	+	+
RNase P RNA	–	+	?	?

[a] *Marchantia polymorpha*.
[b] Genes containing introns.
[c] A minimal set of tRNA genes (sufficient to support mitochondrial protein synthesis) is not
     encoded in any of the angiosperm mtDNAs characterized to date.
[d] Present but designated as a pseudogene because of stop codons within reading frame.
[e] Genes present in the mitochondria of some, but not all, flowering plants.
[f] ORFs (open reading frames encoding specified number of amino acids) are named
     according to the liverwort mitochondrial nomenclature (Oda *et al.*, 1992).

wickerhamii, 55 kbp), although small fungal and protist mtDNAs are also found (e.g. *Schizosaccharomyces pombe*, 17–25 kbp, and *Chlamydomonas reinhardtii*, 16 kbp, respectively). Land plants have relatively enormous mtDNAs, which range from 187 kbp in *M. polymorpha*, to at least 2400 kbp in some cucurbits; the latter approximates the size of a bacterial genome (e.g. 4700 kbp in *E. coli*). Even within a single family of plants (the Cucurbitaceae), mtDNA is seen to vary over a seven-fold size range (compare watermelon and muskmelon, Table 12.2). There appears to be no correlation between the sizes of the nuclear and mitochondrial genomes in plants: *Arabidopsis thaliana* has one of the smallest nuclear genomes, but a moderate-sized mitochondrial genome of about 372 kbp.

Physical and genetic mapping data indicate that most mtDNAs are circular. Supercoiled, covalently closed circular DNA can be isolated in high yield from animal mitochondria. On the other hand, very few intact circular molecules approximating the estimated size of the *S. cerevisiae* mitochondrial genome have ever actually been visualized by electron microscopy. A few mtDNAs are linear, the best characterized being those of the ciliated protozoa *Paramecium aurelia* and *Tetrahymena pyriformis*, as well as that of the unicellular green alga, *C. reinhardtii*. Circular maps have been proposed for a number of flowering plant mtDNAs; however, as discussed below, most plant mtDNAs are physically heterogeneous and display an unusual structural complexity. This contrasts with the liverwort mitochondrial genome which is predicted to be a simple circular molecule of 187 kbp based on restriction mapping and electron microscopic data.

Given the variation in mtDNA size, it is not surprising that there is also marked diversity in the way in which genes in mtDNA are arranged and expressed. The human and yeast mitochondrial genomes represent two very different organizational patterns. Human mtDNA is a model of genetic economy: it has tightly packed genes that are expressed as long polycistronic transcripts, which are then clipped into monocistronic RNAs at the positions of tRNAs that are interspersed end-to-end with protein-coding sequences. In the yeast mitochondrial genome, genetic information is arrayed more spaciously. Coding sequences are separated by A+T-rich non-coding spacer sequences and genes such as *cob* and *coxI* are interrupted by introns.

Post-transcriptional removal of intron sequences from primary transcripts requires a complex processing pathway involving intricate intron folding and in some cases the participation of RNA maturases that are encoded within the introns themselves. Yeast

Table 12.2 Size range of mitochondrial DNA[a]

		kbp
Animals		(14–42)
Homo sapiens		16.6
Mus musculus		16.3
Xenopus laevis		17.6
Caenorhabditis elegans		13.8
Placopecten magellanicus		42
Fungi		(17–176)
Saccharomyces cerevisiae		68–85
Schizosaccharomyces pombe		17–25
Protists		(16–75)
Acanthamoeba castellanii		41.6
Chlamydomonas reinhardtii		15.8
Prototheca wickerhamii		55.3
Tetrahymena pyriformis		47.2[b]
Plants		(187–2400)
Arabidopsis thaliana		372[c]
Beta vulgaris	sugar beet	386[c]
Brassica campestris	turnip	218[c]
Brassica hirta	white mustard	208[c]
Brassica napus	rapeseed	221[c]
Brassica nigra	black mustard	231[c]
Brassica oleracea	cauliflower	219[c]
Citrullus vulgaris	watermelon	330
Cucumis melo	muskmelon	2400
Cucumis sativus	cucumber	1500
Cucurbita pepo	zucchini	840
Helianthus annus	sunflower	300[c]
Marchantia polymorpha	liverwort	187[d]
Petunia hybrida	petunia	443[c]
Pisum sativum	pea	360
Oryza sativa	rice	492[c]
Raphanus sativa	radish	242[c]
Spinacia oleracea	spinach	327[c]
Triticum aestivum	wheat	430[c]
Vicia faba	broad bean	285
Zea mays	maize	570[c]

[a] Data are summarized from Wolstenholme and Jeon, 1992, and supplemented with more recent information from the literature (Vasil and Levings, 1995).
[b] Size excluding telomeres.
[c] Size of 'master chromosome' estimated from complete physical mapping of the genome.
[d] Genome completely sequenced (Oda *et al.*, 1992).

mitochondrial genes are organized into a number of separate transcriptional units (approximately 20) and transcription begins within a conserved nonananucleotide motif that represents the essential promoter element. Topics regarding mitochondrial molecular biology in animal, fungal and protist systems are discussed in detail elsewhere (see reviews in Wolstenholme and Jeon, 1992).

The plant mitochondrial genome

Physical complexity and structural fluidity

The mtDNA of flowering plants is not only exceptionally large (Table 12.2), it also displays an unusual physical complexity suggestive of a heterogeneous collection of molecules (reviewed in Lonsdale, 1984; Hanson and Folkerts, 1992; Schuster and Brennicke, 1994). Restriction endonuclease patterns of plant mtDNA are characterized by a large number of fragments, some of which are present in submolar or multimolar amounts. Renaturation kinetic measurements provide little evidence of highly repetitious sequences, as are found in plant nuclear DNA. Mapping and sequencing studies have revealed the presence of repeated sequences, some short and others >10 kbp in length, that occur only a few times in the plant mitochondrial genome. Some of the large repeats comprise a collection of short repeats interspersed with unique sequences. As discussed below, these repeated sequences appear to play an important role in generating much of the heterogeneity that characterizes plant mtDNA.

Over recent years, physical maps of mtDNAs have been constructed for a number of plants (reviewed in Hanson and Folkerts, 1992; Bonen and Brown, 1993; Levings and Vasil, 1995). Among the first were maize (Fig. 12.1; Lonsdale et al., 1984) and several species of Brassica including B. campestris (Palmer and Shields, 1984). In almost all cases, the mitochondrial genome can be represented as a single circular molecule, termed the master chromosome, which contains the entire genetic complexity of the organelle DNA (Fig. 12.1). A feature of the master chromosome is the presence of repeated sequences that appear to mediate reciprocal recombination. A result of such recombination is the dispersion of

genetic information among a number of subgenomic, circular molecules (Fig. 12.2). If a master chromosome contains one pair of repeat elements in the same orientation, then reciprocal recombination between them leads to its resolution into two smaller circular molecules, each containing one copy of the repeat. Such a tripartite structure has been proposed for Brassica campestris mtDNA. In this model, a master chromosome of 218 kbp contains two copies of a 2 kbp direct repeat that participates in intragenomic recombination to reversibly generate subgenomic circles of 135 kbp and 83 kbp.

As the number of direct and/or inverted repeats increases in a circular molecule, so do the possibilities for recombination (Fig. 12.2). Maize mtDNA contains six major sequence reiterations, each present twice, ranging in size from 0.5 to 14 kbp (Fig. 12.1). Five of these repeats are direct, while the 14 kbp repeat is inverted. Recombination between inverted repeats would cause a 'flip–flop' inversion of the single-copy region between them. The master chromosome, which contains one set of each repeat, is estimated to be 570 kbp in size (Lonsdale et al., 1984). All of the repeats except a 10-kbp pair appear to be recombinationally active. As a result, numerous subgenomic circular molecules are postulated to exist in maize mitochondria, in addition to the 570-kbp master circle. A number of the proposed recombination products do, in fact, correspond closely to discrete size classes that had previously been visualized in the population of rare circular molecules isolated from maize mtDNA. Because the postulated master circle and subgenomic circular molecules are very large, they may not easily survive isolation, which could account for the high proportion of heterogeneous linear molecules and low proportion of circular molecules isolated in plant mtDNA.

Electron microscopic analysis of plant mtDNAs isolated from normal tissue has revealed primarily linear molecules with no fixed size distribution, some even longer than the estimated genome size (reviewed in Bendich, 1993). Circular molecules of various sizes have also been found, but their proportion is vanishingly small and none is the full size predicted from restriction maps. For some plants, the proportion of circular molecules is substantially higher in mtDNA prepared from plant cells grown in

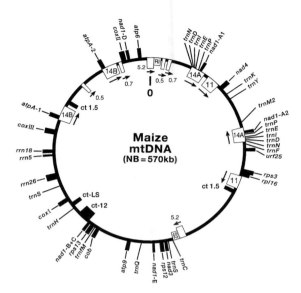

Fig. 12.1 Location of genes on the 570 kbp master circle of the mitochondrial genome of maize (cytotype NB). The approximate sizes (in kbp) and orientations (inner arrows) of repeated DNA sequences are shown (open boxes inside the circle). The black boxes (outside of circle) represent known mitochondrial genes (rRNA and protein-coding) designated as in Table 12.1, as well as individual tRNA (trn) genes (C, cysteine; D, aspartic acid; E, glutamic acid; F, phenylalanine; H, histidine; I, isoleucine; K, lysine; fM methionine (initiator); M methionine (elongator); N, asparagine; P, proline; Q, glutamine; S serine; Y, tyrosine). Sequences homologous to chloroplast DNA are also shown (black boxes labelled ct inside the circle) with approximate sizes in kbp. The ct-LS black box represents sequences homologous to those encoding the large subunit of ribulose 1,5-bisphosphate carboxylase. R1 and S2 correspond to integrated copies of the linear plasmids S1 and S2 found in the maize *cms*-S cytoplasm. Figure updated from Lonsdale *et al.* (1984) and Dawson *et al.* (1986), and kindly provided by Dr. C. M.-R. Fauron.

tissue culture. It is not clear whether this reflects a real difference in the physical form of plant mtDNA in normally grown versus cultured tissue, or is due to a stabilization of the mtDNA during its preparation from tissue culture cells. Based on sensitive hybridization analysis, certain genomic organizations are found to be present at very low levels; these are called 'sublimons' (Small *et al.*, 1989). The amplification of such sub-stoichiometric forms, perhaps in conjunction with recombination events,

may explain changes seen in mtDNA profiles after tissue culture or somatic hybrid production.

The multipartite model of a master chromosome and subgenomic products is attractive because it explains the existence and structural interrelationships of many of the repeated sequences identified in plant mtDNA, and because it can account for much if not most of the observed complexity and physical heterogeneity of this genome. It does, however, raise

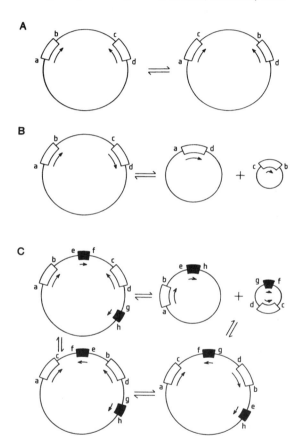

Fig. 12.2 Possible arrangements for repeated sequences in the mitochondrial genome. (A) Single pair of inverted repeats, recombination leads to sequence inversion ('flip–flop'); (B) single pair of direct repeats, recombination leads to two circular products ('loop-out'); (C) a single pair of inverted and direct repeats which are nested. This gives four possible genome configurations. The larger the number of direct and inverted repeats, the more complex the genome organization becomes. Reprinted with permission from Lonsdale (1984).

a number of questions about how plant mtDNA is replicated and segregated in the course of cell division.

Although most flowering plant mtDNAs characterized so far show evidence of high frequency recombination and a multipartite organization, one exception is the mitochondrial genome of *Brassica hirta* which generates restriction maps consistent with a single, circular 208-kbp chromosome (Palmer and Herbon, 1987). Its size is slightly larger than the mitochondrial genome of *M. polymorpha*, which as noted above also has a simple circular structure of 187 kbp (Oda *et al.*, 1992).

Plant mitochondria also contain small circular and linear DNA species that in some cases have sequence homology with the main mitochondrial genome or nuclear DNA. Because these plasmid-like DNAs can in most cases be lost without any phenotypic effect, they seem to have no essential genetic function. RNA plasmids have also been identified in the mitochondria of some plants. The origins of DNA and RNA plasmids are obscure, but they do contribute to the physical heterogeneity of plant mtDNA (reviewed in Hanson and Folkerts, 1992).

Promiscuous DNA in the plant mitochondrial genome

Among the many unique characteristics of flowering plant mtDNA, one of the most intriguing is its ability to incorporate and maintain sequences derived from other genomes: so-called promiscuous DNA. Stern and Lonsdale (1982) first described a 12 kbp DNA sequence common to both the mitochondrial and chloroplast genomes of maize. That sequence contains the chloroplast 16S rRNA gene, so it obviously originated from the maize chloroplast genome. Such promiscuous chloroplast sequences are widely distributed in plant mtDNA, seemingly in random fashion, both with respect to the plant species and the portion of the chloroplast genome that is incorporated into the mitochondrial genome. The finding of promiscuous chloroplast DNA in plant mtDNA was quickly followed by the discovery of mitochondrial and chloroplast sequences in nuclear DNA, and nuclear sequences in the mtDNA, firmly establishing the concept of inter-organellar transfer of

genetic information (reviewed in Hanson and Folkerts, 1992; Brennicke *et al.*, 1993). Although most of this DNA is believed to be non-functional, some transferred chloroplast tRNA genes actually play a role in plant mitochondrial translation (see below). Promiscuous DNA accounts for part of the 'extra' DNA that the plant mitochondrial genome contains, and it illustrates, in graphic fashion, the structural plasticity and genetic adaptibility of the mitochondrial genome in flowering plants. Notably, no chloroplast DNA sequences are evident in the mitochondrial genome of *M. polymorpha*.

Genes in the mitochondrial DNA of flowering plants

The plant mitochondrial genes that have been characterized to date are listed in Table 12.1; some of these (e.g. genes for 5S rRNA, the alpha-subunit of the F_1-ATPase, and homologs of *E. coli* ribosomal proteins and cytochrome *c* biogenesis proteins) have not been found in the mtDNA of animals or fungi. These genes constitute a subset of those present in the completely-sequenced liverwort (*M. polymorpha*) mitochondrial genome. Not all flowering plants have exactly the same set of genes; differences are attributed to the recent transfer of particular genes to the nucleus in certain plant lineages (see below).

Ribosomal RNAs and their genes

Plant mitochondrial ribosomes sediment at about 78S and contain 18S (small subunit) and 26S (large subunit) rRNA species, which are closer in size to the corresponding eukaryotic cytoplasmic rRNAs than to their homologs in bacteria (16S and 23S), and much larger than those in animal mitochondria (12S and 16S). Plant mitochondrial ribosomes also contain a distinctive 5S rRNA (absent in the mitochondrial ribosomes of almost all other eukaryotes).

Plant mitochondrial 26S, 18S and 5S rRNAs are all encoded by the plant mitochondrial genome. The rRNA genes were the first genes to be identified in plant mtDNA, and proved to have a novel arrangement (Bonen and Gray, 1980). The 18S and 5S rRNA genes are closely linked in the same transcriptional orientation, but at a site distant from the 26S rRNA gene in all angiosperm mtDNAs examined. This contrasts with the organization of rRNA genes in eubacterial and chloroplast genomes, where rRNA genes are arrayed and co-transcribed in the order 16S–23S–5S. In *M. polymorpha* mitochondria, the rRNA genes are clustered in the order 18S–5S–26S.

Plant mitochondrial rRNAs show a striking structural resemblance (both in primary structure and potential secondary structure) to their eubacterial homologs in spite of the size differences noted above (Spencer *et al.*, 1984). The degree of structure conservation is much more pronounced than that displayed by the mitochondrial rRNAs of other eukaryotes and this feature has been instrumental in adducing evidence, through rRNA sequence comparisons, for a eubacterial, endosymbiotic origin of all known mitochondrial genomes (reviewed in Gray, 1992). From the wheat mitochondrial 18S rRNA sequence, the origin of mitochondria has been traced to a specific group of eubacteria, the alpha-subdivision of *Proteobacteria*, the so-called non-sulfur purple bacteria. This is precisely the group of eubacteria that had earlier been proposed, on biochemical grounds, to contain the closest contemporary relatives of mitochondria.

Transfer RNAs and their genes

The number of tRNA genes identified in flowering plant mtDNA is inadequate to generate all the different tRNAs required to read the genetic code using the classical 'wobble' pairing rules proposed by Crick. In wheat mitochondria, only sixteen different tRNA genes, including separate ones for initiator and elongator tRNA-Met, have been found. The number and identity of mitochondrially-encoded tRNA genes varies slightly among plant species. In all cases, it is less than in mammalian and fungal mitochondria, where 22–25 tRNA genes are sufficient because of an expanded codon recognition pattern in which a single tRNA species is able to decode all four codons for the same amino acid in a four-codon family. The liverwort (*M. polymorpha*) mitochondrial genome encodes 29 tRNA genes specifying 27 different tRNAs.

The low number of tRNA genes in flowering plant mitochondria is explained at least in part by certain tRNAs being encoded by nuclear genes and the

tRNAs being imported into the mitochondrion (reviewed in Gray, 1992; Marechal-Drouard *et al.*, 1993). The mechanism by which such tRNAs cross the double membrane of the mitochondrion is as yet unknown. It is somewhat surprising that the spacious mitochondrial genome in flowering plants lacks a full complement of tRNA genes and recruits components of the cytosol translational machinery. On the other hand, in certain protists, such as *Trypanosoma*, *Tetrahymena* and *Chlamydomonas*, the majority of mitochondrial tRNAs are nucleus-encoded and imported into the organelle (see reviews in Wolstenholme and Jeon, 1992).

A subset of tRNAs encoded by the plant mitochondrial genome bear an especially striking similarity to their chloroplast homologs. There is strong evidence that many of these genes do in fact give rise to stable tRNAs that participate in mitochondrial protein synthesis; such genes may therefore be viewed as having been recruited from 'promiscuous' chloroplast DNA, with the original mitochondrial gene counterparts being lost. There are plant-specific differences in the presence of particular chloroplast-like tRNA genes in the mitochondrion, suggesting that such transfer events are recent and even ongoing.

Mitochondrial tRNAs in flowering plants therefore have three different genetic origins: 'native' tRNAs encoded by mtDNA, 'chloroplast-like' tRNAs encoded by promiscuous chloroplast DNA that has been integrated into mtDNA, and 'cytosol-like' tRNAs encoded by the nuclear genome and imported from the cytosol into the mitochondrion. In wheat mtDNA, 10 native and 6 chloroplast-like tRNA genes have been identified (Joyce and Gray, 1989); all differ from 15 cytosol-like nucleus-encoded ones (K. Glover and M. W. Gray, unpublished observations). None of the plant mitochondrial tRNAs displays any of the primary or secondary structural aberrations that characterize some fungal and especially animal mitochondrial tRNAs.

The locations of mitochondrial tRNA genes on the master chromosome vary among plants. Some are physically close to and co-transcribed with other genes; others are scattered and appear to form independent transcriptional units. In wheat mtDNA, an initiator tRNA-fMet gene is closely linked to (in the same transcriptional orientation as) the 18S rRNA gene: the 3′ end of the tRNA gene – which does not encode the 3′ CCA terminus – is only one bp removed from the 5′ end of the 18S rRNA gene (Gray and Spencer, 1983). This is a surprisingly compact organization, considering the large size of the plant mitochondrial genome.

Protein-coding genes

Sequence analysis has provided considerable insight into the structure and organization of protein-coding genes in plant mtDNA. The maize *coxII* gene, the first to be sequenced, proved to have a 794 bp intron (Fox and Leaver, 1981). The *coxII* genes examined in other monocots have closely related introns at the same position, but of differing length because of insertions or deletions. In some dicots (such as Oenothera, pea, soybean), *coxII* is not split; in others it has one intron (petunia, sugar beet) or two introns (carrot) (reviewed in Bonen and Brown, 1993).

To date, over twenty introns have been found in angiosperm mitochondrial genes; the vast majority are located within *nad* genes encoding NADH dehydrogenase subunits (Table 12.1). These introns all have elaborate secondary structural features similar to those found in fungal mitochondrial genes and categorized as 'group II'. Certain members of this class of introns are autocatalytic *in vitro*, although none in plant mitochondria has yet been shown to be self-splicing. The terminal intron in the *nad1* gene in plant mitochondria contains an RNA maturase ORF, having homology to reverse transcriptase, a situation similar to that in certain fungal mitochondrial split genes.

Several of these plant mitochondrial split genes (namely, *nad1*, *nad2*, and *nad5*) have undergone DNA rearrangements within intron sequences so that *nad* coding segments are dispersed in the genome. Remarkably, these fragmented genes remain functional because after transcription the RNA molecules are able to find their partners and interact so that the exons are joined in the correct order and proper reading frame for translation. This process is called 'trans-splicing' (reviewed in Wissinger *et al.*, 1992; Bonen and Brown, 1993) and it illustrates the impact that DNA rearrangements can have on plant mitochondrial gene structure and expression. The *Marchantia polymorpha* mitochondrial genes contain a total of 7 group I and 25 group II introns, none of which requires splicing *in trans*. Only the intron in *nad2* is located at the same position as in flowering plants, consistent with this class of introns behaving as mobile genetic elements.

Sequence comparisons have established that plant mitochondrial protein-coding genes are highly conserved in primary structure, but that similarities disappear abruptly outside of the coding region. Comparing wheat and maize *coxII*, for example, exon and intron sequences are equally well conserved (98.9% and 99.3% nucleotide identity, respectively, even without taking into account RNA editing; see below). Immediately upstream of the AUG initiation codon, there is no significant similarity between the maize and wheat sequences. This pattern of extremely high conservation of coding sequence, with abrupt disappearance of similarity in flanking regions, is a common feature of plant mitochondrial genes, and is attributed to rearrangements that have moved coding regions into new sequence contexts. Thus, plant mtDNA is evolving slowly with respect to nucleotide sequence *per se* but rapidly with respect to sequence organization.

A number of ORFs have been identified in plant mtDNA by sequence analysis; a full inventory awaits the complete sequencing of mtDNAs from several angiosperms and confirmation of the presence of corresponding proteins. Such genes might encode additional subunits of respiratory chain complexes or the mitochondrial expression machinery, or proteins unique to plants, which are rather special in having to coordinate activities of three separate genetic compartments (the nucleus, mitochondrion and chloroplast).

It was originally thought that there was one deviation from the universal genetic code in plant mitochondria, namely the use of CGG to encode tryptophan rather than arginine (Fox and Leaver, 1981); however, it is now known that CGG triplets at such positions are in fact converted to UGG at the RNA level, through a process known as RNA editing (see below), so that the expected tryptophan residues are indeed present in the protein. The UGA triplet is used as a normal termination codon in plant mitochondria, rather than coding for tryptophan as in animal, fungal and some protozoan mitochondria.

Gene maps

With the construction of physical maps of plant mtDNAs has come the localization of genes on these maps. The maize mitochondrial gene map is shown in Fig. 12.1. Genes are distributed around the master chromosome, on both strands; many genes are widely separated while others are closely linked. Mitochondrial gene maps have also been published for a number of other plants including spinach, *Brassica campestris*, and petunia (reviewed in Hanson and Folkerts, 1992; Levings and Vasil, 1995); these maps are almost completely different from each other. The relatively low conservation in gene order from plant to plant presumably reflects the recombinogenic nature of plant mtDNA. Mitochondrial genes that are present as single genomic copies in one plant may be located within recombination repeats in another plant; examples of duplicated copies are *atpA* in maize mtDNA, *coxII* in *B. campestris* mtDNA and the rRNA and *atp6* genes in wheat mtDNA. This contrasts with the organization of genes in *M. polymorpha* mitochondria, where genes are generally more tightly clustered, present in single copy, and in some cases retain an ancestral, bacteria-like operon organization. Many of the liverwort mitochondrial ribosomal protein genes have the same organization as their homologs in *E. coli*, as do the *rps3/rpl16* genes in flowering plant mitochondria.

Recent transfer of mitochondrial genes to the nucleus

One example of the dynamic, mosaic nature of plant mtDNA is the presence of functional chloroplast-like tRNA genes in the mitochondrion (discussed above). There is also compelling evidence that certain genes have recently moved from the mitochondrion to the nucleus in particular plant lineages. A systematic analysis of *coxII*, which has traditionally been considered a 'core' mitochondrial gene, has shown that it differs in location among various legumes (Nugent and Palmer, 1991; Covello and Gray, 1992). In cowpea, the functional gene is located in the nucleus and there are no *coxII*-homologous sequences in the mitochondrion. In soybean and pea, there are copies of *coxII* in both the nucleus and the mitochondrion, but in each case only one of the copies gives rise to stable transcripts: the

mitochondrial copy in pea and the nuclear copy in soybean. These cases presumably reflect intermediate stages in the gene transfer process.

If a mitochondrial gene is to be successfully translocated to the nucleus, it must acquire appropriate expression and targeting sequences to enable the polypeptide to be synthesized in the cytosol and routed back into the mitochondrion. For the coding sequences to be correctly translated by the cytosol machinery, the transferred gene must correspond to an RNA-edited copy (see below). In both cowpea and soybean, the nucleus-located *coxII* coding sequences are linked to regulatory sequences by an intron. In Oenothera, the *rps12* gene, which has similarly been transferred from the mitochondrion to the nucleus, contains no intron (reviewed in Brennicke *et al.*, 1993). Based on Southern hybridization data, a number of other ribosomal protein genes are believed to differ in their location among plants (see Table 12.1).

Expression of the plant mitochondrial genome

As the molecular architecture of the plant mitochondrial genome has unfolded, attention has shifted to questions of expression. Because a given gene may be flanked by totally different sequences in the mtDNAs from different plants, and because gene order as a whole is not conserved, we would obviously like to know what motifs are used to signal the beginning and end of transcription, how these promoter and terminator sequences are arrayed, and what features are important for RNA processing, RNA stability and translation. As well, we would like to know what proportion of the large plant mitochondrial genome is, in fact, transcribed and whether certain genomic regions are differentially expressed in different tissues or during different stages in the life cycle of the plant.

Transcription

The *Brassica campestris* mitochondrial genome, which is relatively simple, was the first for which a complete transcriptional map was determined. Makaroff and Palmer (1987) found 24 abundant and positionally distinct transcripts of size >500 nucleotides, which correspond to approximately 30% of the 218-kbp mitochondrial genome. A number of less adundant transcripts, many of which overlap one another and the major transcripts, were also detected. Although *B. campestris* mtDNA contains a number of promiscuous chloroplast sequences, none of these appears to be transcribed within the mitochondrion.

The results of gene and transcript mapping are consistent with some genes being widely scattered throughout the genome in independent transcriptional units, and others being closely linked and cotranscribed. The total number of transcriptional initiation sites in any plant mitochondrial genome is as yet unknown; at least eighteen have been identified in Oenothera (reviewed in Schuster and Brennicke, 1994). Capping experiments with guanylyltransferase to distinguish nascent termini (carrying a 5′ tri- or diphosphate) from processed termini (carrying a 5′ monophosphate) have led to the identification of a short core motif CRTA within a weak, purine-rich consensus sequence (reviewed in Gray *et al.*, 1992; Schuster and Brennicke, 1994). Support for the view that these sequences represent promoter elements has come from *in vitro* transcriptional analysis (reviewed in Gray *et al.*, 1992) and by *in vitro* mutational analysis using linker scanning and point mutagenesis (Rapp *et al.*, 1993).

At this time, it is still an open question whether plant mitochondria have multiple RNA polymerases, and whether transcriptional activity is modified by specific *trans*-acting factors that differ among genes or classes of genes. Run-on transcription studies using isolated organelles have suggested that the rate of RNA synthesis differs among genes and that differences in steady state levels of transcripts for different protein-coding genes reflect both promoter strength and post-transcriptional processing or RNA stability (reviewed in Gray *et al.*, 1992).

Among different plants, the complexity of the transcript profile for a particular gene can also vary dramatically. In wheat, for example, a *coxII* exon-specific probe detects one abundant transcript of size 1.5 kb in RNA blot analysis, whereas in maize a number of discrete transcripts are seen. These differences undoubtedly relate to the fact (noted previously) that 5′ flanking sequences, which carry transcriptional control signals, are very different in

the two cases. In general, complex transcript patterns can reflect multiple promoter sites, a multiplicity of processing sites, different levels of splicing intermediates, or hybridization signals arising from rearranged duplicated sequences that are transcribed. The precise mapping of transcript termini by primer extension and S1 nuclease protection experiments has not revealed any strong consensus motifs, although sequences in the vicinity of the 3′ termini can in some cases be folded into potential secondary structures similar to chloroplast RNA processing sites (reviewed in Gray *et al.*, 1992). In wheat and Oenothera, *in vitro* assays to study tRNA maturation have demonstrated the presence of specific endonucleolytic cleavages that remove 5′ and 3′ sequences and a tRNA nucleotidyltransferase activity that adds the 3′ terminal -CCA$_{OH}$ sequence (reviewed in Gray *et al.*, 1992). Such experiments suggest strong similarities with the yeast mitochondrial, bacterial and chloroplast systems and the involvement of an RNase P-like activity.

An unusual type of post-transcriptional maturation: RNA editing

The most intriguing form of RNA maturation in plant mitochondria is RNA editing. In 1989 it was discovered that certain specific cytidines encoded in the DNA are converted to uridines in the corresponding RNA molecule (Covello and Gray, 1989; Gualberto *et al.*, 1989; Hiesel *et al.*, 1989). Virtually all protein-coding sequences in flowering plant mitochondria are edited to some extent. Usually, editing occurs within the first or second position of a codon so that the encoded amino acid is changed; the amino acid alterations almost always result in increased similarity to homologous proteins from other organisms. For example, within *coxII*, if one examines the codons corresponding to the positions of highly-conserved cysteine and methionine residues that are believed to be involved in copper binding, it can be seen that the wheat DNA sequence predicts arginine and threonine at those positions (Fig. 12.3). However, at the RNA level, editing converts the corresponding CGU and ACG codons to UGU and AUG, respectively, so that the correct amino acids are specified. In maize mitochondria, only the ACG codon counterpart is edited to AUG,

DNA sequence wheat *coxII*	...	TGC	AGT	GAG	ATT	**c**GT	GGA	ACT	AAT	CAT	GCC	TTT	A**c**G ...
DNA-derived *COXII* protein		C	S	E	I	**R**	G	T	N	H	A	F	**T**
RNA sequence wheat *coxII*	...	UGC	AGU	GAG	AUU	UGU	GGA	ACU	AAU	CAU	GCC	UUU	AUG ...
RNA-derived *COXII* protein		C	S	E	I	**C**	G	T	N	H	A	F	**M**
maize		C	S	E	I	C	G	T	N	H	A	F	**M**
Oenothera		C	S	E	I	C	G	T	N	H	**A**	F	M
bovine		C	S	E	I	C	G	S	N	H	S	F	M
yeast		C	S	E	L	C	G	T	G	H	A	N	M
		*				*				*			*

Fig. 12.3 RNA editing in *coxII* mRNA within the region encoding the putative copper-binding (Cu$_A$) site. The region of DNA shown corresponds to residues 224–235 in wheat or 196–207 in bovine COXII (reviewed in Gray and Covello, 1993). Asterisks denote residues believed to be copper ligands. Editing sites are shown in lower case boldface letters and altered amino acids are also in boldface and underlined.

as the codon for cysteine is already genomically-encoded. In Oenothera, both amino acids are correctly specified by the mtDNA so that RNA editing is not required.

The number of RNA editing sites within a given transcript varies among plants and among genes within the same plant; up to 15% of the codons may be altered. In some cases, initiation or termination codons are created by RNA editing. Relatively few editing sites are located in non-coding regions, although a few have been reported in intron sequences; the latter may be important for proper folding of the intron for splicing. Almost all major groups of land plants, including angiosperms and gymnosperms, exhibit some degree of RNA editing in protein-coding transcripts and, in a few cases, in tRNAs; however, it has not been observed (or predicted) in the liverwort, *M. polymorpha*.

The mechanisms and machinery involved in C-to-U type RNA editing in plant mitochondria are as yet unknown, but data are consistent with a base modification system, such as cytosine deamination, analogous to the RNA editing of mammalian apolipoprotein transcripts. It is not yet known how the specific sites that require editing are recognized; no highly conserved motif at the primary sequence or secondary structural level has been found around editing sites. The analysis of incompletely edited RNA molecules suggests that there is no strict polarity to RNA editing in either the 5' or 3' direction, and that the process is a post-transcriptional event. However, it appears to be a relatively early step in RNA maturation, because spliced transcripts are usually completely edited. The C-to-U type of RNA editing in plant mitochondria appears to be fundamentally different from that in the mitochondria of the kinetoplastid protozoa, such as *Trypanosoma*, where uridines are inserted into (or occasionally deleted from) RNA molecules.

Translation

Relatively little is known about the translational process in plant mitochondria. In view of the strongly eubacterial character of plant mitochondrial rRNAs (Spencer *et al.*, 1984), it is anticipated that the translation system as a whole will exhibit eubacteria-like features. One such property is antibiotic sensitivity, with protein synthesis being strongly inhibited by chloramphenicol but resistant to cycloheximide (Forde *et al.*, 1978). The existence of a eubacteria-like 'Shine–Dalgarno' interaction between plant mitochondrial mRNAs and 18S rRNA has been proposed but not experimentally demonstrated (reviewed in Gray *et al.*, 1992). Plant mitochondrial mRNAs often contain 5' and 3' non-coding extensions more than several hundred nucleotides in length, and it is believed that these extensions may be involved in mRNA stability and/or translational regulation. This is analogous to the situation in yeast mitochondria, where RNA binding proteins play an important role in translational control (reviewed in Gillham *et al.*, 1994).

Unlike the case in bacteria, plant mitochondrial transcripts undergo complex processing events, such as RNA editing (and in some cases RNA splicing), prior to translation. It might be expected that immature transcripts would be excluded from taking part in protein synthesis because incorrect proteins would be made. However, analysis of maize *coxII* and *coxIII* (Yang and Mulligan, 1993) and petunia *atp6* (Lu and Hanson, 1994) transcripts, suggests that both mature and partially processed RNAs are present in the polysomal RNA fraction. On the other hand, no unedited-type ATP6 protein could be detected in petunia mitochondria when antibodies that distinguish between the edited and unedited forms were used (Lu and Hanson, 1994).

Cytoplasmic male sterility

Cytoplasmic male sterility (CMS) is a widely distributed, non-Mendelian trait that prevents the formation of viable pollen. Because CMS is valuable for the production of hybrid seed, the molecular biology of this phenomenon is being studied intensively. Although CMS has been reported in over 150 plant species, its cause has been investigated in only a few; such studies suggest that CMS is a complex phenomenon, with perhaps a number of different etiologies. However, substantial evidence does implicate the mitochondrial genome as the determinant of the CMS trait.

Here we will consider a specific case, involving the Texas (T-) cytoplasm of maize as an example of the alterations in mtDNA and mitochondrial function that are seen to accompany CMS and reversion from it. More comprehensive and detailed discussions of the molecular basis of CMS in this and other systems such as petunia, *Brassica* and sunflower can be found elsewhere (Hanson, 1991; Bonen and Brown, 1993;

Levings, 1993; Levings and Vasil, 1995).

The observation that CMS can result from interspecific crosses first suggested that the trait may be a consequence of some kind of incompatibility between the nuclear genome of one species and the cytoplasm of another. In maize, three types of CMS are recognized (*cms-T*, *cms-C* and *cms-S*), which are uniquely distinguished by nuclear restorer genes that suppress the cytoplasmically-determined CMS phenotype. These three maize types are also distinguished by characteristic differences in mtDNA restriction profiles, and by a characteristic spectrum of variant polypeptides synthesized by isolated mitochondria. In particular, an additional 13 kD polypeptide is made in *cms-T* mitochondria. Nuclear suppression of *cms-T*, resulting in fertility restoration, leads to suppression of the synthesis of the 13 kD polypeptide. Maize plants carrying the *cms-T* cytoplasm are preferentially susceptible to the fungal pathogen *Bipolaris maydis* race T (formerly called *Helminthosporium maydis*), which produces a toxin that selectively disrupts the functioning of *cms-T* mitochondria. It was this property that resulted in the Southern corn leaf blight epidemic in the late 1960s in the United States. Sensitivity to the toxin is strictly correlated with the presence of the variant 13 kD mitochondrial polypeptide: nuclear suppression of *cms-T* results in resistance to the *B. maydis* toxin, at the level of isolated mitochondria as well as whole plants. This linkage suggests that in *cms-T* mitochondria, a single defective gene specifies an altered mitochondrial membrane polypeptide that determines the disease susceptibility-male sterility syndrome.

When Dewey *et al.* (1986) sequenced the region of *cms-T* mtDNA that gives rise to abundant transcripts unique to *cms-T*, they identified an ORF, T-*urf13*, which was subsequently confirmed by immunochemical analysis to encode a 13 kD inner mitochondrial membrane polypeptide. T-*urf13* is a remarkable, mosaic gene that apparently originated from multiple rearrangement events that brought together portions of the flanking and/or coding regions of the maize mitochondrial 26S rRNA gene, the *atp6* gene and a promiscuous chloroplast tRNA-Arg gene. When T-URF13 protein is expressed in *E. coli* or yeast, the cells are sensitive to the fungal toxin and membrane properties are altered. Such studies provided direct evidence that the T-*urf13* gene is responsible for the fungal toxin susceptibility in T-cytoplasm maize (reviewed in Levings, 1993).

In *cms-T* maize plants that are restored to fertility by the joint action of the nuclear restorer genes, *Rf1* and *Rf2*, transcript profiles of T-*urf13* are altered and levels of the 13 kD polypeptide are significantly reduced. These changes are attributed solely to *Rf1*; nevertheless, the presence of *Rf2* is also needed for restoration of pollen fertility (reviewed in Levings, 1993). The *Rf2* nuclear restorer gene is being characterized at the molecular genetic level (Schnable and Wise, 1994). In male-fertile, toxin-resistant *cms-T* revertants obtained from maize plants regenerated after passage through tissue culture, the T-*urf13* region was almost always deleted. In one exception, the revertant had a 5 bp insertion within the T-*urf13* gene that caused a frameshift, resulting in a premature stop codon that truncates the T-URF13 gene product (reviewed in Levings, 1993). These genetic studies provide strong evidence that T-*urf13* is responsible for both male sterility and toxin sensitivity.

Why should a defect in mitochondrial function, apparently only manifested in the presence of a fungal toxin, specifically affect male reproductive development in maize? This fundamental question must still be settled. One speculation is that an anther-specific substance, mimicking the *B. maydis* toxin in its effects on *cms-T* mitochondria, may be produced during normal pollen development. In any event, an emerging theme in CMS is the rearrangement of plant mtDNA with the creation of novel functional regions. The products of nuclear restorer genes may then serve to modify transcripts produced from these novel functional regions, so as to minimize or eliminate their deleterious effects. Further detailed investigations of the expression of these chimeric gene regions seems certain to throw further light on the control of normal gene function in plant mitochondria.

Summary and future prospects

Our current picture of plant mtDNA is one of a structurally fluid genome containing much non-coding DNA and able to exist in many different

sequence contexts without major effects on the expression of individual genes. It is a mosaic of genetic information, and movement of genes from the chloroplast to the mitochondrion, and from the mitochondrion to the nucleus, appears to be an ongoing process. Maturation of RNAs requires complex processing events, including RNA editing and *trans*-splicing, but our knowledge of transcriptional and post-transcriptional processes is as yet rudimentary. Further understanding of these processes will be aided by *in vitro* functional assays. Detailed comparisons of closely and distantly related plant mtDNAs should continue to contribute insights into how plant mitochondrial genomes evolve, and the functional constraints that must temper the seemingly rampant rearrangements that characterize plant mtDNA. Finally, plants offer a unique opportunity to study developmental aspects of mitochondrial biogenesis and gene expression; these topics are currently attracting increased attention in the field.

References

Bendich, A. J. (1993). Reaching for the ring: the study of mitochondrial genome structure. *Curr. Genet.* **24**, 279–90.

Bonen, L. and Brown, G. G. (1993). Genetic plasticity and its consequences: perspectives on gene organization and expression in plant mitochondria. *Can. J. Bot.* **71**, 645–60.

Bonen, L. and Gray, M. W. (1980). Organization and expression of the mitochondrial genome of plants. I. The genes for wheat mitochondrial ribosomal and transfer RNA: evidence for an unusual arrangement. *Nucl. Acids Res.* **8**, 319–35.

Brennicke, A., Grohmann, L., Hiesel, R., Knoop, V. and Schuster, W. (1993). The mitochondrial genome on its way to the nucleus: different stages of gene transfer in higher plants. *FEBS Lett.* **325**, 140–45.

Covello, P. S. and Gray, M. W. (1989). RNA editing in plant mitochondria. *Nature* **341**, 662–6.

Covello, P. S. and Gray, M. W. (1992). Silent mitochondrial and active nuclear genes for subunit 2 of cytochrome *c* oxidase (*cox2*) in soybean: evidence for RNA-mediated gene transfer. *EMBO J.* **11**, 3815–20.

Dawson, A. J., Hodge, T. P., Isaac, P. G., Leaver, C. J. and Lonsdale, D. M. (1986). Location of the genes for cytochrome oxidase subunits I and II, apocytochrome b, alpha-subunit of the F_1 ATPase and the ribosomal RNA genes on the mitochondrial genome of maize (*Zea mays* L.). *Curr. Genet.* **10**, 561–4.

Dewey, R. E., Levings, C. S. III and Timothy, D. H. (1986). Novel recombinations in the maize mitochondrial genome produce a unique transcriptional unit in the Texas male-sterile cytoplasm. *Cell* **44**, 439–49.

Forde, B. G., Oliver, R. J. C. and Leaver, C. J. (1978). Variation in mitochondrial translation products associated with male-sterile cytoplasms in maize. *Proc. Natl Acad. Sci. USA* **75**, 3841–5.

Fox, T. D. and Leaver, C. J. (1981). The *Zea mays* mitochondrial gene coding cytochrome oxidase subunit II has an intervening sequence and does not contain TGA codons. *Cell* **26**, 315–23.

Gillham, N. W., Boynton, J. E. and Hauser, C. R. (1994). Translational regulation of gene expression in chloroplasts and mitochondria. *Ann. Rev. Genet.* **28**, 71–93.

Gray, M. W. (1992). The endosymbiont hypothesis revisited. *Int. Rev. Cytol.* **141**, 233–57.

Gray, M. W. and Covello, P. S. (1993). RNA editing in plant mitochondria and chloroplasts. *FASEB J.* **7**, 64–71.

Gray, M. W. and Spencer, D. F. (1983). Wheat mitochondrial DNA encodes a eubacteria-like initiator methionine transfer RNA. *FEBS Lett.* **161**, 323–7.

Gray, M. W., Hanic-Joyce, P. J. and Covello, P. S. (1992). Transcription, processing and editing in plant mitochondria. *Ann. Rev. Plant Physiol. Plant Mol. Biol.* **43**, 145–75.

Gualberto, J. M., Lamattina, L., Bonnard, G., Weil, J. H. and Grienenberger, J. M. (1989). RNA editing in wheat mitochondria results in the conservation of protein sequences. *Nature* **341**, 660–62.

Hanson, M. R. (1991). Plant mitochondrial mutations and male sterility. *Ann. Rev. Genet.* **25**, 461–86.

Hanson, M. R. and Folkerts, O. (1992). Structure and function of the higher plant mitochondrial genome. *Int. Rev. Cytol.* **141**, 129–72.

Hiesel, R., Wissinger, B., Schuster, W. and Brennicke, A. (1989). RNA editing in plant mitochondria. *Science* **246**, 1632–4.

Joyce, P. B. M. and Gray, M. W. (1989). Chloroplast-like transfer RNA genes expressed in wheat mitochondria. *Nucl. Acids Res.* **17**, 5461–76.

Levings, C. S. III (1993). Thoughts on cytoplasmic male sterility in *cms-T* maize. *Plant Cell* **5**, 1285–90.

Levings, C. S. III and Vasil, I. K. (eds) (1995). *The Molecular Biology of Plant Mitochondria*, Dordrecht, Kluwer Academic Press.

Lonsdale, D. M. (1984). A review of the structure and organization of the mitochondrial genome of higher plants. *Plant Mol. Biol.* **3**, 201–6.

Lonsdale, D. M., Hodge, T. P. and Fauron, C. M.-R. (1984). The physical map and organization of the mitochondrial genome from the fertile cytoplasm of maize. *Nucl. Acids Res.* **12**, 9249–61.

Lu, B. and Hanson, M. R. (1994). A single homogeneous form of ATP6 protein accumulates in petunia

mitochondria despite the presence of differentially edited *atp6* transcripts. *Plant Cell* **6**, 1955–68.

Makaroff, C. A. and Palmer, J. D. (1987). Extensive mitochondrial specific transcription of the *Brassica campestris* mitochondrial genome. *Nucl. Acids Res.* **15**, 5141–56.

Marechal-Drouard, L., Weil, J. H. and Dietrich, A. (1993). Transfer RNAs and transfer RNA genes in plants. *Ann. Rev. Plant Physiol. Plant Mol. Biol.* **44**, 13–32.

Nugent, J. M. and Palmer, J. D. (1991). RNA-mediated transfer of the gene *coxII* from the mitochondrion to the nucleus during flowering plant evolution. *Cell* **66**, 473–81.

Oda, K., Yamato, K., Ohta, E., Nakamura, Y., Takemura, M., Nozato, N., Akashi, K., Kanegae, T., Ogura, Y., Kohchi, T. and Ohyama, K. (1992). Gene organization deduced from the complete sequence of liverwort *Marchantia polymorpha* mitochondrial DNA. *J. Mol. Biol.* **223**, 1–7.

Palmer, J. D. and Herbon, L. A. (1987). Unicircular structure of the *Brassica hirta* mitochondrial genome. *Curr. Genet.* **11**, 565–70.

Palmer, J. D. and Shields, C. R. (1984). Tripartite structure of the *Brassica campestris* mitochondrial genome. *Nature* **307**, 437–40.

Rapp, W. D., Lupold, D. S., Mack, S. and Stern, D. B. (1993). Architecture of the maize mitochondrial *atp1* promoter as determined by linker-scanning and point mutagenesis. *Mol. Cell. Biol.* **13**, 7232–8.

Schnable, P. S. and Wise, R. P. (1994). Recovery of heritable, transposon-induced, mutant alleles of the *rf2* nuclear restorer of T-cytoplasm maize. *Genetics* **136**, 1171–85.

Schuster, W. and Brennicke, A. (1994). The plant mitochondrial genome: physical structure, information content, RNA editing, and gene migration to the nucleus. *Ann. Rev. Plant Physiol. Plant Mol. Biol.* **45**, 61–78.

Small, I., Suffolk, R. and Leaver, C. J. (1989). Evolution of plant mitochondrial genomes via substoichiometric intermediates. *Cell* **58**, 69–76.

Spencer, D. F., Schnare, M. N. and Gray, M. W. (1984). Pronounced structural similarities between the small subunit ribosomal RNA genes of wheat mitochondria and *Escherichia coli*. *Proc. Natl Acad. Sci. USA* **81**, 493–7.

Stern, D. B. and Lonsdale, D. M. (1982). Mitochondrial and chloroplast genomes of maize have a 12-kilobase DNA sequence in common. *Nature* **299**, 698–702.

Wissinger, B., Brennicke, A. and Schuster, W. (1992). Regenerating good sense: RNA editing and trans splicing in plant mitochondria. *Trends Genet.* **8**, 322–8.

Wolstenholme, D. R. and Jeon, K. W. (eds) (1992). Mitochondrial genomes. *Int. Rev. Cytol.* **141**, 1–357.

Yang, A. J. and Mulligan, R. M. (1993). Distribution of maize mitochondrial transcripts in polysomal RNA: evidence for non-selectivity in recruitment of mRNAs. *Curr. Genet.* **23**, 532–6.

13 Carbon metabolism in mitochondria

Steven A. Hill

Introduction

Mitochondrial carbon metabolism in plants, as in other eukaryotes, is dominated by the reactions of the tricarboxylic acid (TCA) cycle. Following the ground-breaking elucidation of the cycle in pigeon muscle by Krebs and co-workers (Krebs and Johnson, 1937), evidence has accumulated that the same reactions occur in plant cells (Beevers, 1961). In the thirty five years since this evidence was reviewed remarkably little progress has been made in establishing either the functions of the TCA cycle in plants, or the mechanisms whereby flux through the cycle is regulated. There are a number of reasons for this lack of progress; in particular, the extensive sub-cellular compartmentation of plant cells (ap Rees, 1987), and the difficulties in extracting enzymes and metabolites from plant tissues (ap Rees, 1980), present problems in studying not only the TCA cycle, but also other important pathways. Notwithstanding these difficulties, progress is now being made, and many opportunities are afforded by application of the techniques of molecular biology to study of the TCA cycle. My aim in this chapter is to review our current knowledge of the TCA cycle in plants. First, I will describe the enzymes of the cycle and their properties, and outline briefly the evidence supporting the operation of the TCA cycle in plants. Next, I will move on to the role of the TCA cycle in plant cells, concentrating on the interaction between the cycle and other metabolic pathways, and the emerging evidence for its key role in photosynthetic metabolism. Finally, I will discuss our current understanding of the control and regulation of the TCA cycle in plants.

The tricarboxylic acid cycle

Enzymes of the TCA cycle

The reactions of the TCA cycle are illustrated in Fig. 13.1. For the purposes of this discussion I have

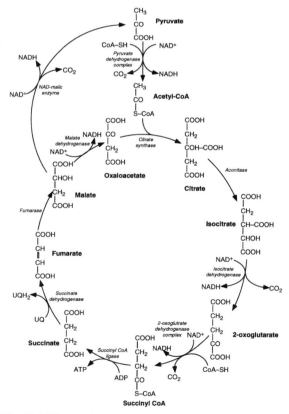

Fig. 13.1 The reactions of the tricarboxylic acid cycle.

also included two reactions that are not strictly involved in the TCA cycle, but are closely associated with it. Pyruvate dehydrogenase complex (PDC) is the major source of acetyl CoA required for TCA cycle operation, and is classically regarded as the entry point of glycolytically-derived carbon into the cycle. NAD-malic enzyme (NAD-ME) may also be significant route for carbon entry into the cycle, particularly when mitochondrial pyruvate uptake rates are insufficient to support respiration, or when malate is the primary respiratory substrate.

Pyruvate dehydrogenase complex

Entry of carbon into the cycle is catalyzed by pyruvate dehydrogenase complex, a large multienzyme complex that catalyzes the oxidative decarboxylation of pyruvate:

$$\text{Pyruvate} + NAD^+ + CoA \rightarrow \text{acetyl CoA} + NADH + H^+ + CO_2$$

The core of the complex consists of three enzymes. Pyruvate dehydrogenase (E1) (EC1.2.4.1) catalyzes the decarboxylation of pyruvate and the transfer of the remaining acetyl moiety to the cofactor thiamine pyrophosphate (TPP). This component consists of two subunits, E1α and E1β. The second component, lipoamide acetyltransferase (E2) (EC2.3.1.12), catalyzes the transfer of the acetyl moiety from TPP to coenzyme A, with the concomitant reduction of a covalently bound lipoamide cofactor. The final component, lipoamide dehydrogenase (E3) (EC1.8.1.4) reoxidizes the lipoamide residue, and converts NAD^+ into its reduced form, via a bound flavin adenine nucleotide cofactor. The reaction scheme for PDC is illustrated in Fig. 13.2.

Our knowledge of the molecular organisation of PDC in plants is by analogy with PDC from other eukaryotes. The bovine complex consists of a central core of 60 E2 subunits, with 20 E1 and 5 E3 subunits arranged around the core (Wieland, 1983). In addition to the E1, E2 and E3 components described above, the complex also contains two other associated enzyme activities, a kinase that phosphorylates the E1α subunit at specific sites (PDC kinase), and a phosphatase that dephosphorylates the protein (phospho-PDC phosphatase). These subunits are involved in the regulation of PDC activity by reversible phosphorylation. PDC from animal mitochondria also contains a protein, termed component X, that is believed to be involved in attaching the E3 component to the E2 core. Whilst the structure of plant PDC has not been determined, all the components have been detected immunologically (Taylor *et al.*, 1992) and cDNA clones have been isolated that encode E1α (Grof *et al.*, 1995), E1β (Luethy *et al.*, 1994) and E2 (Guan *et al.*, 1995). In addition, a cDNA clone encoding the lipoamide dehydrogenase from pea glycine decarboxylase complex has been isolated, and evidence presented that this gene also encodes the E3 subunit of PDC (Bourguignon *et al.*, 1992).

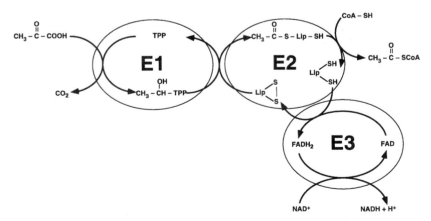

Fig. 13.2 The reactions catalyzed by pyruvate dehydrogenase complex. TPP, thiamine pyrophosphate; Lip, lipoamide residue; CoA-SH, co-enzyme A.

The regulatory properties of PDC from pea leaves have been studied in some detail by Randall and co-workers. PDC is subject to product inhibition – both NADH and acetyl CoA act as strong competitive inhibitors, competing with NAD^+ and CoA, respectively. The K_is for NADH and acetyl CoA are lower than the K_ms for NAD^+ and CoA, so that the rate of the pyruvate dehydrogenase reaction is sensitive to both the $NADH/NAD^+$ and acetyl CoA/CoA ratios. Studies with isolated mitochondria suggest that the effect of the acetyl CoA/CoA ratio is of particular significance (Budde *et al.*, 1991), though we should be cautious in extrapolating observations made *in vitro* to the *in vivo* situation. In common with animal PDC, the mitochondrial complex from plants is regulated by reversible phosphorylation. Phosphorylation of the E1α subunit by PDC-kinase leads to almost complete inactivation of PDC (Budde *et al.*, 1988); the activity is restored by dephosphorylation, catalyzed by phospho-PDC phosphatase (Miernyk and Randall, 1987). These enzymes bring about a dynamic cycle of phosphorylation and dephosphorylation, such that the activation state of PDC is determined by the relative activities of the kinase and phosphatase (Budde and Randall, 1988). PDC-kinase is itself inactivated by pyruvate; this effect is dependent on the presence of the cofactor thiamine pyrophosphate (Budde and Randall, 1987), and enhanced in the presence of ADP (Schuller and Randall, 1990). The activity of PDC-kinase is increased by micromolar concentrations of ammonium (Schuller and Randall, 1989). Phospho-PDC phosphatase requires Mg^{2+} ions for activity, but otherwise has no demonstrated regulatory properties (Miernyk and Randall, 1987). The regulatory properties of PDC are summarized in Fig. 13.3.

In addition to the mitochondrial isoform of PDC, plants also contain a plastidic isoform. The latter has a similar subunit composition to the mitochondrial enzyme, but is immunologically distinct, and is not regulated by reversible phosphorylation. This enzyme is believed to be involved in the provision of acetyl CoA for fatty acid biosynthesis.

NAD-malic enzyme

NAD-ME (EC1.1.1.39) is located exclusively in the mitochondria, and catalyzes the oxidative decarboxylation of malate:

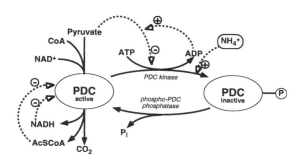

Fig. 13.3 The regulation of pyruvate dehydrogenase complex. The model is based on the work of Randall and coworkers. Dotted arrows indicate regulatory interactions, either positive or negative.

$$Malate + NAD^+ \rightarrow pyruvate + NADH + H^+ + CO_2$$

The enzyme from higher plants consists of two subunits, α and β, and exists in a configuration of (αβ), as either a dimer, tetramer or octamer (Artus and Edwards, 1985). Both the α and β subunits are required for activity (Willeford and Wedding, 1987b). Recently, cDNA clones encoding these subunits have been isolated from potato (Winning *et al.*, 1994), confirming the existence of two separate subunits. The subunits from potato have molecular weights of 59 and 62 kDa, and show 65% identity in their amino acid sequence (Winning *et al.*, 1994).

The kinetic properties of NAD-ME are complex. Activity of the enzyme is dependent on the presence of a divalent cation, either Mn^{2+} or Mg^{2+}. The nature of the divalent cation present affects both the V_{max} and K_m of the enzyme for both malate and NAD^+; for example, in the presence of Mn^{2+} both K_m(malate) and V_{max} are lower than in the presence of Mg^{2+} (Grover *et al.*, 1981). Sulphate ions, CoA and fumarate activate NAD-ME, whilst chloride is an inhibitor (Canellas and Wedding, 1984). Furthermore, these properties are affected by the aggregation state of the enzyme: the tetramer is more active than both the dimer and the octamer (Grover and Wedding, 1982), and *in vitro* the equilibrium position is shifted towards the tetramer in the presence of malate (Grover and Wedding, 1984). However, our understanding of the role of this enzyme is seriously reduced by lack of knowledge of its aggregation state *in vivo*. Kinetic parameters are also modulated by pH (Willeford and Wedding,

1987a), but again the physiological significance of these observations is unclear

Citrate synthase

Citrate synthase (EC4.1.3.7) catalyzes the aldol condensation of acetyl CoA and oxaloacetate to form citroyl CoA, which spontaneously hydrolyzes to form citrate and CoA. The protein has been purified from pea leaves, and consists of a single subunit of 50 kDa (Unger and Vasconcelos, 1989). This subunit, by comparison with citrate synthase from animal sources, is thought to exist as a homodimer of molecular mass 100 kDa. Citrate synthase is confined to the mitochondrial matrix, except in tissues converting fatty acids into sugars, where there is also a glyoxysomal isoform. cDNA clones encoding mitochondrial citrate synthase have been isolated from *Arabidopsis* (Unger *et al.*, 1989) and potato (Landschütze *et al.*, 1995a). Analysis of *Arabidopsis* genomic DNA suggests that there are two genes, but it is not clear whether both are mitochondrial isoforms, or one encodes the glyoxysomal isoform.

Aconitase

Aconitase (EC4.2.1.3) catalyzes the isomerization of citrate into isocitrate, via the bound intermediate cis-aconitate. Aconitase activity is found in both the mitochondrial and cytosolic compartments in plants, with approximately equal activities in each compartment (Brouquisse *et al.*, 1987). The enzyme appears to be absent from peroxisomes, suggesting that cytosolic aconitase is involved in the glyoxylate cycle (Courtois-Verniquet and Douce, 1993). The cytosolic and mitochondrial forms of the enzyme have been separated from *Acer* suspension cells, and have very similar kinetic and physical properties (Brouquisse *et al.*, 1987). Etiolated pumpkin cotyledons contain three isoforms, again with very similar properties, although in this case it is not clear with which compartment each form is associated (De Bellis *et al.*, 1993). Two aconitase isoforms with the same N-terminal amino acid sequence have been isolated from melon cotyledons (Peyret *et al.*, 1995). Using antibodies raised against the proteins from melon, cDNA, and subsequently genomic clones encoding aconitase have been isolated from *Arabidopsis*. This analysis has revealed that in *Arabidopsis* there is only one gene encoding aconitase, raising the possibility that a single gene encodes the mitochondrial and cytosolic isoforms (Peyret *et al.*, 1995). Confirmation that this is the case awaits demonstration that there is more than one isoform of aconitase in *Arabidopsis*. In yeast, there is a single gene that encodes both the cytosolic and mitochondrial isoforms – it has been suggested that import of the gene-product into mitochondria is inefficient, so that some activity remains in the cytosol (Gangloff *et al.*, 1990). Aconitase consists of a single subunit of 90–100 kDa containing an Fe-S cluster in the active site (Verniquet *et al.*, 1991; De Bellis *et al.*, 1993). This cluster has distinct properties from that found in mammalian aconitase and may be either an Fe_3S_4 cluster, or the more normal Fe_4S_4 cluster with an associated Fe centre (Jordanov *et al.*, 1992). Aconitase shows Michaelis–Menton kinetics with respect to citrate: the K_ms for the *Acer* enzymes are 130 and 120 μM for the mitochondrial and cytosolic forms, respectively. In contrast, two of the three forms found in pumpkin cotyledons have K_ms of 480 and 370 μM.

Isocitrate dehydrogenase

Isocitrate dehydrogenase (ICDH) catalyzes the oxidative decarboxylation of isocitrate:

$$Isocitrate + NAD(P)^+ \rightarrow 2\text{-}oxoglutarate + NAD(P)H + H^+ + CO_2$$

Plants contain both NAD^+- and $NADP^+$-specific isoforms of isocitrate dehydrogenase (EC1.1.1.41 and EC1.1.1.42, respectively), with the NAD^+ form being specific to the mitochondria. $NADP^+$-specific forms are found in the plastid and the cytosol, and recent evidence suggests that there is also a mitochondrial isoform. An $NADP^+$-specific form co-purifies with mitochondria isolated from potato tubers and has been shown to be contained within the mitochondrial matrix (Rasmusson and Møller, 1990). This enzyme can be separated from the NAD^+-specific form by ammonium sulphate fractionation, demonstrating that this activity represents a distinct enzyme rather than a lack of specificity of the NAD^+-dependent form. The activity of the $NADP^+$-specific enzyme is around 15% of that of the NAD^+-specific form in potato tuber mitochondria (Rasmusson and Møller,

1990), while mitochondria isolated from pea leaves contain equivalent activities of the NAD^+ and $NADP^+$ forms (McIntosh and Oliver, 1992). It has been suggested that the $NADP^+$-dependent form is involved in the provision of NADPH required for the reduction of glutathione, that protects against oxidative damage in mitochondria (Rasmusson and Møller, 1990).

The kinetics of NAD^+-ICDH are sigmoidal with respect to isocitrate, with $S_{0.5}$ of 0.2 and 0.025 mM for the potato and pea enzymes, respectively. The kinetics with respect to NAD^+ are Michaelis–Menton, and the K_ms are 0.1 mM for the potato enzyme, and between 0.2–0.8 mM for the pea leaf enzyme. The enzyme isolated from pea leaf mitochondria demonstrates product inhibition by NADH ($K_i = 0.2$ mM), and is also inhibited by NADPH ($K_i = 0.3$ mM) (McIntosh and Oliver, 1992). The latter inhibition is not competitive with NAD^+ or NADH, suggesting that NADPH may be an allosteric regulator; the K_i is within the physiological range of mitochondrial NADPH concentrations (Wigge et al., 1993). In contrast to the enzyme from non-plant sources, the activity of which is affected by ATP, ADP and AMP, pea NAD^+-ICDH is unaffected by adenylates (McIntosh and Oliver, 1992). The $NADP^+$-dependent ICDH from potato tuber mitochondria shows Michaelis–Menten kinetics with respect to both $NADP^+$ ($K_m = 5$ μM) and isocitrate ($K_m = 10$ μM) (Rasmusson and Møller, 1990). The K_m for isocitrate is low relative to the $S_{0.5}$ of NAD^+-ICDH, but this may not be of physiological significance; in isolated mitochondria the isocitrate concentration is of the order of 2 mM (MacDougall and ap Rees, 1991), sufficient to saturate both forms of ICDH.

NAD^+-ICDH has been purified to homogeneity from pea leaves (McIntosh and Oliver, 1992). It consists of a single subunit of molecular mass 47 kDa. The freshly isolated enzyme is found in three forms corresponding to an octamer (the major active species), and complexes of two and four octamers. If the preparation is freeze-thawed, NAD^+-ICDH disaggregates forming a tetramer and dimer, which both retain activity. This disaggregation correlates with decreased sigmoidicity of the isocitrate kinetics, and can be reversed by the addition of citrate.

2-oxoglutarate dehydrogenase complex

2-oxoglutarate dehydrogenase (2-OGDC) (formally known as α-ketoglutarate dehydrogenase) catalyzes a reaction that is essentially analogous to that carried out by PDC:

2-oxoglutarate $+ NAD^+ + CoA \rightarrow$ succinyl CoA $+$ NADH $+ H^+ + CO_2$

The complex from animal mitochondria has a similar structure and organization to PDC, differing primarily in the nature of the E1 subunit. In particular, 2-OGDC is not regulated by reversible phosphorylation. As mentioned previously, there is evidence to suggest that the E3 subunit (lipoamide dehydrogenase) is shared between PDC, 2-OGDC and glycine decarboxylase in plants (Bourguignon et al., 1992). The enzyme has been purified from cauliflower florets, and has a native molecular mass of around 2000 kDa (Karam and Bishop, 1989). In common with other TCA cycle dehydrogenases, 2-OGDC is inhibited by NADH, which acts competitively with NAD^+ (Pascal et al., 1990). The enzyme partially purified from potato tubers has a higher K_m for NAD^+ than the K_i for NADH, suggesting that the activity is particularly sensitive to the NADH/NAD ratio.

Succinyl-CoA ligase

Succinyl-CoA ligase (EC6.2.1.5) catalyzes the hydrolysis of succinyl CoA, and the concomitant synthesis of ATP:

Succinyl CoA $+$ ADP $+ P_i \rightarrow$ succinate $+$ ATP

Little is known of the plant enzyme. All succinyl-CoA ligases examined consist of two types of subunit: α, with a molecular mass of 29–34 kDa, and β, with a molecular mass of 41–45 kDa. The eukaryotic enzyme is an αβ dimer. There is now evidence that mammalian cells contain two ligases, one specific for ADP and one for GDP, and that the latter catalyzes the synthesis of succinyl CoA during ketone body metabolism. Thus we should expect plants to contain only the ADP-specific enzyme and this expectation holds for the few plant enzymes studied (Palmer and Wedding, 1966).

Succinate dehydrogenase

Succinate dehydrogenase (EC1.3.99.1) is a component of both the TCA cycle and, as complex II, the mitochondrial electron transport chain. It catalyzes the conversion of succinate into fumarate, with the hydrogen atoms generated being passed to ubiquinone (see Chapter 14 for details).

$$\text{Succinate} + \text{UQ} \rightarrow \text{fumarate} + \text{UQH}_2$$

Electrons and hydrogen ions are passed to ubiquinone via a bound flavin adenine dinucleotide (FAD) cofactor. The enzyme isolated from sweet potato root tissue exists in two forms of identical molecular masses but with distinct charges (Hattori and Asahi, 1982). Both forms consist of two subunits of molecular mass 65 and 26 kDa. The former contains covalently bound FAD. By analogy with the enzyme purified from animal systems, it is likely that a number of smaller subunits are weakly associated with the two large subunits, and that the 26 kDa subunit spans the inner mitochondrial membrane. The K_m for the sweet potato enzyme for succinate is 290 μM.

Fumarase

This enzyme (EC4.2.1.2) catalyzes the addition of water across the double bond of fumarate, forming malate. The addition is stereospecific, so that only the L-form of malate is produced. Little of the enzyme from plants is known, except that it is confined to the mitochondrial matrix. The enzyme from pig heart is a tetramer of identical subunits, each of molecular mass 48.5 kDa.

Malate dehydrogenase

The cycle is completed by malate dehydrogenase (MDH) (EC1.1.1.37), which catalyzes the oxidation of malate to form oxaloacetate. The equilibrium position favours malate and NAD^+, but *in vivo* the removal of oxaloacetate by citrate synthase, and NADH by the respiratory chain causes the reaction to function in the direction of malate oxidation in most tissues. In addition to the mitochondrial forms there are isozymes of NAD^+-MDH found in the cytosol and peroxisomes, and an $NADP^+$-utilizing form found in plastids. The mitochondrial enzyme

isolated from watermelon is a homodimer of 38 kDa subunits (Walk *et al.*, 1977). Genetic studies have indicated that there are three genes encoding mitochondrial subunits of malate dehydrogenase in maize (Newton and Schwartz, 1980). This is supported by the purification of six distinct isoforms of MDH from maize mitochondria; three subunits, with molecular masses of 37, 38 and 39 kDa, have been shown to exist in the three possible homodimers, and the three possible heterodimers (Hayes *et al.*, 1991). The majority of the MDH activity is found in enzymes containing the 39 kDa subunit, suggesting that this subunit plays the major catalytic role *in vivo*. Detailed kinetic analysis of these isoforms has yet to be carried out and it is not yet clear whether the complexity observed in maize is found in other species.

Evidence for the operation of the TCA cycle in plants

There is ample evidence that the TCA cycle operates in plant cells (Beevers, 1961), so only the central pieces of evidence are considered here.

First, it has been demonstrated, for a wide range of plant tissues, that the required enzymes are present at activities well in excess of the observed rates of respiration. There is a lack of quantitative information concerning the maximum catalytic activities of the enzymes within the mitochondria, however, since many of the enzymes have counterparts outside mitochondria and measurement of the total activity does not yield the activity functioning in the TCA cycle. Determination of activities in isolated mitochondrial fractions only gives a measure of the total mitochondrial activity in the tissue if the yield of mitochondria is taken into account. This approach has been applied to extracts and mitochondria isolated from cauliflower florets and the developing club of *Arum maculatum* (MacDougall and ap Rees, 1991). This work has demonstrated that most of the enzymes of the TCA cycle are indeed present at sufficient activities for the observed rate of respiration, but that the observed NAD^+-ICDH activity is only just sufficient, suggesting that this enzyme may be saturated *in vivo*, or that the $NADP^+$-specific form contributes to the flux.

The second line of evidence that the TCA cycle operates in plants is the demonstration that plant tissues contain all the intermediates of the cycle. Once again, there is a lack of adequately authenticated quantitative data, but that which is available indicates that many of the intermediates, particularly malate and citrate, are present in very much larger amounts than are found in animal tissues (MacLennan *et al.*, 1963; ap Rees *et al.*, 1981). These large pools of organic acids are stored in the vacuole, so that the mitochondrial pools are in fact much smaller. This is clearly demonstrated by the classic experiments of MacLennan *et al.* (1963), who supplied ^{14}C-labelled acetate to a range of plant tissues and showed that the specific activity of the CO_2 released exceeded that of the acids extracted from the tissue. Recent studies using *in vivo* ^{13}C nuclear magnetic resonance spectroscopy have reinforced the degree of compartmentation of organic acids in plant tissues (Gout *et al.*, 1993).

The final line of evidence supporting the operation of the TCA cycle in plants is the metabolism of labelled substrates. A wide range of tissues have been supplied with TCA cycle acids, and it has been demonstrated that the correct intermediates are labelled in the expected sequence, and with the predicted intramolecular distribution of label (e.g. Harley and Beevers, 1963).

Thus, for the majority of tissues it is clear that the TCA cycle operates. The exception to this is photosynthetic cells that are being illuminated – the role of the TCA cycle in photosynthesizing leaves remains a matter of some debate.

The substrates for the TCA cycle in plants

Classically, the TCA cycle is considered as a catabolic pathway into which are fed the products of the breakdown of various respiratory substrates, that in animal cells can be derived from proteins, lipids or carbohydrates. All of these potential substrates feed into the cycle and are oxidized. In plants, the situation is very different. Because of their autotrophic nature, plants synthesize their own respiratory substrates, mainly carbohydrates, such as sucrose and starch. As a consequence, the contribution of acetyl CoA and other compounds derived from breakdown of protein and lipid to the TCA cycle is relatively small.

Protein and amino acids

There is no definitive evidence in support of large-scale respiration of amino acids in plants. Indeed when radiolabelled TCA cycle intermediates are supplied to plant tissues there is generally significant incorporation of label into amino acid pools (Fletcher and Beevers, 1971), suggesting that the cycle is involved in the accumulation and interconversion of amino acids rather than their oxidation. In most instances where there is net breakdown of proteins, such as germination or senescence, the amino acids produced are translocated elsewhere rather than being oxidized. An exception to this may be during extreme starvation as is observed when tissue culture cells are deprived of substrate, or root tips are excised from the plant. In this case, oxidation of amino acids occurs when carbohydrate reserves have been depleted (Roby *et al.*, 1987; Brouquisse *et al.*, 1991). Whether this ever occurs naturally has yet to be established.

During germination of many seeds there is significant breakdown of storage proteins, the amino acids from which are required for seedling-growth. In general, the amino acid composition of storage proteins differs from that of newly-synthesized proteins, so that the carbon skeletons of the amino acids are interconverted – this occurs in part via the TCA cycle. During these interconversions, some carbon from the amino acids is inevitably oxidized to CO_2, but this does not represent significant respiration of amino acids. It is merely a necessary result of their metabolism.

There are two instances where there is a significant contribution of amino acid oxidation to respiration. The first is the metabolism of proline. This amino acid accumulates in plant tissues in response to drought or cold stress, then disappears rapidly when the stress is removed (Stewart and Voetberg, 1985). Proline is oxidized via a three-step process that leads ultimately to 2-oxoglutarate that feeds into the TCA cycle. The three enzymes involved, proline oxidase, Δ'-pyrroline 5-carboxylic acid dehydrogenase and NAD^+-dependent glutamate dehydrogenase are localized within the mitochondria. In barley leaves recovering from drought, the oxidation of proline could account for up to 20% of the rate of respiration (Stewart and Voetberg, 1985). The second amino acid that is oxidized at appreciable rates in plant mitochondria is glycine, produced during photorespiration (see Chapter 22 for details). Two molecules of glycine are converted to one of serine, CO_2 and NH_3, with the production of NADH. Although this process does not involve the reactions of the TCA cycle there are likely to be interactions between the oxidation of glycine and the cycle since both share the same pool of NAD^+, and will compete for the NADH-oxidizing capacity of the respiratory chain.

Lipid

In plants, lipid functions as a store of carbon rather than a store of energy. In consequence, when lipid is broken down it is converted to sugars for transport, and is only rarely a respiratory substrate. In lipid storage organs the reactions of β-oxidation are localized not in the mitochondria, but in the peroxisomes, so that the acetyl-CoA produced does not have direct access to the TCA cycle (Gerhardt, 1986). This results in the near quantitative conversion of lipid into sugars during the germination of fat-storing seeds. Although lipid is not oxidized via the TCA cycle, part of the cycle plays a crucial role in the conversion of lipid into sugars via the glyoxylate cycle, as illustrated in Fig. 13.4. The conversion of

succinate to oxaloacetate, required for the operation of the glyoxylate cycle, occurs in mitochondria and is catalyzed by a subset of TCA cycle enzymes (Mettler and Beevers, 1980). Recently, the dogma that lipid does not feed directly into the TCA cycle in plants has been challenged by the demonstration that mitochondria from some tissues do contain the enzymes of the β-oxidation cycle. For example, in germinating pea cotyledons the capacity for β-oxidation is divided equally between the peroxisomes and the mitochondria (Wood et al., 1986). The evidence for this is two-fold. First, careful subcellular fractionation of extracts from a number of tissues has demonstrated that enzymes of the β-oxidation cycle co-purify with mitochondrial markers. That these enzymes are contained within the mitochondria rather than associated with the outer surface is confirmed by their latency and protection from proteolysis. Second, supply of fatty acids to pure preparations of intact mitochondria results in rapid rates of oxygen uptake (Thomas et al., 1988). The mechanism whereby fatty acids cross the mitochondrial membranes remains unclear. The presence of an acyl-carnitine carrier in the inner membrane, analogous to that found in animal mitochondria has been proposed; the main line of evidence to support this is the stimulation of malate

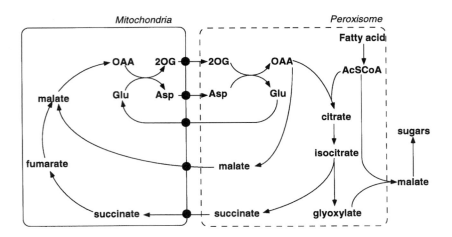

Fig. 13.4 Interactions between the mitochondria and peroxisomes in germinating oilseeds. The reaction scheme is based on that proposed by Mettler and Beevers (1980). The peroxisome membrane is shown as a dotted line, since it is believed that this membrane is freely permeant to small molecules (Heupel et al., 1991). The aconitase reaction is shown occurring in the peroxisome for convenience; in fact this reaction takes place in the cytosol (Courtois-Verniquet and Douce, 1993).

oxidation in isolated mitochondria by the addition of acyl-carnitines. However, the demonstration that acyl-carnitines act to uncouple the inner mitochondrial membrane, thereby bringing about an increase in the rate of respiration, has cast doubt on this evidence (Gerhardt *et al.*, 1995). Consequently, the significance of mitochondrial β-oxidation in plants remains unknown, but two observations suggest that the contribution of lipid to respiration is, at most, minor. These observations are: first, that the respiratory quotient of plant tissues (rate of CO_2 production/rate of O_2 consumption) is generally close to one – a lower figure is expected for the respiration of lipid; and second, that, with the exception of lipid mobilizing seeds, the maximum catalytic activities of the enzymes of β-oxidation are much lower than those of the TCA cycle.

Carbohydrate

It is well established that carbohydrate is the main source of substrate for the TCA cycle. Carbohydrate is metabolized via glycolysis and the oxidative pentose phosphate pathway, ultimately yielding phospho(enol)pyruvate (PEP). In plants this PEP can have one of two fates; either it is converted to pyruvate by the action of pyruvate kinase, or it is carboxylated by the action of PEP carboxylase. The latter reaction yields oxaloacetate, which is converted to malate by cytosolic malate dehydrogenase. Pyruvate and malate are the substrates that enter the mitochondria, but the capacity of the pyruvate transporter is relatively low, and under some circumstances may be insufficient (Hill *et al.*, 1994). The physiological relevance of import of malate into mitochondria is demonstrated by the work of ap Rees *et al.* (1983), who showed that butylmalonate, a specific inhibitor of malate uptake into mitochondria, causes a reduction in the rate of respiration when injected into thermogenic spadicies of *Arum maculatum*. Because of the presence of NAD^+-malic enzyme, plant mitochondria can catalyze full TCA cycle operation with malate as the only substrate. Evidence that malate import can substitute for pyruvate uptake into mitochondria comes from the observation that transgenic tobacco plants without any detectable cytosolic pyruvate kinase are

indistinguishable from the wild-type (Gottlob-McHugh *et al.*, 1992). Thus we should consider the products of glycolysis in plants to be malate and pyruvate, with these acids being the major substrates for the TCA cycle (Fig. 13.5).

The TCA cycle and biosynthesis

Removal of intermediates from the cycle and anaplerotic carbon fixation

Following the supply of labelled pyruvate to plant tissues, much of the label is recovered as CO_2, indicating that this pyruvate is metabolized via the TCA cycle (Neal and Beevers, 1960). However, careful inspection of the results of such experiments reveals that a significant proportion of the label is recovered in other compounds, primarily organic acids and the amino acids that derive from them. It is also regularly observed that plant tissues accumulate organic acids. These data suggest that, far from being an exclusively catabolic pathway, the TCA cycle in plants has an important role in biosynthesis. As well as the production of amino acids and organic acids themselves, these compounds are also the precursors for many biosynthetic pathways. Thus the role of the TCA cycle in plants differs fundamentally from that in animals: in heterotrophic organisms the cycle is the point of *convergence* of catabolic pathways, whereas in autotrophic organisms it is a point of *divergence* of anabolic pathways. A consequence of the anabolic role of the TCA cycle is that intermediates are constantly being removed from it. This poses a problem, since, provided there is a supply of acetyl-CoA, the TCA cycle can catalyze interconversion of its intermediates, but cannot catalyze the net synthesis of any of the intermediates. If intermediates are removed, the cycle will become depleted of organic acids and will ultimately stop through lack of oxaloacetate. There must be a means whereby at least one of the intermediates of the cycle can be replenished through the operation of so-called anaplerotic (literally 'filling up') pathways. In plants this anaplerotic role is provided by the synthesis of oxaloacetate from PEP by PEP carboxylase. The oxaloacetate can be transported into the mitochondria directly, as malate produced by cytosolic malate dehydrogenase, or as aspartate

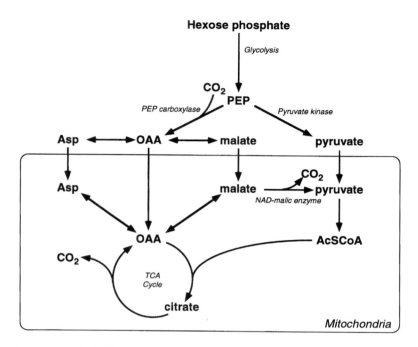

Fig. 13.5 The entry of carbon into the TCA cycle.

following transamination (Fig. 13.5). Evidence that anaplerotic carbon-fixation occurs in plants comes mainly from the demonstration that all plant tissues so far investigated incorporate label from $^{14}CO_2$ into organic acids in the dark. Initially this label is confined to oxaloacetate, malate and aspartate, but further metabolism leads to labelling of all the intermediates of the TCA cycle, and many amino acids (Nesius and Fletcher, 1975). The degree of dark CO_2 fixation correlates well with the demand for carbon skeletons derived from the TCA cycle. For example, in lupin root-nodules the rate of CO_2 fixation increases three-fold at the onset of nitrogen fixation.

Partitioning of C_4 acids

In some cases the C4 acids produced by CO_2 fixation in the cytosol do not enter the TCA cycle. For example, when the malate pools of maize root tips are simultaneously labelled using ^3H-acetate (which labels the mitochondrial pool) and ^{14}C-bicarbonate (which labels the cytosolic pool) the rate of metabolism of

the two pools is significantly different (Lips and Beevers, 1966). Whilst ^3H is lost rapidly from malate, the ^{14}C-malate is much more stable indicating that the latter has moved directly into storage without entering the mitochondria. Similarly, in ripening banana fruit, the labelled malate produced during dark fixation of $^{14}CO_2$ is stable during a subsequent chase in cold CO_2 (Hill and ap Rees, 1994).

Strict separation of cytosolic and mitochondrial organic acid pools does not, however, appear to be the norm. In many tissues citrate is labelled rapidly from $^{14}CO_2$, which can only occur via mitochondrial citrate synthase. In addition, Bryce and ap Rees (1985) demonstrated, for a range of tissues, that much of the $^{14}CO_2$ fixed into C4 acids during dark fixation is lost rapidly as $^{14}CO_2$ during a subsequent chase. From the distribution of label in these experiments, it was concluded that malate entering the mitochondria was decarboxylated via NAD-ME to yield pyruvate. This may represent an important route of respiratory carbon into the cycle. The partitioning of malate between malate dehydrogenase and NAD-ME is of crucial importance, since if malate is metabolized via

NAD-ME this does not result in replenishment of cycle intermediates for biosynthesis. The distribution of malate between these routes would be expected to vary according to tissue type and developmental state, and may be determined by regulation of NAD-ME, the activity of which is potentially modulated via changes in its aggregation state. However, the physiological significance of this is unknown; we do not know the aggregation state *in vivo*, whether this state changes, or what signals may bring about such changes. Demonstration that transgenic potato plants expressing antisense NAD-ME genes appear to grow normally despite having as little as 30% of the wild type NAD-ME activity suggests that this enzyme is subject to regulation *in vivo* (B. M. Winning, C. J. Leaver and S. A. Hill, unpublished).

Synthesis of acetyl-CoA

Acetyl-CoA is the starting point for biosynthesis of fatty acids in the plastids and isoprenoids in the cytosol. There are two hypotheses to explain supply of acetyl-CoA for plastidic fatty acid biosynthesis. At least some of the acetyl-CoA requirements of the plastid are met by the action of plastidic PDC (Fig. 13.6). Pyruvate is either imported from the

cytosol or synthesized within the plastid via glycolysis. This is demonstrated by the ability of isolated plastids to synthesize fatty acids from pyruvate, and, in some cases, hexose phosphates (Kang and Rawsthorne, 1994). The demonstration that plant mitochondria contain an acetyl-CoA hydrolase activity, and that plastids contain an acetyl-CoA synthase activity has led to the suggestion that acetate, which freely crosses membranes, is transferred between these organelles (Fig. 13.6) (Zeiher and Randall, 1990). This hypothesis is weakened, but by no means disproved, by the lack of evidence that free acetate occurs in plant cells.

Supply of acetyl CoA for cytosolic isoprenoid synthesis probably occurs by the export of citrate from the mitochondria (Fig. 13.6). It is converted to acetyl-CoA in the cytosol by the action of the enzyme ATP citrate lyase (Kaethner and ap Rees, 1985):

$$ATP + citrate + CoA \rightarrow acetyl\text{-}CoA + ADP + P_i + oxaloacetate$$

The TCA cycle in illuminated photosynthetic cells

The operation and role of the TCA cycle in illuminated leaves has been, and remains, an area of

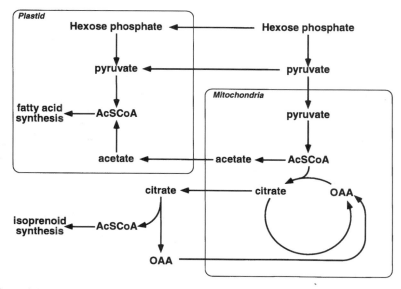

Fig. 13.6 Routes of acetyl-CoA production in plant cells.

considerable controversy (Graham, 1980; Krömer, 1995). The question is of importance because it represents a significant gap in our knowledge of leaf metabolism, and consequently places a serious limitation on our ability to manipulate rationally photosynthesis. Using oligomycin concentrations that inhibit mitochondrial oxidative phosphorylation without affecting chloroplast ATP synthesis, it has been clearly demonstrated that the mitochondrial electron transport chain functions in illuminated leaf cells, providing much of the ATP required for sucrose synthesis (Krömer and Heldt, 1991). The source of reducing equivalents within the mitochondria that support oxidative phosphorylation in the light remains unclear. Reducing equivalents could be provided by oxidation of photorespiratory glycine, oxidation of cytosolic NAD(P)H by the externally-facing mitochondrial dehydrogenases, operation of the TCA cycle or a combination of all three sources.

Support for the suggestion that glycine oxidation provides reducing equivalents for oxidative phosphorylation comes from the demonstration that when photorespiration is inhibited (either using high CO_2 concentrations or aminoacetonitrile (AAN)), there is a fall in both the mitochondrial and cytosolic ATP/ADP ratios (Gardeström and Wigge, 1988). Moreover, Randall and co-workers have shown that on illumination of pea leaves, PDC is rapidly inactivated due to phosphorylation (Budde and Randall, 1990). They have suggested that this is primarily due to the stimulation of PDC-kinase by ammonium ions produced during glycine oxidation, since inhibition of photorespiration (with inhibitors or high CO_2 concentrations) reduces the inactivation of PDC (Gemel and Randall, 1992). However, inhibition of mitochondrial oxidative phosphorylation with oligomycin affects photosynthesis similarly at both high and low CO_2 concentrations (Krömer et al., 1993), which implies that reducing equivalents generated by the TCA cycle can replace those derived from glycine oxidation. Notwithstanding the properties of PDC, the magnitude of the flux through the TCA cycle during photosynthesis under ambient conditions remains an open question. Indeed, it is not clear that light-dependent inactivation of PDC is common to all species (Krömer et al., 1994).

There are two lines of evidence that support operation of the TCA cycle in illuminated leaf cells.

1. McCashin et al. (1988) investigated the metabolism of ^{14}C-succinate by wheat leaf slices in the light and dark. They chose succinate because the content of this TCA cycle acid in leaves is generally low, so that high specific activities in vivo can be achieved, and, unlike most TCA cycle acids in plants there is a single, mitochondrial pool. By examining carefully the fate of supplied ^{14}C they were able to conclude that the TCA cycle between succinate and oxaloacetate operates in the light at a rate approximately 80% of that observed in the dark. They also investigated the metabolism of ^{14}C-acetate, and drew similar conclusions concerning the TCA cycle between acetyl-CoA and succinate.

2. Hanning and Heldt (1993) investigated the TCA cycle flux in isolated spinach mitochondria, taking into account the subcellular contents of mitochondrial substrates and other key metabolites. The latter were determined following non-aqueous fractionation of freeze-stopped leaf material (Gerhardt et al., 1987; Winter et al., 1994). They were able to rule out a significant role for cytosolic NAD(P)H oxidation in mitochondrial oxidative phosphorylation, since the cytosolic concentrations of reduced pyridine nucleotides were insufficient to allow a significant flux through the externally-facing dehydrogenases. Their work confirmed the earlier results of Wiskich et al. (1990) who suggested that some of the reducing equivalents generated by glycine oxidation are exported from the mitochondria as malate; this occurs because of malate dehydrogenase can operate in the reverse direction. Under simulated physiological conditions there is evidence in support of partial TCA cycle activity in isolated mitochondria (Hanning and Heldt, 1993). It is suggested that, during photosynthesis, oxaloacetate is imported into the mitochondria and that citrate is exported to support cytosolic nitrogen assimilation (Fig. 13.7). Some of the reducing equivalents generated within the mitochondria are exported to the cytosol either as malate or citrate, the remainder being used to support oxidative phosphorylation. This hypothesis provides a role for the specific oxaloacetate transporter that is found in the inner membrane of plant mitochondria (Ebbighausen et al., 1985).

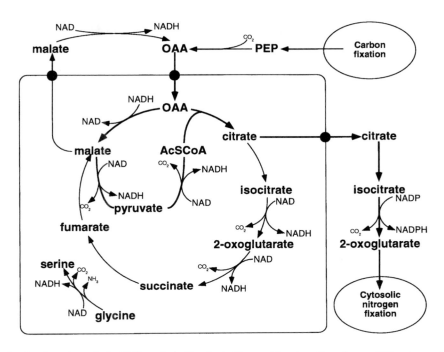

Fig. 13.7 The path of carbon through the TCA cycle in illuminated leaves. Based on the scheme of Hanning and Heldt (1993).

Regulation of flux through the TCA cycle in plants

As can be seen from the previous discussion, the TCA cycle in plants plays a crucial role in both anabolic and catabolic pathways. The role of the cycle varies between tissues and, at least in the case of leaves, can potentially vary diurnally. Thus we might expect that the cycle is tightly regulated, in response to both developmental and environmental signals. Despite this expectation, and the undoubted importance of this pathway, there is remarkably little concrete information concerning either the control or regulation of the TCA cycle in plants. In this section, I will concentrate on three areas: first, I will discuss the growing body of evidence to suggest that TCA cycle flux may be modulated by variation in the maximum catalytic activity of TCA cycle enzymes; second, I will review the little that is known about the control of the TCA cycle; and, third, I will suggest a general hypothesis for regulation of TCA cycle flux.

Modulation of TCA cycle flux by changes in enzyme activities

Many important developmental stages of the plant life cycle are accompanied by increases in the rate of respiration, often associated with increased biosynthesis of cellular constituents. Examples include germination, flower development, fruit ripening and senescence. Many of these are associated with increases in the maximum catalytic activity of respiratory enzymes, including those of the TCA cycle. For example, during the development of the spadix of *Arum maculatum*, the maximum catalytic activity of several TCA cycle enzymes increases in preparation for the massive increase in respiration that accompanies thermogenesis (MacDougall and ap Rees, 1991). These increases are necessary to sustain the respiratory rate that is observed. There is also some evidence to suggest that the TCA cycle is regulated in this way during post-germinative growth, and flower development.

Post-germinative growth

During the post-germinative growth of lipid-storing seeds mitochondria play an important role in the conversion of fatty acids to sugars. This involves the operation of part of the TCA cycle converting succinate to oxaloacetate (see Fig. 13.4). In most cases there is very little loss of carbon as CO_2 during this conversion, suggesting that the remainder of the TCA cycle (oxaloacetate + pyruvate to succinate) operates at a much reduced rate. Studies of the properties of isolated mitochondria and measurement of maximum catalytic activities of TCA cycle enzymes from fat-mobilizing tissues provide evidence that the reduced flux through the decarboxylating portion of the cycle is due to low maximum catalytic activities of the enzymes. Early during post-germinative growth there is an increase in the maximum catalytic activity of enzymes from the non-decarboxylating portion of the cycle, that is not accompanied by increases in activity of enzymes from the decarboxylating portion (Hill et al., 1992). This state persists throughout the development of terminal lipid-storing tissues such as the endosperm of castor bean. In species such as cucumber, however, where the lipid storing cotyledons subsequently become green and photosynthetic, there is a requirement for full TCA cycle activity both for biosynthesis of cellular components, and for the provision of carbon skeletons for nitrogen fixation during photosynthesis (Hill et al., 1992). In this case, the activity of enzymes from the decarboxylating portion of the cycle increases later in development. The maximum catalytic activities of both NAD-ME (S. A. Hill, unpublished) and PDC (Grof et al., 1995) also increase, so that C4 acids derived from the glyoxylate cycle can enter the decarboxylating portion of the cycle. Recently it has been demonstrated that the increases in maximum catalytic activities of fumarase, isocitrate dehydrogenase (D. J. Oliver, unpublished), PDC (Grof et al., 1995) and NAD-ME (B. M. Winning, unpublished) are preceded by increases in the steady state level of mRNA encoding them. This suggests that these changes are regulated either by the rate of transcription or by mRNA stability.

Flower development

Study of cytoplasmic male sterility (CMS) has often led to the assertion that mitochondrial function is of particular importance during flower development (Hanson, 1991). CMS is caused by mutations in the mitochondrial genome, causing production of novel polypeptides, and leads to specific disruption of pollen development. Confirmation that mitochondrial function plays an important role in flower development has been provided by studies of the expression of genes encoding mitochondrial proteins. Increased steady state levels of mRNA encoding NAD-ME (Winning et al., 1994), the E1α subunit of PDC (Grof et al., 1995), citrate synthase (Landschütze et al., 1995a) and aconitase (Peyret et al., 1995) have been detected in developing flowers, which provides circumstantial evidence in support of increased TCA cycle enzyme activities and, consequently, flux at this developmental stage. However, in no case has an increased maximum catalytic activity been demonstrated, and there is no data available that shows particularly high respiratory rates in developing flowers. Recently, the case for increased TCA cycle function being important during flower development has been strengthened by the demonstration that expression of citrate synthase antisense gene in tobacco leads to female sterility (Landschütze et al., 1995b).

Control of flux through the TCA cycle

In discussions of control and regulation we consider pathways as isolated units at our peril. The work of Kacser and colleagues has demonstrated conclusively that control is shared between pathways (Kacser, 1987). Particularly relevant to the present discussion, is the intimate link between the TCA cycle and oxidative phosphorylation, since TCA cycle operation is dependent on the reoxidation of the NADH generated. Consequently, in this section I will consider our current understanding of the control flux through mitochondrial reactions in general, paying particular attention to control and regulation of the TCA cycle (see Chapter 14 for details of the pathway of oxidative phosphorylation in plants).

Much work has been done in recent years applying the techniques of metabolic control analysis to the control of oxidative phosphorylation in plant mitochondria. This approach allows the contribution of particular steps to the control of flux to be quantified in terms of a measurable parameter, the flux control coefficient. The flux control coefficient of a specific enzyme measures the response of flux through a pathway to a small change in the activity of the enzyme; the closer the value of this coefficient to one, the more control is vested in that step. The application of metabolic control analysis to plant metabolism has been reviewed recently (ap Rees and Hill, 1994).

Studies of the control of oxidative phosphorylation have generally minimised the contribution of TCA cycle reactions by analyzing isolated mitochondria oxidizing either succinate or external NADH as respiratory substrate (Padovan *et al.*, 1989; Kesseler *et al.*, 1992; Hill *et al.*, 1993). The picture that emerges from such studies is that much of the control of respiration is vested in the respiratory chain (primarily cytochrome oxidase) at high respiration rates, whilst the passive proton leak across the inner mitochondrial membrane dominates at lower (ADP-limited) rates. The only TCA cycle enzyme studied, succinate dehydrogenase (complex II), makes only a small contribution to the control of flux. When mitochondria oxidizing malate as substrate are analyzed, control shifts away from components of the respiratory chain, perhaps hinting that TCA cycle steps may be important in controlling flux. The control of the TCA cycle in mitochondria isolated from the spadix of *Arum maculatum* has been studied using non-quantitative methods. Determination of the equilibrium position of TCA cycle reactions in mitochondria carrying out the full cycle *in vitro* has identified citrate synthase and isocitrate dehydrogenase as being markedly displaced from equilibrium (MacDougall and ap Rees, 1991). This makes them potentially important controlling steps. However, since work on other pathways has shown that steps catalyzing equilibrium reactions can exert significant control, the possibility that other reactions of the TCA cycle have high flux control coefficients cannot be excluded (Kruckeberg *et al.*, 1989).

The study of the control and regulation of plant metabolism has been revolutionized in recent years by the use of genetic manipulation to modify specific enzyme activities. This approach has been applied to the TCA cycle; transgenic potato plants have been produced where the activity of citrate synthase has been reduced by the expression of the gene in the antisense orientation (Landschütze *et al.*, 1995b). With the exception of the female sterility described above, there is no dramatic effect on the growth or development of the plants. This result implies that citrate synthase is present in excess in most tissues, but a definitive answer concerning the role of this enzyme awaits the measurement of respiratory and other fluxes in these plants. Dramatic reductions in the NAD-ME activity in transgenic potato plants also appear to have little effect on plant growth (B. M. Winning, C. J. Leaver and S. A. Hill, unpublished). Thus, we can conclude that at present we know little about the control of the TCA cycle, but the new technology of genetic manipulation provides an excellent opportunity for such information to be obtained.

Regulation of flux through the TCA cycle

Given the paucity of information concerning the control of TCA cycle flux it is dangerous to speculate in too much detail about potential regulatory mechanisms. As reviewed earlier, there are many well-characterized regulatory properties of TCA cycle enzymes, but it is important that the significance of these properties is critically evaluated in the context of metabolite concentrations that occur *in vivo* and the flux control coefficient of the enzyme concerned. In the framework of metabolic control analysis, the response of an enzyme to a regulator is quantified as the elasticity coefficient. This is a quantitative measure of the sensitivity of an enzyme to an effector, at its physiological concentration, with all other effectors held at their *in vivo* concentrations. The importance of a particular effector acting at a specific enzyme is given by the product of the relevant flux control and elasticity coefficients (Hofmeyer and Cornish-Bowden, 1991). Thus for a particular enzyme to be important in regulation of flux it must have a relatively high flux control coefficient *and* be sensitive to effectors (i.e. have high elasticity coefficients). Therefore demonstration that enzymes respond *in vivo* to effectors, even at physiologically relevant concentrations, is a necessary, but not sufficient condition for a regulatory step.

Based on *in vitro* regulatory properties, the most likely sites for regulation of the flux through the TCA cycle are PDC, and regulation of dehydrogenases by the NADH/NAD ratio (Pascal et al., 1990). In illuminated leaf tissue a potential effector for PDC is photorespiratory ammonium. In non-photosynthetic tissues there is the possibility that the activity of PDC is regulated by the balance between pyruvate and the ADP content, such that the enzyme will be most active when there is a supply of pyruvate, and a high demand for ATP synthesis, indicated by high ADP concentrations. Thus the flux through PDC is potentially regulated by a balance between supply of substrate and demand for ATP. This regulation is reinforced by the sensitivity of PDC and other matrix dehydrogenases to the NADH/NAD ratio, which will tend to balance the flux through the TCA cycle with the rate of electron transport and thus oxidative phosphorylation. However, there appear to be mechanisms that allow the TCA cycle to operate even when demand for ATP is low, since pyruvate accumulation will, in addition to activating PDC, also activate the alternative oxidase (Millar et al., 1993), and high NADH concentrations will lead to the engagement of the rotenone-insensitive NADH dehydrogenase (Møller and Palmer, 1982) (see Chapter 14 for details). The available evidence is thus consistent with a tight but flexible regulation of flux through the TCA cycle, which is perhaps to be expected from the range of metabolic functions performed by the cycle.

Acknowledgements

I thank Dr Susan J. Robertson and Dr Nick J. Kruger for critically reading the manuscript. I am grateful to Dr D. J. Oliver (University of Idaho) and Dr B. M. Winning (University of Oxford) for communicating unpublished data. The version of this chapter that appeared in the First Edition was written by Professor Tom ap Rees, and I am grateful for his permission to use the earlier version as a starting point for writing this chapter.

References

Artus, N. N. and Edwards, G. E. (1985). NAD-malic enzyme from plants. *FEBS Lett.* **182**, 225–33.

ap Rees, T. (1980). Assessment of the contributions of metabolic pathways to plant respiration. In *The Biochemistry of Plants*, Vol. 2, ed. D. D. Davies, Academic Press, New York, pp. 1–29.

ap Rees, T. (1987). Compartmentation of plant metabolism. In *The Biochemistry of Plants*, Vol. 12, ed. D. D. Davies, Academic Press, New York, pp. 87–115.

ap Rees, T., Bryce, J. H., Wilson, P. M. and Green, J. H. (1983). Role and location of NAD-malic enzyme in thermogenic tissues of the Araceae. *Arch. Biochem. Biophys.* **227**, 511–21.

ap Rees, T., Fuller, W. A. and Green, J. H. (1981). Extremely high activities of phosphoenol-pyruvate carboxylase in thermogenic tissues of Araceae. *Planta* **152**, 79–86.

ap Rees, T. and Hill, S. A. (1994). Metabolic control analysis of plant metabolism. *Plant Cell Env.* **17**, 587–99.

Beevers, H. (1961). Respiratory metabolism in plants, Row, Peterson, Evanston.

Bourguignon, J., Macherel, D., Neuberger, M. and Douce, R. (1992). Isolation, characterization, and sequence analysis of a cDNA clone encoding L-protein, the dihydrolipoamide dehydrogenase component of the glycine cleavage system from pea-leaf mitochondria. *Eur. J. Biochem.* **204**, 865–73.

Brouquisse, R., James, F., Raymond, P. and Pradet, A. (1991). Study of glucose starvation in excised maize root tips. *Plant Physiol.* **96**, 619–26.

Brouquisse, R., Nishimura, M., Gaillard, J. and Douce, R. (1987). Characterization of a cytosolic aconitase in higher plant cells. *Plant Physiol.* **84**, 1402–7.

Bryce, J. H. and ap Rees, T. (1985). Rapid decarboxylation of the products of dark fixation of CO_2 in roots of Pisum and Plantago. *Phytochem.* **24**, 1635–8.

Budde, R. J. A., Fang, T. K. and Randall, D. D. (1988). Regulation of the phosphorylation of mitochondrial pyruvate dehydrogenase complex in situ. Effects of respiratory substrates and calcium. *Plant Physiol.* **88**, 1031–6.

Budde, R. J. A., Fang, T. K., Randall, D. D. and Miernyk, J. A. (1991). Acetyl-coenzyme A can regulate activity of the mitochondrial pyruvate dehydrogenase complex in situ. *Plant Physiol.* **95**, 131–6.

Budde, R. J. A. and Randall, D. D. (1987). Regulation of pea mitochondrial pyruvate dehydrogenase complex: inhibition of ATP-dependent inactivation. *Arch. Biochem. Biophys.* **258**, 600–6.

Budde, R. J. A. and Randall, D. D. (1988). Regulation of steady state pyruvate dehydrogenase complex activity in plant mitochondria. Reactivation constraints. *Plant Physiol.* **88**, 1026–30.

Budde, R. J. A. and Randall, D. D. (1990). Pea leaf mitochondrial pyruvate dehydrogenase complex is inactivated in vivo in a light-dependent manner. *Proc. Natl. Acad. Sci. USA* **87**, 673–6.

Canellas, P. F. and Wedding, R. T. (1984). Kinetic properties of NAD malic enzyme from cauliflower. *Arch. Biochem. Biophys.* **229**, 414–25.

Courtois-Verniquet, F. and Douce, R. (1993). Lack of aconitase in glyoxysomes and peroxisomes. *Biochem. J.* **294**, 103–7.

De Bellis, L., Tsugeki, R., Alpi, M. and Nishimura, M. (1993). Purification and characterization of aconitase isoforms from etiolated pumpkin cotyledons. *Physiol. Plant.* **88**, 485–92.

Ebbighausen, H., Jia, C. and Heldt, H. W. (1985). Oxaloacetate translocator in plant mitochondria. *Biochim. Biophys. Acta* **810**, 184–99.

Fletcher, J. S. and Beevers, H. (1971). Influence of cycloheximide on the synthesis and utilization of amino acids in suspension cultures. *Plant Physiol.* **48**, 261–4.

Gardeström, P. and Wigge, B. (1988). Influence of photorespiration on ATP/ADP ratios in the chloroplasts, mitochondria, and cytosol, studied by rapid fractionation of barley (*Hordeum vulgare*) protoplasts. *Plant Physiol.* **88**, 69–76.

Gangloff, S. P., Marguet, D. and Lauquin, G. J.-M. (1990). Molecular cloning of the yeast mitochondrial aconitase gene (ACO1) and evidence of the synergistic regulation of expression by glucose plus glutamate. *Mol. Cell. Biol.* **10**, 3551–61.

Gemel, J. and Randall, D. D. (1992). Light regulation of leaf mitochondrial pyruvate dehydrogenase complex. Role of photorespiratory carbon metabolism. *Plant Physiol.* **100**, 908–14.

Gerhardt, B. (1986). Basic metabolic function of the higher plant peroxisome. *Physiol. Veg.* **24**, 397–410.

Gerhardt, B., Fischer, K. and Maier, U. (1995). Effect of palmitoylcarnitine on mitochondrial activities. *Planta* **196**, 720–26.

Gerhardt, R., Stitt, M. and Heldt, H. W. (1987). Subcellular metabolite levels in spinach leaves. Regulation of sucrose synthesis during diurnal alterations in photosynthetic partitioning. *Plant Physiol.* **83**, 399–407.

Gout, E., Bligny, R., Pascal, N. and Douce, R. (1993). [13]C Nuclear magnetic resonance studies of malate and citrate synthesis and compartmentation in higher plant cells. *J. Biol. Chem.* **268**, 3986–92.

Gottlob-McHugh, S. G., Sangwan, R. S., Blakeley, S. D., Vanlerberghe, G. C., Ko, K., Turpin, D. H., Plaxton, W. C., Miki, B. L. and Dennis, D. T. (1992). Normal growth of transgenic tobacco plants in the absence of cytosolic pyruvate kinase. *Plant Physiol.* **100**, 820–25.

Graham D. (1980). Effects of light on 'dark' respiration. In *The Biochemistry of Plants*, Vol. 2, ed. D. D. Davies, Academic Press, New York, pp. 525–79.

Grof, C. P. L., Winning, B. M., Scaysbrook, T. P., Hill, S. A. and Leaver, C. J. (1995). Mitochondrial pyruvate dehydrogenase. Molecular cloning of the E1α subunit and expression analysis. *Plant Physiol.* **108**, 1623–9.

Grover, S. D., Canellas, P. F. and Wedding, R. T. (1981). Purification of NAD malic enzyme from potato and investigation of some physical and kinetic properties. *Arch. Biochem. Biophys.* **209**, 396–407.

Grover, S. D. and Wedding, R. T. (1982). Kinetic ramifications of the association-dissociation behaviour of NAD malic enzyme. A possible regulatory mechanism. *Plant Physiol.* **70**, 1169–72.

Grover, S. D. and Wedding, R. T. (1984). Modulation of the activity of NAD malic enzyme from *Solanum tuberosum* by changes in oligomeric state. *Arch. Biochem. Biophys.* **234**, 418–25.

Guan, Y., Rawsthorne, S., Scofield, G., Shaw, P. and Doonan, J. (1995). Cloning and characterisation of a dihydrolipoamide acetyltransferase (E2) subunit of the pyruvate dehydrogenase complex from *Arabidopsis thaliana. J. Biol. Chem.* **270**, 5412–17.

Hanning, I. and Heldt, H. W. (1993). On the function of mitochondrial metabolism during photosynthesis in spinach (*Spinacia oleracea* L.) leaves. *Plant Physiol.* **103**, 1147–54.

Hanson, M. R. (1991). Plant mitochondrial mutations and male sterility. *Ann. Rev. Genetics* **25**, 461–86.

Harley, J. L. and Beevers, H. (1963). Acetate utilization by maize roots. *Plant Physiol.* **38**, 117–23.

Hattori, T. and Asahi, T. (1982). The presence of two forms of succinate dehydrogenase in sweet potato root mitochondria. *Plant Cell Physiol.* **23**, 515–23.

Hayes, M. K., Luethy, M. H. and Elthon, T. E. (1991). Mitochondrial malate dehydrogenase from corn. Purification of multiple forms. *Plant Physiol.* **97**, 1381–7.

Heupel, R., Markgraf, T., Robinson, D. G. and Heldt, H. W. (1991). Compartmentation studies on spinach leaf peroxisomes. Evidence for channeling of photorespiratory metabolites in peroxisomes devoid of intact boundary membrane. *Plant Physiol.* **96**, 971–9.

Hill, S. A. and ap Rees, T. (1994). Fluxes of carbohydrate metabolism in ripening bananas. *Planta* **192**, 52–60.

Hill, S. A., Bryce, J. H. and Leaver, C. J. (1993). Control of succinate oxidation by cucumber cotyledon mitochondria. The role of the adenine nucleotide translocator and extra-mitochondrial reactions. *Planta* **190**, 51–7.

Hill, S. A., Bryce, J. H. and Leaver, C. J. (1994). Pyruvate metabolism in mitochondria from cucumber cotyledons during early seedling develoment. *J. Exp. Bot.* **45**, 1489–91.

Hill, S. A., Grof, C. P. L., Bryce, J. H. and Leaver C. J. (1992). The regulation of mitochondrial function and biogenesis in the cotyledons of cucumber (*Cucumis sativus* L.) during early seedling growth. *Plant Physiol.* **99**, 60–66.

Hofmeyer, J.-H. S. and Cornish-Bowden, A. (1991). Quantitative assessment of regulation in metabolic systems. *Eur. J. Biochem.* **200**, 223–36.

Jordanov, J., Courtois-Verniquet, F., Neuburger, M. and Douce, R. (1992). Structural investigations by extended X-ray absorption fine structure spectroscopy of the iron center of mitochondrial aconitase in higher plant cells. *J. Biol. Chem.* **267**, 16775–8.

Kacser, H. (1987). Control of metabolism. In *The Biochemistry of Plants*, Vol. 11, ed. D. D. Davies, Academic Press, New York, pp. 39–67.

Kaethner, T. M. and ap Rees, T. (1985). Intercellular location of ATP citrate lyase in leaves of *Pisum sativum* L. *Planta* **163**, 290–94.

Kang, F. and Rawsthorne, S. (1994). Starch and fatty acid synthesis in plastids from developing embryos of oilseed rape (*Brassica napus* L.). *Plant J.* **6**, 795–805.

Karam, G. A. and Bishop, S. H. (1989). α-ketoglutarate dehydrogenase from cauliflower mitochondria: preparation and reactivity with substrates. *Phytochem.* **28**, 3291–3.

Kesseler, A., Diolez, P., Brinkmann, K. and Brand, M. D. (1992). Characterization of the control of respiration in potato tuber mitochondria using the top-down approach of metabolic control analysis. *Eur. J. Biochem.* **210**, 775–84.

Krebs, H. A. and Johnson, W. A. (1937). The role of citric acid in intermediate metabolism in animal tissues. *Enzymologia* **4**, 148–56.

Krömer, S. (1995). Respiration during photosynthesis. *Ann. Rev. Plant Physiol. Mol. Biol.* **46**, 45–70.

Krömer, S. and Heldt, H. W. (1991). On the role of mitochondrial oxidative phosphorylation in photosynthesis metabolism as studied by the effects of oligomycin on photosynthesis in protoplasts and leaves of barley (*Hordeum vulgare*). *Plant Physiol.* **95**, 1270–76.

Krömer, S., Lernmark, U. and Gardeström, P. (1994). In vivo mitochondrial pyruvate dehydrogenase activity, studied by rapid fractionation of barley leaf protoplasts. *J. Plant Physiol.* **144**, 485–90.

Krömer, S., Malmberg, G. and Gardeström, P. (1993). Mitochondrial contribution to photosynthetic metabolism. A study with barley (*Hordeum vulgare* L.) leaf protoplasts at different light intensities and CO_2 concentrations. *Plant Physiol.* **102**, 947–55.

Kruckeberg, A. L., Neuhaus, H. E., Feil, R., Gottlieb, L. D. and Stitt, M. (1989). Decreased-activity mutants of phosphoglucose isomerase in the cytosol and chloroplast of *Clarkia xantiana*. *Biochem. J.* **261**, 457–67.

Landschütze, V., Müller-Röber, B. and Willmitzer, L. (1995a). Mitochondrial citrate synthase from potato: predominant expression in mature leaves and young flower buds. *Planta* **196**, 756–64.

Landschütze, V., Willmitzer, L. and Müller-Röber, B. (1995b). Inhibition of flower formation by antisense repression of mitochondrial citrate synthase in transgenic potato plants leads to specific disintegration of the ovary tissues of flowers. *EMBO J.* **14**, 660–66.

Lips, S. H. and Beevers, H. (1966). Compartmentation of organic acids in corn roots. *Plant Physiol.* **41**, 709–12.

Luethy, M. H., Miernyk, J. A. and Randall, D. D. (1994). The nucleotide and deduced amino-acidsequences of a cDNA encoding the E1β-subunit of the *Arabidopsis thaliana* mitochondrial pyruvate dehydrogenase complex. *Biochim. Biophys. Acta* **1187**, 95–8.

MacDougall, A. J. and ap Rees, T. (1991). Control of the Krebs Cycle in *Arum spadix*. *J. Plant Physiol.* **137**, 683–90.

MacLennan, D. H., Beevers, H. and Harley, J. L. (1963). 'Compartmentation' of acids in plant tissues. *Biochem. J.* **89**, 316–27.

McCashin, B. G., Cossins, E. A. and Canvin, D. T. (1988). Dark respiration during photosynthesis in wheat leaf slices. *Plant Physiol.* **87**, 155–61.

McIntosh, C. A. and Oliver, D. J. (1992). NAD^+-linked isocitrate dehydrogenase: isolation, purification, and characterization of the protein from pea mitochondria. *Plant Physiol.* **100**, 69–75.

Mettler, I. J. and Beevers, H. (1980). Oxidation of NADH in glyoxysomes by a malate-aspartate shuttle. *Plant Physiol.* **66**, 555–60.

Miernyk, J. A. and Randall, D. D. (1987). Some properties of pea mitochondrial phospho-pyruvate dehydrogenase-phosphatase. *Plant Physiol.* **83**, 311–15.

Millar, A. H., Wiskich, J. T., Whelan, J. and Day, D. A. (1993). Organic acid activation of the alternative oxidase of plant mitochondria. *FEBS Lett.* **329**, 259–62.

Møller, I. M. and Palmer, J. M. (1982). Direct evidence for the presence of a rotenone-resistant NADH dehydrogenase on the inner surface of the inner membrane of plant mitochondria. *Physiol. Plant.* **54**, 267–74.

Neal, G. E. and Beevers, H. (1960). Pyruvate utilization in castor-bean endosperm and other tissues. *Biochem. J.* **74**, 409–16.

Nesius, K. K. and Fletcher, J. S. (1975). Contribution of nonautotrophic CO_2 fixation to protein synthesis in suspension cultures of Paul's scarlet rose. *Plant Physiol.* **55**, 643–5.

Newton, K. J. and Schwartz, D. (1980). Genetic basis of the major malate dehydrogenase isoforms in maize. *Genetics* **95**, 425–42.

Padovan, A. C., Dry, I. B. and Wiskich, J. T. (1989). An analysis of the control of phosphorylation-coupled respiration in isolated plant mitochondria. *Plant Physiol.* **90**, 928–33.

Palmer, J. M. and Wedding, R. T. (1966). Purification and properties of succinyl-CoA synthetase from Jerusalem artichoke mitochondria. *Biochim. Biophys. Acta* **113**, 167–74.

Pascal, N., Dumas, R. and Douce, R. (1990). Comparison of the kinetic behaviour towards pyridine nucleotides of NAD^+-linked dehydrogenases from plant mitochondria. *Plant Physiol.* **94**, 189–93.

Peyret, P., Perez, P. and Alric, M. (1995). Structure, genomic organization, and expression of the *Arabidopsis thaliana* aconitase gene. *J. Biol. Chem.* **270**, 8131–7.

Rasmusson, A. G. and Møller, I. M. (1990). NADP-utilizing enzymes in the matrix of plant mitochondria. *Plant Physiol.* **94**, 1012–18.

Roby, C., Martin, J. B., Bligny, R. and Douce, R. (1987). Biochemical changes during sucrose deprivation in higher plant cells. *J. Biol. Chem.* **262**, 5000–5007.

Schuller, K. A. and Randall, D. D. (1989). Regulation of pea mitochondrial pyruvate dehydrogenase complex. Does photorespiratory ammonium influence mitochondrial carbon metabolism? *Plant Physiol.* **89**, 1207–12.

Schuller, K. A. and Randall, D. D. (1990). Mechanism of pyruvate inhibition of plant pyruvate dehydrogenase kinase and synergism with ADP. *Arch. Biochem. Biophys.* **278**, 211–16.

Stewart, C. R. and Voetberg, G. (1985). Relationship between stress-induced ABA and proline accumulations and ABA-induced proline accumulation in excised barley leaves. *Plant Physiol.* **79**, 24–7.

Taylor, A. E., Cogdell, R. J. and Lindsay, J. G. (1992). Immunological comparison of the pyruvate dehydrogenase complexes from pea mitochondria and chloroplasts. *Planta* **188**, 225–31.

Thomas, D. R., Wood, C. and Masterson, C. (1988). Long chain acyl-CoA synthetase, carnitine and β-oxidation in the pea-seed mitochondrion. *Planta* **173**, 263–6.

Unger, E. A., Hand, J. M., Cashmore, A. R. and Vasconcelos, A. C. (1989). Isolation of a cDNA encoding mitochondrial citrate synthase from *Arabidopsis thaliana*. *Plant Mol. Biol.* **13**, 411–18.

Unger, E. A. and Vasconcelos, A. C. (1989). Purification and characterization of mitochondrial citrate synthase. *Plant Physiol.* **89**, 719–23.

Verniquet, F., Gaillard, J., Neuburger, M and Douce, R. (1991). Rapid inactivation of plant aconitase by hydrogen peroxide. *Biochem. J.* **276**, 643–8.

Walk, R.-A., Michaeli, S. and Hock, B. (1977). Glyoxysomal and mitochondrial malate dehydrogenase of watermelon (*Citrullus vulgaris*) cotyledons. I Molecular properties of the purified isoenzymes. *Planta* **136**, 211–20.

Wieland, O. H. (1983). The mammalian pyruvate dehydrogenase complex: structure and regulation. *Rev. Physiol. Biochem. Pharmacol.* **96**, 123–70.

Wigge, B., Krömer, S. and Gardeström, P. (1993). The redox levels and subcellular distribution of pyridine nucleotides in illuminated barley leaf protoplasts studied by rapid fractionation. *Physiol. Plant.* **88**, 10–18.

Willeford, K. O. and Wedding, R. T. (1987a). pH effects on the activity and regulation of the NAD malic enzyme. *Plant Physiol.* **84**, 1084–7.

Willeford, K. O. and Wedding, R. T. (1987b). Evidence for a multiple subunit composition of plant NAD malic enzyme. *J. Biol. Chem.* **262**, 8423–9.

Winning, B. M., Bourguignon, J. and Leaver, C. J. (1994). Plant mitochondrial NAD$^+$-dependent malic enzyme: cDNA cloning, deduced primary structure of the 59- and 62-kDa subunits, import gene complexity and expression analysis. *J. Biol. Chem.* **269**, 4780–86.

Winter, H., Robinson, D. G. and Heldt, H. W. (1994). Subcellular volumes and metabolite concentrations in spinach leaves. *Planta* **193**, 530–35.

Wiskich, J. T., Bryce, J. H., Day, D. A. and Dry, I. B. (1990). Evidence for metabolic domains within the matrix compartment of pea mitochondria. Implications for photorespiratory metabolism. *Plant Physiol.* **93**, 611–16.

Wood, C., Burgess, N. and Thomas, D. R. (1986). The dual location of β-oxidation enzymes in germinating pea cotyledons. *Planta* **167**, 54–7.

Zeiher, C. A. and Randall, D. D. (1990). Identification and characterization of mitochondrial acetyl-coenzyme A hydrolase from *Pisum sativum* L. seedlings. *Plant Physiol.* **94**, 20–27.

14 Oxidation of mitochondrial NADH and the synthesis of ATP

Hans Lambers

Introduction

Depending on species and growth conditions of a plant, 30–70% of all the carbohydrates fixed in photosynthesis are respired in the same day. A significant part of this respiration proceeds via a non-phosphorylating path, so that energy conservation is less than maximal (Van der Werf *et al.*, 1992). Furthermore, the proportion of carbon lost in respiration is negatively correlated with the plant's growth rate. It is therefore imperative to obtain a good understanding of this important aspect of plant metabolism.

The monograph by Douce (1985), Volume 18 in the *Encyclopedia of Plant Physiology* (Douce and Day, 1985) and the proceedings of the third international meeting on plant respiration (Lambers and Van der Plas, 1992), have been of great help in writing some sections of this chapter. For further information on most of the topics treated here, the reader is referred to the above three works.

This chapter provides firstly the basic information on the structure and functioning of respiratory components, including the biochemistry of substrate oxidation and ATP formation. It then discusses the stoichiometry of substrate consumption, carbon dioxide release and ATP formation *in vivo*. Finally, the problem of the regulation of respiration *in vivo*, both in the light and in the dark is addressed. Notwithstanding the rapid progress made in this field in the last decade, the reader will be confronted with questions which cannot yet be answered.

The location and organization of the mitochondrial electron transport pathways

Respiration involves the transfer of electrons from organic molecules to oxygen. This transfer occurs via an electron transport chain in the inner mitochondrial membrane. The mitochondrial electron transport chain is a complicated arrangement of some 40 redox centers, 50 polypeptides and significant amounts of phospholipids. One single chain has a 'molecular weight' of approximately 1.52×10^6 kDa. The respiratory components of the electron transport pathways are arranged into discrete multiprotein units.

The components of the respiratory chains in plant mitochondria

In mammalian mitochondria four major complexes (I–IV) are associated with electron transfer and one with the production of ATP (complex V). The four electron transport complexes are: complex I (catalyzes the transfer of electrons from internal NADH to ubiquinone), complex II (responsible for the transfer of electrons from succinate to ubiquinone), complex III (generally called the bc_1 complex, transfers electrons from ubiquinol to cytochrome c), and complex IV (also called cytochrome c oxidase, catalyzes the transfer of electrons from cytochrome c to O_2). All of these complexes are located in the inner mitochondrial membrane (Fig. 14.1). Complexes I, III and IV are embedded in the membrane in such a way that they are in contact with both the intermembrane space and the matrix, whereas complex II is only associated with the matrix side. The binding site for NADH of complex I faces the matrix.

Two small redox molecules must be added to the list of electron transferring multiprotein complexes to

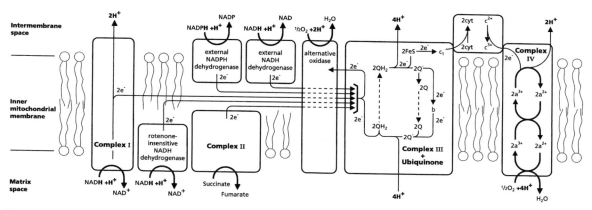

Fig. 14.1 The organization of the four electron transporting complexes (I–IV) of the respiratory chain in higher plant mitochondria. All components are located in the inner mitochondrial membrane. Some of the components are transmembranous, others face the matrix or the space between the inner and the outer mitochondrial membrane (intermembrane space). Q (ubiquinone) is a mobile pool of quinone and quinol molecules, part of which is associated with complex III. Based on information in Douce (1985), Moore and Rich (1985), Lambers and Van der Plas (1992). For further explanations see text.

define the full electron transfer sequence. First, ubiquinone (Q), connects both complex I and complex II to complex III. The quinone and quinol molecules (ubiquinol is the fully reduced form of ubiquinone; the half-reduced form is called ubisemiquinone; Fig. 14.2) operate in the hydrophobic phase of the membrane and are closely associated with complex III. Second, cytochrome c facilitates electron transfer from complex III to complex IV.

In higher plant mitochondria some additional components occur. One of these is a second NADH dehydrogenase, also located in the inner membrane, but with its binding site for NADH facing the intermembrane space. There is also a NADPH dehydrogenase, accepting electrons from external NADPH; this NADPH is mainly produced in the oxidative pentose phosphate pathway. Then there is the cyanide-resistant alternative pathway, also located in the inner membrane. Finally, there is the rotenone-insensitive NAD(P)H dehydrogenase, distinct from complex I. It is not blocked by inhibitors of complex I, such as rotenone. Since this dehydrogenase is reduced by internal NAD(P)H only, it is presumably facing the matrix. Figure 14.1 summarizes the organization of the components and their location in the inner mitochondrial membrane. A further description is given below.

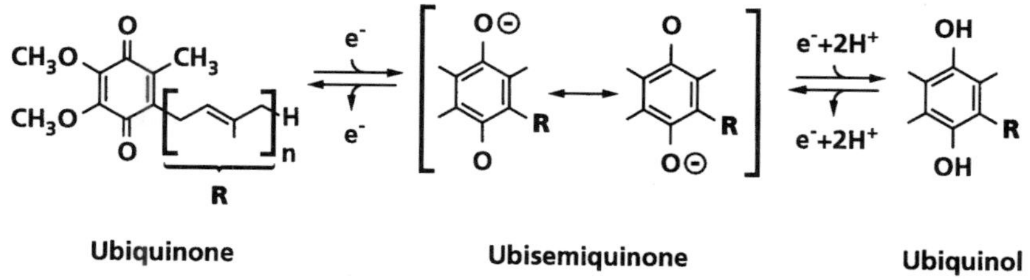

Fig. 14.2 The structure and redox states of ubiquinone (Q). Ubiquinone is a substituted 1,4-benzoquinone; the side chain (R) is composed of 10 isoprenyl units. Reduction of ubiquinone to ubiquinol requires two electrons and two protons. Reduction with only one electron gives ubisemiquinone.

Complex I and the rotenone-insensitive dehydrogenase

Complex I is the main entry point of electrons from 'internal' NADH, that is, NADH produced in the matrix space during the oxidation of malate, pyruvate, oxoglutarate, isocitrate, and glycine (Fig. 14.1). Both a flavin mononucleotide (FMN) and several iron–sulfur proteins are involved in the transfer of electrons from NADH to ubiquinone. This transfer is coupled to the translocation of H^+ from the matrix to the intermembrane space, as further explained below. Complex I is therefore called the 'first coupling site' or 'site 1' of proton extrusion. The rotenone-insensitive NADH dehydrogenase, unlike complex I, is not linked to proton extrusion.

Complex II

Unlike all other intermediates of the TCA cycle, which are oxidized by matrix enzymes, succinate is oxidized by the membrane-bound enzyme succinate dehydrogenase. Electrons from succinate enter the respiratory chain via complex II. Flavin adenine dinucleotide (FAD), several non-heme iron proteins and iron–sulfur proteins are involved in the electron transfer to ubiquinone. Unlike complex I, complex II is not connected with the translocation of H^+ across the inner mitochondrial membrane.

The external NAD(P)H dehydrogenases

External, or cytosolic NAD(P)H feeds its electrons into the chain at the level of ubiquinone, similar to succinate. Unlike complex I, and similar to complex II, the external dehydrogenases are not connected with the translocation of H^+ across the inner mitochondrial membrane.

Complex III

Complex III (cytochrome c reductase) is responsible for the transfer of electrons from ubiquinol to cytochrome c. Ubiquinones are closely associated with complex III, which contains cytochromes b and c_1 and an iron–sulfur center (the Rieske protein). Both the Rieske protein and cytochrome c_1 are exposed to the intermembrane side. The mechanism

of electron transfer, which is coupled to the translocation of four protons per electron pair from the matrix to the intermembrane space, is still poorly understood. It is assumed that a 'Q cycle' allows the observed proton translocation in the following manner. Electrons from the dehydrogenases or complex II are donated to ubisemiquinone (Q^-), which is fairly immobile in the membrane. Ubisemiquinone reduction to ubiquinol (QH_2) requires two protons per electron, which are taken up from the matrix. The ubiquinol then diffuses to the intermembrane side of the inner membrane, where one electron is donated to the Rieske protein, producing ubisemiquinone. Two protons are then released into the intermembrane space. The chemistry of this reaction is outlined in Fig. 14.2. Ubisemiquinone donates an electron to a transmembranous cytochrome b, forming ubiquinone. Q diffuses to the matrix side of the inner membrane where it accepts an electron from cytochrome b so that Q^- is regenerated and the cycle completed (Fig. 14.1). Hence, one electron traverses the electron transport chain and one cycles through cytochrome b. In the process two protons are transported for every electron passing through complex III. Complex III constitutes 'site 2' of proton extrusion.

Complex IV

Complex IV (cytochrom c oxidase) is the terminal oxidase of the cytochrome pathway. It contains cytochromes (cytochrome a and cytochrome a_3) and copper. Cytochrome a, which faces the intermembrane space, accepts electrons from cytochrome c. The electrons are then donated to cytochrome a_3, which reacts with O_2. Cytochrome oxidase generates a proton-motive force by two entirely different processes. Firstly, the reaction is linked to uptake of $1H^+$ from the matrix side per electron donated by cytochrome c at the opposite side to form water from reduced oxygen. Secondly, cytochrome oxidase functions as a proton pump, extruding $1H^+$ per electron to the intermembrane space (Wilkström and Babcock, 1990). This makes complex IV the third coupling site, although only part of the protons are extruded from the matrix as at sites 1 and 2, and part of them are removed via covalently binding to O_2.

The cyanide-resistant path

Respiratory O_2 consumption by most higher plant tissues is not fully inhibited by inhibitors of the cytochrome path such as KCN. O_2 uptake by mitochondria isolated from such tissues also proceeds to some extent in the presence of KCN, showing that the cyanide-resistant O_2 uptake resides in the mitochondria (Lambers *et al.*, 1983). This component of respiration is due to a cyanide-resistant, alternative electron transport pathway. This path consists of one oxidase, with a monomer molecular weight of around 36 kD (McIntosh, 1994). The alternative oxidase molecules are embedded in the inner mitochondrial membrane as homodimers. The oxidase predominantly faces the matrix side (Siedow *et al.*, 1992). When the monomers are covalently linked via disulfide bonds ('oxidized'), the oxidase has a lower activity than when the monomers are non-covalently associated ('reduced'; Umbach and Siedow, 1993). The dimeric nature of the alternative oxidase allows it to be affected allosterically, e.g. by the fraction of its substrate (ubiquinone) which is in its reduced state and by organic acids, as discussed below.

Since the effect of inhibitors of complex IV and complex III (i.e. antimycin; Table 14.1) on mitochondrial O_2 uptake, is the same, the branching point of the alternative path from the cytochrome path must be before complex III. The observation that succinate oxidation, and in some tissues also oxidation of external NAD(P)H, is also cyanide-sensitive leads to the conclusion that the branching point is after complex I. This has implicated ubiquinone as a component common to both pathways (Fig. 14.1). Transport of electrons from ubiquinone to O_2 via the alternative oxidase is not coupled to the extrusion of protons from the matrix to the intermembrane space. This is a major difference in comparison with the cytochrome path.

Substrate oxidation, proton extrusion and oxidative phosphorylation

The major substrates

In mammalian cells, pyruvate is the endproduct of glycolysis and serves as the major substrate for mitochondrial respiration. In addition to pyruvate, plant mitochondria use malate as a major substrate. In this case malate can be viewed as an alternative glycolytic endproduct. It is formed from the carboxylation of phosphoenolpyruvate (PEP) by PEP carboxylase which yields oxaloacetate (OAA). The OAA is then reduced to malate by malate dehydrogenase. The malate formed in the cytosol is transported across the inner membrane of the mitochondria where some may be oxidized in the conventional manner via the TCA cycle and some via malic enzyme, producing pyruvate and CO_2. The pyruvate is then available for subsequent oxidation via pyruvate dehydrogenase and the TCA cycle. In addition to the oxidation of TCA cycle intermediates, glycine produced via the photorespiratory pathway is also oxidized in the mitochondria. Glycine has rapid access to complex I; its oxidation tends to be favoured over that of TCA cycle intermediates.

Figure 14.3 outlines the major pathways of mitochondrial carbon oxidation and the associated production of NADH and $FADH_2$.

The mitochondrial states

Freshly isolated intact mitochondria do not consume an appreciable amount of O_2. Upon addition of one of the substrates included in Fig. 14.3, there is a small increase in O_2 uptake (Fig. 14.4). As soon as ADP is added, a rapid consumption of O_2 can be measured. This 'state' of the mitochondria is called 'state 3'; the two earlier ones are referred to as 'state 1' (mitochondria, but no substrate) and 'state 2' (substrate present), respectively. Upon conversion of all ADP into ATP, the respiration rate of the mitochondria declines again to the rate found before addition of ADP. The state is referred to as 'state 4'. Upon addition of more ADP, the mitochondria go into state 3 again, followed by state 4 upon depletion of ADP. This can be repeated until all O_2 in the cuvette is reduced (Fig. 14.4). Thus the respiratory activity is effectively controlled by the availability of ADP. This phenomenon is called 'respiratory control' and is quantified by the 'respiratory control ratio': the ratio of the rate of O_2 consumption at substrate saturation in the presence of ADP (state 3) to that after ADP has been depleted (state 3) (Fig. 14.4).

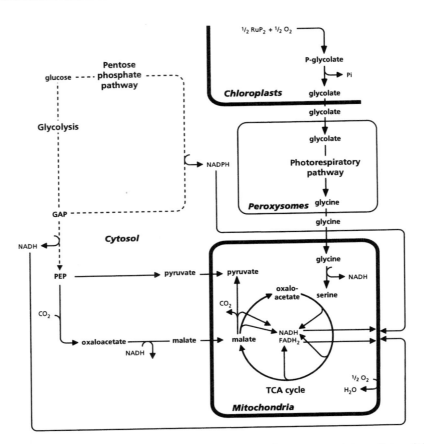

Fig. 14.3 The major substrates for the electron transport pathways are malate, pyruvate, intermediates of the TCA cycle, NADH from glycolysis (oxidized via the external NADH dehydrogenase only), NADPH from the oxidative pentose phosphate pathway (also oxidized via an external dehydrogenase), and glycine (in tissues with an operative photorespiratory pathway).

The O_2 consumed during the phosphorylation of ADP can be related to the total amount of ADP added to calculate the ADP:O ratio. This ratio varies for different substrates. In the absence of inhibitors, it is approximately 3 for NAD-linked substrates and 2 for succinate and external NAD(P)H (Fig. 14.4). The rationale for the observed ADP:O values and effects of inhibitors are discussed below.

Inhibitors, uncouplers and other effectors of respiratory metabolism

Respiratory inhibitors (Table 14.1) have been an important tool in elucidating the organization of the

respiratory pathways, as summarized in Fig. 14.1. First, there are the inhibitors of complex I, of which rotenone is the most specific. There are others, such as amytal, piericidin A, benzyladenine (a synthetic cytokinin) and a range of 'secondary plant compounds', such as glyceollin (from *Glycine max*). However, even when complex I is completely inhibited, internal NADH may still be oxidized. This results from the presence of a bypass, generally called the 'rotenone-insensitive dehydrogenase'. Succinate oxidation is not inhibited by inhibitors of complex I and this is one line of evidence supporting the position of succinate dehydrogenase in Fig. 14.1.

Succinate oxidation is competitively inhibited by malonate, a dicarboxylic acid containing three carbon

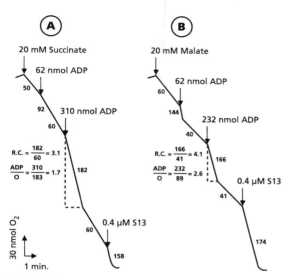

Fig. 14.4 The 'states' of isolated mitochondria. The ADP:O ratio is calculated from the O_2 consumption during the phosphorylation of a known amount of added ADP (state 3 rate). The respiratory control ratio (RC) is the ratio of the state 3 and the state 4 rate of O_2 uptake. State 1 is the respiration rate in the absence of substrate and ADP. Mitochondria were isolated from potato (*Solanum tuberosum*) tubers; rates are expressed as nmol O_2 (mg protein)$^{-1}$ min^{-1}. Unpublished data from A. M. Wagner, Free University of Amsterdam.

atoms as compared with succinate which contains four. Malonate has no effect on the oxidation of NAD-linked substrates or external NADH.

Antimycin binds to cytochrome *b* and inhibits complex III. Cyanide, azide and carbon monoxide bind to cytochrome a_3 and inhibit complex IV. These four compounds have been termed inhibitors of the cytochrome path. When the cytochrome path is blocked, the oxidation of NAD-linked substrates, succinate and external NAD(P)H may not be fully inhibited. This is due to the presence of the alternative path. Both complex III and the alternative path accept electrons from ubiquinone. In the presence of an inhibitor of the cytochrome path and an inhibitor of the alternative path, O_2 uptake by isolated mitochondria is virtually completely inhibited (Fig. 14.4).

Substituted hydroxamic acids (e.g. salicylhydroxamic acid; Table 14.1) are frequently used with isolated mitochondria as inhibitors of the alternative path. They can also be used *in vivo*, but care has to be taken to choose the correct concentration. Disulfiram, though active in lower concentrations, penetrates isolated mitochondria very slowly and is not widely used. It is not an appropriate inhibitor for experiments with intact tissues.

Uncouplers are useful tools in the study of plant respiration. They make membranes, including the inner mitochondrial membrane, permeable to protons, therefore inhibiting oxidative phosphorylation, as explained below. Many compounds belonging to this category are found as 'secondary compounds' in plant tissues.

Valinomycin (in the presence of K^+) indirectly increases the conductance for protons across the inner mitochondrial membrane. Valinomycin allows rapid entry of K^+ ions into the mitochondria where they are then exchanged for protons. Thus, in the presence of K^+, valinomycin dissipates the membrane potential. In the absence of K^+, valinomycin has no effect on the membrane potential, indicating that its effect differs from that of other uncouplers.

Table 14.1 provides a summary of effectors of respiratory metabolism.

Proton extrusion linked with electron transport

During the transfer of electrons from the various substrates to O_2 via the cytochrome electron transport chain, some of the energy is conserved in the form of ATP (Fig. 14.4). However, ATP formation is not directly linked to electron transport, but rather to the generation of a proton-motive force across the inner mitochondrial membrane as described by the 'chemiosmotic theory' (Mitchell, 1966). The transport of protons out of the mitochondrial matrix can be demonstrated by adding a small amount of O_2 to a suspension of intact mitochondria in a lightly buffered medium that contains a suitable substrate. Acidification does not occur when the inner membrane is damaged by detergents. These observations are in agreement with the chemiosmotic theory.

The basic features of this now widely accepted chemiosmotic model are that (1) protons are transported outwards, coupled to the transfer of

Table 14.1 Effectors of respiratory metabolism. Those used most frequently in current research or to be preferred because of their specificity are italicized. (Based on information included in Douce, 1985; Douce and Day, 1985; Lambers and Van der Plas, 1992; Williamson and Metcalf, 1967.) Of those compounds which are naturally found in organisms, (one of) the source(s) is included in the last column.

Inhibitor of	Compound	Source
Complex I	*Rotenone*	*Derris* roots
	Amytal	
	Piericidin A	*Streptomyces mobaraensis*
	Glyceollin	*Glycine max*
Complex II	Malonate	
Complex III	*Antimycin*	
	Myxothiazol	
	2-*n*-heptyl-1-hydroxy-chinoline-N-oxide (HQNO)	
Complex IV	*Cyanide*	
	Azide	
	Carbon monoxide	
Alternative path	Disulfiram	
	Benzhydroxamic acid	
	m-Chlorohydroxamic acid	
	Salicylhydroxamic acid (SHAM)	
	Propyl gallate	
	8-Hydroxyquinoline (8-OHQ)	
Oxidative phosphorylation (uncouplers)	Carbonyl cyanide-*m*-chlorophenyl hydrazone (CCCP)	
	Carbonyl cyanide-*p*-trifluoromethoxy-phenyl-hydrazone (FCCP)	
	5-Chloro-3-*t*-butyl-2′-chloro-4′-nitrosalicylanilide (*S13*)	
	Dinitrophenol (DNP)	
	Pinosylvin monomethylether	*Pinus* sp.
Membrane potential	Valinomycin + K^+	*Streptomyces* sp.
H^+P_i-symporter	N-ethylmaleimide (NEM)	
Complex V (ATP synthetase)	Oligomycin	*Streptomyces diastatochrogenes*
Adenine nucleotide translocator	Carboxy-atractyloside	*Atractylis gummifera*

electrons down the electron transport chain. This gives rise to both a proton gradient (ΔpH) and a membrane potential ($\Delta\Psi$); (2) the inner membrane is impermeable to protons and other ions, except by special transport systems; and (3) there is an ATP synthetase, which transforms the energy of the electrochemical gradient, generated by the proton-extruding system, into ATP.

The pH gradient, ΔpH, and the membrane potential, $\Delta\Psi$, are interconvertible and it is the combination of the two which forms the proton-motive force (Δp), the driving force for ATP synthesis:

$$\Delta p = \Delta\Psi - 2.3\, RT/F \cdot \Delta pH$$

where F is Faraday's number, so that both components in the above equation are expressed in mV.

Δp is estimated by separate determination of $\Delta \Psi$ and ΔpH. This is commonly done by measuring the steady-state distribution of a permeant ion, that is a compound which diffuses freely across the inner membrane. Once the concentration of the permeant ion inside and outside the mitochondria is known, $\Delta \Psi$ can be calculated using the Nernst equation, and ΔpH can be calculated using the Henderson–Hasselbalch equation. In the presence of valinomycin, K^+ ions are freely transported across the inner membrane and thus this system can be used to estimate $\Delta \Psi$ (the system is called a $\Delta \Psi$-probe). Alternatively, synthetic lipophylic cations (e.g. methyltriphenylphosphonium [TPMP$^+$] or tetraphenylphosphonium [TPP$^+$]) can be used as $\Delta \Psi$-probes. Permeant weak acids (e.g. acetate) are used to measure ΔpH (ΔpH-probes). For a description of other techniques used to determine Δp, the reader is referred to Moore and Rich (1985). Higher plant mitochondria in state 4 have Δp values from 153–262 mV. $\Delta \Psi$ contributes 126–250 mV, whereas the ΔpH contributes 12–36 mV (the latter is the equivalent of 0.5 pH units) (Douce, 1985; Moore and Rich, 1985).

The ratio of H^+ extrusion and O_2 uptake has been determined for various substrates (Table 14.2). Since some of the protons flow back in exchange for inorganic phosphate, the H^+/P_i symporter has to be inhibited to allow a proper estimation of the stoichiometry. Unfortunately, inhibition of the H^+/P_i symporter by N-ethylmaleimide (NEM) also reduces the H^+/O stoichiometry when malate is the substrate (Table 14.2).

The determination of the 'sites' of proton extrusion, as indicated in Fig. 14.1, involved the use of different substrates in combination with inhibitors and artificial electron acceptors. Table 14.3 shows that the H^+/O ratio for malate oxidation is 6.96 in the absence of rotenone and 4.88 in the presence of this inhibitor of complex I. Subtracting the value in the presence of rotenone from that in its absence, leads to the conclusion that the H^+/O ratio of site 1 is approximately 2. The H^+/O ratio of succinate oxidation is 6.27 in the absence and 0.0 in the presence of antimycin, an inhibitor of complex III. This indicates that the H^+/O ratio of sites 2 + 3 is 6.27 and that the alternative path is not coupled to proton extrusion. Oxidation of ascorbate, which

Table 14.2 The stoichiometry of proton extrusion and electron transfer associated with oxidation of various substrates in *Phaseolus aureus* (mung bean) and rat liver mitochondria. Values were determined in the absence and precence of *N*-ethylmaleimide (NEM), which inhibits proton uptake associated with inorganic phosphate (H^+/P_i symport). Note that NEM reduced the stoichiometry for NAD-linked substrates; the reason for this is unknown, but it complicates the calculation of the stoichiometry per site (see text and Table 14.3). In the presence of N, N, N', N'-tetramethyl-*p*-phenylene diamine (TMPD), ascorbate donates electrons to cytochrome *c*. (Based on data in Mitchell and Moore, 1984.)

Source	Substrate	Treatment	H^+/O ratio
Mung bean	Malate	−NEM	6.96
	Malate	+NEM	5.87
	Succinate	−NEM	4.58
	Succinate	+NEM	6.27
	NADH	−NEM	4.64
	NADH	+NEM	5.43
Rat liver	Ascorbate + TMPD	−NEM	2.58
	Succinate	−NEM	4.84
	Succinate	+NEM	6.30

donates electrons to complex IV in the presence of tetramethyl-*p*-phenylene diamine (TMPD), gives a H^+/O ratio of 2.58, the stoichiometry for site 3. Combined with the value for the H^+/O ratio of sites 2 + 3, a H^+/O ratio for site 2 of 3.69 is found. The H^+/O ratio for site 1 can also be determined using malate in the presence of an inhibitor of the cytochrome path, e.g. antimycin. Following this approach, a value of 2.57 was found for *Sauromatum guttatum* mitochondria (Table 14.3).

Thus, the H^+/O ratio for site 1 is around 2 (2.08–2.57); the H^+/O ratio for site 2 is close to 4 (3.69); and that of site 3 is close to 3 (2.58) (Table 14.3). For complex I this stoichiometry can be explained by the redox reactions involved. The complex is reduced by NADH (or NADH + H^+, producing NAD$^+$). The electron transfer from NADH to complex I is thus coupled to the removal of two protons from the matrix, which are released in the intermembrane space upon donation of electrons to ubiquinone. The stoichiometry of site 2 can be explained by the operation of a 'Q cycle', described above (Figs 14.1 and 14.2). Reduction of O_2 (concomitantly with uptake of 1H^+ per electron from the matrix side and

Table 14.3 The stoichiometry of H^+/O in mung bean and *Sauromatum guttatum* (last line only) mitochondria, calculated per site, based on information given in Table 14.2. Note that the stoichiometry for site $2 + 3$ cannot be derived from measurements with malate as substrate in the absence of rotenone, since NEM cannot be used in this case so that some protons were allowed to move back to the matrix side with phosphate via the H^+/P_i-symporter). For the same reason the stoichiometry of site 1 may have been somewhat underestimated. (Based on data in Mitchell and Moore, 1984).

Substrate	Treatment	Sites	H^+/O ratio
Malate	Control	$1 + 2 + 3$	6.96
	+ rotenone	$2 + 3$	4.88
	(by difference)	1	2.08
Succinate	Control	$2 + 3$	6.27
	+ antimycin	alt. ox.	0
Ascorbate + TMPD	+ antimycin	3	2.58
	(by difference)	2	3.69
Malate	+ antimycin	1	2.57

pumping of $1H^+$ across the membrane) explains the stoichiometry of site 3. The absence of proton translocation when external NADH is oxidized is due to the fact that the protons donated to ubiquinone are taken up from the intermembrane space (to which they are released again thereafter), rather than from the matrix, as is the case with internal NADH. Similarly, succinate oxidation in complex II is not linked with proton extrusion. Any protons taken up from the matrix during the oxidation of succinate are released to the matrix again upon donation of electrons to ubiquinone.

ATP formation linked with proton re-entry

According to the chemiosmotic theory as explained above, proton extrusion, coupled to electron transport in the respiratory chain, leads to a proton-motive force (Δp). Δp is the driving force for ATP synthesis. ATP synthesis in mitochondria is therefore associated with the re-entry of protons to the matrix, as mediated by complex V (coupling factor or ATP synthetase; Fig. 14.5). This is a reversible reaction, as ATP hydrolysis leads to the extrusion of protons. Although the exact stoichiometry of proton re-entry to ATP formation is still under debate, values around three are widely accepted (Moore and Rich, 1985).

The stoichiometry of H^+/O per site, in combination with the stoichiometry of ATP formation per proton, explains the difference in ADP:O ratio as found for different substrates (Fig. 14.4). For pyruvate oxidation, involving all sites of proton extrusion, an

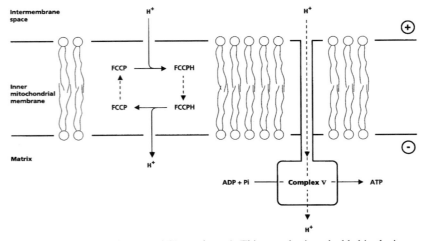

Fig. 14.5 A model of complex V (coupling factor or ATP synthetase). This complex is embedded in the inner mitochondrial membrane. It allows proton re-entry, coupled to the phosphorylation of ADP at the inside surface of the inner membrane. Approximately three protons re-enter per ATP formed. Uncouplers such as FCCP (Table 14.1) inhibit oxidative phosphorylation by allowing proton re-entry without the ATP synthetase.

ADP : O ratio of $(2.57 + 3.69 + 2.58)/3 = 2.94$ is expected. For succinate oxidation, involving site 2 and site 3 only, an ADP : O ratio of $(3.69 + 2.58)/3 = 2.09$ is expected. For a number of reasons the ADP : O values found in isolated mitochondria are often somewhat lower than calculated, but the relative values tend to be 3.0 and 2.0 for internal NADH and succinate, respectively (Fig. 14.4). *In vivo* studies employing ^{31}P NMR have allowed the calculation of ADP : O ratios in intact tissues (Roberts *et al.*, 1984). This work supports the contention that NADH oxidation by complex I is coupled to the production of 3 ATP per O consumed and the oxidation of succinate via complex II is coupled to the production of 2 ATP per O consumed. Observation of ADP : O ratios of 3 for the oxidation of internal NADH and 2 for oxidation of succinate or external NAD(P)H requires that internal NADH oxidation be linked to complex I and that all electron transport occurs via the cytochrome path. Given the presence of both the rotenone-insensitive dehydrogenase and the alternative path, ADP : O ratios may in fact exhibit a high degree of variability.

ATP formation and CO_2 production *in vivo*

From information discussed above the relationship between ATP production and CO_2 release can be calculated, as can the consumption of O_2 during the complete oxidation of hexose and other substrates in the plant cell. This theoretical value of ATP production can then be compared with some experimental data. These calculations assume an ADP : O ratio of 3.0 for internal NADH and 2.0 for succinate or external NAD(P)H.

ATP production coupled to hexose oxidation

Since ADP : O ratios for the oxidation of the TCA cycle intermediates and external NADH have been established, it should be possible to calculate the ATP production per hexose molecule which is fully oxidized to CO_2 and H_2O in the cytosol and the mitochondria. We can carry out these calculations assuming glycolysis ends with either the production

of 2 molecules of pyruvate or 2 molecules of malate per hexose.

If pyruvate is the endproduct the following analysis applies:

$$\text{Hexose} + 2NAD^+ + 2ADP + 2Pi \rightarrow$$
$$2 \text{ pyruvate} + 2NADH + 2H^+ + 2ATP \quad [14.1]$$

If the two molecules of NADH are oxidized via the externally facing dehydrogenase of the electron transport chain, 4ATP will be produced. Glycolysis will therefore yield 6ATP. The subsequent oxidation of pyruvate via the TCA cycle would occur as follows:

$$2 \text{ pyruvate} + 2FAD + 8NAD^+ + 2ADP + 2Pi \rightarrow$$
$$6CO_2 + 2FADH_2 + 8NADH + 8H^+ + 2ATP \quad [14.2]$$

Oxidation of 2 molecules of FAD_2 via complex II will yield 4ATP while oxidation of 8 molecules of NADH via complex I will yield 24ATP. The result is that TCA cycle oxidation of 2 molecules of pyruvate generates 30ATP. Hence, the ATP yield for complete oxidation of a hexose via glycolysis and the TCA cycle is 36ATP/hexose. The net reaction becomes:

$$\text{Hexose} + 6O_2 + 36ADP + 36Pi \rightarrow 6CO_2 + 6H_2O + 36ATP \quad [14.3]$$

If malate is considered the endproduct of glycolysis the analysis differs slightly but the final ATP yield remains unchanged. In this case PEP is converted to OAA via PEP carboxylase, rather than being converted to pyruvate via pyruvate kinase. The OAA is then reduced to malate. The overall reaction for glycolysis becomes:

$$\text{Hexose} + 2CO_2 \rightarrow 2 \text{ malate} \quad [14.4]$$

The reason there is no net ATP production is that pyruvate kinase has been bypassed. The lack of net NADH production results from the consumption of the NADH produced by glyceraldehyde 3-phosphate dehydrogenase in the reduction of OAA to malate. The malate thus produced is imported into the mitochondria where it may be decarboxylated by malic enzyme to form pyruvate.

$$2 \text{ malate} + 2NAD^+ \rightarrow 2 \text{ pyruvate} + 2NADH + 2H^+ + 2CO_2 \quad [14.5]$$

The net effect is that NADH generated in the cytosol during glycolysis is imported into the mitochondrion via the sequential action of three enzymes (PEP carboxylase, malate dehydrogenase and malic enzyme). The oxidation of these 2NADH via complex I yields 6ATP, unlike the 4ATP generated from the oxidation of 2 molecules of cytosolic NADH. This gain of 2ATP offsets exactly the 2ATP lost in bypassing pyruvate kinase. As the pyruvate produced in this pathway is available for TCA cycle oxidation according to eqn [14.2], the final ATP yield for complete oxidation of hexose via 'malate glycolysis' is also 36ATP/hexose (eqn [14.3]).

If the oxidative pentose phosphate pathway, rather than glycolysis, is employed, the ATP yield per hexose is lower. The first part of this oxidation can be described as the cytosolic production of one molecule of glyceraldehyde 3-phosphate (GAP) from hexose:

$$1 \text{ hexose} + 1ATP \rightarrow 1 \text{ hexoseP} + 1ADP \qquad [14.6]$$

$$3 \text{ hexoseP} + 6NADP^+ \rightarrow$$
$$2 \text{ hexoseP} + 1GAP + 6NADPH + 3CO_2 \qquad [14.7]$$

Glyceraldehyde 3-phosphate can be further transformed, starting with the reactions of glycolysis and ending with pyruvate. The sum of these reactions is:

$$1GAP + NAD^+ + ADP \rightarrow 1 \text{ pyruvate} + ATP + NADH \qquad [14.8]$$

Combining [14.6], [14.7] and [14.8] yields:

$$1 \text{ hexose} + 1NAD^+ + 6NADP^+ \rightarrow$$
$$1 \text{ pyruvate} + 1NADH + 6NADPH + 3CO_2 \qquad [14.9]$$

The oxidation of this NADH and NADPH by the externally facing dehydrogenase (ADP:O ratio of 2) yields 14ATP. The subsequent oxidation of the pyruvate according to the stoichiometry of eqn [14.2] would yield 15ATP. As a result, complete oxidation of the hexose via the pentose phosphate pathway and the TCA cycle yields 29ATP. The balanced equation can be written as:

$$\text{Hexose} + 29ADP + 29Pi + 6O_2 \rightarrow 6CO_2 + 29ATP + 6H_2O \qquad [14.10]$$

O_2 consumption and CO_2 release coupled to the oxidation of hexose and other substrates: the respiratory quotient

The molar ratio of CO_2 released to O_2 consumed is called the respiratory quotient (RQ). From reaction eqns [14.3] and [14.10] it is calculated that the RQ for the complete oxidation of hexose to $CO_2 + H_2O$ is 1.0, irrespective of the pathways involved in the

oxidation. Complete oxidation of substrates which are more oxidized than hexose yields an RQ greater than 1. From reaction eqns [14.2] and [14.5] it can be calculated that the complete oxidation of malate yields an RQ of 1.3. The oxidation of more reduced substrates, e.g. fatty acids in some seeds, gives a RQ smaller than 1. Oxidation of fatty acids produces 1NADH per acetyl-CoA, the oxidation of which produces $2CO_2$, 3NADH and $1FADH_2$. Fatty acid oxidation thus gives an RQ of 0.8. Conversely, when a net synthesis of malate occurs in the cell or when fatty acids are synthesised, the RQ is smaller than 1 and greater than 1, respectively. Transfer of electrons to acceptors other than O_2, e.g. nitrate or sulfate, also increases the RQ. In most tissues which do not rapidly metabolize fatty acids or organic acids, the RQ is slightly above 1.0.

Regulation of the partitioning of electrons between the cytochrome and the alternative pathways and between the two internal NADH dehydrogenases

The existence of two respiratory pathways, both transporting electrons from ubiquinol to O_2, raises the question how electron flow is regulated between the two paths. This is particularly relevant, since the cytochrome path and the internal rotenone-sensitive dehydrogenase are coupled to proton extrusion, whereas both the rotenone-insensitive internal dehydrogenase and transport of electrons from ubiquinol to O_2 via the alternative path are not coupled to the generation of a proton-motive force.

The cytochrome path and the alternative path

Bahr and Bonner (1973) concluded that simple competition for electrons between the alternative path and the cytochrome path cannot explain the experimental data. Further evidence has come from 'titration experiments' (Fig. 14.6). In these experiments O_2 uptake is measured at a range of concentrations of an inhibitor of the cytochrome path in both the absence and presence of an inhibitor, which fully blocks the alternative path. If inhibition of the

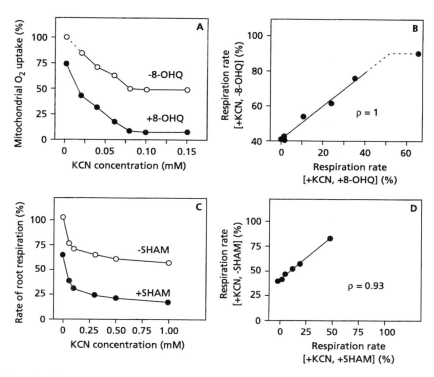

Fig. 14.6 O_2 uptake by mitochondria isolated from callus-forming discs of *Solanum tuberosum* (A) and intact roots of *Carex acutiformis* (C) at a range of concentrations of an inhibitor of the cytochrome path (KCN), in the absence and presence of a concentration of an inhibitor which fully blocks the alternative path. In B and D, the rates obtained in the absence of an inhibitor of the cytochrome path are plotted against those obtained in the presence of an inhibitor blocking the cytochrome path. A straight line with a slope of 1 indicates that the alternative path does not operate until the cytochrome path reaches saturation. In A the applied inhibitor of the alternative path was 8-hydroxyquinoline (8-OHQ); in C it was salicylhydroxamic acid (SHAM). For further explanation, see text. Data on isolated mitochondria: unpublished information provided by A. M. Wagner; data on intact roots: redrawn from information in Van der Werf *et al.* (1988).

alternative path increased the flow of electrons through the cytochrome path, inhibitors of the cytochrome path would be more effective under these conditions than in the absence of an inhibitor of the alternative path. Consequently, the slope of the plot describing the effect of KCN on respiration in the absence of an inhibitor of the alternative path against that in the presence of KCN would be less than 1. When this is not the case, inhibition of the alternative path does not affect the activity of the cytochrome path. In such a situation, the alternative path does not compete for electrons with the cytochrome path, and is active only when the cytochrome path is virtually saturated with electrons. However, it is important to first make titration plots, such as described in Fig. 14.6, before

drawing any conclusions about specific tissues. It has recently been found that the cytochrome path is *not* invariably saturated when the alternative path is active (Atkin *et al.*, 1995). In such a situation there may well be competition for electrons between the two respiratory pathways.

The partitioning of electrons between the two electron transport pathways is explained by the differential response of the cytochrome path and the alternative path to their common substrate, i.e. reduced ubiquinone. The activity of the cytochrome path increases linearly with the fraction of ubiquinone that is in its reduced state. In contrast, the alternative path shows no appreciable activity until a substantial (30–40%) fraction of the ubiquinone is in its reduced state, after which the activity increases very rapidly (Fig. 14.7). The curve describing the activity of the alternative path shifts to the left in the presence of pyruvate, but not because pyruvate is used as a respiratory substrate. Other organic acids have similar effects (Wagner *et al.*, 1989, Millar *et al.*, 1993). Pyruvate is likely to accumulate when its production in glycolysis is not matched by its oxidation by the mitochondria. The response of the alternative oxidase to reduced ubiquinone, in combination with the allosteric effect of pyruvate and the increase in activity when the oxidase becomes 'reduced', allows the alternative path to function as an 'overflow'. However, if effectors shift the curves describing the activity of the respiratory pathways as a function of Q_r/Q_t closer to

each other, competition between the two pathways may well occur.

The application of specific inhibitors has demonstrated that the alternative path contributes to normal respiration, particularly in roots where it may be responsible for up to 40% of all respiration. The results obtained with inhibitors of the alternative path have been confirmed using stable oxygen isotopes. Relative to ^{16}O, ^{18}O is consumed less rapidly by the alternative oxidase than by cytochrome oxidase. This difference in 'isotope discrimination' can be used to quantify the contribution of the two pathways *in vivo* (Guy *et al.*, 1992). However, the results of the two methods do not invariably agree. This is to be expected if the cytochrome path *in vivo* is not fully saturated; in such a situation addition of SHAM leads to an increased activity of the cytochrome path and hence the activity of the alternative path is greater than deduced from the inhibition of respiration by SHAM. The discrepancy between the two methods emphasizes that inhibition of SHAM may greatly underestimate the quantitative significance of the alternative path and that the isotope discrimination technique is to be preferred.

The two internal dehydrogenases

There are two internal dehydrogenases capable of oxidizing NADH. However, only complex I is inhibited by rotenone (Menz *et al.*, 1992). The rotenone-insensitive enzyme also oxidizes NADPH (Rasmusson *et al.*, 1993). The K_m of the internal rotenone-sensitive dehydrogenase is 8 μM, which is an order of magnitude lower than that of the insensitive one (Møller and Palmer, 1982). The lower affinity of the rotenone-insensitive dehydrogenase, and the lack of any proton extrusion, suggest that it is presumably operative when the NADH/NAD ratio is high, or when the availability of ADP is low, such as under state 4 conditions. Figure 14.8 shows that rotenone had very little effect on O_2 uptake of freshly isolated pea leaf mitochondria in state 4 (Day *et al.*, 1987). However, when these mitochondria are deprived of their endogenous NAD, rotenone has a marked effect on the state 4 rate of O_2 uptake. Addition of NAD, which is readily taken up and converted into NADH in the presence of an NAD-linked substrate such as glycine, increases O_2 uptake in the presence of

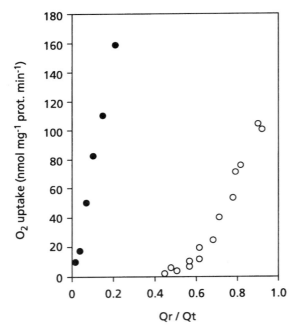

Fig. 14.7 Dependence of mitochondrial oxygen uptake on the fraction of ubiquinone that is in its reduced state (Q_r/Q_t). Filled symbols: total O_2 uptake in state 3; open symbols: O_2 uptake in the presence of an inhibitor of the cytochrome path. Measurements were made on mitochondria isolated from hypocotyls of *Glycine max*, using succinate as a substrate; Q_r/Q_t was varied by inhibiting succinate oxidation to varying degrees with malonate. Redrawn from information in Dry *et al.* (1989).

rotenone. When NAD is added prior to rotenone, this inhibitor has very little effect. The data in Fig. 14.8 lead to the following conclusions: (1) inhibition by rotenone demonstrates that the rotenone-insensitive path has sufficient capacity to take over the role of the sensitive path; and (2) rotenone only has an appreciable effect when the NADH level in the mitochondria is low. Thus it appears that the rotenone-insensitive bypass can only become active in the presence of a high level of NADH in the mitochondria. Note, however, that the results in Fig. 14.8 do not provide information about the operation of the insensitive dehydrogenase in the absence of inhibitors. Furthermore, evidence for the operation of the rotenone-insensitive path *in vivo* is lacking. In the absence of any suitable inhibitor of the rotenone-insensitive bypass such evidence will be hard to obtain.

Regulation of electron transport through the cytochrome path

Respiratory control, as referred to in Fig. 14.4, can be explained in chemiosmotic terms. In the absence of ADP, the proton-motive force increases and restricts the flow of electrons to O_2. Information such as

included in Fig. 14.4 gives the impression that there is an abrupt transition from state 3 to state 4. However, more subtle experimentation with isolated mitochondria has revealed that mitochondria can also operate in a state intermediate between state 3 and state 4. More importantly, it has become apparent that mitochondria mostly operate in this intermediate state *in vivo*.

Isolated mitochondria

As shown in Fig. 14.4, addition of ADP to isolated mitochondria causes a rapid increase in respiration until all ADP is depleted and transition to state 4 occurs. When small doses of ADP are continually added, this does not occur. This can be done with the 'hexokinase–glucose system'. ATP is added to the mitochondrial suspension, in conjunction with hexose and the enzyme hexokinase. This enzyme catalyzes the formation of hexose phosphate, thus releasing ADP. The rate at which ADP is released can be carefully chosen by varying the added activity of hexokinase.

When ADP is added with the 'hexokinase–glucose system', a range of oxidation rates can be obtained (Fig. 14.9). Simultaneously with the decrease in

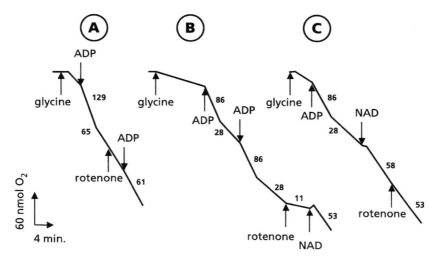

Fig. 14.8 O_2 uptake by mitochondria isolated from *Pisum sativum* leaves. Trace A was obtained with untreated mitochondria, traces B and C with mitochondria depleted of NAD. Added NAD is rapidly taken up by mitochondria and converted into NADH in the matrix, in the presence of NAD-linked substrates, e.g. glycine. Assay conditions were: 10 mM glycine, 0.4 μmol ADP, 0.2 mM NAD, 15 μM rotenone; numbers near the traces give the rate of O_2 uptake in nmol mg^{-1} protein min^{-1}. Data from Day *et al.* (1987).

succinate oxidation the $\Delta\Psi$ increases. This is to be expected, since only in the presence of ADP is the re-entry of protons via complex V possible. When the rate of ADP phosphorylation is controlled by varying degrees of inhibition of ADP import (using carboxyatractyloside), the same relationship between $\Delta\Psi$ and the rate of succinate oxidation is found (Fig. 14.9). Clearly, intermediate states of mitochondrial respiration are possible.

The control of the activity of the cytochrome path in intact tissues

Having established that mitochondrial states, intermediate between state 3 and state 4, can be

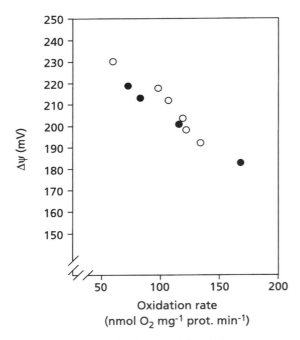

Fig. 14.9 The relationship between $\Delta\Psi$ and the rate of succinate oxidation in mitochondria isolated from potato tubers. The phosphorylation was limited by the rate of supply of ADP (closed symbols), using the 'hexokinase–glucose system'; the enzyme activity was varied from 0–17 nkat. The phosphorylation was also controlled by varying the concentration of carboxyatractyloside from 0 to 0.2 μM (open symbols). This compound inhibits the import of ADP into the mitochondria. $\Delta\Psi$ was measured with the TPP$^+$ method. Data from Diolez and Moreau (1987).

experimentally induced *in vitro*, it is of interest to find out in which state the mitochondria operate *in vivo*. This can be done by using combinations of specific inhibitors, uncouplers and substrates providing electrons for the respiratory chain.

If addition of uncoupler stimulates the rate of respiration in the absence of an inhibitor of the alternative path, respiration must have been limited by adenylates. However, this does not necessarily indicate limitation of oxidative phosphorylation by adenylates, as illustrated for isolated mitochondria (Fig. 14.9). Rather, it may be the substrate supply to the mitochondria which is restricted by adenylates, e.g. via the control of glycolysis. However, if the respiration is first shown to be reduced by an inhibitor of the alternative path (e.g. SHAM), and still stimulated by uncoupler in the presence of SHAM, it must be electron transport *per se* that is limited by ADP, in a manner illustrated in Fig. 14.9. If uncoupler stimulates in the absence but not in the presence of SHAM, adenylates control respiration via the substrate supply to the mitochondria. Comparison of the effects of an exogenous substrate which escapes the control of glycolysis (e.g. glycine oxidation during photorespiration) with one that is oxidized via glycolysis (e.g. sucrose) allows us to discriminate between the two possibilities of control by adenylates.

Table 14.4 shows that indeed adenylates control the rate of O$_2$ uptake *in vivo* and that this control is sometimes exerted via an effect on the substrate supply (Table 14.4a), and sometimes through limitation of the flux of electrons through the respiratory chain (Table 14.4b). The relief of adenylate control of the cytochrome path may actually decrease the flux through the alternative path. However, respiration is not invariably controlled by adenylates. Respiration of *Zea mays* root tips, depleted of sugars, is stimulated by exogenous sugars, but not by DNP. The uncoupler is only effective in the presence of a high endogenous level of sugars or when glucose is supplied (Saglio and Pradet, 1980). A similar observation was made on *Triticum aestivum* leaves, harvested at the end of the night (low carbohydrate levels) and after several hours of photosynthesis (high carbohydrate levels) (Azcón-Bieto et al., 1983; Table 14.4c).

From current information it appears that adenylate control of electron transport and glycolysis often

Table 14.4 Control of respiration of intact tissues: (a) by adenylates, via their effect on glycolysis (Day *et al.*, 1985); (b) by adenylates, via their effect on the cytochrome path (Atkin *et al.*, 1992); and (c) by the substrate supply to the mitochondria (Azcón-Bieto *et al.*, 1983). For further explanation, see text.

(a) Leaves of *Lolium perenne*. Rates are expressed as percentages of the basal rate ($2.0 \, nmol \, g^{-1} \, FW \, s^{-1}$). Applied concentrations: CCCP, $2 \, \mu M$; sucrose and glycine, $40 \, mM$; SHAM, $8 \, mM$; KCN, $0.2 \, mM$.

Addition	O_2 consumption (%)
A. + CCCP	122
+ CCCP + sucrose	136
+ CCCP + sucrose + SHAM	135
+ CCCP + sucrose + SHAM + KCN	26
B. + sucrose	102
+ sucrose + CCCP	123
C. + glycine	119
+ glycine + CCCP	132

(b) Leaves of *Cichorium intybus*. Rates are expressed as percentages of the basal rate ($2.35 \, nmol \, g^{-1} \, FW \, s^{-1}$). Applied concentrations: CCCP, $2 \, \mu M$; SHAM, $10 \, mM$; KCN, $0.5 \, mM$.

Addition	O_2 consumption (%)
+ KCN	13
+ SHAM	95
+ CCCP	120
+ SHAM + CCCP	113

(c) Leaves of *Triticum aestivum*, harvested at the end of the night (A and B) and after 6 hours of photosynthesis (C and D). Rates are expressed as percentages of the basal rate (0.7 and $1.0 \, \mu mol \, m^{-2} \, s^{-1}$, for A and B, and for C and D, respectively). SHAM had no effect on the respiration of control leaves harvested at the end of the night and inhibited 29% when leaves were harvested after 6 hours of photosynthesis. KCN slightly stimulated and inhibited 13%, respectively. Applied concentrations: FCCP, $1 \, \mu M$; sucrose, $60 \, mM$; glycine, $100 \, mM$; SHAM, $5 \, mM$; KCN, $0.3 \, mM$.

Addition	O_2 consumption (%)
A. + sucrose	143
+ sucrose + SHAM	118
+ sucrose + SHAM + KCN	0
B. + FCCP	110
+ FCCP + sucrose	183
C. + sucrose	100
+ sucrose + FCCP	120
+ sucrose + FCCP + SHAM	98
D. + glycine	103
+ glycine + FCCP	130

coincides with a regulation by the availability of substrate. Substrate not only controls respiration in 'starved' cells, but also in tissues going through a normal diel cycle (Table 14.4c). Increased substrate levels tend to increase the activity of the alternative path, rather than that of the cytochrome path (Lambers, 1985).

The physiological function of the alternative path

In trying to understand the physiological role of the cyanide-resistant path, it is important to take into account its non-phosphorylating nature and its operation during high rates of substrate oxidation.

Heat production

Most recent textbooks on plant physiology mention the role of the alternative path in heat production in the male reproductive structures of Araceae and a range of other species. During its 'respiratory crisis' the respiration rate of this reproductive organ approaches the enormously high rate in the flying muscles of humming birds, and the temperature may rise to 10 °C above ambient. Consequently, odoriferous compounds are volatilized and pollinators are attracted. During the respiratory crisis, respiration is largely cyanide-resistant.

Not only the respiration of some reproductive organs, but also that of roots and leaves is (partly) cyanide-resistant. The role of the alternative path in roots and leaves is unlikely to be that of heat production and a number of hypotheses have been put forward to explain the functioning of the alternative path in these organs.

Energy overflow

Experiments as described in Figs 14.6 and 14.7, demonstrate that the alternative path may be operative when the cytochrome path reaches its saturation. Furthermore, increases in carbohydrate supply to cells increase the activity of the alternative path. These observations have led to the 'energy overflow hypothesis' (Lambers, 1985), which states that respiration via the alternative path only proceeds in the presence of high concentrations of respiratory substrate which 'flood' the cytochrome path. It considers the alternative path as a 'coarse control' of carbohydrate metabolism, but certainly not as an alternative to the finer control by adenylates discussed above. Since the alternative path in the roots of many species contributes to respiration without the cytochrome path being saturated, the energy overflow hypothesis as originally proposed is too simple. Allosteric effects on the alternative oxidase also have to be taken into account.

Possibly, the continuous oxidation of part of the substrate via a non-phosphorylating electron transport path allows the plant to increase the availability of carbon for sinks which suddenly arise, e.g. upon a decrease of the water potential in the root environment (Lambers, 1985). It is also likely that the alternative oxidase activity prevents the production of superoxide and/or hydrogen peroxide. Superoxide is produced when electron transport through the cytochrome path is impaired, e.g. due to low temperature or desiccation injury, and this is partly due to a reaction of ubisemiquinone with molecular oxygen (Purvis and Shewfelt, 1993). Superoxide, like other free radicals, may lead to severe metabolic disturbances. However, the various interpretations of the physiological function of an 'energy overflow' remain speculative in the absence of pertinent results with plants lacking the alternative path. Such plants are now available.

Energy overcharge

An increased availability in ADP in cells in which the cytochrome path operates at maximum capacity may lead to increased operation of the alternative path. This was termed the 'energy overcharge model' (De Visser *et al.*, 1986) and is likely to be valid under non-steady-state conditions and/or when the activity of the cytochrome path is no longer controlled by adenylates.

Continuation of respiration in the presence of inhibitors

Naturally occurring inhibitors of the cytochrome path, such as sulfide, carbon dioxide and cyanide, may reach such high concentrations in the tissue that respiration via the cytochrome path is partly or fully inhibited (Palet *et al.*, 1991). The presence of an alternative path, which is unaffected by such inhibitors, may allow continued respiration and ATP production, albeit with low efficiency, under such conditions.

NADH oxidation in the presence of a high energy charge

If cells require a large amount of carbon skeletons (e.g. oxoglutarate or succinate) but do not have a high demand for ATP, the operation of the alternative path could prove useful. However, such a situation is most unlikely *in vivo*. Whenever the rate of carbon skeleton production is high, there tends to be a great need for ATP to further metabolize and incorporate these skeletons. Also, when the carbon skeletons are used for the synthesis of amino acids, significantly more NADH is required for the reduction of nitrate (if this is the source of N) than generated in the production of carbon skeletons.

Cells from soybean root nodules, which are infected with *Rhizobium* bacteroids and rapidly synthesize organic acids, have less alternative path activity than non-infected cells from the same nodules or other tissues of the same plants (Table 14.5). Hence, rapid synthesis of organic acids is by no means invariably associated with greater activity of the alternative path.

Lance *et al.* (1985) have suggested the need for a non-phosphorylating path to allow rapid oxidation of malate in the absence of a large need for ATP. Such a situation occurs during rapid malate decarboxylation in CAM plants. If this decarboxylation occurs in the dark, it is indeed associated with an increased activity

Table 14.5 Cyanide-resistant, SHAM-sensitive respiration (V_{alt} = alternative pathway capacity) in root nodules of *Glycine max*. Measurements were made on infected and uninfected cells (nmol s^{-1}). (Data from Kearns *et al.*, 1992.)

Root nodules	O_2 consumption		Resistance
	Control	V_{alt}	%
Infected cells	1.0	0.0	0
Uninfected cells	0.8	0.4	49

of the alternative path (Robinson *et al.*, 1992). However, malate decarboxylation naturally occurs in the light and it remains to be confirmed that the alternative path plays a vital role in crassulacean acid metabolism.

Dark respiration in green cells during the light

In the light there are some fundamental differences between a photosynthetically active and a heterotrophic cell. A photosynthetic cell has two systems to produce ATP: photophosphorylation and oxidative phosphorylation. For a long time it was believed that photophosphorylation drains all the ADP away from the mitochondria and so restricts oxidative phosphorylation. A number of developments have changed this view. First, the discovery of the non-phosphorylating alternative path: the path would allow electron transport to continue even at low levels of ADP. Second, the level of ADP in the mitochondria is not as low in the light as originally believed. And third, the observation that glycine has preferred access to the respiratory chain also places this hypothesis in doubt. It is no longer a question whether dark respiration continues during the light, but rather, to which extent it does.

Adenylate levels in cell compartments during light and dark

Using a technique for rapid (within 0.1 s) preparation of subcellular fractions, adenine nucleotide levels have been measured in the cytosol, chloroplasts and mitochondria from leaf protoplasts. Levels were measured in protoplasts exposed to light or kept in the dark.

Light does not have a significant effect on the total level of adenylates in the protoplasts. However, it decreases the level of ADP and increases that of ATP in the mitochondria, whereas the levels in the cytosol are virtually unaffected. Moreover, the ATP/ADP ratio in the mitochondria at which electron flow through the cytochrome path becomes restricted exceeds 20 (Wiskich and Dry, 1985). This is considerably higher than that in mitochondria during

photosynthesis (Gardeström and Wigge, 1988). Clearly, mitochondrial electron flow in the light is not fully restricted by the lack of ADP in the cytosol and/ or by the high ATP/ADP ratio in the mitochondria.

Effects of inhibitors of oxidative phosphorylation on photosynthesis

Oligomycin, an inhibitor of oxidative phosphorylation (Table 14.1), inhibits the rate of photosynthetic O_2 evolution in protoplasts as well as intact leaves (Krömer *et al.*, 1992). The same inhibitor has no effect on disrupted protoplasts, because mitochondria and chloroplasts no longer interact in the broken cells. The inhibition of photosynthesis by oligomycin in the intact systems shows that interruption of oxidative phosphorylation inhibits the rate of photosynthesis. This demonstrates the occurrence of oxidative phosphorylation, and thus operation of the cytochrome path, during the light. It also suggests a role for dark respiration in the ATP supply for CO_2 assimilation, presumably in the synthesis of sucrose in the cytosol.

Concluding remarks

In the past several decades, considerable progress has been made in our understanding of the physiology, biochemistry and molecular biology of mitochondrial energy metabolism. In spite of this progress there are still large gaps in our knowledge. Given the role of respiratory rates in determining plant productivity, further insights are required into the problems associated with plant respiration. Only with the combined efforts of plant physiologists, biochemists and molecular biologists will major progress be achieved. This may lead to interesting applications of fundamental knowledge to the rational improvement of crop productivity.

Acknowledgements

I wish to thank Owen Atkin, Ad Borstlap and Ingeborg Scheurwater who greatly contributed to this chapter by their constructive criticism.

Further reading

Douce, R. (1985). *Mitochondria in Higher Plants. Structure, Function, and Biogenesis*, Academic Press, Orlando, FL.

Douce, R. and Day, D. A. (eds) (1985). *Enclyclopedia of Plant Physiology*, New Series, Vol. 18, Springer-Verlag, Berlin.

Lambers, H. and Van der Plas, L. H. W. (eds) (1992). *Plant Respiration. Molecular, Biochemical and Physiological Aspects*, SPB Academic Publishing, The Hage.

Nicholson, D. G. (1982). *Bioenergetics. An Introduction to the Chemiosmotic Theory*, Academic Press, London.

References

Atkin, O. K., Cummins, W. R. and Collier, D. E. (1992). Regulation of electron transport in leaf slices of light and dark-growth belgium endive. In *Plant Respiration. Molecular, Biochemical and Physiological Aspects*, eds H. Lambers and L. H. W. Van der Plas, SPB Academic Publishing, The Hague, pp. 535–40.

Atkin, O. K., Villar, R. and Lambers, H. (1995). Partitioning of electrons between the cytochrome and the alternative pathways in intact roots. *Plant Physiol.*, **108**, 1179–1183.

Azcón-Bieto, J., Day, D. A. and Lambers, H. (1983). The regulation of respiration in the dark in wheat leaf slices. *Plant Sci. Lett.* **32**, 313–20.

Bahr, J. T. and Bonner, W. D. (1973). Cyanide-insensitive respiration. II. Control of the alternate pathway. *J. Biol. Chem.* **248**, 3446–50.

Day, D. A., De Vos, O. C., Wilson, D. and Lambers, H. (1985). The regulation of respiration in the leaves and roots of two *Lolium perenne* populations with contrasting mature leaf respiration rates and yield. *Plant Physiol.* **78**, 678–83.

Day, D. A., Wiskich, J. T. and Dry, I. B. (1987). Regulation of ADP-limited respiration in isolated plant mitochondria. In *Plant Mitochondria. Structural, Functional and Physiological Aspects*, eds A. L. Moore and R. B. Beechey, Plenum Press, New York, pp. 59–66.

De Visser, R., Spreen Brouwer, K. and Posthumus, F. (1986). Alternative path mediated ATP synthesis in roots of *Pisum sativum* upon nitrogen supply. *Plant Physiol.* **80**, 295–300.

Diolez, P. and Moreau, F. (1987). Relationships between membrane potential and oxidation rate in potato mitochondria. In *Plant Mitochondria. Structural, Functional and Physiological Aspects*, eds A. L. Moore and R. B. Beechey, Plenum Press, New York, pp. 17–25.

Dry, I. B., Moore, A. L., Day, D. A. and Wiskich, J. T. (1989). Regulation of alternative pathway activity in plant

mitochondria. Non-linear relationship between electron flux and the redox poise of the quinone pool. *Arch. Biochem. Biophys.* **273**, 148–57.

Guy, R. D., Berry, J. A., Fogel, M. L., Turpin, D. H. and Weger, H. G. (1992). Fractionation of the stable isotopes of oxygen during respiration by plants – the basis of a new technique to estimate partitioning to the alternative path. In *Plant Respiration. Molecular, Biochemical and Physiological Aspects*, eds H. Lambers and L. H. W. Van der Plas, SPB Academic Publishing, The Hague, pp. 443–53.

Kearns, A., Whelan, J., Young, S., Elthon, T. E. and Day, D. A. (1992). Tissue-specific expression of the alternative oxidase in soybean and siratro. *Plant Physiol.* **99**, 712–17.

Krömer, S., Hanning, I. and Heldt, H. W. (1992). On the sources of redox equivalents for mitochondrial oxidative phosphorylation in the light. In *Plant Respiration. Molecular, Biochemical and Physiological Aspects*, eds H. Lambers and L. H. W. Van der Plas, SPB Academic Publishing. The Hague, pp. 167–75.

Lambers, H. (1985). Respiration in intact plants and tissues: Its regulation and dependance on environmental factors, metabolism and invaded organisms. In *Encyclopedia of Plant Physiology*, New Series, eds R. Douce and D. A. Day, Springer-Verlag, Berlin, pp. 418–73.

Lambers, H., Day, D. A. and Azcón-Bieto, J. (1983). Cyanide-resistant respiration in roots and leaves. Measurements with intact tissues and isolated mitochondria. *Physiol. Plant.* **58**, 148–54.

Lance, C., Chauveau, M. and Dizengremel, P. (1985). The cyanide-resistant pathway of plant mitochondria. In *Encyclopedia of Plant Physiology*, New Series, eds R. Douce and D. A. Day, Springer-Verlag, Berlin, pp. 202–247.

McIntosh, L. (1994). Molecular biology of the alternative oxidase. *Plant Physiol.* **105**, 781–6.

Menz, R. I., Griffith, M., Day, D. A. and Wiskich, J. T. (1992). Matrix NADH dehydrogenases of plant mitochondria and sites of quinone reduction by complex I. *Eur. J. Biochem.* **208**, 481–5.

Millar, A. H., Wiskich, J. T., Whelan, J. and Day, D. A. (1993). Organic acid activation of the alternative oxidase of plant mitochondria. *FEBS Lett.* **329**, 259–62.

Mitchell, P. (1966). Chemiosmotic coupling in oxidative and photosynthetic phosphorylation. *Biol. Rev.* **41**, 445–502.

Mitchell, J. A. and Moore, A. L. (1984). Proton stoichiometry of plant mitochondria. *Biochem. Soc. Trans.* **12**, 849–50.

Møller, I. M. and Palmer, J. M. (1982). Direct evidence for the presence of a rotenone-resistant NADH dehydrogenase on the inner surface of the inner membrane of plant nitochondria. *Physiol. Plant* **54**, 267–74.

Moore, A. L. and Rich, P. R. (1985). Organization of the respiratory chain and oxidative phosphorylation.

In *Encyclopedia of Plant Physiology*, New Series, eds R. Douce and D. A. Day, Springer-Verlag, Berlin, pp. 134–72.

Palet, A., Ribas-Carbo, M., Argiles, J. M. and Azcón-Bieto, J. (1991). Short-term effects of carbon dioxide on carnation callus cell respiration. *Plant Physiol.* **96**, 467–72.

Purvis, A. C. and Shewfelt, R. L. (1993). Does the alternative pathway ameliorate chilling injury in sensitive plant tissues? *Physiol. Plant* **88**, 712–18.

Rasmusson, A. G., Fredlund, K. M. and Møller, I. M. (1993). Purification of a rotenone-insensitive NAD(P)H dehydrogenase from the inner surface of the inner membrane of red beetroot mitochondria. *Biochim. Biophys. Acta* **1141**, 107–10.

Roberts, J. K. M., Wemmer, D. and Jardetzky, O. (1984). Measurements of mitochondrial ATP-ase activity in maize root tips by saturation transfer ^{31}P nuclear magnetic resonance. *Plant Physiol.* **74**, 632–9.

Robinson, S. A., Yakir, D., Ribas-Carbo, M., Giles, L., Osmond, C. B., Siedow, J. N. and Berry, J. A. (1992). Measurements of the engagement of cyanide-resistant respiration in the crassulacean acid metabolism plant *Kalanchoe daigremontiana* with the use of on-line oxygen isotope discrimination. *Plant Physiol.* **100**, 1087–91.

Saglio, P. H. and Pradet, A. (1980). Soluble sugars, respiration, and energy charge during aging of excised maize root tips. *Plant Physiol.* **66**, 516–19.

Siedow, J. N., Whelan, J., Kearns, A., Wiskich, J. T. and Day, D. A. (1992). Topology of the alternative oxidase in soybean mitochondria. In *Plant Respiration. Molecular, Biochemical and Physiological Aspects*, eds H. Lambers and L. H. W. Van der Plas, SPB Academic Publishing, The Hague, pp. 19–27.

Gardeström, P. and Wigge, B. (1988). Influence of photorespiration on ATP/ADP ratios in the chloroplasts, mitochondria, and cytosol, studied by rapid fractionation of barley (*Hordeum vulgare*) protoplasts. *Plant Physiol.* **88**, 69–76.

Umbach, A. L. and Siedow, J. N. (1993). Covalent and noncovalent dimers of the cyanide-resistant alternative oxidase protein in higher plant mitochondria and their relationship to enzyme activity. *Plant Physiol.* **103**, 845–54.

Van der Werf, A., Kooijman, A., Welschen, R. and Lambers, H. (1988). Respiratory costs for the maintenance of biomass, for growth and for ion uptake in roots of *Carex diandra* and *Carex acutiformis*. *Physiol. Plant.* **72**, 483–91.

Van der Werf, A., Welschen, R. and Lambers, H. (1992). Respiratory losses increase with decreasing inherent growth rate of a species and with decreasing nitrate supply. A search for explanations for these observations. In *Molecular, Biochemical and Physiological Aspects of Plant Respiration*, eds H. Lambers and L. H. W. Van der Plas, SPB Academic Publishing, The Hague, pp. 421–32.

Wagner, A. M., Kraak, M. H. S., Van Emmerik, W. A. M. and Van der Plas, L. H. W. (1989). Respiration of plant mitochondria with various substrates: Alternative pathway with NADH and TCA cycle derived substrates. *Plant Physiol. Biochem.* **27**, 837–45.

Wilkstrøm, M. and Babcock, G. T. (1990). Catalytic intermediates. *Nature* **348**, 16–17.

Williamson, R. L. and Metcalf, R. L. (1967). Salicylanilides: a new group of active uncouplers of oxidative phosphorylation. *Science* **158**, 1649–95.

Wiskich, J. T. and Dry, I. B. (1985). The tricarboxylic acid in plant mitochondria: Its operation and regulation. In *Encyclopedia of Plant Physiology*, New Series, eds R. Douce and D. A. Day, Springer-Verlag, Berlin, pp. 281–313.

Mitochondrial–Cytosol
Interactions

15 Protein import into the mitochondrion

Karl B. Freeman, Rebecca Hartlen and Debra L. Inglis

Introduction

The formation of new mitochondria occurs by growth and division of pre-existing mitochondria. However, of the hundreds of mitochondrial proteins, only about 60 are coded by plant mitochondrial DNA (mtDNA) and synthesized on mitochondrial ribosomes (see Chapter 12). The rest are coded by nuclear DNA, synthesized on cytosolic ribosomes and imported into mitochondria. To ensure efficient formation of functional mitochondria there must be coordination between the two genetic systems, especially since the respiratory and H^+-ATPase protein complexes of the mitochondrial inner membrane are composed of subunits, some of which are coded by mtDNA and some by nuclear DNA. Although little is known about how this coordination is achieved, there is considerable understanding of the genetic role of mtDNA and on the import of proteins by mitochondria, especially in yeast (*Saccharomyces cerevisiae*), fungus (*Neurospora crassa*) and mammals. The central problem of import was to determine how proteins synthesized on cytosolic polysomes are directed specifically to mitochondria and not to other locations. As will be discussed, specific import reflects the presence of targeting sequences, for example, amino-terminal presequences, on mitochondrial precursor proteins and their recognition by mitochondrial receptors leading to the protein's import, processing and direction to specific suborganellar sites. There are a growing number of studies on protein import into plant mitochondria and these indicate that the basic mechanisms will probably prove to be the same as in other organisms (see Moore *et al.*, 1994 for greater detail). These studies will be considered here, but most of the emphasis will be on results from yeast and *N. crassa*.

General model of import

The current model of import of matrix proteins into mitochondria of yeast, *Neurospora* and mammals is shown in Fig. 15.1. Note that import requires both cytosolic and matrix ATP, likely for the action of chaperones, and the inner membrane potential ($\Delta\Psi$) for translocation of precursors into or across this membrane. The steps in import are:

1. Imported mitochondrial proteins are synthesized on cytosolic free polyribosomes.
2. Most matrix proteins are synthesized with an amino-terminal presequence (see insert in Fig. 15.1). Presequences have been found to contain the information for targeting proteins to mitochondria. Some other details on presequences are given in Fig. 15.1.
3. In the cytosol, precursor proteins interact with chaperones (e.g. hsp70 and its associated proteins), which are thought to maintain precursors in an unfolded state competent for import.
4. The precursor then reacts with receptor proteins on the mitochondrial outer membrane (e.g. MOM19 and MOM72 in *N. crassa*).
5. The precursor then moves to a translocation complex (MOM complex) consisting of at least the outer membrane proteins MOM7, MOM8 and MOM38 in *N. crassa*. This transfer is thought to be facilitated by MOM22.
6. After passage through a pore in the translocation complex, the preprotein, led by its amino terminus, is thought to interact with a mitochondrial inner membrane protein complex, (e.g. MIM17, MIM23 and MIM44 in yeast). This

is facilitated by the $\Delta\Psi$ and might occur at contact sites between the outer and inner membranes.

7. The preprotein enters the matrix and associates with the mitochondrial chaperone mt-hsp70 which may pull the protein into the matrix.

8. The general mitochondrial processing peptidase (MPP) then cleaves the amino-terminal presequence from the protein.

9. It is thought that the chaperonin complex (cpn60/cpn10) in the matrix then allows folding of the protein into its mature conformation and assembly into homo- or hetero-oligomers.

10. Additional signals and mechanisms are necessary to direct proteins to the mitochondrial outer or inner membrane or to the intermembrane space.

Methodology

Import by isolated mitochondria

Detailed studies on the mechanism of import have been possible because isolated mitochondria take up protein precursors synthesized *in vitro*. The approach has been facilitated by *in vitro* translation of cloned cDNAs from plasmid vectors to produce precursors of single proteins (Douglas *et al.*, 1986). A standard import reaction contains mitochondria, precursor, rabbit reticulocyte lysate (used for synthesis of precursors), and energy sources including ATP and often a respiratory substrate. The precursor and mitochondria are usually from the same species. Details of the import mechanism are elucidated by manipulating experimental conditions; for example, by altering the ATP concentration, temperature or mitochondrial membrane potential. There are fewer studies on import by plant mitochondria and it has been suggested that this import is inefficient compared with that in other species (Moore *et al.*, 1994).

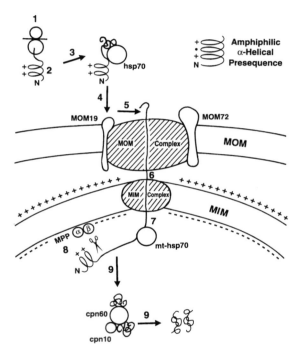

Fig. 15.1 Model of import of matrix proteins into mitochondria. Details are given in the text. An insert diagram shows the amphiphilic α-helical presequence. Presequences have a high content of basic (+) and hydroxyl (·) amino acids, hydrophobic residues which are not present in a continuous stretch, and lack acidic residues. The arrangement of the residues is such that the hydrophilic amino acids are on one side of the α-helix and hydrophobic amino acids on the other side; that is, the helix is amphiphilic (cf. Fig. 15.2). Abbreviations: cpn60 and cpn10, chaperonins of 60 kDa and 10 kDa respectively; hsp70, heat shock protein of 70 kDa; MIM, mitochondrial inner membrane; MOM, mitochondrial outer membrane; MPP, matrix processing protease (both α and β subunits are shown); mt-hsp70, mitochondrial heat shock protein of 70 kDa.

Identification and isolation of components involved in import

An understanding of the detailed mechanism of import can be achieved only by identifying, isolating and characterizing the components involved. This can be accomplished using both biochemical and genetic approaches. Biochemical approaches utilize the specific interaction of precursors with proteins of the import apparatus which allows their cross-linking; mitochondria are incubated with radio-labelled precursors, treated with cross-linking reagents and the import component identified by fractionation and determination of size. Antibodies are also used to identify potential components of the import machinery. If import is inhibited by incubation with monospecifc antibodies, the specific protein is likely involved in the import process. In contrast, the genetic approach requires gene disruption. For example, mutant strains of yeast are observed for growth on a non-fermentable carbon source and their ability to import protein analyzed. If import is defective, mutant proteins are identified by genetic complementation and their structure and role in import determined.

Nature of the precursor, targeting sequences and processing

Site of synthesis

Protein synthesis in the eukaryotic cell cytosol occurs on two classes of polyribosomes, those that are free and those that are bound to membranes, especially of the endoplasmic reticulum. Synthesis on bound ribosomes is termed co-translational since it is linked to translocation of precursor proteins across or into the membrane of the rough endoplasmic reticulum. Co-translational import into mitochondria has also been suggested in yeast since ribosomes synthesizing mitochondrial proteins are found bound to the outer membrane of mitochondria (Douglas *et al.*, 1986; Hartl *et al.*, 1989; Verner *et al.*, 1993), and since both *in vitro* and *in vivo* bulk protein import and synthesis are simultaneously inhibited by cycloheximide (Verner, 1993). However, the demonstration in yeast and *N. crassa* of cytosolic pools of mitochondrial proteins that could be imported when protein synthesis was blocked, pointed to post-translational import *in vivo* (Douglas *et al.*, 1986). Furthermore, this is clearly shown by the ability of isolated mitochondria, including those from plants, to import proteins in the absence of protein synthesis. At present, it appears that import may occur either co-translationally or post-translationally but, aside from in yeast, the evidence favours post-translational import.

Precursors of larger size

The most striking difference between mature and precursor forms of imported mitochondrial proteins is the larger size of most precursors. The precursor's size and the presence of a presequence are determined using sodium dodecyl sulfate–polyacrylamide gel electrophoresis to compare the mobility of the precursor, synthesized *in vitro* with that of the mature protein. Sizes are also determined from cDNA and gene sequences. Detailed lists of the sizes of mature and precursor forms of proteins are available (Hartl *et al.*, 1989). Based on some 60 examples, two general conclusions can be made: (1) proteins destined for the matrix, inner membrane or intermembrane space, but not the outer membrane, are synthesized as larger precursors containing amino-terminal presequences whose lengths vary from about 20 to 80 amino acids and (2) a small number of proteins of the matrix, inner membrane and intermembrane space are not synthesized as precursors of larger size. Although fewer plant precursor sequences are known they do conform to these generalizations (Moore *et al.*, 1994). Interesting variations, of unknown significance, between plant presequences and those from other organisms have been observed (Moore *et al.*, 1994). For example, the inner membrane ADP/ATP carriers of mammals, yeast and *N. crassa* lack presequences but the plant protein precursor contains one.

Characteristics of presequences

Presequences of imported mitochondrial proteins of yeast, *N. crassa* and mammals do not have extensive identity of sequence but do share a number of characteristics (Hartl *et al.*, 1989). These include a high content of positively charged amino acids, particularly arginine and hydroxyl amino acid residues, particularly serine, the presence of hydrophobic residues and a lack of negatively charged residues. These characteristics also apply to plant mitochondrial presequences (Moore *et al.*, 1994), where alanine and leucine are common hydrophobic residues. It has been hypothesized that presequences can form amphiphilic (amphipathic) α-helical structures (Figs 15.1 and 15.2) and that these may be important in import. An example is shown in Fig. 15.2 for the presequence of maize chaperonin 60. A short list of presequences is given in Table 15.1. The table allows comparison of plant and

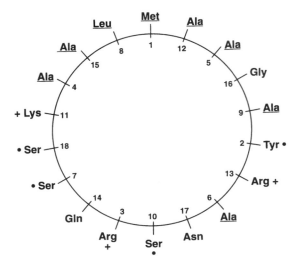

Fig. 15.2 Helical wheel plot of the amino-terminal portion of the presequence of maize mitochondrial chaperonin 60 (Moore *et al.*, 1994). The sequence yields an amphiphilic α-helix with hydrophilic amino acids and hydrophobic amino acids on opposite sides. The basic (+), hydroxyl (·) and more strongly hydrophobic (underlined) residues are shown.

homologous non-plant presequences. A detailed list is given by Hartl *et al.* (1989) and for plant mitochondria by Moore *et al.* (1994). In non-plant species arginine is most commonly found either at position -2 or -10 while in plants this residue occurs frequently at position -3 and less frequently at the -2 and -10 positions (Moore *et al.*, 1994). Other characteristics of plant presequences are discussed by Moore *et al.* (1994).

Targeting role of presequences

The role of presequences was determined directly using recombinant DNA technology. Presequences were joined to non-mitochondrial 'passenger' proteins by constructing expression plasmids containing their coding sequences adjacent to each other (Douglas *et al.*, 1986; Hartl *et al.*, 1989). *In vitro* expression of these fusion proteins and import by isolated mitochondria or import *in vivo* demonstrated that presequences contain the intracellular and

suborganellar targeting information to transport the protein into the mitochondrion (Douglas *et al.*, 1986; Hurt and van Loon, 1986; Hartl *et al.*, 1989; Chaumont *et al.*, 1994).

All of the presequence is not needed for targeting. In yeast, the most amino-terminal portion only is essential (Hurt and van Loon, 1986; Hartl *et al.*, 1989). There are two relevant studies in plants. In the first, a series of 3′-deletions were made in the sequence coding for the first 90 amino acids of the F_1-ATPase β-subunit precursor from *Nicotinia plumbaginifolia* (Chaumont *et al.*, 1994). The remaining 5′-sequences were then fused with the genes of either bacterial chloramphenicol acetyl transferase or β-glucuronidase and introduced into tobacco plants. Import of the fusion proteins was assessed by measuring the activities of these enzymes in mitochondria. The results indicated that the first 23 amino acids of the 54 amino acid presequence were sufficient for specific targeting of the fusion proteins. In a second study, import of maize superoxide dismutase by isolated maize mitochondria was decreased only partly by internal deletions in the presequence again indicating that all of the presequence is not necessary (White and Scandalios, 1989).

In contrast with presequence-containing proteins, the precise targeting sequence of proteins of inner mitochondrial compartments that are not synthesized as larger precursors is still not known. Some of these proteins contain amino termini that resemble presequences (e.g. sweet potato cytochrome *c* oxidase subunit Vc (Moore *et al.*, 1994)), but others do not (e.g. mammalian ADP/ATP carrier). Either the first third or final two thirds of the inner membrane ADP/ATP carrier and uncoupling protein contain targeting information sufficient to direct into mitochondria (Liu *et al.*, 1988). Outer membrane proteins also lack presequences but some, such as the outer membrane 70 kD protein of yeast (Table 15.1), have an amino-terminal sequence similar to the basic portion of presequences. This part of the 70 kD protein sequence can act as a matrix-targeting sequence in a fusion protein but in the normal protein it is not cleaved (Hurt and van Loon, 1986). Such proteins are presumably targeted to and retained in the outer membrane by a sequence (stop-transfer) which follows after the basic sequence.

Table 15.1 Amino-terminal presequences of precursors of some imported mitochondrial proteins.

Matrix

```
    +    ·  ·+ +     ·  ·+     ·+     ·+
1. MYRAAASLASKARQAGNSLATRQVGSRLAWSRNY

   ·+++   ··   +···++· ·+· ·     + ···· ··+ + ·      ·+   −
2. MASRRLLSSLLRSSSRRSVSKSPISNINPKLSSSSPSSKSRASPYGYLLTRAAEY

    +·  ·   +· + ·    ·++   +       +      ·+
3. MFKSGISAFARTARPSFAAASRRAVRPAALNLRAPALSR
```

Inner membrane – unknown location

```
   ·  + ·+    +    ·+·↓   ··+
4. MLSLRQSIRFFKPATRT LCSSRYLL
```

Inner membrane – cytosolic face

```
   ·     ++ +    −   + + ·+     + −·+  ·+ −  +    −    ·  ·+· +        ·   ·   ·
5. MSLGKKIRIGFDGFGRINRFITRGAAQRNDSKLPSRNDALKHGLDGLGSAGSKSFRALAAIGAGVSGLLSFATIAY

   +·  +·+·+·  ·  + +   ++·↓ ·· ··    −· +       ·           ·
6. MLARTCLRSTRTFASAKNGAFKFAKRS ASTQSSGAAAESPLRLNIAAAAATAVAAGSIAWYYHLYGFASA

   ·  ·+   +     +   +·↓ ····
7. MAPVSIVSRAAMRAAAAPARAVRA LTTSTALQ
```

Outer membrane – not a presequence

```
   +· ·+ +·    ·   ··            + ++ ·  +−−+++
8. MKSFITRNKTAILATVAATGTAIGAYYYYNQLQQQQQRGKKNTINKDEKK
```

The sequence given in their single letter codes are taken from Hartl *et al.* (1989) and Moore *et al.* (1994) and are as follows: 1, maize chaperonin 60; 2,3, *Hevea brasiliensis* and *N. crassa* β subunit of F_1ATPase respectively; 4, yeast cytochrome *c* oxidase subunit IV; 5,6, potato and *N. crassa* cytochrome c_1 respectively; 7, *N. crassa* Fe/S protein of the bc_1 complex; 8, yeast 70 kD protein. Basic (+), acidic (−) and hydroxyl (·) amino acyl residues are indicated above the sequences as are points of the first proteolytic cleavage (↓) in proteins cut twice.

Processing

Proteolytic processing of the precursors is clearly necessary to achieve the mature length of the protein. Early studies showed that matrices of mitochondria contain a protease which is dependent on zinc, manganese or cobalt ions for activity. A series of studies in *N. crassa*, yeast, rat and potato tuber led to the identification of this general mitochondrial processing peptidase (MPP) (Moore *et al.*, 1994; Glaser *et al.*, 1994). In all cases, the enzyme has 2 subunits (α and β) although in plants there may be more; in yeast, both subunits are essential for viability. The subunits are homologous and antigenically related to each other and amongst species. They form part of a family of metalloendopeptidases but their submitochondrial location differs as do their roles (Moore *et al.*, 1994; Glaser *et al.*, 1994). In rat and yeast, the subunits

form a soluble heterodimer and both subunits are necessary for activity (Moore *et al.*, 1994; Glaser *et al.*, 1994). In contrast, in *N. crassa*, the β subunit is core II protein of the inner membrane ubiquinol cytochrome *c* oxidoreductase (bc_1-complex) and it serves to enhance the low protease activity of the α subunit. Even more surprising, in plants, all of the subunits are core proteins of the bc_1 complex (Glaser *et al.*, 1994). What features of the presequence are important for the function of the MPP? Given the sequence diversity of targeting presequences, two features appear important: an amphiphilic α-helix and, at least in some cases in yeast and *N. crassa*, the arginine at the -2 position (Arretz *et al.*, 1994). Note that there are other processing peptidases either in the matrix or intermembrane space which are necessary for precursors cleaved in two steps (Moore *et al.*, 1994; Nunnari *et al.*, 1993).

Requirement of the membrane potential and ATP for import

Import of most precursors of matrix or inner membrane proteins but not those of the outer membrane, is blocked by uncouplers of oxidative phosphorylation which collapse the electrochemical potential (Hartl *et al.*, 1989). Under these conditions, the precursors do not even cross the outer membrane. Import is blocked whether or not the precursors have presequences. Of the two energy components of the electrochemical potential, the proton gradient (ΔpH) and membrane potential ($\Delta \Psi$), it is $\Delta \Psi$ which provides energy for import (Hartl *et al.*, 1989). In isolated plant mitochondria, protein import has been shown to occur independent of a proton gradient as long as a $\Delta \Psi$ is present (Moore *et al.*, 1994).

The role of the membrane potential was determined as follows. At a low temperature (e.g. 7 °C) and in the presence of $\Delta \Psi$, import of the *N. crassa* F_1-ATPase β subunit occurred but was stopped short of complete transfer (Hartl *et al.*, 1989). In this situation, the amino-terminal presequence was removed by the matrix protease, but the bulk of the protein was accessible to externally added protease. Import was completed at 25 °C even in the absence of the membrane potential. Thus, the membrane potential (matrix negative) is needed for an initial phase of import; that is, the interaction of the basic presequence with the inner membrane and electrophoretically driven transport of the presequence across it. The latter process also requires matrix ATP for matrix protein precursors (Wachter *et al.*, 1994).

In addition to $\Delta \Psi$, cytosolic ATP is required for import of some precursors (Wachter *et al.*, 1994). This is linked to the ATP requirement for the action of cytosolic chaperones which may keep the precursor in an import-competent conformation. This is considered in the next section. Further, matrix ATP is required for the import of matrix, some inner membrane and perhaps some intermembrane space proteins. Again, this is linked to the role of mitochondrial chaperones and chaperonins and will be discussed below.

The Import Process

Cytosol

Import of proteins into isolated mitochondria requires the presence of reticulocyte lysate suggesting that the lysate may contain cytosolic factors necessary for import. In addition, it is known that proteins having a stable tertiary structure are not imported into mitochondria (Glick and Schatz, 1991). Thus, it may be that the lysate contains proteins necessary to maintain precursors in an unfolded import-competent conformation. From these leads, a number of proteins, including chaperones and two presequence binding proteins, have been suggested to be important for import (Glick and Schatz, 1991; Moore *et al.*, 1994).

The major group of proteins thought to prevent folding or aggregation and to maintain proteins in an import-competent conformation are the 70 kDa heat shock proteins (hsp70s) (Glick and Schatz, 1991; Wachter *et al.*, 1994). It was shown that a yeast strain containing disrupted genes for cytosolic hsp70s could not import proteins into mitochondria (Moore *et al.*, 1994); but when one of the genes was present on a plasmid and expressed, import was restored indicating that cytosolic hsp70 was necessary for import. Further, hsp70s act as ATPases suggesting that the requirement of cytosolic ATP for import might be for hsp70 action (Glick and Schatz, 1991; Wachter *et al.*, 1994). It should be noted that hsp70s act together with another protein, DnaJ (MAS5 in yeast) (Moore *et al.*, 1994). To date, several plant genes encoding cytosolic hsp70s and a cDNA of a plant DnaJ protein of unknown subcellular location have been isolated; preliminary biochemical evidence suggests that the hsp70s may play a role in protein import into plant mitochondria (Moore *et al.*, 1994). When purified cytosolic hsp70 from *Vicia faba* was added to *in vitro* import assays with pea leaf mitochondria, binding of precursor proteins to mitochondria was increased, although the purified hsp70 had little effect on the overall amount of protein imported. These results indicate that cytosolic hsp70 may be required to expose a region, such as the presequence, to allow interaction with outer membrane receptors.

Import receptors

Import of proteins specifically into mitochondria requires not only that imported proteins have a targeting sequence but that this information is recognized by cytosolic factors and/or mitochondria. Attention has been focused on putative specific outer membrane protein receptors for recognizing targeting sequences. Early studies showed that import could be inhibited by protease treatment of isolated mitochondria or with antibodies to outer membrane proteins. Since such treatments affected import of precursors to different extents, it seemed likely that more than one type of receptor was present. Identification of receptors followed quickly in both *N. crassa* and yeast. Söllner, *et al.* (1989) prepared monospecific antibodies and Fab fragments against many of the outer membrane proteins of *N. crassa* mitochondria. Preincubation of *N. crassa* mitochondria with monospecific antibodies and Fab fragments directed against a 19 kDa protein (MOM19, for Mitochondrial Outer Membrane protein), resulted in inhibition of import of a subset of precursor proteins destined for various mitochondrial subcompartments. Further, the observation that this protein could be cross-linked to a surface-arrested precursor, and that the cross-linked product could be identified by immunoprecipitation using specific antibodies suggested that the receptor interacts directly with the precursor. Similar studies led to the discovery of the receptor MOM72 in *N. crassa* (Söllner *et al.*, 1990), and MAS20 and MAS70 (for Mitchondrial Assembly Proteins), the yeast homologues of MOM19 and MOM72 respectively (Ramage *et al.*, 1993). In yeast, a 32 kDa protein, p32, showing sequence identity with the phosphate carrier of the inner membrane, has been proposed to function as an additional import receptor, but this role remains controversial (Kiebler *et al.*, 1993).

Generation of mutants lacking one or both of the receptors gives an indication of the receptors' roles in import as well as whether or not they are essential for viability. For example, temperature-sensitive yeast mutants deficient in MAS20, MAS70, or both receptors were isolated (Ramage *et al.*, 1993). As growth was not completely inhibited in mutants lacking either MAS20 or MAS70, neither of these receptors is itself essential for viability. Analysis of import in these mutants as well as *in vitro* studies with specific antibodies indicate that the two receptors have overlapping specificities in yeast (Kiebler *et al.*, 1993; Hannavy, *et al.*, 1993). In

contrast, similar studies in *N. crassa* showed that mutants lacking MOM19 grew very poorly and protein import in the mutants was decreased by factors of 6 to 30 for most precursors. However, import of the ADP/ATP carrier was unaffected (Harkness *et al.*, 1994). This indicates that the two receptors have distinct precursor specificities with MOM19 specific for presequence-containing precursors and MOM72 specific for import of proteins without presequences, in agreement with earlier *in vitro* studies using specific antibodies (Kiebler *et al.*, 1993).

Passage into mitchondria

Nature of the problem

The hydrophobic portion of the phospholipid bilayer of membranes presents a thermodynamic barrier for the translocation of hydrophilic proteins. The difficulty this presents for the transport of such proteins would be overcome if they were to move through a proteinaceous pore. This is now generally accepted for import and proteins of the putative pore complexes of the outer and inner membranes have been identified (Kiebler *et al.*, 1993). Further, proteins destined for inner compartments must potentially cross the intermembrane space; there may be two mechanisms for this.

Crossing the outer membrane

Insertion and translocation of preproteins across the outer membrane is thought to occur at a site called the general insertion protein or pore (GIP). The GIP was identified as a protease-resistant common pathway for import by competition studies with different precursors (Kiebler *et al.*, 1993). At this site, preprotein intermediates were found which were resistant to externally added proteases and were located partly in the outer membrane and partly in the intermembrane space. Since an import intermediate at the GIP-stage could be extracted by protein denaturants, but not by detergents, it was concluded that the precursor crossed the outer membrane through a hydrophilic pore of the GIP. The proteins forming the GIP were identified by cross-linking to the precursor of the ADP/ATP carrier and by isolation of a high molecular mass complex from the outer membrane, which contained the import receptors MOM19 and MOM72 as well as the GIP (Kiebler *et al.*, 1993). Proteins of the *N. crassa* GIP include MOM7, MOM8, MOM38 and perhaps MOM30; yeast proteins include the MOM38 homologue and essential protein ISP42 (Import Site Protein of 42 kDa) and ISP6 (Hannavy *et al.*, 1993). An additional protein, MOM22, has also been identified in *N. crassa* and found to be complexed with MOM19, MOM72 and MOM38 (Kiebler *et al.*, 1993). MOM22 has an inverted orientation compared to other MOMs; its negatively charged amino terminus is exposed to the cytosol and could thus interact with positively charged presequences and facilitate transfer of preproteins from the receptors into the outer membrane at the GIP site. To date, the only MOM protein identified in plants is a potential homologue of MOM38 (ISP42) called PISP42 from purified pea leaf mitochondria (Moore *et al.*, 1994).

How are the components of the outer membrane import machinery targeted to the mitochondrial outer membrane? As a highly specific and efficient control system is needed to ensure correct localization of these MOM proteins, each component is not responsible for its own import and appears to have an individual pathway leading to interdependent import (Kiebler *et al.*, 1993). The precursor of MOM72 is imported via MOM19. In contrast, MOM19 assembles directly with MOM38, independent of any protease-accessible component. Import of both MOM38 and MOM22 strictly depends on both MOM19 and MOM72 (Kiebler *et al.*, 1993).

Crossing the intermembrane space and inner membrane

A priori, transfer of precursors from the outer to inner membrane could occur via the intermembrane space or at sites of apposition of the two membranes. *In vitro* and *in vivo* studies with a large number of precursors led to the identification of translocation intermediates that spanned both mitochondrial membranes simultaneously, with the amino terminus being in the matrix allowing processing, and the carboxy terminus being outside the mitochondrion. About 50 amino acids were sufficient to span both

membranes. This suggested that the two membranes are in close proximity during import. Such areas of close apposition of the two membranes, also known as contact sites, have been observed using the electron microscope and have been found to be sites of import intermediates (Neupert, 1994). Nevertheless, it has been shown that mitochondria stripped of outer membranes can import precursors suggesting that the inner membrane has a distinct import apparatus (Horst *et al.*, 1993). The function of this apparatus in whole mitochondria was shown using a hybrid protein of cytochrome *c* heme lyase preceded by the F_1-ATPase β-subunit presequence. In the absence of a $\Delta\Psi$, this protein could be imported completely into the intermembrane space due to the lyase moiety, and on reestablishment of a $\Delta\Psi$ it was transferred into the matrix due to the β-subunit presequence (Neupert, 1994). Thus, mitochondria should be viewed as dynamic structures allowing import either via the intermembrane space or contact sites.

Some components of the mitochondrial inner membrane (MIM) import machinery have been identified. The first was ISP45, a protein essential for protein import and yeast viability, and that could be cross-linked to an import intermediate (Horst *et al.*, 1993). Another essential protein (MIM44) has been identified and may or may not be identical to ISP45 (Dekker *et al.*, 1993). Genetic screening was used to identify two other inner membrane proteins (MIM17 and MIM23(MAS6)) required for import (Dekker *et al.*, 1993).

Entry into the matrix

As discussed above, the amino terminus of precursors enters the matrix first. The precursor is then sequentially bound to mitochondrial hsp70 (mt-hsp70), proteolytically processed and transferred to the chaperone chaperonin 60 (cpn60, also called hsp60) which catalyzes its folding and assembly into protein complexes. ATP is necessary for the action of both mt-hsp70 and cpn60. Genetic and biochemical experiments in yeast suggest mt-hsp70 is essential for protein import and viability (Moore *et al.*, 1994). This protein is thought to interact with the amino-terminal region of newly imported proteins in the matrix thereby helping to pull them in. This action is also thought to promote unfolding of precursors outside mitochondria (Neupert, 1994). To date, mt-hsp70 genes have been identified in pea, bean, potato and tomato, and all appear to be homologous to yeast mt-hsp70 (Moore *et al.*, 1994). Two other proteins are thought to be required for mt-hsp70 function: DnaJ-like proteins (MDJ1 or Mdj1p in yeast) and bacterial GrpE homologues (GrpEp or MGE1 in yeast) (Stuart *et al.*, 1994).

Cpn60 and cpn10 (hsp10), the mitochondrial homologues of the bacterial chaperones GroEL and GroES respectively, are the final proteins involved in import of matrix proteins (Moore *et al.*, 1994; Stuart *et al.*, 1993). Cpn60 and cpn10 are essential for viability of yeast, and mutants lacking cpn60 or cpn10 are unable to fold or assemble newly imported proteins (Moore *et al.*, 1994; Stuart *et al.*, 1994). Plant cpn60 genes have been identified in *Arabidopsis thaliana*, *Zea mays*, *Brassica napus* and pumpkin cotyledons and are assumed to have similar functions as their yeast homologues (Moore *et al.*, 1994).

Protein sorting to the inner membrane and intermembrane space

A number of models have been suggested for sorting of proteins to the inner membrane and intermembrane space. One model is that proteins destined for the inner membrane are imported completely into the matrix, are processed and then insert into the inner membrane. This appears to be the case for the inner membrane proteins yeast cytochrome *c* oxidase subunit IV and *N. crassa* ATPase subunit 9 (Pfanner and Neupert, 1990). Further, even the *N. crassa* Fe/S protein of the bc_1 complex, which is on the outer surface of the inner membrane, follows this route (Hartl *et al.*, 1989). In all of these cases, the presequence has two parts, an amino-terminal matrix-targeting region and a region poor in positively charged residues; processing occurs in two steps in the matrix (see Table 15.1). Neupert's group called this mechanism conservative sorting since the *N. crassa* Fe/S protein, in its 're-export' from the matrix, appeared to follow its export path in bacteria (Hartl *et al.*, 1989). A contrasting model holds that proteins insert into the inner membrane during import (Rospert *et al.*, 1994). This appears to be the case for the ADP/ATP carrier which lacks a presequence and yeast cytochrome *c* oxidase subunit

Va which has a presequence that is removed in the matrix (Miller and Cumsky, 1993). This type of mechanism is referred to as stop-transfer (transport) and requires a sequence after the targeting sequence to initiate the integration into the inner membrane. These stop-transfer sequences are usually thought to be hydrophobic and could function by initiating transfer out of a hydrophilic transport pore into the hydrophobic bilayer (Rospert *et al.*, 1994). However, they also have positively charged residues around the hydrophobic sequence and it may be that there is specific recognition of the whole structural motif. An interesting question relates to the ADP/ATP carrier which in plants has a presequence. Will it follow the same route as in mammals, yeast and *N. crassa* where it lacks a presequence? The plant presequence is removed by the matrix protease (Glaser *et al.*, 1994).

The mechanism by which proteins having a matrix-targeting presequence reach the intermembrane space has been controversial. Both conservative sorting, with or without simultaneous import and export, and a stop-transfer mechanism have been suggested for cytochrome b_2, an intermembrane space protein, and cytochrome c_1, which is associated with the inner membrane but mainly exposed to the intermembrane space. Both proteins are processed in two steps, the first in the matrix to remove a matrix-targeting portion and the second in the intermembrane space (see Table 15.1). The conservative sorting model is supported by observations that import is dependent on cpn60 (Ostermann *et al.*, 1989), and that import intermediates are present in the matrix (Hartl *et al.*, 1987). However, others have found opposite results which suggest that intermembrane space proteins do not pass through the matrix during import and therefore support the stop-transfer model (Rospert *et al.*, 1994). Further research is needed to resolve this problem.

The import pathways of the intermembrane space proteins apocytochrome *c* and cytochrome *c* heme lyase (CCHL), which lack presequences, are unique. Apocytochrome *c*, the precursor of cytochrome *c*, crosses the mitochondrial outer membrane independent of surface receptors, ATP, or a membrane potential across the inner membrane (Glick and Schatz, 1991). Apocytochrome *c* is maintained in the intermembrane space either by binding to CCHL or by attachment of heme to it by

CCHL (Moore *et al.*, 1994). Import of CCHL, although dependent on MOM19 and the GIP, does not require an energized inner membrane for translocation across the outer membrane (Lill *et al.*, 1992). This is in contrast to most intermembrane space, inner membrane and matrix proteins whose import across the outer membrane is dependent on a membrane potential.

Import into the plant mitochondrion

Over the past few years there have been a growing number of studies on import by plant mitochondria. The presequences of many plant mitochondrial proteins have been published and several plant proteins have been identified which appear to be homologues of proteins involved in import in yeast and *N. crassa*. One would expect, given these results, that the whole import apparatus of plant mitochondria will be found to be similar if not identical to those of other organisms and that the proteins involved will be homologous. Nevertheless, one should be prepared for novel aspects. For example, import into isolated plant mitochondria is thought to be an inefficient process since much of the precursor form of the protein remains after protease treatment of mitochondria after import (Moore *et al.*, 1994). It has been proposed that the observed inefficiency may be due to limitations in certain components such as mt-hsp70 (Moore *et al.*, 1994). Thus, when studying mitochondrial protein import in plants, it is important to isolate the mitochondria from tissues in which the import components are not limiting and to take into account the stage of development of the plant (Moore *et al.*, 1994).

One critical aspect of import by plant mitochondria is that the targeting sequences of mitochondrial precursors and transit presequences of chloroplast proteins must be specific for the respective organelles. Several studies have examined this problem. For example, Boutry *et al.* (1987) found that although very similar, the presequences of the β-subunit of mitochondrial ATP synthase and of the small subunit of stromal Rubisco, both from *N. plumbaginifolia*, would specifically direct bacterial chloramphenicol acetyltransferase to either the mitochondrion or chloroplast respectively. This is so despite the fact

that both had the overall amino acid characteristics of mitochondrial presequences. However, detailed comparisons of chloroplast transit sequences from stromal and thylakoid lumen proteins with mitochondrial presequences showed clear differences between the two (Brink *et al.*, 1994). For example, transit sequences of stromal proteins lack charged residues in the amino-terminal region whereas mitochondrial presequences contain them. In contrast, transit sequences of two chloroplast envelope inner membrane proteins are similar to mitochondrial presequences and can also form positively charged amphiphilic α-helices. These proteins could be imported at low efficiency and be processed by isolated *V. faba* mitochondria (Brink *et al.*, 1994). The low efficiency of import of these chloroplast proteins into mitochondria may indicate that import is specific *in vivo*, but some missorting cannot be ruled out. Such problems and more detailed aspects of import by plant mitochondria should be forthcoming given the availability of modern molecular biological and genetic techniques.

Note: Recently a uniform nomenclature for mitochondrial membrane proteins involved in protein import has been devised. The proteins are termed translocase of the outer mitochondrial membrane (TOM) and translocase of the inner mitochondrial membrane (TIM). For example MOM72 (MAS70) is termed TOM70, etc. (Pfanner, N., Douglas, M. G., Endo, T., Hoogenraad, N. J., Jensen, R. E., Meijer, M., Neupert, W., Schatz, G., Schmitz, U. K., and Shore, G. C. (1996) Uniform nomenclature for the protein transport machinery of the mitochondrial membranes. *Trends Biochem. Sci.* **21**, 51–52.)

References

Arretz, M., Schneider, H., Guiard, B., Brunner, M. and Neupert, W. (1994). Characterization of the mitochondrial processing peptidase of *Neurospora crassa*. *J. Biol. Chem.* **269**, 4959–67.

Boutry, M., Nagy, F., Poulsen, C., Aoyagi, K. and Chua, N.-H. (1987). Targeting of bacterial chloramphenical acetyltransferase to mitochondria in transgenic plants. *Nature* **328**, 340-42.

Brink, S., Flügge, U.-I., Chaumont, F., Boutry, M., Emmermann, M., Schmitz, U., Becker, K. and Pfanner, N. (1994). Preproteins of chloroplast envelope inner membrane contain targeting information for receptor-dependent import into fungal mitochondria. *J. Biol. Chem.* **269**, 16478–85.

Chaumont, R., de Castro Silva Filho, M., Thomas, D., Leterme, S. and Boutry, M. (1994). Truncated presequences of mitochondrial F1-ATPase *β* subunit from Nicotiana plumbaginifolia transport CAT and GUS proteins into mitochondria of transgenic tobacco. *Plant Mol. Biol.* **24**, 631–41.

Dekker, P. J. T., Keil, P., Rassow, J., Maarse, A. C., Pfanner, N. and Meijer, M. (1993). Identification of MIM23, a putative component of the protein import machinery of the mitochondrial inner membrane. *FEBS Lett.* **330**, 66–70.

Douglas, M. G., McCammon, M. T. and Vassarotti, A. (1986). Targeting proteins into mitochondria. *Microbiol. Rev.* **50**, 166–78.

Glaser, E., Eriksson, A. and Sjöling, S. (1994). Bifunctional role of the bc_1 complex in plants. Mitochondrial bc_1 complex catalyses both electron transport and protein processing. *FEBS Lett.* **346**, 83–7.

Glick, B. and Schatz, G. (1991). Import of proteins into mitochondria. *Ann. Rev. Genetics* **25**, 21–44.

Hannavy, K., Rospert, S. and Schatz, G. (1993). Protein import into mitchondria: a paradigm for the translocation of polypeptides across membranes. *Curr. Opinion Cell Biol.* **5**, 694–700.

Harkness, T. A. A., Nargang, F. E., van der Klei, I., Neupert, W. and Lill, R. (1994). A crucial role of the mitochondrial protein import receptor MOM19 for the biogenesis of mitochondria. *J. Cell Biol.* **124**, 637–48.

Hartl, F.-U., Ostermann, J., Guiard, B. and Neupert, W. (1987). Successive translocation into and out of the mitochondrial matrix: targeting of proteins to the intermembrane space by a bipartite signal peptide. *Cell* **51**, 1027–37.

Hartl, F.-U., Pfanner, N., Nicholson, D. W. and Neupert, W. (1989). Mitochondrial protein import. *Biochim. Biophys. Acta* **988**, 1–45.

Horst, M., Jeno, P., Kronidou, N. G., Bollinger, L., Oppliger, W., Scherer, P., Manning-Krieg, U., Jascur, T. and Schatz, G. (1993). Protein import into yeast mitochondria: the inner membrane import site protein ISP45 is the MPI1 gene product. *EMBO J.* **12**, 3135–41.

Hurt, E. C. and van Loon, A. P. G. M. (1986). How proteins find mitochondria and intramitochondrial compartments. *Trends Biochem. Sci.* **11**, 204–7.

Kiebler, M., Becker, K., Pfanner, N. and Neupert, W. (1993). Mitochondrial protein import: specific recognition and membrane translocation of preproteins. *J. Membrane Biol.* **135**, 191–207.

Lill, R., Stuart, R. A., Drygas, M. E., Nargang, F. E. and Neupert, W. (1992). Import of cytochrome *c* heme lyase into mitochondria: a novel pathway into the intermembrane space. *EMBO J.* **11**, 449–56.

Liu, X., Bell, A. W., Freeman, K. B. and Shore, G. C. (1988). Topogenesis of mitochondrial inner membrane uncoupling protein. Rerouting transmembrane segments to the soluble matrix compartment. *J. Cell Biol.* **107**, 503–9.

Miller, B. R. and Cumsky, M. G. (1993). Intramitochondrial sorting of the precursor to yeast cytochrome *c* oxidase subunit Va. *J. Cell Biol.* **121**, 1021–9.

Moore, A. L., Wood, C. K. and Watts, F. Z. (1994). Protein import into plant mitochondria. *Ann. Rev. Plant Physiol. Plant Mol. Biol.* **45**, 545–75.

Neupert, W. (1994). Transport of proteins across mitochondrial membranes. *Clinical Investigator* **72**, 251–61.

Nunnari, J., Fox, T. D. and Walter, P. (1993). A mitochondrial protease with two catalytic subunits of nonoverlapping specificities. *Science* **262**, 1997–2004.

Ostermann, J., Horwich, A. L., Neupert, W. and Hartl, F.-U. (1989). Protein folding in mitochondria requires complex formation with hsp60 and ATP hydrolysis. *Nature* **341**, 125–30.

Ramage, L., Junne, T., Hahne, K., Lithgow, T. and Schatz, G. (1993). Functional cooperation of mitchondrial protein import receptors in yeast. *EMBO J.* **12**, 4115–23.

Rospert, S., Müller, S., Schatz, G. and Glick, B. S. (1994). Fusion proteins containing the cytochrome b_2 presequence are sorted to the mitochondrial intermembrane space independently of hsp60. *J. Biol. Chem.* **269**, 17279–88.

Söllner, T., Griffiths, G., Pfaller, R., Pfanner, N. and Neupert, W. (1989). MOM19, an import receptor for mitochondrial precursors. *Cell* **59**, 1061–70.

Söllner, T., Pfaller, R., Griffiths, G., Pfanner, N. and Neupert, W. (1990). A mitochondrial import receptor for the ADP/ATP carrier. *Cell* **62**, 107–15.

Stuart, R. A., Cyr, D. M., Craig, E. A. and Neupert, W. (1994). Mitochondrial molecular chaperones: their role in protein translocation. *Trends Biochem. Sci.* **21**, 87–92.

Verner, K. (1993). Co-translational protein import into mitochondria: an alternative view. *Trends Biochem. Sci.* **18**, 366–71.

Wachter, C., Schatz, G. and Glick, B. S. (1994). Protein import into mitochondria: the requirement for external ATP is precursor-specific whereas intramitochondrial ATP is usually needed for translocation into the matrix. *Mol. Biol. Cell* **5**, 465–74.

White, J. A. and Scandalios, J. G. (1989). Deletion analysis of the maize mitochondrial superoxide dismutase transit peptide. *Proc. Nat. Acad. Sci. USA* **86**, 3534–8.

16 Metabolite exchange between the mitochondrion and the cytosol

Roland Douce, Serge Aubert and Michel Neuburger

Introduction

The tricarboxylic acid (TCA) cycle provides reducing equivalents to the electron transport chain for ATP synthesis and also, via ancillary reactions, provides numerous substrates for biosynthetic reactions in the cytoplasm. The relative importance of these roles in the overall plant cell metabolism will depend on (a) the particular tissue; (b) the stage of development (the orderly coordinated sequence of events that attend the change from a single-celled zygote to a multicelled adult); and (c) environmental factors (temperature, light, water deficits, etc.). These mutiple and changing demands necessitate a rather complex detailed regulation of the individual enzymes of the TCA cycle in order to allow a non-uniform flux through the various segments of the cycle. The velocity with which this cycle turns is mainly determined by three quantities: availability of substrates (pyruvate, malate, oxaloacetate, NADH, etc.) for the respiratory chain, availability of substrates (ADP, P_i ($H_2PO_4^-$; HPO_4^{2-}) for the ATP synthase ($F_1F_0ATPase$), and redox state (NADH/NAD$^+$). In addition, these controls are complemented by input and output through the various anion translocators, according to demand and supply of metabolites.

Within a plant cell, the rate of respiration is controlled primarily by the rates of reactions feeding electrons to the respiratory chain and by the rates of reactions consuming or producing ATP. *In vivo*, changes in ATP utilization may be 3- to 5-fold, from a low level of around 4 μmol of ATP used per min per g wet wt (in sycamore cells). In other words, the mitochondria will only speed up if they 'know' that more ATP is consumed in the

cytoplasm. For example, during the course of their studies on the regulation of intracellular pH values in higher plant cells Gout *et al.* (1992) have observed a progressive acceleration in the rate of O_2 consumption as external pH decreased progressively. Indeed, under these conditions, the passive influx of protons through the plasma membrane increased in an exponential manner and the plasma membrane ATPase consumed more and more ATP to reject the invading H$^+$ ions and to maintain cytoplasmic pH at a constant value. Interestingly, when the rate of O_2 consumption was approaching the uncoupled O_2 uptake rate (i.e. at external pH below 4.5), a loss of cytoplasmic pH was observed indicating that at very acidic external pH, the availability of ATP for the plasma membrane ATPase was ultimately responsible for determining the rate of ATP-dependent efflux of H$^+$ ions. In addition, there are several points on the TCA cycle where important intermediates are removed to provide the building blocks for the biosynthesis of proteins, nucleic acids, and a multitude of specialized molecules such as chlorophyll. For mitochondria to function as integral components of the plant cell there must be movement of numerous anions including phosphate ATP^{4-}, ADP^{3-}, P_i, various TCA cycle intermediates and amino acids between the matrix of the mitochondria and the cytosol. In this chapter we will concentrate on the transport systems of plant mitochondria and their importance in plant cell carbon metabolism. Unfortunately, as yet very little work has been done on most of the anion carriers from plant mitochondria. In addition, we know nothing about cation transport activities in mitochondrial membranes.

Mitochondrial outer membrane

The outer membrane of the mitochondrion isolated by rather gentle procedure has been found to be permeable to sucrose, nucleotides, and NAD^+, but not to cytochrome c. In fact, it is now accepted that the outer membrane of mitochondria acts as a molecular sieve which permits the passage of small hydrophilic molecules. This sieving property, which allows uptake of small substrate molecules into the intermembrane space but excludes enzymes, is due to a 31 kDa channel-forming protein called a mitochondrial porin or a voltage-dependent anion-selective channel. All mitochondrial porins investigated so far show a decreasing single channel conductance as a result of increasing voltage (Mannella, 1985). Electron microscope image reconstruction indicates that the mitochondrial pore is a cylinder normal to the membrane plane with a diameter of about 2.5 nm. How many subunits are required to form a pore is not known. Interestingly, it has been shown that the mitochondrial pore can exert a control on the exact exclusion limit for the penetration of the outer membrane (Mannella, 1985). In addition, despite the lack of direct evidence, small 'messenger' molecules may change the aperture of porin (from open- to closed-channel conformations and *vice versa*) thus greatly affecting the permeability properties of the mitochondrion.

Schmid *et al.* (1992) have identified a diffusion channel in the outer membrane of pea mitochondria. This channel was identified as the porin that exhibits single-channel conductances of 1.5 and 3.7 nS in 1 M KCl and an estimated effective diameter of about 1.7 nm. In addition this porin migrated as a single band with an apparent molecular mass of 30 kDa on SDS-PAGE.

Transport systems of plant mitochondria

In contrast with the outer membrane, the hydrophobic bilayers devoid of porin of the inner mitochondrial membrane creates an effective barrier to the passage of charged species including H^+. The high activation energy required to insert an ion into a hydrophobic region is the reason for the extremely low ion permeability of bilayer regions. This overall unspecific impermeability of the inner membrane towards hydrophilic solutes is overcome by specific translocators (Chappell and Croft, 1966). These carriers are essential for the maintenance of a normal mitochondrial function. If anions could freely cross the membrane they would be ejected by the high electric field generated by respiration, and essential substrates would be lost. The kinetic properties of the translocators may vary from species to species and from tissue to tissue in the same species.

On the other hand, the inner membrane is highly permeable to small uncharged molecules such as NH_3, O_2, and H_2O (perhaps the inner mitochondrial membrane contains water selective channels (aquaporins) (Chrispeels and Maurel, 1994) allowing water to pass freely). Available evidence indicates that CO_2 also readily passes through the mitochondrial inner membrane whereas the bicarbonate anion (HCO_3^-) does not. Parenthetically, the inner membrane is permeable to several monocarboxylic acids of low molecular weight with relatively high pK values (such as acetate) which pass, after protonation, through the membrane as undissociated acids (Chappell and Croft, 1966).

Triton-extraction followed by chromatography on hydroxyapatite and celite has been used to purify several carriers from animal mitochondria to homegeneity (Kramer and Palmieri, 1989). However, in plants, none of the mitochondrial metabolite carriers have been purified so far. Likewise, no information is available concerning the spatial and temporal regulation of expression for specific nuclear-encoded plant mitochondrial carriers.

Phosphate and adenine nucleotide translocation in plant mitochondria

The mitochondrial F_1-ATPase, responsible for ATP synthesis in oxidative phosphorylation, is located on the inner surface of the inner mitochondrial membrane (for review see Douce, 1985). Consequently, the transport of P_i (influx), ADP (influx) and ATP (efflux) between the cytosol and the matrix are essential for the continued synthesis of ATP. Under physiological conditions, the non-green plant cell is fully aerobic and over 90% of its ATP is made by mitochondrial oxidative phosphorylation:

$$ADP_{out} + P_{i\,out} + 4H^+_{out} \rightarrow ATP_{out} + 4H^+_{in}$$

The transfer of electrons from substrate to O_2 via the cytochrome electron transport pathway is coupled to an electrogenic translocation of protons across the inner mitochondrial membrane (for review see Douce et al., 1987). Protons have a positive charge and, consequently, proton translocation from the matrix to the cytosol, without cotransport of anions, generates both a proton gradient (ΔpH) and a membrane potential ($\Delta\Psi$) in which the matrix side is negatively charged and the cytoplasmic side positively charged (Fig. 16.1). The two factors are additive, both contributing to a 'protonmotive force' (Δp) differential across the membrane according to the following equation:

$$\Delta\mu H^+/F = \Delta p = \Delta\Psi - 2.303\,RT/F \times \Delta pH$$

where $\Delta\mu H^+$ is the electrochemical potential

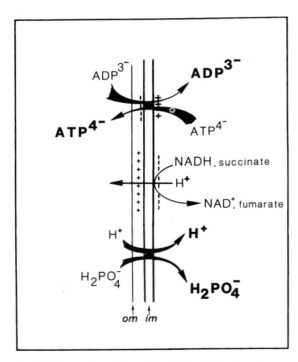

Fig. 16.1 Effect of the electrochemical gradient of protons generated by the electron chain on the electrophoretic asymmetric nucleotide exchange and on the net movement of phosphate (H_2PO4^-) across the inner mitochondrial membrane.

difference between protons in the inner compartment of the mitochondrion and the surrounding bulk solution (or intermembrane space, since the outer mitochondrial membrane is freely permeable to protons); $2.303\,RT/F = 59$ at $25\,^{\circ}C$. Because of convention for membrane potentials (outside − inside), the ΔpH values become negative. Thus a negative sign is included in the equation to allow summation of the two potentials.

The precise mechanism by which the proton flow drives ATP synthesis is not yet known. The Δp provides the driving force for protons to flow back across the membrane through a proton-translocating ATP synthase with concomitant ATP synthesis. According to this delocalized chemiosmotic model, the steady state value of Δp should reflect a steady-state balance of the outward proton flux driven by the respiratory chain and the inward proton flux due to ATP synthase functioning and to a minor extent the passive leakage of protons. A chemiosmotically defined relationship between the protonmotive force and electron transport is summarized by the following equation:

$$J_{ox} . n = L . \Delta p$$

Where J_{ox} is the respiratory rate, n is the ratio of protons translocated per two electrons transported through the respiratory chain to reduce one atom of oxygen (H^+ per O), L is the 'leak' rate of H^+ back across the inner mitochondrial membrane, and Δp the protonmotive force (in mV) across the membrane.

Part of the total energy available is used to move substrates such as adenine nucleotides and phosphate ($H_2PO_4^-$) against concentration gradients between the external medium (or cytosol) and the matrix space (Fig. 16.1). In other words, some of the energy that would otherwise be available for ATP synthesis has been diverted for the translocation of adenine nucleotides and phosphate in intact plant mitochondria. However, further work is required to establish whether $\Delta\mu H^+$ represents the sole and obligatory intermediate driving ATP synthesis. In addition, the major question that remains unresolved is the chemical mechanism of Δp utilization. Very likely, the rate of ATP synthase turnover is modulated in vivo by several mechanisms including: variation in $\Delta\mu H^+$ (via dehydrogenase activation), variation in enzyme capacity via a specific signal (regulatory element), and variation in cytosolic ADP concentrations. For example, the levels of ADP *in vivo* are non-saturating to the ATP synthase; the turnover of the ATP synthase would increase, passively, in response to ADP concentration in the physiological range. In fact, mitochondria *in vivo* are often described as being intermediate between state 4 and 3 since respiration can be increased (to state 3) with uncouplers or decreased (to state 4) by inhibiting ATP production with inhibitors such as oligomycin. Finally there could be a problem with ADP diffusion within the cytosol since it contains a high protein concentration. This could be partially overcome by movement of mitochondria. For example, the cytosolic areas that have a high consumption of ATP might attract mitochondria through microtubule attachment to the outer mitochondrial membrane.

Adenine nucleotide transport

The adenine nucleotide exchange is highly specific for ADP and ATP and is basically electrogenic: free ATP^{4-} and ADP^{3-} are translocated but not their Mg^{2+} complexes ($MgATP^{2-}$) (Fig. 16.1). This is highly significant because it is probable that most ATP molecules in the cytosol and in the mitochondria are chelated with Mg^{2+} and because such a complex is the real substrate of enzymatic reactions involving ATP. A specific system for the removal of Mg^{2+} ions in the vicinity of the carrier could be involved. Since ADP has three and ATP has four negative charges at physiological pH, the ADP^{3-}–ATP^{4-} exchange causes a net movement of charges across the mitochondrial inner membrane. However, since this carrier operates as an antiport it does not change the intramitochondrial concentration of adenine nucleotides (Vignais, 1976). The transmembrane potential is, therefore, a potent driving force for an electrophoretic asymmetric exchange of ATP^{4-} and ADP^{3-} which drives ADP^{3-} inside the mitochondria and ATP^{4-} outside. Accordingly, at equilibrium, the cytosol should have a high ATP/ADP ratio, and this has been found by ^{31}P-NMR spectra obtained *in vivo* from plant cells or tissues (Bligny *et al.*, 1989).

Three specific and powerful inhibitors of the adenine nucleotide transport system are known: atractyloside, carboxyatractyloside (gummiferin, the toxic principles of an Algerian thistle) and bongkrekic acid (an antibiotic formed by a mold in decaying coconut meals). One of the peculiarities of the mitochondrial nucleotide carrier is its ability to bind atractyloside and bongkrekic acid, in an asymmetric manner. The atractyloside bind to the carrier on the cytosolic surface of the inner membrane and bongkrekic acid on the matrix facing surface (Vignais, 1976). The nucleotide carrier has been isolated as a carboxyatractylate protein complex that is probably a dimer, each subunit having an M_r value of 30 000. Interestingly, the nucleotide carrier is the most abundant integral protein in the inner membrane of mitochondria.

Emmermann *et al.* (1991) and Winning *et al.* (1992) reported that the adenine nucleotide translocator of higher plants is synthesized as a large precursor protein, with an N-terminal presequence that is proteolytically processed upon import into isolated mitochondria. The apparent molecular weights of the mature imported translocator polypeptides are identical whether import is carried out with potato or maize mitochondria, indicating that this is a specific processing event, conserved between monocots and dicots. Higher plant mitochondria, therefore, appear to be different to fungal and mammalian mitochondria where targeting of adenine nucleotide translocator to mitochondria is mediated by internal signals with no N-terminal processing.

Phosphate transport

The mitochondrial phosphate carrier catalyzes both P_i–P_i exchange and net movement of the anion across the inner membrane, the latter process being accompanied by the countertransport of OH^-. In fact, it is usually impossible to distinguish between a symport (i.e. a transport process involving the obligatory coupling of two ions in parallel) of a species with H^+ and an antiport of the species with OH^-. The involvement of this transport reaction with ΔpH across the mitochondrial membrane allows this carrier to connect the proton gradient generated by the mitochondrial electron chain to the accumulation of high levels of P_i in the matrix (Fig. 16.1). This transport, which is electroneutral, is sensitive to thiol reagents like *p*-hydroxymercuribenzoate. This carrier protein present in all the mitochondria isolated so far has been isolated, purified and studied in detail in reconstituted liposomes (Wohlrab, 1986). The gene for the protein has been cloned from mammals and more recently from yeasts (Phelps *et al.*, 1991). The mature protein predicted by these sequences has a mol wt of about 33 000. Surprisingly, sequence and structural similarities have been noted between the phosphate transporter and the adenylate transporter. It was suggested, therefore, that these carriers form a family of proteins, probably originating from a common ancestor.

In our laboratory, P_i transport into potato mitochondria has been measured using a rapid filtration technique, which offers fast time resolution (starting from 10 ms), thus allowing for the first time fast kinetic measurements at room temperature (Kathryn Wright, Michel Neuburger, and Roland Douce, unpublished data). The initial rates of P_i accumulation measured within the first 150 ms and at 10 °C give a K_m (total P_i) of 0.6 mM and a maximum velocity (V) equal to 5000–6000 nmol P_i mg protein^{-1} min^{-1} (Fig. 16.2). Potato tuber mitochondria oxidize the substrate succinate at a rate between 400–500 nmol O_2 consumed mg protein^{-1} min^{-1} at 25 °C. Assuming a P/O ratio of 2, this would require that P_i enters the mitochondria at a rate of 1600–2000 nmol P_i mg protein^{-1} min^{-1} if it is not to limit the synthesis of ATP. Furthermore, as the K_m value of P_i for the translocator is approx. 0.6 mM and the cytosolic P_i concentration determined by ^{31}P-NMR is between 1 and 2 mM the translocator could operate at full capacity, and thus in physiological situations, the activity of the phosphate carrier is far higher than is necessary to supply P_i to the H^+-ATPase even at maximum rates of oxidative phosphorylation. Thus, in the case of the phosphate translocator, neither the capacity of the carrier nor the rate of P_i supply to the H^+-ATPase limits the *in vivo* rate of respiration.

The phosphate transporter was solubilized from etiolated pea mitochondrial membranes with triton X-114, purified approx. 500-fold and reconstituted into liposomes that were preloaded with P_i. A number of positively charged amino acid side chains are essential for transport function. Indeed chemical modifiers directed against the basic amino acid residues Arg (*p*-hydroxylphenylglyoxal), His (diethyl pyrocarbonate), and Lys (pyridoxal 5-phosphate) strongly inhibit P_i–P_i exchange. These residues are likely to be involved in binding the phosphate anion

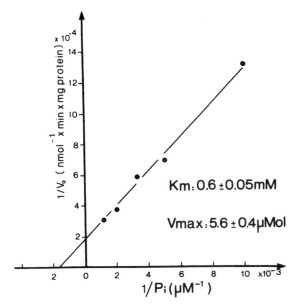

Fig. 16.2 A Lineweaver–Burk plot showing the dependence of rates of the accumulation of internal P_i on external P_i concentration in isolated plant mitochondria. Time courses were obtained at 10 °C using a rapid-filtration technique (Biologic rapid Filtration system, Grenoble, France) using Pi at concentrations of 100 μM, 200 μM, 300 μM and 800 μM. The initial velocity for each concentration was measured from points obtained in the first 150 ms. The kinetic parameters were calculated to be K_m (P_i) = 600–700 μM and V = 5000–6000 nmol mg^{-1} protein min^{-1}.

to the carrier protein (McIntosh and Oliver, 1994). Surprisingly, the pea mitochondrial phosphate transporter is not inhibited by *N*-ethylmaleimide. In fact, the mammal transporters have a reactive Cys[42] residue that results in NEM sensitivity.

Tricarboxylic acid cycle anion translocation in plant mitochondria

The end-product of glycolysis under aerobic conditions in plants is the pyruvate ion (Beevers, 1961) and very likely malate and oxaloacetate ions. Furthermore, the tricarboxylic acid cycle serves as an important source of carbon skeletons for synthetic processes occurring in the cytoplasm and mechanisms for the movement of tricarboxylic cycle anions into and out of mitochondria must occur and have been demonstrated for mitochondria isolated from a wide range of tissues. Mitochondria from different tissues at progressive stages of development vary in their complement of exchange-diffusion carriers (or antiports). The precise carrier composition presumably is a reflection of the function of the tissue from which the mitochondria are isolated. For example, the metabolic pathway for the conversion of fat to carbohydrate in castor bean endosperm or for the conversion of glycolate to glycerate in leaves involves the mitochondria in a number of necessary oxidations (Huang et al., 1983).

The kinetic properties of a tricarboxylic cycle substrate carrier can be examined, free of the constraints of limiting and undefined internal substrate levels, by preloaded mitochondria with a saturating concentration of a transportable substrate (e.g. citrate, malate, succinate, etc.). The exchange reaction is then carried under defined conditions. Indeed, and in contrast to mammalian mitochondria, freshly isolated plant mitochondria appear to contain low levels of endogenous exchangeable metabolites suggesting an inherent 'leakiness' of the inner membrane. For example, depleted mitochondria lose the ability to take up dicarboxylates, an ability which can be restored if a dicarboxylate and an energy source are provided (Douce, 1985).

Pyruvate carrier

The availability or access of pyruvate to the pyruvate dehydrogenase complex in the mitochondrial matrix is of great importance for respiration. Pyruvate, although it is a monocarboxylate, requires a carrier because of its relatively low pK. Pyruvate carrier catalyzes the thermodynamically indistinguishable pyruvate/H^+ symport or pyruvate/OH^- antiport and is inhibited by SH-group poisons, and strongly and specifically by α-cyano-4-hydroxycinnamate with half inhibition occurring at 2 μM (Day and Wiskich, 1984). Consequently, pyruvate transport is electroneutral, and at equilibrium the ratio pyruvate$_{in}$/pyruvate$_{out}$ is a direct function of ΔpH. At high non-physiological concentrations pyruvate is, however, no longer only transported via the translocator, but 'passive' diffusion through the inner membrane also becomes important.

The pyruvate carrier from castor bean mitochondria has been partially purified (Brailsford et al.,1986). Vivekananda and Oliver (1990) have reported the isolation of a monoclonal antibody (generated against mitochondrial proteins) that binds specifically to the pyruvate carrier from pea mitochondria and the use of the antibody to identify this transport protein on Western blots. This transporter which has an apparent molecular mass of 19 kDa appeared to be smaller than the other substrate transporter identified to date. It is possible, however, that this is only one subunit of a multisubunit protein although most transport systems are oligomeric, composed of dimers or higher oligomers.

Dicarboxylate carrier

The dicarboxylate carrier facilitates dicarboxylate transport across the inner mitochondrial membrane. Dicarboxylate transport is mediated by a HPO_4^{2-}–exchange and by a Dicarboxylate^{2-}–Dicarboxylate^{2-} exchange, both of which are inhibited by pentylmalonate and by 2-N-butylmalonate (Day and Wiskich, 1984). Consequently, the matrix phosphate can exchange back out of the mitochondria against the uptake of cytosolic dicarboxylic acids (malate and succinate) on the dicarboxylate transporter. It is clear, therefore, that the phosphate transporter connects the electrochemical gradient generated by electron transport to the uptake of dicarboxylates. The sensitivity of dicarboxylate accumulation to uncouplers and electron transport chain inhibitors is thought to occur by directly affecting P_i uptake, by discharging $\Delta\mu H^+$ and thus lowering the amount of internal P_i available for exchange with external dicarboxylate. Chappell and Beevers (1983) have used a centrifugation method to investigate features of dicarboxylate transport in mitochondria purified from castor bean endosperm. They indicated that the maximum rates of Dicarboxylate^{2-}–Dicarboxylate^{2-} exchange ($V = 250$ nmol mg protein^{-1} min^{-1} at 4 °C) are much greater than those measured for net dicarboxylate accumulation ($V = 15$ nmol mg protein^{-1} min^{-1}) at 4 °C and are sufficient to account for the observed rates of succinate oxidation. This oxidation proceeds to malate, which accumulates in the bathing medium because of the fast succinate^{2-}–malate^{2-} exchange. Hence, initially succinate is taken into the mitochondrion in exchange for P_i. It is then oxidized to malate which, in turn, is exchanged for more external succinate. In general, when plant mitochondria are supplied with a single substrate, its oxidation is frequently limited to a few steps. The resulting intermediates do not remain confined to the mitochondia but are lost and often subsequent re-entry into the mitochondria is observed. ^{13}C-Nuclear magnetic resonance studies of malate synthesis and compartmentation in higher plant cells perfused with $KH^{13}CO_3$ led us to the conclusion that *in vivo* a permanent movement of malate molecules between the mitochondrial matrix and the cytosol occurs via the dicarboxylate carrier of the inner mitochondrial membrane, catalyzing a strict counter exchange of malate–malate (Gout *et al.*, 1993).

Monoclonal antibodies (the hybridoma library was derived from mice that had been immunized with total mitochondrial membranes) which specifically bound to the dicarboxylate carrier of pea leaf mitochondria have been used to identify and characterize that carrier. The plant dicarboxylate carrier has been successfully reconstituted into liposomes and studies of transport in the reconstituted system using different substrates confirmed the specificity of the monoclonal antibodies for the dicarboxylate carrier. This carrier was further identified by Western blotting and immunoprecipitation and has an apparent molecular mass of 26 000 Da (Vivekananda *et al.*, 1988).

α-Ketoglutarate carrier

Intact plant mitochondria that are oxidizing a-ketoglutarate in the presence of ADP excrete malate and there is transient accumulation of succinate in the bathing medium. However, in the presence of malonate, a potent inhibitor of succinate dehydrogenase, mitochondria excrete almost exclusively succinate. During the course of α-ketoglutarate oxidation only that part of the tricarboxylic acid cycle from α-ketoglutarate to malate is operative and the dicarboxylates for the most part pass into the external medium (Douce, 1985). These experiments indicate that α-ketoglutarate transport occurs by a dicarboxylate–α-ketoglutarate exchange.

The α-ketoglutarate carrier from corn shoot mitochondria has been solubilized in triton X-114 and partially purified by chromatography on hydroxyapatite and celite in the presence of diphosphatidylglycerol (cardiolipin). When reconstituted into liposomes, the α-ketoglutarate transport protein catalyzed a phthalonate-sensitive α-ketoglutarate–α-ketoglutarate (succinate, malate) exchange. These exchanges were inhibited by substrate analogs (phenylsuccinate) and by mercurials. In addition, lysine residue(s) may be important for the activity of the α-ketoglutarate carrier because it is inhibited by phenylisothiocyanate and pyridoxal-5′-phosphate (Genchi *et al.*, 1991). Obviously the properties of the reconstituted carrier, i.e. requirement for a couter-anion, substrate specificity, and inhibitor sensitivity, are similar to those of the α-ketoglutarate transport system as characterized in plant and animal mitochondria.

Oxaloacetate carrier

Oxaloacetate has been found to rapidly traverse the inner membrane of all plant mitochondria studied so far (Fig. 16.3). For example, at an extramitochondrial oxaloacetate concentration of 50 μM, the influx of oxaloacetate is so rapid that NAD^+-linked O_2 consumption that is dependent on tricarboxylic acid cycle substrates stops because of the competition for NADH that is used in the reduction of oxaloacetate by malate dehydrogenase (the equilibrium of the malate dehydrogenase reaction lies far towards malate formation; $K_{eq} = 3 \times 10^{-5}$) (Douce, 1985) (Fig. 16.4). This respiratory inhibition is subsequently relieved as the oxaloacetate becomes reduced. While this 'swamping' effect by excess oxaloacetate at best affords only a coarse type of control over electron transport, it amply demonstrates how a metabolic imbalance ensues when those enzyme normally limited by the supply of oxaloacetate are presented with a relative surfeit (Douce, 1985). The rapid phthalonate-sensitive uptake of oxaloacetate is half-saturated at micromolar concentrations of oxaloacetate ($K_{m\ OAA} = 5\ \mu$M; $V_{max} = 2\ \mu$mol mg^{-1} protein min^{-1}; the initial rates of oxaloacetate accumulation were measured using a rapid filtration technique within the first 150 ms and at 10 °C). In some circumstances, intact mitochondria can export oxaloacetate. For example, when malate is being oxidized under state 3 conditions, mitochondria excrete oxaloacetate via the oxaloacetate carrier when the malic enzyme does not operate. As malate oxidation proceeds owing to malate dehydrogenase, the concentration of oxaloacetate in the medium slowly increases up to an equilibrium concentration. When this is achieved, the efflux of oxaloacetate stops and the malate dehydrogenase reaction in the matrix

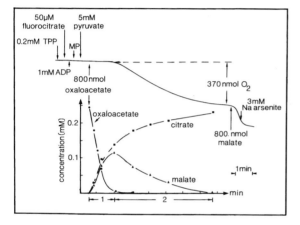

Fig. 16.3 Pyruvate oxidation triggered by oxaloacetate in intact potato tuber mitochondria. Upper trace: O$_2$ consumption during pyruvate oxidation. Lower trace: disappearance of added oxaloacetate, production of malate and production of citrate during pyruvate oxidation. Added oxaloacetate readily penetrates the matrix space through the inner membrane. Part of the oxaloacetate, at least one half, reacts with acetyl-CoA under the control of citrate synthase, as indicated by citrate accumulation. Simultaneously, the other part of the oxaloacetate is converted into malate by malate dehydrogenase, a conversion that depends on NADH generated by the thiamine pyrophosphate (TPP)-linked pyruvate dehydrogenase complex (inhibited by Na arsenite). Under these conditions, O$_2$ consumption by the respiratory chain is inhibited (upper trace). When all the oxaloacetate is consumed, previously accumulated malate re-enters the mitochondria and is rapidly oxidized to form oxaloacetate. Under these conditions, NADH from malate dehydrogenase and pyruvate dehydrogenase is available for oxidation by the respiratory chain (upper trace). Note that during the course of pyruvate oxidation triggered by oxaloacetate, extensive loss of malate and subsequent re-entry into the mitochondria is observed. Fluorocitrate, a potent inhibitor of aconitase, has been added to prevent citrate oxidation. MP: purified mitochondria.

is reversed (Douce, 1985). The excretion of oxaloacetate, which is severely inhibited by phthalonate, occurs under state 3 conditions (i.e. at high NAD$^+$/NADH ratios). Indeed the high concentrations of NAD$^+$ in the matrix space shift the reaction slightly towards oxaloacetate production which facilitates oxaloacetate excretion (Douce, 1985). Likewise during the course of succinate oxidation under state 3 conditions, plant mitochondria also excrete large amounts of oxaloacetate (Douce *et al.*, 1986). Consequently, it is the concentration of oxaloacetate on both sides of the inner mitochondrial membrane that appears to govern the efflux and influx of oxaloacetate. Irrespective of its role, the low K_m of the oxaloacetate carrier for its substrate allows it to compete successfully with cytosolic or matrix malate dehydrogenase.

For the present, the details of oxaloacetate

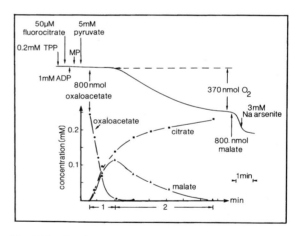

Fig. 16.4 Effect of exogenous oxaloacetate on the oxidation state of endogenous NAD and on pyridine content (%) in purified potato tuber mitochondria oxidizing succinate. In the presence of succinate, NAD reduction (reverse electron transport via ubiquinone pool and complex I) was initiated by addition of ATP. After completion of NADH formation, as indicated by the plateau in the fluorescence trace, addition of small amounts of oxaloacetate caused an immediate partial oxidation of the NADH (decrease in the fluorescence) linked to reduction of the supplied oxaloacetate to malate by mitochondrial malate dehydrogenase. Rapid penetration of plant mitochondria by oxaloacetate is mediated by a unique phthalonate-sensitive oxaloacetate transporter in plant mitochondria.

transport in plant mitochondria remain unknown, but it is thought that the carrier, which is distinct from the phthalonate-sensitive α-ketoglutarate carrier (Genchi *et al.*, 1991), is specific for oxaloacetate. The question whether both malate and oxaloacetate are transported by a single transport protein or by two different ones cannot be answered at present. Interestingly, Zoglowek *et al.* (1988) have provided evidence that a malate–oxaloacetate transport shuttle across the inner mitochondrial membrane is catalyzed by an electrogenic uniport of malate and of oxaloacetate linked to a counter exchange.

The extraordinarily low half-saturation of oxaloacetate transport makes it possible for a very active malate–oxaloacetate transport shuttle to occur between the mitochondria and the cytosolic compartment of a plant cell under physiological conditions, the equivalent of which is not found in mammalian mitochondria. For example, it has been proposed that the NADH generated during glycine oxidation by mitochondria isolated from green leaves (C3 plants) could be reoxidized very rapidly by oxaloacetate, owing to the tremendous excess of malate dehydrogenase located in the matrix space working in the reverse direction. Operation of a rapid malate–oxaloacetate transport shuttle would enable the transfer of reducing equivalents generated in the mitochondria during glycine oxidation to the peroxisomal compartment for the reduction of β-hydroxypyruvate (Krömer *et al.*, 1992) (Fig. 16.5). Indeed, in the presence of oxaloacetate, the glycine decarboxylase reaction in isolated mitochondria is able to operate faster than when the NADH is reoxidized exclusively via the mitochondrial electron transport chain operating under state 3 conditions (Lilley *et al.*, 1987). (A consideration to be taken into account regarding *in vivo* glycine oxidation via the respiratory chain, is that this process is dependent upon the ATP being recycled back to ADP at a rate sufficient to account for the potential rate of glycine-dependent oxidative phosphorylation. However, it is very likely that green cells *in vivo* will generally be respiring between states 4 and 3. If such a limitation were to occur under photorespiratory conditions it would lead to an increase in the matrix NADH/NAD$^+$ ratio and thus reduce the rate of glycine decarboxylase).

Since all the plant mitochondria isolated so far

possess a powerful oxaloacetate carrier, carbon input from the cytosol to the tricarboxylic acid cycle could also occur as oxaloacetate, thus bypassing conventional operation of the tricarboxylic acid cycle. Oxaloacetate arises either from malate oxidation catalyzed by cytosolic malate dehydrogenase and/or β-carboxylation of phospho*enol*pyruvate. The latter reaction is catalyzed by a cytosolic

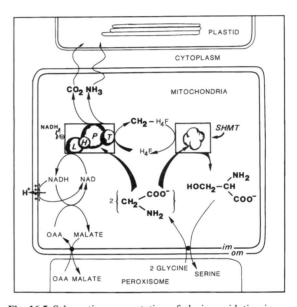

Fig. 16.5 Schematic representation of glycine oxidation in green leaf mitochondria. During photorespiration, glycine is cleaved in the matrix by the glycine cleavage enzyme system (containing four protein components which have been tentatively named P-protein, H-protein, T-protein and L-protein) to CO_2, NH_3 and 5,10-methylene-tetrahydropteroyl glutamate (CH_2-H_4F). The latter compound reacts with a second mole of glycine to form serine and tetrahydropteroyl glutamate (H_4F) in a reaction catalyzed by serine hydroxymethyltransferase (SHMT). NADH produced during the course of glycine oxidation is oxidized either by the respiratory chain or in the conversion of oxaloacetate to malate catalyzed by malate dehydrogenase that is located in the matrix. A rapid malate–oxaloacetate transport shuttle appears to play an important role in the photorespiratory cycle by facilitating the transfer of reducing equivalents generated in the mitochondria during glycine oxidation to the peroxisomal compartment for the reduction of β-hydroxypyruvate. Note the stoichiometry of two glycine molecules entering the mitochondrial matrix in exchange for one serine leaving.

phospho*enol*pyruvate carboxylase present in all plant cells examined so far. This enzyme appears to be a non-biotin containing carboxylase which uses HCO_3^- rather than CO_2 as substrate. Unfortunately, information as to the exact role of phospho*enol*pyruvate carboxylase in the replenishment of tricarboxylic acid cycle intermediates (in the form of oxaloacetate or malate) remains sparse. An alternative source of oxaloacetate is from the degradation of proteins.

Citrate carrier

It is accepted that the transport of citrate by plant mitochondria is by a tricarboxylate carrier similar to the one found in animal mitochondria. This carrier catalyzes an exchange of a tricarboxylate anion (e.g. citrate) with a dicarboxylate anion (Desantis *et al.*, 1976). The tricarboxylate carrier was solubilized from pea mitochondria with triton X-114 and reconstituted into artificial membranes (phospholipid vesicles) (McIntosh and Oliver, 1992). The transporter can exchange external citrate for internal citrate or malate and these exchanges are inhibited by N-butylmalonate and 1,2,3-benzenetricarboxylate. However, the plant citrate carrier differs from the equivalent activity from animal cells in that it did not transport D,L-isocitrate and phospho*enol*pyruvate. In fact intact plant mitochondria supplied with citrate do not excrete isocitrate into the medium.

A fundamental metabolic requirement for the fixation of NH_4^+ involving glutamine synthetase and glutamate synthase is the maintenance of a sufficiently large (and continuously supplied) pool of α-ketoglutarate in the cytoplasm (Miflin and Lea, 1980). The presence of powerful $NADP^+$-linked isocitrate dehydrogenase and aconitase activities in the cytosol may be the means by which citrate is converted to α-ketoglutarate for net ammonia assimilation (Fig. 16.6). The majority (or all) of the extramitochondrial citrate is originally synthesized in the mitochondrion. Malate that is synthesized in the cytosol by phospho*enol*pyruvate carboxylase and malate dehydrogenase can be imported into the mitochondrion and then converted into citrate by the NAD^+-linked malic enzyme and citrate synthase. Hanning and Heldt (1993) have concluded that the mitochondrial oxidation of malate in illuminated leaves produces mainly citrate, which is converted via cytosolic aconitase and NADP-isocitrate dehydrogenase to yield α-ketoglutarate as the precursor for the formation of glutamate and glutamine, which are the main product of photosynthetic nitrate assimilation. There might be endogenous controls polarizing the malate–citrate exchange facilitating citrate efflux from the mitochondrial matrix. The citrate^{3-}–malate^{2-} electrogenic exchange may couple this exchange to the electrochemical gradient and provide directional transport. As suggested by McIntosh and Oliver (1992) the lack of isocitrate exchange suggests that the mitochondrial and cytosolic pools of this intermediate are separate, thereby allowing more independent regulation of the isocitrate dehydrogenase activities in both compartments. These conclusions demonstrate the central importance of the citrate carrier to plant cell metabolism, since the extent to which plant mitochondria are capable of citrate export might dramatically affect the supply of cytosolic α-ketoglutarate available to fast growing cells for amino acid metabolism.

Amino acid translocation in plant mitochondria

The experiments of Leaver and co-workers (for review see Leaver and Gray, 1982) first demonstrated that amino acids were taken up and incorporated into proteins by plant mitochondria. This incorporation is dependent upon the mitochondria being intact and possessing coupled oxidative phosphorylation. This requires that during the biogenesis of mitochondria net influx of amino acids into the matrix space must occur, since almost all amino acids are synthesized inside plastids. Unfortunately, the mechanism by which amino acids are transported into plant mitochondria is not clear.

Glutamate and aspartate transport

Information on the transport of glutamate and aspartate in plant mitochondria is limited. Back exchange experiments with castor bean mitochondria carried out by Millhouse *et al.* (1983) found a freely reversible glutamate–aspartate exchange. This exchange appeared not to be electrogenic, in contrast

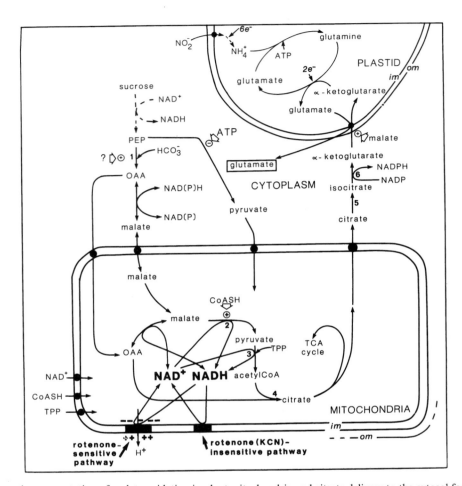

Fig. 16.6 Schematic representation of malate oxidation in plant mitochondria and citrate delivery to the cytosol for the fixation of NH_4^+. This scheme emphasizes the flexibility of plant mitochondria and indicates that there exists a concerted action of malate dehydrogenase, NAD^+-linked malic enzyme (2) and pyruvate dehydrogenase (3) to provide citrate in the anaplerotic function for the TCA cycle. The scheme also indicates that matrix NADH produced by the dehydrogenases, including NAD^+-linked malic enzyme, can be oxidized equally well by either the respiratory chain or by malate dehydrogenase for the reduction of oxaloacetate. Transport of electrons to O_2 can be either phosphorylating (via the rotenone-sensitive pathway) or non phosphorylating (via the rotenone [KCN]-insensitive pathway). The latter mechanism may be important when TCA cycle intermediates such as citrate are exported from the mitochondrion for use elsewhere in the cell (see text). Note that carbon input to the TCA cycle can occur in the form of cytosolic oxaloacetate and malate (produced by the combined operation of phosphoenolpyruvate carboxylase (1) and malate dehydrogenase in the cytosol). Pyruvate can be provided either by the action of pyruvate kinase in the cytosol or by operation of malic enzyme in the matrix utilizing malate generated either in the matrix or the cytosol. Finally, this scheme illustrates that isolated intact plant mitochondria actively accumulate NAD^+, thiamine pyrophosphate (TPP) and coenzyme A (CoA) from the external medium. This leads to a substantial increase in the matrix concentration of these cofactors which would stimulate TCA cycle dehydrogenases, the NAD^+-linked malic enzyme and rotenone [KCN]-insensitive pathway. Other enzymes shown in this diagram are: (5) cytosolic aconitase; (6), cytosolic α-ketoglutarate dehydrogenase; (4) citrate synthase (the first step in the TCA cycle catalyzed by citrate synthase is an aldol condensation of acetyl-CoA and oxaloacetate to give an enzyme-bound thioester intermediate. Hydrolysis of this thioester gives the citrate ion and coenzyme A. The equilibrium position of this reaction is very much in favor of the products because the free energy of hydrolysis of the thioester drives the reaction to completion.)

with mammalian mitochondria. In this latter case, a proton is cotransported on the carrier with glutamate and the exchange becomes virtually unidirectional, glutamate entering the mitochondrial matrix space and aspartate leaving. In other words in mammalian mitochondria, the driving force for the glutamate–aspartate exchange is provided by the electrochemical potential difference of the proton ($\Delta\mu H^+$). Journet *et al.* (1982) presented evidence that plant mitochondria can rapidly exchange external aspartate for internal glutamate. This reaction is electroneutral and therefore readily reversible. Interestingly Vivekamanda and Oliver (1989) reported the generation of a monoclonal antibody that recognizes a single 21 kD polypeptide that was identified as the glutamate–aspartate transporter characterized above. This transporter was partially purified and shown to catalyze the antibody-sensitive transport of glutamate and aspartate in phospholipid vesicles.

A more detailed knowledge of glutamate and aspartate transport in plant mitochondria would be advantageous. However, it is clear that, in contrast with the situation observed in mammals, there is no evidence for a glutamate–OH^- or an electrogenic glutamate–aspartate transporter in plant mitochondria. That plant mitochondria do not exhibit an electrogenic glutamate–aspartate antiporter driving glutamate inside the mitochondria and aspartate outside can be explained by the existence of an NADH dehydrogenase located on the outside of the inner membrane and bypassing complex I and the first site of H^+ translocation. Consequently, a complex exchange system for the transfer of reducing equivalents from cytosol to matrix involving an electrogenic antiport that is driven by the membrane potential across the inner mitochondrial membrane is not required as it is in the case of mammalian mitochondria. Instead, fluxes of aspartate and glutamate will respond to removal of each compound and will be determined by movement of their corresponding ketoacids and also by the local concentration of each throughout the cells.

Serine and glycine transport

In the chloroplasts of higher plants oxygen competes with CO_2 for the active site of ribulose-1,5-bisphosphate carboxylase/oxygenase (Rubisco). When the CO_2

available to chloroplasts becomes progressively limiting in the light (e.g. due to environmental stresses which lead to limitation of CO_2 diffusion through the stomates) Rubisco behaves like an oxygenase to form phosphoglycolate (a two carbon compound) which is rapidly metabolized through a series of reactions (C2 cycle or photorespiration) involving enzymes in the chloroplasts, peroxisomes and mitochondria. In the course of this pathway, two molecules of glycolate (i.e. four carbon atoms) are metabolized to one molecule of 3-phosphoglycerate and CO_2 via glycine formation. The metabolism of photorespiratory glycine, flooding out of the peroxysomes, is carried out by glycine decarboxylase complex which comprises about half of the soluble proteins in the matrix of mitochondria from fully expanded green leaves (Douce and Neuburger, 1989). In the course of glycine decarboxylation and deamination one molecule of serine leaves the mitochondrion, and two molecules of glycine are taken up (Fig. 16.5). Despite the evidence that glycine and serine transporters are common in a number of biological systems, there is no consensus that the much larger flux of glycine across the inner membrane of plant mitochondria is carrier-mediated (Douce, 1985). Indeed, according to several groups, influx of glycine is not related to the energy status of the mitochondrion and appears diffusional because transport rates increase linearly up to very high glycine concentration. In addition, glycine permeates isolated pea leaf mitochondria as the neutral amino acid (Day and Wiskich, 1984). Furthermore, when the initial rates of glycine oxidation by either intact leaf mitochondria or the glycine decarboxylase complex isolated from the matrix space were represented as a double-reciprocal plot, identical apparent K_m (glycine) were obtained: approx. 4–5 mM. This would suggest that leaf mitochondria do not possess a specific carrier for glycine and allow a flexible stoichiometry for the exchange reaction that would accommodate the 2:1 ratio needed for the reaction by which glycine is metabolized (Fig. 16.5).

Cofactor uptake by plant mitochondria

Intact and well coupled plant mitochondria appear capable of the uptake of several important enzyme cofactors. Metabolic pathways that have factor-dependent enzymes localized in the matrix might be subject to modulation in this manner.

Net import of adenine nucleotides

The adenine nucleotide content (ATP + ADP + AMP) of freshly isolated animal mitochondria is fairly constant and in the range of 11–13 nmol per mg of protein. Rapid leakage of nucleotides across the inner mitochondrial membrane does not usually occur. Consequently, addition of ADP to the medium triggers the maximum rate of nucleotide exchange without affecting the overall adenine nucleotide content. In contrast, the total amount of adenine nucleotides present in intact plant mitochondria is very low (1–3 nmol per mg of protein). Consequently, the initial rates of substrate oxidation catalyzed by plant mitochondria in the presence of ADP are very often limited by the internal concentration of adenine nucleotides that can be exchanged by the nucleotide carrier. However, Abou-Khalil and Hanson (1977) have demonstrated the existence in plant mitochondria of a mechanism for net uptake of either ATP or ADP which is insensitive to carboxyatractyloside. This net uptake is strongly accelerated if an electrochemical gradient of protons across the membrane is established. Conversely, this net uptake is strongly inhibited by uncouplers. We can speculate, therefore, that *in vivo* the matrix adenine nucleotide pool is maintained against a concentration gradient by virtue of a one-way movement of adenine nucleotides. This net uptake would also explain how the total adenine nucleotide content is established and maintained during mitochondrial proliferation.

Net import of NAD⁺

Complex I which is the segment of respiratory chain responsible for electron transfer from NADH to ubiquinone, operates in close relationship with all of the NAD^+-linked TCA cycle dehydrogenases and utilizes a common pool of NAD^+ that is present in the matrix. The NADH formed by the TCA cycle dehydrogenases diffuses to the inner membrane where it is oxidized. However, matrix viscosity (0.4 g protein per mL), may limit rates of NADH diffusion during the course of substrate oxidation such that there may

be a steep downward gradient of NADH with higher concentrations in the center of the matrix and much lower concentrations near the periphery. Consequently, the NADH produced by the TCA cycle dehydrogenases located near the periphery is channeled directly to complex I without equilibrium with the bulk 'solution'. To overcome this diffusion problem, the mitochondrial matrix space contains a high concentration of pyridine nucleotides (i.e. above 6–7 nmol per mg protein).

Stimulation of respiration in isolated plant mitochondria by exogenous NAD^+ is well known (Palmer, 1976) and although it occurs during the oxidation of all NAD^+-linked substrates, it is especially the case with malate. However, not all plant mitochondria respond to added NAD^+, and those that do not respond generally have high endogenous NAD^+ contents and rapid rates of respiration (Douce, 1985). Very often, the amount of NAD^+ present in plant mitochondria is sufficient to allow state 3 rates that are approximately 80% of the maximal rate that can be obtained; that is, adding NAD^+ slightly stimulates state 3 rates. However, the rate of O_2 uptake in the presence of rotenone, an inhibitor of electron flow through complex I, is almost completely dependent on exogenous NAD^+ (Douce, 1985). Plant mitochondria have the ability to bypass complex I during oxidation of endogenous NADH. This bypass is coupled to only two sites of proton pumping (complex III and cytochrome oxidase or complex IV) (Palmer, 1976). This rotenone-insensitive internal dehydrogenase has been shown to have a much lower affinity for NADH than its rotenone-sensitive counterpart. Hence, plant mitochondria need to maintain a sufficiently high internal pool of NAD^+ to satisfy the requirements of the internal rotenone-insensitive pathway.

The mitochondrial inner membrane is generally considered to be impermeable to nicotinamide nucleotides (Douce, 1985). The mechanism by which exogenous NAD^+ can stimulate internally located enzymes such as NAD^+-linked malic enzyme and NADH dehydrogenases is therefore not obvious. However, Neuburger and Douce (1983) have shown that isolated intact plant mitochondria actively accumulate NAD^+ from the external medium (Fig. 16.7). This leads to a substantial increase in the matrix concentration of the cofactor which stimulates

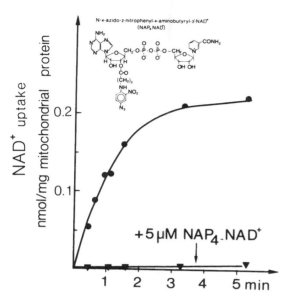

Fig. 16.7 Effect of NAP$_4$-NAD$^+$ on [^{14}C] NAD$^+$ uptake by potato tuber mitochondria. [^{14}C] NAD$^+$ uptake was measured by silicone oil filtration. Labeled NAD$^+$ was added at 10 μM and NAP$_4$-NAD$^+$ at 5 μM. α-ketoglutarate was added as substrate to drive NAD$^+$ uptake. Note that at equilibrium the concentration of [^{14}C] NAD$^+$ in the matrix space (0.2 nmol/ mg mitochondrial protein) was found to be higher than in the medium. Note that NAP$_4$-NAD$^+$ strongly inhibits [^{14}C] NAD$^+$ uptake.

matrix dehydrogenases and electron transport activities, especially the rotenone-resistant respiration. Hence, plant mitochondria appear to possess a specific NAD$^+$ carrier, since NAD$^+$ uptake is concentration-dependent and exhibits Michaelis–Menten kinetics. The maximum velocity of the carrier is strongly affected by the initial concentration of NAD$^+$ in the matrix space. Furthermore, the rate of NAD$^+$ transport is temperature-dependent and the analog N-4-azido-2-nitrophenyl-4-aminobutyryl-3'-NAD$^+$(NAP$_4$-NAD$^+$) inhibits (almost completely) net NAD$^+$ import (Fig. 16.7).

Neuburger and Douce (1983) have also found NAD$^+$ efflux from intact isolated mitochondria. Under these conditions, the concentration of NAD$^+$ in the mitochondrion progressively decreases, and oxidation of NAD$^+$-linked substrates becomes increasingly dependent on added NAD$^+$. We have also demonstrated that warm temperature (40 °C) triggered

a masive and rapid NAD$^+$ efflux from intact, isolated pea leaf mitochondria maintained in a medium that avoids the rupture of the outer membrane, and in the absence of any respiratory substrate. Such a situation leads to an arrest of glycine oxidation and NAD$^+$-dependent processes (Lenne *et al.*, 1993). In the presence of glycine or a respiratory substrate, warm temperatures did not trigger a massive leakage of NAD$^+$ from the mitochondrial matrix. In fact, the proton-motive force built up during the course of substrate oxidation has sufficient driving power to permit retention of NAD$^+$ in the matrix space (Neuburger *et al.*, 1985). The rate of NAD$^+$ efflux from the matrix is dependent upon the intramitochondrial NAD$^+$ concentration and is inhibited by the above analog that inhibits NAD$^+$ uptake, indicating that a protein is required for flux in both directions. These results raise, therefore, the question of the binding of NAD$^+$ to complex I, glycine decarboxylase complex and TCA cycle dehydrogenases. The fact that intact plant mitochondria rapidly lose NAD$^+$ after a short period of warm temperature treatment strongly suggests that most of the NAD$^+$ is not firmly bound to the inner membrane through complex I or to the L-protein, the NAD$^+$-dependent lipoamide dehydrogenase involved in the glycine decarboxylase functioning. Parenthetically, we know that a unique protein utilizing NAD$^+$ as a cofactor is involved in the reoxidation of the dihydrolipoyl moieties of all enzyme complexes containing lipoamide dehydrogenase, including glycine decarboxylase, pyruvate dehydrogenase, and α-ketoglutarate dehydrogenase.

The physiological role of this NAD$^+$ carrier still remains uncertain. Since the intramitochondrial concentration of NAD$^+$ has such a profound influence on the NAD$^+$-linked malic enzyme and O$_2$ uptake via the rotenone-insensitive pathway by isolated mitochondria, it is potentially a very powerful regulator of malate oxidation *in vivo* and could play an important role in the coarse control of metabolism, particularly during the transition from dormancy to active growth and *vice versa*. In this context, the mitochondria from young growing tissues, including mung bean hypocotyls and the shoots of sprouting potato tubers, have much higher matrix NAD$^+$ contents than those from storage tissues. It is also interesting to note that, over the time period of potato tuber storage, the endogenous NAD$^+$ content of the mitochondria declines progressively and thereafter must rise again during sprouting. It is possible that the rates of respiration in these tisssues would be affected by the concentration of NAD^+ in the cytoplasm and this concentration might differ significantly from one tissue to the other, or even between different physiological states of the same tissue. Finally, if NAD^+ rises in the matrix space without a simultaneous requirement for ATP, then the matrix NAD^+ pool will become more reduced. This might increase respiration via the rotenone-insensitive pathway without the need for gross changes in the cytoplasmic phosphorylation status.

Net import of coenzyme A

Although a large number of synthetic and degradative reactions in all tissues depend on coenzyme A (CoA), little is known about the regulation of CoA levels in the cell and even less about the intracellular distribution of CoA between the cytosol and mitochondrial matrix.

Plant mitochondria isolated from a number of tissues have a relatively low endogenous CoA content. These mitochondria are capable of actively accumulating CoA in a manner sensitive to uncouplers and low temperatures. This net uptake is catalyzed by a specific transport system and leads to large increases in the CoA content of the matrix. NAD^+ and CoA transport does not share a common carrier, because NAP_4-NAD^+ has virtually no effect on CoA transport whereas it strongly inhibits NAD^+ influx. This CoA uptake follows saturation kinetics with an apparent K_m of 0.2 mM and a maximum velocity of 4–6.5 nmol mg^{-1} protein min^{-1} (Neuburger et al., 1984).

The physiological function of a mitochondrial system for CoA would be to move the intact CoA molecule from the cytosol where it is synthesized, to the mitochondrial matrix where it is used in the entry into the TCA cycle of all major fuel substrates (pyruvate, malate, α-ketoglutarate). Plant mitochondria, in contrast with mammalian mitochondria, readily oxidize malate without the necessity of removing oxaloacetate because they possess a specific NAD^+-linked malic enzyme. This enzyme has an absolute requirement for Mn^{2+} and is characterized by a low affinity for substrates. It is inhibited by bicarbonate, which accumulates in the matrix space at alkaline pH and this inhibition is fully releaved by CoA (Neuburger and Douce, 1980). Consequently, since the cytosolic pH in vivo is around 7.5, NAD^+-linked malic enzyme will have a low activity unless a rather large pool of CoA is present in the matrix. The CoA transport system might therefore function in vivo to control the activity of the NAD^+-linked malic enzyme. This may be very important in vivo since malic enzyme, which is not very sensitive to high matrix NADH concentrations, can readily pass electrons to the non-phosphorylating rotenone-insensitive NADH dehydrogenase, which would lead to the TCA cycle functioning anaplerotically. In other words, the reaction catalyzed by NAD^+-malic enzyme compensates for the drain on the TCA cycle that occurs when citrate is removed from the cycle for biosynthesis (amino acids).

Net import of thiamine pyrophosphate

There is also evidence that thiamine pyrophosphate (TPP) can enter isolated plant mitochondria. Preparation and purification of plant mitochondria results in the loss of the activities of the pyruvate and α-ketoglutarate complexes (Fig. 16.8). These activities can be restored upon addition of TPP to the mitochondria and possibly this association and dissociation of TPP may provide a mechanism for regulation in vivo: all the plant mitochondria isolated so far appear to be depleted of endogenous TPP (presumably during their isolation, although this has yet to be demonstrated) but rapidly accumulate it when it is provided externally. To determine the kinetics of TPP uptake, mitochondria have been incubated with labeled TPP at concentrations in the range 1–100 μM, and the initial rates of uptake determined by rapid filtration experiments (in the millisecond time range). The K_m value for TPP is 50 μM and the maximum velocity is 0.8 nmol min^{-1} mg^{-1} protein. The extent of TPP accumulation is unaffected by NAP_4-NAD^+ indicating that NAD^+ and TPP do not share a common carrier. Thiamine cannot replace TPP in reconstituting α-ketoglutarate activity nor does it have any competitive activity on reconstitution or uptake (similarly for TMP, pyrimidine pyrophosphate and thiazole phosphate). The carrier is therefore very specific.

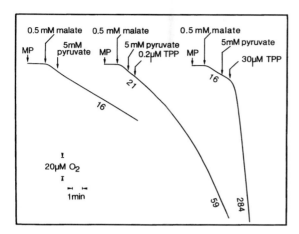

Fig. 16.8 Effect of thiamine pyrophosphate (TPP) on pyruvate oxidation by potato tuber mitochondria. Note that TPP added to the medium even at very low concentration (0.2 μM) triggers, after a lag phase, a rapid rate of O_2 consumption. This experiment demonstrates that TPP molecules penetrate the inner membrane of plant mitochondria.

Finally, during photorespiration, serinehydroxymethyltransferase and the glycine cleavage complex present in the mitochondrial matrix share a common pool of tetrahydropteroyl glutamate with n (4–6) glutamate residues. Oligo-γ-polyglutamate of H_4folate, in contrast with NAD^+, ADP, ATP, TPP and CoA, did not leak out readily from intact isolated mitochondria. The conjugate nature of pea mitochondrial folates may have a role in the retention of these derivatives within the matrix.

Conclusions

The control of plant respiration *in vivo* is likely exerted through the finely regulated supply of substrates to the TCA cycle and phosphorylating system. As shown in this chapter, the mechanisms involved in the regulation of anion transport across the inner mitochondrial membrane, other than adenine nucleotides and P_i, are still poorly defined. Likewise nothing is known about the chemical nature of respiratory substrates (pyruvate, malate, oxaloacetate, NADH, etc.) utilized by the mitochondrion inside the plant cells.

It is now apparent that plant mitochondria possess several distinct carriers for the net uptake of cofactors. These specific carriers would allow regulation of the matrix cofactors pool size *in vivo*, either maintaining them at a constant level or allowing changes in response to specific signals. In this way, metabolic pathways with cofactor-dependent enzymes localized in the matrix might be subject to modulation leading to considerable modification of carbon traffic between the cytosol and matrix. For example, in the presence of a large excess of free CoA in the matrix, NAD^+-linked malic enzyme can operate, allowing the conversion of C4 acids into acetyl-CoA, the normal respiratory substrate, without the necessity of supplying pyruvate from glycolysis. Likewise, an excess of NAD^+ in the matrix space is required to engage the non-phosphorylating rotenone-resistant electron pathway, releasing partially the constraint exerted by the protonmotive force across the inner membrane on internal NADH oxidation. Interestingly, it is only at high matrix NAD^+ concentrations that a functional link between malic enzyme (exhibiting a low affinity for NAD^+) and the non-phosphorylating rotenone-insensitive NADH dehydrogenase (exhibiting a low affinity for NADH) is observed. Under these conditions malic enzyme and the TCA cycle could play a significant role in the metabolism of fast growing cells by providing numerous substrates for biosynthetic reactions in the cytoplasm. There may be an advantage, therefore, in having a non-phosphorylating electron transport chain to enable this role to be fulfilled in the presence of a high protonmotive force across the inner mitochondrial membrane. There is circumstantial evidence that some of these changes in matrix cofactor pool size occur *in vivo*. However, until clearer evidence is forthcoming, their role in the control of respiration in intact cells and tissues cannot be properly evaluated.

Finally, plant mitochondria, very likely, can take up and extrude various cations or inorganic anions to fulfil both general (e.g. pH and volume regulation) and specialized functions. The combination of an efficient expression system (e.g. yeast) with the technique of patch clamping holds promise for the characterization of new ion channels and cation exchangers in plant mitochondria.

References

Abou-Khalil, S. and Hanson, J. B. (1977). Net adenosine diphosphate accumulation in mitochondria. *Arch. Biochem. Biophys.* **183**, 581–7.

Beevers, H. (1961). *Respiratory Metabolism in Plants*, Harpers, New York.

Bligny, R., Roby, C. and Douce, R. (1989). Phosphorus-31 Nuclear Magnetic resonance studies in higher plant cells. In *Nuclear Magnetic Resonance in Agriculture*, eds. P. E. Pfeffer and W. V. Gerasimowicz, CRC Press, Boca Raton, Florida, pp. 71–89.

Brailsford, M. A., Thomson, A. G., Kaderbhai, N. and Beechey, R. B. (1986). Pyruvate metabolism in castor-bean mitochondria. *Biochem. J.* **239**, 355–61.

Chappell, J. and Beevers, H. (1983). Transport of dicarboxylic acids in castor bean mitochondria. *Plant Physiol.* **72**, 434–40.

Chappell, J. and Croft, A. R. (1966). Ion transport and reversible volumes changes of isolated mitochondria. In *Regulation of Metabolic Processes in Mitochondria*, Vol. 17, ed. J. M. Tager, Elsevier, New York, pp. 293–316.

Chrispeels, M. J. and Maurel, C. (1994). Aquaporins: The molecular basis of facilitated water movement through living plant cells? *Plant Physiol.* **105**, 9–14.

Day, D. A. and Wiskich, J. T. (1984). Transport processes of isolated mitochondria. *Physiol. Veg.* **22**, 241–61.

Desantis, A., Arrigoni, O. and Palmieri, F. (1976). Carrier-mediated transport of metabolites in purified bean mitochondria. *Plant Cell Physiol.* **17**, 1221–33.

Douce, R. (1985). *Mitochondria in Higher Plants. Structure, Function, and Biogenesis*, Academic Press, Orlando, FL.

Douce, R. and Neuburger, M. (1989). The uniqueness of plant mitochondria. *Ann. Rev. Plant Physiol. Plant Mol. Biol.* **40**, 371–414.

Douce, R., Neuburger, M. and Givan, C. V. (1986). Regulation of succinate oxidation by NAD$^+$ in mitochondria purified from potato tubers. *Biochim. Biophys. Acta* **850**, 64–71.

Douce, R., Brouquisse, R. and Journet, E.-P. (1987). Electron transfer and oxidative phosphorylation in plant mitochondria. In *The Biochemistry of Plants*, Vol. 11, *Biochemistry of Metabolism*, ed. D. D. Davies, Academic Press, New York, pp. 177–211.

Emmermann, M., Braun, H. P. and Schmitz, U. K. (1991). The ADP/ATP translocator from potato has a long N-terminal extension. *Curr. Gen.* **20**, 405–40.

Genchi, G., De Santis, A., Ponzone, C. and Palmieri, F. (1991). Partial purification and reconstitution of the α-ketoglutarate carrier from corn (*Zea mays* L.) mitochondria. *Plant Physiol.* **96**, 1003–7.

Gout, E., Bligny, R. and Douce, R. (1992). Regulation of intracellular pH values in higher plant cells. *J. Biol. Chem.* **267**, 13903–9.

Gout, E., Bligny, R., Pascal, N. and Douce, R. (1993). ^{13}C Nuclear magnetic resonance studies of malate and citrate synthesis and compartmentation in higher plant cells. *J. Biol. Chem.* **268**, 3986–92.

Hanning, I. and Heldt, H. W. (1993). On the function of mitochondrial metabolism during photosynthesis in spinach (*Spinacia oleracea* L.) leaves. *Plant Physiol.* **103**, 1147–54.

Huang, A. H. C., Trelease, R. N. and Moore, T. S., Jr. (1983). *Plant Peroxisomes*, Academic Press, New York.

Journet, E.-P., Bonner, W. D. and Douce, R. (1982). Glutamate metabolism triggered by oxaloacetate in intact plant mitochondria. *Arch. Biochem. Biophys.* **214**, 336–75.

Kramer, R. and Palmieri, F. (1989). Molecular aspects of isolated and reconstituted carrier proteins from animal mitochondria. *Biochim. Biophys. Acta* **974**, 1–23.

Krömer, S., Hanning, I. and Heldt, H. W. (1992). On the source of redox equivalents for mitochondrial oxidative phosphorylation in the light. In *Molecular, Biochemical and Physiological Aspects of Plant Respiration*, eds H. Lambers and L. H. W. van der Plas, SPB Academic Press, The Hague, The Netherlands, pp. 167–75.

Leaver, C. J. and Gray, M. W. (1982). Mitochondrial genome organization and expression in higher plants. *Ann. Rev. Plant Physiol.* **33**, 373–402.

Lenne, C., Neuburger, M. and Douce, R. (1993). Effect of high physiological temperature on NAD$^+$ content of green leaf mitochondria. *Plant Physiol.* **102**, 1157–62.

Lilley, R. M. C., Ebbighausen, H. and Heldt, H. W. (1987). The simultaneous determination of carbon dioxide release and oxygen uptake in suspensions of plant leaf mitochondria oxidizing glycine. *Plant Physiol.* **83**, 349–53.

McIntosh, C. A. and Oliver, D. (1992). Isolation and characterization of the tricarboxylate transporter from pea mitochondria. *Plant Physiol.* **100**, 2030-4.

McIntosh, C. A. and Oliver, D. J. (1994). The phosphate transporter from pea mitochondria. Isolation and characterization in proteolipid vesicles. *Plant Physiol.* **105**, 47–52.

Mannella, C. A. (1985). The outer membrane of plant mitochondria. In *Encyclopedia of Plant Physiology*, Vol. 18, eds R. Douce and D. Day, Springer-Verlag, Berlin, pp. 106–33.

Miflin, B. J. and Lea, P. J. (1980) Ammonia assimilation. In *The Biochemistry of Plants*, Vol. 5, ed. B. J. Miflin, Academic Press, New York, pp. 169–99.

Millhouse, J., Wiskich, J. T. and Beevers, H. (1983). Metabolite oxidation and transport in mitochondria of germinating castor bean endosperm. *Aust. J. Plant Physiol.* **10**, 167–77.

Neuburger, M. and Douce, R. (1980). Effect of bicarbonate and oxaloacetate on malate oxidation by spinach leaf mitochondria. *Biochim. Biophys. Acta* **589**, 176–89.

Neuburger, M. and Douce, R. (1983). Slow passive diffusion of NAD$^+$ between intact isolated plant mitochondria and the suspending medium. *Biochem. J.* **216**, 443–50.

Neuburger, M., Day, D. A. and Douce, R. (1984). Transport of coenzyme A in plant mitochondria. *Arch. Biochem. Biophys.* **229**, 253–8.

Neuburger, M., Day, D. A. and Douce, R. (1985). Transport of NAD$^+$ in percoll-purified potato tuber mitochondria. Inhibition of NAD$^+$ influx and efflux by *N*-4-azido-2-nitrophenyl-4-aminobutyryl-3′-NAD$^+$. *Plant Physiol.* **78**, 405–10.

Palmer, J. M. (1976). The organization and regulation of electron transport in plant mitochondria. *Ann. Rev. Plant Physiol.* **27**, 133–57.

Phelps, A., Schobert, C. T. and Wohlrab, H. (1991). Cloning and characterization of the mitochondrial phosphate transport protein gene from the yeast Saccharomyces cerevisiae. *Biochemistry* **39**, 248–52.

Schmid, A., Krömer, S., Heldt, H. W. and Benz, R. (1992). Identification of two general diffusion channels in the outer membrane of pea mitochondria. *Biochim. Biophys. Acta* **1112**, 174–80.

Vignais, P. V. (1976). Molecular and physiological aspects of adenine nucleotide transport in mitochondria. *Biochim. Biophys. Acta* **456**, 1–38.

Vivekananda, I. and Oliver, D. J. (1989). Isolation and partial characterization of the glutamate/aspartate-transporter from pea leaf mitochondria using a specific monoclonal antibody. *Plant Physiol.* **91**, 272–77.

Vivekananda, J. and Oliver, D. J. (1990). Detection of the monocarboxylate transporter from pea mitochondria by means of a specific monoclonal antibody. *FEBS Lett.* **260**, 217–19.

Vivekananda, J., Beck, C. F. and Oliver, D. J. (1988). Monoclonal antibodies as tools in membrane biochemistry. Identification and partial characterization of the dicarboxylate transporter from pea leaf mitochondria. *J. Biol. Chem.* **263**, 4782–8.

Winning, B. M., Sarah, C. J., Purdue, P. E., Day, C. D. and Leaver, C. J. (1992). The adenine nucleotide translocator of higher plants is synthesized as a large precursor that is processed upon import into mitochondria. *Plant J.* **2**, 763–73.

Wohlrab, H. (1986). Molecular aspect of inorganic phosphate transport in mitochondria. *Biochim. Biophys. Acta* **853**, 115–34.

Zoglowek, C., Krömer, S. and Heldt, H. W. (1988). Oxaloacetate and malate transport of plant mitochondria. *Plant Physiol.* **87**, 109–15.

Photosynthesis

17 Plastid structure and development
William Newcomb

Introduction

Plastids are self-replicating organelles, surrounded by an envelope comprised of two membranes, that are only found in cells of photosynthetic eukaryotes. With the exception of the generative and sperm cells of certain angiosperms, all plant cells possess plastids which occur in a variety of types, sizes, shapes and colors. It is easiest to categorize plastids on the basis of their color: green, other colors (red, orange and yellow), and colorless. Green-colored plastids are chloroplasts, which contain the chlorophyll pigments and are the site of photosynthesis. Red-, orange- and yellow-colored plastids are chromoplasts which contain carotenoid pigments and are commonly found in flowers, fruits, senescing leaves and certain roots. Chromoplasts are non-photosynthetic and function mainly to attract pollinators and animals, which aid in the distribution of pollen and seeds respectively. Colorless plastids lack chlorophyll and carotenoid pigments and may be referred to by an old term, leucoplast. However, it is more common and precise to name leucoplasts or colorless plastids on the basis of their principal storage products; amyloplasts store starch, while proteins and lipids are stored in proteinoplasts and elaioplasts respectively. Proplastids, which have been regarded as the progenitor for all other types of plastids, are also colorless.

J. Whatley has distinguished three main types of proplastids involved in chloroplast biogenesis, namely eoplasts, amyloplasts and amoeboid plastids. Lack of light can induce the formation of a special type of plastid, etioplasts, which are either colorless or pale yellow in color, form in dark-grown seedlings and are arrested in their development; upon exposure to light etioplasts synthesize chlorophyll and become chloroplasts.

Common features

All plastids have a number of features in common. They are surrounded by an envelope composed of inner and outer membranes which are separated by a 10–20 nm gap. Plastids contain DNA which appears as fibrils in a matrix-free region (the so-called nucleoid) of the stroma, which is the background matrix material of the plastid. The stroma contains numerous ribosomes that are smaller than the cytoplasmic ribosomes. The plastid ribosomes may be arranged singularly or in polyribosomes. Plastoglobuli, small lipid droplets which are not membrane bound, are also present in the stroma of many plastids.

Plastid inheritance

In most angiosperms plastid inheritance is maternal because the sperm cells lack plastids and the egg cell contains numerous plastids. This situation arises during the first pollen mitosis and cytokinesis when all of the plastids go into the vegetative cells resulting in an aplastidic generative cell.

Plastid ultrastructure

Proplastids

Proplastids, the progenitors of other plastids, are found in root and shoot meristems, embryos, endosperm and in young developing leaves. In view of the fact that most plastids may differentiate into another type of plastid, the progenitor function is related to their simple structure which is a primary distinguishing feature. Proplastids are not structurally complex. They consist of an envelope, a nucleoid with DNA fibrils, a small number of ribosomes, some flattened membrane sacs called thylakoids, which may appear to be continuous with the inner membrane of the envelope, and sometimes small starch granules (Fig. 17.1). Proplastids vary in shape and can be ellipsoidal, spherical, cylindrical and branched, with or without bulged or invaginated regions. They are usually only slightly larger than mitochondria. Within meristematic cells, the division of proplastids keeps pace with mitosis and cytokinesis. Proplastids may also divide after cytokinesis in certain root cells where they usually develop into amyloplasts (Fig. 17.2). In light-grown plants, the proplastids in stems and leaves usually differentiate into chloroplasts. Proplastids may also have important biochemical functions such as the synthesis and storage of starch and other metabolites.

Chloroplasts

Chloroplasts are green photosynthetic plastids that have an elaborate arrangement of interconnected membranous sacs in the plastid stroma. Chloroplasts typically measure about $5 \times 2 \times 1$–$2\,\mu m$. The internal membrane system of a typical higher plant chloroplast includes grana (also known as grana thylakoids) which are stacks of flattened disk-shaped membranous sacs. About 50 grana are present in a typical angiosperm chloroplast. Membranous channels, that have been called thylakoids, frets, stroma thylakoids or stroma lamellae by various authors, traverse the stroma and interconnect the grana. The concentration of photosystems I and II differ in the thylakoids from the stroma and grana. The degree of stacking of the thylakoids into grana varies with the physiological requirements of the cell. The chlorophyll pigments and light reactions of photosynthesis are associated with the thylakoid membrane system. The dark reactions of photosynthesis occur in the stroma which also contains numerous ribosomes and large proteinaceous particles. These particles may be ribulose 1,5-bisphosphate carboxylase/oxygenase which makes up more than 50% of the protein in the chloroplast stroma.

Chloroplasts have three compartments, the intermembrane space between the outer and inner membranes, the stroma and thylakoid space. In contrast, mitochondria only possess two compartments – the intermembrane space and a matrix which corresponds to the plastid stroma. In chloroplasts, light reactions cause a proton gradient to develop across the thylakoid membrane. The gradient is discharged during the synthesis of ATP by the ATP synthetase which is located in the membrane. Concomitantly with ATP synthesis, $NADP^+$ is reduced to NADPH and water is oxidized with the liberation of oxygen. In the dark reactions of photosynthesis, the ATP and NADPH synthesized during the light-dependent reactions are used for carbon fixation, the chemical reduction of carbon dioxide to carbohydrate.

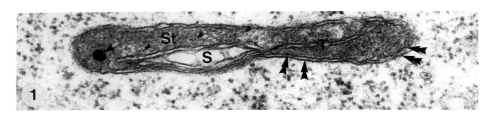

Fig. 17.1 Transmission electron micrograph of a proplastid in an epidermal cell of an embryo of chickpea (*Cicer arietinum* L.). Shown are the two membranes comprising the bounding membrane envelope (double arrowheads), an osmiophilic droplet (D), a starch (S) granule, a thylakoid (T) and ribosomes (single arrowheads) in the stroma (St). $\times\,74\,800$.

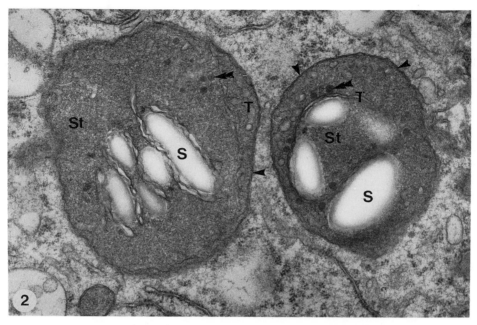

Fig. 17.2 Transmission electron micrograph of two amyloplasts in a pericycle cell adjacent to a root nodule of the garden pea (*Pisum sativum*). Shown within the plastids are the bounding membrane envelope (single arrowheads), osmiophilic droplets (double arrowheads), starch (S) granules and stroma (St). × 33 300.

Chloroplast development

Leaves develop from primordia initiated along the flanks of the shoot apical meristem. These meristematic cells possess proplastids which contain a few thylakoids in their stroma. Some authors have referred to such a plastid as a spherically-shaped eoplast. The eoplast accumulates starch, becoming an amyloplast which subsequently loses the starch deposits and becomes amoeboid. Most authors have not distinguished among these types and have referred to all three of these forms as being proplastids. The amoeboid plastids grow in size and develop stroma lamellae before grana appear. When these plastids mature, the stroma lamellae align and the disk-shaped thylakoids of the grana become stacked. The mature chloroplasts eventually senesce and the stroma lamellae and grana become disorganized. Some of the senescent chloroplasts (termed gerontoplasts by Whatley) may synthesize carotenoid pigments and thus can be termed chromoplasts. In some tissue cultures and evergreen plants, chloroplasts may dedifferentiate and then redifferentiate as chromoplasts. This led Whatley to propose the occurrence of a plastid cycle.

Etioplasts

Etioplasts are found in plants which are grown in continuous darkness, which results in tall etiolated plants that are yellow due to the presence of protochlorophyll in the etioplasts. While etioplasts are regarded commonly as putative chloroplasts whose development has been arrested, it is important to note that etioplasts are not a normal stage in the development of chloroplasts. In some cases, etioplasts represent an interesting developmental alternative. Most seedlings receive a regular diurnal sequence of light and dark periods and etioplasts do not develop under these conditions.

The etioplasts are structurally simple with the most distinctive feature being a paracrystalline arrangement of membranes, termed the prolamellar body. Upon exposure to light the etioplasts differentiate into chloroplasts during which the protochlorophyll becomes converted into chlorophyll and the prolamellar body reorganizes into grana and stroma lamellae. There is considerable variation in plastid development in various species. For example, prolamellar bodies do not form in the etioplasts of dark-grown *Avena* coleoptiles.

Chromoplasts

Chromoplasts are red-, orange- and yellow-colored due to the presence of carotenoid pigments. Chromoplasts are present commonly in flower petals, fruits and certain roots such as carrots. Chromoplasts often develop from chloroplasts (the ripening of bananas and tomatoes) but may also

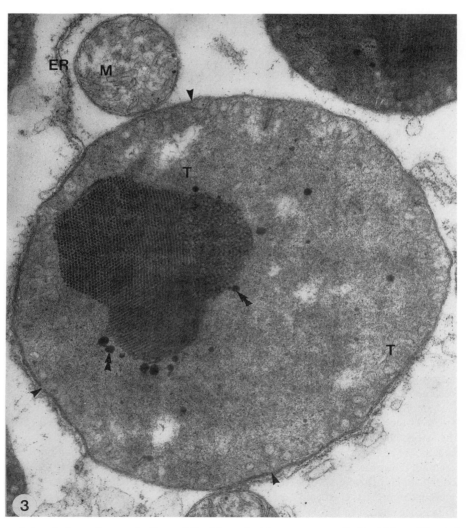

Fig. 17.3 Transmission electron micrograph of a specialized leucoplast in a suspensor cell of a chickpea (*C. arietinum*) embryo. Shown in the plastid are the bounding membrane envelope (single arrowheads), several electron-dense osmiophilic droplets (double arrowheads), several thylakoids (T) and a large region of electron-dense, regularly-arranged, membranous tubules. Shown outside the plastid are a mitochondrion (M) and profiles of endoplasmic reticulum (ER). × 36 400.

originate from proplastids or amyloplasts. During the transition of a chloroplast into a chromoplast, the chlorophyll is destroyed and the grana and stroma lamella become reorganized. Chromoplasts may also revert back to chloroplasts; examples of this include carrot roots and orange fruit skins. Osmiophilic droplets or plastoglobuli, filamentous pigmented bodies, and crystals are normally found in chromoplasts and are the sites of carotenoid synthesis and/or storage.

Leucoplasts

Leucoplasts are colorless plastids which have lost their progenitor function, i.e. are distinct from proplastids. Amyloplasts, elaioplasts and proteinoplasts are colorless and are sites of the synthesis of starch, lipids and proteins respectively. Some leucoplasts have deposits of starch and proteins.

Amyloplasts are characterized by the presence of one or more starch granules which distend and alter the shape of the organelle. Only a few thylakoids are normally found within the stroma. Amyloplasts are most common in root tissues and may be involved in the detection of gravity. Amyloplasts of the root cap will sediment when the orientation of the root is changed. If the root cap is removed, the root's ability to detect gravity is lost. When the root cap reforms, the ability to detect gravity is restored. Amyloplasts of roots may develop into chloroplasts if the roots are exposed to light.

Leucoplasts with specialized functions occur in certain cells and tissues. Some but not all of these plastids have unique ultrastructural features. For example, the cells of some suspensors of angiosperm embryos contain large plastids which have extensive profiles of small diameter membranous tubules (Fig. 17.3). Specialized leucoplasts also occur in sieve elements and the uninfected cells of certain leguminous root nodules (ex. soybean) which export ureides.

Further reading

Alberts, B., Bray, D., Lewis, J., Raff, M., Roberts, K. and Watson, J. D. (1994). *Molecular Biology of the Cell*, 3rd edn, Garland Publishing Co., New York.

Gunning, B. E. S. and Steer, M. W. (1975). *Ultrastructure and the Biology of Plant Cells*, Edward Arnold, London.

Newcomb, E. H. (1967). Fine structure of protein-storing plastids in bean root tips. *J. Cell Biol.* **33**, 143–63.

Possingham, J. V. (1980). Plastid replication and development in the life cycle of higher plants. *Ann. Rev. Plant Physiol.* **31**, 113–29.

Thompson, W. W. and Whatley, J. M. (1980). Development of nongreen plastids. *Ann. Rev. Plant Physiol.* **31**, 375–94.

Whatley, J. B. (1974). Chloroplast development in primary leaves of *Phaseolus vulgaris*. *New Phytol.* **73**, 1097–1100.

Whatley, J. B. (1977). Variations in the basic pathway of chloroplast development. *New Phytol.* **78**, 407–20.

Whatley, J. B. (1978). A suggested cycle of plastid developmental interrelationships. *New Phytol.* **80**, 489–502.

Whatley, J. B. (1979). Plastid development in the primary leaf of *Phaseolus vulgaris*: variations between different types of cells. *New Phytol.* **82**, 1–10.

18 Molecular biology of photosynthesis in higher plants

John E. Mullet

Introduction

This chapter will summarize the organization and regulation of plastid and nuclear genes involved in photosynthesis in higher plants. Not all aspects of plastid gene expression or the expression of nuclear genes encoding proteins involved in photosynthesis will be covered and additional information can be obtained in several books and reviews (see Further reading).

Plastid function

In higher plants, photosynthetic electron transport and CO_2 fixation occur within chloroplasts. Chloroplasts are a specialized type of plastid and are so-named when chlorophyll accumulates within the organelle. Chloroplasts consist of a double outer membrane or envelope, a stromal phase where soluble proteins such as Rubisco are located and an inner membrane, termed the thylakoid. Four major photosynthetic complexes are integrated into the thylakoid: Photosystem I (PSI), Photosystem II (PSII), a cytochrome complex, and an ATP synthetase. These complexes mediate light-dependent vectorial electron flow and the resultant generation of ATP and reducing power. The end products of photosynthetic electron transport are used in CO_2 fixation and other plastid functions. In addition to photosynthesis, plastids carry out numerous other functions including steps in amino acid, nucleic acid, fatty acid, pyrimidine, terpene and tetrapyrrole biosynthesis. This organelle is a site of nitrite and sulfur reduction and some steps in plant growth regulator synthesis. The interdependence of cell and plastid function in photosynthetic as well as non-photosynthetic tissue makes an understanding of the regulation of plastid gene expression inseparable from the larger question of cell and organ biogenesis. Therefore, the special features of genes encoding the photosynthetic apparatus will be dealt with after consideration of plastid origin and the integration of plastid and cell function.

Origin of plastids

Plastids in most higher plants contain a circular genome that contains 120 to 135 genes. However, this is only a fraction of the total number of genes required for photosynthesis. The remaining genes are encoded in nuclear DNA. The presence of DNA in plastids reflects the origin of this organelle from a free-living oxygen-evolving photosynthetic prokaryote (Gray and Doolittle, 1982). Evidence that plastids are derived from prokaryotic endosymbionts includes similarities in gene sequence, gene order and mechanisms for decoding the plastid's genetic information. Plastids, like bacteria, contain 70S ribosomes that translate uncapped, non-polyadenylated RNAs that often contain ribosome binding sites. In contrast, eukaryotic cytoplasms contain 80S ribosomes that translate capped and polyadenylated mRNAs lacking ribosome binding sites. Plastid transcription units are often polycistronic and some plastid operons resemble those found in *E. coli*. In addition, transcription initiation of many plastid genes is specified by DNA sequence elements which are similar to bacterial promoter elements. Overall, plastid gene sequences, gene organization and decoding mechanisms indicate that these organelles arose from free-living photosynthetic prokaryotes.

Integration of plastid and cell function

Plastids isolated from higher plants do not replicate and survive for only a short period of time *in vitro*. These organelles are not autonomous and are dependent on cytoplasmic functions and nuclear gene products. The interdependence of plastids and the cells in which they reside is relatively specific and transfer of plastids into foreign plant cells may result in impaired function and loss of the transferred organelles. Extensive changes have occurred in the photosynthetic endosymbiont during the integration process. Plastids of higher plants do not have peptidylglycan walls, that are found in bacteria. Presumably, the genes required for wall biosynthesis have been lost or inactivated. Moreover, some plastid functions are not found in prokaryotes and must have been acquired during the process of host-endosymbiont coevolution. The involvement of plastids in plant growth regulator biosynthesis is an example of this type of change. Biosynthetic pathways that were present in the host and endosymbiont have become integrated. For example, lipid biosynthesis in higher plants involves the integrated action of enzymes located in plastids and ER membranes.

The integration of cell and plastid function must have involved the transfer of many genes from the plastid to the nucleus. The fact that most of the enzymes involved chlorophyll and carotenoid biosynthesis are encoded by nuclear genes underscores this point. These functions at one time must have been exclusively localized in the plastid genome. In parallel with changes in gene localization, there occurred a restructuring of regulatory circuits. The expression of genes involved in photosynthesis in higher plants is regulated developmentally and shows organ, tissue and cell specificity. Plastid populations in higher plants acquire specialized functions such as starch (amyloplasts), carotenoid (chromoplasts) or lipid (leucoplasts) biosynthesis and storage. Specialized plastids are usually localized in specific organs or tissues such as tubers (amyloplasts) or epidermal layers of fruit and petals (chromoplasts).

Knowledge of the origin of plastids provides a useful perspective when trying to rationalize the organization and regulation of genes encoding plastid functions in higher plants. Moreover, the transfer of genetic information from the plastid to the nucleus raises several questions. Why were some genes transferred but not others? What signals are needed to direct nuclear gene products to the plastid compartment? How are genes in the two genomes coactivated during chloroplast development and what additional mechanisms are used to coordinate gene expression and protein complex assembly? These questions and others will be addressed in the remainder of this chapter.

The plastid genome

Coding capacity

The complete sequence of the plastid genome of tobacco, rice, black pine and liverwort and the identification of genes in numerous other plants indicates that the higher plant plastid genome codes for approximately 130 genes (reviewed by Sugiura, 1992). These genes fall into three categories (Table 18.1). The first category consists of genes that are involved in decoding the plastid genetic information. These genes encode ribosomal RNAs, tRNAs, subunits of an RNA polymerase, ribosomal proteins and proteins which function in translation initiation (IF1). The second category of genes encode proteins involved in photosynthesis. The large subunit of Rubisco and several subunits of each protein complex involved in photosynthetic electron transport (PSI, PSII, cytochrome complex, ATP synthetase) are encoded by plastid DNA. The third group of genes include sequences for an NADH oxidoreductase complex, a protease and open reading frames of unknown function. Genes involved in chlorophyll biosynthesis are present in plastids of *Chlamydomonas*, *Marchantia* and conifers.

There have been many speculations about why the small group of genes described above is still present in plastids whereas most of the endosymbiont's genetic capacity has been lost or transferred to the nucleus. One clue to this puzzle may be found in the functions encoded by plastid genes. The plastid genes identified thus far either participate in decoding the plastid genome or encode proteins involved in photosynthesis. Clearly, these two groups of genes are

Table 18.1 Genes coded by plastid DNA.

Genes	Gene products
I. Genes required to decode plastid DNA	
Transcription	
rpoA, B, C1, C2	RNA polymerase subunits
Translation	
rrn	16S, 23S, 4.5S, 5S ribosomal RNA
trn (30 genes)	tRNAs
rps (12 genes)	30S ribosomal subunits
rpl (7–9 genes)	50S ribosomal subunits
infA	initiation factor 1
II. Genes involved in photosynthesis	
rbcL	Rubisco large subunit
psaA, B, C, I, J	PSI subunits
psbA, B, C, D, E, F, H, I, K, L, M, N	PSII subunits
petA, B, D, G	cytochrome *b6/f* complex
atpA, B, E, F, H, I	ATP synthase subunits
chlN, L	chlorophyll synthesis[a]
III. Other genes	
ndhA, B, C, D, E, F	NADH oxidoreductase
clpP	protease
orfs (15–20 genes)	unidentified

[a] Genes found in *Chlamydomonas*, *Marchantia* and conifers.

coupled, because to express genes involved in photosynthesis, the plastid's transcription and translation apparatus must be synthesized. This latter process is no small undertaking which is shown by the fact that mesophyll leaf cells accumulate between 10^6 and 10^7 plastid ribosomes. Plastid ribosomes and mRNAs are present at much reduced levels in non-photosynthetic plastids such as proplastids, chromoplasts or amyloplasts. Furthermore, functions carried out by these non-photosynthetic plastids (plastid replication, carotenoid biosynthesis, starch accumulation) are not known to require high level plastid gene expression and most likely involve primarily nuclear genes. The localization of these genes in the nucleus means that differentiation of specialized non-photosynthetic plastids does not require concomitant build-up of plastid transcription and translation capacity. The plastid compartment in non-photosynthetic cells can therefore function as other non-DNA containing organelles such as glyoxysomes or microbodies. Perhaps the advantages of this situation contributed to the present distribution of genes in the nucleus and plastid.

The rationale described above poses the question as to why all of the plastid genes have not been relocated to the nucleus. Some genes may have been retained in the plastid because the proteins they encode would be difficult to transport through the chloroplast envelope. This idea, however, does not explain why some ribosomal proteins remain encoded by plastid DNA but not others of similar size and solubility. Likewise, some hydrophobic, chlorophyll-binding proteins are encoded by nuclear genes (*Lhcb*) whereas others are plastid encoded (i.e., *psaA*). This suggests that it is possible to move extant plastid genes to the nucleus without loss of function. Further support for this idea comes from the observation that although EF-Tu is encoded by a plastid gene in many algae, in higher plants it is encoded by a nuclear gene. Similarly, *rpoA* is a nuclear gene in geranium but a plastid gene in other higher plants. This suggests that gene transfer is still occurring although slowly, possibly because gene transfer disrupts complex regulatory circuits that now coordinate plastid and nuclear gene expression.

DNA organization

The plastid genome in most higher plants is circular and varies in size from 120 to 217 kbp. Most of this size variation can be accounted for by the presence or absence of a portion of the plastid genome which has been duplicated and is present in an inverted orientation in the plastid DNA molecule. The location of this inverted repeat is relatively fixed with respect to other genes and it separates a small single copy region from a large single copy DNA region. In most higher plants, the inverted repeat is 22 to 26 kbp within which the rRNA transcription unit is located. In geranium the repeated DNA is larger and genes such as *psbB*, *petB*, *petD*, *petA* and *rbcL* are included in the inverted repeat. Some plastid genomes, such as those in pea and mung bean, lack the inverted repeat. In addition, in rare circumstances such as in the parasitic non-photosynthetic plant *Epifagus*, the plastid genome is smaller and lacks many genes typical of most higher plant genomes including several genes for photosynthesis and the plastid-encoded RNA polymerase.

Plastid gene content in higher plants is relatively constant and many polycistronic transcription units are conserved. Gene pairs such as *psaA–psaB*, *psbD–psbC*, *atpB–atpE*, are cotranscribed in all higher plant plastid genomes examined to date (see Fig. 18.1 for *psbD–psbC* gene organization in barley). The cotranscription of genes may ensure that the synthesis of subunits is stoichiometric or promote protein–protein interactions required for assembly of functional complexes. For example, *psaA* and *psaB* encode polypeptides that are tightly associated in the reaction center of PSI. Some transcription units contain genes encoding the transcription and translation apparatus but not genes for photosynthesis (i.e. *rpoB–rpoC1–rpoC2*, *rrn* operon (16S, 23S, 4.5S, 5S rRNA, *trnI*, *trnA*), *rpl23–rpl2–rps19–rpl22–rps3–rpl16–rpl14–rps8–infA–rpl36–rps11–rpoA*). Other operons or transcription units primarily encode genes for photosynthesis (i.e. *psbB–psbH–petB–petD*, *atpI–atpH–atpF–atpA*, *rbcL*, *atpB–atpE*). The segregation of genes for different functions in separate operons allows groups of related genes to be expressed at similar levels or regulated in a coordinated manner (Rapp *et al.*, 1992).

While the plastid gene content of higher plants is relatively constant, there is variation in gene order resulting primarily from DNA inversions (Palmer, 1990). The DNA inversions have reshuffled plastid genomes such that distances between genes, and relative orientation of transcription units varies in genomes of higher plants. For example, *rps16* is proximal to *trnK* in barley; whereas in pea, *rbcL* occupies this position. The greatest variation in gene order is found in peas (at least 12 rearrangements) perhaps due to the lack of an inverted repeat in this plastid genome which may stabilize the genome.

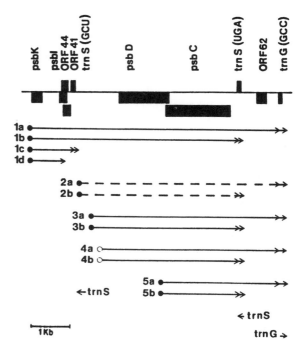

Fig. 18.1 Map of genes and transcripts from a nine kbp region of the barley chloroplast genome that encodes 3 tRNAs and 7 proteins. Gene names are located above the solid boxes that correspond to protein or tRNA coding regions. Arrows indicate the direction of transcription and transcripts synthesized from each DNA region are designated by arrows in the lower part of the figure. Solid circles correspond to RNA 5'-termini and open circles indicate the 5'-termini of transcripts produced from the blue light regulated *psbD–psbC* promoter.

DNA copy number and localization

Multiple copies of the plastid genome are found in each plastid. The amount of DNA per plastid varies with the stage of leaf and chloroplast development. Proplastids with as few as 22 copies of DNA have been reported whereas chloroplasts in general contain 200 to 300 DNA molecules. The polyploid nature of the plastid genome is even more striking when one considers leaf cells. In pea, barley and spinach, each leaf cell contains 9000 to 13 000 copies of plastid DNA that are dispersed in 40 to 120 plastids. This means that *rbcL*, a single copy plastid gene which encodes the large subunit of Rubisco, is present in about 10 000 copies in a mesophyll cell. The small subunit of Rubisco, which accumulates in stoichiometric amounts relative to the large subunit, is encoded by a small gene family (5 to 10 members) within the nucleus. Therefore, the number of *rbcL* genes outnumber *rbcS* genes by 100 to 1, although the gene products accumulate in equal amounts. *In vivo* translation studies show that the large and small subunits are synthesized at approximately equal rates and although the *rbcL* and *rbcS* transcript levels are not known precisely, both mRNAs are abundant in leaves. *RbcL* mRNA is relatively stable with a half life of 5–15 hours. This suggests that *rbcL* genes are transcribed less frequently than *rbcS* genes. The large copy number difference between some plastid genes and their counterparts located in the nucleus raises the question why there are so many copies of plastid DNA per cell. Bendich (1987) has argued that synthesis of rRNA could limit ribosome accumulation during rapid chloroplast development. He proposes that the plastid genome is amplified to provide more copies of rDNA thereby stimulating rRNA and ribosome synthesis.

Plastid DNA is associated with the chloroplast envelope or thylakoid membranes in aggregates of 5 to 20 DNA molecules. In proplastids of wheat, a single aggregate of DNA is observed. As chloroplasts develop, additional DNA accumulates and it becomes arranged first in a ring at the periphery of the plastid and later in a more dispersed arrangement within the organelle. The compaction of plastid DNA can also change during plastid differentiation. Chromoplast DNA is more condensed than chloroplast DNA and this difference is correlated with reduced transcription activity in chromoplasts. The mechanism of plastid DNA compaction is unknown but may be mediated by 'histone-like' proteins found in plastids. The majority of plastid genomes are unit circles but low levels of dimers and trimers have been observed. These higher order structures may represent unresolved replication intermediates.

Decoding the plastid genome

Transcription of plastid genes

The abundance of plastid-encoded proteins varies over 1000-fold with the large subunit of Rubisco being the most abundant protein, followed by proteins involved in electron transport, ribosomal proteins and subunits of the plastid-encoded RNA polymerase. Among plastid genes, mRNA abundance also varies approximately 1000-fold and in general, mRNA abundance parallels protein abundance. A 300-fold range of transcription rates is observed among these same genes, and transcription activity varies in parallel with mRNA level and protein abundance (with some exceptions). Therefore, promoter strength is a primary determinant of plastid gene expression.

Transcription of plastid genes resembles transcription in prokaryotes in many respects. In fact, when *E. coli* is transformed with the plastid genes *rbcL* and *psbA*, transcription is initiated in *E. coli* at sites similar to those used by chloroplast RNA polymerases. This result is consistent with studies showing that the DNA sequence elements that direct transcription initiation of some plastid genes (TTGaca and TAtaaT; located 35 and 10 nucleotides upstream of the site of transcription initiation, respectively) are similar to prokaryotic promoter elements. As in bacteria, these promoter elements precede rRNA, tRNA and protein coding genes. Other plastid genes, such as *rps16* and *rpoB* lack the −35 promoter element. Moreover, transcription of *trnR1* and *trnS1* does not require any upstream sequences and internal promoter sequences may be utilized similar to eukaryotic genes encoding tRNAs.

Variation in plastid gene promoter structure reflects heterogeneity of plastid RNA polymerases and the action of specificity factors that modulate promoter utilization. There are at least two plastid localized RNA polymerases. One RNA polymerase is *E. coli*-like and contains subunits homologous to the bacterial RNA polymerase subunits α, β, and β'. Plastid RNA polymerase subunits α, β, β' and β'' are encoded by the plastid genes *rpoA*, *B*, *C1* and *C2*, respectively (Igloi and Kossel, 1992). The plastid-encoded RNA polymerase is a large complex that transcribes promoters containing –10/–35 promoter elements. In addition to the core subunits, purified RNA polymerase preparations contain several subunits of unknown function. In bacteria, promoter specificity is modulated by sigma factors. Likewise, there is evidence for three different plastid sigma-like factors (Tiller and Link, 1993). Further research will be required to determine if the plastid sigma-like factors are homologous to bacterial sigma factors.

Plastids contain a second RNA polymerase that is encoded by nuclear genes. Evidence for the nuclear-encoded RNA polymerase comes from several sources. In *Epifagus*, a non-photosynthetic parasitic plant, plastids lack *rpoB*, *rpoC1* and *rpoC2*, genes that encode the plastid-encoded RNA polymerase subunits. However, plastid genes that lack –35 promoter elements are retained in the *Epifagus* genome and low levels of rRNA accumulate in plastids. Barley mutants, such as *albostrians*, have been identified that lack plastid ribosomes. Therefore, plastids in these plants cannot synthesize the plastid-encoded RNA polymerase. Even so, *albostrians* plastids are transcriptionally active and accumulate a subset of the normal plastid RNAs (Hess *et al.*, 1993). The nuclear-encoded plastid RNA polymerase has been partially purified. The RNA polymerase preparation contains a 110 kDa catalytic subunit capable of utilizing promoters that lack –35 elements (Lerbs-Mache, 1993).

Plastid transcript 3′-ends often contain sequences with dyad symmetry proximal to RNA termini. In *E. coli*, factor-independent transcription terminators typically contain a GC-rich region of dyad symmetry followed by an AT-rich stretch that contains a run of thymidines. Plastid *in vitro* experiments indicate that DNA sequences with dyad symmetry located at the 3′-end of *rbcL* do not cause transcription termination. Transcription was terminated with low efficiency by 3′-regions of spinach *psbA*, *rrnB* and *petD*. In contrast, *trnS* and *trnH* caused termination with high efficiency (80%) although these genes are not followed by regions of dyad symmetry. These results suggest that the stem–loop structures found at RNA 3′-ends may not play a significant role in factor independent transcription termination. Nevertheless available data do not rule out a role for these sequences in factor-mediated transcription termination.

Transcription from plastid gene promoters is stimulated *in vitro* when supercoiled DNA templates are used instead of relaxed circular or linear templates. Furthermore, changes in DNA conformation altered the relative ratio of transcription of two adjacent promoters (*atpB*, *rbcL*) *in vitro*. Chloroplasts of higher plants contain gyrase and topoisomerase I activity which could alter the superhelicity of plastid DNA *in vivo*. At present, however, no direct test of this possibility has been reported in higher plants. In *Chlamydomonas*, inhibitors of gyrase such as novobiocin change the relative transcription rates of several plastid genes.

RNA processing

tRNAs

Transcripts containing unprocessed tRNA sequences are produced from individual tRNA transcription units, from the ribosomal DNA transcription units [tRNA-Ala(UGC), tRNA-Ile(GAU)] and from transcription units that encode proteins. The production of mature tRNAs involves 5′ and 3′-endonucleolytic cleavage of primary transcripts, addition of CCA to the 3′-end of the processed RNA and base modification. Interestingly, plastid RNAse P, the enzyme responsible for 5′-end cleavage of tRNA precursor RNAs, does not contain an associated RNA. This is in contrast to the situation in other eubacterial RNAse P enzymes that consist of a 377 to 400 nucleotide RNA plus a 14 kD protein. Base modification of chloroplast tRNAs is similar to that found in bacteria. For example, tRNA-Glu contains several modified bases including four pseudouridines, a 5-methylcytosine and a 5-methylaminomethyl-2-thiouridine. This latter modification has been only reported in plastids and prokaryotes providing additional support for the prokaryotic origin of plastids.

Introns

Introns are DNA sequences within genes that disrupt protein coding regions called exons. Introns and exons are cotranscribed and the resultant precursor RNAs are spliced to remove introns. Four major classes of introns have been recognized. One intron type is found in nuclear genes that encode proteins. Introns of this type are characterized by invariant GU and AG dinucleotides at the intron boundaries. A second intron type is found in nuclear tRNA genes. Two additional intron classes, termed Group I and Group II introns, have been distinguished on the basis of conserved sequence elements and potential folding patterns. Group I introns have been found in chloroplasts, mitochondria and ribosomal genes of *Tetrahymena* and Group II introns are present in mitochondria of yeast and chloroplasts. Euglena plastids contain yet another type of small intron.

Euglena plastid protein genes contain numerous introns whereas tRNAs contain none. In contrast, at least 6 tRNA genes and 9 protein coding genes contain introns in higher plants. One plastid gene intron has been identified as a Group I intron (*trnL*(UUA)). In contrast, the gene encoding tRNA-Lys(UUU) contains a 2.5 kbp Group II intron. An open reading frame within this gene's intron encodes a protein homologous to mitochondrial RNA maturases. This raises the interesting possibility that the putative maturase located in *trnK* could facilitate intron processing in plastids. An even more remarkable observation was made with regard to *rps12* (Koller *et al.*, 1987). This gene is encoded by three exons. The 5'-exon is located 28 kbp from the two other 3'-exons and *trans*-splicing is needed to produce functional RNA. *Trans*-splicing has also been reported for *psaA* in *Chlamydomonas* and at least 14 nuclear genes are involved in the *trans*-splicing pathway.

RNA editing and other RNA processing activities

RNA editing, or the post-transcriptional alteration of an RNA's sequence, was first described for transcripts encoded in kinetoplast DNA. In this case, insertions and deletions of U residues were directed by guide RNAs to alter transcript sequences. RNA editing is also extensive in plant mitochondria. In chloroplasts, DNA sequence information has in general been consistent with known protein sequences indicating that RNA editing is not extensive. However, several examples of RNA editing have been uncovered. The initiation codon for *rpl2* is predicted to be ACG based on DNA sequence information. However, analysis of mature RNA revealed that primary transcripts are edited via a C to U transfer to create an AUG initiation codon (Hoch *et al.*, 1991). Internal editing of plastid transcripts has also been documented. The biochemical mechanism involved in plastid RNA editing is unclear and the biological significance of this activity remains a topic of speculation.

Although plastid RNAs are not capped or polyadenylated, RNA maturation pathways can be very complex. In addition to intron removal and editing, primary transcripts are often cleaved to remove a portion of the untranslated RNA proximal to open reading frames. 5'-end RNA processing is likely to alter RNA stability or transcript translatability. This supposition is supported by analysis of jasmonic acid induced cleavage of the 5'-untranslated region of *rbcL* transcripts (Reinbothe *et al.*, 1993). Truncated *rbcL* mRNA was translated at much lower efficiency compared to normal *rbcL* transcripts that accumulate in plastids. This is consistent with the observation that jasmonic acid reduces translation of Rubisco subunits. Polycistronic transcripts are also processed at internal sites. RNA processing results in differential accumulation of subportions of large transcripts.

RNA stability and RNA binding proteins

The stability of several plastid mRNAs have been measured by analyzing the decrease in transcript levels following inhibition of transcription. Half-lives of plastid mRNAs ranged from 6 hours to over 40 hours (Kim *et al.*, 1993). The relatively high stability of plastid transcripts can be contrasted to bacterial mRNAs that have half-lives ranging from 1–10 minutes. Plastid mRNA stability is more similar to eukaryotic mRNA stability that ranges from minutes for unstable mRNAs (e.g. *c-fos*) to hours or days. Significantly, high mRNA stability limits the rate at which an organism can decrease mRNA levels through a change in transcription. *PsbA* mRNA increases in stability during chloroplast development finally reaching a half-life of over 40 hours. The stability of other plastid mRNAs, such as the *atpB* mRNA, did not change during chloroplast development. Differences in plastid mRNA stability are attributed to RNA secondary structure and RNA binding proteins. Deletion of transcript 3′-end sequences that contain stem–loop structures can reduce mRNA stability and it has been proposed that stem-loop structures protect transcripts from exonuclease digestion. In addition, specific proteins that bind to plastid mRNA may modulate RNA stability. RNA binding proteins have been identified in plastids that interact with RNA 3′-termini or 5′-termini. For example, a family of 24–33 kDa RNA binding proteins that bind to plastid transcript 3′-termini were found to contain two RNA binding domains (Li and Sugiura, 1990). These proteins could help stabilize mRNAs or facilitate their correct processing.

Protein translation, secretion and assembly

Transcription, RNA processing and translation all occur within the plastid compartment suggesting that expression of plastid genes could involve coupled transcription and translation. This situation also raises the question how ribosomes distinguish spliced from unspliced transcripts (if they do). Perhaps as has been found in mitochondria, tRNA synthetases or other components of the translation apparatus facilitate the processing of plastid transcripts.

Protein synthesis in plastids is sensitive to chloramphenicol and is carried out by 70S ribosomes. Plastids contain initiation factors with activities similar to bacterial initiation factors (IF1, IF2, IF3). Some plastid transcripts contain ribosome binding sites (GGAGG) just upstream of initiation codons as in bacteria. Toeprint analysis revealed that translation initiation complexes form at the ribosome binding sites. Other transcripts, such as the *psbA* RNA, lack ribosome binding sites immediately upstream from initiation codons. Translation initiation on these transcripts may involve the binding

of 30S ribosomal subunits upstream of initiation sites followed by scanning to the first AUG. Translation initiation in plastids involves the incorporation of formyl-methionine. Translation elongation mediated by EF-TU, EF-G and a pool of 30 different amino-acylated tRNAs continues until a stop codon is reached. The universal code is used to specify amino acids although codon usage shows a clear preference for A or U in the third position. In addition, codon use frequency is correlated with the concentration of isoaccepting tRNAs. Translation can be regulated at the level of initiation or elongation depending on the gene and condition analyzed.

Many of the proteins encoded by plastid genes are localized in the thylakoid membrane. The translocation pathway that facilitates correct localization of these proteins has not been studied in detail but it is known that the proteins can be inserted cotranslationally into the thylakoids. Some membrane proteins (i.e. cytochrome *f*) contain an N-terminal cleavable signal sequence which facilitates their movement through membranes (Rothstein *et al.*, 1985). Other membrane proteins must contain non-cleavable signal domains (i.e. proteins encoded by *psaA–psaB*). In one case, C-terminal cleavage and palmitylation are involved in gene product translation and/or assembly (*psbA* gene product, D1) (Mattoo and Edelman, 1987). In plastids, newly synthesized Rubisco large subunits are found associated with a 60 kD large subunit binding protein (Ellis, 1991). The 60 kDa protein functions as a chaperonin and maintains the large subunit in an assembly competent form during the interim between translation and association with the small subunit of Rubisco which is imported from the cytoplasm.

Nuclear genes encoding plastid proteins

The majority of plastid-localized proteins are encoded by nuclear genes. Nuclear genes that encode abundant plastid proteins involved in photosynthesis are often organized into gene families. For example, there are 16 *Lhcb* genes in Petunia that can be divided into 5 subfamilies. *Lhcb* genes encode Photosystem II antennae proteins that bind chlorophyll *a*, chlorophyll *b* and carotenoids. Similarly, 5 *RbcS* genes that encode the small subunits of Rubisco are present in tomato. Members of multigene families often exhibit different expression patterns and each gene may contribute a large or small fraction of the total output from the gene family (Fluhr *et al.*, 1986).

The steps leading from transcription of nuclear genes to organelle targeting and assembly of the encoded proteins with cofactors and other proteins is shown in Fig. 18.2. Nuclear genes encoding plastid proteins are transcribed by RNA polymerase II. The rate of transcription is modulated by many factors including chromatin structure, DNA methylation, and the action of DNA binding proteins. Deletion

and mutation studies, gel shift and DNase I footprinting assays have shown that proteins interact with specific DNA sequences upstream of transcription initiation sites of nuclear genes that encode plastid proteins (Gilmartin *et al.*, 1990). Some of these DNA binding proteins, such as TFIID, bind to TATA sequences that are required for transcription of a wide variety of genes. Other DNA binding proteins bind to *cis*-elements that specifically mediate light-regulated transcription. This category of DNA binding protein includes GT-1 (binds to GGTTAA) and CA-1 that bind to the promoters of light-regulated *Lhcb*, *RbcS* and *PhyA* genes (Sun *et al.*, 1993). At present, it is not clear how light modifies the action of these proteins to regulate gene expression. The abundance of most of the characterized DNA binding proteins is the same in leaves of dark-grown and illuminated plants. In some cases light-induced changes in protein modification (i.e., phosphorylation) occur that may alter trans-factor activity. Alternatively, the action of DNA binding proteins could be modulated by non-DNA binding proteins that interact with the trans-factors.

Primary transcripts produced by the action of RNA polymerase are spliced, capped and polyadenylated in the nucleus. As in other eukaryotes, specific sequences specify sites of splicing and polyadenylation. Once processed, the mRNA exits the nucleus for translation by 80S ribosomes in the cytoplasm. Proteins destined for plastids are translated in the cytoplasm as precursors that often contain N-terminal targeting sequences called 'transit' sequences. The small subunit of Rubisco and the chlorophyll *a/b* apoproteins, for example, are synthesized in the cytoplasm as precursors containing 33–36 N-terminal amino acids not present in the mature proteins. These additional amino acids form a positively charged 'transit' sequence which directs the import of these proteins into plastids. Transit sequences are important for binding precursor proteins to protein receptors on the chloroplast envelope. Transported proteins are processed in one or two steps by soluble proteases and become localized in the stroma, thylakoid membrane or the thylakoid loculus.

The accumulation of *RbcS* and *Lhcb* gene products is dependent on transcription, translation and protein transport and on the presence of cofactors and other

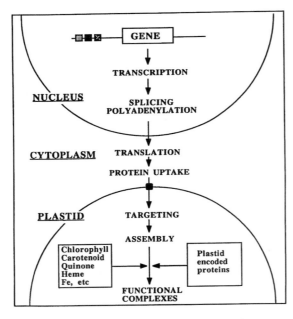

Fig. 18.2 Steps in the expression of nuclear genes that encode plastid localized proteins. A representative nuclear gene is shown at the top of the figure with its multiple *cis*-elements that direct transcription by RNA polymerase II. Transcripts are spliced, capped and polyadenylated in the nucleus. Processed mRNAs that reach the cytoplasm are translated by 80S ribosomes and N-terminal transit sequences direct import of these proteins into plastids. Uptake of nuclear encoded proteins is often followed by multiple protein processing steps and the targeting of the protein to the thylakoid membrane, lumen or soluble phase. Many proteins also must bind to cofactors such as chlorophyll, carotenoid, quinone, heme or metals and assemble with other nuclear and plastid encoded proteins into functional protein complexes.

protein subunits with which they associate in plastids. Mutants that do not synthesize the large subunit of Rubisco, also do not accumulate the small subunit because this protein is unstable when not assembled into the Rubisco holoenzyme. In general, assembly of plastid proteins into protein complexes is important for the stable accumulation of subunits of the complex. Furthermore, protein accumulation is often dependent on the presence of cofactors such as chlorophyll, carotenoid, heme, quinone or metals. The chlorophyll *a/b* apoproteins, for example, do not accumulate in chlorophyll-deficient plants because the apoproteins are proteolyzed in the absence of chlorophyll binding (Cuming and Bennett, 1981). Similarly, the plastid-encoded chlorophyll-binding proteins do not accumulate in dark-grown plants because chlorophyll biosynthesis in most angiosperms is light dependent. Transcripts encoding the plastid-encoded chlorophyll-proteins accumulate in dark-grown plants and are loaded on polysomes. Pulse-labeling studies show that the chlorophyll-apoproteins are synthesized in dark-grown plants but in the absence of chlorophyll, the apoproteins are rapidly degraded (Kim *et al.*, 1994). This mechanism keeps the stoichiometry of chlorophyll and the chlorophyll-apoproteins fixed and prevents the accumulation of free chlorophyll in membranes. This is important because free chlorophyll will mediate the photooxidation of plastid membranes.

Regulation of nuclear gene expression

Nuclear genes encoding plastid-localized proteins involved in photosynthesis are regulated at numerous levels (Fig. 18.3). Genes for photosynthesis are expressed at lower levels in roots and stems relative to leaves. In roots, plastids develop into amyloplasts that are specialized for starch storage rather than photosynthesis. In leaves, expression of nuclear genes for photosynthesis varies depending on cell type. Epidermal cells and vascular cells have much lower levels of the photosynthetic apparatus relative to mesophyll cells and exhibit correspondingly lower levels of photosynthetic gene expression. In C4 plants, expression of nuclear genes for photosynthesis in mesophyll and bundle sheath cells parallels cell specialization. Mesophyll cells in these plants have

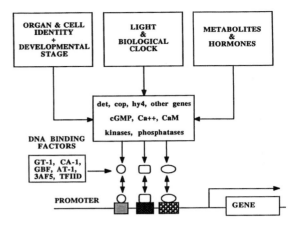

Fig. 18.3 The expression of nuclear genes encoding plastid proteins is regulated by multiple inputs and complex signal transduction pathways. Nuclear genes that encode plastid proteins involved in photosynthesis are expressed in an organ, cell and developmentally specific manner. Expression is regulated by light, the biological clock and metabolites and hormones. A large number of genes are involved in the collection of developmental and environmental signals and the transduction of this information to the gene. Some of the genes involved in the light signal transduction pathways include *det, cop, hy4* and signal propagation involves cGMP, Ca^{++} and calmodulin (CaM), kinases and phosphatases. Gene expression is modulated by a complex set of DNA binding proteins that include GT-1, CA-1, GBF, AT-1, 3AF5 and TFIID depending on the specific gene analyzed.

lower levels of Rubisco and *RbcS* gene expression relative to bundle sheath cells because most of the CO_2 fixation occurs in the bundle sheath cells. Nutrients and metabolites can also dramatically alter the expression of genes encoding the photosynthetic apparatus. Elevated sucrose or limiting nitrogen repress expression of these genes. These metabolic feedback loops help regulate the amount of the photosynthetic apparatus depending on nutrient availability and the need for carbon and reducing power. Similarly, hormones such as cytokinin (stimulatory), abscisic acid and jasmonic acid (inhibitory) modulate expression of genes for photosynthesis and help coordinate gene expression with development and environmental conditions. Expression of genes for photosynthesis also varies as a function of leaf development and in response to light and the biological clock (Anderson *et al.*, 1994).

Leaf and chloroplast development

Proplastids, the progenitors of chloroplasts are inherited maternally by the plant zygote. These non-photosynthetic organelles are 1–2 μm in diameter and present in low numbers in meristematic cells. Leaf and chloroplast development lead to the formation of mesophyll cells that contain 40 to 150 chloroplasts that are 6 to 8 μm in diameter. During chloroplast development, plastid and nuclear genes are activated leading to the accumulation of the photosynthetic apparatus. Light plays a central role in regulating chloroplast development. In many plants, cell division and cell enlargement in primary leaves and the early phases of chloroplast development are light dependent. In contrast, primary foliage leaf development in monocots such as barley is not light dependent and only later stages of chloroplast development require illumination.

Chloroplast development in leaves begins by formation of leaf primordia from cells of the shoot apex. In monocots, meristematic activity soon becomes localized in a basal intercalary meristem and most of the cells of the leaf are derived from this region. Consequently, in developing monocot leaves, older cells are located at the leaf apex and undifferentiated cells at the leaf base. Activation of cell division in the barley leaf basal meristem is paralleled by increased plastid replication and DNA synthesis. When cells stop dividing, cell enlargement begins and cells are displaced apically due to growth of new cells produced by the basal meristem. Plastid division and DNA synthesis are not synchronized with cell division; these processes continue in enlarging mesophyll cells until approximately 60 plastids each with 150 to 200 copies of plastid DNA have accumulated. This phase of chloroplast development is not known to require plastid gene expression but does exhibit cell and organ specificity.

The second phase of chloroplast development, the build-up of the plastid's transcription and translation capacity, requires both plastid and nuclear gene expression. For example, 19 of 52 ribosomal proteins are plastid encoded, the remainder are encoded by nuclear genes. It is not clear how the plastid and nuclear genes that encode the plastids transcription and translation apparatus are co-activated. However,

once the build-up process is initiated, stoichiometric production of subunits could be accomplished through the action of autogenous translational control loops, the turnover of excess protein subunits and feedback loops. For example, transcription of some nuclear genes such as *RbcS* are regulated by a 'plastid factor' produced when plastid transcription and translation are activated (Taylor, 1990).

The accumulation of the chloroplast photosynthetic apparatus begins during build-up of the plastid's transcription and translation capacity. In barley, plastid transcription increases five-fold per plastid (20-fold per cell) soon after cell division stops and cell enlargement begins. Thirty-six to forty-eight hours later a full set of mature chloroplasts is present in mesophyll cells of illuminated plants. Once mature chloroplasts have been produced, the build-up process stops and plastid transcription and translation activity decline.

Role of light

Higher plants contain at least four types of photoreceptors: blue light photoreceptors, and three red light photoreceptors, the phytochromes, protochlorophyllide reductase holochrome and chlorophyll. These four photoreceptors influence nearly all phases of chloroplast development although the extent of each photoreceptor's influence varies with plant species, and gene examined.

The phytochromes are a family of proteins approximately 120 kDa in size that contain a covalently bound open chain tetrapyrrole. In darkness, phytochromes are in their red absorbing form, Pr, which has an absorbance peak at 660 nm. Upon illumination, Pr is converted to Pfr, a form of phytochrome that absorbs maximally in the far-red (760 nm). Pfr can be reconverted to Pr either by slow decay in darkness or by absorption of a photon of far-red light. Conversion of Pr to Pfr stimulates leaf and chloroplast development and the expression of nuclear genes that encode the photosynthetic apparatus. In dark-grown plants, the *PhyA* gene product or 'dark' phytochrome predominates. PHYA levels rapidly decrease when plants are illuminated. In contrast, the *PhyB* gene product or 'green' phytochrome is present at similar levels in dark-grown and illuminated plants. Specific roles for the other phytochrome genes have not been determined.

The nature of the signal transduction pathway that connect formation of Pfr and induction of gene expression is being investigated using pharmacological agents, microinjection of potential signal transduction components into plant cells (Neuhaus et al., 1993) and through the selection and analysis of mutants (reviewed by Chory, 1993). These approaches have demonstrated the involvement of G-proteins, cGMP, Ca++, calmodulin and protein kinases/phosphatases in phytochrome mediated responses. The role of cGMP, a compound involved in animal vision, is particularly intriguing. Furthermore, the cGMP signal transduction pathway and Ca++/ CaM signal transduction pathway interact and either separately or in combination modulate expression of genes involved in photosynthesis (Bowler et al., 1994). Several protein components of the signal transduction pathway have been identified by screening mutants that have altered photomorphogenic responses. The first described was det1, an Arabidopsis mutant that shows derepressed leaf and chloroplast development in darkness (Chory, 1989). This mutant identified a gene involved in red and blue light responses as well as being required for normal repression of photosynthetic gene expression in roots. The interaction of CA-1 and Lhcb gene promoter elements is altered in det1. This provides the first mechanistic connection between phytochrome, det1 and a light-regulated gene. Subsequently, a large number of additional mutants (det, cop mutants) have been identified. In addition, a gene encoding a blue-light photoreceptor has been identified (Ahmad and Cashmore, 1993). This gene, identified by the mutant hy4, encodes a protein with homology to DNA photolyase and probably contains a flavin chromophore. Clearly, additional light signal transduction components will be discovered in the near future allowing the interconnections between the photoreceptors and their target genes to be elucidated.

Maintenance of PSII in mature chloroplasts

The transition from the rapid build-up of the photosynthetic apparatus to maintenance of the accumulated structures presents a special regulatory problem. In mature plastids, proteins which are unstable or damaged need to be turned-over and resynthesized. The proteins which comprise the PSII reaction center (D1, D2) are especially labile in illuminated plants. As a consequence, the rate of synthesis of these proteins in mature chloroplasts remains high in contrast to most other plastid proteins. PsbA (encodes D1) and psbD (encodes D2) mRNA is maintained at relatively high levels in mature chloroplasts even though overall chloroplast transcription activity is low in these plastids relative to developing chloroplasts. High levels of psbA mRNA are maintained by a strong promoter and high RNA stability (half-life greater than 40 hours). Expression of psbD, on the other hand, is maintained in mature barley leaves through the action of a blue light stimulated promoter (Christopher and Mullet, 1994). In addition, D1 translation is regulated in response to light and changes in redox state. This type of regulation is mediated by RNA binding proteins that interact with the 5'-untranslated region of the psbA mRNA (Danon and Mayfield, 1994).

Summary

The acquisition of photosynthetic competence in higher plants requires the coordinated expression of nuclear genes and plastid genes. The nuclear genes that encode plastid proteins are eukaryotic in organization and are regulated in a complex way by environmental signals such as light, as well as by developmental and metabolic signals. Plastid genes, in contrast, are prokaryotic in origin and the mechanism and regulation of expression of these genes differs significantly from nuclear genes. The plastid genome of several plants has been sequenced and most of the genes identified. Plastid gene expression is regulated at the level of transcription, RNA processing and stability, translation and protein turnover.

Further reading

Baker, N. R. and Barber, J. (1984). Chloroplast Biogenesis, Vol. 5, Elsevier/North Holland, Biomedical Press, 380 pp.
Hatch, M. D. and Boardman, N. K. (eds) (1987). Photosynthesis. In The Biochemistry of Plants, Vol. 10, Academic Press, 409 pp.

References

Ahmad, M. and Cashmore, A. R. (1993). *HY4* gene of *A. thaliana* encodes a protein with characteristics of a blue-light photoreceptor. *Nature* **366**, 162–6.

Anderson, S. L., Teakle, G. R., Martino-Catt, S. J. and Kay, S. A. (1994). Circadian clock- and phytochrome-regulated transcription is conferred by a 78 bp *cis*-acting domain of the *Arabidopsis CAB2* promoter. *Plant J.* **6**, 457–70.

Bendich, A. J. (1987). Why do chloroplasts and mitochondria contain so many copies of their genome? *BioEssays* **6**, 279–82.

Berry, J. A., Breiding, D. E. and Klessig, D. F. (1990). Light-mediated control of translational initiation of ribulose-1,5-bisphosphate carboxylase in Amaranth cotyledons. *Plant Cell* **2**, 795–803.

Bowler, C., Yamagata, H., Neuhaus, G. and Chua, N.-H. (1994). Phytochrome signal transduction pathways are regulated by reciprocal control mechanisms. *Genes Develop.* **8**, 2188–2202.

Christopher, D. A. and Mullet, J. E. (1994). Separate photosensory pathways co-regulate blue light/ultraviolet-A-activated *psbD–psbC* transcription and light-induced D2 and CP43 degradation in barley (*Hordeum vulgare*) chloroplasts. *Plant Physiol.* **104**, 1119–29.

Chory, J. (1993). Out of darkness: mutants reveal pathways controlling light-regulated development in plants. *Trends Genetics* **9**, 167–72.

Chory, J., Peto, C., Feinbaum, R., Pratt, L. and Ausubel, F. (1989). *Arabidopsis thaliana* mutant that develops as a light-grown plant in the absence of light. *Cell* **58**, 991–9.

Cuming, A. C. and Bennett, J. (1981). Biosynthesis of the light-harvesting chlorophyll *a/b* protein. Control of messenger RNA activity by light. *Eur. J. Biochem.* **118**, 71–80.

Danon, A. and Mayfield, S. P. (1994). ADP-dependent phosphorylation regulates RNA-binding *in vitro*: implications in light-modulated translation. *EMBO J.* **13**, 2227–35.

dePamphilis, C. W. and Palmer, J. D. (1990). Loss of photosynthetic and chlororespiratory genes from the plastid genome of a parasitic flowering plant. *Nature* **348**, 337–9.

Ellis, R. J. (1991). Chaperone function: cracking the second half of the genetic code. *Plant J.* **1**, 9–13.

Fluhr, R., Moses, P., Morelli, G., Coruzzi, G. and Chua, N.-H. (1986). Expression dynamics of the pea *rbcS* multigene family and organ distribution of the transcripts. *EMBO J.* **5**, 2063–71.

Gilmartin, P. M., Sarokin, L., Memelink, J. and Chua, N.-H. (1990). Molecular light switches for plant genes. *Plant Cell* **2**, 369–78.

Gray, M. W. and Doolittle, W. F. (1982). Has the endosymbiont hypothesis been proven? *Microbiol. Rev.* **46**, 1–42.

Hess, W. R., Prombona, A., Fieder, B., Subramanian, A. R.

and Börner, T. (1993). Chloroplast *rps*15 and the *rpo*B/C1/C2 gene cluster are strongly transcribed in ribosome-deficient plastids: evidence for a functioning non-chloroplast-encoded RNA polymerase. *EMBO J.* **12**, 563–71.

Hoch, B., Maier, R. M., Appel, K., Iglio, G. L. and Kössel, H. (1991). Editing of a chloroplast mRNA by creation of an initiation codon. *Nature* **353**, 178–80.

Iglio, G. L. and Kössel, H. (1992). The transcriptional apparatus of chloroplasts. *Crit. Rev. Plant Sci.* **10**, 525–58.

Kim, J., Eichacker, L. A., Rudiger, W. and Mullet, J. E. (1994). Chlorophyll regulates accumulation of the plastid-encoded chlorophyll proteins P700 and D1 by increasing apoprotein stability. *Plant Physiol.* **104**, 907–16.

Kim, M., Christopher, D. A. and Mullet, J. E. (1993). Direct evidence for selective modulation of *psbA, rpoA, rbcL* and 16S RNA stability during barley chloroplast development. *Plant Mol. Biol.* **22**, 447–63.

Koller, B., Fromm, H., Galun, E. and Edelman, M. (1987). Evidence for *in vivo trans* splicing of pre-mRNAs in tobacco chloroplasts. *Cell* **48**, 111–19.

Lerbs-Mache, S. (1993). The 110-kDa polypeptide of spinach plastid DNA-dependent RNA polymerase: single-subunit enzyme or catalytic core of multimeric enzyme complexes? *Proc. Natl. Acad. Sci. USA* **90**, 5509–13.

Li, Y. and Sugiura, M. (1990). Three distinct ribonucleoproteins from tobacco chloroplasts: each contains a unique amino terminal acidic domain and two ribonucleoprotein consensus motifs. *EMBO J.* **9**, 3059–66.

Mattoo, A. K. and Edelman, M. (1987). Intramembrane translocation and post-translation palmitoylation of the chloroplast 32-kDa herbicide-binding protein. *Proc. Natl. Acad. Sci. USA* **84**, 1497–1501.

Miyamura, S., Nagata, T. and Kuroiwa, T. (1986). Quantitative fluorescence microscopy on dynamic changes of plastid nucleoids during wheat development. *Protoplasma* **133**, 66–72.

Mullet, J. E. (1993). Dynamic regulation of chloroplast transcription. *Plant Physiol.* **103**, 309–13.

Neuhaus, G., Bowler, C., Kern, R. and Chua, N.-H. (1993). Calcium/calmodulin-dependent and -independent phytochrome signal transduction pathways. *Cell* **73**, 937–52.

Palmer, J. D. (1990). Contrasting modes and tempos of genome evolution in land plant organelles. *Trends in Genetics* **6**, 115–20.

Reinbothe, S., Reinbothe, C., Heintzen, C., Seidenbecher, C. and Parthier, B. (1993). A methyl jasmonate-induced shift in the length of the 5′ untranslated region impairs translation of the plastid *rbcL* transcript in barley. *EMBO J.* **12**, 1505–12.

Rothstein, S. J., Gatenby, A. A., Willey, D. L. and Gray, J. C. (1985). Binding of pea cytochrome *f* to the inner membrane of *Escherichia coli* requires the bacterial *secA* gene product. *Proc. Natl. Acad. Sci. USA* **82**, 7955–9.

Sugiura, M. (1992). The chloroplast genome. *Plant Mol. Biol.* **19**, 149–68.

Sun, L., Doxsee, R. A., Harel, E. and Tobin, E. M. (1993). CA-1, a novel phosphoprotein, interacts with the promoter of the *cab*140 gene in Arabidopsis and is undetectable in *det*1 mutant seedlings. *Plant Cell* **5**, 109–21.

Taylor, W. C. (1989). Regulatory interactions between nuclear and plastid genomes. *Annu. Rev. Plant Physiol. Plant Mol. Biol.* **40**, 211–33.

Tiller, K. and Link, G. (1993). Sigma-like transcription factors from mustard (*Sinapis alba* L.) etioplast are similar in size to, but functionally distinct from, their chloroplast counterparts. *Plant Mol. Biol.* **21**, 503–13.

19 The formation of ATP and reducing power in the light

Barbara B. Prézelin and Norman B. Nelson

Introduction

Photosynthesis begins with the absorption of light by pigment molecules within plant cells. Absorbed light energy drives the series of photosynthetic reactions that ultimately lead to the formation of new organic carbon. In this chapter we will introduce the mechanisms by which light is absorbed and its excitation energy funneled into photochemical reactions that result in the formation of chemical energy (ATP) and reducing power (NADPH). The light-dependent formation of ATP and NADPH is generally referred to as the light reactions of photosynthesis, described by the partial reaction:

$$2H_2O + 2NADP^+ + 3ADP^{3-} + 3HPO_4^{2-} + H^+$$
$$\xrightarrow{h\nu} O_2 + 2NADPH + 3ATP^{4-} + 3H_2O$$

The present chapter describes the structural and functional organization of the photosynthetic components which comprise the light reactions and which are found entirely within photosynthetic membranes (Fig. 19.1). Also discussed are the regulatory mechanisms which ensure the efficient use of light energy to photochemically generate the ATP and NADPH required for carbon fixation. Following chapters will address the enzyme systems which catalyze the reduction of inorganic carbon with photochemically produced ATP and NADPH.

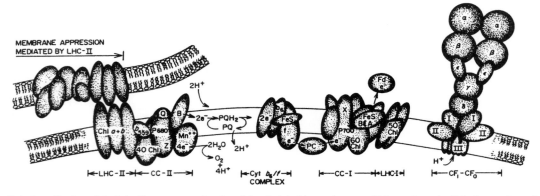

Fig. 19.1 A stylized model of the four structural units of the thylakoid membrane, which catalyze the light-harvesting, electron transport, and energy-coupling of photosynthesis. PSII consists of two parts: a LH pigment–protein complex, (LHCII) and a core complex (CCII) which catalyzes electron transfer from water to plastoquinone (PQ). A cyt. b_6/f complex (also containing a Rieske iron–sulfur (Fe–S) center) oxidizes the plastohydroquinone (PQH_2) and donates electrons to plastocyanin (PC). PSI, consisting of a LH pigment–protein complex (LHCI) and a core complex (CCI) oxidizes plastocyanin and transfers electrons to $NADP^+$ (via ferredoxin). The ATPase, which uses a proton gradient to drive ATP synthesis, consists of a hydrophobic CF_0 and a surface-bound CF_1. By permission, Kaplan and Arntzen (1982).

Light absorption and energy transfer mechanisms

In oxygen-evolving (oxygenic) photosynthesis carried out by higher plants, algae and cyanobacteria, only light in the violet to red part of the visible spectrum (from 400 to 700 nm) can be absorbed by photosynthetic pigments. Light is absorbed in packets of energy known as quanta. The energy of a quantum is directly proportional to its frequency (v) and inversely proportional to its wavelength (λ), the latter being expressed in nanometers (nm). Quanta at shorter wavelengths, near the blue-violet region of the visible spectrum, have higher excitation energy for photosynthesis than quanta at the longer wavelengths found near the red end of the visible spectrum.

Not all visible light reaching a plant will be used for photosynthesis. When light strikes a plant, variable fractions of the photosynthetically available radiation (PAR) are scattered off the plant surface, transmitted through the plant or absorbed by molecular components within the plant. Not all the molecules that absorb light within a plant cell function in photosynthesis. Photosynthetic pigments are localized within the photosynthetic membranes; these can absorb light throughout the visible spectrum and can transfer that absorbed energy with speed and efficiency to the photochemical reaction centers of photosynthesis.

There are three general classes of photosynthetic pigments, including chlorophylls (Chl), phycobilins and carotenoids (Table 19.1). Of the three types of green-colored chlorophylls, only Chl a is found in all plants and cyanobacteria, and is an essential component of the light reactions of photosynthesis. The other two types of chlorophyll, Chl b and c, function as light-harvesting pigments whose absorbed light energy is passed on to Chl a to help drive the photochemical events of photosynthesis. These *accessory* chlorophylls are differentially distributed among plant groups and can be used as chemotaxonomic markers to identify the presence of certain plant groups in natural communities (Table 19.1). There are also three subcategories of light-harvesting phycobilin pigments, which include the red-colored phycoerythrin, the blue-colored phycocyanin, and the violet-colored allophycocyanin. Phycobilins vary in their absorption capabilities and are responsible for the coloration of red algae and cyanobacteria (blue–green algae). While a variety of orange-colored carotenoids are present in all plants, only a small number function as light-harvesting pigments for photosynthesis. Carotenoids can be subdivided into two groups depending on whether they are hydrocarbons (carotenes) or contain some oxygen molecules (xanthophylls). The major photosynthetic carotenoids

Table 19.1 The major plant groups, and their principal photosynthetically active chlorophylls, biliproteins, and carotenoids. A (+) indicates the presence of the pigment in the plant group.

	Chl a	*Chl* b	*Chl* c$_1$	*Chl* c$_2$	*phycoerythrin*	*phycocyanin*	*allophycocyanin*	*major xanthophylls*
Higher plants	+	+						lutein
Green algae	+	+						lutein/siphona-xanthin
Diatoms	+		+	+				fucoxanthin
Brown algae	+		+	+				fucoxanthin
Chrysophytes	+		+	+				butanoyl-fucoxanthin
Prymnesiophytes	+		+	+				hexaoyl-fucoxanthin
Dinoflagellates	+			+				peridinin
Cryptophytes	+		+	+	+			alloxanthin
Red algae	+			+	+	+	+	zeaxanthin
Cyanobacteria (= blue–green algae)	+			+	+	+	+	zeaxanthin, myxoxanthin

are generally the xanthophylls listed in Table 19.1 and can also be used as chemotaxonomic markers for the plant groups in which they occur.

All photosynthetic pigments are conjugated colored molecules and are bound to thylakoid membrane proteins, which are localized in the chloroplast of higher plants and algae or within the cytoplasm of cyanobacteria. Three functional categories of pigment–protein complexes can be recognized. These are the reaction centers of photosystems I and II and the light harvesting complexes. The reaction center of photosystem I (PSI) is the Chl a–protein complex in plants that absorbs at the longest wavelength

(700 nm) (Fig. 19.2a), receiving electrons from a cytochrome carrier via a plastocyanin–protein complex and donating electrons to the enzymatic ferredoxin complex which reduces NADP. The reaction center of photosystem II (PSII) is a shorter wavelength absorbing (680 nm) Chl a–protein complex (Fig. 19.2b) which catalyzes the oxidation of water to oxygen and donates electrons via a plastoquinone pool to PSI. The composition of the reaction center core complexes within PSI and PSII are highly conserved in all plant groups, but only account for a small fraction of total plant chlorophyll a. Most light-harvesting (= accessory) chlorophylls,

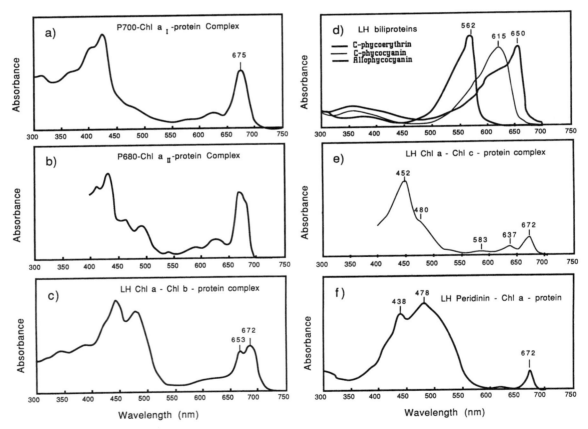

Fig. 19.2 Representative absorption spectra for major pigment–protein complexes in plants, including: (a) P700–Chl a–protein complexes, representing PSI (by permission, Prézelin and Alberte, 1978); (b) P680–Chl a-protein complexes, representing PSII (by permission, Satoh and Butler, 1978); (c) LH Chl a/b–protein complexes (by permission, Thornber and Alberte, 1977); (d) LH phycobiliproteins (by permission, O'Carra and O'hEocha, 1976); (e) LH Chl a/c–protein complexes (by permission, Boczar, 1985); and (f) LH peridinin–chlorophyll a–protein complexes (by permission, Prézelin and Haxo, 1976).

carotenoids and phycobilins are localized in light-harvesting (LH) pigment–protein complexes associated with PSI and PSII. These LH complexes absorb light at wavelengths where PSI and PSII absorption is weakest, and include the Chl *a*–Chl *b*–protein complexes of green algae and higher plants (Fig. 19.2c), the phycobiliprotein complexes of cyanobacteria and red and cryptophyte algae (Fig. 19.2d), and the carotenoid–Chl *a*–Chl *c*–protein complexes of chromophytic (Chl *c*-containing) algae (Fig. 19.2e,f). Their added presence around PSI and PSII core complexes broadens the range of visible light used to drive photosynthesis.

Chlorophylls will be used to illustrate the basic features of how light energy is captured by photosynthetic pigments and efficiently transferred from LH pigment–protein complexes to PSI and II. Chlorophylls are all characterized by a porphyrin ring structure where magnesium is chelated in the center and liganded at four sites to pyrrole nitrogen atoms (Fig. 19.3). It is with this polar head group that most bonding of apoproteins occurs. Only Chl *a* and Chl *b* have phytol, a long-chain lipophilic terpenoid side-group, attached to the porphyrin ring. Phytol may provide additional hydrophobic binding sites for apoproteins (Fenna and Mathews, 1979) or lie to the outside of apoproteins, thereby providing hydrophic bonding with the integral lipophilic components of the thylakoid membrane (Anderson, 1975) (Fig. 19.1). The structure and conformational constraints of the porphyrin ring bound to proteins give different chlorophyll–protein complexes their specific absorption properties (Fig. 19.2).

Porphyrin rings are essentially planar complexes surrounded by dense clouds of pi-electrons which become redistributed when a quanta of light is absorbed, thereby promoting the chlorophyll molecule to a higher excited electronic state (Fig. 19.4). The energized pigment molecule represents an excited *singlet* state whose energy is greater than that of the initial or ground state molecule. The absorbed quanta must be of such a vibrational frequency that the light energy matches possible energy transition states of the pigment molecule. Thus the absorption spectrum of any pigment is a signature of the light energy levels that can be most effectively absorbed (Fig. 19.4A,B) (Sauer, 1975).

Individual pigment molecules can absorb a range

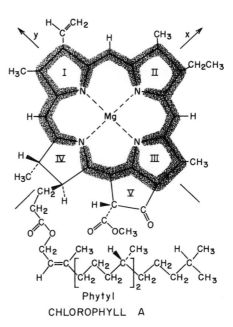

Phytyl
CHLOROPHYLL A

Fig. 19.3 The molecular structure of chlorophyll *a*. The porphyrin pi-electrons are indicated by the shading. Absolute configurations are shown for the assymetrically substituted porphyrin carbon atoms and for phytol. Chlorophyll *b* differs from chlorophyll *a* in the replacement of the methyl group on ring II by a formyl group. Neither chlorophyll *c*$_i$ or *c*$_2$ has phytol, and the propionic group of ring IV is replaced by HC=CH$_2$—COOH. Chlorophyll *a*$_2$ has a vinyl group (—CH=CH$_2$) instead of the ethyl group on ring II of chlorophyll *a* (by permission, Sauer, 1975).

of light energies, but packets of excitation energy (excitons) are transferred to other pigments only from the lowest excited singlet state of the pigment molecule (Fig. 19.4C). For instance, when Chl *a* absorbs the high energy of blue light (440 nm), the molecule is promoted to a higher excited state than it would when excited by absorbed red light (680 nm) (Fig. 19.4). But a blue light excited Chl *a* molecule will quickly relax to its lowest excited state, losing the extra excitation energy through non-radiative dissipative processes (i.e. thermal decay). Thus the energy transferred to another pigment molecule by a blue light or red light excited Chl *a* molecule will be the same.

Resonance transfer of vibrational energies occurs between neighboring pigments when their lowest

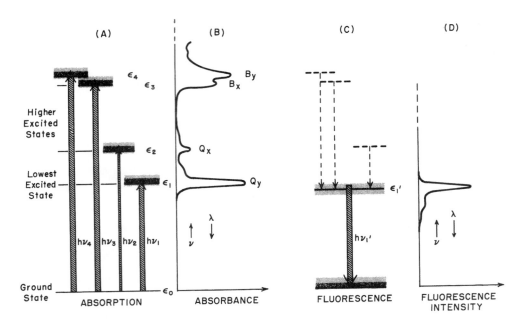

Fig. 19.4 Absorption and fluorescence of chlorphylls. (A) Energy level diagram, showing spectral transitions (vertical arrows). The energy levels are broadened (shading) by vibrational sublevels that are not usually resolved in solution spectra. (B) Absorption spectrum corresponding to energy levels of part (A). This spectrum is turned 90° from the usual orientation in order to show the relationship to the energy levels. Conventions for designating the spectral transitions (Q_y, Q_x, B_x, B_y, where x and y refer to the axes shown in Fig. 19.2) are shown. (C) Diagram showing radiationless relaxation (dashed arrows) and fluorescence (shaded arrow). (D) Fluorescence emission spectrum corresponding to part (C). Note the red shift of fluorescence compared with the corresponding Q_y absorption illustrated in parts (A) and (B). This Stokes shift owes to vibrational relaxation in the excited electronic state prior to fluorescence emission and in the ground electronic state after emission (by permission, Sauer 1975).

energy absorption bands overlap. Exciton transduction from peripheral LH pigment–protein complexes to integrally-bound PSI and PSII is dependent on the unidirectional coupling of lowest excitation states between neighboring pigment molecules. The speed and efficiency of energy transduction depends on the orientation and distance between pigment molecules, which needs to be less than 10 nm to effectively compete with dissipative processes. Due to some energy loss during the time of energy transfer, the excitation energy decreases as excitons are transduced from LH pigments to the reaction centers of PSI and PSII (Fig. 19.5), which represent the lowest energy traps in photosynthesis.

The absorption of quanta and subsequent transduction of excitons to PSI or PSII must be completed within a nanosecond if photochemistry is

to compete successfully against other dissipative processes, i.e. further radiationless (thermal) de-excitation and the luminescing (light-emitting) processes of fluorescence, phosphorescence and delayed light emission. The fraction of excitons funneled into photochemical events or lost through dissipative processes is highly variable and is largely regulated by thylakoid membrane state changes which alter the coupling between LH complexes and the photosystems. Excitons which never reach reaction centers, or reach reaction centers but are not permanently trapped, can migrate back to LH complexes and be dissipated as luminescence. Fluorescence (Fig. 19.4d) is a light-emitting process by which pigments in the excited singlet state return to ground state if their excess energy is not funneled into photochemistry within the fluorescence lifetime

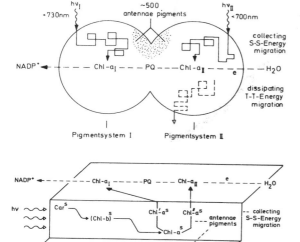

Fig. 19.5 Collecting and dissipating energy migrations in the antenna pigments of photosynthesis (by permission, Witt, 1979).

of the molecule. Fluorescence only occurs from the lowest excited singlet state, so the wavelength of the fluorescence maximum often is a few nanometers longer than the longest absorption maximum of the pigment. For example, chlorophylls absorb light energy throughout the visible spectrum (Fig. 19.4b), but only emit fluoresced light energy at the red end (lowest energy) of the visible spectrum (Fig. 19.4d).

The majority (>90%) of *in vivo* fluorescence at room temperature arises from back reactions of the primary photochemical events occurring in the reaction centers and LH chlorophyll of PSII (Fig. 19.4). In some instances, excitons re-emitted from PSII can be directed to PSI before excitation energy of the absorbed quanta is lost. It is this ability to redirect unused excitation energy that enables PSII to spill over excess excitons to PSI. Spillover is a unidirectional transfer of excitons from PSII to PSI. Excitons are not transferred out of PSI and into PSII because PSI is of a lower energy state than PSII and because it dissipates its excess excitons via radiationless de-excitation. Thus, all excitons reaching PSI have a very low probability of gaining the additional energy required to overcome the energy

barrier which exists between the reaction center and surrounding LH chlorophyll. However, as temperatures are lowered, so are the energy barriers, and fluorescence intensity from PSI in higher plants and some algae increases dramatically at liquid nitrogen temperatures (77 K).

Photochemistry

An exciton is considered trapped when photochemical charge separation of electrons and protons occurs across the thylakoid, allowing transmembrane electron flow to begin. The charge separation reaction can be visualized as the following:

$$DPA \xrightarrow{h\nu} D(P^*A_1) \xrightarrow{5\,ps} D(P^+A_1^-) \xrightarrow{200\,ps} D^+ (PA_1)A_2^-$$

where P is the phototrap chlorphyll *a*, and D and A_1 A_2 represent electron donors and acceptors, respectively. In the excited singlet state (represented by *), phototrap chlorophyll *a* molecules eject an electron which is rapidly transferred (5 ps) to an associated primary electron acceptor (A_1). After 200 ps the primary electron acceptor has passed the electron to a secondary electron acceptor (A_2) and the phototrap chlorophyll has been reduced by the oxidation of an electron donor (D). In both photosystems, the primary electron donors lie near the inner surface of the thylakoid, the reaction centers are buried in the thylakoid membrane, and the electron acceptors are found near the outer stroma surface of the thylakoids (Fig. 19.1). Thus, the photochemical events transfer one electron from each phototrap across the thylakoid membrane. In this manner, light energy is stabilized as stored chemical potential to be used in the formation of oxidizing and reducing compounds, transmembrane electric fields, ion gradients, and membrane conformational changes.

In PSI, the phototrap chlorophyll *a* is termed P700, the electron donor is usually the copper-containing protein plastocyanin (PC) and the electron acceptors are the two iron–sulfur (FeS) proteins ferredoxin A and B (Fig. 19.1). The structure of P700 is not known, but is generally believed to represent a Chl *a* dimer. Other possibilities include an enolized form of a single Chl *a* molecule that is protein-stabilized or a chlorophyll molecule that is structurally distinct from

Chl *a*. Each PSI core complex has a molecular weight of 67 kD and contains about 40 accessory Chl a_I molecules which pass their absorbed light energy exclusively to one P700 reaction center. It is the accessory chlorophylls within the P700–Chl a_I–protein complex which give it the longest wavelength (675–677 nm) red absorption maximum of any Chl–protein complex in plant cells (Fig. 19.2a). The P700–Chl *a*–protein complexes of higher plants are characterized by a photobleaching signal which has a maximum around 700 nm (Fig. 19.6a). This figure represents the difference between the absorbance spectrum of P700 before and after oxidation induced by light or chemical means. The negative signal at around 700 nm and 438 nm arises in less than 20 ns as a result of the charge separation reaction due to photo-oxidation. It has been suggested that when one Chl *a* in the dimer ejects an electron, the nuclear interactions between the two Chl molecules in the dimer are altered and cause a slight shift in the spectral characteristics of the reaction center. The oxidized-minus-reduced signal of P700 has an absorption coefficient of 64 to 70 mequiv cm^{-1} at 700 nm and can be used to quantify the number of PSI reaction centers present in a sample.

In PSII, the phototrap chlorophyll *a* is termed P680, the electron donor is water, an intermediate electron acceptor is a pheophytin molecule and the primary electron acceptor is a specialized protein-embedded plastoquinone molecule (Q) (Fig. 19.1). The PSII reaction center complex ($\geqslant 600$ kD) has been more difficult to isolate than PSI because of its lability in the presence of harsh detergents. The phototrap of PSII is thought to be a special Chl dimer or ligated Chl monomer. It has been termed P680 because of characteristic spectroscopic absorption changes seen in light-minus-dark (oxidized-minus-reduced) difference spectra which are inhibited by specific system II inhibitors (Fig. 19.6b). Each P680-containing core complex of PSII contain one phototrap and 50–60 light-harvesting chlorophyll a_{II} molecules. Unlike P700, the P680 signal is not used to quantify PSII reaction centers as it is not resolved simply, occurring in a part of the visible spectrum where secondary optical signals from P700 complexes and LH chlorophyll *a* molecules interfere. Rather, a diagnostic indicator of PSII is an absorbance change at 550 nm (C-550) which presumably occurs when a secondary electron acceptor for PSII, i.e. a plastoquinone, is reduced and thereby modifies the absorption properties of nearby beta-carotene molecules which are sensitive to the redox changes of the electron acceptor. Two distinct classes of PSII complexes have been identified. The alpha type of PSII are found in appressed regions of thylakoid membranes and have more LH Chl *a* interconnecting them than the beta type of PSII located in non-appressed regions of thylakoid membranes.

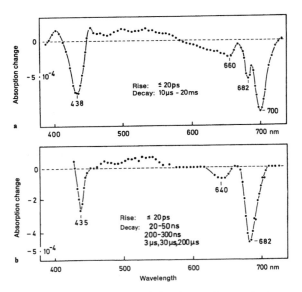

Fig. 19.6 Difference spectra of absorbance changes due to the turnover of Chl a_I (P700) (top) and Chl a_{II} (bottom). The fast rise noted in the figure panel is due to photo-oxidation. The decay is multiphasic and is interpreted as the reduction of reaction centers by electron donors (by permission, Renger, 1983).

Electron transport

Electrons flow from the electron donor side of PSII to the electron acceptor side of PSI, and can be represented by an interpretation of the gross organization of photosynthetic components seen in Fig. 19.1. The two photosystems are linked in series by a transport chain of electron and hydrogen carriers and the flow of electrons from water to

$NADP^+$ is termed 'non-cyclic' electron transport. This pattern of light-driven electron transport is described by the Z-scheme shown in Fig. 19.7, where the ordinate shows the redox potential of the electron carriers. In oxygenic photosynthesis, electrons are transferred from a redox potential sufficiently positive to oxidize water to a redox potential sufficiently negative to reduce $NADP^+$. The electron transport carriers are arranged in an organized manner across the thylakoids, which have a distinct sidedness. While the water-splitting reactions occur on the inner side of the thylakoid which is in contact with an internal aqueous phase (the lumen or intrathylakoid space), $NADP^+$ reduction occurs on the exterior of the thylakoid which is in contact with the surrounding stroma. A $1:1$ ratio of electron carriers per reaction center is generally assumed, with the exception of the plastoquinone (PQ) pool linking PSI and PSII. However, the relative proportion of reaction center components and electron carriers can vary with changes in growth conditions.

The donor side of PSII is represented by a manganese-containing water-splitting enzyme complex (34 kD) which generally donates electrons to P680 within 30 ns. Being slower than the few picoseconds required for the initial charge separation leading to $P680^+$ and reduced pheophytin, the donor side of P680 is the rate-limiting step determining the turnover time of PSII. Facilitated by the enzyme complex, one electron is extracted from two water molecules during each of four successive photochemical acts and eventually results in the release of four protons and one oxygen molecule to the intrathylakoid space.

$$2H_2O \xrightarrow{4h\nu} O_2 + 4H^+ + 4e^-$$

When PSI and PSII are present in equal ratios, the minimum quantum requirement for photosynthesis is eight. For each molecule of oxygen evolved, four photoacts must occur within each paired photosystem to transfer four electrons from water to the carbon dioxide reduction systems.

Between PSI and II are a series of chemical electron carriers. There is some controversy over the order and composition of electron acceptors in the chain, but the most favored simplified pathway is shown in Fig. 19.6. With each photoact, one electron is passed from P680 to pheophytin to a bound quinone, Q, which passes its reducing equivalents to plastoquinone (PQ) via the B protein (Figs 19.1 and 19.6). The secondary acceptor B is a rapidly turned over plastoquinone–protein complex which accumulates two electrons before passing the electron pair on to PQ. It is between Q and B that the photosynthetic inhibitor 3-(3,4-dichlorophenyl)-1,1-dimethylurea (DCMU or diuron) acts to block electron flow. When B accepts two electrons, it also takes up two protons from the stroma, and both electrons and protons are passed along the electron transport chain when B is subsequently oxidized by the neighboring plastoquinone pool (PQ). The transmembrane PQ pool, with five to ten PQ molecules per electron chain, interconnects more than one electron transport chain and thereby regulates electron flow between several photosystems.

When oxidized, PQ donates two electrons to PSI electron acceptors and releases two protons to the inside of the thylakoids into the lumen space. It is the light-driven proton build-up in the lumen that drives ATP formation, thereby linking rates of photophosphorylation to rates of electron transport. Mediating electron flow between the PQ pool and the plastocyanin (PC) molecule, which serves as the primary electron donor to PSI, is a cytochrome b_6/cytochrome f (cyt b/f) complex (Fig. 19.1). The

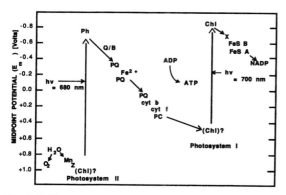

Fig. 19.7 The 'Z' scheme. Electron transfer components and their redox potentials in the photosynthetic systems of green plants. The primary donor species in both PSI and PSII are Chl species. The question marks indicate that their exact nature is not known (by permission, modified from Okamura *et al.*, 1982).

oxidation site for reduced PQH_2 is at an iron–sulfur site (Rieske Fe–S), where protons are released and electrons are passed rapidly to cyt f within the complex. The cyt f reduces PC, which then reduces P700.

On the acceptor side of PSI, electrons from P700 reduce an iron-containing acceptor, X, which passes electrons onto two iron–sulfur centers, A and B (Fig. 19.1). From here electrons are transferred one at a time to ferredoxin (Fd) which in turn reduces $NADP^+$ with the aid of ferrodoxin-NADP oxidoreductase, a FAD-containing flavoprotein. The ferredoxin-NADP reductase is located on the stromal surface of the thylakoid, in protected sites near membrane-bound ATPases, and is light-dependent in its ability to catalyze the two-electron reduction of $NADP^+$ to NADPH.

As $NADP^+$ becomes reduced, the ratio of $NADP^+$ to NADPH is altered in the chloroplast. The $NADP^+ : NADPH$ pool is one of several important feedback mechanisms that regulate the rate of photosynthetic electron transport. When $NADP^+$ concentrations are lowered, then electrons reaching the top of PSI can be recycled back to the PQ pool via ferredoxin. These electrons once more flow to P700 by transfer through the cyt b/f complex and PC, all the while contributing to greater proton transport into the lumen to drive additional ATP formation. This is known as *cyclic electron flow* or *cyclic photophosphorylation*, and is especially important when ATP demand for phosphorylating sugar intermediates is high or when environmental CO_2 limitation slows the rate of carbon fixation and thus slows the rate of NADPH turnover.

Photophosphorylation

The light-driven proton build-up, brought on by both PQ oxidation and water-splitting reactions during electron transport, lowers the intrathylakoid pH significantly. It appears that the internal proton concentration can reach magnitudes 10^4 times as great as the external proton concentration, with the pH of the intrathylakoid space being as low as 4.0. The transmembrane gradient represents a protonmotive force (PMF) which drives transmembrane photophosphorylation reactions.

Thus, the rate of cyclic and non-cyclic electron transport controls the rate of proton build-up, which in turn determines the rate of ATP formation. The phosphorylation of ADP is stimulated by an enzyme complex which couples ATP synthesis to electron transport and is termed 'coupling factor' (CF, or ATPase). CF is a large (325 kD) protein complex consisting of a hydrophobic protein complex (CF_0) which spans the thylakoid membrane and is the binding site for a large hydrophilic protein complex (CF_1) containing the active site for ATP synthesis. CF_1 exists as a particle about 10 nm in diameter, and is attached to the outer, non-appressed stromal surfaces of thylakoid membranes (Fig. 19.8). The entire CF complex is a reversible ATPase. The chemiosmotic coupling theory first proposed by Mitchell (1974) supports the view that when transmembrane gradients in pH are generated during electron transport, thylakoid membrane state changes occur which stimulate proton passage through the proteolipid channel of CF_0. The magnitude of proton movement depends on the light intensity (the driving force), the pH gradient in steady state, and the internal buffer capacity of the membrane. It is

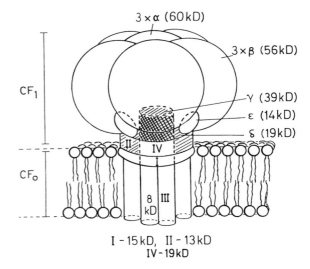

Fig. 19.8 The coupling factor (CF). A schematic model of ATP synthetase (CF_0–CF_1 complex). The shaded components are coded for in the nuclear genome while unshaded polypeptides are chloroplast encoded (by permission, Barber and Baker, 1985).

estimated that between 2.5 and 3.0 protons are pumped for each ADP phosphorylated.

The actual mechanism by which ATP is formed from the dehydration of ADP and the binding of P_i is not known. One hypothesis is that hydroxyl ions and protons are released in ADP dehydration reactions, with the hydroxyl ions drawn toward the proton-rich interior to combine with the protons to form water. As a result, the protons from the dehydration reaction are released in a $1:1$ stoichiometric ratio with each proton hydrated on the interior, and thus a net proton release across the thylakoid membrane results. The reverse process is hypothesized to occur when ATP reserves are mobilized in the stroma, to allow the hydrolysis of ATP to ADP and P_i.

Under ideal conditions of non-cyclic electron transport, when eight quanta are absorbed (four per PSI and four per PSII), eight protons ($4H^+$ from H_2O splitting reactions in PSII and $4H^+$ from electron transport by PQ pool between PSI and PSII) will be deposited in the intrathylakoid space and two NADPH are generated in the stroma. The eight protons in the intrathylakoid space can be used as a protonmotive force to phosphorylate three ADP to three ATP. The 3ATP:2NADPH yielded in non-cyclic electron transport is what is required to fix one molecule of carbon dioxide into a hexose sugar via the reductive pentose phosphate pathway (Chapter 21). Generally, however, the quantum requirement for CO_2 reduction is usually greater than eight, especially when synthetic endproducts of photosynthesis are organic molecules other than simple sugars. Metabolic demands can often require that light-dependent ATP production be increased relative to NADPH reduction.

State 1/State 2 transitions

A balance in exciton distribution between PSI and PSII is required if these interdependent reaction centers are to operate efficiently. However, PSI and PSII do not absorb all wavelengths of light equally. Difference in spectral absorption of PSI and PSII occurs throughout the visible spectrum but is most pronounced at wavelengths greater than 680 nm (far red) (State 1 light), where light is absorbed predominantly by PSI. Under far-red light conditions, there is a decrease in the activity of PSII, termed the red drop. If light of shorter wavelengths (State 2 light) is superimposed on far-red illumination, then there will be a marked *enhancement* of PSII activity.

A number of mechanisms exist to keep a balanced exciton distribution between the two photosystems. For long-term (hours to days) exposure to light environments which favor light absorption by one photosystem over another, plants can adapt by changing the amount or composition of LH pigments servicing each phototrap, by varying the PSI/PSII ratio and/or by altering the density or arrangement of thylakoids within the chloroplast. To accommodate short-term (seconds to hours) biases in the light field, plants can vary the distribution of excitation energy between photosystems in two ways. Plants can redirect some absorbed light out of PSII into PSI (i.e. spillover) and/or rearrange the existing LH pigment–protein complexes among PSI and PSII complexes so as to alter the absorbance cross-section of photosystems. In either instance, exposure to State 2 light will favor absorption by and over-excitation of PSII and induce changes that lead to the redistribution of excitation energy away from PSII and toward PSI. Exposure to State 1 light will favor light absorption and over-excitation of PSI, quickly leading to an inhibition of the transfer of excitation energy from PSII to PSI and/or the inhibition of realignment of PSII light-harvesting complexes with PSI. Molecular mechanisms are now being proposed to explain how changes in environmental light conditions induce an immediate and effective redistribution of excitation energy between PSI and PSII, referred to as State 1/State 2 transitions (Fig. 19.9).

In higher plants and green algae, the light-harvesting capabilities of PSII are determined by the presence of LH Chl *a/b* protein complexes (LHCII) (Fig. 19.2c). These LHCII complexes can be phosphorylated by a kinase enzyme, whose activity is regulated by the redox state of the PQ pool situated between PSI and PSII (Fig. 19.9). Under State 2 light conditions, the PQ pool becomes largely reduced and activates the kinase system, which in turn phosphorylates LHCII complexes. Phosphorylation is thought to increase the net negative charge on surfaces of LHCII complexes which are associated

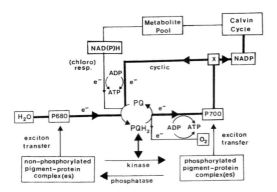

Fig. 19.9 A general model for the control of State 1/State 2 transitions in green plants. A highly reduced PQ pool catalyzes phosphorylation of the LH pigment–protein complexes by the action of a kinase, allowing a spillover of excitons to PSI. Reduction potential in the PQ pool is increased by electron transfer from PSII and from cyclic photophosphorylation and photorespiration, and is reduced by reduction of P700 in PSI (by permission, Williams and Allen, 1987).

with PSII and are located mainly in the appressed regions of grana. As a result of the change in net charge, phosphorylated LHCII complexes are thought to dissociate from PSII within the grana stacks and diffuse to non-appressed regions of stroma lamellae, where they associate with the LH Chl a_I-protein complexes of PSI. If true, the absorption cross-section of PSII is reduced, while that of PSI is increased, thereby leading to a better balance of excitation energy between the two photosystems.

Under State 1 light conditions, the PQ pool should become oxidized, kinase activity decrease, and phosphorylation of LHCII cease. Phosphorylated LHCII complexes may be dephosphorylated by a phosphatase system. The resultant loss of a net negative charge would make it possible for LHCII complexes to return to the stacked regions of the thylakoid and reassociate with PSII. Thus, one model of State 1/State 2 transitions is based on the light-regulated phosphorylation of light-harvesting pigment–protein complexes.

As models develop to describe the control mechanisms underlying state transitions in photosynthesis, additional factors regulating kinase activation via the redox state of the PQ pool are being incorporated (Fig. 19.9). Aside from the initial

distribution of excitation energy between PSI and PSII, excess cyclic electron flow around PSI would also lead to a net reduction of the PQ pool and the phosphorylation of LH pigment–protein complexes. As previously discussed, cyclic electron flow can vary in response to changes in the availability of NADP and ATP, which in turn are influenced by carbon limitation and metabolic demands within the chloroplast. Alternate views regarding mechanisms underlying state transitions do exist, and the pigment – protein phosphorylation model has not been fully accepted for all plant groups in which thylakoid arrangements, composition of LH complexes, and metabolic regulation within chloroplasts can vary broadly.

References

Anderson, J. M. (1975). The molecular organization of chloroplast thylokoids. *Biochim. Biophys. Acta* **416**, 191–235.

Barber, J. and Baker, N. R. (1985). *Photosynthetic Mechanisms and the Environment*. Topics in Photosynthesis, vol. 6, Elsevier, Amsterdam.

Boczar, B. A. (1985). Functional organization of chlorophyll in chlorophyll *c*-containing marine phytoplankton, PhD thesis, University of California, Santa Barbara.

Fenna, R. E. and Mathews, B. W. (1979). Bacterial chlorophyll-proteins from green photosynthetic bacteria. In *The Porphyrins*, Vol. 7, ed. D. Dolphin, Academic Press, New York, pp. 473–94.

Foyer, C. H. (1984). *Photosynthesis*, Wiley-Interscience, New York.

Geacintov, N. E. and Breton, J. (1987). Energy transfer and fluorescence mechanisms in photosynthetic membranes. *CRC Crit. Rev. Plant Sci.* **5**, 1–44.

Govindjee (ed.) (1982). *Photosynthesis: Energy Conversion by Plants and Bacteria*, Vol. 1, Academic Press, New York.

Govindjee, Amesz, J. and Fork, D. C. (1986). *Light Emission by Plants and Bacteria*, Academic Press, New York.

Kaplan, S. and Arntzen, C. J. (1982). Photosynthetic membrane structure and function. In *Photosynthesis: Energy Conversion by Plants and Bacteria*, Vol. 1, ed. Govindjee, Academic Press, New York, pp. 65–140.

Larkum, A. W. D. and Barrett, J. (1983). Light-harvesting processes in algae. In *Advances in Botanical Research*, ed. H. W. Woolhouse, Academic Press, New York, pp. 3-222.

Mitchell, P. (1974). A chemiosmotic molecular mechanism for proton translocating adenosine triphosphatases. *FEBS Lett.* **43**, 189–94.

O'Carra, P. and O'heocha, C. (1976). Algal biliproteins and phycobilins. In *Chemistry and Biochemistry of Plant Pigments*, ed. T. W. Goodwin, Academic Press, London, pp. 328–76.

Okamura, M. Y., Feher, G. and Nelson, N. (1982). Reaction centers. In *Photosynthesis: Energy Conversion by Plants and Bacteria*, Vol. 1, ed. Govindjee, Academic Press, New York, pp. 195–272.

Ort, D. R. and Melandri, B. A. (1982). Mechanism of ATP synthesis. In *Photosynthesis: Energy Conversion by Plants and Bacteria*, Vol. 1, ed. Govindjee, Academic Press, New York, pp. 537–87.

Prézelin, B. B. (1981). Light reactions in photosynthesis. In *Physiological Bases of Phytoplankton Ecology*, ed. T. Platt, Can. Bull. Fish. Aquat. Sci. Vol. 210, pp. 1–43.

Prézelin, B. B. and Alberte, R. S. (1978). Photosynthetic characteristics and organization of chlorophyll in marine dinoflagellates. *Proc. Nat Acad. Sci. USA* **75**, 1801–4.

Prézelin, B. B. and Boczar, B. A. (1986). Molecular bases of cell absorption and fluorescence in phytoplankton: potential applications to studies in optical oceanography. In *Progress in Phycological Research*, Vol. 4, eds F. E. Round and D. J. Chapman, Biopress, Bristol, pp. 349–464.

Prézelin, B. B. and Haxo, F. T. (1976). Purification and characterization of peridinin-chlorophyll *a*-proteins from the marine dinoflagellate *Glenodinium* sp. and *Gonyaulax polyedra*. *Planta* **128**, 133–41.

Raven, J. A. (1984). *Energetics and Transport in Aquatic Plants*, MBL Lectures in Biology, Vol. 4, Alan R. Liss, New York.

Renger, G. (1983). Photosynthesis. In *Biophysics*, eds W. Hoppe, W. Lohmann, H. Markl and H. Ziegler, Springer-Verlag, Berlin, pp. 515–42.

Satoh, K. and Butler, W. L. (1978). Low-temperature spectral properties of subchloroplast fractions purified from spinach. *Plant Physiol.* **61**, 373–9.

Sauer, K. (1975). Primary events and the trapping of energy. In *Bioenergetics of Photosynthesis*, ed. Govindjee, Academic Press, New York, pp. 116–75.

Thornber, J. P. and Alberte, R. S. (1977). The organization of chlorophyll *in vivo*. In *Photosynthesis I: Photosynthetic Electron Transport and Photophosphorylation*, eds A. Trebst and M. Avron, Springer-Verlag, Berlin, pp. 574–82.

Williams, W. P. and Allen, J. F. (1987). State 1/State 2 changes in higher plants and algae. *Photosynth. Res.* **13**, 19–45.

Witt, H. T. (1979). Energy conversion in the functional membrane of photosynthesis. Analysis by light pulse and electric pulse methods: the central role of the electric field. *Biochim. Biophys. Acta* **505**, 355–427.

20 Ribulose 1,5-bisphosphate carboxylase/oxygenase: mechanism, activation and regulation

Keith A. Mott

Introduction

The enzyme ribulose 1,5-bisphosphate carboxylase/oxygenase (Rubisco) catalyzes the addition of atmospheric CO_2 to the 5-carbon-sugar ribulose 1,5-bisphosphate (RuBP), forming two molecules of 3-phosphoglycerate (PGA) (Fig. 20.1). This reaction is part of the photosynthetic reductive pentose phosphate (RPP) pathway. The remainder of the RPP pathway is a complex series of 13 reactions in which the carbon in PGA is either exported to the cytosol or used to regenerate RuBP (see Chapter 21). It may seem inequitable to devote an entire chapter to Rubisco, while the remaining 13 enzymes of the pathway are covered in the same amount of space. The reasons for this apparent injustice lie in the characteristics of Rubisco itself and the reaction it catalyzes.

First, Rubisco catalyzes the only known reaction by which living organisms accrue carbon from atmospheric CO_2. Thus, Rubisco is the portal between the pools of atmospheric carbon and organic carbon, and, as such, its properties are potentially of great importance in determining the flux between these two important global reservoirs of carbon.

Second, as its name implies, Rubisco can also oxygenate RuBP, forming one molecule of PGA and one molecule of phosphoglycolate (Fig. 20.1). The former continues through the RPP cycle, but the latter enters the photorespiratory pathway where 25% of the constituent carbon is lost back to the atmosphere. This wasteful pathway has no known function, and the relative rates of carboxylation and oxygenation by Rubisco are major factors in determining the efficiency of photosynthesis.

Third, Rubisco is a remarkably poor catalyst. The enzyme from higher plants has a turnover number of only about $3.3\,s^{-1}$ for each active site. In addition, the concentration of CO_2 in the stroma of a photosynthesizing leaf is approximately equal to the K_m for CO_2 ($\approx 9\,\mu M$; equivalent to a partial pressure of 27 Pa), so that Rubisco normally functions at only about half of its already-low maximum velocity. The situation is made worse by the fact that the oxygenation reaction competes with the carboxylation reaction, which slows the carboxylation rate by another 28% under current atmospheric conditions.

Thus, to sustain reasonable rates of photosynthetic CO_2 fixation, plants must make an enormous amount of Rubisco. The enzyme may constitute up to 25% of the total nitrogen in a plant, and up to 50% of the protein in the chloroplast stroma. Considering the

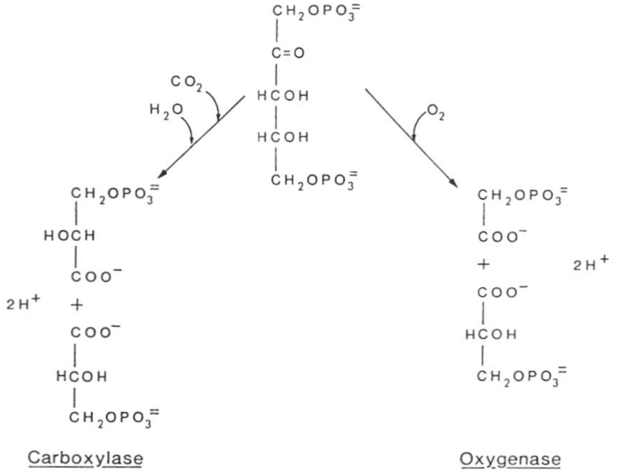

Fig. 20.1 Reactions catalyzed by Rubisco.

quantity of photosynthetic tissue on land and in the ocean, Rubisco is probably the most abundant enzyme in the biosphere. Despite this, the rate of photosynthesis can be substantially limited by the catalytic activity of Rubisco under some circumstances. In addition, the rate of photosynthesis of C_3 plants is usually sensitive to the CO_2 and O_2 concentrations of the atmosphere. This property is of enormous significance with regard to the increasing concentration of CO_2 in the atmosphere.

Rubisco is, therefore, by virtue of its unusual properties and the critical reaction it catalyzes, uniquely situated to have large effects on the rate of photosynthesis. Rubisco from different organisms shows different structures and kinetics, and the catalytic activity of Rubisco in higher plants is regulated by several mechanisms. These aspects of Rubisco and their implications for photosynthesis have been intensively studied and will be explored in this chapter.

Structure

There are two basic structural types of Rubisco, termed form I and form II. All higher plants and most photosynthetic microorganisms have form I Rubisco, which consists of eight large subunits and eight small subunits; often denoted L_8S_8. Some photosynthetic bacteria have form II Rubisco, which is comprised of two or more identical subunits. The subunits of form II Rubisco resemble the large subunits of form I Rubisco and are usually called large subunits. Form II Rubisco from *Rhodospirillum rubrum* has been studied extensively and exists as dimer, L_2, as does form II Rubisco from several other microorganisms (Gutteridge, 1990). Rubisco from *Rhodobacter sphaeroides*, however, varies from L_2 up to at least L_8 depending on pH, so other catalytically competent subunit combinations are apparently possible.

In both forms of Rubisco, the large subunit forms the active site. Although there is low amino acid sequence homology between large subunits of form I and form II enzymes, the three dimensional structures and active site configuration are similar. The large subunit consists of approximately 475 amino acids and has a molecular mass of between 52 and 55 kDa.

It can be divided into an N-terminal domain and a C-terminal domain, which interact very little. The N-terminal domain consists of residues 1–150 and is five-stranded mixed β-sheet with two α-helices on one side (Knight *et al.*, 1990). The C-terminal domain is larger, consisting of residues 151–475, and forms an eight-stranded parallel α/β-barrel structure. This structure is common to a number of enzymes with diverse functions such as glycolate oxidase and triose phosphate isomerase, and consists of eight consecutive $\beta\alpha$-units with the β-strands forming a barrel-shaped β sheet surrounded by the eight helices. The core of the barrel is formed by eight parallel β strands and it is surrounded by eight α helices (Knight *et al.*, 1990).

The active site is formed at the interface of two large subunits, and therefore an L_2 dimer is the minimum functional unit for catalytic activity. This L_2 dimer is roughly ellipsoid with dimensions of $45 \times 70 \times 105\,\text{Å}$ (Knight *et al.*, 1990; Schneider, 1992). The subunits interact with two-fold symmetry, so there are two active sites per dimer. The C-terminal domains of both subunits form the core of the ellipsoid, and the active sites are located in clefts formed by the eight parallel β-strands of the C-terminal domain of one subunit and the N-terminal domain of the other subunit. Residues from both subunits contribute to the active site, and many of the strictly conserved amino acid residues involved in the formation of the active site are located on the loops that connect the β strands to the helices. The N-terminal domain forms a partial lid over the mouth of the barrel formed by the C-terminal domain of the other subunit (Brändén *et al.*, 1991).

Although the L_2 dimer is catalytically active in form II Rubisco, form I Rubisco shows high activity only as a hexadecamer of eight large subunits and eight small subunits having a molecular mass of approximately 550 kDa (Andrews and Lorimer, 1987). The small subunit contains approximately 120 amino acid residues and has a molecular mass of about 13 kDa. It forms a four-stranded antiparallel β sheet with two α helices on one side. Each small subunit interacts extensively with the large subunits and these interactions involve many of the regions that are strictly conserved among small subunits (Knight *et al.*, 1990).

The L_2 dimer is the building block of the

hexadecamer, and the L_8S_8 configuration can be visualized as four ellipsoid L_2 dimers standing on end, roughly forming a cube (Fig. 20.2). The four clefts between dimers at the top of the molecule and the corresponding four clefts at the bottom of the molecule are occupied by the small subunits. (Knight *et al.*, 1990) The four L_2 dimers form a vertical solvent channel through the center of the molecule.

The role of the small subunit is not clear. Since the L_8 core of Rubisco from higher plants is almost completely insoluble and that from cyanobacteria is only slightly soluble, one function of the small subunit is probably to increase the solubility of holoenzyme. The hexadecameric forms of the enzyme show a higher specificity for CO_2 than do the dimeric forms (Gutteridge, 1990), and it is tempting to assign the small subunit a role in this phenomenon. However, the L_8 core of the form I cyanobacterial enzyme retains approximately 1% of the activity of the hexadecamer and shows identical CO_2/O_2 specificity (Andrews and Lorimer, 1987). Thus, the small subunit is apparently not strictly required for catalytic activity and does not directly influence CO_2/O_2 specificity, yet it somehow strongly affects the reactivity of the active site. None of the residues of the small subunit contributes to the active site, so the

small subunit must exert its effect indirectly, possibly by subtly altering in configuration of the active site by interactions with other portions of the large subunit. It has been suggested that the small subunits influence the position of two loop regions of the large subunit that form a portion of the active site (Brändén *et al.*, 1991).

Synthesis and assembly

The large subunit is encoded in the chloroplastic genome and is synthesized on chloroplastic ribosomes. Each chloroplast DNA molecule contains a single copy of the gene, but since there are many DNA molecules per chloroplast and many chloroplasts per cell, there are hundreds to thousands of identical copies of the large subunit gene per cell. The gene does not generally contain introns, although the gene from *Euglena* is a notable exception (Andrews and Lorimer, 1987). Sequence homology for the large subunit is high (more than 80%) among species with the form I enzyme, but there is a segment of about 13 residues at the N terminal that shows relatively weak homology. Overall homology between form I and form II large subunits is low (about 30%) (Gutteridge, 1990), but extensive homology in the areas that form the active site suggests that all Rubisco large subunits had a common heritage.

The small subunit is encoded by a nuclear multigene family. Most higher plants have at least two copies of the small subunit gene, and some may have more than 10. Although these copies are usually distinct at the genomic level, the amino acid sequence of the mature peptide differs very little, if at all (Andrews and Lorimer, 1987). Although the degree of expression of the different genes for the small subunit often varies during development and among different tissues, no significance to these expression patterns has been found. The small subunit is synthesized in the cytoplasm and contains a short leader sequence of about 56 amino acid residues (Gutteridge, 1990) that directs the peptide across the chloroplast membrane via an ATP-dependent transporter. This transit peptide apparently contains all the information necessary for import into the chloroplast because

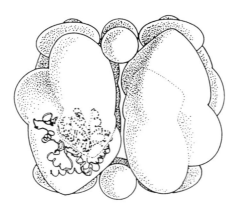

Fig. 20.2 Diagrammatic representation of the structure of the L_8S_8 hexadecameric form of Rubisco. The four large ellipsoids represent large subunit dimers; the eight smaller ellipsoids represent the small subunits (redrawn from Knight *et al.*, 1990, reproduced with the permission of Academic Press).

fusing it to other proteins causes them to be transported into the chloroplast. The peptide is removed after transport by a stromal metalloendopeptidase. In contrast to the large subunit, the small subunit shows much less homology among species, but there are two strongly conserved regions, residues 10–21 and residues 61–76. In addition, many small subunit genes have introns of various lengths, sequences and positions (Andrews and Lorimer, 1987).

Assembly of the holoenzyme occurs in the chloroplast and is mediated by a molecular chaperonin that has been called Rubisco subunit binding protein or chloroplast chaperonin 60 (*cpn60*) (Roy, 1989; Gatenby and Ellis, 1990). This protein is composed of two subunits, each approximately 60 kD, which are encoded in the nucleus and synthesized in the cytoplasm. They are transported into the chloroplast with the aid of a short leader sequence in a manner similar to the small subunit. Fourteen of these subunits – about equal proportions of the two types – form a large doughnut-shaped aggregate and, in the chloroplast, this aggregate binds to newly-synthesized Rubisco large subunits and facilitates the assembly of the holoenzyme (Roy, 1989). The exact function of the chaperonin protein remains unknown; small subunits can be assembled to the L_8 core in the absence of the chaperonin, so this protein probably functions only in the assembly of the L_8 core. ATP is apparently required for the release of large subunits from the chaperonin protein (Hubbs and Roy, 1993). The chloroplast chaperonin shows approximately 50% homology with the bacterial chaperonin encoded by the GroEL gene in *E. coli*. Both proteins form aggregates of 14 subunits, although the bacterial chaperonin is composed of 14 identical subunits. There is some evidence that the bacterial chaperonins can also facilitate the assembly of Rubisco (Gatenby and Ellis, 1990).

Activation

Each active site (i.e. two per L_2 dimer and eight per L_8S_8 hecadecamer) is catalytically competent only after formation of a ternary complex with CO_2 and Mg^{2+}. This reaction, termed activation, begins with the covalent binding of a CO_2 molecule to the

ε-amino group of a lysine residue in the large subunit. The resulting carbamate is then stabilized by the binding of Mg^{2+} (Schneider *et al.*, 1992), making the site catalytically active. The first step of the activation reaction (carbamate formation) is slow and rate-limiting, and its equilibrium is pH dependent. The subsequent binding of Mg^{2+} is rapid.

$$E_{(inactive)} + {}^ACO_2 \overset{slow}{\rightleftharpoons} E \cdot {}^ACO_{2\,(inactive)} + H^+ \qquad [20.1]$$

$$E \cdot {}^ACO_2 + Mg^{2+} \overset{fast}{\rightleftharpoons} E \cdot {}^ACO_2 \cdot Mg_{(active)} \qquad [20.2]$$

The CO_2 molecule that forms the carbamate is termed the activator CO_2 and is distinct from the substrate CO_2 molecule. Other factors, such as the presence of tight-binding inhibitors, can also affect the catalytic activity of an active site once it has complexed with CO_2 and Mg^{2+}. It is important to distinguish these effects from the activation (carbamylation) of the active site.

The lysine residue on which the carbamate is formed (lysine 201 for the spinach enzyme) is the last residue of the second β strand of the β/α barrel and is located at the bottom of the active site (Knight *et al.*, 1990; Brándén *et al.*, 1991). The Mg^{2+} interacts directly with the negative charges of the carbamate and with the side chains of two other residues, Asp203 and Glu204. Other divalent metal ions such as Ni^{2+}, Co^{2+}, Fe^{2+}, Mn^{2+}, and Cu^{2+}, can substitute for Mg^{2+} in stabilizing the carbamate, but the catalytic rates for these forms of the enzyme are generally quite low, and often differ in their ability to discriminate between CO_2 and O_2 (Andrews and Lorimer, 1987). This suggests that in addition to stabilizing the carbamate, the metal ion plays an important role in some aspect of catalysis, and may interact with the gaseous substrate. Crystallographic studies of the enzyme with a transition-state analog CABP (see below) bound to the inactivated (*R. Rubrum*) and the activated (spinach) enzyme suggest that the Mg^{2+} ion may promote the proper binding of the phosphate groups of the substrate RuBP. The Mg^{2+} ion probably also stabilizes the enediol reaction intermediate (see below) (Andrews and Lorimer, 1987; Schneider *et al.*, 1992).

Although there are no large changes in quaternary structure associated with activation (Schneider *et al.*, 1992), there are changes in the chemical and physical properties of the enzyme, presumably caused by modifications of the active site. In contrast, crystallographic studies of the activated enzyme show that binding of CABP causes two loop regions (one from each domain) to fold over the bound CABP molecules. Side chains from each of these loops are directly involved in binding the CABP molecule, and the loops protect the bound substrate from the solvent (Schneider *et al.*, 1992).

Catalysis

Rubisco catalyzes both the carboxylation and the oxygenation of RuBP. The carboxylation reaction has been studied extensively, and there is substantial evidence supporting the reaction mechanism shown in Fig. 20.3 and outlined below. The reaction mechanism is strictly ordered: RuBP binds first and then reacts with either of the gaseous substrates, CO_2 or O_2. After binding of RuBP to an activated site, the first step in catalysis is the deprotonation of C3 to form a 2,3-enediol (1), which creates a nucleophilic center at C2. This intermediate structure is carboxylated to form 2-carboxy 3-keto-arabinitol (ketone form) (2), which is subsequently hydrated to the gem diol form (3). O3 is then deprotonated, and the molecule cleaves to form one molecule of 3-P-glycerate, and one molecule of the C2 carbanion form of 3-P-glycerate (4). The latter is protonated at C2 to produce the second molecule of 3-P-glycerate (5).

The competing oxygenation reaction has been less well-studied, but it is clear that, as with carboxylation, RuBP binds to the carbamylated, active site and undergoes deprotonation of C3 to form an enediol. At this point CO_2 and O_2 compete for the nucleophilic center at C2, with carboxylation producing two molecules of 3PGA, and oxygenation producing one molecule of 3PGA and one molecule of 2-P-glycolate. In the latter reaction, one atom of molecular oxygen is added to C2, which, after cleavage between C2 and C3, forms the carboxyl group of P-glycolate. The other oxygen atom is incorporated into water, and one atom of oxygen from water is incorporated into P-glycolate.

There do not appear to be any binding sites on the enzyme for either CO_2 or O_2. However, the reactivity of the enediol towards these two molecules is different for enzymes from different species and can also be altered by substituting other divalent metals ions for Mg^{2+}. Thus, these substrates may interact with groups on the enzyme.

The study of this reaction mechanism led to the discovery of a transition-state analog, 2-carboxyarabinitol bisphosphate (CABP) (Fig. 20.4). This molecule binds tightly ($K_D = 10$ pM) to activated sites and less tightly to inactivated sites, but it does not undergo catalysis. It has proved valuable in a variety of studies of Rubisco because it causes conformational changes that are presumably similar to those associated with RuBP binding. Also, because

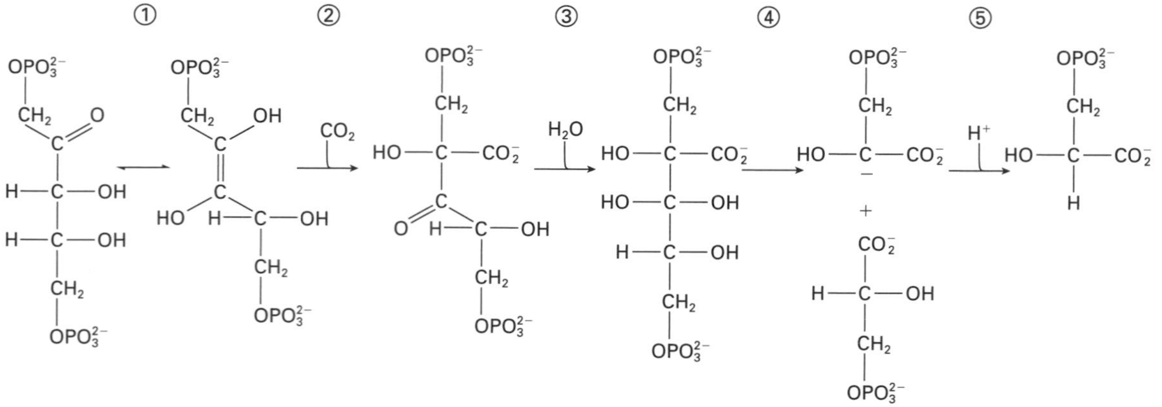

Fig. 20.3 Mechanism of RuBP carboxylation by Rubisco. The numbered steps are described in the text (adapted from Lorimer and Andrews, 1987).

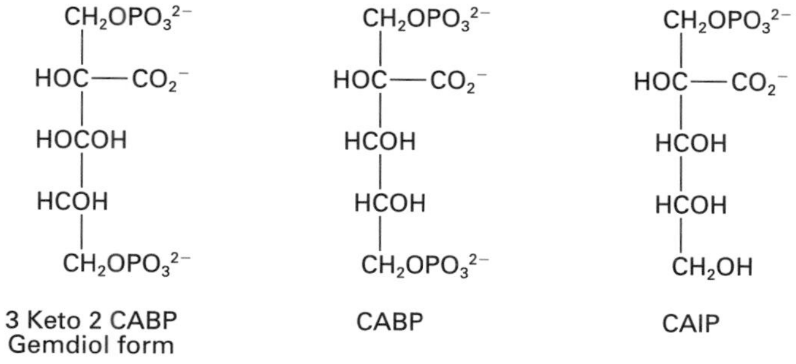

Fig. 20.4 Structures of (1) the six carbon reaction intermediate for the carboxylation of RuBP by Rubisco, 3-keto-2-carboxyarabinitol bisphosphate (3-keto-2CABP); (2) the transition state analog, carboxyarabinitol bisphosphate (CABP); (3) the naturally-occurring inhibitor, carboxyarabinitol 1 phosphate (CA1P).

of the extremely tight binding of this compound to activated sites, radiolabelled forms can be used to determine the number of active sites. A similar compound, 2-carboxyarabinitol 1-phosphate (Ca1P) occurs naturally in some plants and is formed in darkness and low light. CA1P binds somewhat less tightly to activated sites than CABP but still tightly enough to be essentially non-competitive with respect to RuBP. Changes in the catalytic activity of Rubisco caused by the presence of this compound can apparently augment or replace changes in activation state in some plants (see below).

Kinetic properties

The kinetic properties of Rubisco are not easily determined. The reaction rate *in vitro* declines with time, and assays must generally be restricted to short time periods to minimize this problem. This decline in activity has been termed 'fallover'. Although at one time it was attributed to deactivation of the enzyme, fallover is now thought to be caused by the accumulation of tight-binding inhibitors, such as xylulose 1,5-bisphosphate and 3-keto-arabinitol 1,5-bisphosphate (Edmondson, 1990). These may be present as contaminants of the RuBP preparation, but they are also apparently formed by 'misfires' during catalysis. The dual role of CO_2 as an activator and substrate makes it difficult to assess carboxylation rates for the fully activated enzyme at low CO_2 concentrations. It is also difficult to measure

the oxygenation reaction and to control the levels of dissolved CO_2 and O_2 in the reaction mixture, which limits the accuracy with which the Michaelis–Menten constants for CO_2 and O_2 can be determined (Andrews and Lorimer, 1987). For these reasons, the kinetic parameters of Rubisco have been subject to frequent updating in the literature.

The problem of declining reaction rates, when combined with damage to the enzyme during extraction and purification, often results in low activity values. When determined with freshly extracted enzyme and pure RuBP, V_{max} is approximately $3.6 \, \mu mol \, mg^{-1} \, min^{-1}$, and the turnover number (k_{cat}) is approximately $3.3 \, s^{-1} \, site^{-1}$. These values are high enough to account for observed carboxylation rates *in vivo*, but they are rarely achieved with purified preparations of Rubisco and commercially-available RuBP preparations (Woodrow and Berry, 1988).

The competition between CO_2 and O_2 for the intermediate enediol of RuBP results in standard competitive inhibitor kinetics for these gaseous substrates. Thus, the rates of carboxylation (v_c) and oxygenation (v_o) at saturating RuBP concentrations can be expressed as:

$$v_c = \frac{V^c_{max}[CO_2]}{[CO_2] + K_c(1 + [O_2]/K^i_o)} \qquad [20.3]$$

$$v_o = \frac{V^o_{max}[O_2]}{[O_2] + K_o(1 + [CO_2]/K^i_c)} \qquad [20.4]$$

where K_c and K_o are the Michaelis–Menten constants for CO_2 and O_2, respectively, V^c_{max} and V^o_{max} are the maximum velocities of the carboxylation and oxygenation reactions, and K_o^i and K_c^i are the inhibition constants for O_2 and CO_2. Since the reaction mechanism is ordered with RuBP binding first, and CO_2 and O_2 react with the same enediol of RuBP, the K_m values for CO_2 and O_2 are independent of the concentration of RuBP. The effect of RuBP concentration on the reaction rate can therefore be expressed as an effect on the apparent V_{max} in the rate equations above. Similarly, V_{max} is a linear function of the number of activated sites (the activation state), but the K_m values of CO_2 and O_2 are independent of activation state.

From the expressions above it can be shown that the ratio of carboxylation and oxygenation rates (v_c/v_o) is a linear function of the ratio of the concentrations of CO_2 and O_2:

$$\frac{v_c}{v_o} = \left(\frac{V_c/K_c}{V_o/K_o}\right)\frac{[CO_2]}{[O_2]} \qquad [20.5]$$

The first term on the right-hand side of the above equation is often called the relative specificity (s_{rel});

$$s_{rel} = \frac{V_c/K_c}{V_o/K_o} \qquad [20.6]$$

This term is a measure of the reactivities of the enediol reaction intermediate towards CO_2 and O_2. As such, it is major determinant of the efficiency of the enzyme and is often used in describing the catalytic properties of various Rubiscos. It is commonly expressed in molar units, reflecting the dissolved concentrations of CO_2 and O_2 in the solution. Alternatively, it can be expressed in terms of partial pressures, reflecting the composition of the gas in equilibrium with the solution. The latter units are more convenient when relating the kinetic properties of Rubisco to gas-exchange characteristics of whole leaves (Woodrow and Berry, 1988).

Relative specificity varies among Rubiscos from different taxa. Rubiscos from higher plants are the most specific for CO_2, with s_{rel} values between 80 and 100 MM^{-1} (2145–2414 Pa Pa^{-1}). Photosynthetic bacteria with the L_2 form of the enzyme (form II) have s_{rel} values between 15 and 20 MM^{-1} (402–536 Pa Pa^{-1}) (Gutteridge, 1990). To illustrate the consequences of variations in this term, consider an enzyme with specificity factor of 90 MM^{-1}. At the partial pressures of CO_2 and O_2 in the stroma of a

photosynthesizing leaf [*] the ratio of carboxylations to oxygenations will be approximately 2.8. This means that roughly one of every four RuBP molecules is oxygenated rather than carboxylated. For the L_2 form of Rubisco with a specificity factor of 20, the ratio of carboxylations to oxygenations is about 0.6. Since one CO_2 is released for every two oxygenations, this enzyme would barely support a positive carbon flux under these conditions. It is no surprise that the photosynthetic bacteria having Rubiscos with very low values of s_{rel} are anaerobic.

The kinetics for RuBP are more complex. At the low enzyme concentrations used *in vitro*, RuBP follows standard Michealis–Menten kinetics with a K_m of about 28 μM at 5–10 mM Mg^{2+}. Chelation of the active form, $RuBP^{4-}$, by Mg^{2+} complicates the determination of the actual K_m value for RuBP, but since the stroma probably has a Mg^{2+} concentration of between 5 and 10 mM, this value is adequate for predicting kinetic properties *in vivo* (Woodrow and Berry, 1988). The enzyme does not obey a simple rate expression for RuBP *in vivo*, however, because of the extremely high concentration of active sites. This unusual situation, in which the concentrations of both the substrate and the active sites are much higher than the K_m value, results in kinetics resembling a non-rectangular hyperbola. At RuBP concentrations less than the concentration of active sites, the reaction rate is directly proportional to the concentration of RuBP; at RuBP concentrations greater than active sites, the reaction rate is independent of RuBP concentration. In the stroma, the situation is even more complex because total phosphate is conserved by the chloroplast P_i translocater, and any decrease in RuBP must be accompanied by an increase in P_i or phosphorylated intermediates of the RPP cycle, mostly PGA. Since both P_i and PGA are competitive inhibitors of RuBP, a decline in RuBP concentration in the stroma is accompanied by an increase in competitive inhibitors

---

[*] The partial pressure of CO_2 in the atmosphere is approximately 35 Pa, but during photosynthesis diffusional restrictions associated with the boundary layer, stomata, and liquid phase reduce this to 25 Pa or less in the stroma of most plants. The partial pressure of O_2 in the atmosphere is 21 kPa and the fluxes of O_2 are not large enough to alter this significantly in the stroma.

for RuBP. The effective dependence of Rubisco activity on RuBP concentration in the stroma is therefore probably more gradual than would be predicted from a purely kinetic standpoint, and RuBP concentrations several times greater than binding sites concentration may result in only about 80% of the RuBP-saturated rate (Woodrow and Berry, 1988).

Activation state *in vivo*

Activation state *in vivo* is typically quantified by extracting the enzyme rapidly at 0–4 °C and low CO_2 concentrations and then assaying the extract as quickly as possible. These conditions are thought to preserve the *in vivo* activation state of the enzyme. The extract is then incubated with high concentrations of CO_2 and Mg^{2+} and assayed again to determine the activity of the completely activated enzyme. The ratio of these two activities (often called initial and total activities) provides an estimate of the activation state of the enzyme in the leaf, which is essentially the percentage of the total sites that are activated (Perchorowicz *et al.*, 1982).

Studies using these methods have revealed large, reversible changes in activation state in response to several environmental factors. The largest changes in activation state occur in response to light intensity, a phenomenon that has been studied extensively. Rubisco activation state typically varies in parallel with the rate of photosynthesis as light intensity is changed (Perchorowicz *et al.*, 1982), and many plants show almost 100% activation at rate-saturating light intensities. Activation state typically reaches a minimum at very low light intensities, and it is often somewhat higher in darkness than at these low light levels. Changes in activation state in response to light intensity are hysteretic. Following an increase in light intensity, Rubisco activates relatively rapidly, showing a pseudo-first-order rate constant of about $0.5\,min^{-1}$ at atmospheric CO_2 concentration. Deactivation in response to decreased light is much slower and often requires over an hour to reach a steady state (Woodrow and Mott, 1989). Activation state *in vivo* is remarkably insensitive to CO_2 concentrations considering the *in vitro* dependence on CO_2, but some deactivation occurs at CO_2 concentrations greater than ambient and at very low

CO_2 concentrations (Sharkey *et al.*, 1986).

Several aspects of activation state *in vivo* are not consistent with the simple uncatalyzed activation reaction as studied *in vitro*. First, under the conditions thought to exist in the stroma (i.e. $\approx 7\,\mu M$ CO_2 and 5–10 mM Mg, pH 8.0–8.2), the equilibrium of the activation reaction is such that only about 60% of the sites are activated, and the proportion of activated sites varies almost linearly with CO_2 concentration. Complete activation of Rubisco *in vitro* requires very high concentrations of CO_2 ($\approx 150\,\mu M$). The fact that RuBP binds more tightly to inactivated sites ($K_D = 20\,nM$) than to activated sites ($K_m = 28\,\mu M$) results in preferential stabilization of the non-carbamylated form of the enzyme and further reduces the proportion of carbamylated sites. These properties are inconsistent with the high activation states commonly observed *in vivo* at atmospheric CO_2 and high light. In addition, the binding of RuBP to non-carbamylated sites physically blocks the formation of the ternary complex with CO_2 and Mg^{2+} and therefore the activation reaction is very slow in the presence of RuBP. This is inconsistent with the rapid activation observed following an increase in light intensity, when RuBP concentrations in the stroma are very high. Finally, activation state *in vivo* is insensitive to CO_2 concentration over a wide range around ambient concentration, and increases in CO_2 concentration above ambient often causes a *decrease* in activation state. This is not consistent with the equilibrium of the activation reaction *in vitro*.

Rubisco activase

The involvement of another stromal protein, Rubisco activase, explains some aspects of activation *in vivo*. Rubisco activase was discovered in the early 1980s as a result of a nuclear gene mutant of *Arabidopsis thaliana* that showed low levels of Rubisco activation in the light. This mutant was deficient in two polypeptides, 41 and 45 kDa, both of which have been named Rubisco activase (Portis, 1992). These two polypeptides are encoded by a nuclear gene, and are identical except for an extra 37 amino acid residues at the C-terminal end of the 45 kDa form. This may be the result of alternative splicing of an intron near the 3′ end of the gene. Both polypeptides are synthesized on cytoplasmic ribosomes and are transported across the chloroplast envelope with the help of a transit peptide containing approximately 58 amino acid residues. Both forms promote Rubisco activation. The physiological significance of the two forms is unclear.

The reaction that Rubisco activase catalyzes has not yet been elucidated. However, the enzyme promotes carbamylation of Rubisco in the presence of RuBP or other phosphorylated compounds that bind tightly to the uncarbamylated form. This probably explains how Rubisco can activate rapidly following an increase in light intensity despite high RuBP concentrations in the stroma. Furthermore, the full activation of Rubisco at air level CO_2 in the presence of Rubisco activase probably accounts for the high activation states commonly observed *in vivo* at high light intensities, but since an enzyme cannot alter the equilibrium of a reaction, the mechanism for this effect is unclear. Rubisco activase is also an ATPase. Activation of Rubisco in the presence of RuBP requires ATP, and it is possible that the free energy associated with ATP hydrolysis is harnessed to shift the equilibrium of the activation reaction. However, the rates of ATP hydrolysis and Rubisco activation are not always coupled, and Rubisco activase is capable of hydrolyzing ATP in the absence of Rubisco. In addition, Rubisco activase does not appear to actually catalyze the carbamylation reaction; instead it facilitates the removal of RuBP from uncarbamylated sites so that carbamylation can proceed uncatalyzed (Wang and Portis, 1992). This is also not consistent with the apparent role of Rubisco activase in shifting the equilibrium of the activation reaction.

Rubisco activase also facilitates the dissociation of many tight-binding inhibitors from the carbamylated site. Some of these inhibitors, such as xylulose 1,5-bisphosphate and 3-keto-arabinitol, form as 'misfires' during catalysis, and are thought to be the cause of declining reaction rates commonly observed *in vitro* (Portis, 1990). Rubisco activase also probably removes these inhibitors as they form *in vivo*. Another of these inhibitors, CA1P, is formed in darkened leaves of some species where it binds to activated sites and renders them catalytically inactive. It is released and degraded upon illumination, but its role in the regulation of Rubisco activity is somewhat unclear.

Regulation of Rubisco activation state

The processes that regulate activation state *in vivo* are very poorly understood. Some portion of the light-dependent increase in activation is undoubtedly caused by increases in pH and stromal $[Mg^{2+}]$ associated with light-driven electron transport. However, these processes saturate at low light intensities ($100\,\mu mol\ m^{-2}\ s^{-1}$), and activation state varies over a wide range of light intensities (Woodrow and Berry, 1988). Many phosphorylated compounds besides RuBP bind to both carbamylated and uncarbamylated sites of Rubisco (Portis, 1992). These include NADPH, P_i, and many of the intermediates of the reductive pentose phosphate pathway. These compounds usually bind more tightly to either activated or inactivated sites and preferentially stabilize that form. When added to Rubisco preparations, these compounds can, therefore, increase or decrease the activation state of the enzyme. The role of these compounds *in vivo* is not clear, however, because any such effector must be present in concentrations approaching that of the active site to have any substantial effect. Since estimates of active site concentrations in the stroma are around 4 mM, only a few compounds meet this criterion. Also, these effectors bind at the active site and are competitive inhibitors for RuBP binding. Thus, a site cannot be simultaneously stabilized by an effector and catalytically active. Activation of one or two sites on the holoenzyme promotes the activation of the remaining six or seven. Such allosteric cooperativity might allow one site to be stabilized by an effector and thus 'sacrificed' to promote the activation of the remaining sites (Andrews and Lorimer, 1987). There is no evidence that such a system in fact operates *in vivo*, however.

Rubisco activase is probably involved in regulating Rubisco activation *in vivo*, but its role is still uncertain. ADP is a potent inhibitor of Rubisco activase activity, and the activity of Rubisco activase is sensitive to ATP/ADP ratios *in vitro* (Portis, 1990). This property is an attractive possibility for the regulation of activation state *in vivo*. ATP/ADP ratios in the stroma would be expected to change with variations in light and CO_2, and this could explain many of the observed changes in Rubisco activation state *in vivo*. This is consistent with alterations in activation state in isolated chloroplasts by manipulating ATP/ADP ratios inside the chloroplasts. However, ATP/ADP ratios in the chloroplasts of intact leaves do not vary enough to account for the observed changes in activation state. The electron transport reactions of photosynthesis apparently stimulate activation in the presence of Rubisco activase. This effect is not linked to ATP production and appears to be related to photosystem I activity and the development of a *trans*-thylakoid pH gradient via an unknown mechanism (Portis, 1990). Its role in regulating Rubisco activation state is unclear.

CA1P

Carboxyarabinitol 1-phosphate is a naturally occurring tight-binding ($K_D = 32$nM for activated sites) inhibitor of Rubisco (Kobza and Seemann, 1988). In some plants, such as *Phaseolus vulgaris*, it accumulates in darkness and low light and binds to carbamylated sites, therefore reducing the catalytic activity of Rubisco. Upon illumination, CA1P is removed from Rubisco and metabolized, allowing Rubisco activity to increase. The pathway for CA1P synthesis is not known, but it does not appear to form as a 'misfire' in the carboxylation reaction as do some other tight-binding inhibitors. It is degraded in the chloroplast stroma by a specific phosphatase, and the enzyme Rubisco activase facilitates its dissociation from Rubisco (Portis, 1990).

CA1P may regulate Rubisco activity in plants such as *Phaseolus* in much the same way as deactivation regulates Rubisco activity in most other species. Other species, such as *Beta vulgaris*, use a combination of deactivation and CA1P binding to regulate Rubisco activity in response to changes in light intensity. Conflicting data exist, though, and some studies show that deactivation serves as the major regulatory process for Rubisco activity at low light in plants that accumulate CA1P, and that there is a substantial accumulation of CA1P only in darkness (Sage, 1993). The advantages or disadvantages of regulating Rubisco activity with CA1P binding instead of carbamylation are unknown.

Role in determining the rate of photosynthesis

The net CO_2 uptake of a leaf or cell can be expressed as:

$$A = v_c - 0.5 v_o - R_d \qquad [20.7]$$

where R_d is the rate of mitochondrial or 'dark' respiration. This equation illustrates how the reactions catalyzed by Rubisco determine the rate of photosynthesis, but it is more difficult to quantify the role of Rubisco activity. The increase in photosynthesis for most C_3 plants associated with elevated ambient CO_2 concentrations is often taken as evidence that increases in Rubisco activity would also increase productivity. However, much of the increase in photosynthesis associated with elevated CO_2 results from the substitution of carboxylations for oxygenations, a phenomenon that would not occur with an increase in Rubisco activity. The model of C_3 photosynthesis by Farquhar *et al.* (1980) provided one of the first quantitative and testable hypotheses concerning the influence of Rubisco on photosynthesis. Although several aspects of this model have been altered as new data have become available, the basic tenets are still valid and provide a good starting point for this complex issue. The fundamental principle of the model is that photosynthesis can be limited by either (1) the catalytic activity of Rubisco at the prevailing CO_2 and O_2 concentrations, or (2) the capacity to regenerate RuBP by the light reactions and the RPP cycle. The model predicts that at high light intensities and low CO_2 concentrations, photosynthesis should be limited by the catalytic activity of Rubisco. At higher CO_2 concentrations, the capacity to regenerate RuBP becomes limiting.

An important manifestation of this analysis is that at low CO_2 concentrations, the response of photosynthesis to CO_2 and O_2 should be primarily determined by the kinetic parameters of Rubisco. The CO_2 compensation point is an interesting special case. As CO_2 concentration is lowered, carboxylation rate decreases and oxygenation rate increases until the rate of CO_2 uptake exactly equals the rate of CO_2 efflux from photorespiration and mitochondrial respiration. This point, at which there is no net uptake or loss of CO_2, is called the CO_2

compensation point; for C_3 plants it is commonly between 45 and 50 Pa. Equation [20.7] shows that, in the absence of dark respiration, photosynthesis is zero when the oxygenation rate is exactly twice the carboxylation rate. The ratio of carboxylation to oxygenation is given by equation [20.6], and if the effect of dark respiration rate is ignored, the CO_2 compensation point (Γ) can be estimated as:

$$\Gamma \approx \frac{0.5[O_2]}{S_{rel}} \qquad [20.8]$$

For a higher-plant Rubisco with specificity factor around 2200 Pa Pa^{-1}, the estimated CO_2 compensation point at atmospheric O_2 concentration would be about 48 Pa, which is very close to the observed value. Precise measurements of these parameters confirm that the specificity factor of Rubisco determines the CO_2 compensation point, which is evidence of Rubisco's role in determining the rate of photosynthesis at low CO_2 concentrations (Woodrow and Berry, 1988).

Control theory analyses support the hypothesis that photosynthesis is substantially limited by Rubisco at low CO_2 concentrations but becomes more limited by other processes – such as the availability of RuBP – at higher CO_2 concentrations (Woodrow, 1986; Stitt *et al.*, 1991). At atmospheric CO_2, rate control is shared among several processes, suggesting that resources are partitioned such that one single enzyme or process does not substantially limit the overall rate of photosynthesis.

The recent advent of antisense mutants with reduced levels of Rubisco provide yet another method for investigating the role of Rubisco in limiting the rate of photosynthesis. These experiments confirm that decreases in Rubisco activity have a larger effect on photosynthesis at low CO_2 concentrations (Hudson *et al.*, 1992; Stitt *et al.*, 1991). There is, however, some discrepancy between studies concerning the degree to which Rubisco controls the rate of photosynthesis at ambient CO_2 concentrations.

The role of Rubisco activity in limiting photosynthesis at low light intensities is problematic because Rubisco deactivates when light is reduced below rate-saturating intensities. Most data suggest that this deactivation does not reduce photosynthesis below the maximum rate sustainable by the light

reactions, and control theory analyses and Rubisco antisense experiments confirm that Rubisco exerts very little control over photosynthesis rate at low light intensities. However, when light intensity increases following a period of low light, Rubisco must return to a higher activation state necessary to sustain the higher rate of photosynthesis. Under some circumstances, the rate at which photosynthesis achieves steady-state at the higher light intensity can be substantially limited by the rate at which Rubisco activates (Woodrow and Mott, 1989).

If deactivation of Rubisco does not directly influence the rate of photosynthesis at low light intensities, what function does it serve in photosynthetic metabolism? The answer to this question probably involves the regulation of other processes in the chloroplast. In the absence of deactivation, RuBP concentrations would decline and PGA concentrations would increase as light intensity decreased. Deactivation of Rubisco maintains the concentrations of RuBP and PGA at approximately constant values despite changes in photosynthesis rate caused by variations in light intensity. This maintains the concentrations of P_i and RPP cycle intermediates relatively constant and may have important ramifications for the regulation of other processes such as the partitioning of carbon between starch and sucrose (Woodrow and Berry, 1988). Also, the formation of PGA from RuBP and CO_2 produces two protons, which are consumed by subsequent reactions in the RPP pathway. Therefore, maintaining the concentrations of PGA and RuBP relatively constant might help stabilize the pH of the stroma (Andrews and Lorimer, 1987).

Prospects for improving Rubisco

Considerable effort has been devoted to producing a 'better' Rubisco in hopes of increasing photosynthetic rate and agricultural productivity. Since Rubisco is at least partially rate-limiting under many conditions, an increase in Rubisco activity would be expected to result in an increase in photosynthesis. Also, a more catalytically active form of Rubisco would allow a plant to maintain lower stomatal apertures and invest less nitrogen in Rubisco without decreasing photosynthetic rate, thereby improving the efficiency of photosynthesis with respect to water loss rates and nitrogen use (Andrews and Lorimer, 1987). However, since most studies show that rate control of photosynthesis is shared by several components of the system under ambient conditions, simply increasing the catalytic activity of Rubisco would probably not result in a proportional increase in photosynthesis.

Another possibility for improving Rubisco is to increase the specificity of the enzyme for CO_2. Although benefits from photorespiration have been proposed, it seems more likely that the oxygenation reaction is a consequence of the fact that Rubisco evolved before there was a substantial amount of oxygen in the atmosphere. Improving the CO_2/O_2 specificity of Rubisco without reducing the catalytic activity of the enzyme would almost certainly improve photosynthetic efficiency. In addition to the benefits associated with nitrogen and water use, decreased oxygenations would reduce the energy loss that is incurred each time an RuBP is oxygenated rather than carboxylated. The specificity for CO_2 does vary among Rubiscos from different species; it is higher in L_8S_8 forms than in L_2 forms and is highest in Rubisco from higher plants. However, considering the intense selection pressure over millions of years, it is somewhat surprising that a more discriminant form of the enzyme has not evolved. In addition, large-scale efforts to select mutants with improved CO_2/O_2 specificity have failed. Thus, it seems likely that modifications to eliminate or substantially reduce oxygenations involve multiple changes in the enzyme, which would not be favored by evolutionary mechanisms and would be unlikely to occur in mutagenesis experiments (Andrews and Lorimer, 1987).

There is no theoretical reason why the specificity of Rubisco for CO_2 cannot be increased, and efforts to improve Rubisco continue. Much of this work involves the use of mutants with Rubiscos of reduced specificity to gain a better understanding of how CO_2 and O_2 interact with enzyme (Spreitzer, 1993).

Global significance

No chapter on Rubisco would be complete without brief consideration of the role this enzyme plays in global carbon balance. As the CO_2 is added to the atmosphere by the burning of fossil fuels, fluxes of CO_2 into other carbon pools will increase and 'buffer' the increase in atmospheric CO_2 to some extent. One of the more important of these fluxes is the CO_2 assimilation rate of C_3 plants. The magnitude of the increase in photosynthetic rate associated with increases in CO_2 depends on the kinetic parameters of Rubisco and the degree to which Rubisco limits the overall rate of CO_2 assimilation. Although it is clear that increases in atmospheric CO_2 can substantially increase photosynthetic rate in the short term, longer-term gains are often less substantial (Bowes, 1993). Repartitioning of carbon and nitrogen among photosynthetic enzymes as well as limitations imposed by plant growth form and water and nutrient availability complicate the prediction of how much extra carbon will be sequestered by C_3 plants. Intensive study of these processes and the role of Rubisco in the global carbon budget is ongoing.

Further reading and references

Andrews, J. T. and Lorimer, G. H. (1987). Rubisco: structure, mechanisms, and prospects for improvement. In *The Biochemistry of Plants*, eds M. D. Hatch and N. K. Boardman, Academic Press, New York.

Brändén, C. I., Lindqvist, Y. and Schneider, G. (1991). Protein engineering of rubisco. *Acta Crystallographica* **B47**, 824–35.

Bowes, G. (1993). Facing the inevitable: plants and increasing atmospheric CO_2. *Ann. Rev. Plant Physiol. Plant Mol. Biol.* **44**, 309–32.

Edmondson, D. L., Badger, M. R. and Andrews, T. J. (1990). Slow inactivation of ribulosebisphosphate carboxylase during catalysis is caused by accumulation of a slow, tight-binding inhibitor at the catalytic site. *Plant Physiol.* **93**, 1390–97.

Farquhar, G. D. (1979). Models describing the kinetics of ribulosebisphosphate carboxylase–oxygenase. *Arch. Biochem. Biophys.* **193**, 456–68.

Farquhar, G. D., Caemmerer, Sv. and Berry, J. A. (1980). A biochemical model of photosynthetic CO_2 fixation in leaves of C_3 plants. *Planta* **149**, 78–90.

Gatenby, A. A. and Ellis, R. J. (1990). Chaperone function: the assembly of ribulose bisphosphate carboxyalase–oxygenase. *Ann. Rev. Cell Biol.* **6**, 125–49.

Gutteridge, S. (1990). Limitations of the primary events of CO_2 fixation in photosynthetic organisms: the structure and mechanism of rubisco. *Acta Biochim. Biophys.* **1015**, 1–14.

Hubbs, A. E. and Roy, H. (1993). Assembly of *in vitro* synthesized large subunits into ribulose-bisphosphate carboxylase/oxygenase. Formation and discharge of an L_8-like species. *J. Biol. Chem.* **268**, 13519–25.

Hudson, G. S., Evans, J. R., von Caemmerer, S., Arvidsson, Y. B. C. and Andrews, T. J. (1992). Reduction of ribulose 1,5-bisphosphate carboxylase/oxygenase content by antisense RNA reduces photosynthesis in transgenic tobacco plant. *Plant Physiol.* **98**, 294–302.

Knight, S., Andersson, I. and Brändén, C. I. (1990). Crystallographic analysis of ribulose 1,5-bisphosphate carboxylase from spinach at 2.4 Å resolution. Subunit interactions and active site. *J. Mol. Biol.* **215**, 113–60.

Kobza, J. and Seemann, J. R. (1988). Mechanisms for light-dependent regulation of ribulose 1,5-bisphosphate carboxylase activity and photosynthesis in intact leaves. *Proc. Nat. Acad. Sci.* **85**, 3815–19.

Perchorowicz, J. T., Raynes, D. A. and Jensen, R. G. (1982). Measurement and preservation of \em *in vivo* activation of ribulose 1,5-bisphosphate carboxylase in leaf extracts. *Plant Physiol.* **65**, 902–5.

Portis, A. R. (1990). Rubisco activase. *Biochim. Biophys. Acta* **1015**, 15–28.

Portis, A. R. (1992). Regulation of ribulose 1,5-bisphosphate carboxylase/oxygenase activity. *Ann. Rev. Plant Physiol. Plant Mol. Biol.* **43**, 415–37.

Roy, H. (1989). Rubisco assembly: a model system for studying the mechanism of chaperonin action. *Plant Cell* **1**, 1035–42.

Sage, R. F. (1993). Light-dependent modulation of ribulose 1,5-bisphosphate carboxylase/oxygenase activity in the genus. *Phaseolus, Photosynthesis Res.* **35**, 219–26.

Schneider, G., Lindqvist, Y. and Brändén, C. (1992). Rubisco: structure and mechanism. *Ann. Rev. Biophys. Biomol. Structure* **21**, 119–43.

Sharkey, T. D., Seemann, J. R. and Berry, J. A. (1986). Regulation of ribulose 1,5-bisphosphate carboxylase activity in response to changing partial pressure of O_2 and light in *Phaseolus vulgaris. Plant Physiol.* **81**, 788–91.

Spreitzer, R. J. (1993). Genetic dissection of rubisco structure and function. *Ann. Rev. Plant Physiol. Plant Mol. Biol.* **44**, 411–34.

Stitt, M., Quick, W. P., Schurr, U., Schulze, E.-D., Rodermel, S. R. and Bogorad, L. (1991). Decreased ribulose 1,5-bisphosphate carboxylase–oxygenase in transgenic tobacco transformed with 'antisense' *rbcS*. II. Flux-control coefficients for photosynthesis in varying light, CO_2, and air humidity. *Planta* **183**, 555–66.

Wang, Z. Y. and Portis, A. R. (1992). Dissociation of ribulose 1,5-bisphosphate bound to ribulose 1,5-bisphosphate carboxylase/oxygenase and its enhancement by ribulose 1,5-bisphosphate carboxylase/oxygenase activase-mediated hydrolysis of ATP, *Plant Physiol.* **99**, 1348–53.

Woodrow, I. E. (1986). Control of the rate of photosynthetic carbon dioxide fixation. *Biochim. Biophys. Acta* **851**, 181–92.

Woodrow, I. E. and Berry, J. A. (1988). Enzymatic regulation of photosynthetic CO_2 fixation in C_3 plants. *Ann. Rev. Plant Physiol. Mol. Biol.* **39**, 533–94.

Woodrow, I. E. and Mott, K. A. (1989). Rate limitation of non-steady-state photosynthesis by ribulose 1,5-bisphosphate carboxylase in spinach. *Australian J. Plant Physiol.* **16**, 487–500.

21 The reductive pentose phosphate pathway and its regulation

Fraser D. Macdonald and Bob B. Buchanan

Introduction

Life on our planet obtains its substance and energy through the process of photosynthesis, a grand device by which photosynthetic organisms use the electromagnetic energy of sunlight to synthesize carbohydrates (CH_2O) and other cellular constituents from carbon dioxide and water.

$$CO_2 + 2H_2O^* \xrightarrow{\text{light}} (CH_2O) + O_2^* + H_2O$$

Photosynthesis may be broadly divided into two phases: a light phase, in which the electromagnetic energy of sunlight is trapped and converted into ATP and NADPH (see Chapter 19), and a synthetic phase, in which the ATP and NADPH generated by the light phase are used, in part, for biosynthetic carbon reduction. As described below, light also functions in the regulation of the synthetic or carbon reduction phase of photosynthesis and in associated biochemical processes of chloroplasts.

In most plants, the major products of photosynthesis are starch (formed in chloroplasts), and sucrose (formed in the cytosol). Both of these products (collectively called photosynthate) are synthesized from photosynthetically generated dihydroxyacetone phosphate (DHAP) via pathways that in some respects are similar to the gluconeogenic pathway of animal cells. In the first case, DHAP is converted to hexose phosphates, which, in turn, are converted to starch within the chloroplast. In sucrose synthesis, DHAP (or a derivative) is transported to the cytosol and is there converted to sucrose. Some plant species use DHAP to form transport compounds other than sucrose, for example, sorbitol.

All oxygenic (oxygen-evolving) organisms from the simplest prokaryotic cyanobacteria to the most complicated land plants have a common pathway for the reduction of CO_2 to sugar phosphates. This pathway is known as the reductive pentose phosphate (RPP), Calvin–Benson or C3 cycle (Calvin and Bassham, 1962; Bassham and Buchanan, 1982).

Although the RPP cycle is the fundamental carboxylating mechanism, a number of plants have evolved adaptations in which CO_2 is first fixed by a supplementary pathway and then released in the cells in which the RPP cycle operates. One of these supplementary pathways, the C4 pathway, involves special leaf anatomy and a division of biochemical labor between cell types. Plants endowed with the C4 pathway are able to flourish under conditions of high light intensity and elevated temperature. A second supplementary pathway was found in species of the Crassulaceae and is called Crassulacean acid metabolism (CAM). These plants are often found in dry areas and fix CO_2 at night into C4 acids. During the day, the leaves can close their stomata to conserve water while CO_2 released from the C4 acids is converted to sugar phosphates by the RPP cycle using absorbed light energy (see Chapter 23).

CO_2 fixation is also found in many bacteria, both photosynthetic and non-photosynthetic. However, unlike plants, anoxygenic photosynthetic bacteria employ pathways other than the RPP cycle. The type first found to utilize another pathway, the photosynthetic green sulfur bacteria, use widely occurring ferredoxin-linked carboxylases in a pathway known as the reductive carboxylic acid or reverse Krebs cycle (Bassham and Buchanan, 1982; Buchanan and Arnon, 1990). It has more recently been recognized that the other types of photosynthetic bacteria are also able to assimilate CO_2 independently of the RPP cycle. In some of these organisms, the RPP cycle is not present (e.g. green non-sulfur bacteria for which a new cycle was described, see Eisenreich *et al.*, 1993) whereas in others the RPP cycle is present (e.g. purple non-sulfur bacteria which have the reductive carboxylic acid cycle, see Bassham and Buchanan, 1982; Buchanan and Arnon, 1990). In organisms such as the purple non-sulfur bacteria it is not clear to what extent the two pathways contribute to overall photosynthesis.

In the following sections the RPP cycle is described by first following the cycle itself, then its regulation, and finally the way in which its activity is coordinated with the utilization of photosynthate.

The reductive pentose phosphate cycle

The RPP cycle is the primary carboxylating mechanism in plants. The enzymes which catalyze steps in the RPP cycle are water soluble and are located in the soluble portion of the chloroplast (the stroma). Elucidation of the pathway was chiefly the work of Calvin, Benson, Bassham and coworkers, although there were important contributions by others. In their experiments, the group of Calvin used green algae, *Chlorella* and *Scenedesmus*, but their results have since been confirmed in a wide variety of organisms, including many higher plants.

The crux of the pathway (Fig. 21.1) is the carboxylation of ribulose 1,5-bisphosphate (Rbu-1,5-P_2) to produce two molecules of 3-phosphoglycerate (3-PGA). The next reactions constitute the reductive phase of the cycle in which the ATP and NADPH produced by the light reactions of photosynthesis are consumed in the reduction of 3-PGA to glyceraldehyde 3-phosphate (G3P). The cycle is completed by the regeneration phase: intermediates formed from G3P are utilized via a series of isomerizations, condensations and rearrangements resulting in the conversion of five molecules of triose phosphate to three of pentose phosphate, eventually ribulose 5-phosphate (Rbu-5-P). Phosphorylation of Rbu-5-P by ATP regenerates the original carbon acceptor Rbu-1,5-P_2, thus completing the cycle.

The three phases of the RPP cycle will now be described in greater detail. The individual reactions and the characteristics of the enzymes which catalyze each step have been reviewed in more detail elsewhere (Robinson and Walker, 1981).

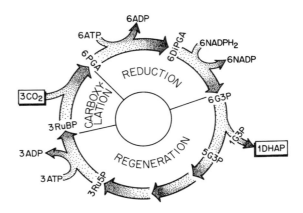

Fig. 21.1 The reductive pentose phosphate (RPP) cycle. Abbreviations: RuBP, ribulose 1,5-bisphosphate; PGA, 3-phosphoglycerate; DiPGA, 1,3-bisphosphoglycerate; Ru5P, ribulose 5-phosphate; G3P, glyceraldehyde 3-phosphate.

Carboxylation phase (3CO_2 + 3Rbu-1,5-P_2 → 6PGA)

$$
\begin{array}{c}
CH_2O\!\!-\!\!P \\
| \\
C\!\!=\!\!O \\
| \\
3\ HCOH + 3^*CO_2 + 3H_2O \\
| \\
HCOH \\
| \\
CH_2OH\!\!-\!\!P
\end{array}
\quad
\xrightarrow[\text{carboxylase/oxygenase}]{\substack{\text{Ribulose} \\ \text{1,5-bisphosphate}}}
\quad
\begin{array}{c}
^*COOH \\
| \\
6\ HCOH \\
| \\
CH_2O\!\!-\!\!P
\end{array}
$$

Ribulose 1,5-bisphosphate   3-Phosphoglycerate
(Rbu-1,5-P_2)   (3-PGA)

[21.1]

Carboxylation is catalyzed by ribulose 1,5-bisphosphate carboxylase/oxygenase, Rubisco, probably the most abundant protein on earth. Carboxylation of Rbu-1,5-P_2 at the C-2 carbon gives rise to a short-lived, six carbon intermediate which breaks down to give two molecules of 3-PGA (eqn [21.1]). A stable analog of this intermediate (carboxyarabinitol 1,5-bisphosphate) is a very tight binding inhibitor of Rubisco that has been of importance in analyzing the enzyme reaction. A similar inhibitory analog is now known to occur naturally in some plants (see later). Note that, in following carbon through the cycle, we begin with three molecules of CO_2 to give eventually a net synthesis of one molecule of triose phosphate.

Reduction phase
(6PGA + 6ATP + 6NADPH → 6G3P + 6ADP + 6NADP$^+$ + 6P$_i$)

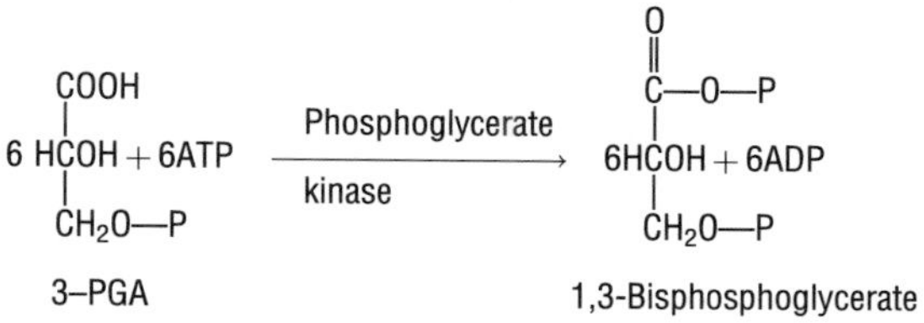

```
  COOH                                            O
   |                                              ‖
6 HCOH + 6ATP   ──Phosphoglycerate──→    6 HCOH   C—O—P
   |                 kinase                   |
  CH₂O—P                                  6 HCOH + 6ADP
                                              |
                                             CH₂O—P
  3–PGA                                1,3-Bisphosphoglycerate
```
[21.2]

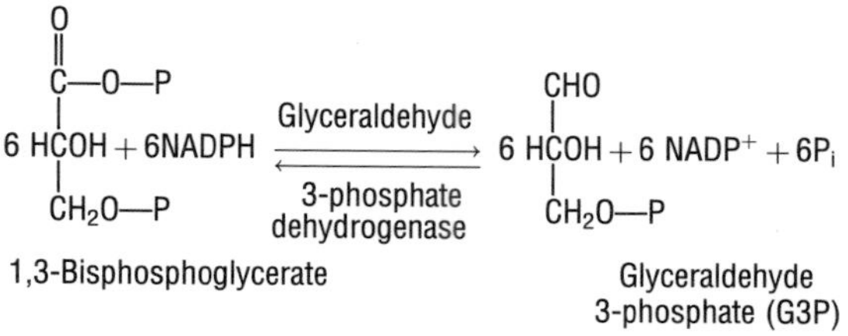

The two reactions of the reduction phase (eqns [21.2] and [21.3]) utilize ATP and NADPH produced by the light reactions of photosynthesis to synthesize G3P from 3-PGA. these steps resemble part of the gluconeogenic pathway in the cytosol, except that glyceraldehyde 3-phosphate dehydrogenase (G3PDH) in chloroplasts is specific for NADPH rather than NADH.

```
  O
  ‖
  C—O—P                                      CHO
   |          Glyceraldehyde                  |
6 HCOH + 6NADPH ⇌ 3-phosphate    6 HCOH + 6 NADP⁺ + 6Pᵢ
   |          dehydrogenase                   |
  CH₂O—P                                     CH₂O—P

1,3-Bisphosphoglycerate                   Glyceraldehyde
                                       3-phosphate (G3P)
```
[21.3]

Regeneration phase

(5G3P + 3ATP → 3Rbu-1,5-P$_2$ + 3ADP + 2P$_i$)

Of the six G3P molecules produced above, one represents a net synthesis of fixed carbon and the other five must be used to regenerate Rbu-1,5-P$_2$ via eqns [21.4] to [21.12].

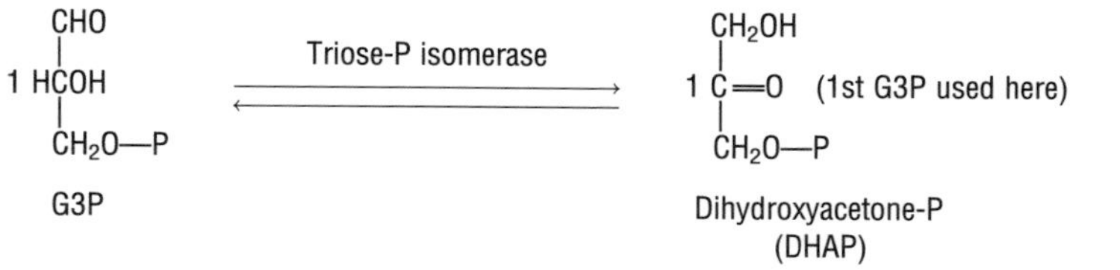

```
   CHO                                      CH₂OH
    |          Triose-P isomerase            |
1 HCOH         ───────────────⇌       1 C=O   (1st G3P used here)
    |                                        |
   CH₂O—P                                   CH₂O—P

   G3P                                 Dihydroxyacetone-P
                                            (DHAP)
```
[21.4]

```
   CHO        CH₂OH                          CH₂O—P
    |          |            Aldolase          |
1 HCOH  +  1 C=O        ───────────⇌         C=O
    |          |                              |
   CH₂O—P     CH₂O—P                    HO—C—H   (2nd G3P used here)
                                             |
   G3P        DHAP                          HCOH
                                             |
  (from reaction [21.4])                    HCOH
                                             |
                                            CH₂O—P

                                    Fructose-1,6-bisphosphate
                                          (Fru-1,6-P₂)
```
[21.5]

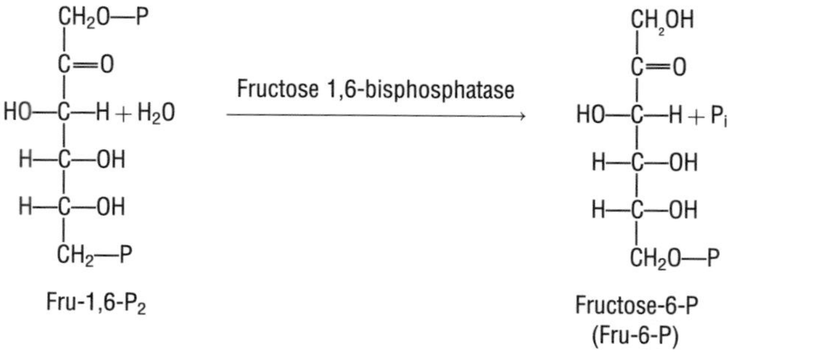

```
   CH₂O—P                                    CH₂OH
    |                                         |
    C=O                                       C=O
    |         Fructose 1,6-bisphosphatase     |
 HO—C—H + H₂O  ──────────────────→       HO—C—H + Pᵢ
    |                                         |
  H—C—OH                                   H—C—OH
    |                                         |
  H—C—OH                                   H—C—OH
    |                                         |
   CH₂—P                                     CH₂O—P

   Fru-1,6-P₂                           Fructose-6-P
                                          (Fru-6-P)
```
[21,6]

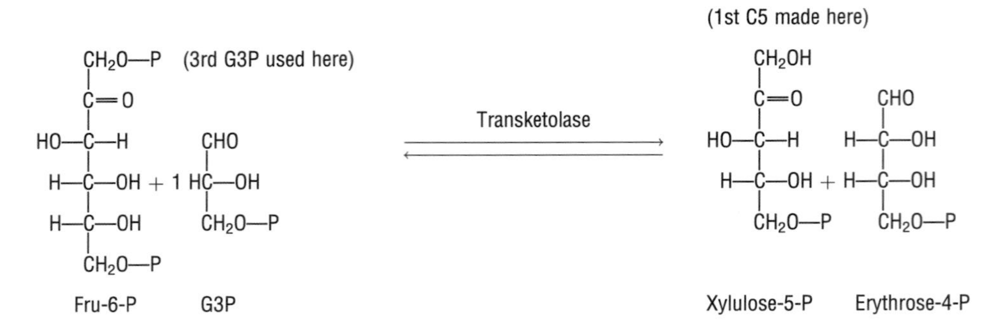

(1st C5 made here)

(3rd G3P used here)

$$\text{Fru-6-P} + \text{G3P} \xrightarrow{\text{Transketolase}} \text{Xylulose-5-P} + \text{Erythrose-4-P}$$

[21.7]

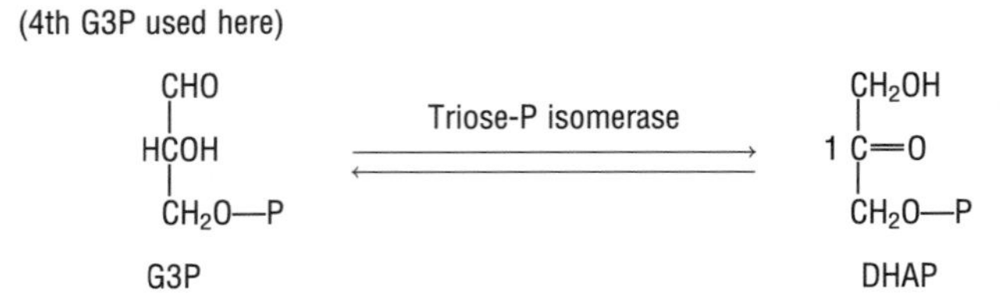

(4th G3P used here)

$$\text{G3P} \xrightarrow{\text{Triose-P isomerase}} \text{DHAP}$$

[21.8a]

$$\text{DHAP (from reaction 21.8a)} + \text{Erythrose-4-P (from reaction [21.7])} \xrightarrow{\text{Aldolase}} \text{Sedoheptulose-1,7-bisphosphate}$$

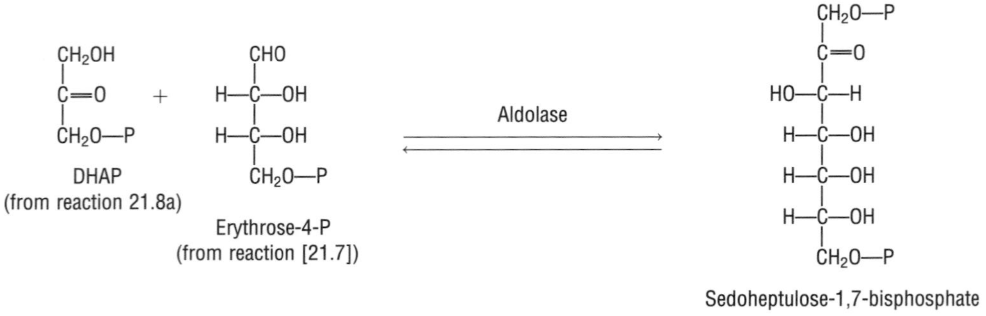

[21.8b]

$$\text{Sedoheptulose-1,7-bisphosphate} + H_2O \xrightarrow{\text{Sedoheptulose 1,7-bisphosphatase}} \text{Sedoheptulose-7-phosphate} + P_i$$

[21.9]

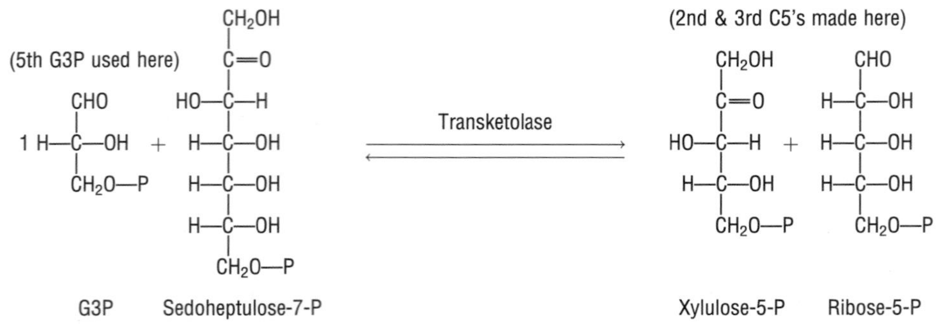

(5th G3P used here)

(2nd & 3rd C5's made here)

$$\text{G3P} + \text{Sedoheptulose-7-P} \xrightarrow{\text{Transketolase}} \text{Xylulose-5-P} + \text{Ribose-5-P}$$

[21.10]

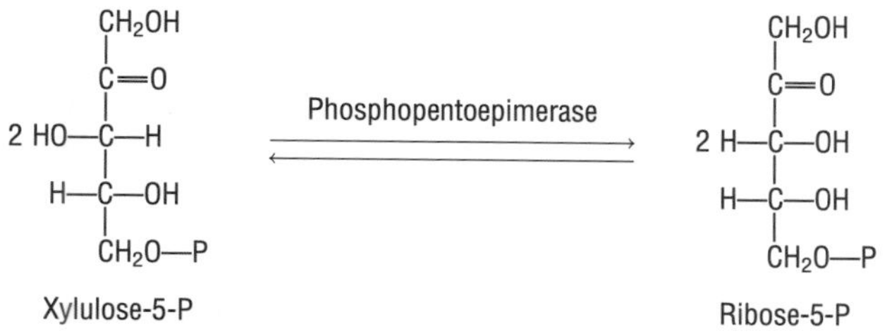

$$
\begin{array}{ccc}
\text{CH}_2\text{OH} & & \text{CH}_2\text{OH} \\
| & & | \\
\text{C}=\text{O} & \xrightarrow{\text{Phosphopentoepimerase}} & \text{C}=\text{O} \\
| & & | \\
2\ \text{HO}-\text{C}-\text{H} & \rightleftharpoons & 2\ \text{H}-\text{C}-\text{OH} \\
| & & | \\
\text{H}-\text{C}-\text{OH} & & \text{H}-\text{C}-\text{OH} \\
| & & | \\
\text{CH}_2\text{O}-\text{P} & & \text{CH}_2\text{O}-\text{P} \\
\text{Xylulose-5-P} & & \text{Ribose-5-P}
\end{array}
$$

(One xylulose-5-P is from reaction [21.7]; the other is from reaction [21.10].)

[21.11]

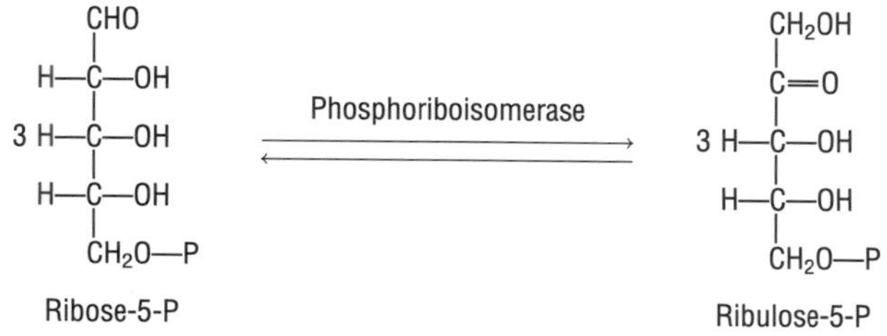

$$
\begin{array}{ccc}
\text{CHO} & & \text{CH}_2\text{OH} \\
| & & | \\
\text{H}-\text{C}-\text{OH} & & \text{C}=\text{O} \\
| & \xrightarrow{\text{Phosphoriboisomerase}} & | \\
3\ \text{H}-\text{C}-\text{OH} & \rightleftharpoons & 3\ \text{H}-\text{C}-\text{OH} \\
| & & | \\
\text{H}-\text{C}-\text{OH} & & \text{H}-\text{C}-\text{OH} \\
| & & | \\
\text{CH}_2\text{O}-\text{P} & & \text{CH}_2\text{O}-\text{P} \\
\text{Ribose-5-P} & & \text{Ribulose-5-P}
\end{array}
$$

[21.12]

Of the three ribose-5-P used in this reaction [21.12] one is from reaction [21.10] and two are from reaction [21.11]. The three ribulose-5-P formed comprise the three used below in reaction [21.13].

$$
3\text{Ribulose-5-P} + 3\text{ATP} \xrightarrow{\text{Phosphoribulokinase}} 3\text{Ribulose 1,5-bisphosphate} + 3\text{ADP}
$$

[21.13]

Overall reaction

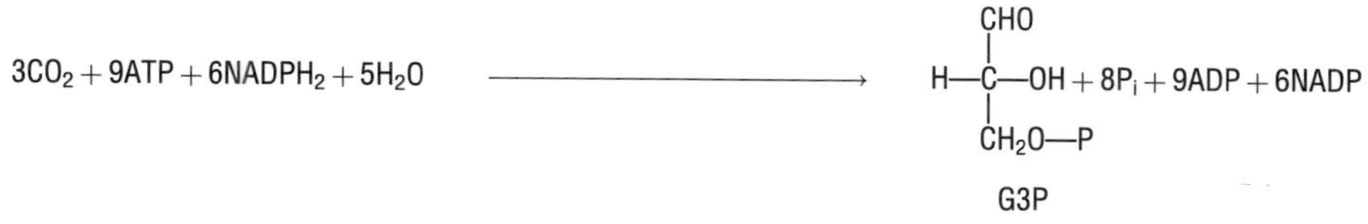

$$
3\text{CO}_2 + 9\text{ATP} + 6\text{NADPH}_2 + 5\text{H}_2\text{O} \longrightarrow
\begin{array}{c}
\text{CHO} \\
| \\
\text{H}-\text{C}-\text{OH} + 8\text{P}_i + 9\text{ADP} + 6\text{NADP} \\
| \\
\text{CH}_2\text{O}-\text{P} \\
\text{G3P}
\end{array}
$$

[21.14]

The RPP cycle displays four features which are necessary for its role as a fundamental carboxylating system (Robinson and Walker, 1981): (1) the Rubisco reaction has a highly favorable equilibrium in the direction of PGA synthesis ($\Delta G' = -35.1$ kJ); (2) Rubisco has a high affinity for CO_2 which occurs at a relatively low concentration in air, $\sim$350 ppm, or in an air saturated aqueous medium, $\sim$10 μM; (3) there is a cyclic regeneration of the CO_2 acceptor, Rbu-1,5-P_2, from the products of the carboxylation reaction, thus enabling the continued operation of the cycle; and (4) the cycle is capable of the net fixation of carbon and yields triose phosphate as net product. For every three turns of the cycle during which six molecules of triose phosphate are formed, five molecules must be utilized to reform three molecules of Rbu-1,5-P_2 while, as noted above, the sixth triose phosphate molecule is available as an end product (photosynthate) for biosynthetic reactions, predominantly starch synthesis in the chloroplast and sucrose synthesis in the cytosol.

In addition to its carboxylase activity, Rubisco can

act as an oxygenase (see Chapter 20). In this reaction, molecular O_2 is bound and reacts with Rbu-1,5-P_2 to give 3-PGA and 2-phosphoglycolate. 2-Phosphoglycolate cannot be utilized in the RPP cycle and thus represents a loss of fixed carbon. This loss is partly compensated by the process of photorespiration during which three-quarters of the lost carbon is returned to the chloroplast as 3-PGA (see Chapter 22). The oxygenase reaction is suppressed by lowered O_2 or elevated CO_2 pressure (compared to air levels), and hence photorespiration is greatly reduced in C4 plants, CAM plants, algae and cyanobacteria, all of which possess CO_2-concentrating mechanisms. The oxygenase activity of Rubisco may be necessary to protect the chloroplast against photo-oxidation damage when CO_2 is limiting. Alternatively, it has been suggested that Rbu-1,5-P_2 oxygenation is an inevitable consequence of the carboxylation mechanism of Rubisco (see Chapters 20 and 22).

Regulation of the reductive pentose phosphate cycle

The principal and ultimate regulator of chloroplast carbohydrate metabolism is light. In fulfilling its regulatory role, light is absorbed by chlorophyll and is then converted to regulatory signals that modulate selected enzymes. Such regulation is essential because enzymes for degrading carbohydrates coexist in chloroplasts with enzymes of carbohydrate synthesis. Selected biosynthetic enzymes are light-activated, whereas degradative enzymes are light-deactivated. In this way chloroplasts minimize the concurrent functioning of enzymes or pathways that operate in opposing directions ('futile cycling') and thereby maximize the efficiency of temporally disparate metabolic processes. In C3 plants, the regulatory function of light maintains 'enzyme order' by assuring that carbon dioxide assimilation takes place during the day, and carbohydrate degradation occurs primarily at night (Buchanan, 1980; 1991; 1992; Buchanan *et al.*, 1994; Cséke and Buchanan, 1986; Scheibe, 1991; Wolosiuk *et al.*, 1993). Through the provision of DHAP formed either from newly fixed carbon or the breakdown of stored starch, chloroplasts are

able to supply carbon for the cytosolic synthesis of sucrose – the transport carbohydrate in most plants (see Chapters 7 and 25) – and thereby meet the energy needs of non-photosynthetic (heterotrophic) tissues at all times.

Identification of the sites of regulation

The sensitivity of a metabolic pathway to regulation residues principally in only a small number of the total steps in the pathway (ap Rees, 1980). Such regulatory steps characteristically have large, negative free energy changes (ΔG) and thus are essentially irreversible. The physiological free energy changes (ΔG^S) for the reactions of the RPP cycle were calculated by Bassham and Krause (1969) from measurements of the steady-state levels of radioactive compounds in photosynthesizing *Chlorella*. The reactions shown to be substantially displaced from equilibrium and therefore potential sites of metabolic regulation were those catalyzed by Rubisco, fructose 1,6-bisphosphatase (FBPase), sedoheptulose 1,7-bisphosphatase (SBPase) and phosphoribulokinase (PRK) (equations [21.1], [21.6], 21.9] and [21.13], respectively).

Further evidence on the importance of these sites in the regulation of the pathway comes from the analysis of light–dark and dark–light transient changes in levels of metabolites. It would be expected that increasing flux through a regulated step of the pathway (in the light) would lead to depletion of the substrate for that step, and decreasing flux (in the dark) would lead to a rise in the concentration of that substrate. The kinetic analysis of such experiments is complicated by the cyclic nature of the pathway, since the production of substrate for one reaction may be affected by the regulation of a subsequent step. Nevertheless, analysis confirms that the reactions catalyzed by Rubisco, FBPase, SBPase and PRK are of greatest significance in controlling the flux through the RPP pathway (see Bassham and Buchanan, 1982 for review). As recalled from Fig. 21.1, Rubisco is a member of the carboxylation phase and FBPase, SBPase and PRK are members of the regeneration phase of the cycle.

In contrast, the enzymes involved in the reduction

of 3-PGA to triose phosphate (the reduction phase) together catalyze a freely reversible oxidation/reduction the direction of which, *in vivo*, is largely determined by the levels of ATP and ADP, NADPH and NADP$^+$. In the light, with high levels of ATP and NADPH, the reactions proceed in the direction of triose phosphate because of sustained production of 3-PGA and consumption of triose phosphate. In steady-state photosynthesis this balance provides for a coordination of the activity of other parts of the cycle. Any component tending to increase the activity of PRK, for example, will cause the consumption of ATP and production of ADP. The resulting deficiency of ATP in turn will slow the rate of 3-PGA reduction, leading to decreased synthesis of Rbu-5-P and bringing the cycle back into balance. It should be noted, however, that one enzyme of the reduction phase, NADP-G3PDH (equation [21.3]), is also regulated directly by light (see below).

Mechanisms of regulation

Regulation of ribulose 1,5-bisphosphate carboxylase/oxygenase

The capacity of Rubisco to carboxylate Rbu-1,5P$_2$ is determined by the concentration of substrates (Rbu-1,5-P$_2$, CO$_2$, and O$_2$), in addition to the amount and activity of enzyme. Under conditions of low CO$_2$ and high light, it is possible to show a direct correlation between Rubisco content and CO$_2$ fixation of spinach leaves (Seeman, 1986). During short-term changes in the rate of photosynthesis, however, modulation of the activity of the enzyme occurs (Perchorowicz *et al.*, 1981). Activation of the enzyme involves the formation of a complex with CO$_2$ and the subsequent addition of a divalent metal ion (Mg^{2+}, *in vivo*) to form the activated ternary complex (Miziorko and Lorimer, 1983). The equilibrium of this reaction is sensitive to pH, and low pH in the stroma would be expected to lead to deactivation. Upon illumination, protons move rapidly from the stromal compartment into the thylakoids, causing an increase in stromal pH from about 7.0 to 8.0. The efflux of H$^+$ is countered by an influx of other cations, possibly including Mg^{2+} and Ca^{2+} (Barber, 1976), and thus both the pH and cation concentration in the stroma upon illumination

have been proposed as being favorable for Rubisco activation in the presence of CO$_2$. Two other mechanisms for the activation/deactivation of Rubisco have been described more recently (see Chapter 20 for more details). Work with a mutant of *Arabidopsis* that requires a high CO$_2$ concentration for growth has led to the discovery of a mechanism whereby Rubisco is activated by light (Ogren *et al.*, 1986). The mechanism, which presently appears to be unrelated to other systems of enzyme regulation, involves a protein, Rubisco activase, that links light to enzyme activity. While details of the activation mechanism are not yet established, it currently appears that ATP, perhaps in concert with light-induced changes in the electrochemical potential of thylakoid membranes, facilitates the combination of CO$_2$ with an epsilon group of a specific lysine of Rubisco thereby yielding active enzyme (Portis, 1992; Wolosiuk *et al.*, 1993). Rubisco activase also appears to act through the removal of inhibitors from the active site of Rubisco. By these means, Rubisco activase is able both to enhance and sustain the activity of Rubisco during photosynthesis. Such a light-dependent mechanism of Rubisco regulation explains results obtained over the years by a number of different laboratories.

A second novel mechanism of Rubisco regulation involves a phosphorylated inhibitor of catalysis which can occupy the catalytic site of the enzyme. The discovery of this inhibitor in some but not all plant species followed the observation that Rubisco extracted from illuminated *Phaseolus* leaves was significantly more active than from darkened leaves, despite optimal *in vitro* activation of the enzyme with CO$_2$ and Mg^{2+}. Several studies have shown that phosphorylated compounds can be inhibitors of Rubisco *in vitro*. The results with *Phaseolus*, however, were the first to document the importance *in vivo* of a compound which is light modulated and present in sufficient amounts to reduce the activity of Rubisco to close to zero in the dark (Seeman, 1986). The inhibitor was identified as carboxyarabinitol 1-phosphate, an analog of the six carbon intermediate formed during catalysis (Berry *et al.*, 1986; Gutteridge *et al.*, 1986). Carboxyarabinitol 1-phosphate is representative of the inhibitors that, as noted above, may be removed by Rubisco activase in the light.

The ferredoxin/thioredoxin system

The ferredoxin/thioredoxin system, involving ferredoxin, ferredoxin–thioredoxin reductase (FTR) and a thioredoxin (Buchanan, 1980; 1991; 1992; Buchanan *et al.*, 1994; Cséke and Buchanan, 1986; Scheibe, 1991; Wolosiuk, *et al.*, 1993) is important among the more general mechanisms of light-mediated enzyme regulation. As described in Chapter 19, ferredoxin is an iron–sulfur protein that functions as the first soluble carrier of electrons from Photosystem I. As such, ferredoxin provides electrons for reduction of a number of components central to metabolism – for example, NADP, nitrite and, as described below, thioredoxin. Thioredoxins are proteins, typically with a molecular weight of 12 kD, that are widely, if not universally distributed in the animal, plant, and bacterial kingdoms. Thioredoxins undergo reversible reduction and oxidation through changes in a disulfide group ($S–S \rightarrow 2SH$) that is located in a strongly conserved sequence, Cys–Gly–Pro–Cys. In the ferredoxin/thioredoxin system, a thioredoxin (Trx) is reduced via the iron–sulfur protein FTR, by ferredoxin (Frx), which itself is reduced by the chlorophyll system of illuminated chloroplast thylakoid membranes (equations [21.15] and [21.16].

$$4Frx_{ox} + 2H_2O \xrightarrow{\text{Light}} 4Fr_{red} + O_2 + 4H^+ \quad [21.15]$$

$$2Frx_{red} + 2H^+ + Trx_{ox} \xrightarrow{\text{FTR}} 2Frx_{ox} + Trx_{red} \quad [21.16]$$

Two different thioredoxins, designated *f* and *m*, are a part of the ferredoxin/thioredoxin system in higher oxygenic photosynthetic organisms. (Cyanobacteria contain thioredoxin *m* and a novel thioredoxin, but not an *f* type.) In the reduced state, the two thioredoxins selectively activate enzymes of carbohydrate biosynthesis, including FBPase, SBPase and PRK. In addition, thioredoxins have been shown to activate NADP-G3PDH and deactivate glucose 6-phosphate dehydrogenase (G6PDH), a regulatory enzyme of the oxidative pentose phosphate pathway, a route of carbohydrate degradation alternate to glycolysis. The ferredoxin/thioredoxin system also functions in chloroplasts in regulating other enzymes such as NADP-malate dehydrogenase (NADP-MDH) (Scheibe, 1991; Buchanan *et al.*, 1994), the 'coupling

factor' (CF_1-ATPase) (Ort and Oxborough, 1992), and the glutamine synthetase of green algae (Florencio *et al.*, 1993). The type of thioredoxin interacting with each of these chloroplast enzymes is shown in Fig. 21.2. Cyanobacteria, C3, C4 and Crassulacean acid metabolism (CAM) plants have been shown to utilize the ferredoxin/thioredoxin system of enzyme regulation. In general, the enzymes identified in Fig. 21.2 play identical functions in the different types of photosynthetic organisms. NADP-MDH, however, is an exception in that its major function in C4 plants is to trap CO_2, whereas in C3 plants it is needed to transport excess reducing equivalents from chloroplasts to cytosol.

The ferredoxin/thioredoxin system functions by changing the sulfhydryl status of target enzymes. FBPase, which is a key regulatory enzyme of the RPP cycle, is activated by a net transfer of reducing equivalents (hydrogen) from reduced thioredoxin *f* to enzyme disulfide ($S–S$) groups, thereby yielding oxidized thioredoxin *f* and reduced (SH) enzyme. It is thought that deactivation of FBPase takes place in the dark through an oxidation by the $S–S$ group of oxidized thioredoxin. The thioredoxin has been

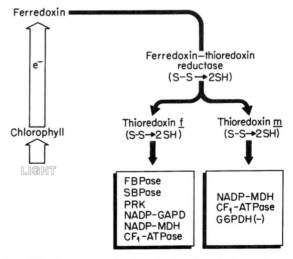

Fig. 21.2 Enzymes regulated by the ferredoxin/thioredoxin system. Role of thioredoxins in regulating phosphoglycerate kinase of C4 mesophyll cells is not indicated (cf. Cséke and Buchanan, 1986). Cyanobacteria appear to lack thioredoxin *f* and, in contrast to plants, utilize thioredoxin m for regulating enzymes.

previously oxidized by molecular oxygen. This light-dependent reduction mechanism also functions in the activation of NADP-MDH and PRK. The role of thioredoxin *m* in the activation of NADP-MDH has been studied intensively (Issakidis *et al.*, 1994). Recently, interesting models have been proposed to explain how thioredoxin modulates this activity as well as that of its other target enzymes of chloroplasts (Li *et al.*, 1994).

In summary, current evidence is consistent with the view that the ferredoxin/thioredoxin system functions in photosynthetically diverse types of organisms as a master switch to restrict the activity of degradatory enzymes and activate biosynthetic enzymes in the light. In this way, the organisms control the processes identified in Fig. 21.3. It is significant to note that, aside from NADP-G3PDH, RPP cycle enzymes controlled by the ferredoxin/thioredoxin system (FBPase, SBPase, and PRK) function in the regenerative phase of the reductive pentose phosphate cycle needed to sustain its continued operation, i.e. to regenerate the carbon dioxide acceptor, Rbu-1,5-P_2, from newly formed 3-PGA. It seems likely that one of these thioredoxin-linked enzymes limits the regeneration of Rbu-1,5-P_2.

Coordinate regulation of photosynthetic enzymes

Biochemical processes are generally regulated not by one, but by several interacting systems of regulation. From early work, it was concluded that the ferredoxin/thioredoxin system acts jointly with other light-actuated systems in achieving a particular regulatory effect, e.g. light-induced shifts in concentration of metabolite effectors and pH (Buchanan, 1980). Since those early studies, results from a number of laboratories support such a coordinate function of the different regulatory systems. Noteworthy among the metabolite effector studies are the demonstration of the inhibition of thioredoxin-linked NADP-MDH activation by $NADP^+$, the inhibition of PRK by compounds such as 6-phosphogluconate, and the enhancement of thioredoxin-linked FBPase and SBPase activation by substrate (sugar bisphosphate) and divalent cations (Ca^{2+}, Mg^{2+}). In short, it appears that the ferredoxin/thioredoxin system functions jointly with light-dependent mechanisms promoting shifts in ions and metabolites in the regulation of a number of chloroplast enzymes.

Considerable debate has centered on the question of whether it is the activity of Rubisco or the rate of regeneration of Rbu-1,5-P_2 (governed by the regulatory steps of the rest of the cycle, FBPase, SBPase, and PRK and by the supply of ATP and NADPH from the light reactions) that primarily sets the rate of the RPP cycle and CO_2 fixation *in vivo*. Recent results suggest that during rapid changes from high to low light, the rate of photosynthesis at subsaturating light intensity in certain plants is determined by the rate of Rbu-1,5-P_2 regeneration. Many workers believe that under steady-state conditions Rubisco activity is the most important limitation on the rate of photosynthesis, even under saturating CO_2 (Dietz and Heber, 1984a,b). Recent experiments with transgenic plants having decreased Rubisco activity tend to bear this out, although the results depend on experimental conditions and the situation appears not to be simple (Stitt and Schulze, 1994).

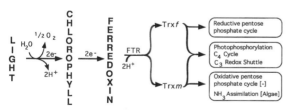

Fig. 21.3 Role of the ferredoxin/thioredoxin system in the regulation of enzymes of photosynthesis and associated processes. While the reductive and oxidative pentose phosphate cycles appear to be regulated in all oxygenic photosynthetic organisms by the ferredoxin/thioredoxin system (see Fig. 21.2), the regulation of the other processes listed above differs among the major groups of organisms – e.g. cyanobacteria, eukaryotic algae, higher plants.

Compartmentation and triose phosphate export

Because the RPP cycle supplies fixed carbon for export, the biochemical regulation of the cycle appears insufficient to prevent the intermediates from being consumed by other metabolic processes. Thus, in addition to the biochemical controls discussed above, there is also metabolic compartmentation within the cell.

The chloroplast is encircled by a double membrane called the envelope. Of the two membranes, the inner is practically impermeable to hydrophilic compounds, such as P_i, phosphate esters, dicarboxylates, glucose and sucrose. Transport of certain of these metabolites is accomplished by carrier proteins, specific for groups of compounds. Individual carriers have been shown to facilitate the transport of P_i and phosphate esters, dicarboxylates, ATP and ADP, and glucose.

The carrier protein facilitating P_i and phosphate ester transport is of particular interest in leaves in connection with carbon processing, i.e. the synthesis, transport, and degradation of carbohydrate, all of which occur in the cytosol (Cséke et al., 1984; Cséke and Buchanan, 1986). This metabolite carrier, called the phosphate translocator, is a polypeptide with a molecular weight of 29 000 and is a major component of the inner envelope membrane (Flügge and Heldt, 1984). The phosphate translocator mediates the counter transport of 3-PGA, DHAP and P_i. The rate of P_i transport by itself is three orders of magnitude lower than with simultaneous DHAP or 3-PGA countertransport. Consequently, operation of the phosphate translocator keeps the total amount of esterified phosphate and P_i constant inside the chloroplast.

During photosynthesis, chloroplasts convert CO_2, water and P_i to triose phosphates that are exported to the cytosol (see equation [21.14] above). As phosphate is a substrate of this process, the continued operation of the RPP cycle is dependent on the utilization of triose phosphate for the synthesis of starch (in the chloroplast) and sucrose (in the cytosol). These synthesis processes release P_i, preventing the level of free P_i in the cell from falling to a concentration at which photosynthesis may be limited by its availability. Such a limitation of photosynthesis is observed when mannose (which sequesters cytosolic P_i as mannose phosphate) is fed to a leaf (Robinson and Walker, 1981), and is suggested by the increase in CO_2 fixation detected on feeding P_i via the transpiration stream to a cut leaf (Sivak and Walker, 1986). It has long been known that isolated chloroplasts require a continuous supply of P_i in order to sustain photosynthesis.

A further ramification of the translocator-mediated exchange of exported triose phosphate and imported P_i pertains to starch synthesis. When cytosolic metabolism and P_i availability are limited, thereby leading to a high 3-PGA/P_i ratio in the chloroplast, starch synthesis will be stimulated. This occurs because ADP-glucose pyrophosphorylase, the major regulatory enzyme in starch synthesis, is strongly activated by 3-PGA and inhibited by P_i (Preiss, 1982; Preiss et al., 1991). Starch synthesis from triose phosphate will release P_i, relieving to some extent the P_i limitation of CO_2 fixation described above.

Coordination of CO_2 fixation and sucrose synthesis

The requirement of chloroplast photosynthesis for P_i and the release of P_i by sucrose synthesis in the cytosol means that these two processes must be closely coordinated. This coordination is essential if photosynthesis is to continue, since a large fraction (five-sixths) of the triose phosphate produced in chloroplasts must be used to regenerate Rbu-1,5-P_2 to allow the continued function of the RPP cycle. Part of this coordination, as explained above, lies in the characteristics of the triose phosphate translocator and the supply of cytosolic P_i to the chloroplast (Robinson and Walker, 1981). Results obtained in the last fifteen years have led to the identification of a regulatory metabolite that also serves this function. Fructose 2,6-bisphosphate (Fru-2,6-P_2) coordinates the metabolism of sucrose, starch and CO_2 fixation and, in so doing, links metabolic processes of the chloroplast to those of the cytosol.

Fructose 2,6-bisphosphate

Fru-2,6-P_2 occurs in plant tissues and exerts its effects on metabolism through the modulation of cytosolic FBPase and pyrophosphate, fructose 6-phosphate, 1-phosphotransferase (PFP) (see Chapters 7 and 25). PFP catalyzes the reversible phosphorylation of fructose 6-phosphate (Fru-6-P) by pyrophosphate and is believed to function in the regulation of glycolysis and gluconeogenesis (sucrose synthesis) in plant tissues. The essentiality of PFP has, however, recently been called into question by experiments showing that a 70 to 90% inactivation of its gene results in no change in visible phenotype and only little change in metabolism (Hajirezaei *et al.*, 1994). A series of biochemical analyses revealed some ways by which the transgenic plants compensated for the loss in PFP activity. These experiments represent a prime example of the adaptability of higher plants – an adaptability that seems to align them metabolically with bacteria (prokaryotes).

Studies with spinach leaves revealed that: Fru-2,6-P_2 is present in the cytosolic fraction of photosynthetic (leaf parenchyma) cells; a PFP strongly activated by Fru-2,6-P_2 is present in the cytosol; and Fru-2,6-P_2 strongly inhibits cytosolic FBPase, an important regulatory enzyme of sucrose synthesis. Fru-2,6-P_2 can affect sucrose metabolism by inhibiting cytosolic FBPase and by activating PFP, an enzyme that, because of the reversibility of the reaction it catalyzes, can potentially function in both sucrose synthesis and breakdown. Fru-2,6-P_2 has little effect on phosphofructokinase (PFK) in plants, in contrast to its effect in animal cells. Instead, the cytosolic PFK of plants is regulated by changes in pivotal metabolites that alter enzyme activity either directly through activation or inhibition or indirectly through association or disassociation (Wong *et al.*, 1987).

In the initial studies on plants (for review see Cséke *et al.*, 1984) a substrate-specific fructose 6-phosphate, 2-kinase (Fru-6-P,2K) was identified in leaves, specifically in the cytosol fraction. Experiments revealed that the synthesis of Fru-2,6-P_2 by leaf Fru-6-P,2K was regulated by metabolite effectors: P_i and Fru-6-P served as activators and 3-PGA and DHAP as inhibitors. Also, an enzyme was partially purified from spinach leaves that selectively hydrolyzed Fru-2,6-P_2 to Fru-6-P and P_i (Macdonald *et al.*, 1987). The enzyme, designated fructose 2,6-bisphosphatase (Fru-2,6-P_2ase), was strongly inhibited by its products, Fru-6-P and P_i. Thus the regulation of Fru-2,6-P_2ase by metabolites was found to be opposite to the regulation of Fru-6-P,2K which, as noted above, is activated by the same metabolites (Fig. 21.4).

It should be noted that Fru-6-P,2K and Fru-2,6-P_2ase of animal tissues are regulated by phosphorylation via a cAMP-dependent protein kinase that, in turn, is regulated hormonally. So far, there is no conclusive evidence that Fru-6-P,2K and

Metabolite	Fru-6-P,2K	Fru-2,6-P_2ase
Pi	Activator	Inhibitor
Fru-6-P	Activator	Inhibitor
DHAP	Inhibitor	No effect
3-PGA	Inhibitor	No effect
PPi	Inhibitor	No effect
Mg^{2+}	Cofactor	Inhibitor

Fig. 21.4 Principal metabolites regulating Fru-6-P,2K and Fru-2,6-P_2ase.

Fru-2,6-P_2ase in plants are regulated by phosphorylation physiologically. There is, however, evidence that plant Fru-6-P,2K is regulated covalently in addition to its regulation by metabolites (Stitt, 1987), perhaps through an ATP-dependent modification (Huber, 1990).

As discussed above, the P_i released in sucrose synthesis is recycled to the chloroplast via the phosphate translocator, in strict counter-exchange for tiose phosphate (Flügge and Heldt, 1984). 3-Phosphoglycerate can also be transported by this same carrier, though its export from the chloroplast in the light is restricted by a pH-dependent change in the charge on the molecule.

It is thus obvious that the metabolites modulating Fru-6-P,2K and Fru-2,6-P_2ase occupy strategic positions in the pathway of sucrose synthesis in leaves. Extensive export of triose phosphates by chloroplasts into the cytosolic C3 pool would lower the Fru-2,6-P_2 concentration (by inhibiting Fru-6-P,2K) and thereby promote the use of photosynthate for sucrose synthesis by relieving the Fru-2,6-P_2-linked inhibition of cytosolic FBPase. On the other hand, elevated levels of P_i (e.g. in the dark) or Fru-6-P (e.g. as sucrose accumulated in the leaf) would tend to raise the Fru-2,6-P_2 concentration, and thus restrict sucrose synthesis or favor sucrose degradation. The role of chloroplasts in controlling the content of Fru-2,6-P_2 in the cytosol via export and import of central metabolites is illustrated diagrammatically in Fig. 21.5.

Changes in DHAP concentration and the accompanying alteration of the Fru-2,6-P_2 concentration thus provide a mechanism to coordinate sucrose synthesis in the cytosol with the rate of carbon dioxide fixation in chloroplasts. This coordination is necessary to prevent the inhibition of

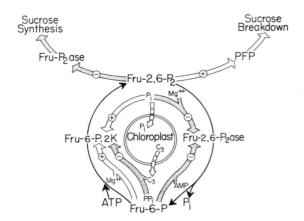

Fig. 21.5 Role of chloroplasts and effector metabolites in Fru-2,6-P$_2$-linked control of cytosolic sucrose transformations in spinach leaves. Fru-P$_2$ase is equivalent to FBPase. Regulation of Fru-6-P,2K and Fru-2,6-P$_2$ase is indicated by '+' for activation and '−' for inhibition.

photosynthesis that would result from either an excessive drain of metabolites from the chloroplast or from an inadequate release of P$_i$ in the cytosol to sustain photophosphorylation in the chloroplast.

Other end products of CO$_2$ assimilation

Ultimately, all the fixed carbon that makes up plant cells is derived from CO$_2$ that is reduced via the RPP cycle. As stated above, the major end products of CO$_2$ fixation in mature leaves from most plant species are starch and sucrose. A number of other compounds, however, are derived directly from intermediates of the cycle so that their synthesis may be light dependent. These compounds may account for a significant percentage of the fixed carbon in growing and developing tissues.

In unicellular green algae such as *Chlorella* (Calvin and Bassham, 1962), as much as one-third of the carbon fixed during a one hour exposure to ^{14}CO$_2$ is found in amino acids, and not in sucrose as is the case for higher plants (which produce sucrose as an easily transported carbohydrate). In the algae, the major amino acids labeled are glycine and serine. In addition, the ^{14}C label appears in alanine (from pyruvate) and metabolites produced by the tricarboxylic acid cycle: aspartate, asparagine (from

oxaolacetate) and glutamate, glutamine (from 2-oxoglutarate). These compounds originate from the cytosolic metabolism of triose phosphates exported from the chloroplast. Though quantitatively insignificant in most C3 plants (with the exception of photorespiratory production of glycine and serine), such synthesis is clearly of great importance and is probably tightly controlled. There is evidence of feedback inhibition of the pathways of amino acid production.

In addition to the cytosolic metabolism described above, a number of products may be produced wholly within chloroplasts. These include the aromatic amino acids (tryptophan, tyrosine and phenylalanine) and plastoquinone derived from the shikimic acid pathway, the pyruvate-derived amino acids, fatty acids and terpenes produced from acetyl-CoA and nucleic acids (Schulze-Siebert *et al.*, 1984). As with cytosolic metabolism, such products are quantitatively insignificant in comparison to export to the cytosol and starch production in the plastid, but are nonetheless of vital importance to the plant.

The existence in chloroplasts of complete pathways leading from the RPP cycle to the above mentioned compounds continues to be a matter of debate and their synthesis at least in some plant species may require the cooperation of the cytosol and plastids. Pea chloroplasts, for example, appear not to contain phosphoglyceromutase activity which would be required for the production of PEP and pyruvate from PGA generated photosynthetically (ap Rees, 1985). Spinach chloroplasts, which apparently do contain a complete pathway leading from 3-PGA to PEP, pyruvate and acetyl-CoA, presumably face a problem of control so as to prevent metabolites being drained from the cycle during photosynthesis. Little is known of the control of this or other biosynthetic pathways in the chloroplast.

Concluding remarks

The mechanism of carbon dioxide assimilation by the RPP cycle has been known for four decades. During this interval, it has been established that light functions not only to produce ATP and NADPH$_2$ to drive the cycle, but also to regulate selected enzymes. In oxygen-evolving systems (chloroplasts and cyanobacteria), light absorbed by chlorophyll is converted to several different regulatory signals – changes in pH, cations, metabolite effectors, and sulfhydryl groups – that collectively interact to 'inform' selected enzymes that the light is on or off and that their activities should be altered accordingly. In the case of sulfhydryl changes, the light signal is carried from chlorophyll containing thylakoid membranes via ferredoxin to thioredoxins, which, through redox changes in their own sulfhydryl groups, bring about reversible changes in the sulfhydryl status of target enzymes. In this way, light alters key activities and directs major biosynthetic and degradatory pathways in the appropriate direction. With certain enzymes, the light-produced alkalization of the chloroplast stroma and increase in the concentration of cations and selected metabolite effectors enhance the sulfhydryl changes. By linking these regulatory changes to light, the cell is in command of its biosynthetic and degradatory capabilities at all times and can direct available resources to increase growth and survival under a wide range of environmental conditions. It is significant that photosynthetic bacteria (anaerobic photosynthetic organisms that lack the ability to evolve oxygen) seemingly do not regulate metabolic processes in this manner.

In higher plants, which utilize photosynthetically fixed carbon to form transportable sugars such as sucrose, the light-directed systems of enzyme regulation of chloroplasts interact with a metabolite-directed system of the cytosol (Fig. 21.6). Here, $Fru-2,6-P_2$ plays a key role. In leaves, $Fru-2,6-P_2$ acts as a regulatory link between chloroplasts and the cytosol, thus allowing metabolic communication between these compartments, and signaling changes in environmental conditions. In this way, carbon processing, i.e. the synthesis, degradation and transport of carbohydrate in the cytosol, can be adjusted in accordance with needs of the plant. In performing its function, $Fru-2,6-P_2$ acts at several levels, i.e. sucrose synthesis (FBPase), sucrose degradation (PFP regulation), and the related process of carbon partitioning (accumulation of photosynthetically fixed carbon as sucrose versus starch).

The evidence at hand is thus in accord with the view that the $Fru-2,6-P_2$ system connects cytosolic

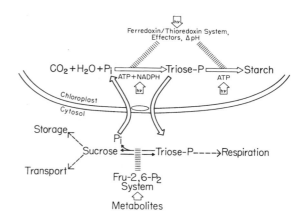

Fig. 21.6 Relation of carbon processing in the cytosol to photosynthetic CO_2 assimilation in chloroplasts. Note the dual function of light in supplying ATP and $NADPH_2$ as well as in enzyme regulation.

carbohydrate metabolism with the light-directed regulatory mechanisms of chloroplasts, and with other regulatory signals significantly altering cytosolic metabolite status. The role of $Fru-2,6-P_2$ as an environmental sensor enables plants to make effective use of available energy for processes taking place either in leaves or in distal sink tissues.

Acknowledgements

Work from the authors' laboratory was supported by grants from the National Science Foundation, the U.S. Department of Agriculture, National Aeronautics and Space Administration and private gifts.

References

ap Rees, T. (1980). Integration of pathways of synthesis and degradation of hexose phosphates. In *The Biochemistry of Plants*, Vol. 3, ed. J. Preiss, Academic Press, New York, pp. 1–42.

ap Rees, T. (1985). The organization of glycolysis and the oxidative pentose phosphate pathway in plants. In *Encyclopedia of Plant Physiology*, Vol. 18, eds R. Douce and D. Day, Springer-Verlag, Berlin, pp. 391–417.

Barber, J. (1976). Cation control in photosynthesis. *Trends Biochem. Sci.* **1**, 33–6.

Bassham, J. A. and Buchanan, B. B. (1982). Carbon dioxide fixation pathways in plants and bacteria. In *Photosynthesis*, ed. Govindjee, Vol. II, Academic Press, New York, pp. 141–89.

Bassham, J. A. and Krause, G. H. (1969). Free energy changes and metabolic regulation in steady-state photosynthetic carbon reduction. *Biochim. Biophys. Acta* **189**, 207–21.

Berry, J. A., Lorimer, G. H., Pierce, J., Seeman, J. R., Meek, J. and Freas, S. (1986). Isolation, identification and synthesis of 2-carboxyarabinitol 1-phosphate, a diurnal regulator of ribulose-bisphosphate activity. *Proc. Natl. Acad. Sci. USA* **84**, 734–8.

Buchanan, B. B. (1980). Role of light in the regulation of chloroplast enzymes. *Ann. Rev. Plant Physiol.* **31**, 341–74.

Buchanan, B. B. (1991). Regulation of CO_2 assimilation in oxygenic photosynthesis: the ferredoxin/thioredoxin system. *Arch. Biochem. Biophys.* **288**, 1–9.

Buchanan, B. B. (1992). Carbon dioxide assimilation in oxygenic and anoxygenic photosynthesis. *Photosyn. Res.* **33**, 147–62.

Buchanan, B. B. and Arnon, D. I. (1990). A reverse KREBS cycle in photosynthesis: consensus at last. *Photosyn. Res.* **24**, 47–53.

Buchanan, B. B., Schürmann, P. and Jacquot, J.-P. (1994). Thioredoxin and metabolic regulation. In *Seminars in Cell Biology*, ed. A. R. Grossman, Academic Press, London, **5**, 285–293.

Calvin, M. and Bassham, J. A. (1962). *The Photosynthesis of Carbon Compounds*, Benjamin, New York.

Cséke, C. and Buchanan, B. B. (1986). Regulation of the formation and utilization of photosynthate in leaves. *Biochim. Biophys. Acta* **853**, 43–64.

Cséke, C., Balogh, A., Wong, J. H., Buchanan, B. B., Stitt, M., Herzog, B. and Heldt, H. W. (1984). Fructose-2,-6-bisphosphate: a regulator of carbon processing in leaves. *Trends Biochem. Sci.* **9**, 533–5.

Dietz, K. J. and Heber, U. (1984a). Rate-limiting factors in leaf photosynthesis 1. Carbon fluxes in the Calvin cycle. *Biochim. Biophys. Acta* **767**, 432–43.

Dietz, K. J. and Heber, U. (1984b). Rate-limiting factors in leaf photosynthesis 2. Electron transport. *Biochim. Biophys. Acta* **767**, 444–50.

Eisenreich, W., Strauss, G., Werz, U., Fuchs, G. and Bacher, A. (1993). Retrobiosynthetic analysis of carbon fixation in the phototrophic eubacterium *Chloroflexus aurantiacus*. *Eur. J. Biochem.* **215**, 619–32.

Florencio, F. J., Gadal, P. and Buchanan, B. B. (1993). Thioredoxin-linked activation of the chloroplast and cytosolic forms of *Chlamydomonas reinhardtii* glutamine synthetase. *Plant Physiol. Biochem.* **31**, 649–55.

Flügge, U. I. and Heldt, H. W. (1984). The phosphate-triose phosphate-phosphoglycerate translocator of the chloroplast. *Trends Biochem. Sci.* **9**, 530–33.

Gutteridge, S., Parry, M. A. R., Burton, S., Keys, A. J.,

Mudd, A., Feeny, J., Servaites, J. C. and Pierce, J. (1986). A nocturnal inhibitor of carboxylation in leaves. *Nature* **324**, 274–6.

Hajirezaei, M., Sonnewalt, U., Viola, R., Carlisle, S., Dennis, D. and Stitt, M. (1994). Transgenic potato plants with strongly decreased expression of pyrophosphate-fructose-6-phosphate phosphotransferase show no visible phenotype and only minor changes in metabolic fluxes in their tubers. *Planta* **192**, 16–30.

Huber, S. C. (1990). On the nature of fructose-2,6-P_2 metabolizing enzymes in plants. In *Fructose-2,6-Bisphosphate*, ed. S. J. Pilkis, CRC Press, Boca Raton, FL, pp. 211–18.

Issakidis, E., Saarinen, M., Decottignies, P., Jacquot, J. P., Cretin, C., Gadal, P. and Miginiac-Maslow, M. (1994). Identification and characterization of the second regulatory disulfide bridge of recombinant sorghum leaf NADP-malate dehydrogenase. *J. Biol. Chem.* **269**, 3511–17.

Li, D., Steven, F. J., Schiffer, M. and Anderson, L. E. (1994). Mechanism of light modulation: Identification of potential redox-sensitive cysteines distal to catalytic site in light-activated chloroplast enzymes. *Biophys. J.* **67**, 1–7.

Macdonald, F. D., Cséke, C., Chou, Q. and Buchanan, B. B. (1987). Activities synthesizing and degrading fructose-2,6-bisphosphate in spinach leaves reside on different proteins. *Proc. Natl. Acad. Sci. USA* **84**, 2742–6.

Miziorko, H. M. and Lorimer, G. H. (1983). Ribulose-1,5-bisphosphate carboxylase–oxygenase. *Ann. Rev. Biochem.* **52**, 507–35.

Mott, K. A., Jensen, R. G. and Berry, J. A. (1986). Limitation of photosynthesis by RuBP regeneration rate. In *Biological Control of Photosynthesis*, eds R. Marcelle, H. Clijsters and M. Van Pouke, Martinus Nijhoff, Dordrecht, pp. 33–43.

Ogren, W., Salvucci, M. and Portis, A. (1986). The regulation of Rubisco activity. *Phil. Trans. R. Soc. London B* **313**, 337–46.

Ort, D. R. and Oxborough, K. (1992). *In situ* regulation of chloroplast coupling factor. *Annu. Rev. Plant Physiol.* **43**, 269–91.

Perchorowicz, J. T., Raynes, D. A. and Jensen, R. G. (1981). Light limitation of photosynthesis and activation of ribulose bisphosphate carboxylase in wheat seedlings. *Proc. Natl. Acad. Sci. USA* **78**, 2985–9.

Portis, A. R. (1992). Regulation of ribulose 1,5-bisphosphate carboxylase oxygenase activity. *Ann. Rev. Plant Physiol.* **43**, 415–37.

Preiss, J. (1982). Regulation of biosynthesis and degradation of starch. *Ann. Rev. Plant Physiol.* **33**, 431–54.

Preiss, J., Ball, K., Smith-White, B., Iglesias, A., Kakefuda, G. and Li, L. (1991). Starch biosynthesis and its regulation. *Biochem. Soc. Transactions* **19**, 539–47.

Robinson, S. P. and Walker, D. A. (1981). Photosynthetic carbon reduction cycle. In *The Biochemistry of Plants*, Vol. 8, Academic Press, New York, pp. 194–236.

Scheibe, R. (1991). Redox-modulation of chloroplast enzymes. A common principle for individual control. *Plant Physiol.* **96**, 1–3.

Schulze-Siebert, D., Heinke, D., Scharf, H. and Schultz, G. (1984). Pyruvate-derived amino acids in spinach chloroplasts. *Plant Physiol.* **76**, 465–71.

Seeman, F. R. (1986). Mechanisms for the regulation of CO_2 fixation by ribulose-1,5-bisphosphate carboxylase. In *Biological Control of Photosynthesis*, eds R. Marcelle, H. Clijsters and M. Van Poucke, Martinus Nijhoff, Dordrecht, pp. 71–82.

Sivak, M. N. and Walker, D. A. (1986). Summing up: measuring photosynthesis *in vivo*. In *Biological Control of Photosynthesis*, eds R. Marcelle, H. Clijsters and M. Van Poucke, Martinus Nijhoff, Dordrecht, pp. 1–31.

Stitt, M. (1987). Fructose-2,6-bisphosphate and plant carbohydrate metabolism. *Plant Physiol.* **84**, 201–4.

Stitt, M. and Schulze, D. (1994). Does rubisco control the rate of photosynthesis and plant growth – an exercise in molecular ecophysiology. *Plant Cell. Environ.* **17**, 465–87.

Wolosiuk, R. A., Ballicora, M. A. and Hagelin, K. (1993). The reductive pentose phosphate cycle for photosynthetic CO_2 assimilation: enzyme modulation. *FASEB J.* **7**, 622–37.

Wong, J., Yee, B. C. and Buchanan, B. B. (1987). A novel type of phosphofructokinase from plants. *Biol. Chem.* **262**, 3185–91.

22 Photorespiration and CO₂-concentrating mechanisms

David T. Canvin and Christophe Salon

Introduction

Photorespiration, which is light dependent CO_2 evolution and O_2 uptake, occurs concurrently with photosynthesis in green leaves of C3 plants. Due to the existence of photorespiration, photosynthesis is inhibited by oxygen and when the release of CO_2 by photorespiration equals CO_2 fixation by photosynthesis, a concentration of CO_2 is reached which is called the CO_2 compensation point. When the light is extinguished, continued CO_2 production from photorespiration results in a post-illumination burst (PIB). Warburg observed the inhibition of photosynthesis by oxygen in 1920, the post-illumination burst was described in 1955, and the competition between O_2 and CO_2, as expressed in the direct dependence of the compensation point on O_2 concentration, was observed in 1966.

During the 1950s and 1960s, the extensive synthesis of glycolate by leaves in high O_2 concentrations or low CO_2 concentrations was demonstrated and the enzymatic pathways for CO_2 production, now called the photosynthetic carbon oxidation cycle (PCO), was elucidated (Fig. 22.1). The origin of the glycolate was established and the gas exchange results were reconciled with the biochemistry of glycolate metabolism by the discovery (Bowes *et al.*, 1971) that the CO_2 fixing enzyme, ribulose 1,5-bisphosphate carboxylase (Rubisco) also displays oxygenase activity. CO_2 and O_2 are competitive substrates for the enzyme (see Chapter 20). The addition of CO_2 to ribulose 1,5-bisphosphate (RuBP) by the enzyme yields two molecules of 3-phosphoglyceric acid (3-PGA), whereas the addition of O_2 to RuBP yields a molecule of 3-PGA and a molecule of phosphoglycolate. Glycolate is formed by the removal

of the phosphate. Subsequent studies on the incorporation of oxygen into glycolate and other PCO cycle intermediates and the isolation of mutants deficient in PCO cycle enzymes definitively established the mechanics of CO_2 production by the PCO cycle and the release and reassimilation of NH_3 in the photorespiratory nitrogen cycle.

It was also apparent, however, that certain plants, such as CAM plants and C4 plants, had developed mechanisms to avoid oxygenase activity or the effect of oxygen on photosynthesis. In these plants, elaborate biochemical and anatomical developments have provided a means for elevating CO_2 concentration in the vicinity of Rubisco, thereby decreasing oxygenase activity. Microalgae and cyanobacteria have adopted a more direct approach by developing systems for the active transport of both CO_2 and HCO_3^- which also provide higher concentrations of CO_2 in the vicinity of the enzyme. These 'CO₂-concentrating' mechanisms seem to be nature's primary way of decreasing photorespiration, which appears to not have an essential function in the life of the plant.

Inorganic carbon chemistry

Before proceeding further some understanding of inorganic carbon chemistry may be helpful. The main source of inorganic carbon for plants is CO_2 which is present at 0.036% (360 parts per million or $360\,\mu l\,l^{-1}$) in the air. CO_2 can dissolve in water and the maximum concentration (which is independent of pH) in equilibrium with the CO_2 in air is about $12\,\mu M$. When CO_2 dissolves in water, however, the carbon atom of CO_2 is susceptible to nucleophilic

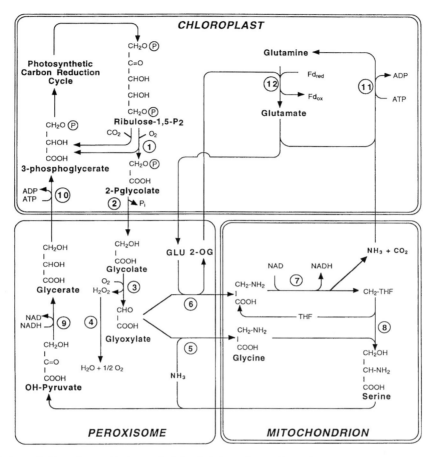

Fig. 22.1 The photosynthetic carbon oxidation cycle. The circled numbers refer to the enzyme or reaction: (1) ribulose 1,5-bisphosphate carboxylase/oxygenase; (2) phosphoglycolate phosphatase; (3) glycolate oxidase; (4) catalase; (5) serine : glyoxylate aminotransferase; (6) glutamate : glyoxylate aminotransferase; (7) glycine decarboxylase; (8) serine hydroxymethyltransferase; (9) hydroxypyruvate reductase; (10) glycerate kinase; (11) glutamine synthetase; (12) glutamate synthase.

attack by either H$_2$O, yielding in a slow reaction H$_2$CO$_3$ (carbonic acid) which in turn dissociates instantaneously to HCO$_3^-$ and H$^+$, or by OH$^-$, yielding HCO$_3^-$. HCO$_3^-$ can further dissociate to CO$_3^{2-}$ + H$^+$. As a consequence dissolved inorganic carbon is composed of four different forms (H$_2$CO$_3$, CO$_2$, HCO$_3^-$ and CO$_3^{2-}$) in varying concentrations according to the pH. At low pH, CO$_2$ predominates while at pH higher than 6.3 HCO$_3^-$ is the main species and, above pH 10.3, CO$_3^{2-}$ predominates. Above pH 6.0, H$_2$CO$_3$ is negligible. At pH 7.0 dissolved inorganic carbon in equilibrium with the CO$_2$ in air

will consist of 12 μM CO$_2$ and 75 μM HCO$_3^-$ whereas at pH 8.0 it will consist of 12 μM CO$_2$, 734 μM HCO$_3^-$ and 4 μM CO$_3^{2-}$. Whatever the pH, the rate of interconversion between CO$_2$ and HCO$_3^-$ is relatively slow. As all carboxylating enzymes use only one form of inorganic carbon, the supply of this form is assured in those enzyme locations by the presence of carbonic anhydrase (CA), an enzyme that catalyzes the interconversion of the various forms and keeps all forms available in the concentrations that will be determined by the prevailing pH.

In all terrestrial plants CO$_2$ diffuses through the

stomata or pores to the intercellular spaces. As the stomata constitute a resistance, the CO_2 concentration in the intercellular spaces ($270 \, \mu l \, l^{-1}$) is less than that in air. The intercellular CO_2 will dissolve in the cytosol (mostly water) and the maximum concentration will be about $8.5 \, \mu M$ in the cell.

In C3 plants this source of CO_2 is close to the affinity (K_m (CO_2) $\approx 10 \, \mu M$) of Rubisco for its substrate, CO_2. In the alkaline stroma (pH 8.0) of the chloroplast HCO_3^- will predominate and CA is present to rapidly convert the HCO_3^- to CO_2.

In C4 plants CA is confined to the cytosol of mesophyll cells together with PEP carboxylase whose substrate is HCO_3^-. CA operates here by increasing the supply of HCO_3^- for its subsequent fixation in C4 acids which then move to the bundle sheath cells where they are rapidly decarboxylated. The absence of CA and the slow rate of CO_2 hydration to HCO_3^- ensure a sufficient supply of CO_2 to Rubisco in the bundle sheath cells.

Most aquatic plants fix CO_2 via Rubisco. CO_2 diffusion in water is much slower (10^4 times) than in air and aquatic plants may take up other forms of inorganic carbon. Such uptake may be passive, by diffusion, or active, requiring the expenditure of metabolic energy. In some cases, the plants excrete carbonic anhydrase into the cell wall region (external CA) which ensures that an equilibrium concentration of all forms of inorganic carbon will be available. Carbonic anhydrase will also be present in the chloroplasts and in the cells of cyanobacteria (internal CA) to facilitate the fixation of the internal inorganic carbon.

The photosynthetic carbon oxidation cycle

The photosynthetic carbon oxidation (PCO) cycle involves the coordinated activity of reactions in three cellular organelles, the chloroplast, the peroxisome and the mitochondrion, as well as transport of the compounds through the cytoplasm between these organelles (Fig. 22.1). The oxygenation of RuBP (Reaction 1, Fig. 22.1) produces phosphoglycolate in the chloroplast. For every four carbons (two molecules) of phosphoglycolate metabolized in the PCO cycle, one carbon is released as CO_2 (Reaction

7, Fig. 22.1) and the other three carbons are returned to the photosynthetic carbon reduction (PCR) cycle (Reaction 10, Fig. 22.1). Hence, two molecules of O_2 are consumed in the oxygenase reaction for every CO_2 released, or the ratio between CO_2 evolution and oxygenation in photorespiration is 0.5. If total O_2 uptake, including that consumed by the oxidation of two glycolate molecules (Reaction 3, Fig. 22.1) is considered then three O_2 will be taken up for every CO_2 released.

This complex cycle, which can only function completely in intact cells, has been elucidated and formulated through a variety of approaches, including isolation and characterization of the various enzymes and isotope experiments with leaves (see Lorimer and Andrews, 1981) and algae (De Veau and Burris, 1989a). About 1950, glycolate was identified as an early product of photosynthesis and shown to be uniformly labeled during photosynthesis in $^{14}CO_2$. Uniform labeling is consistent with label distribution in the intermediates of the PCR cycle and the origin of glycolate from carbons 1 and 2 of RuBP through the oxygenation reaction. By using specifically labeled intermediates (e.g. 1-^{14}C-glycolate, 2-^{14}C-glycolate, 2-^{14}C-glycine, etc.) supplied to leaves in the light, it was shown that glycolate was directly converted to glycine. Carbon 1 of glycine was released as CO_2 and serine was derived from one complete molecule of glycine plus carbon 2 of another glycine molecule. When oxygenation occurs in the presence of $^{18}O_2$, one atom of ^{18}O ends up as water but one atom of ^{18}O is incorporated into the carboxyl group of phosphoglycolate and is retained in the glycine, serine and glycerate produced in subsequent reactions. By using intact isolated chloroplasts, the measured ^{18}O enrichment in glycolate was over 90% of the ^{18}O content supplied, showing that all, or essentially all, the glycolate was derived from the oxygenase reaction of Rubisco.

Glycolate production

The initial reaction of photorespiration and the PCO cycle is the oxygenase reaction of Rubisco which produces 2-phosphoglycolate and 3-phosphoglycerate with O_2 as the substrate (Reaction 1, Fig. 22.1). whereas two molecules of 3-phosphoglycerate are produced with CO_2 as the substrate. CO_2 and O_2 are competitive substrates for the enzyme and the relative amounts of each that react will depend upon the relative concentrations of each gas at the active site of the enzyme and the kinetic properties of the enzyme. A more detailed treatment of Rubisco can be found in Chapter 20 and more will be said about the rates of carboxylation and oxygenation later.

Phosphoglycolate is converted to glycolate and inorganic phosphate in the chloroplast by phosphoglycolate phosphatase (Reaction 2, Fig. 22.1). The enzyme has been found in a number of C3 and C4 plants and algae and has been purified from several sources (see Husic *et al.*, 1987). The molecular weight of the enzyme from spinach is 93 kD; it has a broad pH optimum (6.0–8.5) and requires a divalent cation and monovalent anion for activity. The enzyme has a high degree of specificity for phosphoglycolate and does not hydrolyze other phosphate esters that are normally present in the chloroplast. Definitive evidence for the importance of this enzyme in the PCO cycle was provided by Somerville and Ogren who isolated a mutant of *Arabidopsis thaliana* deficient in this enzyme (see Artus *et al.*, 1986). Under non-oxygenating conditions (e.g. 1% CO_2) the mutant grew normally and the enzyme was not essential, but under oxygenating conditions (e.g. 0.03% CO_2) phosphoglycolate accumulated. Photosynthesis was inhibited, presumably because phosphoglycolate inhibits triose phosphate isomerase.

Glycolate oxidation

The glycolate that is formed moves to the cytoplasm and enters the peroxisome. As the cell organelles move about in the living cell and as peroxisomes are frequently seen appressed to the chloroplasts, there may also be the possibility of direct transfer of glycolate across the appressed chloroplast and peroxisomal membranes without transit through the cytoplasm. There is a glycolate/D-glycerate translocator on the inner membrane of the chloroplast which carries out exchange transport as well as proton symport with either substrate (Young and McCarty, 1993). Specific metabolite transporters are not present in the single membrane of the peroxisomes and glycolate presumably diffuses through pores in this membrane (Heupel and Heldt,

1994). In microalgae some glycolate may be excreted into the medium (see De Veau and Burris, 1989a).

In the peroxisome, glycolate is oxidized by glycolate oxidase (Reaction 3, Fig. 22.1). A molecule of O_2 is taken up and the products of the reaction are glyoxylate and H_2O_2. The H_2O_2 is broken down by catalase (Reaction 4, Fig. 22.1) to H_2O and 0.5 O_2 so that the net uptake of O_2 in the oxidation of one molecule of glycolate is 0.5 O_2. Glycolate oxidase and catalase are tightly associated in the peroxisome (Heupel and Heldt, 1994). A catalase deficient mutant of barley grew poorly in normal air with death of the older leaves, but grew well in 0.2% CO_2 (Kendall *et al.*, 1983). The reported molecular weights of glycolate oxidase from various sources range from 93 to 199 kD, the pH optimum ranges from 7.8 to 8.5, the K_m (O_2) ranges from 76 to 130 μM and the k_m (glycolate) is around 0.25 to 0.4 mM. Glycolate oxidase contains flavin mononucleotide (FMN) which, in the reduced form, is oxidized by molecular O_2. The enzyme will also oxidize other α-hydroxy acids, including L-lactate. The enzyme is not inhibited by CN^- but is inhibited by a number of compounds, of which α-hydroxypyridine methane sulfonate (α-HPMS) and 2-hydroxy-3-butynoate (HBA) have been mostly widely used. The results of the use of these compounds for *in vivo* studies, however, should be interpreted with caution as they seem to inhibit a number of reactions.

Cyanobacteria and microalgae have, instead of the higher plant glycolate oxidase described above, a glycolate dehydrogenase that is inhibited by CN^- and oxidizes D-lactate (not L-lactate) in addition to glycolate. The natural electron acceptor is not known and the enzyme is assayed with artificial electron acceptors such as 2,6-dichlorophenol-indophenol (DCIP) or phenazine methosulfate (PMS). In the green algae the enzyme appears to be localized in the mitochondria; in the cyanobacteria it appears to be associated with the thylakoids. The oxidation of glycolate in these organisms would appear to be linked to the respiratory electron transport chain.

Glycine formation

The glyoxylate that is produced from the oxidation of glycolate is aminated by two aminotransferases, serine : glyoxylate aminotransferase (Reaction 5, Fig.

22.1) and glutamate : glyoxylate aminotransferase (Reaction 6, Fig. 22.1). The enzymes are essentially irreversible and both are necessary as two molecules of glycine must be produced to form one molecule of serine. From *in vitro* studies where both aminotransferases are present, serine is the preferred amino donor and this will ensure the production of hydroxypyruvate and the continued movement of carbon through the PCO cycle. The serine : glyoxylate aminotransferase is relatively specific for serine, although asparagine may function as a donor in pea. The glutamate : glyoxylate aminotransferase will also use alanine in addition to glutamate. In a Fd-GOGAT mutant of *Arabidopsis*, alanine is also used up along with glutamate under photorespiratory conditions, but the relative use of alanine versus glutamate in normal conditions is not known.

That the serine : glyoxylate aminotransferase is essential for the operation of the PCO cycle was demonstrated by the isolation of several mutants of *Arabidopsis* deficient in this enzyme. The plants grew under high CO_2 conditions but died under photorespiratory conditions. Glycine and serine accumulated as the glutamate : glyoxylate aminotransferase was still functional but little metabolism of serine could be demonstrated by a chase experiment with $^{12}CO_2$ after $^{14}CO_2$ fixation.

The aminotransferases are poorly inhibited with isonicotinyl hydrazide (INH) but strongly inhibited with hydroxylamine and aminooxyacetate (AOA). In algae, inhibition with AOA results in glycolate excretion but it has not been very useful for studies in higher plants.

In vitro, glyoxylate reacts non-enzymatically with H_2O_2 to produce formate and CO_2, and formate could be converted to CO_2 by an NAD-formate dehydrogenase in the mitochondria. It has been suggested that these reactions could produce at least part of the CO_2 observed in photorespiration. Indeed, a mutant of *Arabidopsis* deficient in serine hydroxymethyltransferase (SHMT) activity and hence unable to metabolize glycine did not accumulate glyoxylate but apparently metabolized it to CO_2 after all the amino donors in the cell were exhausted. However, if NH_3 (and serine) was supplied, the CO_2 evolution was prevented, which demonstrates that if amino donors are available the glyoxylate is aminated to glycine rather than being oxidized to CO_2. This

view is further supported by results with catalase-deficient barley (Kendall *et al.*, 1983) where much higher concentrations of H_2O_2 than normal did not result in increased CO_2 production. Hence, with the possible exception of the situation in which there are no amino donors, glyoxylate is not oxidized but converted to glycine and hence does not contribute to photorespiratory CO_2.

Glycolate oxidase, catalase, the two aminotransferases and hydroxypyruvate reductase form a multi-enzyme complex that remains intact after the peroxisome membrane is removed by osmotic shock (Heupel and Heldt, 1994). This close association of these enzymes ensures the orderly movement of metabolites through these reactions and reduces the possibility that glyoxylate or H_2O_2 can exit into the cytoplasm where they may have detrimental effects.

Glycine oxidation and CO_2 and serine production

The glycine generated in the peroxisome moves to the mitochondria and is taken up via a glycine/serine exchange protein in the mitochondrial inner membrane. In the mitochondria, glycine is metabolized by two enzymes, glycine decarboxylase (Reaction 7, Fig. 22.1) and serine hydroxymethyltransferase (Reaction 8, Fig. 22.1) in an overall reaction that converts two molecules of glycine to one molecule of serine and one molecule each of NADH, CO_2 and NH_3. There now seems to be little disagreement that this reaction produces the CO_2 that is observed in photorespiration. It also releases NH_3 that must be reassimilated for the continued operation of the PCO cycle.

Glycine decarboxylase (GDC) consists of four different component enzymes named P-protein, H-protein, T-protein and L-protein (Oliver, 1994). P-protein decarboxylates glycine and H-protein, which contains lipoic acid, acts as a carrier of the methylamine and electrons. The T-protein transfers the methyl group to tetrahydrofolic acid (THF) with the release of NH_3 to form N^5,N^{10}-methylene THF. The L-protein transfers the electrons to NAD^+ to form NADH. The enzyme can be inhibited by INH and aminoacetonitrile. Glycine decarboxylase has largely been studied in intact mitochondria but the individual components have been isolated and their cDNA clones characterized from pea leaves (Douce *et al.*, 1994).

Serine hydroxymethyltransferase (SHMT) activity results in the formation of serine through the transfer of the C1 unit from N^5,N^{10}-methylene THF to another molecule of glycine. In addition to the mitochondrial enzyme, there is some evidence that an isozyme is present in the chloroplast. A SHMT mutant of *Arabidopsis*, under non-photorespiratory conditions, did not require serine for growth or another source of C1 units for other biosynthetic reactions. SHMT is reversible but in leaves appears to function largely in the direction of serine synthesis.

Mutants deficient in GDC or SHMT have been isolated in *Arabidopsis* and barley. In both cases, under photorespiratory conditions glycine accumulates and photosynthesis is inhibited for reasons which are not yet clear.

Only low levels of GDC and SHMT are present in non-green tissues but the amounts increase in leaves on exposure to light and in mature leaves represent nearly half of the soluble protein of the mitochondria (Douce *et al.*, 1994). Glycine is the preferred substrate and the oxidation capacity of glycine in these mitochondria exceeds the capacity for oxidation of other tricarboxylic acid cycle intermediates. Serine and NADH inhibit GDC and are removed by the serine/glycine translocator and the oxidation of NADH. NADH may be oxidized by the respiratory electron transport chain with the production of ATP or alternatively it may be used to reduce oxaloacetate (OAA) to malate in the mitochondria via malate dehydrogenase. The oxidation of this malate in the peroxisome or elsewhere in the cell will be discussed later.

Glycerate formation

The serine that was produced in the mitochondria by the action of SHMT moves to the peroxisome where the amino group is removed by the serine : glyoxylate aminotransferase (Reaction 5, Fig. 22.1) and hydroxypyruvate is produced. The hydroxypyruvate is reduced by hydroxypyruvate reductase to glycerate using NADH as the electron donor (Reaction 9, Fig. 22.1). Hydroxypyruvate reductase has been isolated from several plants and microalgae but no

activity is found in cyanobacteria. The molecular weight of the enzyme is 91 to 97 kD, the pH optimum is 6.0 to 7.4 and the K_m (hydroxypyruvate) for the spinach enzyme is 120 μM. The enzyme will also reduce glyoxylate. NADPH dependent hydroxypyruvate reductase and glyoxylate reductase are present in the cytosol of plant cells and may be important in reducing and recovering any hydroxypyruvate or glyoxylate that escapes from the peroxisome (Givan and Kleckzowski, 1992).

Hydroxypyruvate reduction

Reducing equivalents must be imported into the peroxisome for the reduction of hydroxypyruvate, as a reaction generating NADH is not part of the major metabolic reactions of this organelle. Peroxisomes do contain an isozyme of malate dehydrogenase and reducing equivalents could be imported as malate (Heupel and Heldt, 1994). Malate would be oxidized to OAA with the production of NADH, and the OAA would leave the peroxisome.

The malate that is required for the above system could be generated in the mitochondria using the NADH produced from glycine decarboxylation to reduce OAA. Addition of OAA to mitochondria metabolizing glycine severely inhibits oxygen uptake and greatly stimulates glycine oxidation (McC. Lilley *et al.*, 1987). The mitochondria possess a high affinity ($K_m = 7 \mu$M) transporter for OAA, and the observed effects are presumably due to the reduction of OAA to malate by malate dehydrogenase in the mitochondria and the reoxidation of NADH by that means rather than by the transfer of the electrons to O_2. The inhibition of the glycine decarboxylase is relieved by the lowering of the NADH level. It does not seem that the reoxidation of NADH *per se* by the respiratory electron transport chain is limiting, even in the presence of glycine oxidation, since the addition of another tricarboxylic acid cycle intermediate can stimulate O_2 uptake. The malate that is produced could leave the mitochondria and move through the cytosol to the peroxisome where it could be used to generate the NADH for the reduction of hydroxypyruvate to glycerate. This malate/OAA shuttle seems to be a feasible mechanism by which the production of NADH for the reduction of hydroxypyruvate is linked to the oxidation of glycine, another intermediate in the PCO cycle (Hanning and Heldt, 1993). Chloroplasts also possess a high affinity OAA transporter, and a malate/OAA shuttle between the chloroplast and the peroxisome is another means by which reducing equivalents could be supplied to the peroxisome (see Hanning and Heldt, 1993). Such a reaction would, of course, affect the amount of NADH from glycine oxidation that could be reoxidized by the malate/OAA shuttle between the mitochondria and peroxisomes.

In vivo, it is not possible to determine the source of NADH for the reduction of hydroxypyruvate. The malate/OAA shuttle between the mitochondria and peroxisome conveniently balances the production and utilization of reducing equivalents in the cycle. Both the malate/OAA shuttle and oxidation through the respiratory electron transport chain may contribute to the reoxidation of the NADH produced in glycine decarboxylation. If NADH from glycine decarboxylation is reoxidized by O_2, then NADH must be supplied to the peroxisome from some other source to satisfy the requirements for the reduction of hydroxypyruvate.

The glycerate that is produced leaves the peroxisome and is transported into the chloroplast via the glycerate/glycolate transporter. In the chloroplast, glycerate kinase catalyses the phosphorylation of glycerate (Reaction 10, Fig. 22.1) to 3-phosphoglycerate, and the carbon re-enters the phosphorylated carbon pool of the photosynthetic carbon reduction cycle. Glycerate kinase is entirely located in the chloroplast and in spinach has a molecular weight of about 40 kD, a pH optimum of 6.5 to 8.5 and a K_m (D-glycerate) of about 0.2 mM.

Photorespiratory nitrogen cycle

During the decarboxylation of glycine (Reaction 7, Fig. 22.1), NH_3 is produced in stoichiometric amounts with the photorespiratory CO_2. In C3 plants, as the rate of this NH_3 production can be 5 to 10 times the rate of primary nitrate assimilation (Hanning and Heldt, 1993), the plant could certainly not afford to lose this NH_3 as a volatile gas but must have an effective mechanism for its reassimilation. The reassimilation of this NH_3 occurs through the concerted action of glutamine synthetase (GS) (Reaction 11, Fig. 22.1) and glutamate synthase (Reaction 12, Fig. 22.1); the latter enzyme is also called glutamine : oxoglutarate aminotransferase (GOGAT) and the system is referred to as the GS/GOGAT system.

Two isozymes of GOGAT are located in chloroplasts or plastids. The major enzyme of the chloroplast is a ferredoxin-linked GOGAT (Fd-GOGAT) and the minor enzyme, which occurs at 3% or less of the total activity of the former enzyme, uses NAD(P)H in place of ferredoxin as the electron donor. GOGAT catalyzes the formation of two molecules of glutamate according to the following equation:

$$\text{glutamine} + \text{2-oxoglutarate} + Fd_{red} \rightarrow \text{2 glutamate} + Fd_{ox}$$

The Fd-GOGAT from *Vicia faba* is reported to have a molecular weight of 150 kD, a pH optimum of 7.4, a K_m(Fd) of 2 μM, a K_m(glutamine) of 300 μM and a K_m(2-oxoglutarate) of 150 μM. Azaserine and 6-diazo-5-oxo-L-norvaline (DON), which are glutamine analogs, strongly inhibit the enzyme.

Fd-GOGAT is essential for the reassimilation of photorespiratory NH_3, as mutants of barley or *Arabidopsis* lacking this enzyme accumulate large quantities of NH_3 under photorespiratory conditions and eventually die. Under non-photorespiratory conditions the activity of the NAD(P)H GOGAT would seem to be sufficient for the assimilation of NH_3 produced from the reduction of nitrate.

Although there are some exceptions (e.g. spinach), two isozymes of glutamine synthetase are present in higher plants, one cytoplasmic, the other chloroplastic, as shown by histochemical localization experiments (Edwards *et al.*, 1990). It has been demonstrated with transgenic tobacco plants that the different isozymes perform separate non-overlapping metabolic functions (Edwards *et al.*, 1990). The chloroplast isozyme is the predominant activity. Glutamine synthetase catalyzes the formation of glutamine according to the following equation:

$$\text{glutamate} + NH_3 + ATP \rightarrow \text{glutamine} + ADP + P_i$$

Glutamine synthetase, which requires Mg^{2+}, has a molecular weight in the 350–400 kD range, a pH optimum of 8.0, a K_m(glutamate) of 2–13 mM, a K_m(ATP) of 0.1–1.5 mM and a strikingly high affinity for ammonia, K_m(10–20 μM).

The cytosolic glutamine synthetase is not expressed in leaf mesophyll cells but in the phloem elements and is unlikely to play any role in the reassimilation of the photorespiratory NH$_3$ that is released in the mesophyll cells (Edwards et al., 1990).

Mutants of barley lacking the chloroplastic GS and Fd-GOGAT have been isolated and a double mutant produced (Joy et al., 1992). In photorespiratory conditions ammonia accumulated in the mutants and photosynthesis was inhibited. In non-photorespiratory conditions (0.7% CO$_2$) growth of the mutants was normal, showing that the activities of root enzymes and residual leaf enzymes were sufficient for the assimilation of nitrogen for growth but were inadequate to cope with the increased ammonia flux from photorespiration.

Although mitochondria possess glutamate dehydrogenase, this enzyme would appear to play no role in the assimilation of the NH$_3$. The enzyme has a low affinity for ammonia, $K_m(NH_3) \approx 5\text{--}70$ mM, and when 2-oxoglutarate is supplied to mitochondria that are actively decarboxylating glycine, little if any glutamate is produced. When GS is inhibited with L-methionine sulfoximine (MSO), little, if any, NH$_3$ is assimilated in the leaves. Mutants lacking GDH are not lethal under photorespiratory conditions.

The reassimilation of the NH$_3$ in the chloroplast requires the import of 2-oxoglutarate and the export of glutamate, the substrate for the glutamate : glyoxylate aminotransferase in the peroxisome. There are two dicarboxylate transporters in the chloroplast membrane (Woo et al., 1987). The 2-oxoglutarate (2-OG) translocator has been purified from spinach chloroplasts (Menzslaff and Flugge, 1993) and transports 2-OG, malate, succinate, fumarate and glutarate. The other dicarboxylate transporter transports all these acids plus glutamate and aspartate. Yet another transporter for glutamine may be present. The two dicarboxylate transporters are thought to work in concert. The 2-OG translocator transports 2-OG into the chloroplast in exchange for malate and the dicarboxylate transporter transports glutamate out of the chloroplast in exchange for malate. Since the transporters are exchange transporters, the movement of malate in the opposite direction to 2-OG or glutamate is essential, but no net movement of malate into or out of the chloroplast occurs. The importance of the dicarboxylate transporter has been demonstrated by the isolation of a mutant of Arabidopsis defective in this protein. In non-photorespiratory conditions, growth of the mutant was normal, whereas the plants died under photorespiratory conditions.

Photorespiration and leaf photosynthesis

Because O$_2$ is a competitive inhibitor with CO$_2$ for Rubisco, higher O$_2$ concentrations produce striking inhibitions of photosynthesis in C3 leaves. Photosynthesis in C3 leaves is stimulated 40 to 60% when the oxygen concentration is decreased from 21% to 2% (Fig. 22.2A and 22.2E) and photosynthesis is inhibited a similar amount when the oxygen concentration is increased from 21% to 50% (Fig. 22.2E).

The CO$_2$ concentration where release of CO$_2$ by photorespiration is equal to fixation of CO$_2$ by photosynthesis, called the CO$_2$ compensation point, is normally 40–60 μl CO$_2$. As O$_2$ fixation is required for CO$_2$ release, the compensation point is close to zero at 2% O$_2$ and increases linearly with the O$_2$ concentration in C3 leaves (Fig. 22.2C). Above the light compensation point (i.e. that light intensity where gross CO$_2$ fixation equals gross CO$_2$ release and net CO$_2$ exchange is zero) the CO$_2$ compensation point does not change with light intensity.

Just as photorespiration will vary with the O$_2$ concentration, it will also vary with the CO$_2$ concentration, and glycine formation is largely suppressed at CO$_2$ concentrations above 0.2% (Somerville and Somerville, 1986). O$_2$ uptake (which is largely due to oxygenase activity of Rubisco) in C3 leaves at low light intensities is also inhibited by increasing CO$_2$ concentrations (Fig. 22.2F). More complex patterns of O$_2$ uptake occur as a function of CO$_2$ concentration at high light intensities (Badger and Canvin, 1981), but in all cases higher CO$_2$ concentrations inhibit O$_2$ uptake. In C3 leaves, the ratio of photorespiration to photosynthesis increases with leaf temperature. In part, this is due to the kinetics of Rubisco as increased temperature reduces the affinity of the enzyme for CO$_2$ and favors oxygenation. It is also a result of the increased solubility of O$_2$ and the decreased solubility of CO$_2$

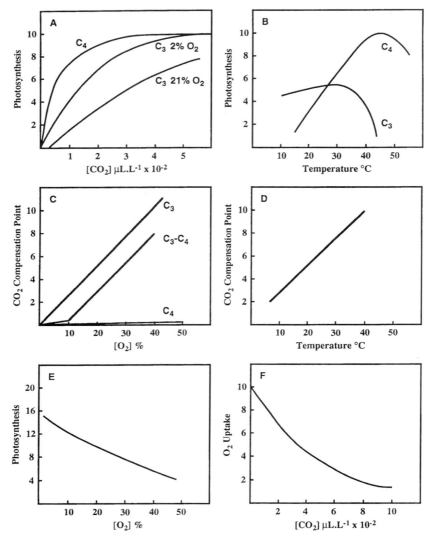

Fig. 22.2 Typical responses of some features of gas exchange in plant leaves. All rates are relative as detailed for individual panels. (A) Net photosynthesis of C3 leaves in 21% or 2% O_2 and C4 leaves in either 21% or 2% O_2 in response to the external CO_2 concentration. Rate of C3 leaf photosynthesis expressed relative to the maximum rate observed in the C3 leaf in 2% O_2. Maximum rate of C3 leaf photosynthesis in 2% O_2 and maximum rate of C4 leaf photosynthesis arbitrarily set to 10. Temperature about 25 °C and about one-half full sunlight. (B) Effect of temperature on photosynthesis of C3 and C4 leaves. Rate of photosynthesis expressed relative to the maximum rate in the C4 leaf. Air and about one-half full sunlight. (C) Effect of O_2 concentration on the CO_2 compensation point of a C3 leaf, a C4 leaf and a C3–C4 intermediate leaf. Compensation points expressed relative to the compensation point in the C3 leaf at 20% O_2 being arbitrarily set to 5. Temperature about 25 °C and one-half full sunlight. (D) Effect of temperature on the CO_2 compensation point of a C3 leaf. Compensation point at 20 °C arbitrarily set equal to 5. O_2 (21%) and about one-half full sunlight. (E) Net photosynthesis of a C3 leaf as a function of O_2 concentration. Rate of photosynthesis in 20% O_2 arbitrarily set equal to 10. CO_2 (0.036%), temperature about 25 °C and one-half full sunlight. (F) The effect of CO_2 concentration on O_2 uptake of a C3 leaf. Maximum rate of O_2 uptake arbitrarily set equal to 10. O_2 (21%), temperature about 25 ° and one-quarter full sunlight.

that occurs at higher temperatures (see Chapter 20). As expected, the CO_2 compensation point of C3 plants increases with temperature (Fig. 22.2D) and the quantum yield decreases.

The increasing losses due to photorespiration with increasing temperature in C3 plants might be expected to limit the temperatures under which these plants can grow. In general, this is true, as the quantum yield in C3 plants is decreased by over 50% at temperatures above 35 °C. The quantum yield in C4 plants, which have developed a mechanism for suppressing photorespiration (see later), is not affected by such temperatures. The optimal temperature for photosynthesis of C4 leaves is markedly higher than for C3 leaves (Fig. 22.2B), and results in C4 plants being found more frequently in environments with high temperatures.

Measurement and rate of photorespiration

Photorespiration, when measured as light-dependent O_2 uptake and CO_2 evolution, involves the same gases as photosynthesis but the fluxes are in the opposite direction. It is relatively easy to qualitatively detect photorespiration but, because photosynthesis and photorespiration occur simultaneously, it is exceedingly difficult to measure photorespiration quantitatively (Sharkey, 1988).

In C3 plants, when the light is extinguished, there is a rapid evolution of CO_2 which greatly exceeds the rate of dark respiration and which has been called the post-illumination burst (PIB). The PIB is thought to arise from the continued metabolism of intermediates of the PCO cycle up to and including glycine (Cournac *et al.*, 1993) that are present after the synthesis of phosphoglycolate has ceased. The PIB is then a qualitative indicator of photorespiration whose size depends on the amount of accumulated intermediates and the technical ability of the measuring system to measure the 'burst' with the minimum amount of dispersion (Peterson, 1987).

If the inward CO_2 flux of photosynthesis is reduced by depriving the leaf of the substrate (i.e. placing the leaf in CO_2-free air) the outward flux of CO_2 from photorespiration can be measured. This will be an underestimate because some of the CO_2 released in photorespiration will be refixed in photosynthesis

before it can exit the leaf. The change in CO_2 concentration at the site of Rubisco could presumably also alter the amount of substrate produced for the PCO cycle. The method provides a quick and relatively easily obtained estimate of photorespiration, but it cannot be considered quantitative.

Ogren and his coworkers (Ogren, 1984) have developed the following equation to determine the ratio of the carboxylation to oxygenation reactions catalyzed by Rubisco:

$$v_c/v_o = (V_c K_o/V_o K_c) ([CO_2]/[O_2])$$

where v_c and v_o are the velocity of carboxylation and oxygenation respectively, V_c and V_o the maximal velocities of these reactions, and K_c and K_o the K_m values of the enzyme for CO_2 and O_2 respectively. The term $(V_c K_o/V_o K_c)$ determines the relative rate at any given CO_2 and O_2 concentration and has been called the specificity factor (see Chapter 20). The specificity factors for higher C3 and C4 plants are about 80. They decrease to about 50 to 60 for cyanobacteria and unicellular algae and are as low as 10 to 15 for some photosynthetic bacteria. Higher specificity factors indicate a greater affinity of the enzyme for CO_2.

As one CO_2 is released in photorespiration for every two O_2 fixed in the oxygenase reaction, the ratio of CO_2 evolution to oxygenation is 0.5 (see above). The velocity of the relative oxygenase activity $(V_o K_c)$ can then be expressed in terms of CO_2 released as $0.5 V_o K_c$. Hence, the specificity factor for carbon fixation in photosynthesis and carbon release in photorespiration is $V_c K_o/0.5 V_o K_c$. As Ogren states, this is the 'essence' of photorespiration (in terms of CO_2) as it will determine the relative rate of photorespiration in relation to photosynthesis.

With a specificity factor of 80 and assuming concentrations of CO_2 and O_2 at the enzyme site of 7.5 μM and 250 μM under normal ambient conditions, the ratio of carboxylation to oxygenation is 2.4. Photorespiration as CO_2 release will be 0.5 or about 20% of the rate of photosynthesis. From a similar consideration of the enzyme kinetics Sharkey (1988) estimates photorespiratory CO_2 release as 26% of net photosynthesis. Since four carbons must traverse the PCO cycle for every CO_2 released in photorespiration, the rate of carbon flux through the

PCO cycle will be 80% of the rate of CO_2 fixation or photosynthesis.

The measurement of opposite movements of the same component can be studied by altering the ratio of isotopes of the element in one compartment. For CO_2 uptake the external gas can be enriched in the carbon-14 isotope, and for O_2 uptake the gas can be enriched in the oxygen-18 isotope.

Photorespiration as measured by the dual carbon ($^{14}CO_2/^{12}CO_2$) method yields rates of carbon flux through the PCO cycle of 0.65 to 0.9 the rate of CO_2 fixation, and hence, CO_2 evolution equal to 16–22% of the rate of photosynthesis under normal conditions. As some of the CO_2 that is produced from photorespiration must be refixed before it can exit the leaf, the obtained rates represent minimum estimates of photorespiration. The amount of refixation has been estimated (Gerbaud and André, 1987) as 16% under normal conditions, and application of this correction factor would yield rates of photorespiration as CO_2 evolution equal to 18–25% of the rate of photosynthesis.

Using oxygen-18 and the mass spectrometer the measurement of O_2 uptake is not complicated by the problem of extensive recycling (Gerbaud and André, 1987). It is, however, affected by O_2 uptake that occurs through reactions other than the Rubisco oxygenase reaction and hence tends to overestimate photorespiration. When allowance is made for oxygen uptake by other means the oxygenase/carboxylase ratio is about 0.55 under normal conditions (Badger and Canvin, 1981). This would yield a rate of carbon flux through the PCO cycle that is 110% of the rate of CO_2 fixation and result in a rate of photorespiration as CO_2 evolution that is 27% of the rate of photosynthesis.

The rate of photorespiration can also be determined by measuring the rate of carbon flux through glycine and serine, assuming a stoichiometry of one CO_2 released for every four carbons of traffic. These measurements indicate a flux of carbon through the PCO cycle about equal to the rate of apparent CO_2 fixation and a rate of photorespiration as CO_2 evolution equal to 20–25% of the rate of photosynthesis.

The rate of photorespiration has also been estimated by an elegant method using a mutant of *Arabidopsis* lacking in serine

hydroxymethyltransferase (Somerville and Somerville, 1986). If the mutant is supplied with serine and NH_3, the carbon that enters the PCO cycle accumulates as glycine. Under normal conditions 0.53 molecules of glycine accumulates for every CO_2 fixed. Hence, oxygenation would be 50% of the rate of carboxylation, carbon flow through the PCO cycle would be about equal to the rate of CO_2 fixation and photorespiration as CO_2 evolution would be about 25% of the rate of photosynthesis.

Costs, function and regulation of the PCO cycle

As described above, under normal conditions, the flux of carbon through the PCO cycle is 0.65 to 1.1 times the rate of CO_2 fixation, and CO_2 release in photorespiration is 16–27% of the CO_2 fixed. In addition to this direct loss of CO_2, the rate of CO_2 fixation is also decreased because of the competition between O_2 and CO_2 at the active site of Rubisco. For every O_2 that is fixed by the oxygenase reaction a CO_2 cannot be fixed.

The inhibition of fixation of CO_2 by O_2 under normal conditions must be twice the rate of photorespiratory CO_2 release because two O_2 must be fixed to produce the two glycine required for the production of one CO_2. If photorespiratory CO_2 release is 25% of the true rate of CO_2 fixation, the true rate of CO_2 fixation would increase by 50% (equivalent to the two O_2) in the absence of O_2 and apparent photosynthesis (true CO_2 fixation minus photorespiratory CO_2 release) should theoretically increase by 75%.

Experimentally, increases of 50 to 70% in both apparent (Fig. 22.2E) and true CO_2 fixation rates are observed when the oxygen concentration is decreased from 21% to 1%. This is in the range of increase expected for true CO_2 fixation but is somewhat lower than that expected for apparent CO_2 fixation. The failure to observe the expected increase in CO_2 fixation when the O_2 concentration is decreased may be related to the reduction in photosynthetic electron transport or true photosynthetic capacity that occurs when the O_2 concentration is decreased (Badger and Canvin, 1981).

The costs of oxygenase activity and

photorespiration to the plant can also be determined from the energy requirements of the PCR and PCO cycles. These are shown in Table 22.1. The cost of fixing one molecule of CO$_2$ in the PCR cycle is nine ATP equivalents. The cost of fixing one molecule of O$_2$ in the oxygenase reaction is 9.5 ATP equivalents. However 0.5 CO$_2$ is lost and the cost to fix this originally would have been 4.5 ATP equivalents. Hence the total cost of the oxygenase reaction would be 14 ATP per O$_2$ fixed or 28 ATP per CO$_2$ released. As oxygenase activity is approximately one-half the carboxylase activity under normal conditions, the utilization of all the energy for CO$_2$ fixation could support a rate of CO$_2$ fixation at least 50% higher.

All Rubisco's that have been studied, even those from chemolithotrophic bacteria and photoautotrophic bacteria such as *Rhodospirillum rubrum* that grow under anaerobic conditions, have oxygenase activity. The ubiquity of the oxygenase activity has led to the suggestion that oxygenase activity is an inevitable consequence of the reaction chemistry required for CO$_2$ fixation by the enzyme

(see Chapter 20). If this is indeed the case, the PCO cycle is a means whereby 75% of the carbon that is diverted to phosphoglycolate can be recovered into products useful to the plant. This recovery or 'salvage' role may be a sufficient and only function for the PCO cycle.

Because of the magnitude of carbon transit through the PCO cycle and the substantial expenditure of energy associated with it, however, one may continue to have a lingering feeling that there should be a more essential function for the cycle. Such a case can certainly not be demonstrated for the *Arabidopsis* or barley mutants that lack essential enzymes of the PCO cycle, because they grow normally under conditions that prevent oxygenase activity and that in other respects are non-stressful. But is this also the case when plants are exposed to extreme stress, such as water deficiency or cold temperatures under high light conditions? In such situations, because CO$_2$ may be limiting when the stomata are closed or because the activity of carboxylation may be low, the capacity to dissipate

Table 22.1 Energy requirements for CO$_2$ and O$_2$ fixation by Rubisco.

CO$_2$ fixation

$$3 \text{ RuBP} + 3\text{CO}_2 + 2\text{H}_2\text{O} \rightarrow 6 \text{ (3-PGA)}$$
$$6 \text{ (3-PGA)} + 6 \text{ ATP} + 6 \text{ (2H}^+ + 2 \text{ e}^-) \rightarrow 6 \text{ triose phosphate} + 6 \text{ ADP} + 6 \text{ P}_i$$
$$5 \text{ triose phosphate} \rightarrow 3 \text{ R-5-P} + 2 \text{ P}_i$$
$$3 \text{ R-5-P} + 3 \text{ ATP} \rightarrow 3 \text{ RuBP} + 3 \text{ ADP}$$

Sum:
$$3 \text{ CO}_2 + 3 \text{ H}_2\text{O} + 9 \text{ ATP} + 6 \text{ (2H}^+ + 2 \text{ e}^-) \rightarrow \text{ triose phosphate} + 9 \text{ ADP} + 8 \text{ P}_i$$

If (2 H$^+$ + 2 e$^-$) is assumed to be equivalent to 3 ATP, the cost for each CO$_2$ fixed would be 9 ATP equivalents.

O$_2$ fixation

$$10 \text{ R-5-P} + 10 \text{ ATP} \rightarrow 10 \text{ RuBP} + 10 \text{ ADP}$$
$$10 \text{ RuBP} + 10 \text{ O}_2 \rightarrow 10 \text{ (3-PGA)} + 10 \text{ phosphoglycolate}$$
$$10 \text{ phosphoglycolate} \rightarrow 10 \text{ glycolate} + 10 \text{ P}_i$$
$$10 \text{ glycolate} + 5 \text{ O}_2 \rightarrow 10 \text{ glyoxylate} + 10 \text{ H}_2\text{O}$$
$$10 \text{ glyoxylate} + 10 \text{ glutamate} \rightarrow 10 \text{ glycine} + 10 \text{ (2-oxoglutarate)}$$
$$10 \text{ glycine} \rightarrow 5 \text{ serine} + 5 \text{ CO}_2 + 5 \text{ NH}_3 + 5 \text{ (2H}^+ + 2 \text{ e}^-)$$
$$5 \text{ serine} + 5 \text{ (2-oxoglutarate)} \rightarrow 5 \text{ hydroxypyruvate} + 5 \text{ glutamate}$$
$$5 \text{ hydroxypyruvate} + 5 \text{ (2H}^+ + 2 \text{ e}^-) \rightarrow 5 \text{ glycerate}$$
$$5 \text{ glycerate} + 5 \text{ ATP} \rightarrow 5 \text{ (3-PGA)} + 5 \text{ ADP}$$
$$5 \text{ NH}_3 + 5 \text{ ATP} + 5 \text{ glutamate} \rightarrow 5 \text{ glutamine} + 5 \text{ ADP} + 5 \text{ P}_i$$
$$5 \text{ glutamine} + 5 \text{ (2-oxoglutarate)} + 5 \text{ (2H}^+ + 2 \text{ e}^-) \rightarrow 10 \text{ glutamate}$$
$$15 \text{ (3-PGA)} + 15 \text{ ATP} + 15 \text{ (2H}^+ + 2 \text{ e}^- \rightarrow 15 \text{ triose phosphate} + 15 \text{ ADP} + 15 \text{ P}_i$$
$$15 \text{ triose phosphate} \rightarrow 9 \text{ R-5-P} + 6 \text{ P}_i$$

Sum:
$$\text{R-5-P} + 1.5 \text{ O}_2 + 3.5 \text{ ATP} + 2 \text{ (2H}^+ + 2 \text{ e}^-) \rightarrow 0.9 \text{ (R-5-P)} + 3.5 \text{ ADP} + 3.6 \text{ P}_i + 0.5 \text{ CO}_2 + \text{H}_2\text{O}$$

The cost for each O$_2$ fixed by oxygenase would be 9.5 ATP equivalents.

photochemical energy may be limited. The high energy state that would result in the photochemical system leads to a damage of the system which has been called photoinhibition (Powles, 1984). Photoinhibition has been shown to occur when leaves are exposed to high light in the absence of both CO_2 and O_2, conditions where both the PCR and PCO cycles could not operate. With oxygen and oxygenase activity, CO_2 would be maintained at the compensation point and both O_2 and CO_2 fixation could contribute to using the photochemical energy. In C3 plants, either O_2 or CO_2 reduced photoinhibition to some degree, whereas in C4 plants only CO_2, and not O_2, was effective.

Any role of photorespiration in offering some protection against photoinhibition (Osmond, 1981) would seem to be entirely fortuitous. As noted above, all Rubisco's have oxygenase activity, so oxygenase activity was not fixed in the enzyme by evolutionary pressure in plants that routinely encountered stress conditions. There is no evidence that oxygenase activity is greater in plants that grow in regions where stress is frequent as compared with plants that grow in regions where stress is never encountered. C4 plants, in which oxygenase activity is reduced due to a 'CO$_2$-concentrating' mechanism and lower amounts of Rubisco, do not seem to be any more susceptible to photoinhibition than C3 plants. No doubt, this possible role of photorespiration will continue to be considered. One cannot escape from the fact, however, that if the carbon from phosphoglycolate is not returned to the PCR cycle, i.e. the 'salvage' function, the death of the plant follows, as shown in *Arabidopsis* and barley mutants deficient in PCO cycle enzymes.

Cyanobacteria and green algae, which are evolutionary older than C3 terrestrial plants, do not seem to have a complete effectively functioning PCO cycle. They do, however, possess active transport systems for both CO_2 and HCO_3^- that function as a 'CO$_2$-concentrating' mechanism to elevate the internal CO_2 concentration and suppress oxygenase activity. Aquatic macrophytes seem to retain, at least in some plants, some 'CO$_2$-concentrating' activity, but they also have developed a PCO cycle. C3 terrestrial plants have apparently lost all 'CO$_2$-concentrating' capability. With the progressive loss of the 'CO$_2$-concentrating' capability it was imperative for

the plants to develop an efficient PCO cycle to salvage the carbon that was diverted to phosphoglycolate and, hence, the PCO cycle is essential but the oxygenase reaction is merely unavoidable.

As a salvage pathway, it must have the capability of metabolizing all the substrate, i.e. the phosphoglycolate, that is produced. One should not expect to find that the operation of the pathway is regulated by any other factor than the availability of substrate. While all the enzymes of the PCO cycle have not been examined extensively, there is little evidence for regulation of them at the present time. The capacity of the total PCO cycle, however, may be able to moderate the activity of the PCR cycle. There is little doubt that in the extreme case where the PCO cycle cannot operate, such as in the PCO cycle enzyme-deficient mutants, one of the consequences is that the PCR cycle is inhibited. It is possible that where substrate supply exceeds the capacity of the PCO cycle, i.e. the PCO cycle is overloaded, the accumulation of intermediates would moderate the PCR cycle to again bring substrate production into line with the PCO cycle capacity.

Mechanisms leading to reduced photorespiration

C4 plants

In C4 plants two distinct types of photosynthetic cells are observed in the leaves. The mesophyll cells are arranged throughout the leaf lamina while the bundle sheath cells are arranged in a ring, called Kranz (wreath-like) anatomy, surrounding the vascular strands. The bundle sheath cells have thickened walls and prominent starch-filled chloroplasts, whereas the mesophyll cells have thinner walls and smaller chloroplasts.

The distinctive anatomy in C4 plants is accompanied by a specialized biochemistry of photosynthesis. Rubisco and the PCR cycle are confined to the bundle sheath. CO_2 fixation in the mesophyll cells consists of the addition of HCO_3^- to phospho*enol*pyruvate (PEP) by PEP carboxylase to form oxaloacetic acid (OAA). The OAA is converted to either malate or aspartate and transported to the bundle sheath cells. In the bundle sheath cells the four carbon acid (malate or aspartate) is decarboxylated, depending on the plant, by NADP or NAD malic enzyme or PEP carboxykinase to provide CO_2 for Rubisco. The three carbon acid remaining (pyruvate or PEP) is transported back to the mesophyll cells where it serves as the acceptor for further CO_2 fixation (see Chapter 23).

In terms of whole-leaf CO_2 exchange, C4 photosynthesis is insensitive to the O_2 concentration (Fig. 22.2A) and the CO_2 compensation point is less than $5 \mu l\ CO_2\ l^{-1}$ (Fig. 22.2C). O_2 uptake is observed in C4 plants but it is largely insensitive to changes in the CO_2 concentration. These characteristics of the O_2 and CO_2 exchange in C4 plants suggest that oxygenase activity and photorespiration are suppressed or absent in C4 plants.

As Rubisco is confined to the bundle sheath cells, oxygenase activity would be similarly limited and the PCO cycle should also be confined to these cells. Photosynthesis of isolated bundle sheath cells is sensitive to O_2 concentration and while present in lower amounts than in C3 plants, many of the enzymes of the PCO cycle are present in the bundle sheath cells. Many of the enzymes, however, are also present in the mesophyll cells and, surprisingly, glycerate kinase seems to be localized in these cells. Peroxisomes are present in both cell types at 10 to 50% of the frequency observed in C3 plants, but they are usually several times more numerous in the bundle sheath cells. C4 leaves or bundle sheath strands metabolize glycolate or glycine to CO_2, and glycine and serine become labeled when C4 leaves photosynthesize in $^{14}CO_2$. The general picture that emerges is that oxygenase activity occurs under some circumstances in C4 leaves but this activity is much reduced. For example, photorespiratory CO_2 release has been estimated as 6 to 8% (De Veau and Burris, 1989b; Dai et al., 1993) of net photosynthesis in maize seedlings and as 2% in older maize leaves (De Veau and Burris, 1989b). The operation of the PCO cycle may be more complex, involving both the bundle sheath and mesophyll cells, and any CO_2 that is released is effectively refixed in the mesophyll cells before it could exit to the surroundings.

The suppression of oxygenase activity and, hence, phosphoglycolate production in C4 plants is attributed to a high CO_2 concentration in the bundle

sheath cells (Jenkins et al., 1989). Bundle sheath cells have a low permeability to CO_2 (Jenkins et al., 1989; Furbank et al., 1989) and a high CO_2 concentration is formed and maintained in the bundle sheath cells by the rapid decarboxylation of the C4 acids that were formed in the initial carboxylation in the mesophyll cells and transported to the bundle sheath cells. Only at elevated CO_2 concentrations will the carboxylation of RuBP become equal to the rates of decarboxylation to maintain the system in steady state.

Estimates of the [CO_2] in bundle sheath cells range from $27 \mu M$ (Dai et al., 1993) to $70 \mu M$ (Jenkins et al., 1989). These concentrations of CO_2 are about 4 to 9 times the normal CO_2 concentration ($7.5 \mu M$) in mesophyll cells of C3 plants and would be sufficient to markedly suppress oxygenase activity.

CAM plants

CAM plants are characterized by a marked diurnal variation in titratable acidity, mostly malate. This is due to extensive CO_2 fixation at night by PEP carboxylase to produce malate which is stored in the vacuole. Stomata are open at night when the evaporative demand is low and PEP is produced from stored carbohydrate. During the day, when evaporative demand is high, the stomata close. The malate produced during the preceding night is decarboxylated by malic enzyme or PEP carboxykinase, depending on the plant, to produce CO_2 which is fixed by the normal PCR cycle and a three carbon molecule which is converted to carbohydrate. Because the stomata are closed, the CO_2 concentration in the leaf can reach high levels which are sufficient to suppress or inhibit the oxygenase reaction and hence photorespiration. Sampling of the internal gas phase of a number of CAM plants revealed CO_2 concentrations ranging from 0.08% to as high as 2.5% (Cockburn et al., 1979; Spalding et al., 1979).

At the end of the day when de-acidification is complete, CAM plants, growing under reasonable moisture levels, carry out normal C3 photosynthesis. During this phase, photosynthesis is sensitive to O_2 and a post-illumination burst can be observed. In these conditions CAM plants would seem to be similar to C3 plants in the potential capacity for photorespiration, but this capacity is normally suppressed due to the elevated CO_2 concentration that prevails in the light during photosynthesis.

C3-C4 intermediates

C3-C4 intermediates are characterized by a CO_2 compensation point of $7-28 \mu l \ CO_2 l^{-1}$ and reduced sensitivity of photosynthesis to the O_2 concentration (Fig. 22.2C). The lower CO_2 compensation point in C3–C4 intermediates compared with C3 plants is consistent with reduced photorespiration in these plants. C3–C4 intermediates also have a leaf anatomy intermediate between that of C3 plants and C4 plants. A bundle sheath can be observed, but it is not as well developed or defined as in C4 plants.

C3–C4 intermediates have been extensively studied because they seem to represent natural instances of reduced photorespiration, and if the mechanism can be discovered, some guidance as to how photorespiration in C3 plants can be reduced may be provided. From these studies two proposals have emerged. The C4-like anatomy and the presence of a number of C4 photosynthesis enzymes has suggested that the C3–C4 intermediates are capable of a limited amount of C4 photosynthesis. Alternatively, the C3–C4 intermediates may possess a more efficient refixation system for any CO_2 that may be released. Support for this proposal rests largely on the fact that the CO_2 compensation point of C3–C4 intermediates, which at low light intensity is equal to closely related C3 plants, decreases as the light inensity is increased while for C3 plants it does not change above the light compensation point. Obviously, increased refixation is only possible at higher light intensities that would provide the energy for this process. Other explanations have been presented to account for reduced photorespiration in these plants, but none are entirely satisfactory (see Holaday and Chollet, 1984; Monson *et al.*, 1984; Schuster and Monson, 1990).

Aquatic macrophytes

Aquatic macrophytes include angiosperms, macroalgae and non-vascular higher plants. Plants may be completely submerged or may be amphibious with a portion of the plant emergent. The emergent leaves of aquatic macrophytes are similar to C3 terrestrial plants in reference to their photosynthesis and photorespiration characteristics. Submerged plants, however, display a variety of photosynthetic characteristics and also a plasticity whereby these characteristics are modified in response to environmental conditions (Bowes and Salvucci, 1989).

Many submerged plants rely exclusively on CO_2 diffusion to supply the substrate for photosynthesis and while photosynthetic rates are low and often limited by CO_2 diffusion, the characteristics of photosynthesis and photorespiration are similar to those in C3 terrestrial plants. Other submerged plants, however, have developed mechanisms to improve the efficiency of carbon utilization and to suppress photorespiration. In cooler temperatures and short days the plants exist in the high-photorespiration (PR) form mentioned above but in warmer temperatures and longer photoperiods the plants change to a low-PR form. The CO_2 compensation point and the sensitivity of photosynthesis to oxygen is greatly reduced compared to those observed in the high-PR plants. The mechanisms by which the low-PR condition is achieved are associated with either an increased ability for carbon uptake through HCO_3^- use or a C4/CAM metabolism.

HCO_3^- concentrations are often many times higher than CO_2 concentrations in water and the ability to use this source of carbon in addition to the diffusion of CO_2 has been shown in a wide variety of aquatic macrophytes (Madsen and Sand-Jensen, 1991). Although active transport of HCO_3^- has been shown in only a few species, active transport of HCO_3^- is widely assumed, resulting in elevated CO_2 levels in the vicinity of Rubisco and a suppression of oxygenase activity. Because an external carbonic anhydrase (CA) has been found in a few species, several authors have proposed its involvement in an increased supply of either CO_2 or HCO_3^- to the cells. To date no clear formulation has been reached and the role of this extracellular CA is still obscure.

The development of the low-PR condition may be accompanied by an increase in the amount of PEP carboxylase, PEP carboxykinase, Pyuvate-P_i-dikinase and the enzymes associated with the C4 pathway (Bowes and Salvucci, 1989). Label from $^{14}CO_2$

incorporated into malate and aspartate rapidly flush from these acids during a chase period but major changes in anatomy or in the activities of PCR or PCO cycle enzymes are not observed. Increased dark fixation, similar to that in CAM plants, may also occur. The operation of C4 fixation in these plants is not fully understood but it has been postulated that C4 acids are formed in the cytosol, transported to the chloroplast where decarboxylation and CO_2 is made available to Rubisco.

Direct evidence for increased intracellular levels of inorganic carbon by the above mechanisms is sparse. A 100–500 fold concentration of inorganic carbon has been estimated in a fresh-water angiosperm and 6–80 fold concentration have been observed in macroalgae. Indirect evidence (low CO_2 compensation points, reduced sensitivity of photosynthesis to O_2, low CO_2 evolution in the light or reduced labeling of PCO cycle intermediates from $^{14}CO_2$) however, indicates that photorespiration is reduced by the above mechanisms, presumably due to elevated CO_2 in the vicinity of Rubisco.

Aquatic microorganisms

The photosynthetic properties of microalgae and cyanobacteria vary widely with the dissolved inorganic carbon (C_i) concentration on which they are grown. Most experimental work, however, is done with cells grown either at high [C_i] (>1 mM), called high C_i cells or at low [C_i] (<0.1 mM), called low C_i cells. Standing cultures where the level of [C_i] is lower than 0.03 mM have also been used.

High C_i cells have a higher growth rate and maximum photosynthetic rate than low C_i cells. Their $K_{1/2}(C_i)$ varies with the measured pH; it is low (<50 μM) at pH 5.0, and rises to high values (>5 mM) at pH 10.0. At pH 8.0, where most of the studies are done, it is usually in the range of 1–2 mM. Under high C_i conditions, photosynthesis of high C_i cells is not inhibited by O_2. However, when placed under low C_i conditions, high C_i cells are not able to effectively use the low C_i or CO_2 that is present and their photosynthetic rates are very low. Under low C_i conditions, photosynthesis of high C_i cells becomes sensitive to O_2, the CO_2 compensation point at pH 8.0 is usually greater than 60 μl $CO_2 l^{-1}$, and apparent

photorespiration can be observed. Glycolate synthesis can be demonstrated and can be greatly increased in microalgae if the incubation is done in the presence of aminooxyacetate (AOA), an inhibitor of the glyoxylate aminotransferase. While the cells have some ability to metabolize glycolate due to the presence of glycolate dehydrogenase, most of the glycolate is excreted into the medium.

Low C_i cells of unicellular algae and cyanobacteria have a $K_{1/2}(C_i)$ for photosynthesis in the range 10–40 μM which is essentially independent of the pH. In low C_i cells, photosynthesis is not sensitive to the O_2 concentration, the CO_2 compensation point is less than 5 μl $CO_2 l^{-1}$ and little photorespiration can be demonstrated.

Both unicellular algae and cyanobacteria fix CO_2 by the normal PCR cycle and are C3 plants in that respect. Photosynthesis of low C_i cells, however, exhibits a much higher affinity for CO_2 (< 1μM) than the $K_m(CO_2)$ of Rubisco which is 30–60 μM in green unicellular algae and 100–200 μM in cyanobacteria. The marked discrepancy between the apparent affinity of the primary CO_2-fixing enzyme and the ability of the low C_i cells of unicellular algae and cyanobacteria to use the CO_2 in the medium was reconciled when it was discovered that they possess mechanisms for the active uptake of inorganic carbon which results in an internal inorganic carbon concentration that can be 1000 fold the external concentration. If the intracellular CO_2 and HCO_3^- are in equilibrium with the prevailing internal pH, the internal CO_2 concentration can be more than 17 000 times the external CO_2 concentration. The rate of photosynthesis is directly correlated with the size of the internal pool and agents that block pool formation result in a reduced rate of photosynthesis (Miller *et al.*, 1990). Hence it is the high internal concentration of inorganic carbon that enhances photosynthesis and suppresses oxygenase activity in these organisms.

Cyanobacteria

The generation and use of high internal C_i concentration in cyanobacteria require: (1) active (energy dependent) transport of inorganic carbon (2) a resistance to leakage of inorganic carbon from the cell (3) an internal CA to maintain the CO_2 supply to Rubisco, as at an internal pH of 8.0 the inorganic carbon will be over 98% in the form of HCO_3^-. The mechanisms of C_i concentration have been investigated by a variety of physiological techniques (see Miller, 1990) and by the use of high CO_2 requiring mutants (see Badger and Price, 1992; Kaplan *et al.*, 1994).

Low C_i cells   *Transport systems.*   In low C_i grown cells or standing culture cells of the cyanobacterium *Synechococcus*, CO_2 and HCO_3^- can be transported simultaneously and the transport requires light and is energy dependent (Badger and Andrews, 1982; Espie *et al.*, 1988a). A single transporter for both inorganic carbon species has been proposed (Volokita *et al.*, 1984; Badger and Price, 1992) but there does not appear to be any interaction between the transport of CO_2 and HCO_3^- and other evidence also suggests (see Fig. 22.3) that the CO_2 and HCO_3^- transporters are separate and independent (Miller *et al.*, 1990).

In low C_i cells at pH 8.0, HCO_3^- transport requires millimolar concentrations of Na^+ (Espie *et al.*, 1988a) and is inhibited by Li^+ and monensin (Espie *et al.*, 1988b; Espie and Kandasamy, 1994). This 'Na$^+$-dependent' HCO_3^- transport could be due to a Na^+/HCO_3^- symport but an Na^+/H^+ antiporter cannot be excluded (Espie and Kandasamy, 1994). In standing culture cells there is, in addition to the Na^+-dependent HCO_3^- transport, an Na^+-independent HCO_3^- transport which requires micromolar concentrations of Na^+, is not inhibited by Li^+ and only transiently inhibited by monensin (Espie and Kandasamy, 1992, 1994). CO_2 transport is present in both low C_i cells and standing culture cells but is not affected by Li^+ and monensin and requires micromolar concentrations of Na^+ (Espie *et al.*, 1988b; Espie and Kandasamy, 1994; Miller *et al.*, 1990). CO_2 transport is prevented by suitable concentrations of COS or H_2S (Miller *et al.*, 1990), treatments which do not impair HCO_3^- transport. HCO_3^- and CO_2 are not structural analogues and although $^{12}CO_2$ proves to be an efficient competitive inhibitor of $^{13}CO_2$ uptake, $H^{13}CO_3^-$ is not a competitive inhibitor of CO_2 transport (Espie *et al.*, 1991). Strong evidence for distinct and independent uptake routes is also provided by the isolation of mutants impaired in the transport of only CO_2 or only HCO_3^-. A *Synechocystis* mutant (mutant SC), isolated by Ogawa *et al.* (1993) was able to grow normally at an air level of CO_2

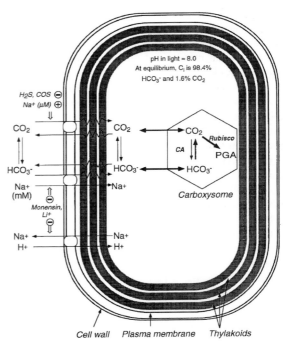

Cell wall   Plasma membrane   Thylakoids

Fig. 22.3 Schematic representation of an air-grown cyanobacterial cell and of the inorganic carbon fluxes involved in the C_i concentrating mechanism. The scheme depicts separate CO_2 and HCO_3^- transport systems and their regulation. The CO_2 transport system requires micromolar amounts of Na^+ and is inhibited by H_2S and COS. The HCO_3^- transport system is fully activated by millimolar amounts of Na^+ and is inhibited by Li^+ and monensin. As monensin collapses the Na^+ gradient through an electroneutral exchange with H^+, the possibility of a Na^+/H^+ antiport is also mentioned. The CO_2 and HCO_3^- are actively transported through the plasma membrane, then cross the thylakoid layers and are delivered to the cytoplasm. It is not known how the C_i species cross the thylakoid layers (zig-zag arrow). In the cytoplasm the HCO_3^- or CO_2 species are allowed to diffuse freely through the thin proteic shell (less than 10 nm) of the carboxysome in either direction. The presence of carbonic anhydrase (CA) in the carboxysome increases the interconversion of the CO_2 and HCO_3^- species, ensuring an efficient utilization of CO_2 by Rubisco. The CO_2 and HCO_3^- can leak back to the cytosol, where either they are hydrated or dehydrated, respectively, or escape to the surrounding medium. By convenience only one carboxysome has been drawn and the relative proportions of the membranes and compartments are not respected. Rubisco and CA are confined to the carboxysomes. At an average intracellular pH of 8.0 in the light, the inorganic carbon (C_i) pool at equilibrium will consist of 98.4% of HCO_3^- and 1.6% CO_2.

(0.04%) but eventually died when transferred to very low CO$_2$ conditions (0.008%). CO$_2$ transport but not HCO$_3^-$ transport was impaired. Mutants that are impaired in HCO$_3^-$ transport but not in CO$_2$ transport have been reported (Kaplan *et al.*, 1994).

The kinetic characteristics of the CO$_2$ and Na$^+$-dependent HCO$_3^-$ transport systems have been determined (Espie *et al.*, 1991; Yu *et al.*, 1994b; Salon *et al.*, 1996a). CO$_2$ transport has a very high affinity for its substrate, with a $K_{1/2}$(CO$_2$) of 0.4–0.6 μM, and a maximum rate in the range 400–700 μmol mg^{-1} Chl h^{-1}. The $K_{1/2}$(HCO$_3^-$) and V_{max} of the HCO$_3^-$ transport system are 17 μM and about 220 μmol mg^{-1} Chl h^{-1} respectively (Yu *et al.*, 1994b; Salon *et al.*, 1996a).

The nature of the CO$_2$ or HCO$_3^-$ transporters is unknown but several suggestions have been made (Badger and Price, 1992; Miller *et al.*, 1991). In *Ulva lactuca* (a green macroalgae), HCO$_3^-$ transport was inhibited by 4,4'-diisothiocyanostilbene-2,2'-disulfonate (DIDS), suggesting that transport may be by an anion exchange carrier (Axelsson *et al.*, 1995).

The capacity of either transport system is sufficient to sustain maximum rates of photosynthesis if the concentration of the appropriate substrate is available. At alkaline pH and low C$_i$, HCO$_3^-$ transport provides most of the carbon necessary for the maintenance of photosynthesis and CO$_2$ transport is primarily involved in scavenging CO$_2$ which has leaked from the cell. The death of the mutant TM17 (Yu *et al.*, 1994a) at very low C$_i$ conditions was correlated with a large decrease in the affinity of the HCO$_3^-$ transport system. At neutral or acid pH or high C$_i$, CO$_2$ transport provides most of the carbon for photosynthesis and the HCO$_3^-$ transporter may be largely involved in a scavenging role.

The use of the internal C$_i$ pool.   Accumulation of C$_i$ into the cell must result in an increase of the CO$_2$ concentration at the active site of Rubisco. In cyanobacteria, Rubisco is mainly located in polyhedral bodies, called carboxysomes, regardless of the [C$_i$] in the growth medium (MacKay *et al.*, 1993). Use of mass spectrometric measurement of ^{18}O depletion from labeled C^{18}O$_2$ provides evidence for the presence of an intracellular CA activity in cyanobacteria. In *Synechococcus* an intracellular CA activity (which requires Mg^{2+}, is inhibited by DTT and sensitive to

EZA) was associated with Rubisco and the carboxysomes (Price *et al.*, 1992). Yu *et al.* (1992) isolated a DNA fragment which complemented high C$_i$ mutants defective in intracellular C$_i$ utilization but not in transport (Type II mutants). The properties of the CA activity which was expressed was similar to that associated with carboxysomes (Price *et al.*, 1992). Upon transfer of high C$_i$ cells to low C$_i$ conditions intracellular CA activity is slightly enhanced concomitantly with the C$_i$ concentrating mechanism. The close association of Rubisco and CA in a properly constituted carboxysome would seem to be required for the utilization of the internal pool as mutants lacking CA or properly constituted and organized carboxysomes are not able to use the internal C$_i$ pool (Price and Badger, 1991; Badger and Price, 1992; Kaplan *et al.*, 1994).

The retention of internal C$_i$.   Plasma membrane of most cells are highly permeable to CO$_2$ (permeability coefficient $\approx 10^{-4}$ m s^{-1}) and the rate of CO$_2$ leakage through such membranes would prevent any significant internal accumulation of C$_i$. C$_i$ leakage, however, is limited from cyanobacteria and calculated permeability coefficients for CO$_2$ are 10^{-7} to 10^{-8} m s^{-1} and for HCO$_3^-$ 10^{-9} m s^{-1} (Price and Badger, 1991; Salon *et al.*, 1996b). Two views, which are still under debate, of C$_i$ flux in cyanobacteria have been suggested.

The carboxysome model.   In the 'carboxysome model' suggested by Reinhold *et al.* (1987, 1989, 1991), the inorganic carbon is delivered into the cell in the form of HCO$_3^-$ and remains largely as HCO$_3^-$ except in the carboxysome where CA maintains the HCO$_3^-$/CO$_2$ equilibrium (Badger and Price, 1992). CO$_2$ leakage out of the carboxysome is prevented either through (1) a special property of the carboxysome shell (Reinhold *et al.*, 1989) or (2) by a spatial arrangement of CA and Rubisco in the carboxysomes (Reinhold *et al.*, 1991) which dictates that the CO$_2$ produced is efficiently fixed before it can escape to the cytosol. This picture was attractive in that no special diffusion resistance for CO$_2$ or O$_2$ would be needed at the level of the plasmalemma membrane and O$_2$ could freely escape from the cells but could not (like CO$_2$) enter the carboxysomes, providing extra insurance for a decrease in oxygenase activity. The special properties of the carboxysomal shell (selective permeability for HCO_3^- not for CO_2) were originally assigned to its proteic nature but to date no experimental evidence has been produced in support of such properties. To date the only support for this hypothesis comes from a mutant (pCA mutant) where human CA was expressed in the cytosol (Price and Badger, 1989a). This was expected to equilibrate CO_2 and HCO_3^- throughout the cell and reduce the ability to concentrate C_i internally, due to 'extensive CO_2 leakage'. The expression of human CA in the cytosol did result in a high CO_2 requiring phenotype but corroborating evidence of 'extensive CO_2 leakage' has not been presented.

In 'Type I' mutants, the mistargeting of CA to the cytosol (Price and Badger, 1989b) instead of to the carboxysomes strongly impaired CO_2 fixation but these cells were still able to accumulate internal C_i pools of normal size (40 mM). The presence of both CA and a consistent C_i pool in the cytosol of these cells should preclude the very low level of cytosolic CO_2 postulated for the wild type and the barrier of CO_2 leakage, supposed to be located at the level of the carboxysome, should no longer be relevant. As a consequence, the permeability coefficient of the cell for CO_2 should be close to that of a plasma membrane and CO_2 efflux would then be substantial for a reported internal C_i pool size of 40 mM (Price and Badger, 1989b). This turns out not to be the case and the permeability coefficient for CO_2 of these cells was only three fold that of the wild type but still two orders of magnitude lower than that of plasma membranes. Similarly, in the deletion mutant PN (Price and Badger, 1991), where no carboxysomes are detected electron microscopically and the Rubisco and CA activities are found to be principally soluble, the permeability coefficient for CO_2 of these cells was similar to that of the wild type (Price and Badger, 1991). The carboxysomeless mutants M3 (Marco et al., 1993) and EK6 (Lieman-Hurwitz et al., 1990) were also not impaired in the accumulation of C_i internally but were unable to use it. An alteration in the small subunit of Rubisco in the mutant EK6 suggested its involvement in the correct assembly of carboxysomes. The association of CA and Rubisco and the correct organization of the carboxysomes (Reinhold et al., 1991) might be responsible for the efficient supply of CO_2 at the active site of Rubisco

but carboxysome integrity does not seem to be associated with the efficient retention of C_i internally.

Kaplan et al. (1994) propose an extension of the carboxysome scheme in the form of an energy-dependent CA-like activity which would be located at the level of the plasma membrane. The CO_2 molecules escaping from the carboxysomes and leaking from the cytosol to the exterior would be hydrated to HCO_3^- by this CA like activity. However, again, no direct evidence is consistent with this scheme and the energy-dependent mechanism driving this CA-like activity against the already suggested tremendous disequilibrium between HCO_3^- and CO_2 in the cytosol is unknown.

The whole-cell equilibrium state.   In this proposal of the C_i fluxes in cyanobacteria (Espie et al., 1989), either CO_2 or HCO_3^- could be delivered to the interior of the cell and they would be kept in equilibrium by CA in the carboxysomes. Simply because the intracellular pH in the light is about 8.0 the C_i would be over 98% in the form of HCO_3^-. The carboxysome would be permeable to both HCO_3^- and CO_2 and while CO_2 fixation in the carboxysomes could contribute to a reduction in CO_2 leakage other components of the cell would also be involved. This suggestion is compromised by the inability to identify the other cell components that contribute a resistance to C_i leakage but considerable evidence is consistent with the maintenance of the HCO_3^-/CO_2 equilibrium in the cell.

When C_i accumulates internally the concentration of both CO_2 and HCO_3^- would rise (Fig. 22.3) and both CO_2 and HCO_3^- efflux would consequently increase. The rapid loss of ^{18}O from $C^{18}O_2$ that is observed when CO_2 efflux is caused by the addition of COS is consistent with rapid interconversion of CO_2 and HCO_3^-. Initial CO_2 uptake is the same but subsequent CO_2 uptake and the size of the internal pool are significantly different when only CO_2 transport is allowed and when CO_2 and HCO_3^- transport is allowed (Fig. 22.4). When only CO_2 transport is active only a small internal pool of C_i develops and CO_2 efflux cannot be detected in the light. However, when the HCO_3^- transport system is also allowed to proceed, a characteristic increase in the extracellular $[CO_2]$ trace, only explainable by an increased CO_2 leakage due to the higher size of the intracellular C_i pool, has always been observed (Fig.

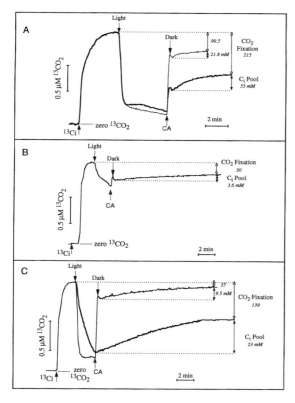

Fig. 22.4 Typical changes in the extracellular CO_2 concentration recorded by a mass spectrometer during incubation of air-grown cells of *Synechococcus* (A), *Selenastrum minutum* (B) or *Chlamydomonas reindardtii* (C) in the presence of 100 μM inorganic carbon. During the incubation of *Selenastrum minutum* and *Chlamydomonas reinhardtii* only contaminant levels of Na^+ were present in the reaction medium while *Synechococcus* was incubated in the presence of either 100 μM NaCl (narrow line, A) or 25 mM NaCl (broad line, A). The inorganic carbon in the form of $KH^{13}CO_3$ was introduced in the dark (arrow $^{13}C_i$) and allowed to equilibrate. Acetozolamide (100 μM) was injected into the reaction medium in order to inhibit the external CA of *Chlamydomonas* (narrow line, C). Upon illumination the cells deplete the extracellular C_i through their active CO_2 and HCO_3^- transport systems. In the extracellular medium and in the absence of CA externally, the rate of CO_2 production from the spontaneous dehydration of HCO_3^- is slower than the rate of active CO_2 transport, and a rapid decrease in the extracellular $[CO_2]$ is observed (A, B and C, narrow line). When CO_2 and HCO_3^- are kept in equilibrium by the extracellular CA of *Chlamydomonas* (broad line, C) the rapid drop in the extracellular $[CO_2]$ is not observed because the rate of CO_2

uptake is slower than the catalyzed rate of HCO_3^- dehydration (broad line, C). After a steady rate of CO_2 depletion is observed, an injection of carbonic anhydrase (CA) in the reaction medium (arrow CA) causes the CO_2 and HCO_3^- to return to their equilibrium values (A, B, and C narrow line). In the presence of CA, the CO_2 trace represents the total extracellular $[C_i]$ and it is obvious, that before addition of CA, the CO_2 and HCO_3^- were being kept out of equilibrium by rapid CO_2 transport. In *Chlamydomonas* cells which were not treated with acetozolamide, the C_i species were already maintained at equilibrium by the external CA and the addition of CA does not change the trace. When the lights are turned off the internal inorganic carbon (C_i pool) leaks back to the medium (double filled arrows). The amount of C_i fixed photosynthetically during the light period is given by the difference between the initial and the final extracellular C_i (double empty arrows) once all the intracellular inorganic carbon has leaked back into the medium in the dark. For *Synechococcus* when both the CO_2 and HCO_3^- transport systems are allowed to proceed (25 mM NaCl, broad line) the size of the internal C_i pool and the resulting amount of C_i fixed photosynthetically are much higher than when only the CO_2 transport system is operative (100 μM NaCl, narrow line). During the light period the characteristic shape of the curve which is always observed when the HCO_3^- transport system is allowed to proceed (broad line) is a consequence of the increase of the CO_2 efflux resulting from the enhanced internal C_i pool. For *Selenastrum minutum* and *Chlamydomonas reinhardtii* treated with acetozolamide the rate of net CO_2 transport just before the addition of CA represents 3 and 30%, respectively, of the rate of CO_2 fixation. For *Chlamydomonas* incubated in the absence of acetozolamide, both the internal C_i pool and CO_2 fixation are much higher than when the CO_2 supply was limited by the absence of an external CA. CO_2 fixation is given in units of μmol mg^{-1} Chl h^{-1}. All incubations were conducted in 25 mM BTP/HCl buffer, pH 8.0 except for *Chlamydomonas* which was incubated in 50 mM HEPES, pH 8.0. Temperature and light intensity was 30 °C and 220 μmol photons m^{-2} s^{-1} for *Synechococcus*, 25 °C and 500 μmol photons m^{-2} s^{-1} for *Selenastrum* and 25 °C and 400 μmol photons m^{-2} s^{-1} for *Chlamydomonas*.

22.4; Badger and Andrews, 1982; Salon *et al.*, 1995a). The measurement of light and dark efflux rates of CO_2 and HCO_3^- are consistent with an increase of 0.6 units in the intracellular pH upon transfer of the cells from darkness to light (Salon *et al.*, 1996b). This suggests that the HCO_3^-/CO_2 equilibrium is rapidly attained.

At pH 8.0 when the extracellular $[C_i]$ is high, CO_2 efflux represents only a small fraction of the maximum rate of the CO_2 transport system but the HCO_3^- efflux represents more than 45% of the maximum rate of the

HCO_3^- transport system (Salon *et al.*, 1996b). The CO_2 and HCO_3^- transport systems may act as scavengers of the leaking corresponding inorganic carbon species to increase the overall efficiency of the C_i concentrating mechanism under conditions where C_i leakage is inevitable (Salon *et al.*, 1996b).

High C_i cells   When high C_i cells are illuminated, the internal C_i concentration can reach values much higher than the extracellular C_i, so there is little doubt that high C_i cells of cyanobacteria also possess an active inorganic carbon transport mechanism. It differs in its properties, however, from that of low C_i cells or standing culture cells. While the HCO_3^- transport system can be shown to occur in cells grown on low levels of C_i when the rate of photosynthesis exceeds the maximum production of CO_2 from the spontaneous dehydration of HCO_3^-, the rate of photosynthesis of high C_i cells does not exceed the rate of spontaneous dehydration of HCO_3^- to CO_2. Thus high C_i cells appear to have retained only the CO_2 transport system. In high C_i cells, the $K_{1/2}(CO_2)$ and $K_{1/2}(C_i)$ are similar at pH 6.0 (about 15–30 μM) but when the pH is increased to 9.0 the $K_{1/2}(CO_2)$ remains relatively constant whereas the $K_{1/2}(C_i)$ increases by 200 fold. During disequilibrium experiments, when CO_2 is supplied to the cells the size of the internal C_i pool and the photosynthetic rate rapidly increase without any lag whereas when HCO_3^- is the supplied species, the size of the internal pool and the photosynthetic rate are much smaller and photosynthesis proceeds with a lag.

The transport systems for CO_2 may not be the same, however, in low C_i cells and high C_i cells. Although the requirements for Na^+ are similar and Li^+ does not inhibit the CO_2 transport system either in low or in high C_i cells, the affinity for CO_2 is about 10 to 50 times higher in low C_i cells than the $K_{1/2}(CO_2)$ determined for photosynthesis in high C_i cells.

Microalgae

There is, as yet, no direct evidence for an active inorganic carbon concentrating mechanism in high C_i-grown unicellular algae. In *Chlamydomonas* the $K_m(CO_2)$ for Rubisco has been reported as 29–57 μM, whereas the $K_{1/2}(CO_2)$ for whole cell photosynthesis

has been reported to be 20–30 μM so that little, if any, concentration of the CO_2 would be required to reconcile the affinity of Rubisco for CO_2 with the apparent affinity of the cell for CO_2.

In contrast, it is well established that when high C_i cells of unicellular algae are transferred to low C_i conditions they develop an active inorganic carbon transport mechanism which results in intracellular concentrations many times higher than the extracellular concentrations (Aizawa and Miyachi, 1986). It has been suggested that an intermediate of the photorespiratory pathway (phosphoglycolate was proposed) might be an effector of the C_i concentrating mechanism (Marcus *et al.*, 1983; Ramazanov and Cardenas, 1992), but definitive evidence is still lacking.

Although the accumulation of inorganic carbon in microalgae and cyanobacteria was demonstrated at about the same time, much less is known about the carbon concentrating mechanism in microalgae. The eukaryotic structure requires that the carbon cross the chloroplast membrane as well as the plasmalemma. Evidence has been presented for a transporter(s) on the chloroplast membrane with CO_2 reaching that site by diffusion (Moroney *et al.*, 1987). Evidence has also been presented for the absence of any transporters on the chloroplast membrane (Williams and Turpin, 1987; Ramazanov and Cardenas, 1992; Sültemeyer *et al.*, 1990; Rotatore and Colman, 1991) which would suggest that the transporters would be present on the plasma membrane. Depending on the location of the transporters the internal inorganic carbon accumulation may be in the whole cell or only in the chloroplast.

All microalgal cells with carbon concentrating mechanisms actively transport CO_2. In low C_i microalgae which lack external CA this may be the only source of inorganic carbon (Aizawa and Miyachi, 1986). *Chlamydomonas* and some other microalgae can also directly transport HCO_3^- (Sültemeyer *et al.*, 1990, 1991; Palmqvist *et al.*, 1994). In *Chlamydomonas* and a number of other microalgae, however, the development of the low C_i state is accompanied by a large increase in extracellular CA (Moroney *et al.*, 1985; Spalding *et al.*, 1983; Husic *et al.*, 1989; Sültemeyer *et al.*, 1990; Coleman *et al.*, 1991; Palmqvist *et al.*, 1994). In

Chlamydomonas reinhardtii two periplasmic forms of extracellular CA have been extensively characterized and both are under transcriptional control. Most of the activity of the extracellular CA (90%) has been found by immunoelectron microscopy to be associated with the inner layer of the cell wall (Coleman *et al.*, 1991). This CA activity is coded by the CAH-1 gene and induced at growth under-limiting C_i (Bailly and Colman, 1988). The remaining extracellular CA activity is coded by the CAH-2 gene and expressed at high extracellular C_i but repressed at low extracellular C_i (Tachiki *et al.*, 1992). In some microalgae the use of HCO_3^- is mediated through its conversion to CO_2 by this extracellular CA (Aizawa and Miyachi, 1986). In others, such as *Chlamydomonas* however, the presence of the extracellular CA seems to ensure that both forms of inorganic carbon (CO_2 and HCO_3^-) are available for transport.

Evidence for soluble cytoplasmic and insoluble chloroplastic CA has been found in *Chlamydomonas reinhardtii*, *Chlorella* and *Dunaliella* (Sültemeyer *et al.*, 1990; Coleman *et al.*, 1991; Ramazanov and Cardenas, 1992; Husic and Marcus, 1994). Intracellular CAs are present both in low and high C_i cells but their activities increase in low C_i cells. These internal CA activities seem to be primarily involved in the CO_2 supply for fixation by Rubisco. In mutants deficient in intracellular CA activity, or when the internal CA is inhibited, although an internal inorganic carbon pool develop, the cells are unable to use it efficiently for photosynthesis. This suggests that most of the C_i in the cell or specially in the chloroplast or in the pyrenoid, where Rubisco is found, is in the form of HCO_3^- due presumably to the prevailing pH. Although some models have been proposed (Sültemeyer *et al.*, 1990; Ramazanov and Cardenas, 1992) the function of the soluble cytoplasmic CA and the involvement of the chloroplastic CA in the supply of CO_2 to Rubisco needs further investigations.

Energy for transport The source of energy for driving internal C_i accumulation against a strong electrochemical gradient is still under investigation. A low level of PSII activity seems to be necessary for the activation of the C_i concentrating mechanism (Kaplan *et al.*, 1987) but PSI provides the energy for

sustained transport (Ogawa *et al.*, 1985). A strong correlation between quenching of the fluorescence of chlorophyll *a* and the activity of C_i transport demonstrated the occurrence of a link between photosynthetic electron transport and C_i accumulation (Miller *et al.*, 1991; Espie *et al.*, 1991). In cyanobacteria, the transport system can be studied in the absence of CO_2 fixation by inhibiting it with iodoacetamide, glycoladehyde (Miller *et al.*, 1989) or glyceraldehyde. In both inhibited cells or non-inhibited cells the provision of inorganic carbon causes a large increase in the photoreduction of oxygen. For *Chlamydomonas reinhardtii*, Sültemeyer *et al.* (1993) recently suggested pseudocyclic photophosphorylation as a source of the energy requirement of the C_i concentrating mechanism. These authors found that low C_i cells displayed higher O_2 photoreduction than high C_i cells at high pH (8.0) but not at low pH (5.5). This enhancement of O_2 photoreduction was concomitant with the development of the C_i concentrating mechanism and a 4 to 20 fold increase in the activity of superoxide-radical-degrading enzymes was observed in low C_i cells at pH 8.0. The authors concluded that pseudocyclic photophosphorylation could supply at least 70% of the ATP requirement of the C_i concentrating mechanism. However, for cyanobacteria, removing the O_2 from the medium does not preclude but enhances the formation of the intracellular C_i pool (Mir *et al.*, 1995). Under such conditions pseudocyclic photophosphorylation could not account for the source of energy necessary for the inorganic carbon transport system. As cyanobacteria lack non-photochemical quenching of chlorophyll *a* fluorescence a possible role for O_2 photoreduction in cyanobacteria could be protection against photoinhibition, as previously hypothesized for photorespiration. Energy for the C_i concentrating mechanism must then be provided by cyclic and/or non-cyclic photophosphorylation. Inactivation of genes encoding various subunits of NADH dehydrogenase resulted in high CO_2 requiring mutants impaired in C_i accumulation (Ogawa, 1992) but not in internal C_i utilization. So NADH dehydrogenase could be involved in the path of electron flow. When cyanobacterial cells are exposed to far red light which is absorbed by PSI, electrons flow toward P700. Monitoring the oxidation state of

P700, Mi *et al.* (1992a,b) recently demonstrated that a cyclic electron flow through PSI, mediated via pyridine nucleotide to the plastoquinone pool, occurred and hence might provide a link between photosynthesis and respiration after the Q_B site (Mi *et al.*, 1992a,b). This was supported by studies with a mutant in which inactivation of a gene coding for NADH dehydrogenase resulted in the inability to accumulate the C_i pool. Hence C_i transport could obtain its energy requirement from cyclic electron flow and according to the authors, such a process would produce an adjustment of the ATP/NAD(P)H ratio under varying cellular and environmental conditions. However, much work still needs to be done to establish firmly the mechanisms of active C_i transport. Additional studies are also required concerning the distribution of such transport systems in the plant kingdom. Molecular biology has already provided substantial advances concerning photorespiration and also promises to provide considerable insights into the intrinsic mechanisms of active C_i accumulation in cyanobacteria and microalgae.

Concluding remarks

In C3 plants, which constitute the bulk of economically important plants, photorespiration results in considerable loss of fixed carbon and utilization of energy, which subsequently decreases the productivity and yield, without any apparent beneficial function. Much research will continue to be devoted towards finding a means of suppressing photorespiration. Whether or not this can be done remains an open question and additional work is also necessary in order to understand the significance of photorespiration.

A possible protective role of photorespiration in higher plants under severe stress needs further confirmation and research. For example, this role has not been shown under natural conditions. Moreover, oxygenase activity, the first reaction of photorespiration, occurs in the Rubisco of all organisms and may well be a chemical consequence of the carboxylation mechanism, photorespiration being then a salvage operation. Although oxygenase activity is ubiquitous, the increased specificity of Rubisco for CO_2 in C3 higher plants compared with that of the Rubisco of microorganisms shows that the carboxylase/oxygenase ratio is not inflexible but can be modified. To date many attempts have been made in order to increase the specificity factor of Rubisco but while some success was obtained, the carboxylation activity decreased concomitantly. The substitution of Mn^{2+} for Mg^{2+} as the activating metal for Rubisco also alters the carboxylase/oxygenase ratio. These changes, while they are no doubt significant, are relatively small compared with the changes that are observed from the other strategy nature has adopted to cope with oxygenase activity, namely, that of CO_2 concentration.

The CO_2 concentrating mechanisms, under normal conditions, virtually eliminate oxygenase activity as no effect of oxygen on photosynthesis can be observed. CAM plants and C4 plants have developed a supplementary system of fixation of CO_2 into C4 acids that acts as the front-end to the carboxylation reaction of Rubisco. In both plants, the decarboxylation of these C4 acids elevates the CO_2 concentration in the vicinity of Rubisco to enhance carboxylation and suppress oxygenation. In CAM plants there is a temporal separation between the initial CO_2 fixation and the final CO_2 fixation. In C4 plants there is an anatomical separation between these events, with Rubisco specifically localized only in a certain portion of the chlorophyll-containing cells. In both types of plants an elaborate complex biochemical system has had to develop for the complete pathway of CO_2 fixation to operate. Microalgae and cyanobacteria tackled the problem more directly by developing systems that concentrate CO_2 through the active transport of CO_2 as well as HCO_3^-. In cyanobacteria these transport systems are located in the plasma membrane, but in the microalgae they may also be present in the chloroplast envelope in addition to the plasma membrane. There is no evidence for active CO_2 transport in the chloroplast of higher plants. Why have these mechanisms for CO_2 concentration been lost in the evolution of algae into higher plants? This question and many others will continue to provide fertile ground for research for years to come, although the increasing CO_2 concentration of the atmosphere may diminish photorespiration and make CO_2-concentrating mechanisms less important.

Further reading and references

Aizawa, K. and Miyachi, S. (1986). Carbonic anhydrase and CO_2 concentrating mechanisms in microalgae and cyanobacteria. *Fed. Eur. Microbiol. Soc. Microbiol. Rev.* **39**, 215–33.

Artus, N. N., Somerville, S. C. and Somerville, C. R. (1986). The biochemistry and cell biology of photorespiration. *CRC Crit. Rev. Plant Sci.* **4**, 121–47.

Axelsson, L., Ryberg, H. and Beer, S. (1995). Two modes of bicarbonate utilization in the marine green macroalga *Ulva lactuca. Plant Cell and Environ.* **18**, 439–45.

Badger, M. R. (1985). Photosynthetic oxygen exchange. *Ann. Rev. Plant Physiol.* **36**, 27–53.

Badger, M. R. and Andrews, T. J. (1982). Photosynthesis and inorganic carbon usage by the marine cyanobacterium, *Synechococcus* sp. *Plant Physiol.* **70**, 517–23.

Badger, M. R. and Canvin, D. T. (1981). Oxygen uptake during photosynthesis in C3 and C4 plants. In *Photosynthesis IV. Regulation of Carbon Metabolism*, ed. G. Akoyunoglou, Balaban International Science Services, Philadelphia, pp. 151–61.

Badger, M. R., Price, G. D. and Yu, J. W. (1991). Selection and analysis of mutants of the CO_2 concentrating mechanism in cyanobacteria. *Can. J. Bot.* **69**, 974–83.

Badger, M. R. and Price, G. D. (1992). The CO_2 concentrating mechanism in cyanobacteria and green algae. *Physiol. Plant.* **84**, 606–15.

Bailly, J. and Coleman, J. R. (1988). Effect of CO_2 concentration on protein biosynthesis and carbonic anhydrase expression in *Chlamydomonas reinhardtii. Plant Physiol.* **87**, 833–40.

Blackwell, R. D., Murray, A. J. S., Lea, P. J., Kendall, A. C., Hall, N. P., Turner, J. C. and Wallsgrove, R. M. (1988). The value of mutants unable to carry out photorespiration. *Photosynthesis Res.* **16**, 155–76.

Bowes, G., Ogren, W. L. and Hageman, R. H. (1971). Phosphoglycolate production catalyzed by ribulose diphosphate carboxylase. *Biochem. Biophys. Res. Commun.* **45**, 716–22.

Bowes, G. and Salvucci, M. E. (1989). Plasticity in the photosynthetic carbon metabolism of submersed aquatic macrophytes. *Aquatic Bot.* **34**, 233–66.

Cockburn, W., Ting, I. P. and Sternberg, L. O. (1979). Relationships between stomatal behavior and internal carbon dioxide concentration in Crassulacean acid metabolism plants. *Plant Physiol.* **63**, 1029–32.

Coleman, J. R., Rotatore, C., Williams, T. G. and Colman, B. (1991). Identification and localization of carbonic anhydrase in two *Chlorella* species. *Plant Physiol.* **95**, 331–43.

Colman, B. (1989). Photosynthetic carbon assimilation and the suppression of photorespiration in the cyanobacteria. *Aquatic Bot.* **34**, 211–31.

Cournac, L., Dimon, B. and Peltier, G. (1993). Evidence for O-18 labeling of photorespiratory CO_2 in photoautotrophic cell cultures of higher plants illuminated in the presence of O-18. *Planta* **190**, 407–14.

Dai, Z. Y., Ku, M. S. B. and Edwards, G. E. (1993). C4 Photosynthesis – the CO_2-concentrating mechanism and photorespiration. *Plant Physiol.* **103**, 83–90.

De Veau, E. J. and Burris, J. E. (1989a). Glycolate metabolism in low and high CO_2-grown *Chlorella pyrenoidosa* and *Pavlova lutheri* as determined by ^{18}O-labeling. *Plant Physiol.* **91**, 1085–93.

De Veau, E. J. and Burris, J. E. (1989b). Photorespiratory rates in wheat and maize as determined by ^{18}O-labeling. *Plant Physiol.* **90**, 500–11.

Douce, R., Bourguignon, J., Macherel, D. and Neuburger, M. (1994). The glycine decarboxylase system in higher plant mitochondria – Structure, function and biogenesis. *Biochem. Soc. Trans.* **22**, 184–8.

Edwards, J. W., Walker, E. L. and Coruzzi, G. M. (1990). Cell specific expression in transgenic plants reveals nonoverlapping roles for chloroplast and cytosolic glutamine synthetase. *Proc. Nat. Acad. Sci. USA* **87**, 3459–63.

Espie, G. S. and Kandasamy, R. A. (1992). Na$^+$ independent HCO$_3^-$ transport and accumulation in the cyanobacterium *Synechococcus* UTEX 625. *Plant Physiol.* **98**, 560–68.

Espie, G. S. and Kandasamy, R. A. (1994). Monensin inhibition of Na$^+$ dependent HCO$_3^-$ transport distinguishes it from Na$^+$ independent HCO$_3^-$ transport and provides evidence for Na$^+$/HCO$_3^-$ symport in the cyanobacterium *Synechococcus* UTEX 625. *Plant Physiol.* **104**, 1419–28.

Espie, G. S., Miller, A. G., Birch, D. G. and Canvin, D. T. (1988a). Simultaneous transport of CO_2 and HCO$_3^-$ by the cyanobacterium *Synechococcus* UTEX 625. *Plant Physiol.* **87**, 551–4.

Espie, G. A., Miller, A. G. and Canvin, D. T. (1988b). Characterization of the Na$^+$ requirement in cyanobacterial photosynthesis. *Plant Physiol.* **88**, 757–62.

Espie, G. S., Miller, A. G. and Canvin, D. T. (1989). The selective and reversible inhibition of active CO_2 transport by hydrogen sulfide in a cyanobacterium. *Plant Physiol.* **91**, 387–94.

Espie, G. S., Miller, A. G. and Canvin, D. T. (1991). High affinity transport of CO_2 in the cyanobacterium *Synechococcus* UTEX 625. *Plant Physiol.* **97**, 943–53.

Furbank, R. T., Jenkins, C. L. D. and Hatch, M. D. (1989). CO_2 concentrating mechanism of C4 photosynthesis. Permeability of isolated bundle sheath cells to inorganic carbon. *Plant Physiol.* **91**, 1364–71.

Gerbaud, A. and André, M. (1987). An evaluation of the recycling in measurements of photorespiration. *Plant Physiol.* **83**, 933–7.

Givan, C. V. and Kleckzowski, L. A. (1992). The enzymic

reduction of glyoxylate and hydroxypyruvate in leaves of higher plants. *Plant Physiol.* **100**, 552–6.

Hanning, I. and Heldt, H. W. (1993). On the function of mitochondrial metabolism during photosynthesis in spinach (*Spinacia oleracea* L.) leaves – Partitioning between respiration and export of redox equivalents and precursors for nitrate assimilation products. *Plant Physiol.* **103**, 1147–54.

Heupel, R. and Heldt, H. W. (1994). Protein organization in the matrix of leaf peroxisomes – A multi-enzyme complex involved in photorespiratory metabolism. *Eur. J. Biochem.* **220**, 165–72.

Holaday, A. S. and Chollet, R. (1984). Photosynthetic/photorespiratory characteristics of C3–C4 intermediate species. *Photosynthesis Res.* **5**, 307–23.

Husic, D. W., Husic, H. D. and Tolbert, N. E. (1987). The oxidative photosynthetic carbon cycle or C2 cycle. *CRC Crit. Rev. Plant Sci.* **5**, 45–100.

Husic, H. D. and Marcus, C. A. (1994). Identification of intracellular carbonic anhydrase in *Chlamydomonas reinhardtii* with a carbonic anhydrase-directed photoaffinity label. *Plant Physiol.* **105**, 133–9.

Jenkins, C. L. D., Furbank, R. T. and Hatch, M. D. (1989). Inorganic carbon diffusion between C4 mesophyll and bundle sheath cells. Direct bundle sheath CO_2 assimilation in intact leaves in the presence of an inhibitor of the C4 pathway. *Plant Physiol.* **91**, 1356–63.

Joy, K. W., Blackwell, R. D. and Lea, P. J. (1992). Assimilation of nitrogen in mutants lacking enzymes of the glutamate synthase cycle. *J. Exp. Bot.* **43**, 139–45.

Kaplan, A., Zenvirth, D., Marcus, Y., Omata, T. and Ogawa, T. (1987). Energization and activation of inorganic carbon uptake by light in cyanobacteria. *Plant Physiol.* **84**, 210–13.

Kaplan, A., Schwarz, R., Lieman-Hurwitz, J., Ronen-Tarazi, M. and Reinhold, L. (1994). Physiological and molecular studies on the response of cyanobacteria to changes in the ambient inorganic carbon concentration. In *The Molecular Biology of Cyanobacteria*, ed. D. A. Bryant, Kluwer Academic Publishers, Dordrecht, pp. 469–85.

Kendall, A. C., Keys, A. J., Turner, J. C., Lea, P. J. and Miflin, B. J. (1983). The isolation and characterization of a catalase-deficient mutant of barley (*Hordeum vulgare* L.). *Planta* **159**, 505–11.

Lieman-Hurwitz, J., Schwarz, R., Martinez, F., Maor, Z., Reinhold, L. and Kaplan, A. (1990). Molecular analysis of high-CO_2 requiring mutants indicates that genes in the region of *rbc* are involved in the ability of cyanobacteria to grow under low CO_2. *Can. J. Bot.* **69**, 645–50.

Lorimer, G. H. (1981). The carboxylation and oxygenation of ribulose 1,5-bisphosphate: The primary events in photosynthesis and photorespiration. *Ann. Rev. Plant Physiol.* **32**, 349–83.

Lorimer, G. H. and Andrews, T. J. (1981). The C2 chemo-

and photorespiratory carbon oxidation cycle. In *The Biochemistry of Plants*, Vol. 8, eds M. D. Hatch and N. K. Boardman, Academic Press, New York, pp. 329–74.

McKay, R. M. L., Gibbs, S. P. and Espie, G. S. (1993). Effect of dissolved inorganic carbon on the expression of carboxysomes, localization of Rubisco and the mode of inorganic carbon transport in cells of the cyanobacterium *Synechococcus* UTEX 625. *Arch. Microbiol.* **159**, 21–9.

McLilley, R. M., Ebbighausen, H. and Heldt, H. W. (1987). The simultaneous determination of carbon dioxide release and oxygen uptake in suspensions of plant leaf mitochondria oxidizing glycine. *Plant Physiol.* **83**, 349–53.

Madsen, T. V. and Sandjensen, K. (1991). Photosynthetic carbon assimilation in aquatic macrophytes. *Aquatic Bot.* **41**, 5–40.

Marco, E., Ohad, N., Schwarz, R., Lieman-Hurwitz, J., Gabay, C. and Kaplan, A. (1993). High CO_2 concentration alleviates the block in photosynthetic electron transport in an *ndhB*-inactivated mutant of *Synechococcus* sp. PCC 7942. *Plant Physiol.* **101**, 1047–53.

Marcus, Y., Harel, E. and Kaplan, A. (1983). Adaptation of the cyanobacterium *Anabaena variabilis* to low CO_2 concentration in their environment. *Plant Physiol.* **71**, 208–10.

Menzlaff, E. and Flugge, U. I. (1993). Purification and functional reconstitution of the 2-oxoglutarate/malate translocator from spinach chloroplasts. *Biochim. Biophys. Acta* **1147**, 13–18.

Mi, H. L., Endo, T., Schreiber, U. and Asada, K. (1992a). Donation of electrons from cytosolic components to the intersystem chain in the cyanobacterium *Synechococcus* Sp PCC-7002 as determined by the reduction of P700$^+$. *Plant Cell Physiol.* **33**, 1099–105.

Mi, H. L., Endo, T., Schreiber, U., Ogawa, T. and Asada, K. (1992b). Electron donation from cyclic and respiratory flows to the photosynthetic intersystem chain is mediated by pyridine nucleotide dehydrogenase in the cyanobacterium *Synechocystis* PCC 6803. *Plant Cell Physiol.* **33**, 1233–7.

Miller, A. G., Espie, G. S. and Canvin, D. T. (1989). Glycoladehyde inhibits CO_2 fixation in the cyanobacterium *Synechococcus* UTEX 625 without inhibiting the accumulation of inorganic carbon or the associated quenching of chlorophyll *a* fluorescence. *Plant Physiol.* **91**, 1044–9.

Miller, A. G. (1990). Inorganic carbon transport and accumulation in cyanobacteria. In *Autotrophic Microbiology and One-carbon Metabolism*, eds G. A. Codd, L. Dijkhuuizen and F. R. Tabita. Kluwer Academic Publishers, Dordrecht, pp. 25–53.

Miller, A. G., Espie, G. S. and Canvin, D. T. (1990). Physiological aspects of CO_2 and HCO_3^- transport by cyanobacteria: a review. *Can. J. Bot.* **68**, 1291–302.

Miller, A. G., Espie, G. S. and Canvin, D. T. (1991). Active CO_2 transport in cyanobacteria. *Can. J. Bot.* **69**, 925–35.

Mir, N. M., Salon, C. and Canvin, D. T. (1995). Inorganic carbon stimulated nitrite reduction in *Synechococcus*. *Plant Physiol.*, **108**, 313–18.

Monson, R. K., Edwards, G. E. and Ku, M. S. D. (1984). C3–C4 intermediate photosynthesis in plants. *Bioscience* **34**, 563–74.

Moroney, J. V., Husic, H. D. and Tolbert, N. E. (1985). Effect of carbonic anhydrase on inorganic carbon accumulation by *Chlamydomonas reinhardtii*. *Plant Physiol.* **79**, 177–83.

Moroney, J. V., Kitayama, M., Togasaki, R. K. and Tolbert, N. E. (1987). Evidence for inorganic carbon transport by intact chloroplasts of *Chlamydomonas reinhardtii*. *Plant Physiol.* **83**, 460–63.

Ogawa, T., Kaneda, T. and Omata, T. (1985). Photosystem-I-driven inorganic carbon transport in the cyanobacterium *Anacystis nidulans*. *Biochim. Biophys. Acta* **808**, 77–84.

Ogawa, T. (1992). Identification and characterization of the *ictA/ndhL* gene product essential to inorganic carbon transport of *Synechocystis* PCC 6803. *Plant Physiol.* **99**, 1604–8.

Ogawa, T. (1993). Molecular analysis of the CO₂-concentrating mechanism in cyanobacteria. In *Photosynthetic Responses to the Environment*, eds H. Yamamoto and C. Smith, American Society of Plant Physiology, pp. 113–25.

Ogren, W. L. (1984). Photorespiration: pathways, regulation, and modification. *Ann. Rev. Plant Physiol.* **35**, 415–42.

Oliver, D. J. (1994). Glycine decarboxylase. *Ann. Rev. Plant Physiol. Plant Mol. Biol.* **45**, 577–607.

Osmond, C. B. (1981). Photorespiration and photoinhibition. Some implications for the energetics of photosynthesis. *Biochim. Biophys. Acta* **639**, 77–98.

Palmqvist, K., Yu, J. W. and Badger, M. R. (1994). Carbonic anhydrase activity and inorganic carbon fluxes in low-C_i and high-C_i cells of *Chlamydomonas reinhardtii* and *Scenedesmus obliquus*. *Physiol. Plant.* **90**, 537–47.

Peterson, R. B. (1987). Quantitation of the O₂-dependent, CO₂-reversible component of the postillumination CO₂ exchange transient in tobacco and maize Leaves. *Plant Physiol.* **84**, 862–7.

Powles, S. B. (1984). Photoinhibition of photosynthesis induced by visible light. *Ann. Rev. Plant Physiol.* **35**, 14–44.

Price, G. D. and Badger, M. R. (1989a). Expression of human carbonic anhydrase in the cyanobacterium *Synechococcus* PCC 7942 creates a high CO₂ requiring phenotype: evidence for a central role for carboxysomes in the CO₂ concentrating mechanism. *Plant Physiol.* **91**, 505–13.

Price, G. D. and Badger, M. R. (1989b). Isolation and characterization of high CO₂ requiring mutants of the cyanobacterium *Synechococcus* PCC 7942. *Plant Physiol.* **91**, 514–25.

Price, G. D. and Badger, M. R. (1991). Evidence for the role of carboxysomes in the cyanobacterial CO₂-concentrating mechanism. *Can. J. Bot.* **69**, 963–73.

Price, G. D., Coleman, J. R. and Badger, M. R. (1992). Association of carbonic anhydrase activity with carboxysomes isolated from the cyanobacterium *Synechococcus* PCC 7942. *Plant Physiol.* **100**, 784–93.

Ramazanov, Z. and Cardenas, J. (1992). Inorganic carbon transport across cell compartments of the halotolerant alga *Dunaliella salina*. *Physiologia Plant.* **85**, 121–28.

Reinhold, L., Zviman, M. and Kaplan, A. (1987). Inorganic carbon fluxes and photosynthesis in cyanobacteria – A quantitative model. In *Progress in Photosynthetic Research*, Vol. IV, ed. J. Biggins, Martinus Nijhoff Publishers, Dordrecht, pp. 289–96.

Reinhold, L., Zviman, M. and Kaplan, A. (1989). A quantitative model for inorganic carbon fluxes and photosynthesis in cyanobacteria. *Plant Physiol. Biochem.* **27**, 945–54.

Reinhold, L., Kosloff, R. and Kaplan, A. (1991). A model for inorganic carbon fluxes and photosynthesis in cyanobacterial carboxysomes. *Can. J. Bot.* **69**, 984–8.

Reiskind, J. B., Seamon, P. T. and Bowes, G. (1988). Alternative methods of photosynthetic carbon assimilation in marine macroalgae. *Plant Physiol.* **87**, 686–92.

Rotatore, C. and Colman, B. (1991). The acquisition and accumulation of inorganic carbon by the unicellular green alga *Chlorella ellipsoida*. *Plant Cell Environ.* **14**, 377–82.

Salon, C., Mir, N. A. and Canvin, D. T. (1996a). Influx and efflux of inorganic carbon in air-grown cells of *Synechococcus* UTEX 625. *Plant Cell Environ.* **19**, 247–59.

Salon, C. N., Mir, A. and Canvin, D. T. (1996b). HCO₃⁻ and CO₂ leakage from *Synechococcus* UTEX 625. *Plant Cell Environ.* **19**, 260–79.

Schuster, W. S. and Monson, R. K. (1990). An examination of the advantages of C3–C4 intermediate photosynthesis in warm environments. *Plant Cell Environ.* **13**, 903–12.

Sharkey, T. D. (1988). Estimating the rate of photorespiration in leaves. *Physiol. Plant.* **73**, 147–52.

Somerville, S. C. and Somerville, C. R. (1986). Analysis of photosynthesis with mutants of higher plants and algae. *Ann. Rev. Plant Physiol.* **37**, 467–507.

Spalding, M. H., Stumpf, D. K., Ku, M. S. B., Burris, R. H. and Edwards, G. E. (1979). Crassulacean acid metabolism and diurnal variations of internal CO₂ and O₂ concentrations in *Sedum praealtum* DC. *Australian J. Plant Physiol.* **6**, 557–67.

Spalding, M. H., Spreitzer, R. J. and Ogren, W. L. (1983). Genetic and physiological analysis of the CO₂-concentrating system of *Chlamydomonas reinhardtii*. *Planta* **159**, 261–6.

Sültemeyer, D. F., Fock, H. P. and Canvin, D. T. (1990). Mass spectrometric measurement of intracellular carbonic anhydrase activity in high and low C_i cells of *Chlamydomonas*. *Plant Physiol.* **94**, 1250–57.

Sültemeyer, D. F., Fock, H. P. and Canvin, D. T. (1991). Active uptake of inorganic carbon by *Chlamydomonas reinhardtii* – Evidence for simultaneous transport of HCO_3^- and CO_2 and characterization of active CO_2 transport. *Can. J. Bot.* **69**, 995–1002.

Sültemeyer, D. F., Biehler, K. and Fock, H. P. (1993). Evidence for the contribution of pseudocyclic photophosphorylation to the energy requirement of the mechanism for concentrating inorganic carbon in *Chlamydomonas*. *Planta* **189**, 235–42.

Tachiki, A., Fukuzawa, H. and Miyachi, S. (1992). Characterization of carbonic anhydrase isozyme CA2, which is the CAH2 gene product, in *Chlamydomonas reinhardtii*. *Biosci. Biotechnol. Biochem.* **56**, 794–8.

Volokita, M., Zenvirth, D., Kaplan, A. and Reinhold, L. (1984). Nature of the inorganic carbon species actively taken up by the cyanobacterium *Anabaena variabilis*. *Plant Physiol.* **76**, 599–602.

Wallsgrove, R. M., Keys, A. J., Lea, P. J. and Miflin, B. J. (1983). Photosynthesis, photorespiration and nitrogen metabolism. *Plant Cell Environ.* **6**, 301–9.

Williams, T. G. and Turpin, D. H. (1987). The role of external anhydrase in inorganic carbon aquisition by *Chlamydomonas reinhardtii* at alkaline pH. *Plant Physiol.* **83**, 92–9.

Woo, K. C., Flugge, U. I. and Heldt, H. W. (1987). A two-translocator model for the transport of 2-oxoglutarate and glutamate in chloroplasts during ammonia assimilation in the light. *Physiol. Plant.* **84**, 624–32.

Young, X. K. and McCarty, R. E. (1993). Assay of proton-coupled glycolate and D-glycerate transport into chloroplast inner envelope membrane vesicles by stopped-flow fluorescence. *Plant Physiol.* **101**, 793–9.

Yu, J. W., Price, G. D., Song, L. and Badger, M. R. (1992). Isolation of a putative carboxysomal carbonic anhydrase gene from the cyanobacterium *Synechococcus* PCC7942. *Plant Physiol.* **100**, 794–800.

Yu, J. W., Price, G. D. and Badger, M. R. (1994a). A mutant isolated from the cyanobacterium *Synechococcus* PCC 7942 is unable to adapt to low inorganic carbon conditions. *Plant Physiol.* **104**, 605–11.

Yu, J. W., Price, G. D. and Badger, M. R. (1994b). Characterization of CO_2 and HCO_3^- uptake during steady state photosynthesis in the cyanobacterium *Synechococcus* PCC 7942. *Australian J. Plant Physiol.* **21**, 185–95.

Zoglowek, C., Kromer, S. and Heldt, H. W. (1988). Oxaloacetate and malate transport by plant mitochondria. *Plant Physiol.* **87**, 109–15.

23 Metabolite transport and photosynthetic regulation in C4 and CAM plants

R. C. Leegood, S. von Caemmerer and C. B. Osmond

Introduction

Plants with the C4 dicarboxylic acid pathway of photosynthesis (C4 plants) and those with Crassulacean acid metabolism (CAM) plants can be distinguished from others by the initial products of $^{14}CO_2$ fixation and by leaf anatomy. These two features belie a complex of different metabolite transport processes which serve a common physiological function, a CO_2-concentrating mechanism which mitigates the oxygenase activity of Rubisco (Andrews and Lorimer, 1987). This potentially improves the efficiency of carboxylation, of water use and of nitrogen use in photosynthesis, with a number of important implications for plant performance and survival (Osmond et al., 1982; Edwards and Walker, 1983; Winter, 1985; Nobel, 1988; Henderson et al., 1994).

Our appreciation of the significance of these leaf anatomical differences in relation to photosynthetic processes is quite recent. Although Haberlandt (1884) drew attention to the peculiar 'Kranz' (German for wreath) organization of photosynthetic cells which calls for intercellular metabolite transport in what we now know as C4 plants (Fig. 23.1a), it remained for Burr et al. (1957) and Karpilov (1960) to highlight the labeling of C4 acids as early products of $^{14}CO_2$ fixation in sugar cane and maize, respectively. The pathway was subsequently named after these initial products of carboxylation and early developments of this research are well documented by Hatch (1992). Although the unusual capacity for dark CO_2 fixation (De Saussure, 1804) and nocturnal acidification and diurnal de-acidification (Heyne, 1815) in succulent plants was established in the 19th century, the importance of intracellular transport of malic acid into and out of the large vacuole (Fig. 23.1c) was not appreciated. The photosynthetic implications were not recognized until Kunitake and Saltman (1958) showed that malic acid labeled in the dark with $^{14}CO_2$ was conserved and converted to photosynthetic products in the light. Well before this, the metabolic pathway had been named after the family Crassulaceae, which contains many genera and species which show this acid metabolism. It is fair to say that interpretation of CAM as a photosynthetic process was stimulated by analogies with C4 photosynthesis (cf. Ranson and Thomas, 1960; Osmond, 1978).

The carboxylation and decarboxylation events which drive the CO_2-concentrating mechanisms of C4 and CAM plants are similar, but they operate on different anatomical, physiological and biochemical principles. They are initiated by the fixation of HCO_3^- via PEPcase, leading to the synthesis of malate and aspartate, and the subsequent decarboxylation of these C4 acids to generate CO_2 for Rubisco. In C4 plants the PEPcase and Rubisco are spatially separated in outer mesophyll and inner bundle sheath cells of the Kranz complex (Fig. 23.1a,b). Simultaneous bidirectional fluxes of C3 and C4 metabolites in the symplasm connecting these cells are required to sustain C4 photosynthesis, and the 4-C carboxyl carbon of malate and aspartate serves as a short-lived reservoir of CO_2 during transport. Small amounts of C4 acids (a few μmol per mg chlorophyll) are involved, and this transport pool turns over in about 10 s. In CAM plants, on the other hand, both PEPcase and Rubisco are present in all chloroplast-containing cells. The activities of these enzymes are thus regulated in time, rather than in space, with PEPcase active in the dark but inactive for much of the light period, and Rubisco active only in the light.

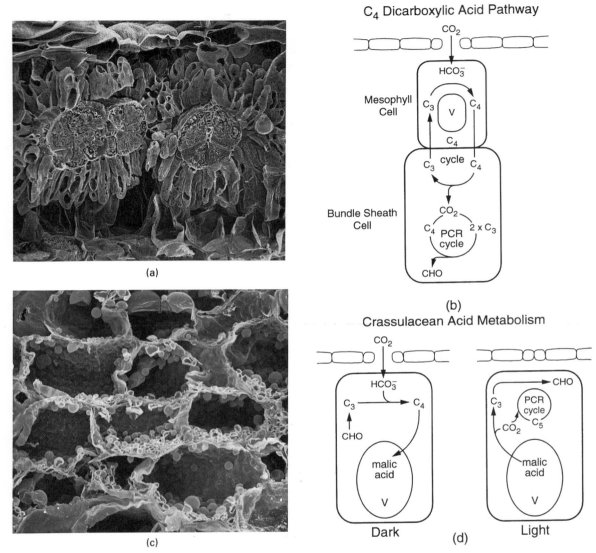

Fig. 23.1 Spatial (C4 plants) and temporal (CAM plants) separation of components of the C4 cycle and photosynthetic carbon reduction (PCR) cycle which underlie metabolite transport in C4 and CAM plants. (a) SEM of a transverse section of the leaf of *Atriplex spongiosa*, a C4 plant, showing the radial arrangement of outer mesophyll (M) and inner bundle sheath cells (BS) (courtesy of J. H. Troughton). (b) A generalized schematic representation of the spatial separation of the carboxylation step in the C4 cycle from decarboxylation and the PCR cycle in C4 plants. (c) SEM of part of a *Kalanchoe daigremontiana* leaf showing the large cell vacuoles (V) (courtesy R. A. Balsamo and E. G. Uribe). (d) A generalized schematic representation of the temporal separation of the carboxylation step of the C4 cycle from decarboxylation and the PCR cycle in CAM plants.

In CAM plants the 4-C carboxyl of malate serves as a longer-lived, much larger temporary store of CO_2. Large amounts of malic acid (several hundreds of μmoles per mg chlorophyll) may accumulate in the large cell vacuoles of CAM plants (Fig. 23.1c,d), and may have a turnover time of thousands of seconds.

The tonoplast fluxes of malic acid, rather than symplastic transport, thus play the key role in the CO_2-concentrating mechanism of CAM, and also in the day–night regulation of enzyme activity.

In general, none of the reactions, regulations or transport processes in C4 or CAM plants is unique. However, they are coordinated in space and time in several unique ways which permit biochemical, physiological and anatomical differentiation of subtypes of C4 plants (Hatch and Osmond, 1976; Hatch 1992, Hattersley 1992) and CAM plants (Osmond, 1978). In addition, it is now recognized that there are many intermediate forms between C4 and C3 metabolism, and between CAM and C3 metabolism. In early studies, artificial hybrids between C3 and C4 plants in the genus *Atriplex* were produced to explore many of the features of C4 photosynthesis (Osmond *et al.*, 1980). However, it is now known that what may be natural hybrids in the genus *Flaveria* display genetically stable, but intermediate physiology and biochemistry (Edwards and Ku, 1987). Artificial interspecific hybridization has been very successful in this genus and is currently being used to further explore the genetics and physiology of the C4 syndrome (Brown and Bouton 1993). Similarly, there are many genetically-determined variants of CAM, and the environmental control of this pathway leads to many manifestations intermediate between strict CAM and C3 photosynthesis (Ting, 1985; Winter, 1985). It should also be noted that many of the physiological and biochemical features (but rarely the anatomical features) of photosynthesis in terrestrial C4 and CAM plants have been observed during photosynthesis in aquatic macrophytes (Cockburn, 1985; Bowes and Salvucci, 1989). We cannot deal with these variations in detail here, but emphasize the utility of these 'intermediate' organisms in evaluating the real functional significance, and hence the selective advantage and probable evolutionary relationships, of the different photosynthetic pathways.

Intercellular metabolite transport during C4 photosynthesis

The dramatic specialization of photosynthetic functions between the mesophyll and bundle sheath cells of C4 plants has few parallels. Perhaps the best

analog is the cooperative function of heterocysts and vegetative cells in the N_2-fixing cyanobacteria (see Chapter 31). In its most sophisticated form, C4 photosynthesis in tropical grasses such as sugar cane and maize depends on mesophyll cell chloroplasts which lack most of the enzymes of the photosynthetic carbon reduction (PCR) cycle, and bundle sheath cells with poorly developed PSII activities. This means that photosynthetic CO_2 fixation in C4 plants depends on inter- and intracellular regulation of metabolite transport within and between cells. Karpilov (1970) introduced the term 'cooperative photosynthesis' to describe the interdependencies of mesophyll and bundle sheath cells. The developmental anatomy of the 'Kranz' sheath in different types of C4 plants has been explored by Dengler *et al.* (1985). The genome of the chloroplasts from the two cell types is the same (Walbot, 1977), but differential expression of the genes for the enzymes peculiar to the C4-cycle seems to involve light-modulation of transcript pools in the two cell types. Nelson and Langdale (1992) conclude that the first step is the light-dependent repression of Rubisco expression in mesophyll cells no more than three cells distant from veins. They suggest a major role for differentiating vascular tissue in the initial processes of photosynthetic differentiation, and that methylation plays a role in subsequent stabilization of the patterns of gene expression.

Extensive surveys of enzyme distribution within and between cells in different C4 plants led to the identification of three decarboxylase enzymes in the bundle sheath cells: $NADP^+$-malic enzyme, NAD^+-malic enzyme and PEP carboxykinase (PEPCK) (Hatch and Osmond, 1976; Edwards and Walker, 1983) (Fig. 23.2). Apart from the different patterns of intercellular movement of metabolites that these subtypes imply, there are patterns of regulation in C4 plants which are unique. One of the most striking examples is the way in which 3-PGA reduction is shared between the PCR cycle in bundle sheath chloroplasts and the mesophyll chloroplasts.

In the majority of C4 plants the photosynthetic cells are organized in two concentric cylinders. Thin-walled mesophyll cells with large intercellular spaces radiate from the thick-walled bundle sheath cells (Fig. 23.1a). In some C4 plants, the cell walls between the bundle sheath and mesophyll contain suberin, but it seems likely that the thickened cell wall is itself a major barrier to the diffusion of solutes and gases in all C4 plants (see Hattersley, 1992). Extensive pit-fields on the wall between mesophyll and bundle sheath cells with plasmodesmata provide symplastic connections between cells (Botha and Evert, 1988). The plasmodesmata evidently exclude large molecules, such as the cytoplasmic enzyme PEPcase which is restricted to mesophyll cells, and the size exclusion limit seems to be about 900 daltons (Burnell, 1988). However, plasmodesmata generally permit the rapid exchange of solutes between cells and slower exchange of gases in solution (Lucas *et al.*, 1993).

Using an analogy based on symplastic transport of solutes across the concentric cylinders of cortical and vascular tissue in roots, Osmond (1971) concluded that metabolite transport in C4 photosynthesis could be sustained by diffusion, driven by gradients in the concentration of specific metabolites between the source cells and the sink cells. Thus, in the case of malate or aspartate, Hatch and Osmond (1976) estimated that a gradient with the concentration in mesophyll source cells 10 to 30 mM higher than in bundle sheath sink cells would be needed to sustain observed rates of photosynthesis in *Zea* (maize) (NADP$^+$-ME) and *Amaranthus* (NAD$^+$-ME), respectively. *Zea*, which has centrifugally arranged bundle sheath chloroplasts and a short diffusion path, and *Amaranthus,* which has centripetally arranged chloroplasts and a longer diffusion path, are representative of a wide range of C4 plants (Fig. 23.2).

Many of the assumptions on which these conclusions were based have been supported by subsequent evidence. For example, Hatch and Osmond (1976) assumed an effective surface area of only 3% of the mesophyll/bundle sheath interface, a value consistent with observation. Plasmodesmata are frequent only in primary pit-fields at the area of contact between mesophyll and bundle sheath cells, and in the C4 grass *Themeda,* 57% of all the plasmodesmata in vascular bundles are at this interface (Botha and Evert, 1988), and the cross-sectional area of the sphincter in plasmodesmata occupies 1.5–3% of the pit-field of the surface between mesophyll and bundle sheath cells. Although resolution of the biochemical and ultrastructural complexity of plasmodesmata has

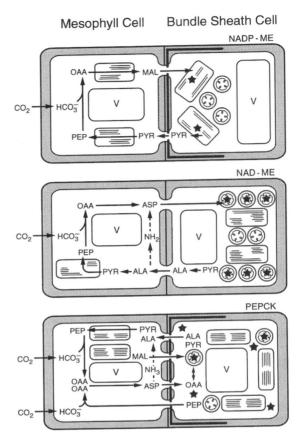

Fig. 23.2 Generalized schematic representation of biochemical and structural bases for metabolite transport and the differentiation of subtypes of C4 photosynthesis (based on Hatch and Osmond, 1976; Hattersely, 1987; Burnell and Hatch, 1988). The sites of C4 acid decarboxylation are indicated by the * and the suberin lamella by the solid black line.

expanded greatly, it remains likely that small molecules diffuse through micro channels between the globular proteins in the appressed ER-protein complex (desmotubule). The path is tortuous, but the distances are small (1 μm or less) and intercellular metabolite exchange may be complete in 50 to 100 ms (Lucas *et al.*, 1993). Perhaps the greatest uncertainty stems from the need to sustain bidirectional metabolite fluxes, when there is a net flux of sucrose from mesophyll to bundle sheath coincident with water flux in the other direction.

Intercellular transport and leaf anatomy

The necessity for metabolite transport between the two types of cells requires intimate contact between cells and therefore sets limits on the amount of mesophyll tissue which can be functionally associated with bundle sheath tissue. For this reason, the leaf thickness is limited in C4 plants, and there are significant differences between the various C4 decarboxylation types (see Hattersley, 1992). For example, in C4 plants the maximum number of chlorenchymatous mesophyll cells which intervene between the photosynthetic bundle sheath cells is between two and four. Hence, the interveinal distance is usually smaller than in the leaves of C3 plants. The chloroplast position and the presence of other cells between the mesophyll and the bundle sheath will affect the distance over which metabolites must be transported during C4 photosynthesis and also need to be considered in relation to translocation of sucrose from the mesophyll cells to the vascular tissue. The occurrence of a suberized lamella varies; it is absent in dicotyledonous species, and in grasses is present only in species with either an uneven bundle sheath outline or with centrifugally located chloroplasts. In those species with uneven cell outlines the suberized lamella may be important in restricting CO_2 leakage through the high surface area of the bundle sheath/mesophyll interface (Hattersley, 1992).

A number of structural features, such as the position of chloroplasts within the bundle sheath cells, have been used to predict the different decarboxylation type (Fig. 23.2). Others include the presence of cells which intervene between the metaxylem vessel elements and the chlorenchymatous bundle sheath cells and the occurrence of an even or uneven outline to the bundle sheath cell walls, which influences the area of contact between the two cell types (see Hattersley, 1992). However, the evidence suggests that such associations between biochemical type and structure may not be wholly reliable (e.g. PEPCK and NAD$^+$-ME type *Panicum* species are anatomically indistinguishable; Prendergast *et al.*, 1987). At the subcellular level, other differences that are related to metabolic pathways arise. In NADP$^+$-ME plants, the bundle sheath chloroplasts contain few grana and show a deficiency of PSII proteins and

function (Hatch and Osmond, 1976; Höfer *et al.*, 1992). In NAD$^+$-ME and PEPCK plants which decarboxylate malate in the bundle sheath mitochondria, the mitochondrial frequency is between two and four times that in the bundle sheath of NADP$^+$-malic enzyme type C4 species (Fig. 23.2).

Intercellular movement of metabolites

Evidence of rapid metabolite movement may be inferred from the rapid transfer of ^{14}C from C4 acids (labeled in the mesophyll) to 3-PGA and products (labeled in the bundle sheath) which occurs when $^{14}CO_2$ is supplied to leaves of C4 plants (Hatch, 1971). It can be visualized by microautoradiography in leaves during in pulse-chase experiments. (Osmond, 1971). After a 2 s pulse of $^{14}CO_2$ (when the majority of the label is in C4 acids), the cytoplasm of the mesophyll cells is clearly labeled. However, a considerable amount of label is already found in bundle sheath cells, and the majority of the label is transferred to the bundle sheath during a 10 s chase.

Direct measurements of the gradients of metabolites in leaves of *Zea* are shown in Fig. 23.3 (Leegood, 1985; Stitt and Heldt, 1985b). The gradient of malate between the mesophyll and the bundle sheath is sufficient to generate a flux between the two cell types (7.2 μmol m^{-2} s^{-1}) which is much larger than the photosynthetic flux (0.5 μmol m^{-2} s^{-1}) estimated by Hatch and Osmond (1976). The gradient of malate in Fig. 23.3, and hence the flux, is likely to be a considerable overestimate, since much of the malate pool is present in the vacuole or in non-photosynthetic leaf cells, and thus does not participate in photosynthesis. An anomalous feature shown in Fig. 23.3 is the predominance of pyruvate in the mesophyll compartment, so that the gradient of pyruvate appears to lie in the opposite direction to the expected flux, but this is a result of the intracellular transport of pyruvate within mesophyll cells (see below). The amounts of 3-PGA and triose-P in leaves of maize are typically 10 to 20 times higher than in the leaves of C3 plants. This reflects the opposing intercellular concentration gradients of these metabolites (Fig. 23.3) which are adequate to generate fluxes required by the assimilation rates. Non-aqueous fractionation of maize leaves has

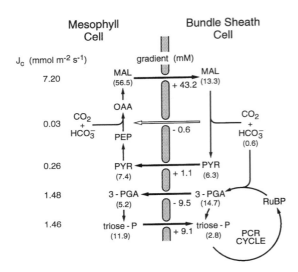

Fig. 23.3 Metabolite concentrations (mM) in mesophyll and bundle sheath cells which drive metabolite transport in *Zea mays* (mean values from Leegood (1985) and Stitt and Heldt (1985a) are shown in parentheses). The gradient in concentration (relative to the bundle sheath cells) is shown across the plasmodesmata, and the maximum flux which could be sustained by the gradient (J_c) is calculated using the assumptions of Hatch and Osmond (1976). These authors calculated that a flux of about $0.5\,\mu\mathrm{mol\,m^{-2}\,s^{-1}}$ on a bundle sheath surface area would support the observed rate of photosynthesis.

subsequently shown that these gradients of 3-PGA and triose-P exist between the cytosols of the mesophyll and bundle-sheath.

Although most other enzymes of the PCR cycle are absent from the mesophyll cells, all C4 species possess the enzymes for 3-PGA reduction in the mesophyll chloroplasts (Hatch and Osmond, 1976). The triose-P which is formed from 3-PGA in the mesophyll has two fates. The first is its conversion to sucrose or starch in the mesophyll cells (see below), while the second is its return to the bundle sheath to regenerate RuBP. In plants such as maize, in which 50% of the 3-PGA generated in the bundle sheath may be reduced in the mesophyll, a minimum of two-thirds of the triose-P must be returned to the bundle sheath to maintain pools of PCR cycle intermediates. In NADP$^+$-ME plants this, and NADPH generation from malic enzyme, compensates for the low PSII activity in the bundle

sheath. Reduced O_2 evolution in the bundle sheath of these plants augments the elevated CO_2 concentration and further favors the carboxylation of RuBP over its oxygenation. Another role for 3-PGA reduction in the mesophyll could be coordination of PCR cycle and C4 cycle turnover (see below), and also it may be a means of ensuring H$^+$ transport, and hence charge balance, between the two cell types. The reduction of 3-PGA to triose-P in the mesophyll consumes a proton, after which the triose-P is transported to the bundle sheath. The consumption of a proton in this reaction is necessary because a proton is released when CO_2 is hydrated and the resulting HCO_3^- is fixed by PEPcase.

Leaves of NAD$^+$-ME plants such as *Amaranthus edulis* contain amounts of triose-P and 3-PGA similar to those in maize, suggesting that extensive intercellular transport of these metabolites occurs. No direct measurements have been made of metabolite gradients in leaves of C4 plants other than maize. In leaves of *Amaranthus edulis*, an NAD$^+$-ME enzyme species which is believed to transfer aspartate from the mesophyll to the bundle sheath and return alanine to the mesophyll cells (Fig. 23.2), leaf pools of aspartate and alanine are sufficient to account for diffusion-driven transport of these compounds between the mesophyll and bundle sheath cells under many different flux conditions (Leegood and von Caemmerer, 1988). In PEPCK plants, transport is likely to be rather more complex, because both malate and aspartate must be transferred from the mesophyll to the bundle sheath, and PEP and alanine must be returned to the mesophyll (Fig. 23.2).

Intercellular movement of inorganic carbon

One of the principal selective pressures leading to the evolution of the above complexes of intercellular metabolite transport is believed to be the gain in photosynthetic rate which can be achieved by concentrating CO_2 in the vicinity of Rubisco and largely eliminating its oxygenase activity. Direct measurements of the pool of $HCO_3^- + CO_2$ during radiotracer studies indicate that the total pool of inorganic carbon is about 0.6 mM in the bundle sheath of maize (Hatch, 1971), corresponding to concentrations of CO_2 some 2 to 5 times ambient. Analysis of CO_2-concentrating mechanisms in the three different decarboxylation types (Furbank and Hatch, 1987) indicates that the absence of carbonic anhydrase in bundle sheath cells is of critical importance. The symplastic pathway of metabolite movement also permits movement of inorganic carbon out of the bundle sheath. Detailed modeling of the compartmentation of the inorganic carbon pool in mesophyll and bundle sheath show that the efflux of HCO_3^- via plasmodesmata is insignificant compared to the C4 acid flux (Fig. 23.4). Thus leakage of HCO_3^- via the plasmodesmata is not likely to be a serious problem, nor is the leakage of CO_2, because the diffusion coefficients of gases in solution are 10^4 times less than in air.

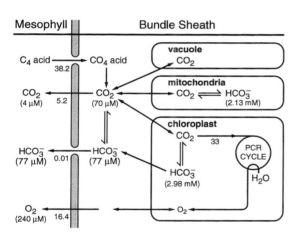

Fig. 23.4 Schematic description of inorganic carbon fluxes and concentrations in C4 leaves during steady state photosynthesis (adapted from Jenkins et al., 1989). Double headed arrows indicate assumed equilibration between CO_2 and O_2 between compartments. Fluxes were converted to a leaf area basis from the original data by assuming 310 mg Chl m^{-2}. A bundle sheath to leaf area ratio of 2.5 can be used to convert fluxes to bundle sheath area basis. Key values are: A net photosynthetic rate of 6.4 μmol min^{-1} (mg Chl^{-1}) (33 μmol m^{-2} s^{-1}), a total bundle sheath inorganic carbon pool of 55 nmol (mg Chl^{-1}); a bundle sheath cell volume 19% of leaf volume. Intercellular compartment volumes (as percent of cell volume) and pH for these compartments were: cytosol, 15% (pH 7.4), chloroplast 17% (pH 7.8), mitochondria 2% (pH 7.6), and vacuole, 51% (pH < 5.5).

Intracellular metabolite transport in C4 plants

The primary carboxylase of C4 photosynthesis is a cytoplasmic enzyme, PEPcase, which draws on the light-dependent production of its substrate, (PEP), in the chloroplast and in many cases the light-dependent reduction of its product, (OAA), also occurs in the chloroplast (Fig. 23.2). Consequently, intracellular transport of metabolites between the cytoplasm and organelles is a key component of C4 photosynthesis in both mesophyll and bundle sheath cells. Leaf organelles in C4 plants share the translocators of organelles in leaves of C3 plants (Heldt and Flügge, 1992), but also contain translocators with unique or greatly altered kinetic properties.

Mesophyll chloroplasts

We have seen that measurements of intercellular compartmentation of pyruvate suggest that it is transported within cells. Transport of pyruvate on a specific carrier occurs in chloroplasts of both C3 and C4 plants, but the translocator is much more active and is light-dependent in mesophyll chloroplasts of C4 plants (Flügge et al., 1985). Pyruvate transport appears to be driven by an H^+ gradient in NADP$^+$-malic enzyme species and a Na$^+$ gradient, or perhaps Na$^+$ activation of the transporter (Murata et al., 1993), in NAD$^+$-malic enzyme and PEP carboxykinase species (Heldt and Flügge, 1992; Ohnishi and Kanai, 1990).

The dicarboxylates, malate, oxaloacetate, 2-oxoglutrarate, aspartate and glutamate are transported through the mesophyll chloroplast envelope in a carrier-mediated mode (Fig. 23.5). These compounds undergo counter-exchange on the dicarboxylate translocator and the K_m for uptake of a particular dicarboxylic acid is similar to the K_i for the inhibition of uptake by other dicarboxylates. In maize, for example, the K_i (0.3 mM) for oxaloacetate inhibition of malate transport is comparable to the K_m (0.5 mM) for malate uptake (Day and Hatch, 1981). Such a system is unsuitable for catalyzing oxaloacetate uptake where oxaloacetate concentrations are several orders of magnitude less than malate concentrations, as occurs in NADP$^+$-ME plants such as maize, in which OAA concentrations

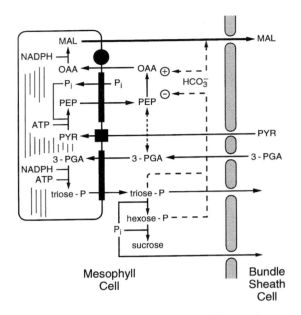

Fig. 23.5 Schematic representation of intercellular and mesophyll chloroplast metabolite transport in a NADP⁺ME-type C4 plant such as *Zea mays*. The dicarboxylate shuttle is shown by a circle, the pyruvate translocator by a square and the P_i translocator by the rectangular symbols on the chloroplast envelope. The equilibrium between 3-PGA and PEP in the cytoplasm is indicated by the dotted line. Feedback (−) and feedforward (+) modulation of cytoplasmic PEPcase by cytoplasmic malate and triose-P/hexose-P respectively, is indicated by the dashed lines.

are probably less than $50 \, \mu M$. Not surprisingly, Hatch *et al.* (1984) have provided evidence for a very active oxaloacetate carrier in maize (K_m $45 \, \mu M$) which is little affected by malate (K_i $7.5 \, mM$).

One feature which distinguishes chloroplastic transport of phosphorylated intermediates in C4 plants from that in C3 plants is the direction of transport across the chloroplast envelope. During photosynthesis in C4 plants, the mesophyll chloroplasts *import* 3-PGA and *export* triose-P, and the bundle sheath chloroplasts *export* 3-PGA and *import* triose-P which has been reduced by the mesophyll chloroplasts (Fig. 23.5). In addition to the exchange of 3-PGA, triose-P and P_i, the mesophyll chloroplasts also catalyze the export of PEP, formed in the chloroplast by the action of pyruvate P_i dikinase, in exchange for P_i to sustain

PEPcase in the cytoplasm. Exchange of PEP, P_i, 3-PGA and triose-P occurs on a common translocator in the chloroplast envelope of C4 plants. The C4 mesophyll translocator is very similar to the C3 type phosphate translocator, with between 83 and 94% identity in amino acid residues. Only minor changes in amino acid sequence have occurred to extend the substrate specificity of the C3 phosphate translocator to recognize PEP in C4 plants (Fischer *et al.*, 1994).

In terms of metabolic regulation, the practical consequences of having intermediates of the C3 and C4 pathways carried on a common translocator cannot easily be predicted since we are entirely ignorant of the subcellular compartmentation of these metabolites under varying environmental conditions.

Bundle sheath chloroplasts

Although transport processes across the envelope of C4 mesophyll chloroplasts are now adequately characterized, little is known of transport into the bundle sheath chloroplasts, largely because these have yet to be isolated intact and in appreciable quantities from any C4 plants. Bundle sheath chloroplasts are unusual in catalyzing 3-PGA export at high rates (Fig. 23.5). In C3 plants, 3-PGA is not exported by chloroplasts to any great extent because 3-PGA is transported by the translocator as the 3-PGA^{2-} ion, whereas 3-PGA^{3-} is the form that predominates at the pH which occurs in the illuminated stroma (Heldt and Flügge, 1992). It is not known whether the stromal pH is lower in bundle sheath chloroplasts, whether 3-PGA^{3-} is specifically exported, or whether 3-PGA within the chloroplast reaches such high concentrations that transport of 3-PGA^{2-} from the chloroplast is inevitable. Little is known about transport of dicarboxylates into bundle sheath chloroplasts, but aspartate has been shown to stimulate malate decarboxylation by bundle sheath chloroplasts of maize. It has therefore been suggested that a carrier specific for malate uptake is present which depends upon the presence of aspartate for maximum activity (Boag and Jenkins, 1985). In contrast to mesophyll chloroplasts, the bundle sheath chloroplasts of *Panicum miliaceum* show a slow carrier-mediated uptake of pyruvate which has very similar characteristics to the carrier in wheat and pea chloroplasts and which is not light-stimulated.

Leaf mitochondria and C4 photosynthesis

Since the mesophyll cell chloroplasts lack Rubisco, and because Rubisco in bundle sheath chloroplasts functions at high CO_2 concentrations, the flux of carbon through photorespiration is low in C4 plants. Leaf mitochondria in these plants contain low activities of the glycine decarboxylase complex, compared with C3 plants (Woo and Osmond, 1977; Ohnishi and Kanai, 1983). However, in both the NAD^+-ME and PEPCK subtypes of the C4 pathway, bundle sheath cell mitochondria are major sites of C4 acid decarboxylation (Fig. 23.2).

The carbon fluxes through the mitochondria in the above C4 sub-types are equivalent to the rate of photosynthesis, i.e. several-fold greater than those during photorespiration in C3 plants and 10 to 20-fold greater than respiratory carbon fluxes in leaves. However, even these fluxes are an order of magnitude slower than those involved in thermogenesis in the *Arum* spadix (see Chapter 14). Both in the *Arum* spadix and in NAD^+-ME-type C4 plants it seems that these high carbon fluxes are freed of respiratory control by engagement of the alternative, cyanide-insensitive pathway of respiration (Gardeström and Edwards, 1985). In PEPCK-type C4 plants, on the other hand, a much more complicated interaction between malate decarboxylation in mitochondria and OAA decarboxylation in the cytoplasm is indicated. Burnell and Hatch (1988) suggest that the ATP required for PEPCK in the cytoplasm may arise from the mitochondrial oxidation of malate and oxidative phosphorylation, and Carnal *et al.* (1993) suggest possible mechanisms which might regulate partitioning of C4 acid decarboxylation between PEP carboxykinase in the cytosol and NAD malic enzyme in the mitochondria. Here there is an interesting possible link between sucrose synthesis and decarboxylation because P_i is required not only for ATP synthesis but possibly also for malate transport into the mitochondria. These complex intracellular metabolite transport patterns are reminiscent of those which appear to operate during de-acidification in CAM plants. In none of these systems have the mitochondrial metabolite translocators been examined, chiefly because of difficulties in extraction of intact mitochondria, in quantity, from bundle sheath cells.

Regulation of C4 photosynthesis as a consequence of metabolite transport

The consequences of intercellular transport in C4 plants for the regulation of photosynthetic carbon metabolism and electron transport are far-reaching. For example, the export of reductant from the mesophyll as malate means that mitochondrial respiration may generate the ATP required for decarboxylation in the bundle sheath cytoplasm of PEPCK-type plants. The operation of the 3-PGA/triose-P shuttle has important consequence for the regulation of starch and sucrose synthesis, regulation of the C4 cycle and regulation of electron transport. Strict regulation of metabolism is also required if metabolite gradients (Fig. 23.3) are not to collapse. In maize, interchange of carbon between PEP and 3-PGA must be curtailed in the bundle sheath in order to prevent the collapse of the gradient of 3-PGA, as must the overall conversion of pyruvate to triose phosphate and of 3-PGA to malate in the mesophyll.

Mechanisms of control within the C4 cycle include control by metabolites, by the thioredoxin system ($NADP^+$-malate dehydrogenase), which is a feature shared with the PCR cycle, and also complex control of two enzymes, PEP carboxylase and pyruvate, P_i dikinase, by phosphorylation.

Coordination of the PCR and C4 cycles

Coordination of the rate at which the PCR and C4 cycles fix CO_2 is necessary if photosynthesis is to proceed efficiently under different environmental conditions. The breakdown of such coordination during light flecks, for example, has been shown to result in inefficient CO_2 assimilation (Krall and Pearcy, 1993). Coordination could occur in a variety of ways (Fig. 23.5), all of which are linked to metabolite fluxes between the bundle sheath and the mesophyll cells:

(1) In $NADP^+$-malic enzyme species such as maize, the C4 cycle is obligatorily coupled to the PCR cycle, because NADPH generated by $NADP^+$-malic enzyme is reoxidized by the reduction of 3-PGA in bundle sheath chloroplasts.

(2) Electron transport in the mesophyll chloroplasts not only powers conversion of pyruvate to malate

in the C4 cycle, but also drives 3-PGA reduction in all C4-types.

(3) Interconversion of 3-PGA and PEP, catalyzed by phosphoglycerate mutase and enolase in the mesophyll cytoplasm, provides metabolic communication between the C4 cycle and the PCR cycle.

(4) Products of 3-PGA exported from the PCR cycle, such as triose-P, and ultimately hexose-P, act as positive effectors of PEPcase, and relieve inhibition by malate.

Mathematical models which relate the underlying biochemistry of the C4 photosynthetic pathway to the CO_2 exchange characteristics of leaves provide a quantitative framework to examine the connectivity of bundle sheath and mesophyll cells (Berry and Farquhar 1978, Collatz et al., 1992). These models link C3 photosynthesis in the bundle sheath with a CO_2 pump driven by the activity of PEPcase in the mesophyll and the steady state balance of these processes can be described by two equations (Berry and Farquhar, 1978):

$$A = V_p - L,$$ [23.1]

where A is the net rate of CO_2 assimilation, V_p is the rate of PEP carboxylation and L is the leakage from the bundle sheath, and

$$A = V_c(1 - \Gamma_*/c_s),$$ [23.2]

which describes C3 photosynthesis in the bundle sheath. The rate of Rubisco carboxylation is given by V_c, c_s the bundle sheath CO_2 concentration and Γ_* the CO_2 concentration at which rate of photorespiratory CO_2 release equals V_c (Farquhar and von Caemmerer, 1982). Indeed, $V_c\Gamma_*/c_s$ quantifies the rate of photorespiration occurring in the bundle sheath.

Analytical solutions to these equations allow the calculation of parameters such as bundle sheath CO_2 and O_2 concentration and permit the exploration of interdependencies of the various components such as the capacities of PEPcase and Rubisco and the effect bundle sheath conductance has on the overall rate (Collatz et al., 1992) Two such examples are given below. Fig 23.6 shows a typical CO_2 response of assimilation for a C4 leaf. At low intercellular CO_2 concentration, CO_2 assimilation rate is determined by the rate of PEP carboxylation. At high intercellular

CO_2 concentration the abrupt saturation of the CO_2 response curve occurs when the capacity of Rubisco

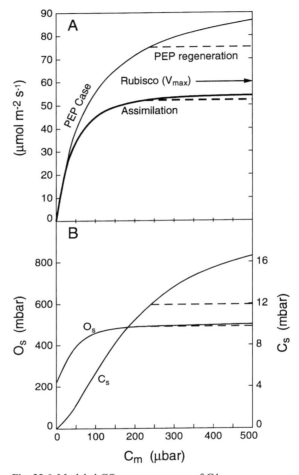

Fig. 23.6 Modeled CO_2 response curve of C4 photosynthesis. (A) At low intercellular CO_2 partial pressure, c_i, CO_2 assimilation rate, A, is limited by PEP carboxylation rate, V_p, at high c_i, A is limited either by the maximum activity of Rubisco, V_{cmax} or by the rate of PEP regeneration. (B) The resultant partial pressures of bundle sheath CO_2, (c_s) and O_2, (o_s). It was assumed that half of the net O_2 evolution occurred in the bundle sheath. The curves were calculated from the model of Berry and Farquhar (1978) with a maximum PEP carboxylase activity of $101 \mu mol\, m^{-2}\, s^{-1}$ and a $K_m(CO_2) = 80 \mu bar$, a maximum Rubisco activity of $60 \mu mol\, m^{-2}\, s^{-1}$; $K_m(CO_2) = 1015 \mu bar$ a $K(O_2) = 675\, mbar$ and $\Gamma_* = 69 \mu bar$ at $200\, mbar$ O_2. Bundle sheath conductance to CO_2, g_s, was $2\, mmol\, m^{-2}\, s^{-1}$ and the conductance to O_2 was $0.045\, g_s$ (Farquhar, 1983).

(or the regeneration of its substrate RuBP) or the regeneration of PEP become rate limiting steps. The bundle sheath CO_2 concentration is determined by the relative capacities of the C4-cycle and Rubisco, and the physical conductance of the bundle sheath to CO_2 diffusion. It is evident that if PEP carboxylase is not curtailed through metabolic regulation at high intercellular CO_2 concentration once Rubisco capacity has been exceeded, bundle sheath CO_2 concentration would continue to increase and consequently leakage. The O_2 concentration in the bundle sheath depends on the amount of PSII activity and O_2 evolution in the bundle sheath varies amongst the different C4 decarboxylation types, with $NADP^+$-ME types having little or none. In Fig. 23.6 net O_2 evolution in the bundle sheath is assumed to be 50% of net CO_2 assimilation so bundle sheath O_2 concentration saturates at approximately twice the ambient concentration.

The leakiness of the C4 pathway is an important measure of the coordination between the C4 and the PCR cycle. The leak rate of CO_2 from the bundle sheath cannot be measured directly but estimates have been made in various ways. Hatch and Osmond (1976) estimated that on diffusional grounds the back flux of CO_2 would be only 10% of the C4 acid flux, a value which was confirmed by recent studies on inorganic carbon pools in C4 plants (Furbank and Hatch, 1987, Jenkins et al., 1989, Fig. 23.4). Leakage of CO_2 from the bundle sheath occurs because of the gradient of CO_2 between the two compartments generated by the pump, so that

$$L = g_s(c_s - c_i) \qquad [23.3]$$

where c_i is the intercellular CO_2 concentration and g_s the conductance of the bundle sheath to CO_2. Leakiness (ϕ), which defines leakage as a fraction of the C4 acid flux,

$$\phi = L/V_p, \qquad [23.4]$$

also describes the efficiency of the C4 cycle.

Jenkins et al. (1989) and Furbank et al. (1990) use a related term, overcycling, to define leakage as a fraction of CO_2 assimilation, which gives the proportion by which the flux through the C4 acid cycle has to exceed net CO_2 assimilation rate.

$$L/A = (V_p - A)/A, \qquad [23.5]$$

Leakiness of C4 photosynthesis should also correlate with carbon isotope discrimination, since the latter is an index of the extent to which Rubisco functions in a closed compartment carboxylating all the CO_2 previously fixed by PEPcase (O'Leary, 1981; Farquhar, 1983). That is, ^{13}C discrimination is greater the more leaky the system. Hattersley (see Hattersley, 1992) noted significant differences in dry matter ^{13}C composition between different C4 subtypes, and concluded that $NADP^+$-ME type was less leaky than NAD^+-ME and PCK types. Brown and Byrd (1993), who estimated bundle sheath conductances on a leaf area basis through CO_2 exchange measurements on leaves in which PEPcase had been chemically inhibited, obtained estimates in the range from $1–2.5\,\mu mol\,m^{-2}\,s^{-1}$, which ranked the same way. Recent direct measurements of ^{13}C discrimination during CO_2 assimilation have so far failed to detect differences in leakiness between the various decarboxylation types (Henderson et al., 1992). They measured a leakiness of approximately 20% in a wide variety of C4 species and under a range of environmental conditions and concluded that, once the bundle sheath conductance is sufficiently low, leakiness may largely be determined by the relative capacities of the PCR cycle and the C4 cycle. This is illustrated in Fig. 23.7, which calculates the amount of Rubisco required at different bundle sheath conductances to obtain a CO_2 assimilation rate of $45\,\mu mol\,m^{-2}\,s^{-1}$ at a leakiness of 20%. Thus, bundle sheath conductances which are related to leaf anatomy (such as the presence or absence of a suberin lamella (23.2)) can be compensated for with altered biochemical capacities. Slightly more Rubisco activity is required if O_2 evolution takes place in the bundle sheath.

The regulation of PEPcase in the mesophyll

Coordination of the C4 cycle with the PCR cycle during photosynthesis is partly achieved by metabolite modulation of PEPcase in the mesophyll cytoplasm. Leaves of C4 plants actually contain both C3 and C4-type PEPcase, although the activity of the C4-isoenzyme predominates (Huber et al., 1994). PEPcase activity is modulated by a wide range of effectors such as phosphorylated intermediates

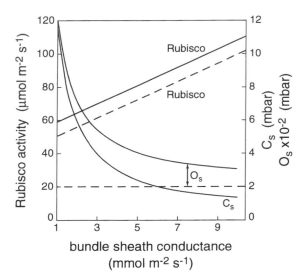

Fig. 23.7 The amount of Rubisco required in the bundle sheath at different bundle sheath conductances for a CO_2 assimilation rate of $45 \mu mol\, m^{-2}\, s^{-1}$ and a leakiness of 20%. Shown are the conditions where no O_2 evolution occurs in the bundle sheath (dashed line) or when the net O_2 evolution in the bundle sheath is $45 \mu mol\, m^{-2}\, s^{-1}$ (solid line). The bundle-sheath CO_2 concentration is the same for both cases. The O_2 sensitivity of CO_2 assimilation is always below 5% (data not shown).

(triose-P and hexose-P), amino acids and organic acids. Malate is a strong inhibitor of PEP carboxylases from all plants. Following elucidation of the role of malate in regulation of CAM plant PEP-case (Winter, 1982), the enzyme in C4 plants was found to be more sensitive to malate when extracted from darkened leaves than when extracted from illuminated leaves (Huber *et al.*, 1986; Doncaster and Leegood 1987) (i.e. the converse of the diurnal regulation in CAM plants). Figure 23.8 shows how the two forms of PEPcase, from darkened and from illuminated leaves, respond to the concentration of PEP in the presence of concentrations of malate and triose-P which might be expected to occur in the mesophyll cytoplasm (Figure 23.3). The enzyme from darkened leaves is virtually inactive at physiological concentrations of PEP (indicated by the arrow), whereas the enzyme from illuminated leaves shows a strong substrate dependence and a high activity. As in CAM plants, this change in malate sensitivity was

subsequently shown to be associated with a change in the phosphorylation state of the enzyme (Jiao and Chollet, 1991; Huber *et al.*, 1994), such that the phosphorylated (active) enzyme is less sensitive to inhibition by malate than the dephosphorylated enzyme (K_i increases from *ca.* 0.2 mM to 1 mM malate) (McNaughton *et al.*, 1989). PEP carboxylase is phosphorylated on a serine residue (Ser[8] in sorghum, Ser[15] in maize). Activation occurs rather slowly *in vivo* (30–60 min, or more, by comparison with 5–10 min for thioredoxin-activated enzymes or pyruvate, P_i dikinase).

The protein kinase which phosphorylates PEP carboxylase has been partially purified and has been shown to be Ca^{2+}-insensitive, but sensitive to inhibition by malate (McNaughton *et al.*, 1991; Wang

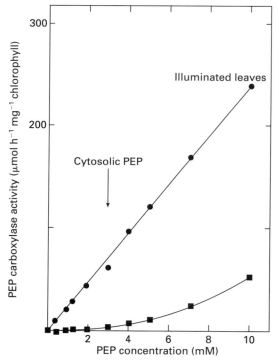

Fig. 23.8 Measured activity of PEPcase in extracts of illuminated and darkened leaves assayed in a reconstituted cytosol (20 mM malate, 15 mM triose-P) at different PEP concentrations, demonstrating the importance of interactions shown in Fig. 23.5 (Redrawn from Doncaster and Leegood, 1987).

and Chollet, 1993). A type 2A protein phosphatase leads to dephosphorylation of PEPcase. However, light affects the activity of the kinase, although not that of the phosphatase. This modulation of the activity of the kinase could come about either by covalent modification or by protein turnover, since cycloheximide inhibits light-dependent increases in kinase activity (Bakrim *et al.*, 1992). Inhibitors of photosynthesis, such as DCMU, gramicidin and DL-glyceraldehyde, prevent phosphorylation of PEPcase in leaves, which may indicate the involvement of both electron transport and carbon metabolism in the signal transduction chain which leads to activation of the kinase (Bakrim *et al.*, 1992; Jiao and Chollet, 1992). The relative importance of effects of protein synthesis, metabolites and other factors in modulating kinase activity have yet to be determined, but our knowledge of these factors is summarized in Fig. 23.9, in which the parallels between regulation in C4 and CAM plants can be seen.

Evidence of coordination between the C3 and C4 cycles is to be seen in the relationship between the assimilation rate and photosynthetic intermediates in *Amaranthus edulis* and maize in response to the

intercellular concentration of CO_2 (Fig. 23.10). As the rate of photosynthesis increases with increasing CO_2, the RuBP pool falls, but the amounts of PEP and metabolites of the C4 cycle increase (Leegood and von Caemmerer, 1988, 1989). This behavior suggests that a feedback loop from the PCR cycle is operative *in vivo*. The simplest explanation is that the amount of PEP declines when the assimilation rate falls because the amount of 3-PGA exported to the

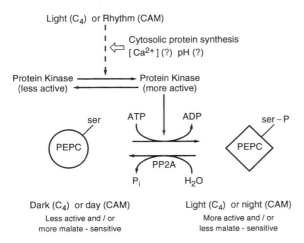

Fig. 23.9 Schematic representation of the regulation of PEP carboxylase in C4 plants by light/dark signals and in CAM plants by an endogenous rhythm. Phosphorylation of a single N-terminus seryl residue (Ser^{-15} in maize; Ser-8 in sorghum, or Ser^{-11} in *M. crystallimum*) induces a putative conformational change (represented by circles and squares) that affects the catalytic activity and or malate sensitivity. Adapted from Jiao and Chollet (1991).

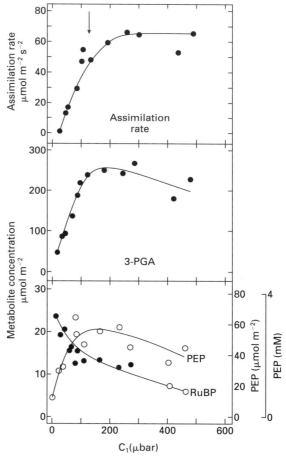

Fig. 23.10 Changes in steady-state photosynthesis and amounts of metabolites in leaves of *Zea mays* as a function of the intercellular partial pressure of CO_2 (c_i). The PEP concentration is estimated on the basis of a mesophyll cell cytoplasmic volume of $25 \, \mu l \, mg^{-1}$ chlorophyll. The arrow indicates c_i at ambient CO_2. (Leegood and von Caemmerer, 1989).

mesophyll also declines (Fig. 23.5).

Activation of PEPcase by triose-P and hexose-P is important in determining the response to the supply of 3-PGA (the metabolite 'message' from the PCR cycle), the rate of triose-P utilization by sucrose synthesis in the mesophyll cytosol, and to triose-P consumption in the PCR cycle. In leaves of maize and *Amaranthus edulis*, the amount of triose-P is always closely related to the assimilation rate whether the flux is changed by alterations in irradiance, CO_2 or temperature. Sucrose synthesis is regulated by fructose-2,6-P_2 (to which PEPcase is unresponsive) which ensures that the formation of sucrose is highly responsive to the supply of triose-P because the synthesis of hexose-P is triggered by an appropriate threshold concentration of triose-P (see Chapter 25). Increasing the concentration of hexose-P then increases the activity of PEPcase. On the other hand, the tendency for diversion of 3-PGA to malate by excessive rates of PEP carboxylation will be ameliorated both by accumulation of malate and aspartate, inhibitors of PEPcase, and by decreases in the levels of the activators, triose-P and hexose-P (Fig. 23.5). In these ways, rates of initial carboxylation in the C4 cycle can be linked to the rates of PCR cycle turnover and to rates of product synthesis, despite the complexity and compartmentation of the pathway.

The regulation of other enzymes of the C4 cycle

Like PEPcase, pyruvate, P_i dikinase is also regulated by phosphorylation, but this enzyme is unusual in that regulation is part of the catalytic mechanism of the enzyme (Edwards *et al.*, 1985; Huber *et al.*, 1994). Regulation is accomplished by a protein (Smith *et al.*, 1994) that is a bifunctional enzyme catalyzing the interconversion between an active form phosphorylated on a catalytic site histidine and an inactive form phosphorylated on a regulatory threonine residue. The mechanism allows response to changing irradiance. The importance of the role of adenylates, pyruvate and PEP in the light activation and dark inactivation of pyruvate, P_i dikinase is open to question (Roeske and Chollet, 1989). In the mesophyll NADP-malate dehydrogenase is regulated

by the thioredoxin system in a manner that allows it to respond to a range of irradiance. The activation of $NADP^+$-malate dehydrogenase (Edwards *et al.*, 1985) is also inhibited by NADP, but there is no evidence that other metabolites play any direct role in the regulation of this enzyme.

Regulation of decarboxylation in the bundle-sheath is much less well-characterized than that of other enzymes of the C4-cycle. All three decarboxylases are inhibited by CO_2 within the physiological range (Ashton *et al.*, 1990). As we have seen, decarboxylation by $NADP^+$-ME enzyme is tightly coupled to the availability of NADP and hence the rate of reduction of 3-PGA. Activity of NAD^+-ME enzyme shows a sigmoidal response to malate concentration and regulation by effectors, including CoA, acetyl CoA and fructose-1,6-bisphosphate. Activity of PEPCK is inhibited by 3-PGA, fructose-6-P, fructose-1,6-bisphosphate and dihydroxyacetone phosphate (Ashton *et al.*, 1990). This enzyme in particular would be expected to be switched off in the dark by some mechanism (Carnal *et al.*, 1993). Although these observations point to further interactions between C3 metabolites and the C4 pathway, there is insufficient information on the changes in effect or concentrations within various subcellular compartments to be able to evaluate the physiological significance of these properties.

The regulation of sucrose and starch synthesis in C4 plants

Although in a species such as maize the synthesis of sucrose appears to occur largely in the mesophyll cells, and the synthesis of starch occurs mainly in the bundle sheath cells, there is considerable flexibility both within maize and between C4 plants in general. For example, while the mesophyll tissue of maize grown under normal conditions contains no detectable starch, growth of plants in continuous light or at low temperatures induces starch formation in the mesophyll. On the other hand, *Digitaria* spp., which are also $NADP^+$-ME plants, synthesize both sucrose and starch in the mesophyll compartment. During $^{14}CO_2$ fixation in maize, labelled sucrose appears first in the mesophyll cells (Furbank *et al.*, 1985). The majority of the sucrose-phosphate synthetase, fructose-6-P,2-kinase, fructose-2,6-P_2ase and fructose-2,6-P_2 itself is present in the mesophyll cells of maize leaves (Stitt and Heldt, 1985a). However, at least under some conditions, sucrose-phosphate synthetase activity is also present in the bundle sheath cells. The major function of bundle sheath cell sucrose-phosphate synthetase may be sucrose synthesis following starch degradation (Ohsugi and Huber, 1987).

When triose-P is exported from the chloroplast of a C3 plant and is converted to sucrose, the P_i released re-enters the chloroplast in exchange for more triose-P. This provides a direct link between the provision of triose-P and its utilization in sucrose synthesis. Fine control of the level of fructose-2,6-P_2 and of the enzymes of sucrose synthesis by triose-P, hexose-P and P_i then allows a sensitive adjustment of the rate of sucrose synthesis to the availability of fixed carbon (see Chapter 25). The situation in maize is rather more complicated. The triose-P which is available for sucrose synthesis in the mesophyll contains P_i which was incorporated into 3-PGA in the bundle sheath. Hence, P_i released in sucrose synthesis in the mesophyll must be transported back to the bundle sheath (Fig. 23.5). However, the P_i will nevertheless play a part in regulating the transport of 3-PGA, triose-P and PEP across the mesophyll chloroplast envelope by the P_i translocator.

We have seen that triose-P is present in the mesophyll of C4 plants at far higher concentrations than in C3 plants. Consequently, the amount of fructose-1,6-P_2 formed through the action of aldolase is also higher in the mesophyll cytoplasm. In maize the cytosolic fructose 1,6-bisphosphatase (which is the first step in the synthesis of sucrose, shows a much higher K_m for FBP (20 mM in maize compared with 3 μM in spinach and, in the presence of fructose-2,6-P_2, 3–5 mM in maize and 20 μM in spinach; Stitt and Heldt, 1985a). These properties prevent discharge of the gradient of triose-P by sucrose synthesis. In addition, higher concentrations of triose-P and 3-PGA are needed to inhibit the fructose 6-phosphate, 2-kinase from maize, and the enzyme is also inhibited by PEP and OAA. Thus, fructose-2,6-P_2 will build up and inhibit sucrose synthesis when both C3 and C4 metabolites are low.

Studies have shown that the activities of starch synthase, of the branching enzyme and of ADP-glucose pyrophosphorylase are higher in the bundle sheath of maize than in the mesophyll. On the other hand, the enzymes of starch degradation, starch phosphylase and amylase, are evenly distributed and may be slightly higher in the mesophyll (Spilatro and Preiss, 1987). The regulation of starch synthesis has also been modified in C4 plants. ADP-glucose pyrophosphorylase from spinach chloroplasts is activated by 3-PGA and inhibited by P_i, with ratios of 3-PGA/P_i for half-maximal activation typically being less than 1.5. ADP-glucose pyrophosphorylase from maize leaves requires a ratio of 3-PGA to P_i of between 7 and 10 for half-maximal activation in the bundle sheath and even higher ratios in the mesophyll (Spilatro and Preiss, 1987). The metabolite gradients which develop during photosynthesis lead to lower 3-PGA/P_i ratios in the mesophyll than in the bundle sheath which, together with the relatively low activities of enzymes of starch synthesis, would appear to limit synthesis of starch in the mesophyll relative to the bundle sheath.

Intercellular and intra-cellular metabolite transport in C3–C4 intermediates

Over twenty species of plants exhibit photosynthetic characteristics that are intermediate between C3 and C4 plants in that they show reduced photorespiratory CO_2 release and CO_2 compensation points in the range of 7–30 μl l^{-1} (Edwards and Ku, 1987; Rawsthorne, 1992), compared with typical values in C3 plants of 50 μl l^{-1} and in C4 plants of less than 5 μl l^{-1}. Although all show a degree of Kranz anatomy, mesophyll cells are arranged as in leaves of C3 species, where interveinal distances are much greater than in C4 species. However, bundle sheath cells contain large numbers of organelles, and mitochondria and peroxisomes are particularly prominent (see Rawsthorne, 1992, and Brown and Bouton, 1993). Different biochemical variants give rise to the syndrome of C3–C4 intermediacy. In all species both mesophyll and bundle sheath cells contain Rubisco but C4 pathway enzymes are generally low (Edwards and Ku, 1987). In several types of intermediates (e.g. *Panicum milioides, Neurachne minor* and *Moricandia arvensis*) there is little evidence for a functional C4 acid cycle which donates CO_2 from C4 acids to the PCR cycle. The intermediate character with respect to the compensation point appears to be due to efficient refixation of photorespiratory CO_2 in the bundle-sheath cells. The mesophyll cell mitochondria of *M. arvensis* and many other intermediate plants, like true C4 plants, have low activities of glycine decarboxylase. The loss of glycine decarboxylase activity in the mesophyll of *M. arvensis* is due to the loss of a single subunit of the glycine decarboxylase protein complex, whereas none of the four subunits of the glycine decarboxylase complex was detectable in intermediates of *Flaveria* and *Panicum* species examined (Morgan *et al.*, 1993). Studies in Rawsthorne's laboratory have shown that transformation of *M. arvensis* with the P subunit of the glycine decarboxylase complex restores the C3 phenotype. Thus photosynthesis in leaves of these intermediates (Hylton *et al.*, 1988) might involve shuttling of photorespiratory intermediates such as glycine from mesophyll to bundle sheath and the return of serine to the mesophyll (Fig. 23.11).

Although the operation of a glycine shuttle has yet to be demonstrated directly, the quantitative feasibility of such localized photorespiratory CO_2 release accounting for the observed CO_2 exchange characteristics of C3–C4 intermediate plants is

demonstrated by the model of von Caemmerer (1989). Recycling of photorespiratory CO_2 cannot result in a large increase in photosynthetic rate at ambient CO_2 concentrations (and at most 15% of leaf Rubisco might be allocated to the bundle sheath for such recycling purposes), but may be advantageous when stomatal restrictions on CO_2 supply, or high leaf temperatures increase photorespiratory rates (Monson, 1989).

In *M. arvensis* pool sizes of glycine and serine are large enough to suggest that a shuttle of glycine from the mesophyll to the bundle sheath and the subsequent return of serine to the mesophyll could occur down a concentration gradient (Leegood and von Caemmerer, 1994). Decarboxylation of glycine in the bundle sheath also confines photorespiratory ammonia release to the bundle sheath, and high maximum catalytic activity of the glutamine synthase/ glutamate synthase cycle (GS/GOGAT) in the bundle sheath of *M. arvensis* implies that it can be reassimilated there (see Rawsthorne, 1992). This also places a greater demand on the bundle sheath for redox equivalents and ATP. Once ammonia is assimilated there is likely to be a demand for return of this nitrogen to the mesophyll to avoid imbalances in transamination reactions. How this nitrogen is recycled is unknown. Large pools of alanine and aspartate are observed in *M. arvensis* and it may be that these amino acids are involved in the maintenance of a nitrogen balance (Leegood and von Caemmerer, 1994).

In C3–C4 intermediate species of *Flaveria* appreciable activities of C4 pathway enzymes PEPcase, pyruvate P_i dikinase and $NADP^+$-malic enzyme are found, and it has been suggested that these plants represent genuine evolutionary intermediates between C3 and C4 modes of photosynthesis (Edwards and Ku, 1987; Brown and Bouton, 1993). These plants show varying capacities to fix ^{14}C into C4 acids during short-term exposure to [14]CO_2, and to transfer this to products of the PCR cycle (Edwards and Ku, 1987). Furthermore, studies with a PEPcase inhibitor have shown that CO_2 fixation via PEPcase contributes to overall CO_2 fixation in *Flaveria* intermediate species (Brown *et al.*, 1991) suggesting a limited capacity for the C4 pathway. The complete compartmentation of glycine decarboxylase (but not Rubisco) to the bundle sheath

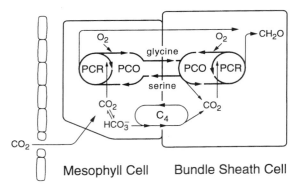

Fig. 23.11 Schematic representation of carbon metabolism in C3–C4 intermediates. CO_2 fixation by the PCR-cycle occurs in mesophyll and bundle sheath cells. CO_2 is supplied to the bundle sheath via decarboxylation of glycine from both mesophyll and bundle sheath PCO (photosynthetic carbon oxidation)-cycles, and in some species additionally via a C4 cycle.

in these species (Fig. 23.11) suggests that they also use a glycine shuttle to refix photorespiratory CO_2 (Hylton *et al.*, 1988; Morgan *et al.*, 1993). Comparison of quantum yields show that the *Flaveria* intermediates have a higher energy requirement than C3 or C4 plants, presumably because the C4 pathway activity is largely futile (Edwards and Ku, 1987). The evidence that the CO_2 compensation points are light dependent in most of the intermediate species is also an indication that the mechanisms of C3–C4 intermediate photosynthesis is limited by the supply of ATP and reductant at low CO_2.

Intracellular metabolite transport in CAM plants

The distinctive features of succulent plants that have CAM metabolism are the nocturnal fixation of CO_2 via PEPcase, accumulation of malic acid in the vacuole and subsequent diurnal de-acidification in which CO_2 released from malic acid is fixed via Rubisco during photosynthesis. This day/night pattern occurs within all chloroplast-containing cells of CAM plants, and the expression of its many components is very sensitive to biological and environmental factors. The essentials of intracellular

transport of malic acid and the analogies in metabolic regulation between C4 and CAM plants are thus best discussed by dissecting the patterns of metabolism into four phases (Osmond, 1978; Fig. 23.12).

Proof of the participation of these differing patterns of carboxylase activity throughout the daily cycle of CAM *in-vivo* is found in the changed patterns of ^{13}C-isotope discrimination as revealed by on-line measurements (Griffiths *et al.*, 1990). Phase I embraces dark acidification, involving the stoichiometric conversion of carbohydrates (glucans or soluble sugars) to PEP, carboxylation via PEPcase, reduction to malic acid and acid accumulation in the vacuole. Upon illumination, CO_2 fixation often increases transiently, involving both PEPcase and Rubisco, before malic acid efflux, de-acidification and stomatal closure. Designated phase II (Fig. 23.12), the complex mechanism of this phase will not concern us further here. Phase III is the stoichiometric antithesis of phase I, in which de-acidification can be accomplished in 4 to 8 hours, depending on light and temperature. During phase III, the CO_2 which is released by decarboxylation of malic acid accumulates to high concentrations and is refixed by Rubisco. To sustain CAM it is essential that the C3 residue from malic acid decarboxylation should be conserved during phase III. In phase IV, stomata open again

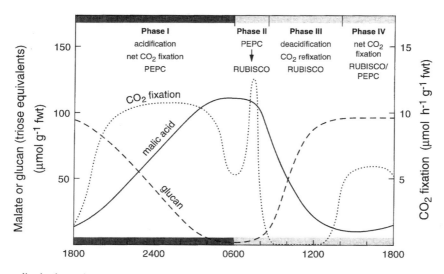

Fig. 23.12 Generalized schematic representation of malic acid and glucan levels, and rates of net CO_2 fixation in air, used to identify four phases of metabolism in ME-type CAM plants (redrawn from Osmond (1978) and other data).

and CO_2 fixation by Rubisco will predominate, although some CO_2 fixation via PEPcase also occurs.

Like C4 plants, CAM plants can be divided into major subtypes on the basis of their decarboxylation pathways during phase III (Osmond, 1978). However, unlike C4 plants, the malic enzyme type (ME-CAM plants) have a cytoplasmic $NADP^+$-ME which, with mitochondrial NAD^+-ME, participates in decarboxylation. Gluconeogenic recovery of the C3 residue from decarboxylation requires pyruvate P_i dikinase in the chloroplast. Unlike C4 plants, CAM plants seem not to be further differentiated into $NADP^+$-ME or NAD^+-ME types. Unlike C4 plants, PEPCK-type CAM plants have very low malic enzyme activities and no pyruvate P_i dikinase, but high activities of cytoplasmic PEPCK. The implications for metabolic transport between organelles in ME-type CAM plants are easily discerned, but in PEPCK-type CAM plants they are not (Osmond and Holtum, 1981).

All aspects of metabolic regulation in CAM plants are dominated by intracellular transport of malic acid (Smith and Bryce, 1992). In constitutive CAM plants these processes have been studied by manipulation of the environment to control the physiological expression of different events, such as acid accumulation in phase I. Experiments with inducible CAM plants, such as day length inducible *Kalanchoe blossfeldiana* (Brulfert and Quieroz, 1982) and salinity/water stress inducible *Mesembryanthemum crystallinum* (Winter *et al.*, 1982) have also been essential to the elucidation of these processes. Induction of CAM in these plants involves the whole 'package' of enzyme activities and organelle translocators required to sustain CAM. This 'package' can be expressed in response to salinity stress or ABA in pre-existing leaf cells of mature plants with C3 metabolism, making the phenomenon one of the most interesting cases of regulation of gene expression in response to development and physiological stress (Ostrem *et al.*, 1987; De Rocher and Bohnert 1993; Vernon *et al.*, 1993).

Malic acid influx across the tonoplast

Considerations of volume, pH, and osmotic pressure imply that the massive amounts of H^+ malic acid that may be accumulated in CAM plants during phase I must be transported to the cell vacuole. Electrophysiological considerations indicate that active transport of malic acid to the vacuole must occur (Lüttge and Ball, 1979), and many studies have demonstrated both an ATP-dependent, and PP_i-dependent H^+ transport in tonoplast membranes and vesicles of CAM plants (Lüttge, 1987). Nishida and Tominga (1987) showed Mg^{2+} ATP^{2-}-dependent malic acid uptake in isolated vacuoles, and Balsamo and Uribe (1988) demonstrated an ATPase on the cytoplasmic side of tonoplast vesicles. Although the rapid accumulation of such vast quantities of malic acid is unique to the vacuole of CAM plants, the proton-ATPase transport system shares many similarities with proton pumps in other systems. (Lüttge, 1987). Accumulation of malic acid at a rate of $10 \, \mu mol \, g^{-1}$ fresh wt. h^{-1} in a 5 mm thick leaf of *Kalanchoe* corresponds to a rate of CO_2 fixation of about $12 \, \mu mol \, m^{-2} \, s^{-1}$ (tonoplast area), 2.5×10^3 times slower than the flux across the symplast interface in C4 photosynthesis. The much larger surface area of the tonoplast (approx. $160 \, cm^2 \, g^{-1}$ fresh wt.), as compared with the area of the plasmodesmata (approx. $1 \, cm^2 \, g^{-1}$ fresh wt.), compensates for the slower rate.

Although the H^+-ATPase and H^+-PP_iase drive the active transport process (Fig. 23.13), the associated $malate^{2-}$ 'carrier' has been identified as a voltage-regulated ion channel (White and Smith, 1989). The strongly rectifying system favors $malate^{2-}$ selective ion channels, and activity is adequate to support the maximum rates of malic acid accumulation (Smith and Bryce, 1992). The energy requirements for malic acid transport to the vacuole dominate the nocturnal respiratory activity of CAM plants. Malic acid synthesis is energetically self-contained (Osmond, 1978) and, depending on the carbohydrate utilized, may even lead to net ATP production which can be used for transport. Thus, in ME-CAM plants such as *Kalanchoe*, glycolysis of glucans is initiated by hexose-P production via phosphorylase, and the overall equation for glycolysis, carboxylation and reduction to malate in these plants is

$$Glucan + P_i \rightarrow hexose\text{-}P$$
$$Hexose\text{-}P + ADP + 2\,CO_2 \rightarrow 2\,malate + ATP$$

In PEPCK-type CAM plants, such as the bromeliaed *Aechmea,* soluble hexoses are the principle carbohydrate sources for glycolysis. In these, the additional ATP required to synthesize hexose-P means that the overall equation for glycolysis, carboxylation and malate reduction is simply

$$Hexose + CO_2 \rightarrow 2\ malate$$

Assuming a similar stoichiometry of the proton ATPase in both *Kalanchoe* and *Aechmea*, one can predict that the ATP demands on respiration for malic acid transport will be about twice as large in the latter as in the former. Lüttge and Ball (1987) confirmed these predictions, showing molar ratios of O_2 uptake to malic acid of about 6 in *Kalanchoe* and 12 in *Aechmea*. In pineapple, which can use both glucan and soluble sugar in glycolysis (Kenyon *et al.*, 1985), the ratio was intermediate.

The stoichiometries of the proton ATPase, and the distinctly different enzyme, the H^+-PP_iase, have been explored by Lüttge (1987; 1988) (see Bremberger and Lüttge 1992). The relative contributions of the H^+-ATPase and the H^+-PP_iase during CAM induction in *M. crystallinum* have been clarified, and the activity of the tonoplast H^+ATPase increases, whereas that of the H^+-PP_iase declines (Bremberger and Lüttge, 1992). In several CAM plants, at the highest malic acid concentrations, the calculated electrochemical potential gradient (which presumably drives the tonoplast ATPase) is equivalent to a Gibbs free energy of 20–30 kJ mol^{-1}. This corresponds to almost half the free energy of ATP hydrolysis, suggesting that the proton transport system involved in malic acid accumulation in CAM plants is a $2H^+$/ATPase. Many CAM plants also accumulate extraordinary concentrations of citric acid nocturnally (Franco *et al.*, 1992), and Lüttge (1988) has outlined possible bioenergetic advantage of this pathway. However, little is known of the biosynthetic pathways and metabolite transport events in these circumstances.

Malic acid efflux from the vacuole

In contrast to the active transport of protons and malate into the vacuole, thermodynamic arguments suggest the return passage of free malic acid during phase II is a passive process (Smith and Bryce, 1992). During nocturnal malate accumulation in the vacuole, the dissociation equilibrium in the vacuole

shifts towards the undissociated acid (Lüttge and Smith, 1984).

$$malate^{2-} \underset{-H^+}{\overset{+H^+}{\rightleftharpoons}} H\ malate^{-1} \underset{-H^+}{\overset{+H^+}{\rightleftharpoons}} H_2\ malate$$

$$pK_a\ 4.25 \qquad pK_a\ 3.18$$

At high malic acid contents, efflux across the tonoplast is likely to occur via a passive 'lipid solution' mechanism driven by the high permeability of lipid membranes to the undissociated acid (Fig. 23.13). Passive efflux towards the end of phase I would also cause cytoplasmic malic acid levels to increase, and the rate of malic acid synthesis to decrease due to malate inhibition of PEPcase (Winter, 1985). Under natural conditions, this situation is often observed at the end of a 10 to 14 hour dark period (Fig. 23.12). Although this process could set an upper limit to vacuolar malic acid concentration attained in the dark, kinetic regulation of efflux is likely if futile cycling is to be avoided (Smith and Bryce, 1992). Efflux is presumably amplified following the activation of decarboxylase enzymes during

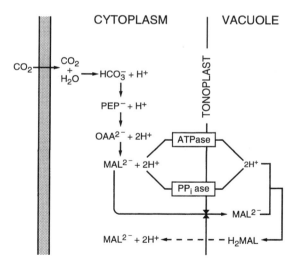

Fig. 23.13 Schematic representation of charge balance and malic acid transport during synthesis and accumulation of malate in CAM plants (redrawn from Lüttge, 1987). A tonoplast proton pump driven by a $2H^+$/ATPase, a H^+-pyrophosphatase (Bremberger and Lüttge 1992) and a malate^{2-} selective ion channel (Smith and Bryce, 1992) are indicated. The efflux of the undissociated acid is also shown as a passive flux, denoted by broken lines.

illumination in phase II. Net efflux is sustained throughout phase III by malic acid decarboxylation in the cytoplasm or mitochondria, and the sustained malate inhibition of malate synthesis va PEPcase (Kluge, 1971; Osmond *et al.*, 1988).

Several authors have suggested that the malic acid efflux in phase III might be triggered by light, and may also respond to the increased turgor generated by malic acid in the vacuole. Evidence for a light-dependent trigger is poor, and the circadian rhythms of CO_2 fixation via PEPcase in CAM plants in continuous light are consistent with an active malate influx, passive efflux model. The notion that a critical turgor might precipitate a change in membrane properties and facilitate malic acid efflux is contradicted by sustained diel malate influx and efflux in wilted leaves (Rygol *et al.*, 1987).

Organelle transport processes in CAM plants

As might be expected on the basis of the analogous metabolic pathways, CAM plants seem to share the same organelle translocator systems as C3 and C4 plants, but they tend to be deployed differently. Thus the energetic self-sufficiency of malic acid synthesis in those CAM plants which metabolize glucans requires that the chloroplast P_i translocator and the dicarboxylic acid shuttle operate in phase I and that the chloroplast P_i translocators have a high affinity for PEP (Neuhaus *et al.*, 1988). This translocator evidently works in the opposite direction in phase III (Fig. 23.14). The gluconeogenic recovery of pyruvate via chloroplastic pyruvate P_i dikinase during phase III presumably depends on an electrogenic pyruvate

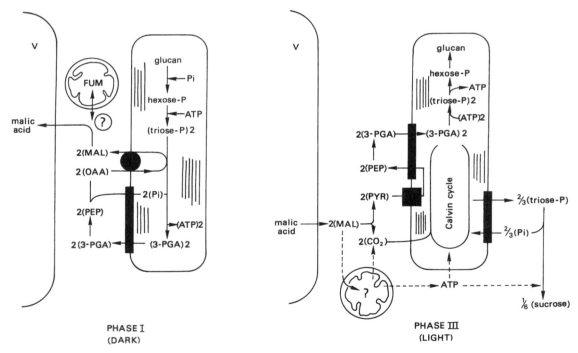

PHASE I
(DARK)

PHASE III
(LIGHT)

Fig. 23.14 Schematic representation of intracellular and chloroplast metabolite transport in a ME-CAM plant such as *Kalanchoe daigremontiana*. The organelle translocators are identified by the following symbols: dicarboxylate shuttle, a circle, the pyruvate translocator, a square, and the P_i translocator by rectangular symbols on the chloroplast envelope. During cytoplasmic malic acid synthesis in phase I, the translocators ensure that chloroplast glycolysis is energetically self-contained, but regulation of malic acid metabolism in mitochondria is not understood. In phase III the translocators ensure recovery of glucan in the chloroplast and sucrose synthesis in the cytosol, but once again the extent of mitochondrial metabolism is unclear.

translocator (Demmig and Winter, 1983) which is analogous to that in mesophyll chloroplasts of C4 plants (cf. Figs 23.5 and 23.14). Subsequently, in phase IV, PCR cycle CO_2 fixation predominates and chloroplasts which share the properties of mesophyll chloroplasts from C4 plants then behave as those of a C3 plant. Whether all chloroplasts in CAM plants have these shared properties, or whether a mixed population of specialized chloroplasts is present, is not known.

The intricacies of the organelle transport systems in other CAM plants which utilize soluble sugars, or decarboxylate malic acid via PEPCK rather than ME, remain to be explored. Little is known about metabolite exchanges in mitochondria of CAM plants. However, a large part of the total malate flux in phase I equilibrates with fumarase, as indicated by randomization of C-1 and C-4 carboxyl carbons (Cockburn and McAuley, 1975), implying rapid exchange of C4 acids into and out of the mitochondria. Kalt *et al.* (1990) observed that the rate of ^{13}C- randomization between carboxyls C-1 and C-4 was greatest at the beginning and end of phase I, at least when the rate of net CO_2 fixation and acid accumulation was greatest. These observations fit with the balance of net malic acid fluxes to the vacuole described above.

The proportion of malic acid decarboxylated by NAD^+-ME in mitochondria of ME-type CAM plants during de-acidification is very much dependent on pH (Day, 1980). Although mitochondrially-generated ATP might be used to make up the ATP requirements of pyruvate P_i dikinase during gluconeogenesis in phase III, Rustin and Quieroz-Claret (1985) found that respiration was more insensitive to cyanide during de-acidification, and presumably less tightly linked to ATP synthesis. This has been confirmed by direct, *in vivo* assessment of the engagement of the alternative oxidase by means of oxygen isotope discrimination (Robinson *et al.*, 1992). depending on whether malate oxidation proceeds to CO_2, or only as far as pyruvate and CO_2, these authors conclude 14 to 58% of malic acid could be oxidized via the mitochondria in phase III.

Regulation of CAM as a consequence of metabolite transport

Because Rubisco, PEPcase and several decarboxylase enzymes are present at high activities in all

chloroplast-containing cells of CAM plants, the regulation of these activities by metabolites is essential if CAM is to proceed. Thus for malic acid to accumulate in phase I, PEPcase must be active, and Rubisco and the decarboxylase systems must be inactive. If de-acidification and the PCR cycle are to proceed in phase III, decarboxylases and Rubisco must be active and PEPcase inactive. At the same time, glycolysis and glucoeogenesis must be regulated to ensure needed carbohydrate substrate supply to restore the cyclical carbon fluxes in CAM. It is evident that the above transport events mediate many of the key regulatory processes.

Regulation of futile carboxylation, decarboxylation, and competitive carboxylation in CAM

Regulation of PEPcase in CAM plants has been extensively studied and there is general agreement as to the mechanism involved. Although early studies implied coarse control, by means of changes in the amount of enzyme, this level of control is only observed during induction of CAM (DeRocher and Bohnert, 1993). The amount of PEPcase does not vary throughout the day–night cycle (Fig. 23.15). As in C3 and C4 plants, malic acid is a potent inhibitor of PEPcase and early models of CAM proposed this to be the key regulator of the cycle of malic acid synthesis and consumption (Kluge, 1971). The PEPcase of CAM plants has a higher affinity for PEP, and a higher sensitivity to malic acid, than the enzyme from C4 plants. Malic acid synthesis continues in phase I because this inhibitory end product of PEPcase is efficiently transported from cytoplasm to vacuole, and then declines as flux equilibrium is approached. In the light, at least during phase III, futile cycles of malic acid decarboxylation and resynthesis are prevent because PEPcase is inhibited by high malic acid concentration in the cytoplasm during de-acidification. Direct evidence that PEPcase is inhibited during de-acidification in phase III comes from studies of ^{13}C label patterns in malate. The ^{13}C distribution among 4-^{13}C and 1-^{13}C labeled malate remains unchanged during de-acidification; futile cycling, if it were to occur, would yield doubly labeled malic acid (Osmond *et al.*, 1988).

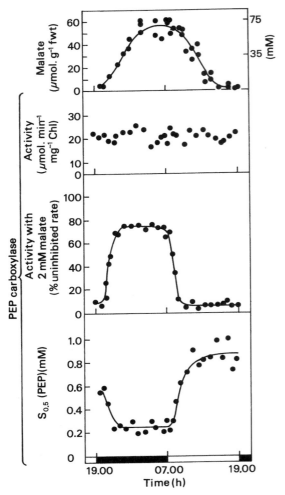

Fig. 23.15 Changes in the properties of PEPcase extracted from leaves of *Mesembryanthemum crystallinum* at different times of the diel cycle of CAM (redrawn from Winter, 1982). The concentration of malic acid in the vacuole was calculated assuming 85% water content. Although the total activity of PEP carboxylase remains unchanged, the decreased affinity for PEP and the increased sensitivity to malate ensure that the enzyme is inhibited during de-acidification in the light.

This hypothesis has been augmented by evidence that PEPcase is sensitized to malate inhibition in the light, thus greatly amplifying the feedback control (Winter, 1982). When PEPcase is extracted from CAM plants in the dark, it shows a high affinity for

PEP, is susceptible to effectors such as glucose 6-P stimulation, and has a low sensitivity to malic acid. If extracted from the same tissues during the light period, the enzyme has a low affinity for its substrate, low sensitivity to effectors and is much more sensitive to malic acid (Fig. 23.15). Nimmo *et al.* (1987) reported that the malic acid insensitive form of the enzyme isolated in the dark is phosphorylated, and that this covalent modification of the protein is responsible for changes in kinetic properties. The protein kinase responsible is under circadian control (Carter *et al.*, 1991). This is analogous to phosphorylation of PEPcase in C4 plants (see above and Fig. 23.9), except that in CAM plants PEP carboxylase is activated (phosphorylated) at night and inactivated during the day (Huber *et al.*, 1994). As in C4 plants, a serine residue near the N-terminus (Ser[11]) appears to be the target for the phosphorylation in *Mesembryanthemum crystallinum* (Baur *et al.*, 1992). Phosphorylation by the partially purified kinase leads to a 5 to 10-fold change in malate sensitivity of PEPcase in *Bryophyllum fedtschenkoi*. The activity of the kinase is inhibited by PEP, glucose-6-P and malate, and the kinase undergoes diurnal changes in activity, appearing 4–5 h after the onset of darkness and disappearing 2–3 h before the end of the dark period. As in C4 plants, the activity of the type 2A phosphatase, which dephosphorylates PEPcase, shows no day/night variation. The appearance of protein kinase activity is blocked by inhibitors of protein synthesis under the control of an endogenous circadian rhythm, rather than light as in C4 plants (Carter *et al.*, 1991; Wilkins, 1992).

The regulation of decarboxylase enzymes in relation to the day–night cycle is not so well understood. It is often observed that the rate of CO_2 fixation in the dark declines as the malic acid content of the vacuole reaches a maximum level, towards the end of phase I (Fig. 23.12). Commonly attributed to feedback inhibition due to malate, this could also reflect the onset of decarboxylation. Control of the cytoplasmic decarboxylases ($NADP^+$-ME or PEPCK) in the dark is not easily explained, and whether futile cycles of carboxylation and decarboxylation occur in the dark probably depends on the residence time of malic acid in the cytosol, and on the balance of metabolite fluxes to the vacuole (Kalt *et al.*, 1990).

There is evidence that mitochondrial NAD^+-ME is inactive in the dark in some CAM plants, but not in others (Wedding and Black, 1983; Artus and Edwards, 1985). In these circumstances, inhibition of mitochondrial NAD^+-ME could prevent futile cycling.

It is now clear that Rubisco is inactive during dark CO_2 fixation via PEPcase (cf. Ranson and Thomas, 1960), and there is evidence in some CAM plants that specific inhibitors, such as carboxyarabinitol-1P may be involved (Chapter 24). Both Rubisco and the decarboxylase enzymes are active in the light. After malic acid is consumed in photosynthesis, the relief of PEPcase inhibition permits parallel CO_2 fixation via both carboxylases in phase IV (Ritz *et al.*, 1986; Osmond *et al.*, 1988). However, because net CO_2 fixation in phase IV shows the O_2 sensitivity normally associated with Rubisco, it seems that PEPcase activity is quantitatively unimportant (Osmond, 1978). Not surprisingly, CAM plants display the full suite of enzymes of the photorespiratory carbon oxidation cycle (Whitehouse *et al.*, 1991) and engage these in CO_2 recycling and photoprotective functions if stomata remain closed following de-acidification under water stress (Adams and Osmond, 1988).

Regulation of carbohydrate metabolism in CAM

The pattern of malic acid synthesis and transport characteristic of CAM can only be sustained by stoichiometric degradation and resynthesis of carbohydrates as substrates and products of malic acid synthesis and breakdown. There is no analogy of this stoichiometry in other photosynthetic systems, and very little is known about the ways in which carbohydrate pools dedicated to malic acid synthesis (which are conserved during photosynthesis in phase III) are distinguished from photosynthetic end products produced and presumably translocated in phase IV. Yet there is tantalizing evidence for the isolation of the glycolytic and gluconeogenic pathways associated with CAM (Deleens and Quieroz, 1984). The carbohydrate demands of carbon cycling in CAM, appear to have priority over those of translocation. Whereas CAM can be impaired by reduced sink strength following girdling, increased

sink strength following shading of all other leaves does not impair CAM (Mayoral *et al.*, 1991).

Enzyme compartmentation and organelle metabolite translocators play key roles in the regulation of carbohydrate metabolism in CAM plants. These roles can be identified with some confidence in the glucan-utilizing ME-CAM plants, but are quite uncertain in PEPCK-CAM plants which convert soluble carbohydrates to malic acid and vice versa. In ME-CAM plants decarboxylation of malic acid produces pyruvate and CO_2. Gluconeogenic recovery of pyruvate can only take place in the chloroplast after conversion of pyruvate to PEP. Chloroplasts of CAM plants, like mesophyll chloroplasts of C4 plants, evidently contain a specific pyruvate translocator (Demmig and Winter, 1983) which could ensure the stoichiometric transport of pyruvate into the chloroplast during de-acidification. However, gluconeogenic recovery of PEP involves further transport of PEP from, and 3-PGA into, the chloroplasts (Fig. 23.14). A specific form of the phosphate translocator in CAM chloroplasts which exchanges PEP and 3-PGA at high rates (Neuhaus *et al.*, 1988) thus has a potentially significant role in restoring chloroplast glucan levels.

At the same time as these organelle metabolite transport events are taking place, CO_2 released from malic acid is fixed via the PCR cycle and sucrose synthesis proceeds in the cytoplasm. This presumably involves metabolite and P_i exchanges across the chloroplast envelope analogous to those in C3 plants and similar regulatory interactions. Thus levels of fructose-2, $6-P_2$ are lowest during de-acidification in *Bryophyllum tubiflorum*, consistent with sustained cytoplasmic sucrose synthesis in the face of concurrent starch synthesis in the chloroplast (Fahrendorf *et al.*, 1987). However, in a PEPCK-CAM plant (*Ananas comusus*) which derives only part of the carbon for nocturnal malic acid synthesis from glucans, these authors found 20 to 50-fold higher levels of fructose-2,$6-P_2$ throughout the 24 h cycle. Although these high levels of effector could inhibit sucrose synthesis in the cytosol and promote some glucan synthesis in the chloroplast during de-acidification, there is a paradox so far as sucrose accumulation is concerned. Fahrendorf *et al.* (1987) suggested that the high activity of pyrophosphate-dependent phosphofructokinase (PFP) in *Ananas,* which is promoted by fructose-2,6-P_2 (see Chapter 25), could serve as a path for cytoplasmic sucrose synthesis in the presence of high concentrations of fructose-2,6-P_2. The overall effect of high fructose-2,6-P_2 levels and high PFP activity may be to promote the recovery of both glucan and soluble carbohydrates during de-acidification. Osmond and Holtum (1981) deduced that, for many reasons, the ratio of 3-PGA to P_i is likely to be very high during de-acidification and hence to favor glucan synthesis via ADP-glucose pyrophosphorylase. Together with the extraordinary effect of 3-PGA on lowering the $S_{0.5}$ (glucose-1-P), and the fact that P_i does not completely inhibit the enzyme, ADP-glucose pyrophosphorylase from CAM plants seems uniquely suited to ensure glucan synthesis in the chloroplast (Singh *et al.*, 1984).

Conclusions

Metabolic fluxes between and within chloroplast-containing cells underlie almost every aspect of photosynthetic metabolism in C4 and CAM plants. None of these processes is unique to these photosynthetic systems, but the activities and regulatory properties are markedly different from those in C3 plants. The broad distinction between the spatial and temporal separation of the carboxylation phase of the C4-cycle from the decarboxylation phase and the PCR cycle, in C4 and CAM plants respectively, leads to further differentiation in respect of mitochondrial transport and regulation. Many details of the symplastic transport of photosynthetic metabolites in C4 plants (the most rapid of all symplast processes) remain to be resolved, but simple models based on concentration gradients and diffusion are adequate and consistent with fundamental geometric, structural and biochemical properties of the system. Variants of the three major subtypes of C4 photosynthesis, and the C3–C4 intermediates, offer great scope for further testing of the principles of symplast transport in C4 photosynthesis. By the same token, progress in the evaluation of malic acid transport across the tonoplast of CAM plants, evidently driven by a 2H$^+$ ATPase or pyrophosphatase, opens many avenues for the study of the most rapid cell membrane transport

processes in plants. The overwhelming influence of malic acid compartmentation on photosynthetic metabolism in CAM plants has yet to be explored in detail, especially in PEPCK-type CAM plants, and especially so far as leaf mitochondria are concerned. Understanding the regulation of carbohydrate metabolism in both C4 and CAM plants remains a major challenge. Overall, a remarkable complementarity in organelle translocator properties and in enzyme regulatory mechanisms, especially those involving protein photophorylation and allostric effectors, is evident in these two physiologically distinctive photosynthetic CO_2 concentrating systems.

Further reading and references

Adams III, W. W. and Osmond, C. B. (1988). Internal CO_2 supply during photosynthesis of sun and shade grown CAM plants in relation to photoinhibition. *Plant Physiol.* **86**, 117–23.

Andrews, T. J. and Lorimer, G. H. (1987). Rubisco: structure, mechanisms and prospects for improvement. In *The Biochemistry of Plants*, Vol. 10, eds M. D. Hatch and N. K. Boardman, Academic Press, New York, pp. 131–218.

Artus, N.N. and Edwards, G. E. (1985). Properties of leaf NAD-malic enzyme from the inducible Crassulacean acid metabolism species *Mesembryanthemum crystallinum*. *Plant Cell Physiol.* **26**, 341–50.

Ashton, A. R., Burnell, J. N., Furbank, R. T., Jenkins, C. L. D. and Hatch, M. D. (1990). Enzymes of C4 photosynthesis. In *Methods in Plant Biochemistry*, ed. P. J. Lea, Vol. 3, Academic Press, London. pp. 39–72.

Bakrim, N., Echevarria, C., Cretin, C., Arrio-Dupont, M., Pierre, J. N., Vidal, J., Chollet, R. and Gadal, P. (1992). Regulatory phosphorylation of *Sorghum* leaf phospho*enol*pyruvate carboxylase. Identification of the protein-serine kinase and some elements of the signal transduction cascade. *Eur. J. Biochem.* **204**, 821–30.

Balsamo, R. A. and Uribe, E. G. (1988). Plasmalemma and tonoplast ATPase activity in mesophyll protoplasts, vacuoles and microsomes of the Crassulacean-acid metabolism plant *Kalanchoe daigremontiana*. *Planta* **173**, 190–96.

Baur, B., Dietz, K.-J. and Winter, K. (1992). Regulatory protein phosphorylation of phosphoenolpyruvate carboxylase in the facultative crassulacean-acid-metabolism plant *Mesembryanthemum crystallinum* L. *Eur. J. Biochem.* **209**, 95–101.

Berry, J. A. and Farquhar, G. D. (1978). The CO_2

concentration function of C4 photosynthesis: a biochemical model. In *Proceedings of the 4th International Congress on Photosynthesis,* eds D. Hall, J. Coombs and T. Goodwin, Biochemical Society, London, pp. 119–31.

Boag, S. and Jenkins C. L. D. (1985). CO_2 assimilation and malate decarboxylation by isolated bundle sheath chloroplasts from *Zea mays. Plant Physiol.* **79**, 165–70.

Botha, C. E. J. and Evert, R. F. (1988). Plasmodesmatal distribution and frequency in vascular bundles and contiguous tissues of the leaf of *Themeda triandra. Planta* **173**, 433–41.

Bowes, G. and Salvucci, M. E. (1989). Plasticity in the photosynthetic carbon metabolism of submersed aquatic macrophytes. *Aquat Bot.* **34**, 233–66.

Bremberger, C. and Lüttge, U. (1992). Dynamics of tonoplast proton pumps and other tonoplast proteins of *Mesembryanthemum crystallinum* L. during induction of Crassulacean acid metabolism. *Planta* **188**, 575–80.

Brown, R. H. and Bouton, J. H. (1993). Physiology and genetics of interspecific hybrids between photosynthetic types. *Annu. Rev. Plant Physiol. Plant Mol. Biol.* **44**, 435–56.

Brown, R. H., Byrd, G. T., Black, C. C. (1991). Assessing the degree of C4 photosynthesis in C3–C4 species using an inhibitor of phosphoenolpyruvate carboxylase. *Plant Physiol.* **97**, 985–89.

Brown, R. H. and Byrd, T. G. (1993). Estimation of bundle sheath cell conductance in C4 species and O_2 sensitivity of photosynthesis. *Plant Physiol.* **103**, 1183–88.

Brulfelt, J. and Quieroz, O. (1982). Photoperiodism and Crassulacean acid metabolism. III. Different characteristics of photoperiod-sensitive and nonsensitive isoforms of phosphoenolpyruvate carboxylase and Crassulacean acid metabolism operation. *Planta* **154**, 3399–43.

Burnell, J. N. (1988). An enzymic method for measuring the molecular exclusion limit of plasmodesmata of bundle sheath cells of C4 plants. *J. Exp. Bot.* **39**, 1575–80.

Burnell, J. N. and Hatch, M. D. (1985). Light-dark regulation of pyruvate, P_i dikinase. *Trends Biochem. Sci.* **10**, 288–91.

Burnell, J. N. and Hatch, M. D. (1988). Photosynthesis in phosphoenolpyruvate carboxykinase-type C4 plants: pathways of C4 acid decarboxylation in bundle sheath cells of *Urochloa panicoides. Arch. Biochem. Biophys.* **260**, 187–99.

Burr, G. O., Hartt, C. E., Brodie, H. W., Tanimoto, T., Kortschak, H. P., Takahashi, D., Ashton, F. M. and Coleman, R. E. (1957). The sugar cane plant. *Ann Rev. Plant Physiol.* **8**, 275–98.

Carnal, N. W., Agostino, A. and Hatch, M. D. (1993). Photosynthesis in phospho*enol*pyruvate carboxykinase-type C4 plants: Mechanism and regulation of C4 acid decarboxylation in bundle sheath cells. *Arch. Biochem. Biophys.* **306**, 360–367.

Carter, P. J., Nimmo, H. G., Fewson, C. A. and Wilkins, M. B. (1991). Circadian rhythms in the activity of a plant protein kinase. *EMBO J.* **10**, 2063–68.

Cockburn, W. (1985). Variation in photosynthetic acid metabolism in vascular plants: CAM and related phenomena. *New Phytol.* **101**, 3–25.

Cockburn, W. and McAuley, A. (1975). Pathway of dark CO_2 fixation in CAM plants. *Plant Physiol.* **55**, 87–89.

Collatz, G. J., Ribas-Carbo, M. and Berry, J. A. (1992). Coupled photosynthesis-stomatal model for leaves of C4 plants. *Aust. J. Plant. Physiol.* **19**, 519–38.

Day, D. A. (1980). Malate decarboxylation by *Kalanchoe daigremontiana* mitochondria and its role in crassulacean acid metabolism. *Plant Physiol.* **65**, 675–79.

Day, D. A. and Hatch, M. D. (1981). Dicarboxylate transport in maize mesophyll chloroplasts. *Arch. Biochem. Biophys* **211**, 738–42.

De Saussure, T. (1804). *Recherches Chimiques sur la Vegetation,* Nyon, Paris.

Deleens, E. and Quieroz, O. (1984). Effects of photoperiod and ageing on the carbon isotope composition of *Bryophyllum diagremontiana* Berger. *Plant Cell Environ.* **7**, 279–83.

Demmig, B. and Winter, K. (1983). Photosynthetic characteristics of chloroplasts isolated from *Mesembryanthemum crystallinum*; a halopyte plant capable of crassulacean acid metabolism. *Planta* **159**, 66–76.

Dengler, E. G., Dengler, R. F. and Hattersley, P. W. (1985). Differing ontogenetic origins of PCR ('Kranz') sheaths in leaf blades of C4 grasses (Poaceae). *Am. J. Bot.* **72**, 284–302.

De Rocher, E. J. and Bohnert, H. J. (1993). Development and environmental stress employ different mechanisms in the expression of a plant gene family. *Plant Cell* **5**, 1611–25.

Doncaster, H. D. and Leegood, R. C. (1987). Regulation of phosphoenolpyruvate carboxylase activity in maize leaves. *Plant Physiol.* **84**, 82–87.

Edwards, G. E. and Walker, D. A. (1983). C3, C4: *Mechanisms, and Cellular and Environmental Regulation of Photosynthesis,* Blackwell, Oxford.

Edwards, G. E. and Ku, M. S. B. (1987). Biochemistry of C3–C4 intermediates. In *The Biochemistry of Plants,* Vol. 10, eds M. D. Hatch and N. K. Boardman, Academic Press, New York, pp. 275–325.

Edwards, G. E., Nakamoto, H., Burnell, J. N. and Hatch, M. D. (1985). Pyruvate, P_i dikinase and NADP-malate dehydrogenase in C4 photosynthesis: Properties and mechanism of light/dark regulation. *Ann. Rev. Plant Physiol.* **36**, 255–86.

Fahrendorf, T., Holtum, J. A. M., Muckerjee, U. and Latzko, E. (1987). Fructose-2,6-bisphosphate, carbohydrate partitioning and crassulacean acid metabolism. *Plant Physiol.* **84**, 182–87.

Farquhar, G. D. and von Caemmerer, S. (1982). Modeling

of photosynthetic response to environmental conditions. In *Physiological Plant Ecology II. Water Relations and Carbon Assimilation,* eds O. L. Lange, P. S. Nobel, C. B. Osmond and H. Ziegler. *Encyclopedia of Plant Physiology.* Vol. 12B, Springer-Verlag, Berlin pp. 550–87.

Farquhar, G. D. (1983). On the nature of carbon isotope discrimination in C4 species. *Aust. J. Plant Physiol.* **10**, 205–26.

Fischer, K., Arbinger, B., Kammerer, B., Busch, C., Brink, S., Wallmeier, H., Sauer, N., Eckerskorn, C. and Flügge, U.-I. (1994). Cloning and *in vivo* expression of functional triose phosphate/phosphate translocators from C3- and C4-plants: evidence for the putative participation of specific amino acid residues in the recognition of phosphoenolpyruvate. *Plant J.* **5**, 215–26.

Flügge, U. I., Stitt, M. and Heldt, H. W. (1985). Light-driven uptake of pyruvate into mesophyll chloroplasts from maize. *FEBS Lett.* **183**, 335–39.

Franco, A. C., Ball, E. and Lüttge, U. (1992). Differential effects of drought and light levels on accumulation of citric and malic acids during CAM in *Clusia. Plant Cell Environ.* **15**, 821–29.

Furbank, R. T. and Hatch, M. D. (1987). Mechanism of C4 photosynthesis. The size and composition of the inorganic carbon pool in bundle-sheath cells. *Plant Physiol.* **85**, 958–64.

Furbank, R. T., Stitt, M. and Foyer, C. H. (1985). Intercellular compartmentation of sucrose synthesis in leaves of *Zea mays. Planta* **164**, 172–87.

Furbank, R. T., Jenkins, C. L. D. and Hatch, M. D. (1990). C4 photosynthesis: quantum requirements, C4 acid overcycling and Q-cycle involvement. *Aust. J. Plant Physiol.* **17**, 1–7.

Gardenström, P. and Edwards, G. E. (1985). Leaf mitochondria (C3 + C4 + CAM). In *Encyclopedia of Plant Physiology,* New Series, Vol. 18, eds R. Douce and D. A. Day, Springer-Verlag, Berlin, pp. 314–46.

Griffiths, H., Broadmeadow, M. S. J., Borland, A. M. and Hetherington, C. S. (1990). Short-term changes in carbon isotope discrimination identify transitions between C3 and C4 carboxylation during crassulacean acid metabolism. *Planta* **181**, 604–610.

Haberlandt, G. (1884). *Physiological Plant Anatomy,* (transl. M. Drummond), Macmillan, London.

Hatch, M. D. (1971). The C4 pathway of photosynthesis. Evidence for an intermediate pool of carbon dioxide and the identity of the donor C4 acid. *Biochem. J.* **125**, 425–32.

Hatch, M. D. and Osmond, C. B. (1976). Compartmentation and transport in C4 photosynthesis. In *Encyclopedia of Plant Physiology,* New Series Vol. 3, eds C. R. Stocking and U. Heber, Springer-Verlag, Berlin, pp. 144–84.

Hatch, M. D., Dröuscher, L., Flügge, U. I. and Heldt, H. W. (1984). A specific translocator for oxaloacetate transport in chloroplasts. *FEBS Lett.* **178**, 15–19.

Hatch, M. D. (1992). I can't believe my luck. *Photosynth. Res.* **33**, 1–14.

Hattersley, P. W. (1992). C4 photosynthetic pathway variation in grasses (Poaceae): Its significance for arid and semi-arid lands. In *Desertified Grassland: Their Biology and Managements.* Linnean Society Symposium Series 13, ed. G. P. Chapman, Academic Press, London, pp. 181–212.

Heldt, H. W. and Flügge, U. I. (1986). Transport of metabolites across the chloroplast envelope. *Methods Enzymol.* **125**, 705–16.

Heldt, H. W. and Flügge, U. I. (1992). Metabolite transport in plant cells. In *Plant Organelles,* ed. A. K. Tobin, Cambridge University Press, Cambridge, pp. 21–47.

Henderson, S. A., von Caemmerer, S., Farquhar, G. D. (1992). Short-term measurements of carbon isotope discrimination in several C4 species. *Aust. J. Plant Physiol.* **19**, 263–285.

Henderson, S., Hattersley, P., von Caemmerer, S. and Osmond, C. B. (1994). Are C4 pathway plants threatened by global climatic change? In *Ecophysiology of Photosynthesis.* Ecological Studies Vol. 100, eds E-D. Schulze and M. M. Caldwell. Springer-Verlag, Berlin, pp. 529–49.

Heyne, B. (1815). On the deoxidation of the leaves of *Cotyledon calycina. Trans. Linn. Soc. Lond.* **11**, 213–15.

Höfer, M. U., Santore, U. J., Westhoff, P. (1992). Differential accumulation of the 10-, 16- and 23-kDa peripheral components of the water-splitting complex of photosystem II in mesophyll and bundle sheath chloroplasts of dictyledonous C4 plant *Flaveria trinervia* (Spreng.) C.Mohr. *Planta* **186**, 304–12.

Holtum, J. A. M. and Osmond, C. B. (1981). The gluconeogenic metabolism of pyruvate during deacidification in plants with crassulacean acid metabolism. *Aust. J. Plant Physiol.* **8**, 31–44.

Huber, S. C., Sugiyama, T. and Akazawa, T. (1986). Light modulation of maize leaf phosph*enol*pyruvate carboxylase. *Plant Physiol.* **82**, 550–54.

Huber, S. C., Huber, J. L. and McMichael, R. W. (1994). Control of plant enzyme activity by reversible protein phosphorylation. *Int. Rev. Cytol.* **149**, 47–98.

Hylton, C. M., Rawsthorne, S., Smith, A. M., Jones, D. A. and Woolhouse, H. W. (1988). Glycine decarboxylase is confined to the bundle sheath cells of C3–C4 intermediate species. *Planta* **175**, 452–59.

Jenkins, C. L. D., Furbank, T. R. and Hatch, M. D. (1989). Mechanisms of C4 photosynthesis – a model describing the inorganic carbon pool in bundle sheath cells. *Plant Physiol.* **91**, 1372–81.

Jiao, J. A. and Chollet, R. (1991). Post translational regulation of phosphoenolpyruvate carboxylase in C4 and CAM plants. *Plant Physiol.* **95**, 981–85.

Jiao, J. and Chollet, R. (1992). Light activation of maize phospho*enol*pyruvate carboxylase protein-serine kinase activity is inhibited by mesophyll and bundle sheath directed photosynthesis inhibitors. *Plant Physiol.* **98**, 152–56.

Kalt, W., Osmond, C. B. and Siedow, J. N. (1990). Malate metabolism in the dark after $^{13}CO_2$ fixation in the crassulacean plant *Kalanchoe tubiflora*. *Plant Physiol.* **94**, 826–32.

Karpilov, Y. S. (1960). The distribution of radioactivity of ^{14}C among the products of photosynthesis in maize. *Trans. Kazan Agric. Inst.* **41**, 15–24 (in Russian).

Karpilov, Y. S. (1970). Cooperatve photosynthesis in zerophytes. *Proc. Moldavian Inst. Irrigation and Agric. Res.* **11**, 3–66 (in Russian).

Kenyon, W. H., Severson, R. F. and Black, C. C. (1985). Maintenance carbon cycle in Crassulacean acid metabolism plant leaves. Source and compartmentation of carbon for nocturnal malate synthesis. *Plant Physiol.* **77**, 183–89.

Kluge, M. (1971). Studies on CO_2 fixation by succulent plants in the light. In *Photosynthesis and Photorespiration,* eds M. D. Hatch, C. B. Osmond and R. O. Slatyer, Wiley-Interscience, New York, pp. 283–87.

Krall, J. K. and Pearcy R. W. (1993). Concurrent measurements of oxygen and carbon dioxide exchange during lightflecks in maize (*Zea mays* L.). *Plant Physiol.* **103**, 823–8.

Kunitake, G. and Saltman, P. (1958). Dark CO_2 fixation by succulent leaves: conservation of the dark fixed CO_2 under diurnal conditions. *Plant Physiol.* **34**, 123–27.

Leegood, R. C. (1985). The intercellular compartmentation of metabolites in leaves of *Zea mays*. *Planta* **164**, 163–71.

Leegood, R. C. and von Caemmerer, S. (1988). The relationship between contents of photosynthetic intermediates and the rate of photosynthetic carbon assimilation in leaves of *Amaranthus edulis* L. *Planta* **174**, 253–62.

Leegood, R. C. and von Caemmerer, S. (1989). Some relationships between the contents of photosynthetic intermediates and the rate of photosynthetic carbon assimilation in leaves of *Zea mays* L. *Planta* **178**, 258–66.

Leegood, R. C. and von Caemmerer, S. (1994). Regulation of photosynthetic carbon assimilation in leaves of C3–C4 intermediate species of *Moricandia* and *Flaveria*. *Planta* **192**, 232–38.

Lucas, W. J., Ding, B. and van der Schoot, C. (1993). Plasmodesmata and the supracellular nature of plants. *New Phytol.* **125**, 435–76.

Lüttge, U. (1987). Carbon dioxide and water demand: Crassulacean acid metabolism (CAM), a versatile ecological adaptation exemplifying the need for integration in ecophysiological work. *New Phytol.* **106**, 593–629.

Lüttge, U. (1988). Day–night changes in citric acid levels in crassulacean acid metabolism phenomenon and ecophysiological significance. *Plant Cell Environ.* **11**, 445–51.

Lüttge, U. and Smith, J. A. C. (1984). Mechanism of passive malic-acid efflux from vacuoles of the CAM plant *Kalanchoe daigremontiana*. *J. Membrane Biol.* **81**, 149–58.

Lüttge, U. and Ball, E. (1987). Dark respiration of CAM plants. *Plant Physiol. Biochem.* **25**, 3–10.

Mayoral, M. L., Medina, E. and Garcia, V. (1991). Effect of source-sink manipulations on the crassulacean acid metabolism of *Kalanchoe pinnata*. *J. Exp. Bot.* **42**, 1123–29.

McNaughton, G. A. L., Fewson, C. A., Wilkins, M. B. and Nimmo, H. G. (1989). Purification, oligomerization state and malate sensitivity of maize leaf phosphoenolpyruvate carboxylase. *Biochem. J.* **261**, 349–55.

McNaughton, G. A. L., MacKintosh, C., Fewson, C. A., Wilkins, M. B. and Nimmo, H. G. (1991). Illumination increases the phosphorylation state of maize leaf phospho*enol*pyruvate carboxylase by causing an increase in the activity of a protein kinase. *Biochim. Biophys. Acta* **1093**, 189–95.

Monson, K. R. (1989). The relative contributions of reduced photorespiration, and improved water- and nitrogen-use efficiencies, to the advantages of C3–C4 photosynthesis in *Flaveria*. *Oecologia* **80**, 215–21.

Morgan, C. L., Turner, S. R., Rawsthorne, S. (1993). Coordination of cell-specific distribution of the four subunits of glycine decarboxylase and of serine hydroxymethyltransferase in leaves of C3–C4 intermediate species from different genera. *Planta* **190**, 468–73.

Murata, S., Kobayashi, M., Matoh, T. and Sekiya, J. (1992). Sodium stimulates regeneration of phospho*enol*pyruvate in mesophyll chloroplasts of *Amaranthus tricolor*. *Plant Cell Physiol.* **33**, 1247–50.

Nelson, T. and Langdale, J. A. (1992). Developmental genetics of C4 photosynthesis. *Ann. Rev. Plant Physiol. Plant Mol. Biol.* **43**, 25–47.

Neuhaus, H. E., Holtum, J. A. M. and Latzko, E. (1988). Transport of phosphoenolpyruvate by chloroplasts from *Mesembryanthemum crystallinum* L. exhibiting crassulacean acid metabolism. *Plant Physiol.* **87**, 64–68.

Nimmo, G. A., Wilkins, M. B., Fewson, C. A. and Nimmo, H. G. (1987a). Persistent circadian rhythms in the phosphorylation state of phosphoenolpyruvate carboxylase from *Bryophyllum fedtschenkoi* leaves and in its sensitivity to inhibition by malate. *Planta* **170**, 408–15.

Nishida, K. and Tominga, O. (1987). Energy-dependent uptake of malate into vacuoles isolated from a CAM plant, *Kalanchoe daigremontiana*. *J. Plant Physiol.* **127**, 385–93.

Nobel, P. S. (1988). *The Environmental Biology of Agaves and Cacti,* Cambridge University Press, New York.

O'Leary, M. H. (1981). Carbon isotope discrimination in plants. *Phytochemistry* **20**, 553–67.

Ohnishi, J. and Kanai, R. (1983). Differentiation of photorespiratory activity between mesophyll and bundle sheath cells of C4 plants. I. Glycine oxidation by mitochondria. *Plant Cell Physiol.* **24**, 1411–20.

Ohnishi, J. and Kanai, R. (1990). Pyruvate uptake induced by a pH jump in mesophyll chloroplasts of maize and sorghum, NADP-malic enzyme type C4 species. *FEBS Lett.* **269**, 122–4.

Ohnishi, J., Flügge, U. I., Heldt, H. W. and Kanai, R. (1990). Involvement of Na^+ in active uptake of pyruvate in mesophyll chloroplasts of some C4 plants: Na^+/pyruvate transport. *Plant Physiol.* **94**, 950–95.

Ohsugi, R. and Huber, S. C. (1987). Light modulation and localization of sucrose phosphate synthase activity between mesophyll cells and bundle sheath cells in C4 species. *Plant Physiol.* **94**, 1096–101.

Osmond, C. B. (1971). Metabolite transport in C4 photosynthesis. *Aust. J. Biol. Sci.* **24**, 159–63.

Osmond, C. B. (1978). Crassulacean acid metabolism, a curiosity in context. *Ann. Rev. Plant Physiol.* **29**, 379–414.

Osmond, C. B., Björkman, O. and Anderson, D. J. (1980). *Physiological Processes in Plant Ecology,* Ecological Studies, Vol. 36, Springer-Verlag, Heidelberg.

Osmond, C. B. and Holtum, J. A. M. (1981). Crassulacean acid metabolism. In *The Biochemistry of Plants,* Vol. 8, eds M. D. Hatch and N. K. Boardman, Academic Press, New York, pp. 283–328.

Osmond, C. B., Holtum, J. A. M., O'Leary, M. H., Roeske, C., Summons, R. E., Wong, O. C. and Avadhani, P. N. (1988). Regulation of malic-acid metabolism in Crassulacean-acid-metabolism plants in the dark and light: *In vitro* evidence from ^{13}C labeling patterns after $^{13}CO_2$ fixation. *Planta* **175**, 184–92.

Osmond, C. B., Winter, K. and Ziegler, H. (1982). Functional significance of different pathways of CO_2 fixation in photosynthesis. In *Encyclopedia of Plant Physiology,* New Series Vol. 12B., eds O. L. Lange, P. S. Nobel, C. B. Osmond and H. Ziegler, Springer-Verlag, Berlin, pp. 479–547.

Ostrem, J. A., Olson, S. W., Schmitt, J. M. and Bohnert, H. J. (1987). Salt stress increases the level of translatable mRNA for phosphoenolpyruvate carboxylase in *Mesembryanthemum crystallinum. Plant Physiol.* **84**, 1270–75.

Prendergast, H. D. V., Hattersley, P. W. and Stone, N. E. (1987). New structural/biochemical associations in leaf blades of C4 grasses (Poaceae). *Aust. J. Plant Physiol.* **14**, 403–20.

Ranson, S. L. and Thomas, M. (1960). Crassulacean acid metabolism. *Ann. Rev. Plant Physiol.* **11**, 81–110.

Rawsthorne, S. (1992). C3–C4 intermediate photosynthesis: linking physiology to gene expression. *Plant J.* **2**, 267–74.

Ritz, D., Kluge, M. and Veith, H. J. (1986). Mass-spectrometric evidence for double-carboxylation pathway of malate synthesis in plants in light. *Planta* **167**, 284–91.

Robinson, S. A., Yakir, D., Ribas-Carbo, M., Giles L., Osmond, C. B., Siedow, J. N. and Berry, J. A. (1992). Measurements of the engagement of cyanide-resistant respiration in the crassulacean acid metabolism plant *Kalanchoe daigremontiana* with the use of on-line oxygen isotope discrimination. *Plant Physiol.* **100**, 1087–91.

Roeske, C. A. and Chollet, R. (1989). Role of metabolites in the reversible light activation of pyruvate, orthophosphate dikinase in *Zea mays* mesophyll cells *in vivo. Plant Physiol.* **90**, 330–37.

Rustin, P. and Queiroz-Claret, C. (1985). Changes in oxidative properties of *Kalanchoe blossfeldiana* leaf mitochondria during development of crassulacean acid metabolism. *Planta* **164**, 415–22.

Rygol, J., Winter, K. and Zimmerman, U. (1987). The relationship between turgor pressure and titratable acidity in the mesophyll cells of intact leaves of a Crassulacean acid metabolism plant, *Kalanchoe daigremontiana* Hamet et Perr. *Planta* **172**, 487–93.

Singh, B. K., Greenberg, E. and Preiss, J. (1984). ADP glucose pyrophosphorylase from the CAM plants *Hoya carnosa* and *Xerosicyos danguyi. Plant Physiol.* **74**, 711–16.

Smith, C. M., Duff, S. M. G. and Chollet, R. (1994). Partial purification and characterization of maize-leaf pyruvate, orthophosphate dikinase regulatory protein: a low abundance, mesophyll chloroplast stromal protein. *Arch. Biochem. Biophys.* **308**, 200–206.

Smith, J. A. C. and Bryce, J. H. (1992). Metabolite compartmentation and transport in CAM plants. In *Plant Organelles,* ed. A. K. Tobin, Cambridge Univ. Press, Cambridge. pp. 141–67.

Spilatro, S. R. and Preiss, J. (1987). Regulation of starch synthesis in the bundle sheath and mesophyll of *Zea mays* L. Intercellular compartmentation of enzymes of starch metabolism and the properties of the ADPglucose pyrophosphorylases. *Plant Physiol.* **83**, 621–27.

Stitt, M. and Heldt, H. W. (1985a). Control of photosynthetic sucrose synthesis by fructose-2,6-bisphosphate. Intercellular metabolite distribution and properties of the cytosolic fructose bisphosphatase in leaves of *Zea mays* L. *Planta* **164**, 179–88.

Stitt, M. and Heldt, H. W. (1985b). Generation and maintenance of concentration gradients between the mesophyll and bundle sheath in maize leaves. *Biochim. Biophys. Acta* **808**, 400–14.

Stitt, M., Huber, S. C. and Kerr, P. (1987). Control of photosynthetic sucrose formation. In *The Biochemistry of Plants,* Vol. 10, eds. M. D. Hatch and N. K. Boardman, Academic Press, New York, pp. 327–409.

Ting, I. P. (1985). Crassulacean acid metabolism. *Ann. Rev. Plant Physiol.* **36**, 595–622.

Vernon, D. M., Ostrem, J. A. and Bohnert, H. J. (1993). Stress perception and response in a facultative halophyte. The regulation of salinity-induced genes in *Mesembryanthemum crystallinum. Plant Cell Environ.* **16**, 437–44.

von Caemmerer, S. (1989). Biochemical models of photosynthetic CO_2 assimilation in leaves of C3–C4 intermediates and the associated carbon isotope discrimination. I. A model based on a glycine shuttle between mesophyll and bundle sheath cells. *Planta* **178**, 463–74.

Walbot, V. (1977). The dimorphic chloroplasts of the C4 plant *Panicum maximum* contain identical genomes. *Cell* **11**, 729–37.

Wang, Y.-H. and Chollet, R. (1993). Partial purification and characterization of phosphoenolpyruvate carboxylase protein-serine kinase from illuminated maize leaves. *Arch. Biochem. Biophys.* **304**, 496–502.

Wedding, R. T. and Black, C. C. (1983). Physical and kinetic properties and regulation of the NAD malic enzyme purified from leaves of *Crassula argentea*. *Plant Physiol.* **72**, 1021–28.

Weiner, H. and Heldt, H. W. (1992). Inter- and intracellular distribution of amino acids and other metabolites in maize (*Zea mays* L.) leaves. *Planta* **187**, 242–46.

White, P. J. and Smith, J. A. C. (1989). Proton and anion transport in crassulacean acid metabolism plants: specificity of the malate-influx system in *Kalanchoe daigremontiana*. *Planta* **179**, 265–74.

Whitehouse, A. G., Rogers, W. J. and Tobin, A. K. (1991). Photorespiratory enzyme activities in C3 and CAM forms of the facultative CAM plant *Mesembryanthemum crystallinum* L. *J. Exp. Bot.* **42**, 485–92.

Wilkins, M. B. (1992). Circadian rhythms: their origin and control. *New Phytol.* **121**, 347–75.

Winter, K. (1982). Properties of phosphoenolpyruvate carboxylase in rapidly prepared, desalted leaf extracts of the Crassulacean acid metabolism plant *Mesembryanthemum crystallinum* L. *Planta* **154**, 298–308.

Winter, K. (1985). Crassulacean acid metabolism. In *Photosynthetic Mechanisms and the Environment*. Topics in Photosynthesis, Vol. 6, eds J. Barber and N. R. Baker, Elsevier, Amsterdam, pp. 329–87.

Winter, K., Foster, J. G., Edwards, G. E. and Holtum, J. A. M. (1982). Intracellular localization of carbon metabolism in *Mesembryanthemum crystallinum* exhibiting photosynthetic characteristics of either a C3 or Crassulacean acid metabolism plant. *Plant Physiol.* **69**, 300–307.

Woo, K. C. and Osmond, C. B. (1977). Participation of leaf mitochondria in the photorespiratory carbon oxidation cycle: glycine decarboxylation activity in leaf mitochondria from different species and its intramitochondral location. In *Photosynthetic Organelles*, eds S. Mujachi, S. Katoh, Y. Fujita and K. Shibata. *Plant Cell Physiol. Special Issue 3*. Japanese Society of Plant Physiologists, Kyoto. pp. 315–23.

VI

Chloroplast–Cytosol
Interactions

24 Transport of proteins into chloroplasts
Kenneth Keegstra

Introduction and overview

Chloroplasts are functionally complex organelles that perform a diverse array of metabolic processes in addition to their well-known role in photosynthesis. Consistent with their functional complexity, chloroplasts are structurally complex organelles. They possess three different lipid bilayer membranes enclosing three different aqueous compartments (Fig. 24.1). Each membrane and each aqueous compartment has a unique set of proteins and enzyme activities that reflect their respective functions.

Because of the limited coding capacity of the plastid genome, most of these chloroplastic proteins are encoded by nuclear genes, synthesized in the cytoplasm as precursors and transported into chloroplasts via a post-translational process (Fig. 24.1). Understanding how these precursor proteins are targeted from the cytoplasm to their proper location within chloroplasts is a major challenge; progress in achieving this objective is described in this chapter. This subject has been the topic of several reviews in recent years and readers should consult these reviews for further details (de

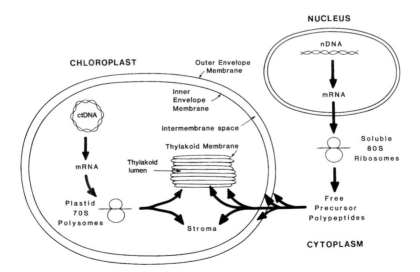

Fig. 24.1 Schematic diagram depicting the structural complexity of chloroplasts and the biogenetic origins of chloroplastic proteins. Cytoplasmically synthesized proteins are directed to all known compartments. Proteins synthesized within chloroplasts are directed mainly to the thylakoid membrane or the stromal compartment, although it is possible that some envelope proteins are synthesized inside the organelle.

Boer and Weisbeek, 1991; Gray and Row, 1995; Keegstra *et al.*, 1989; Schmidt and Mishkind, 1986; Soll and Alefsen, 1993; Theg and Scott, 1993). Moreover, many cytoplasmically synthesized proteins are targeted into mitochondria (see Chapter 15). Transport of proteins into these two organelles share some similarities, but also have significant differences that account for the specific targeting of proteins to their respective organelle.

Two different methods have been used to investigate the transport of proteins into chloroplasts. The first, and most widely used, is an *in vitro* assay in which the transport of radioactive precursor proteins into isolated intact chloroplasts is measured. In this case, precursor proteins are usually synthesized from cloned genes either via sequential *in vitro* transcription and translation reactions or via expression in *E. coli* or some other convenient protein expression system. Radiolabeled or chemically purified precursor proteins are then incubated with isolated intact chloroplasts and the extent of protein import can be quantitatively measured. The second method is *in vivo* analysis where the intracellular localization of a protein is examined and information about the targeting process is deduced from the localization data.

Both *in vitro* and *in vivo* approaches have benefited greatly from the application of recombinant DNA techniques that allow the production of altered or chimeric precursor genes. *In vivo* studies with chimeric genes have allowed 'foreign' proteins to be targeted into chloroplasts in transgenic plants. Such efforts have important implications in biotechnology (della-Cioppa *et al.*, 1987). *In vitro* assays have allowed the transport process to be dissected and analyzed in ways that are not possible *in vivo*. Such studies have demonstrated that the import process can be divided into discrete steps (Fig. 24.2) that often can be investigated independently.

The first step in the transport process, the binding of precursor proteins to the organelle surface (step 1 in Fig. 24.2), can be detected when precursor translocation is prevented. Generally this is accomplished by limiting ATP levels. Translocation of bound precursors across the two membranes of the chloroplastic envelope (step 2 in Fig. 24.2) occurs when ATP levels are adequate. During or immediately after transport, a stromal protease removes the portion of the precursor responsible for

targeting across the envelope membranes (step 3 in Fig. 24.2). The processed protein is either assembled in the stroma or further targeted to or across the thylakoid membrane or to the inner envelope membrane (steps 4a, 4b or 4c in Fig. 24.2). The targeting of proteins to the outer membrane of the chloroplastic envelope (step 5 in Fig. 24.2) is dramatically different from the targeting of proteins to internal locations. Further details regarding each of these steps is provided below. First, a brief review of precursor structure will be presented.

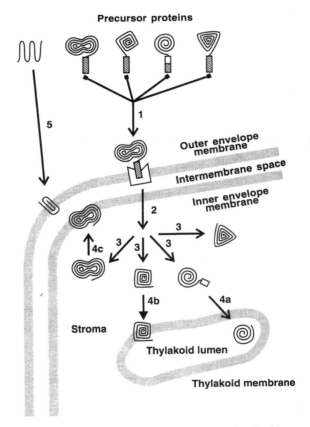

Fig. 24.2 Schematic representation of the steps involved in the targeting of proteins to the various compartments. Transit peptides are shown as cross-hatched boxes attached to the precursor proteins. The bipartite transit peptide of lumenal proteins is shown with a cross-hatched and open box. No attempt is made to depict the components of the transport apparatus other than the schematic representation of the putative receptor involved in precursor binding.

Precursor structure

Most chloroplastic proteins are synthesized as higher molecular weight precursors. The extra amino acids are present at the amino terminus of a precursor and are referred to as a transit peptide because of their importance in directing precursors into chloroplasts. Experiments from several laboratories have demonstrated that a transit peptide is both necessary and sufficient for the proper targeting of precursors (Keegstra *et al.*, 1989). Deletion of the transit peptide produces a protein that is not capable of being targeted into chloroplasts. Even more important was the observation that it is possible to redirect 'foreign' proteins into chloroplasts by the addition of a transit peptide.

Efforts to define essential regions within the primary structure of transit peptides have not been successful. Transit peptides have a distinctive amino acid composition, being rich in hydroxylated, basic and small hydrophobic amino acids. However, sequence comparisons have failed to identify conserved motifs present in all transit peptides (Keegstra *et al.*, 1989; von Heijne *et al.*, 1989). Given the variability among the sequences of transit peptides, it is unlikely that the essential features of transit peptides reside in the primary structure. Rather it seems likely that the essential features lie with secondary structure, as has been postulated for mitochondrial presequences (von Heijne *et al.*, 1989). However, to date, the essential features that determine the functionality of chloroplastic transit peptides have not been determined (von Heijne and Nishikawa, 1991).

One important conclusion regarding transit peptides is that they are responsible for organelle specificity. Because plant cells contain both mitochondria and chloroplasts, the question of how organelle specificity is determined is important. Boutry *et al.* (1987) examined this question with a series of *in vivo* experiments. They compared the targeting of chimeric precursors containing a chloroplastic transit peptide or a mitochondrial presequence fused to the same passenger protein, bacterial chloramphenicol acetyltransferase (CAT). These chimeric constructs were introduced into plants by Ti-mediated transformation. Upon expression of the chimeric genes, the chloroplastic transit sequence directed CAT into chloroplasts, but not mitochondria. On the other hand, the mitochondrial presequence directed CAT to mitochondria, but not chloroplasts. The results provide convincing evidence that organelle specificity resides in the respective targeting sequences.

Transport into chloroplasts

All proteins that enter the chloroplast stroma (step 2 in Fig. 24.2) follow a similar pathway early in the transport process. Several lines of evidence support the conclusion that a single translocation apparatus mediates the transport of all internal proteins (Fig. 24.2 and Fig. 24.3). The most compelling evidence comes from competition studies where excess unlabeled precursor proteins block the binding and translocation of different radiolabeled precursors (Cline *et al.*, 1993; Oblong and Lamppa, 1992; Perry *et al.*, 1991; Schnell *et al.*, 1991). A second system for transport across the envelope membranes may exist and some evidence points in this direction (Reinbothe *et al.*, 1995), but further work will be needed to define and characterize it.

Transport across the envelope membranes can be divided into two discrete stages and each can be studied separately. The first step is the binding of precursor proteins to the chloroplastic surface (step 1 in Fig 24.2; see also Fig. 24.3). Binding is mediated, at least in part, by a receptor protein, but it is likely that lipid–protein interactions have an important role (Fig. 24.3) (van't Hof *et al.*, 1991; van't Hof *et al.*, 1993). Specific binding can be observed only if translocation is prevented; otherwise specifically bound precursors are transported across the envelope. Inhibition of translocation can be accomplished by limiting the levels of ATP because binding requires approximately 0.1 mM ATP whereas translocation requires greater than 1 mM ATP. The ATP needed for binding is utilized in the intermembrane space (Olsen and Keegstra, 1992) whereas the ATP needed for translocation is utilized in the stroma (Theg *et al.*, 1989) (Fig. 24.3). Specifically bound precursors are irreversibly attached to the surface of chloroplasts and cannot be removed by washing in the absence of ATP. The precise topology of a bound precursor is not clear, but two possibilities are shown in Fig. 24.3.

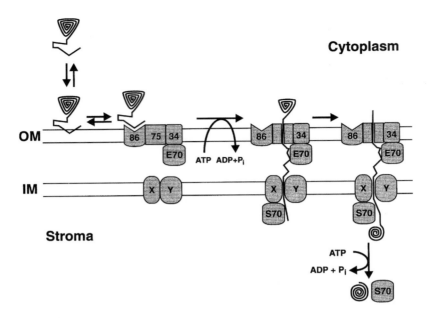

Fig. 24.3 More detailed schematic representation of protein transport across the envelope membranes. Both the binding and translocation steps are depicted. Those components of the translocation apparatus that have been identified are designated by numbers that represent their approximate molecular mass in kD. The stromal and envelope located versions of hsp70 are labeled S70 and E70 respectively. Those components that are suspected to occur, but have not yet been identified, are labeled with letters. The functions shown for the various proteins are still speculative, however the sites of ATP utilization are based on experimental evidence.

Translocation across the envelope membranes is thought to involve unfolding of a precursor protein so that it can pass through a putative channel (Fig. 24.3). It is generally thought that translocation across the two envelope membranes occurs simultaneously, with transport occurring at contact sites where the two membranes are present in close physical proximity (Schnell and Blobel, 1993). Each envelope membrane has its own translocation apparatus, although it is unclear whether the two systems can operate independently or whether they must act in concert to accomplish the transport of precursors from the cytoplasm to the stroma. It has been observed that ATP hydrolysis is needed inside chloroplasts to support the import of precursor proteins (Theg *et al.*, 1989). This observation is usually interpreted in terms of a general model for transport into organelles whereby an internal molecular chaperone of the 70 kD heat shock protein (hsp70) family provides the driving force for internalization by binding to unfolded proteins (Fig. 24.3). ATP is required to release an hsp70 from its interaction with an unfolded protein (Craig *et al.*, 1993). It has not yet been established whether this model applies to chloroplasts, but it has been established that chloroplasts contain hsp70s (Marshall *et al.*, 1990) and a transient interaction between hsp70 and recently imported proteins has been reported (Tsugeki and Nishimura, 1993).

Several laboratories have made progress recently in identifying the envelope membrane polypeptides that mediate the binding and transport of precursor proteins (see review by Gray and Row, 1995). Two laboratories employed a chemical cross-linking strategy to identify envelope components that are in close physical proximity to bound precursor proteins (Perry and Keegstra, 1994; Wu *et al.*, 1994). Two other laboratories used detergent solubilization to isolate complexes formed between envelope membrane proteins and precursors that were either bound to the translocation complex or were trapped as intermediates in the process of transport across the envelope membranes (Schnell and Blobel, 1993; Soll and Waegemann, 1992). Despite the different approaches, there is general agreement regarding the envelope membrane proteins identified by the different groups. An 86 kD Outer Envelope-membrane Protein (OEP86) was identified as a putative receptor that mediates the early interactions with precursor proteins (Fig. 24.3) (Hirsch et al., 1994; Perry and Keegstra, 1994). Bound precursors later form a complex that includes outer envelope polypeptides of 75 and 34 kD (OEP75 and OEP34) as well as E70, an hsp70 associated with the outer envelope membrane (Schnell et al., 1994). Both OEP86 and OEP34 have nucleotide binding domains and may be involved in using the ATP that is required for binding (Kessler et al., 1994; Seedorf et al., 1995). On the other hand, it is possible that the nucleoside triphosphates needed for binding are utilized by E70, the hsp70 associated with the outer envelope membrane. OEP75 is thought to form a channel that mediates translocation across the outer envelope membrane (Schnell et al., 1994; Tranel et al., 1995). cDNA clones have been isolated and characterized for OEP85, OEP75, OEP34 (Hirsch et al., 1994; Kessler et al., 1994; Schnell et al., 1994; Seedorf et al., 1995; Tranel et al., 1995) as well as S70 (Marshall and Keegstra, 1992). However, a functional translocation complex almost certainly includes other polypeptides, both from the outer and inner envelope membrane. Further work will be required to identify and characterize these proteins and to isolate cDNA clones encoding the transport components. In addition, a great deal of work will be required to confirm the function of each component and determine the sequence of molecular events that occurs during protein translocation.

During or immediately after transport across the envelope membranes, part or all of the transit peptide is removed by a stromal processing protease (step 3 in Fig. 24.2). Current evidence supports the conclusion that a single protease cleaves the many different precursor proteins that enter chloroplasts. The protease has been partially purified and some of its properties have been described (Oblong and Lamppa, 1992). Recently, a cDNA clone encoding the protease has been reported (VanderVere et al., 1995). The

features of the transit peptide that are recognized by the protease have not been identified, but it seems unlikely that it recognizes motifs in the primary structure of precursors (see discussion earlier under precursor structure).

Targeting within chloroplasts

Once a precursor protein is in the stromal space and has been processed, several different pathways are available depending upon the final destination of the mature protein. Stromal proteins need to be folded and assembled into their mature form. For the small subunit of Ribulose bisphosphate carboxylase (Rubisco) this means assembly with the chloroplast-encoded large subunit to form the holoenzyme. In this case, the chaperonin from the 60 kD heat shock protein family is required for proper assembly of Rubisco holoenzyme (Gatenby and Viitanen, 1994). Other proteins are active as monomers and only need to refold to be active. Still others require further modification, such as ferredoxin, to which iron-sulfur centers must be added (Li et al., 1990).

Transport to the thylakoid lumen

One of the most complicated targeting pathways is that followed by proteins destined for the thylakoid lumen. These proteins must cross three different biological membranes between their site of synthesis in the cytoplasm and their final destination in the thylakoid lumen. Consistent with this complicated targeting pathway, precursor proteins destined for the thylakoid lumen have complicated transit peptides (de Boer and Weisbeek, 1991). Lumenal precursors have a bipartite transit peptide with two distinct domains. The first directs the precursor across the envelope membranes (steps 1 and 2 in Fig. 24.2), whereas the second domain directs the protein across the thylakoid membrane (step 4a in Fig. 24.2). The first domain has features similar to those of other stromal targeting transit peptides and, indeed, can be replaced by the transit peptide of stromal proteins (de Boer and Weisbeek, 1991). The second domain is similar to bacterial signal sequences that direct proteins into or across the membranes of bacteria. It has been postulated that the thylakoid transfer system, including the signal peptides of lumenal precursors, was derived from the bacteria that gave rise to chloroplasts via the endosymbiotic events (Yuan *et al.*, 1994).

The transport of proteins across the thylakoid membrane can be reconstituted *in vitro* using isolated thylakoid membranes and either lumenal precursors or the intermediate form that is generated by the stromal processing protease. In contrast to the envelope membranes, where a single transport system is thought to be responsible for transporting all precursor proteins, thylakoid membranes appear to have at least two different systems for transporting proteins to the lumen (Cline *et al.*, 1993; Robinson *et al.*, 1994). One system, which transports a distinct subset of lumenal proteins, requires ATP as an energy source for translocation and requires stromal proteins as part of the transport apparatus. One of these proteins has recently been identified as a chloroplastic equivalent of bacterial secA (Yuan *et al.*, 1994). The second system, which transports a different subset of lumenal proteins, requires only a ΔpH across the thylakoid membrane as the source of energy for transport and does not need any stromal proteins for transport (Cline *et al.*, 1993; Robinson *et al.*, 1994). It is unclear why chloroplasts have evolved two different systems for transporting the relatively few proteins to the thylakoid lumen whereas many different proteins are transported across the envelope membranes by a single system.

Insertion into the thylakoid membrane

Proteins destined for the thylakoid membrane need to be inserted into the thylakoid membrane (step 4b in Fig. 24.2) before their assembly can be completed. The best studied example of a thylakoid membrane protein is the chlorophyll *a*/*b* binding (Cab) protein that forms the antenna complex for photosystem II. The transit peptide of the Cab precursor can be replaced by the transit peptide of a stromal protein, indicating that the transit peptide functions as a stromal targeting domain (Hand *et al.*, 1989). The information directing Cab to the thylakoid membrane resides within the mature region of the protein (Auchincloss *et al.*, 1992).

Insertion of Cab into the thylakoid membranes can be reconstituted *in vitro* using isolated thylakoid membranes and either the precursor or the processed form of Cab (Cline, 1988). The insertion of Cab into isolated thylakoid membranes requires both ATP and a ΔpH across the thylakoid membrane. Insertion of Cab also requires the presence of a soluble stromal protein (Payan and Cline, 1991). This protein has recently been identified as a chloroplastic homolog of SRP54 (Li *et al.*, 1995). The cytoplasmic version of SRP54 is an integral part of the signal recognition particle that is needed for the initiation of protein translocation across the endoplasmic reticulum membrane. It is not yet clear whether the chloroplastic version of SRP54 is used only for Cab insertion or whether it is required for the insertion of other thylakoid membrane proteins.

Once Cab has been inserted into the thylakoid membrane, additional assembly steps are required. The Cab apoprotein needs to combine with the chlorophyll molecules that are found in the mature protein and this complex needs to assemble with other Cab monomers to form the trimer that is functional in the light-harvesting antenna complex.

Targeting to the inner envelope membrane

The targeting of proteins to the inner membrane of the chloroplastic envelope is poorly understood. Indeed, even the pathway shown in Fig. 24.2 is uncertain; this pathway is one of two equally plausible possibilities. The one depicted in Fig. 24.2 is known as the conservative sorting pathway because the protein is first targeted to stromal space (steps 1 and 2 in Fig. 24.2) and then directed back to the inner envelope membrane (step 4c in Fig. 24.2) by the ancestral mechanisms that existed in the free-living bacterium that presumably gave rise to chloroplasts. The second alternative is the stop-transfer, or non-conservative sorting, pathway whereby a stop-transfer signal within the mature protein halts translocation across the inner envelope membrane causing the protein to be anchored in this membrane. Relatively few inner membrane proteins have been studied in detail and it is possible that both pathways operate, with some inner membrane proteins following one and other inner membrane proteins following the other pathway.

In a study of one inner membrane protein, Li *et al.* (1992) conducted a series of transit peptide swapping experiments and came to the conclusion that the transit peptide serves as a stromal targeting domain and that the targeting information directing the protein to the inner envelope membrane resides within the mature protein. Furthermore, targeting to the inner membrane appears to utilize the same apparatus used for stromal and thylakoid proteins (Li *et al.*, 1992). However, these observations do not distinguish between the two hypotheses for the pathway of targeting to the inner envelope membrane and are consistent with either pathway.

Targeting to the outer envelope membrane.

The targeting of proteins to the outer envelope membrane is distinctly different from the pathways and mechanisms described thus far (Fischer *et al.*, 1994; Ko *et al.*, 1992; Li *et al.*, 1991; Salomon *et al.*, 1990). With the exception of novel targeting pathways used by two components of the transport apparatus (described below), targeting of proteins to the outer envelope membrane is different in almost all respects from proteins destined for the inner compartments of chloroplasts. In contrast to internal proteins, which are synthesized as larger precursors containing a transit peptide, the precursors for outer membrane proteins are the same size as the mature form. The proteins appear to undergo a conformational change during insertion into the outer membrane, as depicted in schematic form in Fig. 24.2 (step 5). Other differences from the targeting of internal proteins are that proteins destined for the outer envelope membrane do not need ATP for their insertion and they do not utilize OEP 86 as a receptor. Indeed, it is not clear that they utilize protein receptors at all. As noted above, these features have been observed for several outer membrane proteins (Fischer *et al.*, 1994; Ko *et al.*, 1992; Li *et al.*, 1991; Salomon *et al.*, 1990). Moreover, the targeting of proteins to the outer mitochondrial membrane is distinctly different from the targeting of most internal mitochondrial proteins and has many similarities to the targeting of chloroplastic outer membrane proteins.

Recently two outer membrane proteins have been described that have different targeting pathways, i.e. the precursors to OEP86 and OEP75 (Hirsch *et al.*, 1994; Tranel *et al.*, 1995). Both precursors are made as a larger form with a cleavable transit peptide and both require ATP for translocation to the outer membrane. It has been demonstrated that the precursor to the small subunit of Rubisco competes for translocation sites with the precursor to OEP75, leading to the conclusion that this precursor utilizes at least part of the same import pathway as internal proteins (Tranel *et al.*, 1995). The reason for the novel targeting pathway for these two proteins is unclear, but one possibility is that it constitutes a safeguard to prevent the mistargeting of these components of the transport apparatus to the wrong organelle. If they were mistargeted they might cause further mistargeting of other precursors that serve as substrates for the chloroplastic protein transport apparatus.

Summary

Considerable progress has been made in recent years in describing the process of protein transport into chloroplasts. However, many questions remain unanswered and many aspects of the problem remain to be resolved. The importance of a transit peptide in determining organelle specificity has been established, however, the essential features of a transit peptide that carry the specific targeting information remain to be identified. The pathways of targeting to the several locations within chloroplasts have been described, however, the mechanistic details of how this targeting is accomplished remain to be revealed. Some of the components of the transport apparatus have been identified, however, additional components need to be identified, and even for those whose identity is known, much remains to be learned about their molecular level functions. The final goal is a molecular level description of the interactions between precursors and components of the transport apparatus that lead to the proper localization of chloroplastic proteins.

References

Auchincloss, A. H., Alexander, A. and Kohorn, B. D. (1992). Requirement for three membrane-spanning α-helices in the post-translational insertion of a thylakoid membrane protein: *J. Biol. Chem.* **267**, 10439–46.

Boutry, M., Nagy, F., Poulsen, C., Aoyagi, K. and Chua, N. H. (1987). Targeting of bacterial chloramphenicol acetyltransferase to mitochondria in transgenic plants: *Nature* **328**, 340–42.

Cline, K. (1988). Light-harvesting chlorophyll *a/b* protein. Membrane insertion, proteolytic processing, assembly into LHC II, and localization to appressed membranes occurs in chloroplast lysates: *Plant Physiol.* **86**, 1120–26.

Cline, K., Henry, R., Li, C. and Yuan, J. (1993). Multiple pathways for protein transport into or across the thylakoid membrane. *EMBO J.* **12**, 4105–14.

Craig, E. A., Gambill, B. D. and Nelson, R. J. (1993). Heat shock proteins: Molecular chaperones of protein biogenesis *Microbiol. Rev.* **57**, 402–14.

de Boer, A. D. and Weisbeek, P. J. (1991). Chloroplast protein topogenesis: Import, sorting and assembly. *Biochim. Biophys. Acta Rev. Biomembr.* **1071**, 221–53.

della-Cioppa, G., Bauer, S. C., Taylor, M. L., Rochester, D. E., Klein, B. K., Shah, D. M., Fraley, R. T. and Kishore, G. M. (1987). Targeting a herbicide-resistant enzyme from *Escherichia coli* to chloroplasts of higher plants. *Bio/Technology* **5**, 579–84.

Fischer, K., Weber, A., Arbinger, B., Brink, S., Eckerkorn, C. and Flügge, U. (1994). The 24 kDa outer envelope membrane protein from spinach chloroplasts: molecular cloning, in vivo expression and import pathway of a protein with unusual properties. *Plant Molecular Biology* **25**, 167–77.

Gatenby, A. A. and Viitanen, P. V. (1994). Structural and functional aspects of chaperonin-mediated protein folding. *Annu. Rev. Plant Physiol. Plant Mol. Biol.* **45**, 469–91.

Gray, J. C. and Row, P. E. (1995). Protein translocation across chloroplast envelope membranes. *Trends in Cell Biology* **5**, 243–7.

Hand, J. M., Szabo, L. J., Vasconcelos, A. C. and Cashmore, A. R. (1989). The transit peptide of a chloroplast thylakoid membrane protein is functionally equivalent to a stromal targeting sequence. *EMBO J.* **8**, 3195–206.

Hirsch, S., Muckel, E., Heemeyer, F., von Heijne, G. and Soll, J. (1994). A receptor component of the chloroplast protein translocation machinery. *Science* **266**, 1989–92.

Keegstra, K., Olsen, L. J. and Theg, S. M. (1989). Chloroplastic precursors and their transport across the envelope membranes. *Annu. Rev. Plant Physiol. Plant Mol. Biol.* **40**, 471–501.

Kessler, F., Blobel, G., Patel, H. A. and Schnell, D. J. (1994). Identification of two GTP-binding proteins in the chloroplast protein import machinery. *Science* **266**, 1035–9.

Ko, K., Bornemisza, O., Kourtz, L., Ko, Z. W., Plaxton, W. C. and Cashmore, A. R. (1992). Isolation and characterization of a cDNA clone encoding a cognate 70-kDa heat shock protein of the chloroplast envelope. *J. Biol. Chem.* **267**, 2986–93.

Li, H., Sullivan, T. D. and Keegstra, K. (1992). Information for targeting to the chloroplastic inner envelope membrane is contained in the mature region of the maize *Bt1*-encoded protein. *J. Biol. Chem.* **267**, 18999–9004.

Li, H., Theg, S. M., Bauerle, C. M. and Keegstra, K. (1990). Metal-ion-center assembly of ferredoxin and plastocyanin in isolated chloroplasts. *Proc. Natl. Acad. Sci. USA* **87**, 6748–52.

Li, H-m., Moore, T. and Keegstra, K. (1991). Targeting of proteins to the outer envelope membrane uses a different pathway than transport into chloroplasts. *The Plant Cell* **3**, 709–17.

Li, X., Henry, R., Yuan, J., Cline, K. and Hoffman, N. E. (1995). A chloroplast homologue of the signal recognition particle subunit SRP54 is involved in the post-translational integration of a protein into thylakoid membranes. *Proc. Natl. Acad. Sci. USA* **92**, 3789–93.

Marshall, J. S., DeRocher, A. E., Keegstra, K. and Vierling, E. (1990). Identification of heat shock protein hsp70 homologues in chloroplasts. *Proc. Natl. Acad. Sci. USA* **87**, 374–8.

Marshall, J. S. and Keegstra, K. (1992). Isolation and characterization of a cDNA clone encoding the major Hsp70 of the pea chloroplastic stroma. *Plant Physiol.* **100**, 1048–54.

Oblong, J. E. and Lamppa, G. K. (1992). Precursor for the light-harvesting chlorophyll *a/b*-binding protein synthesized in *Escherichia coli* blocks import of the small subunit of ribulose-1,5-bisphosphate carboxylase/oxygenase *J. Biol. Chem.* **267**, 14328–34.

Oblong, J. E. and Lamppa, G. K. (1992). Identification of two structurally related proteins involved in proteolytic processing of precursors targeted to the chloroplast. *EMBO J.* **11**, 4401–9.

Olsen, L. J. and Keegstra, K. (1992). The binding of precursor proteins to chloroplasts requires nucleoside triphosphates in the intermembrane space. *J. Biol. Chem.* **267**, 433–9.

Payan, L. A. and Cline, K. (1991). A stromal protein factor maintains the solubility and insertion competence of an imported thylakoid membrane protein. *J. Cell Biol.* **112**, 603–13.

Perry, S. E., Buvinger, W. E., Bennett, J. and Keegstra, K. (1991). Synthetic analogues of a transit peptide inhibit binding or translocation of chloroplastic precursor proteins. *J. Biol. Chem.* **266**, 11882–9.

Perry, S. E. and Keegstra, K. (1994). Envelope membrane proteins that interact with chloroplastic precursor proteins. *Plant Cell* **6**, 93–105.

Reinbothe, S., Runge, S., Reinbothe, C., Van Cleve, B. and

Apel, K. (1995). Substrate-dependent transport of the NADPH : protochlorophyllide oxidoreductase into isolated plastids. *Plant Cell* **7**, 161–72.

Robinson, C., Cai, D., Hulford, A., Brock, I. W., Michl, D., Hazell, L., Schmidt, I., Herrmann, H. G. and Klösgen, R. B. (1994). The presequence of a chimeric construct dictates which of two mechanisms are utilized for translocation across the thylakoid membrane: Evidence for the existence of two distinct translocation systems *EMBO J.* **13**, 279–85.

Salomon, M., Fischer, K., Flügge, U.-I. and Soll, J. (1990). Sequence analysis and protein import studies of an outer chloroplast envelope polypeptide. *Proc. Natl. Acad. Sci. USA* **87**, 5778–82.

Schmidt, G. W. and Mishkind, M. L. (1986). The transport of proteins into chloroplasts. *Ann. Rev. Biochem.* **55**, 879–912.

Schnell, D. J. and Blobel, G. (1993). Identification of intermediates in the pathway of protein import into chloroplasts and their localization to envelope contact sites. *J. Cell Biol.* **120**, 103–15.

Schnell, D. J., Blobel, G. and Pain, D. (1991). Signal peptide analogs derived from two chloroplast precursors interact with the signal recognition system of the chloroplast envelope. *J. Biol. Chem.* **266**, 3335–42.

Schnell, D. J., Kessler, F. and Blobel, G. (1994). Isolation of components of the chloroplast protein import machinery. *Science* **266**, 1007–12.

Seedorf, M., Waegemann, K. and Soll, J. (1995). A constituent of the chloroplast import complex represents a new type of GTP-binding protein. *Plant J.* **7**, 401–11.

Soll, J. and Alefsen, H. (1993). The protein import apparatus of chloroplasts. *Physiol. Plant.* **87**, 433–40.

Soll, J. and Waegemann, K. (1992). A functionally active protein import complex from chloroplasts. *Plant J.* **2**, 253–6.

Theg, S. M., Bauerle, C., Olsen, L. J., Selman, B. R. and Keegstra, K. (1989). Internal ATP is the only energy requirement for the translocation of precursor proteins across chloroplastic membranes *J. Biol. Chem.* **264**, 6730–36.

Theg, S. M. and Scott, S. V. (1993). Protein import into chloroplasts. *Trends Cell Biol.* **3**, 186–90.

Tsugeki, R. and Nishimura, M. (1993). Interaction of homologues of Hsp70 and Cpn60 with ferredoxin-NADP$^+$ reductase upon its import into chloroplasts. *FEBS Lett.* **320**, 198–202.

Tranel, P., Froehlich, J., Goyal, A. and Keegstra, K. (1995). A component of the chloroplastic protein import apparatus is targeted to the outer envelope membrane via a novel pathway. *EMBO J.* **14**, 2436–46.

VanderVere, P. S., Bennett, T. M., Oblong, J. E. and Lamppa, G. K. (1995). A chloroplast processing enzyme involved in precursor maturation shares a zinc-binding motif with a recently recognized family of metalloendopeptidases. *Proc. Natl. Acad. Sci. USA* **92**, 7177–81.

Van't Hof, R., Demel, R. A., Keegstra, K. and de Kruijff, B. (1991). Lipid-peptide interactions between fragments of the transit peptide of ribulose-1,5-bisphosphate carboxylase/oxygenase and chloroplast membrane lipids. *FEBS Lett.* **291**, 350–54.

Van't Hof, R., Van Klompenburg, W., Pilon, M., Kozubek, A., De Korte-Kool, G., Demel, R. A., Weisbeek, P. J. and de Kruijff, B. (1993). The transit sequence mediates the specific interaction of the precursor of ferredoxin with chloroplast envelope membrane lipids. *J. Biol. Chem.* **268**, 4037–42.

von Heijne, G. and Nishikawa, K. (1991). Chloroplast transit peptides: The perfect random coil. *FEBS Lett.* **278**, 1–3.

von Heijne, G., Steppuhn, J. and Herrmann, R. (1989). Domain structure of mitochondrial and chloroplast targeting peptides. *Eur. J. Biochem.* **180**, 535–45.

Wu, C., Seibert, F. S. and Ko, K. (1994). Identification of chloroplast envelope proteins in close physical proximity to a partially translocated chimeric precursor protein. *J. Biol. Chem.* **269**, 32264–71.

Yuan, J., Henry, R., McCaffery, M. and Cline, K. (1994). SecA homolog in protein transport within chloroplasts: Evidence for endosymbiont-derived sorting. *Science* **266**, 796–8.

25 The flux of carbon between the chloroplast and cytoplasm

Mark Stitt

Introduction

In plants, there is a compartmentation of metabolism between the cytosol and the plastid. The major storage carbohydrate, starch, is restricted to the plastid while the major transport carbohydrate, sucrose, is metabolized in the cytosol. This strict compartmentation is found in all plant tissues, from photosynthetic leaves to storage tissues like wheat grains or potato tubers. Clearly, processes which facilitate and control movement across the plastid envelope membrane are crucial in determining how carbon is distributed between sucrose, starch and respiratory metabolism.

Furthering our understanding of carbon flow between the plastid and the cytosol is dependent upon advances in two areas. First, the transport proteins present in the envelope membrane must be identified and characterized. Obviously, until we know the form of carbon that crosses the plastid membrane, the biochemical pathways for starch and sucrose turnover will be unknown. Second, the mechanisms regulating the rate of transport must be elucidated. These might include direct regulation of the transport proteins, and could also involve regulation of metabolism in the interlinked subcellular compartments. Research using traditional biochemical methods are now being extended by genetic and molecular approaches. These use mutants or transgenic plants with altered expression of a chosen enzyme or transporter to directly test its role and importance.

In this chapter, the known transport proteins will be described, and one example of chloroplast-cytosol interactions will be considered in detail, namely, the control of carbon fluxes across the chloroplast envelope membrane during photosynthetic starch and sucrose synthesis. Then the routes and regulation of carbon fluxes between starch and sucrose during respiratory metabolism will be introduced.

Transport in chloroplasts

Study of transport

Many experiments have attempted to characterize transport by incubating isolated organelles with a potential precursor (usually radioactive) and investigating its incorporation into a product within the organelle. Such studies only provide imprecise and indirect information, and can be very misleading. For example, there is a strong possibility that the added compound is modified prior to uptake. Alternatively, the rate and kinetics of incorporation into products may be more dependent on the metabolism of the compound after it has been taken up, than on the transport step itself. For these reasons, the study of transport requires direct measurements of the initial rate of uptake into the organelle before other factors such as metabolism or back exchange can lead to an apparent change in the transport kinetics.

This has been achieved for isolated chloroplasts by the silicone oil centrifugation method, pioneered by Heldt and coworkers (Fig. 25.1). A suspension of isolated chloroplasts is pipetted into a $400 \, \mu l$ microcentrifuge tube which already contains $70 \, \mu l$ 10% (v/v) perchloric acid and, above this, $20 \, \mu l$ of silicone oil. Silicone oil is immiscible with aqueous solutions, and is used at a density which is heavier than the suspension medium, but lighter than the chloroplasts themselves or the perchloric acid.

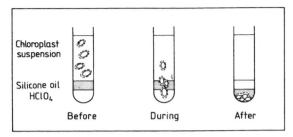

Fig. 25.1 Silicone oil centrifugation of chloroplasts.

Consequently, it provides a barrier between the suspension of chloroplasts and the perchloric acid. To measure transport, a small amount of radioactively-labeled substrate is mixed into the chloroplast suspension and a few seconds later uptake is terminated by centrifuging the chloroplasts out of the medium through the silicone oil into the perchloric acid. This separates the chloroplasts from the incubation medium, and stops further uptake or metabolism. The amount of radioactivity in the perchloric acid is a measure of its uptake into the chloroplast, although (see below) control experiments must also be carried out to correct for the small amounts of medium which are carried through the silicone oil by the organelles.

This technique allows Km and V_{max} values to be measured for the transport of a substrate, just as if the transporter were a soluble enzyme. Commonly, transporters mediate movement of more than one substrate. In this case, the alternative substrates may compete for binding and transport, and act as competitive inhibitors to one another. By comparing the K_m and K_i values of the various substrates, it is possible to decide which compounds are transported on the same protein (see, for example, Fliege *et al.*, 1978).

In an important advance, Flügge and Weber (1994) developed a technique to reconstitute transport proteins into liposomes from a crude extract. Using this method, Schunemann and Borchert (1996) isolated envelope membrane and then reconstituted their transport proteins into liposomes, thereby allowing studies of membrane transport without previous isolation of the organelle. This approach will greatly fascilitate the study of transporters in organelles which are difficult to isolate because they are fragile, or are difficult to separate from other contaminating organelles, or in cases where the silicone oil centrifugation technique does not work.

Further characterization and isolation of the protein usually depends upon finding a compound that will bind specifically and irreversibly to the protein, providing a 'tag' by which the protein can be identified. This is essential because the isolation of a protein requires that its presence during the various isolation steps can be monitored. The activity of a transport protein can obviously not be assayed once it has been removed from the membrane across which it performs the transport. The only exception to this are transport proteins that have a side reaction which is retained after solubilization, (e.g. ATP-linked ion pumps that retain an ion-stimulated ATPase activity). Once a protein has been 'tagged', it can be solubilized with detergents, purified, and then reconstituted into artificial membranes or liposomes. Its kinetic properties and regulation can then be studied without other reactions or processes interfering.

Following purification of the transporter protein, the corresponding gene can be cloned. One strategy involves screening of libraries with antibodies. However, antibodies are often difficult to raise and use against integral membrane proteins. An alternative is partial sequencing and subsequent screening with a collection of oligonucleotides (Flügge *et al.*, 1991; 1992). Recent advances have allowed cloning of the transporters without the need to purify the protein. One strategy is to raise polyclonal antibodies against total proteins in a purified membrane fraction. They are then used to screen a library, and the clones are sequenced to identify putative genes for transporters. Another strategy is to use functional complementation or activity assays in a heterologous system like yeast (Frommer, 1995).

The various transport systems found in the envelope membrane of chloroplasts will now be considered. One of these, the triose-phosphate translocator (TPT), has been unambiguously characterized in the strict sense of having been isolated and reconstituted, cloned, and its expression altered *in vivo*. Research into other transporters is not so far advanced.

The inner membrane as the site of specific transport

The chloroplast envelope consists of two functionally and physically distinct membranes. Low molecular weight compounds can freely pass through the outer membrane, because it contains proteins called *porins*. Similar proteins, which are found in the outer membrane of mitochondria and Gram-negative bacteria, form pores in the membrane, and permit free movement of molecules up to a molecular weight of 10 000 (Flügge and Benz, 1984).

The inner membrane is the site at which specific transport of metabolites occurs. This can be shown by incubating chloroplasts in increasing concentrations of a non-permeating osmoticum such as sorbitol. As the water is removed from the chloroplast, the stromal volume decreases and the inner impermeable membrane contracts away from the outer membrane so that the volume of the space between these two membranes increases. This can be directly visualized in the electron microscope. The volume of the intermembrane space can be measured using the technique of silicone oil centrifugation. Chloroplasts are suspended in medium that contains tritiated water and ^{14}C-sorbitol. Tritiated water will readily permeate both envelope membranes, but ^{14}C-sorbitol only equilibrates with the intermembrane space because the inner membrane is impermeable to sorbitol. By comparing the amounts of ^{3}H and ^{14}C in the perchloric acid after centrifuging the chloroplasts through silicone oil, it is possible to quantitatively estimate the volumes of the stroma and intermembrane spaces. This approach allows a correction to be made for the radioactivity which is carried through the silicone oil in the space between the two membranes when transport is being studied (see below).

The triose-phosphate translocator

The triose-phosphate translocator (TPT) plays an essential role during photosynthesis (Flügge and Heldt, 1991; Heldt and Flügge, 1992). CO_2 fixation by isolated chloroplasts requires a carefully controlled supply of phosphate (P_i) (Edwards and Walker, 1983). The P_i enters the chloroplast in exchange for triose-P and glycerate-3-P, which are exported to the medium (Lilley *et al.*, 1976). In intact leaves and protoplasts, newly fixed ^{14}C appears in the cytosol initially as triose-P and glycerate-3-P (Heber and Willenbrink, 1964; Stitt *et al.*, 1987b). These are then converted to sucrose, releasing P_i which re-enters the chloroplast (Fig. 25.2). These fluxes of P_i and carbon metabolites are tightly coupled. The discovery and characterization of the TPT by Heldt, Flügge and coworkers revealed how this coupled movement of metabolites between the chloroplast and cytosol is achieved.

The TPT has been purified and incorporated into liposomes (Flügge *et al.*, 1983). The kinetic properties, which were first investigated in isolated chloroplasts (Fliege *et al.*, 1978), could then be confirmed and extended in a purified system. The favoured substrates for the spinach leaf phosphate translocator are P_i, dihydroxyacetone-P,

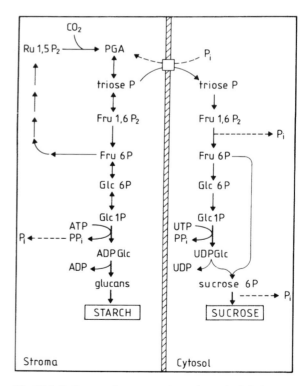

Fig. 25.2 Pathways of sucrose and starch synthesis in leaves of C3 plants.

glyceraldehyde-3-P and glycerate-3-P, all of which have a K_m between 0.1–0.4 mM, and comparable V_{max} activities (Table 25.1). All these substrates are transported as the divalent anion. At physiological pH, most of the P_i and triose-P are present as the divalent ion. However, the pK_a for the reaction $PGA^{2-} \leftrightarrow PGA^{3-} + H^+$ is 7.2 which means that only half of the total PGA is present as the transported form at pH 7.2, and the proportion will fall as the pH rises. At pH 8, less than 10% of the glycerate-3-P is in the transported form. The phosphate translocator also transports other compounds. For example, erythrose-4-P, glycerol-1-P, phosphoenolpyruvate and glycerate-2-P are translocated with 2 to 5-fold lower affinities whereas pentose-P and hexose-P are poor substrates, with K_m values of 40 mM or more.

Each substrate inhibits the transport of the other substrates competitively. Hence, a compound with a poor K_m will be transported at low rates *in vivo*, if other high-affinity substrates are present at comparable concentrations. The high-affinity substrates will also compete with each other for transport. For example, it is unlikely that glyceraldehyde-3-P will be important as a substrate for transport because its concentration is 10 to 20-fold lower than that of dihydroxyacetone-P concentration as a result of the equilibrium position of triose phosphate isomerase.

The TPT catalyzes a *passive counterexchange*; that is, for each molecule transported in one direction, one molecule is transported in the opposite direction. This counterexchange is strictly coupled, the unidirectional transport of P_i being at least 1000-fold slower than

the rate of counterexchange. This strict stoichiometry is important because it ensures that every molecule of phosphoester exported from the chloroplast is counterbalanced by the uptake into the chloroplast of a molecule of P_i.

The molecular mechanism which ensures that transport only occurs by counterexchange may be similar to that of the ADP/ATP translocator of mitochondria, which has some similarities with the plastid TPT. Both these transport proteins are dimers, with subunit molecular weights of about 30 000, have only one binding site, and catalyze a strict counterexchange (Flügge, 1987). A 'gated' pore mechanism has been suggested for the ADP/ATP translocator by Klingenberg (1981). The binding site of this translocator is formed by the two subunits, and is directed at any given time point to one side of the membrane. After a substrate is bound, the dimer undergoes a conformational change, the substrate is moved across the membrane and, simultaneously, the binding site is directed towards the other side of the membrane. In this mechanism, a strict coupling of transport in the two directions is ensured because the conformational change that reorientates the transporter can only occur when a ligand is bound in the pore.

Sequence analysis has revealed that the TPT is a transmembrane protein with 6 transmembrane helices per subunit (Flügge *et al.*, 1992; Heldt and Flügge, 1992). These transmembrane helices have a typical hepta motive, which allows them to build an α-helix which is hydrophobic on one side, and hydrophilic on the other side. The 12 α-helices (2 dimers, each with 6 α-helices) of a dimer are thought to be arranged in a circle. The hydrophobic side of each helix is orientated outward, and interacts with the lipid phase of the membrane. The hydrophilic sides are orientated inwards and combine to form a hydrophilic pore through the membrane. Two amino acid residues (lysine 273 and arginine 274) which have been identified by chemical studies (Gross *et al.*, 1990) to be involved in substrate binding (presumably interacting with the negative P_i group of the substrate) are located midway through this transmembrane channel. This molecular structure is in agreement with the 'gated pore' model. The location of the putative binding site in the middle of the transmembrane pore allows it to bind and release substrates to both sides without this requiring any

Table 25.1 Kinetic constants for the phosphate translocator of spinach chloroplasts

	K_m (mM)	V_{max} ($mol\,mg\,Chl^{-1}\,h^{-1}$)
P_i	0.3	57
Dihydroxyacetone-P	0.13	51
Glyceraldehyde-P	0.08	41
Glycerate-3-P	0.14	36
Glycerate-2-P	2.8	24
Glycerol-1-P	1.1	59

The results are from Fliege *et al.* (1978) and were obtained at 4 °C in the dark.

large (energy-requiring) changes of conformation. The identification of amino acids which are involved in opening and blocking the substrate site to inside and outside (gating) should be possible in the future by using site specific mutagenesis or, possibly, functional selection of mutant proteins in heterologous systems.

The affinity of the TPT varies in different species and tissues (Heldt *et al.*, 1991). For example, in C4 species the affinity for phosphoenolpyruvate increased in mesophyll chloroplasts where this metabolite has to be exchanged with P_i during photosynthesis. In guard cells and non-photosynthetic tissues, hexose-P can also be transported (see below for more details). There is evidently a multi-gene family encoding various TPT proteins. The cloning and sequencing of these genes is providing information about which amino acids participate in the active site to define substrate size (Fischer *et al.*, 1993).

The hexose transporter

The chloroplast envelope can also transport free sugars, including D-glucose, D-fructose and D-ribose (Schäfer *et al.*, 1977). The disaccharide maltose is also transported by this system (Beck, 1985). The uptake is selective, shows saturation kinetics towards the substrates with K_m values of about 20 mM, and can be inhibited by phloretin which is known to inhibit hexose uptake in other systems, including mammalian red blood cells.

The maximum velocity of the hexose carrier (5 μmol hexose mg Chl^{-1} h^{-1}) is only one tenth of the phosphate translocator in spinach chloroplasts. However, the fluxes of carbon across the envelope during respiratory metabolism are much lower than those needed during photosynthesis. As will be described later, hexose transport is rapid enough to cope with a significant part of the flux of carbon in leaves in the dark when starch is being degraded. A mutant of *Arabidopsis* which lacks the glucose transporter shows strongly impaired starch degradation (Trethewey and ap Rees, 1994).

Organic acid transport

Transport of organic acids across the envelope plays an essential role in photorespiration, and in facilitating the flux of carbon that is associated with nitrogen metabolism. Many of these transporters are also involved in the accessory pathways of CO_2 fixation during C4 or CAM photosynthesis; indeed, they are often first described in these specialized systems. In addition, it is now becoming apparent that the transport of dicarboxylates may allow redox groups to be transferred between the chloroplast and the cytosol (Heldt and Flügge, 1992). Chloroplasts resemble mitochondria in having several transport systems for dicarboxylates, carboxylates, and related organic acids. However, the actual transport systems differ between the two organelles.

The dicarboxylate transporters

Chloroplasts transport L-malate, 2-oxoglutarate, L-aspartate, L-glutamate and L-glutamine. The transport is mediated by a counterexchange mechanism, with unidirectional transport being about 100-fold slower than exchange transport. The K_m values for organic acids are in the 1–2 mM range. It is likely that the transport is mediated by translocators with different but overlapping substrate specifities. This is suggested by detailed studies of uptake kinetics in spinach chloroplasts (Lehner and Heldt, 1978) and by studies of *Arabidopsis* mutants, which revealed the presence of two separate translocators. One of them transported dicarboxylates but not glutamine, whereas the other transported glutamine in preference to dicarboxylates (Somerville and Somerville, 1985).

The oxaloacetate translocator

Oxaloacetate can also be transported by the dicarboxylate transporter but this route is unlikely to be of any significance *in vivo*. The dicarboxylate transporter has similar K_m values for malate and oxaloacetate but the *in vivo* concentration of oxaloacetate is 1000-fold lower than the malate concentration, because of the equilibrium position of the reaction catalyzed by malate dehydrogenase. Complex schemes in which oxaloacetate was aminated and deaminated to allow it to be moved across the membrane as aspartate were devised to overcome this problem. These schemes are no longer needed, because a high affinity ($K_m = 50$ μM) and very specific transporter for oxaloacetate has been found in maize mesophyll chloroplasts (Hatch *et al.*, 1984) where it is necessary for the carbon fluxes during C4 photosynthesis (see Chapter 23). It is also present in spinach chloroplasts, where it cooperates with the malate transporter to export reducing equivalents from the chloroplast during photosynthesis (see Heldt and Flügge, 1987; 1992).

Pyruvate

The pyruvate transporter was also first discovered in mesophyll chloroplasts isolated from C4 plants. In species that use NADP-malic enzyme, the transporter is needed to facilitate the entry of the pyruvate returning from the bundle sheath into mesophyll chloroplasts where pyruvate P_i dikinase is located (see Chapter 23). The uptake of pyruvate is driven by a light-dependent cation gradient across the envelope (Flügge *et al.*, 1985). It allows accumulation to occur against a concentration gradient. The pyruvate transporter may be present at lower activities in other plastids, and could provide a source of pyruvate for the plastid pyruvate dehydrogenase complex. This implies that biosynthesis of fatty acids may not be entirely dependent on glycolysis within the plastid.

Glycerate and glycolate transport

During photorespiration, glycolate has to be rapidly exported from, and glycerate returned to, the chloroplast. Two glycolate molecules are exported for every glycerate that returns (Chapter 22). This transport occurs via a single transporter which binds and transports glycolate, glyoxylate, D-glycerate and D-lactate, all with a similar affinity (Howitz and McCarty, 1985). The transport is driven by either a proton symport or a hydroxyl antiport.

Exchange of ATP and ADP

Chloroplasts and mitochondria both contain a translocator in their inner membrane that exchanges ATP and ADP, but they have very different properties (Heldt and Flügge, 1987). The mitochondrial ADP/ATP translocator is extremely active and catalyzes a strict counterexchange of ATP and ADP, which is energized by the electrical gradient across the inner membrane. This allows a selective uptake of ADP and export of ATP, and is

obviously well suited to export energy to the cytosol so that a high phosphorylation potential can be maintained. In contrast, in chloroplasts the exchange of ATP and ADP is relatively slow, being 10 to 100-fold lower than that of the phosphate translocator. Chloroplasts also import ATP in preference to ADP. This, and the observation that the activity is higher in young immature pea leaves, suggests the ATP translocator operates to import ATP into chloroplasts in the dark, or into plastids in non-photosynthetic cells (see below).

Control of transport during photosynthesis

The operation and regulation of these transporters *in vivo* will now be discussed using the TPT during photosynthesis as an example. The need for regulation of carbon exchange between the stroma and cytosol can be illustrated by considering the influence of the P_i concentration in the medium on photosynthesis by isolated chloroplasts (Fig. 25.3). In this simplified system, the rate of exchange across the translocator can be varied by altering the concentration of P_i outside the chloroplast. When P_i is low, the phosphate translocator activity is restricted, which leads to an accumulation of metabolites in the chloroplast and a depletion of P_i in the stroma (Table 25.2). The low P_i concentration results in a decrease in ATP, which restricts Calvin cycle activity and, hence, inhibits photosynthesis. On the other hand, if too much P_i is present in the medium, photosynthesis is also inhibited. This results from the fact that CO_2 can only be fixed if the

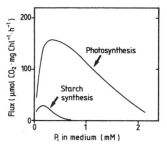

Fig. 25.3 Influence of the phosphate concentration in the medium on the rate of photosynthesis by isolated chloroplasts.

Table 25.2   Influence of the P_i concentration in the medium on photosynthesis, starch synthesis and stromal metabolites in isolated spinach chloroplasts

| P_i in medium (mM) | Fluxes | | P_i | Stromal metabolites | | | |
	CO_2 fixation (μmol mg Chl^{-1} h^{-1})	Starch synthesis (μmol hexose mg Chl^{-1} h^{-1})		Glycerate-3-P	Triose-P (mM)	Hexose-P	Ru-1,5-P$_2$
2.0	10	–	16.0	tr	tr	0.1	0.06
0.5	107	1.4	7.0	6.0	0.33	4.1	0.28
0.1	52	7.9	2.2	8.9	0.25	4.2	0.57

tr = trace (50 μM).
The results are from Heldt *et al.* (1977) and Flügge *et al* (1980).

acceptor, Ru-1,5-P$_2$ can be regenerated. For every three CO_2 molecules fixed, six molecules of triose-P are formed. One of the triose-P molecules can be exported, but the remaining five must be retained in the chloroplast and used to regenerate three molecules of Ru-1,5-P$_2$ again. If triose-P is withdrawn too rapidly, the Calvin cycle metabolite pools would be depleted, and photosynthesis would be inhibited. In other words, high rates of photosynthesis depend upon a controlled exchange of triose-P and P_i, so that there is a compromise between adequate P_i for ATP synthesis, and a sufficient concentration of stromal phosphorylated metabolites for the turnover of the Calvin cycle to regenerate Ru-1,5-P$_2$.

In a leaf, this balance must be achieved under more complex circumstances. The cytosol contains triose-P and glycerate-3-P, as well as P_i, all competing for entry into the chloroplast. This means the rate and direction of the resulting flux depends upon the relative concentrations of the substrates in the cytosol, and in the stroma. It also means that only a small fraction of the translocator capacity is actually involved in mediating the required export of triose-P and import of P_i. A considerable proportion will be involved in exchanges between other substrate pairs, or even so-called homologous exchanges between like partners (Flügge *et al.*, 1983). Hence, although the phosphate translocator is clearly in excess in isolated chloroplasts and has to be restrained by using a subsaturating P_i concentration in the surrounding medium, it may not be in excess *in vivo* (see below).

Another important difference between leaves and *in vitro* systems is that in leaves the chloroplasts are not surrounded by a large volume of medium with a fixed P_i concentration, acting as an 'infinite' source and sink for P_i. Instead, triose-P is continually removed by being synthesized into sucrose. This pathway regenerates P_i which returns to the chloroplast (see Section II). The pools of metabolites in the cytosol are small and turn over every few seconds (Stitt *et al.*, 1987b; Stitt, 1996). This means that photosynthesis will be inhibited unless the rate of sucrose synthesis and, hence, the supply of P_i is continually adjusted so that it is equal to the rate of CO_2 fixation. If sucrose synthesis is too rapid, CO_2 fixation will be inhibited because the high cytosolic P_i concentration will deplete the stromal metabolites and Ru-1,5-P$_2$ will no longer be regenerated at an adequate rate. If sucrose synthesis is too slow, the P_i concentration within the chloroplast will be too low, and ATP synthesis will be inhibited. The rate of transport therefore depends on metabolic events occuring in the cytosol.

To study the regulation of transport, complementary approaches must be used. Isolated chloroplasts, or preferably reconstituted systems in liposomes (Heldt and Flügge, 1987) can be used to study the regulatory properties of the proteins. It is also necessary to investigate how substrate concentrations and fluxes vary *in vivo*. Several techniques can be used to measure the subcellular metabolite distribution in leaves or in protoplasts (Stitt *et al.*, 1989). These measurements are dependent on the metabolism being rapidly quenched. They also depend upon adequate techniques being available for the extraction and assay of metabolites (ap Rees, 1980). The reliability of the data should therefore always be assessed by carrying out control experiments, in which the recovery of representative amounts of metabolites added to the plant material in the killing mixture is determined. Data presented without this kind of authentication have to be regarded as potentially unreliable. Studies of metabolites and fluxes only provide correlative evidence for the importance of control sites and regulation mechanisms. It is therefore necessary also to use genetic or molecular techniques to manipulate expression of proteins and directly test their *in vivo* contribution to control (Stitt, 1995; Stitt and Sonnewald, 1995).

In principle, the rate of transport could be controlled by one or more of the following mechanisms: (1) Regulation of the amount and/or type of TPT protein expressed in the cell; (2) regulation of the activity of the TPT protein; and (3) regulation of the reactions which produce and utilize its substrates.

Direct proof of the importance of TPT *in vivo*

In recent experiments, expression of TPT was decreased by transforming potato plants with the homologous gene in 'antisense' configuration (Riesmeier *et al.*, 1993). The transformants contained 30% less TPT than wild-type plants. Their leaves contained lower sucrose and higher starch. Measurements of sucrose, starch and photosynthesis during a 24 h diurnal period indicated that sucrose synthesis was inhibited by 25–50% during the day (Heineke *et al.*, 1994). Subcellular metabolite measurements showed that the chloroplast stroma in the leaves of the transformants contained high glycerate-3-P levels and a low ATP/ADP ratio. This resembles isolated chloroplasts in a medium with very low P_i to restrict activity of the TPT. Clearly, a change in the amount of TPT protein will lead to a change in the rate of export of carbon from the chloroplast. Research is still needed, however, to investigate if and when the expression of TPT actually changes in wildtype plants. It is already clear that expression of the gene-encoding chloroplasts TPT in potato is light-dependent (Schulz *et al.*, 1992).

It has usually been assumed that passive transporters like TPT are present in excess, and fully equilibrate substrates between the two compartments they link. However, there may be limits on the amounts of protein that can be packed into a membrane (TPT already represents 20–30% of the protein in the envelope membrane, Flügge and Heldt, 1991). Much of the capacity of a passive counterexchange transporter will anyway be catalyzing homologous exchanges *in vivo*, rather than the exchange required for photosynthetic sucrose synthesis (see above). Measurements of subcellular metabolites show that its substrates are not equilibrated between the cytosol and stroma (Stitt *et al.*, 1983; Gerhardt *et al.*, 1983). Since (see below for further discussion) triose-P, glycerate-3-P and P_i regulate the activity of key enzymes of starch and sucrose metabolism, relatively small changes in their concentration in the stroma and cytosol will have a large 'knock-on' effect on the rate of sucrose and starch synthesis.

Regulation of the activity of TPT

There is no evidence for direct regulation of TPT activity via post-translational modification, or regulator metabolites. However, light does exert an indirect regulatory effect on its activity. In the light, the stroma becomes more alkaline, and most of the glycerate-3-P is present in the trivalent form. As discussed previously, the phosphate translocator transports divalent anions. This means glycerate-3-P is retained in the chloroplast in the light, where it is reduced to triose-P using the ATP and NADPH that are generated by photosynthetic electron transport and photophosphorylation in the thylakoids. The triose-P can preferentially be exported in exchange for P_i under these conditions.

Fru-2,6-P_2 as a regulatory molecule to coordinate chloroplast and cytosol metabolism

The rate of carbon export via TPT also depends on the rate of triose-P formation (and P_i consumption) in the chloroplast, and on the rate at which triose-P are converted to sucrose (and P_i is released) in the cytosol. These pathways are described in detail in Chapters 8 and 21, Stitt *et al.* (1987b) and Stitt (1995). The following sections discuss how the rate of triose-P export from the chloroplast is coupled to metabolic events and requirements in the chloroplast and cytosol.

The first reaction leading to sucrose synthesis is catalyzed by the cytosolic Fru-1,6-Pase. No metabolites were known that affected the activity of this enzyme when they were present at concentrations equivalent to those found in the cytosol, until work on liver metabolism in the early 1980s led to the discovery of a regulator metabolite termed Fru-2,6-P_2. In animals, Fru-2,6-P_2 activates PFK and inhibits Fru-1,6-Pase. It plays an important role in regulating glycolysis and gluconeogenesis. Fru-2,6-P_2 is synthesized and degraded by specific enzymes termed Fru-6-P,2-kinase and Fru-2,6-P_2ase. In liver, both of these activities are present on a single bifunctional protein. They are regulated by a series of metabolites, often acting in an opposite manner on the synthesis and breakdown of Fru-2,6-P_2. They are also regulated in a reciprocal manner by cAMP-dependent protein kinase, which inactivates Fru-6-P,2-kinase and activates Fru-2,6-P_2ase. In heart muscle and yeast, kinase and phosphatase activities can be separated and Fru-6-P,2-kinase is activated by phosphorylation.

Fru-2,6-P_2 plays an important role in the regulation of sucrose synthesis (Stitt, 1990). Fru-2,6-P_2 is present in the cytosol but not in the chloroplast, and is a potent inhibitor of the cytosolic Fru-1,6-Pase. It exerts its effect by decreasing the substrate saturation kinetics. It also increases the enzyme's sensitivity to other inhibitors, such as AMP and P_i. Many plant tissues, including leaves, contain Fru-6-P,2-kinase, and Fru-2,6-P_2ase. Fru-6-P,2-kinase and Fru-2,6-P_2ase can be separated, similar to the situation in heart muscle and yeast (MacDonald *et al.*, 1989). They are regulated by the metabolites that are transported by the TPT (Fig. 25.4). Fru-6-P,2-kinase is stimulated by P_i, and is inhibited by glycerate-3-P and dihydroxyacetone-P. In contrast, Fru-2,6-P_2ase is inhibited by P_i. They are also regulated by the product of the Fru-1,6-Pase reaction. Fru-6-P is an activator of Fru-6-P,2-kinase and an inhibitor of Fru-2,6-P_2ase.

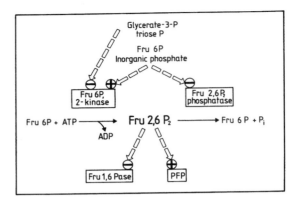

Fig. 25.4 The fructose 2,6-bisphosphate system.

Feedforward control

When the rate of photosynthesis increases, the concentration of dihydroxyacetone-P and the glycerate-3-P : P_i ratio will rise. As a result of this, there will be an inhibition of Fru-6-P,2-kinase, which will lead to a decrease of Fru-2,6-P_2. Hence, as the light intensity or the CO_2 concentration is raised, the increasing rate of photosynthesis is coupled to a progressive decrease of Fru-2,6-P_2 (Fig. 25.5A). Low concentrations of methylviologen can be used to prevent the increase of glycerate-3-P and amplify the rise of triose-P in the light (Neuhaus and Stitt, 1989). The light-dependent decrease of Fru-2,6-P_2 is abolished, indicating that rising glycerate-3-P : P_i ratio is the crucial signal linking photosynthesis to Fru-2,6-P_2. The decrease of Fru-2,6-P_2 will relieve the inhibition of the cytosolic Fru-1,6-P_2ase, and sucrose synthesis will be stimulated. In this way, Fru-2,6-P_2 signals how much photosynthate is available to make sucrose.

The activity of the cytosolic Fru-1,6-Pase is *not* just affected by Fru-2,6-P_2. Like other important enzymes that are regulated, activity depends upon an interaction between several factors. As the rate of photosynthesis increases, triose-P is exported from the chloroplast and converted to Fru-1,6-P_2 by triose phosphate isomerase and aldolase. These reactions are near equilibrium and, since two triose-P molecules are needed to produce one Fru-1,6-P_2 molecule, the Fru-1,6-P_2 concentration will increase as a square of the triose-P concentration. This means that there is a

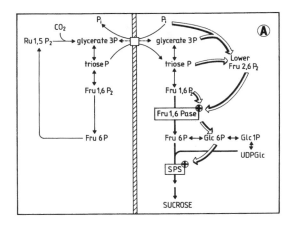

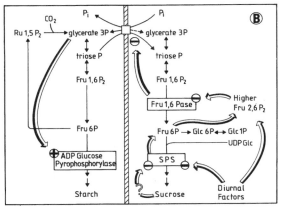

Fig. 25.5 Regulation of photosynthetic sucrose synthesis. (A) Feedforward control, (B) feedback control.

simultaneous increase of the substrate (Fru-1,6-P$_2$) and, because of the inhibition of Fru-6-P,2-kinase by three-carbon metabolites, a decrease of the concentration of the inhibitor (Fru-2,6-P$_2$) as photosynthesis increases (Fig. 25.6A). This interaction will exert a powerful regulation of the cytosolic Fru-1,6-Pase in response to a changing rate of photosynthesis. An empirical model can be developed to illustrate how the cytosolic fluxes are regulated in response to the requirements of the chloroplast (Fig. 25.6B). This approach relates the estimated Fru-1,6-Pase activity and the measured rate of photosynthesis to the triose-P content of the leaf (Stitt *et al.*, 1987a). Up to a threshold level of triose-P, the cytosolic Fru-1,6-Pase is effectively inactive.

Obviously, when triose-P is low and there are inadequate levels of metabolites in the stroma for turnover of the Calvin cycle, the plant cannot afford to make sucrose. Once this 'threshold' is exceeded, there are adequate levels of metabolites for Calvin cycle turnover, and the 'surplus' may be removed for synthesis of sucrose. There is now a strong activation of the Fru-1,6-Pase in response to small increments of triose-P. This highly sensitive activation as triose-P rise above the 'threshold' is important: it ensures sucrose synthesis is stimulated and will remove triose-P and regenerate P$_i$ before phosphorylated intermediates accumulate to the point where P$_i$ becomes limiting for photosynthesis.

The response of the cytosolic Fru-1,6-Pase may also be modified by other factors (Stitt *et al.*, 1987b). Other *metabolites*, such as AMP, modulate this enzyme. If, the cytosolic ATP/ADP ratio decreases, a concomitant increase in AMP will lead to an

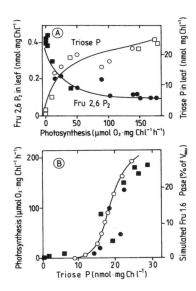

Fig. 25.6 Regulation of photosynthetic sucrose synthesis. (A) Changes of Fru-2,6-P$_2$ and triose-P as photosynthesis increases. (B) A model for the activation of the cytosolic Fru-1,6-Pase as availability of triose-P increases: (○) modeled Fru-1,6-Pase activity based on empirical measurements of Fru-2,6-P$_2$ and Fru-1,6-P$_2$ concentration estimated from the triose-P concentration assuming the reactions catalyzed by aldolase and triose phosphate isomerase are at equilibrium. The measured rate of photosynthesis is also shown (●, ■).

inhibition of sucrose synthesis. The response of the cytosolic Fru-1,6-Pase is also modified by *temperature*. As the temperature is lowered this enzyme becomes increasingly sensitive to inhibition by Fru-2,6-P_2 and AMP. In effect, the 'threshold' for activating sucrose synthesis is shifted upwards at low temperatures, and higher concentrations of triose-P and other metabolites can be maintained. This may be an adaptation which promotes photosynthesis at low temperatures since the higher levels of metabolites in the chloroplast may partially compensate for a decreased turnover rate of the Calvin cycle enzymes. The response of the cytosolic Fru-1,6-Pase also depends on the *species*. For example, cytosolic Fru-1,6-Pase from the mesophyll cells of maize has a 10-fold lower substrate affinity than the enzyme from C3 species. This change is functionally important. In maize, glycerate-3-P diffuses from the Calvin cycle in the bundle sheath to the mesophyll cells, where much of the glycerate-3-P reduction takes place. Triose-P has then to diffuse back to the bundle sheath cells. The low substrate affinity of the mesophyll Fru-1,6-Pase allows very high concentrations of triose-P to be maintained in the mesophyll. This provides a driving force for their diffusion back to the bundle sheath.

The rate of sucrose synthesis is also regulated at later steps in the pathway (Stitt *et al.*, 1987b; Stitt, 1995), including sucrose phosphate synthase (SPS). SPS is modulating by several mechanisms including (a) allosteric regulation involving activation by glucose-6-P and inhibition by P_i and (b) inactivation of the enzyme due to phosphorylation of a specific serine residue (Huber and Huber, 1992; Huber *et al.*, 1992). Allosteric activation by a rising Glc-6-P/P_i serves to stimulate SPS in response to increased Fru-1,6-Pase activity. The phosphorylation state of SPS is regulated at several levels. SPS kinase and phosphatase are inhibited by Glc-6-P and P_i respectively (Huber *et al.*, 1992). An increase of the Glc-6-P/P_i ratio will therefore lead to decreased phosphorylation (activation) of SPS. This provides a second link between carbon export and SPS activity. In addition, there is evidence that SPS is regulated via a minicascade which involves changes in the activity of SPS phosphatase and depends on the *de novo* protein synthesis. The details of this cascade have not yet been elucidated.

Feedback control

So far, we have considered how the use of triose-P for sucrose synthesis is adjusted to the requirements of the chloroplast. However, the export of triose-P from the chloroplast is also regulated by 'demand'. Although sucrose accumulates in leaves when photosynthesis is faster than the rate of sucrose export, in most plants, a point is reached where sucrose accumulation slows down or stops. At this point, the 'surplus' photosynthate is retained in the chloroplast and converted to starch.

Experimentally, the sucrose content of leaves can be increased by supplying sucrose exogenously, or by girdling or detaching leaves to prevent export. In spinach and soybean, these treatments lead to an increase of Fru-2,6-P_2 (Stitt *et al.*, 1984; Kerr and Huber, 1987). The increased Fru-2,6-P_2 inhibits Fru-1,6-Pase, which leads to an increase of triose-P and restricts the recycling of P_i to the chloroplast. Feedback regulation can also be observed in intact plants. In soybean, there are endogenous rhythms of sucrose phosphate synthase activity (Kerr and Huber, 1987) which are accompanied by lower rates of sucrose synthesis, increases in Fru-2,6-P_2 and increased starch synthesis. In spinach, there is a gradual shift towards starch synthesis during the day. This is again correlated with decreased sucrose phosphate synthase activity (Stitt *et al.*, 1988, Fig.

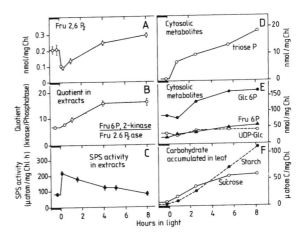

Fig. 25.7 Diurnal rhythms of enzyme activities and cytosolic metabolite levels in spinach leaves.

25.7). The reduced sucrose phosphate synthase activity leads to an accumulation of Fru-6-P in the cytosol. This, in turn, would activate Fru-6-P,2-kinase and inhibit Fru-2,6-P$_2$ase, and result in an increase of Fru-2,6-P$_2$. The increased Fru-2,6-P$_2$ would then inhibit cytosolic Fru-1,6-Pase activity and result in an increase in the concentration of triose-P in the cytosol. The resulting restriction on the recycling of P$_i$ to the chloroplast then leads to a stimulation of starch accumulation.

Experiments with isolated chloroplasts reveal how a decreased rate of sucrose synthesis (equivalent to decreased recycling of P$_i$) leads to a stimulation of starch synthesis. The key enzyme is ADP-glucose pyrophosphorylase. Preiss and coworkers (Preiss, 1988; 1991) have shown that ADP-glucose pyrophosphorylase is activated by glycerate-3-P and inhibited by P$_i$. Heldt et al. (1977) demonstrated that glycerate-3-P accumulates in chloroplasts when P$_i$ becomes limiting, probably because phosphoglycerate kinase is particularly sensitive to inhibition by the falling concentrations of ATP which are found in these conditions. An increase of the glycerate-3-P : P$_i$ ratio in the stroma is therefore a signal indicating that the supply of P$_i$ has decreased. This results in a stimulation of starch synthesis. By allowing phosphorylated intermediates to be converted to starch within the chloroplast, it facilitates the recycling of P$_i$ within the stroma. An analogous sequence of events is thought to occur in leaves when sucrose synthesis is restricted (see Neuhaus et al., 1989; Heineke et al., 1994).

The mechanisms involved in the feedback control of sucrose synthesis are complex, and are not yet fully understood. It has often been suggested that sucrose directly inhibits the enzymes that are responsible for its synthesis, but the available evidence is not convincing. High levels of sucrose are needed to inhibit the enzymes involved in sucrose biosynthesis. However, sucrose levels do not change much after export is inhibited. In several species hexose increase more than sucrose (Goldschmidt and Huber, 1994; Stitt and Sonnewald, 1995): It has been proposed that recyling of hexose may lead to a decreased net rate of sucrose synthesis and raised metabolite levels. Another explanation could be that sucrose acts indirectly by triggering protein modification or turnover. Studies of diurnal rhythms in leaves have

provided some evidence for this possibility (Stitt et al., 1987b) but the mechanisms involved still have to be clarified (Huber et al., 1992).

In summary, the rate of carbon export from the chloroplast depends on a balance between feedforward mechanisms which decrease Fru-2,6-P$_2$ and also activate sucrose phosphate synthase, and feedback mechanisms which deactivate sucrose phosphate synthase and increase Fru-2,6-P$_2$. A balance is achieved which allows the leaf to adjust the partitioning of photosynthate between various end-products, while simultaneously maintaining levels of stromal metabolites and P$_i$ which allow rapid photosynthesis to continue. The mechanisms involved in this regulation are complex, because many interacting factors make a contribution, and regulation occurs at several enzymes. This complexity is, however, probably unavoidable if the transport of triose-P and P$_i$ across the envelope membrane is to be adjusted to the wide variety of environmental and physiological conditions that are experienced by a leaf.

Use of genetically manipulated plants to assess the significance of regulation *in vivo*

The complexity of this regulation network leads to a new problem: how can the crucial sites be identified at which flux is regulated. This problem can be approached via studies of mutants or genetically manipulated plants, in which the expression of one gene is specifically altered (Stitt, 1994; Stitt and Sonnewald, 1995). For example, mutants of *Clarkia xantiana* have been used to confirm that increasing Fru-2,6-P$_2$ inhibits sucrose synthesis and stimulates starch synthesis in leaves. Genotypes of this species can be generated which have decreased cytosolic phosphoglucose isomerase activity. This enzyme is needed to convert Fru-6-P to Glc-6-P and, hence, these mutants contain more Fru-6-P than the wild type (Table 25.3). The higher Fru-6-P activates Fru-6-P,2-kinase and inhibits Fru-2,6-P$_2$ase which causes an increase of Fru-2,6-P$_2$. These plants allow an *in vivo* quantification of (a) the effect of Fru-6-P on Fru-2,6-P$_2$ levels, and (b) the effect of Fru-2,6-P$_2$ on the Fru-1,6-Pase (Neuhaus et al., 1989). Recently Scott et al. (1995) increased Fru-2,6-P$_2$ levels by expressing the

Table 25.3 Stimulation of starch synthesis by increased Fru-2,6-P$_2$ in a mutant of *Clarkia xantiana* with reduced phosphoglucose isomerase activity

	Metabolite levels			Synthesis rates	
	Fru-6-P	Fru-2,6-P$_2$ (nmol mg Chl^{-1})	PGA	Sucrose (μmol hexose mg Chl^{-1} h^{-1})	Starch
Wild type	137	0.16	282	38	40
18% Mutant	176	0.25	360	34	56

The mutant line had 18% of the wild-type complement of phosphoglucose isomerase. The results were measured in leaves illuminated at 125 μmol m^{-2} s^{-1} in saturating CO$_2$, and are taken from Neuhaus *et al.* (1989).

rat liver bifunctional Fru-6-P,2-kinase/phosphatase in tobacco and showed that this altered the rate of sucrose synthesis. An analogous approach is currently being used to quantify the importance of SPS. Overexpression of maize leaf SPS in tomato leaves leads to an increased rate of sucrose synthesis (Galthier *et al.*, 1983; T. Sharkey, pers. comm.). Decreasing SPS by 'antisense' in potato leaves leads to a proportionately large inhibition of sucrose synthesis (Stitt and Sonnewald, 1995). Evidence that AGPase plays a key role in regulating starch synthesis in leaves has been provided by studies of *Arabidopsis* mutants (Neuhaus and Stitt, 1990) and 'antisense' potato transformants (Müller-Röber *et al.*, 1992). Relatively small changes in AGPase expression lead to a change in the rate of starch synthesis. Clearly, the rate of carbon export from the plastid during photosynthesis depends on an interaction between the TPT itself, Fru-1,6-Pase and SPS in the cytosol, and AGPase in the plastid.

Control of transport during respiratory metabolism

Less is known about the routes and regulation of carbon fluxes across the plastid envelope membrane during respiratory metabolism. Non-photosynthetic plastids are fragile and difficult to isolate in an intact and functional state. Plastids with large starch grains are especially difficult to prepare, because the starch grains disrupt the plastids during centrifugation. Study of metabolism is also more difficult in these tissues. For example, it is much easier to measure the flow of carbon from CO$_2$ to sucrose or starch than it is to measure the flux between them. It is also difficult

to quantify the absolute rate of carbohydrate breakdown for respiration and biosynthesis.

In this section, starch degradation and carbon fluxes in leaves in the dark will be discussed first. The pathways of starch–sucrose interconversions in non-photosynthetic tissues will then be considered. Further information on the enzymes involved in starch and sucrose turnover can be found in Chapter 7, and in reviews by Preiss (1988; 1991), ap Rees (1980), or Stitt and Steup (1985).

Starch mobilization in leaves in the dark

Intact chloroplasts containing large quantities of starch can be isolated from spinach leaves. When these chloroplasts are incubated in the dark, the starch is degraded and a mixture of glucose, maltose, triose-P and glycerate-3-P (Fig. 25.8), are exported to medium (Stitt and Heldt, 1981). The distribution between these products depends on the conditions during the incubation (Table 25.4).

When the concentration of P$_i$ in the medium is high, phosphorylated products and CO$_2$ account for up to half of the products. In these conditions, starch is degraded phosphorolytically, giving rise to hexose phosphates in the stroma. These are metabolized via the stromal phosphofructokinase and the oxidative pentose phosphate pathway to triose-P and glycerate-3-P (Fig. 25.8), which are exported via the phosphate translocator to the cytosol. This export occurs in counterexchange for P$_i$, which is required for the continued mobilization of starch by this route. The relation between metabolism in the chloroplast and cytosol during phosphorolytic starch breakdown could be very similar to the interaction that occurs

Table 25.4 Influence of P_i in the medium on the products of starch breakdown in intact, starch-loaded spinach chloroplasts.

P_i in medium (mM)	Starch breakdown	Accumulation of products		
		Neutral sugars	Phosphorylated intermediates ($\mu atom\ C\ mg\ Chl^{-1}\ h^{-1}$)	CO_2
0.05	9.0	6.3	0.5	1.6
5.0	10.4	5.0	4.6	0.8

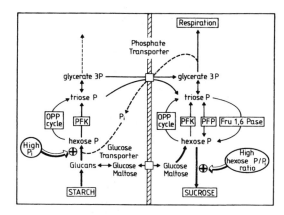

Fig. 25.8 Routes of starch mobilization in chloroplasts from spinach leaves. The possible fate of the starch degraded via the different routes is also indicated.

during photosynthesis. It can be envisaged how increasing demand for respiratory substrates in the cytosol will be associated with a decrease of triose-P and glycerate-3-P, and an increase of P_i. This would increase the exchange of cytosolic P_i and chloroplast metabolites via the phosphate translocator and would simultaneously stimulate phosphorolysis.

However, this is not the only route that is available for starch mobilization in leaves (Fig. 25.8). Even when P_i is present, glucose and maltose account for half of the starch degradation in isolated chloroplasts (Table 25.4). They are produced even more rapidly when P_i is omitted from the medium. Clearly, starch is being degraded hydrolytically, and maltose and glucose are then being released via the hexose transporter. Definitive evidence for the importance of this route *in vivo* has been provided by Trethewey and ap Rees (1994). They showed that an *Arabidopsis* mutant which lacks the envelope transporter has

strongly impaired rates of starch breakdown. Further *in vivo* evidence implicating the hexose transporter in leaf starch mobilization has been provided by studies of antisense TPT plants (Heineke *et al.*, 1994). The starch that these plants accumulate during the day is rapidly converted to sucrose during the night, presumably via the hexose transporter. These plants also illustrate that the presence of alternative transport routes allows flexibility (see below for further treatment of this idea).

Export of hexoses allows degradation of starch and export of carbon from the chloroplast when metabolite levels are high and P_i is low. This could be important during the conversion of starch to sucrose, because sucrose phosphate synthase is stimulated by high concentration of hexose phosphates and low P_i. This route would also avoid having to convert triose-P back to hexose-P in the cytosol in the dark. As already discussed, the cytosolic Fru-1,6-Pase is inactive in the dark because leaves contain high Fru-2,6-P_2 and negligible Fru1,6-P_2. The cytosolic Fru-1,6-Pase could be bypassed by pyrophosphate-fructose-6-phosphate phosphotransferase (PFP). However, this enzyme has a low activity in mature leaves. At least in *Arabidopsis*, a route via phosphorylase, the TPT and PFP cannot substitute for a missing hexose transporter.

Starch degradation in non-photosynthetic tissues

Starch fulfils several different roles in non-photosynthetic tissues. In some tissues, starch is accumulated when sucrose import exceeds the immediate needs of the tissue. This starch is later degraded to support respiration and growth. Changes of P_i and metabolites could link the mobilization of starch and export from the plastid, as in leaves. Rebeille *et al.* (1984) used ^{31}P-NMR to monitor the concentrations of P_i and metabolites in the cytoplasm of sycamore cell suspension cultures. When the supply of exogenous carbohydrate is removed, reserve carbohydrates in the cells are remobilized, a process that starts with sucrose. As sucrose is exhausted, there is a fall in the concentrations of phosphorylated metabolites, P_i increases, and then starch mobilization commences.

In other tissues, that we would normally designate as 'storage' tissues, starch is deposited in large quantities in specialized plastids (amyloplasts) as the seed or tuber develops. This starch is remobilized during germination or sprouting, when it provides a source of carbon from which sucrose can be synthesized for export. Two different strategies exist for moving carbon across the envelope membrane. In seeds in which starch is accumulated in an endosperm, the envelope and the remainder of the cell disintegrate when the seed matures. Upon germination, the starch is degraded by hydrolytic enzymes and the glucose is taken up into specialized cells where it is converted to sucrose. Many seeds and tubers, however, retain their plastid envelope during starch remobilization. The route and regulation of carbon fluxes during starch mobilization in non-photosynthetic plastids could be quite flexible and varied.

Plastids from a wide variety of plant tissues contain PFK, and the enzymes needed for the oxidative pentose phosphate pathway and the oxidation of triose-P (Chapter 6). They are usually able to convert glycerate-3-P to pyruvate, even though this section of glycolysis is always more active in the cytosol. In analogy with chloroplasts, they could contain transport systems allowing them to transport hexoses, three-carbon phosphoesters, pyruvate, dicarboxylates and adenylates. Two important differences to mature leaves should be noted. Firstly (see below), the TPT of non-photosynthetic plastids transports hexose-phosphates as well as three-carbon metabolites. Secondly, plant tissues often contain high activities of pyrophosphate : fructose-6-phosphate phosphotransferase (PFP) in the cytosol (see Chapter 8). This enzyme catalyzes a reversible reaction and could substitute for PFK or Fru-1,6Pase especially when glycolytic metabolites are high, P_i is low and

Fru-2,6-P_2 is high (Stitt, 1990). This means, that the hydrolytic and phosphorolytic routes of starch degradation could both supply six-carbon and three-carbon intermediates for the cytosol. Both of these pools could be used for respiration or for carbohydrate synthesis (Fig. 25.8). These alternative routes could reflect a genuine redundancy. It is also possible that different routes are used in different cell types, developmental stages or conditions.

Conversion of sucrose to starch

Several lines of evidence show that the conversion of sucrose to starch involves degradation to hexose-phosphate in the cytosol (see Chapter 7) followed by uptake of hexose-phosphates into the plastid and conversion to starch in the plastid via ADP-glucose pyrophosphorylase. Transport measurements in plastids from pea roots (Borchert *et al.*, 1989; 1993) and cauliflower buds (Neuhaus *et al.*, 1993) reveal that the TPT in these tissues transports Glc-6-P as well as P_i and three-carbon phosphoesters. In isolated plastids from many tissues including pea roots (Bowscher *et al.*, 1989), pea endosperm (Hill and Smith, 1993) and cauliflower buds (Batz *et al.*, 1994) the best precursor for starch is ^{14}C-glucose-6-P. Triose-P are a poor precursor because non-photosynthetic plastids usually lack a stromal Fru-

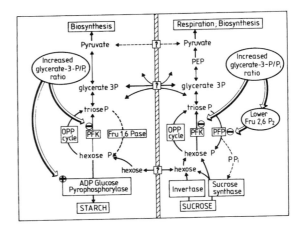

Fig. 25.9 Possible routes for starch accumulation in storage tissues.

1,6-Pase (Escheverria *et al.*, 1988, Entwhistle and ap Rees, 1990; Borchert *et al.*, 1993). Indirect evidence that hexose-phosphates are exported into plastids *in vivo* is provided by studies of the randomization of asymmetrically labelled glucose in wheat endosperm (Keeling *et al.*, 1988), potato tubers, maize endosperm and cell cultures (Hatzfeld and Stitt, 1989).

It is possible that glucose-1-P is transported instead of glucose-6-P in some species. ^{14}C-Glucose-1-P was the preferred substrate for starch synthesis in isolated wheat endosperm plastids (Tyson and ap Rees, 1988), and was also taken up more rapidly in transport assays (M. Emes, unpubl.). A similar picture is emerging in soybean cell culture and tomato fruit. Proof of these differences will require isolation and characterization of the corresponding protein and/or the cloning of the gene and functional studies of the gene product in heterologous systems.

There have been reports that other ^{14}C-labelled compounds including triose-P, glucose and ADP-glucose can act as substrates for starch synthesis. In evaluating such results it is important to critically evaluate the specifity and rate of incorporation. For example, triose-P lead to heavy labelling of *insoluble* compounds in plastids, but this is due to lipid synthesis, not starch (Tyson and ap Rees, 1988). Glucose and ADP-glucose will both label starch in cauliflower floret plastids, but the rates are very low compared to those obtained with glucose-6-P, and are almost totally suppressed when physiological concentrations of glucose-6-P are added to the medium (Batz *et al.*, 1994). Although it would be premature to exclude the possibility that import of other compounds may also play a role in some species (especially crop plants where breeders have selected for yield and starch content and might have accidently recruited or amplified novel or minor pathways), evidence at present indicates that starch synthesis occurs primarily via import of hexose-phosphates.

Studies of pea mutants and antisense potato plants have provided definitive evidence that hexose-phosphates are then converted to starch via ADP-glucose pyrophosphorylase. The storage tissues of these plants contain negligible amounts of starch (Smith and Denyer, 1992; Müller-Röber *et al.*, 1992). The enzyme is regulated via the glycerate-3-P : P_i ratio in analogy to leaves (Preiss, 1988). Glycerate-3-P

stimulates starch synthesis in isolated plastids (Neuhaus *et al.*, 1993). Starch accumulation can be increased in potato tubers when a mutant non-regulated bacterial ADP-glucose phosphorylase is expressed, but not when the wild-type protein is expressed. This implies that AGPase is normally down-regulated by metabolites *in vivo* (Stark *et al.*, 1992). The TPT therefore has two roles during starch synthesis in non-photosynthetic plastids. It allows the entry of substrate, and also allows regulatory metabolites to be transferred between the cytosol and plastid (Fig. 25.8). It should be noted, however, that storage synthesis may also be regulated by additional factors including turgor and interactions with other aspects of cell growth. It is not known how this information is sensed, or transmitted to the plastid.

Starch synthesis requires ATP. Some ATP could be regenerated via plastid glycolysis, provided mechanisms are available to remove the NADH which is produced by glyceraldehyde-3-P dehydrogenase. Alternatively, ATP might be imported in exchange for ADP via the ATP/ADP transporter. Indirect evidence for the latter is provided by the observation that exogenous ATP leads to a strong stimulation of starch synthesis in isolated plastids (Hill and Smith, 1993; Neuhaus *et al.*, 1993). Rapid operation of the ATP/ADP transporter will, however, lead to a problem. Entry of ATP in exchange for ADP would lead to a net transfer of phosphate into the plastid. This would disrupt metabolism, unless P_i can be recycled out of the stroma. More research is needed to show how the plastid can be provided with ATP without this disrupting the subcellular distribution of phosphate.

Concluding remarks

In this chapter the importance of understanding the transport of carbon between the chloroplast and cytosol has been emphasized. In leaves, considerable progress has been made in clarifying routes and regulation, and research in the future will probably be aimed at understanding protein modification and turnover, probing the structure and function of key proteins, and in manipulating the levels of individual enzymes and transporter proteins. These investigations have already opened the way to a rational manipulation of export and storage strategies in leaves.

However, plant productivity depends upon the distribution of photosynthate to growing tissues, and its efficient utilization or storage. Recent research is starting to fill serious gaps in our knowledge of the routes and regulation of starch-sucrose interconversions in non-photosynthetic tissues and the first successful manipulation of these pathways have been reported. Continuing research at biochemical and molecular level is needed, however, to provide an adequate understanding of their developmental and environmental regulation, and generate a broader theoretical basis for rational manipulation of sucrose and starch accumulation.

Further reading and references

ap Rees, T. (1980). Integration of pathways of synthesis and degradation of hexose phosphates. In *The Biochemistry of Plants*, Vol. 3, ed. J. Preiss, Academic Press, New York, pp. 1–42.

ap Rees, T. (1985). The organisation of glycolysis and the oxidative pentose phosphate pathway in plants. In *Encyclopedia of Plant Physiology*, Vol 8, eds R. Douce and D. A. Day, Springer-Verlag, Berlin, pp. 347–90.

Batz, O., Maaß. U., Henrichs, G., Scheibe, R. and Neuhaus, H. E. (1994). Glucose and ADP-glucose dependent starch synthesis in isolated cauliflower bud amyloplasts. Analysis of the interaction of various potential precursors. *Biochim. Biophys. Acta* **1200**, 148–54.

Beck, E. (1985). The degradation of transitory starch in plastids. In *Regulation of Carbohydrate Partitioning in Photosynthetic Tissues*, eds R. L. Heath and J. Preiss, American Society of Plant Physiologists, Rockville, Maryland, pp. 27–40.

Borchert, S., Grosse, H. and Heldt, H. W. (1989). Specific transport of inorganic phosphate, glucose-6-phosphate, dihydroxyacetone phosphate and 3-phosphoglycerate into amyloplasts from pea roots. *FEBS Lett.* **253**, 183–86.

Borchert, S., Harborth, J., Schünemann, D., Hoferichter, P. and Heldt, H. W. (1993). Studies of the enzymic capacities and transport properties of pea root plastids. *Plant Physiol.* **101**, 301–12.

Bowscher, C. G., Huckelsby, D. B. and Emes, M. J. (1989). Nitrate reduction and carbohydrate metabolism in plastids purified from roots of *Pisum sativum*. *Planta* 177, 359–66.

Edwards, G. E. and Walker, D. A. (1983). C_3 C_4: *Mechanisms, and cellular and environmental regulation of photosynthesis*, Blackwell Scientific Publications, Oxford, pp. 1–542.

Entwhistle, G. and ap Rees, T. (1990). Lack of fructose-1,6-bisphosphatase in a range of higher plants that store starch. *Biochem. J.* **271**, 467–72.

Echeverria, E., Boyer, C. D., Thomas, P. A., Liu, K.-C. and Shannon, J. C. (1988). Enzyme activities associated with maize endosperm amyloplasts. *Plant Physiol.* **86**, 786–92.

Fischer, K., Arbinger, B., Kammerer, B., Busch, C., Brink, S., Wallmeier, H., Sauer, N., Eckerstrom, C. and Flügge, U.-I. Cloning and *in vivo* expression of functional triose-phosphate/phosphate translocator from C_3 and C_4 plants: evidence for the putative participation of specific amino acids in the recognition of phosphoenolpyruvate. *Plant J.*, in press.

Fliege, R., Flügge, U.-I., Werdan, K., and Heldt, H. W. (1978). Specific transport of inorganic phosphate, 3-phosphoglycerate and triose phosphate across the inner membrane of the envelope in spinach chloroplasts. *Biochim. Biophys. Acta* **502**, 232–47.

Flügge, U.-I. (1987). Physiological function and physical characteristics of the chloroplast phosphate translocator. In *Progress in Photosynthesis Research*, Vol. III, ed J. Biggins, Martinus Nijhoff, Dordrecht, pp. 739–47.

Flügge, U.-I. and Benz, R. (1984). Pore forming activity in the outer membrane of the chloroplast envelope. *FEBS Lett.* **169**, 85–89.

Flügge, U.-I. and Heldt, H. W. (1991). Metabolite translocator of the chloroplast envelope. *Ann. Rev. Plant. Physiol. Mol. Biol.* **42**, 129–44.

Flügge, U.-I. and Weber, A. (1994). A rapid method for measuring organelle-specific substrate transport in homogenates from plant tissues. *Planta* 194, 181–85.

Flügge, U.-I., Freisl, M. and Heldt, H. W. (1980). Balance between metabolite accumulation and transport on relation to photosynthesis by isolated spinach chloroplasts. *Plant Physiol.* **65**, 574–77.

Flügge, U.-I., Gerber, J. and Heldt, H. W. (1983). Regulation of the reconstituted chloroplast phosphate translocator by an H^+ gradient. *Biochim. Biophys. Acta* **725**, 229–37.

Flügge, U.-I., Weber, A, Fischer, K., Löddenkötter, B. and Wallmeier, H. (1992). Structure and function of the chloroplast triose-phosphate/phosphate translocator. In *Research in Photosynthesis*, Vol. 3, ed N. Murata, Kluwer Academic Publishers, Dordrecht, pp. 667–74.

Flügge, U.-I., Stitt, M. and Heldt, H. W. (1985). Light driven uptake of pyruvate in mesophyll chloroplasts from maize. *FEBS Lett.* **183**, 335–39.

Gerhardt, R., Stitt, M. and Heldt, H. W. (1987). Subcellular metabolite levels in spinach leaves. *Plant Physiol.* **87**, 399–7.

Goldschmidt, E. E. and Huber, S. C. (1992). Regulation of photosynthesis by end-product accumulation in leaves storing starch, sucrose and hexose sugars. *Plant Physiol.* **99**, 1443–48.

Gross, A., Brückner, G., Heldt, H. W. and Flügge, U.-I. (1990). Comparison of the kinetic properties, inhibition and labelling of the phosphate carriers from maize and spinach chloroplasts. *Planta* **180**, 262–71.

Hatch, M. D., Dröscher, L., Flügge, U.-I. and Heldt, H. W. (1984). A specific translocator for oxaloacetate transports in chloroplasts. *FEBS Lett.* **178**, 15–19.

Hatzfeld, W. D. and Stitt, M. (1989). A study of the rate of recycling of triose-phosphates in heterothropic *Chenopodium rubrum* cells, potato tubers and maize endosperm. *Planta* **180**, 198–204.

Heber, W. and Willenbrink, J. (1964). Sites of synthesis and transport of photosynthetic products within the leaf cell. *Z. Naturforschung* **25b**, 710–28.

Heineke, D., Kruse, A., Flügge, U.-I., Frommer, W. B., Riesmeier, J., Willmitzer, L. and Heldt, H. W. (1994). Effect of 'antisense' repression of the chloroplast triose-phosphate translocator on photosynthesis metabolism in transgenic potato plants. *Planta* **193**, 174–80.

Heldt, H. W. and Flügge, U.-I. (1987). Subcellular transport of metabolites in a plant cell. In *The Biochemistry of Plants*, Vol. 12, Academic Press, New York, pp. 49–85.

Heldt, H. W. and Flügge, U.-I. (1992). Metabolite transport in plant cells. In *Society for Experimental Biology Seminar Series 50, Plant Organelles*, ed A. K. Tobin, Cambridge University Press, Cambridge.

Heldt, H. W:, Chon, C. J., Maronde, D., Herold, A., Stankovic, Z. S., Walker, B. A., Kraminer, A., Kirk, M. R. and Heber, U. (1977). Role of orthophosphate and other factors in the regulation of starch formation in leaves and isolated chloroplasts. *Plant Physiol.* **59**, 1146–55.

Heldt, H. W., Flügge, U.-I. and Borchert, S. (1991). Diversity of specifity and function of phosphate translocators in various plastids. *Plant Physiol.* **95**, 341–43.

Hill, L. M. and Smith, A. M. (1993). Evidence that glucose-6-phosphate is imported as the substrate for starch synthesis by the plastids of developing pea embryos. *Planta* **185**, 91–96.

Howitz, K. T. and McCarthy, R. E: (1985). Substrate specifity of the pea chloroplast glycolate transporter. *Biochemistry* 24, 3645–50.

Huber, S. C. and Huber, J. L. A. (1992). Role of sucrose phosphate synthase in sucrose metabolism in leaves. *Plant Physiol.* **99**, 1275–78.

Huber, S. C., Huber, J. L. A. and McMichael, R. W. Jr. (1991). The regulation of sucrose phosphate synthase in leaves. In *Carbon Partitioning Within and Between Organisms*, eds C. T. Pollock, J. F. Farrar and A. J. Gorden, Bios. Scientific Publ. Oxford, pp. 1–26.

Keeling, P. L., Wood, J. T., Tyson, R. H. and Bridges, I. G. (1988). Starch biosynthesis in developing wheat grain. Evidence against the direct involvement of triose phosphates in the metabolic pathway. *Plant Physiol.* **87**, 311–19.

Kerr, P. S. and Huber, S. C. (1987). Coordinate control of sucrose formation in soybean leaves by sucrose phosphate synthase and fructose-2,6-bisphosphate. *Planta* 170, 197–94.

Klingenberg, M. (1981). The mitochondrial ATP/ADP translocator. *Nature* 290, 449–54.

Kruckeberg, A. L., Neuhaus, E., Feil, R., Gottlieb, L. D. and Stitt, M. (1989). Reduced activity mutants of phosphoglucose isomerase in the chloroplast and cytosol of *Clarkia xantiana*. *Biochem. J.* 261, 457–67.

Lehner, K. and Heldt, H. W. (1978). Dicarboxylate transport across the inner membrane of the chloroplast envelope. *Biochim. Biophys. Acta* **501**, 531–44.

Lilley, R. McC., Chon, C.J, Mosbach, A. M. and Heldt, H. W. (1976). The distribution of metabolites between spinach chloroplasts and medium during photosynthesis *in vitro*. *Biochim. Biophys. Acta* **460**, 259–72.

MacDonald, F. D., Chou, Q., Buchanan, B. B, and Stitt, M. (1989). Purification and characterization of fructose-2,6-bisphosphate. A substrate-specific cytosolic enzyme from leaves. *J. Biol. Chem.* **264**, 5540–44.

Müller-Röber, B. T., Sonnewald, U. and Willmitzer, L. (1992). Inhibition of ADP-glucose pyrophosphorylase in transgenic potatoes leads to sugar-storing tubers and influences tuber formation and expression of tuber storage proteins. *EMBO J.* **11**, 1229–38.

Neuhaus, H. E. and Stitt, M. (1989). Peturbation of photosynthesis in spinach leaf discs by low concentrations of methylviologen. *Planta* 179, 151–61.

Neuhaus, H. E., Kruckeberg, A. L., Feil, R and Stitt, M. (1989). Reduced activity mutants of phosphoglucose isomerase in the cytosol and chloroplast of *Clarkia xantiana*. II Study of the mechanisms which regulate photosynthate partitioning. *Planta* 178, 10–22.

Neuhaus, H. E., Henrichs, G. and Scheibe, R. (1993). Characterization of glucose-6-phosphate incorporation into starch by isolated intact cauliflower bud plastids. *Plant Physiol.* **101**, 573–78.

Rebeille, F., Bligny, R., Martin, J.-P. and Douce, R. (1984). Effect of sucrose starvation on sycamore (*Acer pseudoplatanus*) cell carbohydrate and P_i status. *Biochem. J.* **226**, 679–84.

Riesmeier, J. W., Flügge, U.-I., Schulz, B., Heineke, D., Heldt, H. W., Willmitzer, L., and Frommer W. D. (1993). Antisense repression of the chloroplast triose-phosphate translocator effects carbon partitioning in transgenic potato plants. *Proc. Natl. Acad. Sci. USA* 90, 6160–64.

Preiss, J. (1988). Biosynthesis of starch and its degradation. In *Biochemistry of Plants*, Vol. 13, ed J. Preiss, Academic Press, pp. 181–254.

Preiss, J. (1991). Biology and molecular biology of starch synthesis and its regulation. In *Oxford Surveys of Plant Molecular and Cell Biology*, Vol. 7, ed B. Miflin, Oxford University Press, pp. 59–114.

Schäfer, G., Heber, U. and Heldt, H. W. (1977). Glucose transport into spinach chloroplasts. *Plant Physiol.* **60**, 286–89.

Schulz, B., Frommer, W. B., Flügge, U.-I., Hummel, S., Fischer, K. and Willmitzer, L. (1993). Expression of the triose-phosphate translocator gene from potato is light-dependent and restricted to green tissues. *Mol. Gen. Genet.* **238**, 357–61.

Schunemann, D., Borchert, S. (1996) Specific transport of inorganic phosphate and C3 and C6 sugar phosphates across the envelope membranes of tomato leaf-chloroplasts, tomato-fruit chloroplasts and fruit chromoplasts. *Biochim. Biophys. Acta, in* press.

Scott, P., Lange, A., Pilkis, S. J. and Kruger, N.J. (1995). Carbon metabolism in leaves of transgenic tobacco plants containing elevated fructose-2,6-bisphosphate levels. *Plant J, 7,* 461–69.

Smith, A. M. and Denyer, K. (1992). Starch synthesis in developing pea embryos. *New Phytol.* **122**, 21–33.

Somerville, S. C. and Somerville, C. R. (1985). A mutant of *Arabidopsis* deficient in chloroplast dicarboxylase transport is missing an envelope protein. *Plant Sci. Lett.* **37**, 217–20.

Stark, D. M., Timmermar, K. P., Barry, G. F., Preiss, J. and Kishose, G. M. (1992). Regulation of the amount of starch in plant tissues by ADP-glucose pyrophosphorylase. *Science* **258**, 287–92.

Stitt, M. (1990). Fructose-2,6-bisphosphate as a regulatory metabolite in plants. *Ann. Rev. Plant Physiol. Mol. Biol.* **41**, 153–85.

Stitt, M. (1996). Metabolic Regulation of photosynthesis. In *Advances in Photosynthesis*, Vol. 3, Environmental Stress and Photosynthesis, ed N. K. Baker, Academic Press, in press.

Stitt, M. and Heldt, H. W. (1981). Physiological rates of starch breakdown in isolated intact spinach chloroplasts. *Plant Physiol.* **68**, 755–61.

Stitt, M. and Sonnewald, U. (1995). Regulation of metabolism in transgenic plants. *Ann. Rev. Plant Physiol. Mol. Biol.*, in press.

Stitt, M. and Steup, M. (1985). Starch and sucrose degradation. In *Encyclopedia of Plant Physiol.*, Vol. 18, eds R. Douce and D. A. Day, Springer-Verlag, Heidelberg, pp. 347–90.

Stitt, M., Gerhardt, R., Wilke, I. and Heldt, H. W. (1987a). The contribution of fructose-2,6-bisphosphate to the regulation of sucrose synthesis during photosynthesis. *Physiol. Plantarum* **69**, 377–86.

Stitt, M., Huber, S. C. and Kerr, P. (1987b). Control of photosynthetic sucrose synthesis. In *The Biochemistry of Plants*, Vol. 10, eds M. D. Hatch and N. K. Boardman, Academic Press, New York, pp. 327–409.

Stitt, M., Kürzel, B. and Heldt, H. W. (1984). Control of photosynthetic sucrose synthesis by fructose-2,6-bisphosphate. II Partitioning between sucrose and starch. *Plant Physiol.* **75**, 554–60.

Stitt, M., Lilley, R. McC., Gerhardt, R. and Heldt, H. W. (1989). Determination of metabolite levels in specific cells and subcellular compartments in plant leaves. *Methods Enzymol.* **174**, 518–52.

Stitt, M., Wilke, I., Gerhardt, R. and Heldt, H. W. (1988). Coarse control of sucrose phosphate synthase in leaves. Alterations of the kinetic properties in response to the rate of photosynthesis and the accumulation of sucrose. *Planta* **174**, 217–30.

Trethewey, R. N. and ap Rees, T. (1994). A mutant of *Arabidopsis* lacking the ability to transport glucose across the chloroplast envelope. *Biochem. J.* **301**, 449–54.

Tyson, R. H. and ap Rees, T. (1988). Starch synthesis by isolated amyloplasts from wheat endosperm. *Planta* **175**, 33–8.

The Formation and
Breakdown of Lipids

26 The structure and formation of microbodies

Alison Baker, Claire Halpin and J. Michael Lord

Introduction

Microbodies were first described in mouse kidney cells in the early 1950s by Rhodin, a Swedish electron microscopist. They are small organelles about half a micrometer in diameter bounded by a single membrane and containing a fine granular matrix. Since morphologically similar organelles have now been discovered in a wide range of plant and animal tissues, in fungi and in unicellular organisms such as protozoa and yeasts, microbodies are considered to be ubiquitously present in eukaryotic cells. Almost without exception, microbodies can be biochemically characterized by the possession of a primitive respiratory chain involving hydrogen peroxide-producing oxidases and hydrogen peroxide-decomposing catalase, and as such are more commonly referred to as peroxisomes. In addition, an inducible fatty acid β-oxidation system is present in the peroxisomes of all organisms analyzed so far, although the activity of this system varies between tissues. In plants, β-oxidation enzymes are found predominantly in the peroxisome and are induced to high levels of activity during germination of fatty seeds, but show little activity during other developmental stages. In mammals, the bulk of fatty acid oxidation occurs in mitochondria, but enzymes present in peroxisomes are essential for the initial shortening of long chain fatty acids. In rat liver, high levels of activity of these enzymes can be induced by high fat diets or administration of drugs that are known to lower serum lipid levels. The importance of peroxisomes is emphasized by the finding that a number of severe inherited diseases in man are due to mutations which interfere with the biogenesis of peroxisomes (van den Bosch et al., 1992).

Apart from these morphological and biochemical similarities, peroxisomes from different organisms and tissues can contain very different complements of enzymes, reflecting the diverse metabolic functions that these organelles can perform. For example, one very specialized type of plant peroxisome contains enzymes of the glyoxylate cycle and has been given the special name glyoxysome. However, although peroxisomes in several algae and yeast strains metabolize certain specific compounds, they have neither been given names nor have been classified separately.

Morphology and biochemistry of plant microbodies

Electron microscopic examination of microbodies has revealed them to be small organelles 0.5–1.5 μm in diameter, which exhibit spherical, elongate, or pleiomorphic profiles in section. They are bounded by a single membrane which encloses an amorphous or granular matrix. The matrix may also contain electron-dense, often semicrystalline, material known as a core or nucleoid. Microbodies do not contain any internal membranes, ribosomes, or nucleic acid. On sucrose density gradients, microbodies band at an equilibrium density of 1.23–1.26 g cm^{-3}.

Three specialized types of peroxisome have so far been identified in plant tissues: peroxisomes in photosynthetic leaves, glyoxysomes in the fat-storing cells of oil seeds, and peroxisomes found in the uninfected nodule cells of certain legumes. In addition, most other plant tissues contain some microbodies whose particular function is unknown, and these organelles are known as unspecialized peroxisomes. A more detailed survey of the biochemical pathways associated with peroxisomes can be found in Tolbert (1981).

Leaf peroxisomes

Peroxisomes in photosynthetic green leaves play an important role in photorespiration (see Chapter 22), a process by which carbon dioxide, newly fixed in the plant cell during photosynthesis, is liberated back into the atmosphere. Glycolate (produced in the chloroplast) is oxidized in the peroxisome to glyoxylate with the liberation of hydrogen peroxide, which is detoxified by catalase. Glyoxylate is then transaminated to yield glycine which is transported to the mitochondrion where is it used to produce serine and carbon dioxide. The serine can be converted in the peroxisome to glyceric acid which can be used for carbohydrate formation in the chloroplast. Thus leaf peroxisomes display a metabolic function common to most plant and lower eukaryote peroxisomes; that is, they play a role in gluconeogenesis (the new formation of carbohydrate).

At an early stage of leaf development, as cotyledons are converted into green photosynthetic tissue, the enzymes associated with the photorespiratory glycolate pathway accumulate in leaf peroxisomes in a light-dependent manner. The leaves of plants with high levels of photorespiration have much larger and more numerous peroxisomes than those of plants with low photorespiration, consistent with the difference in biochemical activities between the two types of plant (Newcomb, 1982). The peroxisomes may reach $1.5\,\mu m$ in diameter and often contain large crystalline inclusions of the enzyme catalase (see Fig. 26.1).

Glyoxysomes

Glyoxysome is the name given to a specialized type of peroxisome that possesses enzymes of the glyoxylate cycle. In higher plants, glyoxysomes are found in the endosperm or cotyledons of oil seeds during the early stages of growth after germination, when conversion of stored oil to sucrose is the major metabolic process. This metabolic activity serves to provide developing shoots with energy in the form of

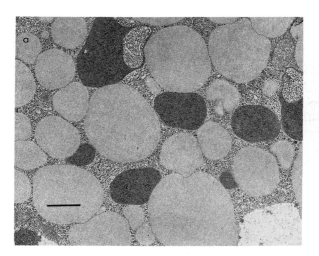

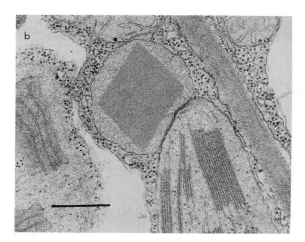

Fig. 26.1 Microbody morphology. (a) Glyoxysomes (in close association with spherosomes) in a tomato cotyledonary cell; (b) a crystalloid-containing spinach leaf peroxisome. Bar represents $1\,\mu m$ (micrographs courtesy of Dr. E. H. Newcomb).

carbohydrate, at the expense of the seed's stored oil, until the first green leaves appear and photosynthesis begins. The compartmentation of β-oxidation enzymes and those of the glyoxylate cycle together within the glyoxysome is crucial to the efficiency of this process. Acetyl-CoA generated by β-oxidation of fatty acids avoids the oxidative decarboxylations of the mitochondrial citric acid cycle, and instead is converted exclusively to succinate via the glyoxylate cycle. Thus, up to 75% of the fatty acid carbon can be ultimately recovered in the cytoplasm as sucrose. Because of the magnitude of gluconeogenesis during this time, glyoxysomes may account for up to 20% of the total particulate protein in tissues such as castor bean endosperm when at the peak of their activity.

In oilseed species where oil reserves are stored in the cotyledons (e.g. cucumber, sunflower, pumpkin, cotton), relatively small, spherical glyoxysomes present at germination are thought to have been synthesized during seed formation (Trelease, 1984). Activities of glyoxysomal enzymes do not appear coordinately, but low levels of activity of certain enzymes notably catalase, β-oxidation enzymes, and the glyoxylate cycle enzyme malate synthase, have been detected during seed maturation. Following germination, the activity of all glyoxysomal enzymes increases dramatically during the first five days of growth (in seeds germinated at a constant 25–30 °C), peaks, and then declines again (Huang et al., 1983). During this period of intense activity, the glyoxysomes themselves become more elongate and highly pleiomorphic, increasing in volume nearly seven-fold.

As the cotyledons emerge and are exposed to light (about day 3–4), the enzymes associated with the photorespiratory glycolate pathway, e.g. hydroxypyruvate reductase, serine : glyoxylate aminotransferase and glycolate oxidase, are induced and begin to accumulate in microbodies, while the glyoxylate cycle enzymes continue to decline. The nature of this transition from glyoxysomal to peroxisomal function in cotyledonary microbodies remains poorly understood and will be discussed later.

In oilseed species where oil reserves are stored in the endosperm (e.g. castor bean), this transition does not occur. Glyoxysomes and their constituent enzymes are rapidly synthesized during the first five days of growth (see Fig. 26.2). After this time the organelles, and indeed the whole endosperm tissue, senesces and effectively has disappeared by day nine of growth at 30 °C (Lord and Roberts, 1983).

Glyoxysomes are also found in algae, yeasts and other fungi and protozoa which have been grown on acetate or compounds initially converted into acetyl units. In these cases, however, only two of the five glyoxylate cycle enzymes, malate synthase and isocitrate lyase, are present. For this reason some

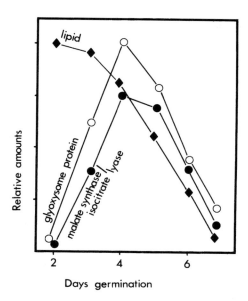

Fig. 26.2 Induction of glyoxysomes and glyoxylate cycle enzymes during germination of castor bean seeds.

authors prefer to refer to these organelles as 'glyoxysome-like'.

Root nodule peroxisomes

A third specialized, though not very well characterized, type of plant peroxisome is found in certain cells of the root nodules of some legumes, e.g. soybean and cowpea. Early in nodule development, only a few relatively small peroxisomes are seen either in infected or uninfected cells. As the nodule develops, however, peroxisomes in the uninfected cells become larger and more numerous, and in soybean the induction of at least one peroxisomal enzyme, a uricase, has been demonstrated (Nguyen et al., 1985). Such legumes principally produce and transport fixed nitrogen as the ureides allantoin and allantoid acid. These compounds are formed via purine biosynthesis and oxidation. It has been suggested that the enzymes involved in purine biosynthesis are located in the proplastids, whereas the oxidation of purines to ureides occurs in the peroxisomes. In this oxidation the purine xanthine is first converted to uric acid which is then converted to allantoin by uricase. This second reaction liberates hydrogen peroxide whose destruction requires the action of catalase. The localization of both uricase and catalase in the peroxisomes of uninfected nodule cells provides good evidence for the organelles' involvement in at least one reaction during the metabolism of compounds of recently fixed nitrogen. It has been suggested that such compounds may even induce peroxisome proliferation and uricase activity in this tissue (Newcomb, 1982).

Glyoxysome–peroxisome transition

Microbodies in the cotyledonary cells of oil seeds perform two successive but distinct functions. During germination when fat is being converted to carbohydrate, large numbers of glyoxysomes containing active β-oxidation and glyoxylate cycle enzymes are present; whereas later, after greening, the cotyledons contain abundant peroxisomes which carry out photorespiratory glycolate metabolism. During the transition between these states, the activity of the glyoxysomal glyoxylate cycle enzymes declines while that of the peroxisomal photorespiratory enzymes increases. Thus, for a short time during transition, both sets of enzyme activity are present in the same cell.

The nature of the glyoxysome–peroxisome transition has been investigated by a number of researchers over the past 10–15 years and a number of models have been proposed to explain it. Beevers (1979) suggested the two population model proposing that at transition two biochemically distinct sets of microbody exist, one containing the glyoxylate cycle enzymes and the other containing the glycolate pathway enzymes. In contrast, Trelease et al. (1971) proposed that the change in enzyme composition occurs within one population of homogeneous microbodies by gradual replacement of glyoxysomal enzymes by peroxisomal enzymes. A variation of this 'one population' model proposes that transition occurs by the continual synthesis and degradation of microbodies, which as development proceeds will contain changing patterns of enzymes (Schopfer et al., 1976).

Only recently have the experimental techniques evolved which make it possible to differentiate between these models. Double-label immunoelectron microscopy has been employed by Titus and Becker

(1985) to determine the localization of glyoxysomal and peroxisomal enzymes in transition-stage cotyledons of cucumber. Using two sizes of protein A-gold, these workers have unequivocally demonstrated the coexistence of a glyoxysomal marker enzyme (isocitrate lyase) and a peroxisomal marker enzyme (serine : glyoxylate aminotransferase) within the same microbody at transition, suggesting that only one population of microbodies exists which contains both sets of enzymes.

Recently glyoxysomal isocitrate lyase (ICL) has been expressed in green leaf cells and in roots of transgenic plants, and was correctly targeted to the resident microbody population (Olsen et al., 1983; Onyeocha et al., 1993). Double labelling experiments showed that in leaves of the transformed plants all peroxisomes contained ICL (Marrison et al., 1993). These results further support the notion that only one population of microbodies exists within a cell and demonstrate that all types of microbodies use the same type of targeting mechanisms for their proteins. The various microbodies should perhaps best be considered to be a family of related organelles (in the same way as plastids are) which perform different functions in different tissues.

Reverse transition during senescence

The microbodies of epigeous species undergo a second transition as the cotyledons senesce. During senescence enzymes characteristic of photorespiration are lost and the enzymes of β-oxidation and the glyoxylate cycle reappear (Gut and Matile, 1988). It seems likely that this results in the salvage of carbon and nitrogen from the breakdown of membrane lipids and possibly nucleic acids in the senescing cells. Double label immuno-electron microscopy demonstrated that in this case too the enzyme complement of the existing microbodies is turned over and replaced with the new activities as senescence occurs (Nishimura et al., 1993).

Plant microbody gene expression

Over the past decade a number of genes which encode plant microbody proteins have been cloned. These include isocitrate lyase, malate synthase, catalase, thiolase, glyoxysomal malate dehydrogenase, glycolate oxidase and hydroxypyruvate reductase. Studies of mRNA accumulation using the relevant genes as probes have shown that control of gene expression is primarily at the level of transcription, with changes in enzyme activity generally following and mimicking changes in steady state levels of mRNA. The case of catalase is, however, more complicated where post-transcriptional processes seem to play an important role (Ni and Trelease, 1991).

The expression of the malate synthase (MS) gene has been studied in cucumber cotyledons. The MS gene is expressed in early post germinative growth, is repressed as the cotyledons become photosynthetic and subsequently reactivated as the cotyledons senesce (Graham *et al.*, 1992). The malate synthase promoter has been shown to direct faithful expression of a reporter gene, β-glucuronidase (GUS) in transgenic plants opening up the way for detailed studies on the mechanism of regulation of this gene (Graham *et al.*, 1990).

In the case of leaf peroxisomal enzymes, transcription has been shown to be activated by light, as might be expected for enzymes involved in photorespiration. The promoter of the hydroxypyruvate reductase gene has been analyzed by studying the activity of various mutated versions of the promoter in transgenic plants and the light responsive elements identified. These are similar to light responsive elements found in nuclear encoded chloroplast genes (Sloan *et al.*, 1993).

Biogenesis of microbody proteins

Elucidation of the pathway of microbody biogenesis has lagged behind that of nearly all other major cellular organelles. Over the past 15–20 years some controversy has developed between investigators supporting opposing models, and much conflicting and contradictory data has been produced, confusing the issue still further. Many technical difficulties have beset this area of investigation, making microbody biogenesis inherently more difficult to study than that of, say, chloroplasts or mitochondria. In recent years, however, the advent of modern molecular biology allied with improved biochemical techniques has

overcome many of these problems, and all the recent evidence supports a post-translational mechanism for the incorporation of proteins into microbodies. The key findings are summarised below, but for a more extensive review see Subramani (1993) and Baker (1994).

Microbody matrix proteins

As microbodies have no DNA, all their constituent proteins are encoded by nuclear genes. The mRNA produced is translated on free as opposed to membrane bound polysomes and the proteins are imported post translationally into the organelle. This was shown initially by pulse-chase experiments where microbody matrix proteins were detected in the soluble cytosolic fraction before their incorporation into the microbody fraction. Subsequently, reconstitution of the import process *in vitro* has also demonstrated that it is post translational and requires ATP (Imanaka *et al.*, 1987; Behari and Baker, 1993).

Most matrix proteins are synthesized at their final size and are not proteolytically processed upon or after import. Notable exceptions to this are thiolase (an enzyme of the β-oxidation pathway) and glyoxysomal malate dehydrogenase from watermelon. There is now good evidence, which will be summarized later, that these proteins follow a different import pathway from most of the other matrix proteins.

Recent work indicates that there are multiple pathways for the import of matrix proteins and two types of targeting signal have been characterized. The first to be discovered, now called Peroxisomal Targeting Signal 1 (PTS-1), consists of a conserved tripeptide at the extreme carboxyl-terminus of a number of matrix proteins. In animal cells, the consensus is serine or cysteine or alanine in the 3rd last position, lysine, arginine, or histidine in the penultimate position and leucine in the final position. This is often referred to as the SKL motif from the single letter code for serine, lysine, leucine; the first PTS-1 motif to be characterized in the firefly peroxisomal enzyme luciferase (Gould *et al.*, 1989). In lower eukaryotes such as various yeasts, more degenerate forms of this tripeptide motif appear to be functional such as AKI and NKL. Many peroxisomal enzymes end with a tripeptide which matches this consensus and this includes several plant microbody proteins. However, only in two cases has the role of this tripeptide in microbody targeting in plants been tested.

Glycolate oxidase is an enzyme of the photorespiratory glycolate pathway. The spinach enzyme ends with the tripeptide ARL and the last 6 amino acids RAVARL when fused to the carboxyl terminus of a reporter protein, β-glucuronidase, could redirect it to peroxisomes (Volikita, 1991). The glyoxylate cycle enzyme isocitrate lyase (ICL) has been studied by two groups who have reached different conclusions about the role of the carboxyl terminus of the protein in microbody targeting. Olsen *et al.* (1993) studied the localization of mutant versions of the *Brassica* ICL in transgenic *Arabidopsis* plants and concluded that the C-terminus played an important role in microbody targeting, whereas Behari and Baker (1993) studying the import of the castor bean protein in an *in vitro* import system could find no requirement for the C-terminal tripeptide.

A second type of targeting signal (PTS-2) is found in the thiolases and glyoxysomal malate dehydrogenase. The signal is short, is located close to the amino terminus of the protein and is removed after import. Mutagenesis experiments point to the conserved motif R.L/I.X.X.X.X.H.L (where X is any amino acid) being essential for the function of PTS-2 (Gietl *et al.*, 1994). Some matrix proteins appear to lack either of these signals, raising the possibility of a third route for import of matrix enzymes.

The biochemical evidence for at least two import pathways is corroborated by the phenotypes of yeast mutants defective in peroxisomal assembly (pas mutants) and in cells from human patients with a variety of genetic diseases which result in defects in peroxisome biogenesis (Fig. 26.3). The mutant pas 10 in the yeast *Saccharomyces cerevisiae* is defective in the import of PTS-1 containing proteins but not those which use the PTS-2 pathway. Pas 7 has the converse phenotype; PTS-2 containing proteins such as thiolase are imported but PTS-1 containing proteins are not. The homologue of pas 10 from *Pichia pastoris* (confusingly called pas 8) has been shown to be a peroxisomal membrane protein and to bind to peptides containing the SKL motif and is therefore a good candidate for the PTS-1 receptor (McCollum *et al.*,

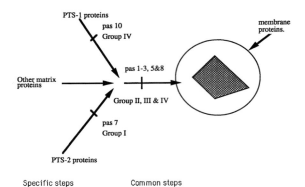

Specific steps   Common steps

Fig. 26.3 Cartoon illustrating the effects of different peroxisomal assembly mutants in yeast and mammalian cells. The numbering of pas mutants is as for *S. cerevisae*. Similar mutants have been described in other yeasts. Groups I–IV refer to the human peroxisome disorder complementation groups studied by Motley *et al.* (1994). While it is possible that mutations which block common steps act downstream of the mutations affecting the import of specific subsets of proteins, this has not been demonstrated and remains speculative.

1993). Other mutants appear to block the import of all peroxisomal proteins tested. Analogous mutants are now recognized in cells from humans with peroxisomal disorders (Motley *et al.*, 1994).

Microbody membrane proteins

Like matrix proteins, membrane proteins are also synthesized on free ribosomes and imported post-translationally. Several lines of evidence suggest that proteins are inserted into the peroxisome membrane by quite different route(s) to the matrix enzymes. Firstly, none of the integral membrane proteins characterized so far have either PTS-1 or 2. Secondly, in many of the yeast and mammalian mutant cells described in the previous section, membrane proteins are inserted to produce peroxisomal 'ghosts', membrane sacs with little or no matrix content. Only one membrane protein has so far been subjected to extensive analysis; peroxisomal membrane protein 47 (PMP 47) from the yeast *Candida boidinii*. In this case peroxisomal targeting information appears to reside in two of the transmembrane segments of the protein (McCammon *et al.*, 1993).

Proliferation of microbodies

If both microbody matrix and integral membrane proteins are synthesized on free polysomes and enter the organelle post-translationally, where does the importing microbody originally come from? Do microbodies, like chloroplasts and mitochondria, arise by division of pre-existing organelles, or, conversely, might there still be a role for the ER in the production by vesiculation of a 'pre-microbody' vesicle into which matrix and membrane proteins might insert post-translationally? This question has still not been conclusively answered and various lines of evidence exist that would seem to support either theory.

ER Vesiculation

In the early days of microbody research, it was expected that microbodies arose directly from the ER by vesiculation. Ultrastructural studies provided some evidence for this theory, as did early reports concerning the phospholipid composition of microbody membranes and the detection of glycoproteins in them. This theory, however, suggested that microbody proteins should be synthesized on bound polysomes and cotranslationally inserted into the ER, which conflicts with the growing body of biochemical and molecular evidence. A re-evaluation of the data has therefore been necessary which has left many of the earlier interpretations in doubt. In accordance with this, some investigators support a modified ER vesiculation model, i.e. that microbodies are formed by vesiculation from the ER and that matrix and possibly membrane proteins are added post-translationally during formation and/or following detachment from the ER (Trelease, 1984). The evidence supporting involvement of the ER in microbody biogenesis is as follows.

In ultrastructural studies, microbodies have often been seen in very close proximity to sections of the ER, sometimes being entirely surrounded by it. Regions where the ER and microbody are seen to touch have been interpreted as demonstrating direct membrane continuity between the two organelles (e.g. Novikoff and Shin, 1964). In many cases the membranes in these micrographs are broken or fuzzy around the putative connecting region.

Unambiguous views of direct membrane and luminal connections between ER and microbodies are rare, if they exist at all. Several workers have looked for connections in rat liver, pneumocytes, and bean leaves, and failed to observe them. Other workers have found that connections are only apparent when the microsomal or ER membranes are sectioned tangentially. Connections could therefore be artefactual and no definitive evidence of their existence has been found.

Many workers have investigated the existence of glycoproteins in the peroxisomal matrix and membrane. As N-glycosylation is a cotranslational modification acquired during synthesis on the rough ER, identification of glycoproteins in peroxisomes would demonstrate ER involvement in their biogenesis. Efforts to detect such glycoproteins have been inconclusive. Malate synthase from castor bean was initially thought to be glycosylated but subsequent analyses have shown it to lack sugar residues. Reports conflict as to whether cucumber isocitrate lyase is glycosylated; the castor bean enzyme is not. Bergner and Tanner (1981) examined isolated glyoxysomes after incorporation of radioactive sugars, and found label in the membrane but not the matrix fraction. Membrane glycoproteins were also detected by Lord and Roberts (1983) in similar experiments. Lazarow and Fujiki (1985) suggest that the use of radioactive sugars in such experiments is dangerous because the method is so sensitive it permits the detection of trace contaminant glycoproteins in the peroxisome fractions.

There has therefore been no totally convincing demonstration of glycoproteins in either the peroxisomal matrix or membrane.

Growth and division

As much of the evidence for the ER vesiculation theory is now in doubt, the model is being increasingly abandoned, and some investigators (e.g. Lazarow and Fujiki, 1985) have adopted an alternative one. As microbodies develop like mitochondria and chloroplasts by post-translational incorporation of proteins, might they not also be capable of division, as mitochondria and chloroplasts are? Some evidence supports this.

Studies of the morphological changes in the

microbodies of yeasts following induction suggest that the rapid proliferation of microbodies that occurs is due to growth and division of pre-existing microbodies. When peroxisome proliferation is induced in *Candida tropicalis* by growth on alkanes the size of the peroxisomal compartment increases twelve-fold (Osumi *et al.*, 1975). At the start of induction one or two very small peroxisomes (approximately 0.1 μm) can be seen per cell. By 6–8 h after induction both the number and the size of peroxisomes has increased five times and larger oblate peroxisomes, some containing septa which divide them unequally, are also obvious. After 46 h the cytoplasm is full of peroxisomes 0.3–0.6 μm in diameter.

Similar changes in peroxisome morphology have been observed when *Hansenula polymorpha* cells, grown to stationary phase on methanol, are diluted into fresh methanol medium (Veenhuis *et al.*, 1978). After three hours the yeast cells begin to bud, and the peroxisomes inside them also appear to divide unequally, producing small new peroxisomes that apparently migrate into the yeast bud. After yeast division, small new yeast cells contain one or two small peroxisomes while the mother yeast cells are filled with large peroxisomes. During a second growth phase, the small peroxisomes in the daughter cells themselves enlarge and divide.

Both these studies suggest that microbodies, in yeast at least, arise by division from pre-existing microbodies. Direct evidence for such division in higher plants is lacking, but glyoxysome and peroxisome 'preforms' have been documented that might be considered analogous to the 'small' peroxisomes found in yeast cells prior to induction. In suspension cultures of de-differentiated anise cells growing in sucrose, glyoxysome 'preforms', which contain three β-oxidation enzymes and catalase but which are less dense than mature organelles (equilibrating at 1.13 g ml^{-1} on sucrose gradients), have been described (Lutzenberger and Theimer, 1986). When glyoxysome proliferation is induced in these cells by growth on acetate for 48 h, six glyoxysomal marker enzymes accumulate in organelles equilibrating at 1.23 g ml^{-1} sucrose, indicating a development of functional glyoxysomes from the less dense organelles. The presence of small glyoxysomes in the cotyledonary cells of cotton and cucumber seeds that are detectable during seed development (inferred from appropriate

enzyme activities) and that enlarge on germination, has already been mentioned. Might these organelles also be considered as 'pre' or progenitor glyoxysomes from which glyoxysome proliferation is effected on germination purely by growth and fission? The growth and division model is also in keeping with our recent understanding of the glyoxysome–proxisome transition and explains the existence of small or pre-peroxisomes in cells when no specific microbody activity or proliferation is occurring. In fact, the growth and division model predicts that as microbodies never arise *de novo* but only through division of pre-existing organelles, most cells, especially germ cells, must contain at least one. In the case of yeast at least, this prediction has been confirmed, and yeast cells, grown on glucose which suppresses peroxisome formation, still contain one or two small peroxisomes.

Conclusions

The validity of the models for microbody proliferation, by ER vesiculation or by growth and division from pre-existing organelles, cannot yet be properly assessed. Much of the evidence suggesting ER involvement is now in doubt, although persistent claims that glycoproteins can be detected in the microbody membrane and matrix cannot be easily dismissed. On the other hand, evidence for the growth and division model is still largely circumstantial but is conceptually persuasive in the light of all the recent data which supports post-translational incorporation of proteins into microbodies. If the ER does play a role in microbody biogenesis, why is it that every well-characterized microbody protein investigated to date is synthesized on free ribosomes and post-translationally imported, while not a single microbody protein that will cotranslationally insert into ER microsomes has been documented? Clearly the signals for microbody targeting which have been characterized are quite distinct from ER signal sequences.

The simplest model in keeping with virtually all the recent evidence is that microbodies grow by the post-translational incorporation of new matrix and membrane proteins into pre-existing microbodies which then divide to form daughter microbodies. In this respect, microbodies are like mitochondria and chloroplasts. However, microbody proteins, unlike mitochondria and chloroplast proteins, are not initially synthesized as precursors. A role for the ER in this biogenetic process is still supported by many investigators and cannot be discounted, but as yet there is no unequivoval evidence for it.

Perspectives

Since the first edition of this book was published there has been rapid progress in the understanding of microbody protein targeting and in the regulation of expression of microbody proteins. However, a number of important questions still remain.

Mechanism of protein import

We still have little idea about the molecular mechanisms involved in targeting proteins to the microbodies and integrating them into or transporting them across the membrane. Progress is likely to be quite rapid as there are now powerful experimental systems available to address just these questions.

(1) *Better* in vitro *import systems*

In vitro systems are very powerful because they allow the experimentalist (almost) complete control over the reaction conditions. Components can be added or removed at will or the amounts carefully titrated. The ultimate goal of *in vitro* systems would be to reconstitute the process from defined components, as has recently become possible for ER protein translocation. However, *in vitro* systems have limitations in that they are often much less efficient that the *in vivo* process and may be technically difficult to work with; both of these are certainly true with the microbody systems. This in turn leads to concern about the physiological relevance of the results.

(2) *Permeabilized cell systems*

Recently peroxisomal protein transport has been demonstrated in permeabilized mammalian cells (Rapp *et al.*, 1993; Wendland and Subramani, 1993).

This is a kind of half-way house between *in vivo* and *in vitro* studies.

(3) *Mutants*

A large number of mutants defective in peroxisome formation or proliferation have been generated in the yeasts *Saccharomyces cerevisiae*, *Pichia pastoris*, *Yarrowa lipolytica* and *Hansenula polymorpha*. Several cell lines from patients with a variety of peroxisomal disorders have also been established and characterized as have some similar mutants in cultured Chinese Hamster Ovary cells (CHO cells). A dozen or more genes have been cloned by complementation of the corresponding mutants, but in most cases the function of the gene product in peroxisome biogenesis remains unknown. Finding the answers will require the convergence of the biochemical and genetic approaches, for example, testing the functions of the gene products uncovered by the genetic approach in the *in vitro* or permeabilized cell systems.

Proliferation and the role of the ER

While it is clear that all microbodies are competent to import proteins and do so by specific routes which do not require the participation of the endoplasmic reticulum, it is still difficult to exclude any role of the ER in the formation of microbodies. The origin of microbody lipids is poorly understood, but it is unlikely that the organelles can synthesize all their own requirement. The ER is therefore a likely source for some or all microbody lipids. In germinating oilseeds it was demonstrated that lipids were recruited into glyoxysomes from oil bodies (Chapman and Trelease, 1991), which are themselves formed from the ER during seed development, but this question has not been extensively investigated for other types of microbody.

References

Baker, A. (1996). Biogenesis of Plant Peroxisomes. In *Membranes: Specialised Functions in Plant Cells,* eds D. Bowles, J. P.Knox and M. Smallwood, Bios Scientific Publishers, Oxford.

Beevers, H. (1979). Microbodies in higher plants. *Ann. Rev. Plant Physiol.* **30**, 159–93.

Behari, R. and Baker, A. (1993). The carboxyl terminus of isocitrate lyase is not essential for import into glyoxysomes in an *in vitro* system. *J. Biol. Chem.* **268**, 7315–22.

Bergner, U. and Tanner, W. (1981). Occurrence of several glycoproteins in glyoxysomal membranes of castor beans. *FEBS Lett.* **131**, 68–72.

Chapman, K. D. and Trelease, R. N. (1991). Acquisition of membrane lipids by differentiating glyoxysomes: role of lipid bodies. *J. Cell Biol.* **115**, 995–1007.

Fujiki, Y. and Lazarow, P. B. (1985). Post-translational import of fatty acyl CoA oxidase and catalase into peroxisomes of rat liver *in vitro*. *J. Biol. Chem.* **260**, 5603–9.

Gietl, C., Faber, K. N., van der Klei, I. J. and Veenhuis, M. (1994). Mutational analysis of the N terminal topogenic signal of watermelon glyoxysomal malate dehydrogenase using the heterologous host *Hansenula polymorpha*. *Proc. Natl. Acad. Sci. (USA)* **91**, 3151–55.

Gould, S. J., Keller, G. A., Hosken, N., Wilkinson, J. and Subramani, S. (1989). A conserved tripeptide sorts proteins to peroxisomes. *J. Cell Biol.* **108**, 1657–64.

Graham, I. A., Smith, L. M., Leaver, C. J. and Smith, S. M. (1990). Developmental regulation of expression of the malate synthase gene in transgenic plants. *Plant Mol. Biol.* **15**, 539–49.

Graham, I. A., Leaver, C. J. and Smith, S. M. (1992). Induction of malate synthase gene expression in senescent and detached organs of cucumber. *The Plant Cell* **4**, 349–57.

Gut, H. and Matile, P. (1988). Apparent induction of key enzymes of the glyoxylic acid cycle in senescent barley leaves. *Planta* **176**, 548–50.

Huang, A. H. C., Trelease, R. N. and Moore, T. S. Jr. (1983). *Plant Peroxisomes*, Academic Press, New York.

Imanaka, T., Small, G. M. and Lazarow, P. B. (1987). Translocation of acylCoA oxidase into peroxisomes requires ATP hydrolysis but not a membrane potential. *J. Cell Biol.* **105**, 2915–22.

Lazarow, P. B. and Fujiki, Y. (1985). Biogenesis of peroxisomes. *Ann. Rev. Cell Biol.* **1**, 489–530.

Lord, J. M. and Roberts, M. R. (1983). Formation of glyoxysomes. In *Aspects of Cell Regulation*, ed. J. F. Danielli, International Review of Cytology Suppl. 15, pp. 115–56.

Lutzenberger, A. and Theimer, R. R. (1986). Nutrient dependent induction of formation and degradation of glyoxysomal isocitrate lyase in cell suspension cultures of anise (*Pinpinella anisum* L.). *Eur. J. Cell Biol.* **41**, 28–33.

Marrison, J. L., Onyeocha, I., Baker, A. and Leech, R. M. (1993). Recognition of peroxisomes by immunofluoresence in transformed and untransformed tobacco cells. *Plant Physiol.* **103**, 1055–59.

McCammon, M. T., McNew, J. A., Willy, P. J. and Goodman, J. M. (1994). An internal region of the peroxisomal membrane protein PMP47 is essential for sorting to peroxisomes. *J. Cell Biol.* **124**, 915–25.

McCollum, D., Monosov, E. and Subramani, S. (1993). The pas 8 mutant of *Pichia pastoris* exhibits the peroxisomal protein import deficiencies of Zellweger syndrome cells – the pas 8 protein binds to the COOH-terminal tripeptide peroxisomal targeting signal and is a member of the TPR protein family. *J. Cell Biol.* **121**, 761–74.

Motley, A., Hettema, E., Distel, B. and Tabak, H. (1994). Differential protein import deficiencies in human peroxisomal assembly disorders. *J. Cell Biol.* **125**, 755–67.

Newcomb, E. H. (1982). Ultrastructure and cytochemistry of plant peroxisomes and glyoxysomes. In *Peroxisomes and Glyoxysomes*, eds H. Kindl and P. B. Lazarow, Annals of the New York Academy of Sciences, Vol. 386, pp. 228–41.

Nguyen, T., Zelechowska, M., Foster, V., Bergmann, H. and Verma, D. P. S. (1985). Primary structure of the soybean nodulin-35 gene encoding uricase 11 localized in the peroxisomes of uninfected cells of nodules. *Proc. Natl. Acad. Sci. U.S.A.* **82**, 5040–44.

Ni, W. and Trelease, R. N. (1991). Post transcriptional regulation of catalase isozyme expression in cotton seeds. *The Plant Cell* **3**, 737–44.

Nishimura, M., Takeguchi, Y., DeBellis, L. and Hara-Nishimura, I. (1993). Leaf peroxisomes are transformed directly into glyoxysomes during senescence of pumpkin cotyledons. *Protoplasma* **175**, 131–37.

Novikoff, A. B. and Shin, W-Y. (1964). The endoplasmic reticulum in the Golgi zone and its relation to microbodies, Golgi apparatus and autophagic vacuoles in rat liver cells. *J. Micros. Oxford* **3**, 187–206.

Olsen, L. J., Ettinger, W. F., Damsz, B., Matsudaira, K., Webb, A. and Harada, J. J. (1993). Targeting of glyoxysomal proteins in leaves and roots of a higher plant. *The Plant Cell* **5**, 941–52.

Onyeocha, I., Behari, R., Hill, D. and Baker, A. (1993). Targeting of castor bean glyoxysomal isocitrate lyase to tobacco leaf peroxisomes. *Plant Mol. Biol.* **22**, 385–96.

Osumi, M., Fukuzumi, T., Teranishi, Y., Tanaka, A. and Fukui, S. (1975). Development of microbodies in *Candida tropicalis* during incubation in *n*-alkane medium. *Arch. Microbiol.* **103**, 1–11.

Rapp, S., Soto, U. and Just, W. W. (1993). Import of firefly luciferase into peroxisomes of permeabilized chinese hamster ovary cells; a model system to study peroxisomal protein import *in vitro*. *Exp. Cell Res.* **205**, 59–65.

Schopfer, P., Bajracharya, D., Bergfeld, R. and Falk, H. (1976). Phytochrome-mediated transformation of glyoxysomes into peroxisomes in the cotyledons of mustard seedlings. *Planta* **133**, 73–80.

Sloan, J. S., Schwartz, B. W. and Becker, W. M. (1993). Promoter analysis of a light regulated gene encoding hydroxypyruvate reductase, an enzyme of the photorespiratory glycolate pathway. *The Plant J.* **3**, 867–74.

Subramani, S. (1993). Protein import into peroxisomes and biogenesis of the organelle. *Annu. Rev. Cell. Biol.* **9**, 445–78.

Titus, D. E. and Becker, W. M. (1985). Investigation of the

glyoxysome-peroxisome transition in germinating cucumber cotyledons using double-label immunoelectron microscopy. *J. Cell Biol.* **101**, 1288–99.

Tolbert, N. E. (1981). Metabolic pathways in glyoxysomes and peroxisomes. *Ann. Rev. Biochem.* **50**, 133–57.

Trelease, R. N. (1984). Biogenesis of glyoxysomes. *Ann. Rev. Plant Physiol.* **35**, 321–47.

Trelease, R. N., Becker, W. M., Gruber, P. J. and Newcomb, E. H. (1971). Microbodies (glyoxysomes and peroxisomes) in cucumber cotyledons. Correlative biochemical and ultrastructural study in light- and dark-grown seedlings. *Plant Physiol.* **48**, 461–75.

van den Bosch, H., Schutgens, R. B. H., Wanders, R. J. A. and Tager, J. M. (1992). Biochemistry of peroxisomes. *Ann. Rev. Biochem.* **61**, 157–97.

Veenhuis, M., Van Dijken, J. P., Pilon, S. A. F. and Harder, W. (1978). Development of crystalline peroxisome in methanol-grown cells of the yeast *Hansenula polymorpha* and its relation to environmental conditions. *Arch. Microbiol.* **117**, 153–63.

Volokita, M. (1991). The carboxy-terminal end of glycolate oxidase directs a foreign protein into tobacco leaf peroxisomes. *The Plant J.* **1**, 361–66.

Wendland, M. and Subramai, S. (1993). Cytosol dependent peroxisomal protein import in a permeabilized cell system. *J. Cell Biol.* **120**, 675–85.

27 Fatty acid and lipid biosynthesis and degradation

Katherine Schmid, Jaen Andrews and John Ohlrogge

Structures and functions of plant lipids

Lipids are a diverse group of chemicals which perform several major functions in plants. Phospholipids, galactolipids and sterols provide the hydrophobic barrier of cell membranes. Cuticular lipids and wax esters form a coating on the aerial surface of plants which both prevents water loss and protects the tissues from environmental and biological stress. An analogous material, suberin, is synthesized by underground organs or in response to wounding. Energy rich triacylglycerols (Fig. 27.6) are a prime form of carbon storage in most seeds, and a major source of calories for human consumption. Lipids also may serve a variety of less well defined functions such as hormones, second messengers, insect attractants, and defense chemicals (phytoalexins).

The most abundant plant membrane lipids contain two long chain fatty acids esterified to the *sn* 1 and 2 positions of glycerol. These fatty acids are almost always 16 or 18 carbons in length and contain from 0 to 3 *cis* double bonds (Fig. 27.1). Attached to the third position on the glycerol backbone is a polar head group. The combination of the non-polar fatty acyl chains and the polar head group leads to the amphipathic properties of membrane lipids. Most such lipids spontaneously form bilayer or micellar structures when mixed with water, and in biological systems they are organized into the classic fluid bilayer.

The composition of plant membrane lipids depends on a number of factors including the tissue type, subcellular localization and environmental influences (Harwood, 1980). The lipid composition of plant leaves is dominated by the abundant membranes of the chloroplast thylakoids. These membranes contain high levels of monogalactosyl- and digalactosyldiacylglycerol. These galactolipids are absent or present in low levels in other subcellular organelles or in non-photosynthetic tissues (Table 22.1). The composition of non-photosynthetic membranes in plants is not substantially different from that of animals or other eukaryotic organisms in that phosphatidylcholine, phosphatidylethanolamine and phosphatidylinositol are major components. Plant plasma membranes are also known to contain significant amounts of sphingolipid. The dominant sterols of plant membranes vary from species to species, although as a rule sterols are absent from the thylakoids.

Fatty acid biosynthesis

The hydrophobic properties of lipids are determined by their long chain non-polar fatty acid constituents. The synthesis of fatty acids is conducted through sequential two carbon additions which are catalyzed by a group of at least 6 soluble enzymes referred to collectively as the fatty acid synthase (FAS). Because soluble proteins are easier to purify and study, much more is known about the biochemistry of fatty acid synthesis than about the subsequent membrane-bound reactions of fatty acid desaturation and glycerolipid assembly. Most of the enzymes for the early steps of plant lipid metabolism have been purified, and recently cDNA and genomic clones have been obtained for almost all of them (Ohlrogge *et al.*, 1993; Töpfer and Martini, 1994).

Subcellular localization of fatty acid synthesis

In animals, yeast and many other eukaryotic organisms, fatty acid synthesis occurs in the

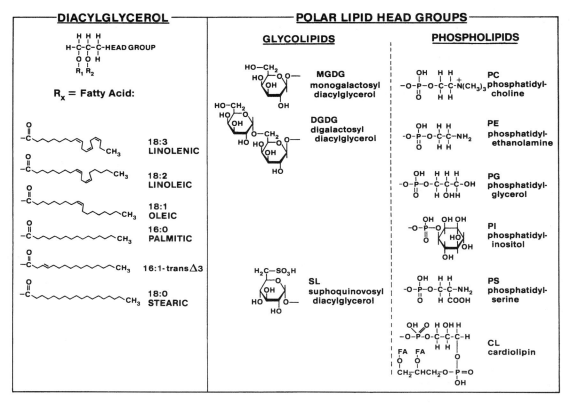

Fig. 27.1 Glycerolipids of plant cell membranes. Note that the fatty acids are referred to by the number of carbon atoms followed by the number of double bonds after the colon.

cytoplasm. In plants, however, there is a fundamentally different subcellular organization. As early as 1960, it was found that isolated chloroplasts have the capacity to synthesize fatty acids. Despite an expectation that a more animal-like, cytoplasmic FAS activity would be found to provide fatty acids for membranes outside the plastids, it now appears that, in plant mesophyll cells, *de novo* fatty acid synthesis occurs almost exclusively in the chloroplasts. This organelle therefore must not only serve its own needs but also supply membrane fatty acids for the entire plant cell.

Although the data are clearest for leaf tissues, whose subcellular components can be fractionated with relative ease, plastids of other tissues are clearly capable of fatty acid synthesis. For example, the proplastids of developing seeds

appear to play the major role in synthesis of fatty acids for triacylglycerols. However, it is not yet certain if the plastid is the sole site of fatty acid synthesis in all tissues. Recently acyl carrier protein (ACP), an important constituent of FAS, has been discovered to occur also in the mitochondria. Although the amount of ACP in this organelle is considerably less than in plastids, it is now known that mitochondria possess a complete fatty acid biosynthetic pathway.

Role of acyl carrier protein (ACP)

ACP is a small acidic protein that has a phosphopantetheine prosthetic group attached to a serine residue near the middle of the protein

(Fig. 27.2) (Ohlrogge, 1987). The prosthetic group is similar in structure to coenzyme A and serves a similar function. A sulfhydryl group at the end of the pantetheine can be joined with the carboxyl carbon of a fatty acid to form a thioester. Formation of the thioester bond requires energy (supplied by ATP) and results in the activation of the carbonyl carbon of the acyl group so that it can participate in several key reactions of lipid metabolism. In general, acyl groups are not directly esterified to ACP but are first linked to coenzyme A and then transferred from coenzyme A to ACP by the action of a transacylase.

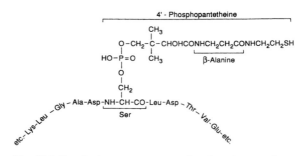

Fig. 27.2 Prosthetic group structure of acyl carrier protein. The 4'-phosphopantetheine group is attached to a serine residue near the middle of the polypeptide chain.

The enzymes of fatty acid synthesis

The fatty acid synthesis pathway uses acetyl-CoA as the building block for assembly of long chain (C16 and C18) fatty acids. Acetyl-CoA is probably supplied to the pathway primarily by the action of the pyruvate dehydrogenase reactions. Although this enzyme occurs both in the mitochondria and in plastids, it is not yet certain if the plastid isozyme has sufficient activity to account for *in vivo* rates of fatty acid synthesis. In non-photosynthesizing tissues, pyruvate is provided from the glycolytic pathway. In chloroplasts, pyruvate can be derived from 3-phosphoglycerate produced by the Calvin cycle reactions. Plastids also contain a very active acetyl-CoA synthetase, and isolated chloroplasts or developing seed plastids rapidly incorporate radiolabeled acetate into fatty acids. Therefore, it appears that acetyl-CoA for fatty acid biosynthesis can derive either from the pyruvate dehydrogenase reaction inside the plastid or from extraplastidial production of acetate (possibly by mitochondrial pyruvate dehydrogenase) followed by its activation inside the plastid by acetyl-CoA synthetase.

The assembly of an eighteen carbon fatty acid requires the condensation of nine two-carbon units. All of these units are derived from acetyl-CoA. However, their actual assembly first requires that the acetyl group be further 'activated'. This activation is accomplished when acetyl-CoA is carboxylated to form malonyl-CoA by the action of acetyl-CoA carboxylase (ACCase), an enzyme which catalyzes the following two partial reactions:

$$ATP + HCO_3^- + BCCP \xrightarrow[\substack{\text{Biotin}\\\text{carboxylase}}]{Mg^{2+}} ADP + P_i + BCCP\text{-}CO_2$$

$$BCCP\text{-}CO_2 + Acetyl\text{-}CoA \xrightarrow[\substack{\text{Trans}\\\text{carboxylase}}]{} BCCP + Malonyl\text{-}CoA$$

Recent evidence indicates that this enzyme occurs in two forms in plants (Konishi and Sasaki, 1994; Alban *et al.*, 1994). In one type, the complete acetyl-CoA carboxylase consists of a very large (> 200 kDa) multifunctional polypeptide with three functional domains. Biotin is attached covalently to the biotin carboxyl carrier protein domain (BCCP). CO_2 is activated and attached to biotin by the action of the biotin carboxylase domain. After formation of the carboxyl-biotin, the CO_2 is transferred to acetyl-CoA by the acetyl-CoA : malonyl-CoA transcarboxylase domain. In the second type of structure, the acetyl-CoA carboxylase is made up of several smaller subunits. The biotin carboxylase resides on a 50–52 kDa subunit, the BCCP is 34–38 kDa, and the transcarboxylase is encoded by a separate 65–75 kDa subunit. These subunits are organized into a 700 kDa complex which probably involves other, yet unidentified subunits. Which type of acetyl-CoA carboxylase structure is used depends on the subcellular localization and the type of plant. Dicots appear to have the multisubunit form in their plastids whereas the multifunctional structure occurs in the cytoplasm. The grasses differ in that both the plastid and cytosolic forms are large, multifunctional polypeptides (Konishi and Sasaki, 1994). Although the plastidial isozyme of ACCase is involved primarily in fatty acid synthesis, the cytosolic isozyme of ACCase is needed to supply malonyl-CoA for a variety of pathways including flavonoid biosynthesis and fatty acid elongation for cuticular lipid production.

There is now clear evidence that ACCase is a regulatory enzyme in leaf fatty acid synthesis (Post-Beittenmiller *et al.*, 1991), and its activity may be a major determinant of the flux of carbon into lipids in other tissues as well. In animals and yeast, ACCase activity is tightly controlled by a variety of mechanisms including feedback regulation, phosphorylation, and transcriptional control over its expression. Although feedback regulation also occurs on plant ACCase, the mechanism of this or other controls on ACCase activity is not yet known.

After the formation of malonyl-CoA, all subsequent steps of plant fatty acid synthesis require acyl carrier protein (ACP). Malonyl-CoA produced by acetyl-CoA carboxylase is transferred to ACP by the action of a transacylase.

Malonyl-CoA + ACP → Malonyl-ACP + CoA

The next step in fatty acid synthesis involves the condensation of acetyl and malonyl groups by the enzyme 3-ketoacyl-ACP synthase to form the four carbon intermediate, acetoacetyl-ACP.

Acetyl-CoA + Malonyl-ACP → Acetoacetyl-ACP + CO_2

This key reaction results in the formation of a carbon–carbon bond and in the release of the CO_2 which was added by the acetyl-CoA carboxylase reaction. The removal of CO_2 in this reaction helps to drive this reaction in the forward direction, making it essentially irreversible.

The acetoacetyl-ACP is next reduced at the carbonyl group by the enzyme 3-ketoacyl-ACP reductase, which uses NADPH as the electron donor.

Acetoacetyl-ACP + NADPH + H^+ → D-3-Hydroxybutyryl-ACP + $NADP^+$

The third reaction in the fatty acid synthesis cycle, which is catalyzed by 3-hydroxyacyl-ACP dehydratase, removes a water molecule from hydroxy-ACP to yield *trans*-2-acyl-ACP.

D-3-Hydroxybutyryl-ACP → *trans*-2-Butenoyl-ACP + H_2O

One round of fatty acid synthesis is then completed by the enzyme enoyl-ACP reductase, which uses NADH or NADPH to reduce the *trans*-2 double bond to form a saturated fatty acid.

trans-2-Butenoyl-ACP + NAD(P)H + H^+ Butyryl-ACP + $NAD(P)^+$

The combined action of these four reactions leads to the lengthening of the two carbon acetic acid to the four carbon butyric acid, which remains attached to ACP as a thioester. The condensation reaction is then repeated with additional malonyl-ACP followed again by the keto reduction, dehydration, and enoyl reduction steps. This four-step cycle continues until the fatty acid chain length is 16 or 18 carbons long. Although single isozymes of the reductases and dehydrase are apparently sufficient to act on acyl-ACP substrates of all chain lengths, at least three separate condensing enzymes (or 3-ketoacyl-ACP synthases) are required to produce an 18 carbon chain. The first condensation to form a four carbon product is carried out by 3-ketoacyl-ACP synthase III. Chain lengths from 6 to 16 carbons are produced primarily by 3-ketoacyl-ACP synthase I. Finally, the elongation of 16 carbon palmitoyl-ACP requires a separate condensing enzyme known as 3-ketoacyl-ACP synthase II and yields stearoyl-ACP.

Acyl-ACP utilization in the plastid

The 18 carbon saturated fatty acid, stearate, is the substrate for the introduction of the first *cis* double bond in plant fatty acids. Almost all aerobic fatty acid desaturation in nature is catalyzed by membrane-bound enzymes. However, in plants, the enzyme stearoyl-ACP desaturase is a soluble component of the chloroplast stroma. This enzyme requires O_2 and an electron donor such as reduced ferredoxin. A *cis* double bond is introduced exactly in the middle of the acyl chain between carbons 9 and 10.

Both palmitoyl-ACP and the oleoyl-ACP produced by the stearoyl-ACP desaturase are potential substrates for two additional branch point enzymes in chloroplast fatty acid metabolism. These fatty acids may be transferred from ACP to glycerolipids or they may be released as free fatty acids from ACP by the action of soluble acyl-ACP thioesterases. If the fatty acids from acyl-ACP are used immediately in glycerolipid synthesis, they remain in the plastid. On the other hand, if they are released as free fatty acids they may be exported for use by other cellular membranes. Thus, the different fates of oleate and palmitate in these two reactions determine the allocation of fatty acids between retention in or export from the chloroplast.

Source of NADPH and ATP

The ATP and reducing power needed for the fatty acid biosynthesis pathway is provided differently in different tissues. In leaves, the photosynthetic electron transport chain provides ATP and NADPH. In developing seeds, energy must be derived from sucrose imported from the leaves. Therefore, the pentose phosphate pathway in the plastid functions specifically to provide reducing power for fatty acid synthesis. ATP required for acetyl-CoA carboxylase can be derived from the glycolytic pathway which occurs both in plastids and in the cytosol of developing seeds (Dennis and Miernyk, 1982).

Genes for the plastidial pathway

Although the fatty acid synthesis pathway is localized inside the plastid, it is now clear that the genes coding for most of the enzymes of the pathway are in the nuclear genome. This was first suggested by the observation that mutants which lack plastid protein synthesis are still capable of producing characteristic plastid lipids. More recently, cDNA clones for acyl carrier protein, acyl-ACP thioesterases, and other enzymes of fatty acid metabolism have been found to encode transit peptide extensions comparable to those described in Chapter 24 (Ohlrogge et al., 1991). Thus, most of the proteins involved in producing fatty acids are synthesized in the cytoplasm as precursors and cut to their mature size during transport into the plastids. One exception to this rule is the transcarboxylase subunit of acetyl-CoA carboxylase. It has recently been discovered that this subunit is encoded in the chloroplast genome of a number of plants (Sasaki et al.,1993). Therefore, assembly of an active enzyme complex requires participation of both nuclear and plastidial gene products.

Occurrence of isozymes

A number of the proteins involved in plant fatty acid synthesis occur in different forms that are encoded by separate genes. In some cases, the use of such multiple isoforms may permit temporal or spatial control of the same reactions. For example, seeds storing oil or epidermal cells manufacturing cuticle would be expected to require different amounts of fatty acid than cells that are merely maintaining membranes. At the other extreme, there exist enzymes like the three keto-acyl synthases described earlier. These are closely related proteins with such different substrate specificities that they are considered distinct enzymes. Still other proteins exhibit subtle differences in substrate specificity that may or may not be significant in the plant. Acyl carrier proteins are currently the best studied examples of isoforms in lipid biosynthesis (Ohlrogge, 1987). In all plants examined, members of the ACP gene family are expressed quite differently in different tissues. There is evidence in vitro that oleoyl-ACP thioesterase and the first enzyme of plastid glycerolipid synthesis prefer acyl groups attached to different ACP isoforms. If this kind of discrimination occurs in vivo in plants, ACP isoforms could help to control the branch point between fatty acid use and export by plastids. However, all isoforms tested can be used by both enzymes, and the spinach ACP expected to promote the plastidial pathway is actually the main isoform in export-dominated tissues. The isoforms therefore may be more significant for the fine control of ACP concentration.

Two pathways of glycerolipid synthesis

As described above, acyl groups synthesized in the plastids are either used for glycerolipid synthesis in situ or exported for glycerolipid synthesis in the endoplasmic reticulum and elsewhere. Glycerolipids made by the 'prokaryotic pathway' of the plastids and the 'eukaryotic pathway' of the endoplasmic reticulum can easily be distinguished (Roughan and Slack, 1982; Browse and Somerville, 1991). In lipids synthesized by the plastids, position 2 of the glycerol backbone is occupied almost exclusively by 16 carbon fatty acids (Fig. 27.3). Both 16 and 18 carbon fatty

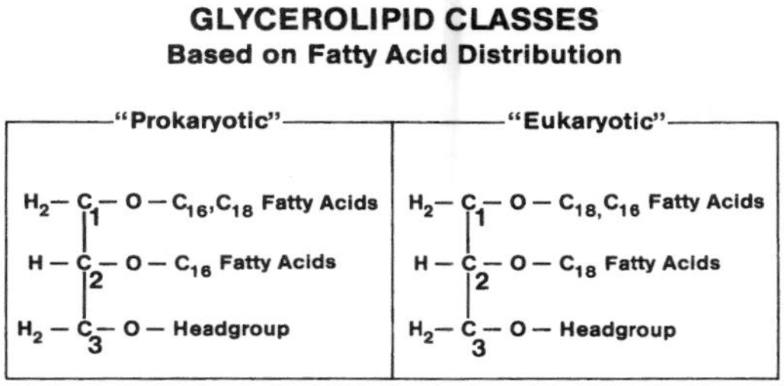

GLYCEROLIPID CLASSES
Based on Fatty Acid Distribution

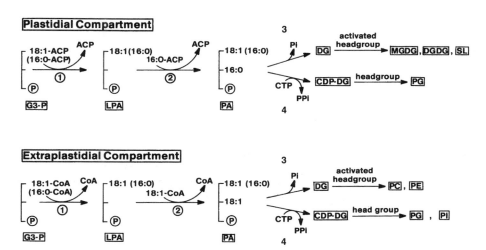

Fig. 27.3 Glycerolipid classes based on fatty acid distribution. The 'prokaryotic' and 'eukaryotic' lipid classes are based on the distribution of the fatty acids within the diacylglycerol moiety; the major difference occurs at position 2 of the glycerol backbone.

acids are found at position 1. This 'prokaryotic' fatty acid distribution is also characteristic of lipids from photosynthetic prokaryotes. Glycerolipids synthesized by the 'eukaryotic' pathway of the plant endoplasmic reticulum resemble phospholipids from mammals and yeast. Position 2 is occupied primarily by 18 carbon unsaturated fatty acids, while position 1 again contains either 18 carbon or 16 carbon fatty acids. Lipids synthesized by extraplastidial organelles other than endoplasmic reticulum are probably also eukaryotic.

Both prokaryotic and eukaryotic glycerolipids are formed by sequential transfer of acyl groups to positions 1 and 2 of glycerol-3-phosphate. Glycerol-3-phosphate is synthesized by the enzyme dihydroxyacetone-phosphate reductase. Several isozymes of the reductase occur, including at least one in the chloroplast and one in the cytoplasm (Kirsch *et al.*, 1992). Thus, both the plastidial and the extraplastidial compartments can produce the backbone for glycerolipid synthesis. The 'prokaryotic' and 'eukaryotic' arrangements of fatty acids on this backbone result from the different specificities of plastidial and endoplasmic reticulum acyltransferases.

In chloroplasts, the first acyltransferase moves an acyl group from ACP to position 1 on glycerol-3-P, forming lysophosphatidic acid (LPA, Fig. 27.4) (Frentzen, 1986). The enzyme is found in the chloroplast stroma and shows some preference for oleoyl ACP. The second acyltransferase is located in the inner membrane of the chloroplast envelope, and transfers almost exclusively 16:0 from ACP to position 2 of LPA, resulting in the formation of 'prokaryotic' phosphatidic acid (PA, Fig. 27.4) (Frentzen, 1986).

In the 'eukaryotic' pathway, fatty acyl donors are acyl-CoAs rather than the acyl-ACPs used in the

Fig. 27.4 Acyltransferases involved in glycerolipid synthesis. The sequential acylation of glycerol 3-P (G3P) by (1) acyl-ACP (or acyl-CoA): glycerol 3-P acyltransferase and (2) acyl-ACP (or acyl-CoA): lysophosphatidic acid acyltransferase gives rise to PA in both the plastid and ER. PA may then be either dephosphorylated to DG by (3) phosphatase or (4) activated by the addition of CTP to CDP-DG by phosphatidate cytidyltransferase in both compartments.

plastids. The free fatty acids released by plastidial thioesterases are apparently added to CoA by an acyl-CoA synthetase in the outer envelope of the plastids (Fig. 27.5) (Joyard and Stumpf, 1981). The high energy acyl-CoAs are probably bound to an acyl-CoA binding protein, allowing the fatty acids to move through the cytoplasm to the endoplasmic reticulum and other cell membranes for glycerolipid synthesis. At the endoplasmic reticulum, the glycerol-3-phosphate : acyl-CoA acyltransferase can use either 16:0-CoA or 18:1-CoA at position 1 to form LPA.

The second acyltransferase employs 18:1-CoA almost exclusively, resulting in newly-synthesized PA with a 'eukaryotic' fatty acid pattern (Frentzen, 1986; Browse and Somerville, 1991).

A central role for phosphatidic acid

Whether phosphatidic acid (PA) is produced in plastids or endoplasmic reticulum, it is a central intermediate in glycerolipid synthesis. In either

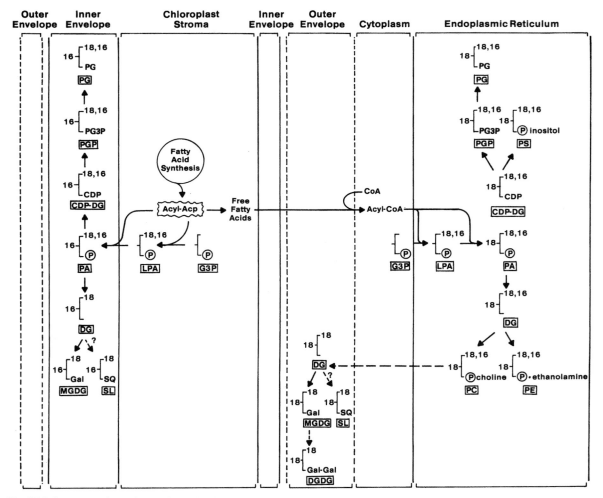

Fig. 27.5 Summary of membrane glycerolipid synthesis. Lipid synthesis in the plastid and ER is interdependent for the supply of fatty acids and the supply of DG moieties. Within the plastid, compartmentalization of lipid synthesis suggests that lipids must be moved from their site of synthesis in the envelope to the thylakoids.

compartment, it is directed into a variety of lipids by first entering one of two branching reactions (Fig. 27.4) (Browse and Somerville, 1991). In one branch, PA is dephosphorylated to yield diacylglycerol (DG), which is a precursor for those lipids that are synthesized by the addition of head groups activated by a high energy phosphate bond. In the other, the DG portion of PA is itself activated by the addition of CTP to form CDP-DG. The lipid head group is then added to this high-energy intermediate.

Plastidial lipid synthesis: phosphatidylglycerol and glycolipids

Since the phosphatidylglycerol (PG) of plastid membranes is 'prokaryotic' in every case examined to date, it is presumably assembled by plastid enzymes. PG is formed from CDP-DG by the sequential action of two enzymes (Fig. 27.4 and 27.5) (Mudd *et al.*, 1987). The first adds glycerol 3-P to CDP-DG to form phosphatidylglycerol 3-P (PGP). PGP is then rapidly dephosphorylated to PG by the second enzyme. In fact, the only way to detect the intermediate PGP is to inhibit the activity of the second enzyme. Both are located in the inner membrane of the chloroplast envelope. A curious characteristic of plastidial PG is the occurrence of *trans*-Δ^3-16:1 at the 2-position. This unusual fatty acid is apparently produced by desaturation of 16:0 on the PG itself (Ohnishi and Thompson, 1991). Although a mutant of *Arabidopsis* lacking *trans*-Δ^3-16:1 survives without difficulty, there is some evidence that the presence of this fatty acid is correlated with cold resistance (Gibson *et al.*, 1994).

As mentioned earlier, chloroplasts are rich in glycolipids, specifically monogalactosyl diacylglycerol (MGDG), digalactosyl diacylglycerol (DGDG), and sulfoquinovosyl diacylglycerol (SQDG) (Fig. 27.1). DG provides the glycerol and fatty acids for all three (Joyard and Douce, 1987; Joyard *et al.*, 1994). A phosphatase that releases DG from PA has been identified on the inner chloroplast envelope, and this is the probable origin of 'prokaryotic' glycolipid backbones (Figs. 27.4 and 27.6). MGDG is produced by transfer of galactose from UDP-galactose to DG. This reaction is catalyzed by a UDP-galactose: diacylglycerol galactosyltransferase in the inner

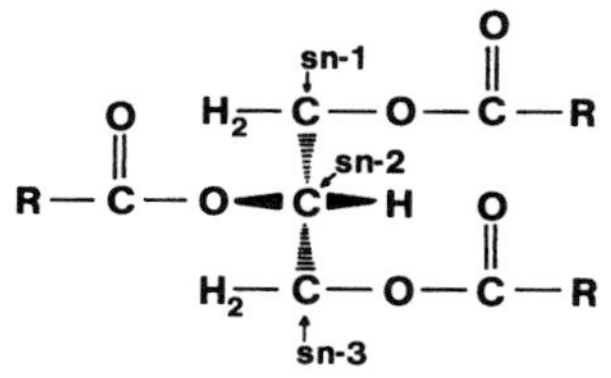

Fig. 27.6 Structure of triacylglycerol.

chloroplast envelope of spinach, although outer envelope participation has not been ruled out in other plants (Browse and Somerville, 1991; Joyard *et al.*, 1993). The source of DGDG has been a more controversial issue. An outer envelope enzyme that catalyzes donation of a second galactose from one MGDG to another is currently the best candidate for DGDG synthesis. Finally, envelopes from spinach chloroplasts contain an enzyme that will transfer sulfoquinovose from UDP-sulfoquinovose to DG, resulting in the synthesis of SQDG (Seifert and Heinz, 1992). The pathway by which adenosine phosphosulfate is incorporated into the sulfolipid head group remains an enigma, although recent advances in the molecular biology of SQDG synthesis in bacteria are expected to advance the field immensely.

Although the fatty acid synthesis pathway results primarily in 16:0 and 18:1-ACPs, glycolipids are extremely rich in triunsaturated fatty acids. Current research with desaturation mutants indicates that plastids contain desaturases active on fatty acids after they have been incorporated into glycolipids. One enzyme is able to desaturate the prokaryotic 16:0 at the 2-position of MGDG to 16:1. Two other enzymes working in sequence appear to desaturate either 16:1 or 18:1 on any of the three glycolipid classes to di- and then triunsaturated fatty acids (Browse and Somerville, 1991).

Extraplastidial lipid synthesis: endoplasmic reticulum

The assembly of phospholipids is known to occur in both the endoplasmic reticulum (ER) and the mitochondria. Whether it also occurs in other cell membranes is under investigation. Currently, it is thought that the phospholipids synthesized in the ER are also transported to other cellular membranes. The recipient membranes may then assemble those lipids that are unique to them.

As in the plastid, PA is the precursor for all phospholipids assembled in the ER. Again, phospholipid synthesis requires that PA either be hydrolyzed to DG or activated to CDP-DG (Figs. 27.4 and 27.5). The DG produced by the dephosphorylation pathway serves as the precursor to phosphatidylcholine (PC) or phosphatidylethanolamine (PE). The head groups for these lipids are first activated by the addition of CTP, resulting in CDP-choline (for PC) or CDP-ethanolamine (for PE). They are then added to DG to form PC or PE. Thus, the synthesis of these lipids is analogous to the synthesis of the glycolipids in the plastid. In fact, PC and PE are the major lipids found in the ER membranes (Table 22.1).

The CDP-DG produced by reaction of CTP and PA serves as the precursor for PG, phosphatidylserine (PS) and phosphatidylinositol (PI), all of which are relatively minor components of the ER membrane (Table 22.1). PG is synthesized by the same steps which occur in the plastid: glycerol 3-P is added to CDP-DG to form PGP, which is then dephosphorylated to PG. The reaction of CDP-DG and serine to form PS is also well documented in plants, although in some tissues PS is generated primarily by exchange of serine with the ethanolamine of PE (Kinney, 1993). Finally, the addition of inositol to CDP-DG results in the formation of PI. It has recently been shown that plants also convert a small proportion of their PI to phosphatidylinositol phosphates. The possible roles of these compounds in plant signal transduction are under intense scrutiny (Coté and Crain, 1993).

The plastidial-endoplasmic reticulum connection

The only phospholipid synthesized by the plastid appears to be PG. Yet the plastid membranes also contain PC, and in much smaller amounts PS and PI as well (Table 22.1). Since these lipids possess eukaryotic fatty acid distributions, it is thought that they are imported from their site of synthesis in the ER to the plastid. In fact, PC may play a special role.

Not only is it present as a structural component in the plastid membranes, but it may also serve as a precursor to some of the plastidial glycolipids as well.

As mentioned earlier, the chloroplast can synthesize the glycolipids MGDG, DGDG and SQDG. If only DG from PA made in the plastids were incorporated into these glycolipids, we would expect them to be entirely prokaryotic in fatty acid distribution. However, in some plants, only about half the glycolipids are prokaryotic, while in other plants almost none of them are; instead, they are eukaryotic in nature. The results of many experiments suggest that PC synthesized in the ER is the precursor for these eukaryotic glycolipids. After its assembly in the ER, at least the diacylglycerol portion from PC is transported to the plastid (Fig. 27.5). It may be that the entire PC molecule is moved by a phospholipid transport protein, although there is no direct evidence to support this idea. At some point, the phosphocholine head group is removed. The resulting DG can then enter the glycolipid synthesis pathway.

What determines the proportion of prokaryotic and eukaryotic glycolipids that a particular plant will synthesize? If the sole source of prokaryotic glycolipids is DG derived from plastidial PA, the plastidial enzyme that dephosphorylates PA to make that DG is an obvious candidate. In fact, it has been shown that those plants with high proportions of prokaryotic glycolipids have high levels of the plastidial phosphatase, while plants with low proportions of prokaryotic glycolipids have low levels of phosphatase activity (Mudd et al., 1987).

Additional regulation may be exerted at the point at which fatty acids are either used by the plastid for lipid synthesis, or hydrolyzed for export to other cell membranes. In those plants with a small proportion of prokaryotic glycolipids, the predominant lipid synthesized from PA by the plastidial pathway is PG. Since PG is a minor lipid component of the plastid membranes compared to glycolipids, it seems reasonable that the plastids in which eukaryotic glycolipids dominate would synthesize relatively little PA. They could then export unused fatty acids for synthesis of eukaryotic PC, which would in turn donate eukaryotic backbones back to the plastids for glycolipid synthesis. Recent research indicates that this may in fact be the case. Moreover, plants mutated so that the plastidial pathway is largely disabled are able to compensate by shunting fatty acids through the eukaryotic pathway into glycolipids (Browse and Somerville, 1991; Gibson *et al.*, 1994).

Thus it can be seen that lipid metabolism may be regulated at several points within one organelle, and that it must be coordinated among several different organelles. Alternative pathways and interconversions between phospholipids add complexity to the overall picture (Kinney, 1994). Many other factors also come into play. A plant must be able to keep pace with demands that vary during its development. For example, rapidly expanding leaves require increased lipid synthesis for production of thylakoids and other membranes, while developing seeds specialize in triacylglycerol synthesis. In addition, environmental conditions can alter both proportions and fatty acid compositions of plant lipids.

Extraplastidial lipid synthesis: mitochondria

Lipid biosynthetic activities elsewhere have been less well characterized than those in the chloroplast and ER. One difficulty has been the challenge of obtaining cellular subfractions uncontaminated by fragments of ER. Another is preparing subfractions which are still active, and yet another is determining the appropriate conditions under which to investigate lipid synthetic activities.

Indications that plant mitochondria may possess their own fatty acyl synthase were discussed earlier. In addition, mitochondria are able to synthesize PA by the sequential acylation of glycerol 3-P in much the same manner as in the ER (Frentzen, 1994). This synthesis appears to occur in both the inner and outer membranes of the mitochondria, and the resulting PA is eukaryotic in nature. It may then be used in the synthesis of PG, which apparently occurs only in the inner mitochondrial membrane, by the same set of steps described earlier. There is evidence that cardiolipin, a lipid unique to the mitochondria, is derived from PG by reaction with CDP-DG in mitochondria. Finally, in addition to the work described here, there have been reports of PC and PE synthetic enzymes in both mitochondria and Golgi bodies (Moore, 1982). As the technology for studying membrane-bound enzymes improves, it is possible

that glycerolipid synthesis will be confirmed in the Golgi and perhaps even in other subcellular fractions.

Glycerolipid desaturation

As described earlier, 16:0 and 18:1 are the main fatty acids used for glycerolipid synthesis in both plastids and ER. In an earlier section, the desaturation of 16:0 and 18:1 on glycolipid substrates was discussed. Plants in which the prokaryotic pathway is most active contain relatively high levels of 16:3 in their MGDG due to this pathway, and are sometimes called '16:3 plants' for this reason. When the eukaryotic pathway is the primary source of glycolipid backbones, only 18-carbon fatty acids can be fully desaturated after their incorporation into galactolipids, and the plants are known as '18:3 plants'.

The plastids are not the only organelle in which glycerolipids may be desaturated. The ER of leaves, roots and developing seeds can desaturate oleate (18:1) to linoleate (18:2) and linolenate (18:3). Single mutations in ER desaturation affect the fatty acid compositions of all ER phospholipids, particularly PC, and labeling studies indicate that the substrates for desaturation are the phospholipids themselves. Insertion of each double bond requires NADH and O_2, and it is thought that electrons are passed from the NADH to the desaturase via cytochrome b_5 and cytochrome b_5 reductase in a manner analogous to that observed in animal systems (Jaworski, 1987; Heinz, 1993).

There has been considerable discussion concerning the substrate specificity of ER desaturases. One popular model (Browse and Somerville, 1991) postulates relatively nonspecific 18:1 and 18:2 desaturases that act on their respective fatty acyl groups at either position of any available phospholipid. Another scenario calls for enzymes that work only on 18:1 or 18:2 at the 2-position of PC. A recent paper suggests that the ER 18:2 desaturase may desaturate at either the 1- or the 2-position, but that its productivity is greater at the 2-position (Stymne *et al.*, 1992). Whatever the specificities of the desaturases themselves, an integrated view of glycerolipid synthesis must account for the fact that different classes of lipids and different membranes have their own characteristic fatty acid compositions. In addition, fatty acid compositions change in response to changing environmental conditions, and perhaps during plant growth and development as well. There are a number of other ways in which the fatty acid profiles of membrane lipids can be altered. New lipids may be synthesized to replace or supplement preexisting lipids. Such a mechanism could account for slow changes observed during growth or adjustment of membrane unsaturation to seasonal fluctuations in temperature. However, some changes occur too rapidly to be accounted for by *de novo* synthesis. One way to quickly alter membrane lipid composition is to remove fatty acids from lipids while they remain in the membrane, and replace them with new fatty acids. This type of 'acyl exchange' may be accomplished by a combination of the reverse and forward reactions of acyltransferases, where the reverse reaction removes a fatty acid by attaching it to CoA and the forward reaction replaces it with a new fatty acid from another acyl-CoA (Fig. 27.8) (Stymne and Stobart, 1987). Very little is known about these acyltransferases, although they appear to differ from those involved in *de novo* lipid synthesis. Another possibility is removal of fatty acids by specific lipases followed by insertion of new ones via acyltransferases. Even less is known about such lipases, although there are recent reports that at least some of these activities do exist.

Triacylglycerol synthesis

Unlike the glycerolipids found in membranes, triacylglycerols do not perform a structural role but instead serve primarily as a storage form of carbon. Although small quantities of TG are found even in lettuce leaves, and certain fruits such as olive and avocado accumulate substantial TG, the presence of these storage oils in seeds has received the most attention in recent years. The amount of oil in different species may vary, from as little as 1–2% to as much as 60% of the total dry weight of the seed. Plants which produce seeds containing oil of economic importance for either food or industrial use include soybean, palm, rapeseed, sunflower, safflower, cotton, peanut, and coconut. Study of the mechanism and control of TG deposition in seeds has been

impeded by fact that only developing seeds are suitable experimental material. In fact, very rapid lipid synthesis usually occurs only during a limited period when the seed is rapidly gaining weight, and it essentially stops as the seed begins to mature and dehisce.

Triacylglycerol consists of a glycerol backbone with fatty acids esterified to all 3 carbons (Fig. 27.6). Its synthesis, at least in some species, is relatively straightforward (Fig. 27.7). Once again, fatty acids are sequentially transferred to positions 1 and 2 of glycerol 3-phosphate, resulting in the formation of the central metabolite, PA. As in the synthesis of glycolipids, PC, and PE, dephosphorylation of PA releases DG. In the unique step of TG synthesis, a third fatty acid is transferred to the vacant position 3 of the DG. This step is catalyzed by diacylglycerol acyltransferase, apparently in the endoplasmic reticulum.

Is PA the only source of DG for triacylglycerol synthesis? In many seeds, there is clear evidence that at least some of the DG used in TG synthesis passes through PC. As discussed above, incorporation into PC gives fatty acids an important opportunity for desaturation. Since most seed oils used for human (and plant) consumption are polyunsaturated, the PC pathway is of major significance.

Fatty acids from PC may become available for TG synthesis by one of two mechanisms. In the first, a

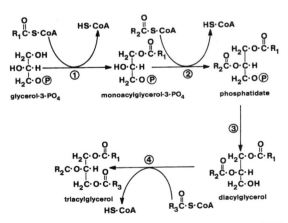

Fig. 27.7 Synthesis of triacylglycerol. The DG moiety of TG may be derived from either PA by the action of phosphatase (3) or PC by the reverse reaction of CDP-choline : DG choline transferase.

fatty acid attached to CoA and a fatty acid on PC may essentially trade places. Such an 'acyl exchange' probably occurs by the combined reverse and forward reactions of an acyl-CoA : PC acyltransferase (Fig. 27.8) (Stymne and Stobart, 1987). The resulting acyl-CoA may then be used as an acyl donor in triacylglycerol synthesis. The exchange reaction allows 18:1 newly produced and exported from the plastid to enter PC while desaturated or otherwise modified fatty acids depart for TG or other lipids. The second mechanism by which PC can participate in TG synthesis is by donation of its entire DG portion (Stymne and Stobart, 1987). In some plants, the synthesis of PC from DG and CDP-choline appears to be rapidly reversible. This would allow the DG moiety of PC, including any modified fatty acids, to become available for TG synthesis.

Storage of triacylglycerols in oil bodies

In the mature seed, TG is stored in densely packed lipid bodies, which are roughly spherical in shape with an average diameter of 1 μm (Stymne and Stobart, 1987; Tzen and Huang, 1992). This size does not change during seed development, and accumulation of oil is accompanied by an increase in the number of lipid bodies. These organelles appear

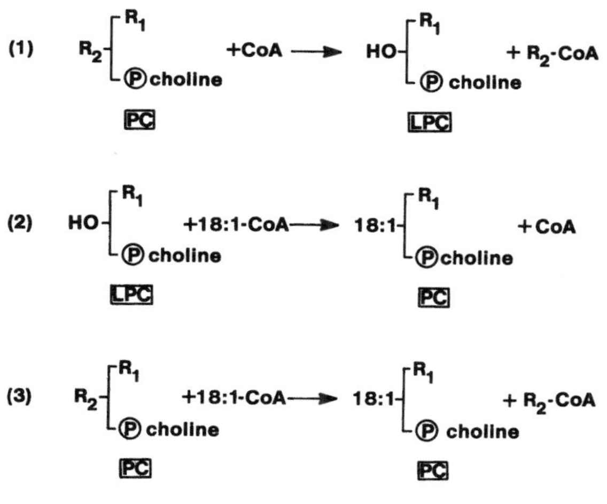

Fig. 27.8 Acyl exchange in PC. The combined reverse (1) and forward (2) reactions of acyl-CoA : lysophosphatidylcholine acyltransferase will result in acyl exchange between acyl-CoA and phosphatidylcholine (3).

to be surrounded by a type of membrane which is only half of a normal bilayer membrane. The polar headgroups of this half unit membrane face the cytoplasm, while the nonpolar acyl groups are associated with the nonpolar TG within. The membranes of isolated lipid bodies contain both phospholipids and proteins.

The synthesis of TG is located in the ER, as these membranes contain all three acyltransferase activities. In addition, the reactions by which fatty acids are modified or elongated also appear to occur in the ER. How triacylglycerols are moved from their site of synthesis in the ER to oil bodies is unknown. In fact, how oil bodies are generated is poorly understood. A popular model shows TG accumulation inside one bilayer of an ER cisterna. As the oil accumulates, it pushes the halves of the bilayer apart. Eventually, this bubble of oil buds off of the ER, complete with half unit membrane. When first synthesized, these oil bodies coalesce easily, as do oil droplets synthesized by ER vesicles *in vitro* (Stymne and Stobart, 1987; Murphy, 1993). However, seeds that will undergo desiccation synthesize oleosins, small proteins with a hydrophobic central domain. These oleosins appear to become associated with lipid bodies after their release from the ER (Murphy, 1993). Oleosins are believed to preserve individual lipid bodies as discrete entities. Lipid body-like structures can be reconstituted by sonication of TG with appropriate amounts of phospholipid and oleosin, but not if one of these ingredients is left out (Tzen and Huang, 1992).

Seed oils and unusual fatty acids

The fatty acid composition of storage oils varies much more than that of membrane glycerolipids. While the structural glycerolipids of all plants contain predominantly 6 fatty acids (18:1, 18:2, 18:3, 16:0, 16:1 3-*trans*, and in some plants 16:3), there are more than 300 different fatty acids known to occur in seed TG (Harwood, 1980). The reason for this diversity is unknown, but the special properties of some of these 'unusual' fatty acids are utilized commercially. The medium chain lauric acid (12:0) derived from coconut or palm kernel oil is used as a detergent in shampoos and toothpastes, while the longer chain erucic acid

(22:1) from rapeseed oil is an excellent lubricating oil at high temperature. The fatty acid composition of seed oil is species specific, and the characteristics of any particular oil are dependent upon both the types of fatty acids present and the positions these fatty acids occupy in TG. For example, TG from cocoa oil contains 16:0 (or 18:0), 18:1, and 18:0 at positions 1, 2 and 3 of the glycerol backbone, respectively; this gives it the characteristic of 'melting in your mouth' which is so important for chocolate. Sunflower oil, on the other hand, may contain 18:1 esterified to all three positions, resulting in an oil which is liquid at room temperature. In general, edible oils contain predominantly the saturated fatty acids 16:0 and 18:0 and the unsaturated fatty acids 18:1, 18:2 and 18:3.

The variations observed in fatty acids may be roughly classified as differences in chain length and modification of the fatty acid itself, such as the addition of double bonds or functional groups (Harwood, 1980).

Economically significant examples of fatty acids longer than 18:1 or shorter than 16:0 have already been described. Elongation of fatty acids from 18:1 to 20:1 or 22:1 (or even longer) occurs outside the plastid, at least partly in the ER, using a membrane bound fatty acid elongation system (Slack and Browse, 1984; Stymne and Stobart, 1987). Two carbon units are sequentially transferred from malonyl-CoA to 18:1 in much the same manner as during fatty acid synthesis. Plants that synthesize medium chain fatty acids appear to contain a special medium chain acyl-ACP thioesterase in their plastids in addition to thioesterases that release 16:0 and 18:1 from ACP. When a gene for medium-chain thioesterase is inserted into a plant that does not normally make laurate, fatty acid is in fact released from ACP when it is only 12 carbons long, resulting in genetic engineering of laurate accumulation (Voelker et al., 1992).

Fatty acids with double bonds in unusual positions are another class of modified fatty acids. The mechanisms responsible for alternative desaturation patterns vary. For example, the plastids of members of the carrot family contain an extra acyl-ACP desaturase that inserts a double bond four carbons from the esterified end of a 16:0. In the borage family, on the other hand, one double bond is placed in an unusual position while its parent fatty acid is

attached to PC. PC has also been implicated as a substrate in formation of fatty acids with hydroxy or epoxy groups. Recent advances in the molecular biology of these and other unusual fatty acids may well lead to the marketing of common oilseed crops engineered to produce useful fatty acids previously unavailable in practical quantities (Ohlrogge, 1994).

Very little is known about the regulation of the fatty acid content of TG. The fact that TG and membrane lipids often have widely different fatty acid compositions continues to provoke a great deal of study. One possibility is that fatty acids incompatible with membrane structure are either edited out of or not accepted by enzymes of membrane lipid synthesis. This would leave the rejected fatty acids to form the less structurally demanding TG. Another possibility is that TG are synthesized from fatty acids associated with localized pools of DG or earlier precursors. The final composition of the TG probably depends on both the selectivities and specificities of the three acyltransferases. Which acyl-CoAs these enzymes preferentially select to transfer and how quickly they transfer any particular acyl-CoA do vary from species to species. In some cases, they may help to channel specific fatty acids into TG, while in other situations they may accept whatever fatty acyl-CoAs are available.

Release of fatty acids from acyl lipids

When oil storing seeds germinate, a massive breakdown of the triacylglycerol reserves is initiated. This process is begun by the action of lipases which catalyze the hydrolysis of fatty acids from the glycerol backbone (Huang, 1987). Lipase activity is typically absent in ungerminated seeds but rises rapidly shortly after imbibition. In some cases, lipases appear to become embedded in the half unit membrane of lipid bodies; other TG lipases are easily stripped from this membrane. In still another permutation, removal of the last fatty acid from storage lipid glycerol may take place in the glyoxysomes. Glyoxysomes are vital constituents of all oilseeds, since they house both fatty acid oxidation and the glyoxylate cycle that enables plants to convert carbon in fatty acid to the more mobile sucrose.

Fatty acids may also be released from other

glycerolipids. At least some of the array of phospholipases and nonspecific esterases that plague biochemists attempting to analyze plant lipids probably participate in membrane lipid turnover.

Fatty acid oxidation for energy and sugar production

Although plants are known to remove single carbons from either the carboxyl end (α-oxidation) or the acyl end (ω-oxidation) of fatty acids, fatty acyl moieties derived from both membrane lipid turnover and storage lipid mobilization are typically broken down by β-oxidation (Gerhardt, 1992, 1993). During β-oxidation, two-carbon units are cleaved from a fatty

acyl-CoA beginning at its carboxyl end. As shown in Fig. 27.9, each two-carbon unit proceeds through a series of oxidation steps culminating in its release as acetyl-Coenzyme A. In leaves and other plant organs, the β-oxidation apparatus has been identified in the peroxisomes. Oilseeds, on the other hand, house the multifunctional protein responsible for β-oxidation in their glyoxysomes. Since, as described in the previous chapter, glyoxysomes are the precursors of peroxisomes in greening oilseed cotyledons, this is not surprising. In both peroxisomes and glyoxysomes, oxidation of acyl-CoA to enoyl-CoA results in the formation of hydrogen peroxide that must be detoxified by catalase, an enzyme diagnostic for the microbody family to which these organelles belong (Kindl, 1993; Gerhardt, 1992). There is thus less

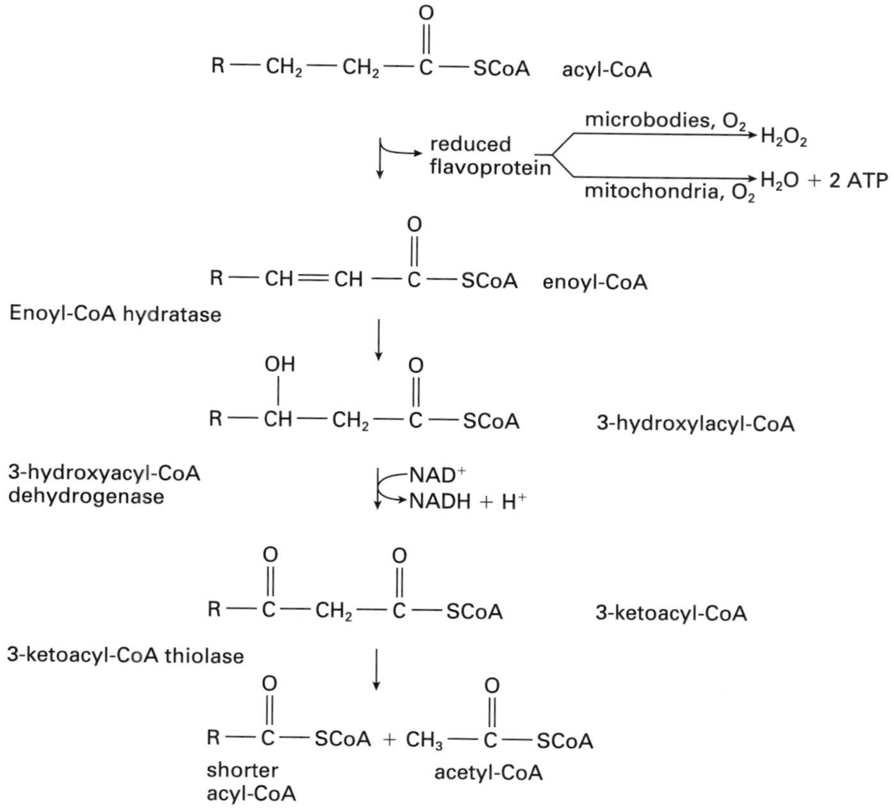

Fig. 27.9 The β-oxidation pathway breaks down fatty acids to produce acetyl-CoA. Four reactions are required to remove two carbons in each cycle. In microbodies (peroxisomes and glyoxysomes) acyl-CoA oxidase catalyzes the first reaction and produces H_2O_2 whereas in mitochondria, acyl-CoA dehydrogenase catalyzes the first reaction.

energy released by β-oxidation in microbodies than in the more classical mitochondrial pathway, in which the $FADH_2$ generated during the corresponding step feeds into the mitochondrial electron transport pathway (Fig. 27.9). Do plants also possess a mitochondrial β-oxidation system? Although the subject has been a contentious one, there is recent evidence that at least some plants have both systems (Miernyk *et. al.,* 1991; Dieuaide *et al.,* 1993).

There is little doubt that a portion of the acetyl-CoA generated during β-oxidation is further oxidized in the mitochondrion to release CO_2, H_2O and energy. However, plants, unlike animals, have the ability to use acetyl-CoA for carbohydrate synthesis. In germinating oilseeds, the main role of stored triacylglycerol is to provide carbon for the growing seedling, a role that requires conversion of insoluble lipid to the more easily transported sucrose. A comparable mobilization of fatty acids into carbohydrate may take place in senescing leaves, in which peroxisomes appear to metamorphose into glyoxysomes (Pistelli *et. al.,* 1992). In either case, the synthesis of carbohydrate from lipid is made possible by the glyoxylate cycle, which converts acetyl-CoA to four-carbon compounds. Since this cycle is housed in the glyoxysomes, acetyl-CoA produced by glyoxysomal β-oxidation can feed directly into the cycle (Fig. 27.10). As in the citric acid cycle (Chapter 8), acetyl-CoA condenses with oxaloacetate to form citrate, which is subsequently isomerized to isocitrate. Two enzymes unique to the glyoxylate cycle then intervene. Isocitrate lyase converts isocitrate to succinate and glyoxylate. In a reaction catalyzed by malic enzyme, the glyoxylate then combines with a second acetyl-CoA to generate malate. Finally, oxaloacetate is regenerated from malate as in the citric acid cycle.

The succinate generated from isocitrate in the glyoxysome is further metabolized in the mitochondrion by citric acid cycle enzymes to produce oxaloacetate. Oxaloacetate then leaves the mitochondrion for the cytosol, where phospho-*enol*-pyruvate (PEP) is produced and formation of sucrose occurs by way of gluconeogenesis. Thus, the conversion of triacylglycerol to sucrose is a dramatic example of the interrelationships between several intracellular compartments in plant metabolism.

Further reading and references

Alban, C., Baldet, P. and Douce, R. (1994). Localization and characterization of two structurally different forms of acetyl-CoA carboxylase in young pea leaves, of which one is sensitive to aryloxyphenoxypropionate herbicides. *Biochem. J.* **300**, 557–65.

Browse. J. and Somerville, C. (1991). Glycerolipid synthesis: Biochemistry and regulation. *Annu. Rev. Plant Physiol. Plant Mol. Biol.* **42**, 467-506.

Coté, G. G. and Crain, R. C. (1993). Biochemistry of phosphoinositides. *Ann. Rev. Plant Physiol. Plant Molec. Biol.* **33**, 333–56.

Dennis, D. T. and Miernyk, J. A. (1982). Compartmentation of nonphotosynthetic carbohydrate metabolism. *Ann. Rev. Plant Physiol.* **33**, 27–50.

Dieuaide, M., Couee, I., Pradet, A. and Raymond, P. (1993). Effects of glucose starvation on the oxidation of fatty-acids by maize root-tip mitochondria and peroxisomes – Evidence for mitochondrial fatty acid beta-oxidation and acyl-CoA dehydrogenase activity in a higher plant. *Biochem. J.* **296**, 199–207.

Frentzen, M. (1993). Acyltransferases and triacylglycerols. In *Lipid Metabolism in Plants*, ed. T.S. Moore, Jr., CRC Press, Boca Raton, pp 195–230.

Frentzen, M. (1986). Biosynthesis and desaturation of the different diacylglycerol moieties in higher plants. *J. Plant Physiol.* **124**, 193–209.

Gerhardt, B. (1992). Fatty acid degradation in plants. *Prog. Lipid Res.* **31**, 417–46.

Gerhardt, B. (1993). Catabolism of fatty acids (α- and β-Oxidation). In *Lipid Metabolism in Plants*, ed. T. S. Moore, Jr., CRC Press, Boca Raton, 527–65.

Gibson, S., Falcone, D. L., Browse, J. and Somerville, C. (1994). Use of transgenic plants and mutants to study the regulation and function of lipid composition. *Plant Cell Env.* **17**, 627–637.

Harwood, J. L. (1980). Plant acyl lipids: Structure, distribution and analysis. In *The Biochemistry of Plants*, Vol. 4, ed. P. K. Stumpf, Academic Press, New York, pp. 2–56.

Heinz, E. (1993). Biosynthesis of polyunsaturated fatty acids. In *Lipid Metabolism in Plants*, ed. T. S. Moore, Jr., CRC Press, Boca Raton, pp. 33–90.

Huang, A. H. C. (1987). Lipases. In *The Biochemistry of Plants*, Vol. 9, eds P. K. Stumpf and E. E. Conn, Academic Press, New York, pp. 91–116.

Jaworski, J. G. (1987). Biosynthesis of monoenoic and polyenoic fatty acids. In *The Biochemistry of Plants*, Vol. 9, eds P. K. Stumpf and E. E. Conn, Academic Press, New York, pp. 159–75.

Joyard, J., Block, M. A., Malherbe, A., Maréchal, E. and Douce, R. (1994). Origin and synthesis of galactolipid and sulfolipid head groups. In *Lipid Metabolism in Plants*, ed. T. S. Moore, Jr., CRC Press, Boca Raton, pp 231–58.

Joyard, J. and Douce, R. (1987). Galactolipid synthesis. In *The Biochemistry of Plants*, Vol. 9, eds P. K. Stumpf and E. E. Conn, Academic Press, New York, pp. 215–75.

Joyard, J. and Stumpf, P. K. (1981). Synthesis of long-chain acyl-CoA in chloroplast envelope membranes. *Plant Physiol.* **67**, 250–56.

Kindl, H. (1993). Fatty acid degradation in plant peroxisomes – Function and biosynthesis of the enzymes involved. *Biochimie* **75**, 225–30.

Kinney, A. J. (1993). Phospholipid head groups. In *Lipid Metabolism in Plants*, ed. T. S. Moore, Jr., CRC Press, Boca Raton, pp 259–84.

Kirsch, T., Gerber, D. W., Byerrum, R. U. and Tolbert, N. E. (1992). Plant dihydroxyacetone phosphate reductases. Purification, characterization and localization. *Plant Physiol.* **100**, 352–59.

Kolattukudy, P. E. (1984). Biochemistry and function of cutin and suberin. *Can. J. Bot.* **62**, 2918–33.

Konishi, T. and Sasaki, Y. (1994). Compartmentalization of two forms of acetyl-CoA carboxylase in plants and the origin of their tolerance toward herbicides. *Proc. Natl. Acad. Sci. USA* **91**, 3598–3601.

Miernyk, J. A., Thomas, D. R., and Wood, C. (1991). Partial purification and characterization of the mitochondrial and peroxisomal isozymes of enoyl-coenzyme A hydratase from germinating pea seedlings. *Plant Physiol.* **95**, 564–69.

Moore, T. S., Jr. (1982). Phospholipid biosynthesis. *Ann. Rev. Plant Physiol.* **33**, 235–59.

Mudd, J. B. (1980). Phospholipid biosynthesis. In *The Biochemistry of Plants*, Vol. 4, ed. P. K. Stumpf, Academic Press, New York, pp. 249–82.

Mudd, J. B., Andrews, J. E. and Sparace, S. A. (1987). Phosphatidylglycerol synthesis in chloroplast membranes. *Methods Enzymol.* **148**, 338–45.

Murphy, D. J. (1993). Structure, function and biogenesis of storage lipid bodies and oleosins in plants. *Prog. Lipid Res.* **32**, 247–280.

Ohlrogge, J. B., Browse, J. and Somerville C. R. (1991). The genetics of plants lipids. *Biochim. Biophys. Acta.* **1082**, 1–26.

Ohlrogge, J. B., Jaworksi, J. G., and Post-Beittenmiller, D. (1993). *De Novo* fatty acid biosynthesis. In *Lipid Metabolism in Plants*, ed. T. S. Moore, Jr., CRC Press, Boca Raton, pp. 3–32.

Ohlrogge, J. B. (1987). Biochemistry of plant acyl carrier proteins. In *The Biochemistry of Plants*, Vol. 9, eds P. K. Stumpf and E. E. Conn, Academic Press, New York, pp. 137–57.

Ohlrogge, J. B. (1994). Design of new plant products: engineering of fatty acid metabolism. *Plant Physiol.* **104**, 821–26.

Ohnishi, M. and Thompson, G. A., Jr. (1991). Biosynthesis of the unique *trans*-Δ^3-hexadecenoic acid component of chloroplast phosphatidylglycerol: Evidence concerning its site and mechanism of formation. *Arch. Biochem. Biophys.* **288**, 591–99.

Pistelli, L., Perata, P. and Alpi, A. (1992). Effect of leaf senescence on glyoxylate cycle enzyme activities. *Aust. J. Plant Physiol.* **19**, 723–29.

Post-Beittenmiller D, Jaworski J. G., Ohlrogge J. B. (1991). *In vivo* pools of free and acylated acyl carrier proteins in spinach: Evidence for sites of regulation of fatty acid biosynthesis. *J. Biol. Chem.* **266**, 1858–65.

Roughan, P. G. and Slack, C. R. (1982). Cellular organization of glycerolipid metabolism. *Ann. Rev. Plant Physiol.* **33**, 97–132.

Sasaki, Y., Hakamada, K., Suama, Y., Nagano, Y., Furusawa, I. and Matsuno, R. (1993). Chloroplast-encoded protein as a subunit of acetyl-CoA carboxylase in pea plant. *J. Biol. Chem.* **268**, 25118–23

Seifert, U. and Heinz, E. (1992). Enzymatic characteristics of UDP-sulfoquinovose: diacylglycerol sulfoquinovosyltransferase from chloroplast envelopes. *Bot. Acta* **105**, 197–205.

Stumpf, P. K. (1987). The biosynthesis of saturated fatty acids. In *The Biochemistry of Plants*, Vol. 9. eds P. K. Stumpf and E. E. Conn, Academic Press, New York, pp. 121–34.

Stymne, S. and Stobart, A. K. (1987). Triacylglycerol biosynthesis. In *The Biochemistry of Plants*, Vol. 9, eds P. K. Stumpf and E. E. Conn, Academic Press, New York, pp. 175–214.

Stymne, S., Tonnet, M. L. and Green, A. G. (1992). Biosynthesis of linolenate in developing embryos and cell-free preparations of high-linolenate linseed (*Linum usitatissimum*) and low-linolenate mutants. *Arch. Biochem. Biophys.* **294**, 557–63.

Töpfer, R. and Martini, N. (1994). Molecular cloning of cDNAs or genes encoding proteins involved in *de novo* fatty acid biosynthesis in plants. *J. Plant Physiol.* **143**, 416–25.

Tzen, J. T. C. and Huang, A. H. C. (1992). Surface structure and properties of plant seed oil bodies. *J. Cell Biol.* **117**, 327–35.

Voelker, T. A., Worrell, A. C., Anderson, L., Bleibaum, J., Fan, C., Hawkins, D. J., Radke, S. E., Davies, H. M. (1992). Fatty acid biosynthesis redirected to medium chains in transgenic oilseed plants. *Science* **257**, 72–4.

von Wettstein-Knowles, P. M. (1993) Waxes, cutin and suberin. In *Lipid Metabolism in Plants*, ed. T. S. Moore, Jr., CRC Press, Boca Raton, pp 127–66.

28 Terpene biosynthesis and metabolism

Charles A. West

Introduction

It was recognized early that the structures of a large, diverse array of natural products could be rationalized as covalently linked, branched C_5 units related to isoprene (methylbutadiene). The name terpene was derived from one group of these so-called isoprenoid substances that are derived from turpentine and shown to contain two C_5 units. In time, terpene became a generic name for all classes of substances composed of isopentenoid units irrespective of the number present per molecule. The major classes of terpenes are: monoterpenes (C_{10}), sesquiterpenes (C_{15}), diterpenes (C_{20}), sesterterpenes (C_{25}), triterpenes (C_{30}), tetraterpenes (C_{40}) and polyterpenes ($> C_{40}$). Higher plants produce members of all these classes of terpenes. In addition, natural products are known in which a terpenoid moiety is covalently linked with another moiety of different biogenetic origin. Examples of such mixed terpenoids from higher plants include chlorophyll, with its diterpenoid phytyl side chain esterified to the cyclic tetrapyrrole nucleus, and plastoquinone, with a polyprenyl side chain substituted by alkylation on the benzoquinone ring.

The existence of common structural features in all terpenoid substances implies that their biosynthetic origins also share common features. Ruzicka (1953) formulated the 'biogenetic isoprene rule' which predicted the biosynthetic relationships between the various classes of terpenes. Subsequent investigations of the biosynthetic pathways have established the essential correctness of these ideas and have led to the generalized scheme illustrated in Fig. 28.1. A common pathway leads from acetyl-CoA via mevalonate to the central intermediate isopentenyl diphosphate (IPP). IPP is isomerized to dimethylallyl diphosphate (DMAPP) and also participates in the series of chain elongation steps to generate the acyclic precursors of the various classes of terpenes, geranyl diphosphate (GPP) (C_{10}), farnesyl diphosphate (FPP) (C_{15}) and geranylgeranyl diphosphate (GGPP) (C_{20}). These acyclic precursors (prenyl diphosphates) are converted to the various members of the monoterpene, sesquiterpene and diterpene classes by an initial cyclization, often followed by further, mostly oxidative, transformations of the cyclic hydrocarbons. In some instances, rearrangements occur and carbons are lost or gained, resulting in structures in which the regular isoprenoid skeleton is no longer intact. Reductive coupling of two FPP molecules in a 'head-to-head' fashion generates the triterpene, squalene, that in turn can be modified by cyclization and further transformations to the family of triterpenes and related sterols. In analogous fashion, GGPP is the precursor of the tetraterpenes that include the carotenes and xanthophylls. The polyterpenes result from the further elongation of GGPP, or in some cases FPP or GPP more directly, by the addition of units from IPP. Based on considerations evident from Fig. 28.1, terpenes can be defined as those substances that are derived biosynthetically from IPP.

Higher plants produce a diverse array of isoprenoid substances. Some found in most or all plants have a role in fundamental processes of energy metabolism and in macromolecular assemblages required for growth. Sterols, which are required membrane components, and carotenes, which serve as accessory photosynthetic pigments, are examples of such primary terpenoid metabolites. Many other isoprenoid substances are of a more specialized

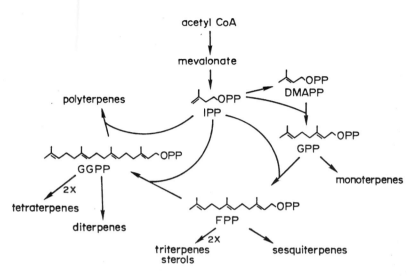

Fig. 28.1 Biosynthetic relationships among the various classes of terpenes. IPP, isopentenyl diphosphate; DMAPP, dimethylallyl diphosphate; GPP, geranyl diphosphate; FPP, farnesyl diphosphate; GGPP, geranylgeranyl diphosphate. PP, diphosphate or pyrophosphate; both names are used in the literature.

nature in that they are found in only a limited range of plants. In cases where the function of such secondary metabolites is known or suspected, they serve a regulatory role in permitting the plant that produces them to adapt to changes in the environment or biological stresses. For example, the sesquiterpenoid and diterpenoid phytoalexins produced by some plants are thought to participate in resistance to infection by microbial pathogens. Isoprenoid growth regulators, such as the gibberellins and abscisic acid, fall somewhere in between these two groups in that they are produced by most or all higher plants and function as regulatory agents for normal growth and development as well as in adaptation to stresses. Many isoprenoid substances in plants, like secondary metabolites in general, are of unknown function.

Table 28.1 lists the major classes of higher plant terpenes and some of their general characteristics.

The biosynthesis of isoprenoid substances is much better understood than their degradative metabolism, although it is clear that many isoprenoid substances do turn over in plant tissues. Also, we know that the metabolism of terpenoid substances in plants must be subject to regulation at the enzyme level, at the level of gene expression, and by features of

compartmentation; however, our understanding of these features of terpene biosynthesis and catabolism are still very fragmentary. This chapter will attempt to summarize with selected examples the current status of understanding of these problems, and will try to show how terpene metabolism integrates with other aspects of plant metabolism.

Biosynthesis of isopentenyl diphosphate

Pathway

The pathway for the conversion of acetyl-CoA to IPP is illustrated in Fig. 28.2. This pathway and the participating enzymes have been studied in detail from yeast and mammalian tissues (Qureshi and Porter, 1981). Two molecules of acetyl-CoA are condensed to form acetoacetyl-CoA in a reaction catalyzed by acetoacetyl-CoA thiolase (step 1). This reaction is made favorable under physiological conditions through its coupling with the strongly favorable synthesis of 3-hydroxy-3-methylglutaryl-CoA (HMG-CoA) from acetyl-CoA and acetoacetyl-CoA catalyzed by HMG-CoA synthase (step 2). Mevalonate is then formed from HMG-CoA in the

Table 28.1 Characteristics of Major Classes of Higher Plant Terpenes

Class	Precursor	Sites of synthesis	Some suggested functions	General references
Cyclic monoterpenes	GPP	Epidermal oil glands Plastids	Attractants Defensive agents C and energy source	Croteau (1987) Croteau (1981)
Cyclic sequiterpenes	FPP	Epidermal oil glands Cytoplasm/ER[a]	Defensive agents Phytoalexins Anti-feeding agents	Cane (1981)
Cyclic diterpenes	GGPP	Cytoplasm/ER[a] Plastids	Defensive agents Phytoalexins Plant hormones- gibberellins	West (1981)
Cyclic triterpenes (sterols)	$2 \times$ FPP	Cytoplasm/ER[a]	Membrane components Defensive agents Plant hormones- brassinosteroids	Goodwin (1981)
Tetraterpenes (carotenes, xanthophylls)	$2 \times$ GGPP	Plastids	Photosynthesis- photoreception and photoprotection Source of plant hormone- abscisic acid	Spurgeon and Porter (1983) Britten (1988)
Prenylated pigments Ubiquionones	FPP or GGPP	Mitochondria	Mitochondrial electron transport	Pennock and Threlfall (1983)
Plastoquinones Tocopherols Phylloquinone Chlorophyll	GGPP	Plastids	Photosynthesis Photosynthetic electron Transport	Pennock and Threlfall (1983)
Polyprenols Dolichols	FPP or GGPP	Mitochondria, Golgi	Biosynthesis of glyconcon- jugates	Hemming (1983)
Rubber	FPP or GGPP	–	–	Benedict (1983)

[a]ER = Endoplasmic reticulum

essentially irreversible reaction involving successive NADPH-dependent reductions at the same catalytic site of 3-hydroxy-3-methylglutaryl-CoA reductase (HMGR; step 3). Two successive phosphoryl transfers from ATP catalyzed by separate kinases (steps 4 and 5) generates 5-diphosphomevalonate. And finally, IPP is generated from 5-diphosphomevalonate by a decarboxylation coupled with dehydration and the obligate hydrolysis of ATP (step 6). The overall stoichiometry of this sequence emphasizes the substrates that must be supplied for IPP synthesis.

$$3 \text{ acetyl CoA} + 2\text{NADPH} + 2\text{H}^+ + 3\text{ATP} + \text{H}_2\text{O} \rightarrow$$
$$\text{IPP} + \text{CO}_2 + 3 \text{ CoASH} + 2\text{NADP}^+ + 3\text{ADP} + \text{P}_i$$

The enzymes responsible for the formation of IPP in higher plants have not been as extensively studied as those from yeast, but the pathway is thought to be the same as that illustrated in Fig. 28.2. A membrane-associated complex of enzymes capable of converting acetyl-CoA to HMG-CoA (steps 1 and 2) has been identified in radish (Bach *et al.*, 1991a). HMGR activity for the synthesis of mevalonate from HMG-CoA (step 3) has been demonstrated in extracts from plant sources, and the primary structures for a

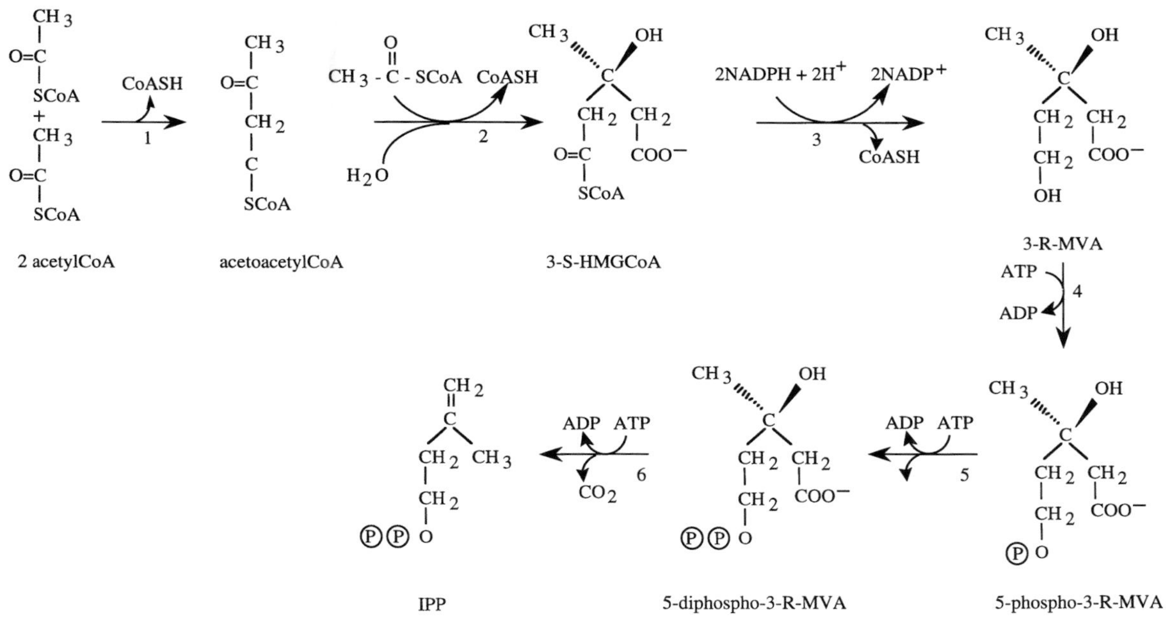

Fig. 28.2 Pathway for the conversion of acetyl-CoA to isopentenyl diphosphate. Catalysts are: step 1, acetoacetyl-CoA thiolase [acetyl-CoA acetyl transferase (EC 2.3.1.9)]: step 2, 3-hydroxy-3-methylglutaryl-CoA synthase (EC 4.1.2.5); step 3, 3-hydroxy-3-methylglutaryl-CoA reductase [mevalonate: NADP$^+$ oxidoreductase acetylating CoA (EC 1.1.1.34)]; step 4, mevalonate kinase [ATP: mevalonate 5-phosphotransferase (EC 2.7.1.36)]; step 5, phosphomevalonate kinase [ATP: 5-phosphomevalonate phosphotransferase (EC 2.7.4.2)]; step 6, diphosphomevalonate decarboxylase [ATP: 5-diphosphomevalonate decarboxylase (dehydrating) (EC 4.1.1.33)]. P, phosphoryl group.

number of plant HMGR species have been deduced from cDNA clones. Analysis of these primary structures indicates the presence of two membrane-spanning domains in the N-terminal region and a highly conserved catalytic domain in the C-terminal region (Bach *et al.*, 1991a). These plant HMGRs resemble closely those from other eukaryotic sources except for the presence of only two membrane-spanning domains instead of seven. The enzymes responsible for the conversion of mevalonate to IPP (steps 4 through 6) have also been shown to be present in many plants.

Compartmentation of the pathway in plants

Early experiments of Goodwin and his associates suggested the possibility that the biosynthesis of IPP and isoprenoid products formed from it were localized in at least two separate compartments in

plant cells (Goodwin and Mercer, 1963). Dark-grown maize seedlings excised from their roots were fed DL-[2-^{14}C]-mevalonate and illuminated for 24 h. Radioactivity was readily detected in squalene, phytosterols, β-amyrin (a pentacyclic triterpene synthesized in plants) and ubiquinone, but not in β-carotene, the phytyl side chain of chlorophyll or plastoquinone, all of which are plastid constituents being rapidly synthesized under these conditions. Conversely, when the precursor supplied in these experiments was ^{14}CO$_2$ instead of [2-^{14}C]-mevalonate, the plastid pigments were readily labeled while squalene, sterol precursors and β-amyrin were not. These and other results were interpreted to suggest the existence of two separate isoprenoid biosynthetic pathways, one in the chloroplast and the other in the cytoplasm/endoplasmic reticulum where sterol synthesis occurs. They speculated that exogenously supplied mevalonate failed to penetrate the chloroplast membranes.

Several lines of evidence that are consistent with the existence of an autonomous pathway for IPP synthesis in plastid compartments were developed in a number of laboratories in ensuing years. The pathway generally proposed to account for IPP synthesis in plastids involves the conversion of triose phosphate generated directly or indirectly from photosynthesis to acetyl-CoA through the actions of glycolytic enzymes and pyruvate dehydrogenase, followed by the transformation of acetyl-CoA to IPP via the reactions depicted in Fig. 28.2. Some of the evidence obtained in support of this pathway includes: (1) the demonstrated presence in plastid preparations of the six enzymes required for the conversion of acetyl-CoA to IPP according to the scheme in Fig. 28.2; (2) significant incorporations of radioactivity from proposed intermediates, $^{14}CO_2$, [^{14}C]-3-phosphoglycerate, [^{14}C]-acetate and [^{14}C]-mevalonate, into β-carotene and plastoquinone-9 in seemingly intact isolated chloroplasts from spinach leaves; and (3) the observation that mevinolin, a potent inhibitor of HMGR, inhibited sterol accumulation in radish seedlings, whereas plastidic prenyl lipid accumulation was unaffected. Leidvogel (1986) presents a more detailed discussion of these and other results and the relevant references. In general, these findings were consistent with the proposal that plastid compartments possess a pathway for IPP synthesis from acetyl-CoA that is distinct from the pathway in the cytoplasm/endoplasmic reticulum compartment.

Not all of the observations, however, were consistent with this view. Kleinig and his associates reported that seemingly intact chloroplasts from spinach and chromoplasts from daffodil readily utilize exogenously supplied IPP for plastid prenyl lipid synthesis, whereas these same organelle preparations are unable to use mevalonate, 5-phosphomevalonate or 5-diphosphomevalonate as precursors. Similar results were obtained with either intact or disrupted etioplasts from mustard seedlings grown under continuous red light. These investigators were unable to detect activities for mevalonate kinase, 5-phosphomevalonate kinase or 5-diphosphomevalonate decarboxylase in isolated etioplasts, whereas these enzyme activities were readily detected in the cytoplasm. To explain these results, it was proposed that these plastids utilize IPP

imported from the cytoplasm as a precursor for plastidic prenyl lipid synthesis because they lack the capacity for IPP synthesis from endogenous precursors within the plastid (Kreuz and Kleinig, 1981; Kleinig, 1989). Further support for this model has come from the report that intact plastids isolated from non-photosynthetic cell suspensions of *Vitis vinifera* are capable of a carrier-dependent facilitated uptake of IPP against a concentration gradient (Soler *et al.*, 1993).

An explanation for these seemingly contradictory sets of observations is suggested by the results of Heintze *et al.* (1990). The activity of sections of barley leaf for the synthesis of chloroplast isoprenoid lipids from CO_2 via plastid-produced mevalonate varies from high in sections from the base of the leaf, which is rich in developing chloroplasts, to quite low in sections from the leaf tip, which contains mature chloroplasts of high photosynthetic efficiency. Conversely, the conversion of supplied [2-^{14}C]-mevalonate to isoprenoid lipids was greatest in leaf sections from the tip region and lowest in sections from the base of the leaf. Apparently, [^{14}C]-IPP (formed from [2-^{14}C]-mevalonate in the cytosol) can be most readily taken up and utilized by plastids which have a low capacity for the endogenous synthesis of mevalonate and IPP. The molecular basis for these differences is not known, although it was ascertained that the pyruvate decarboxylase/dehydrogenase activity is greatly decreased in mature chloroplasts in comparison with its activity in young developing chloroplasts. These findings suggest that the extent to which plastids utilize IPP derived from the cytoplasm/endoplasmic reticulum versus IPP synthesized *de novo* in the plastid may vary depending on the physiological state of the plastids or their source. Young developing chloroplasts would appear to be at one extreme and non-photosynthetic plastids at the other.

Sources of precursors for isopentenyl diphosphate biosynthesis

The summary equation for IPP biosynthesis indicates that three substrates – acetyl CoA, NADPH and ATP – must be available in any cellular compartment where this process is occurring. Mechanisms for maintaining NADPH and ATP levels in the cytoplasm, chloroplasts, and mitochondria are well known. However, the source of acetyl-CoA for use in the cytoplasm/endoplasmic reticulum and in plastids, two compartments considered the most likely sites for IPP synthesis, is less well established. Reviews by Givan (1983) and Leidvogel (1986) discuss this problem and summarize the evidence from the original literature.

The source of acetyl-CoA for use as a substrate for IPP synthesis in the cytoplasm is unclear. Pyruvate dehydrogenase complexes that generate acetyl-CoA appear to be restricted to organelles – mitochondria and in some cases chloroplasts. Since organellar membranes are thought not to be permeable to acetyl-CoA, an indirect transport mechanism would presumably be necessary to provide acetyl-CoA in the cytoplasm from that formed by pyruvate dehydrogenase action within the organelles. There is some evidence for a cytosolic ATP-dependent citrate lyase. The presence of this enzyme in the cytoplasm would present the possibility for citrate generated from acetyl-CoA in mitochondria and transported to the cytoplasm to serve as a source of cytoplasmic acetyl-CoA. An alternative possibility would be to activate acetate to form acetyl-CoA through the action of acetyl-CoA synthetase in the cytoplasm. Although the presence of acetyl-CoA synthetase in spinach leaves, pea seedlings and other plant sources has been documented, it appears to be restricted to the stromal compartment of plastids in the cases examined. Thus, the source of the acetyl-CoA required in the cytoplasmic compartment for IPP synthesis remains unclear.

Two sources of acetyl-CoA for the plastid compartment have been considered most likely. The first route was proposed by Stumpf (1984). A number of investigators had shown that [^{14}C]- acetate is efficiently incorporated into fatty acids in the chloroplast. Stumpf and his associates demonstrated acetyl-CoA hydrolase activity in the mitochondrial matrix and an acetyl-CoA synthetase activity in the stromal compartment of the chloroplast. The pathway they suggested is illustrated in Fig. 28.3A. Acetyl-CoA produced from pyruvate in the mitochondrial matrix is hydrolyzed to free acetate.

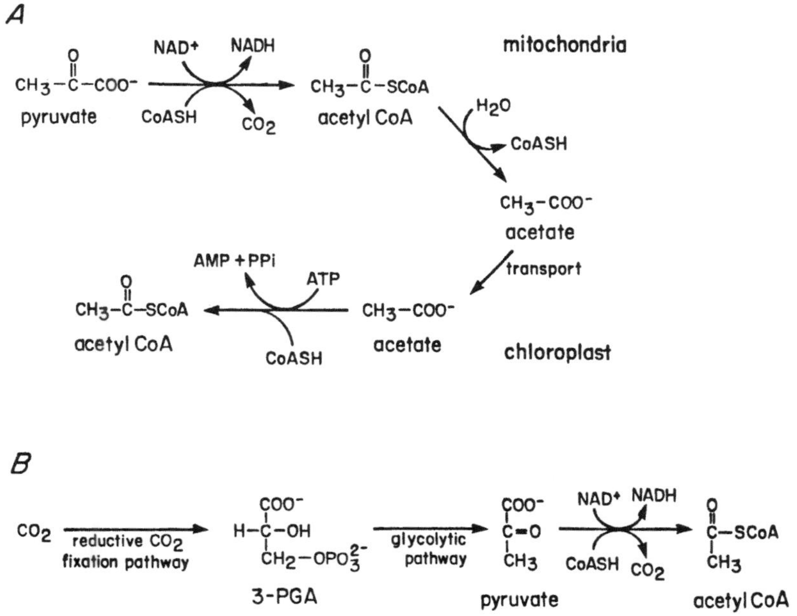

Fig. 28.3 Proposed pathways for acetyl-CoA formation in chloroplasts. (A) Pathway via acetate. (B) Pathway utilizing pyruvate dehydrogenase.

The acetate is then transported into the stromal compartment of the plastid where it is reactivated by acetyl-CoA synthetase to provide acetyl-CoA.

The second proposal (Fig. 28.3B) relates acetyl-CoA formation to photosynthetic CO_2 fixation. As pictured, 3-phosphoglycerate formed by CO_2 fixation in the Calvin–Benson–Basham cycle is directed through a series of glycolytic steps to pyruvate, which then undergoes oxidative decarboxylation through the action of the stromal pyruvate dehydrogenase to yield acetyl-CoA. Evidence has been presented for the presence of all of the required glycolytic enzymes in the chloroplast except for phosphoglyceromutase, which has not been detected to date. Also, it should be noted that the pyruvate dehydrogenase complex has been found in chloroplasts from a number of plants, including pea, but efforts to detect this enzyme in spinach leaf chloroplasts have failed. Obviously, the absence of either of these enzymes in chloroplasts would preclude the operation of the intraplastidic pathway shown in Fig. 28.3B. The very low rates of incorporation of $^{14}CO_2$ into fatty acids in isolated chloroplasts (Givan, 1983) might suggest that this pathway operates to only a limited extent as shown, if at all. An alternative to the scheme shown in Fig. 28.3B might be more likely in view of the questions that have been raised. Triose phosphate that is transported from the chloroplast to the cytoplasm could be transformed via glycolytic enzymes in that compartment into pyruvate. If the pyruvate then re-entered the chloroplast, it could serve as a substrate for chloroplast pyruvate dehydrogenase to yield acetyl-CoA.

The relative importance of the two pathways for the formation of acetyl-CoA in plastids probably differs from plant to plant, and may well differ within a plant, depending on the circumstances. For example, spinach chloroplasts appear to lack a pyruvate dehydrogenase complex and therefore may require the pathway depicted in Fig. 28.3A, whereas pea chloroplasts possess both pyruvate dehydrogenase and acetyl-CoA synthetase, so both pathways may be operational. As noted earlier, developing barley leaf chloroplasts possess pyruvate dehydrogenase activity, whereas this activity is greatly reduced in mature chloroplasts (Heintze et al., 1990). However, the ability to incorporate [^{14}C]-acetate into fatty acids was much the same for barley leaf

chloroplasts at all stages of development. Thus, it is possible that contributions of the two pathways for the synthesis of stromal acetyl-CoA could vary within an individual leaf as a function of the physiological status of the chloroplasts. The extent to which the two pathways are used could depend on the plant species, on the stage of plastid development, and on the relative acetate and pyruvate levels available in the plastid. However, a frequently stated view holds that acetate is quantitatively the most important precursor of acetyl-CoA for most plant plastids.

Regulation of isopentenyl diphosphate biosynthesis

Since the pathway pictured in Fig. 28.2 is the starting point for the biosynthesis of all isoprenoid end-products, one might expect to find regulation of both the activity and levels of regulatory enzymes of the pathway in response to the plant's momentary needs for various end-products. Furthermore, if the pathway operates in different subcellular compartments, as has been suggested, one would expect that the enzymes in each compartment would have their own set of regulatory influences. Unfortunately, the information about regulation of these pathways in plants is still quite fragmentary, so a coherent picture of the control of IPP biosynthesis is not yet available. Some factors that may have regulatory significance are described briefly in this section.

HMGR has received a lot of attention in animal and fungal systems, particularly in connection with the regulation of cholesterol biosynthesis in animals. Not surprisingly, this enzyme has also received most of the attention as a potential regulatory enzyme for isoprenoid biosynthesis in plants, although it is not clear that the reductase occupies the same position of importance as a primary target for regulation in plants. The catalytic activity of plant HMGR does not appear to be regulated by isoprenoid end-products such as the gibberellins or abscisic acid. There have been reports that reductase preparations are inactivated by treatment with ATP, and then reactivated by incubation with phosphatases in a manner suggesting regulation by phosphorylation/dephosphorylation reminiscent of that seen with rat liver reductase. However, this phenomenon has not been examined in detail, nor has its physiological significance been evaluated. Thus, there is little indication that plant reductases are regulated by modulation of their catalytic activities.

Animal HMGR, which participates primarily in sterol biosynthesis, is clearly associated with membranes of the endoplasmic reticulum. The situation in higher plants is more complicated. Evidence was obtained a number of years ago for the association of HMGR activity with chloroplasts and mitochondria a well as with the cytoplasm/endoplasmic reticulum (Brooker and Russell, 1975). Some questions were raised about the validity of these observations because of the low levels of activity seen in the organellar preparations, particularly in mitochondria, the known instability of the catalytic activity *in vitro*, and the possibilities for contamination of the organelle preparations with cytoplasm/endoplasmic reticulum. At the same time, the idea was attractive because of the known role of these organelles, especially plastids, in isoprenoid biosynthesis. Support for the concept has come from more recent studies of HMGR genes and their expression in plants. It is now established for a number of plants that their HMGR genes exist in multigene families. Furthermore, it has been seen that the multiple genes are expressed in a differential fashion during development. The pathways in which the individual gene products participate and their tissue and subcellular compartment localizations have not yet been established. But it seems likely that the different isoforms will have distinctive distributions and specific functions.

A number of environmental factors can influence HMGR activity in plants, including light, growth regulators and phytohormones, herbicides, wounding, infections with pathogens or treatment with fungal elicitors (Bach *et al.*, 1991b). Stimulation by one of these factors in at least some of these cases leads to the selective accumulation of one or more of the HMGR mRNA species in differential fashion. This raises the possibility that HMGR activities in different cell compartments are being regulated at the transcriptional level by specific signals. This progress is encouraging, but many unanswered questions remain to be addressed before a coherent picture of the role of HMGR in the regulation of isoprenoid biosynthesis will emerge.

Two observations point to 5-diphosphomevalonate decarboxylase (Fig. 28.2, step 6) as a possible site of regulation between mevalonate and IPP. In one study, the activity of this enzyme was found to be rate-limiting and was seen to increase most markedly after induction of furanosesquiterpenoid phytoalexin synthesis in sweet potato root tissue by treatment with a fungus or toxic chemical (Oba *et al.*, 1976). In another study, adenylate energy charge was found to regulate the activity of this enzyme in a cell-free enzyme preparation from immature *Marah macrocarpus* seeds that was catalyzing the conversion of mevalonate to the diterpene hydrocarbon *ent*-kaurene (Knotz *et al.*, 1977). The response was typical of that seen for regulatory enzymes in a biosynthetic sequence. No other step was affected by adenylate energy charge. These indications that 5-diphosphomevalonate decarboxylase is rate-limiting and may be regulated are surprising because of its position in a part of the pathway removed from a branch point.

Utilization of isopentenyl diphosphate for the synthesis of prenyl diphosphates.

Isomerase and prenyl transferases

The scheme portrayed in Fig. 28.1 emphasizes the central importance of IPP as the precursor of all classes of terpenes. In Fig. 28.4 are illustrated the reactions responsible for the conversion of IPP to the family of four prenyl diphosphates that serve in turn as precursors of all of the major classes of terpenes.

Dimethylallyl diphosphate is formed from IPP by an isomerization reaction (Fig. 28.4, reaction 1). This involves the addition of a proton to C-4 of IPP coupled with the sterospecific elimination of a proton from C-2 and the resulting shift of the double bond from the 3–4 to the 2-3 position. Isomerase activity is readily detected in plant tissues, but the enzyme has not been very thoroughly characterized from these sources.

Prenyl transferase activities (Fig. 28.4, reactions 2, 3, 4) are responsible for transferring the prenyl unit of a prenyl diphosphate donor to isopentenyl diphosphate to generate a new prenyl diphosphate donor with one additional C_5-prenyl unit in the

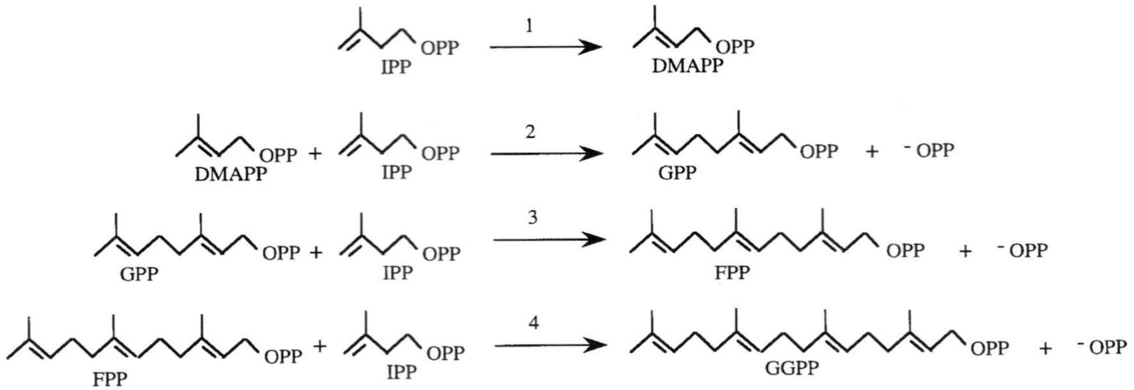

Fig. 28.4 Reactions utilizing IPP in the formation of prenyl diphosphates. Catalysts: reaction 1, isopentenyl diphosphate: dimethylalyl diphosphate isomerase; reaction 2 -dimethylallyl transferase; reaction 3, geranyl transferase; reaction 4 - farnesyl transferase.

chain. The reaction involves the electrophilic addition of a carbonium ion generated at C-1 of the prenyl donor by elimination of diphosphate, to the electron-rich C-4 position of the acceptor IPP molecule. A proton elimination from the adduct establishes the 2–3 double bond of the new prenyl diphosphate product to complete the process.

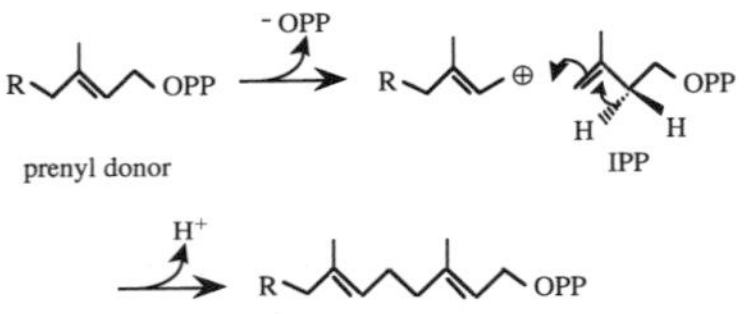

A detailed mechanism that serves as a model for all prenyl transferase reactions has been developed by Poulter and Rilling (1981) on the basis of extensive supporting evidence from studies of farnesyl diphosphate synthase.

GPP synthase, FPP synthase and GGPP synthase have all been identified in plants. GPP synthase accepts only DMAPP at physiological concentrations as a prenyl donor and thus possesses only a single prenyl transferase activity. FPP synthase and GGPP synthase, on the other hand, may possess more than one prenyl transferase activity. For example, liver FPP synthase has been shown to catalyze both dimethylallyl and geranyl transferase reactions at the same catalytic site. No appreciable GPP accumulates with this enzyme during the synthesis of FPP from DMAPP and IPP. The substrate and product

specificities of plant prenyl transferases have been examined in a number of cases. Purified FPP synthases from pumpkin seedlings and castor bean seedlings produce all-*trans*-FPP from either DMAPP or GPP as the prenyl donor with similar efficiencies. In a similar manner, purified GGPP synthases from carrot root, pumpkin seedlings and tomato fruit plastids produce all-*trans*-GGPP from DMAPP, GPP or FPP with similar efficiencies. However, the purified farnesyl transferase from castor bean seedlings infected with fungus used only FPP at physiological substrate concentrations as a prenyl donor (Dudley *et al.*, 1986). It was therefore proposed that this enzyme must function in conjunction with isomerase and FPP synthase to produce GGPP from IPP *in vivo*.

Compartmentation and regulation

GPP synthase has been found in the plastid compartment. The FPP synthases in plant cells are believed to be localized predominantly in the cytoplasm/endoplasmic reticulum region where sterol biosynthesis occurs. GGPP synthase, on the other hand, appears to reside predominantly in plastids. This is consistent with the observed production of the major diterpene-containing pigment, chlorophyll, as well as carotenes and xanthophylls, in plastid compartments. It is possible that smaller amounts of these prenyl transferases are present in other compartments as well. Localization of prenyl transferases in subcellular compartments is doubtless an important feature in the regulation of terpene biosynthesis in higher plants.

There is little at present to suggest that the enzymes participating in the conversion of IPP to prenyl diphosphates are physiologically important sites of regulation.

Pathways for the biosynthesis of terpenoid products from prenyl diphosphates

In spite of their structural and functional diversity, the family of terpenes share many biogenetic features in common. The pathway from acetyl-CoA to IPP is common to all classes of terpenes. The next stage involves a combination of isomerase and prenyl transferases to generate four prenyl diphosphate substrates: DMAPP (C_5), GPP (C_{10}), FPP (C_{15}), and GGAPP (C_{20}). It seems likely that each of the diverse array of terpenoid substances that are found in nature is derived from one of these four prenyl diphosphates plus, in the case of polyprenyl compounds, additional IPP.

The primary reactions for the utilization of prenyl diphosphates for the synthesis of this array of terpenoid products are of four general types

1. *Cyclization reactions.* In the case of the monoterpenes, sesquiterpenes and diterpenes, most of which are carbocyclic compounds, the primary reaction is an intramolecular cyclization using the appropriate prenyl diphosphate as a substrate. Some of these cyclization reactions utilize a proton addition to a carbon–carbon double bond to promote the electrocyclization, whereas others are initiated by the elimination of an allylic diphosphoryl substituent as diphosphate anion.

2. *Coupling reactions.* The primary step in the biosynthesis of triterpenes and tetraterpenes involves the 'head-to-head' covalent coupling of two molecules of the appropriate prenyl diphosphate. The reductive coupling of two FPP molecules yields squalene, the acyclic C_{30} precursor of the triterpenes and sterols. The non-redox coupling of two GGPP molecules yields phytoene, the acyclic C_{40} precursor of the family of carotenes and xanthophylls.

3. *Prenyl transfer reactions.* The synthesis of long chain polyprenyl compounds involves a chain elongating prenyl transferase that catalyzes the processive addition of IPP units to a prenyl-donating prenyl diphosphate primer that can be either GPP or FPP or GGPP, depending on the case.

4. *Prenylation (alkylation) reactions.* This type of reaction is involved in the synthesis of mixed terpenoids in which a prenyl chain is linked to a moiety of different biosynthetic origin, often an aromatic or heterocyclic ring compound. In these cases, a prenyl diphosphate acts as a prenyl donor to alkylate (prenylate) the acceptor. These reactions resemble a prenyl transferase reaction: a prenyl carbocation is first formed from the ionization of the prenyl diphosphate followed by electrophilic attack of the prenyl carbocation on an electron-rich aromatic ring or heteroatom. Ubiquinone, plastoquinine, phylloquinone, tocopherol and cytokinins are examples of mixed terpenoid compounds which require a prenylation step in their biosynthesis. Recently it has been discovered that selected proteins also can be prenylated by the transfer of a prenyl group from either FPP or GGPP to the thiol group of a specific cysteine residue of the protein to form a thioether linkage.

It is noteworthy that none of the reactions involved in the biosynthesis of terpenes from acetyl-CoA through the primary steps of prenyl diphosphate utilization described above are oxidative in nature or require molecular oxygen as a substrate. On the other hand, the end-products of terpene biosynthetic pathways frequently contain oxygenated functional groups and other structural features not represented in the primary products of prenyl diphosphate utilization. It is clear that a secondary phase of most terpene biosynthetic pathways is required to introduce these structural features. The reactions involved are frequently oxidative in nature, including those for the more hydrophobic substrates by mono- and dioxygenases requiring molecular oxygen and others catalyzed by dehydrogenases. Other secondary reactions include cyclizations and rearrangements of the carbon skeletons, desaturation, isomerization and alkylation.

A great deal is now known about the pathways leading to terpenoid products. In some cases, the nature of the enzyme catalysts involved has been investigated, although much remains to be done in this area. A detailed consideration of this complex subject is not possible here. The following sections will elaborate on the very general summary just presented with a few selected examples. It will be necessary to consult the references and monographs cited for a more detailed consideration of the biosynthetic pathways of the major classes of terpenes.

Cyclic diterpene biosynthesis

An outline of the biosynthetic pathway leading to a plant growth regulating gibberellin is shown in Fig. 28.5 to illustrate the general features of biosynthesis of a polycyclic diterpenoid compound. A more detailed consideration of the characteristics of biosynthesis of the gibberellins will be found in the chapter by Jones and MacMillan (1984) and the review by Graebe (1987). All-*trans*-GGPP is the acyclic precursor of the entire family of diterpenes, including the gibberellins. GGPP is cyclized to *ent*-kaurene (steps 1a and 1b) by the successive action of two catalysts, which are known collectively as kaurene synthase. Step 1a involves the proton-initiated cyclization of GGPP to *ent*-copalyl diphosphate (CPP), as shown by the curved arrows to indicate the direction of electron migrations to establish new bonds (Fig. 28.5). CPP is further cyclized in step 1b to form *ent*-kaurene; this reaction involves the elimination of diphosphate anion coupled with a further cyclization and a rearrangement of the carbon skeleton to generate the C and D rings of *ent*-kaurene. Kaurene synthase from the endosperm of immature *Marah macrocarpus* fruit is composed of two separable polypeptides which must function as a complex to catalyze the efficient accumulation of *ent*-kaurene from GGPP without significant accumulation of free CPP (Duncan and West, 1981). This enzyme complex is specific for the production of *ent*-kaurene among the diterpene hydrocarbons. However, CPP can serve with enzymes from other plant sources as a precursor of additional polycyclic diterpenes with different carbon skeletons.

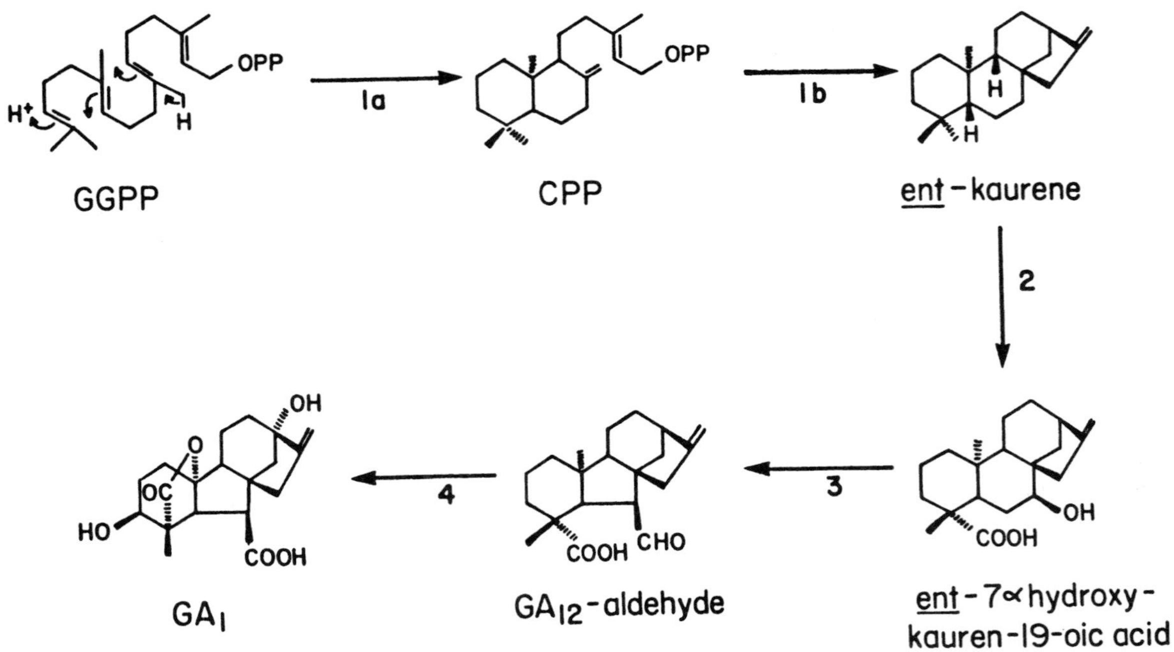

Fig. 28.5 Outline of a gibberellin biosynthesis pathway. CPP, *ent*-copalylPP; GA₁, gibberellin A₁; GA₁₂-aldehyde, gibberellin A₁₂-aldehyde.

The further transformation of *ent*-kaurene to *ent*-7α-hydroxykauren-19-oic acid (Fig. 28.5, step 2) is catalyzed by a series of specific cytochrome P450-dependent, membrane-bound mixed function monooxygenases requiring NADPH and O_2 as co-substrates. The interesting contraction of the B-ring (step 3) producing gibberellin A_{12}-aldehyde, the first intermediate with a gibberellane skeleton, is a reaction of the same type. However, investigations of the oxygenation enzymes catalyzing the interconversions of the gibberellins as part of the complex group of reactions represented by step 4 indicate that for the most part these are soluble dioxygenases requiring Fe^{2+} and 2-oxoglutarate as co-substrates with O_2.

This sequence is generally characteristic of the biosynthesis of cyclic monoterpenes, sesquiterpenes and diterpenes, where an initial intramolecular cyclization of an acyclic prenyl diphosphate is followed by secondary transformations, with an emphasis on oxygenation steps.

Biosynthesis of carotenes

The initial step utilizing prenyl diphosphates in the synthesis of triterpenes and tetraterpenes involves the coupling of two molecules of either FPP or GGPP, respectively, to form a symmetrical coupled precursor. The reaction involved in tetraterpene synthesis is illustrated as the first step in the outline of a tetraterpene pathway in a higher plant (Fig. 28.6). The coupling reaction involves the initial formation of a discrete cyclopropyl carbinyl diphosphate intermediate, called prephytoene diphosphate, which is further rearranged to form *cis*-phytoene, in which the newly synthesized central double bond has the *cis* configuration. The coupling reaction between two FPP molecules that is a part of triterpene synthesis occurs in an analogous manner except that the intermediate presqualene diphosphate undergoes reductive rearrangement in the presence of NADPH to form squalene, which lacks the central double bond.

Further transformations of *cis*-phytoene lead to the tetraterpenes that accumulate in higher plants, including β-carotene and violaxanthin. A series of steps, including the isomerization of the central *cis* double bond to the *trans* isomer and the introduction

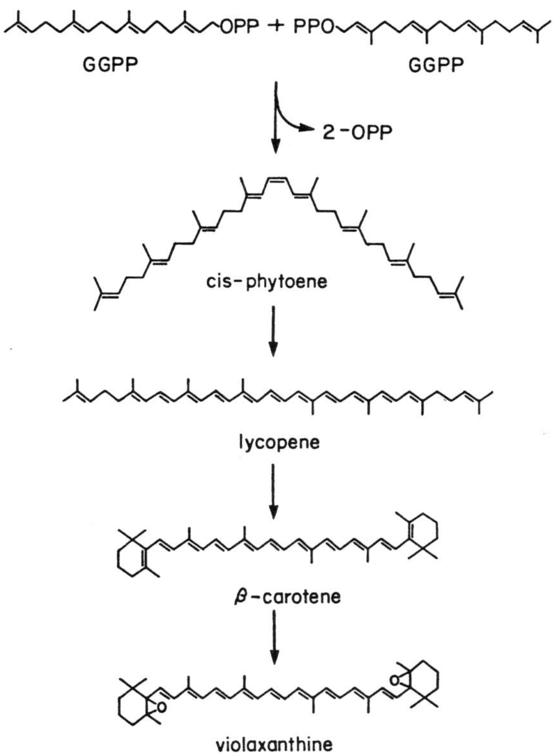

Fig. 28.6 Outline of a tetraterpene biosynthesis pathway.

by two flavoprotein desaturases of four sites of unsaturation in conjugation, is involved in the transformation of *cis*-phytoene to the acyclic tetraterpene, lycopene. Cyclization enzymes introduce the carbocyclic rings at the ends of the chain to produce β-carotene, and O_2-requiring oxygenases catalyze the further conversion of β-carotene to violaxanthin. Variations on these latter steps produce the other carotenes and xanthophylls present in the plastids of higher plants. More detailed accounts of the biochemistry and molecular biology of carotene biosynthesis are given in the reviews by Spurgeon and Porter (1983), Sandmann (1991) and Bartley *et al.*, (1994). The natural plant growth regulator, abscisic acid, is biosynthesized in higher plants by the oxidative cleavage of the C-11,12 double bond of epoxide-containing xanthophylls, such as violaxanthin, to release a C_{15} product (xanthoxin) that is further transformed to abscisic acid.

Rubber biosynthesis

Rubber is a *cis*-polyisoprene polymer ($M_r = 10^5$ to 4×10^6) that accumulates in rubber particles dispersed in the latex of *Hevea brasiliensis* plants or in stem and root cortical parenchyma cells of guayule (*Parthenium argentatum*), and to a lesser extent in other plants. Recently, the presence of a small number of *trans* double bonds have been detected by spectral means in natural rubber in the midst of the predominant *cis* double bonds. This observation prompted a reexamination of the requirements for rubber biosynthesis in isolated systems and the discovery that the rate of IPP incorporation into rubber is greatly stimulated by the inclusion of an all-*trans* prenyl diphosphate, FPP or GGPP, in the incubation mixture. A revised model for rubber biosynthesis has been developed based on these observations (Benedict *et al.*, 1992*)*. In this model, a combination of isomerase and either FPP synthase or GGPP synthase generates a pool of 'initiator' all-*trans* prenyl diphosphate from IPP. Then, another prenyl transferase catalyzes a reaction in which the 'initiator' prenyl diphosphate donates its prenyl moiety to IPP to generate a product with a new prenyl unit added in the *cis* configuration. This prenyl transferase continues to add more IPP molecules in processive fashion until a long *cis* polymer is generated with the 'trans' double bonds of the initiator molecule incorporated at the distal end. The mechanism for termination of polymer growth has not been elucidated. These enzymes of rubber biosynthesis are associated with rubber particles.

According to this view, rubber biosynthesis fits into the pattern for other types of terpenes in that it requires one of the four standard prenyl diphosphates to initiate the synthesis. This is an example of the use of a prenyl transferase to form a polyprenyl compound.

Isoprene biosynthesis

The smallest member of the isoprenoid family is the C_5 metabolite isoprene (2-methylbuta-1,3-diene). Isoprene is a major component of the gaseous hydrocarbon emissions to the atmosphere from a large number of plants. It has been estimated that on the order of $350-400 \times 10^9$ kg of isoprene are emitted globally from plants per year. Although the production of isoprene by plants has been recognized for well over thirty years, its physiological significance for the plants producing it has remained uncertain. This was also true of the biosynthetic origins of isoprene until recently when Silver and Fall (1991) were able to purify and partially characterize an enzyme from alder extracts that catalyzes the conversion of DMAPP to isoprene. The reaction catalyzed by this enzyme involves the elimination of diphosphate anion from DMAPP coupled with the loss of a proton to generate the conjugated diene.

This is analogous to an acid-catalyzed elimination reaction, but the biological process clearly requires a specific enzyme catalyst.

Regulation of terpene biosynthesis

Relatively little is known about the regulation of terpene biosynthetic pathways in higher plants. Since terpenoid compounds serve a wide variety of functions, it is anticipated that many different factors could be involved in regulating terpene biosynthesis. In spite of this, few systematic studies of regulation have been undertaken to clarify the specific mechanisms involved. A critical review of the status of investigations of control of isoprenoid biosynthesis in higher plants has been published (Gray, 1987).

The initial cyclization enzymes appear to be rate-limiting steps in the monoterpene biosynthetic pathways investigated by Croteau and his associates. This suggests the possibility of regulation by metabolic effectors at these steps. However, extensive studies have failed to identify good candidates for intracellular modifier metabolites of these reactions (Croteau, 1987). Also, attempts to identify feed-back inhibitors or other natural effector metabolites for kaurene synthase, the cyclization enzyme complex catalyzing the initial step of gibberellin biosynthesis from GGPP (Fig. 28.5) were similarly negative (Frost and West, 1977). Although these are negative results and are limited to two systems, they suggest that regulation of preformed enzyme by intracellular effectors may not be of such general importance in these types of terpene biosynthetic pathways.

On the other hand, transcriptional activation of genes may be a more important regulatory feature for some terpene biosynthetic pathways. Casbene, a macrocyclic diterpene synthesized from GGPP through the action of a single enzyme, casbene synthase, is produced in appreciable quantities in castor bean seedling extracts only after the seedlings have been in contact with microbial pathogens or elicitor substances derived from them. Casbene has antimicrobial properties and thus has been considered to serve the castor bean plant as a phytoalexin. Transient increases in hybridizable and translatable casbene synthase mRNA precede the accumulation of active enzyme after treatment of the seedlings with an elicitor (Moesta and West, 1986). These results coupled with the results of nuclear run-on experiments, strongly support the idea that the appearance of active casbene synthase, and the potential for casbene biosynthesis, is regulated by elicitor at the level of transcription of the casbene synthase gene (Lois and West, 1990). It seems likely that activation of gene transcription and translation will prove to be an important means of regulation of pathways for the production of secondary terpenoid metabolites like casbene that are produced in response to environmental or biological stresses.

Another example of the regulation of terpene accumulation by induction of biosynthetic enzymes, in this case in response to wounding, is illustrated by the studies of oleoresinosus carried out by Croteau and his associates. Grand fir (*Abies grandis* Lindl) stems respond to wounding by producing oleoresin (pitch), which is a defensive secretion made up of monoterpene olefins (turpentine) and diterpene resin acids (rosin). Funk *et al.* (1994) have reported that the activities of the monoterpene and diterpene cyclases and the two cytochrome P450-dependent diterpenoid hydroxylases involved in the synthesis of the principal resin acid, (–)-abietic acid, are coordinately induced by wounding grand fir stems. As a consequence, an oleoresin with antibiotic properties composed of monoterpene olefins as the solvent in which the water-insoluble diterpenoid resin

acids are dissolved is produced in the resin duct and transported to the sites of wounding by insects. This cleanses the wounded area and, as the volatile turpentine evaporates, leaves a rosin barrier which seals the wound and protects the tree. This interesting defense phenomenon illustrates both the role of induction of terpene biosynthetic enzymes in a response to an external stimulus as a means of regulation and the combined use of two classes of terpenes with different physical and chemical properties in a defensive secretion.

The activity of some terpene biosynthetic pathways is regulated by environmental factors. For example, the synthesis of prenyl lipids in chloroplasts and other plastids – chlorophyll, carotenoids, plastoquinone, α-tocopherol and phylloquinone (vitamin K_1) – are regulated by phytochrome and light treatment (Mohr, 1981). Also, it has been observed that the periodic low temperature treatment at night in field-grown guayule plants strikingly stimulates the formation of rubber while inhibiting the accumulation of oleoresins (Goss *et al.*, 1984). In neither of these cases have the regulatory mechanisms been elucidated at the molecular level. Clearly much remains to be done to develop understanding of the regulation of terpene metabolism.

Catabolism of terpenes

It is evident that at least some end-products of terpene biosynthesis do undergo further metabolic transformations in the plant that produces them. Carotenoids have been shown to undergo chemical changes during leaf senescence, some of which appear to be enzyme-catalyzed (Spurgeon and Porter, 1983). Physiologically active gibberellins are converted to physiologically inactive forms by hydroxylation in the 2β-position, and the 2β-hydroxy derivatives are then subject to further oxidative degradations to unidentified catabolites (Jones and MacMillan, 1984). In many cases, the further metabolic fates of terpenes accumulated in the plant have not been examined. One notable exception involves the cyclic monoterpenes that accumulate in oil glands. Croteau (1987) has shown that these monoterpenes are in a state of metabolic flux, and has undertaken a detailed examination of the catabolic pathways involved in

Salvia and *Mentha* species. This long-term turnover results in a net decrease in the monoterpene content at a late stage in development during which the oil gland is undergoing ultrastructural changes characteristic of senescence. (−)-Menthone, the major terpene component in the leaf, is converted by reduction to (−)-menthol and (+)neomenthol. These two products are conjugated to form (−)-menthyl acetate and (+)-neomenthyl-β-D-glucoside in the leaf. It was further demonstrated that the glucoside is transported in large part to the rhizome where it undergoes catabolism via a pathway that has been elucidated. The glycoside is first hydrolyzed by a glucosidase, the resulting (+)-neomenthol is oxidized to (−)-menthone, and the latter is oxidized to menthone lactone, which is subjected to a β-oxidation scheme yielding acetyl-CoA as the end-product. This acetyl-CoA is available for reincorporation into fatty acids and lipids, or for oxidative catabolism to yield energy and other biosynthetic intermediates. Thus, this might be viewed as a salvage pathway during leaf senescence in which a portion of the carbon that had been stored in abundant monoterpenes of the leaf can be recovered in a form for recycling by a non-senescing tissue. An analogous pathway seems to operate for the recovery of camphor carbon from oil glands in sage.

Structures of enzymes of plant terpene metabolism

It has become possible to deduce primary structures for proteins in cases where the cDNA sequence can be obtained. This approach is being utilized to learn the structures of enzymes of plant terpene metabolism, but only a limited number of examples are available at this time. The general similarity of structures of plant HMGR species to HMGR enzymes from other eukaryotes has been noted earlier. A comparison of the deduced primary structures for three cyclization enzymes involved in terpene biosyntheses – limonene synthase from spearmint (Colby *et al.*, 1993), *epi*-aristolochene synthase from tobacco (Facchini and Chappell, 1992), and casbene synthase from castor bean (Mau and West, 1994) – has revealed a striking level of conservation in the amino acids found in comparable positions of the aligned structures.

Limonene is a monoterpene, *epi*-aristolochene is a sesquiterpene, and casbene is a diterpene. Even though their substrates differ in size, all three enzymes are thought to catalyze mechanistically analogous cyclization reactions in which diphosphate anion elimination is accompanied by the formation of a cyclic intermediate with the participation of electrons from the distal double bond of the substrate. Since these enzymes come from plants in three different plant families, the homology between them suggests the possibility that they arose by divergent evolution from an ancestral gene that existed prior to angiosperm speciation.

References

Bach, T. J., Boronat, A., Caelles, C., Ferrer, A., Weber, T. and Wettstein, A. (1991a). Aspects related to mevalonate biosynthesis in plants. *Lipids* **26**, 637–48.

Bach, T. J., Wettstein, A., Boronat, A., Ferrer, A., Enjuto, M., Gruissem, W. and Narita, J. O. (1991b). Properties and molecular cloning of plant HMG-CoA reductase. In *Physiology and Biochemistry of Sterols*, eds G. W. Patterson and W. D. Nes, American Oil Chemists Society, Champaign, IL, pp. 29–49.

Bartley, G. E., Scolnik, P. A. and Giuliano, G. (1994). Molecular biology of carotenoid biosynthesis in plants. *Annu. Rev. Plant Physiol. Plant Mol. Biol.* **45**, 287–301.

Benedict, C. R. (1983). Biosynthesis of rubber. In *Biosynthesis of Isoprenoid Compounds*, Vol. 2, eds J. W. Porter and S. L. Spurgeon, John Wiley and Sons, New York, pp. 355–69.

Benedict C. R., Madhavan, S., Greenblatt, G. A., Venkatachalem, K. V. and Foster, M. A. (1992). The enzymatic synthesis of rubber polymer in *Parthenium argentatum* Gray. *Plant. Physiol.* **92**, 816–21.

Britten, G. (1988). Biosynthesis of carotenoids. In *Plant Pigments*, ed. T. W. Goodwin, Academic Press, London, pp. 133–82.

Brooker, J. D. and Russell, D. W. (1975). Subcellular localization of 3-hydroxy-3-methylglutaryl Coenzyme A reductase in *Pisum sativum* seedlings. *Arch. Biochem. Biophys.* **167**, 730–37.

Cane, D. E. (1981). Biosynthesis of sesquiterpenes. In *Biosynthesis of Isoprenoid Compounds*, Vol. 1, eds J. W. Porter and S. Spurgeon, John Wiley and Sons, New York, pp. 283–374.

Colby, S. M., Alonso, W. R., Katahira, E., McGarvey, D. J. and Croteau, R. (1993). 4-*S*-Limonene synthase from the oil glands of spearmint (*Mentha spicata*). *J. Biol. Chem.* **268**, 23016–24.

Croteau, R. (1981). *Biosynthesis of Isoprenoid Compounds,*

Vol. 1, eds J. W. Porter and S. Spurgeon, John Wiley and Sons, New York, pp. 225–82.

Croteau, R. (1987). Biosynthesis and catabolism of monoterpenoids. *Chem. Rev.* **87**, 929–54.

Dudley, M. W., Green, T. R., and West, C. A. (1986). Biosynthesis of the macrocyclic diterpene casbene in castor bean (*Ricinus communis* L.) seedlings. The purification and properties of farnesyl transferase from elicited seedlings. *Plant Physiol.* **81**, 343–48.

Duncan, J. D. and West C. A. (1981). Properties of kaurene synthetase from *Marah macrocarpus* endosperm: evidence for the participation of separate but interacting enzymes. *Plant Physiol.* **68**, 1128–34.

Facchini, P. J. and Chappell, J. (1992). Gene family for an elicitor-induced sesquiterpene cyclase in tobacco. *Proc. Nat. Acad. Sci. USA* **89**, 11088–92.

Frost, R. G. and West C. A. (1977). Properties of kaurene synthetase from *Marah macrocarpus*. *Plant Physiol.* **59**, 22–29.

Funk, C., Lewinsohn, E., Vogel, B. S., Steele, C. L. and Croteau, R. (1994). Regulation of oleoresinosus in Grand Fir (*Abies grandis*): Coordinate induction of monoterpene and diterpene cyclases and two cytochrome P450-dependent diterpenoid hydroxylases by stem wounding. *Plant Physiol.* **106**, 999–1005.

Givan, C. V. (1983). The source of acetyl coenzyme A in higher plants. *Physiologica Plantarum* **57**, 311–16.

Goodwin, T. W. (1981). Biosynthesis of plant sterols and other triterpenoids. In *Biosynthesis of Isoprenoid Compounds*, Vol. 1, eds J. W. Porter and S. Spurgeon, John Wiley and Sons, New York, pp. 443–80.

Goodwin, T. W. and Mercer, E. I. (1963). The regulation of sterol and carotene metabolism in germinating seedlings. *Biochemical Society Symposia*, **24**, 37–41.

Goss, R. A., Benedict, C. R., Keithley, J. H., Nessler, C. L., Madhavan, S. and Stipanovic, R. D. (1984). *cis*-Polyisoprene synthesis in guayule (*Parthenium argentatum* Gray) exposed to low, non-freezing temperatures. *Plant Physiol.* **74**, 534–7.

Graebe, J. E. (1987). Gibberellin biosynthesis and control. *Annu. Rev. Plant Physiol.* **38**, 419–65.

Gray, J. C. (1987). Control of isoprenoid biosynthesis in higher plants. *Adv. Botan. Res.* **14**, 25–91.

Heintze, A., Görlach, J., Leuschner, C., Hoppe, P., Hagelstein, P., Schulze-Siebert, D. and Schultz, G. (1990). Plastidic isoprenoid synthesis during chloroplast development. Change from metabolic autonomy to division-of-labor stage. *Plant Physiol.* **93**, 1121–7.

Hemming, F. W. (1983). Biosynthesis of dolichols and related compounds. In *Biosynthesis of Isoprenoid Compounds*, Vol. 2, eds J. W. Porter and S. L. Spurgeon, John Wiley and Sons, New York, pp. 305–54.

Jones, R. L. and MacMillan, J. (1984). Gibberellins. In *Advanced Plant Physiology*, ed. M. Wilkins, Pitman Publishing Ltd., London, pp. 21–52.

Kleinig, H. (1989). The role of plastids in isoprenoid biosynthesis. *Annu. Rev. Plant Physiol. Plant Mol. Biol.* **40**, 39–59.

Knotz, J., Coolbaugh, R. C. and West. C. A. (1977). Regulation of the biosynthesis of *ent*-kaurene from mevalonate in the endosperm of immature *Marah macrocarpus* seeds by adenylate energy charge. *Plant Physiol.* 60, 81–85.

Kreuz, K. and Kleinig, H. (1981). On the compartmentation of isopentenyl diphosphate synthesis and utilization in plant cells. *Planta* 153, 578–81.

Liedvogel, B. (1986). Acetyl Coenzyme A and isopentenyl pyrophosphate as lipid precursors in plant cells – biosynthesis and compartmentation. *J. Plant Physiol.* **124**, 211–22.

Lois, A. F. and West, C. A. (1990). Regulation of expression of the casbene synthase gene during elicitation of castor bean seedlings with pectic fragments. *Arch. Biochem. Biophys.* **276**, 270–77.

Mau, C. J. D. and West, C. A. (1994). Cloning of casbene synthase cDNA: Evidence for conserved structural features among terpenoid cyclases in plants. *Proc. Nat. Acad. Sci. USA* **91**, 8497–501.

Moesta, P. and West, C. A. (1986). Casbene synthetase: regulation of phytoalexin biosynthesis in *Ricinus communis* L. seedlings. Purification of casbene synthetase and regulation of its biosynthesis during elicitation. *Arch. Biochem. Biophys.* **238**, 325–33.

Mohr, H. (1981). Control of chloroplast development by light – some recent aspects. In *Photosynthesis V. Chloroplast Development*, ed. G. Akoyunglow, Balaban International Science Services, Philadelphia, PA, pp. 869–83.

Oba, K., Tatematsu, H., Yamashita, K. and Uritani, I. (1976). Induction of furano-terpene production and formation of the enzyme system from mevalonate to isopentenyl pyrophosphate in sweet potato tissue injured by *Ceratocystis fimbriata* and by toxic chemicals. *Plant Physiol.* **58**, 51–6.

Pennock, J. F. and Threlfall, D. R. (1983). Biosynthesis of ubiquinone and related compounds. In *Biosynthesis of Isoprenoid Compounds*, Vol. 2, eds J. W. Porter and S. L. Spurgeon, John Wiley and Sons, New York, pp. 191–303.

Poulter, C. D. Rilling, H. C. (1981). Prenyl transferases and isomerase. In *Biosynthesis of Isoprenoid Compounds* Vol. 1, eds J. W. Porter and S. L. Spurgeon, John Wiley and Sons, New York, pp. 161–224.

Qureshi, N. and Porter J. W. (1981). Conversion of acetyl-Coenzyme A to isopentenyl pyrophosphate. In *Biosynthesis of Isoprenoid Compounds,* Vol. 1, eds J. W. Porter and S. Spurgeon, John Wiley and Sons, New York, pp. 47–94.

Ruzicka, L. (1953). The isoprene rule and the biogenesis of terpenic compounds. *Experientia* **9**, 357–67.

Sandmann, G. (1991). Biosynthesis of cyclic carotenoids: biochemistry and molecular genetics of the reaction sequence. *Physiologica Plantarum* **83**, 186–93.

Silver, G. M. and Fall, R. (1991). Enzymatic synthesis of isoprene from dimethylallyl diphosphate in aspen leaf extracts. *Plant Physiol.* **97**, 1588–91.

Soler, E., Clastre, M., Bantignies, B., Marigo, G. and Ambid, C. (1993). Uptake of isopentenyl diphosphate by plastids isolated from *Vitis vinifera* L. cell suspensions. *Planta* **191**, 324–29.

Spurgeon, S. L. and Porter, J. W. (1983). Biosynthesis of carotenoids. In *Biosynthesis of Isoprenoid Compounds*, Vol. 2, eds J. W. Porter and S. L. Spurgeon, John Wiley and Sons, New York, pp. 1–122.

Stumpf, P. K. (1984). Fatty acid biosynthesis in higher plants. In *Fatty Acid Metabolism and Its Regulation,* ed. S. Numa, Elsevier Science Publishers, Amsterdam, New York, Oxford, pp. 155–179.

West, C. A. (1981). Biosynthesis of diterpenes. In *Biosynthesis of Isoprenoid Compounds,* Vol. 1., eds J. W. Porter and S. L. Spurgeon, John Wiley and Sons, New York, pp. 375–411.

Nitrogen Metabolism

29 The molecular biology of N metabolism*

C. P. Vance

Introduction

Nitrogen is the major limiting nutrient for most plant species. Acquisition and assimilation of nitrogen is second in importance only to photosynthesis for plant growth and development. Seed viability and germination are directly related to nitrogen content. Production of high-quality, protein-rich food is extremely dependent upon availability of sufficient nitrogen. Clearly, the crucial role that nitrogen plays in plant growth requires that physiologists understand the biochemical and molecular events that regulate nitrogen metabolism.

Plants acquire nitrogen from two principal sources: (1) the soil, through commercial fertilizer, manure, and/or mineralization of indigenous organic matter; and (2) the atmosphere, through symbiotic N_2 fixation. Soil-derived nitrogen, generally in the form of nitrate (NO_3^-), and atmospheric N_2 must be reduced to ammonia (NH_4^+) to become available for amino acid and protein synthesis. Nitrate is reduced to NH_4^+ by the plant enzymes nitrate reductase (NR) and nitrite reductase (NiR) while atmospheric N_2 is reduced by the microbial enzyme nitrogenase.

$$NO_3^- + NAD(P)H + H^+ \rightarrow NO_2^- + NAD + H_2O$$
(Nitrate reductase) [29.1]

$$NO_2^- + 6e^- + 8H^+ \rightarrow NH_4^+ + 2H_2O$$
(Nitrite reductase) [29.2]

$$N_2 + 16ATP + 8e^- + 10H^+ \rightarrow 2NH_4^+ + 16$$
$$ADP + 16P_i + H_2 \text{ (Nitrogenase)} \qquad [29.3]$$

Irrespective of the source, in higher plants the reduced form of nitrogen ultimately available for direct assimilation is NH_4^+ (Lea *et al.*, 1990). Although the predominant sources of NH_4^+ are from NO_3^-

reduction and symbiotic N_2 fixation, significant amounts are also derived from photorespiration and phenylpropanoid biosynthesis. The rate of NH_4^+ assimilation in tissues varies as a function of organ development, environmental conditions, nutritional status, and genotype or species. Since NH_4^+ is generally toxic to plant cells at high concentrations, it must be rapidly assimilated into non-toxic amino acids.

Over the past 25 years NH_4^+ assimilation has been the focus of intense study. It is now generally agreed that the primary assimilation of NH_4^+ into amino acids occurs via the concerted action of four enzymes (Fig. 29.1). The initial reaction involves the ATP-dependent amination of glutamate to glutamine by glutamine synthetase (GS). The next and collaborative step, catalyzed by ferredoxin (Fd-) or NADH-glutamate synthase (GOGAT), involves the reductive transfer of the amide-amino group of glutamine to α-ketoglutarate to yield two glutamates. These two reactions are collectively referred to as the GS/GOGAT cycle. Glutamate at this point can either be used to replenish the pool of glutamate available for GS or to donate its N to form other amino acids and nitrogenous compounds such as alkaloids, ureides, and polyamines. A second tier of control for NH_4^+ assimilation is represented by the enzymes aspartate aminotransferase (AAT) and asparagine synthetase (AS), which control the flow of carbon (C) between

*Joint contribution of the United States Department of Agriculture-Agricultural Research Service and the Minnesota Agricultural Experiment Station. This work was supported in part by United States Department of Agriculture CRGO grant #94-37305-0575, and NSF grant#JBN-9206890. Paper 21,678 Scientific Journal Series, Minnesota Agricultural Experiment Station.

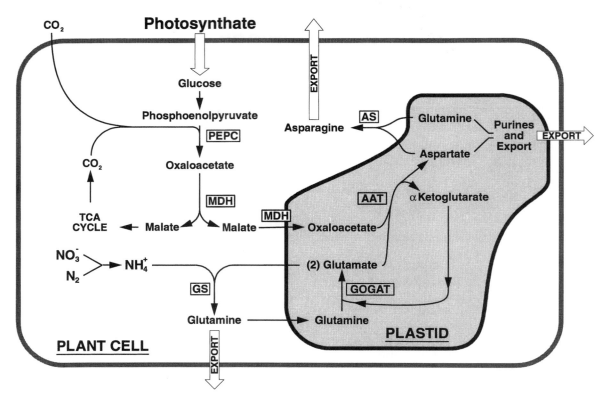

Fig. 29.1 A general scheme for nitrogen assimilation in higher plants and the enzymes involved. In this particular scheme, glutamine, asparagine, and ureides derived from purines are the primary nitrogenous compounds transported to other cells and plant organs. Photosynthate is used via glycolysis and the TCA cycle to generate carbon skeletons for amino acid biosynthesis. Substantial carbon for amino acids can also be derived from nonphotosynthetic CO_2 fixation via phosphoenolpyruvate. Enzymes involved are within boxes: AAT, aspartate aminotransferase; AS, asparagine synthetase; GOGAT, glutamate synthase; GS, glutamine synthetase; MDH, malate dehydrogenase, PEPC, phosphoenolpyruvate carboxylase.

amino and organic acids and catalyze the synthesis of the key amino acids aspartate and asparagine. Glutamate dehydrogenase (GDH) offers an alternative route for the assimilation of NH_4^+ equation [29.4]. However, it now appears that this ubiquitous enzyme functions in glutamate catabolism during growth and/or senescence rather than N assimilation.

$$\alpha\text{-ketoglutarate} + NH_4^+ + NAD(P)H \leftrightarrow$$
$$\text{glutamate} + NAD(P) + H_2O \qquad [29.4]$$

The C skeletons required for the initial assimilation of NH_4^+ are derived from the tricarboxylic acids α-ketoglutarate and oxaloacetate. Hence, nitrogen assimilation is inextricably linked to C metabolism.

Phosphoenolpyruvate carboxylase (PEPC) catalyzing the carboxylation of PEP to oxaloacetate, provides a substantial amount of C to replenish the organic acid pool and for the synthesis of aspartate (Day and Copeland, 1991). Any consideration of NH_4^+ assimilation must include this key enzyme (Fig. 29.1).

This chapter will attempt to familiarize the reader with molecular aspects relating to nitrogen metabolism. The aim is not to present a comprehensive review, but more to provide a working knowledge of the current status of the topic. Because some steps in nitrogen metabolism have received more attention than others, this chapter will of necessity not devote equal space to each reaction.

Nitrate reductase

Characteristics

Nitrate reductase (NR, equation [29.1]) catalyzes the first step in the reduction of NO_3^- to NH_4^+. The enzyme occurs as three forms NADH-NR, NAD(P)H-NR, and NADPH-NR, with NADH-NR as the predominate form in green tissues (Pelsy and Caboche, 1992). By comparison NAD(P)H-NR is found in all tissues, particularly roots. Fungi contain NADPH-NR. The NR enzyme is a large, complex protein containing FAD, cytochrome b_{557} and molybdenum (MoCo) as prosthetic groups. Several plant and fungal NRs have been purified and have characteristics in common. The most well characterized plant NRs have been isolated from barley (*Hordeum vulgare* L.), squash (*Curcurbita maxima* L.), spinach (*Spinacia oleracea* L.), and tobacco (*Nicotiana tabacum* L.). Consensus indicates that the enzyme is a homodimer with a native molecular mass of 200–230 kD and a subunit mass of 100–115 kD (Warner and Kleinhofs, 1992). The NR holoenzyme contains two each of FAD, cytochrome b_{557}, and MoCo (Fig. 29.2). Most immunogold localization studies indicate that the enzyme is located in the cytosol (Hoff *et al.*, 1992).

The NR mechanism of action is proposed to be a two site ping-pong model. The prosthetic groups, in actuality redox centers, occur structurally as three

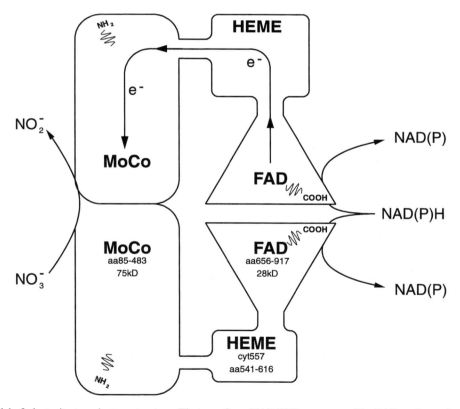

Fig. 29.2 Model of plant nitrate reductase structure. Electrons from NAD(P)H are accepted by FAD and transferred directly to cyt.$_{b557}$ and then used at the molybdenum cofactor (MoCo) site to reduce NO. $FMNH_2$ can also donate electrons for the direct reduction of NO_3^-. The enzyme is composed of two identical subunits of approximately 110 kD MW, two FAD, two cyt$_{b557}$, and two MoCo. The molybdenum cofactor (MoCo) is found at the amino-terminus of the polypeptide (aa #85-483) while the FAD binding portion of the polypeptide is found at the carboxy-terminus (aa #656-916).

distinct domains and function as an electron transport system. Reduced NAD(P)H donates two electrons to the initial acceptor, FAD. Electrons then flow through the cytochrome b_{557} site to the MoCo domain, where they are donated to NO_3^- reducing it to NO_2^-. NR also has partial activities that involve one or more of the prosthetic groups. One partial reaction involves the dehydrogenase activity in which NAD(P)H acts as the electron donor for the reduction of ferricyanide and cytochrome C, while the other involves the reduction of NO_3^- to NO_2^- using electron donors such as methylviologen and bromphenol blue. Although these partial reactions can occur *in vitro*, they are probably of limited importance in the plant cell.

Isolation of cDNAs and genomic clones for NR has provided significant insight into the primary structure of the enzyme (Crawford and Arst, 1993). The FAD region is approximately 28 kD in molecular mass and is found at the carboxy-terminus of the deduced amino acid sequence. The heme prosthetic domain is approximately 14 kD in size and occupies amino acid positions 541–620 between the FAD site and MoCo. The MoCo domain is approximately 75 kD in molecular mass and is found at the amino-terminus of the protein (Fig. 29.2). The deduced amino acid sequence identity among plant NRs varies from about 60 to 90%.

Nitrate reductase genes have been cloned and sequenced from tomato, barley, tobacco, rice, and *Arabidopsis* (Crawford and Arst, 1993; Hoff *et al.*, 1992; Pelsy and Caboche, 1992). Genomic Southern blots indicate that there are one to three NR genes per haploid genome depending upon the species. The structural organization of the NR genes is well conserved among plants. Generally, NR genes have three introns and four exons. The introns are found only in the MoCo domain of the gene and vary in size from 50 to 1950 bp. Studies with NR promoter-reporter gene constructs have shown that the elements controlling the NO_3^- and light response are found in the 5′ flanking region upstream from the transcription start site.

The MoCo is required for full NR activity. Mutants either defective in or lacking MoCo lack NR activity. Activity from such mutants can be restored through *in vitro* complementation with MoCo from other sources. For example, when barley *nar2*

mutants, which are defective in MoCo and lack NR activity, are used as the apoprotein source and complemented *in vitro* with MoCo from xanthine oxidase, over 70% of the NR activity is restored (Warner and Kleinhofs, 1992). All molybdoenzymes tested to date, except nitrogenase, have MoCos that are active in the *in vitro* complementation test.

Genetic regulation

Expression and genetic regulation of NR have been evaluated through: (1) selection of plant mutants lacking NR activity; (2) use of antibodies to assess NR polypeptide expression both spatially and temporally; (3) characterization of NR transcript levels using cDNAs; and (4) NR gene promoter-GUS fusions. Each approach provides unique information about NR and how the enzyme is regulated.

Nitrate reductase deficient mutants have been isolated in *Hordeum*, *Arabidopsis*, *Nicotiana*, *Glycine*, *Pisum*, *Datura*, *Rosa*, *Petunia*, and *Hyoscyamus*. These mutants have provided a wealth of information on biochemical and genetic control of NR. Both apoprotein and MoCo mutants have been isolated. In general, mutational analysis indicates that most plants have two or more functional NR structural genes (Pelsy and Caboche, 1992; Warner and Kleinhofs, 1992). Mutation analysis in barley and *Arabidopsis* has shown that two loci control the synthesis of NR apoenzyme. The barley *nar1* locus codes for the NADH-NR structural gene, which is expressed primarily in shoots. Mutations in *nar1* reduce NR activity by 90% yet there is little effect on barley growth (Crawford and Arst, 1993; Hoff *et al.*, 1992). The *nar1* mutation has little to no effect on barley growth because the NAD(P)H-NR bispecific enzyme that is usually expressed in roots appears to be expressed at elevated levels in leaves of the *nar1* mutant. The barley NAD(P)H-NR is encoded by the *nar7* locus. Double mutations involving *nar1* and *nar7* have no NR activity and grow poorly on NO_3^-. Similarly, in *Arabidopsis*, mutations in the *nia2* gene that encodes a NADH-NR reduces NR activity by 90% yet the plants grow normally. This is because another NR structural gene designated *nia1* that usually comprises only 10% of total NR activity in *Arabidopsis* could compensate for mutations in *nia2*.

A *nai1 nai2* double mutant had almost no NR activity and grew poorly on NO_3^-. These data suggest that there is a 10-fold excess of NR activity in normal plants and that the long-held dogma that NR is the limiting enzyme in NO_3^- assimilation is inaccurate. This excess NR activity is probably the reason that agronomic selection for NR activity has not led to improved plant performance.

Synthesis of the MoCo cofactor in barley and *N. plumbaginifolia* is controlled by six separate loci (Pelsy and Caboche, 1992). Six loci have also been identified in fungi that affect MoCo synthesis. Although the functions of these loci are unknown, they could be involved in assembly of MoCo into the enzyme, molybdopterin synthesis or other aspects of MoCo formation.

In all plants, NR activity increases dramatically in response to applied NO_3^- (Crawford and Arst, 1993; Lillo, 1994). It is one of the few truly inducible plant enzymes. As NO_3^- is depleted or removed, NR activity diminishes. Light is also required for the induction of high levels of NR in leaves of higher plants. NR activity is induced to low levels in dark treated and/or etiolated plants exposed to NO_3^-. Upon exposure to light, NO_3^- treated plants accumulate maximum amounts of NR activity. As NO_3^- is depleted or removed, NR activity diminishes. Enhanced NR activity results from increased synthesis of NR mRNA and protein. Several studies have shown that NR mRNA is low to non-detectable prior to addition of NO_3^-. However, shortly after addition of NO_3^-, NR mRNA increases several-fold, as the result of increased transcription.

Western blotting and immunoprecipitation of *in vitro* translation products further confirm that NO_3^- induction of NR activity is due to synthesis of NR protein and mRNA. When RNA is isolated from seedings grown with and without NO_3^- and translated *in vitro*, immunoprecipitates using NR antibodies showed a striking increase in the 110 kD NR band in NO_3^--induced as compared to uninduced seedlings. Likewise, proteins that cross-react with NR antibodies are not detectable in roots or shoots of seedlings grown without NO_3^- (Lillo, 1994; Warner and Kleinhofs, 1992). Polypeptides which cross-react with NR antibodies appear within 10 h after exposure to NO_3^- and increased NR activity is directly related to the appearance of NR polypeptides. Then as NO_3^- is depleted, NR polypeptides decline and NR activity is reduced. Taken inclusively, these data indicate that NR is, in part, regulated by *de novo* synthesis of NR mRNA and protein.

The effect of light on NR regulation is complex and may involve several effectors. As mentioned previously, in photosynthetic tissues light is required for NO_3^- induced synthesis of NR mRNA and activity. Superimposed upon this aspect, light is also involved in the circadian rhythm patterns observed for NR mRNA and activity. Steady state levels of NR mRNA increase during the night reaching a peak either near the end of the dark period or at the beginning of the light period and then decline precipitously throughout the day to nearly undetectable levels at the end of the light period (Hoff *et al.*, 1992; Lillo, 1994). By comparison, NR activity increases sharply upon illumination and is maintained at relatively high levels until late in the light cycle. Upon darkening NR activity declines rapidly. Thus the diurnal cycle of NR mRNA transcript level does not strictly correspond to NR activity and protein, which suggests a post-translational form of control involving photoactivation of existing NR enzyme.

Recently the light/dark modulation of NR activity in spinach has been shown to involve reversible phosphorylation (Huber *et al.*, 1992). Similar to sucrose-phosphate synthase (SPS), NR is partially dephosphorylated upon illumination resulting in activation of the protein. Okadaic acid and microcystin prevent the rapid light-induced activation of NR suggesting that protein dephosphorylation is catalyzed by a protein phosphatase 2A. In darkness, NR is phosphorylated at multiple seryl residues by a protein kinase. The resultant phosphorylated enzyme is covalently modified, much more sensitive to Mg^{++} inhibition, and has greatly reduced activity. Whether the same phosphatase and kinase which act on NR also act on SPS is unknown. However, since N and C metabolism are so tightly coupled, it has been speculated that such regulation would be convenient for plant control of metabolism.

Metabolites of CO_2 fixation and N assimilation also appear to play a role in regulation of NR gene expression (Cheng *et al.*, 1992; Vincentz *et al.*, 1993). Sucrose can replace light in the induction of NR mRNA in *Arabidopsis*. When 16-day-old etiolated plants grown in the presence of NO_3^- were supplemented with sucrose, high levels of NR mRNA accumulated. In plants not supplemented with sucrose, little mRNA accumulation occurred. Also with the use of a reporter gene transformed into *Arabidopsis* it was shown that the sucrose and light response domain of NR was located in the 2.7 kb 5' flanking sequence of the gene. Likewise, addition of sucrose, glucose, or fructose to dark-adapted *N. plumbaginifolia* leaves resulted in induction of NR mRNA and activity. Also glucose and light induction were controlled by a 1.35 kb region of the 5' flanking sequence of the *Nicotiana* NR gene. A further role for sugars in NR regulation was shown by the fact that glucose abolished the circadian rhythm fluctuations of NR transcription in *N. plumbaginifolia*. These data indicate that carbon constituents play a crucial role in NR expression and further exemplify the tight link between carbon and nitrogen metabolism.

Reduced nitrogen in the form of glutamine and glutamate appear to have a negative effect on NR expression and activity. Glutamine inhibits NR gene expression in both soybean and *N. plumbaginifolia* (Lillo, 1994; Vincentz *et al.*, 1993). Sucrose and glucose supplements can partially overcome this inhibitory effect. Glutamate applications can also inhibit NR mRNA induction in *Nicotiana* and squash. The effect that amino acids have on down regulation of NR gene expression is further confirmed by the fact that treatments which inhibit amino acid biosynthesis result in increased synthesis of NR mRNA. For example, plants treated with inhibitors of glutamine synthetase (GS) and plants in which nitrite reductase (NiR) is inhibited through antisense RNA have higher amounts of NR mRNA than control plants.

The foregoing discussion shows that NR activity in plants is controlled by numerous mechanisms including: (1) transcriptional and translational regulation; (2) post-translational modification; and (3) negative and positive effectors. The steps in signal transduction for NO_3^- induction of NR gene expression and activity remain to be understood and will prove to be exciting areas for future research.

Nitrite reductase

The six electron reduction of NO_2^- to NH_4^+ equation [29.2] is catalyzed by nitrite reductase (NiR) a plastid localized enzyme. Reduced ferredoxin serves as the electron donor with NADPH acting as the source of reducing power through the action of a ferredoxin-NADP oxidoreductase. NiR in plants is a monomer having a molecular mass of about 63 kD. It contains a [4Fe4S] center and siroheme as prosthetic groups. Similar to NR, NiR requires nitrate and light for induction. Over the last few years, both NiR cDNAs and genomic clones have been characterized and thereby significantly extend our understanding of this enzyme (Wray, 1993).

The plant NiR cDNAs cloned to date are very similar (Friemann *et al.*, 1992). The *Betula* (birch) NiR cDNA is described herein as a typical example. The 2.4 kb birch NiR cDNA contains a 1752 bp open reading frame, which encodes a protein of 583 amino acids. The deduced molecular mass of the protein is 65 kD, while the actual mass of the *in vivo* protein is 63 kD. This discrepancy in mass is due to the presence of a targeting sequence that is processed upon import into plastids. The birch protein has a putative 22 amino acid presequence, while a 32 amino acid presequence has been confirmed for spinach. Both the birch and spinach presequences are high in serine and threonine which is very characteristic of plastid targeting transit sequences. The transit peptides of NiR are among the shortest identified in plants. The deduced amino acid sequence of birch NiR is greater than 76% similar to other higher plant NiRs, indicating that the primary amino acid sequence of plant NiRs is quite highly conserved.

Sequence comparisons with the related redox proteins, sulphite reductase and ferredoxin-NADP reductase, show that birch NiR has a ferredoxin binding site at positions 95–119 in the N-terminal half of the protein. By comparison, the [4Fe4S] cluster and siroheme which form the active center of the enzyme are found in the C-terminus of the protein. Four cysteines at positions 443, 449, 484, and 488 appear to be involved in binding of the cofactors. Other plant NiRs have cysteines conserved at comparable positions within their primary structure.

Southern blot analysis indicates that NiR is encoded by a single gene in birch, spinach, maize, and barley, while at least four NiR genes occur in tobacco (Friemann *et al.*, 1992; Vincentz *et al.*, 1993; Wray, 1993). Isolation of a barley mutant designated *nir1* which lacks NiR enzyme activity and protein in both roots and leaves confirms that NiR is controlled by a single gene in this species. Moreover, this mutation is a conditional-lethal unless plants are grown on reduced nitrogen. In spinach a single NiR gene has been cloned and characterized. The spinach NiR gene has three introns of differing lengths (90–500 bp). A 3.1 kb element of the spinach gene upstream from the transcriptional start site was found to control expression in both leaves and roots. This element was also responsive to NO_3^- addition. In contrast, tobacco has at least two isozymes of NiR, one expressed in leaves and the other in roots. Isolation of several distinct cDNAs confirm that tobacco NiR is encoded by several genes. Transcripts corresponding to the *nir-1* cDNA were expressed only in leaves, while those corresponding to the *nir-2* cDNA were expressed in roots.

Synthesis of NiR message, protein, and enzyme activity are rapidly induced *de novo* by the addition of NO_3^- but not NH_4^+. Induction occurs in both shoots and roots, with roots being induced slightly sooner than shoots. For example, in NO_3^- starved maize NiR message is induced in roots within 45 min of NO_3^- addition. Interestingly, NiR mRNA appears to have a short half-life of approximately 30 min to 2 h depending upon the species. NiR protein, however, has a half-life substantially longer reaching 24 h in some species. Although NO_3^- appears to be the primary effector of NiR expression, light also plays an important role. Light, mediated through phytochrome, can stimulate in some instances mRNA synthesis and in others enzyme activity. Enzyme activity may be stimulated through light activation/deactivation of existing enzyme. Reduced N in the form of amino acids also appears to affect NiR expression by exerting a negative feedback control on synthesis of NiR mRNA and protein. It is increasingly apparent that the regulation of NiR is very similar to that of NR.

Nitrogenase

Characteristics

The second major process by which plants acquire nitrogen is through the microbial enzyme nitrogenase expressed by *Rhizobium* and *Bradyrhizobium* during symbiotic N_2 fixation equation [29.3]. It should be noted that nitrogenase is also present in certain free-living bacteria such as *Azotobacter*, *Clostridium*, *Klebsiella*, as well as cyanobacteria such as *Anabaena*. Amino acid analyses have shown that the enzyme is highly conserved between symbiotic and free-living bacteria (Newton, 1993). Therefore, the characteristics described in this section are derived from experiments with both types of bacteria. For molecular analysis of *nif* genes the *Klebsiella pneumoniae* system will be described and then, superimposed, the *Rhizobium* system will be considered.

Nitrogenase is comprised of two easily separable proteins designated the iron protein (Fe protein) or component II and the molybdenum–iron protein (MoFe protein) or component I (Dean and Jacobsen, 1992; Merrick, 1992). The Fe protein (encoded by the *nifH* gene) is a homodimer with a native molecular mass of 60 to 64 kD and a subunit molecular mass of 30 to 32 kD. The Fe protein contains approximately 4 g-atoms of Fe and S per mole of preparation, which form a single [4Fe–4S] cluster which is bound between the subunits. The Fe protein has two Mg ATP binding sites. As ATP binds to these sites, the potential of electrons present at the [4Fe–4S] cluster is reduced allowing the Fe protein to donate electrons to the MoFe protein. Mutational analysis of the Fe-protein suggests that the amino acid residues Arg-101 and Glu-113 are the contact surface for the transfer of electrons from the Fe-protein to the MoFe-protein. The Fe-protein also plays a role in the assembly of MoFe-cofactor but this is currently ill defined.

The MoFe protein is a tetramer ($\alpha 2\ \beta 2$) of approximately 220 kD molecular weight. The α subunit has a molecular mass of about 56 kD and is encoded by the *nifD* gene while the β subunit has a molecular mass of approximately 60 kD and is encoded by the *nifK* gene. The MoFe protein contains two atoms of Mo and 24 to 32 atoms of Fe and S per molecule. There appear to be two to four [4Fe–4S] clusters and two [MoFe$_6$S$_8$] clusters. The two [MoFe$_6$S$_8$] clusters comprise the MoFe cofactor. The role of the MoFe protein is to transfer electrons to N$_2$ and H$^+$.

In the last few years it has become apparent that some bacterial species contain Mo-independent nitrogenases (Newton, 1993). In these species, the Mo in component I is replaced by either V or Fe and the enzymes are referred to as V-nitrogenase and Fe-only nitrogenase. The V- and Fe-only nitrogenases are synthesized when Mo is limiting. While all N$_2$ fixing microbes have the standard Mo-nitrogenase, the distribution of alternative nitrogenases is less uniform. For example, *Rhizobium* does not contain either alternative nitrogenase while *Azotobacter vinelandii* contains both and *Rhodobacter capsulatus* contains the Fe-only form. The VFe or FeFe component I proteins of the alternative nitrogenases are hexameric composed of an a$_2$β$_2$d$_2$ arrangement with the D protein having a subunit molecular mass of 10 kD. The synthesis of the V- and Fe-only nitrogenases are under control of the *vnfDGK* and *vnfDGK* genes. Synthesis of the alternative and standard nitrogenases are transcriptionally controlled. In the presence of adequate Mo, only Mo-dependent component I is produced, whereas when Mo is limiting and V is available, V-dependent component I is synthesized. However, when both Mo and V are limiting, only the Fe-dependent component I is produced. The V- and Fe-dependent nitrogenases have lower specific activities than their Mo-dependent counterpart. Also, a greater proportion of the electron flow through nitrogenase is directed toward H$_2$ in the V- and Fe-dependent nitrogenases as compared to the Mo enzyme.

Genetic regulation

The genes required for nitrogen fixation have been most clearly defined in the free-living bacterium *K. pneumoniae*. The organization of the nitrogen-fixing (*nif*) gene cluster of *K. pneumoniae* is shown in Fig. 29.3. Some 20 genes are transcribed in eight adjacent operons which occupy 25 kb of the genome (Dean and Jacobsen, 1992). The *nif* gene functions can be grouped into several categories: (1) *nifH*, *nifD*, *nifK*-structural proteins for nitrogenase; (2) *nifF*, *nifJ*-

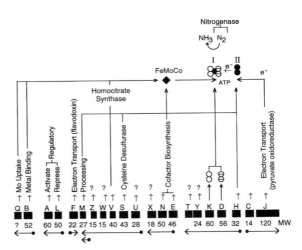

Fig. 29.3 Map of the *nif* gene cluster in *Klebsiella pneumoniae*. The function of each gene is denoted by arrows above the gene letter designation and the molecular weight of the gene product is given below each gene. The size of the individual operons and the direction of transcription are noted by arrows below the gene product molecular mass in kD (modified from Vance, 1990).

flavodoxin electron transport proteins; (3) *nifQ*, *nifB*, *nifN*, *nifE*, *nifV*, *nifS*-proteins involved in MoFe-cofactor and 4Fe4S cluster synthesis; (4) *nifM*, *nifY*-proteins required for processing NifH and MoFe-cofactor insertion into NifDK, respectively; (5) *nifA*-positive regulator, *nifL*-negative regulator; (6) *nifW*, *nifT*, *nifZ*, *nifU*-unknown but may be related to molecular chaperonin activity. All 19 to 20 genes have been cloned and deduced amino acid sequences have been determined. Regulation of the *nif* genes in *K. pneumoniae* is complex and at present not completely understood. Only the salient features of this regulation will be covered here.

Nitrogenase is synthesized when *K. pneumoniae* is grown under anaerobic, nitrogen-limiting conditions. This is not surprising since nitrogenase is irreversibly denatured in the presence of oxygen and is not required when alternative sources of reduced nitrogen are available. Regulation of nitrogenase is controlled by the NifA and NifL proteins in the *nifLA* operon. The *nifLA* operon is functionally regulated by the prokaryotic universal nitrogen control system designated *ntr*. The primary components of the *ntr* system are *ntrA*, *ntrB*, and *ntrC*. Under anaerobic, nitrogen-limiting conditions, *ntrA*, which encodes the sigma factor (σ^{54}), and *ntrC* gene product, a transcriptional activator, activate transcription of the *nifLA* operon. The NifA protein then activates transcription of all other *nif* operons. Since activation of all *nif* genes except *nifA* require the same gene product (NifA), it seems reasonable that the promoter region of the *nif* genes would contain similar recognition elements, which is in fact the case. The promoter region of all *K. pneumoniae nif* genes contains a conserved region at −24 to −12, which is the binding site for $\Sigma\sigma^{54}$, and about 100 bp upstream from the −24 to −12 region is an upstream activator sequence, which is the binding site for NifA (Merrick, 1992). Activation of these common regulatory regions under appropriate environmental and nutritional conditions results in a cascade effect leading to synthesis and assembly of functional nitrogenase.

Since *nif* gene expression is positively controlled by transcriptional activators and requires the *nifA* gene product, repression of nitrogenase synthesis in the presence of excess nitrogen and/or oxygen involves inactivation of these positive controlling elements. In the presence of oxygen and/or excess nitrogen the *nifL* gene product is altered. The altered *nifL* gene product in some, as yet unknown, fashion inactivates the *nifA* gene product resulting in lack of expression of the other *nif* operons. The *ntrB* gene product is also thought to be involved in sensing of excess nitrogen and repression of *nif* genes. Evidence for the involvement of *nifL* and *ntrB* in repression of nitrogenase has been obtained through *nifL* and *ntrB* mutants which synthesize nitrogenase in the presence of oxygen and/or excess nitrogen. Understanding the organization and regulation of the *K. pneumoniae nif* genes has provided the framework and tools to more fully ascertain how symbiotic N_2 fixation is controlled.

Symbiotic N_2 fixation

Rhizobium-legume symbiosis

Symbiotic N_2 fixation results from the complex interaction between the host plant and micro-organism (Franssen *et al.*, 1992; Hirsch, 1992). The host plant provides the microorganism with a source of energy for growth and function and a specialized ecological niche.

The microorganism fixes atmospheric N_2 and provides the plant with a source of reduced nitrogen in the form of NH_4^+. While the *Rhizobium/Bradyrhizobium*-legume symbiosis typifies such an interaction, symbiotic N_2 fixation also occurs in non-legumes such as *Parasponia* in symbiosis with *Rhizobium*, actinorhizal systems including *Alnus*, *Casuarina*, and *Elaeagnus* in association with the actinobacterium *Frankia*, and in cycads and the water fern *Azolla* in symbiosis with cyanobacteria. Since genetic regulation of the *Rhizobium*-legume has been the most fully characterized, that system will be covered in detail.

The ecological niche for *Rhizobium/Bradyrhizobium*-legume symbioses is the root nodule. Root nodules are highly organized, hyperplastic tissue masses derived from root cortical cells (Vance *et al.*, 1988; Hirsch, 1992). Nodules are generally divided into two major groupings characterized by shape, meristematic activity, and fixed N transport products: (1) nodules that are elongate-cylindrical with indeterminate apical meristematic activity that transport fixed N as amides such as alfalfa (*Medicago sativa* L.), pea (*Pisum sativum* L.) and clover (*Trifolium*); and (2) nodules that are spherical with determinate internal meristematic activity that transport fixed N as ureides, such as soybean (*Glycine max* L. Merr) and common bean (*Phaseolus vulgaris* L.) (Fig. 29.4). The complex series of events leading from bacterial colonization of the legume rhizosphere to fixation of N_2 and export of that fixed N requires controlled coordinated expression of both bacterial and host plant genes. The contribution of these genes to symbiosis can be grouped into several functions including: recognition, root hair invasion, infection thread growth, nodule differentiation, carbon assimilation, organic acid metabolism, ammonia assimilation and possibly, suppression of host plant defense responses. In the last 10 years, substantial progress has been made in understanding molecular aspects of both the bacterial and host plant role in symbiosis.

Rhizobium symbiotic genes

Nod genes

Symbiotic genes of rhizobia are categorized as those affecting nodulation (*nod* genes), those controlling nitrogenase based on their homology to *K.*

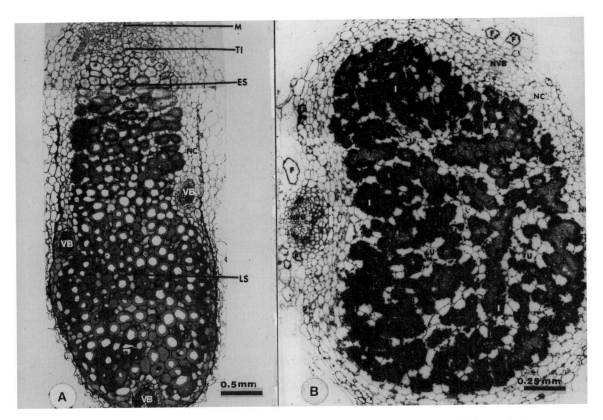

Fig. 29.4 Median longitudinal sections of mature, pink, N₂-fixing nodules of (A) alfalfa and (B) birdsfoot trefoil. (A) Indeterminate alfalfa nodule about 30 days after inoculation. Zones illustrated are meristem (M), thread invasion (TI), early symbiotic (ES), late symbiotic (LS), and vascular bundles (VB). (B) Determinate nodule of trefoil is composed of a central zone with infected (I) cells containing bacteroids and uninfected cells (U). Nodule vascular bundles (NVB) are embedded in cortex (NC) and some cortical cells contain flavolans (F).

pneumoniae (*nif* genes), and others that affect symbiotic N₂ fixation, but share no homology to known *K. pneumoniae* genes (*fix* genes) (Denarie *et al.*, 1992; Fisher and Long, 1992). For the most part these genes are highly conserved in all rhizobia, however, their location in the genome may vary. For example, the symbiotic genes of *R. meliloti* and *R. leguminosarum* are clustered within an approximately 60 kb region on indigenous plasmids while those in *Bradyrhizobium* are more scattered and located on the chromosome. Irrespective of the organism and the location in the genome, the organization and regulation of symbiotic genes is similar. The symbiotic genes of *R. meliloti* and *R. leguminosarum*

have been analyzed in great detail and serve as the foundation of our understanding of the molecular control of N₂ fixation in the microbial symbiont.

A region approximately 20 kb in size containing about 12 genes comprises the *nod* gene cluster of the *R. leguminosarum* symbiotic plasmid (Denarie *et al.*, 1992; Fisher and Long, 1992). Two small clusters 6 kb apart make up the analogous *nod* gene region of the *R. meliloti* symbiotic (*sym*) plasmid (Fig. 29.5). Transfer of these regions to *Agrobacterium* and other rhizobia cured of their sym plasmid confers the ability to nodulate pea or alfalfa. Those nodules are, however, ineffective (non-N₂ fixing), indicating that, although the heterologous genes carry information

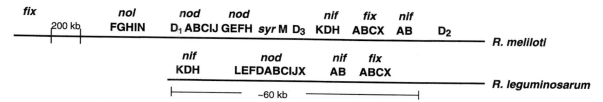

Fig. 29.5 Identification and organization of the symbiotic genes on plasmids of *Rhizobium meliloti* and *Rhizobium leguminosarum*. Note the difference in proximity of *nif* K,D,H to *fix* A,B,C,X in *R. meliloti* as compared to *R. leguminosarum*. Other features of interest include three copies of the *nod*D regulatory gene found in *R. meliloti* and a second *fix* region located some 150 to 200 kb away from the 60 kb symbiotic gene cluster of *R. meliloti*.

necessary for nodule development, other genes are required for effective N_2 fixation. Mutations in the *nodA,B,C* operon block root hair curling, infection, and nodulation indicating that these genes are involved in the earliest steps of the symbiotic interaction. The *nodA,B,C,I,J* genes are conserved in most rhizobia and are functionally identical as shown in experiments where the *nodA,B,C,I,J* genes from one *Rhizobium* or *Bradyrhizobium* species could complement mutations in homologous genes of other species. The *nodA,B,C,I,J* genes are frequently referred to as the 'common *nod* genes'.

Although all *Rhizobium* and *Bradyrhizobium* species contain one or more copies of the *nodD* gene, unique features of individual *nodD* genes indicate that it is not a common *nod* gene (Schlaman *et al.*, 1992). Mutations in *nodD* of *R. leguminosarum* prevent nodulation. There is only one copy of *nodD* in *R. leguminosarum*. However, mutations in *nodD* of *R. meliloti* only delay nodulation because *R. meliloti* has three functional copies of *nodD* and another *nodD*-like gene designated *syrM*. Although in some instances *nodD* genes from one species may partially rescue *nodD* mutant from another, host specificity is usually altered. This is exemplified by the fact that when *R. meliloti* *nodD* mutants are transformed with *nodD* from the wide host range *Rhizobium* NGR234, the transconjugants have a broader host range. Moreover, isogenic strains of *Rhizobium* which vary only in the source of their *nodD* gene differed in response to a variety of inducers and in host specificity. Lastly, the *NodD* gene product plays a regulatory role in signal perception and functions essentially as a transcriptional activator.

Nodulation by rhizobia is very host specific. For example, *R. meliloti* will nodulate alfalfa, but not clover or soybean. Similarly, *B. japonicum* will nodulate soybean but not pea, alfalfa, etc. This specificity is affected through a set of *nod* genes designated the host specificity genes (*hsn*). In *R. meliloti* host specificity is regulated by the *nodE,F,G,H,P,Q* genes (Figs. 29.5 and 29.7). The corresponding host specificity genes in *R. leguminosarum* are *nodE,F,L,M,N,O,T* (Fisher and Long, 1992). Mutations in these genes cannot be complemented by similar genes from other species. Other interesting host specific genes occur within strains of some *Rhizobium* species. The *R. leguminosarum* strain designated TOM nodulates primitive pea plants from Afghanistan, but not adapted European pea cultivars. A gene designated *nodX* has been shown to be required for *R. leguminosarum* TOM to nodulate primitive pea. European strains of *R. leguminosarum* have no homologous *nodX* gene. The common and host specific *nod* genes reside in two to four operons covering approximately 50 kb of the *sym* plasmid in *R. meliloti* and *R. leguminosarum*. While these genes are found on the chromosome of *B. japonicum*.

Regulation of host specificity, signal molecule synthesis, and nodule initiation have been particularly exciting areas of study in recent years because the mechanisms controlling these processes have largely been elucidated in the *Rhizobium*–legume symbiosis, thus providing a molecular paradigm for host–microbe interactions (Fisher and Long, 1992; Schlaman *et al.*, 1992). Interdisciplinary approaches using microbial genetics, natural product chemistry, and developmental biology have shown that activation of *Rhizobium nod* genes and host specificity of nodule induction occurs through an exquisite signaling process (Hirsch, 1992). This process is initiated as host legume roots grow through the soil and secondary plant products are released in exudates which activate transcription of *Rhizobium nod* genes (Denarie *et al.*, 1992). This activation is affected by the interaction of the *Nod*D gene product with compounds in root exudates. The *nodD* gene is constitutively expressed in free-living cultures of *Rhizobium* and *Bradyrhizobium*. However, the other *nod* genes are transcribed little, if any, except in the presence of root exudates. Induction of all of the other *nod* genes, both common and host-specific genes, is dependent upon the presence of NodD protein and phenolic compounds in legume root exudates (Fig. 29.6). The NodD protein is modified, in an unknown fashion, by the active factors in root exudates to become functional as a transcriptional activator. The NodD protein binds to a 35 to 45 bp conserved element (called the *nod* box) in the promoter region of the *nod* operons and activates transcription of the *nod* genes. Activation of *nod* gene expression in *Rhizobium/Bradyrhizobium* leads to the synthesis of chitin-like lipo-oligosaccharides which act as signals to the legume plant for the initiation of the first committed phases of nodule development (Figs. 29.6 and 29.7).

The phenolic compounds in root exudates responsible for induction of *nod* gene transcription are flavones, flavanones, and isoflavones (Peters *et al.*, 1986). These compounds are derived from the condensation of phenolic cinnamic acid derivatives and malonate units (Fig. 29.6). The initial inducing compounds from alfalfa, pea, clover, and soybean seed exudates are luteolin, hesperitin, 7,4'-dihydroxyflavone, and 4',7-dihydroxyisoflavone (daidzein), respectively. The most effective inducer compounds have hydroxyl groups substituted at the 3' or 4' position on the B ring and a hydroxyl or glucoside linkage at position 7 of the A ring. While isoflavonoid compounds induce *nod* gene expression in *B. japonicum* and *R. fredii*, strains nodulating soybean, these compounds act as antagonists of *nod* gene expression in *R. meliloti*, *R. trifolii*, and *R. leguminosarum*. Isoflavonoids not only play a role in legume-*Rhizobium* symbiosis, but also in plant disease resistance in legumes. Microbial infection of legumes frequently induces the accumulation of certain isoflavonoids that act as antibiotics (phytoalexins) which limit the growth of invading organisms. Therefore subtle differences in secondary plant products may regulate whether an interaction results in symbiosis or pathogenesis (Vance, 1990).

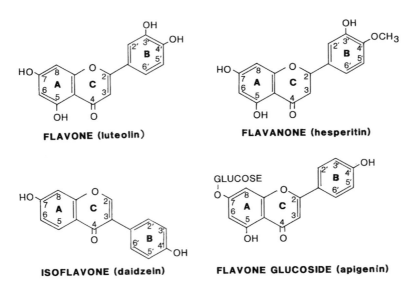

Fig. 29.6 Flavonoid and isoflavonoid compounds exuded from legume roots that activate and/or inhibit transcription of the *nod* genes in *Rhizobium* and *Bradyrhizobium*. The B ring is derived from phenylalanine, while the A and C rings are derived from malonate.

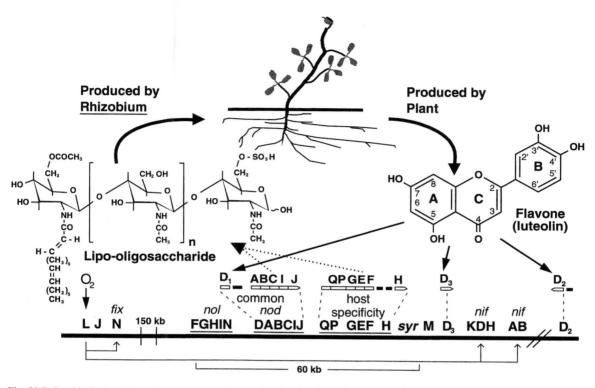

Fig. 29.7 Symbiotic signaling between legume plant and *Rhizobium* bacteria. Phenolic flavonoids and isoflavonoids released from plant roots bind to rhizobial NodD gene product which in turn activates transcription of other *nod* genes. Nod gene proteins catalyze synthesis of chitin-like lipo-oligosaccharides which act to induce the first steps (root hair curling and nodule cell division) in root nodule initiation.

The *nodD* gene was initially thought to be a common *nod* gene because mutations in *nodD* of one *Rhizobium* sp could be complemented, in part, by the *nodD* gene of another species. In addition, coding regions of *nodD* genes of various *Rhizobium* sp. share significant homology. However, studies showing that isogenic strains of *Rhizobium* which vary only in the source of their *nodD* gene differed in host specificity and in response to a variety of phenolic inducers and root exudates indicate that *nodD* is not common and is involved in mediating host specificity (Schlaman *et al.*, 1992). Further support for *nodD* involvement in species-specific nodulation was demonstrated in *R. leguminosarum* bv. *trifolii* by replacement of the *R. l.* bv *trifolii nodD* gene with that of the promiscuous strain *Rhizobium* NGR234. Although the *nod* genes of *R.l.* bv. *trifolii* are usually induced by 7,4'-dihydroxy-

flavone and flavanone, the transconjugant containing the NGR234 *nodD* was responsive to numerous flavonoids and other phenolics, reflecting the promiscuous nodulation capability of this particular strain. It is safe to say that *nodD* is a major determinant of host specificity through its role in recognition of flavonoid *nod* gene inducer molecules.

The function of the *nod* genes in synthesis of chitin-like lipo-oligosaccharide *nod* factors has been clarified through identification of the products they encode (Table 29.1) and how site directed mutagenesis of *nod* genes affects *nod* factor production (Lerouge *et al.*, 1990; Spaink *et al.*, 1991). The *nodA,B,C* genes are involved in the synthesis of the chitin-like oligosaccharide backbone of the *nod* factors. NodC has been identified as an *N*-acetylglucosaminyltransferase (chitin synthase), while

Table 29.1 *Rhizobium* nodulation (*nod*) genes and their proposed or determined functions

Gene*	Enzyme or function[a]
nodA	*N*-acylation of deacetylated nonreducing end of glucosamine oligosaccharide
nodB	Deacetylase – deacetylation of non-reducing end of glucosamine oligosaccharide
nodC	*N*-acetylglucosaminyltransferase – synthesis of β-1,4-*N*-acetylglucosamine oligosaccharide
nodD	DNA binding protein – transcriptional activator of other *nod* genes
nodE	β-ketoacylsynthase – postulated to be involved in Nod factor acyl chain synthesis
nodF	Acyl carrier protein – involved in Nod factor acyl chain synthesis
nodG	Similar to alcohol dehydrogenases and 3-oxoacyl reductase, may be involved in modifying fatty acyl side chain
nodH	Sulfotransferase – involved in transfer of activated sulphate to reducing end of Nod factor
nodI	ATP-binding protein – proposed to form membrane complex with nod J
nodJ	Hydrophobic transmembrane protein
nodL	Acetyl transferase – involved in addition of *O*-acetyl group to nonreducing end (position 6) of Nod factor
nodM	Glucosamine synthetase – may aid synthesis of Nod factor subunits
nodP,Q	ATP sulfurylase and ATS kinase – provide activated sulfur for sulfated Nod factor synthesis
nodV,W	Sensor, regulator in two component regulatory system
nodX	*O* acetyltransferase – specifically *O*-acetylates the C-6 of the reducing sugar of the penta-N-acetylglucosamine of *R. leguminosarum* TOM (which nodulates Afghanistan pea)

*Information adapted from Denarie *et al.* (1992), Hirsch (1992), and Fisher and Long (1992).

nodA and *B* are involved in *N*-acylation and deacetylation, respectively, of the nonreducing end of the glucosamine oligosaccharide. The chitin oligosaccharide backbone occurs as a 3-to-5-mer and is common to all *Rhizobium/Bradyrhizobium* species, thus the common *nod* gene designation for *nodA,B,C*. Specificity, however, resides in the decorations on the common chitin-like oligosaccharide backbone. The *R. meliloti nodE,F,G* genes affect synthesis of the fatty acyl chain located at the non-reducing end of the molecule, with *nodH,P,Q* controlling synthesis of the sulphated molecule found at the reducing end of the chitin oligosaccharide. The *nod L* gene has homology to an acetyl transferase and controls the addition of the *O*-acetyl group at position 6 of the nonreducing end of the polysaccharide. Interestingly, mutations in either of the *nodQ* or *H* genes of *R. meliloti* results in the production of an unsulphated *nod* factor which is inactive on alfalfa, the usual host for *R. meliloti*, but is active on *Vicia sativa* (vetch). It should be pointed out that *nod* factors produced by *R. meliloti* cause curling of root hairs and nodule meristem induction on alfalfa but not other legume species. Likewise, *nod* factors produced by *B. japonicum* cause root hair curling and nodule meristem induction on soybean but not other legume species.

Specificity in the *Rhizobium/Bradyrhizobium–* legume symbiosis is therefore controlled by release of host specific flavonoid-isoflavonoid molecules from the roots of the legume, which in turn activate transcription of the *nod* genes of the compatible *Rhizobium* species (Fig. 29.7). *Nod* genes then synthesize lipooligosaccharides which induce root hair curling and nodule meristem induction on the compatible host. Since all plants: (1) contain flavonoids and isoflavonoids; (2) have chitin degrading enzymes; and (3) release potential *nod* gene inducing compounds at various rates during growth and decomposition, specificity of nodulation must involve a finely tuned balance between inducers and antagonists and probably as yet other unidentified genes.

Nif and fix genes

Using *K. pneumoniae nif* genes as probes, DNA elements homologous to *nifK,D,H,A,B,* and *N* have been identified in all *Rhizobium* and *Bradyrhizobium* species. These corresponding genes are located on plasmids in fast-growing *R. meliloti* and *R. leguminosarum*, and on the chromosome in slow-growing *Bradyrhizobium*. The *nifK,D,H,A,B* and *N* genes have the same functions in rhizobia as in *Klebsiella*. Although a *nifL* gene has not yet been identified in *Rhizobium*, regulation of the *nif* operons in *Rhizobium/Bradyrhizobium* is similar to that in *Klebsiella*. The nifA gene product is a transcriptional activator for other *nif* operons. Mutations in *nifA* block symbiotic N_2 fixation and such mutants do not synthesize nitrogenase polypeptides. In addition, the promoter sequences of *Rhizobium nif* genes which bind nifA are similar to those in *Klebsiella*.

While the control of nitrogenase in symbiotic root nodules of *Rhizobium/Bradyrhizobium* is a two-component system like that in *Klebsiella*, regulation of the symbiotic system differs from that of the free-living system. The expression of *nifA* in rhizobial systems is not autoregulatory nor is it under control of the global *ntr* system. Instead *nifA* is oxygen regulated (Fig. 29.7). Two genes designated *fixL* and *fixJ*, which have no homologs in free-living N_2-fixing organisms, act as a sensor-transducer of low oxygen potential in root nodules and to activate transcription of *nifA*, which in turn activates transcription of the other *nif* operons. The FixL product is a transmembrane heme-containing protein which perceives low oxygen and becomes auto-phosphorylated. FixL then phosphorylates FixJ which activates nifA transcription. In *R. meliloti* these genes are located on the *sym* plasmid about 200 kb away from *nod* and *nif* genes.

In addition to the *nif* and *fixL,J* genes, another group of genes essential for symbiotic N_2 fixation, designated *fix* genes, have been identified in *Rhizobium/Bradyrhizobium*. Mutations in *fix* genes result in nodules with a Fix⁻ phenotype (nodules form but they do not fix N_2). Several have been identified (*fixA,B,C,N,K,X*). These genes are for the most part not found in free-living diazotrophs. As our knowledge of the biochemistry and physiology of symbiosis grows, undoubtedly more bacterial genes affecting symbiosis will be identified. Identification and manipulation of these genes may allow for improvement of *Rhizobium/Bradyrhizobium*–legume symbiosis.

Plant genes in symbiosis

A complete description of host plant genes involved in N_2 fixation is beyond the scope of this chapter. However, a brief overview of the available plant nodulation and N_2 fixation mutants will provide some perspective of the progress in this area. The capacity for symbiotic N_2 fixation is acquired through the coordinated expression of both plant and bacterial genes, which give rise to a unique organ, the root nodule, in which bacteria reduce N_2 gas to NH_4^+ in exchange for carbon-rich nutrients from the plant (Day and Copeland, 1991; Vance, 1990). The host contribution to symbiosis can be grouped into several functions including recognition, invasion, infection thread formation, nodule differentiation (e.g. meristems, vascular bundles, O_2 diffusion barrier), carbon metabolism, organic acid production, NH assimilation, senescence, and possibly suppression of host defense responses required for compatibility. While substantial progress has been made in identifying microbial genes and gene products contributing to nodulation and N_2 fixation, comparatively less progress has been made in understanding the contribution of plant genes to symbiosis. Plant genetic control of symbiosis has been documented through classical genetic studies and by induced mutagenesis in loci affecting nodulation, yet the molecular and biochemical manifestations of these genes are poorly understood.

To date, some 60 genes across 11 legume species have been identified as affecting nodulation and N_2 fixation (Table 29.2). Most are inherited as recessive traits and involve a single gene. The identified genes condition four major phenotypes: (1) non-nodulation (resistant to infection or infection occurs and nodules do not develop); (2) ineffective Fix⁻ tumor-like nodules with few bacteria present in nodules; (3) ineffective early-senescent nodules containing bacteria; and (4) supernodulation and/or nodulation in the presence of applied nitrogen fertilizer (Caetano-Anolles and Gresshoff, 1991; Phillips and Teuber, 1992).

Although a broad array of plants having altered nodulation and symbiotic traits now exist, other than structural descriptions, few have received further attention. The most well-characterized appear to be the ineffectively nodulating pea, alfalfa, and faba bean, and the supernodulating soybean mutants. Molecular tagging of nodulation and N assimilation genes in many legumes is difficult because of the large genome size and outcrossing is required for seed set. However, numerous laboratories are now attempting

Table 29.2 Host plant genes affecting nodulation and N_2 fixation[a]

Species	Number of genes	Comments
Trifolium pratense L.	7	Naturally occurring; condition non-nodulation and ineffective nodulation. Nodules vary from early senescencing to tumor-like
Pisum sativum L.	15	Naturally occurring and EMS mutagenesis; condition non-nodulation, ineffective, and super-nodulation, and nodulation in presence of NO_3^-. Some traits temperature sensitive
Medicago sativa L.	7	Naturally occurring; condition non-nodulation and ineffective nodulation. Nodules vary from early senescencing to tumor-like
Glycine max L. Merr.	8	Naturally occurring and EMS mutagenesis; condition non-nodulation, ineffective, and super-nodulation, and nodulation in the presence of NO_3^-.
Trifolium incarnatum L.	1	Naturally occurring; conditions ineffective nodulation
Arachis hypogaea L.	2	Naturally occurring; conditions non-nodulation
Cicer arietinum L.	5	Derived by γ irradiation; condition non-nodulation, reduced nodulation, and ineffective nodulation. Some traits temperature sensitive
Vicia faba L.	2	Naturally occurring; conditions ineffective nodulation
Phaseolus vulgaris L.	5	Derived by EMS mutagenesis; condition non-nodulation, super-nodulation, and nodulation in the presence of NO_3^-.
Melilotus alba Desr.	5	Derived by EMS and neutron radiation; condition non-nodulation
Medicago truncatula L.	4	Derived by transformation with T-DNA and EMS mutagenesis, non-nodulation and ineffective nodulation

[a]Derived from Vance, 1990.

to use self-compatible, small seeded species with a small genome size such as *Lotus japonicus* L. and *Medicago truncatula* L. for tagging purposes. Plant mutants are invaluable sources through which to gain an understanding of host plant genetic control of N_2 fixation, NH_4^+ assimilation, and carbon–nitrogen relationships. Such studies require multidisciplinary approaches involving plant breeders, plant biologists, and molecular geneticists.

Control of ammonia assimilation

General

Primary assimilation of NH_4^+ involves complex intermingling with C metabolites. Integrated functioning of both cytosolic and organelle-associated enzymes is required to link NH_4^+ assimilation with C metabolism (Lea *et al.*, 1990; Lam *et al.*, 1995). Efforts to understand and improve N metabolism require knowledge of how the enzymes of NH_4^+ assimilation are regulated, which, in turn necessitates fundamental insight into the plant genes encoding the

enzymes involved in this process. Until the last few years, most of our knowledge of plant genes involved in primary NH_4^+ assimilation focused on GS. Now, however, other than GDH whose function in primary NH_4^+ assimilation is questionable, genes encoding the remaining enzymatic steps in the pathway have been cloned and characterized. Because many of the enzymes of primary NH_4^+ assimilation are highly expressed in root nodules (Fig. 29.8), genetic regulation of the pathway has been frequently most closely scrutinized in these organs.

Glutamine synthetase (GS)

GS, the enzyme catalyzing the initial step in NH_4^+ assimilation can comprise up to 1 to 2% of the total soluble protein in organs actively assimilating N. GS activity increases dramatically during legume root nodule development (Fig. 29.8), in etiolated leaves exposed to light, and in leaves and roots of plants grown in NH_4^+ or NO_3^- (Cullimore and Bennett, 1992; McGrath and Coruzzi, 1991). The holoenzyme ranges in molecular mass from 320 to 380 kD and is

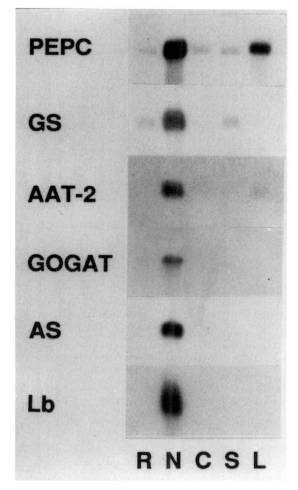

Fig. 29.8 Steady state levels of mRNAs involved in primary assimilation of NH_4^+ in alfalfa. Lanes correspond to: R, roots; N, nodules; C, cotyledons; S, stems; and L, leaves. All lanes contain 1 μg polyA$^+$ RNA. Enzyme designations are: AAT, aspartate aminotransferase isozyme 2; AS, asparagine synthetase; GOGAT, NADH-glutamate synthase; GS, glutamine synthetase; and PEPC, phosphoenolpyruvate carboxylase. Lb refers to the oxygen binding protein leghemoglobin. The AS cDNA used as a probe for the AS RNA blot was kindly provided by Gloria Coruzzi.

comprised of eight subunits ranging from 38 to 46 kD in mass. Cytosolic and chloroplastidic isoforms have been identified in both monocot and dicot species. The cytosolic and plastid forms were originally designated as GS_1 and GS_2, respectively. However, it

is now evident that these designations are insufficient because up to three cytosolic forms have been characterized. Changes in GS activity during organ development and in response to environmental signals are the result of differential expression of isozymes. In leaves GS activity is comprised primarily of the plastid form of the enzyme, although cytosolic GS can be expressed to high levels. Conversely, in roots cytosolic GS is the major form expressed. In legume root nodules two cytosolic forms of GS are expressed and in some instances one of those is nodule specific. Plastid GS can also be expressed in root nodules. Cytosolic and plastid forms of GS are immunologically and kinetically distinct. Antibodies to cytosolic GS from one species will usually recognize cytosolic GSs from other species but not plastid forms. The converse is true of plastid GS antibodies, which do not generally recognize cytosolic GS polypeptides.

The molecular basis for the generation of isozymes and differential expression of GS resides in the fact that different subunits are encoded by different genes (Lam *et al.*, 1995). In most species GS belongs to a small gene family. GS has been purified from various organs of *Phaseolus vulgaris* and antibodies prepared to root nodule cytosolic GS (Cullimore and Bennett, 1992; Forde *et al.*, 1989; Gebhardt *et al.*, 1986). The cytosolic enzyme in each organ was comprised of various proportions of three isoelectric variants α, β, and γ. Chloroplast GS was comprised of another subunit variant designated δ. These subunits are encoded by four distinct genes designated gln-α, gln-β, gln-γ, and gln-δ. The cytosolic GS genes encode proteins that are about 80–90% identical, but are highly divergent in their 5′- and 3′- untranslated regions. Cytosolic GS activity in leaves involves expression of all three genes with preferential enhancement of that encoding gln-α. Similarly, gln-β shows preferentially enhanced expression in roots, while gln-γ is preferentially expressed in nodules. The gln-γ polypeptide and gene were originally thought to be nodule specific. Recently, however, gln-γ mRNAs have been detected in other plant organs. Nodule GS, originally thought to be comprised of equal amounts of the γ and β subunits, may occur as other combinations of these polypeptides. Isolation of genomic clones for *Phaseolus* gln-β and gln-γ have allowed more precise studies of GS regulation.

Promoter analysis through chimeric gene fusions in transgenic *Lotus corniculatus* have shown: (1) gln-γ is preferentially expressed in rhizobial infected cells of nodules; and (2) gln-β directs high expression to roots. The gln-β promoter also directs expression to the nodule cortex and infected cell zone of very young nodules but is restricted to vascular bundles of older nodules.

In pea, four nuclear genes encode GS polypeptides (Brears *et al.*, 1991; McGrath and Coruzzi, 1991; Tingey *et al.*, 1988). The GS_2 gene encodes the plastid form of GS. The primary translation product of this gene is processed from 49 kD to 44 kD upon transport into pea leaf chloroplasts. The GS_2 gene is expressed in leaves in a light-dependent fashion. Leaf, root, and root nodule GS polypeptides are encoded by three distinct genes. A 37 kD cytosolic GS polypeptide is encoded by two genes which are nearly identical, now termed twin genes GS_3A and GS_3B. By comparison a separate and distinct 38 kD polypeptide is encoded by GS_1. These genes are expressed in most organs and tissues. However, there is selective preferential enhancement of specific GS genes, for example, cytosolic GS_3A and GS_3B are highly expressed in root nodules and in cotyledons of germinating seeds. However, GS_3A is consistently expressed at higher levels than GS_3B. The predominant form in roots is GS_1. GS_1 is also expressed in nodules but at much lower levels than GS_3A. Promoter analysis through chimeric gene fusions in transgenic plants has also been instrumental in more accurately defining the functions of pea GS genes. In transgenic tobacco the promoter of GS_2 directs expression to photosynthetic cell types, while the GS_3A promoter directs expression to the phloem. These data suggest GS_2 is functional in reassimilation of NH_4^+ derived from photorespiration and GS_3A may be involved in glutamine transport. In alfalfa the GS_3A promoter also directs expression to the vascular bundles but highest expression from the GS_3A promoter was directed to nodules, further confirming its role in assimilation of symbiotically fixed N.

In both maize and rice GS is also encoded by a small gene family (Kozaki *et al.*, 1992; Li *et al.*, 1993; Sakamoto *et al.*, 1989; Snustad *et al.*, 1988). At least three genes are found in these species, one which encodes the plastidic form and two which encode the cytosolic GS polypeptides. Exposure to light stimulates the expression of the plastid GS genes. It should be noted that plastid GS gene expression in response to light in both monocots and dicots is transcriptionally regulated. Light induces an increase in plastid GS mRNA and protein synthesis. Neither dicot or monocot cytosolic GS genes respond to light. Although maize and rice cytosolic GS genes are expressed at varying degrees in all tissues, there appears to be preferential expression of specific cytosolic GS genes in leaves versus roots. The cytosolic genes of maize are quite similar within the coding regions but show significant divergence in the 3'-untranslated regions.

The importance of plastid GS in photorespiration was conclusively demonstrated by the molecular analysis of barley photorespiratory mutants which lack GS activity (Freeman *et al.*, 1990). These mutants grow normally under conditions which suppress photorespiration, but they are dramatically impaired in growth under normal growth conditions conducive to photorespiration. Evaluation of chloroplast GS mRNA and protein expression showed three classes of mutants. Class I mutants lack GS mRNA and protein. Class II mutants had normal amounts of mRNA but little chloroplast GS protein and class III mutants contained significant amounts of GS protein but no GS activity. Class I mutants could arise from transcriptional or post-transcriptional lesions, while class II and III mutants may arise from translational and post-translational lesions.

The role of NH_4^+ and NO_3^- in affecting the expression of GS is variable. In some species such as soybean and rice (Marsolier *et al.*, 1993; Sakakibara *et al.*, 1992) transcription and translation of selected forms of cytosolic GS genes are enhanced and/or induced by either NO_3^- or NH_4^+. However, in other species such as *Phaseolus* and *Medicago*, cytosolic GS gene expression appears to require neither N source for initial expression. In these species, however, NH_4^+ and NO_3^- usually stimulate cytosolic GS expression to maximum levels. The data regarding the effects of NO_3^- and NH_4^+ on plastidic GS_2 gene expression are also variable (Redinbaugh and Campbell, 1993; Yamaya *et al.*, 1992). In maize leaves GS_2 plastidic gene expression is not elevated by either NO_3^- or NH_4^+. By comparison, expression of the maize root plastidic form of GS is stimulated by NO_3^- and NH_4^+. In rice, by contrast, NO_3^- and NH_4^+ stimulate the expression of plastidic GS in both leaves and roots. Clarification of the role of NH_4^+ and NO_3^- in GS gene induction requires further evaluation in other species. It is abundantly clear, however, that striking species to species variation exists.

Thus, in both monocots and dicots the genetic basis of multiple GS isozymes and subunits is well established. GS isoforms are encoded by a small gene family giving rise to several polypeptides. The genes encoding these polypeptides are differentially expressed as a function of age, organ, tissue, and N status. Future exciting research will focus on whether modification of GS expression can improve plant N use efficiency.

Glutamate synthase (GOGAT)

GOGAT catalyzes the reductive transfer of the amido group of glutamine to the α-keto position of 2-oxoglutarate (α-ketoglutarate), resulting in the formation of two molecules of glutamate. GOGAT, in collaboration with GS, functions to maintain a cyclic flow (GS/GOGAT cycle) of N from NH_4^+ into glutamine and glutamate. Nitrogen from glutamine and glutamate is then used for a number of aminotransferase reactions in the synthesis of other amino acids.

In higher plants GOGAT occurs as two distinct forms that differ in molecular mass, kinetics, location within the plant, and reductant specificity (Gregerson et al., 1993; Sakakibara et al., 1991): NADH-GOGAT and ferredoxin-GOGAT (Fd-GOGAT). Another form of GOGAT (NADPH-GOGAT) occurs in bacteria but there is little evidence for its occurrence in plants. Fd-GOGAT and NADH-GOGAT can be readily separated by exclusion and ion exchange chromatography. Antibodies to Fd-GOGAT do not recognize NADH-GOGAT, nor do NADH-GOGAT antibodies recognize Fd-GOGAT. Such data indicate that Fd- and NADH-GOGAT are distinct proteins. This has been unequivocally proven by recent studies in which the N-terminal amino acid and cDNA sequences have been determined for both forms of GOGAT.

Fd-GOGAT is predominantly localized in chloroplasts and is involved in the assimilation of NH_4^+ derived from light-dependent reduction of NO_3^- and from photorespiration (Sakakibara et al., 1991). Fd-GOGAT has also been reported in maize roots where it is implicated in assimilation of NH_4^+ derived from NO_3^- reduction. Legume root nodules are also reported to contain Fd-GOGAT, however, its role is not understood. Both Fd-GOGAT enzyme activity and protein increase in etiolated seedlings exposed to light. The light-induced increase in Fd-GOGAT activity and protein is preceded by an increase in Fd-GOGAT mRNA suggesting transcriptional control of the Fd-GOGAT gene in response to light. In maize roots but not leaves Fd-GOGAT protein and mRNA increase in response to applied NO_3^-, this also suggests increased transcription of Fd-GOGAT in response to an environmental signal (Redinbaugh and Campbell, 1993).

A complete cDNA encoding Fd-GOGAT has been isolated and characterized from maize and incomplete cDNAs for the enzyme have been isolated from tobacco and barley (Avila et al., 1993; Sakakibara et al., 1991; Zehnacker et al., 1992). The maize leaf Fd-GOGAT is 5.6 kb in length and hybridizes to a 5.7 kb mRNA that increases in abundance upon illumination. The maize cDNA encodes a 1616-amino acid protein which includes a 97 amino acid presequence (Fig. 29.9). The maize Fd-GOGAT presequence is rich in serine and arginine residues and has a net positive charge, all very characteristic of a plastid transit sequence. The deduced amino acid sequence of maize Fd-GOGAT is about 42% identical to that of the large subunit of *Escherichia coli* NADPH-GOGAT; however, the maize cDNA has no sequence corresponding to the small subunit of *E. coli* NADPH-GOGAT. Regions within the deduced amino acid sequence of maize Fd-GOGAT show similarity to the *E. coli* enzyme for iron–sulfur cluster binding and FMN binding sites. The barley and tobacco Fd-GOGAT cDNAs show greater than 85% similarity to the maize enzyme at both the deduced amino acid and nucleotide sequence levels. Southern blot analysis showed that Fd-GOGAT exists as a single gene in maize and barley but tobacco appeared to contain two Fd-GOGAT genes.

Further understanding for the role of Fd-GOGAT in photorespiration has been derived from recent

Amino Acid Identity			
	Maize Fd-GOGAT	**E.coli L**	**E.coli S**
Alfalfa	48%	46%	38%
Maize	--	42%	--

Fig. 29.9 Diagrammatic comparison of known GOGAT deduced protein sequences and the amino acid identity between various portions of the maize Fd-GOGAT, *E. coli* NADPH-GOGAT, and alfalfa NADH-GOGAT. Each polypeptide includes a presequence of varying length. These presequences occur at the N-terminus and begin with the first amino acid residue. Amino acids 102–1638 of alfalfa NADH-GOGAT and 98-1616 of the maize Fd-GOGAT polypeptides show significant sequence identity to the large subunit (L) of the *E. coli* NADPH-GOGAT. In comparison, amino acid residues 1695–2194 of alfalfa NADH-GOGAT show significant sequence identity to the small subunit(s) of *E. coli* NADPH-GOGAT. Maize Fd-GOGAT does not contain a region corresponding to the *E. coli* small subunit. See text for other details.

studies of barley photorespiratory mutants which lack Fd-GOGAT activity (Avila *et al.*, 1993). Fd-GOGAT enzyme protein and mRNA amounts were determined in five mutants lacking enzyme activity. Two classes of mutants were found, one of which was similar to Class II GS photorespiratory mutants. This class of mutants contained no enzyme protein but had normal amounts of Fd-GOGAT mRNA. The other class of Fd-GOGAT mutants did not resemble either class I or III GS mutants. This new class, designated IV, was characterized by having normal Fd-GOGAT mRNA replaced by mRNAs of different size. The class II mutants were probably affected at a post-translational level such as incorrect folding of the protein or defective targeting. The class IV mutant may be affected in processing of the primary transcript or initiation/termination of transcription.

Like Fd-GOGAT, NADH-GOGAT is a flavoprotein containing an iron–sulfur cluster. The enzyme is found primarily in non-green tissue. Plant NADH-GOGAT has been well characterized from legume root nodules and rice cell cultures (Gregerson *et al.*, 1993; Yamaya *et*

al., 1992). The enzyme exists as a monomer with a native and subunit mass in excess of 200 kD. Antibodies to nodule NADH-GOGAT will recognize comparable 200 kD polypeptides in other tissues. Rice antibodies to NADH-GOGAT recognize a single 200 kD polypeptide in seeds, roots, and young non-green leaves. The amount of NADH-GOGAT polypeptide in tissues is generally an order of magnitude less than that of GS and/or Fd-GOGAT. Root nodule NADH-GOGAT increases markedly during the development of effective root nodules and this activity increase is accompanied by an increase in enzyme protein and mRNA. By contrast, ineffective non N_2-fixing root nodules have little to no NADH-GOGAT enzyme activity, protein, and mRNA. The root nodule data suggest that nodule NADH-GOGAT expression is controlled at the transcriptional level. Little to no NADH-GOGAT mRNA can be detected in leaves, stems, roots, and cotyledons (Fig. 29.8).

Recently a 7.2 kb cDNA encoding the complete NADH-GOGAT enzyme has been isolated from alfalfa root nodules (Gregerson *et al.*, 1993). This 7.2 kb cDNA contains a single long open reading frame of 2194-amino acids that corresponds to a 240 kD protein (Fig. 29.9). Amino acid sequence determination of the N-terminus of the protein showed that the mature protein resulted from a processing event at position 101. Thus, the primary translation product is synthesized containing a 101 amino acid presequence and the mature protein has a processed molecular mass of 229 kD. Whether the NADH-GOGAT presequence is a plastid transit peptide is unclear. The deduced amino acid sequence of the presequence more resembles a mitochondrial than a plastid targeting presequence. However, most root nodule organelle isolation studies show the major portion of NADH-GOGAT activity resides in the plastid fraction. Interestingly, *E. coli* NADPH-GOGAT contains a presequence of 43-amino acid yet the function of this domain is unknown.

When the deduced amino acid sequence of alfalfa NADH-GOGAT was aligned with plant Fd-GOGATs and *E. coli* NADPH-GOGAT interesting similarities and unique differences were noted. The alfalfa protein contains domains that correspond to both the large (46% identity) and small (38% identity) subunits of the *E. coli* protein (Fig. 29.9). In the alfalfa NADH-GOGAT there is a 61-amino acid highly charged, hydrophilic domain that serves as a connector for the region corresponding to the prokaryotic large subunit C-terminus to the small subunit N-terminus. The deduced amino acid sequence of the alfalfa NADH-GOGAT shows about 48% identity to Fd-GOGATs. Interestingly, the processing site for all three GOGATs occurs at a cysteine residue and the first seven amino acids in all three mature proteins are very similar. The deduced amino acid sequence of these proteins suggest that the genes encoding these enzymes share a common evolutionary origin. Bacterial and yeast GOGAT genes are comprised of two different subunits. By contrast, higher plant GOGATs are single polypeptides. Whether the GOGAT gene arrangement resulted from the combining of two GOGATs subunit-encoding genes or from the split of a single ancestral progenitor gene into separate coding regions is unknown.

While no genomic clones for either Fd- or NADH-GOGAT have been reported, my laboratory has recently isolated and characterized an alfalfa NADH-GOGAT gene (Vance *et al.*, 1995). The gene is 14 kb in size and is comprised of 22 exons interrupted by 21 introns. The 5′-flanking region of the gene targets *GUS* reporter gene expression to root nodules. The alfalfa NADH-GOGAT gene shows little expression in tissues other than nodules. Effective nodules are required for NADH-GOGAT expression. We propose that NADH-GOGAT could be the rate limiting step in assimilation of symbiotically fixed N. Future studies of both Fd- and NADH-GOGAT will involve in-depth promoter analysis for the genes and efforts to use sense and antisense constructs to modify *in vivo* GOGAT activity.

Aspartate aminotransferase (AAT)

AAT is a pyridoxal-5′-phosphate dependent enzyme that catalyzes an essential step in the biosynthesis and degradation of aspartate in all species (Lea *et al.*, 1990). AAT catalyzes the reversible reaction: glutamate + oxaloacetate ↔ aspartate + α-ketoglutarate. In plants AAT plays an essential role in several metabolic pathways including the transfer of fixed C from mesophyll cells to bundle sheath cells in C_4 plants, a malate–aspartate shuttle that distributes reducing equivalents to chloroplasts, mitochondria, and peroxisomes, and the assimilation of NH_4^+ into aspartate and asparagine. Cytosolic, mitochondrial, plastid, and glyoxysomal forms of the enzyme have been documented and characterized. Independent segregation of soluble, mitochondrial, plastid, and glyoxysomal forms indicate that these distinct forms are controlled by separate genes. Many forms of the enzyme have been purified from several species including carrot, alfalfa, *Panicum*, soybean, and lupin. Within the last three years, cDNAs encoding the plastid, cytosolic, and mitochondrial forms of AAT have been characterized (Gantt *et al.*, 1992; Reynolds *et al.*, 1992; Taniguchi *et al.*, 1992; Udvardi and Kahn, 1991; Wadsworth *et al.*, 1993). AAT is a homodimer with a native mass ranging from 80 to 94 kD and subunit masses ranging in mass from 40 to 47 kD. Antibodies have been raised against cytosolic, plastid, and mitochondria forms of AAT. These antibodies are immunological distinct. They show no cross-reaction with heterologous forms of the enzyme. Immunoprecipitations and western blots have shown that plastid and mitochondrial AATs are synthesized as preproteins and are processed upon import into organelles.

Alfalfa serves as a model for the molecular basis of differential expression of AAT genes in plants. In alfalfa, two AAT isozymes have been observed. AAT-1 is expressed in leaves, stems, roots, and nodules at similar levels and consists of predominantly a single isoform. AAT-2, by comparison, is expressed at low levels in leaves, stems, and roots, but is induced to very high levels as effective root nodules develop, and occurs as three allozymes AAT2a, AAT2b, and AAT2c. The three AAT-2 allozymes that can be resolved into distinct bands by native PAGE result from the expression of two alleles, AAT-2A and AAT-2C at the AAT-2 locus. These two alleles differ in net charge depending upon the amino acid at position 296, being either valine or glutamic acid.

AAT-1 and AAT-2 cDNAs were cloned from alfalfa leaf and root nodule cDNA libraries, respectively (Gantt *et al.*, 1992; Udvardi and Kahn, 1991). Both cDNAs could rescue an *E. coli* aspartate auxotroph. The AAT-1 cDNA is 1.5 kb in length and encodes a 424-amino acid protein with a molecular mass of 46 kD. The AAT-1 cDNA hybridizes to a 1.5 kb mRNA that is expressed relatively uniformly in root, stems, leaves, and nodules. By comparison, the AAT-2 cDNA is 1.7 kb in length and encodes a 465-amino acid with a deduced mass of 50 kD. The mass of purified AAT-2 is about 40 kD as determined by SDS-PAGE. Complementation of the *E. coli* aspartate auxotroph with the AAT-2 cDNA resulted in the synthesis of a processed AAT-2 protein of 40 kD in mass. The discrepancy between the deduced and observed mass of the AAT-2 protein suggests AAT-2 is synthesized as a preprotein with a presequence of approximately 6 to 10 kd mass. AAT-2 cDNA hybridizes to a 1.7 kb mRNA that is highly expressed in root nodules with slight expression in leaves and cotyledons and little to no expression in roots and stems (Fig. 29.8). Nodules appear to contain 15- to 20-fold more AAT-2 mRNA and protein than do other tissues. AAT-1 and AAT-2 are approximately 50% similar at the deduced amino acid sequence level. The N-terminal domain of AAT-2 (amino acids 1-59) shares many characteristics of plastid targeting peptides. It contains just two acidic amino acids, is positively charged, and contains a high proportion of serine and threonine residues (25%). cDNAs bearing striking similarity, greater than 85%, to AAT-2 have been isolated from lupin and soybean (Reynolds *et al.*, 1992; Wadsworth *et al.*, 1993). The N-terminal domain of the lupine and soybean AAT-2 like cDNAs show approximately 66% similarity to the alfalfa AAT-2 N-terminal domain and contain equivalent percentages of serine residues.

Recent immunogold studies (Robinson *et al.*, 1994) confirm that alfalfa AAT-2 is localized to plastids. Using monospecific AAT-2 antibodies this form of AAT was shown to be localized to amyloplasts of the infected cell zone of alfalfa nodules. Amyloplast staining was highly specific and concentrated in the stroma surrounding starch grains. The density of gold label in plastids of infected cells was 4-fold that of plastids in uninfected cells. Gold labeling was significantly reduced (65%) over plastids of ineffective as compared to effective nodules. It should be noted that alfalfa AAT-2 antibodies also stain leaf chloroplasts but the density of label is much less than that in nodule amyloplasts. These data unequivocally show that AAT-2, the nodule-enhanced form of AAT, occurs primarily in the plastids of infected cells. Thus plastids are not only important in nodule C metabolism, as evidenced by starch accumulation and depletion, but also have a central function in primary

assimilation of N since they are an important site for aspartate biosynthesis. Furthermore, the formation of α-ketoglutarate by AAT in plastids provides a necessary substrate for GOGAT in the initial incorporation of NH_4^+ into amino acids.

In contrast to AAT-2, AAT-1 in alfalfa appears to be the cytosolic form of the enzyme. Analogous cDNAs have been isolated from carrot and *Panicum* and are greater than 80% identical to alfalfa AAT-1 (Taniguchi *et al.*, 1992; Udvardi and Kahn, 1991). The alfalfa AAT-1 cDNA and corresponding ones from other species lack any targeting-transit sequences in the N-terminal domain of the deduced amino acid sequence. Initial immunogold localization studies with alfalfa indicate AAT-1 antibody labels the cytosol.

Although there have been no definitive studies to assess the effect of light, development, NO_3^-, and NH_4^+ on the expression of most forms of AAT, we do know that both root nodule development and effectiveness interact to modulate expression of alfalfa nodule AAT-2 (Fig. 29.8). As root nodules develop, prior to detectable nitrogenase activity, AAT-2 activity, protein, and mRNA are expressed. Then as bacterial nitrogenase activity is expressed, AAT-2 activity, protein, and mRNA are further enhanced to maximum levels. In ineffective plants, AAT-2 mRNA, protein, and enzyme activity show the initial increase correlated with nodule initiation but fail to show the further increase in expression associated with effective nitrogenase activity. Therefore one signal for enhanced AAT-2 expression is related to nodule development while another is associated with nodule effectiveness and NH_4^+ production. We have also evaluated expression of AAT-2 in alfalfa leaves as affected by light. Etiolated leaves show a basal level of AAT-2 enzyme protein and mRNA which increases about 50% on exposure to light (Robinson and Vance, unpublished). A similar increase in plastid AAT has been noted upon exposure of soybean and *Panicum* leaves to light (Taniguchi *et al.*, 1992; Wadsworth *et al.*, 1993).

We have recently isolated alfalfa genes encoding AAT-1 and AAT-2 (Gregerson *et al.*, 1994). The *AAT-1* gene contains 12 exons, and the corresponding mRNAs are polyadenylated at multiple sites differing in location by more than 250 nucleotides. By contrast, *AAT-2* contains 11 exons and its transcripts are polyadenylated at a very limited range of sites.

The transit peptide of AAT-2 is encoded by the first two and part of the third exon. The alfalfa *AAT-2* gene has five intron positions identical to those found in animal AAT genes, and the alfalfa *AAT-1* gene has one such conserved intron position. A comparison of the region upstream of the ATG translational initiation codon of *AAT-2* with that of *AAT-1* showed that *AAT-2* contains two sequence motifs (CTCTT and AAAGAT) that are conserved in several nodule-enhanced genes. These sequences have been implicated as *cis*-acting elements necessary for appropriate expression of leghemoglobin and N-23 genes from soybean. Further evidence for the differences in the regulation of gene expression between *AAT-1* and *AAT-2* is shown by the fact that *AAT-1* lacks these conserved sequences. *AAT-2*-promoter/GUS fusions transformed into birdsfoot trefoil and alfalfa show that this promoter enhances transcription of downstream sequences in nodules. We are currently preparing several deletion derivatives of the *AAT-1* and *AAT-2* promoters to evaluate the *cis*-elements contributing to control of these genes. These constructs will enable us to conclusively assess the role of light and other effectors on the expression of *AAT-1* and *AAT-2*.

Asparagine synthetase (AS)

AS catalyzes the glutamine-dependent amidation of aspartate with the simultaneous hydrolysis of ATP to AMP and pyrophosphate. Studies using inhibitors of GS have shown that NH_4^+ may directly serve as a substrate for AS in some instances. The enzyme has been recalcitrant to study due to its instability, low abundance, and presence of endogenous inhibitors. Until the studies of alfalfa nodule AS currently in progress in the laboratory of Scott Twary and Pat Unkefer (personal communication), the plant enzyme had not been purified to homogeneity nor had antibodies been produced to it. (These studies will be discussed later.) Significant advances made in the understanding of pea AS clearly show the power of molecular biology in the resolution of intractable biological problems.

Using the strategy of designing a molecular probe from the human AS enzyme, Tsai and Coruzzi (1990, 1991) isolated and characterized cDNAs encoding pea AS and evaluated factors contributing to control of AS gene expression. Two AS cDNAs, *AS1* and *AS2*, of approximately 2.2 kb were isolated. *AS1* encodes a 586-amino acid deduced protein of 66.3 kD mass, while *AS2* encodes a 583-amino acid deduced protein with a mass of 65.6 kD. Both cDNAs are greater than 80% similar at the nucleotide and deduced amino acid sequence level but the 3'-untranslated regions are highly divergent. Both appear to be encoded by a single gene. *AS1* was more highly expressed in leaves and nodules than *AS2*. Light represses the transcription of both genes in leaves and roots. When plants were grown in continuous light or normal photoperiods and then transferred to dark, *AS1* and *AS2* mRNAs increased some 20- and 5-fold, respectively. Nuclear run-on experiments showed that within 20 min of exposure to light AS transcriptional rates decreased. Exposure of pea plants to continuous light, dark, and red-far red light treatments demonstrated that AS expression was mediated through phytochrome. The authors also showed light-dark regulation of AS in tobacco with companion experiments. The dark stimulation of AS directly contrasts with the light stimulation of plastid GS. The enhanced expression of AS in the dark is consistent with reports showing enhanced transport of asparagine in darkened tissues.

In efforts to modify amino acid metabolism in tobacco, the pea AS cDNA was fused to the cauliflower mosaic virus 35S promoter (Brears *et al.*, 1993). This construct was transformed into tobacco resulting in constitutive expression of AS at high levels in all tissues in both the dark and light. The resultant transgenic tobacco showed asparagine levels 10- to 100-fold greater than the control. Moreover, when the region corresponding to the glutamine binding site was deleted from the AS cDNA, there was a 3- to 19-fold increase in asparagine levels. These data show the feasibility of modifying amino acid metabolism through direct over-expression of enzymes of primary amino acid metabolism. They also show the possibility of designing an NH_4^+-dependent AS activity into plants.

In contrast to pea, expression of AS in asparagus does not appear to be affected by dark (Davies and King, 1993). Isolation of an asparagus AS cDNA, which is highly conserved relative to *AS1* of pea, and evaluation of AS expression in harvested asparagus spears showed that AS mRNA was induced within

2 h of harvest and message continued to increase for up to 12 h after harvest. The amount of AS message was not affected by light and little AS could be detected in roots, ferns, or spears before harvest. The increase in AS in harvested asparagus spears is consistent with observations in many species that asparagine increases under stress conditions. The increased expression of AS in these instances may be related to prevention of ammonia toxicity.

Lastly and mentioned earlier, AS is currently being evaluated in alfalfa. Alfalfa nodule AS activity has been sufficiently stabilized to allow purification of the enzyme to homogeneity (Scott Twary and Pat Unkefer, personal communication). There are apparently two forms in nodules both are homodimers comprised of subunits with molecular masses of 66 and 62 kD, respectively. High titer antibodies which immunoprecipitate AS activity have been prepared to a complex of both forms of the enzyme. Western blots show that the 62 kD form is strikingly enhanced in nodules. The AS antibodies were used to screen an alfalfa nodule cDNA expression library and a 2.2 kb alfalfa AS cDNA showing high similarity to pea AS1 has been isolated (S. Twary, P. Unkefer, R. Gregerson, C. Vance, S. Gantt, unpublished). RNA blots show that the alfalfa AS mRNA increased some 15-fold during the development of effective nodules and is low to non-existent in ineffective nodules. Evaluation of poly A^+RNA from various alfalfa tissues showed high expression in nodules and almost no expression in leaves, stems, roots, or cotyledons (Fig. 29.8). Whether light affects expression of the alfalfa AS has not been evaluated.

The above studies show at least three variants of AS may be important in plants: a leaf form affected by light–dark exposure, a post-harvest stress induced form, and a form associated with symbiotic N_2 fixation. Further characterization of the various AS genes and their promoters will provide insight into the signals regulating AS activity and expression (Lam et al., 1995). The continuing question of whether plants contain an NH_4^+-dependent form of AS should also be resolved.

Phosphoenolpyruvate carboxylase (PEPC)

PEPC catalyzes the irreversible carboxylation of phosphoenolpyruvate to oxaloacetate. The enzyme is widely distributed in plants and microbes, playing a crucial role in C and N metabolism (Latzko and Kelly, 1983). In plants, most studies have focused on the role of PEPC in C_4 and CAM species where the enzyme provides an effective mechanism of concentrating CO_2 within leaves. Numerous studies have documented physical and biochemical properties of C_4 and CAM PEPCs. Both cDNAs and genomic clones have been isolated which encode the enzyme in these species (Cretin et al., 1991; Cushman et al., 1989; Hudspeth and Grula, 1989). The enzyme is a tetramer comprised of subunits of 100 to 110 kD and several isoforms of PEPC have been detected in both C_4 and CAM plants. Regulation of the enzyme's activity is achieved through both transcriptional and post-translational events. Light induces an increase in the amount of PEPC mRNA and enzyme protein in maize and sorghum while salt stress increases the amount of mRNA and enzyme protein in ice plant. Post-translational regulation of PEPC is achieved through phosphorylation of the protein, oligomerization as affected by PEP, malate and glucose-6-PO_4 and protein turnover (Jiao and Chollet 1991; Nimmo, 1992). In both C_4 and CAM species, PEPC is encoded in the nucleus by a small gene family. The gene encoding maize leaf C_4 PEPC is expressed in leaf sheaths, tassels, and husks but not roots or seeds. A separate gene encodes the maize root form of PEPC and this gene is only slightly expressed in green tissues. PEPC expression in ice plant is a function of at least three genes, a non-autotrophic root form and two leaf forms, the expression of one of which is increased about 30-fold by salt stress. Regulation and genetic control of PEPC in CAM and C_4 species is considered in detail elsewhere in this volume.

Although Latzko and Kelly (1983) hypothesized several functions for PEPC in C_3 plants, the most clearly defined role is related to primary assimilation of NH_4^+ in legume nodules. CO_2 fixation in root nodules via PEPC is integrally associated with N_2 fixation and NH_4^+ assimilation (Fig. 29.1). In fact, up to 30% of the carbon required for nitrogenase activity and primary assimilation of NH_4^+ can be supplied by PEPC. Similar to C_4 and CAM species, nodule PEPC can comprise 1% to 2% of the total soluble protein of nodules (Egli et al., 1989). Furthermore the specific activity of nodule PEPC is comparable to that of C_4 and CAM PEPCs. Ineffective nodules, whether induced by ineffective rhizobia or controlled by plant genes, have strikingly reduced amounts of PEPC activity and enzyme protein. Also treatments that inhibit nitrogenase activity also cause a dramatic reduction in PEPC activity. The growing awareness that PEPC plays a critical role in NH_4^+ assimilation has further been confirmed in studies with *Selenastrum* and maize which show striking increases in PEPC activity and mRNA in response to NH_4^+ additions. Rapid incorporation of ^{14}C, derived from fixation of $^{14}CO_2$ through PEPC, into newly synthesized amino acid in NH_4^+ treated plants also exemplifies the interconnectedness of N and C metabolism.

Along with its instrumental role in NH_4^+ assimilation, PEPC plays another salient, but less obvious, role in providing C in the form of malate and perhaps aspartate to fuel root nodule bacteroid growth and nitrogenase (Day and Copeland, 1991; Vance and Heichel, 1991). Root nodule bacteroids acquire malate, and perhaps aspartate, by a pathway that is nearly identical to that by which the bundle sheath cells of C_4 plants acquire these compounds (Fig. 29.10). A side by side comparison of C acquisition by legume nodules and C_4 leaf cells shows provocative similarities. In mesophyll cells of C_4 species phosphoenolpyruvate derived from sucrose is carboxylated by PEPC to form oxaloacetic acid, which is then converted to malate and aspartate by malate dehydrogenase (MDH) and AAT, respectively. Malate and aspartate are then transported to bundle sheath cells where they are decarboxylated and the resulting CO_2 is used by RUBISCO in chloroplasts for production of PGA. An almost identical set of reactions occurs in root nodules to form malate and aspartate. These compounds then are transported into bacteroids and used as energy sources for reduction of N_2 to NH_4^+. In light of the fact that these pathways are so similar and that some symbiotic bacteria are capable of not only N_2 fixation but also photosynthesis, it is

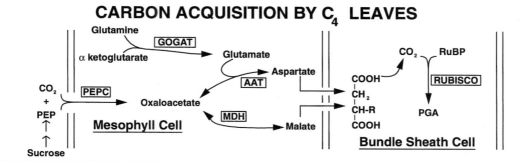

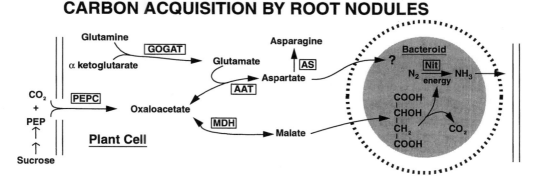

Fig. 29.10 Schematic representation of carbon acquisition by C_4 leaves and root nodules of C_3 legumes. Note that the two systems are functionally analogous in many respects. Enzyme acronyms are in blocks and described in the text.

intriguing to consider whether incorporation of such bacteria into plants had an evolutionary role in the development of symbiosis or C_4 metabolism.

Nodule PEPC has been purified from several species and documented to occur as multiple forms in soybean, alfalfa, and lupine (Schuller and Werner, 1993; Vance, et al., 1994). While such reports suggest the presence of nodule-specific forms of PEPC, studies to date have not confirmed this hypothesis. To date only alfalfa nodule PEPC has been thoroughly examined at the mRNA and protein level (Pathirana et al., 1992). Using monospecific antibodies to alfalfa nodule PEPC, an alfalfa nodule PEPC cDNA of 3.6 kb was isolated. This cDNA contains a single long open reading frame of 966-amino acids with a deduced protein molecular mass of 111 kD. The deduced amino acid sequence of the alfalfa nodule PEPC shows greater than 80% identity to both C_4 and CAM PEPCs. Characterization of PEPC mRNA by RNA gel blot analysis of polyA$^+$ RNA extracted from roots, nodules, cotyledons, stems, and leaves showed that although expression was greatest in nodules, PEPC mRNA could be detected in all organs with substantial amounts in leaves (Fig. 29.8). Expression of PEPC in nodules was 10- to 15-fold greater than in roots, cotyledons, and stems. The 10- to 15-fold increase in PEPC activity which occurs during effective nodule development is accompanied by a similar increase in PEPC mRNA and protein. In ineffective nodules PEPC activity and protein are low to non-detectable yet substantial PEPC mRNA is present. These data suggest nodule effectiveness is not required for PEPC mRNA synthesis and that PEPC is in part regulated at the translational or post-translational level.

Evidence from soybean (Schuller and Werner, 1993) and alfalfa (Vance et al., 1994) indicate that similar to C_4 and CAM PEPCs, root nodule PEPC activity is modulated through phosphorylation. Alignment of the N-terminal first 20 amino acids residues of alfalfa PEPC with those of the maize – C_4 and ice plant – CAM PEPCs shows that the phosphorylation site, serine residues 15, and 11, respectively, of C_4 and CAM PEPCs have been conserved in alfalfa nodule PEPC. In vitro phosphorylation experiments show that nodule PEPC can be phosphorylated and that phosphorylation reduces the sensitivity of the enzyme to malate. To function effectively, nodule PEPC requires such a mechanism to buffer against the high malate concentrations found in nodules.

Immunogold localization studies (Robinson et al., 1994) show that alfalfa nodule PEPC is localized to the cytosol of infected and uninfected cells. Uninfected cells appeared to contain as much if not more gold label than did infected cells. Moreover, dense labeling was also found in vascular bundle transfer cells. These data indicate a function for PEPC in three different cell types and are consistent with the several roles proposed for nodule PEPC. Day and Copeland (1991) proposed that uninfected cells play a major role in nodule metabolism, particularly in malic acid synthesis. Localization of PEPC to uninfected cells provides support for this hypothesis.

An alfalfa PEPC gene corresponding to the alfalfa nodule PEPC cDNA has been isolated (S. Pathirana, S. Gantt, C. Vance, unpublished). The transcribed region of the PEPC gene has eleven exons separated by ten introns. Nine of the ten introns are located in the same position as those found in C_4 and CAM PEPC genes. Promoter-GUS fusion constructs transformed into Lotus and alfalfa target expression to root nodules. Deletion analysis of the promoter element is currently underway. These deletions fused to GUS and transformed into alfalfa and Lotus will be useful in identifying the signals which affect PEPC expression and the genetic elements controling nodule expression of PEPC.

Synopsis

Striking progress has been made in the isolation and characterization of the genes involved in primary assimilation of N in plants. Antibodies have been prepared to and both cDNAs and genomic clones have been isolated for NR, NiR, GS, AS, AAT, GOGAT, and PEPC. Furthermore, significant progress has been made in identifying cis-acting elements in promoter regions of these genes. The expression of many of the genes involved in the primary assimilation of N is closely associated with C metabolism. In fact NiR, GS, AAT, and GOGAT have plastid targeted forms that are essential for N metabolism. Thus, plastids not only play a major role in C metabolism as evidenced by the location for starch biosynthesis and photosynthesis, but they also have primary functions associated with N assimilation. Light is also integrally involved in regulating the expression of many genes of N assimilation as evidenced by its positive effect on NR and NiR and its negative effect on AS. Because there is little information on the signal transduction pathway involved in activation of the genes involved in primary N assimilation, this field will be a central target for future research. Additionally, direct genetic manipulation of N assimilation by over- and/or under-expression of enzyme activity through sense and antisense technology will show whether plant N metabolism and quality can be improved or altered.

Fundamental progress has also been made in the molecular genetics of N_2 fixation for both free-living and symbiotic systems. The signal transduction mechanisms involved in symbiotic N_2 fixation have been identified and offer a model paradigm for plant–microbe interactions. Future research will focus on whether similar signals are involved in other systems and whether these signals can be used to improve root nodulation. Many plant mutants altered in N_2 fixation and nodulation have been identified and characterized at the structural and biochemical levels. However, no plant genes controling these processes have been identified; future studies will pursue this exciting field.

Even though substantial progress has been made in the molecular genetics of primary acquisition and assimilation of N, we still do not know which genes are most crucial to enhancing N use efficiency in plants. It seems apparent that a multidisciplinary approach utilizing both conventional and molecular methods should focus on improvement of N metabolism in plants. This approach must not fail to consider how C metabolism impacts N assimilation.

References

Avila, C., Marquez, A. J., Pajuelo, P., Canell, M. E., Wallsgrove, R. M. and Forde B. G. (1993). Cloning and sequence analysis of a cDNA for barley ferredoxin-dependent glutamate synthase and molecular analysis of photorespiratory mutants deficient in the enzyme. *Planta* **189**, 475–83.

Brears, T., Walker, E. and Coruzzi, G. (1991). A promoter sequence involved in cell-specific expression of the pea glutamine synthetase GS3A gene in organs of transgenic tobacco and alfalfa. *Plant J.* **1**, 235–44.

Brears, T., Christopher, L., Knight, T. J. and Coruzzi, G. M. (1993). Ecotopic overexpression of asparagine synthetase in transgenic tobacco. *Plant Physiol.* **103**, 1285–90.

Caetano-Anolles, G. and Gresshoff, P. M. (1991). Plant genetic control of nodulation. *Annu. Rev. Microbiol.* **45**, 345–82.

Cheng, C. L., Acedo, G. N., Cristinsin, M. and Conkling, M. A. (1992). Sucrose mimics the light induction of *Arabidopsis* nitrate reductase gene transcription. *Proc. Natl. Acad. Sci. USA* **89**, 1861–64.

Crawford, N. M. and Arst, H. N. (1993). The molecular genetics of nitrate assimilation in fungi and plants. *Annu. Rev. Genet.* **27**, 115–46.

Cretin, C., Sonti, S., Keryer, E., Lepiniec, L., Tagu, D., Vidal, J. and Gadal, P. (1991). The phosphoenolpyruvate carboxylase gene family of Sorghum: promoter structures, amino acid sequences and expression of genes. *Gene* **99**, 87–94.

Cullimore, J. V. and Bennett, M. J. (1992). Nitrogen assimilation in the legume root nodule: current status of the molecular biology of the plant enzymes. *Can. J. Microbiol.* **38**, 461–66.

Cushman, J. C., Meyer, G., Michalowski, C. B., Schmitt, J. M. and Bohnert, H. J. (1989). Salt stress leads to differential expression of two isogenes of phosphoenolpyruvate carboxylase during crassulacean acid metabolism induction in the common ice plant. *Plant Cell* **1**, 715–25.

Davies, K. M. and King, G. A. (1993). Isolation and characterization of a cDNA clone for a harvest-induced *asparagine synthetase* L. *Plant Physiol.* **102**, 1337–40.

Day, D. A. and Copeland, L. (1991). Carbon metabolism and compartmentation in nitrogen-fixing legume nodules. *Plant Physiol. Biochem.* **29**, 185–201.

Dean, D. R. and Jacobsen, M. R. (1992). Biochemical genetics of nitrogenase. In *Biological Nitrogen Fixation* eds G. Stacy, R. H. Burris and H. J. Evans, Chapman-Hall, New York, pp. 763–831.

Denarie, J., Debelle, F. and Rosenberg, C. (1992). Signaling and host range variation in nodulation. *Annu. Rev. Microbiol.* **46**, 497–531.

Egli, M. A., Griffith, S. M., Miller, S. S., Anderson, M. P. and Vance, C. P. (1989). Nitrogen assimilating enzyme activities and enzyme protein during development and senescence of effective and ineffective alfalfa nodules. *Plant Physiol.* **91**, 898–904.

Fisher, R. F. and Long, S. R. (1992). *Rhizobium*-plant signal exchange. *Nature* **357**, 655–60.

Forde, B. G., Day, H. M., Turton, J. F., Wen-jun, S., Cullimore, J. V. and Oliver, J. E. (1989). Two glutamine synthetase genes from *Phaseolus vulgaris* L. display contrasting developmental and spatial patterns of expression in transgenic *Lotus corniculatus*. *Plant Cell* **1**, 391–401.

Franssen, H. J., Vijn, I., Yang, W. C. and Bisseling, T. (1992). Developmental aspects of the *Rhizobium*-legume symbiosis. *Plant Mol. Biol.* **19**, 89–107.

Freeman, J., Marquez, A., Wallsgrove, R., Saarelainen, R. and Forde, B. (1990). Molecular analysis of barley mutants deficient in chloroplast glutamine synthetase. *Plant Mol. Biol.* **14**, 297–311.

Friemann, A., Brinkmann, K. and Hachtel, W. (1992). Sequence of a cDNA encoding nitrite reductase from the tree *Betula pendula* and identification of conserved protein regions. *Mol. Gen. Genet.* **231**, 411–16.

Gantt, J. S., Larson, R. J., Farnham, M. W., Pathirana, S. M., Miller, S. S. and Vance, C. P. (1992). Aspartate amino transferase in effective and ineffective alfalfa nodules: cloning of a cDNA and determination of enzyme activity, protein and mRNA levels. *Plant Physiol.* **98**, 868–78.

Gebhardt, C., Oliver, J. E., Forde, B. G., Saarelainen, R. and Miflin, B. J. (1986). Primary structure and differential expression of glutamine synthetase genes in nodules, roots and leaves of *Phaseolus vulgaris*. *EMBO J* **5**, 1429–35.

Gregerson, R. G., Miller, S. S., Petrowski, M., Gantt, J. S. and Vance, C. P. (1994). Genomic structure, expression, and evolution of the aspartate aminotransferase genes of alfalfa. *Plant Mol. Biol.* **25**, 387–99.

Gregerson, R. G., Miller, S. S., Twary, S. N., Gantt, J. S. and Vance, C. P. (1993). Molecular characterization of NADH-dependent glutamate synthase from alfalfa nodules. *Plant Cell* **5**, 215–26.

Hirsch, A. M. (1992). Developmental biology of legume nodulation. *New Phytol.* **122**, 211–37.

Hoff, T., Stummann, B. M. and Henningsen, K. W. (1992). Structure, function, and regulation of nitrate reductase in higher plants. *Physiol. Plant.* **84**, 616–24.

Huber, J. L., Huber, S. C., Campbell, W. H. and Redinbaugh, M. G. (1992). Reversible light/dark modulation of spinach leaf nitrate reductase activity involves protein phosphorylation. *Arch. Biochem. Biophys.* **296**, 58–65.

Hudspeth, R. L. and Grula, J. W. (1989). Structure and expression of the maize gene encoding the phosphoenolpyruvate carboxylase isozyme involved in C_4 photosynthesis. *Plant Mol. Biol.* **12**, 579–89.

Jiao, J. A. and Chollet, R. (1991). Post-translational regulation of phosphoenolpyruvate carboxylase in C_4 and CAM plants. *Plant Physiol.* **95**, 981–85.

Kozaki, A., Sakamoto, A. and Takeba, G. (1992). The promoter of the gene for plastidic GS_2 from rice is developmentally regulated and exhibits substrate-induced expression in trangenic tobacco. *Plant Cell Physiol.* **33**, 233–38.

Lam, H.-M., Cochigano, K., Schultz, C., Melo-Oliveria, R., Tjaden, G., Oliveria, I., Nagi, N. Hsieh, M.-H. and Ceruzzi, G. (1995). Use of *Arabidopsis* mutants and genes to study amide amino acid biosynthesis. *Plant Cell.* **7**, 887–98.

Latzko, E. and Kelly, G. J. (1983). The many-faceted functions of phosphoenolpyruvate carboxylase in C_3 plants. *Physiol. Veg.* **21**, 805–15.

Lea, P. J., Robinson, S. A. and Stewart, G. R. (1990). The enzymology and metabolism of glutamine, glutamate, and asparagine. In *The Biochemistry of Plants: An Advanced Treatise*. Vol. 16, eds. P. K. Stumpf and E. E. Conn, Academic Press, San Diego, pp. 121–60.

Lerouge, P., Roch, P., Faucher, C., Maillet, F., Truchet, G., Prome, J. C. and Denarie, J. (1990). Symbiotic host specificity is determined by a sulphated and acylated glucosamine oligosaccharide signal. *Nature* **344**, 781–84.

Li, M.-G., Villemur, R., Hussey, P. J., Silflow, C. D., Gantt, J. S. and Snustad, D. P. (1993). Differential expression of six glutamine synthetase genes in *Zea mays*. *Plant Mol. Biol.* **23**, 401–407.

Lillo, C. (1994). Light regulation of nitrate reductase in green leaves of higher plants. *Physiol. Plant.* **90**, 616–20.

Marsolier, M. C., Carrayol, E. and Hirel, B. (1993). Multiple functions of promoter sequences involved in organ-specific expression and ammonia regulation of a cytosolic soybean glutamine synthetase gene in transgenic *Lotus corniculatus*. *Plant Journal* **3**, 405–14.

McGrath, R. B. and Coruzzi, G. M. (1991). A gene network controling glutamine and asparagine biosynthesis in plants. *Plant Journal* **1**, 275–80.

Merrick, M. J. (1992). Regulation of nitrogen fixation genes in free-living and symbiotic bacteria. In *Biological Nitrogen Fixation*, eds G. Stacy, R. H. Burris, H. J. Evans, Chapman-Hall, New York, pp. 835–76.

Newton, W. E. (1993). Nitrogenases: distribution, composition, structure and function. In *New Horizons in Nitrogen Fixation*, eds R. Palacios, J. Mora, and W. Newton, Kluwer, Dordrecht, pp. 5–17.

Nimmo, H. G. (1992). The regulation of phosphoenolpyruvate carboxylase by reversible phosphorylation. In *Post-translational Modification in Plants*, eds N. H. Battey, H. G. Dickinson, and A. M. Hetherington, Cambridge Univ. Press, pp. 161–70.

Pathirana, S. M., Vance, C. P., Miller, S. S. and Gantt, J. S. (1992). Alfalfa root nodule phosphoenolpyruvate carboxylase: characterization of the cDNA and expression in effective and plant-controlled ineffective nodules. *Plant Mol. Biol.* **20**, 437–50.

Pelsy, F. and Caboche, M. (1992). Molecular genetics of nitrate reductase in higher plants. *Adv. Genet.* **30**, 1–39.

Peters, N. K., Frost, J. W. and Long, S. R. (1986). A plant flavone, luteolin, induces expression of *Rhizobium meliloti* nodulation genes. *Science* **233**, 977–80.

Phillips, D. A. and Teuber, L. R. (1992). Plant genetics of symbiotic nitrogen fixation. In *Biological Nitrogen Fixation*, eds G. Stacy, R. H. Burris, H. J. Evans. Chapman-Hall, New York, pp. 625–45.

Redinbaugh, M. G. and Campbell, W. H. (1993). Glutamine synthetase and ferredoxin-dependent glutamate synthase expression in the maize (*Zea mays*) root primary response to nitrate. *Plant Physiol.* **101**, 1249–55.

Reynolds, P. H. S., Smith, L. A., Dickson, J. M. J., Jones, W. T., Jones, S. D., Rodber, K. A., Carne, A. and Liddane, C. P. (1992). Molecular cloning of a cDNA encoding aspartate aminotransferase-P_2 from lupin root nodules. *Plant Mol. Biol.* **19**, 465–72.

Robinson, D. L., Kahn, M. L. and Vance, C. P. (1994). Cellular localization of nodule-enhanced aspartate aminotransferase in *Medicago sativa* L. *Planta* **192**, 202–10.

Robinson, D. L., Pathirana, S. M., Gantt, J. S. and Vance, C. P. (1996). Immunogold localization of nodule-enhanced phosphoenolpyruvate carboxylase in alfalfa. *Plant Cell Environ.*, **19**, 602–608.

Sakakibara, H., Kawabata, S., Takahashi, H., Hase, T. and Sugiyama, T. (1992). Molecular cloning of the family of glutamine synthetase genes from maize: expression of genes for glutamine synthetase and ferredoxin-dependent glutamate synthase in photosynthetic and non-photosynthetic tissues. *Plant Cell Physiol.* **33**, 49–58.

Sakakibara, H., Watanabe, M., Hase, T., and Sugiyama, T. (1991). Molecular cloning and characterization of complementary DNA encoding for ferredoxin-dependent glutamate synthase in maize leaf. *J. Biol. Chem.* **266**, 2028–35.

Sakamoto, A., Ogawa, M., Masumura, T., Shibata, D., Takeba, G., Tanaka, K. and Fujii, S. (1989). Three cDNA sequences coding for glutamine synthetase polypeptides in *Oryza sativa* L. *Plant Mol. Biol.* **13**, 611–14.

Schlaman, H. R. M., Okker, R. J. H. and Lugtenberg, B. J. J. (1992). Regulation of nodulation gene expression by *nodD* in rhizobia. *J. Bacteriol.* **174**, 5177–82.

Schuller, K. A. and Werner, D. (1993). Phosphorylation of soybean (*Glycine max* L.) nodule phosphoenolpyruvate carboxylase *in vitro* decreases sensitivity to inhibition by L-malate. *Plant Physiol.* **101**, 1267–73.

Snustad, D. P., Hunsperger, J. P., Chereskin, B. M. and Messing, J. (1988). Maize glutamine synthetase cDNAs: isolation by direct genetic selection in *Escherichia coli*. *Genetics* **120**, 1111–24.

Spaink, H. P., Sheeley, D. M., van Brussel, A., Glushka, J., York, W., Tak, T., Geiger, O., Kennedy, E., Reinhold, V. and Lugtenberg, B. (1991). A novel highly unsaturated fatty acid moiety of lipooligosaccharide signals determines host specificity of *Rhizobium*. *Nature* **354**, 125–30.

Taniguchi, M., Sawaki, H., Sasakawa, H., Hase, T. and Sugiyama, T. (1992). Cloning and sequence analysis of cDNA encoding aspartate aminotransferase isozymes from *Panicum miliaceum* L., a C_4 plant. *Eur. J. Biochem.* **204**, 611–20.

Tingey, S. V., Tsai, F. Y., Edwards, J. W., Walker, E. L. and Coruzzi, G. M. (1988). Chloroplast and cytosolic glutamine synthetase are encoded by homologous nuclear genes which are differentially expressed *in vivo*. *J. Biol. Chem.* **263**, 9651–57.

Tsai, F. Y. and Coruzzi, G. M. (1990). Dark-induced and organ-specific expression of two asparagine synthetase genes in *Pisum sativum*. *EMBO J.* **9**, 323–32.

Tsai, F. Y. and Coruzzi, G. (1991). Light represses transcription of asparagine synthetase genes in photosynthetic and nonphotosynthetic organs of plants. *Mol. Cell. Biol.* **11**, 4966–72.

Udvardi, M. K. and Kahn, M. L. (1991). Isolation and analysis of a cDNA clone that encodes an alfalfa (*Medicago sativa*) aspartate aminotransferase. *Mol. Gen. Genet.* **231**, 97–105.

Vance, C. P. (1990). Symbiotic nitrogen fixation: recent genetic advances. In *The Biochemistry of Plants: An Advanced Treatise*. Vol. 16. eds. P. K. Stumpf and E. E. Conn. Academic Press, San Diego, pp. 43–88.

Vance, C. P., Egli, M. A., Griffith, S. M. and Miller, S. S. (1988). Plant regulated aspects of nodulation and N_2 fixation. *Plant Cell Environ.* **11**, 413–27.

Vance, C. P., Gregerson, R. G., Robinson, D. L., Miller, S. S. and Gantt, J. S. (1994). Primary assimilation of nitrogen in alfalfa nodules: molecular features of the enzymes involved. *Plant. Sci.* **101**, 51–64.

Vance, C. P. and Heichel, G. H. (1991). Carbon in N_2 fixation: limitation or exquisite adaptation? *Annu. Rev. Plant Physiol. Plant Mol. Biol.* **42**, 373–92.

Vance, C. P., Miller, S. S., Gregerson, R. G., Samac, D. A., Robinson, D. L. and Gantt, J. S. (1995). Alfalfa NADH-dependent glutamate synthase: structure of the gene and importance in symbiotic N_2 fixation. *Plant Journal* **8**, 345–58.

Vincentz, M., Moureaux, T., Leydecker, M. T., Vaucheret, H. and Caboche, M. (1993). Regulation of nitrate and nitrite reductase expression in *Nicotiana plumbaginifolia* leaves by nitrogen and carbon metabolites. *Plant J.* **3**, 315–24.

Wadsworth, G. J., Marmaras, S. M. and Matthews, B. F. (1993). Isolation and characterization of a soybean cDNA clone encoding the plastid form of aspartate aminotransferase. *Plant Mol. Biol.* **21**, 993–1009.

Warner, R. L. and Kleinhofs, A. (1992). Genetics and molecular biology of nitrate metabolism in higher plants. *Physiol. Plant.* **85**, 245–52.

Wray, J. L. (1993). Molecular biology, genetics and regulation of nitrite reduction in higher plants. *Physiol. Plant.* **89**, 607–12.

Yamaya, T., Hayakawa, T., Tanasawa, K., Kamachi, K., Mae, T. and Ojima, K. (1992). Tissue distribution of glutamate synthase and glutamine synthetase in rice leaves. *Plant Physiol.* **100**, 1427–32.

Zehnacker, C., Becker, T. W., Suzuki, A., Carrayol, E., Caboche, M. and Hirel, B. (1992). Purification and properties of tobacco ferredoxin-dependent glutamate synthase, and isolation of corresponding cDNA clones. *Planta* **187**, 266–74.

30 Amino acid and ureide biosynthesis

Robert Ireland

Introduction

Amino acids serve a wide range of functions in plants. They are involved in transporting nitrogen between roots, leaves, fruits etc., and are precursors in the syntheses of many nitrogen-containing compounds, such as chlorophyll and the enzyme cofactors thiamine pyrophosphate (from alanine and methionine) and coenzyme A (from valine, aspartate and cysteine). Amino acids are the structural units from which proteins are made, and also serve as the carbon and nitrogen source for the production of most 'secondary', or 'natural' products, such as alkaloids, phenolic acids and cyanogenic compounds, which are responsible for plant colour, taste, smell and toxicity. Despite the obvious importance of the amino acids, we know far less about their metabolism than we do about that of many other pathways. In several cases, the pathway that is generally accepted for the synthesis of a particular amino acid has been assumed to be the same as that in bacterial or animal systems, and it is only relatively recently that these have been confirmed or novel pathways demonstrated. The production of mutant plants that are deficient in specific enzymes has added much to our knowledge of these pathways, as have the advances in molecular genetic techniques. A number of amino acids can be made by more than one synthetic pathway, depending on, for example, the tissue concerned, time of day, stage of growth, etc. The synthesis of some amino acids in plants remains unclear, as is the case with histidine and many of the non-protein amino acids.

The branched and interwoven nature of amino acid synthetic pathways requires precise control. The nitrogen and carbon flowing through aspartate does not all end up in a single amino acid, but is distributed to several different amino compounds, in a defined manner, according to the needs of the cell at that particular time. The consumption of amino acids for synthetic processes requires the re-establishment and maintenance of pool levels. The conversion of phenylalanine to secondary products, for example, requires increased activity of the shikimate pathway to restore phenylalanine levels, without affecting tyrosine levels. Other metabolic conditions may require the synthesis of phenylalanine and tyrosine simultaneously, and thus these pathways have to be under strict control. The enzymes involved in amino acid synthesis are regulated to different degrees: at one end of the spectrum are enzymes like the transaminases, whose activities appear to be solely dependent on pH and substrate/product concentrations. At the other end of the scale are enzymes like aspartate kinase, that are highly regulated by several metabolites. As with many metabolic pathways, end product inhibition appears to be the principal mechanism by which much amino acid synthesis is regulated.

Amino acids can be grouped together into 'families', each of which are derived from a single 'head' amino acid. For example, the 'aspartate family' is comprised of asparagine, homoserine, threonine, isoleucine and lysine, all synthesized from aspartic acid. Because more than one amino acid may be involved in the synthesis of another, a single amino acid may be assigned to more than one family, and it is not unusual for different authors to differ in their assignments. This chapter will largely be confined to the 'protein' amino acids, so-called because they are those commonly found in proteins. Plants do, however, synthesize hundreds of other amino acids, the 'non-protein' amino acids, which will be dealt with briefly towards the end of the chapter.

Ammonia assimilation: glutamine, glutamate and asparagine

Because of the high affinity of glutamine synthetase (GS) for ammonia, nearly all plant nitrogen is first assimilated into organic form as glutamine, (Lea *et al.*, 1990). This enzyme works in concert with glutamate synthase (GOGAT) in what is generally referred to as the glutamate synthase or GS/GOGAT cycle (see Chapter 29). The net effect of this is the amination of α-ketoglutarate to glutamate. A similar result is achieved by glutamate dehydrogenase (GDH), which also produces glutamate from α-ketoglutarate and ammonia, but it is generally believed that this enzyme is primarily involved in glutamate oxidation, and contributes little to ammonia assimilation (Lea *et al.*, 1992). Other pathways have also been implicated in ammonia assimilation in plants. Alanine dehydrogenase, which has been purified from the blue–green alga, *Anabaena* (Rowell and Stewart, 1976), catalyzes the amination of pyruvate to alanine:

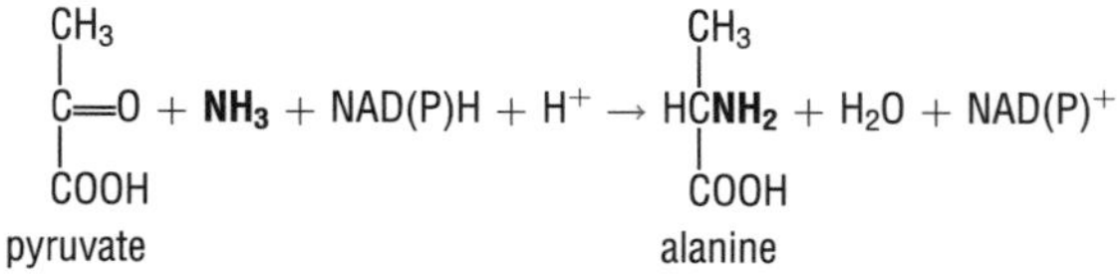

$$CH_3-C(=O)-COOH + NH_3 + NAD(P)H + H^+ \rightarrow CH_3-HCNH_2-COOH + H_2O + NAD(P)^+$$

pyruvate → alanine

In a similar reaction, aspartate dehydrogenase uses ammonia to aminate oxaloacetate to aspartate. Aspartate can also be produced by the enzyme, aspartase, which catalyzes the addition of ammonia to fumarate. Although these three enzymes have been detected in a few plants, they are not thought to make significant contributions to the assimilation of ammonia in plant tissues (Miflin and Lea, 1982). Asparagine synthetase (AS), on the other hand, is present in many plants, and catalyzes the amidation of aspartate to asparagine, using ammonia or glutamine as the amino donor (Fig 30.1). In most cases, it appears that the *in vivo* substrate for this enzyme is glutamine, not ammonia, and hence asparagine synthetase does not usually constitute a route for ammonia assimilation. Oaks and Hirel

(a)

(b)

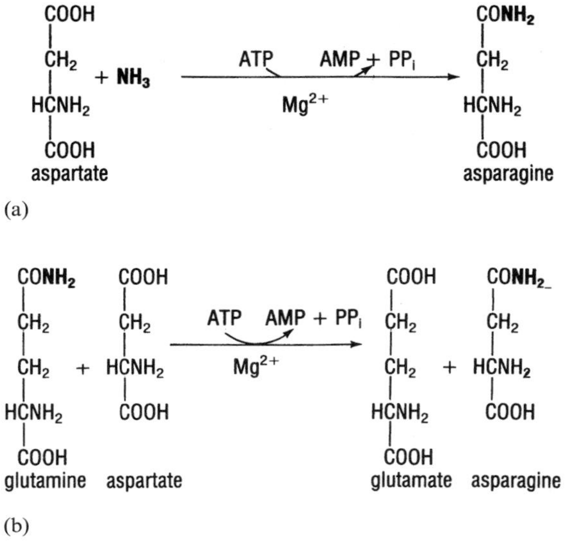

Fig. 30.1. The synthesis of asparagine from aspartate by asparagine synthetases, with either ammonia (a) or glutamine (b) providing the amide nitrogen.

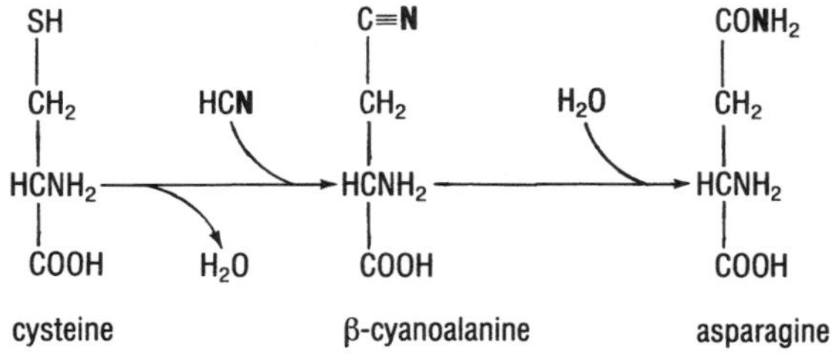

(1985) report that in some tissues, ammonia is the preferred substrate, and asparagine synthetase may indeed represent an entry point for ammonia into organic metabolism in some plants. Asparagine may also be synthesized from the hydrolysis of β-cyanoalanine, which is formed from hydrogen cyanide (HCN) and cysteine (Fig. 30.2).
This pathway has been demonstrated in various plants, including lupins, sorghum, sweet pea and asparagus, but relies on the supply of HCN, which is thought to be limited in most plants. Recent work on the synthesis of methionine and ethylene, however (see below), suggests that HCN synthesis may be

cysteine → β-cyanoalanine → asparagine

Fig. 30.2. The synthesis of asparagine via β-cyanoalanine by the actions of β-cyanoalanine synthase (a) and β-cyanoalanine hydrolase (b).

much more widespread than is currently thought, and the β-cyanolalanine pathway of HCN detoxification may be functioning in many, if not most plants.

Asparagine synthesis is particularly important in the root nodules of legumes, where much of the nitrogen fixed by the bacteria is rapidly transferred to asparagine through the joint activities of glutamine synthetase and asparagine synthetase. Thus a lot of the nitrogen transported in the xylem, away from a nodule, is in the form of asparagine (Sieciechowicz *et al.*, 1988). Asparagine levels in plant tissues vary diurnally, and often increase under stress conditions, such as mineral deficiencies, salt stress, or drought (Stewart and Larher, 1980). The significance of such increases has not been established but may be a means of storing nitrogen when protein synthesis is limited by the stress that the plant is experiencing. The biochemistry and molecular biology of the enzyme, asparagine synthetase is discussed in greater detail in Chapter 29.

For most plants, the data are in agreement with ammonia being assimilated via the actions of the GS/GOGAT cycle. When nitrate labeled with the heavy isotope of nitrogen,[15]N, is fed to plants, the label rapidly appears in both glutamine and glutamate, which then act as the precursors for the synthesis of all of the other amino acids. Nitrate reduction is not the only source of ammonia production in plant tissues, in fact most of the ammonia assimilated by glutamine synthetase in green tissues comes from photorespiration. Other sources of ammonia include reactions catalyzed by enzymes such as arginase, asparaginase and other deamidases (Lea *et al.*, 1992).

As well as being the entry point for nitrogen into plant metabolism, glutamine also serves as a nitrogen transport compound, as does asparagine. These two amides differ only in chain length, but exhibit considerable differences in their biochemical activities and roles in the cell: asparagine is more soluble and less reactive than glutamine and is thus more suited to its role as a transport and storage compound (Sieciechowicz *et al.*, 1988).

Nitrogen flow

The route that nitrogen flow takes in a plant cell depends on the tissue and plant species involved.

Developing wheat leaves, for example, receive most of their nitrogen as nitrate, transported in the xylem from the roots, and thus their starting point for amino acid biosynthesis is glutamine/glutamate as described above. In other plants, such as legumes, much of the nitrate is reduced and converted to organic form in the roots prior to transport in the xylem. Thus, the nitrogen arrives at the leaves in different forms, often as asparagine and glutamine. Similarly, developing seeds or fruits, which are very active in amino acid biosynthesis, will receive most of their nitrogen in the form of amino acids supplied by the phloem. In both cases, these transport compounds are subsequently metabolized to the other amino acids (Lea *et al.*, 1990). The age of the tissue also affects nitrogen flow: young leaves consume all of the incoming nitrogen for growth, whereas mature leaves re-export (in the phloem) most of the nitrogen they receive to the growing apex or developing fruit, as do senescing leaves, which also convert a lot of their proteins and other nitrogenous molecules to transport compounds for export. Diurnal variations in nitrogen flow are also seen. The transport and partitioning of N in plants is discussed in more detail in Chapter 33.

Carbon flow

Three metabolites of glycolysis and the citric acid cycle serve as the major withdrawal points for organic carbon in the syntheses of the amino acids; pyruvate, oxaloacetate, and α-keto-glutarate (Fig. 30.3). Large amounts of phosphoenolpyruvate (PEP) and acetyl coenzyme A are also consumed for amino acid synthesis, and the pentose phosphate pathways (reductive and oxidative) yield ribose 5′-phosphate and erythrose 4′-phosphate for the synthesis of the aromatic amino acids. The Calvin cycle provides the carbon skeletons for the photorespiratory synthesis of glycine and serine, following the oxidation of ribulose 1′,5′-bisphosphate to yield phosphoglycolate. Considerable quantities of α-ketoglutarate are required by the ammonia-assimilating activities of the glutamate synthase cycle, which represents the greatest drain on these carbon pathways. Consumption of amino acids for protein or other

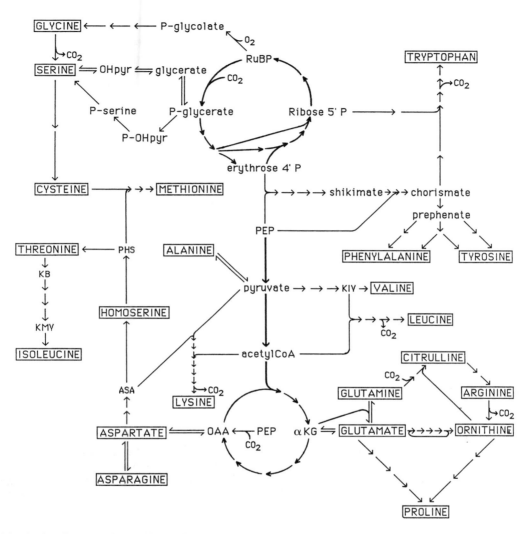

Fig. 30.3. Carbon flow in amino acid biosynthesis. Each arrow represents a separate reaction: for clarity some reactions in the oxidative/reductive pentose phosphate pathway have been omitted, as have the links between it and glycolysis. Some of the intermediates are included using abbreviated names: ketoisovalerate (KIV), ketobutyrate (KB), ketomethylvalerate (KMV), aspartate semialdehyde (ASA), α–ketoglutarate (αKG), oxaloacetic acid (OAA), phosphoenol pyruvate (PEP), rubulose bisphosphate (RuBP), hydroxy-pyruvate (OH-pyr).

syntheses requires that carbon compounds be withdrawn from their pathways to replenish the amino acids and maintain their concentrations in the 'amino acid pools'. Increased glycolytic and anaplerotic activities (such as PEP carboxylase) are then required to restore the levels of these carbon compounds (see Chapters 13 and 32).

Demand for carbon skeletons varies considerably according to many factors – plant species, tissue, age, time of day, stress level etc. In wounded or infected plant tissues there is an increase in the activity of glycolytic and pentose phosphate pathway enzymes, in order to increase production of carbon skeletons for the synthesis of the aromatic amino acids. These are the precursors of so-called 'secondary' metabolites, such as phenolic acids and flavonoids, which are used in defense reactions. The relationship between carbon and nitrogen flow can clearly be seen when nitrogen-deficient plants are transferred to a nitrogen-rich environment. This results in an increase in cellular respiration, because of the increased demand placed on respiratory pathways to supply carbon skeletons needed for the incorporation of nitrogen into organic form (see Chapter 32). The consumption of α-ketoglutarate by the nitrogen-assimilating activities of the GS/GOGAT cycle causes a depletion of citric acid cycle intermediates. Responses to this include increased glycolytic activity which serves to restore the levels of citric acid cycle metabolites, as does increased PEP carboxylase activity. Similar responses are seen when oxaloacetate is withdrawn for aspartate synthesis and when pyruvate is withdrawn from glycolysis for transamination to alanine.

Transamination

It is difficult to discuss the biosynthesis of amino acids without considering their breakdown, since the synthesis of one amino acid often involves the degradation of another. Such is the case in transamination reactions, which are central to amino acid metabolism, distributing nitrogen from glutamate to other amino acids in all cells and most subcellular compartments. Transaminases catalyze the transfer of an amino group from the α-carbon of an amino acid to the α-carbon of a keto acid, producing a new amino and a new keto acid. Less commonly, amines may act as amino donors and aldehydes as amino acceptors in these reactions. Transaminases, also known as aminotransferases, each have a tightly-bound coenzyme; pyridoxal 5'-phosphate. This coenzyme accepts an amino group from the amino acid substrate (A), becoming aminated to pyridoxamine phosphate. The keto acid thus produced (A) is released and the aminated form of the coenzyme then undergoes a reversal of the process, giving up its newly-acquired amino group to the keto acid substrate (B) to produce the amino acid product (B).

$$\begin{array}{cccc} R_1 & R_2 & R_1 & R_2 \\ | & | & | & | \\ HC NH_2 + C=O & \rightleftharpoons & C=O + HC NH_2 \\ | & | & | & | \\ COOH & COOH & COOH & COOH \\ \text{amino} & \text{keto} & \text{keto} & \text{amino} \\ \text{acid A} & \text{acid B} & \text{acid A} & \text{acid B} \end{array}$$

Enzymes have been isolated from plant tissues that can catalyze the transamination of all of the common, or protein amino acids, except proline, which is not an amino acid but an imino acid, and thus has no primary amino group available for transamination.

Transaminases play a key role in the synthesis of amino acids. Since they are usually freely reversible, it is theoretically possible for all amino acids to be synthesized by transamination, but in some cases, the only source of the required keto acid appears to be the transamination reaction itself, so no net synthesis of the amino acid can occur by this route. Where there is another pathway able to provide the keto acid, transamination can contribute significantly to amino acid synthesis, and in fact transaminases catalyze the final step in the synthesis of many amino acids. An example of this is seen in the complex pathway leading to the synthesis of phenylalanine, where the final step is the transamination of phenylpyruvate. Similarly, tyrosine and leucine are produced by the amination of hydroxyphenylpyruvate and ketoisocaproate, respectively (Ireland and Joy, 1985).

Unlike other metabolites, many amino acids are often present in millimolar concentrations in plant cells, not just in the vacuole, which seems to act as a repository for high concentrations of many compounds, but also in the cytosol, chloroplasts and other organelles. Corn leaf chloroplasts, for example, have been shown to contain $16 \, \text{mM}$ asparagine, $10 \, \text{mM}$ serine and alanine, and $5 \, \text{mM}$ glycine and aspartate (Chapman and Leech, 1979). Not surprisingly, the kinetic constants for the enzymes metabolizing these compounds also seem to be high when compared with, for example, the enzymes of the glycolytic pathway: it is common to see K_m values for amino acid substrates in the $2–6 \, \text{mM}$ range. Transaminases are not highly specific, and will react with a number of amino acid or keto acid substrates. Thus a transaminase may be designated 'aspartate transaminase', but also be able to use glutamate as an amino acid substrate, and either α-ketoglutarate or oxaloacetate as keto acid substrates (Ireland and Joy, 1985). Given that nearly all plant nitrogen has to pass through glutamate, it is not surprising that many transaminases can use glutamate or, to a lesser extent, aspartate as amino donors in the synthesis of a wide range of amino acids such as glycine, phenylalanine, tyrosine, serine etc. The extensive use of glutamate as an amino donor results in the production of large quantities of α-ketoglutarate, which is probably reaminated by the GS/GOGAT cycle during ammonia assimilation.

Following the assimilation of nitrogen into glutamine and glutamate, transaminases serve to redistribute the nitrogen to a range of other amino acids and they contribute to the maintenance of relatively stable amino acid pools. Several transaminases have functions that are unique to plants, such as those involved in carbon assimilation and other processes requiring the 'shuttling' of metabolites. Carbon shuttling in C4 plants makes extensive use of transaminases to produce aspartate and alanine, which are used as transport compounds for the movement of fixed carbon between the mesophyll and bundle sheath cells (Ireland and Joy, 1985) (see Chapter 23). Organelle membranes are largely impermeable to pyridine nucleotides, and so shuttle mechanisms involving transaminases have to be used to transfer reducing power across chloroplast and mitochondrial membranes (see Chapters 19 and 25).

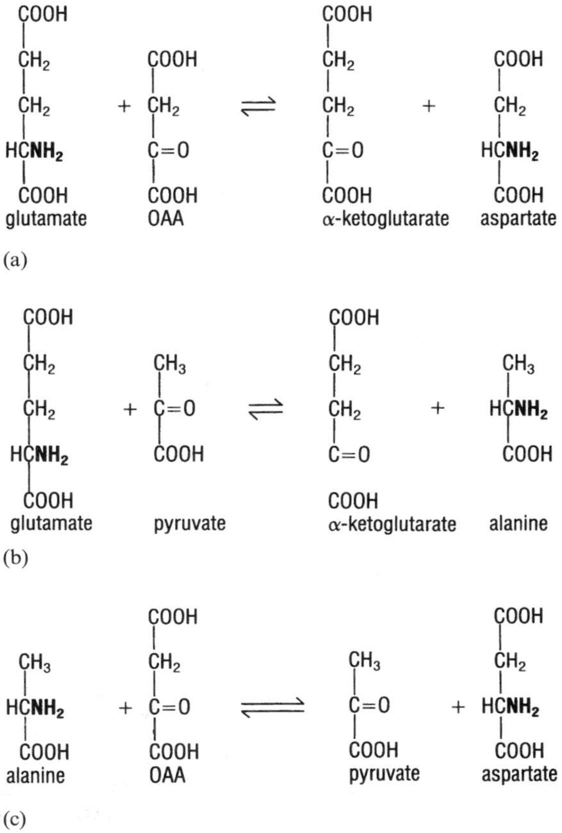

Aspartate and alanine

After incorporation into glutamate, nitrogen is quickly distributed to other amino acids, much of it going directly to alanine and aspartate via transamination (Fig. 30.4, a and b).
In both of these reactions, α-ketoglutarate is produced, which can be reused in the GS/GOGAT cycle. Transamination between alanine and oxaloacetate (Fig. 30.4c) completes the aspartate–alanine–glutamate triangle seen in Fig. 30.5, transferring amino groups between alanine and aspartate. The transaminases involved are all present in both the chloroplast (Kirk and Leech, 1972) and the cytosol, and the first two are also found in mitochondria (Ireland and Joy, 1985).

Fig. 30.4. The major transamination reactions that occur in plant tissues. (a) glutamate : oxaloacetate aminotransferase (b) glutamate : pyruvate aminotransferase (c) alanine : oxaloacetate aminotransferase

Another enzyme responsible for aspartate synthesis is asparaginase, a cytosolic enzyme that catalyzes the hydrolysis of asparagine to aspartate and ammonia. This is a major source of aspartate in developing legume leaves and fruits/seeds which often receive much of their nitrogen in the form of asparagine (from the xylem or phloem). Asparaginase activity varies during the development of legume seeds and leaves, and also varies diurnally, increasing in light, and decreasing in darkness. It appears that the diurnal variation is due to a relatively rapid turnover of the enzyme, with synthesis occurring in the light, and degradation by proteolysis in the dark. Since the enzyme synthesises ammonia, which can be harmful

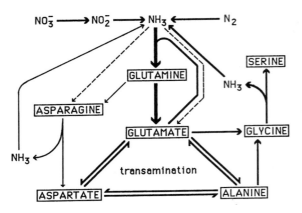

Fig. 30.5. Primary routes of nitrogen flow in amino acid synthesis. Ammonia is assimilated into the amino acids shown, which then serve as precursors and amino donors in the synthesis of the other amino acids.

to plant tissues, it is not surprising that it is regulated in this way. The enzyme is only functional in the light when there is adequate ATP and reducing power (from photosynthetic electron transport) to allow the GS/GOGAT cycle to assimilate the ammonia produced. In the dark, ATP is produced from the respiration of storage carbohydrates. Hence, reducing ammonia production by decreasing the level of asparaginase activity will serve to alleviate demand on this limited ATP supply. Furthermore, in darkness, protein synthesis is reduced because of a lower concentration of ATP, and hence there will be a reduced demand for aspartate, which is a precursor in the synthesis of many of the other amino acids. (Sieciechowicz *et al.*, 1988)

Asparaginase has been cloned from lupin (*Lupinus angustifolius*) seeds, and the genomic sequence established (Dickson *et al.*, 1992). Analysis of the upstream region revealed sequences associated with seed-specific and nodule-specific expression. Other regulatory sequences, such as might be involved in the light–dark switching described above, have yet to be identified.

Regardless of the form in which nitrogen arrives at a tissue, be it nitrate, asparagine, glutamate or other amino acids, it is quickly distributed within pools of amino acids in different subcellular compartments, within which aspartate, glutamate and alanine are often the major components.

Glycine and Serine

Most of the work on glycine and serine metabolism in plant tissues has been concerned with photorespiration. In this process, glycolate, produced in the chloroplast by the oxygenase activity of RUBISCO (see Chapter 18, Fig. 18.1), is transported via the cytosol into the peroxisome where it is oxidized to glyoxylate. It is then transaminated to glycine, using glutamate, alanine, serine or asparagine as amino donors (Ta *et al.*, 1985). In pea leaves, it is estimated that 38% of the glycine amino nitrogen comes from serine, 28% from alanine, 23% from glutamate, and 7% from asparagine (Ta and Joy, 1986). Following transport into the mitochondrion, two glycine molecules are converted to one molecule of serine by the actions of the enzymes, glycine decarboxylase and serine hydroxymethyltransferase (see Chapter 22, Fig. 22.1). This serine enters the amino acid pools where some is used for protein synthesis, some for the synthesis of other amino acids, but most of it moves to the peroxisome where it is transaminated with glyoxylate (see above) yielding glycine and the keto analog of serine, hydroxypyruvate. The latter is reduced to glycerate and subsequently converted to sugars.

Regulation of photorespiratory production of glycine and serine is largely dependent on the relative concentrations of CO_2 and O_2, which compete for the active site of RUBISCO (see Chapter 18). The rate of glycine production during photorespiration is very high – as much as 50% of the photosynthetic rate. It is therefore not surprising that collectively glycine decarboxylase and serine hydroxymethyltransferase account for half of the soluble matrix proteins of green leaf mitochondria (Oliver *et al.*, 1990).

Glycine decarboxylase is a complex composed of four different subunits, in an uneven stoichiometry. Serine hydroxymethyltransferase exists in leaf mitochondria as a homotetramer. Each of the subunits of the two enzymes have been isolated and cloned from peas. All of these proteins are nuclear-encoded, and synthesized in the cytosol with a characteristic N-terminal leader sequence that is removed following uptake by the mitochondria. The expression of all of these proteins is light-dependent and tissue-specific: expression is at a much higher level in leaves than in stems or roots (see Rawsthorne *et al.*, 1995).

Serine can also be synthesized by two other routes, not involving glycine. Both of these pathways use phosphoglycerate (derived from either the Calvin cycle or the glycolytic pathway) as the starting point. Phosphoglycerate phosphatase and glycerate dehydrogenase can convert phosphoglycerate first to glycerate and then to hydroxypyruvate, which can be transaminated to serine. The other route from phosphoglycerate is via phosphohydroxypyruvate and phosphoserine (Fig. 30.3). The enzymes for this pathway have been found in plants, but the contribution it makes to serine synthesis is not clear (Ireland and Hiltz, 1995).

Lysine, isoleucine and threonine

The relationship amongst the amino acids in the 'aspartate family' is shown in Fig. 30.3. Interest in these pathways remains high, partly because this group contains the 'essential' amino acids, lysine and methionine, whose levels in many crop plants are not as high as we would like (Last, 1992). The key reactions in the synthesis of these amino acids are the two involved in the synthesis of aspartic semialdehyde (ASA) from aspartate, catalyzed by the enzymes aspartate kinase and aspartate semialdehyde dehydrogenase (Fig. 30.6). Aspartate semialdehyde is at an important branch point in amino acid synthesis, since it can either be reduced to homoserine or condensed with pyruvate to give dihydropicolinic acid, which subsequently undergoes a series of reactions to produce lysine (Fig. 30.3). The synthesis of lysine involves six steps, including acylation (from acetyl-CoA), transamination and decarboxylation, but few of the enzymes responsible have been isolated from plant tissues. The pathway has been deduced from feeding studies, isolation of intermediates, and

comparison with the bacterial pathway (see Bryan, 1990). Homoserine is an amino acid not found in proteins and hence is not usually present in appreciable concentrations in plants, with the exception of peas, where it can constitute 70% of the soluble nitrogen in 1-week old seedlings (Mitchell and Bidwell, 1970). In most plants nearly all of their homoserine is phosphorylated to phosphohomoserine. This metabolite is also at a metabolic branch point, since it can be converted to methionine (see below) or rearranged in a single step to threonine by the enzyme threonine synthase. Threonine can either be used for protein synthesis or metabolized to isoleucine, the synthesis of which begins with the deamination of threonine to ketobutyrate. This is then converted by a series of reactions to ketomethylvalerate, which is subsequently transaminated to isoleucine (Bryan, 1990).

The branched nature of the pathway requires that the regulation of the synthesis of the aspartate family amino acids must occur at several points. The first enzyme showing regulatory properties is aspartate kinase, which occurs in most plants as two distinct isozymes, both of which are found in the chloroplast (Wallsgrove *et al.*, 1983). Several metabolites affect the activity of these isozymes, but the most effective are lysine and threonine, each of which inhibits only one of the isozymes. Thus there are lysine-sensitive and threonine-sensitive aspartate kinase isozymes in plant cells. Hence the presence of high levels of only one of these compounds is insufficient to shut the pathway down since the other isozyme will still allow the pathway to operate. Lysine is also a major regulator of another key enzyme, dihydropicolinate synthase, which catalyzes the first step in the sequence of reactions leading from aspartate semialdehyde to lysine. Most of this inhibition data comes from work on isolated enzymes, but is supported both by feeding studies and by work with mutants (see below).

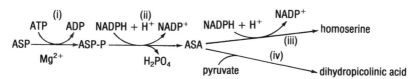

Fig. 30.6. The synthesis of homoserine and dihydropicolinic acid by the actions of aspartate kinase (i), aspartate semialdehyde dehydrogenase (ii), homoserine dehydrogenase (iii) and dihydropicolinate synthase (iv).

Efforts to increase the levels of lysine (and threonine) in crop plants have led to the development of mutants where the lysine feedback inhibition described above is diminished. A mutant of *Nicotiana sylvestris* has been developed which contains an Aspartate kinase isozyme that is completely unaffected by lysine. As a result, levels of free lysine, methionine, and particularly threonine, increased significantly in the plant (Frankard *et al.*, 1992). Another mutant of *Nicotiana sylvestris* has a form of dihydropicolinate synthase that is also insensitive to lysine. Further work on these two key enzymes has produced cDNA clones from a variety of plants encoding the different isoforms of aspartate kinase and dihydropicolinate synthase (see Jacobs *et al.*, 1995). These cDNA clones have facilitated the isolation of genomic clones and an examination of upstream regulatory elements is now in progress.

As well as being a substrate for dihydropicolinate synthase, aspartate semialdehyde is also a substrate for homoserine dehydrogenase, which is also regulated by feedback inhibition. Two forms of homoserine dehydrogenase exist in plant tissues, one is sensitive to inhibition by threonine, the other is not. The threonine-sensitive form can exist in a variety of conformational states, each of which exhibit different catalytic and regulatory properties (Krishnaswamy and Bryan, 1983).

The metabolism of phosphohomoserine is also regulated, since it is the common precursor of threonine, isoleucine and methionine. In barley, the methionine derivative, *S*-adenosylmethionine, has been shown to stimulate the activity of threonine synthase up to 20-fold (Giovanelli *et al.*, 1984). This would divert carbon and nitrogen towards threonine synthesis when levels of methionine are high. In radish and sugar beet, threonine synthesis is inhibited by cysteine. Cysteine reacts with phosphohomoserine as the first step in methionine synthesis, so that when cysteine levels are high, threonine synthesis is decreased to make more phosphohomoserine available to react with cysteine. The first step in isoleucine synthesis, catalyzed by threonine dehydratase, is subject to feedback inhibition by isoleucine (Giovanelli *et al.*, 1988). Multiple forms of threonine dehydratase have been found in Maize, but show similar properties to each other (Bryan, 1990).

Most of the enzymes involved in aspartate family synthesis have been found in the chloroplast, and it is likely that the whole pathway occurs in this organelle (see Bryan, 1990). Illuminated chloroplasts will convert ^{14}C-aspartate or malate to lysine, threonine and isoleucine (Mills *et al.*, 1980), and analysis of cDNA sequences from all aspartate family enzymes cloned so far has shown them all to have the N-terminal transit peptide characteristic of nuclear-encoded chloroplastic enzymes (see Jacobs *et al.*, 1995).

Valine and leucine

Pyruvate and acetyl-CoA provide the carbon skeletons for these two 'branched-chain' amino acids. Isoleucine is also a branched chain amino acid, similar in structure to leucine, but derives its carbon from aspartate (see above). These three amino acids are often grouped together, not only because of structural considerations, but also because they share several common enzymes in their syntheses. The same enzymes that convert ketobutyrate to isoleucine also convert pyruvate to valine in a parallel but distinct pathway, with no sharing of intermediates. A branch point in the valine pathway is at α-ketoisovalerate (KIV), which can be considered equivalent to α-ketobutyrate in the isoleucine pathway. α-ketoisovalerate can either be transaminated directly to valine, or condensed with the methyl group of acetyl coenzyme A to give isopropylmalate, which is isomerized and decarboxylated to α-ketoisocaproate (see Fig. 30.3), then transaminated to leucine (Bryan, 1990). As with other branched pathways, close regulation is necessary to produce the desired levels of the different products, and again feedback inhibition is in effect. Valine and leucine both inhibit the enzyme, acetolactate (acetohydroxyacid) synthase, which catalyzes both the conversion of α-ketobutyrate to acetohydroxybutyrate in the synthesis of isoleucine, and the conversion of pyruvate to acetolactate in the synthesis of valine and leucine. Leucine inhibits the synthesis of isopropylmalate from ketoisovalerate (isopropylmalate synthase) (Bryan, 1990).

The elucidation of these pathways has been driven, in part, by the success of the imidazolinone and sulfonylurea herbicides, which inhibit acetohydroxyacid synthase (Shaner and Singh, 1992). These pathways appear to be in the chloroplast since isolated chloroplasts can synthesize valine from $^{14}CO_2$, and several enzymes of the pathway have been found in isolated chloroplasts (Bryan, 1990).

The sulfur amino acids: cysteine and methionine

Methionine is also a member of the aspartate family, since one of its precursors is phosphohomoserine. However, it is often put in a separate group with the other sulfur amino acid, cysteine, which is another precursor in methionine synthesis.

Sulfur is usually taken up by plants as sulfate, which is transported in the xylem to leaf tissue and reduced by a series of reactions to protein-bound sulfide, then incorporated into amino acids. Nearly all sulfate reduction occurs in photosynthetic tissues, specifically in chloroplasts, where it accounts for a significant amount of the ATP and reduced ferredoxin that is consumed. Most of the sulfide is incorporated into cysteine by reaction with o-acetylserine, also in the chloroplast (Fig. 30.7). The control mechanisms in cysteine synthesis appear to involve feedback inhibition by cysteine at several steps in the reduction of sulfate to sulfide, and at sulfate uptake itself. There has been some discussion, but little conclusive evidence, about the regulation of cysteine synthase and serine acetyltransferase (see Anderson, 1990).

Cysteine and phosphohomoserine react to produce cystathionine, which is subsequently cleaved to give homocysteine. The final step in methionine synthesis is the methylation of homocysteine, which involves the polyglutamyl derivative of folic acid (Cossins, 1987). These reactions have been demonstrated in many plants, but the specific nature and location of the enzymes involved has not been established. There is some evidence that cystathionine synthase is in the chloroplast, and that the subsequent reactions leading to methionine occur in the mitochondria.

Another pathway involved in methionine synthesis has been demonstrated in plants. This is a cycle which starts with the ribose moiety of ATP being used in the conversion of methionine to S-adenosylmethionine, an important one-carbon donor in plant metabolism. The S-adenosylmethionine is cleaved to give methylthioadenosine and the non-protein amino acid, aminocyclopropane carboxylate, which is subsequently broken down to give CO_2, hydrogen cyanide, and the plant growth regulator, ethylene (Kushad et al., 1983; Miyazaki and Yang, 1987). The methylthioadenosine is converted through a series of reactions back to methionine. This is thus a route for ethylene synthesis from ATP in which methionine is turned over but results in no net methionine synthesis. The control and integration of this pathway with the homoserine route of methionine synthesis has yet to be determined.

As well as being used in protein and ethylene synthesis, methionine is also withdrawn from amino acid pools for the synthesis of the enzyme cofactors spermine and spermidine, as well as various other compounds. Methionine is known to inhibit its own synthesis, but the nature of this control is not understood (see Anderson, 1990).

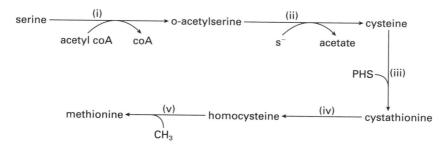

Fig. 30.7. The synthesis of cysteine and methionine. Enzymes: (i) serine acetyltransferase, (ii) cysteine synthase, (iii) cystathionine synthase, (iv) cystathionine lyase, (v) methionine synthase.

Arginine and proline

Glutamate is the precursor of glutamine, arginine and proline. In the synthesis of arginine, glutamate is first metabolized to the nonprotein amino acid, ornithine, via a series of acetylated intermediates (Fig. 30.8). The subsequent metabolism of ornithine appears to be by the same series of reactions that occur in animal tissues, the ornithine cycle (Ugalde *et al.*, 1995).

Little is known about the subcellular localization of these reactions in plant cells, but some of the reactions have been shown to be chloroplastic (Taylor and Stewart, 1981). The use of acetylated intermediates in this pathway probably serves to prevent competition between the syntheses of arginine and proline, which are similar in several ways. Arginine has been shown to inhibit the phosphorylation of acetylglutamate, another example of an end product inhibiting an early reaction in its synthesis. Proline levels are often observed to increase dramatically (up to 200 fold) under stress conditions such as increased salinity, drought, or high temperatures (Stewart and Larher, 1980; Pahlich, 1992).

In the first step of proline synthesis, glutamate is reduced to glutamyl-5'-semialdehyde (Fig. 30.9), via an enzyme-bound intermediate, glutamyl 5'-phosphate (similar to the production of aspartate semialdehyde from aspartate). The semialdehyde

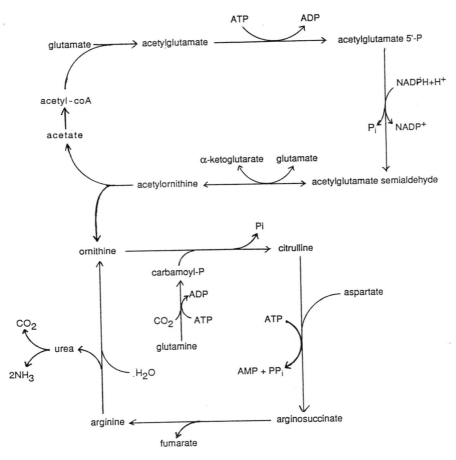

Fig. 30.8. The synthesis of ornithine, citrulline and arginine.

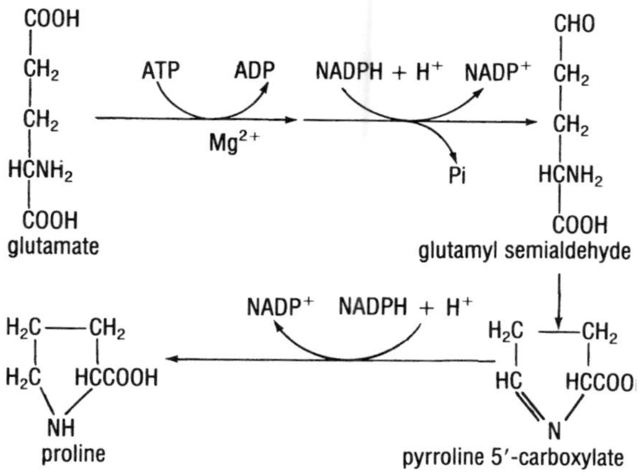

Fig. 30.9. The production of proline from glutamate.

spontaneously cyclizes to give pyrroline 5'-carboxylic acid, which is then reduced to proline (Thompson, 1980).

Proline appears to feed-back inhibit its own synthesis, which probably occurs in the cytosol, but the same series of reactions apparently occurs in the reverse direction in the mitochondrion, providing another route for glutamate synthesis. This pathway is not normally of great significance, but its importance may increase following stress conditions when it also appears that proline and arginine metabolism become linked: ornithine can be converted to glutamate semialdehyde either directly by transamination, or through a series of reactions via proline and the reverse of the proline synthetic pathway. These complex interconversions have been demonstrated by feeding experiments: radioactive label introduced into plant tissue via ornithine or arginine is transferred to proline, and label from proline is transferred to pyrroline 5'-carboxylate, glutamyl 5'-semialdehyde and glutamate (with prolonged feedings also to other amino acids and citric acid cycle intermediates) (Thompson, 1980). Glutamyl semialdehyde can also be produced by transamination of ornithine. The transaminase responsible for this is present in appreciable quantities in many plant tissues, including wheat endosperm, where it is believed that all proline is synthesised from ornithine rather than via glutamyl phosphate (Ugalde et al., 1995).

The aromatic amino acids: phenylalanine, tyrosine and tryptophan

These three amino acids appear to be synthesized exclusively by the 'shikimic acid pathway', the first step of which is the condensation of erythrose 4'-phosphate (derived from the oxidative pentose phosphate pathway or the Calvin cycle) with phosphoenolpyruvate (from glycolysis) to produce 3'-deoxy D-arabino heptulosonic acid 7'-phosphate (DAHP). This undergoes a series of reactions, including loss of a phosphate, ring closure and a reduction to give shikimic acid which is then phosphorylated by shikimate kinase (Fig. 30.3). Shikimate phosphate is combined with a further molecule of PEP to give 3-enolpyruvylshikimate 5-phosphate (EPSP). This latter reaction is catalyzed by the enzyme EPSP synthase, which has received considerable attention because it is inhibited by the herbicide, glyphosate. To have crop plants resistant to this herbicide with weeds being susceptible, is of obvious commercial interest. Glyphosate-tolerance has been induced in transgenic plants by having the EPSP synthase gene over-expressed. EPSP is converted to chorismic acid, which is at a branch point in this pathway, and can undergo two different reactions, one leading to tryptophan, and the other to phenylalanine and tyrosine.

The synthesis of tryptophan from chorismate begins with the reaction of chorismate with the amide group of glutamine to produce anthranilic acid, which subsequently condenses with phosphoribosyl pyrophosphate (derived from ribose 5'-phosphate) to give phosphoribosyl anthranilate. This molecule undergoes a further series of reactions to produce indole glycerol phosphate, which then reacts with serine to produce tryptophan (catalyzed by tryptophan synthetase) (see Bryan, 1990).

The synthesis of phenylalanine and tyrosine starts with the rearrangement of chorismate by chorismate mutase to prephenic acid, whose further metabolism has been the subject of some debate (Fig. 30.10). For some time the synthesis of phenylalanine and tyrosine from prephenate in plants was assumed to be the same as in bacteria, where the prephenate is either dehydrated to phenylpyruvate (prephenate dehydratase) or oxdatively decarboxylated to

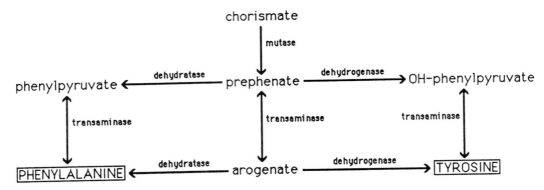

Fig. 30.10. Phenylalanine and tyrosine synthesis via different routes from chorismate.

hydroxyphenylpyruvate (prephenate dehydrogenase). Both of these keto acids are subsequently aminated by transaminases, the former to phenylalanine, and the latter to tyrosine. Although some of the enzymes involved in this route have been found in plants, there is a growing body of evidence which suggests that another route is either also, or in some plants solely, in operation (see Rubin and Jensen, 1979). Formerly called the 'pretyrosine' pathway, it is now generally referred to as the 'arogenate pathway', and involves the transamination of prephenate to arogenate which is then directly converted to either phenylalanine (arogenate dehydratase) or tyrosine (arogenate dehydrogenase). Arogenate dehydratase has been purified from sorghum and its activity shown to be inhibited by phenylalanine and stimulated by tyrosine, as might be expected from its position in the pathway (Siehl and Conn, 1988). Arogenate dehydrogenase has also been purified from sorghum and characterized, and is strongly inhibited by tyrosine, but unaffected by phenylalanine or other metabolites of the pathway (Connelly and Conn, 1985). There is good evidence that the arogenate pathway is functioning in a number of plants, but further work is required to determine the contribution made by the two pathways to aromatic amino acid synthesis in other plants.

Many of the enzymes of tryptophan synthesis have been found in the chloroplast, and labeling studies with $^{14}CO_2$ have shown that chloroplasts contain the complete pathways for the synthesis of the aromatic amino acids. It is believed that these pathways also exist in the cytosol, and perhaps other subcellular compartments. As might be expected, feedback inhibition by tryptophan affects the synthesis of anthranilate from chorismate. Phenylalanine and tyrosine also inhibit their own synthesis, but it is not clear how this occurs. Two isoforms of chorismate mutase exist in a variety of plants, one being sensitive to inhibition by phenylalanine, tyrosine and tryptophan, whereas the other is not. The inhibition, however, is very much dependent on assay conditions, and is not yet well-defined (see Bryan, 1990).

When plant tissues are wounded or infected by parasites, respiration increases, along with associated carbohydrate metabolism. This is principally to provide the precursors for the synthesis of defense and wound-repair compounds, such as phenolic acids, suberin and lignin. Most of these compounds are synthesized from phenylalanine, tyrosine or components of the shikimate pathway, all of which are synthesized from phosphoenolpyruvate (from glycolysis) and erythrose 4'-phosphate (from the pentose phosphate pathways). Under such circumstances, many of the enzymes of glycolysis, such as phosphofructokinase and pyruvate kinase show increased activity, and in some cases, such as pyruvate kinase, there is increased synthesis of the enzyme. The net result is a dramatic increase in glycolytic capacity. Pentose phosphate pathway activity increases even more than glycolysis in response to wounding or infection, largely as a result of increased synthesis of glucose 6'-phosphate dehydrogenase and

6'-phosphogluconate dehydrogenase. Thus the supply of shikimate pathway precursors is enhanced, facilitating an increase in the synthesis of aromatic amino acids, which are then used in the synthesis of defense or wound-repair compounds (Uritani and Asahi, 1980).

Non-protein amino acids

Many plants channel large amounts of nitrogen into amino acids that are not usually constituents of proteins. These non-protein amino acids comprise a very diverse and often complex group of compounds: several hundred of them have been found in plants, most often in seeds (especially those of the legumes), where they can accumulate to high levels. They are found in all plant tissues as intermediates in the synthesis of protein amino acids, and in a more restricted range of plants as metabolic 'end products' (Fowden, 1980). Examples of the former category have already been encountered in this chapter: homoserine is a nonprotein amino acid intermediate in the synthesis of threonine, isoleucine and methionine, as is diaminopimelic acid in the synthesis of lysine. In the case of the nonprotein amino acids which are end products, their function is often unclear, but they are frequently toxic to animals and found in seeds where they appear to serve both as a storage reserve and as a feeding-deterrant to herbivores. The mode of their toxicity varies, but is usually based on interference with regulation, transport or protein synthesis, where enzymes and other proteins are not able to distinguish between the amino acid they normally deal with, and the anolog which presents a very similar structure. One nonprotein amino acid fairly common in legumes is canavanine, which can account for up to 6% of the fresh weight of the seeds of the Jackbean (*Canavalia ensiformis*). It is very similar in structure to arginine and is thus able to interfere with arginine metabolism in animals that ingest the seeds. When ingested it causes a variety of toxic effects, including pupal malformation in insects, and immune dysfunction in vertebrates (see D'Mello, 1995). Other toxic arginine analogs include homoarginine and indospicine, also found in legumes.

canavanine   homoarginine   indospicine   arginine

As with most non-protein amino acids, the pathway by which canavanine is synthesized is not clear, but seems to be similar to the ornithine cycle which produces arginine (Rosenthal, 1982). The levels of non-protein amino acids often increase under stress conditions; for example, the levels of γ-aminobutyric acid, the decarboxylation product of glutamic acid, often increase in response to stresses such as hypoxia (Stewart and Larher, 1980).

Ureides

The ureides are compounds such as citrulline, allantoin and allantoic acid, which are related to urea and used for nitrogen transport in a variety of plants, including many legumes. Citrulline synthesis has been mentioned above, and so this section will briefly describe the synthesis of allantoin and allantoic acid. An excellent and comprehensive discussion of this subject can be found in the review by Schubert and Boland (1990).

Many legumes which normally use the amides, glutamine and asparagine for nitrogen transport from their roots switch to ureide synthesis when they are nodulated. These plants are sometimes referred to as the 'tropical' legumes and include soybeans, cowpeas and mungbeans. The 'temperate' legumes such as peas, lupins and alfalfa, continue to synthesize amides regardless of whether they are nodulated or not. The rationale for this switch from amide to ureide synthesis in the tropical legumes involves the economy of carbon use: the relative metabolic costs of using ureides or amides has been the subject of some debate, but it does appear that the ureide producers use less organic carbon to transport the same amount of nitrogen as do the amides (Pate *et al.*, 1981). Catabolic costs are difficult to estimate, since the pathways have not been completely elucidated, but catabolism of ureides appears to be less efficient than catabolism of amides, since some of the ureide carbon is lost as CO_2. This may not be particularly significant in the light since there is presumably an excess of ATP under such conditions (Schubert and Boland, 1990).

Ureides are synthesized by the oxidation of purines, which are themselves derived from glutamine, glycine, aspartate and ribose 5'-phosphate. Root nodules contain both infected (with the nitrogen-fixing bacteroids) and uninfected cells, both of which are involved in ureide synthesis. In the bacteroids, dinitrogen gas is reduced to ammonia, which is excreted into the cytosol of the infected host cell, then assimilated into glutamine by glutamine synthetase. Purine synthesis from this glutamine then appears to occur in the plastids of these infected cells (Fig. 30.11). Feeding experiments with ^{14}C glycine have shown that *de novo* synthesis is far more important than the salvage pathways from the nucleic acids in providing purines for ureide synthesis (Boland and Schubert, 1982).

In legume nodules, the purine, xanthine, is exported from the plastid to the cytosol, where it is oxidized to uric acid by xanthine dehydrogenase (Fig. 30.12). There has been some debate about the location of this reaction in the nodule. The use of sucrose density gradients indicated that xanthine dehydrogenase is in the cytosol of both infected and uninfected cells (Shelp *et al.*, 1983), whereas immunochemical studies have assigned it solely to infected cells (Triplett, 1985), or to the plastids of uninfected cells (Nguyen, 1986). The next step occurs in the peroxisomes, apparently in the uninfected cells, and involves the action of uricase on

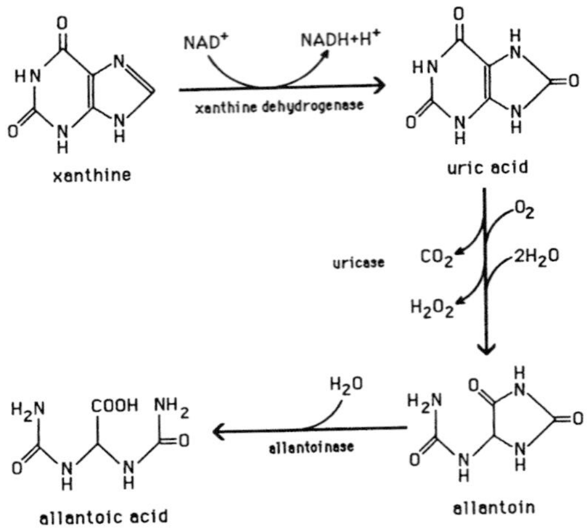

Fig. 30.12. Synthesis of allantoin and allantoic acid from the purine, xanthine.

uric acid, producing allantoin, CO_2 and H_2O_2. Depending on the plant species concerned, differing proportions of the allantoin are used directly for transport in the xylem, or first converted to allantoic acid in the endoplasmic reticulum by the enzyme allantoinase. Little is known about the regulation of ureide synthesis in plant tissues. Supplying exogenous nitrate to nodulated plants will decrease ureide production and increase amide synthesis, but the mechanism behind this switch is unclear.

Concluding Remarks

Research on the biosynthesis of amino acids is entering a new and fascinating stage. Prior to 1980, pathways elucidation followed from extensive enzyme isolation and characterization, tied in with feeding studies. With the development of specific mutants, with either diminished or enhanced levels of specific enzymes, putative pathways have been confirmed or, in some cases, proved to be incorrect. Most recently, extensive activity on the molecular genetics front has led to the availability of cDNA clones for the enzymes involved in amino acid biosynthesis, which, in turn, have allowed us to examine regulation of these pathways at the gene expression level. Our attention is now turning to using these clones to develop transgenic plants which may lead to direct

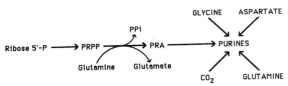

Fig. 30.11. Biosynthesis of purines in root nodules: 5-phosphoribosylpyrophosphate (PRPP), 5-phosphoribosylamine (PRA).

economic benefits, perhaps by increasing levels of essential amino acids in some of our crop plants, or by producing new herbicide-resistant plants or new herbicide strategies. Of more direct relevance to the theme of this chapter will be the use of such transgenic plants to further elucidate the synthetic pathways, to determine the regulatory steps and to determine the relative contributions made by enzymes found in more than one subcellular location.

Further reading

Bryan, J. K. (1990). Advances in the Biochemistry of Amino Acid Biosynthesis. In *The Biochemistry of Plants*, Vol. 16, eds B. J. Miflin and P. J. Lea, Academic Press, New York, pp. 161–195.

Ireland, R. J. and Joy, K. W. (1985). Plant Transaminases. In *Transaminases*, eds P. Christen and D. E. Metzler, John Wiley & Sons, New York, pp. 376–84.

Last, R. L. (1992). The genetics of nitrogen assimilation and amino acid biosynthesis in flowering plants: progress and prospects. *International Review of Cytology* **143**, 297–330.

Lea, P. J., Robinson, S. A. and Stewart, G. R. (1990). The Enzymology and Metabolism of Glutamine, Glutamate and Asparagine. In *The Biochemistry of Plants*, Vol. 16, eds B. J. Miflin and P. J. Lea, Academic Press, New York, pp. 121–159.

Rosenthal, G. A. (1982). *Plant Nonprotein Amino and Imino Acids*, Academic Press, New York.

Schubert, K. R. and Boland, M. J. (1990). The Ureides. In *The Biochemistry of Plants* Vol. 16, eds B. J. Miflin and P. J. Lea, Academic Press, New York, pp. 197–282.

Singh, B. K., Flores, H. E. and Shannon, J. C. (1992) *Biosynthesis and Molecular Regulation of Amino Acids in Plants*, eds B. Singh, H. Flores and J. Shannon, American Society of Plant Physiologists.

References

Anderson, J. W. (1990). Sulfur Metabolism in Plants. In *The Biochemistry of Plants*, Vol. 16, ed. B. J. Miflin and P. J. Lea, Academic Press, New York, pp. 327–81.

Boland, M. and Schubert, K. R. (1982). Purine biosynthesis and catabolism in soybean root nodules: incorporation of ^{14}C from $^{14}CO_2$ into xanthine. *Arch. Biochem. Biophys.* **213**, 486–91.

Chapman, D. J. and Leech, R. M. (1979). Changes in Pool Sizes of Free Amino Acids and Amides in Leaves and Plastids of *Zea Mays* during leaf development. *Plant Physiol.* **63**, 567–72.

Connelly, J. A. and Conn, E. E. (1986). Tyrosine

biosynthesis in *Sorghum bicolor*: isolation and regulatory properties of arogenate dehydrogenase. *Z. Naturforsch.* **41c**, 69-78.

Cossins, E. A. (1987). Folate biochemistry and the metabolism of one-carbon units. In *The Biochemistry of Plants* Vol. 11, ed. D. D. Davies, Academic Press, New York, pp. 317–53.

D'Mello, J. P. F. (1995). Toxicity of non-protein amino acids from plants. In *Amino Acids and Their Derivatives in Higher Plants*, ed R. M. Wallsgrove, Cambridge University Press, Cambridge, pp. 145–53.

Dickson, J. M. J. J., Vincze, E., Grant, M. R., Smith, L. A., Rodber, K. A., Farnden, K. J. F. and Reynolds, P. H. S. (1992). Molecular cloning of the gene encoding developing seed asparaginase from *Lupinus angustifolius*. *Plant Mol. Biol.* **20**, 333–36.

Fowden, L. (1981). Nonprotein amino acids. In *The Biochemistry of Plants*, Vol. 7, ed. E. E. Conn, Academic Press, New York, pp. 215–47.

Frankard, V., Ghislain, M. and Jacobs, M. (1992). Two feedback-insensitive enzymes of the aspartate pathway in *Nicotiana sylvestris*. *Plant Physiol.* **99**, 1285–93.

Giovanelli, J., Veluthambi, K., Thompson, G. A., and Mudd, S. H. (1984). Threonine synthase of *Lemna paucicostata* Hegelm. 6746. *Plant Physiol.* **76**, 285–92.

Giovanelli, J., Mudd, S. H. and Datko, A. H. (1988). *In vivo* regulation of threonine and isoleucine biosynthesis in *Lemna paucicostata* Hegelm. 6746. *Plant Physiol.* **86**, 369–77.

Ireland, R. J. and Hiltz, D. A. (1995). Glycine and serine synthesis in non-photosynthetic tissues. In *Amino Acids and Their Derivatives in Higher Plants*, ed. R. M. Wallsgrove, Cambridge University Press, Cambridge, pp. 111–18.

Jacobs, M., Frankard, V., Ghislain, M. and Vauterin, M. (1995). The genetics of aspartate derived amino acids in higher plants. In *Amino Acids and Their Derivatives in Higher Plants*, ed. R. M. Wallsgrove, Cambridge University Press, Cambridge, pp. 29–50.

Kirk, P. R. and Leech, R. M. (1972). Amino acid biosynthesis by isolated chloroplasts during photosynthesis. *Plant Physiol.* **50**, 228–34.

Krishnaswamy, S. and Bryan, J. K. (1983). Ligand-induced interconversions of maize homoserine dehydrogenase. *Arch. Biochem. Biophys.* **222**, 449–63.

Kushad, M. M., Richardson, D. G. and Ferro, A. J. (1983). Intermediates in the recycling of 5-methylribose to methionine in fruits. *Plant Physiol.* **73**, 257–61.

Lam, H.-M., Peng, S. S.-Y. and Coruzzi, G. M. (1994). Metabolic regulation of the gene encoding glutamine-dependent asparagine synthetase in *Arabidopsis thaliana*. *Plant Physiol.* **106**, 1347–57.

Lea, P. J., Blackwell, R. D. and Joy, K. W. (1992). Ammonia assimilation in higher plants. In *Nitrogen Metabolism of Plants*, eds K. Mengel and B. J. Pilbeam, Oxford University Press, pp. 154–86.

Miflin, B. J. and Lea, P. J. (1982). Ammonia assimilation and amino acid metabolism. In *Encyclopaedia of Plant Physiology*, New Series, Vol. 14A, eds D. Boulter and D. Parthier, Springer-Verlag, Berlin, pp. 5–64.

Mills, W. R., Lea, P. J. and Miflin, B. J. (1980). Photosynthetic formation of the aspartate family of amino acids in isolated chloroplasts. *Plant Physiol.* **65**, 1166–72.

Mitchell, D. J. and Bidwell, R. G. S. (1970). Compartments of organic acids in the synthesis of asparagine and homoserine in pea roots. *Can. J. Bot.* **48**, 2001–7.

Miyazaki, J. H. and Yang, S. F. (1987). Metabolism of 5-methylribose to methionine. *Plant Physiol.* **84**, 277–81.

Nguyen, J., Machal, L., Vidal, J., Perrot-Rechenmann, C. and Gadal, P. (1986). Immunochemical studies on xanthine dehydrogenase of soybean root nodules. *Planta* **167**, 190–95.

Oaks, A. and Hirel, B. (1985). Nitrogen metabolism in roots. *Ann. Rev. Plant Physiol.* **36**, 345–65.

Oliver, D. J., Neuburger, M., Bourguignon, J. and Douce, R. (1990). Interaction between the component enzymes of the glycine decarboxylase multienzyme complex. *Plant Physiol.* **94**, 833–39.

Pahlich, E. (1992). Environmentally induced adaptive mechanisms of plant amino acid metabolism. In *Nitrogen Metabolism of Plants*, eds K. Mengel and B. J. Pilbeam, Oxford University Press, pp. 187–200.

Pate, J. S., Atkins C. A. and Rainbird, R. M. (1981). Theoretical and experimental costing of nitrogen fixation and related processes in nodules of legumes. In *Current Perspectives in Nitrogen Fixation*, eds A. H. Gibson and W. E. Newton, Aust. Acad. Sci., Canberra, pp. 105–16.

Rawsthorne, S., Douce, R. and Oliver, D. (1995). The glycine decarboxylase complex in higher plant mitochondria: structure, function and biogenesis. In *Amino Acids and Their Derivatives in Higher Plants*, ed. R. M. Wallsgrove, Cambridge University Press, Cambridge, pp. 87–110.

Rowell, P. and Stewart, W. D. P. (1976). Alanine dehydrogenase of the N_2-fixing blue–green alga *Anabaena cylindrica*. *Arch. Microbiol.* **107**, 115–24.

Rubin, J. L. and Jensen, R. A. (1979). Enzymology of L-tyrosine biosynthesis in mung bean (*Vigna radiata* [L.] Wilczek). *Plant Physiol.* **64**, 727–34.

Schubert, K. R. (1986). Products of biological nitrogen fixation in higher plants: synthesis, transport and metabolism. *Ann. Rev. Plant Physiol.* **37**, 539–74.

Shaner, D. and Singh, B. K. (1992). How does inhibition of amino acid biosynthesis kill plants? In *Biosynthesis and Molecular Regulation of Amino Acids in Plants*. eds B. Singh, H. Flores and J. Shannon. American Society of Plant Physiologists. pp. 174–183.

Shelp, B. J., Atkins, C. A., Storer, P. J. and Canvin, D. T. (1983). Cellular and subcellular organization of pathways of ammonia assimilation and ureide synthesis in nodules of cowpea (*Vigna unguiculata* L. Walp.). *Archiv. Biochem. Biophys.* **224**, 429–41.

Sieciechowicz, K. A., Joy, K. W. and Ireland, R. J. (1988). The metabolism of asparagine in plants. *Phytochemistry* **27**, 663–71.

Siehl, D. L. and Conn, E. E. (1988). Kinetic and regulatory properties of arogenate dehydratase in seedlings of *Sorghum bicolor* (L.) Moench. *Archiv. Biochem. Biophys.* **260**, 822–29.

Stewart, G. R. and Larher, F. (1980). Accumulation of amino acids and related compounds in relation to environmental stress. In *The Biochemistry of Plants*, Vol. 5, ed. B.J. Miflin, Academic Press, New York, pp. 609–30.

Ta, T. C, Joy, K. W. and Ireland, R. J. (1985). Role of asparagine in the photorespiratory nitrogen metabolism of pea leaves. *Plant Physiol.* **78**, 334–37.

Ta, T. C and Joy, K. W. (1986). Metabolism of some amino acids in relation to the photorespiratory nitrogen cycle of pea leaves. *Planta* **169**, 117–22.

Taylor, A. A. and Stewart, G. G. (1981). Tissue and subcellular localization of enzymes of arginine metabolism in *Pisum sativum*. *Biochem. Biophys. Res. Commun.* **101**, 1281–89.

Thompson, J. F. (1980). Arginine synthesis, proline synthesis, and related processes. In *The Biochemistry of Plants*, Vol. 5, ed. B. J. Miflin, Academic Press, New York, pp. 375–402.

Triplett, E. W. (1985). Intercellular nodule localization and nodule specificity of xanthine dehydrogenase in soybean. *Plant Physiol.* **77**, 1004–9.

Tsai, F.-Y. and Coruzzi, G. M. (1991). Dark-induced and organ-specific expression of two asparagine synthetase genes in *Pisum sativum*. *EMBO J.* **9**, 323–32.

Tsai, F.-Y. and Coruzzi, G. M. (1992). Light represses the transcription of asparagine synthetase genes in photosynthetic and non-photosynthetic organs of plants. *Mol. Cell Biol.* **11**, 4966–72.

Ugalde, T. D., Maher, S. E., Nardella, N. E. and Wallsgrove, R. M. (1995). Amino acid metabolism and protein deposition in the endosperm of wheat; synthesis of proline via ornithine. In *Amino Acids and Their Derivatives in Higher Plants*, ed. R.M. Wallsgrove, Cambridge University Press, Cambridge, pp. 77–86.

Uritani, I. and Asahi, T. (1980). Respiration and related metabolic activity in wounded and infected tissues. In *The Biochemistry of Plants*, Vol. 2, ed. D. D. Davies, Academic Press, New York, pp. 463–85.

Urquhart, A. A. and Joy, K. W. (1981). Use of phloem exudate technique in the study of amino acid transport in pea plants. *Plant Physiol.* **68**, 750–54.

Wallsgrove, R. M., Lea, P. J. and Miflin, B. J. (1983). Intracellular localization of aspartate kinase and the enzymes of threonine and methionine synthesis in green leaves. *Plant Physiol.* **71**, 780–84.

31

The physiology and biochemistry of legume N₂ fixation

David B. Layzell and Craig A. Atkins

Introduction

Elemental nitrogen is one of the most important nutrients required for plant growth. It is a key constituent of protein, nucleic acids and other cellular components and its availability to plants frequently limits their growth and yield. Most plants acquire N from the soil solution as either nitrate (NO_3^-) or ammonium (NH_4^+) ions. In addition, some plants can utilize the atmospheric N_2 pool through symbiotic associations with species of bacteria, cyanobacteria or actinomycetes that contain the N_2 fixing enzyme, nitrogenase.

This chapter will summarize our understanding of the biochemistry and physiology of nitrogen fixation in the legume–rhizobia symbiotic association. The original work on the reactions of nitrogenase, and on the nitrogenase mechanism was not done with the bacteria that infect legume nodules, but with free-living bacteria such as *Klebsiella* and *Clostridium*. However, there is good evidence that nitrogenase is similar in the microsymbionts of legumes (*Rhizobium* and *Bradyrhizobium* species).

Nodule structure

Much is known about the signaling that occurs between the bacteria and the legume plant, and the developmental and molecular processes that lead to the establishment of the legume nodule. These matters will be considered in another chapter in this volume (Chapter 29) and have been discussed in recent reviews on the subject (van Rhijn and Vanderleyden, 1995; Vijn *et al.*, 1993; Fischer, 1994;

Sprent and Sprent, 1990). This section will summarize the general structure of a legume nodule as a background for the subsequent discussion of nodule metabolism.

In functional nodules, the bacteria-infected *central zone* of the nodule is surrounded by layers of uninfected cells that occupy a region of the nodule referred to as the *nodule cortex* (Fig. 31.1). The vascular tissue within the nodule cortex contains phloem and xylem surrounded by a vascular endodermis. These tissues are continuous with similar tissues in the subtending root. In some nodules, especially those that produce ureides as the end product of N_2 fixation (see Chapter 30), the central zone contains both infected and uninfected cells. The uninfected cells are generally smaller than the infected cells, are distributed throughout the entire central zone (at least one in contact with each infected cell), often occur in radially-oriented wedges and, in contrast to the infected cells, are highly vacuolated. As will be discussed later, the uninfected cells are thought to play a role in ureides synthesis.

Within the infected cells, the symbiotic bacteria (called *bacteroids*) occupy enclosures (*symbiosomes*) surrounded by a plant-derived membrane called a *peribacteroid* or *symbiosome membrane* (Fig. 31.1). Depending on the plant–bacterial symbiosis, one to four, or more, bacteroids may be found within each symbiosome. The bacteroids differ from the free-living bacteria in that they are larger, frequently lobed and express a complement of genes that are not expressed in the free-living form. Plant organelles, including the mitochondria, plastids and peroxisomes, tend to be localized near gas-filled *intercellular spaces* that form a network throughout the entire central zone (Fig. 31.1). These spaces are thought to have a

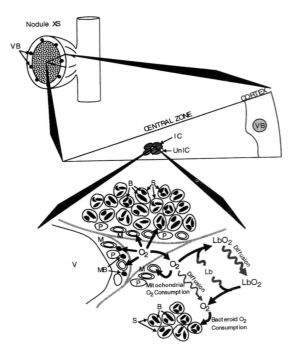

Fig. 31.1 Diagrammatic representation of a cross section of a legume nodule (in this case, a ureide producing nodule such as soybean) showing the nodule cortex with vascular tissue surrounding the central, infected zone. In the exploded view of the nodule, both infected (IC) and uninfected (UnIC) cells are depicted within the central zone, and a vascular bundle is shown within the nodule cortex. At the bottom of the figure, a gas filled space is shown along with portions of two infected cells and one uninfected cell. The uninfected cell contains a large central vacuole (V), plastids (P), mitochondria (M) and peroxisomes or microbodies (MB). The bacteria-infected cells lack microbodies, but contain large numbers of symbiosomes (S) that comprise bacteroids (B) enclosed within a plant membrane. Minimal subcellular details are depicted in one of the infected cells to show the path of leghemoglobin (Lb)-facilitated O_2 diffusion into the cell. Note that the plant organelles (mitochondria, plastids, peroxisomes) tend to be clustered around the gas-filled intercellular space (Millar *et al.*, 1995).

role in providing a low-resistance diffusion pathway for O_2 supply to, and H_2 and CO_2 removal from, the metabolically active cells within the central zone. The cytoplasm of the infected cells contains high concentrations (up to 3mM) of an O_2-binding heme-protein called *leghemoglobin*. Leghemoglobin

facilitates the diffusion of O_2 from the surface of the cell at the intercellular spaces to the O_2-consuming symbiosomes within the center of the infected cell (Fig. 31.1).

Nitrogenase catalyzed reactions and the electron allocation coefficient

Nitrogenase is a prokaryotic enzyme that catalyzes the transfer of electrons to a number of substrates having a triple bond. The products formed by the enzyme depend on the substrates available to it. For example, in an atmosphere containing N_2 gas (e.g. air), nitrogenase catalyzes the reduction of both N_2 to NH_4^+ (N_2 fixation, equation [31.1a]) and protons to H_2 gas (H_2 production, equation [31.1b]):

$$N_2 + 12ATP + 6e^- + 8H^+ \rightarrow 2NH_4^+ + 12ADP + 12P_i \text{ (N}_2\text{ fixation)} \qquad [31.1a]$$

$$4ATP + 2e^- + 2H^+ \rightarrow H_2 + 4ADP + 4P_i \text{ (H}_2\text{ production)} \qquad [31.1b]$$

The production of H_2 by nitrogenase seems to be associated with the binding of N_2 to the enzyme, hence there is a ratio of at least 1 H_2 produced per N_2 fixed (Simpson, 1987). In many N_2 fixing systems, more than one H_2 is produced per N_2 fixed, and the term *electron allocation coefficient* (EAC) is used to describe the proportion of total electron flow through the enzyme which is allocated to N_2 fixation rather than H_2 production. Since N_2 fixation consumes three electron pairs (i.e. $6e^-$) per N_2 reduced, and H_2 production requires only one electron pair, the maximum EAC is 0.75. In most legume symbioses, EAC values typically range from 0.60 to 0.70. The reasons for these less-than-maximal EAC values will be discussed later in this chapter.

The rate of H_2 production in air has been termed the *Apparent Nitrogenase Activity* (ANA, Table 31.1) since H_2 production represents only that proportion of electron flux through nitrogenase directed to proton reduction. In the absence of N_2 gas (i.e. in an atmosphere where N_2 is replaced by an inert gas such as Ar or He), all electron flow through nitrogenase is used to reduce protons to H_2 as shown in equation [31.2]:

$$16ATP + 8e^- + 8H^+ \rightarrow 4H_2 + 16ADP + 16P_i \text{ (H}_2\text{ production in Ar : O}_2) \qquad [31.2]$$

Table 31.1. Aspects of nitrogenase activity and how they can be measured in plant symbioses

Type of nitrogenase activity	Measured as:
1. Apparent nitrogenase activity (ANA)	H$_2$ evolution in air (N$_2$:O$_2$)
2. Total nitrogenase activity (TNA)	H$_2$ evolution in Ar:O$_2$ or C$_2$H$_4$ production in ca. 10 kPa C$_2$H$_2$
3. Potential nitrogenase activity (PNA)	Max. H$_2$ evolution in Ar:O$_2$ (or C$_2$H$_4$ production in C$_2$H$_2$) attained as pO$_2$ around nodules is increased.
4. N$_2$ fixation rate	^{15}N methods[a], N increment methods[b] or (TNA-ANA) ÷ 3

[a]Includes a number of techniques that rely on ^{15}N-enriched soils or atmospheres, or the difference in the natural abundance of ^{15}N in atmosphere (^{15}N$_2$) and soil (^{15}NO$_3^-$ or ^{15}NH$_4^+$). The dominant N isotope in our atmosphere and soil is ^{14}N; ^{15}N is a stable isotope that accounts for about 0.365% of the N in the atmosphere. See Hansen (1994) for further details.
[b]Includes methods that measure the increase in reduced plant N content in N$_2$ fixing plants compared with non-N$_2$ fixing control plants.

A stoichiometry showing the production of 4 H$_2$ has been chosen here so that the total electron flow through the enzyme (8e$^-$ or 4 electron pairs) is similar to that in the combined eqns of [31.1a] and [31.1b].

In the presence of saturating concentrations of acetylene (e.g. 10 kPa C$_2$H$_2$), virtually all of the electron flow through nitrogenase is used in the reduction of C$_2$H$_2$ to ethylene (C$_2$H$_4$) (eqn [31.3]) and few, if any, of the electrons are allocated to N$_2$ fixation or H$_2$ production:

$$4C_2H_2 + 16ATP + 8e\text{-} + 8H^+ \rightarrow 4C_2H_4 + 16ADP + 16P_i$$
$$\text{(C}_2\text{H}_2 \text{ reduction)} \quad [31.3]$$

C$_2$H$_2$ is a non-competitive inhibitor of N$_2$ fixation and H$_2$ production, and acts at a site which is separate from the N$_2$ binding site, but competing for electrons from the same pool as N$_2$ fixation and H$_2$ production (Ludden and Burris, 1986). Neither N$_2$-free atmospheres (e.g. Ar:O$_2$) or atmospheres with 10% C$_2$H$_2$ are present in the natural environment of N$_2$ fixing bacteria. However, these treatments are commonly used by researchers to obtain an estimate of the rate of total electron flow through the nitrogenase enzyme. The rate of C$_2$H$_2$ production in 10% C$_2$H$_2$ or H$_2$ production in Ar:O$_2$ is termed *total nitrogenase activity* (TNA, Table 31.1).

Like C$_2$H$_2$, CO is a non-competitive inhibitor of N$_2$ fixation, but unlike C$_2$H$_2$, CO only binds to the enzyme and is not reduced. At saturating levels of CO, N$_2$ fixation is blocked and all electron flow through nitrogenase is diverted to H$^+$ reduction to H$_2$.

Although CO has been used by researchers studying the properties of purified nitrogenase, it cannot be used with legume symbioses since the CO has a very high affinity for the nodule leghemoglobin, and therefore inhibits O$_2$ transport within the infected cell.

Finally, H$_2$ is not only a product of the nitrogenase reaction, but is also a competitive inhibitor of N$_2$ reduction ($K_i(H_2) = 3$ to 4 kPa H$_2$), blocking N$_2$ reduction and diverting electron flow to the reduction of protons and the release of additional H$_2$ gas (Burris, 1985). Studies with D$_2$ (i.e. deuterium gas or ^{2}H$_2$) have shown that D$_2$ inhibition of N$_2$ fixation results in the production of two molecules of HD for every D$_2$ reacting at the active site. Therefore, by stimulating its own production, H$_2$ could be considered an 'autocatalytic' competitive inhibitor. This H$_2$ inhibition of N$_2$ fixation can account for why legume nodules typically have EAC values that are less than the theoretical maximum value of 0.75 (Moloney et al., 1994).

Uptake hydrogenase activity and relative efficiency

In many N$_2$ fixing bacteria, the H$_2$ produced by nitrogenase is lost to the surrounding environment. However, some N$_2$ fixers have an enzyme called an uptake hydrogenase (HUP) that can recover some or all of the H$_2$ produced by the nitrogenase. HUP is a membrane bound enzyme that operates in only one direction:

$$H_2 \rightarrow 2e^- + 2H^+ \quad \text{(Uptake Hydrogenase)} \qquad [31.4]$$

Like other hydrogenases, it is an Fe–S protein, and in rhizobia and some other bacteria, Ni is an integral component of the active enzyme.

There is evidence that the electrons generated from H_2 oxidation can be passed to a respiratory pathway to produce ATP and consume O_2 (Maier, 1986). Presumably, the ATP produced can be utilized by the bacteria, thereby improving the efficiency of energy use by the bacteria or symbiotic association. However, in legume nodules there is evidence that O_2 may be a major limiting substrate in the infected cell. In this case, H_2 oxidation may compete with more efficient respiratory pathways in the generation of the ATP required for plant and bacteroid metabolism (Layzell and Moloney, 1994). Perhaps these considerations may account for the fact that there is no consistent evidence that HUP expression in legume nodules is correlated with higher rates of N_2 fixation or improved crop yield (O'Brian and Maier, 1988).

Although some N_2 fixing bacteria lack the gene that codes for HUP, in those with the gene, HUP expression is regulated by a variety of factors. For example, its expression is induced by H_2, but repressed by O_2 and C substrates (Maier, 1986; Batut and Boistard, 1994). In certain legume symbioses, a host effect on HUP expression has also been reported (Phillips *et al.*, 1985).

The term 'relative efficiency' (RE) has been used to define the proportion of total electron flow through nitrogenase that is not lost as H_2 evolution to the environment. Hence, RE differs from EAC in that it takes into consideration the net H_2 exchange resulting from the activities of both nitrogenase and HUP (Saari and Ludden, 1987). In an N_2 fixing system lacking HUP activity, RE = EAC.

Nitrogenase mechanism

As described in Chapter 29, the nitrogenase enzyme is comprised of two separate proteins. The Fe protein (or component II) contains an Fe–S center and two Mg-ATP binding sites, whereas the MoFe Protein has two atoms of Mo and a number of Fe–S centers. The active site of nitrogenase is within the MoFe cofactor of the MoFe protein, and is made up of two

$[MoFe_6S_8]$ clusters. The role of the Fe protein is to accept electrons from either ferredoxin or flavoredoxin (dithionite can be used in *in vitro* studies), and pass them to the MoFe protein where they are used to reduce N_2 or H^+.

Thorneley and Lowe (1985) have proposed a model that attempts to describe the mechanism of nitrogenase action in the fixation of N_2 and the production of H_2. Although new information in the past decade has been used to fine-tune their model (e.g. Leigh, 1995), its basic principles are still widely accepted and will be described here.

The nitrogenase mechanism is thought to be separated into two cycles: the Fe protein cycle, and the MoFe protein cycle. In the Fe protein cycle (Fig. 31.2), electrons are passed first to the Fe protein, and then from the Fe protein to the MoFe protein. Two MgATP must be bound to the reduced Fe protein (k_4, Fig. 31.2) before this molecule can form a reversible complex with the MoFe protein (k_1 and k_{-1}, Fig. 31.2). Once complexed, the reduced Fe protein can transfer the electron to the MoFe Protein (k_2, Fig. 31.2). The oxidation–reduction step is coupled to MgATP hydrolysis, and is effectively irreversible. The Fe protein:MgATP:reduced MoFe protein complex can undergo a reversible dissociation (k_{-3} and k_3, Fig. 31.2). The k_{-3} reaction is the rate-limiting step in the catalytic cycle for substrate reduction, and is responsible for making nitrogenase one of the slowest enzymes found in bacteria (Thorneley and Lowe, 1985). The significance of this rate-limiting reaction will be discussed in more detail below.

The oxidized Fe protein:MgADP complex can be reduced and the ADP exchanged for ATP in reaction k_4 (Fig 31.2) whereas the reduced MoFe protein can return to acquire additional electrons from the Fe protein through the Fe protein cycle. In total, 8 electron transfers to the MoFe protein are required to reduce $1 N_2$ and produce $1 H_2$, and between each electron transfer, the Fe protein:MoFe protein complex dissociates completely. This cycle of 8 electron transfers has been called the MoFe protein cycle and is summarized in Fig. 31.3. Note that the Fe protein cycle forms the basic unit for the MoFe cycle.

When the MoFe protein has received 3 or 4 electrons (i.e. E_3H_3 or E_4H_4), it is thought to be in a

The Fe Protein Cycle

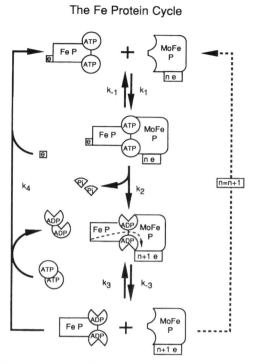

Fig. 31.2 Model of the Fe protein cycle of nitrogenase as proposed by Thorneley and Lowe (1985). The binding sites for the MgATP (ATP) and Mg ADP (ADP) molecules are shown to be localized between the Fe protein (Fe P) and MoFe protein (MoFe P) components. In fact, little is known of the precise location of the MgATP binding sites. The MoFe component shown here represents one of the two independently functioning halves of the tetrameric protein. The MoFe component receives from the Fe protein one of the eight electrons (e) that it requires to reduce N_2 or protons to NH_4^+ or H_2, respectively. As each electron is passed to the MoFe component, the number of electrons it holds (n) is incremented by one (i.e. n = n + 1). Note that the protein dissociation step (reaction k_{-3}) is the slow rate-determining step in the cycle. The MoFe protein cycle of Fig. 31.3 shows how the 8 cycles of the Fe protein cycle may be organized to bring about N_2 fixation and H_2 production.

The MoFe Protein Cycle

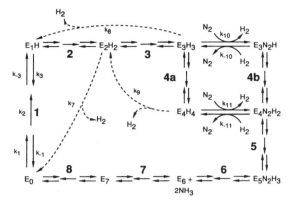

Fig. 31.3 Model of the MoFe protein cycle of nitrogenase as proposed by Thorneley and Lowe (1985). Note that reactions k_1 to k_3 (and k_{-1} to k_{-3}) of the Fe protein cycle (Fig. 31.2) form the basic unit for the MoFe cycle. The large numbers (1 to 8) associated with each set of arrows represent the sequential transfer of electrons from the Fe protein to the MoFe protein. The species E_n represents the MoFe protein of Fig. 31.2, where the subscript n denotes the total number of electrons which have been passed to that protein. The H_n and N_n associated with E refers to the number (n) of H and N atoms bound to the enzyme respectively. N_2 binding to the enzyme and the H_2 production associated with that binding is depicted in reactions k_{10} and k_{11}, while reactions k_{-10} and k_{-11} show a possible mechanism for the competitive inhibition of H_2 on N_2 fixation. The dashed lines (k_7 to k_9) show possible pathways of H_2 production not associated directly with N_2 binding to the enzyme. The protons involved in the reaction are assumed to be derived from pools within the bacterial cytosol, and the precise stage at which ammonia is released from the enzyme is not known.

state which is receptive to N_2 binding (k_{10} and k_{11}, Fig 31.3). The binding of N_2 to the enzyme is thought to displace H_2 gas, thereby accounting for the prerequisite one H_2 evolved per N_2 fixed, discussed previously. In this model, additional H_2 can be evolved from the dissociation of the E_2H_2, E_3H_3 and E_4H_4 intermediates (reactions k_7, k_8 and k_9,

Fig. 31.3). Therefore, these intermediates are substrates for reactions leading to both N_2 binding/ fixation and wasteful H_2 production. Based on a kinetic analysis of the competing reactions, Thorneley and Lowe (1985) predicted that minimal rates of H_2 production by nitrogenase would result if (1) the *in vivo* concentration of nitrogenase is as high as possible, and (2) the pool sizes of the E_2H_2, E_3H_3 and E_4H_4 intermediates are very low. Herein lie explanations both for the high *in vivo* concentration of nitrogenase and for the slow rate of catalysis of reaction k_{-3} (Fig. 31.2), the rate-determining step in the nitrogenase mechanism. If the dissociation of the

Fe protein:MgADP:MoFe protein complex is slow, pool sizes of the reduced MoFe protein will be very low, and N_2 binding will be favored over H_2 production. In the absence of N_2 gas (i.e. eqn [31.2]), reactions k_7, k_8 and k_9, (Fig. 31.3) would be favored and, in effect, all electron flow through nitrogenase would go to H_2 production.

The binding of N_2 and the release of H_2 seems to be a reversible reaction, such that H_2 can displace N_2 from the active site. This portion of the model can account for the observation that, in the presence of D_2, N_2 fixation is inhibited and HD is produced, presumably through dissociation of intermediates E_3HD_2 and $E_4H_2D_2$ (Fig 31.3).

In the presence of N_2 and reasonably low H_2, the MoFe protein will continue to receive electrons, and after receiving a total of 5, 6 or 7 electrons (the precise number is not known), two NH_3 are released from the enzyme. Since the reduction of the N_2 to ammonia requires a total of 8 electrons (assuming 1 H_2 produced), the MoFe protein may be 'oxidized' following the release of the NH_3, and additional electrons must be received from the Fe protein to bring it to the E_0 state (Fig. 31.3).

Oxygen and nitrogen fixation

Legume nodules have a much higher rate of respiratory O_2 consumption than most other plant tissues; about four fold higher than that observed in a similar biomass of root tissue. This O_2 is required to meet the large demands for ATP within the bacteroids (e.g. 16–20 ATP are consumed per N_2 fixed, eqn [31.1a and b]), and the ATP requirements for the synthesis of the exported organic solutes of N, in addition to the other needs for cell growth and maintenance. However, O_2 is a very potent, irreversible inhibitor of the nitrogenase enzyme. Therefore, legume nodules must provide a very high flux of O_2 to the bacteria-infected cells, but at an extremely low O_2 concentration.

O_2 concentrations in the infected cells (O_i) of legume nodules are among the lowest that have been measured in aerobic cells of any organism (Fig. 31.4). At 4 to 70 nM O_2, the concentration is about 0.0001 times that expected to occur in cells that are in equilibrium with air (250 μM). In other words, the outer cells in the

nodule cortex could have an intercellular O_2 concentration 10 000 times higher than that in the infected cells, only a mm away within the central zone. If the O_i rises above approximately 60 to 70 nM O_2, nitrogenase activity is inhibited. The inhibition is reversible if nitrogenase exposure to elevated O_2 is brief (sec to 5 min) and the O_2 concentration does not approach that in the atmosphere. However, longer exposures to higher O_i levels (e.g. above 100 μM), will cause irreversible inhibition of nitrogenase protein. Nodule activity may recover if O_2 is returned to normal levels, but this require 24 h or more for *de novo* synthesis of new enzyme.

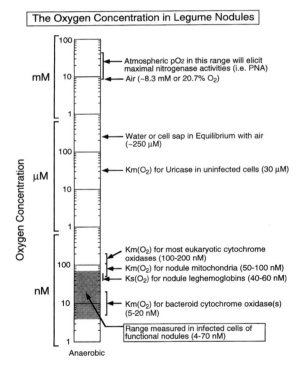

Fig. 31.4 A logarithmic scale of O_2 concentrations showing the level in the infected cells of legume nodules (shaded regions) relative to the K_m (O_2) and K_S(O_2) values of cellular proteins, and concentrations typical of the natural environment. The infected cell O_2 concentration was measured by a spectrophometric technique based on differences in light absorption between deoxygenated leghemoglobin (Lb) and oxygenated leghemoglobin (LbO₂) within the cytoplasm of the infected cell. (Kuzma *et al.*, 1993).

We do not fully understand how legume nodules provide such a precise level of control over O_i, but studies to date indicate that there are two key components to the regulation:

(a) nodule metabolism is O_2 limited, thereby providing at least some capacity to consume additional O_2, should it diffuse into the nodule, and

(b) nodules are able to adjust their resistance to O_2 diffusion.

Fick's First Law of Diffusion explains the relationships among nodule respiration, nodule resistance and O_i:

$$O_e - O_i = F \times R \quad \text{(Fick's First Law)} \qquad \text{[Eqn 31.5]}$$

This law states that under steady-state conditions, the difference between the O_2 concentration external to the nodule (O_e, mole/m^3) and in the infected cell (O_i, moles/m^3) is equal to the product of the flux of O_2 into the nodule (i.e. the respiration rate) (F, moles/m^2/s) and the resistance to O_2 diffusion (R, s/m). Therefore, if O_e were to rise (due, for example, to an increase in atmospheric pressure), increasing the O_2 differential ($O_e - O_i$), a small change in Oi may relieve O_2-limited respiration (i.e. F increases) and stabilize a new equilibrium. However, if the nodule has insufficient capacity to increase flux (F), it may increase resistance (R) to offset the larger O_2 differential and maintain control over O_i.

There are many examples of treatments that are known to cause a change in nodule resistance (Hunt and Layzell, 1993):

(1) In many symbioses, nodule exposure to O_2 concentrations of 25 kPa or more (ambient air is *ca.* 20.7 kPa), causes a rapid increase in O_i, and an inhibition of nodule respiration and nitrogenase activity. However, within 5 to 10 min., O_i returns to its 'normal' range, and over the subsequent 30 min, nitrogenase activity and respiration recover even though the external O_2 remains at the higher level. This is because R increases to compensate for the greater O_2 differential.

(2) Within 2 h of cutting off phloem sap supply to nodules (e.g. by stem girdling or shoot removal), nodule respiration (i.e. F in equation [31.5]) and nitrogenase activity are greatly reduced. Surprisingly, O_i is also reduced, and R is greatly increased to compensate for both the lower O_i and the lower F. It is interesting to note that increases in O_e can recover up to 70% of the initial nodule activity, providing evidence that the increased resistance has down-regulated metabolism by making it more O_2-limited.

(3) Nitrate fertilization of legumes for 2 to 4 days, or exposure of nodules to an Ar:O$_2$ atmosphere for 40 min also cause an increased R, decreased O_i and a greater O_2-limitation of nodule respiration and nitrogenase activity.

(4) Exposing nodulated roots to low soil temperatures, or to drought, causes declines in nodule respiration and nitrogenase activity, yet O_i levels change little, since nodule R increases.

The mechanism used by nodules to regulate their resistance to O_2 diffusion is not completely understood. However, we do know that the innermost region of the nodule cortex plays an important role. Here, the cells are tightly packed and have few intercellular spaces and long term (i.e. days to weeks) exposure of nodulated roots to high or low O_2 concentrations, increase or decrease, respectively, the number of cell layers in this region. Fewer, and smaller spaces will result in less gas-phase and more liquid phase O_2 diffusion through this cell layer, and since O_2 solubility and O_2 diffusion in water are 0.03 and 0.0001, respectively, times that in the gas phase, there will be a greater resistance to O_2 movement to the central infected zone.

Legume nodules are also able to regulate their resistance in response to short-term (minutes to days) changes in environmental and physiological conditions such as those described in point form above. This time scale would not permit changes in the cell structure of the nodule cortex similar to those occurring with longer ter O_2 treatments. It has been suggested that short-term regulation of R may be achieved by displacement of the gas-filled spaces of the nodule inner cortex with water or a glycoprotein (Iannetta *et al.*, 1995). In addition, recent studies have provided evidence that the infected cells themselves may have the ability to exercise some level of control over their own intercellular O_2 concentration. For example, Thumfort *et al.* (1994) have shown that if the O_2 concentration in the gas spaces of the central zone rises to too high a level, the leghemoglobin adjacent to the gas-filled spaces will become O_2 saturated, thereby destroying the LbO_2 gradient in this region and greatly reducing the effectiveness of Lb-facilitated diffusion. In addition, it has been suggested that the mitochondria that are clustered around the gas spaces may provide a form of respiratory protection by consuming much of the O_2 before it can diffuse to the symbiosomes within the center of the infected cells (Millar *et al.*, 1995). Both mechanisms would serve to stabilize O_i in situations when there may be relatively large variations in the central zone O_2 concentration. However, these infected cell mechanisms require increases in the rate of O_2 consumption to deal with higher space O_2 concentrations, and many of the short-term, physiologically-regulated mechanisms (stem girdling, etc.) cause decreases in respiration when nodule resistance increases. Hence, these mechanisms cannot account for all short term regulatory responses, and there is still a need for direct control over nodule resistance such as that which could be provided by changes in the spaces of the nodule cortex.

Carbon, nitrogen and oxygen metabolism in legume nodules

In legumes, the phloem provides the nodule with reduced carbon in the form of sucrose whereas the xylem removes the products of nitrogen fixation as either the ureides, allantoin (ALLN) and allantoic acid (ALLA), or as the amide, asparagine (ASN) (Fig. 31.5) (Atkins, 1991). Tropical or subtropical legumes such as cowpea and soybean typically export ureides, whereas temperate or Mediterranean legumes like alfalfa, lupin, clover and pea are typically amide exporters. The sucrose provided to the nodules is not only a source of C for growth, but provides C for the deposition of starch reserves in plastids of uninfected cells, for polyhydroxybutyrate storage in bacteroids, and for the C skeletons required in the synthesis of ASN and ureides. In addition, sucrose is the source of

oxidizable substrates needed for plant and bacteroid respiration.

A wealth of biochemical and genetic evidence supports the idea that C4 acids, principally malate, are synthesized in the cytosol of the infected cell and

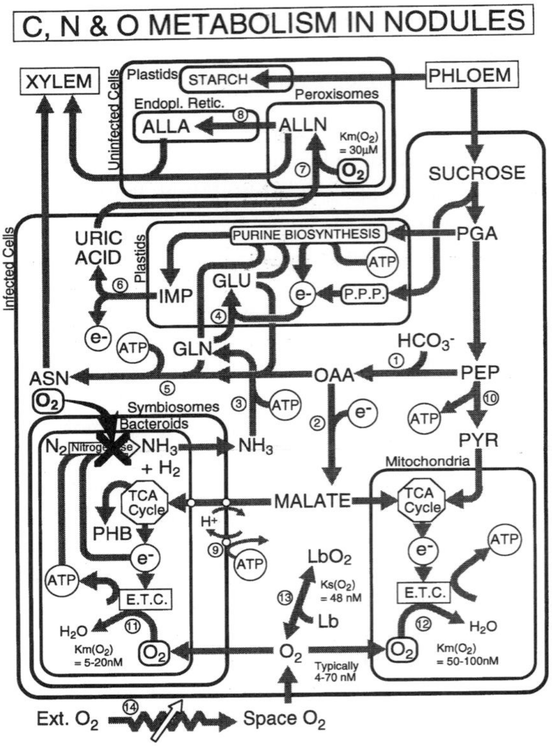

Fig. 31.5 A summary of key pathways of C, N and O metabolism and their location in the infected and uninfected cell of a legume nodule. Key reactions or process are depicted as Items 1, PEP carboxylase; 2, malate dehydrogenase (MDH); 3, glutamine synthetase (GS); 4, glutamate synthase (GOGAT); 5, asparagine synthetase (AS); 6, xanthine oxidoreductase; 7, uricase; 8, allantoinase; 9, ATPase; 10, pyruvate kinase; 11, bacteroid terminal oxidase; 12, mitochondrial terminal oxidase; 13, leghemoglobin equilibrium reaction; 14, variable cortical diffusion barrier. ALLA, allantoic acid; ALLN, allantoin; ASN, asparagine; ATP, adenosine triphosphate; e⁻, reductant; E.T.C. electron transport chain; GLN, glutamine; GLU, glutamate; IMP, inosine monophosphate; Lb, leghemoglobin; OAA, oxaloacetic acid, PEP, phosphoenolpyruvate; PGA, phosphoglycerate; PHB, polyhydroxybutyric acid; P.P.P., pentose phosphate pathway; PYR, pyrivate; TCA, Tricarboxylic Acid Cycle.

transported across the plant symbiosome membrane and the bacterial plasma membrane where they are metabolized in a bacterial TCA cycle (Fig. 31.5). A proton-ATPase (Item 9, Fig. 31.5) is present in the plant membrane of the symbiosome where it is thought to provide the proton gradient needed for malate transport to the bacteroid (Day and Udvardi, 1993). Phosphoenolpyruvate carboxylase (Item 1, Fig. 31.5, see also Chapter 29) is a key enzyme in the infected cell. Together with malate dehydrogenase (Item 2, Fig. 31.5) it generates the malate for the bacteria, and in nodules that export ASN, it provides the 4-C skeleton for ASN synthesis.

Although the initial step of NH$_3$ assimilation is catalyzed by a nodule-specific isozyme of glutamine synthetase (GS, Item 3, Fig. 31.5) located in the cytosol of the infected cell (Chapter 29), the transfer of amide N to form an alpha-amino group is catalyzed by glutamate synthase (GOGAT, Item 4, Fig. 31.5), an enzyme localized within the plastids of these cells. This is true of nodules which export asparagine as well as those which export ureides, so that in both cases, the assimilation of NH$_3$ requires the presence of reductant in the plastids and ATP in the cytosol. The location of asparagine synthetase (Item 5, Fig. 31.5) is not well established and the ATP requirement for the final reaction of asparagine synthesis could be in the plastid or the cytosol.

In legume nodules that synthesize ureides, present evidence indicates that the whole of the *de novo* purine pathway, and its supporting reactions of glycine synthesis and synthesis of C1 folate derivatives, occurs within the plastid. The initial nucleotide product of the pathway, IMP, is thought to be exported to the cytosol where it is oxidized to xanthine and uric acid. Interestingly, the oxidation of xanthine is linked to NAD reduction by xanthine oxidoreductase (Item 6, Fig. 31.5) rather than to O$_2$ by a xanthine oxidase. However, further oxidation of uric acid by the enzyme uricase (Item 7, Fig. 31.5) is linked to molecular O$_2$, and is localized within microbodies (peroxisomes) in the adjacent uninfected cells. Therefore, uric acid must be transported from one cell type to another, in much the same sort of cooperative way that C4 acids are transported in the C4 leaf. This pathway is further compartmentalized within the uninfected cells, with allantoin hydrolysis being localized to the endoplasmic reticulum (Item 8,

Fig. 31.5). Uninfected cells are also found in many nodules that export ASN, but in these nodules their role in N$_2$ fixation pathways is not clear. They probably are not essential since in some ASN producing nodules (e.g. lupin), there are no uninfected cells in the central zone.

The large ATP demands of nitrogenase are provided by bacteroid oxidative phosphorylation, whereas mitochondrial oxidative phosphorylation generates most of the ATP needed for NH$_3$ assimilation, and for maintaining the proton gradient across the symbiosome membrane (Item 9, Fig. 31.5), as well as the needs for basic cell growth and maintenance. Substrate level phosphorylation, such as that provided by pyruvate kinase (Item 10, Fig. 31.5), are probably of limited importance in meeting the ATP demands within the infected plant cell. To maintain these high rates of oxidative phosphorylation, the bacteria and mitochondria (Items 11 and 12, Fig. 31.5, respectively) have the capacity for high rates of O$_2$ consumption. However, since the O$_2$ concentration in the infected cell must be stringently regulated to protect nitrogenase from irreversible inhibition, it is maintained at a level that limits the rate of oxidative phosphorylation, and therefore the availability of ATP in support of cellular processes. As mentioned previously, infected cell O$_2$ concentration is regulated by the combined effects of respiratory O$_2$ consumption and the action of a variable barrier to O$_2$ diffusion (Item 13, Fig. 31.5).

Physiological data showing that small increases in external pO$_2$ stimulate respiration and nitrogenase activity in legumes (Hunt and Layzell 1993), and measurements of the nodule adenylate energy charge (AEC = [ATP + 1/2 ADP]/[ATP + ADP + AMP]) provide evidence for O$_2$ and ATP-limited metabolism in nodules. In aerobic tissues, AEC values are typically 0.80 or greater, whereas the central zone of legume nodules have AEC values (0.60 to 0.72) typical of hypoxic metabolism. Moreover, treatments which cause a greater O$_2$ limitation of nodule metabolism and nitrogen fixation generally reduce the AEC, whereas exposure to high pO$_2$ increases AEC (deLima *et al.*, 1994).

The site of O$_2$ limitation of nodule metabolism is not known. An O$_2$ limitation within the bacteroid could easily account for the decline in nitrogenase activity as the infected cell O$_2$ concentration is reduced. However, the $K_m(O_2)$ for the bacteroid terminal oxidase is very low (5–20 nM; Item 11, Fig. 31.5) compared with that in the mitochondria (50–100 nM O_2; Item 12, Fig. 31.5), and mitochondrial respiration is therefore likely to be more O_2-limited than bacteroid respiration at the O_2 concentration measured within infected cells (4–70 nM O_2). If this is the case, the lower nitrogenase activity in more O_2-limited nodules may be associated with the availability of ATP needed to maintain the proton gradient across the symbiosome membrane (Item 9, Fig. 31.5) that is required for malate transport to the bacteroids. Perhaps then, an O_2-limitation within the mitochondria may impose a carbohydrate limitation within the bacteroids. Further studies are required to test this hypothesis.

Further studies are also required to determine how the uninfected cells can maintain an O_2 pool sufficient to support uricase activity when the $K_m(O_2)$ for that enzyme (30 μM; Item 7, Fig. 31.5) is 1000 times higher than that measured in the adjacent infected cell. Either the measured $K_m(O_2)$ for isolated uricase (Rainbird and Atkins, 1981) is incorrect, or the infected and uninfected cells have an ability to exercise some degree of control over their own intercellular O_2 concentration, rather than all of the control being localized in the nodule cortex.

The presence of a mechanism to maintain O_2 limitation of C and N metabolism in legume nodules makes these structures very different from other plant organs, and this mechanism is likely to be essential for the existence of the symbiotic relationship. Recent research indicates that the metabolic pathways in nodules are governed by the rate of synthesis and utilization of reductant and ATP which, in turn, are a function of O_2 supply. An effective symbiosis requires that the infected cell O_2 concentration is maintained in a very low, but precise range, not only to protect nitrogenase from irreversible inhibition, but to provide the conditions necessary to supply reduced C to the bacteroid, to synthesize the end products of N_2 fixation, and to maintain healthy plant and bacterial cells. The complexity of the control mechanisms required to sustain the interchange of metabolites between the plant host and bacteroid endosymbiont may account for the fact that N_2 fixing symbioses are restricted to relatively few organisms within the plant world.

Acknowledgment

Special thanks to Dr Stephen Hunt for his useful comments on this manuscript.

References

Atkins, C. A. (1991). Ammonia assimilation and export of nitrogen from the legume nodule. In *Biology and Biochemistry of Nitrogen Fixation*, eds M. Dilworth and A. Glenn, Elsevier, Amsterdam, pp. 293–319.

Batut, J. and Boistard, P. (1994). Oxygen control in *Rhizobium. Antonie van Leeuwenhoek* **66**, 129–50.

Burris, R. H. (1985). H_2 as an inhibitor of N_2 fixation, *Physiologie Vegetale* **23**, 843–48.

Day, D. A. and Udvardi, M. K. (1993). Metabolite exchange across Symbiosome membranes, Symbiosis **14**, 175–89.

deLima, M. L., Oresnick, I. J., Fernando, S. M., Hunt, S., Smith, R., Turpin, D. H. and Layzell, D. B. (1994). The relationship between nodule adenylates and the regulation of nitrogenase activity by O_2 in soybean. *Physiol. Plant.* **91**, 687–95.

Fisher, H.-M. (1994). Genetic regulation of nitrogen fixation in Rhizobia. *Microbiol. Rev.* **58** (3), 352–86.

Hansen, A. P. (1994). *Symbiotic N_2 Fixation of Crop Legumes: Achievements and Perspectives*, Margraf Verlag, Weikersheim, Germany.

Hunt, S. and Layzell, D. B. (1993). Gas exchange of legume nodules and the regulation of nitrogenase activity, *Ann. Rev. of Plant Physiol. and Plant Mol. Biol.* **44**, 483–511.

Iannetta, P. P. M., James, E. K., Sprent, J. I., Minchin, F. R. (1995). Time-course changes involved in the operation of the oxygen diffusion barrier in white lupin nodules. *J. Exp. Biol.* **46**, 565–75.

Kuzma, M. M., Hunt S. and Layzell, D. B. (1993). The role of oxygen in the limitation and inhibition of nitrogenase activity and respiration rate in individual soybean nodules. *Plant Physiol.* **101**, 161–69.

Layzell, D. B. and Moloney, A. H. (1994). Dinitrogen Fixation. In *Physiology and Determination of Crop Yield*, eds K. J. Boote, J. M. Bennett, T. R. Sinclair and G. M. Paulsen, American Society of Agronomy, Madison, pp. 311–35.

Leigh, G. J. (1995). The mechanism of dinitrogen reduction by molybdenum nitrogenases, *Eur. J. Biochem.* **220**, 14–20.

Ludden, P. K. and Burris, R. H. (1986). Nitrogenase: properties and regulation. In *Biochemical Basis of Plant Breeding*, ed. C. A. Neyra, Vol II, CRC Press, Boca Raton, pp.41–58.

Maier, R. J. (1986). Biochemistry, regulation and genetics of hydrogen oxidation in *Rhizobium. CRC Crit. Rev. Biotechnology* **3**, 17–38.

Moloney, A. H., Guy, R. D. and Layzell, D. B. (1994). A model of the regulation of nitrogenase electron allocation in legume nodules–II Comparison of empiricial and theoretical studies in soybean. *Plant Physiol.* **104**, 541–50.

Millar, A. H., Day, D. A. and Bergersen, F. J. (1995). Microaerobic respiration and oxidative phosphorylation by soybean nodule mitochondria: implications for nitrogen fixation. *Plant Cell Environ.* **18**, 715–26.

O'Brian, M. R. and Maier, R. J. (1988). Hydrogen metabolism in *Rhizobium*: energetics, regulation, enzymology and genetics. *Adv. Microbial Physiol.* **29**, 1–52.

Phillips, D. A., Bedmar, E. J., Qualset, C. O. and Teuber, L. R. (1985). Host legume control of *Rhizobium* function. In *Nitrogen Fixation and CO$_2$ Metabolism*, eds P. W. Ludden and J. E. Burris, Elsevier Publishing Company, New York, pp. 203–12.

Rainbird, R. M. and Atkins, C. A. (1981). Purificaton and some properities of urate oxidase from nitrogen fixing nodules of cowpea. *Biochim. Biophys. Acta* **659**, 132–40.

Saari, L. L. and Ludden, P. W. (1987). The energetics and energy cost of symbiotic nitrogen fixation. In *Plant Microbe Interactions,* eds T. Kosuge and E. W. Nester, Vol. 2, Macmillan Publishing Company, New York, pp. 147–93.

Simpson, F. B. (1987). The hydrogen reactions of nitrogenase. *Physiol. Plant.* **69**, 187–90.

Sprent, J. I. and Sprent, P. (1990). Nitrogen Fixing Organisms. Pure and Applied Aspects, Chapman and Hall, Publ., New York, NY.

Thorneley, R. N. F. and Lowe, D. J. (1985). Kinetics and mechanisms of the nitrogenase enzyme system. In *Molybdenum Enzymes,* ed T. G. Spiro, J Wiley and Sons, New York, pp. 220–84.

Thumfort, P. P., Atkins, C. A. and Layzell, D. B. (1995). A re-evaluation of the role of the infected cell in the control of O$_2$ diffusion in legume nodules. *Plant Physiol.* **105**, 1321–33.

van Rhijn, P. and Vanderleyden, J. (1995). The *Rhizobium*-Plant Symbiosis. *Microbiol. Rev.* **59** (1), 124–42.

Vijn I., das Neves, L., van Kammen, A., Franssen, H. and Bisseling, T. (1993). Nod factors and nodulation in plants. *Science* **260**, 1764–5.

Assimilate Partitioning and Storage

32 Interactions between photosynthesis, respiration and nitrogen assimilation

David H. Turpin, Harold G. Weger and Heather C. Huppe

Introduction

Photosynthesis, respiration and N assimilation are interrelated processes (Fig. 32.1) (Turpin *et al.*, 1988). Many of these processes have been addressed separately in considerable detail in other chapters of this book. The purpose of this chapter is to integrate these processes by providing an overview of the interactions involved.

In simple terms, photosynthesis and respiration appear to be opposing processes. Photosynthesis involves the light-driven oxidation of H_2O, resulting in the production of O_2. This oxidation is coupled to the reduction of CO_2 or other physiological electron acceptors (e.g. NO_2^-, SO_4^-, O_2). The electron transfer between H_2O and ferredoxin (Fd) is mediated by the photosynthetic electron transport chain (ETC), and is coupled to ATP production (photophosphorylation, Chapter 19). Subsequently, Fd donates its electrons to $NADP^+$, to form NADPH, that is then used to reduce CO_2 via the reductive pentose phosphate (RPP or Calvin) cycle. Carbon fixation also consumes ATP produced by photophosphorylation. Conversely, respiration results in the oxidation of reduced carbon compounds to CO_2. Usually, electrons removed during the carbon oxidation are transferred to O_2 in the mitochondrion, resulting in respiratory O_2 consumption, although other respiratory acceptors may also be used. As in photosynthesis, respiratory electron transfer may be coupled to ATP production (oxidative phosphorylation, Chapter 14). Fig. 32.2 compares and contrasts these processes at this fundamental level. Despite the opposite direction of carbon flow, i.e. CO_2 to carbohydrate via photosynthesis and back to CO_2 via respiration, futile cycling between these pathways is apparently absent.

In fact, respiration plays an important role in the growth and metabolism of photosynthetic organisms in the light. Given the lack of futile cycling, one would predict that regulatory mechanisms must exist which function to coordinate between photosynthetic and respiratory carbon metabolism.

The process of inorganic N assimilation interacts with both photosynthesis and respiration. The ATP and reductant required to assimilate N may be provided by photosynthesis, respiration, or both. As well, N assimilation requires carbon skeletons that are provided by respiration. Under some circumstances, enhancement of respiration for N assimilation diverts carbon from the RPP pathway and this is correlated with a decrease in the rate of CO_2 fixation.

The effects of photosynthesis on respiration

Both photosynthesis and respiration are vital parts of plant metabolism; however, there is still debate about the magnitude of mitochondrial respiration during photosynthesis. Mitochondrial respiration can be divided into two distinct but interrelated processes. In the first process, organic carbon is oxidized via glycolysis, the oxidative pentose phosphate (OPP) pathway and the tricarboxylic acid (TCA) cycle which produce CO_2, reducing equivalents (NAD(P)H and $FADH_2$) and carbon skeletons. The carbon skeletons produced may be used in biosynthesis or recycled to support continued TCA cycle activity. In the second part of mitochondrial respiration, the mitochondrial electron transport chain (ETC) oxidizes TCA cycle-generated reductant resulting in the reduction of O_2 to H_2O and the production of ATP.

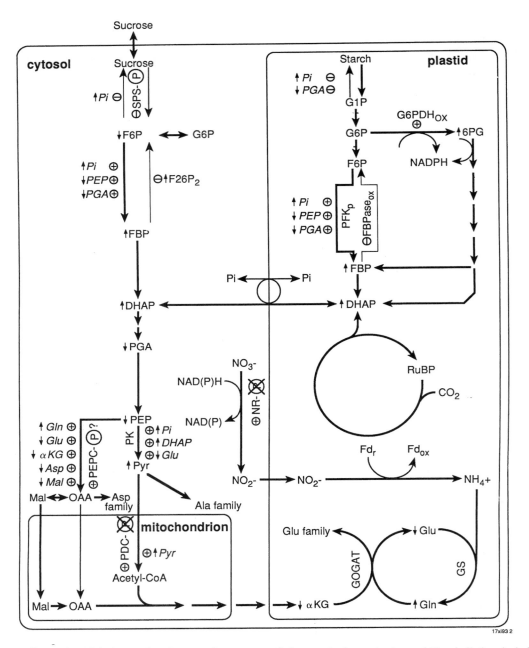

Fig. 32.1 An illustration of the interactions between the processes of photosynthesis, respiration and N assimilation, including the regulatory integration of these processes via metabolite effectors and/or covalent modification of the enzymes involved. Carbon is provided from starch, sucrose and recent photosynthate to amino acid synthesis depending on the tissue and its physiological status. Small arrows preceding a metabolite indicates whether there is an increase ($\uparrow$) or decrease ($\downarrow$) in the metabolite level at the onset of nitrogen assimilation. Abbreviations are as in text, symbols are as follows: activator, $\oplus$; inhibitors, $\ominus$; phosphorylated, $\textcircled{P}$; not phosphorylated; $\cancel{\bigotimes}$.

PHOTOSYNTHESIS

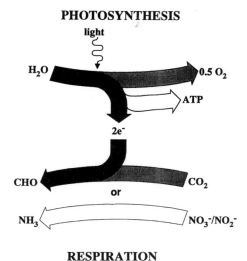

RESPIRATION

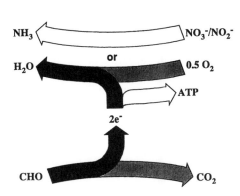

Fig. 32.2 A simplified representation of photosynthesis and respiration illustrating the opposing nature of these two processes. Photosynthesis is the oxidation of H_2O and the reduction of some physiological electron acceptor such as CO_2 or NO_3^-/NO_2^-. Respiration is the oxidation of carbohydrate (CHO) and the reduction of O_2 to H_2O.

Although TCA cycle activity and mitochondrial electron transport are considered interdependent, these processes can operate in isolation of one another. The NAD(P)H and $FADH_2$ produced by respiratory carbon flow must be reoxidized continually to ensure a supply of $NAD(P)^+$ and FAD to maintain the carbon oxidation. We commonly think of the mitochondrial ETC serving this role, but many other reactions may use NAD(P)H and produce $NAD(P)^+$ (e.g. OAA reduction to malate,

NO_3^- reduction to NO_2^-). As well, the mitochondrial ETC may oxidize NAD(P)H produced in reactions other than respiration, e.g. oxidation of NADH produced by glycine decarboxylase during photorespiration (see Chapter 22). Dehydrogenases located on the outer surface of the mitochondrial membrane may oxidize cytosolic NADH (from glycolysis) and NADPH (from the OPP pathway) and feed electrons to the mitochondrial ETC (Douce and Neuberger, 1989). Consequently, it is important to distinguish between the pathways of carbon respiration and the mitochondrial ETC when discussing interactions between respiration and other metabolic processes.

Evaluation of respiratory carbon flow during photosynthesis has been facilitated by two methods. The first method employs feeding ^{14}C carbon compounds to cells. Tracking radiolabeled glycolytic and TCA cycle intermediates has provided evidence that respiratory carbon metabolism continues during photosynthesis (e.g. Marsh *et al.*, 1965; Chapman and Graham, 1974; McCashin *et al.*, 1988; Zubkova *et al.*, 1988, Mamushina *et al.*, 1991; Mamushina and Zubkova, 1992). The pathway of radiolabeled carbon provided as [14]CO_2 is also consistent with maintenance of at least some movement of label through respiratory pathways during photosynthesis (e.g. Chapman and Graham, 1974; Scherer *et al.*, 1984; Elrifi and Turpin, 1987; Maslov *et al.*, 1993). These experiments have been extremely useful in providing evidence for respiratory carbon flow during photosynthesis, but they have failed to provide a consistent picture about the actual rates of respiratory carbon flow during the light versus dark.

The second method used to measure respiratory carbon flow in the light involves examining the gas exchange characteristics of photosynthesizing tissues. This approach has inherent difficulties because photosynthetic CO_2 assimilation can mask the low rates of CO_2 released by the TCA cycle and the OPP pathway. Respiratory CO_2 production may be unmasked by using a mass spectrometer to measure photosynthetic and respiratory CO_2 exchange. In the mass spectrometer, stable carbon isotopes (^{12}C, ^{13}C) are used to distinguish between respiratory and photosynthetic CO_2 exchange in the light. Rates measured this way can be biased by intracellular refixation of respired CO_2, and may be confounded by photorespiratory CO_2 release (Cournac et al., 1993). The effects of photorespiration can be minimized in studies with unicellular algae which possess a CO_2-concentrating mechanism, or in higher plant studies conducted at elevated CO_2 and/or reduced O_2 concentrations. In such cases, respiratory CO_2 release has been observed during photosynthesis, although at a reduced rate compared to the dark. The lower rate of CO_2 release in the light has been attributed both to photosynthetic re-fixation of respired CO_2 (Weger et al., 1988), and to light-inhibition of mitochondrial substrate decarboxylations (Avelange et al., 1991).

In the light, one might assume that photophosphorylation fulfills cellular ATP requirements and so there is little need for oxidative phosphorylation by the mitochondria. The rate of direct transport of ATP from the chloroplast to the cytosol is very low and it is thought that photosynthetic ATP energy is usually stored in the form triosephosphate sugars (TP) for transport to the cytosol (Heldt and Flügge, 1987). The conversion of TP back to 3-phosphoglycerate (PGA) releases ATP to the cytosol, and PGA can reenter the chloroplast as part of the TP-PGA shuttle. TP could also be converted to ATP energy by respiration via glycolysis, the TCA cycle and, if the mitochondrial ETC were operative, oxidative phosphorylation. In addition, glycine, a product of photorespiration, is known to be oxidized at a high rate in the mitochondrion during photosynthesis by C_3 plants and this could produce ATP via oxidative phosphorylation (Gardeström and Wigge, 1988). It is generally agreed that some respiratory carbon flow must be maintained during photosynthesis to supply the large demand for glycolytic and TCA carbon skeletons created by the high rate of biosynthesis in the light (Graham, 1980). Continued operation of the mitochondrial electron transport during photosynthesis is more controversial. There is evidence from experiments using inhibitors of the mitochondrial ETC that mitochondria contribute to the ATP supply in the cytosol even in the light (Krömer and Heldt, 1991).

Until recently it was believed that an increase in the ATP/ADP ratio during photosynthesis would prevent operation of the mitochondrial ETC, thereby removing one potential sink for electrons generated

by the TCA cycle. Several reports have provided evidence that ATP/ADP ratios, as influenced by photosynthesis, are probably not an important regulatory mechanism for mitochondrial ETC activity. First, cellular (especially cytosolic) ATP/ADP ratios are relatively constant, and not greatly affected by light/dark transitions (Goller et al., 1982; Hampp et al., 1982; 1985; Stitt et al., 1982). Second, in vitro evidence has shown that extremely high ATP/ADP is necessary to inhibit mitochondrial ETC activity (Dry and Wiskich 1982), and such conditions are unlikely during steady-state photosynthesis. Indeed, in vivo ETC activity is probably much more sensitive to the absolute ADP concentration than to the ATP/ADP ratio. Therefore, any biosynthetic process which consumes ATP and produces ADP could conceivably stimulate ETC activity. Finally, the existence of alternative pathway respiration would minimize ATP/ADP effects on ETC activity, especially if electron flow through the alternative pathway is completely uncoupled from ATP production through utilization of the rotenone-resistant bypass (Chapter 14).

Measurement of O_2 consumption in the light by mass spectrometry has been used to evaluate mitochondrial electron transport during photosynthesis. Photosynthetic O_2 evolution is monitored by the appearance of $^{16}O_2$, and respiratory ETC activity is monitored by the disappearance of $^{18}O_2$. Although there have been numerous studies of O_2 exchange in the light, there are also a variety of interpretations of these results. A clear appreciation of the role of mitochondrial ETC activity during photosynthesis has yet to emerge. This may be due, in part, to a number of reactions which could consume O_2 photosynthesis, including photorespiration, O_2 photoreduction (Mehler reaction) and mitochondrial respiration (Badger, 1985). Under conditions where photorespiration is suppressed (e.g. high CO_2), a high rate of O_2 consumption at high irradiance has often been demonstrated and attributed to photoreduction (Glidewell and Raven, 1975; Radmer and Kok, 1976; Marsho et al., 1979; Canvin et al., 1980; Furbank et al., 1982; Ishii and Schmid, 1982; Brechignac and Furbank, 1987; Sültemeyer and Fock, 1986; Shiraiwa et al., 1988; Sültemeyer et al., 1993). Many of these latter experiments have also suggested that illumination results in a decrease in the rate of mitochondrial O_2 consumption. A decline in mitochondrial ETC activity may provide an explanation for the higher apparent quantum yield at low light compared to high light ('Kok effect') (Kok, 1949; Sharp *et al.*, 1984). In contrast, some experiments have indicated that O_2 consumption decreases in the light, and does not respond to increasing irradiance (Bate *et al.*, 1988). Still other work has demonstrated that O_2 consumption at all levels of light occurs at rates comparable to those in the dark (Gerbaud and André, 1980; Peltier and Thibault, 1985; Weger *et al.*, 1988). There is also evidence for transient O_2 photoreduction upon initiation of illumination (Radmer and Kok, 1976; Avelange and Rébeillé, 1991). Finally, it has been suggested that low levels of irradiance may result in partial inhibition of mitochondrial ETC activity while high irradiance results in enhancement (Nespoulous *et al.*, 1989).

The photosynthetic ETC inhibitor 3-(3,4-dichloro)-1,1-dimethylurea (DCMU) has been used in attempts to separate O_2 consumption by the Mehler reaction and photorespiration from mitochondrial respiration. DCMU inhibits non-cyclic electron flow and so maintenance of O_2 consumption in the presence of DCMU rules out photorespiration and the Mehler reaction. In the green algae, mass spectrometry has indicated that O_2 consumption in the light occurred at high rates and was completely unaffected by DCMU. This indicated little change in mitochondrial ETC activity between the light and the dark and suggested that, in this case, rates of both photorespiratory and Mehler O_2 consumption are low during photosynthesis (Peltier and Thibault, 1985; Weger *et al.*, 1988). In contradiction to these data, earlier studies using O_2 electrodes to monitor O_2 exchange have provided evidence that illumination results in a decrease in respiratory O_2 consumption in green algal cells in the presence of DCMU (Kowallik, 1969; Sargent and Taylor, 1972).

Finally, there is evidence for enhanced mitochondrial respiration following periods of illumination. Cells which have been recently darkened may contain increased levels of substrates, such as TP from both the Calvin cycle and/or the breakdown of recently synthesized starch, that could contribute to the respiratory rate (Azcón-Bieto and Osmond, 1983; Weger *et al.*, 1989; Reddy *et al.*,

1991). Part of the post-illumination stimulation of respiration may be associated with photorespiration (Jackson and Volk, 1970), however; it is still observed under conditions in which photorespiration is suppressed (Heichel, 1970). The enhanced rate of respiration may sometimes be associated with engagement of the alternative respiratory pathway (Chapter 14; Azcón-Bieto *et al.*, 1983). Some authors have argued that the enhanced rate of post-illumination mitochondrial respiration is representative of the rate in the preceding light period (Weger *et al.*, 1989) while others have suggested that the enhanced respiration occurs strictly post-illumination (Bate *et al.*, 1988).

Much attention has been paid to the potential for light regulation of the pyruvate dehydrogenase complex (PDC). This enzyme complex, found in the mitochondrial matrix, converts pyruvate to acetyl-CoA and is a potentially important site for regulating carbon flow to the TCA cycle. Leaf mitochondrial PDC is inactivated *in vivo* by phosphorylation in the light, and the inactivation may be prevented by the presence of DCMU (Budde and Randall, 1990; Gemel and Randall, 1992). It appears that inactivation of PDC in the light may be related to photorespiratory metabolism because products of photorespiration, such as ammonium (NH_4^+) and glycine, inhibit PDC in the dark, while inhibitors of photorespiration prevent inactivation in the light (Schuller and Randall, 1989; Gemel and Randall, 1992).

Another relevant issue concerning mitochondrial respiration in the light is the potential for chloroplast respiration ('chlororespiration'). A 'chlororespiratory' ETC, which shares some common electron carriers with the photosynthetic ETC, has been postulated to exist in chloroplasts from algae and higher plants (Bennoun, 1982; Caron *et al.*, 1987; Garab *et al.*, 1989; Wilhelm and Duval, 1990). Chlororespiratory O_2 consumption is suggested to be mediated by an unknown terminal oxidase which accepts electrons from the plastoquinone pool (Ravenel and Peltier, 1991). Light inhibits chlororespiration (Peltier *et al.*, 1987) and it is suggested that the reaction functions to supply reducing equivalents and ATP from starch breakdown for chloroplast metabolism in the dark (Gfeller and Gibbs 1985; Singh *et al.*, 1992). However, chloroplasts isolated from *Chlamydomonas*, an algae in which chlororespiration is well studied, showed no O_2 requirement for starch breakdown (Klöck et al., 1989). As well, the rate of chlororespiratory electron flow is still not clearly established. Based on results to date, one can assume that any contribution of chlororespiration to respiration in the light is minor.

Clearly, many possible relationships between the rates of respiration and photosynthesis exist. Currently, it seems unlikely that a single point of view will be valid under all conditions in every organism. We should expect there to be conditions where photosynthesis enhances respiration, others in which it causes its inhibition, and still others where there is little effect.

The effects of respiration on photosynthesis

The majority of studies examining the interactions between photosynthesis and respiration have focused on how photosynthesis might modulate the rate of respiration. Recently, it has been suggested that mitochondrial respiration can modulate photosynthesis (Krömer et al., 1988; Vani et al., 1990).

Selective inhibition of mitochondrial ATP synthetase activity by low concentrations of oligomycin decreased the rate of photosynthesis in barley protoplasts, but not chloroplasts, suggesting that this inhibition of photosynthesis is mediated by the mitochondrion (Krömer et al., 1988) and that the mitochondrion may serve to oxidize excess photosynthetic energy (Krömer and Heldt, 1991). Inhibition of mitochondrial ETC activity may lead to increased susceptibility to photoinhibition of photosynthesis (Saradadevi and Raghavendra, 1992), or to a decline in photosynthesis due to feedback inhibition as a result of altered cytosolic sucrose metabolism (Hanson, 1992; Krömer et al., 1993).

Krömer et al. (1988) have suggested that oxidative phosphorylation may be more efficient than cyclic photophosphorylation in producing ATP from photogenerated reductant. According to currently accepted stoichiometries, cyclic electron flow around Photosystem I would result in the formation of 1.5 ATP per electron pair. Non-cyclic photosynthetic electron flow would result in the formation of 1.5 ATP and 1 molecule of NADPH per electron pair. Oxidation of photogenerated NADPH by the mitochondrial ETC would then result in the additional formation of 1.5 to 2.5 ATP, depending upon whether the oxidation proceeded via Complex I or the externally facing NADPH dehydrogenase (see Chapter 14). This may represent a mechanism to supply the cytosolic ATP requirements during photosynthesis. Intriguing as this possibility might be, at present it represents only 'paper chemistry'. Whether this is an important in vivo mechanism is unknown. The characterization of a Chlamydomonas mutant deficient in chloroplast ATP synthetase that is able to grow photoautotrophically (Lemaire et al., 1988), however, implies that oxidative phosphorylation in the mitochondrion can supply the ATP demands of the RPP cycle in the chloroplast.

Mass spectrometry has frequently shown that light causes a large decrease in the rate of CO_2 production, but has a smaller effect on the rate of respiratory O_2 consumption. Given the interdependency of respiration and photosynthesis, one could speculate that in these instances light (photosynthesis) causes a decrease in the rate of respiratory carbon flow, while respiratory O_2 consumption is maintained via oxidation of photo-generated reductant. Cytosolic ATP requirements could be met by oxidative phosphorylation powered by photogenerated reductant (Krömer et al., 1988), while at the same time PDC activity is inhibited (Budde and Randall, 1990). A variety of schemes could be drawn to suggest possible routes of carbon and reductant flow between cellular compartments during photosynthesis.

While the specific interactions between photosynthesis and respiration are not yet resolved, the information available to date points to some generalities. Photosynthesis and respiration are clearly interdependent processes. Reductant, ATP and reduced carbon compounds produced in one cellular compartment may be used in a different compartment. Photosynthesis and respiration are linked via common metabolites. Given these interconnections, it is not surprising that a change in the rate of one of the processes has consequences for the rate of the other process.

Interactions between N assimilation and photosynthesis

The assimilation of inorganic N can be considered a photosynthetic process. Assimilation of nitrate (NO_3^-) or NH_4^+ requires reductant and ATP, both of which may be supplied by the light reactions of photosynthesis. The required carbon skeletons are produced by the oxidation of reduced carbon compounds that, ultimately, are produced by photosynthesis (Fig. 32.1). Inorganic N assimilation may be summarized by the following four reactions:

Nitrate reductase: $NO_3^- + 2e^- \longrightarrow NO_2^-$ [32.1]

Nitrite reductase: $NO_2^- + 6e^- \longrightarrow NH_4^+$ [32.2]

GS: $NH_4^+ + \text{glutamate} + ATP \longrightarrow \text{glutamine} + ADP + P_i$ [32.3]

GOGAT: $\text{glutamine} + \alpha KG + 2e^- \longrightarrow 2 \text{ glutamate}$ [32.4]

These reactions are discussed in more detail in Chapter 31.

The light reactions

The reduction of NO_3^- to NH_4^+ and subsequent assimilation into glutamate requires 10 electrons and, consequently, 2.5 moles of O_2 would be evolved by photosynthesis per NO_3^- assimilated (Fig. 32.3) (Grant and Turner, 1969; Ullrich and Eisele, 1977; Larsson et al., 1982). Although several reductive processes use photogenerated electrons, nitrogen reduction is second only to carbon assimilation in its total requirement for photosynthetic energy (Syrett, 1981). CO_2 and NO_3^- reduction not only draw their energy from a similar source, but 8 of the 10 electrons required to reduce NO_3^- are added in the chloroplast (reactions [32.2] and [32.4]), the site of CO_2 assimilation in plant cells. The proximity of NO_2^- and CO_2 reduction requires that the partitioning of electrons between these reactions be controlled. Reports of NO_3^--dependent increases in O_2 evolution by algae and higher plants indicate that in some cases non-cyclic electron transport increases to meet the extra demands of NO_3^- reduction (Bloom et al., 1989; de la Torre et al., 1991). Oxygen evolution is greater if the nitrogen is supplied as NO_3^- rather than NH_4^+ which further supports the idea that electron flow increases to meet energy demand.

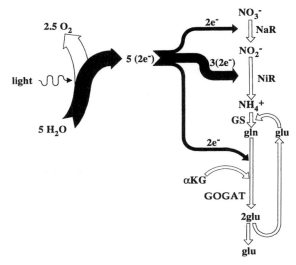

Fig. 32.3 A diagrammatic representation of the use of photosynthetic electron flow in the reduction and assimilation of NO_3^- to glutamate. NaR, nitrate reductase; NiR, nitrite reductase; GS, glutamine synthetase; GOGAT, glutamine:oxoglutarate aminotransferase.

Interestingly, although N-limited algae may increase O_2 evolution during NO_3^- resupply, this is frequently accompanied by a decline in CO_2 fixation (Turpin et al., 1988; Huppe et al., 1994). Evidently, the capacity to increase non-cyclic electron transport is not adequate to supply all of the electrons required for NO_3^- reduction in these cells. It appears that both mitochondrial respiration and an activation of the oxidative pentose phosphate pathway of the chloroplast are involved in supplying electrons under this circumstance (Weger et al., 1988; Weger and Turpin, 1989; Huppe et al., 1992; Huppe et al., 1994). A capacity to use the TCA cycle and OPP pathway reductant rather than photosynthetic electrons for the assimilation of NO_3^- is clearly demonstrated by three circumstances: NO_3^- assimilation occurs in roots; NO_3^- assimilation occurs in darkened photosynthetic tissues of both higher plants and algae; and the dark inhibition of nitrogen assimilation can be overcome in some circumstances by an exogenous supply of carbon (for review see Huppe and Turpin, 1994).

In addition to supplying reductant, the light reactions of photosynthesis may provide ATP for N assimilation via cyclic and non-cyclic photophosphorylation. In fact, under some circumstances, NH_4^+ assimilation, which requires a higher ATP/electron supply than the assimilation of NO_3^-, may result in a state transition of the thylakoid membranes (see Chapter 19) thereby favoring ATP production by cyclic photophosphorylation (Turpin and Bruce, 1990).

Photosynthetic carbon fixation

Addition of NO_3^- or NH_4^+ to N-sufficient photosynthetic cells has little or no effect on photosynthetic CO_2 fixation (Larsson et al., 1982; see Turpin et al., 1988). In contrast, there is a striking effect of CO_2 availability on the ability of photosynthetic cells to assimilate NO_3^- and NH_4^+ (Larsson et al., 1985; Lara et al., 1987). Work with the green alga Selenastrum minutum has demonstrated that both N-sufficient and N-limited cells require CO_2 in order for NH_4^+ assimilation to occur. These cell types differ, however, in the nature of the CO_2 requiring process (Amory et al., 1991). NH_4^+ assimilation by N-sufficient cells is strictly dependent on photosynthetic CO_2 fixation, presumably because recent photosynthate is required to produce carbon skeletons. N-limited cells, however, assimilate NH_4^+ in darkness indicating that recent photosynthate is not required. Removal of CO_2 prevents NH_4^+ assimilation by these cells in the dark (Amory et al., 1991). Presumably, the absence of CO_2 restricts PEP carboxylase (PEPcase) activity in these N-limited cells which compromises anapleurotic replenishment of the TCA cycle. The role of PEPcase in the provision of carbon for amino acid synthesis will be explored in more detail later.

The effects of N assimilation on photosynthetic carbon fixation in N-sufficient tissues are relatively minor. In contrast, when algae are grown under N-limitation the capacity for N assimilation relative to carbon assimilation is increased. At the onset of N assimilation in these cells there is a large demand for both energy and carbon skeletons which in some cases leads to a suppression in photosynthetic CO_2 fixation (see Turpin et al., 1988). At saturating light intensities, addition of NH_4^+ or NO_3^- has been found to cause a major suppression of photosynthetic carbon fixation in S. minutum whereas in another

green alga, Chlorella pyrenidosa, there was no suppression during NH_4^+ assimilation until the light intensity was lowered (Elrifi et al., 1988).

Photosynthetic CO_2 fixation is primarily controlled by the quantity and activation of Rubisco and the availability of this enzyme's substrates: CO_2, O_2, and ribulose bisphosphate (RuBP) (Jones, 1973; Walker, 1976; Collatz, 1977; Collatz et al., 1979; Perchorowicz, et al., 1981; von Caemmerer et al., 1983; Sharkey, 1985; Salvucci et al., 1987). The suppression in photosynthetic carbon fixation in S. minutum was shown to coincide with a decrease in RuBP levels (Elrifi and Turpin 1986) and a resulting RuBP limitation of Rubisco (Elrifi et al., 1988). The mechanism by which N assimilation causes a decrease in RuBP regeneration is uncertain. Measurement of changes in the RPP cycle metabolites immediately following NH_4^+ addition to N-limited S. minutum showed a rapid increase in both TP and FBP (Smith et al., 1989) suggesting that regulation occurs in the regeneration steps of the cycle rather than during the carboxylation or reduction steps (see Chapter 21).

Redox regulation is a key control mechanism of chloroplast metabolism (see Chapter 21). During photosynthesis, reduction via the ferredoxin–thioredoxin system activates several enzymes of the RPP cycle and inactivates the regulatory enzyme of the chloroplastic OPP pathway, glucose 6-phosphate dehydrogenase (G6PDH). Oxidation in the dark shuts off the RPP cycle and activates the OPP pathway. This system acts to prevent futile cycling of energy during the light and the dark within the chloroplast (see Chapter 21). Interestingly, during photosynthetic NO_3^- assimilation by N-limited Chlamydomonas reinhardtii the activation of at least one of the ferredoxin–thioredoxin system target enzymes, phosphoribulosekinase, is decreased by oxidation while the G6PDH is activated (Huppe et al., 1994) (Fig. 32.1). Although redox regulation may play an important role in the suppression of CO_2 fixation during NO_3^- assimilation, NH_4^+ does not effect a similar response.

The changes in the RPP cycle activity which occur during N assimilation by N-limited algae should not be viewed in isolation from the rest of cell metabolism. This transient suppression of carbon assimilation occurs simultaneously with a dramatic increase in demand for carbon skeletons for amino acid synthesis. The demands of amino acid synthesis requires increased carbon export from the chloroplast to provide substrate for respiratory production of carbon skeletons.

Interactions between N assimilation and respiration

The supply of carbon skeletons for amino acid biosynthesis

The largest portion of N assimilated by a C3 plant cell comes from the NH_4^+ released during photorespiration (see Chapter 22). This process, however, does not result in a net production of amino nitrogen. Primary N assimilation converts inorganic nitrogen obtained from the environment into amino acids. This net synthesis of amino acids requires the provision of new carbon skeletons in the form of keto acids.

Figures 32.1 and 32.4 illustrate the role of glycolysis and the TCA cycle in providing αKG for net glutamate synthesis. The synthesis of the other amino acids also requires the provision of carbon skeletons, most of which are also intermediates in respiratory pathways (Chapter 30). The implication of this increased demand for carbon is that the assimilation of new nitrogen must be met by an increased supply of keto acids, which, in turn, requires an increased flow of carbon through respiratory pathways. A simple demonstration of this requirement is the stimulation of respiratory CO_2 efflux upon supply of inorganic nitrogen in both the dark and during photosynthesis (Fig. 32.5).

The TCA cycle intermediates used for amino acid synthesis must be replenished because organic acids removed from the cycle become unavailable for the regeneration of OAA. OAA is a substrate of citrate synthase, the enzyme responsible for incorporation of acetyl-CoA into the TCA cycle. Depletion of OAA would cause a decrease in TCA cycle activity and compromise the provision of carbon skeletons for amino acid synthesis. The most common reaction serving this anaplerotic function is the carboxylation of PEP resulting in the production of OAA by the

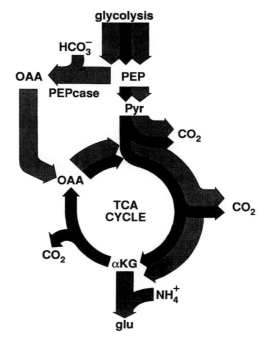

Fig. 32.4 A diagrammatic representation of the stoichiometry required to maintain both catabolic TCA cycle activity (black arrows) or anabolic TCA activity (light arrows) for the net synthesis of glutamate. The widths of the arrows are proportional to the rates of carbon flow. This figure illustrates the requirement for anapleurotic reactions (e.g. PEPcase) to replenish OAA during biosynthetic consumption of TCA cycle intermediates.

enzyme PEPcase. Bicarbonate (HCO_3^-) is the substrate for this enzyme.

$$HCO_3^- + PEP \xrightarrow{\text{PEPcase}} OAA$$

$$OAA + NADH \xleftrightarrow{\text{MDH}} \text{Malate} + NAD^+$$

PEPcase is a cytosolic enzyme that converts PEP from glycolysis into OAA. The OAA produced is rapidly reduced to malate via cytosolic malate dehydrogenase (MDH) and then imported into the mitochondrion where it enters the TCA cycle. The significance of anapleurotic PEP carboxylation to supply keto acid synthesis for N assimilation is supported by reports of increased PEPcase activity during N assimilation by algae and higher plants (Guy et al., 1989; Vanlerberghe et al., 1990, 1992; Le

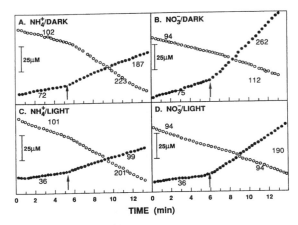

Fig. 32.5 The interactions between respiration and nitrogen assimilation both the light and dark in the N-limited green alga *Selenastrum minutum*. Respiratory O_2 consumption (○) and CO_2 release (●) were measured in quadruple isotope experiments with mass-spectrometry which enables the determination of respiratory gas exchange during photosynthesis. The rates reported are in units of μmol (O_2 or CO_2) mg^{-1} Chl h^{-1}. (A) The effects of the onset of ammonium assimilation on CO_2 release (●) and O_2 consumption (○) in the dark; (B) the effects of NO_3^- in the dark; (C) the effects of NH_4^+ during photosynthesis; (D) the effects of NO_3^- during photosynthesis. Arrows indicate the addition of nitrogen.

Van Quy *et al.*, 1991a). In N-limited cells of *S. minutum*, PEPcase activity increased 20–40 fold after NH_4^+ addition in both the light and dark (Elrifi and Turpin, 1986; Guy *et al.*, 1989). In fact, the rate of dark carbon fixation is a linear function of the NH_4^+ assimilation rate (Vanlerberghe *et al.*, 1990). Figure 32.4 illustrates a simplified stoichiometry of carbon flow necessary for the TCA cycle to operate simultaneously in both catabolic and anabolic (biosynthetic) processes, assuming that glutamate is the only amino acid being produced. On balance, the removal of one molecule of αKG for use in net glutamate synthesis requires the production of one OAA via PEP carboxylation and the entry of one additional molecule of acetyl-CoA. Hence, net glutamate synthesis requires two additional molecules of PEP to enter the TCA cycle; one via PEPcase or another anapleurotic route, the other via pyruvate kinase and the pyruvate dehydrogenase complex (Fig. 32.4).

The main routes of entry of carbon into the TCA cycle are via PEPcase and pyruvate kinase (PK); therefore, it is not surprising that substantial evidence has accumulated indicating that both PEPcase and PK are key regulatory enzymes governing the increase in anapleurotic carbon flow to the TCA cycle during N assimilation (Smith *et al.*, 1989; Turpin *et al.*, 1990). In N-limited cells of *S. minutum*, the initiation of N assimilation causes a large increase in the Pyr/PEP and malate/PEP ratios which are consistent with the activation of PK and PEPcase, respectively (Fig. 32.1). Similarly, an activation of PK and PEPcase is indicated by a drop in PEP and PGA during N assimilation by the developing cotyledon of *Ricinus communis* (Geigenberger and Stitt, 1991).

Kinetic properties of cytosolic PK(PKc) and PEPcase suggest regulatory mechanisms that may activate these enzymes (Fig. 32.1). Glutamate inhibits PKc from algae and from *R. communis* cotyledons (Lin *et al.*, 1989, Wu & Turpin, 1992; Podestá and Plaxton, 1994; see Chapter 8 and Fig. 32.4). There have been several reports of a correlation between the *in vivo* activity of PEPcase and the level of glutamine and glutamate (Vanlerberghe *et al.*, 1990; Thi Mahn *et al.*, 1993; Foyer *et al.*, 1994). *In vitro*, glutamate inhibits PEPcase from *S. minutum* and *R. communis* (Schuller *et al.*, 1990; Podestá and Plaxton, 1994) and glutamine is an activator of *S. minutum* PEPcase. Dihydroxyactetone phosphate (DHAP) activates both PKc and PEPcase from this alga (Lin *et al.*, 1989; Schuller *et al.*, 1990). Interestingly, initiation of NH_4^+ assimilation in *S. minutum* causes a transient decrease in glutamate due to the action of GS and an increase in DHAP from the activation of phosphofructokinase (PFK) (Smith *et al.*, 1989; Turpin *et al.*, 1990). As a result, a rapid activation of PKc could be expected at the onset of NH_4^+ assimilation. Simultaneously, activation of *S. minutum* PEPcase could be mediated via the drop in glutamate and increased level of glutamine that occurs at the onset of NH_4^+ assimilation. Metabolic events associated with the activation of respiration events during dark NH_4^+ assimilation are summarized in Fig. 32.6. During NO_3^- assimilation by *S. minutum*, similar metabolite fluxes occur, but more gradually. Respiratory carbon flow activates to provide carbon skeletons to NO_3^- assimilation, but the energy requirement for the reduction of the N-source must be met first (Vanlerberghe *et al.*, 1992).

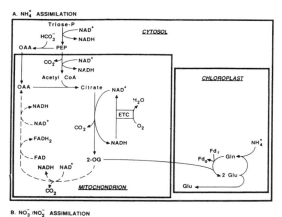

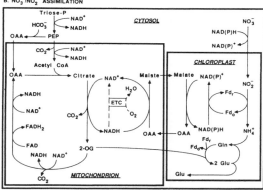

Fig. 32.6 Proposed scheme for electron and carbon flow during transient assimilation of NH_4^+ or NO_3^-. Triose phosphate would be provided by the chloroplast from either the RPP cycle or starch degradation. The potential for exchange of reductant, produced by photosynthetic electron transport, activation of the OPP pathway for starch degradation, or within the mitochondria, across organellar membranes is illustrated by the inclusion of the malate/ oxaloacetate shuttle. The pathways for generating reduced Fd and $NADP^+$ in the chloroplast are not illustrated. Abbreviations: αOG, 2-oxoglutarate; OAA, oxaloacetate; Fd_r, reduced ferredoxin; Fd_o, oxidized ferredoxin (adapted from Weger and Turpin, 1989).

Whether N assimilation is occurring in photosynthesizing or non-photosynthesizing tissues determines the ultimate source of respiratory substrate for these anapleurotic reactions (Fig. 32.1). In the absence of photosynthesis, starch or sucrose must supply the carbon. In the light, under N-sufficient conditions, recent photosynthate is the

likely source of carbon for steady-state amino acid synthesis. In some cases, however, high rates of N assimilation during photosynthesis may actually result in net starch breakdown during photosynthesis (Smith *et al.*, 1989). In higher plants the rate of sucrose synthesis has been shown to decrease if NO_3^- assimilation is increased (Le Van Quy *et al.*, 1991b; Thi Mahn *et al.*, 1993; Foyer *et al.*, 1994). While our understanding of the regulatory mechanisms controlling carbon partitioning in response to N assimilation is still incomplete, it is obvious that N assimilation depends upon increased respiratory carbon flow. Although the regulation is complex, activation of PK, PEPcase, PFK and starch breakdown are all involved in determining the carbon flow to nitrogen metabolism. During photosynthetic N assimilation by N-limited cells there is a diversion of TP from the RPP cycle to respiratory metabolism (Fig. 32.1, Fig. 32.6).

Oxidation of TCA cycle reductant during N assimilation

NAD(P)H and $FADH_2$ must be continually oxidized to maintain TCA cycle carbon flow during N assimilation. In the case of NH_4^+ assimilation, enhancement of O_2 consumption implicates the mitochondrial ETC in this reoxidation. During NH_4^+ assimilation by N-limited *S. minutum* this large stimulation of respiratory O_2 consumption occurs in both light and dark (Fig. 32.5) (Weger *et al.*, 1988; Weger and Turpin, 1989). If mitochondrial ETC activity is inhibited there is a decrease in the rate of NH_4^+ assimilation. The situation during NO_3^- assimilation is quite different. Although the increase in respiratory CO_2 production can be even greater for NO_3^- than for NH_4^+ assimilation, there is a much smaller increase in mitochondrial O_2 consumption both in light and darkness (Fig. 32.5; Weger and Turpin, 1989). The difference is probably due to the fact that NO_3^- is a more oxidized N source than NH_4^+ (8 electrons are required to reduce NO_3^- to NH_4^+). Synthesis of carbon skeletons requires respiration, which, in turn, generates reductant. If the reduction of NO_3^- reoxidizes the reductant produced then the mitochondrial ETC activity would be less essential in maintaining high rates of NO_3^-

assimilation. Calculations indicate that under some circumstances the rate of photosynthetic reductant production is inadequate to account for the observed rates of NO_3^- assimilation and additional reductant must be supplied by either TCA cycle or OPP pathway activity (Weger and Turpin, 1989; Huppe et al., 1992). The observation that respiratory reducing power may be required during photosynthetic NO_3^- assimilation serves to further highlight the potential complexity of the interactions between photosynthesis, respiration and nitrogen assimilation.

Provision of carbon skeletons – mitochondrial or cytosolic?

Operation of the GS/GOGAT pathway requires large amounts of αKG, which is produced by isocitrate dehydrogenase (IDH). Three potential sites for the production of αKG exist: the mitochondria, the cytosol and the chloroplast. Two arguments favor cytosolic αKG production: higher activity of cytosolic NADP-IDH is relative to IDH activity in other compartments (Chen et al., 1988; Martínez-Rivas and Vega, 1993); and the kinetics of αKG and citrate transport by the chloroplast and mitochondria (Chen and Gadal, 1990; Hanning and Heldt, 1993). It has been suggested, therefore, that citrate produced by the TCA cycle is exported to the cytosol in support of N assimilation, and that αKG is synthesized in the cytosol from cytosolic aconitase and NADP-IDH and then transported into the chloroplast.

Summary

The interactions between photosynthesis, respiration and nitrogen assimilation are complex. For simplicity, these processes have usually been studied in isolation; however, to view these processes as isolated pathways is to ignore the integration of metabolism. Organic carbon compounds are the product of photosynthesis, and are the ultimate source of carbon skeletons needed for biosynthesis. The light reactions of photosynthesis are capable of producing ATP and reductant for biosynthesis, but synthesis of carbon skeletons requires the participation of respiratory pathways. Therefore, high rates of biosynthesis in the light require high rates of respiration in the light. The rate of respiration is modulated by biosynthetic demands for carbon skeletons, and by the supply of respiratory substrate, which is produced by photosynthesis. Recent studies suggest that photosynthetic rates are also modulated by respiration.

References

Amory, A. M, Vanlerberghe, G. C. and Turpin, D. H. (1991). Demonstration of both a photosynthetic and a non-photosynthetic CO_2 requirement for NH_4^+ assimilation in the green alga *Selenastrum minutum*. *Plant Physiol.* **95**, 192–6.

Avelange, M.-H. and Rébeillé, F. (1991). Mass spectrometric determination of O_2 exchange during a dark-to-light transition in higher-plant cells. Evidence for two individual O_2-uptake components. *Planta* **183**, 158–63.

Avelange, M.-H., Thiéry, J. M., Sarrey, F., Gans, P. and Rébeillé, F. (1991). Mass-spectrometric determination of O_2 and CO_2 gas exchange in illuminated higher-plant cells. Evidence for light-inhibition of substrate decarboxylations. *Planta* **183**, 150–57.

Azcón-Bieto, J. and Osmond, C. B. (1983). Relationship between photosynthesis and respiration: the effect of carbohydrate status on the rate of CO_2 production by respiration in darkened and illuminated wheat leaves. *Plant Physiol.* **71**, 574–81.

Azcón-Bieto, J., Lambers, H. and Day, D. A. (1983). Effect of photosynthesis and carbohydrate status on respiratory rates and the involvement of the alternative pathway in leaf respiration. *Plant Physiol.* **72**, 598–603.

Badger, M. R. (1985). Photosynthetic oxygen exchange. *Ann. Rev. Plant Physiol.* **36**, 27–53.

Bate, G. C., Sültemeyer, D. F. and Fock H. P. (1988). $^{16}O_2/^{18}O_2$ analysis of oxygen exchange in *Dunaliella tertiolecta*. Evidence for the inhibition of mitochondrial respiration in the light. *Photosyn. Res.* **16**, 219–31.

Bennoun, P. (1982). Evidence for a respiratory chain in the chloroplast. *Proc. Natl Acad. Sci. USA* **79**, 4352–6.

Bloom, A. J., Caldwell, R. M., Finazzo, J., Warner, R. L. and Weissbart, J. (1989). Oxygen and carbon dioxide fluxes from barley shoots depend on nitrate assimilation. *Plant Physiol.* **91**, 352–6.

Brechignac, F. and Furbank, R. T. (1987). On the nature of the oxygen uptake in the light by *Chondrus crispus*. Effects of inhibitors, temperature and light intensity. *Photosyn. Res.* **11**, 45–59.

Budde, R. J. A. and Randall, D. D. (1990). Pea leaf mitochondrial dehydrogenase complex is inactivated *in vivo* in a light-dependent manner. *Proc. Natl. Acad. Sci. USA* **87**, 673–6.

Canvin, D. T., Berry, J. A., Badger, M. R., Fock, H. and Osmond, C. B. (1980). Oxygen exchange in leaves in the light. *Plant Physiol.* 66, 302–7.

Chapman, E. A. and Graham, D. (1974). The effects of light on the tricarboxylic acid cycle in green leaves. I. Relative rates of the cycle in the dark and the light. *Plant Physiol.* 53, 879–85.

Chen, R. D. and Gadal, P. (1990). Do the mitochondria provide the 2-oxoglutarate needed for glutamate synthesis in higher plants? *Plant Physiol. Biochem.* 28, 141–6.

Chen, R. D., Le Maréchal, P., Vidal, J., Jacquot, J. P. and Gadal, P. (1988). Purification and comparative properties of the cytosolic isocitrate dehydrogenases (NADP) from pea (*Pisum sativum*) roots and green leaves. *Eur. J. Biochem.* 175, 565–72.

Collatz, G. J. (1977). The interaction between photosynthesis and ribulose-P_2 concentration – effects of light, CO_2. *Carnegie Inst. Washington Year Book, 1977,* 248–51.

Collatz, G. J., Badger, M., Smith, C. and Berry, J. A. (1979). A radioimmune assay for RuP_2 carboxylase protein. *Carnegie Inst. Washington Year Book, 1979,* 171–5.

Cournac, L., Bimon, B. and Peltier, G. (1993). Evidence for ^{18}O labeling of photorespiratory CO_2 in photoautotrophic cell cultures of higher plants illuminated in the presence of $^{18}O_2$. *Planta* 190, 407–14.

de la Torre, A., Delgado, B. and Lara, C. (1991). Nitrate-dependent O_2 evolution in intact leaves. *Plant Physiol.* 69, 1196–9.

Douce, R. and Neuberger, M. (1989). The uniqueness of plant mitochondria. *Annu. Rev. Plant Physiol. Plant Mol. Biol.* 40, 371–414.

Dry, I. B. and Wiskich, J. T. (1982). Role of external adenosine triphosphate/adenosine diphosphate ratio in the control of plant mitochondrial respiration. *Arch. Biochem. Biophys.* 217, 72–9.

Elrifi, I. R. and Turpin, D. H. (1986). Nitrate and ammonium induced photosynthetic suppression in N-limited *Selenastrum minutum*. *Plant Physiol.* 81, 273–9.

Elrifi, I. R., and Turpin, D. H. (1987). The path of carbon flow during NO_3^- induced photosynthetic suppression in N-limited *Selenastrum minutum*. *Plant Physiol.* 83, 97–104.

Elrifi, I. R., Holmes, J. J., Weger, H. G., Mayo, W. P. and Turpin, D. H. (1988). RuBP limitation of photosynthetic carbon fixation during NH_3 assimilation: Interaction between photosynthesis, respiration and ammonium assimilation in N-limited green algae. *Plant Physiol.* 87, 395–401.

Foyer, C. H., Noctor, G., Lelandais, M., Lescure, J. C., Valadier, M. H., Boutin, J. P. and Horton, P. (1994). Short-term effects of nitrate, nitrite and ammonium assimilation on photosynthesis, carbon partitioning and protein phosphorylation in maize. *Planta* 192, 211–20.

Furbank, R. T., Badger, M. R. and Osmond, C. B. (1982).

Photosynthetic oxygen exchange in isolated cells and chloroplasts of C_3 plants. *Plant Physiol.* 70, 927–31.

Garab, G., Lajkó, F., Mustárdy, L. and Márton, L. (1989). Respiratory control over photosynthetic electron transport in chloroplasts of higher-plant cells: evidence for chlororespiration. *Planta* 179, 349–58.

Gardeström, P, and Wigge, B. (1988). Influence of photorespiration on ATP/ADP ratios in the chloroplasts, mitochondria and cytosol, studied by the rapid fractionation of barley (*Hordeum vulgare*) protoplasts. *Plant Physiol.* 88, 69–76.

Geigenberger, P. and Stitt, M. (1991). Regulation of carbon partitioning between sucrose and nitrogen assimilation in cotyledons of germinating *Ricinus communis* L. seedings. *Planta* 185, 563–8.

Gemel, J. and Randall, D. D. (1992). Light regulation of leaf mitochondrial dehydrogenase complex. Role of photorespiratory carbon metabolism. *Plant Physiol.* 100, 908–14.

Gerbaud, A. and André, M. (1980). Effect of CO_2, O_2 and light on photosynthesis and photorespiration in wheat. *Plant Physiol.* 66, 1032–6.

Gfeller, R. P. and Gibbs, M. (1985). Fermentative metabolism of *Chlamydomonas reinhardtii*. II. Role of plastoquinone. *Plant Physiol.* 77, 509–11.

Goller, M., Hampp, R. and Ziegler, H. (1982). Regulation of the cytosolic adenylate ratio as determined by rapid fractionation of mesophyll protoplasts of oat. Effect of electron transport inhibitors and uncouplers. *Planta* 156, 255–63.

Glidewell, S. M. and Raven, J. A. (1975). Measurement of simultaneous oxygen evolution and uptake in *Hydrodictyon africanum*. *J. Exp. Bot.* 26, 479–88.

Graham, D. (1980). Effects of light on 'dark' respiration. In *The Biochemistry of Plants*, Vol. 2, ed. D. D. Davies, Academic Press, New York, pp. 525–79.

Grant, B. R. and Turner, I. M. (1969). Light-stimulated nitrate and nitrite assimilation in several species of algae. *Comp. Biochem. Physiol. A* 29, 995–1004.

Guy, R. D., Vanlerberghe, G. C. and Turpin, D. H. (1989). Significance of phosphoenolpyruvate carboxylase during ammonium assimilation: Carbon isotope discrimination in photosynthesis and respiration by the N-limited green alga *Selenastrum minutum*. *Plant Physiol.* 89, 1150–57.

Hampp, R., Goller, M. and Ziegler, H. (1982). Adenylate levels, energy charge, and phosphorylation potential during dark/light and light/dark transition in chloroplasts, mitochondria, and cytosol of mesophyll protoplasts from *Avena sativa* L. *Plant Physiol.* 69, 448–55.

Hampp, R., Goller, M., Fullgraf, H. and Eberle, I. (1985). Pyridine and adenine nucleotide status, and pool sizes for a range of metabolites in chloroplasts, mitochondria, and the cytosol/vacuole of *Avena* mesophyll protoplasts during dark/light transition: Effect of pyridoxal phosphate. *Plant Cell Physiol.* 26, 99–108.

Hanning, I. and Heldt, H. W. (1992). Provision of carbon skeletons for nitrate assimilation by spinach leaf mitochondria. In *Molecular, Biochemical and Physiological Aspects of Plant Respiration*, eds H. Lambers and L. H. W. van der Plas, SPD Academic, The Hague, pp. 249–53.

Hanson, K. R. (1992). Evidence for mitochondrial regulation of photosynthesis by a starchless mutant of *Nicotiana sylvestris*. *Plant Physiol.* **99**, 276–83.

Heldt, H. W. and Flügge, U. I. (1987). Subcellular transport of metabolites in plant cells. In *Physiology of Metabolism*, ed. D. D. Davis, Academic Press, San Diego, pp. 49–85.

Heichel, G. H. (1970). Prior illumination and respiration of maize leaves in the dark. *Plant Physiol.* **46**, 359–62.

Huppe, H. C., Vanlerberghe, G. C. and Turpin, D. H. (1992). Evidence for activation of the oxidative pentose phosphate pathway during photosynthetic assimilation of NO_3^- but not NH_4^+ by a green alga. *Plant Physiol.* **100**, 2096–9.

Huppe, H. C. and Turpin, D. H. (1994). Integration of carbon and nitrogen metabolism in plant and algal cells. *Annu. Rev. Plant Physiol. Plant Mol. Biol.* **45**, 577–607.

Huppe, H. C., Farr T. J. and Turpin, D. H. (1994). Coordination of chloroplastic metabolism in N-limited Chlamydomonas reinhardtii by redox modulation. II. Redox modulation activates the oxidative pentose phosphate pathway during photosynthetic nitrate assimilation. *Plant Physiol.* **105**, 1043–8.

Ishii, R. and Schmid, G. H. (1982). Studies on $^{18}O_2$ uptake in the light by entire plants of different tobacco mutants. *Z. Naturforsch.* **37c**, 93–101.

Jackson, W. A. and Volk, R. J. (1970). Photorespiration. *Annu. Rev. Plant Physiol.* **21**, 385–432.

Jones, H. G. (1973). Limiting factors in photosynthesis. *New Phytol.* **62**, 1089–94.

Klöck, G., Sültemeyer, D. F., Fock, H. P. and Kreuzberg, K. (1989). Gas exchange in intact isolated chloroplasts from *Chlamydomonas reinhardtii* during starch degradation in the dark. *Physiol. Plant.* **75**, 109–13.

Kok, B. (1949). On the interrelation of respiration and photosynthesis in green plants. *Biochim. Biophys. Acta* **3**, 625–31.

Kowallik, W. (1969). Der einfluß von licht auf die atmung von *Chlorella* bei gehemmter photosynthese. *Planta* **86**, 50–62.

Krömer, S. and Heldt, H. W. (1991). On the role of mitochondrial oxidative phosphorylation in photosynthesis metabolism as studied by the effect of oligomycin on photosynthesis in protoplasts and leaves of barley (*Hordeum vulgare*). *Plant Physiol.* **95**, 1270–76.

Krömer, S., Malmberg, G. and Gardeström, P. (1993). Mitochondrial contribution to photosynthetic metabolism. *Plant Physiol.* **102**, 947–55.

Krömer, S., Stitt, M. and Heldt, H. W. (1988). Mitochondrial oxidative phosphorylation participating in photosynthetic metabolism of a leaf cell. *FEBS Lett* **226**, 352–6.

Lara, C., Romero, J. M., Coronil, T. and Guerrero, M. G. (1987). Interactions between photosynthetic nitrate assimilation and CO_2 fixation in cyanobacteria. In *Inorganic Nitrogen Metabolism*, eds W. R. Ullrich, P. J. Aparicio, P. J. Syrett and F. Castillo, Springer-Verlag, New York, pp. 45–52.

Larsson, M., Ingemarsson, B. and Larsson, C.-M. (1982). Photosynthetic energy supply for NO_3^- assimilation in *Scenedesmus*. *Physiol. Plant.* **55**, 301–8.

Larsson, M., Olsson, T. and Larsson, C.-M. (1985). Distribution of reducing power between photosynthetic carbon and nitrogen assimilation in *Scenedesmus*. *Planta* **164**, 246–53.

Le Van Quy, Foyer, C. and Champigny, M.-L. (1991a). Effect of light and NO_3^- on wheat leaf phosphoenolpyruvate carboxylase activity. *Plant Physiol.* **97**, 1476–82.

Le Van Quy, Lamaze, T. and Champigny, M.-L. (1991b). Short-term effects of nitrate on sucrose synthesis in wheat leaves. *Planta* **185**, 53–7.

Lemaire, C., Wollman, F.-A. and Bennoun, P. (1988). Restoration of phototrophic growth in a mutant of *Chlamydomonas reinhardtii* in which the chloroplast at pB gene of ATP synthase has a deletion: An example of mitochondria-dependent photosynthesis. *Proc. Natl Acad. Sci. USA* **85**, 1344–8.

Lin, M., Turpin, D. H. and Plaxton, W. (1989). Pyruvate kinase isozymes from the green alga *Selenastrum minutum* II. Kinetic and regulatory properties. *Arch. Biochem. Biophys.* **269**, 228–38.

Mamushina, N. S. and Zubkova, E. K. (1992). Functioning of the Krebs cycle in autotrophic leaf tissues of C_3 plants under light at natural CO_2 concentration. *Sov. Plant Physiol. (Engl. Tr.)* **39**, 447–53.

Mamushina, N. S., Zubkova, E. K., Ivanova, T. L., Yudina, O. S. and Filippova, L. A. (1991). Functioning of the main stages of dark respiration under light in the early-spring ephemeroid *Ficaria verna*. *Sov. Plant Physiol. (Engl. Tr.)* **38**, 346– 53.

Martínez-Rivas, J. M. and Vega, J. M. (1993). Effect of culture conditions on the isocitrate dehydrogenase and isocitrate lyase activities in *Chlamydomonas reinhardtii*. *Physiol. Plant.* **88**, 599–603.

Marsh, H. V. Jr., Galmiche, J. M. and Gibbs, M. (1965). Effect of light on tricarboxylic acid cycle in *Scenedesmus*. *Plant Physiol.* **40**, 1913–22.

Marsho, T. V., Behrens, P. W. and Radmer, R. J. (1979). Photosynthetic oxygen reduction in isolated intact chloroplasts and cells from spinach. *Plant Physiol.* **64**, 656–9.

Maslov, A. I., Zubova, S. V., Kuz'min, A. N. and Romanova, A. K. (1993) Dynamics of distribution of ^{14}C-labeled photosynthetic products during exposure of pea leaf protoplasts under light and dark. *Russ. Plant Physiol. (Engl. Tr.)* **40**, 313–8.

McCashin, B. G., Cossin, E. A. and Canvin, D. T. (1988). Dark respiration during photosynthesis in wheat leaf slices. *Plant Physiol.* **87**, 155–61.

Nespoulous, C., Peltier, G. and Gans, P. (1989). Photosynthetic, photorespiratory and respiratory gas exchange in *Lemna minor*. *Plant Physiol. Biochem.* **27**, 863–71.

Peltier, G. and Thibault, P. (1985). O_2 uptake in the light in *Chlamydomonas*: evidence for persistent mitochondrial respiration. *Plant Physiol.* **79**, 225–30.

Peltier, G., Ravenel, J. and Verméglio, A. (1987). Inhibition of respiratory activity by short saturating flashes in *Chlamydomonas*: evidence for chlororespiration. *Biochim. Biophys. Acta* **893**, 83–90.

Perchorowicz, J. T., Raynes D. A. and Jensen, R. G. (1981). Light limitation of photosynthesis and activation of ribulose bisphosphate carboxylase in wheat seedlings. *Proc. Natl Acad. Sci. USA* **78**, 2895–989.

Podestá, F. E. and Plaxton, W. C. (1994). Regulation of cytosolic carbon metabolism in germinating *Ricinus communis* cotyledons. II. Properties of phosphoenolpyruvate carboxylase and cytosolic pyruvate kinase associated with the regulation of glycolysis and nitrogen assimilation. *Planta* **194**, 381–7.

Radmer, R. J. and Kok, B. (1976). Photoreduction of O_2 primes and replaces CO_2 assimilation. *Plant Physiol.* **58**, 336–40.

Ravenel, J. and Peltier, G. (1991). Inhibition of chlororespiration by myxathiazol and antimycin A in *Chlamydomonas reinhardtii*. *Photosyn. Res.* **28**, 141–8.

Reddy, M. M, Vani, T. and Raghavendra, A. S. (1991). Light-enhanced dark respiration in mesophyll protoplasts from leaves of pea. *Plant Physiol.* **96**, 1368–71.

Salvucci, M. E., Werneke, J. M., Ogren, W. L. and Portis, A. R. (1987). Purification and species distribution of Rubisco activase. *Plant Physiol.* **84**, 930–36.

Saradadevi, K. and Raghavendra, A. S. (1992). Dark respiration protects photosynthesis against photoinhibition in mesophyll protoplasts of pea (*Pisum sativum*). *Plant Physiol.* **99**, 1232–7.

Sargent, D. F. and Taylor, C. P. S. (1972). Light-induced inhibition of respiration in DCMU-poisoned *Chlorella* caused by photosystem I activity. *Can. J. Bot.* **50**, 13–21.

Scherer, S., Stürzel, E. and Böger, P. (1984). Photoinhibition of respiratory CO_2 release in the green alga *Scenedesmus*. *Physiol. Plant.* **60**, 557–60.

Schuller, K. A. and Randall, D. D. (1989). Regulation of the mitochondrial dehydrogenase complex. Does photorespiratory ammonium influence mitochondrial carbon metabolism? *Plant Physiol.* **89**, 1207–12.

Schuller, K. A., Plaxton, W. C. and Turpin, D. H. (1990). Regulation of phospho*enol*pyruvate carboxylase from the green alga *Selenastrum minutum*. *Plant Physiol.* **93**, 1303–11.

Semikhatova, O. A. (1992). Relations between chloroplasts and mitochondria in the dark. *Sov. Plant Physiol. (Engl. Tr.)* **39**, 391–5.

Singh, K. K., Chen, C. and Gibbs, M. (1992). Characterization of an electron transport pathway associated with glucose and fructose respiration in the intact chloroplasts of *Chlamydomonas reinhardtii* and spinach. *Plant Physiol.* **100**, 327–33.

Sharkey, T. D. (1985). Photosynthesis in intact leaves of C_3 plants: physics, physiology and rate limitations. *Bot. Rev.* **51**, 53–105.

Sharp, R. E., Mathews, M. A. and Boyer, J. S. (1984). Kok effect and the quantum yield of photosynthesis. Light partially inhibits dark respiration. *Plant Physiol.* **75**, 95–101.

Shiraiwa, Y., Bader, K. P. and Schmid, G. H. (1988). Mass spectrometric analysis of oxygen exchange in high and low-CO_2 cells of *Chlorella vulgaris*. *Z. Naturforsch.* **43C**, 709–16.

Smith, R. G., Vanlerberghe, G. C., Stitt, M. and Turpin, D. H. (1989). Short-term metabolite changes during transient ammonium assimilation by the N-limited green alga *Selenastrum minutum*. *Plant Physiol.,* **91**, 749–55.

Stitt, M., Lilley, R. McC. and Heldt, H. W. (1982). Adenine nucleotide levels in the cytosol, chloroplasts and mitochondria of wheat leaf protoplast. *Plant Physiol.* **70**, 971–7.

Sültemeyer, D. F. and Fock, H. P. (1986). Effect of photon fluence rate on oxygen evolution and uptake by *Chlamydomonas reinhardtii* suspensions grown in ambient and CO_2 enriched air. *Plant Physiol.* **81**, 372–5.

Sültemeyer, D. F., Biehler, K. and Fock, H. P. (1993). Evidence for the contribution of pseudocyclic photophosphorylation to the energy requirement of the mechanism for concentrating inorganic carbon in *Chlamydomonas*. *Planta* **189**, 235–42.

Syrett, P. J. (1981). Nitrogen metabolism of microalgae. *Can. Bull. Fish. Aquat. Sci.* **219**, 182–210.

Thi Mahn, C., Bismuth, E., Boutin, J.-P., Provot, M. and Champigny, M.-L. (1993). Metabolite effectors for short-term nitrogen-dependent enhancement of phosphoenolcarboxylase activity and decrease of net sucrose synthesis in wheat leaves. *Physiol. Plant.* **89**, 460–66.

Turpin, D. H. and Bruce, D. (1990). Regulation of photosynthetic light harvesting by nitrogen assimilation in the green alga *Selenastrum minutum*. *FEBS Lett.* **263**, 99–103.

Turpin, D. H., Elrifi, I. R., Birch, D. G., Weger, H. G. and Holmes, J. J. (1988). Interactions between photosynthesis, respiration, and nitrogen assimilation in microalgae. *Can. J. Bot.* **6**, 2083–97.

Turpin, D. H., Botha, F. C., Smith, R. G., Feil, R., Horsey, A. K. and Vanlerberghe, G. C. (1990). Regulation of carbon partitioning to respiration during dark ammonium assimilation by the green alga *Selenastrum minutum*. *Plant Physiol.* **93**, 166–75.

Ullrich, W. R. and Eisele, R. (1977). Relations between nitrate uptake and nitrate reduction in *Ankistrodesmus braunii*. In *Transmembrane Ionic Exchange in Plants*, eds M. Thellier, A. Monnier, M. Demarty and J. Dainty, Colloque du CNRS, Rouen, pp 307–13.

Vani, T., Reddy, M. M. and Raghavendra, A. S. (1990). Beneficial interaction between photosynthesis and respiration in mesophyll protoplasts of pea during short light-dark cycles. *Physiol. Plant.* **80**, 467–71.

Vanlerberghe, G. C., Huppe, H. C., Vlossak, K. D. and Turpin, D. H. (1992). Activation of respiration to support dark NO_3^- and NH_4^+ assimilation in the green alga *Selenastrum minutum*. *Plant Physiol.* **99**, 495–500.

Vanlerberghe, G. C., Schuller, K. A., Smith, R. G., Feil, R., Plaxton, W. C. and Turpin, D. H. (1990). Relationship between NH_4^+ assimilation rate and *in vivo* phospho*enol*pyruvate carboxylase activity. *Plant Physiol.* **94**, 284–90.

von Caemmerer, S., Coleman, J. R. and Berry, J. A. (1983). Control of photosynthesis by RuBP concentration: studies with high- and low-CO_2 adapted cells of *Chlamydomonas reinhardtii*. *Carnegie Inst. Washington Year Book, 1982*, 91–5.

Walker, D. A. (1976). Regulatory mechanisms in photosynthetic carbon metabolism. *Curr. Top. Cell. Regul.* **11**, 203–41.

Weger, H. G. and Turpin, D. H. (1989). Mitochondrial respiration can support NO_3^- and NO_2^- reduction during photosynthesis: Interactions between photosynthesis, respiration, and N assimilation in the N-limited green alga *Selenastrum minutum*. *Plant Physiol.* **89**, 409–15.

Weger, H. G., Birch, D. G., Elrifi, I. R. and Turpin, D. H. (1988). Ammonium assimilation requires mitochondrial respiration in the light: A study with the green alga *Selenastrum minutum*. *Plant Physiol.* **86**, 688–92.

Weger, H. G., Herzig, R., Falkowski, P. G. and Turpin, D. H. (1989). Respiratory losses in the light in a marine diatom: Measurements by short-term mass-spectrometry. *Limnol. Oceanogr.* **34**, 1153–61.

Wilhelm, C. and Duval, J.-C. (1990). Fluorescence induction kinetics as a tool to detect a chlororespiratory activity in the prasinophycean alga, *Mantoniella squamatal. Biochim. Biophys. Acta* **1016**, 197–202.

Wu, H. B. and Turpin, D. H. (1992). Purification and characterization of pyruvate kinase from the green alga, *Chlamydomonas reinhardtii. J. Phycol.* **28**, 472–81.

Zubkova, E. K., Filippova, I. S., Mamushina, N. S. and Chupakhina, G. N. (1988). Effects of light on dark respiration of albino and green regions of the barley leaf. *Sov. Plant Physiol. (Engl. Tr.)* **35**, 254–9.

33 Regulation of the transport of nitrogen and carbon in higher plants

Mark B. Peoples and Roger M. Gifford

INTRODUCTION

A poorly understood aspect of plant function is the regulation of the interactions between sources (organs which supply assimilates) and sinks (sites of storage or growth) for nutrients and the role that such interactions play in determining assimilate partitioning between plant parts. Nonetheless, overall partitioning of nitrogen (N) and carbon (C) assimilates over a growing season is a major determinant of crop yield and product quality (Gifford, 1986; Gifford and Evans, 1981). Furthermore, the factors determining the relationship between C and N partitioning also contribute to the determination of C:N ratio of tissues. The C:N ratio of tissues not only affects its nutritional quality but the C:N ratio of crop residues and of litter influence the rate of their decomposition. Decomposition is in turn critical for ecosystem processes like nutrient mineralization, organic matter build-up, and net carbon sequestration. These are important issues for sustainable development of agricultural systems and for an analysis of the input of 'global change' (Gifford, 1994). Thus, it is important to consider the mechanisms governing assimilate export, translocation, import and utilization, and the factors that may influence N and C allocation and distribution within the plant on both a whole-season basis, and in the short term.

The broad overview presented here draws substantially on reviews devoted to aspects of N (Pate, 1980; Raven and Smith, 1976; Simpson, 1986) and C assimilation and transport (Delrot and Bonnemain, 1985) and assimilate loading and unloading (Thorne, 1985; Wolswinkel, 1985; van Bel, 1993) or partitioning (Gifford and Evans, 1981; Pate *et al.*, 1988).

Translocation of N and C

Requirements, characteristics and constraints

The specific requirements of each plant part for C and N compounds changes throughout development. With fixed C originating in leaves while N compounds (and minerals) come from roots, higher plants have faced a challenging task to move these substrates simultaneously in opposite directions over long distances to meet these changing demands.

To achieve upwards transport of nutrients from the roots, plants capitalize on the large flow of water from root to shoot which is an inevitable consequence of the open stomata required to ventilate the leaves with CO_2 for photosynthesis. The driving force for upward movement of solutes in the xylem is therefore principally 'suction' generated by transpiration. It is a wick effect relying on capillary continuity in the xylem vessel network, which is a 'super-apoplast of dead cells within the living plant axis' (Raven, 1977). 'Root pressure', a positive hydrostatic pressure in the xylem (when transpiration rate is very low) arising from osmotic uptake of soil water following active accumulation of ions (or in special cases, organic solutes), plays only a minor part by day. It may, however, serve to re-establish capillary continuity at night after this has been lost in some xylem vessels by cavitation due to excessive tension during rapid daytime transpiration.

Export of C compounds from the leaves to areas of growth and storage needs a different mechanism since the flow is against an overall water potential gradient throughout most of the plant. Therefore, the C transport system needs (1) to use minimal water;

(2) to be compartmentalized away from the xylem counter-flow, and to minimize uncontrolled leakage of C compounds into the fast-flowing, dilute super-apoplast; (3) have an active driving force in contrast to the 'free-rider' method for xylem solutes. The solution to this problem has been that the phloem's sieve-tubes, forming a network of 'super-symplastic' tubes (Raven, 1977) throughout the plant, are (together with companion cells) powerful active scavengers of sugars from the neighbouring apoplast. The translocated sugars, usually sucrose, having favourable solubility/concentration/viscosity relationships (Lang, 1978), can accumulate to concentrations as high as $500–1000 \, mol \, m^{-3}$ and still move by mass solution flow. In contrast, xylem sap concentration is only approximately $1–50 \, mol \, m^{-3}$, depending on transpiration rate.

The most favoured hypothesis for the phloem transport mechanism – that of mass flow in the sieve tube lumen ('Münch pressure flow') – requires a hydrostatic pressure gradient along the lumen of the sieve tube. This is achieved by active loading of carbohydrates at source, causing water uptake osmotically, and unloading of the solutes and water at the sink. Assuming open sieve plate pores, the required gradient in osmolarity to drive the observed flow is only a small part of the total osmolarity of sieve tube sap.

While the apoplastic xylem sap flows under tension, requiring vessel re-enforcement by encircling lignocellulosic bands to prevent collapse (Raven, 1977), the symplastic sieve tube sap is driven under pressure. These two conduits are in close proximity in vascular strands throughout the plant. This may seem to provide a risk of inefficiency by 'short-circuiting'. In practice it offers an opportunity for improved efficiency via regulated solute exchanges between xylem and phloem at strategic sites. The need for such regulated exchange arises, in part, because N in the xylem would be transported directly to the sites of transpiration, not to the growing points in the plants where the N demand is greatest. On the other hand, solutes and water transported to growing points in phloem sap may not all be utilized (or transpired) by the growing points. The unused molecules can then be carried away in the xylem back to the leaves. Whereas loading and unloading of carbohydrates induces the very driving force needed to move this

material to where it is required from where it is produced, N compounds have to 'hitchhike' first to where the transpiration stream takes them, and then (after phloem loading in the leaves) to where the carbohydrate stream takes them. Transfer of N compounds between xylem and phloem *en route* could potentially adjust the sieve tube sap composition to avoid delivery of unneeded compounds to final sinks.

There is a substantial C requirement in the roots for growth and respiration. So it would seem, at first sight, most efficient for N to move out of the roots in the form in which it is absorbed from the soil (as NH_4^+ and/or NO_3^-), not as organic N compounds, and for phloem not to transport N compounds into the root system. This rarely happens. Linked constraints of pH regulation, charge balance, avoidance of osmotic stress, and of energetics at cell, tissue and whole plant levels influence the molecules transported and cause close co-ordination between xylem and phloem transport (Raven and Smith, 1976).

When NO_3^- is assimilated to the redox level of protein, excess OH^- is produced in the cytoplasm. When NH_4^+ is assimilated, excess H^+ are produced. Movement of such excess protons and hydroxyls into and out of vacuoles and apoplast, in association with organic or inorganic anions and cations for charge balance, would allow short term buffering of cytoplasmic pH. However, this would involve substantial pH and osmotic shifts in the vacuole and cell walls. In the long term, overall plant pH stability might be achievable by three methods: (1) uptake and assimilation of NH_4^+, NO_3^- (and N_2 in legumes) in such ratios that OH^- production everywhere balances H^+ production; (2) assimilation of NH_4^+ and NO_3^- solely in the roots so that H^+ and OH^- (respectively) can be excreted into the soil water; or (3) if N assimilation occurs in the shoot, utilization of a biochemical 'pH-stat' and phloem transport to shift H^+ or OH^- equivalents to the roots for excretion. Method 1 (pH-balancing NH_4^+/NO_3^- uptake) would be excessively restrictive since the proportions of these ions in the soil varies widely. In practice, free NH_4^+ is never found in more than trace amounts in the xylem. Possibly it is toxic. All the assimilation of both NO_3^- and NH_4^+ to organic N would have to occur entirely in the roots to maintain pH constant:

this would require extra C import by the roots. If NH_4^+ and NO_3^- were assimilated solely in the roots, then the availability of the soil water as a sink for excreted NH_4^+ and/or NO_3^- would mean that the two N forms did not have to be taken up in a pH-balancing ratio anyway (i.e. method 2 could be used). However, the import of carbohydrate into the roots would still be required to provide C skeletons for organic N export to the shoot. A further cost of NO_3^- reduction in the roots, instead of in the chloroplasts, is that it must be energized wholly by respiration of phloem-transported carbohydrate. The opportunity to utilize photophosphorylation energy surplus to the needs of CO_2 reduction in the chloroplast is then foregone.

To take advantage of photophosphorylation energy for NO_3^- assimilation in the shoot (involving method 3) requires the operation of a biochemical 'pH-stat' to neutralize the surplus OH^-. This is considered to involve the carboxylation of neutral photosynthetic precursors to organic acids like malic, succinic, or citric acids. Salts of these organic acids (e.g. K^+ malate) must then either be stored in shoot vacuoles or be phloem-translocated to the root where they can be decarboxylated releasing CO_2 and OH^- to the soil in exchange for NO_3^- uptake. This suggested route has the disadvantage of transporting K^+ out of the root system against its net flux, but the products of decarboxylation, such as pyruvate, would be readily available for root respiration to energize further NO_3^- uptake, and to provide C skeletons to carry the N assimilated from NH_4^+ up the xylem. Thus, where NO_3^- is available, its uptake and assimilation in the shoot appears to confer overall advantages for plant function and efficiency.

The NO_3^- uptake capacity of root systems usually greatly exceeds the actual rate of NO_3^- uptake, and this rate appears to be under feedback regulation by the plant's N demand as determined by photosynthetically-paced growth. The mechanism of this regulation of NO_3^- uptake by whole plant N demand is unclear, but it appears to involve a coordinated suppression in response to amino-acid levels and stimulation in response to malate levels in the phloem (Imsande and Touraine, 1994).

In summary, the constraints of pH, charge balance, and osmotic regulation together with the 'hitchhiker' mechanism of main-channel N transport, result in the apparent inefficiencies of transporting C out of the roots and N and minerals into the roots. However, the resultant circulation of materials around the plant provides flexibility for growing points to extract what solutes they need, when they need them from the circulating mixed-composition saps. Different species have adopted different ways of coping with the various conflicting requirements as we next document. A common feature, however, is that the N compounds transported are often those with relatively low C:N ratios and a large number of N atoms per molecule.

Identity of transported solutes

It is relatively easy to examine the solutes of xylem fluids by vacuum extracting 'tracheal' sap from freshly harvested shoot segments, or by collecting sap exuding from the cut xylem of roots or stems passively under root pressure or as the result of applying pneumatic pressure to the root system. However, the delicacy of phloem tissue and the ease with which sieve tubes become blocked with slime and callose create difficulties in collecting phloem contents. Phloem sap exudes spontaneously from shallow incisions for only relatively few woody and non-woody species. Techniques involving severing the stylets of phloem-feeding insects (Fisher and Frame, 1984), the use of aqueous solutions of chelating agents to maintain incision exudation (Simpson and Dalling, 1981), or 'cryopuncturing' vascular strands of legume pods (Peoples et al., 1985b) have been employed to sample phloem translocates.

Xylem saps are mildly acidic (e.g. pH 6–6.5) while phloem saps are alkaline (pH 7–8.5). Nitrogenous compounds frequently comprise the main solute component of xylem while carbohydrates are the major constituent of phloem sap (although high levels of N solutes may also be found). K^+ is the main inorganic ion in phloem. The relative importance of N and C in the two transport streams is reflected in their C:N weight ratios which can range between 1.5 to 6 in xylem exudate, and 10 to 200 in phloem sap. For comparison, a typical range in C:N ratio of whole herbaceous plants is 15 to 100, while in soil organic matter it typically ranges from 10 to 20.

In plants such as *Xanthium*, where virtually no

nitrate is reduced in the root, over 95% of the xylem's soluble-N can consist of free nitrate. The roots of most other plants, however, exhibit some capacity to reduce nitrate and a proportion of xylem N is transported in an organic form (Fig. 33.1). Nitrate is essentially absent from phloem saps (Fig. 33.1). Generally one N-rich molecule, characteristic of the species, dominates the spectrum of organic N compounds present in xylem and

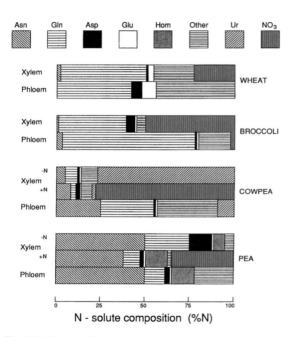

Fig. 33.1 Composition of nitrogenous fractions of xylem exudates and phloem saps collected from wheat (*Triticum aestivum* L.), broccoli (*Brassica oleracea* var. *italica*), cowpea (*Vigna unguiculata* (L.) Walp.), and pea (*Pisum sativum* L.). The major N-solutes in xylem exudates from the legumes cowpea and pea are depicted for plants grown under two different N-regimes: −N, plants totally dependent upon an effective symbiosis; or +N, plants fed with culture solution containing nitrate. Phloem sap constituents were similar for each legume regardless of whether grown under −N or +N regimes. Data for wheat were derived from Simpson and Dalling (1981) and Simpson *et al.* (1982); broccoli from Shelp (1987); cowpea from Peoples *et al.* (1985a); and pea from Peoples *et al.* (1987) and unpublished data. Asn, asparagine; Gln, glutamine; Asp, Aspartate; Glu, glutamate; Hom, homoserine; Other, all other amino compounds detailed in above references; Ur, ureides (allantoin plus allantoic acid); NO_3, nitrate.

phloem. While the advantage of N being transported in N-rich molecules in the xylem is evident, it is not so clear why this should be so in phloem. The amide glutamine for instance, is commonly a major N compound detected in both xylem and phloem saps collected from non-leguminous crops (e.g. Fig. 33.1). In legumes, glutamine is only rarely the main N solute transported from the nodule in xylem (Peoples *et al.*, 1987) despite its synthesis as the initial product of ammonia assimilation during N_2 fixation. Most nodulated legumes tend to export either the ureides, allantoin and allantoic acid (tropical species e.g. cowpea, soybean, mungbean, see Peoples and Herridge, 1990), or the amide asparagine (mainly temperate species e.g. lupin, pea, clovers) as the principal forms of symbiotically fixed N in xylem exudate (Fig. 33.1). The ureides, having four N atoms per molecule (C:N = 1:1), have an advantage over asparagine with only two N atoms per molecule (C:N = 2:1) in terms of N export and efficient C use. On the other hand the low solubility of ureides may limit the amount of N that can be transported in this form in temperate conditions. Both tropical and temperate legumes appear to translocate a mixture of asparagine and glutamine in phloem (Fig. 33.1).

In ureide-producing species, the ureides dominate xylem composition only so long as the legume derives a high proportion of its N requirements for growth from atmospheric N_2 and this can be used as a means of monitoring N_2 fixation (Peoples and Herridge, 1990). With amide-producing legumes, asparagine is the principal exported organic product of both root nitrate reduction and N_2 fixation and continues to be the major form of reduced-N despite changes in N nutrition (Fig. 33.1). The principal N solutes of broccoli (a non-legume), cowpea and pea phloem exudates, depicted in Fig. 33.1, have also been shown to remain relatively unchanged regardless of whether the roots are assimilating soil NO_3^-, or NH_4^+, or fixing atmospheric N_2. In view of the different constraints relating to the regulation of pH imposed by the assimilation of varying forms of N, this invariance is perplexing. Other compounds, such as arginine, can be important xylem components in some deciduous perennial trees, and high levels of citrulline have been found in *Alnus* xylem fluid. Significant amounts of N in certain species are carried as alkaloids, while the non-protein amino acids homoserine and γ-methylene glutamine carry N in other species (Pate, 1980; Fig. 33.1).

The C skeletons of the main organic N constituents often form the bulk of xylem-C (Fig. 33.2). Sugars and sugar alcohols are generally not the major C component in xylem of herbaceous plants (e.g. Fig. 33.2) but they may occur in quantity in deciduous trees such as maple in early spring, or apple during winter dormancy. These are times of year when the lack of transpiring leaves means that the mechanism of xylem transport must be analogous to that of phloem – osmotically driven pressure flow rather than suction flow, i.e. 'root pressure'. In the phloem on the other hand, soluble carbohydrates commonly account for over 80% of total C (Fig. 33.2). Sucrose is most commonly the predominant phloem component, especially in

herbaceous species. The oligosaccharides raffinose, stachyose and verbascose are present in phloem of some species in a few genera and families, while sugar alcohols like sorbitol and mannitol occur in others. Various organic acids detected in phloem translocates are minor carriers of C, whereas in xylem saps, malate, malonate, succinate, citrate and tartrate, can represent between 3 and 27% of total xylem C depending on species, N nutrition and place of sap collection (Fig. 33.2). In legumes there can be quantitative and qualitative changes in xylem organic acid contents associated with specific organic acid requirements for nodule functioning and with the requirements of pH regulation and charge balancing during NO_3^- uptake (Fig. 33.2).

The adaptive rationale for the wide diversity of molecules in xylem and phloem among species in terms of the optimization of growth and partitioning in each species is elusive in detail. Explanations can be sought in terms of constraints of the type described above.

Uptake, loading, transport and unloading

Nitrogen

After exhaustion of the seed protein reserves, plants are solely dependent upon the roots for a continued N input for growth and development. Crop plants may differ in their N source preferences, but most grow faster on NO_3^-. This is believed to be the commonest form of N taken up from agricultural soils, although this may not be true for species adapted to acid soils or anaerobic conditions. The uptake of NO_3^- by root cells is the net result of an active influx causing concentrations several hundredfold greater in the cells than in the soil solution, and of a passive efflux (leakage) of nitrate back into the rooting medium. Nitrate influx is thought to include ATPase-facilitated transport across the plasmalemma and from cytoplasmic nitrate pools within the root to other pools (e.g. the vacuole). Uptake of NH_4^+ by roots also appears to involve a simultaneous influx and efflux (Morgan and Jackson, 1989). Kinetic studies indicate that there may be two transport systems controlling influx: a saturable high affinity system acting at low external NH_4^+ concentrations, and a low affinity, linear transport system for high

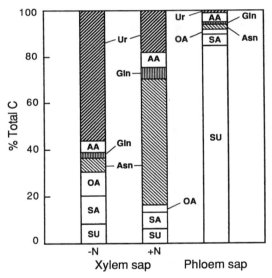

Fig. 33.2 The relative contribution of sugars (Su), sugar alcohols (SA), organic acids (OA), the amides asparagine (Asn) and glutamine (Gln), amino acids (AA), and the ureides, allantoin and allantoic acid (Ur) to the carbon composition of xylem and phloem exudates collected from effectively nodulated (−N) or nitrate fed (+N) soybean (*Glycine max* [L.] Merrill.). Data derived from Layzell and LaRue (1982). The major sugars were fructose, glucose and sucrose in xylem saps, but over 90% of the sugar was as sucrose in phloem. Pinitol was the principal sugar alcohol detected. Malonate and malate were the main organic acids in −N xylem and phloem exudates while malonate and citrate dominated +N xylem sap.

NH_4^+ concentrations (Wang *et al.*, 1993). Evidence suggests that NH_4^+ uptake is under negative feedback control by amino acid end-products of NH_4^+ assimilation. Efflux of NO_3^- back out of the root, which may be as high as 40% of the influx rate, provides another means of regulating net uptake. However, a proportion of the NH_4^+ efflux seems to originate from endogenous metabolism and is not directly related to free NH_4^+ pools within the root.

The enzyme responsible for catalyzing the first step of NO_3^- reduction, is localized almost entirely in epidermal cells at the root periphery at low external NO_3^- concentrations. However, additional NO_3^- reductase activity can also be detected in cortical and stelar cells at high NO_3^- concentrations (Rufty *et al.*, 1986). The root cortex also contains high activities of many of the enzymes involved in subsequent glutamine synthesis. Therefore, these tissues appear to be the principal sites of N assimilation in many plants, although further N metabolism can occur in the vascular bundle region of the stele. Transport of NO_3^- and newly-synthesized amino compounds across the root cortex to the xylem is probably symplastic, but there is little information on the mechanism of release into the xylem. There is also little known in legumes about the route or mechanisms involved in transfer of N solutes from their site of synthesis to the vascular elements which carry them out of the nodule to the host plant. However, specialized pericycle transfer cells, which could facilitate vein loading of a symplastic flux of N solutes have been found adjacent to the vascular bundles in nodules of amide-exporting legumes (Pate, 1980).

While movement of N solutes in the xylem is by mass flow in the transpiration stream, the rate of movement may also be influenced to some extent by adsorption to the walls of the vessels and exchange with the surrounding apoplast, by xylem to phloem transfer *en route*, or by diffusion down a potential gradient within the flowing system. The importance of transpirational flow for export of N from roots is especially evident in legume nodules. Nitrogenous solutes tend to accumulate within nodules during the night, but in the early morning this pool is rapidly depleted as the night's backlog of fixation products is swept from the nodule with the onset of transpiration.

Since xylem flow is towards the points of evaporative water loss, most of the xylem-borne N enters the leaf apoplast close to those sites. Phloem transport is then required to re-distribute N compounds to weakly transpiring growth centres. Nevertheless, transfer cells in leaf vascular tissue may also assist in regulating the supply of N. Some N solutes (e.g. glutamine, asparagine) may be transferred preferentially from the xylem directly into the phloem, whereas several amino acids (e.g. glutamate, aspartate, glycine, arginine), nitrate, and ureides require metabolic conversion to specific amino compounds in mesophyll tissue. These are then exported in the phloem. Loading of amino acids into the phloem may be accomplished by co-transport with protons in a manner akin to that suggested for sucrose transport in phloem (see below). Translocation speeds of amino compounds in phloem have been found to be similar to those of sugars, probably because they flow with the sugars.

Carbon

The regulation of C movement from sources to sinks involves different processes on various time-scales. Seasonally, phenological and developmental processes are of paramount importance to understanding the cumulative partitioning of C into the various plant parts. On a diurnal scale, the regulation of temporary storage in leaf and stem assumes significance, as does the daily photosynthetic integral, the daily sink growth and the feedback interaction between sources and sinks. On a minute or second time-scale, regulation can be considered in terms of enzymic and carrier-mediated transport processes, transmembrane movement, phloem loading in the leaves, and phloem unloading in the sinks. Most of the partitioning literature relates to this very short-term regulation (principally within the leaf) and is difficult or impossible to relate to long-term C partitioning between organs.

In the leaf, the complex fine regulation of partitioning of photosynthetically-produced triose phosphate between chloroplast starch formation and sucrose synthesis in the cytosol or stored in the mesophyll vacuoles is described elsewhere in this volume. The mechanism involves the regulatory compound fructose 2,6-bisphosphate which may also play a role in relating photosynthesis and sucrose synthesis to sink demand for assimilate. The net effect of this fine metabolic control is that over a diurnal cycle or longer, approximately all the photosynthetic assimilate produced by a fully expanded leaf is loaded into phloem and exported while high photosynthetic efficiency is sustained. The starch and vacuolar pools act as buffers smoothing out sucrose availability for loading during the light/dark cycle. Under some circumstances the level of sink demand can feedback through translocation rate, loading, and sucrose synthesis hence modifying photosynthesis rate to some extent. Just how important such feedback is in field photosynthesis is still unclear as are the details of its mechanism. Discussion often seems to imply that the amount of sucrose available for export is determined by partitioning between pools within the leaves. Since the sizes of these pools do not grow or shrink continually, it seems more likely that on longer time-scales, the amount available is usually the amount of photosynthetic C fixed according to environmental conditions, as temporarily modified by sink demand feedback under extreme circumstances. Thus manipulating the metabolism of C partitioning within the leaf seems unlikely to offer a route to increasing seasonal partitioning of C into sinks having economic importance.

The mechanism of sucrose movement from mesophyll to sieve tube is unresolved and keenly debated (van Bel, 1993). The favoured hypothesis for most investigated species, despite shaky evidence, is that sucrose diffuses symplastically from cell to cell towards the phloem. Any leakage *en route* would presumably flow away from the vascular strand in the apoplastic transpiration drift between the mesophyll cells and be re-absorbed by the 'general [sucrose] accumulating system' (Maynard and Lucas, 1982) of those cells. Sucrose unloaded or leaked from mesophyll cells in close proximity to phloem – in a 'backwater' from the transpirational drift – would be retrieved by the powerful retrieval system of companion cells and/or sieve-tubes. The loading from acidic apoplast to alkaline sieve tube is considered to be via a proton-cotransport mechanism driven by a plasmalemma-bound ATPase that exudes protons into the apoplast (Giaquinta, 1983; van Bel, 1993).

Although functional plasmodesmata connecting the cytosol of the sieve tube-companion cell complex to that of the phloem parenchyma exist, no mechanism has been identified whereby purely symplastic transport from mesophyll through plasmodesmata to sieve tubes could operate against the very high concentration gradient that exists between them viz. $1-20 \, mol \, m^{-3}$ in mesophyll cytoplasm to $500-1000 \, mol \, m^{-3}$ in the sieve tubes of the fine leaf veins. Only weak and confusing evidence has been offered favouring symplastic vein loading in leaves and then only for oligosaccharide loading in cucurbits (Turgeon and Beebe, 1991). There is the suggestion that symplastic loading may be the more ancient form of phloem loading and be confined to species from the humid tropics (van Bel, 1993).

The high osmotic concentrations in the minor veins of source leaves generates, by osmotic water uptake, the high turgor pressure necessary to drive mass flow towards sinks where sucrose is unloaded. At the sink, solutes are unloaded from the sieve tubes. Transfer of sugars (and amino acids) from parent plant to developing embryo or endosperm certainly involves apoplastic transport since embryo and endosperm do not have symplastic continuity with the parent plant. But in roots, stems and expanding leaves phloem unloading might be wholly symplastic and directly into the growing cells. However, the mechanism of such symplastic transport from cell to cell via plasmodesmata is not clear. Calculations based on our current understanding of plasmodesmatal frequency, dimensions, structure and conductivity, together with the cytosolic distances from root protophloem termini to growing root cells, indicated that the sucrose concentration gradient required to drive the necessary flux to satisfy growth, respiration and mucilage formation in the primary maize root tip was well beyond physiological possibility. Thus symplastic diffusion alone seemed indaquate to explain phloem to sink transport in that system (Bret-Harte and Silk, 1994). In no tissue (meristem, storage tissue or seed) is the mechanism of phloem unloading understood, but most progress has been made with legume seeds because their embryos can be removed and replaced with solutions into which phloem contents continue to be released. These solutions can subsequently be collected and analyzed (Thorne, 1985; Patrick, 1994).

In developing seeds transport from sieve tube–companion cell complex to the apoplastic interface with the filial sink is down a steep diffusion gradient through parenchymatous maternal cells of the seedcoat (in legume seeds) or funicular-chalazal and nucellar tissue (in temperate cereals). There is bi-directional equilibration of sugars between symplast and apoplast of these transfer tissues (in soybean at least, Gifford, 1986), so it is not necessarily meaningful to ask precisely where translocated solutes leave the symplast and enter the apoplast – net transfer to apoplast could be dispersed along the parenchymatous path from sieve tube to sink. However, in fruits where water imported through the phloem is partly removed in the xylem (Peoples *et al.*, 1985b) there is presumably a zone around the xylem where efflux of solutes from the seedcoat symplast is minimal otherwise they would appear in the xylem export stream.

During the passage of translocated substances across the seedcoat, they can be transformed. A range of amino acids is synthesized in seed coats, ureides are withdrawn, and sucrose may be partially hydrolyzed for example. The compounds secreted from the seedcoat inner surface are therefore derived from, but not identical to, the material unloaded from the sieve tubes into the seedcoat tissue. The mechanisms of control over sieve tube unloading and over secretion from the maternal seedcoat or pericarp tissue seem different, having different response to inhibitors and other exogenous agents, but details are unclear. Despite the steep solute gradients from sieve tubes to final sinks the transfer processes appear to involve respiration and to be metabolically controlled, possibly involving phloem turgor relations.

Sink cells maintain the sucrose concentration gradient needed to sustain continued unloading by taking up the unloaded solutes for conversion to other soluble products (e.g. hydrolysis of sucrose), or vacuolar storage (e.g. in sugarbeet roots), or conversion to insoluble products (e.g. starch in seeds). Most of the source-to-sink gradient in solute concentration appears to occur across the unloading zone. This places phloem unloading strategically for involvement in the regulation of sink growth and partitioning.

Redistribution and utilization of N and C

Nitrogen

Partitioning of N between the roots and shoot is influenced by environmental factors and by the physiological and nutritional status of the plant. For example, plants tend to invest more N in roots relative to the shoot when growing in soils of low N status. Roots may preferentially utilize the N they absorb to satisfy their own requirements before translocating the remainder to the shoot. However, roots have also been reported to be partly dependent for their growth on N cycled through the shoot. In some cases, the movement of N to the roots via the phloem is in excess of the root's total N requirements and in this case, some N is re-exported in the xylem (Simpson *et al.*, 1982). There is a possibility of some metabolic conversion of this N during the movement from phloem to xylem.

Nitrogen translocated from roots in the xylem stream is not partitioned to leaves and growing points of the shoot in strict proportion to their transpirational activity. Older leaves acquire less N per volume of water transpired than do expanding leaves, apices, fruits and stems. It has been suggested that the stem plays a significant role in partitioning N to leaves and apices partially through storage in stem parenchyma cells and partially through a succession of xylem-to-xylem and xylem-to-phloem transfers of N solutes (Pate, 1980; Simpson, 1986). It is proposed that there is lateral withdrawal by stem nodal tissue of N initially directed in the transpiration stream towards lower leaves and then a subsequent re-routing of this N to xylem streams moving further up the shoot, and a transfer of amino acids and amides to the phloem. The importance of such N exchanges are illustrated in the models of N flow prepared for the legume, lupin (Pate *et al.*, 1988). Empirical models of whole lupin plant partitioning of N indicate that of the total requirements of the developing lateral apices and terminal inflorescence: (1) 41% was attracted directly through xylem in accordance with transpiration loss; (2) 40% was donated through phloem from leaves; and (3) 19% was obtained through xylem to phloem transfer in upper regions of the stem.

Bearing in mind that these three sources of N all

involved xylem-to-xylem transfer low down the shoot, and that (2) and (3) were implemented by xylem-to-phloem transfer, the importance of active short distance exchanges between conducting elements in coping with limitations imposed by whole plant pH-regulation and the passive mechanism of long-distance N transport in both xylem and phloem, is apparent.

The N content of many dryland crops reaches a maximum during early reproductive phase. This is usually because of limitations to further N uptake caused by water stress, by depletion of available N in the root zone or, in annual legumes, by nodule senescence. Seed filling in these circumstances is usually dependent on extensive redistribution of N assimilated and invested in the vegetative organs prior to anthesis (Peoples and Dalling, 1988). The various vegetative organs differ significantly from one another in the extent to which they contribute by N mobilization to the final yield of grain-N. The differences are a reflection of the absolute N content of the organ at anthesis and the extent of N removal from the organ. Mobilization of N from leaves usually contributes most to the seed's N requirements (from 16–40% of seed N), but fruit parts (e.g. glumes in cereals and grasses, legume pod walls, the burr in cotton, or the husk and cob of maize) are also capable of N redistribution to seeds (Peoples and Dalling, 1988). In general, leaves, stem and fruit parts are characterized by a high efficiency of N removal (>65%). Roots re-allocate N less readily (<30%). It has been suggested that these differences are a reflection of an 'ordered priority system' within the plant whereby there is a need to balance integrity of organ function during senescence against any inherent tendency towards a high efficiency of N removal. The 'cost' of this compromise is less N redistribution than would otherwise occur, especially in the roots which seem to be the last organ to senesce fully as befits their role of keeping the shoot hydrated and physically supported.

Although some crops accumulate large amounts of soluble N in storage organs or stems, the majority of the N redistributed within the plant arises from the enzymatic degradation of protein by peptide hydrolases and the metabolic interconversion of breakdown products. It is likely that, through turnover, much of the N entering the plant early in its life passes through several age groups and types of vegetative structures, and is incorporated into several generations of proteins and compounds before being finally stored in developing seed. In species having the C_3 photosynthetic pathway, the chloroplast enzyme ribulose-1,5-bisphosphate carboxylase is likely to represent the single largest remobilizable reserve of protein-N since it generally accounts for 30 to 60% of the total soluble protein and between 20 to 30% of total leaf N (Peoples and Dalling, 1988).

The reallocation of N is regarded as occurring principally via the phloem, but there is evidence that N can also be redistributed via the xylem if phloem transport is disrupted. In wheat, less than 50% of the N redistributed from leaves may be exported directly to the grain via the phloem. The rest appears to be transported to the roots and re-exported to the shoot via the xylem, and transferred again to the phloem for export to grain. Studies with legumes suggest that plants have the ability to manipulate solute composition via metabolic conversions and by xylem-to-xylem and xylem-to-phloem transfers in transit to reproductive structures so that fruits receive a uniform spectrum of N compounds in the xylem or phloem regardless of whether they are utilizing current assimilates from the roots or N redistributed from senescing tissue for seed growth (Peoples *et al.*, 1985a).

Carbon

During vegetative development, photoassimilate tends to flow preferentially to the sink closest to the assimilating leaf. Photoassimilate from the uppermost fully expanded leaf, for instance, moves predominantly to the shoot apex and newly emerging leaves, while from lower leaves photoassimilate moves to the roots (and tillers). In legumes, N_2 fixation by nodules may consume up to 18% of the net photoassimilate produced, with total phloem translocation of C to the nodulated root representing over half of the shoot's photoassimilate supply. Some 16% of the C directed to the nodulated roots may be returned to the shoot in the xylem stream as the C skeletons of N transport compounds and as organic acids, and as much as 53% can be lost through respiration in some legumes (Pate *et al.*, 1988). This large respiratory release of C, however, is likely to represent only a net respiratory loss since part of the CO_2 respired will have been refixed by the active phosphoenolpyruvate carboxylase (PEPCase) enzymes in legume nodules and roots. Such non-photosynthetic fixation of CO_2 by below-ground parts appears to play a key role in the synthesis of C skeletons for amino acid biosynthesis, in the provision of respiratory substrates and as an energy source for root assimilation of NH_4^+ (Cramer et al., 1993), in pH and ion-balance regulation, and in the de novo synthesis of purines for ureide biogenesis in legumes such as soybean. Dark CO_2 fixation rates of the nodulated roots of alfalfa have been estimated as averaging 26% of the gross respiratory rate, with nodule PEPCase contributing as much as 25% of the C required for the assimilation of fixed N (Anderson et al., 1987). In contrast to legumes, relatively less of the C allocated to the roots of cereals appears to be lost through root respiration (32% in young wheat plants), but more (39% of the total C translocated to roots) may be cycled through the roots to be exported back to the shoot associated with amino acids (Lambers et al., 1982).

There appears to be some capacity for C to be exchanged between translocation streams within the stem by mechanisms similar to those described above for N. It seems that the phloem transport pathway is to some extent leaky, with passive unloading into, and active reloading from, the apoplast of legume stems. Continual phloem unloading (leakage) and reloading within the stem could provide a sucrose pool in the apoplast buffering the sieve tubes against sudden changes in phloem sucrose concentration. The significance of phloem-to-phloem or phloem-to-xylem C transfer to shoot nutrition has not been fully evaluated (Minchin and McNaughton, 1987).

Towards the end of the vegetative phase of growth, the upper leaves transport progressively more of their photoassimilate to the stem and the developing inflorescence until anthesis and early flowering. There is commonly thereafter an abrupt decline in the flow of photoassimilate to below-ground parts (and tillers), a gradual decrease in source leaf photosynthetic rate, and an increasing monopolization of photosynthate by reproductive organs. Temporary storage of non-structural carbohydrates in upper stem internodes of cereals during this phase, and subsequent movement

to the ear during grain-filling, has been reported in many studies. Fructan levels can decline in wheat from around 20% of the dry weight of the second internode below the ear after ear emergence to zero at the end of grain-filling. Quantitatively, however, such mobilization of C reserves in temperate cereals contributes less than 10% to the final ear weight except when the crop becomes droughted during grain filling. In this respect, the time-course of C assimilation and partitioning appears very different from that of N. Whereas a major portion of the seed-N yield is derived from the re-allocation of N from vegetative organs to the seed, more than 75% of the final seed C usually comes from direct transfer of current photosynthate from source leaves. Even in cowpea, a legume noted for its rapid fruit growth and early senescence of leaves, reallocation of C from vegetative parts contributes less than 20% to the plant's post-anthesis C requirements, even though up to 50% of leaflet, stem, petiole and peduncle C may be mobilized during the period of fruit development. Appreciable amounts of extractable starch and sugars can be lost from vegetative tissues between anthesis and seed maturity in cowpea; however, these alone could account for no more than 75% of the total C loss from any one organ, indicating that C from structural components may also be mobilized in addition to non-structural carbohydrates to help satisfy the C demand of fruiting plants (Pate et al., 1983).

N and C interrelationships of individual organs

The following case studies for the leaves and fruits of legumes exemplify the dynamic interrelationships between xylem and phloem transport in the distribution of C and N during growth and development.

The leaf

The net turnover of C and N in a white lupin leaf during its life is summarized in Fig. 33.3a, and the time course of exchanges made with the rest of the plant through xylem and phloem are shown in Fig. 33.3b and c.

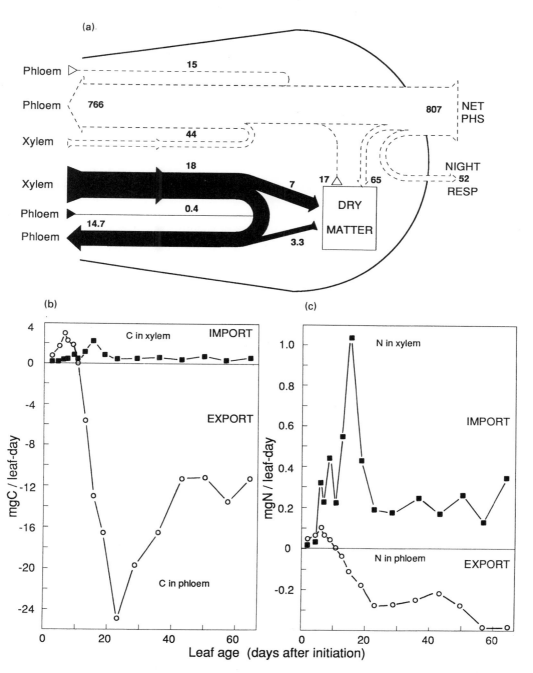

Fig. 33.3 Carbon and nitrogen economies of the uppermost main stem leaf of lupin (*Lupinus albus* L.). Data are expressed as (a) net total exchanges of C and N between plant and leaf (mg C and N) and as daily rates of export and import of (b) C and (C) N by the leaf through phloem (○) and xylem (■). Adapted from Pate *et al.* (1988). NET PHS, net photosynthetic C input; NIGHT RESP, night-time respiratory C loss.

Initially, both C and N were imported by the very young leaf mainly via the phloem (Fig. 33.3b and c). As the developing leaf began to transpire, the xylem became an increasingly important conduit for the import of N, contributing more than 80% of the N utilized for leaf growth by the time the leaf began to export C (i.e. at between one-third to one-half of its fully expanded size), and 94% of the leaf's final dry matter N requirements (Fig. 33.3a). Rapid leaf growth and accumulation of N continued in the period after the leaf's transition from C sink to C source until maximum N content was reached just prior to full leaf expansion. Except for the early phloem importing phase, the leaf continuously cycled N (i.e. simultaneously receiving and exporting N) by xylem to phloem transfer in the minor leaf veins (Fig. 33.3a). Of the N that was received through the xylem during the leaf's life, two-thirds were cycled back to the plant more or less immediately by xylem to phloem exchange, and a further one-sixth was translocated out of the leaf, following protein breakdown at the approach of leaf senescence. The time courses of these transfers are depicted in Fig. 33.3c.

The budget for C (Fig. 33.3a) records a net photoassimilate production of 807 mg C of which 766 mg C were exported from the leaf. Only 65 mg C were left behind in the leaf as non-retrievable dry matter, and 52 mg C were lost in night respiration. Fig. 33.3b shows a sharp peak in export of C shortly after the leaf achieved maximum size and N content, and a gradual lowering of the C:N ratio of the phloem export stream due to declining photosynthesis relative to N cycling and, later, to increasing mobilization of N from the leaf.

The legume fruit

The fruits on a modern cultivar of grain legume contain up to 40% of the plant's final content of C and 60 to 85% of its final N. To meet this especially large demand for N and C during seed filling a heavy premium is placed on mechanisms to allocate these to the developing fruit. This applies particularly to fast maturing, determinant cultivars in which near-synchronous maturation of fruits generates a climacteric in demand for C and N

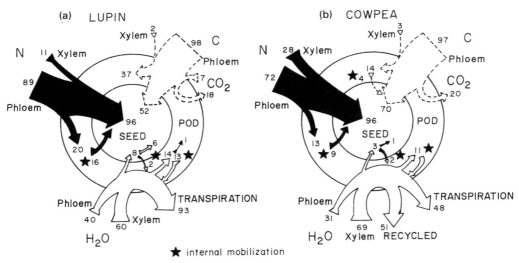

Fig. 33.4 Proportional intakes of C, N and water through xylem and phloem during development of (a) white lupin (*Lupinus albus* L.) and (b) cowpea (*Vigna unguiculata* [L.] Walp.). All components of a fruit's budget are expressed relative to a net intake of 100 units of C, N or water through xylem and phloem. Ratios of absolute amounts by weight of C, N and water consumed by lupin fruits are given in Pate *et al.* (1977) and by cowpea in Peoples *et al.* (1985b) from which the developmental budgets were derived. The hypothetical 'recycled' component of cowpea's water budget is shown in (b). Values for mobilization C, N, or water are indicated by an asterisk (*).

during the final weeks of plant growth. The following discussion compares the strategies developed by a legume with a long reproductive phase (lupin) with one characterized by rapid rates of seed-filling (cowpea) to effectively supply C and N for fruit growth.

Budgets for C and N have been provided for both the lupin fruit (Fig. 33.4a) and the cowpea fruit (Fig. 33.4b). Both species conserved C by refixing some or all CO_2 respired by day in green tissues of the pod wall (cowpea) and pod and seed (lupin). Mobilization of pod N during seed-filling provided a significant proportion (17% in lupin, 9% in cowpea) of the seeds' N requirements (Fig. 33.4).

In lupin the C, N and H_2O budgets for the fruit balanced closely over its 12 week period of growth. Phloem supplied 89% of the N and 98% of the C (Fig. 33.4a). Transpiration and tissue water accumulation matched water input via xylem and phloem. While the phloem supplied a similar proportion of the C (97%) and slightly less N (72%) in the cowpea fruit, the water accompanying the C and N inputs in xylem and phloem vastly exceeded the fruit's tissue requirements for water plus the transpiration losses during the 3 weeks between anthesis and maturation. It appears that xylem intake by the cowpea fruit was mostly at night. However, during the day intake of phloem-borne water associated with the high rates of dry matter gain by pod and seed exceeded fruit transpiration. It is surmised that the water potential gradient must have been in the direction for water to flow out of the pod in the xylem when there is this surplus of phloem-delivered water. This would account for the discrepancy in the water budget. Whether such diurnally reversible xylem exchange of water carries solutes, including 'fruit induced leaf senescence' hormones, out of the seed is worthy of investigation (see Pate *et al.*, 1988).

CONCLUSIONS

It seems miraculous that higher plants manage to cope with all the conflicting and ever-changing boundary conditions involved in taking up N from the soil and C from the air and moving them to where they are needed for integrated operation of the whole growing plant. In achieving the miracle, there are apparent inefficiencies, like the transfer of C from the roots to the leaves and N from the shoots to the roots. But the closer one looks the more one realizes that the inefficiencies may be inevitable means to identifiable ends. Diverse N and C compounds are found in the xylem and phloem of different species and one wonders 'Why so many?'. It is not always apparent. In view of the complexity of the constraints that plants are coping with in diverse environments it would seem to be a useful working hypothesis to assume that each compound is specifically needed where and when it is found in the resolution of a specific developmental, evolutionary, or environmental conflict. Further exploration of the system on that basis might reveal situations where the hypothesis is wrong and provide opportunity to improve plant performance by plant breeding.

References

Anderson, M. P., Heichel, G. H. and Vance, C. P. (1987). Nonphotosynthetic CO_2 fixation by alfalfa (*Medicago sativa* L.) roots and nodules. *Plant Physiol.* **85**, 183–89.

Bret-Harte, M. S. and Silk, W. K. (1994). Nonvascular, symplasmic diffusion of sucrose cannot satisfy the carbon demands of growth in the primary root tip of *Zea mays* L. *Plant Physiol.* **105**, 19–33.

Cramer, M. D., Lewis, O. A. M. and Lips, S. H. (1993). Inorganic carbon fixation and metabolism in maize roots as affected by nitrate and ammonium nutrition. *Physiol. Plant.* **89**, 632–9.

Delrot, S. and Bonnemain, J.-L. (1985). Mechanism and control of phloem transport. *Physiol. Vég.* **23**, 199–220.

Fisher, D. B. and Frame, J. M. (1984). A guide to the use of the exuding-stylet technique in phloem physiology. *Planta* **161**, 385–93.

Giaquinta, R. T. (1983). Phloem loading of sucrose. *Ann. Rev. Plant Physiol.* **34**, 347–87.

Gifford, R. M. (1986). Partitioning of photoassimilate in the development of crop yield. In *Phloem Transport*, eds J. Cronshaw, W. J. Lucas, R. T. Giaquinta, Alan R. Liss Inc., New York, pp. 535–49.

Gifford, R. M. (1994). The global carbon cycle: A viewpoint on the missing sink. *Aust. J. Plant Physiol.* **21**, 1–15.

Gifford, R. M. and Evans, L. T. (1981). Photosynthesis, carbon partitioning, and yield. *Ann. Rev. Plant Physiol.* **32**, 485–509.

Imsande, J. and Touraine, B. (1994) N demand and the regulation of nitrate uptake. *Plant Physiol.* **105**, 3–7.

Lambers, H., Simpson, R. J., Beilharz, V. C. and Dalling, M. J. (1982). Translocation and utilization of carbon in wheat (*Triticum aestivum*). *Physiol. Plant* **56**, 18–22.

Lang, A. (1978). A model of mass flow in phloem. *Aust. J. Plant Physiol.* **5**, 535–46.

Layzell, D. B. and LaRue, T. A. (1982). Modelling C and N transport to developing soybean fruits. *Plant Physiol.* **70**, 1290–98.

Maynard, J. W. and Lucas, W. (1982). Sucrose and glucose uptake into *Beta vulgaris* leaf tissue: a case for general (apoplastic) retrieval systems. *Plant Physiol.* **70**, 1436–43.

Minchin, P. E. H. and McNaughton, G. S. (1987). Xylem transport of recently fixed carbon within lupin. *Aust. J. Plant Physiol.* **14**, 325–29.

Morgan, M. A. and Jackson, W. A. (1989). Reciprocal ammonium transport into and out of plant roots: Modifications by plant nitrogen status and elevated root ammonium concentration. *J. Exp. Bot.* **40**, 207–14.

Pate, J. S. (1980). Transport and partitioning of nitrogenous solutes. *Ann. Rev. Plant Physiol.* **31**, 313–40.

Pate, J. S., Atkins, C. A., Peoples, M. B. and Herridge, D. F. (1988). Partition of carbon and nitrogen in the nodulated grain legume: Principles, processes and regulation. In *World Crops: Cool Season Food Legumes*, ed. R. J. Summerfield, Kluwer Academic Publ., Dordrecht, pp. 751–65.

Pate, J. S., Peoples, M. B. and Atkins, C. A. (1983). Post-anthesis economy of carbon in a cultivar of cowpea. *J. Exp. Bot.* **34**, 544–62.

Pate, J. S., Sharkey, P. J. and Atkins, C. A. (1977). Nutrition of a developing legume fruit. Functional economy in terms of carbon, nitrogen, water. *Plant Physiol.* **59**, 506–10.

Patrick, J. W. (1994). Turgor-dependent unloading of assimilates from coats of developing legume seed. Assessment of the significance of the phenomenon in the whole plant. *Physiol. Plant.* **90**, 645–54.

Peoples, M. B. and Dalling, M. J. (1988). The inter-play between proteolysis and amino acid metabolism during senescence and nitrogen re-allocation. In *Senescence and Aging in Plants*, eds L. D. Noodén and A. C. Leopold, Academic Press, New York, pp. 181–217.

Peoples, M. B. and Herridge, D. F. (1990). Nitrogen fixation by legumes in tropical and subtropical agriculture. *Adv. Agron.* **44**, 155–223.

Peoples, M. B., Pate, J. S. and Atkins, C. A. (1985a). The effect of nitrogen source on transport and metabolism of nitrogen in fruiting plants of cowpea (*Vigna unguiculata* (L.) Walp.). *J. Exp. Bot.* **36**, 567–82.

Peoples, M. B., Pate, J. S., Atkins, C. A. and Murray, D. R. (1985b). Economy of water, carbon, and nitrogen in the developing cowpea fruit. *Plant Physiol.* **77**, 142–47.

Peoples, M. B., Sudin, M. N. and Herridge, D. F. (1987). Translocation of nitrogenous compounds in symbiotic and nitrate-fed amide-exporting legumes. *J. Exp. Bot.* **38**, 567–79.

Raven, J. A. (1977). The evolution of vascular land plants in relation to supracellular transport processes. *Advances in Bot. Res.* **5**, 154–219.

Raven, J. A. and Smith, F. A. (1976). Nitrogen assimilation and transport in vascular land plants in relation to intracellular pH regulation. *New Phytol.* **76**, 415–31.

Rufty, T. W., Thomas, J. F., Remmier, J. L., Campbell, W. H. and Volk, R. J. (1986). Intercellular localization of nitrate reductase in roots. *Plant Physiol.* **82**, 675–80.

Shelp, B. J. (1987). The composition of phloem exudate and xylem sap from broccoli (*Brassica oleracea* var. *italica*) supplied with NH_4^+, NO_3^-, $NH_4 NO_3$. *J. Exp. Bot.* **38**, 1619–36.

Simpson, R. J. (1986). Translocation and metabolism of nitrogen: whole plant aspects. In *Fundamental, Ecological and Agricultural Aspects of Nitrogen Metabolism in Higher Plants*, eds H. Lambers, J. J. Neetsson and K. Stulen, Martinus Nijhoff Publ., Dordrecht, pp. 71–96.

Simpson, R. J. and Dalling, M. J. (1981). Nitrogen redistribution during grain growth in wheat (*Triticum aestivum* L.) III Enzymology and transport of amino acids from senescing flag leaves. *Planta* **151**, 447–56.

Simpson, R. J., Lambers, H. and Dalling, M. J. (1982). Translocation of nitrogen in a vegetative wheat plant. *Physiol. Plant* **56**, 11–17.

Thorne, J. H. (1985). Phloem unloading of C and N assimilates in developing seeds. *Annu. Rev. Plant Physiol.* **36**, 317–43.

Turgeon R. and Beebe, D. U. (1991). The evidence for symplastic phloem loading. *Plant Physiol.* **96**, 349–54.

van Bel, A. J. E. (1993). Strategies for phloem loading. *Annu. Rev. Plant Physiol. Plant Molec. Biol.* **44**, 253–81.

Wang, M. Y., Siddiqi, M. Y., Ruth, T. J. and Glass, A. D. M. (1993). Ammonium uptake by rice. II. Kinetics of $^{13}NH_4^+$ influx across the plasalemma. *Plant Physiol.* **103**, 1259–67.

34 Protein storage and utilization in seeds

Rod Casey, J. Derek Bewley and John S. Greenwood

Introduction

The economic importance of seed proteins derives from the fact that about 70% of human intake of protein comes directly from seeds; cereal and legume grains also provide a significant component of the diet of non-ruminant animals, which in turn contribute appreciably to the remaining 30% of our requirements for protein. The accumulation of proteins during seed development, in cereals and grain legumes in particular, is therefore one of the most significant processes in plants with respect to their usefulness to people. This provides the justification for much of the research on seed proteins, one long-term objective of which is to obtain the basic understanding of structure, synthesis and utilization required to improve the quality and quantity of the proteins for human and animal nutrition.

Seeds characteristically contain a few major proteins known as storage proteins, and many minor proteins that are required for a plethora of functions. The storage proteins are sequestered in storage tissues of the developing seed (cotyledons in grain legumes, endosperm in many cereals) and stored in the dry seed; they are subsequently hydrolyzed after germination to amino acids, which are used as a source of reduced nitrogen for early seedling growth.

Storage protein structure

Legume seed proteins

The major grain legumes in which seed proteins have been extensively studied are pea (*Pisum sativum*),

French ('navy', 'kidney') bean (*Phaseolus vulgaris*), soybean (*Glycine max*) and broad ('faba', 'field') bean (*Vicia faba*). There is remarkable conservation of the structures of the storage proteins between species, although the relative proportions of particular proteins can vary greatly between species and also between genotypes within a species. The major legume storage proteins are soluble in dilute salt solution, but insoluble in water, and are therefore biochemically classified as globulins; they account for up to 70% of the total seed protein. They contain very little of the sulfur-containing amino acids, methionine and cysteine, and consequently legume seed protein generally is sulfur-deficient.

The storage globulins from legumes are historically known as legumins and vicilins, but several trivial names have gained preference in the literature (see Casey *et al.*, 1986): French bean vicilin is known as phaseolin and soybean legumin and vicilin are known as glycinin and β-conglycinin, respectively. Native legumins exist as hexamers ($M_r \sim 350\,000–400\,000$) and vicilins as trimers ($M_r \sim 150\,000–200\,000$), having sedimentation coefficients of 11–12S and 7–8S, respectively. The terms legumin/11S and vicilin/7S often are used interchangeably.

Each monomer of hexameric 11S proteins consists of an acidic (α- or A-) polypeptide of M_r 35 000–40 000 linked through a disulfide bond to a basic (β- or B-) polypeptide of $M_r \sim 20\,000$. There are several of each in a given 11S protein preparation, with individual A and B polypeptides being specifically paired; this is a consequence of their synthesis as covalently linked AB precursors (see later). The amino acid sequence of the A- and B-polypeptides of 11S proteins from soybean, pea and broadbean indicates the existence of multiple 11S genes; in peas, for instance, there are at least ten legumin genes located in at least three genetic loci.

The 7S proteins are more heterogeneous than 11S proteins, comprising many polypeptides ranging in size from $M_r \sim 12\,000$ to $\sim 80\,000$, depending on the species. In all cases, some of the polypeptide heterogeneity arises from the fact that 7S proteins are encoded by small gene families with divergent members. Pea has a complex family of vicilins; there are at least 24 vicilin genes located in at least six genetic loci (Ellis *et al.*, 1992). Further complexity arises from differential glycosylation (see later) and differential proteolytic processing. Despite this apparently great diversity, most 7S proteins have highly homologous primary (and probably tertiary) structures and may all have evolved from a common precursor gene.

Cereal seed proteins

The major storage proteins of most cereal seeds are found in the endosperm, are mostly soluble in alcohol–water mixtures and are compositionally deficient in lysine and tryptophan. They are classified as prolamins, a name that reflects their high content of proline and glutamine, and have a more varied structure from one species to another compared to legume seed globulins. Prolamins generally account for 50–60% of the total seed protein and are found only in monocots, whereas globulins are found in dicots and monocots. Rice and oats are exceptions amongst the cereals because their prolamin fractions account for only about 10% of the seed protein; their major storage proteins are globulins. The prolamins of wheat (*Triticum aestivum*), rye (*Secale cereale*), barley (*Hordeum vulgare*), oats (*Avena sativa*) and rice (*Orysa sativa*) (Festucoideae) have some similarities, but differ from those of maize (*Zea mays*), sorghum (*Sorghum bicolor*) and millet (*Pennisetum americanum*) (Panicoideae). The prolamins of wheat and maize are used here as examples of each group.

Wheat prolamins can be subdivided into sulfur-rich and sulfur-poor gliadins, and high- and low-molecular weight (HMW and LMW) subunits of glutenin. The S-rich gliadins are the most abundant and diverse, comprising the α-, β- and γ-gliadins; these collectively have a methionine/cystene content

of $\sim 3\%$. They contain several repeats that are rich in proline and glutamine and have M_r 32 000–44 000. The S-poor gliadins (ω-gliadins) have M_r 50 000–70 000, also contain peptide repeats, and 80% of their sequence consists of proline, glutamine and phenylalanine. The LMW-glutenins have a similar overall structure to the γ-gliadins (see Fig. 34.1). The HMW-glutenins range in M_r from 80 000 to 150 000, and are cross-lined by disulfide bonds. They too contain a region of peptide repeats, unrelated to those in the other wheat prolamins, and it is this combination of repeat elements and disulfide bonds that imparts the physical properties that make HMW glutenins important in breadmaking (see Flavell *et al.*, 1989 for a review).

The prolamins of maize – the zeins – have been classified into α-, β-, γ- and δ-zeins on the basis of solubility and comparative sequence analysis. α-Zeins, which account for 70–80% of the total in common inbred maize lines, have a central region of nine or ten tandemly repeated peptides, each of about 20 amino acids, within polypeptides of M_r 19 000 and 22 000. The β-zeins have smaller polypeptides (M_r 14 000), do not have repeats, and have less proline and glutamine, but more cysteine and methionine, than α-zeins. The γ-zeins have two major polypeptides of M_r 16 000 and 27 000, are made up of eight distinct regions, including three (16 000 M_r) or eight (27 000 M_r) copies of Pro–Pro–Pro–Val–His–Leu at the N terminus, contain 25% proline and are also cysteine-rich. δ-Zeins are minor proteins of

(a)

(b)

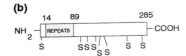

(c)

Fig. 34.1 Diagrammatic structure of (a) γ-gliadin, (b) LMW-glutenin and (c) HMW-glutenin (subunit 12) polypeptides. S, cysteine residues. After Shewry and Gutteridge (1992).

$M_r \sim 10\,000$ that are very rich in cysteine and methionine. None of the α-zeins analyzed to date has been found to encode a lysine residue and in only one case is tryptophan present (see Kriz and Larkins, 1991 for references); this is largely the reason for the lysine/tryptophan deficiency of maize seed protein.

The genes encoding wheat gliadins and maize α-zeins are members of large, multigene families of up to 100 members, whereas wheat glutenins and maize β-, γ- and δ-zeins are encoded by genes present in only a few, possibly single, copies (see Shotwell and Larkins, 1991). There are no obvious reasons for this wide range in gene copy number.

Storage protein synthesis, modification and sequestering in protein bodies

The subcellular repositories for the majority of storage proteins in reserve tissues of mature seeds are the protein bodies or protein storage vacuoles. These unit-membrane bound organelles are usually spherical with a diameter of 0.1 to $25\,\mu m$. Protein body size and the presence of proteinaceous crystalloid and/or phytin-containing globoid inclusions (see Fig. 34.2) within the proteinaceous matrix are species- and tissue-dependent (Lott, 1980).

Protein body formation in dicotyledonous genera. General scheme

Microscopic studies of developing legume seeds revealed that protein bodies within the cotyledonary storage parenchyma cells are of vacuolar origin, and the presence of vacuolar enzymes within protein bodies supports this view. The cotyledonary cells generally contain one or two large vacuoles at the conclusion of the cell expansion phase. At the onset of storage protein accumulation, deposits of protein collect on the luminal side of the vacuolar membrane. The membrane then surrounds these deposits and smaller, virtually mature protein bodies are pinched off from the central vacuole. With continued development, the vacuole itself divides and is gradually replaced by numerous smaller vacuoles and protein bodies. Filling of the smaller vacuoles with storage protein continues until the maturation-drying

phase of development. By maturation, vacuolar subdivision and filling results in a large number ($> 150\,000$ per cell in *Pisum sativum*) of 1–$5\,\mu m$ diameter, relatively homogeneous, protein bodies (Craig *et al.*, 1980; Chrispeels, 1985) (Fig. 34.2).

This general pattern of vacuolar subdivision and filling in protein body ontogeny is common to the cells of the storage tissues (cotyledon or endosperm) in seeds of the dicotyledonous species studied to date. Minor variations exist in the general scheme, primarily in the timing and degree of vacuolar division. In developing castor bean endosperm, for example, major accumulation of storage proteins occurs after the large central vacuole divides. Continued subdivision of the protein bodies is not evident, with the result that castor bean endosperm cells have fewer, larger protein bodies compared to storage parenchyma cells of legumes (Greenwood and Bewley, 1985) (Fig. 34.2).

The movement of storage proteins from site of synthesis to the vacuolar/protein body compartment is outlined in Fig. 34.3. Following the vectoral insertion of the translation products into the rough endoplasmic reticulum (RER) lumen, and co- and post-translational modification and assembly (detailed below), the storage polypeptides are competent for transport. Transport of the storage polypeptides to the Golgi apparatus most likely takes place va direct tubular ER connections between the RER and the *cis*-Golgi stack (Harris, 1986). Sorting of the storage protein polypeptides and modification of oligosaccharide side-chains, if present, take place during transport through the Golgi stacks. The storage polypeptides are packaged into vesicles in the *trans*-Golgi stack for transport to the vacuolar compartment. Fusion of the vesicular membranes with the vacuolar membrane discharges the storage proteins into the vacuolar lumen. Vacuolar subdivision and continued filling of the protein bodies completes the process.

The fusion of Golgi-derived vesicles with the vacuolar membrane results in an increase in the membrane surface area of the vacuolar/protein body compartment (Craig *et al.*, 1980). The infolding of the vacuolar membrane and subdivision of the vacuole from a large compartment with a small surface-to-volume ratio to numerous smaller compartments each having a much larger surface-to-volume ratio

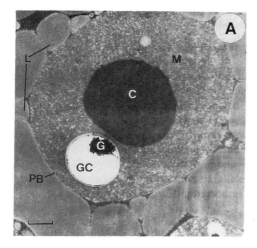

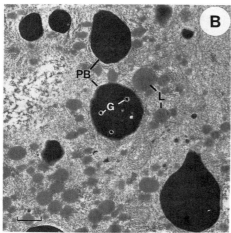

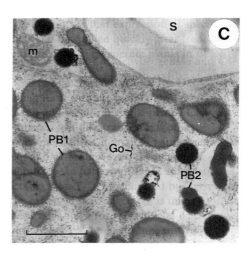

Fig. 34.2 Types of protein bodies found in storage parenchyma cells of seeds. (A) A protein body with protein crystalloid (C) and phytin-containing globoid (G) inclusions embedded in the proteinaceous matrix (M). A portion of the globoid has been lost due to sectioning (GC) (*Ricinus communis* endosperm). (B) Protein bodies typical of those found in mature legume seed cotyledon cells. Crystalloid inclusions are absent, numerous small globoids may be found embedded in the matrix protein, as illustrated here. (*Medicago sativa* cotyledon). (C) Protein bodies of cereal grains. Spherical (PB1) and crystalline (PB2) protein bodies as found in the starchy endosperm are illustrated. Matrix protein in both types of protein bodies is usually free of inclusions (*Zea mays* starchy endosperm). L, lipid; m, mitochondrion; Go, Golgi; S, starch.

compensates for this increased surface area. Thus, the protein bodies may also store preassembled membrane components that can be readily used for cell expansion during imbition and following germination.

Translation of storage protein messages, and co- and post-translational processing

The RER, Golgi complex and vacuolar/protein body compartment each have discrete functions in the synthesis, assembly, movement and sequestering of storage proteins in dicots.

The synthesis and assembly of storage proteins is illustrated here with reference to legumin (glycinin) in soybean cotyledons. The steps involved are illustrated in Fig. 34.4, and commence with the transcription of their genes; the resultant mRNAs are then transported from the nucleus to the protein synthesizing site within the cytoplasm, the RER. Processing of the message (e.g. splicing out of introns) occurs in the nucleus, and then it is recruited into the translational machinery, the polysomes. The ribosomes become attached at the translation start site toward the 5' end of the message, and translation commences. The initial translation product contains a sequence at the amino-terminus of the protein, the

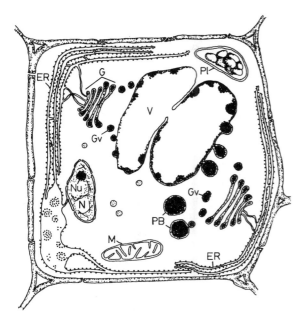

Fig. 34.3 Diagrammatic representation of the events involved in protein body formation in storage parenchyma cells of legume seeds. Storage proteins are synthesized on the rough endoplasmic reticulum and transported to the Golgi apparatus via tubular smooth ER connections. The storage proteins are sorted and packaged into Golgi-derived vesicles and transported to the vacuole/protein body compartment. Fusion of the vesicles with the vacuolar membrane, illustrated on the left of the central vacuole, discharges the storage protein into the vacuolar lumen. Evagination of the vacuolar membrane around concentrations of storage proteins, seen on the right of the vacuole, results in the formation of virtually mature protein bodies. Continued vacuolar subdivision gives rise to numerous protein bodies. ER, endoplasmic reticulum; G, Golgi apparatus; Gv, Golgi-derived vesicle; M, mitochondrion; N, nucleus; Nu, nucleolus; Pl, plastid, V, vacuole.

signal or transit peptide, which allows the protein to pass into the lumen of the RER (Blobel *et al.*, 1979). Once the protein has passed through a specific site within the membrane, the signal peptide is enzymically removed.

The subunits of glycinin, which contain an acidic and a basic polypeptide, are synthesized sequentially on the same mRNA template (Fig. 34.4). The primary translation product (preprolegumin) contains the signal peptide at the amino-terminal end, the acidic and basic polypeptides joined by a short linker

sequence of amino acids, and a short (penta-) peptide at the carboxy-terminus. The glycinin is now further processed in the lumen of the RER; following removal of the signal peptide co-translationally, the carboxy-terminal peptide is removed post-

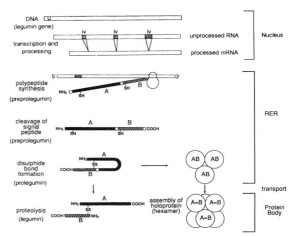

Fig. 34.4 A representation of the synthesis of the legumin storage protein (glycinin) in soybean cotyledons. The gene for legumin possesses an upstream sequence (u) of several hundred nucleotides which controls its tissue-specific and temporal expression. The first transcriptional product is an unprocessed mRNA which contains nontranslated intron regions (iv), about 1150 bases altogether, and which are spliced out to produce the mature processed legumin mRNA, which is 1455 bases long. (The mRNA is polyadenylated at the 3′ end). This mRNA is transported from the nucleus to the rough endoplasmic reticulum (RER) where it is translated. The mRNAs for the acidic (A) and basic (B) polypeptides are joined by codons for a 4-amino-acid linker sequence on the primary translation product (preprolegumin). The mRNA also contains a code for the signal peptide at the amino-terminal end and a pentapeptide at the carboxy-terminal end. The primary product (prolegumin) is processed to yield the A and B polypeptides joined by the linker sequence and disulfide bonds (S–S); the signal peptide and the pentapeptide are cleaved off. In the final step of processing the mature protein is formed as the linker sequence is removed and the acidic and basic polypeptides are joined only by the disulfide bonding. The mature acidic polypeptide contains 278 amino acids (approx. 40 000 M_r) and the basic polypeptide 180 amino aids (approx. 20 000 M_r). Assembly of the joined acidic and basic polypeptides, the subunits, within the protein bodies yields the mature hexameric legumin protein. Based on Krochko and Bewley (1989). See also Dickinson *et al.* (1989), Bednarek and Raikhel (1992), and Bewley and Black (1994).

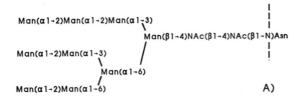

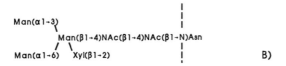

Fig. 34.5 Probable structures of typical high-mannose (A) and complex oligosaccharide side-chains (B) of seed storage glycoproteins (after Johnson and Chrispeels, 1987).

translationally, i.e. after the protein has become dissociated from the ribosome. The protein begins to fold and assemble within the RER into its correct three-dimensional (tertiary and quaternary) structure, an event which is required before the protein can be targeted and transported to the protein body. Proteins within the ER membrane facilitate this process, including a binding protein (BiP) which promotes the correct folding of the storage polypepties, and protein disulfide isomerase (PDI) which forms disulfide bonds between the cysteine residues of the acidic and basic polypeptides. The resultant product is prelegumin which is translocated to the protein body for the final stage of maturation, the assembly of hexamers. This occurs after the removal in the protein body of the pentapeptide linking the acidic and basic polypeptides by a thiol-containing proteinase. The six subunits which compose the glycinin holoprotein can now assemble (Fig. 34.4), and they become associated into their correct configuration by surface charge interactions.

Many of the storage proteins in mature seeds are glycoproteins; one or more oligosaccharide-side chains are covalently linked to the constituent polypeptides. Examples include the vicilins of legume seeds (e.g. β-conglycinin of soybean) and water-soluble carbohydrate-binding storage proteins (lectins) of common bean (phytohemagglutinin) and soybean (soyin). The oligosaccharide side-chains present on the storage proteins fall into two major categories. Simple or high-mannose oligosaccharides are composed exclusively of mannose (Man) and N-acetylglycosamine (NAc) residues, usually in a 5–9:2 ratio. Complex, or modified, oligosaccharides have xylose (Xyl) and/or fucose (Fuc) in addition to Man and NAc (Chrispeels, 1984, 1991; Faye et al., 1989) (Fig. 34.5).

The initial glycosylation of storage protein polypeptides is a co-translational event. High-Man chains are assembled on carrier molecules of dolichol pyrophosphate (a lipid) in the RER membrane via RER-associated glycosyltransferases. Construction up to dolichol-Nac$_2$Man$_5$ takes place on the cytosolic side of the membrane. The nascent dolichol-oligosaccharide chain then flips in the membrane, the oligosaccharide moiety being presented to the luminal side of the RER membrane. Additions of Man and glucose (Glc) to the chain are made via dolichol-single sugar intermediates, the final oligosaccharide chain having the composition Glc$_3$Man$_9$NAc$_2$. These chains are then transferred *en bloc* to asparagine (Asn) residues occurring in the sequence –Asn–(any amino acid)–threonine (or serine) on nascent polypeptide chains. The three terminal Glc are subsequently cleaved by glucosidases resident in the RER, leaving mature high Man chains attached to the polypeptide (see Faye et al., 1989; Johnson and Chrispeels, 1987).

Although there is considerable variation in the structure and composition of complex oligosaccharide side-chains, all are post-translationally derived from the enzymatic modification of high-Man chains, as outlined in Fig. 34.6. The modification begins with the obligatory trimming of Man$_9$NAc$_2$-Asn to Man$_5$NAc$_2$-Asn, the smallest high-Man chain. In the case of the complex side-chain of phytohemagglutinin, transfer of an additional NAc to the terminal α1,3-linked Man occurs in the Golgi apparatus and is necessary for continued processing. Fuc may be added to the β1,N-linked core NAc at any stage following the addition of the terminal NAc. The addition of xylose, however, requires the prior removal of the remaining two terminal Man. Transfer of an additional terminal NAc occurs prior to transport from the Golgi, the final removal of the two terminal NAc occurring in the protein body (Johnson and Chrispeels, 1987).

Processing of high-Man to complex oligosaccharide

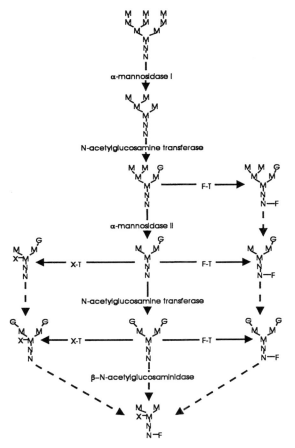

Fig. 34.6 Steps involved in the processing of high-mannose to complex asparagine-linked oligosaccharide side-chains of *Phaseolus vulgaris* storage proteins. Slashed arrows indicate probable steps in the processing. Either xylose or fucose or both are found on the complex side-chain in the final conformation. M, mannose; N, *N*-acetyl glucosamine; X, xylose; F, fucose; F-T, fucosyl transferase; X-T, xylosyl transferase. Modified from Johnson and Chrispeels (1987).

side chains does not always occur. The lack of processing appears to be a function of the accessibility of the high-Man chains to the Golgi-localized α-mannosidases and glycosyltransferases. If the high-Man side chain is hidden within the tertiary structure of the polypeptide then it will not be modified (Faye *et al.*, 1986).

Finally, although we have a good understanding of the events leading to glycosylation and modification

of oligosaccharide side-chains on storage proteins, the necessity for glycosylation is not clear. It is doubtful that oligosaccharide side-chains contain information responsible for sorting and directing storage proteins to the vacuole, since not all storage proteins destined for vacuoles are glycosylated. Further, tunicamycin, an inhibitor of glycosylation, does not interfere with subunit association or transport of most storage proteins to the vacuolar compartment. It has been suggested that the oligosaccharide side-chains offer some protection against proteolysis in the RER, or may promote (but are not absolutely required for) protein folding and oligomer association in the RER (Chrispeels, 1991).

Sorting and targeting of storage proteins to the protein bodies

The accumulation of relatively massive quantities of a limited number of storage proteins in the vacuole/protein bodies of developing storage parenchyma cells offers an excellent model system for studying the mechanisms underlying protein sorting and targeting in plants. By analogy to animal systems, the Golgi complex should be the site where vacuolar proteins are sorted from secretory proteins, each then being packaged into vesicles destined for transport to the appropriate site. Some element in the Golgi is able to distinguish between vacuolar and secretory proteins.

In animal cells, receptors embedded in the membranes of the *trans*-Golgi stack recognize and bind to a specific targeting signal, a terminal Man phosphorylated in the sixth position, on the oligosaccharide side-chains of hydrolases destined for the lysosomal (vacuolar) compartment. The sorted proteins are then packaged into vesicles and delivered to the lysosome. If the terminal Man-6-P is not present, the proteins are transported to the plasmalemma via Golgi-derived vesicles and secreted by default.

The Man-6-P signal: receptor mechanism of protein sorting and targeting does not appear to operate in plant cells. Man-6-P residues do not occur on seed storage proteins and Man-6-P receptors have not been detected (Gaudreault and Beevers, 1984; Vitale and Chrispeels, 1992). Many storage proteins that are targeted to protein bodies are never glycosylated, legumins being prime examples. The basic concept that the sorting of vacuolar proteins is via the recognition and binding of a signal region(s) in the protein to membrane-embedded receptors in the Golgi remains valid, however, and is presently an area of intense research (Chrispeels, 1991; Bednarek and Raikhel, 1992; Vitale and Chrispeels, 1992).

Genetic engineering and transgenic approaches are essential for elucidating targeting determinants in vacuolar storage (and other) proteins. Suspected targeting domains encoded in the genes can be subjected to mutational analysis by truncation, deletion, point mutation(s), or excision and fusion to genes encoding proteins not normally transported to the vacuolar compartment. The newly engineered genes can then be introduced into an heterologous organism (usually plants, plant cells or yeast) that does not have genes encoding the protein or fusion protein of interest.

Application of the above technology to the elucidation of vacuolar targeting determinants in plant proteins has revealed that there are many variations on the theme. These signals exist as contiguous propeptide amino acid sequences at either the carboxy- (barley lectin, vacuolar chitinases) or amino- (sporamin, barley aleurain, potato M_r 22 000 protein) terminus of the polypeptides, the propeptides being cleaved upon entry into the vacuolar compartment, or as non-continuous amino acid domains located in non-terminal portions of the protein (legumin, phytohemmaglutinin). Although some, very minor, amino acid sequence homology is conserved in the targeting determinants within each of the former two groups, no homology exists between groups (see Bednarek and Raikhel, 1992; Chrispeels, 1991; Vitale and Chrispeels, 1992). This suggests that there are multiple mechanisms for sorting and targeting of vacuolar proteins in plants.

The presence of targeting determinants on plant vacuolar proteins, including storage proteins, immediately suggests that there are receptors located in the Golgi apparatus which recognize and bind to these signals. A potential vacuolar targeting receptor has been isolated recently from Golgi-derived vesicles from developing pea seeds (Kirsch et al., 1994). The M_r 80 000 protein has a transmembrane domain, co-sediments with Golgi apparatus membranes, and has a strong affinity for the amino-terminal sorting

domain of barley proaleurain, a vacuolar, but not storage, protein. Sporamin, also having an amino-terminal sorting signal, weakly competes for the binding site. It remains to be shown that this potential receptor is actually involved in sorting and targeting of pea storage proteins. If this is the case, the possibility of multiple mechanisms of protein sorting and targeting can be tested by ascertaining the binding affinity of this receptor for proteins having demonstrated carboxy-terminal or internal vacuolar sorting determinants.

Protein body formation in the cereals

In the seeds of cereals, protein bodies in embryonic cells and in the cells of the aleurone layer of the endosperm are also of vacuolar origin. Formation and filling of these protein bodies proceeds as described above. Cells of the starchy endosperm, however, contain as many as three distinct protein body types: large (1–3 μm) and small (0.1–1 μm) diameter spherical, and angular, 1–5 μm diameter 'crystalline' protein bodies (Krishnan et al., 1986). In rice, the 'crystalline' protein bodies are the sites of accumulation of 11S legumin-like globulin storage proteins. Immunocytochemical evidence strongly suggests that these protein bodies have a vacuolar origin and that transport of the storage proteins from the RER to the protein body proceeds via the Golgi apparatus as described above (Krishnan et al., 1986; see Fig. 34.2). In rice, maize and sorghum, the large and small spherical protein bodies of the starchy endosperm are the sites of accumulation of the aqueous-insoluble prolamins, as evidenced by immunocytochemical localization (Krishnan et al., 1986; Lending and Larkins, 1989; Taylor et al., 1984). During development, the protein body membranes are continuous with those of the RER, and polysomes are attached to the outer surface (Larkins and Hurkman, 1978). Isolated spherical protein bodies are able to direct the synthesis of prolamin storage proteins in vitro. Thus the spherical prolamin-storing protein bodies form as localized dilations of the RER (see Higgins, 1984), and this mechanism of protein body formation may be an adaptation for sequestering highly insoluble proteins.

Although a different origin of different types

protein bodies in maize and rice starchy endosperm cells seems evident and seems to reflect relative solubilities of the different storage proteins, this is not the case for all cereals. Prolamins are initially sequestered in RER dilations in the starchy endosperm cells of wheat, but are later seen to accumulate in vacuoles, and there is some evidence that related prolamins may have different mechanisms of transport (Kim *et al.*, 1988; Simon *et al.*, 1990).

The targeting of prolamin storage proteins to the RER-derived protein bodies initially seems to be simple. The signal sequence on the nascent polypeptide chain targets the protein to the ER and directs its insertion into the lumen, where it remains. The retention of proteins in the ER is usually dependent on the presence of a carboxy-terminal histidine (or lysine)–aspartic acid–glutamic acid–leucine (HDEL or KDEL) domain, but cereal prolamins do not have this ER-retention signal. The aqueous insolubility of the prolamins may prevent their transport out of the ER. Alternatively, associations of prolamin polypeptides with other proteins which have an ER-retention signal, e.g. BiP, may cause the prolamins to remain in the ER and may promote formation of prolamin-containing protein bodies by dilation of the ER (Li *et al.*, 1993).

The switch from protein deposition to protein mobilization

One of the striking differences between the metabolic events occurring during seed development and those during post-germinative seedling growth is that during the former there is an extensive deposition of reserves, and during the latter, their mobilization. What then, causes this switch in direction of metabolism from an anabolic to a catabolic mode? The terminal event in seed development is maturation drying, during which protein synthesis gradually declines, the ribosomes become dissociated from the mRNAs, and the latter are degraded and are present in only low amounts in the dry seed. Upon subsequent rehydration of the seed, as germination proceeds, the residual mRNAs for storage proteins continue to be degraded; in addition, their synthesis is completely suppressed. A new set of mRNAs arise specifically after drying, however, which include

proteins essential for the completion of germination and for the mobilization of stored reserves, including proteolytic enzymes. Thus drying is an important controlling event for suppressing the synthesis of developmentally and related proteins, and activating the synthesis of those essential for germination and growth (Kermode, 1990). The mode of action of drying in causing the switch is not known.

Mobilization of stored protein reserves

Hydrolysis of storage proteins to the constituent amino acids requires the presence of proteinases. Some of these, the endopeptidases, can effect extensive hydrolysis, to produce small poly- or oligo-peptides which must be degraded further by peptidases. Amino- and carboxy-peptidases sequentially cleave the terminal amino acid from the free amino- or carboxy-end, respectively, of the polypeptide chain. The liberated amino acids may be reutilized for protein synthesis or be deaminated to provide carbon skeletons for respiration, for example. The sequence of events involved in the mobilization of a dicot, storage protein, such as glycinin, is outlined in Fig. 34.7.

A group of proteinases ('proteinase A'), usually acidic endopeptidases, increase in activity due to their *de novo* synthesis and insertion into the protein bodies; they commence the hydrolysis of the major storage reserves. Short chain oligopeptides are cleaved from the native proteins, markedly increasing their solubility and susceptibility to further proteolytic activity. 'Proteinase B'-type activity, again due to endopeptidases, now accomplishes extensive degradation of the proteins modified by proteinase A. Proteinase B enzymes also increase in activity due to their *de novo* synthesis; they are inactive against unmodified native proteins and must await their modification by proteinase A activity. carboxypeptidases in the protein body also contribute to the hydrolysis of the products of proteinase A activity as new carboxy-terminal sequences arise during proteolysis. Aminopeptidases may not play a role in protein mobilization within the protein bodies, for their optimum pH (6.5–8) is such that they are active in the more alkaline environment of the cytosol. They, along with peptide hydrolases, cleave the di- and tri-peptides released from the protein bodies.

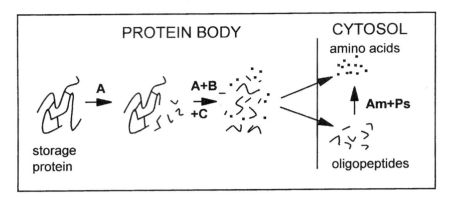

Fig. 34.7 A generalized pathway for the mobilization of storage proteins in the protein bodies of dicot seeds. The native storage protein, here a disulfide-linked insoluble legumin, is initially trimmed by proteinase A (endopeptidase) activity (A) to render it more soluble as small oligopeptides are released. Further proteinase A activity, and that of proteinase B endopeptidases (A and B) result in hydrolysis of the protein to amino acids and small peptides which are transported into the cytosol. Hydrolysis of the polypeptides in the protein bodies is aided by carboxypeptidases (C). The released oligopeptides are degraded further by aminopeptidases (Am) and peptidases (di and tripeptidases, Ps) within the cytosol to yield amino acids. Based on Wilson (1986). See also Shutov and Vaintraub (1987).

The cellular changes that precede and accompany proteolysis have been studied most thoroughly in the cotyledons of mung bean (*Vigna radiata*), and are summarized in Fig. 34.8. Here, the major storage protein is a vicilin, and an endopeptidase, vicilin peptidohydrolase (VP), is responsible for its hydrolysis, with some participation by a carboxypeptidase. VP is synthesized on the cisternal (rough) endoplasmic reticulum (CER), inserted into the ER lumen and packaged into vesicles. These vesicles are transported to the protein bodies where they fuse with its surrounding membrane, releasing the VP inside so that it comes into contact with the storage protein matrix. Eventually, the protein bodies come to contain other hydrolytic enzymes, including α-mannosidase and a glucosaminidase, for hydrolysis of the mannose and glucosamine residues associated with the vicilin, which is a glycoprotein. Hydrolases which are not proteinases are also inserted into the protein body following their *de novo* synthesis, e.g., acid phosphatase, ribonuclease, phosphodiesterase and phospholipase D. As protein digestion proceeds, the emptying protein bodies fuse to form a single large vacuole containing the many hydrolases, and becomes analogous in function to a lysosome. Digestion of cell contents by the enzymes within the vacuole is

achieved when vesicles are internalized by an autophagic process in which a portion of the cytoplasm is engulfed and sealed off by the protein body membrane. Hence destruction of the cotyledons is achieved by cellular autolysis.

Within different seeds, there are variations in this general pattern of protein hydrolysis. Within the germinated cereal grains, for example, there are three major sites of proteolytic activity: (1) the aleurone layer, where proteinases are synthesized to mobilize storage proteins within the cells of this layer, and for secretion into the starchy endosperm; (2) the starchy endosperm, into which proteinases are secreted by the aleurone layer, and in which proteinases may exist in the dry seed, to be activated following germination; (3) the embryo, particularly the scutellum, which contains peptidases to hydrolyze the small oligopeptides absorbed from the starchy endosperm following digestion of the major protein reserves stored therein.

The products of proteolysis, the amino acids, are redistributed within the growing seedling via its vascular system. Since the favored form of transportation of amino acid nitrogen is as the amides, glutamine and asparagine, enzymes which catalyze their synthesis frequently arise in storage tissues at the time of storage protein mobilization.

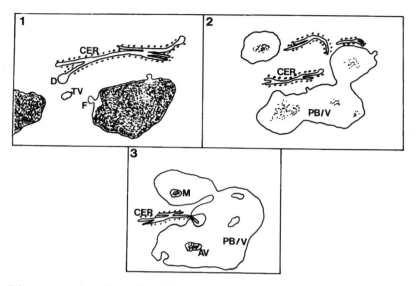

Fig. 34.8 Part of a cell from a cotyledon of mung bean illustrating the changes undergone by the protein body and ER during reserve hydrolysis and cell autolysis. (1) Three to five days from start of imbibition. Vicilin peptidohydrolase (VP) is synthesized on polysomes attached to newly formed cisternal endoplasmic reticulum (CER) and inserted into the CER lumen. Dilations (D) of the cisternae form containing the enzymes; these break off as transport vesicles (TV) and carry the peptidohydrolase to the protein bodies, with which they fuse (F). The VP commences hydrolysis of the vicilin. Other enzymes, e.g. ribonuclease, start to be inserted into the protein body. (2) As proteins are hydrolyzed the protein bodies coalesce to form large vacuoles (PB/V) and other hydrolytic enzymes are inserted. (3) Autophagic vacuoles (AV) form, engulfing cell contents such as the CER and mitochondria (M). More protein bodies fuse to form a large central vacuole, with autolytic enzymes. Note that this is an illustrative scheme of events and is not drawn to scale. Moreover, organelles and cell structures other than the ER and protein bodies are deliberately omitted for clarity. See Chrispeels and Jones (1980/81), and Herman *et al.* (1981).

Developing seeds and the regulation of storage protein synthesis

Timing and location of storage protein synthesis

Seed development has proved an especially attractive system for the study of gene expression. This is in part due to the availability of large amounts of material, the ability to determine a fixed starting point for the process (fertilization) and the recognition that many seed products, including the storage proteins, generally are absent from all other parts of the plant.

One of the striking features of seed development is the large increase in amounts of DNA during the course of embryogenesis/seed formation. Pea cotyledon cells may contain in excess of 64C (copies), and DNA in maize endosperm nuclei can increase to as much as 690C (Lopes and Larkins, 1993). Early speculation that DNA endoreduplication allowed selective amplification and consequent high expression of storage protein genes proved unfounded (Millerd, 1975; Goldberg *et al.*, 1981a), and the function of such large DNA amounts during seed development is still unclear.

Owing to their abundancy the mRNAs encoding seed storage proteins were amongst the first plant nucleic acids to be cloned (as cDNAs), providing DNA probes to study the expression of specific genes during seed development. Consequently a great deal of detail is known of the expression of storage protein genes in developing seeds.

At any given stage of cotton or soybean seed development, 15 000–30 000 genes are expressed (Goldberg, 1986), greater than 90% of which continue to be expressed throughout the life of the plant (Goldberg *et al.*, 1989). mRNA complexity increases slightly as embryos develop (Dure, 1985). There are, however large *quantitative* changes in gene expression and the composition of the mRNA population during seed development. By soybean seed mid-maturation, the small number of superabundant mRNAs that encode glycinin and β-conglycinin comprise approximately 50% of the total mRNA mass (Goldberg *et al.*, 1981b). These mRNA molecules are undetectable (< 1 molecule/20 cells) in leaf tissue, demonstrating the precision of temporal and spatial regulation.

The spatial distribution of cells containing particular storage protein mRNAs has been demonstrated using *in situ* hybridization, in which radioactively-labeled single-stranded probes are hybridized to tissue sections at various stages of seed development and the distribution of mRNA-probe hybrids analyzed by microscopy. Such studies have shown that the mRNA for soybean β-conglycinin accumulates progressively in a wave-like pattern across the cotyledon as embryogenesis proceeds (Perez-Grau and Goldberg, 1989). A similar observation was made by Hauxwell *et al.* (1990) for pea vicilin, except that the progression of the wave of mRNA accumulation was in the opposite direction.

The temporal pattern of gene expression during seed development is similar for all storage protein genes in the sense that mRNAs appear at early- to mid-development accumulate over time and finally decay before seed maturation and desiccation (Fig. 34.9); the specific timing of this process may, however, vary both between and within gene classes (see Higgins, 1984).

Temporal changes in and post-transcriptional regulation of storage protein gene expression

Vicilins and legumins in pea

The synthesis of the majority of vicilin proteins precedes that of legumin in the few pea genotypes examined so far (for references see Casey *et al.*, 1993), reflecting the relative abundance of the major vicilin and legumin mRNAs. However, as described earlier, both types of protein are the products of multiple, non-identical genes at several genetic loci, with consequent heterogeneity not only in protein

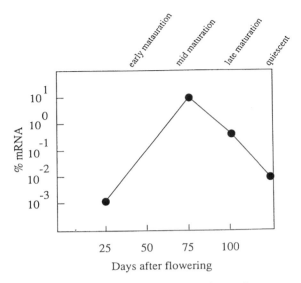

Fig. 34.9 Changes in the amount of mRNA encoding soybean β-conglycinin during seed development. After Goldberg *et al.* (1981b).

structure, but also in the patterns of gene expression within a protein class; at least two types of mRNA encoding 7S (vicilin) proteins appear relatively late in seed development, for instance. Similarly, different types of legumin gene are expressed at different times, and to different extents, during seed development (Domoney and Casey, 1987). The relative timing of the appearance of cytoplasmic mRNA appears to reflect transcriptional activity (Thompson *et al.*, 1989), but the final amounts of mRNA from the different gene families are subject to appreciable post-transcriptional modification as a consequence of genotype and nutrient effects.

β-Conglycinins in soybean

The α- and α'-polypeptides of soybean β-conglycinin accumulate earlier in seed development than the β-polypeptides and also are mobilized deferentially (Meinke *et al.*, 1981). In contrast, their corresponding genes are transcriptionally activated and repressed at the same stages of seed development (Goldberg *et al.*, 1989), suggesting that transcriptional and post-transcriptional processes are important in the regulation of β-conglycinin accumulation. β-polypeptide mRNA is present in cotyledon cells some

7–10 days before the appearance of the protein, suggestive of control of synthesis at the level of translation; alternatively, the β-polypeptides might be hydrolyzed by proteolytic enzymes at early, but not later, stages of seed development.

Prolamins in barley

Horden, the prolamin of barley, is conventionally separated into groups by polyacrylamide gel electrophoresis in the presence of sodium dodecyl sulfate, the B and C groups comprising 95% of the total hordein. During barley seed development, the various polypeptides of B and C hordein contribute differentially to the total hordein at any particular developmental stage; the kinetics of accumulation of individual polypeptides also varies within the B hordein as a function of developmental stage (see Kreis *et al.*, 1985). Thus, even within a class of proteins or subgroup of polypeptides, individual members are differentially regulated with respect both to timing of synthesis and to the amount of product.

Globulins and prolamins in oats

Approximately 70–85% of oat seed protein is a globulin that is structurally similar to legumin, and 10–15% is a prolamin called avenin. Although there are about 50 globulin genes compared to about 25 avenin genes per haploid oat genome, the amounts of avenin and globulin mRNAs are about equal throughout seed development. In contrast, about 10 times as much globulin as avenin protein, on a molar basis, is synthesized; this is a consequence of differences of an order of magnitude in rate of synthesis from approximately equal amounts of mRNA. Initiation of protein synthesis seems not to be rate-limiting for the translation of either protein, and it appears that translation elongation and/or termination is a critical regulatory step in controlling storage protein accumulation in oat endosperm (Boyer *et al.*, 1992).

Thus, the overall regulation of storage protein synthesis is a function of several transcriptional and post-transcriptional processes that can vary from species to species, between genotypes within a species, and also as a consequence of environmental conditions.

The regulation of storage protein gene expression

Expression of seed storage proteins is mostly organ-specific, occurring in cotyledons or endosperm of developing seeds but never in mature vegetative tissues. This regulated expression has made seed storage proteins especially useful as models of gene expression. Our understanding of the factors that are responsible for this strict regulation in time and space have come largely from the analysis of the behaviour of storage protein genes, or of fusions between reporter genes and storage protein gene promoters, in transgenic plants, particularly in tobacco and petunia (because of the ease of their transformation and regeneration into seed-bearing plants) (see Chapter 35).

The first storage protein gene to be introduced into the genome of another plant species (tobacco) was that for β-phaseolin from French bean (Sengupta-Gopalan *et al.*, 1985). The introduced DNA fragment contained 863 bp of 5'-, and 1226 bp of 3'-flanking sequence in addition to the transcribed sequence of ∼ 1700 bp. The β-phaseolin gene was fully developmentally regulated in the transgenic plants, being expressed specifically in seeds at the appropriate stage of development, suggesting that the signals for the regulation of gene expression reside within the introduced 3.8 kb fragment. Once this had been established, a number of systems were exploited to examine the role of various DNA sequences in seed-specific developmental regulation, special attention being paid to 5'-flanking ('upstream', promoter) sequences. Some elements within such sequences are often referred to as *cis*-acting sequences, and these are assumed to play a part in transcriptional activation of the gene through an interaction with so-called *trans*-acting factors in the nucleus. Analysis of such sequences has generally employed a bacterial reporter gene – usually chloramphenicol acetyl transferase or β-glucuronidase – fused to the promoter, or its mutated derivative(s). The fusion is introduced into transgenic plants and the presence of the appropriate enzyme activity taken as evidence of the activity of the promoter fragment in a given organ or tissue.

Sequence analysis of seed protein promoters from a variety of species identified several conserved elements within them, including A/T rich regions, AACCCA in 7S protein promoters, the so-called 'legumin box' of 11S protein promoters, which contains a conserved 'RY repeat' (CATGCATG; Dickinson et al., 1988), and the highly conserved –300 element ('endosperm box') in a number of cereal seed protein promoters. Progressive deletion of promoters, or selective alteration of their sequences, provided identification of both positive and negative regulatory elements (Bustos et al., 1991; Chen et al., 1988; Matzke et al., 1990, for example) and have indicated that a range of sequence elements are responsible for the regulation of seed storage protein gene transcription in a temporal, spatial and cell-specific manner.

Several protein (trans-acting) factors that bind to the promoter regions of seed storage protein genes have been detected in nuclear extracts from many species. Such proteins can be detected by their ability to bind to and alter the mobility of specific promoter DNA fragments on agarose gel electrophoresis (gel-shift assays) or protect the DNA from degradation by chemicals or DNase ('footprinting'). In many cases such in vitro assays recognized the sequence motifs that had been identified through sequence comparisons and transgenic expression studies (see e.g. Yunes et al., 1994). In vivo 'footprinting' assays have also been used to analyze the developmentally-regulated binding of nuclear factors to the endosperm box of a low-molecular-weight glutenin gene during wheat seed development (Hammond-Kosack et al., 1993). Several studies have shown that specific sequence elements play key roles in seed-specific gene regulation only in coordination, via trans-acting factors, with other cis-acting elements (e.g., Bäumlein et al., 1992; Chamberland et al., 1992; Fujiwara and Beachy, 1994); more information is required before we have a detailed understanding of how the interactions between trans- and cis-acting factors regulate seed storage protein gene expression.

The molecular effects of abscisic acid

Changes in the endogenous content of abscisic acid (ABA) during seed development are consistent with it playing a role in the inhibition of precocious germination and the promotion of embryo maturation; concentrations of ABA are maximal at the stage of development when embryos first become competent to germinate and decrease as development proceeds (for references see Quatrano, 1986). ABA influences the accumulation of embryo-specific mRNAs, including those for storage proteins and the so-called late-embryogenesis-abundant (Lea) proteins (see Thomas, 1993). Many seed-expressed genes, including those for phaseolin and the 11S protein from sunflower (helianthinin), contain ABA-responsible elements (ABREs) in their 5'-flanking sequence. The core of the ABRE consensus sequence from a number of plant genes (Thomas, 1993) is the ACGT motif that has been associated with binding to bZiP transcription factors, but it is clear that motifs that deviate from the ACGT core also can confer ABA-responsive gene expression.

Developmental mutants and storage protein synthesis

A number of variant lines have been identified that are defective in the production of specific storage protein subunits, or classes of storage proteins; these can be a consequence of mutation in the structural gene encoding the storage protein, an alteration in a regulatory gene, or a biochemical change during seed development. A very wide range of structural gene mutations can have the effect of disrupting storage protein gene expression, including deletion of the genes (hordein in barley Risø 56; Kreis et al., 1983), deletion of the 5' region and most of the structural gene (β-conglycinin α' subunit; Ladin et al., 1984), chromosomal inversion (glycinin; Cho et al., 1989) and mutation of the initiation codon (glycinin; Scallon et al., 1987, and legumin; Thompson et al., 1991). Such mutations provide useful information on specific aspects of the expression of seed storage protein genes and establish the molecular basis of the lesion.

Opaque-2 in maize

Several mutations are known to affect the amounts of zeins in maize endosperm, one of which, opaque-2 (*o2*), has a major effect on the rate of transcription of the genes encoding the 22 000 M_r α-zein polypeptides. Because the 22 000 M_r α-zein polypeptides are lysine-deficient, one consequence of the *o2* mutation is an enhancement of the lysine content of the protein in *o2* kernels, but other phenotypic characteristics (low yield, low seed protein content, soft kernels that are susceptible to damage and spoilage) have prevented *o2* from being adopted in breeding programmes. The wild-type *o2* gene has been cloned by transposon tagging and the protein it encodes is a bZiP binding factor that interacts with the sequence TCCACGTAGA in the promoters of the 22 000 M_r α-zein genes, to activate their transcription (Schmidt *et al.*, 1992). The *opaque-2* mutation is pleiotropic; transcription of the genes encoding β-, γ- and δ-zeins is also reduced, the free amino acid pool is increased, and the amounts of mRNAs for several non-zein proteins, including RNase, trypsin inhibitors and catalase, are elevated compared to normal wild-type maize kernels. It has been suggested that many of the changes in gene expression in *opaque-2* endosperm are in response to a stress induced by the mutation (Habben *et al.*, 1993).

Floury-2 in maize

Floury-2 (*fl2*), like *opaque-2*, increases the lysine content of maize seed protein and decreases zein accumulation; unlike *opaque-2*, it equally affects the synthesis of all classes of zeins. Protein bodies in *fl2* endosperms are asymmetrical compared to those of normal endosperms. The mutant phenotype is associated with the presence of a 24 000 M_r α-zein containing an intact signal peptide; sequence analysis of the gene corresponding to this polypeptide shows a change from alanine to valine at the C terminus of the signal peptide. It has been proposed that this change leads to retention of the signal peptide, a consequence of which is that the 24 000 M_r α-zein remains anchored to the endoplasmic reticulum. Since normal protein bodies in maize have α-zeins within a shell of β- and α-zeins, such anchorage would perturb normal protein body production, possibly blocking import sites for other α-zeins into the protein body (B. Larkins, pers. comm.).

The *r* mutation in peas

The *r* (for *rugosus* = wrinkled or shrivelled) mutation in peas leads to a wrinkled appearance of seeds in homozygous recessive (rr) lines; the wild type dominant allele produces round seeds in the homozygous (RR) or heterozygous (Rr) state. The inheritance of round/wrinkled seed shape was examined by Mendel in his seminal studies and *r* is, by definition, a single Mendelian locus. The *r* mutation, however, has pleiotropic effects, including changes in starch quantity and quality, sugar and lipid content and storage protein composition (see Casey *et al.*, 1993; Wang and Hedley, 1993). The alterations in starch and sugar consequent to mutation at the *r* locus can be attributed to the nature of the mutation; mutant embryos have very low activities of starch-branching enzyme (SBE I) as a result of the disruption of the coding sequence of the SBE I gene (at the *r* locus) by a transposon-like insertion (Bhattacharyya *et al.*, 1990). Lack of starch-branching activity leads to a low content of relatively unbranched starch and an accumulation of sugar; the latter results in a high osmotic potential in cotyledon cells, a relatively large accumulation of water and increase in cotyledon size during seed development, and a subsequent wrinkling of the seed as it dries out at maturation. It is, however, not clear how the nature of the allele at the *r* locus affects seed protein composition. Round-seeded genotypes have higher legumin/vicilin ratios than corresponding wrinkled-seeded lines, which appear to result from decreased stability of legumin, relative to vicilin, mRNAs in the high-sugar/high osmotic potential environment of the developing seeds of the rr genotypes. This effect on mRNA accumulation can be mimicked by growing RR genotypes in culture in high sucrose concentrations (Turner *et al.*, 1990).

Challenges and prospects

There are several features of seed storage proteins that may be improved. These include better amino acid composition (increased methionine/cysteine in legume globulins and increased lysine/tryptophan in cereal prolamins), increased digestibility, reduced antigenicity, and desirable physical properties in foodstuffs (for example, in relation to bread-making, gelatinization and emulsification). Significant progress has been made in defining those wheat HMW glutenin polypeptides that are desirable in breadmaking and ascribing a molecular basis to their better physical properties (see Flavell *et al.*, 1989). The relationship between structure and physical behaviour is less clear for legume seed storage proteins, but it is known that 11S and 7S globulins have different physicochemical properties (Wright and Bumstead, 1984).

Attempts to improve the amino acid composition of cereal and grain legume seed protein by conventional plant breeding have so far met with only modest success. The use of molecular biology and genetic engineering therefore has been considered as a means of developing seeds with a better balance of essential amino acids. In principle, recombinant protein technology (protein engineering) offers the opportunity to introduce into plants modified seed proteins with improved properties (see Shotwell and Larkins, 1991; Shewry and Gutteridge, 1992). Several potential constraints, however, need to be considered before modifications are made to improve the chemical or physical properties of storage proteins. Any alteration designed to increase the content of a given amino acid must avoid associated changes in other properties. For example, alterations to solubility may affect deposition in protein bodies; modification of critical residues might affect post-translational processing and assembly, decrease protein stability, or disrupt targeting; introduction of cysteine residues could lead to inappropriate aggregation and affect deposition in protein bodies; and alteration of structure also may lead to detrimental effects in the germinating seedling.

Identification of areas that are more likely to tolerate structural change has come from three-dimensional structure and comparative sequence analyses. The polar region at the C terminus of the A-polypeptide of 11S globulins, for instance, varies greatly in length from one 11S protein to another, is situated at the exterior of the assembled hexamer and is a good candidate area for the introduction of additional methionine residues; by modifying this region, Dickinson *et al.* (1990) have improved significantly the methionine content of a specific glycinin polypeptide without perturbing subunit assembly. Although it will prove possible to engineer storage protein genes to improve amino acid composition, the modified gene will need to be introduced into transgenic crop plants, where it must be expressed in seeds in sufficient amounts to overcome the diluting effects of expression of the endogenous genes; Lending *et al.* (1992) have calculated that if two lysine residues were introduced into an α-zein sequence and the modified sequence expressed at the same level as the resident, unmodified, α-zein genes, the amount of seed lysine would increase by only 0.01%. In theory, this could be overcome by repressing the endogenous genes. Since, however, many of the storage protein genes are members of moderate sized gene families (see earlier), this requires much more understanding of the *cis* and *trans*-acting factors that regulate high-level, seed-specific expression of seed protein genes, the ways in which they interact and to what extent they can be manipulated to increase (or decrease, as desirable) expression.

An alternative approach to modification of existing proteins/genes is to express a chemically synthesized gene for a high quality polypeptide or a heterologous protein from another source (e.g., the 2S high-sulfur albumins from Brazil nut or sunflowers, for legume seed protein improvement) in transgenic plants (see, for example, Saalbach *et al.*, 1994). Expression of high-quality proteins in seeds and their protection by sequestration into protein bodies undoubtedly has great potential, exemplified by the production of the peptide pharmaceutical Leu-enkaphalin in transgenic rapeseed (Vandekerckhove *et al.*, 1989). It is also possible that an increase in the amounts of minor, non-storage, protein components of desirable amino acid composition could improve the overall quality of seed protein.

At present, genetic engineering of seed protein is in its infancy (see Shewry and Gutteridge, 1992) and is unlikely by itself to be sufficient to dramatically improve the nutritional quality of seed proteins (Habben and Larkins, 1994). A combination of protein chemistry, protein engineering, plant transformation and the use of genetic variants such as *opaque-2* should eventually lead to improvements in and diversification of the quality of seed protein, either in the form of directed alterations in amino acid composition or changes in protein physical properties.

References

Bäumlein, H., Nagy, I., Villarroel, R., Inzé, D. and Wobus, U. (1992). *Cis*-analysis of a seed protein gene promoter: the conservative RY repeat CATGCATG within the legumin box is essential for tissue-specific expression of a legumin gene. *Plant J*. **2**, 233–39.

Bednerak, S. Y. and Raikhel, N. V. (1992). Intracellular trafficking of secretory proteins. *Plant Mol. Biol*. **20**, 133–50.

Bewley, J. D. and Black, M. (1994). *Seeds. Physiology of Development and Germination*, Second Edition. Plenum Press, New York.

Bhattacharyya, M. K., Smith, A. M., Ellis, T. H. N., Hedley, C. and Martin, C. (1990). The wrinkled-seed character of pea described by Mendel is caused by a transposon-like insertion in a gene encoding starch-branching enzyme. *Cell* **60**, 115–22.

Blobel, G., Walker, P., Chang, C. N., Goldman, B. M., Erickson, A. H. and Lingappa, V. R. (1979). Translocation of proteins across membranes: The signal hypothesis and beyond. In *Secretory Mechanisms, The Society of Experiment Biology Symposium 33*, eds C. R. Hopkins and C. J. Duncan. Cambridge University Press, Cambridge, pp. 9–36.

Boyer, S. K., Shotwell, M. A. and Larkins, B. A. (1992). Evidence for the translational control of storage protein gene expression in oat seeds. *J. Biol. Chem*. **267**, 17449–57.

Bustos, M. M., Begum, D., Kalkan, F. A., Battraw, M. J. and Hall, T. C. (1991). Positive and negative *cis*-acting DNA domains are required for spatial and temporal regulation of gene expression by a seed storage promoter. *EMBO J*. **10**, 1469–79.

Casey, R., Domoney, C. and Ellis, T. H. N. (1986). Legume storage proteins and their genes. *Oxford Surv. Plant Mol. Cell Biol*. **3**, 1–95.

Casey, R., Domoney, C. and Smith, A. M. (1993). Biochemistry and molecular biology of seed products. In *Peas: Genetics, Molecular Biology and Biotechnology*, eds R. Casey ad D. R. Davies, CAB International, pp. 121–163.

Chamberland, S., Daigle, N. and Bernier, F. (1992). The legumin boxes and the part of a soybean β-conglycinin promoter are involved in seed gene expression in transgenic tobacco plants. *Plant Mol. Biol*. **19**, 937–49.

Chen, Z.-L., Pan, N.-S. and Beachy, R. N. (1988). A DNA sequence element that confers seed-specific enhancement to a constitutive promoter. *EMBO J*. **7**, 297–302.

Cho, T.-J., Davies, C. S., Fischer, R. L., Turner, N. E., Goldberg, R. B. and Nielsen, N. C. (1989). Molecular characterization of an aberrant allele for the Gy_3 glycinin gene: a chromosomal rearrangement. *Plant Cell* **1**, 339–50.

Chrispeels, M. J. (1984). Biosynthesis, processing and transport of storage proteins and lectins in cotyledons of developing legume seeds. *Phil. Trans. Royal Soc. London B*. **304**, 309–22.

Chrispeels, M. J. (1985). The role of the Golgi apparatus in the transport and post-translational modification of vacuolar (protein body) proteins. *Oxford Surv. Plant Mol. Cell Biol*. **2**, 43–68.

Chrispeels, M. J. (1991). Sorting of proteins in the secretory system. *Ann. Rev. Plant Physiol. Plant Mol. Biol*. **42**, 21–53.

Chrispeels, M. J. and Jones, R. L. (1980/81). The role of the endoplasmic reticulum in the mobilization of reserve macromolecules during seedling growth. *Israel J. Bot*. **29**, 225–45.

Craig, S., Goodchild, D. J. and Miller, C. (1980). Structural aspects of protein accumulation in developing pea cotyledons. II. Three-dimensional reconstructions of vacuoles and protein bodies from serial sections. *Aust. J. Plant Physiol*. **7**, 329–37.

Dickinson, C. D., Evans, R. P. and Nielsen, N. C. (1988). RY repeats are conserved in the 5' flanking regions of legume seed protein genes. *Nucl. Acids Res*. **16**, 371.

Dickinson, C. D., Hussein, E. H. A. and Nielsen, N. C. (1989). Role of post-translational cleavage in glycinin assembly. *Plant Cell* **1**, 459–69.

Dickinson, C. D., Scott, M. P., Hussein, E. H. A., Argos, P. and Nielsen, N. C. (1990). Effect of structural modifications on the assembly of a glycinin subunit. *Plant Cell* **2**, 403–13.

Domoney, C. and Casey, R. (1987). Changes in legumin messenger RNAs throughout seed development in *Pisum sativum* L. *Planta* **170**, 562–66.

Dure, L. (1985). Embryogenesis and gene expression during seed formation. *Oxford Surv. Plant Mol. Cell Biol*. **2**, 179–97.

Ellis, T. H. N., Turner, L., Hellens, R. P., Lee, D., Harker, C. L., Enard, C., Domoney, C. and Davies, D. R. (1992). Linkage maps in pea. *Genetics* **130**, 649–64.

Faye, L., Johnson, K. D. and Chrispeels, M. J. (1986). Oligosaccharide side-chains of glycoproteins that remain in the high-mannose form are not accessible to glycosidases. *Plant Physiol*. **81**, 206–11.

Faye, L., Johnson, K. D., Sturm, A. and Chrispeels, M. J. (1989). Structure, biosynthesis, and function of asparagine-linked glycans on plant glycoproteins. *Physiol. Plant*. **75**, 309–14.

Flavell, R. B., Goldsbrough, A. P., Robert, L. S., Schnick, D. and Thompson, R. D. (1989). Genetic variation in wheat HMW glutenin subunits and the molecular basis of bread-baking quality. *Bio/Technology,* **7**, 1281–86.

Fujiwara, T. and Beachy, R. N. (1994). Tissue-specific and temporal regulation of a β-conglycinin gene: roles of the RY repeat and other *cis*-acting elements. *Plant Mol. Biol*. **24**, 261–72.

Gaudreault, P.-R. and Beevers, L. (1984). Protein bodies and vacuoles as lysosomes. Investigations into the role of mannose-6-phosphate in intracellular transport of glycosidases in pea cotyledons. *Plant Physiol*. **76**, 228–32.

Goldberg, R. B. (1986). Regulation of plant gene expression. *Phil. Trans. R. Soc. Lond. B.* **314**, 343–53.

Goldberg, R. B., Barker, S. J. and Perez-Grau, L. (1989). Regulation of gene expression during plant embryogenesis. *Cell* **56**, 149–60.

Goldberg, R. B., Hoschek, G., Ditta, G. S. and Breidenbach, R. W. (1981a). Developmental regulation of cloned superabundant embryo mRNAs in soybean. *Develop. Biol.* **83**, 218–31.

Goldberg, R. B., Hoschek, G., Tam, S. H., Ditta, G. S. and Breidenbach, R. W. (1981b). Abundance, diversity, and regulation of mRNA sequence sets in soybean embryogenesis. *Develop. Biol.* **83**, 201–17.

Greenwood, J. S. and Bewley, J. D. (1985). Seed development in *Ricinus communis* cv. Hale (castor bean). III. Pattern of storage protein and phytin accumulation in the endosperm. *Can. J. Bot.* **63**, 2121–28.

Habben, J. E., Kirleis, A. W. and Larkins, B. A. (1993). The origin of lysine-containing proteins in *opaque-2* maize endosperm. *Plant Mol. Biol.* **23**, 825–38.

Habben, J. E. and Larkins, B. A. (1994). Improving protein quality in seeds. In *Seed Development and Germination*, eds J. Kigel, M. Nebgi and G. Galili. Marcel Dekker Inc., pp. 791–810.

Hammon-Kosack, M. C. U., Holdsworth, M. J. and Bevan, M. W. (1993). *In vivo* footprinting of a low molecular weight glutenin gene (LMWG-1D1) in wheat endosperm. *EMBO J.* **12**, 545–54.

Harris, N. (1986). Organization of the endomembrane system. *Ann. Rev. Plant Physiol.* **37**, 73–92.

Hauxwell, A. J., Corke, F. M. K., Hedley, C. L. and Wang, T. L. (1990). Storage protein gene expression is localized to regions lacking mitotic activity in developing pea embryos. An analysis of seed development in *Pisum sativum* L. XIV. *Development* **110**, 283–89.

Herman, E. M., Baumgartner, B. and Chrispeels, M. J. (1981). Uptake and apparent digestion of cytoplasmic organelles by protein bodies (protein storage vacuoles) in mung bean cotyledons. *Eur. J. Cell Biol.* **24**, 226–35.

Higgins, T. J. V. (1984). Synthesis and regulation of major proteins in seeds. *Ann. Rev. Plant Physiol.* **35**, 191–221.

Johnson, K. D. and Chrispeels, M. J. (1987). Substrate specificities of *N*-acetylglycosaminyl-, fucosyl-, and xylosyltransferases that modify glycoproteins in the Golgi apparatus of bean cotyledons. *Plant Physiol.* **84**, 1301–8.

Kermode, A. R. (1990). Regulatory mechanism involved in the transition from seed development to germination. *CRC Crit. Rev. Plant Sci.* **9**, 155–95.

Kim, W. T., Franceschi, V. R., Krishnan, H. B. and Okita, T. W. (1988). Formation of wheat protein bodies; involvement of the Golgi apparatus in gliadin transport. *Planta* **176**, 173–82.

Kirsch, T., Paris, N., Butler, J. M., Beevers, L. and Rogers, J. C. (1994). Purification and initial characterization of a potential plant vacuolar targeting receptor. *Proc. Nat. Acad. Sci. USA*, **91**, 3403–37.

Kreis, M., Shewry, P. R., Forde, B. G., Rahman, S. and Miflin, B. J. (1983). Molecular analysis of a mutation conferring the high-lysine phenotype on the grain of barley (*Hordeum vulgare*). *Cell* **34**, 161–67.

Kreis, M., Shewry, P. R., Forde, B. G., Forde, J. and Miflin, B. J. (1985). Structure and evolution of seed storage proteins and their genes with particular reference to those of wheat, barley and rye. *Oxford Surv. Plant Mol. Cell Biol.* **2**, 253–317.

Krishnan, H. B., Franceschi, V. R. and Okita, T. W. (1986). Immunocytochemical studies on the role of the Golgi complex in protein body formation in rice seeds. *Planta* **169**, 471–80.

Kriz, A. L. and Larkins, B. A. (1991). Biotechnology of seed crops: genetic engineering of seed storage proteins. *HortScience* **26**, 1036–41.

Krochko, J. E. and Bewley, J. D. (1988). Use of electrophoretic techniques in determining the compositon of seed storage proteins in alfalfa. *Electrophoresis* **9**, 751–63.

Ladin, B. F., Doyle, J. J. and Beachy, R. N. (1984). Molecular characterization of a letion mutation affecting the α'-subunit of β-conglycinin of soybean. *J. Mol. Appl. Genet.* **2**, 372–89.

Larkins, B. A. and Hurkman, W. J. (1978). Synthesis and deposition of zein in protein bodies of maize endosperm. *Plant Physiol.* **62**, 256–63.

Lending, C. R. and Larkins, B. A. (1989). Changes in the zein composition of protein bodies during maize endosperm development. *Plant Cell* **1**, 1011–23.

Lending, C. R., Wallace, J. C. and Larkins, B. A. (1992). Synthesis of zeins and their potential for amino acid modification. In *Plant Protein Engineering*, eds P. R. Shewry and S. Gutteridge. Cambridge University Press, Cambridge. pp. 209–18.

Li, X., Wu, Y., Zhang, D.-J., Gillikin, J. W., Boston, R. S., Franceschi, V. R. and Okita, T. W. (1993). Rice prolamin protein body biogenesis: A BiP-mediated process. *Science* **262**, 1054–56.

Lopes, M. A. and Larkins, B. A. (1993). Endosperm origin, development, and function. *Plant Cell* **5**, 1383–99.

Lott, J. N. A. (1980). Protein bodies. In *The Biochemistry of Plants*, Vol. 1, ed. N. E. Tolbert, Academic Press, New York, pp. 589–623.

Matzke, A. J. M., Stöger, E. M., Schernthaner, J. P. and Matzke, M. A. (1990). Deletion analysis of a zein gene promoter in transgenic tobacco plants. *Plant Mol. Biol.* **14**, 323–32.

Meinke, D. W., Chen, J. and Beachy, R. N. (1981). Expression of storage protein genes during soybean seed development. *Planta* **153**, 130–39.

Millerd, A. (1975). Biochemistry of legume seed proteins. *Ann. Rev. Plant Physiol.* **26**, 53–72.

Perez-Grau, L. and Goldberg, R. B. (1989). Soybean seed protein genes are regulated spatially during embryogenesis. *Plant Cell* **1**, 1095–1109.

Quatrano, R. S. (1986). Regulation of gene expression by abscisic acid during angiosperm embryo development. *Oxford Surv. Plant Mol. Cell Biol.* **3**, 467–77.

Saalbach, I. Pickardt, T., Waddell, D. R., Schieder, O. and Müntz, K. (1995). The sulfur-rich Brazil nut 2S albumin is specifically formed in transgenic seeds of the grain legume *Vicia narbonensis. Euphytica,* **85**, 181.

Scallon, B. J., Dickinson, C. D. and Nielsen, N. C. (1987). Characterization of the null allele for the *Gy4* glycinin gene from soybean. *Mol. Gen. Genet.* **208**, 107–13.

Schmidt, R. J., Ketudat, M., Aukerman, M. J. and Hoschek, G. (1992). Opaque-2 is a transcriptional activator that recognizes a specific target site in 22-kD zein genes. *Plant Cell* **4**, 689–700.

Sengupta-Gopalan, C., Reichert, N. A., Barker, R. F., Hall, T. C. and Kemp, J. D. (1985). Developmentally regulated expression of the bean β-phaseolin gene in tobacco seed. *Proc. Natl. Acad. Sci. USA* **82**, 3320–24.

Shewry, P. R. and Gutteridge, S. (eds) (1992). *Plant Protein Engineering.* Cambridge University Press, Cambridge.

Shotwell, M. A. and Larkins, B. A. (1991). Improvement of the protein quality of seeds by genetic engineering. In *Molecular Approaches to Crop Improvement*, eds E. S. Dennis and D. J. Llewellyn. Springer-Verlag, Berlin, pp. 33–61.

Shutov, A. D. and Vaintraub, I. A. (1987). Degradation of storage proteins in germinating seeds. *Phytochem.* **26**, 1557–66.

Simon, R., Altschuler, Y., Rubin, R. and Galili, G. (1990). Two closely-related wheat storage proteins follow a markedly different subcellular route in *Xenopus laevis* oocytes. *Proc. Nat. Acad. Sci. USA.* **88**, 834–38.

Taylor, J. R. N., Schussler, L. and Leibenberg, N.v.d. W. (1984). Protein body formation in starchy endosperm of *Sorghum bicolor* (L.) Moench seeds. *S. Afr. J. Bot.* **51**, 35–40.

Thomas, T. L. (1993). Gene expression during plant embryogenesis and germination: an overview. *Plant Cell* **5**, 1401–10.

Thompson, A. J., Evans, I. M., Boulter, D., Croy, R. R. D. and Gatehouse, J. A. (1989). Transcriptional and post-transcriptional regulation of seed storage-protein gene expression in pea (*Pisum sativum* L). *Planta* **179**, 279–87.

Thompson, A. J., Bown, D., Yaish, S. and Gatehouse, J. A. (1991). Differential expression of seed storage protein genes in the pea *legJ* subfamily; sequence of gene *legK. Biochem. Physiol. Pflanzen* **187**, 1–12.

Turner, S. R., Barratt, D. H. P. and Casey, R. (1990). The effect of different alleles at the *r* locus on the synthesis of seed storage proteins in *Pisum sativum. Plant Mol. Biol.* **14**, 793–803.

Vandekerckhove, J., Van Damme, J., Van Lijsenbettens, M. V., Botterman, J., De Block, M., Vandewiele, M., De Clercq, A., Leemans, J., Van Montagu, M. and Krebbers, E. (1989). Enkephalins produced in transgenic plants using modified 2S seed storage proteins. *Bio/Technology* **7**, 929–32.

Vitale, A. and Chrispeels, M. J. (1992). Sorting of proteins to the vacuoles of plant cells. *Bioessays* **14**, 151–60.

Wang, T. L. and Hedley, C. L. (1993). Genetic and developmental analysis of the seed. In *Peas: Genetics, Molecular Biology and Biotechnology*, eds R. Casey and D. R. Davies, CAB International, pp. 83–120.

Wilson, K. A. (1986). Role of proteolytic enzymes in the mobilization of protein reserves. In *Plant Proteolytic Enzymes*, Vol. 2, ed. M. J. Dalling, CRC Press, Boca Raton, FL. pp. 19–47.

Wright, D. J. and Bumstead, M. R. (1984). Legume proteins in food technology. *Phil. Trans. R. Soc. Lond. B.* **304**, 381–93.

Yunes, J. A., Cord Neto, G., da Silva, M. J., Leite, A., Ottoboni, L. M. M. and Arruda, P. (1994). The transcriptional activator opaque 2 recognizes two different target sequences in the 22-kD-like α-prolamin genes. *Plant Cell* **6**, 237–49.

Prospects for
Plant Improvement

35 Fundamentals of gene transfer in plants

Brian L. A. Miki and V. N. Iyer

Introduction

We now have the ability to isolate and clone genes from plant cells or their compartments, to experimentally alter them and to assess the effect of such manipulation *in vitro*. Such studies will, however, be incomplete unless they are complemented by dependable techniques that will transmit these manipulated genes back into the organism. It is in this manner that we can best correlate the experimental alterations to the gene with specific biochemical and physiological responses in specific cells of the whole plant. This need arises even more in cases where the gene to be installed in the plant is derived from a foreign source. In this chapter, we describe and compare several methods that have been developed recently for such transfers and discuss their relevant advantages, present limitations and promise.

This chapter begins with a consideration of the growth and differentiation of plant cells in culture and of marker genes used for the identification of transformed tissues. Practical considerations mandate the need for genes which can be used for the selection of those few cells in a population that have been genetically transformed. This is followed by sections considering transfer vectors and modes of transfer. The chapter concludes with sections considering the demonstrated or potential utility of these techniques for increasing our understanding of plant gene regulation and in obtaining transgenic plants with potential for the agricultural industry.

Plant cell and tissue culture

Great experimental opportunities exist for the genetic manipulation of plant cells and tissues in culture.

Plants are unique among higher organisms in their capacity to regenerate whole organisms from a variety of cell types as well as from a fertilized ovum. Genetic transformation or the successful introduction and integration of exogenous DNA into these cells can give rise to transgenic plants. Fig. 35.1 illustrates potential pathways for growth and differentiation of plant cells and tissues in culture. The regeneration of plantlets can occur by two different processes. Organogenesis usually involves the differentiation of shoots followed by roots. Embryogenesis in culture follows similar stages of maturation to zygotic embryos in seeds but originates from different cells. For species such as tobacco, the culture conditions have been determined for most of these steps. Tobacco is frequently used as an experimental plant and as a model system for studying organogenesis in culture. Similar manipulations can be performed with some crop species such as alfalfa or rapeseed but with lower efficiencies. These particular species are excellent for studying the process of embryogenesis in culture. The regeneration of cereals from protoplasts has been very difficult to demonstrate, but the recent achievement with rice suggests that it is possible. Cells at any of the stages shown in Fig. 35.1 can be used as recipients for transformation with isolated DNA. Indeed successful transformation has been demonstrated for most of these using a variety of DNA transfer techniques.

Marker genes

A critical step in the development and evaluation of transformation strategies for plants was the construction of vectors with genes that could act as

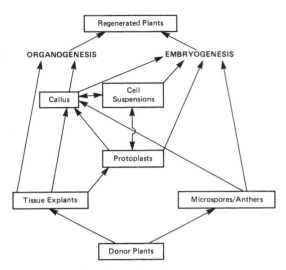

Fig. 35.1 Potential pathways for the growth and differentiation of plant cells and tissues in culture. The regeneration of plantlets in culture may occur by organogenesis or embryogenesis. This diagram illustrates some of the manipulations that have been performed with plant cells in culture. The experimental capabilities that exist for a given plant species vary greatly.

dominant selectable markers. These genes provide a growth advantage to cells that integrate the vectors and express the marker gene. An example is *nptII* from the bacterial transposon Tn5 which codes for the enzyme neomycin phosphotransferase. This enzyme catalyzes the transfer of a phosphate moiety from ATP to a number of aminoglycoside antibiotics, including kanamycin (Km), thereby detoxifying them. In the presence of kanamycin transformed cells can grow and differentiate into plantlets in culture whereas normal cells and tissues of most plant species cannot.

Among recipient cells, transformation events may occur at very low frequencies, such as one cell in a hundred thousand cells. Without an efficient selection system, the transformants would be difficult to detect and could not be recovered or separated from the untransformed cells. The transformation frequencies or proportion of cells that are transformed may be estimated from the proportion of resistant cells that are recovered in an experiment. This provides a standard for comparing the efficiencies of different transformation techniques and for manipulating

conditions to optimize a system. Genetic analyses of transgenic plants using selectable markers is relatively straightforward. Figure 35.2 illustrates the segregation of kanamycin resistance among tobacco seedlings germinated on culture media supplemented with kanamycin. The ratio of resistant, green seedlings to sensitive, white seedlings is approximately 3 : 1 in this illustration. This is expected for the segregation of a dominant marker among progeny derived from self-fertilization of a transgenic plant with a single integration.

A number of selectable marker genes have been developed which provide resistance to a range of antibiotics and chemicals (Table 35.1). This variety is important because plant species and particular tissues of a plant may differ widely in their sensitivities to different selective agents. Furthermore, the choice of several markers increases the range of genetic manipulations and analyses that can be performed with transgenic plants.

Most of the selectable markers are not of plant origin. To achieve constitutive expression in culture and in the various tissues of the plants, the coding region of the genes have been fused to promoters or other regulatory sequences known to function in plants. These include regulatory sequences from the nopaline synthase (*nos*) or octopine synthase (*ocs*)

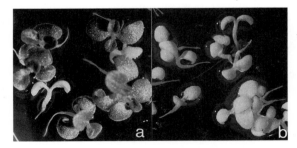

Fig. 35.2 Segregation of kanamycin resistance among seedlings of transgenic (a) and normal (b) tobacco plants. Seedlings were germinated on media with 200 ug/ml kanamycin. Under these conditions seedlings of normal plants (b) were white and eventually died. In self-pollinated transgenic plants, kanamycin resistance segregated as a single dominant marker, i.e. approximately three green resistant seedlings were recovered for every white sensitive seedling. (Photographs courtesy of Jérôme Gabard and Pierre Charest, Agriculture Canada and Carleton University).

Table 35.1 Examples of selectable markers

Chimeric gene	Source	Selective agent	
nos-nptII-nos	Tn5	kanamycin	Herrera-Estrella et al. (1983)
ocs-aphIV-nos	E. coli	hygromycin	Waldron et al. (1985)
35S-ble-nos	Tn5	bleomycin	Hille et al. (1986)
35S-dhfr-nos	mouse	methotrexate	Eichholtz et al. (1987)

genes of the Ti (tumour inducing) plasmids found in pathogenic *Agrobacterium tumefaciens* strains and from the cauliflower mosaic virus (CaMV) 35S or 19S transcript. The 35S promoter ensures high levels of transcription. Although various 3′ termination signals have been used, that taken from *nos* is common.

Reporter genes are genes that are not selectable but code for enzymes that can be detected with great ease and sensitivity. These have been important for monitoring transformation events or for the rapid detection of promoter activity. The opine synthase genes of the Ti plasmids, *nos* and *ocs* and the chloramphenicol acetyltransferase (*cat*) gene from bacteria are early examples. *Ocs* and *nos* catalyze principally the reductive condensation of arginine with α-ketoglutarate and pyruvate, respectively. These reactions can be performed with cell-free extracts of transformed tissues and the opines detected by staining or fluorography after paper electrophoresis (Fig. 35.3a). In many early studies this provided clear evidence for transformation of plants with Ti plasmids of *Agrobacterium tumefaciens* (Kemp, 1982). It provides a generally useful detection system; however, comigrating compounds or loss of activity may interfere with analyses of samples. The acetylation of chloramphenicol catalyzed by cat can be easily detected with great sensitivity by autoradiography (Fig. 35.3b). This system is valuable for monitoring transient expression and the regulation of transcription by promoters in transgenic plants (Hauptmann *et al.*, 1987). However, some species such as those of the genus *Brassica*, have a very high level of endogenous plant cat activity (Fig. 35.3b) and inhibitors of bacterial cat which obscures the activity of the transformed bacterial gene.

Recently, more versatile reporter genes have been developed. The firefly luciferase gene (Ow *et al.*,

1986) and the *E. coli* β-glucuronidase (GUS) gene (*uid A* locus; Jefferson *et al.*, 1987) in particular allow great sensitivity of detection. Luciferase catalyzes the oxidative decarboxylation of luciferin and emits light that can be detected in cell-free extracts by luminometry. Alternatively, the activities in tissues or whole plants can be examined by the exposure of X-ray films (autoluminography). β-Glucuronidase is a

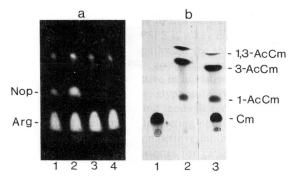

Fig. 35.3 (a) Detection of nopaline synthase activity in broccoli crown gall tumour tissues transformed with nopaline Ti plasmid C58. Extracts from transformed tissues in culture revealed small amounts of nopaline (lane 1) whereas normal tissues did not (lane 4). Following standard assay protocols, nopaline was synthesized in cell-free extracts of transformed tissues (lane 2) but not with normal tissues (lane 3). Nopaline (Nop) was separated from the substrate arginine (Arg) by paper electrophoresis and detected by staining of quanidines with a fluorescent dye. (Reproduced from Holbrook *et al.*, 1986, with permission).
(b) Detection of chloramphenicol acetyl transferase (cat) activity in transgenic tobacco. [14C]-chloramphenicol (Cm) was acetylated (Ac) in cell free extracts; separated into 1-AcCm, 3-AcCm and 1,3-AcCm forms by thin-layer chromatography; and detected by autoradiography. Cat activity can be detected in transgenic tobacco (lane 2) but not in normal tobacco (lane 1). High levels of cat activity are normally present in *Brassica* species (Lane 3). (Photograph courtesy of Pierre Charest, Carleton University).

hydrolase that cleaves a variety of β-glucuronides including fluorogenic substrates that can be detected by fluorometry. β-Glucuronidase activity can be detected histochemically in cells using substrates such as X-gluc (5-bromo-4-chloro-3-indolyl glucoronide) which stain a blue color. The histochemical detection of gene expression in transgenic plants offers greater precision in identifying the specific cells and tissues in which the promoters are active. GUS, in particular, is widely used as a reporter gene in plants and retains activity when fused to other proteins at the N terminus. This capability allows for the construction of bifunctional proteins (Fig. 35.4) and for gene tagging by insertion of GUS genes.

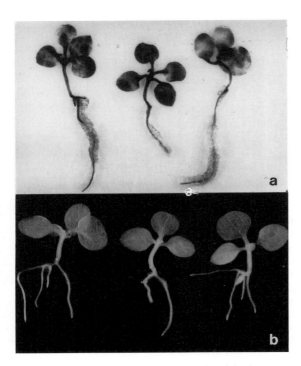

Fig. 35.4 Histochemical detection of GUS activity in transgenic tobacco seedlings. The GUS gene was fused to the mammalian metallothionin gene and expressed in tobacco under the control of the 35S promoter and *nos* 3′ end. A bifunctional fusion protein was expressed that retained GUS activity and tissues stained dark blue with x-gluc (panel a) whereas untransformed control seedlings did not stain (panel b). (Photographs courtesy of Hélène Labbé and Jiro Hattori, Agriculture Canada).

Agrobacterium-mediated gene transfer

Bacteria of the genus *Agrobacterium* are free-living but opportunistic soil bacteria which evolved the unique capacity to interact genetically with susceptible plants. The interaction results in the stable insertion of part of the genome of the bacterium into the genome of the plant. It is this natural ability of the bacterium to genetically transform the plant tissue that is the basis of all gene transfer technology that attempts to exploit this system. How does part of the bacterial DNA end up inserted in plant chromosomal DNA? How and why did this capacity evolve? We do not have satisfying answers. Circumstantial evidence suggests that DNA transfer occurs by a process akin to interbacterial mating followed or accompanied by a form of DNA recombination. The molecular biology of the *Agrobacterium*-plant interaction has been periodically reviewed (Zambryski, 1988; Ream, 1989; Winans, 1992; Charles *et al.*, 1992). These reviews can provide the student with entry points into the primary research literature. The outline that follows emphasizes aspects, the understanding of which is most relevant to the utilization of the system.

Natural history and molecular basis

The definition of 'species' within the genus *Agrobacterium* is a matter of convenience and has no phylogenetic implications. All isolates have been recognized on the basis of the symptoms they produce on susceptible plants. Thus *A. tumefaciens* induces tumours and *A. rhizogenes* induces root proliferation at the site of infection on appropriate plants. These symptoms do not reflect basic differences in the gene transfer process but rather the genetic determinants contained in the DNA segment that is transferred to the plant. This DNA segment called the T-DNA (Transferred DNA) is initially present in the bacterium as part of a bacterial plasmid which is self-transmissible by inter-bacterial mating and called the Ti (Tumour inducing) plasmid. Thus it is feasible to interconvert one 'species' of *Agrobacterium* into another simply by exchanging its Ti plasmid. This observation does not imply that genes present on the chromosome of *Agrobacterium* are irrelevant. A number of such genes have been identified and are shown in Table 35.2.

Table 35.2 Chromosomal genes of *Agrobacterium* that participate in the transmission of DNA into the plant cell[a]

Gene	Probable role
*chv*B, *chv*A	Synthesis and secretion of cyclic β-1,2-glucan that aids binding to plant cells
*exo*C	Synthesis of exopolysaccharides
cel	Cellulose biosynthesis that assists aggregation and colonization
ros	Regulation of some of the plasmid *vir* genes
*chv*E	Sugar-binding and promoting positive chemotax

[a] Mutations in some of these genes have been found to have pleiotrophic effects.

Figure 35.5 is a diagrammatic summary of what we know concerning the infection and transformation process as it is believed to occur in nature. Infection is initiated by accidental wounding of plant tissue. The wound site releases a mixture of several compounds the nature of which will vary with the plant. One class of compounds is phenolic and another consists of monosaccharides derived from plant cell walls. These compounds diffuse from the wound and at low concentrations some of them exert positive chemotaxis on the bacterium. Chemotaxis contributes to the colonization of the wound site. Other factors may also contribute to this colonization. They include (1) the extracellular formation of cellulose microfibrils by the bacteria upon contact or close exposure with plant cells;

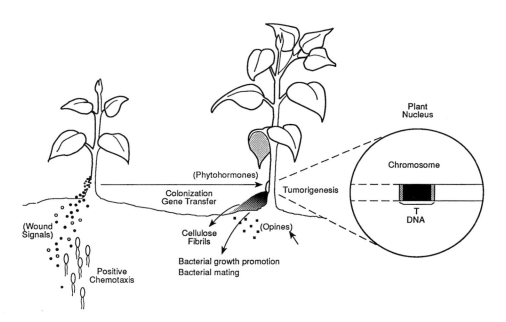

Fig. 35.5 Outline of events in the genetic transformation of susceptible plant tissue by *Agrobacterium*. Phenolic compounds diffusing from a plant wound site serve as chemotactic attractants for *Agrobacterium*. At higher concentrations, such compounds also induce the expression of several bacterial genes some of which are essential for plant cell transformation (see Fig. 35.6). Bacteria attach to and colonize cells in the wound site. This colonization is aided by the induction of a mesh of cellulose fibrils by the colonizing bacteria. A region of the Ti plasmid DNA defined and bordered by 25 base pair directly repeated sequences is found to be transferred and inserted into the plant nuclear genome. The molecular intermediates and mechanisms of these processes are not yet firmly established. Genes within the T-DNA that are expressed by the recipient plant cells include (1) genes for the synthesis of rare amino acids (opines) that are a second colonizing factor as they promote both selective bacterial growth and interbacterial conjugation, and (2) genes for novel and unregulated pathways of biosynthesis of auxin and/or cytokinin that results in the localized tumours. Not shown in this diagram is that substantial parts of the T-DNA (except one or both borders) can be deleted and replaced by other desired and functional genes and that regenerant transgenic plants can be obtained from transformed tissue.

(2) the synthesis and release by the transformed plant tissue of a class of amino acid-derived metabolites called the opines. These can be catabolized by the infecting bacterium and can also induce inter-bacterial mating so that a larger fraction of the population in this ecological niche can use the opine. In laboratory experiments, colonization steps can be by-passed by artificially infecting wound sites. The colonization pathways are thus probably important in nature but not essential to the transformation process. One role of wound exudate compounds is to induce certain bacterial genes, the expression of which is essential for T-DNA transfer.

There are only two regions on the Ti plasmid that are essential to this transmission, the T-DNA region and another region called the *vir* (virulence) region. Although linked in all native plasmids, these two regions will also function when they are cloned in separate plasmids, a circumstance which facilitates the manipulation of the T-DNA-bearing vector in so-called binary vector systems (see section on Gene Vectors).

The T-DNA

As shown within the enlarged circle in Fig. 35.5, the T-DNA is the portion of the Ti plasmid DNA that is transmitted to the plant DNA. Under natural conditions, it has within its boundaries, phytohormone genes, the activity of which leads to the morphology of the plant tumors and genes determining the synthesis and secretion of opines. These genes can be replaced by non-tumorogenic marker genes and/or desired plant or other expressible foreign genes which will then be transferred into the plant DNA. T-DNA vectors in which these tumorogenic genes have been removed are called disarmed vectors. The only parts of the T-DNA that are usually sufficient for the transfer process are the 25 bp region shown as left and right borders (LB and RB) and in some cases the overdrive sequence (Fig. 35.6). This figure also shows the genes associated naturally with the T-DNA of a number of plasmids that have been analyzed.

The *vir* region

The only other region on the plasmid that is needed for T-DNA transfer is the *vir* (virulence) region

(Fig. 35.6). It is a large region containing eight regulatable transcriptional units and several proteins. Of these, *vir*A and *vir*G constitute a two-component system in which the VirA protein can sense the environmental wound-induced signal(s) and transmit the signal to VirG (Winans, 1992). Evidence suggests that these signals lead to the autophosphorylation of VirA followed by transfer of the phosphate to a VirG residue (Fig. 35.7A). This transfer is apparently needed to activate VirG but the details of this sensory-transduction mechanism needs to be better understood. When this understanding is secured, it is likely to extend our ability to use the system in productive ways because VirG, thus phosphorylated is an activator of all the other *vir* genes and the eventual transmission into plant DNA of natural or artificial T-DNA constructs (Fig. 35.7A). The *vir*B region is large and specifies 11 proteins most of which span the bacterial cell surface (Kuldau *et al.*, 1990; Berger and Christie, 1994). It is believed that these proteins are involved in the formation of a transmembrane conduit to enable polarized DNA transfer from the bacterium to the plant cell nucleus (Kado, 1991, 1994) (see Fig. 35.7A). This polarized DNA transfer, by analogy with bacterial conjugation systems, is believed to require the participation of at least a few different proteins that process the DNA for transfer (analogous to the mobilization or *mob* proteins). At the least, this processing must involve DNA strand unwinding, strand-nicking at or near the right border, stabilization of single strands and polarized strand transfer accompanied by complementary strand synthesis (Fig. 35.7B). Potential Vir protein candidates for all of these steps have been identified although there are some Vir proteins whose functions are not known. Thus the VirC proteins are believed to bind to DNA at the right border or close to it in the overdrive region before or after binding to each other (Peralta *et al.*, 1986; Toro *et al.*, 1989). Some of the VirD proteins are involved in strand nicking with Vir D2 attached covalently to the 5' end of the transmitted strand to serve as a 'pilot' protein (Fig. 35.7B) (Herrera-Estrella *et al.*, 1988; Howard *et al.*, 1989; Pansegrau *et al.*, 1993; Filickin and Gelvin, 1993). The Vir E2 protein has been shown to be a single strand DNA-binding protein (Sen *et al.*, 1992) and also to be a nuclear localizing protein (Citovsky *et al.*, 1992). The

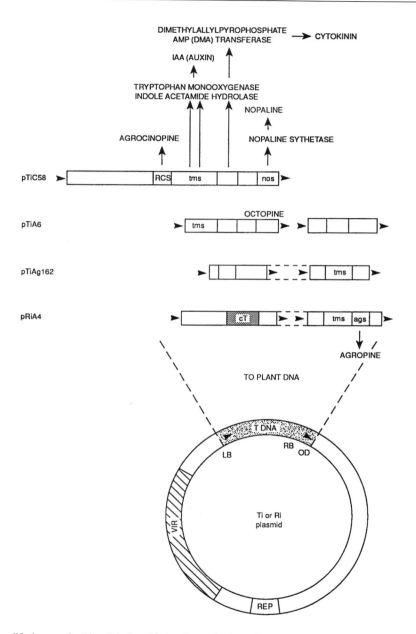

Fig. 35.6 A simplified map of a Ti or Ri plasmid showing only the regions necessary for T strand transfer and examples of the structures of T-DNA in plant tissue after transformation by *Agrobacterium* strains carrying different Ti (or Ri) plasmids (top). OD is the overdrive sequence. Vir is the large Vir region that is not found stably in the nucleus and which is discussed further. Rep is the Ti plasmid replicon. Regions not directly related to T-DNA transfer are now shown. pTiC58 and pTiA6 are the intensively-studied plasmids that incite tumors producing nopaline or octopine, respectively. pTiAg162 is a plasmid from a grapevine isolate now called *A. vitis*. It confers a limited host-range. pRiA4 is a plasmid that incites tumors with a profusion of roots (hairy root disease). Shaded arrow indicate the border regions that signal the extent of DNA that has been transmitted. cT is a DNA region that has been found to hybridize with the genomic region of some untransformed plant species.

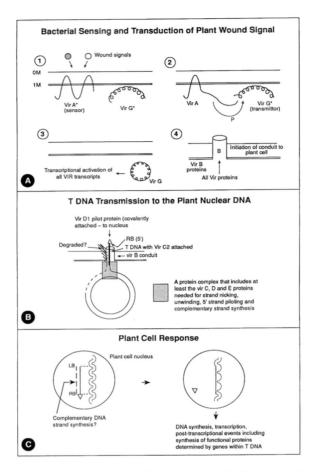

Fig. 35.7 A simplified 'thought diagram' intended only to aid the student in thinking about the possible way by which *Agrobacterium* transmits its T-DNA into plant nuclear DNA. This scenario is favoured currently and the evidence for it is discussed in the text. It has been influenced by what is known about DNA transfer by bacterial mating.

In [A], ● and ○ are signals from the environment (plant wounds) such as phenolic compounds and sugars which the VirA* protein located in the bacterial periplasm senses and becomes activated to VirA by autophorphorylation. The phosphate group is then transferred to the inactive and cytoplasmically located VirG* protein. This converts VirG* to VirG which is a transcriptional activator that activates all Vir transcripts by interaction at its promoters. The number of proteins coded by each transcript is large but the VirB transcript is the largest coding for 11 proteins most of which are exported to the cell surface. It is speculated that they form a conduit for the transport of the T-DNA.

The possible transport pathway is shown in [B]. A structure only loosely similar in function to the relaxosome formed in the origin of transfer region of conjugative plasmids is believed to be formed at or near the right border of T-DNA. Such a structure would include as a minimum, the VirC, D and E proteins. Of these the VirD2 protein is known to attach covalently to an end of one strand of T-DNA generated by a specific nick and the VirE2 protein to coat the transferred single strand. Both have plant nuclear localization signals which is consistent with their role in transmitting the DNA to the nucleus via the VirB conduit. The VirC protein may be involved in the synthesis in the bacterium of the strand complementory to the transferred strand. It is not known whether a complement exists in the plant cytoplasm or nucleus.

Information on stage [C] is scant. However, the T-DNA is inserted in any one of nuclear DNA regions by a mechanism that has been described as illegitimate recombination (presumably, double-stranded DNA will be required for this purpose). Plant gene products are likely to be involved although VirD2 could also participate. Following this recombinational insertion, the genes within the insert, whether natural or artificial are expressed in the same way as any other plant nuclear gene. Introns are not present.

role of the VirF and VirH proteins is not yet determined unambiguously. Evidence that the Vir D2 pilot protein and the Vir E2 single strand DNA binding proteins have a nuclear organization signal is consistent with their role in guiding T-DNA to plant nuclear DNA. Plant extra-nuclear proteins may also be involved in this targeted passage (Citovsky and Zambryski, 1993). What happens thereafter is obscure (Fig. 35.7C). We have little or no information on mechanisms of transfer termination and its relationship to integration into the plant DNA. However, since T-DNA is eventually found to be inserted in one or more of several regions in the plant nuclear DNA, some recombination events must intervene. The players in this process have not been identified. Our best understanding comes only from a few cases where the plant target sites have been sequenced both before and after T-DNA integration (Gheysen *et al.*, 1987; Mayerhofer *et al.*, 1991). These studies indicate that in the inserted T-DNA, the right border is nearly precise but the left is not. Furthermore, the target sequences are often observed to suffer rearrangements. Fig. 35.7 is an outline of the possible events that occur between the binding of bacteria to wound sites and the integration of T-DNA.

T-DNA genes determining phytohormone production

It will have been noted (Figs 35.5, 35.6) that in all naturally occurring Ti (or Ri) plasmids, the genes present in their respective T-DNA have evolved to function in plants and not in the bacterium. Prominent among these functions are those that result in the production of auxins, cytokinins or both. This results in the disease symptoms of tumorigenesis and of the phytohormone independence of transformed cell growth in culture, phenotypes that are associated with the prevention of plant regeneration from the transformed tissue. In early studies, a correlation was observed between the ability of *A. tumefaciens*-transformed tobacco cell cultures to regenerate into plants and the loss of a central portion of the T-DNA. This portion codes for auxin and cytokinin biosynthesis pathways novel to the plant. Since an important objective in using the

Agrobacterium system is to obtain transgenic plants, it is the general practice to 'disarm' natural or derived Ti plasmid systems by deleting from their T-DNA these genes that appear to interfere with normal plant morphogenesis. Examples of such 'disarmed' vectors will be considered in the next section.

Gene vectors

All systems involving plant genetic engineering with *Agrobacterium* rely on strategies for manipulating the bacterium and its plasmid(s) which in turn engineer the plant cells or plant. Since the first demonstration of the potential usefulness of this system, these strategies have become progressively simpler. The development of expression and selection systems for plant cells has also kept pace with this. These systems are summarized here in a practical rather than in a historical context. In general, an *Agrobacterium* strain carrying a disarmed and engineered Ti plasmid or one that carries the binary system composed of a deleted Ti plasmid providing the *vir* functions and a separate vector providing the engineered T-DNA are used. Figure 35.8 illustrates the basic attributes of *Agrobacterium* strains that provide either of these two systems. In binary systems, the *vir* region is *in trans* to the T-DNA region and is usually provided conveniently as a deleted derivative of a Ti plasmid. This acts essentially as the helper for the transmission of T-DNA and two such helper plasmids are pAL4404 which is derived from the octopine-inducing plasmid pTiAch5 (closely related to pTiA6) and pMP90RK derived from the nopaline-inducing plasmid pTiC58. Although each of the *vir* genes from one of the parental plasmids is replaceable by a corresponding gene from the other plasmid, there is evidence suggesting that the efficiencies of expression of each *vir* region are different. It may therefore be advantageous to use a particular helper for a specific purpose. A large number of existing plasmid vectors provide the T-DNA component as a binary system. As a minimum, these vectors contain (1) one or both the right and left borders; (2) a selectable dominant plant cell marker under the control of plant transcription signals; (3) a single replicon and markers for maintenance and selection in *Agrobacterium* and *E. coli* and (4) restriction sites suitably positioned to allow the cloning of desired plant sequences and their detection.

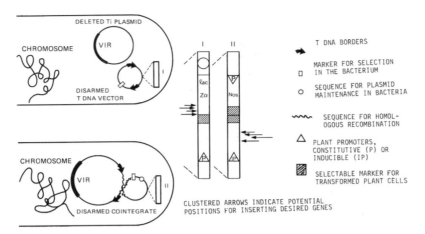

Fig. 35.8 General attributes of *Agrobacterium* plasmid constructs that are used in the genetic engineering of plants. The top figure of the bacterium illustrates a binary system, one in which the *vir* functions are provided *in trans* to the plasmid carrying the engineered T-DNA. In the bottom figure, these two essential regions are on the same molecule, usually the modified Ti plasmid that is refered to as the *cointegrate* because of the way in which it is usually constructed. In either situation, at least one border 25 bp sequence is essential (shown as thick shaded arrows). I and II illustrate two of several possible kinds of cassettes that can be engineered to be contained within the border sequence (and therefore within the T-DNA). The various symbols used are as follows:

thick shaded arrows:	T-DNA border sequences
small open rectangle:	marker that permits selection for stable maintenance of the construct in the bacterium (often *E. coli* and *Agrobacterium* species)
open circle:	sequence that permits replication of the plasmid in the bacterium. Note that in the binary system (top figure), this sequence is shown as part of the cassette I while in the cointegrate system (bottom figure) it is now within cassette II but still within the T-DNA borders. Both such systems can be used. The inclusion of such sequences within the T-DNA borders can allow recovery and recloning in bacteria of these regions from the transformed plant cells.
wavy line:	region of sequence similarity that permits cointegrate formation by homologous recombination
open triangles:	promoters functional in plants (P, constitutive; IP, inducible)
hatched rectangle:	dominant marker that is selectable for the transformed plant tissue
clustered arrows:	positions for inserting a desired gene (by *in vitro* methods)
*lac*Z:	portion of the *β*-galactosidase gene that permits assay by alpha-complementation
NOS:	Nopaline synthase (assayable in transformed plant tissue)

Additionally it is useful to incorporate into these vectors *ori*T regions (origins of transfer) that allow the facile transfer of the vectors across bacterial species boundaries and lambda *cos* (cohesive end) sites that enable the packaging of lambda DNA into lambda virions. The replicons in these vectors have been derived usually from the Incompatibility (Inc)P or IncQ plasmid groups of *E. coli* which have a broad bacterial host range. Virtually any plasmid marker conferring resistance to a bacterial antibiotic or inhibitor can be used as a selection marker to ensure plasmid maintenance in *Agrobacterium* before plant cell

transformation. For selection of plant cell transformants, several resistance markers have been used successfully but as indicated above in the section on Marker genes, the most appropriate tend to vary with the plant species. Advantages with the T-DNA vectors of binary systems is that they are manipulated relatively more easily, that they have a higher copy number and that they can be transmitted between bacteria more efficiently. It has also been observed that in some instances the binary system can be used effectively without eliminating or disarming indigenous Ti plasmids native to the *Agrobacterium* host. These observations and the availability of a range of selection systems make binary vectors the systems of choice at the present time.

Unlike vectors of the binary type, those of the co-integrate type incorporate a replicon that is not stably maintained in *Agrobacterium* and are therefore rescued by integration into the resident Ti plasmid (Fig. 35.8). The principle is to first engineer, using *in vitro* methods, a cassette containing desired genes and sequences into a well-characterized *E. coli* plasmid like pBR322. The acceptor Ti plasmid for the integration of this pBR322 derived intermediate is a disarmed one in which most of the central region of its T-DNA has been replaced by pBR322 sequences. Since the intermediate vector cannot replicate in *Agrobacterium*, it is rescued by a single recombination event within the pBR322 sequences to give rise to the co-integrate plasmid in which the genes to be transmitted into the plant are now contained within the right and left borders. The relative efficiency of cointegrate systems is less than that of binary systems but with the availability of good selection systems this has not been a serious disadvantage. With cointegrate systems, plasmid sequences outside those that are delimited by the T-DNA borders tend to be less often transmitted to the plant.

With both kinds of vector systems the simplest way to obtain transgenic plants is to infect tissue explants such as leaf discs with the manipulated *Agrobacterium* and to select for transformants in culture. *Agrobacterium*-mediated transformation has also been achieved with a variety of cell types including protoplasts, callus tissues and cells from suspension cultures. Generally the recipient cell or tissue source employed depends largely on their capabilities for plant regeneration which varies greatly among species. Although these strategies have been successful for a diverse array of plant species, T-DNA transfer may be extremely inefficient for other species, such as the cereals. As discussed in the following sections, a variety of alternative methods have therefore been developed.

Virus-mediated gene transfer

Plant viral genomes or their cDNAs can be cloned in bacterial plasmids and deliberately altered by genetic engineering. Host plants can be inoculated artificially without the invertebrate vectors that usually transmit the viruses in nature and extrachromosomal replication may generate high copy numbers in the infected plants. For these reasons, viruses are interesting candidates as vectors but strategies to employ them effectively must be developed. At this time a critical assessment of the utility of viral vectors is difficult. An enormous experimental effort is needed to understand the complex viral life cycles in relation to the organization of their genomes and the genes they encode (see Hohn and Schell, 1987 for further readings). Most plant viruses possess RNA genomes which makes genetic engineering more complicated. Despite this it has been shown that expression of the *cat* gene can be achieved in barley protoplasts from RNA prepared from a cDNA of brome mosaic virus (BMV). The caulimoviruses and geminiviruses are the two groups of plant DNA viruses. Among the caulimoviruses which have circular ds DNA genomes, the cauliflower mosaic virus has been most extensively explored as a vector. Replacement of a non-essential gene with the coding regions from bacterial dihydrofolate reductase or mammalian metallothionein have yielded functional expression of these proteins in turnip. Major limitations are the small size of DNA inserts that can be stably maintained when the recombinant genome is packaged as a virus particle and the narrow host-range of CaMV which is limited to the Cruciferae and some Solanaceae. The geminiviruses possess small circular ssDNA genomes. Collectively these viruses have a much wider host range which include a number of economically important monocotyledonous species; however, their use as vectors have yet to be demonstrated.

The integration of *Agrobacterium*-mediated transformation technology has greatly expanded the experimental opportunities for studying viruses. Complete and partial copies of viral genomes have been introduced into Ti plasmid vectors. The Ti plasmid-encoded mechanisms for transferring T-DNA into plant cells also mediates the efficient release of infectious viruses within plants if viral oligomers are used. This process can be achieved by simply inoculating plants with the bacteria. Alternatively, transgenic plants can also be produced with the same bacteria with virus being released from the integrated T-DNA. In both cases virulent Ti plasmids are required to transfer the viral genomes into plants. The term 'agroinfection' has been adopted to describe these phenomena.

Agroinfection studies have been employed widely. Studies of *Agrobacterium* host-range and Ti plasmid vir genes have been facilitated by capitalizing on virus amplification as a sensitive indicator of T-DNA transfer. The construction of transgenic plants that express specific genome components chromosomally permits detailed analyses by complementation and may generate new strategies for employing viral vectors. A number of agricultural applications are feasible and will be discussed in a later section.

DNA transfer without vectors

Over the years, several approaches have been used for introducing isolated DNA directly into plant cells without biological vectors. In most cases, evidence for transformation was ambiguous. Recently, clear evidence for transformation has been obtained with a variety of recipient cell types. Successful approaches include direct DNA uptake into protoplasts stimulated by chemical and/or electrical treatments; intranuclear microinjection of protoplasts having partially reformed cell walls and proembryos; fusion of protoplasts with liposomes that encapsulate DNA; bombardment of cells with gold or tungsten particles that are coated with DNA. Each of these procedures are designed to overcome the cell wall and plasma membrane as barriers to DNA entry and to protect the DNA from degradation by nucleases. Since these are essentially artificial mechanisms, they are not biologically restricted to specific species. The major limitation is the prerequisite for a system of plant regeneration from the transformed cells.

Microprojectile-mediated DNA transfer

This approach is complementary to *Agrobacterium*-mediated approaches and is perhaps the most versatile method for transferring nucleic acids into plant cells (Sanford, 1990). Due to the simplicity of the method it has recently become favored over other methods that do not employ *Agrobacterium*. Nucleic acids are first coated onto gold or tungsten particles

of 1–4 μm diameter by precipitation with $CaCl_2$, spermidine or polyethylene glycol. A variety of devices have been constructed which employ gun powder, compressed gases or electrical discharge to propel the particles at velocities sufficient to penetrate plant cell walls. Transformation of nuclear, chloroplast, and mitochondrial genomes have been achieved in this fashion. As the cell wall does not constitute a barrier to DNA transfer, a wide variety of regenerable cell types can be transformed. Consequently, the range of species for which transgenic plants can be produced has increased to include several members of the cereals and gymnosperms. When combined with sensitive reporter genes such as the GUS (*uid*A) gene, transient expression analyses can be rapidly performed with a wide variety of tissue and cell types, such as pollen, leaf or cultured cells. The major drawback of the method is the experimental variability that is inherent to the process.

Direct DNA uptake

Among the artificial mechanisms for protoplast transformation with isolated DNA, direct DNA uptake has been the most developed (Potrykus *et al.*, 1987). The uptake and integration of free or precipitated DNA is mediated by a number of factors. Often protoplast fusion treatments such as polyethylene glycol, polyvinyl alcohol and calcium combined with high pH enhances the transfer of DNA across the plasma membrane. Other factors such as high temperature treatments and electric pulses (electroporation) contribute significantly to this process. Generally, the kind and concentration of divalent cations are important presumably by mediating interactions with membranes and DNA or by protecting DNA from nucleases. Some protocols present the protoplasts with DNA packaged as a calcium phosphate coprecipitate before inducing uptake. Other factors that eventually affect the transformation frequencies include the use of carrier DNA and the DNA concentrations. The condition of the protoplasts is an important factor in these experiments. The cell cycle stage has been suggested as another factor. When the factors that have been identified so far are combined optimally for tobacco protoplasts, transformation frequencies may exceed 2% in the absence of selection. The optimal conditions differ with the source of the protoplasts and procedures adopted among laboratories vary widely.

The list of protoplast systems that have been transformed by these techniques is growing rapidly. Generally, the transformation frequencies are lower than reported for tobacco. Other Solanaceae include *Nicotiana plumbaginofolia*, *Petunia hybrida* and *Hyocyamus muticus*. Among the Brassiceae, *Brassica rapa* and *Brassica napus* protoplasts have been transformed using different procedures. Transformed cereal calli have been recovered by transformation of protoplasts of the Graminae species, *Lolium multiflorum*, *Triticum monococcum*, *Zea mays* and *Oryza sativa*. Clearly the transfer of DNA directly into protoplasts using a variety of procedures for stimulating uptake is generally applicable.

The integration of input DNA in the plant genome has been examined in tobacco tissues generated by the uptake of both Ti plasmids and bacterial vectors carrying selectable marker genes. In both cases, significant rearrangements occur presumably due to nuclease activity before integration. Concatenation of smaller plasmids is commonly observed; however, the number of copies is not correlated with the level of gene expression or antibiotic resistance conveyed by the marker genes. The integration into the plant genome appears to occur randomly by illegitimate or non-homologous recombination mechanisms governed by the plant cell. Certainly the virulence genes and T-DNA borders associated with Ti plasmids (see section on *Agrobacterium*-mediated gene transfer) do not contribute to integration when it is delivered as free DNA. Generally the DNA is integrated stably and it is maintained during meiosis. The chromosomal location appears to be random. Frequently the marker gene is inherited as a single Mendelian factor and in a few cases *in situ* hybridization has confirmed integration at a single locus on metaphase chromosomes. In studies with Ti plasmid DNA and calf thymus carrier DNA it was shown that different fragments could integrate at separate locations and segregate independently. Generally, however, cotransformation which is the transformation of nonselected DNA along with DNA carrying selectable marker genes occurs at a relatively high frequency.

Direct DNA uptake procedures have been used to demonstrate the transformation of tobacco with total genomic DNA taken from transgenic plants. Selection for kanamycin resistance yielded transformants at low frequencies of 10^{-6}. These experiments demonstrate clearly the feasibility of using these techniques for transferring genes or gene families that have not been isolated from their genomes.

Microinjection of DNA

Conceptually, microinjection is the most direct approach for introducing DNA into plant cells. The technology is unique in that it offers precise control over the form of the DNA that reaches the target site and the specific cells in culture that receive DNA (Miki *et al.*, 1987). Selectable marker genes are not essential for the recovery of transformants. Preliminary experiments with regenerating alfalfa and tobacco protoplasts indicate that intranuclear microinjection yield extremely high frequencies of transformation (14%); whereas, cytoplasmic injections are very inefficient. This observation parallels those made with animal cells in culture and implies that the transfer of free DNA from the cytoplasm to the nucleus is a significant barrier in the transformation process. As with direct DNA uptake, the microinjected DNA that is integrated may be rearranged; however, the extent to which this occurs may be less severe. Presumably delivery to the nucleus with glass capillaries protects the DNA from the extracellular nucleases and the complex chemical interactions that may alter the form of the DNA extensively before integration.

Among laboratories the technologies differ greatly depending on the systems employed for holding the protoplasts or cells, visualizing the intracellular target sites, as well as the equipment and methods for the identification and culture of microinjected cells. A major objective in the design of these systems is to obtain the maximum rates of injection because only small numbers of cells (200–300) can be manipulated in an experiment. A major limitation of this technology is that it can only be applied to cells for which efficient culture conditions have been determined.

Liposome-mediated DNA transfer

The encapsulation of DNA in artificial membrane-enclosed vacuoles called liposomes followed by transfer to protoplasts by polyethyleneglycol-induced fusion have yielded low transformation frequencies (10^{-5}; Deshayes *et al.*, 1985). Tandemly repeated copies of plasmid DNA carrying a selectable marker gene for kanamycin resistance appeared to integrate at a single locus in transgenic tobacco plants and segregated as a dominant genetic marker among progeny. The analyses of transformation paralleled those obtained by direct DNA uptake procedures.

Advantages and limitations

Among the different gene transfer systems that are now available for plants, that mediated by *Agrobacterium* and its Ti plasmids are relatively the best understood. Especially important have been the significant insights that have been obtained on the molecular biology of relevant genes and events in *Agrobacterium*. It is likely that these insights can be brought to bear on fundamental questions that remain outstanding and which have limited the full exploitation of this system. One can now confidently predict that this transfer system will be increasingly exploited to address fundamental problems in plant molecular biology. Although *Agrobacterium* is known to transfer genes to over a thousand plant species, a practical limitation that has to be considered is the apparent recalcitrance of some major crop species, especially the cereals. It is likely that to overcome this limitation a molecular understanding of factors that limit or extend host-range will be needed.

For the genetic engineering of plants, viral vectors have several limitations that must be considered. Viruses are generally poorly transmited by pollen or seed to progeny, therefore this approach is not practical for crop variety development. Distribution throughout a host plant may not be uniform. It may not be possible to dissociate viral disease symptoms from the vectors. Relatively high rates of recombination and error accumulation may generate instability. The advantage that transduction with viral vectors has over approaches for producing stable transgenic plants is that expression of specific gene products can be studied quickly after transduction without the need to regenerate plants in tissue culture.

A variety of artificial methods for DNA transfer and integration without vectors have now been developed and yield transgenic plants. Compared with tobacco leaf disc transformation by *Agrobacterium*, these approaches are generally less efficient and/or require special equipment. Furthermore, the pattern of DNA integration is less predictable than T-DNA delivered by *Agrobacterium* and the extent of rearrangement is much more severe. The clear advantage for the artificial delivery systems is that they bypass the host range restrictions of *Agrobacterium*. Currently, microprojectile-mediated delivery of DNA is very popular. This is largely due to the versatility and simplicity of the methods and the demonstrated potential to transform the mitochondria and chloroplast genomes as well as the nuclear genome.

Recently electroporation of protoplasts and microprojectile-mediated delivery to diverse tissues have been used to study transient expression of non-integrated DNA (Fromm and Walbot, 1987; Hauptmann *et al.*, 1987). The DNA is eventually lost after several days; however, the extent of DNA transfer is sufficient for the detection of reporter gene activity. Since viability but not cell division is a prerequisite for analyses of transient expression, the range of species that can be employed is greatly expanded over those that can be stably transformed. Transient expression can be used to analyze gene promotors in homologous plant systems for which regeneration technologies have not been developed yet.

Applications for basic studies

A direct consequence of the development of vectors and protocols for the transfer of genes into plants is that it is enabling us to address important questions concerning the organization of DNA sequences that determine varied aspects of plant gene expression and development. This has involved the construction of novel second generation vectors tailored to specific purposes. In the relatively short history of their development, they have been used to delve into questions that were not otherwise easily accessible to investigation. Only some examples illustrative of different applications are provided here.

Gene regulation

A common tactic is to reconstruct vectors that are composed of DNA sequences that are suspected to have a role in regulation with other reporter genes from which the regulatory sequences have been deleted and the expression of which can be reliably detected or measured. Such constructs may also contain the native gene sequences whose regulation is under scrutiny. In one early example the transcriptionally-active regulatory sequence of about 1 kbp upstream of the nuclear gene (*ss*) for the small subunit of ribulose 1,5-bisphosphate carboxylase of pea was fused to the coding sequence of the bacterial gene for chloramphenicol acetyl transferase (*cat*). The vector containing this artificial *ss*:*cat* fusion was then exploited to dissect and assess the complexity of this upstream *cis*-acting regulatory sequence (Fluhr *et al.*, 1986). This was done by using the vector to transmit the hybrid gene to heterologous plant cells (tobacco or petunia) which could then be regenerated into plants. The tissues and organelles of the transgenic plants could then be assayed for *cat* activity both with and without light induction. A similar strategy has been used to study regulatory sequences for the chlorophyll binding protein and for chalcone synthase, a key enzyme in flavanoid biosynthesis (Kaulen *et al.*, 1986; Sommer and Saedler, 1986). Other examples where this kind of approach is proving useful are for the analysis of organ-specific and wound-inducible genes (Sanchez-Serrano *et al.*, 1987; Thornburg *et al.*, 1987), and genes inducible by symbiotic root nodulating bacteria (Jensen *et al.*, 1985).

Intracellular trafficking

A second set of questions for which vector constructs are proving to be of value relates to the rules of protein localization and the targeting of proteins into organelles such as chloroplasts. Such import is usually associated with the presence of a hydrophobic transit peptide at the amino terminal end of a protein, the transit peptide being usually cleaved during import. Vectors in which the transit peptide alone or along with portions of the mature proteins have been fused to the bacterial neomycin phosphotransferase gene II are used to study aspects of targeting and translocation both *in vivo* and *in vitro* (Van den Broeck *et al.*, 1985; Smeekens *et al.*, 1987).

T-DNA tagging

A unique application of the T-DNA of the Ti plasmid of *Agrobacterium* is in its use both as an insertion element to uncover plant genes by mutation and also to create transcriptional, and translational fusions with plant genes (Koncz *et al.*, 1992). Promoter- and enhancer-trap experiments have been conducted with T-DNA vectors that contain reporter genes which lack promoters or enhancers essential for activity. Once inserted into the plant genome the reporter gene may be activated by neighboring regulatory sequences within the plant DNA. This event appears to occur at high frequencies indicating that T-DNA may preferentially integrate within transcribed chromatin of plants such as *Arabidopsis* or tobacco. Novel promoters have been isolated through such approaches. In *Arabidopsis*, T-DNA insertions have also generated a wide range of mutants and have led to the isolation of novel genes such as those involved in the regulation of development. Another approach involves the introduction of transcriptional enhancers alone into the plant genome. Insertion can activate neighbouring genes and through selection specific classes of genes can be identified. T-DNA tagging has become an important method for isolating novel plant genes or their regulatory sequences and for studying the organization of the plant genome.

Gene targeting

In plants, the integration of introduced DNA is predominantly through the mechanisms of illegitimate recombination, therefore, long stretches of homologous DNA are not required. Homologous recombination is known to occur in plants but it has not been characterized as well as in other organisms, such as mammals (Puchta *et al.*, 1994). For example, targeted gene replacement in mouse using cloned homologous genes is being used to create mutants thereby greatly accelerating the study of mammalian molecular genetics (Capecchi, 1994). In tobacco, a mutant acetolactate synthase was targeted to a resident gene using a T-DNA vector (Lee *et al.*, 1990) demonstrating the potential for similar experiments in plants. The technology, however, is not easily implemented at this time and most experiments have been conducted with selectable marker genes. It seems that the frequency of gene targeting relative to illegitimate recombination is very low (10^{-4}) but it can be achieved by both direct DNA uptake and *Agrobacterium*-mediated transformation. Advances in these technologies will have a dramatic impact in determining the functions of the many plant genes that have been cloned but as yet remain poorly understood.

A popular alternative for generating mutants with cloned genes is the random insertion of constructs in which the coding region of the gene is placed in the antisense configuration. Transcription of antisense RNA may reduce the mRNA levels dramatically and generate phenotypes related to deficiency of the gene product.

Novel genes

Another particular application exploits the normal presence within the T-DNA, of genes for novel pathways for auxin and cytokinin biosynthesis that are not normally present in plants. For example, it has been illustrated that by using mutations of appropriate phytohormone genes of the T-DNA, it is possible to obtain transgenic tobacco and petunia plants with altered levels of auxin in their tissues (Klee *et al.*, 1987). If the implications of these recent observations are fully realized through further experimentation and predictable technology, they could have a major impact on understanding the role of these phytohormones in the control of plant development.

All applications that were considered here involve the stable integration of transferred DNA into the plant nuclear genome. Recently techniques have also become available for studying the transient expression of transferred genes over periods in which they have not been stably integrated into the genomes of plant cells (Fromm and Walbot, 1987). In the near future, it is anticipated that the experimental capabilities described here will form the basis for more advanced genetic manipulations. Cloning of certain plant genes in plant cells is reasonable to expect. Deliberate transformation of specific organelles other than the nucleus has been achieved. These and other technological advances will have a profound effect on our understanding of basic genetic control mechanisms in plants.

Applications for agriculture

Many examples of transgenic plants that appear to be normal in all respects have been documented. Among these the number of crop species is expanding rapidly; therefore, it is feasible to incorporate transgenic plants into breeding programs for crop improvements. The next challenge is the identification and isolation of new genes that will have an impact on the quality and productivity of a particular crop. Current technological capabilities do not limit the source of these genes to plants or to natural origins. A comprehensive examination of agricultural applications is beyond the scope of this chapter, therefore, only examples that demonstrate some fundamental principles will be discussed.

Herbicide resistance

Significant progress has been made in the genetic engineering for resistance to broad spectrum herbicides. Compared with other agronomically important characteristics, the biochemical basis for herbicide sensitivity and resistance has been well studied and the number of genes that are involved may be limited. Generally, the herbicides inhibit key biological processes or specific enzymes in metabolic pathways unique to plants. At least three basic genetic principles that confer resistance have been documented for nuclear genes. These are gene amplifications resulting in over-production of target enzymes; mutations that alter herbicide binding to enzymes; and expression of genes that code for enzymes which detoxify herbicides. Experimental exploitation of these principles in transgenic plants has yielded resistance to a number of herbicide classes. For instance, mutant genes from bacteria and other plants have been used to substitute for resident plant genes coding for enzymes in key metabolic pathways. By this means, resistance to the sulfonylurea herbicides and glyphosate has been achieved. Bacterial genes have been used to provide a mechanism for the detoxification of another herbicide, phosphinothricin, in transgenic plants.

Pest resistance

Similar strategies are being examined for the protection of crops from insect damage. Preliminary analyses have been performed with a gene coding for the toxic region of the insecticidal protein of the bacterium *Bacillus thuringiensis*. Preparations of the bacterium have been used as commercial insecticides for the control of insect larvae and a number of strains exist with selective toxicity to Lepidopteran, Dipteran and Coleopteran species. Under controlled conditions, expression of the Lepidopteran protein in tobacco and tomato was toxic to feeding larvae thereby protecting the transgenic plants from damage (Fischhoff *et al.*, 1987). Other approaches include the use of genes coding for proteins such as trypsin inhibitor that will interfere with insect digestive enzymes.

Disease resistance

Mechanisms for the protection of transgenic plants from viral diseases are being examined. The phenomenon of viral cross protection in which inoculation with mild strains prevent damage from more virulent strains can be mimicked by the expression of cloned viral coat protein genes in transgenic plants such as tobacco and tomato (Powell *et al.*, 1986). Studies with tobacco mosaic virus (TMV) and alfalfa mosaic virus (AMV) have shown that the appearance of disease symptoms can be significantly delayed or inhibited by this

approach. A separate mechanism that suppresses disease symptoms is the expression in transgenic plants of the satellite RNA of viruses such as cucumber mosaic virus (CMV). Genetic transformation provides a tool for studying the principles responsible for these phenomenon. As knowledge is acquired the strategies for imparting viral protection will be improved.

The applications of gene transfer technology to agriculture is limited only by our understanding of the fundamental genetic principles responsible for crop quality and production. The examples offered above serve to illustrate a few of the many genetic manipulations that have been shown to be feasible. An examination of the chapters preceding and following this one reveal other areas in which genetic transformation technology will serve to acquire knowledge and eventually for genetic manipulation for agricultural objectives. The efficiencies and sophistication of the technologies will undoubtedly improve rapidly as these experiments progress since this field of research is still in the early stages of development.

Acknowledgements

We are grateful to colleagues and students for thoughtful comments on this chapter. Our research is supported by Agriculture Canada and the Natural Science and Engineering Research Council of Canada.

Further reading

Evans, D. A., Sharp, W. R. and Ammirato, P. V. (1986). *Handbook of Plant Cell Culture*, Vols 1–5, Macmillan Publishing Company, New York; Collier Macmillan Publishers, London.

Hohn, T. and Schell, J. (1987). *Plant DNA Infectious Agents*, Springer-Verlag, New York, Vienna.

Kahl, G. and Schell, J. (1982). *Molecular Biology of Plant Tumors*, Academic Press, New York.

Ream, W. (1989). *Agrobacterium tumefaciens* and interkingdom genetic exchange. *Ann. Rev. Phytopathol.* **27**, 583–618.

Wu, R. and Grossman, L. (1987). *Recombinant DNA* Part D *Methods in Enzymology*, Vol. 153, Academic Press, London.

References

Berger, B. R. and Christie, P. J. (1994). Genetic complementation analysis of the *Agrobacterium tumefaciens vir*B operon: *vir*B2 through *vir*B11 are essential virulence genes. *J. Bacteriol.* **176**, 3646–60.

Capecchi, M. R. (1994). Targeted gene replacement. *Sci. Am.* March, 52–9.

Charles, T. C., Shouguang, Jin and Nester, E. W. (1992). Two-component sensory transduction systems in Phytobacteria. *Ann. Rev. Phytopathol.* **30**, 463–84.

Citovsky, V. and Zambryski, P. (1993). Transport of nucleic acids through membrane channels: snaking through small holes. *Ann. Rev. Microbiol.* **47**, 167–97.

Citovsky, V., Zupan, J., Warnick, D. and Zambryski, P. (1992). Nuclear localization of *Agrobacterium* VirE2 protein in plant cells. *Science* **256**, 1803–5.

Deshayes, A., Herrera-Estrella, L. and Caboche, M. (1985). Liposome-mediated transformation of tobacco mesophyll protoplasts by an *Escherichia coli* plasmid. *EMBO J.* **4**, 2731–7.

Eichholtz, D. A., Rogers, S. G., Horsch, R. B., Klee, H. J., Hayford, M., Hoffman, N. L., Braford, S. B., Fink, C., Flick, J., O'Connell, K. M. and Fraley, R. T. (1987). Expression of mouse dihydrofolate reductase gene confers methotrexate resistance in transgenic petunia plants. *Somat. Cell Molec. Genet.* **13**, 67–76.

Filichkin, S. A. and Gelvin, S. B. (1993). Formation of a putative relaxation intermediate during T-DNA processing directed by the *Agrobacterium tumefaciens* VirD1, D2 endonuclease. *Mol. Microbiol.* **8**, 915–26.

Fischhoff, D. A., Bowdish, K. S., Perlak, F. J., Marrone, P. G., McCormick, S. M., Niedermeyer, J. G., Dean, D. A., Kusano-Kretzmer, K., Mayer, E. J., Rochester, D. E., Rogers, S. G. and Fraley, R. T. (1987). Insect tolerant transgenic tomato plants. *Bio/Technology* **5**, 807–13.

Fluhr, R., Kuhlemeier, C., Nagy, F. and Chua, N.-H. (1986). Organ-specific and light-induced expression of plant genes. *Science* **232**, 1106–12.

Fromm, M. and Walbot, V. (1987). Transient expression of DNA in plant cells. In *Plant DNA Infectious Agents*, eds T. Hohn and J. Schell, Springer-Verlag, Vienna, pp. 303–10.

Gheysen, G., Van Montagu, M. and Zambryski, P. (1987). Integration of *Agrobacterium tumefaciens* transfer DNA (T-DNA) involves rearrangements of target plant DNA sequences. *Proc. Natl. Acad. Sci. USA* **84**, 6169–73.

Hauptmann, R. M., Ozias-Akins, P., Vasil, V., Tabaeizadeh, Z., Rogers, S. G., Horsch, R. B., Vasil, I. K. and Fraley, R. T. (1987). Transient expression of electroporated DNA in monocotyledonous and dicotyledonous species. *Plant Cell Rep.* **6**, 265–70.

Herrera-Estrella, L., De Block, M., Messens, E.,

Hernalsteens, J.-P., Van Montagu, M. and Schell, J. (1983). Chimeric genes as dominant selectable markers in plant cells. *EMBO J.* **2**, 987–95.

Herrera-Estrella, A., Chen, Z., VanMontagu, M. and Wang, K. (1988). Vir D proteins of *Agrobacterium tumefaciens* are required for the formation of a covalent DNA-protein complex at the 5' terminus of t-strand molecules. *EMBO J.* **7**, 4055–62.

Hille, J., Verheggen, F., Roelvink, P., Franssen, H., van Kammen, A. and Zabel, P. (1986). Bleomycin resistance: a new dominant selectable marker for plant cell transformation. *Plant Mol. Biol.* **7**, 171–6.

Holbrook, L. A., Haffner, M. and Miki, B. L. (1986). A sensitive fluorographic method for the detection of nopaline and octopine synthase activities in *Brassica* crown gall tissues. *Biochem. Cell Biol.* **64**, 126–32.

Howard, E. A., Winsor, B. A., DeVos, G. and Zambryski, P. (1989). Activation of the T-DNA transfer process in *Agrobacterium* results in the generation of a T-strand-protein complex: Tight association of Vir D2 with the 5' ends of T-strands. *Proc. Natl. Acad. Sci. USA* **86**, 4017–21.

Jefferson, R. A., Kavanagh, T. A. and Bevan, M. W. (1987). Gus fusions: β-glucoronidase as a sensitive and versatile gene fusion marker in higher plants. *EMBO J.* **6**, 3901–7.

Jensen, J. S., Marcker, K. A., Otten, L. and Schell, J. (1986). Nodule-specific expression of a chimaeric soyabean leghaemoglobin gene in transgenic *Lotus corniculatus*. *Nature* **321**, 669–74.

Kado, C. I. (1991). Molecular mechanisms of crown gall tumorogenesis. *Crit. Rev. Plant Sci.* **10**, 1–32.

Kado, C. I. (1994). Promiscuous DNA transfer system of *Agrobacterium tumefaciens*: role of the *vir*B operon in sex pilus assembly and synthesis. *Mol. Microbiol.* **12**, 17–22.

Kaulen, H., Schell, J. and Kreuzaler, F. (1986). Light-induced expression of the chimeric chalcone synthase-NPTII gene in tobacco cells. *EMBO J.* **5**, 1–8.

Kemp, J. D. (1982). Enzymes in octopine and nopaline metabolism. In *Molecular Biology of Plant Tumors*, eds G. Kahl and J. S. Schell, Academic Press, New York, pp. 461–74.

Klee, H., Horsch, R. B., Hinchee, M. A., Hein, M. B. and Hoffman, N. L. (1987). The effects of overproduction of two *Agrobacterium tumefaciens* T-DNA auxin biosynthetic gene products in transgenic petunia plants. *Genes and Devel.* **1**, 86–96.

Koncz, C., Németh, K., Rédei, G. P. and Schell, J. (1992). T-DNA insertional mutagenesis in *Arabidopsis*. *Plant Mol. Biol.* **20**, 963–76.

Koukolikova-Nicola, Z., Albright, L. and Hohn, B. (1987). The mechanism of T-DNA transfer from *Agrobacterium tumefaciens* to the plant cell. In *Plant DNA Infectious Agents*, eds T. Hohn and J. Schell, Springer-Verlag, Vienna, pp. 109–48.

Kuldau, G. A., DeVos, G., Owen, J., McCaffrey, G. and Zambryski, P. (1990). The *vir*B operon of *Agrobacterium* *tumefaciens* pTiC58 encodes 11 open reading frames. *Mol. Gen. Genet.* **221**, 256–66.

Lee, K. Y., Lund, P., Lowe, K. and Dunsmuir P. (1990). Homologous recombination in plant cells after *Agrobacterium*-mediated transformation. *Plant Cell* **2**, 415–25.

Mayerhofer, R., Koncz-Kalman, Z., Nawrath, C., Bakkeren, G., Crameri, A., Angelis, K., Redei, G. P., Schell, J., Hohn, B. and Koncz, C. (1991). T-DNA integration: a mode of illegitimate recombination in plants. *EMBO J.* **10**, 697–704.

Miki, B. L. A., Reich, T. J. and Iyer, V. N. (1987). Microinjection: An experimental tool for studying and modifying plant cells. In *Plant DNA infectious agents*, eds T. Hohn and J. Schell. Springer-Verlag, Vienna, pp. 248–65.

Ow, D. W., Wood, K. V., DeLuca, M., DeWet, J. R., Helinski, D. R. and Howell, S. H. (1986). Transient and stable expression of the firefly luciferase gene in plant cells and transgenic plants. *Science* **234**, 856–9.

Pansegrau, W., Schoumacher, F., Hohn, B. and Lanka, E. (1993). Site-specific cleavage and joining of single-stranded DNA by VirD2 protein of *Agrobacterium tumefaciens* Ti plasmids: Analogy to bacterial conjugation. *Proc. Natl. Acad. Sci. USA* **90**, 11538–42.

Potrykus, I., Paszkowski, J., Shillito, R. D. and Saul, M. W. (1987). Direct gene transfer to plants. In *Plant DNA Infectious agents*, eds T. Hohn and J. Schell, Springer-Verlag, Vienna, pp. 229–47.

Peralta, E. G., Helmiss, R. and Ream, L. W. (1986). Overdrive, a T-DNA transmission enhancer on the *A. tumefaciens* tumor-inducing plasmid. *EMBO J.* **5**, 1137–42.

Powell Abel, P., Nelson, R. S., De, B., Hoffman, N., Rogers, S. G., Fraley, R. T. and Beachy, R. N. (1986). Delay of disease development in transgenic plants that express the tobacco mosaic virus coat protein gene. *Science* **232**, 738–43.

Puchta, H., Swoboda, P. and Hohn, B. (1994). Homologous recombination in plants. *Experientia* **50**, 277–84.

Ream, W. (1989). *Agrobacterium tumefaciens* and interkingdom genetic exchange. *Ann. Rev. Phytopathol.* **27**, 583–618.

Sanchez-Serrano, J. J., Keil, M., O'Connor, A., Schell, J. and Willmitzer, L. (1987). Wound-induced expression of a potato proteinase inhibitor II gene in transgenic tobacco plants. *EMBO J.* **6**, 303–6.

Sanford, J. C. (1990). Biolistic plant transformation. *Physiol. Plant* **79**, 206–9.

Smeekens, S., van Steeg, H., Bauerle, C., Bettenbroek, H., Keegstra, K. and Weisbeek, P. (1987). Import into chloroplasts of a yeast mitochondrial protein directed by ferredoxin and plastocyanin transit peptides. *Plant Molec. Biol.* **9**, 377–88.

Sommer, H. and Saedler, H. (1986). Structure of the Chalcone synthase gene of *Antirrhinum majus*. *Mol. Gen. Genet.* **202**, 429–34.

Sen, P., Pazour, G. J., Anderson, D. and Das, A. (1989). Cooperative binding of *Agrobacterium tumefaciens* Vir E2 protein to single-stranded DNA. *J. Bacteriol.* **171**, 2573–80.

Thornburg, R. W., An, G., Cleveland, T. E., Johnson, R. and Ryan, C. A. (1987). Wound-inducible expression of a potato inhibitor II-chloramphenicol acetyl transferase gene fusion in transgenic plants. *Proc. Natl. Acad. Sci. USA* **84**, 744–8.

Toro, N., Dalta, A., Carmi, O. A., Young, C., Prusti, P. K. and Nester, E. W. (1989). The *Agrobacterium tumefaciens virC1* gene product binds to overdrive, a T-DNA transfer enhancer. *J. Bacteriol.* **171**, 6845–9.

Van den Broeck, G., Timko, M. P., Kausch, A. P., Cashmore, A. R., Van Montagu, M. and Herrera-Estrella, L. (1985). Targeting of a foreign protein to chloroplasts by fusion to the transit peptide from the small subunit of ribulose 1,5-bisphosphate carboxylase. *Nature* **313**, 358–63.

Waldron, C., Murphy, E. B., Roberts, J. L., Gustafson, G. D., Armour, S. L. and Malcolm, S. K. (1985). Resistance to hygromycin B. A new marker for plant transformation studies. *Plant Molec. Biol.* **5**, 103–8.

Winans, S. C. (1992). Two-way chemical signaling in *Agrobacterium* – plant interactions. *Microbiol. Revs.* **56**, 12–21.

Zambryski, P. (1988). Basic processes underlying *Agrobacterium* – mediated DNA transfer to plant cells. *Ann. Rev. Genet.* **22**, 1–30.

36 The manipulation of resource allocation in plants

Stephen Blakeley

Introduction

Crop plants accumulate carbohydrates, lipids and proteins in a variety of storage organs including seeds, roots and tubers. The relative quantity of these storage compounds varies considerably and these differences determine, to a large extent, the utility of the crop (Table 36.1). Accumulation of storage material in these sink tissues depends upon photosynthetic carbon fixation in the leaf, or source tissue, the rate of translocation of photoassimilate and the simultaneous assimilation and delivery of nitrogen containing compounds. Various metabolic pathways within the developing seed, fruit or tuber will then compete for these substrates leading, ultimately, to the deposition of a characteristic array of storage materials. Photosynthesis, nitrogen assimilation and the long distance transport of nitrogen and carbon have been covered in detail in other chapters of this book. Here, recent attempts to manipulate resource allocation in plants will be described and the potential for enhancement of the nutritional or economic value of crop species will be discussed.

Fundamentals of plant metabolism

Before attempting to manipulate plant metabolism, it is necessary to review the fundamental pathways through which the majority of storage compounds are produced. Glycolysis, gluconeogenesis and the pentose phosphate and shikimate pathways are at the core of the allocation of carbon to the various biosynthetic pathways (Figs 36.1, 36.2). Since several of the intermediates in these pathways are also utilized for amino acid biosynthesis, they play a key role in determining the distribution of photoassimilate between the carbon-based storage compounds, such as starch and oil, and the nitrogen-containing storage proteins.

A feature of plants that distinguishes them from other organisms is the sequestration of many biosynthetic pathways into plastids (Chapter 6; Dennis and Miernyk, 1982). These organelles are present in all plant tissues but their functions vary considerably with one activity being dominant, depending on the specialized role of the tissue (Kirk and Tilney-Bassett, 1978). Thus, in leaves, chloroplasts are specialized to perform photosynthesis and contain the enzymes of the Calvin cycle. In tubers and cereal grains, amyloplasts are responsible for the biosynthesis of starch whereas oilseeds contain leucoplasts, the site of fatty acid synthesis (Fig. 36.2) (Blakeley and Dennis, 1993; Dennis and Blakeley, 1993). Despite these differences, all plastids within a plant are derived from the proplastid and contain the

Table 36.1 Chemical composition of some seeds of economic importance.

| | % Dry weight | | |
	Carbohydrate	Protein	Lipid
Barley (*Hordeum vulgare*)	80	9	1
Corn (*Zea mays*)	84	10	5
Oat (*Avena sativa*)	77	13	5
Peanut (*Arachis hypogaea*)	27	27	45
Rape (*Brassica napus*)	25	23	48
Rice (*Oryza sativa*)	88	8	2
Soybean (*Glycine max*)	38	38	20
Wheat (*Triticum esculentum*)	82	14	2

From Sinclair and deWit, 1975.

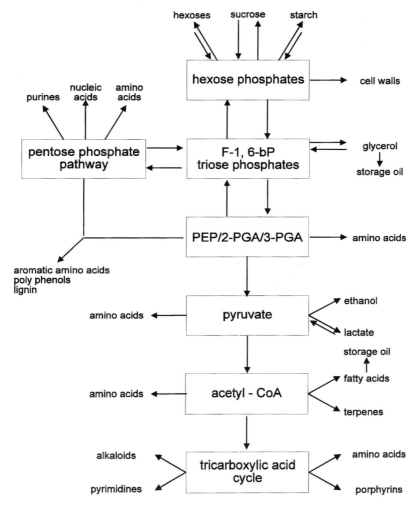

Fig. 36.1 Schematic representation of the glycolytic and pentose phosphate pathways to illustrate their central role in the allocation of carbon for biosynthesis. Abbreviations: F1,6bP, fructose 1,6 bisphosphate; PEP, phospho*enol*pyruvate; 2-PGA, 2-phosphoglycerate; 3-PGA, 3-phosphoglycerate. Reproduced in modified form from Blakeley and Dennis (1993) with permission from the *Canadian Journal of Botany*.

same genome (Chapter 17; Kirk and Tilney-Bassett, 1978). Most amino acids are also synthesized in plastids and the flow of carbon and nitrogen into storage compounds therefore requires interaction between different tissues and also between the subcellular compartments within them.

If Figs 36.1 and 36.2 are studied together, it can be seen that there are certain key points in glycolysis at which intermediates are withdrawn for the synthesis

of storage material or to enter other metabolic pathways such as the oxidative pentose phosphate pathway or the citric acid cycle. Competition between the different biosynthetic pathways will ultimately define the composition of storage material deposited in a particular tissue. In addition, the energy and cofactor requirements of biosynthetic pathways differ and the availability of ATP and reductant will also influence the final composition of the storage organ.

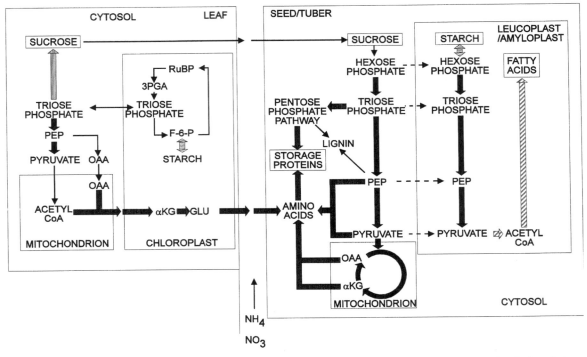

Fig. 36.2 Schematic representation of the major metabolic and transport pathways in plants. Pathways are glycolysis and the TCA cycle ➡️, and those leading to starch ⇨, fatty acids ⬈, amino acids ⇨ and sucrose ➡️.

In tubers and storage grains, for example, sucrose, translocated in the phloem, is converted into hexose phosphates which subsequently enter the amyloplast and are converted into starch through the action of ADP-glucose pyrophosphorylase. Starch biosynthesis requires only one molecule of ATP per hexose (6-carbon) and is thus a relatively inexpensive way for plants to produce storage material (Fig. 36.3). In oilseeds, sucrose can be metabolized by either cytosolic or plastid glycolysis. Fatty acid biosynthesis in leucoplasts requires a supply of carbon and also energy, in the form of ATP, and reducing power as NAD(P)H. In fact, for every 3 acetate molecules (6-carbon equivalents) converted into oil, 6 ATP and 6 NAD(P)H are required. Fatty acids are, therefore, expensive to produce and oil biosynthesis requires the operation of the glycolytic pathway within the leucoplast to maintain a supply of cofactors within the organelle (Dennis and Blakeley, 1993).

The metabolic expense incurred during the

production of different storage compounds can, perhaps, be related to the composition of some seeds at maturity. For example, barley and corn deposit approximately 80% of their storage reserves as starch and contain very little protein or oil (Table 36.1). Since starch biosynthesis requires a modest input of energy, it is possible to utilize most of the photoassimilate arriving at the seed for the production of this single storage compound. In oilseeds, such as peanut and rape, however, only 50% of storage material is produced as oil, and the remaining carbon is stored as starch, or in association with nitrogen, in storage proteins. Since oil biosynthesis requires a large input of energy and reducing power it is possible that the developing seed cannot generate sufficient quantities of these commodities to support further fatty acid biosynthesis. Residual photoassimilate in these seeds may then be deposited as starch or in storage proteins. The presence of some starch in oilseeds also

STARCH SYNTHESIS

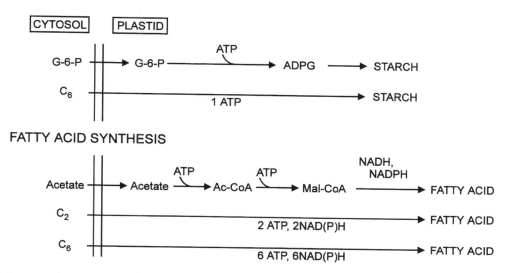

Fig. 36.3 Energy and co-factor requirements for starch and long chain fatty acid biosynthesis within plastids in plants. Reproduced from Dennis and Blakeley (1993) by permission of Oxford University Press.

provides the seed with a readily accessible source of energy during germination.

Plant metabolism in developing storage organs can, therefore, be viewed as a competition between central pathways for certain key metabolic intermediates. For example, during storage protein biosynthesis in seeds, *de novo* synthesis of amino acids within the seed will both complement the supply of amino acids from the phloem and, at the same time, drain resources from glycolysis and the pentose phosphate pathway. In addition, the production of storage protein also has a large requirement for ATP so the intermediates of glycolysis will be further depleted as pyruvate enters the mitochondrion. As a consequence, fewer carbon-based metabolic intermediates will be available for starch and oil biosynthesis. The final distribution of carbon and nitrogen derived from a variety of compounds transported in the xylem and phloem will depend on factors such as the affinity of the enzymes of these pathways for their substrates, the concentration of enzyme and substrate and the ability of metabolic intermediates to travel between the different cellular compartments.

Before attempting to manipulate the distribution of storage compounds by modification of the above

pathways, a point in metabolism must be selected at which an alteration in the regulatory properties of an enzyme, or an increase or decrease in it's activity, will cause the desired effect. Initially, this may seem like a relatively straightforward task since biochemists have identified the key enzymes in metabolism and one would presume that manipulation of the level of these enzymes would have a profound effect on the resulting allocation of resources. In practice, this is not always the case. These experiments have, however, led to both a greater understanding of the differences between plant metabolism and that of other organisms and to a number of guidelines to consider prior to embarking on a genetic approach to manipulation of resource allocation in plants.

Manipulation of metabolism

Details of the methods available for the transformation of plants are available in Chapter 35. These methods allow us to introduce foreign genetic material into plant cells and these cells can subsequently be regenerated into reproductively competent plants. For the purposes of metabolic manipulation, these methods allow the increase or decrease of a specific enzyme activity or, alternately, the introduction of an enzyme or isozyme in a compartment in which it is not normally found. The genes to be introduced, or transgenes, are often coupled to segments of DNA encoding peptides that target the product of the transgene to a specific subcellular location. A promoter is also chosen to ensure that the transgene is expressed at the appropriate developmental stage.

For the transgene to influence metabolism it must be expressed and, despite the attachment of appropriate regulatory elements, it is not always easy to predict the extent to which this will occur. Firstly, integration of the transferred DNA into the host genome is random and there may be rearrangement of the inserted DNA. The expression of the transgene can also be influenced by where it has integrated into the host genome. This depends, for example, on whether the insertion occurs in a region of the chromosome that is constitutively transcribed or one in which the genes are regulated during development. If the latter occurs it is possible that the transgene will be expressed in an unpredictable way as the plant regenerates. In some cases novel phenomena, notably co-suppression, have been observed (Gottlob-McHugh *et al.*, 1992; Jorgensen, 1990). In these studies, transgenic plants have been generated in which expression of both the transgene and the corresponding host gene are inhibited simultaneously.

In addition to these technical considerations, it will often be necessary to target the product of a transgene to a specific subcellular location. Plant cells contain distinct isozymes in different compartments. These isozymes have similar primary structures and catalyze identical reactions but often differ in their regulatory or kinetic properties. It may, therefore, be necessary to identify exactly which gene has been cloned, and where the product of this gene is localized *in vivo*. Since the majority of proteins found within a plastid are encoded in the nuclear genome and imported into the organelle from the cytosol (Chapter 24; Keegstra, 1989) it will be necessary to test the efficiency of this process *in vitro* to ensure that any planned intervention in plastid metabolism does not fail simply because the product of the transgene is incorrectly targeted. Selection of the gene for the correct isozyme is very important when the

goal is to inactivate a specific enzyme activity in a specific compartment using an antisense strategy. On the other hand, when an increase in activity is desired, it may be advantageous to express the gene for an isozyme not normally found in that compartment, or even a non-plant enzyme, thereby bypassing the regulatory properties of the resident enzyme and reducing the possibility of cosuppression (Jorgenson, 1990).

The choice of enzyme may not always be obvious. Metabolic pathways have both highly regulated steps and those that are essentially at equilibrium. Since the activity of some enzymes is regulated by metabolic intermediates within the pathway in which they function, the artificial perturbation of one step may have an undesirable effect on another part of the pathway. In addition, the control analysis theory (Chapter 37; Kacser and Burns, 1973), reviewed in ap Rees and Hill (1994) indicates that all enzymes in a pathway may actually exert some control on flux through the pathway. Indeed, recent studies have indicated that the level of enzymes which were thought to be present in great excess, and at equilibrium, can influence the mass action ratio of substrate to product of the reaction they catalyze (Kruckeberg *et al.*, 1989; Neuhaus and Stitt, 1990; Stitt *et al.*, 1991). This indicates that, *in vivo,* these enzymes are not in excess but at a concentration sufficient to maintain an adequate flux of metabolites through the pathway. The results of these studies demonstrate that, to affect metabolism, it is not necessarily sufficient to manipulate the levels of enzymes that have the most profound regulatory properties *in vitro.*

Since plants are immobile they cannot avoid stresses such as nutrient deficiency, salinity, temperature or drought, and have evolved a highly flexible metabolism. Hence, there are several seemingly redundant enzymes which have probably been retained to bypass more conventional metabolic processes when conditions are unfavourable. Thus, it has been possible to completely eliminate the activity of several enzymes of primary carbon metabolism without having a discernable effect on the plant (Gottlob-McHugh *et al.*, 1992; Hajirezaei *et al.*, 1994). These studies have demonstrated that it may be wise to select enzymes for manipulation at steps in pathways at which there are no obvious bypass reactions.

Thus, although the basic framework of carbon metabolism in plants resembles that of other organisms, there are also significant differences. Several of these differences have been identified unexpectedly through the creation of transgenic plants in which the metabolic modification has been tolerated even though it would be fatal in other organisms. In many cases the plants show no apparent phenotype. In the remainder of this chapter, some recent achievements in the manipulation of plant metabolism will be reviewed and the future potential for the creation of improved crop plants will be discussed.

Glycolysis

Pools of metabolites

Glycolysis is at the centre of the allocation of carbon to storage compounds and because of this has been targeted as a pathway to manipulate. The intermediates of glycolysis can be divided into pools of metabolites (Fig. 36.1) (Dennis, 1987). The first of these pools, the hexose-phosphates, are at equilibrium through the action of hexose phosphate isomerase and phosphoglucomutase. Metabolites enter this pool through sucrose breakdown, phosphorylation of free hexoses, or via gluconeogenesis from trioses. Alternatively, starch or sucrose synthesis and glycolysis will deplete this pool. The demonstration, at least in amyloplasts, that hexose phosphates can cross the plastid envelope (Keeling *et al.*, 1988; Hill and Smith, 1991) suggests that the hexose phosphate pools in the cytosol and plastid are not entirely separate. The second pool, the intermediates of the oxidative pentose phosphate pathway are in equilibrium with the trioses, dihydroxyacetone phosphate and glyceraldehyde-3-phosphate. This equilibrium also serves to interconnect glycolysis and the pentose phosphate pathway. Depletion of this pool occurs through biosynthesis of glycerol for storage oils and phospholipids, nucleic acids, aromatic amino acids and through the lower part of glycolysis. The phosphate translocator in the plastid envelope serves to equilibrate triose-phosphates and phosphate in the cytosol and plastid. The lower reactions of the glycolytic pathway generate the third pool of metabolites, between glyceraldehyde-3-phosphate and PEP. In this pool, 3-phosphoglycerate is the precursor for serine biosynthesis and PEP is involved in aromatic amino acid biosynthesis. Several of these compounds can also cross the plastid envelope. The final step of glycolysis generates pyruvate. This, the precursor for acetyl-CoA and fatty acid biosynthesis in leucoplasts, is the direct precursor for alanine biosynthesis and is utilized by the mitochondrion for energy production via the TCA cycle.

Glycolysis is present in the cytosol of all plant cells and is also present, at least partially, in all plastids. In developing oilseeds a complete glycolytic pathway is present in the leucoplast where it is thought to provide both carbon, in the form of pyruvate, and cofactors in the form of ATP and NAD(P)H, for the biosynthesis of fatty acids (Fig. 36.4) (Dennis and Blakeley, 1993). Hence, it is anticipated that the genetic modification of one or both of these pathways will have a significant impact upon the partitioning of carbon between the cytosol and plastid in these seeds and could, perhaps, lead to an increase in the yield of oil.

Manipulation of glycolytic enzymes

The first enzyme of glycolysis to be manipulated by genetic engineering was pyruvate kinase (PK) which occurs as isozymes in both the plastid and cytosol. This enzyme catalyzes the irreversible conversion of PEP to pyruvate and was chosen because, in many organisms, it is a key regulatory enzyme (Muirhead, 1990). An attempt was therefore made to perturb glycolytic flux through over-expression of PK (Gottlob-McHugh *et al.*, 1992). As has been alluded to earlier, the results of this manipulation were not as expected. Firstly, over-expression was not achieved. Instead, a complete inactivation of cytosolic PK activity was observed in leaves as the result of cosuppression. Despite the absence of PK, the plants appeared normal. Whether or not PK activity is essential in other parts of the plant is not known but leaves, clearly, can tolerate the absence of the cytosolic isozyme. This study exemplifies some of the problems discussed earlier and highlights the patience that will be required for the successful manipulation of plant metabolism.

Several genes for plastid isozymes of PK have now been cloned (Blakeley *et al.*, 1991, 1995) and it should, therefore, be possible to manipulate glycolysis in this compartment. Studies such as this will

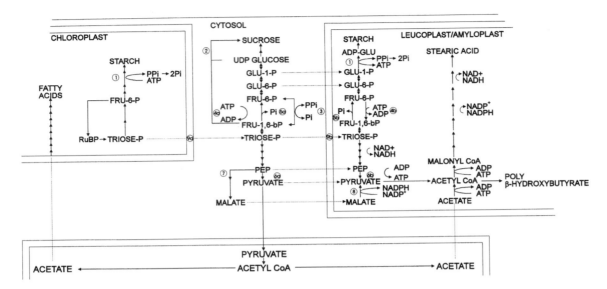

Fig. 36.4 Summary of the pathways and enzymes involved in the biosynthesis of starch, sucrose and fatty acids in the cytosolic, plastid and mitochondrial compartments in plant cells. The enzymes are as follows: (1) ADP-glucose pyrophosphorylase; (2) sucrose phosphate synthase; (3) pyrophosphate dependent phosphofructokinase; (4a,b) cytosolic and plastid ATP-dependent phosphofructokinase; (5a,b) cytosolic and plastid fructose 1,6 bisphosphatase; (6a,b) cytosolic and plastid pyruvate kinase; (7) PEP-carboxylase; (8) NADP-dependent malic enzyme; (9a,b) plastid envelope phosphate translocators. Abbreviations: Suc-6-P, sucrose-6-phosphate; UDP-Glu, UDP-glucose; ADP-Glu, ADP-glucose; Glu-1-P, glucose-1-phosphate; Glu-6-P, Glucose-6-phosphate; Fru-6-P, fructose-6-phosphate; F-1,6-bP, fructose 1,6-bisphosphate; 3-PGA, 3-phosphoglycerate; PEP, phospho*enol*pyruvate; RuBP, ribulose 1,5-bisphosphate. Reproduced in modified form from Blakeley and Dennis (1993) with permission from the *Canadian Journal of Botany.*

determine whether glycolysis in the leucoplast is, indeed, supplying carbon substrates and cofactors for use during fatty acid biosynthesis. In addition, it will now be possible to assess whether elevating carbon flow through the glycolytic pathway in leucoplasts will have an impact on the level of fatty acid synthesis in these organelles.

It is probably naive to expect that a process as complex as oil biosynthesis can be improved through the manipulation of the activity of a single enzyme. As has been discussed, most oilseeds also contain some starch and storage protein and an improvement in oil yield may only be achieved if the production of these other storage compounds can be reduced thus enhancing the availability of carbon compounds and cofactors for use in the production of fatty acids. In addition, malate has been shown to be an excellent substrate for utilization in the biosynthesis of fatty acids (Smith *et al.*, 1992), perhaps because conversion

of malate to pyruvate in the leucoplast generates NADPH within the organelle (Fig. 36.4). Thus, enhancing the availability of malate in the cytosol might also lead to increased fatty acid biosynthesis. This scenario provides an example of how it may be possible to combine genetic engineering with conventional breeding. For example, one plant can be engineered to have reduced starch biosynthesis through manipulation of ADP-glucose pyrophosphorylase (see later). A second plant could be produced with elevated PEP-carboxylase activity in the cytosol thereby draining PEP from glycolysis and generating malate while at the same time leading to a reduced availability of precursors for amino acid and storage protein synthesis. The effects of each individual manipulation can then be investigated in detail and, if deemed successful, the two characteristics can be combined into a single plant by conventional breeding. In this way, the construction of complex vectors for transformation can be avoided and individual manipulations assessed independently. Since fatty acid synthesis is a complex process, several such manipulations can be envisioned and it is anticipated that the combination of the more effective of these will ultimately lead to an improvement in oil yield.

The demonstration that plants lacking cytosolic PK activity in their leaves can survive is an illustration of the flexibility of plant metabolism. Another example of this flexibility is the presence, in plants, of two enzymes in the cytosol that phosphorylate fructose-6-phosphate (F-6-P) to fructose 1,6-bisphosphate (F-1,6-bP). In animals, this reaction is catalyzed by the enzyme phosphofructokinase (PFK) which utilizes ATP as the phosphoryl donor. PFK is present in plants but there is also a second enzyme, pyrophosphate dependent phosphofructokinase (PFP) catalyzing the same reaction but, instead, utilizing pyrophosphate as the phosphoryl donor (Black et al., 1987; Carlisle et al., 1990). Unlike PFK, the reaction catalyzed by PFP is freely reversible and the enzyme can therefore function in either the glycolytic or gluconeogenic direction (Dennis and Greyson, 1987). The function of PFP in plants is still a subject of speculation but it has been proposed to be a bypass for PFK at times of phosphate limitation (Theodorou and Plaxton, 1993, 1994) and also to operate, at least in part, to equilibrate the hexose- and triose-phosphate pools (Dennis and Greyson, 1987). Because PFP is an enzyme that catalyzes a highly regulated step in glycolysis, several studies have been performed in which the activity of this enzyme has been manipulated. It was found that the virtual elimination of the activity of PFP had no profound effect on the growth or development of transgenic potatoes (Hajirezaei et al., 1994). It is assumed that other enzymes catalyzing the same reactions as PFP, namely PFK and fructose-1,6-bisphosphatase (Fig. 36.4) are compensating for it's absence.

As an alternative to the elimination of PFP activity outlined above, another recent study described the effects of forcing the reversible reaction catalyzed by this enzyme to operate in only one direction. This was achieved by introducing a bacterial pyrophosphatase into the cytosol of transgenic plants thus utilizing pyrophosphate in this compartment and reducing the ability of PFP to operate in the glycolytic direction and generate F-1,6-bP (Sonnewald, 1992). The results of

this study indicate that glycolysis in these plants was no longer able to operate effectively and, as a consequence, sucrose accumulated in the leaves. This is an elegant demonstration that it is possible to manipulate the allocation of carbon to end products by modifying the activity of a single enzyme and that attempts to alter the partitioning of photoassimilate are not necessarily in vain. Using these plants as a model, it will now be possible to determine whether an increase in sucrose available for translocation to developing storage organs will lead to an increase in yield.

Other enzymes

Sucrose phosphate synthase

Manipulation of the activity of PFP is not the only way in which source tissue metabolism has been perturbed. Carbon, assimilated by photosynthesis, is exported from the chloroplast as triose phosphate. Sucrose, for export from the leaf, is then synthesized in the cytosol through the actions of several enzymes including sucrose phosphate synthase (SPS). This enzyme catalyzes the formation of sucrose-6-phosphate from UDP-glucose and F-6-P. There are several lines of evidence indicating that SPS is a site for the regulation of sucrose synthesis. Firstly, SPS activity is modulated by both protein kinase mediated phosphorylation and allosteric regulation by glucose-6-phosphate (Stitt and Quick, 1989; Huber et al., 1991). In addition, extractable SPS activity in different maize genotypes can be correlated with growth rate (Rocher et al., 1989).

SPS has been introduced into transgenic plants under the control of the Rubisco small subunit promotor to assess the effects of elevated SPS activity on leaf metabolism (Micallef et al., 1995; Worrell et al., 1991). In the most recent study, elevated SPS activity in tomato leaves actually led to an increase in yield although it is not yet known whether this effect will be repeated when the plants are grown under conditions that are less favourable than those provided in laboratory growth chambers. Worrell et al. (1991) used an enzyme that was not subject to the regulatory mechanisms controlling the resident enzyme and SPS activity in the leaves of these plants was increased sixfold, again resulting in an increase in the level of sucrose in this tissue. Both of these studies demonstrated that it is feasible to manipulate source tissue metabolism and, at least under ideal growth conditions, that this can lead to an increase in yield.

Invertase

Assimilate partitioning and yield in plants does not only depend upon competition between the various metabolic pathways in source and sink tissues. Before any storage material can be accumulated, photoassimilate, generally sucrose, must be transported from the leaves to the developing sink organ. The studies described above should allow us to determine whether the presence of an elevated level of sucrose in the leaf will be sufficient to lead to an increase in the translocation of this compound to other tissues or whether it will also be necessary to manipulate the transport process itself. The latter possibility has been addressed recently through the expression of yeast invertase in the apoplasm of source tissues, thus removing sucrose from this compartment and effectively blocking phloem loading (Von Schaewen et al., 1990; Dickenson et al., 1991; Sonnewald et al., 1991). The resulting accumulation of sucrose and starch in these leaves actually led to an inhibition of photosynthesis and ultimately to plants with stunted growth and suppressed root formation. These results are not unexpected but have demonstrated that there is an apoplastic component to the process of phloem loading and have, therefore, contributed to our understanding of plant metabolism. It will be interesting to see whether expression of invertase in sink tissues will lead to an increase in phloem unloading. If this occurs, the combination of elevated sucrose in leaves through manipulation of PFP or SPS along with increased translocation of this compound through expression of invertase in the developing storage organs could prove to be a valuable combination.

ADP-glucose pyrophosphorylase (AGPase)

Perhaps the most spectacular demonstration of the feasibility of manipulating plant carbohydrate metabolism, and an alternative to invertase for increasing demand from the sink, have been studies involving the enzyme AGPase. This enzyme is considered to be the pacemaker in the pathway from hexose-phosphate to starch. The importance of AGPase in starch biosynthesis was illustrated by starch deficient mutants of *Arabidopsis* that contained only residual activity of this enzyme (Lin et al., 1988; Smith et al., 1989). Like PK, PFP and SPS, AGPase activity in plants is regulated. This regulation ensures that starch biosynthesis does not place an excessive demand on cellular resources. In a recent report, a bacterial AGPase, lacking these regulatory properties, was expressed in potato tubers. This resulted in a large increase in the starch content of these organs (Stark et al., 1992). Another significant observation in this study was that the leaves of these plants could keep pace with the increased demand for photoassimilate. This augers well for the combination of plants in which sucrose translocation is facilitated and, at the same time, in which there is increased demand for this compound in the sink.

In another study, AGPase activity was inhibited in transgenic potatoes resulting in a reduction in starch synthesis. These tubers contained an elevated level of sucrose (Muller-Rober et al., 1992). This demonstrates that it may be possible to manipulate the composition of carbon containing storage compounds. This, in theory, could be employed to manipulate sweetness in crops such as grapes. It is interesting to note that the inhibition of starch biosynthesis in these tubers also resulted in a reduction in the level of storage proteins demonstrating that modification of one pathway can have unpredictable effects on another. It is likely that the study of these secondary effects will be crucial for our understanding of the interrelations between different biosynthetic pathways in plants.

Plastid translocators

The movement of metabolic intermediates across the plastid envelope will have a profound effect on the flux of carbon through the various pathways in both the cytosol and plastid. Recent work has demonstrated that the phosphate translocator in the plastid envelope can transport a wider variety of metabolic intermediates than was originally thought (Gross et al., 1990). Phosphate translocators have also been identified in non-green plastids, such as those in pea roots (Emes and Traska, 1987), and there is probably a similar translocator in leucoplasts in developing oilseeds. The pea root phosphate translocator has a wider specificity than that of the chloroplast and can transport glucose-6-phosphate and PEP in addition to triose phosphates and 2- and 3-phosphoglycerate (Borchert et al., 1993). The gene for the chloroplast phosphate/triose-phosphate translocator has been cloned and it is likely that genes encoding other transporters will soon be isolated and available for expression in transgenic plants. Incorporation of the pea root translocator in the leucoplast might enhance the range of metabolites available for the biosynthesis of storage oil since it has affinity for metabolites all the way from the hexose phosphate pool to PEP. No doubt other transporters, with different affinities, will be identified and cloned and the expression of these genes in tissues with different endogenous transporters will be used to manipulate plant metabolism.

In vitro mutagenesis

Several of the key enzymes at the heart of carbon metabolism are regulated to ensure that a balance is maintained between carbon assimilation, utilization and transport. Carbon metabolism is also tightly linked to that of nitrogen ensuring, amongst other things, an adequate supply of amino and nucleic acids. Many of these enzymes have now been cloned in plants and their amino acid sequences compared with the equivalent non-plant enzymes. This type of analysis, along with an understanding of the regulatory properties of these enzymes in plants and other organisms, will allow the identification of residues involved in enzyme regulation. These residues can then be altered by *in vitro* mutagenesis and the modified enzymes subsequently re-introduced into transgenic plants. Possibilities for this approach might include deregulation of glycolysis through the modification of PFK. PFK is subject to feedback inhibition by PEP and the elimination of this regulation would allow glycolysis to proceed unchecked. It should also be possible to modify the regulatory properties of PK, SPS or PFP and, perhaps, to influence the specificity of plastid

envelope translocators. This approach will be somewhat more sophisticated than the introduction of heterologous or non-plant genes since it should allow manipulation of specific regulatory properties without completely changing the kinetic characteristics of the enzyme involved.

Future prospects

In addition to the manipulation of production of naturally occurring plant products, future goals also include the synthesis of raw materials not normally produced by plants. That this is possible has been demonstrated by the synthesis of the biodegradable plastic, poly-β-hydroxybutyrate (PHB), in plants (Chapter 37; Poirier et al., 1992). The PHB biosynthetic pathway in bacteria consists of three enzymes and begins with the condensation of two molecules of acetyl-CoA to form acetoacetyl-CoA. Acetyl-CoA is also the direct precursor of fatty acid biosynthesis in leucoplasts and it is anticipated that the expression of all three enzymes, targeted to the leucoplast with appropriate transit peptides during seed development, will lead to significant production of PHB. Other than plastics, it is anticipated that plants could also be used as bioreactors for the synthesis of drugs and other high value commodities.

Another potential target for genetic manipulation is the pathway leading to lignin. Lignin, either in the seed coat or as a structural element in the plant, contributes to the indigestibility of forage crops and to the ease with which seeds can be processed. In addition, alterations to the structure and composition of lignin in trees could have a significant economic impact on the pulp and paper industry. Lignin biosynthesis begins with the utilization of PEP and erythrose-4-phosphate, intermediates of the glycolytic and pentose phosphate pathways, respectively (Fig. 36.2). The biosynthesis of lignin thus depletes resources from both of these pathways and, therefore, may also compromise productivity. It has been demonstrated that genetic approaches can be utilized to manipulate both the quantity and composition of lignin (reviewed in Whetton and Sederoff, 1995) and these methods will be used in the future to modify lignin biosynthesis in several different plant tissues.

In summary, enormous progress has been made in defining the factors controlling the allocation of photoassimilate in plants. The creation of transgenic plants expressing invertase in the apoplasm, AGPase in the tuber and pyrophosphatase in the leaf, along with those lacking PK and PFP activity have all contributed to our understanding of plant biochemistry and physiology. Furthermore, these studies have demonstrated the clear potential for rational manipulation of primary plant metabolism and productivity. It is only six years since the characterization of starchless *Arabidopsis* mutants confirmed the critical role played by the enzyme AGPase (Lin *et al.*, 1988) and this enzyme has already been successfully manipulated in transgenic plants. More and more mutants are being isolated and characterised and some, such as *fus3* and *lec* (Keith *et al.*, 1994; Meinke *et al.*, 1994), profoundly affect the composition of storage material in the seeds, changing from mostly oil to mostly starch. Analysis of the metabolism and anatomy of these mutants should allow us to identify why this change has taken place and the information gathered from these, and other, mutants will contribute to our ability to engineer high yielding and economically viable crops.

References

ap Rees, T. and Hill, S. A. (1994). Metabolic control analysis of plant metabolism. *Plant Cell Env.* **17**, 587–99.

Black, C. C., Mustardy, L., Sung, S. S., Kormanic, P. P., Xu, W.-P. and Paz, N. (1987). Regulation and roles for alternative pathways of hexose metabolism in plants. *Physiol. Plant* **69**, 387–94.

Blakeley, S. D. and Dennis, D. T. (1993). Molecular approaches to the manipulation of carbon allocation in plants. *Can. J. Bot.* **71**, 765–78.

Blakeley, S. D., Plaxton, W. C. and Dennis, D. T. (1991). Relationship between the subunits of leucoplast pyruvate kinase from *Ricinus communis* and a comparison with the enzyme from other sources. *Plant Physiol.* **96**, 1283–8.

Blakeley, S. D., Gottlob-McHugh, S., Wan, J., Crews, L., Ko, K., Miki, B. and Dennis, D. T. (1995). Molecular analysis of plastid pyruvate kinase in castor and tobacco. *Plant Mol. Biol.* **27**, 79–89.

Borchert, S., Harborth, J., Schunemann, D., Hoferichter, P. and Heldt, H. W. (1992). Studies on the enzymic capacities and transport properties of pea root plastids. *Plant Physiol.* **101**, 303–12.

Carlisle, S. M., Blakeley, S. D., Hemmingsen, S. M., Trevanion, S. J., Hiyoshi, T., Kruger, N. J. and Dennis, D. T. (1990). Pyrophosphate dependent phosphofructokinase. Conservation of protein sequence between the α- and β-subunits and with the ATP dependent phosphofructokinase. *J. Biol. Chem.* **265**, 18366–71.

Dennis, D. T. (1987). *The Biochemistry of Energy Utilisation by Plants*, Blackie and Sons, London, Glasgow.

Dennis, D. T. and Blakeley, S. D. (1993). Carbon and cofactor partitioning in oilseeds. In *Seed Storage Compounds*, eds P. R Shewry and K Stobart, Proceedings of the Phytochemical Society of Europe.

Dennis, D. T. and Greyson, M. F. (1987). Fructose 6-phosphate metabolism in plants. *Physiol. Plant* **69**, 394–404.

Dennis, D. T. and Miernyk, J. A. (1982). Compartmentation of non-photosynthetic carbohydrate metabolism. *Ann. Rev. Plant Physiol.* **33**, 27–50.

Dickinson, C. D., Altabella, T. and Crispeels, M. J. (1991). Slow growth phenotype of transgenic tomato expressing apoplastic invertase. *Plant Physiol.* **95**, 420–25.

Emes, M. J. and Traska, A. (1987). Uptake of inorganic phosphate by plastids purified from the roots of *Pisum sativum* L. *J. Exp. Bot.* **38**, 1781–8.

Gottlob-McHugh, S. G., Sangwan, R. S., Blakeley, S. D., Vanlerberghe, G. C., Ko, K., Turpin, D. H., Plaxton, W. C. Miki, B. L. and Dennis, D. T. (1992). Normal growth of transgenic tobacco plants in the absence of cytosolic pyruvate kinase. *Plant Physiol.* **100**, 820–25.

Gross, A., Bruckner, G., Heldt, H. W. and Flugge, U.-I. (1990). Comparison of the kinetic properties, inhibition and labelling of the phosphate translocators from maize and spinach mesophyll chloroplasts. *Planta* **180**, 262–71.

Hajirezaei, M., Sonnewald, S., Viola, R., Carlisle, S., Dennis, D. T. and Stitt, M. (1994). Transgenic potato plants with strongly decreased expression of pyrophosphate: fructose-6-phosphate phosphotransferase show no visible phenotype and only minor changes in metabolic fluxes in their tubers. *Planta* **192**, 16–30.

Hill, L. M. and Smith, A. M. (1991). Evidence that glucose-6-phosphate is imported as the substrate for starch synthesis by plastids of developing pea embryos. *Planta* **185**, 91–6.

Huber, J. L., Hite, D. R. C., Outlaw, W. H. and Huber, S. C. (1991). Inactivation of highly activated spinach leaf sucrose-phosphate-synthase by dephosphorylation. *Plant Physiol.* **95**, 291–7.

Jorgensen, R. (1990). Altered gene expression in plants due to trans interactions between homologous genes. *Trends in Biotechnology* **8**, 340–44.

Kacser, H. and Burns, J. A. (1973). The control of flux. *Symp. Soc. Exp. Biol.* **27**, 65–107.

Keegstra, K. (1989). Transport and routeing of proteins into chloroplasts. *Cell* **56**, 247–53.

Keeling, P. L., Wood, J. R., Tyson, R. H., Bridges, I. G. (1988). Starch biosynthesis in developing wheat grain. Evidence against the direct involvement of triose phosphates in the metabolic pathway. *Plant Physiol.* **87**, 311–19.

Keith, K., Kraml, M., Dengler, N. G. and McCourt, P. (1994). *Fusca3*: A heterochronic mutation affecting late embryo development in Arabidopsis. *Plant Cell.* **6**, 589–600.

Kirk, J. T. O. and Tilney-Bassett, R. A. E. (1978). *The plastids. Their Chemistry, Structure, Growth and Inheritance*, Elsevier/North Holland Biomedical Press, New York.

Kruckeberg, A. L., Neuhaus, H. E., Gottlieb, L. D. and Stitt, M. (1989). Decreased activity mutants of phosphoglucose isomerase in the cytosol and chloroplast of *Clarkia xantiana*. Impact on mass action ratios on fluxes to sucrose and starch, and estimation of flux control coefficients and elasticity coefficients. *Biochem. J.* **261**, 457–67

Lin, T. P., Caspar, T., Somerville, C. and Preiss, J. (1988). Isolation and characterisation of a starchless mutant of *Arabidopsis thaliana* (L.) Heynh lacking ADP-glucose pyrophosphorylase activity. *Plant Physiol.* **86**, 1131–5.

Micallef, B. J., Haskins, K. A., Vanderveer, P. J., Kwang-Soo, Roh, Shewmaker, C. K. and Sharkey, T. D. (1995). Altered photosynthesis, flowering and fruiting, in transgenic tomato plants that have an increased capacity for sucrose synthesis. *Planta.* **196**, 327–34.

Meinke, D. W., Franzmann, L. H., Nickle, T. C. and Yeung, E. C. (1994). *Leafy cotyledon* mutants of Arabidopsis. *Plant Cell.* **6**, 1049–64.

Muirhead, H. (1990). Isozymes of pyruvate kinase. *Biochem. Soc. Trans.* **18**, 193–6.

Muller-Rober, B., Sonnewald, U. and Willmitzer, L. (1992). Inhibition of ADP-glucose pyrophosphorylase in transgenic potatoes leads to sugar storing tubers and influences tuber formation and expression of tuber storage protein genes. *EMBO. J.* **11**, 1229–38.

Neuhaus, E. H. and Stitt, M. (1990). Control analysis of photosynthate partitioning: Impact of reduced activity of ADP-glucose pyrophosphorylase or plastid phosphoglucomutase on the fluxes to starch and sucrose in *Arabidopsis thaliana* (L.) Heynh. *Planta* **182**, 445–54.

Poirier, Y., Dennis, D. E., Klomparens, K. E. and Somerville, C. (1992). Polyhydroxybutyrate, a biodegradable thermoplastic, produced in transgenic plants. *Science* **256**, 520–23.

Rocher, J. P., Prioul, J. L., Lecharny, A., Reyss, A. and Joussaume, M. (1989). Genetic variability in carbon fixation, sucrose-P-synthase and ADP-glucose pyrophosphorylase in maize plants of differing growth rate. *Plant Physiol.* **89**, 416–20.

Sinclair, T. W. and Dewit, C. T. (1975). Photosynthate and nitrogen requirements for seed production by various crop species. *Science.* **189**, 565–7.

Smith, A. M., Bettey, M. and Bedford, I. M. (1989). Evidence that the rb locus alters the starch content of developing embryos through an effect on ADP-glucose pyrophosphorylase. *Plant Physiol.* **89**, 1279–84.

Smith, R. G., Gauthier, D. A., Dennis, D. T. and Turpin, D. H. (1992). Malate- and pyruvate- dependent fatty acid biosynthesis in leucoplasts from developing castor endosperm. *Plant Physiol.* **98**, 1233–8.

Sonnewald, U. (1992). Expression of *E.coli* inorganic pyrophosphatase in transgenic plants alters photoassimilate partitioning. *Plant J.* **2**, 571–81.

Sonnewald, U., Brauer, M., Von Schaewen, A., Stitt, M. and Willmitzer, L. (1991). Transgenic tobacco plants expressing yeast derived invertase in either the cytosol, vacuole or apoplast: a powerful tool for studying sucrose metabolism and sink/source interactions. *Plant J.* **1**, 95–106.

Stark, D. M., Timmerman, K. P., Barry, G. F., Preiss, J. and Kishore, G. M. (1992). Regulation of the amount of starch in plant tissues by ADP glucose pyrophosphorylase. *Science.* **258**, 287–92.

Stitt, M. and Quick, W. P. (1989). Photosynthetic carbon partitioning: its regulation and possibilities for manipulation. *Physiol. Plant.* **77**, 633–41.

Stitt, M., Quick, W. P., Schurr, U., Schulze, E.-D., Rodermel, S. R. and Bogorad, L. (1991). Decreased ribulose 1,5-bisphosphate carboxylase–oxygenase in transgenic tobacco transformed with 'antisense' rbcs. II. Flux control coefficients for photosynthesis in varying light, CO_2 and air humidity. *Planta* **183**, 555–66.

Theodorou, M. and Plaxton, W. C. (1993). Metabolic adaptations of plant respiration to nutritional phosphate deprivation. *Plant Physiol.* **101**, 339–44.

Theodorou, M. and Plaxton, W. C. (1994). Adaptations of plant respiratory metabolism to nutritional phosphate deprivation. In *Environment and Plant Metabolism*, ed. N. Smirnoff, Bios Scientific Publishers, Oxford.

Von Schaewen, A., Stitt, M., Schmidt, R. Sonnewald, U. and Willmitzer, L. (1990). Expression of a yeast-derived invertase in the cell wall of tobacco and *Arabidopsis* plants leads to accumulation of carbohydrate and inhibition of photosynthesis and strongly influences growth and phenotype of transgenic tobacco plants. *EMBO. J.* **9**, 3033–44.

Whetton, R. and Sederoff, R. (1995). Lignin biosynthesis. *Plant Cell* **7**, 1001–13.

Worrell, A. C., Bruneau, J.-M., Summerfelt, K., Boersig, M. and Voelker, T. A. (1991). Expression of a maize sucrose phosphate synthase in tomato alters leaf carbohydrate partitioning. *Plant Cell* **3**, 1121–30.

37 The biochemical basis for crop improvement

William D. Hitz and John W. Pierce

The theories, knowledge and methodologies of molecular biology and biochemistry have developed to the point where numerous alterations to plants can be made through direct genetic manipulation of their genomes. Indeed, the ability to manipulate plant metabolism in ways designed to enhance either the absolute productivity or the economic productivity of crops is one of the practical outcomes of the disciplines of plant molecular biology and plant biochemistry. The tools required to put these directed manipulations into practice include transgenic expression along with plant transformation and the use of molecular analysis and genetic mapping to facilitate the use of natural or induced mutations. At the present time, large scale, commercial production of transgenic crop plants is underway for the first time, and many additional transgenic crops are in various stages of commercial development. These results presage an era in which direct intervention in the underlying biochemistry and biology of plants will become a common means for enhancing plant productivity and varying the types and qualities of materials produced by plants.

There are many reviews of the development of our capacities to modify plants through transgenesis, of the likely routes for commercialization of these plants and of the techniques that can be used to create transgenic plants (Burrell *et al.*, 1993; Bennet, 1993; Riazudin, 1994). In this article we will discuss how the underlying biology and biochemistry of plant metabolism create both opportunities and difficulties for the production of crop plants that have been improved through intervention in their existing metabolism. We will use a discussion of the types of alterations in plant metabolism that have been successful to point out some general considerations in using currently available technology and indicate developments in technology and understanding of quantitative plant biochemistry that are likely to be required to achieve further goals in crop improvement.

Introduction

When seeking to alter plant metabolism via transgenesis, one would ideally want to know the qualitative and quantitative relationships between all the steps in a metabolic pathway in order to design strategies for redirecting metabolites in that pathway to new or alternate products. The ideal situation is seldom the starting point given our current knowledge base of metabolism at the quantitative level. A much more likely scenario is that a good, qualitative outline of all or a portion of a metabolic pathway will be available from some combination of biochemistry and genetics and the initial attempts at modification of the existing pathway will begin to provide the quantitative relationships between steps in the path. The experience gained from the first alterations must then provide the information for improved schemes for reaching the original goal. The formal methods for quantitating control of flux in multistep biochemical pathways are expressed in flux control theory (Kacser and Porteous, 1987). Measurements of flux control coefficients are made in practice by altering flux by producing specific and measurable changes in the capacity at specific steps. In fact the variety of enzyme expression levels and the specificity of alteration often obtained in transgenic manipulations makes the technique ideal for determination of control coefficients (Stitt and Schulze, 1994).

Inherent in this type of analysis is the concept that control is seldom localized to a single metabolic step. Certain steps may exert a majority of the total control in some situations, but in general control is achieved by fractional limitation to total flux at many individual reactions. The amount of control at a given step may not be the same at all fluxes and may therefore vary with environment and development.

Fig. 37.1 is a diagrammatic simplification of metabolism which may help explain the strategies used and requirements placed on the altering step in achieving different goals in altering plant composition. An existing metabolic pathway which produces two main products, A and B in unequal proportions is shown in panel 1. One of the goals in altering plant metabolism is to produce an entirely new product, not previously made by the plant, but made from intermediates that already exist. In principle this is a simple modification and is depicted in panel 2. All that is required is the introduction of the molecular machinery to make the new product 'C'. The simplest form of this modification is the production of the detoxifying enzymes used as selectable markers. The transgenic introduction of the gene for the enzyme diverts amino acids from the production of endogenous proteins to the production of a foreign protein using the existing protein synthesis pathway. Provided that the efficacy of the

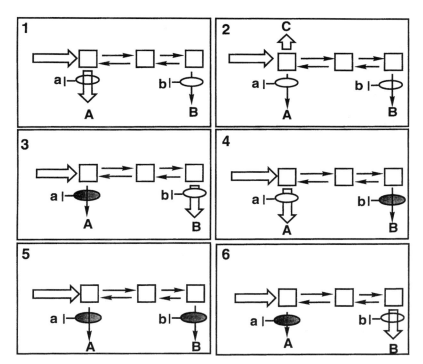

Fig. 37.1 A schematic representation of a multistep biosynthetic pathway which produces two products, 'A' and 'B' in unequal amounts. Single steps which exert a strong control over the amount of an end product made are shown as shaded ovals and those which exert only a small amount of the total flux control are shown as unshaded ovals. Panel 1 represents the situation in an unaltered cell while panels 2 through 6 represent the changes made and outcomes expected from: (2) addition of an alternative fate for an intermediate of both 'A' and 'B'. (3) Reduction of the total capacity of a metabolic step leading to 'A' such that flux control is increased at that step. (4) Reduction of the total capacity of a metabolic step leading to 'B' such that flux control is increased at that step. (5) Reduction of the total capacity of a metabolic step leading to 'B' such that flux control is increased at that step in a situation where flux to 'A' is already strongly limited at a single synthesis step. (6) Alteration of the situation in panel 4 by simultaneously limiting flux to 'A' and increasing the total flux capacity to 'B'.

detoxifying enzyme is very high, the amount of transcript required from the transgene is small, the quantity of amino acid diverted to that use is of little consequence, and the desired goal of metabolizing the selection toxin can be met with existing transgene technology and without additional effect on the whole plant.

The other extreme in this type of composition change is exemplified by attempts to produce entirely new storage products. In such cases it may be necessary to divert nearly all of the existing flux to the new product 'C' if the goal is economically viable production. In this case the transgene expression requirement may be very high since its gene product must succeed in diverting metabolite from the existing pathway by competition with the elements of that pathway. In many cases the transcription requirements may challenge the limits of transgene technology, and if the metabolic intervention is highly successful, the cell expressing the change will be faced with both the storage of a new product and the loss of the old products.

A second type of alteration of composition is changing the proportion of endogenous products in a mixture of products coming from a common metabolite. In practice the mixture of end-products to be altered might be as similar as isomers of a single compound or have no more in common than being products of carbon metabolism, yet the principles underlying the change in composition remain the same. If in our pathway in panel 1, there are no single steps in the paths leading to 'A' or 'B' that exert strong control over their production such that the proportion of 'A' or 'B' produced is the product of the control at several steps, the introduction of a strongly rate controlling step can effectively switch the product mix to favor 'B' (panel 3). Such changes might be effected by mutations that reduce either message level or enzyme catalytic capacity or by transgene strategies that accomplish the same things. Observations coming from this type of intervention strategy which show that product composition can be altered without alteration of the total amount of product made indicate that overall metabolic control is accomplished in nature with multiple, low control steps.

Examples to the contrary are also known however, and that scenario is shown in panel 4. If the overall flux to 'B' is strongly controlled at a single, or even a very few steps, the strategy of simply inhibiting the formation of 'A' by the introduction of a new limiting step in its synthesis path may not produce the ideal result. As shown in panel 5, the ratio of 'A' to 'B' may change, but the amount of the change may be less that expected and the total amount of 'A' + 'B' produced may be decreased. The desired 'A' to 'B' ratio may require release of the strong control over the synthesis of 'B' and the placement of a strong control point in the pathway to 'A' simultaneously (panel 6). Since over expression is a very rare occurrence, modifications like the one depicted in panel 6 will probably require transgenic modification, or the combination of transgenic modification and mutagenesis.

Many layers of complexity may be placed on the simple diagram shown in Fig. 37.1. Among the first that should be considered is the control of regulatory steps themselves. While the enzymatic steps are shown as on/off valves (i.e. simple enzyme products of genes controlled by developmental regulation), in many cases the controlling step may be the result of an allosterically controlled enzyme or a gene whose transcription or translation is under metabolic control. While these complexities do not change the possibility of introducing new controls over metabolism, they do influence the strategies employed.

In the sections that follow we will use examples of plants with altered composition that have been created using the strategies outlined in Fig. 37.1. These examples are not intended to be an all inclusive survey, but were chosen to illustrate problems that have been encountered in attempts to alter plant metabolism and ways in which those problems have been addressed.

Engineering of new products of metabolism

Herbicide resistant crops

The antibiotic detoxifying enzymes usually originating in bacteria and expressed in plant cells as selectable markers for transformation are the simplest and earliest examples of the transgenic production of a new product in plant cells. The antibiotics used for selection were, at least in part chosen because the amount required to eliminate the growth of plant cells that were not stably transformed was low in relation to the amount of detoxifying catalytic capacity that could be expressed by the transgene promoters available. Since the usual goal of selection is to recover all transformants and still allow only a small percentage of escapes, it is a system designed to function with the production of a low level of transgene product. The exact amount of detoxifying protein required is dependent upon the level of antibiotic required to eliminate growth of non-transformed cells and the efficacy of the enzyme in converting the antibiotic to a non-toxic metabolite in transformed cells. Even in this system it is often the inability of available promoters to provide sufficient levels of transcription in the tissue most critical to the survival of transformed cells as differentiation begins to occur that limits the application of certain antibiotic/detoxification systems to many transformation and plant regeneration systems.

This same strategy has been successful in synthesizing enough catalytic capacity to convert herbicides to non-toxic metabolites at a rate sufficient to produce plants that are resistant to a specific herbicide at levels required for tolerance of field applications. For a relatively non-mobile herbicide whose mode of action is inhibition of photosynthesis, production of a nitrile hydrolase isolated from a soil bacteria at levels comprising less than 0.01% of the total leaf protein was sufficient to produce tobacco plants that were resistant to more than four times the normal herbicide application rate (Stalker *et al.*, 1988).

A slightly different strategy, production of an enzyme which is resistant to a specific herbicide's normal target, but still functional in its metabolic capacity has been successful in producing herbicide tolerance in crop plants also. Both mutational and transgenic approaches have been successful in producing herbicide resistant plants using this strategy (Chaleff and Ray, 1984; Haughn *et al.*, 1988). Sulfonyl urea and imidazolinone herbicides have as their target acetolactate synthase, a key enzyme in the biosynthesis of branched chain amino acids. Several mutant forms of the enzyme have

been described that have varying degrees of resistance to members of both herbicide classes and that are catalyticly very similar to the wild-type enzyme (Sarri and Mauvis 1995). In practice, these mutant enzymes need only provide about one half of wild type specific activity at normally herbicidal concentrations of their inhibitor to produce plants with useful herbicide resistance (Haughn *et al.*, 1988). Depending upon the mutant form of the enzyme and the amount of tolerance required, this level of expression can be met by a variety of promoters, including the actual acetolactate gene promoter (Sarri and Mauvis, 1995).

In contrast to these rather simple expression systems used to produce commercially useful resistance, several characteristics of both the herbicide glyphosate and its target, 5-*enol*-pyruvylshikimate-3-phosphate synthase (EPSPS) have made mutant and transgenic approaches to achieving useful resistance to this herbicide difficult (Kishore *et al.*, 1992). EPSPS is a key enzyme in a comparatively high flux pathway, the shikimate pathway, leading to many aromatic compounds. The enzyme is most highly active in meristematic regions of plants and its inhibitor, glyphosate, is highly mobile within the plant and tends to accumulate in regions that are actively importing photosynthate. Glyphosate is a strong competitive inhibitor with respect to one of the substrates for EPSPS, phospho*enol*pyruvate (PEP). All of these factors make this a difficult system to detoxify. The series of steps that have been taken to create commercially useful tolerance has involved a great deal of careful research, much of it unpublished, which provides a useful perspective on the types of issues that must be dealt with in creating commercially useful transgenic plants.

The first glyphosate resistant mutants were selections from petunia tissue culture that overexpress the wild-type form of the enzyme. This strategy requires that the inhibitor free enzyme become a large enough term in the equilibrium equation that it can handle the flux through the pathway. In practice, expression sufficient to prevent killing of plants was achievable, but expression sufficient to avoid growth retardation was not.

Several glyphosate resistant forms of EPSPS have been created through mutagenesis and used in the preparation of transgenic plants. Decreased herbicide binding in these forms correlates with decreased substrate binding (increased K_m) due to an overlap between the glyphosate and PEP binding sites. While these mutations have allowed an increase in the ratio of binding constants K_i/K_m to produce enzymes that are relatively more tolerant to the herbicide, these enzymes also tend to have a lower catalytic capacity. This decreased capacity per unit of enzyme must be overcome by an increase in total enzyme production. The best of these tolerant enzymes, when expressed behind the cauliflower mosaic virus (CaMV) 35S promoter, produced canola plants that were highly tolerant of field applications of glyphosate, but which were sufficiently retarded in growth by the herbicide application as to be agronomically compromised.

An additional enhancement involved the isolation and use of a promoter from cauliflower that provided strong expression of the mutant EPSPS in the meristematic regions of the plant (G. Kishore, pers comm). It is in these regions that the demand for detoxification is greatest and in these regions also that the consequences of enzyme inhibition are most damaging to the long-term performance of the plant. Provision of more resistant enzyme to these regions produced plants with very high levels of herbicide tolerance.

Even though the efforts taken above resulted in plants with very high levels of herbicide tolerance and with demonstrated efficacy in field trials, further developments were thought necessary to ensure a margin of safety for herbicide use under commercial conditions, e.g., to allow for inadvertent overspraying of herbicide, tolerance under adverse weather and soil conditions, etc. (G. Kishore, pers comm).

Bacterial populations were screened for their ability to grow in the presence of glyphosate, and one particular *Agrobacterium* isolate was found to possess an EPSPS enzyme with the most optimal balance of resistance and enzymic activity found to date. Expression of this enzyme in soybean using the CaMV 35S promoter has afforded full, commercially useful glyphosate tolerance to soybeans, and these plants have undergone extensive field testing. However, this promoter/enzyme pair does not confer adequate resistance to canola. Again, from the bacterial populations that grew in the presence of glyphosate, one strain was found which harbored an

enzyme, glyphosate oxidoreductase, which converted glyphosate to the much less toxic glyoxylate and amino-methyl phosphonate. This provided the researchers with an additional mode for detoxification, namely metabolism of the herbicide to inactive forms. When the oxidoreductase and the resistant EPSPS mutant was co-expressed in canola behind the 35S promoter, commercially useful levels of herbicide tolerance were achieved.

These examples serve to show that current transgene technology is able to produce enzymes in the growing portions of plants at levels sufficient to carry the flux of prominent metabolic pathways. When the flux is rather low and the enzyme highly efficient, as in the example of sulfonyl urea resistance, the expression task is a simple one. When the flux is major and the introduced enzyme of low catalytic capability, as in the case of glyphosate resistance, sufficient protein expression can severely test the upper limits of transgene expression.

The examples of detoxification systems also serve to point out one method for overcoming the limits to the amount of product produced from transgenes. Those systems that produce a catalyst to amplify the effect of the transgenic transcript in carrying out the required function greatly reduce the burden on the strength of the transgene.

Insect deterring proteins in leaves

The systems described above achieve success when the amount of protein produced on a whole leaf basis is still very small (as little as 0.01% or less). A comparison of this level of expression to a study of expression of the insect control protein of *Bacillus thuringiensis* in potato leaves (Perlak *et al.*, 1993) is informative as to the actual levels of steady-state protein concentration achievable in a system that has been optimized for expression using a strong promoter active in a wide range of tissues. To achieve protein levels between 0.1 and 0.3% of total leaf protein in about 50% of the transformants, the CaMV 35S promoter with a duplication of the enhancer region (Kay *et al.*, 1987) was used to drive expression of the bacterial gene which had been modified to remove sequences known to destabilize mRNAs in eukaryotes (Perlak *et al.*, 1991).

Production of a new storage product

While it has been speculated that plants might be modified to produce entirely new storage products (panel 2 of Fig. 37.1, in which a large diversion of an existing metabolite to a new end-product occurs) there are as yet few examples where this has been demonstrated. One such work in progress is the production of the bacterial carbon storage polymer polyhydroxybutyrate in plants. This polymer is produced by some bacteria under conditions in which growth is limited by some nutrient other than carbon. The synthesis requires three reactions starting from acetyl-CoA. 3-Ketolase catalyzes the formation of acetoacetyl-CoA from two acetyl-CoA molecules, the acetoacetyl-CoA is reduced to 3-hydroxybutyryl-CoA by acetoacetyl-CoA reductase and the 3-hydroxy product is polymerized by polyhydroxybutyrate synthase. Acetyl-CoA is a key intermediate in plant cells and 3-ketolase is expressed in the cytoplasm to provide acetoacetyl-CoA for the mevalonate pathway. Acetoacetyl-CoA might thus serve as the intermediate for polyhydroxybutyrate synthesis if the remaining enzymes in the pathway existed in plants. The feasibility of the biochemical pathway was demonstrated by the transgenic expression of acetoacetyl-CoA reductase and polyhydroxybutyrate synthase in the cytoplasm of leaves of *Arabidopsis* (Poirier *et al.*, 1992). Low levels of product formation were obtained and plants expressing either the reductase or both activities had very low growth rates. Even at the low levels of accumulation obtained it was calculated that as much as 50% of the normal flux to mevalonate could have been diverted to polyhydroxybutyrate (Poirier *et al.*, 1992). Plants expressing both enzymes were more severely inhibited in growth so there is a possibility that cytoplasmic storage of the polymer is also toxic.

The same experiment, modified to express 3-ketolase and the enzymes of the polyhydroxybutyrate pathway in plastids greatly improved polymer yield and the apparent health of the plant (Nawrath *et al.*, 1994). This tactic accomplishes two things: first the polymer is isolated in the plastid, and secondly, isolating the acetoacetyl-CoA reductase from the enzymes of the entirely cytoplasmic mevalonate pathway prevents competition for acetoacetyl-CoA. The result also suggests that sufficient acetyl-CoA

could be produced to provide carbon for essential cellular processes that rely on that intermediate while at least a modest amount of polyhydroxybutyrate was made.

These examples illustrate both the need to quantitatively understand the metabolic paths the introduced modifications must interact with and the iterative nature of plant modification as it is now practised.

Alteration of existing metabolic pathways

Altering fatty acid metabolism

Two manipulations in the fatty acid biosynthetic pathway that produces the triacylglyceride in storage organs illustrate the type of manipulations shown in panels 3 through 6 in the schematic representations of general metabolism in Fig. 37.1. Fatty acid biosynthesis, fatty acid modification and complex lipid assembly are compartmentalized into separate metabolic compartments in all plant organs. Fatty acid synthesis occurs in the plastid. In photosynthetic organs, the complex lipids that make up the thylakoid membrane system are also assembled in the chloroplast, while the lipids of the remaining membranes are assembled in the cytoplasm from fatty acid exported from the chloroplast. In storage organs that synthesize large quantities of triacylglyceride, very little lipid assembly occurs in the plastid and the pathway can be simplified as shown in Fig. 37.2.

This branched and interconnected pathway leads to the production of a heterogeneous product. Storage triacylglycerides from any given species are composed of from as few as two or three main fatty acids to as many as seven to ten. The relative proportion of these fatty acids varies between species and within species in response to environment. In addition, unusual fatty acids that are unique to storage lipids may be found as major components of the triacylglyceride but not found in the membrane lipids of the same tissue. This naturally occurring plasticity implies that variation in storage lipid composition imposed by mutation or by genetic engineering might be tolerated by crop plants.

The incentive for making modifications in the lipid biosynthetic pathway in temperate oilseeds has been the production of vegetable oils with improved health characteristics and with improved functional characteristics for the various uses of oils in foods. To those ends three fatty acid profiles have been sought: (1) oils with lower levels of multiply unsaturated fatty acids to improve stability at high temperature and to improve storage life; (2) oils with lower amounts of saturated fatty acids to meet the needs of an overall reduction in saturated fat in the diet; and (3) oils with increased saturated fat to supply solid fats without cholesterol and with less or no hydrogenation.

Fig. 37.2 is drawn to emphasize the branch points in the pathway which might be modified to achieve these goals. Reduction in the relative level of the 18:3 acid should occur if the activity of the 18:2 desaturating enzymes were decreased sufficiently to limit flux at that point and if the acyltransferases on the path to triacylglyceride retain capacity to carry the increased flux of 18:2 and 18:1 fatty acids. The same is true for the 18:1 desaturating enzymes. Increased levels of saturated fatty acids might be

obtained by decreasing the capacity of the last elongating activity, ketoacyl synthase II or of the first desaturase, stearoyl-ACP desaturase. This increase can only take place without decreasing the total flux through the pathway if sufficient activity exists in the acyl-ACP hydrolyzing enzymes capable of hydrolyzing the saturated fatty acids to compensate for the loss of flow through to the mono-unsaturated fatty acid. Alternatively, the capacity to hydrolyze saturated fatty acids could be increased independently of the capacity to hydrolyze 18:1-ACP.

In assessing the likelihood of success in modifying flux through these branch points it is useful to note the fatty acid phenotypes that have been achieved through mutagenesis and the likely flux controlling steps in the pathway.

The first enzyme in the fatty acid biosynthetic pathway, acetyl-CoA carboxylase (ACCase) has often been speculated to be a major control point in the fatty acid biosynthesis pathway. This is due in part to its location at the beginning of a very energy intensive biosynthetic process and to experiments in

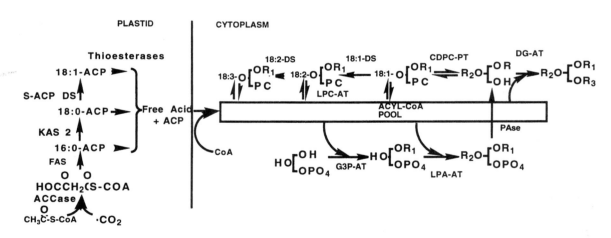

Fig. 37.2 The flux and acyl chain length controlling steps in fatty acid biosynthesis and the steps in triacylglyceride assembly. The diagram is based on a review of glycerolipid synthesis (Browse and Somerville, 1991) but is simplified to show only the very dominant eukaryotic assembly and desaturation pathway which accounts for most of the triacylglyceride synthesis in developing oilseeds. The multiple steps of the chain elongation cycle are shown only as fatty acid synthesis (FAS). The abbreviations for enzymes are as follows beginning with the start of the pathway: ACCase; acetyl-Coenzyme A carboxylase, KAS 2 keto-acyl-ACP synthase 2 (16:0 elongating), 18:0-ACP DS; stearoyl-ACP desaturase, thioesterases; the set of enzymes which hydrolyze acyl-ACPs, G3P-AT; glycerol-3-phosphate:acyl-CoA acyltransferase, LPA-AT; lysospatidic acid acyltransferase, LPAse; lysophosphatic acid phosphatase, 18:1-DS and 18:2-DS; oleate and linoleate desaturases, LPC-AT; lyosphosphatidyl choline:acyl-CoA acyltransferase, CDPC-PT; CDP-choline phosphotransferase, DG-AT; diacylglycerol:acyl-CoA acyltransferase.

chloroplasts which indicate that the flux through the enzyme is controlled by light (Post-Beittenmiller, 1991). Recently the flux control coefficient of ACCase in chloroplasts of barley and corn leaves has been estimated (Page *et al.*, 1994). Of the more than twenty steps in progressing from acetyl-CoA to the simplest glycerolipid, acetyl-CoA carboxylation to malonyl-CoA accounted for 45 to 61% of the control over flux in chloroplasts of these two species. This estimate leaves room for other major controlling steps, but it does make it less likely that other steps that are highly limiting are present in the normally functioning pathway. This assessment bodes well for the possibility of reducing flux through one branch of the pathway while maintaining flux through other branches. It also suggests, however, that if many of the activities are present in large excess to the normal flux, large changes in expressed enzyme activity must be made in order to substantially change flux at that step. For example, doubling the activity of an enzyme which is capable of functioning at a rate ten times greater than the usual flux will provide only a very small increase in flux through that step. Similarly, decreasing activities which are present in large excess will have little effect on flux, and therefore on end metabolite phenotype until the activity has been sufficiently reduced as to become a dominant part of the overall flux control.

This point is illustrated by experiments aimed at increasing the relative 18:0 content in rapeseed oil using a transgenic antisense approach (Knutzon *et al.*, 1992). Firstly, in both of the plants for which detailed analysis was given by these authors, a very large reduction in stearoyl-ACP desaturase activity (to levels which could not be detected with the available *in vitro* assay) was required to decrease flux through the first desaturation from about 90% in non-transformed controls to 55% in maximally altered plant lines. This suggests that in the wild-type, developing *Brassica* seeds, the enzymatic capacity for desaturating 18:0-ACP to 18:1-ACP greatly exceeds flux and that the converse experiment, increasing the 18:0-ACP desaturase capacity to decrease the 18:0 content of the triacylglyceride end product, will produce only small phenotypic changes.

The second point illustrated by this example is that unless the total flux through a pathway is maintained when one branch is blocked, secondary alterations are likely. In the *Brassica rapa* line described by Knutzon and coworkers (Knutzon *et al.*, 1992) which was most reduced in its 18:0-ACP desaturase activity, the total amount of triglyceride produced during seed filling was greatly decreased. In other, less inhibited lines and in the *Brassica napus* line described, there was no obvious decrease in total lipid synthesis, however, the levels of 18:2 and 18:3 fatty acids were increased relative to non-transformed control lines in addition to the increase in 18:0. Similar increases in unsaturated fatty acids which occur at low temperatures have been attributed in part to the flux through the rate controlling steps decreasing more than the flux through the desaturating enzymes in response to temperature (Browse and Slack 1983). Knutzon and colleagues suggested a similar explanation for their result. If the capacity to release 18:0 from ACP by hydrolysis were insufficient to maintain normal flux of fatty acid out of the plastid, the overall rate of fatty acid biosynthesis may be slowed and as a result the normally limited capacity for 18:3 desaturation may be sufficient to produce a higher level of 18:3 in the end product triacylglyceride. In our metabolism schemes shown in Fig. 37.1 this result is the one expected in proceeding from the condition in panel 4 to that in panel 5.

A similar increase in the 18:0 content of canola seeds was accomplished using the strategy depicted in moving from panel 4 to panel 6 of Fig. 37.1 (Hitz and Rieter, unpublished data). The 18:1 preferring thioesterases are capable of hydrolyzing 18:0 and 16:0-ACPs at about 10% of the rate at which they hydrolyze 18:1-ACP (Loader *et al.*, 1993; Yadav *et al.*, 1993). Transgenic over-expression of this enzyme therefore alters the ratio of activities acting on the 16:0-ACP and 18:0-ACP pools to produce higher levels of 18:0 and 16:0 in the seed oil. Using this approach, a line producing 18:0 levels about 5 fold higher than wild-type canola was obtained (Yadav *et al.*, 1993). The activity of the 18:0-ACP desaturase was decreased in a separate canola line by cosuppression to give a line that contained 11 to 13% 18:0 in its seed lipid. Genetic crossing of these two lines and isolation of the double homozygote produced a line with 38 to 41% 18:0 at normal levels of 18:3.

In several oilseed crops mutation breeding followed by screening seeds for lower levels of 18:2 and 18:3

has been successful in producing lines with decreased polyene fatty acids in their seed triacylglyceride. A nearly complete block in the synthesis of 18:2 that is specific to the cotyledonary tissues of the seed was obtained in sunflower and has been used to produce a commercial variety which produces oil with very high 18:1 content. Thus the pathway in at least some oilseeds species, does have the capacity to use less unsaturated fatty acids for the acylations required to produce triacylglyceride and still retain normal overall flux.

Similar strategies have produced soybean lines with elevated levels of 18:1, however, the extent of the blockage of 18:2 synthesis is not as complete as in the case of sunflower seeds. Further, the high 18:1 mutant soybean lines have oil phenotypes that are variable with environment. Temperature during the seed filling period influences the relative proportion of 18:1 to (18:2 + 18:3) fatty acids with high temperatures favoring 18:1 synthesis. The mutant lines are much more sensitive to temperature during seed fill than wild-type lines (Martin *et al.*, 1986). Both the inability of mutation breeding to obtain complete blockage of the pathway between 18:1 and 18:2 and the temperature sensitive nature of the wild-type and mutant pathways suggest that multiple genes encoding the same activity are present in developing soybean seeds.

The observation that antisense expression or cosuppression due to over expression of one member of a gene family can provide inhibition of expression in highly related members may provide a workable solution to the problem of controlling multiple isozymes. The cloning of the cytosolic and plastid forms of the fatty acid desaturases has made tests of these transgenic modifications of lipid biosynthesis possible. While the typical 18:1, 18:2 and 18:3 contents of soybean seed triacylglyceride are about 23%, 54% and 7% of the total fatty acids respectively, soybeans transformed with chimeric genes designed to give seed specific antisense expression of the 18:1 desaturase and of the 18:2 desaturase produced seeds with 80% 18:1 and 1.4% 18:3 respectively (Hitz *et al.*, 1994). Both of the introduced transgenes show simple, single gene inheritance and initial indications are that the fatty acid contents of the lines are less sensitive to alteration with temperature.

Since the change in the ratio of 18:1 to (18:2 + 18:3) is the only observed change in these seeds, this is an example of the simple manipulation shown in going from panel 1 to panel 3 of Fig. 37.1. Blockage of the 18:1 desaturating capacity is all that is required to shift from the production of 18:2 + 18:3 as the dominant product to 18:1. This implies that the increased flux of 18:1 into triacylglyceride can be accommodated by the catalytic capacity in the existing enzymes in that portion of the pathway.

This same strategy, using transgenes to make alterations in a metabolic pathway may be used to provide tissue specificity in cases where the genes whose expression needs to be modified are widely expressed. This has proven advantageous in fatty acid modifications in developing rapeseeds. While there is good evidence that the fatty acid desaturases in developing seeds of some crops such as soybean and sunflower are relatively seed specific isoforms, this does not appear to be the case in oilseed rape. Mutant lines have been reported in which the seed 18:1 content is greatly elevated. Development of agonomically viable lines from these mutants has proven difficult however. Analysis of three mutant lines with high 18:1 content in seeds (and therefore lower relative levels of 18:2 and 18:3) has shown that the relative 18:1 to (18:2 + 18:3) level in cytoplasmicly synthesized lipids in all parts of the plant are modified. While it is difficult to prove that these whole plant lipid modifications are the cause of the observed decrease in plant vigor associated with the high 18:1 phenotype, the two phenomena always co-segregate.

Cosuppression of the endogenous 18:1 desaturases due to seed-specific over-expression of one of the two seed expressed copies of the rapeseed 18:1 desaturase elevated the seed 18:1 content nearly as high as the 18:1 desaturase mutant lines, but total root lipid and leaf phosphatidyl choline fatty acids were unchanged (Hitz, unpublished).

The experiences in modifying even an essential pathway such as lipid biosynthesis have taught that there are distinctly different patterns of gene expression between species. This holds true even though the overall pathway is biochemically identical between species and even though the coding regions of homologous genes in the pathway are very similar in sequence.

Alteration of existing products: free amino acids

It is well known that the ratio of amino acids provided by the storage proteins of most seeds is less than optimal in providing the essential amino acids required in animal diets. Mutations which alter the ratio of proteins normally produced by seeds have been exploited to improve nutritional quality. In these mutants classes of proteins that have higher content of one or more of the essential amino acids dominate. In commerce, free amino acids, produced mainly by microbial fermentation, are used to supplement animal feed rations. Production of high levels of free amino acids in seed storage tissue might serve as a third alternative strategy.

An increase in the existing, small pool of a given free amino acid might be achieved by either decreasing its catabolism or increasing its rate of synthesis to exceed its possible metabolic fates. Lysine is one of the essential amino acids that is not produced in sufficient quantity by most plant seeds for animal feed. A schematic diagram of its synthesis and breakdown pathways is shown in Fig. 37.3. Two of the enzymes in the pathway are allosterically inhibited by lysine itself (Bryan, 1980). One is aspartyl kinase (AK), the enzyme which initiates conversion of aspartate to other amino acids. The other enzyme is dihydrodipicolinate synthase (DHPS) that initiates branching to

produce lysine. Bacteria which produce mutant, non-regulated forms of either of these enzymes produce large quantities of free lysine. Transgenic tobacco plants expressing these non-regulated, bacterial forms of either AK alone or in combination with a non-regulated form of DHPS produced free lysine in their leaves (Shaul and Galili, 1993).

When this same manipulation was applied to tobacco using a seed specific promoter, however, little or no increase in free lysine was observed. Instead, in one study slight accumulation of the intermediates of lysine catabolism were observed (Falco et al., 1995) and in another elevated levels of the first enzyme in lysine catabolism, lysine-ketoglutarate reductase (LKR), were observed (Hagai et al., 1994). In contrast to the experience with tobacco, seed specific expression of a non-regulated DHPS in canola resulted in very large increases in free lysine (Falco et al., 1995). Seed specific expression of non-regulated forms of both AK and DHPS was required to achieve high levels of free lysine accumulation in soybean seed (Falco et al., 1995). In both species, sufficient free lysine was accumulated to more than double the total seed lysine content when compared to the non-transformed seed from the same species.

Despite these increases in free lysine, intermediates in the catabolic pathway leading from lysine back to a common intermediate in carbon metabolism accumulated in both canola and soybean (Falco et

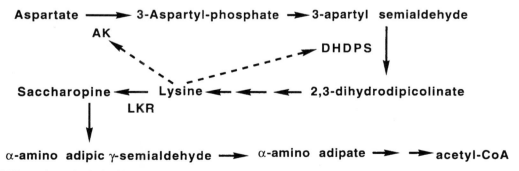

Fig. 37.3 The pathway for lysine biosynthesis and catabolism in plants with the steps that are strongly rate controlling in synthesis due to product inhibition of enzymes shown. Both aspartate kinase (AK) and dihydrodipicolinate synthase (DHDPS) are inhibited by the end product of the pathway, lysine. The first enzyme in the catabolic pathway, lysine-ketoglutarate reductase (LKR) appears to control flux at the low free lysine concentrations found in wild-type plants, but not in plants in which high levels of free lysine accumulate.

al., 1995). In canola it is the third intermediate, α-amino adipate, that shows the most significant increase, while in soybean it is the first intermediate saccharophine, that accumulates to levels exceeding that of most free amino acids.

These intermediates are known to be on the normal catabolic pathway, but are usually present in only very minor amounts. Their accumulation indicates either that only a portion of the whole catabolic path exists in these tissues during the period of lysine synthesis or that the normal path is operating at a flux which causes the build up of intermediates at strongly limiting points. This pattern of intermediate build up also suggests that either flux down the catabolic path from lysine is controlled by the capacity of LKR when it functions at low lysine concentration or that LKR activity increases when the concentration of free lysine increases. In Fig. 37.1 this situation would be depicted by uncovering a minor branch off the intermediate pool as that pool size is increased.

Alteration of existing products: starch

Starch is used as the insoluble, transient, carbon and energy store in many plant tissues and in most plant species. In addition to its role as a major storage product and metabolite, starch is one of the primary economic outputs of many crop plants such as corn and potato. In that regard, alteration of the amount of starch produced, either increased or decreased, as a per cent of the existing dry matter in storage tissues might produce a crop with added economic value.

The positioning of starch biosynthesis in the overall scheme of large scale carbon metabolism is shown diagrammatically in Fig. 37.4. While the committed steps to starch are only those following the production of ADP-glucose, the flux of carbon through to starch from all the preceding steps can be measured and control coefficients determined if starch synthesis is measured in an intact system. When such experiments were done on leaf tissue using *Arabidopsis thaliana* mutants with reduced activities

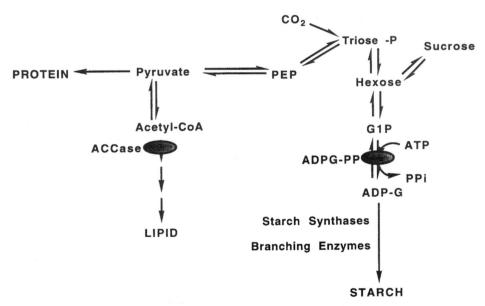

Fig. 37.4 The metabolic relationship between three carbon storage products and the apparent control points in storage lipid and starch synthesis. Abbreviations: triose-P; dihydroxy acetone phosphate and glycerol-3-phosphate, PEP; phosphoenol pyruvate; G1P, glucose-1-phosphate, ADP-G; ADP-glucose, ADPG-PP; ADP-glucose pyrophosphate, ACCase; acetyl-CoA carboxylase, ACCase and ADPG-PP. While the data is most complete for non-storage tissues, both ACCase and ADPG-PP appear to be strong control points for the flux of carbon into lipids and starch respectively in many plant tissues. In both cases it appears to be allosteric control of the enzyme that is responsible for flux limitation at that point in the pathway.

in several enzymes of carbohydrate metabolism including ADP-glucose pyrophosphorylase (ADPG-PP), to generate variation in activity at individual steps, the control coefficient for ADPG-PP was estimated at 0.28; over ten fold greater than any of the other measured control coefficients (Neuhaus and Stitt, 1990).

Since these measurements are of net starch synthesis from CO_2, the number of steps in the pathway is very large and yet the total control remaining among the steps other than ADP-glucose formation must only sum only to about 0.7. The analysis leaves the possibility of another major controlling step prior to ADPG-PP, but at the same time suggests that of the reactions that are exclusively involved in starch biosynthesis in leaves, ADPG-PP exerts greatest control.

ADPG-PP from bacteria and all plant sources where it has been studied is strongly regulated by allosteric effectors which are themselves metabolites in carbohydrate metabolism. The plant enzyme is inhibited by inorganic phosphate and activated by several sugar phosphates, primarily 3-phophosphoglycerate (Preiss, 1988). Flux analysis modeling and measurements of photosynthetic carbon metabolism pools have shown that it is this allosteric regulation of ADPG-PP that confers its dominant regulatory role in chloroplast starch synthesis. In some tissues where the *in vitro* activities of all three of the enzymes on the starch biosynthesis branch have been measured under optimal conditions of activity, starch synthase activity is some twenty fold less than ADPG-PP activity. If ADPG-PP activity strongly controls the rate of starch biosynthesis in these tissues, it must be due to its allosteric down regulation.

Potato tubers have been genetically modified to produce slightly higher amounts of starch (Stark *et al.*, 1992). In assessing the metabolic changes required to effect this increase, Stark and coworkers were aware that ADPG-PP might be strongly limiting although this was unknown for storage tissues. They were also aware that allosteric regulation might be important but again no data was available for storage tissues. The location of the ADPG-PP that is important for starch biosynthesis in amyloplasts was also uncertain. In devising a test to deal with each of these unknowns, Stark *et al.*, used a mutant form of

the *E. coli* ADPG-PP which is not affected by allosteric regulators. By targeting the bacterial ADPG-PP either to the cytoplasm or to the amyloplast in transgenic plants, they produced sets of potato plants with tuber specific expression of three different forms of the *E. coli* ADPG-PP: (1) the mutant enzyme localized to the plastid; (2) the wild-type enzyme localized to the plastid and (3) the mutant enzyme in the cytoplasm. Tubers from plants transformed with construct 1, the non-allosterically controlled enzyme targeted to the plastid produced about 35% more starch as a fraction of fresh weight than controls or either of the other two constructions.

These results show that in potato tubers ADPG-PP has a strong influence over the long-term rate of starch synthesis and that it is the allosteric regulation of enzymatic activity rather than the absolute amount of enzyme which exerts the control.

The starch content of canola seeds was increased using the same strategy and promoters appropriate for canola seed expression. Starch synthesis in this seed, which primarily stores triacylglyceride, was sufficiently increased to divert carbon from lipid biosynthesis (Barry and Kishore, 1995).

The result in canola seeds is an interesting one in that it demonstrates that *in vitro* determined relationships between the common intermediates in carbon metabolism function *in vivo*. The equivalent change in Fig. 37.1 is moving from panel 3 to panel 1. In panel 3, product A is produced at a low level from intermediates common to the dominant product B due to a single, highly restrictive step in the pathway to A. Relief of that restriction both increases the production of A and decreases the production of B without additional alterations.

Summary

It is evident that the tools of molecular biology will find increasing use in practical agriculture through the development of transgenic plants with a wide variety of traits. Our present state of knowledge of the underlying biochemistry, metabolism and gene expression in plants is such that the use of the term 'genetic engineering' to describe our attempts to create transgenic crops implies a precision that has not yet been reached. For example, even when the intention of the researcher is clearly a practical one, more often than not, an experiment is performed that produces, not the desired product, but rather a very precise perturbation of plant biochemistry and metabolism which, when analyzed, forms the basis for proceeding in an iterative fashion towards the desired goal.

Even with these limitations, the progress in the field in the past ten years has been substantial, and numerous examples of transgenic crops have been produced and are being commercialized. The methodologies at hand are becoming more powerful and precise, and the use of these tools is providing a much more thorough, quantitative understanding of plant biochemistry. Taken together, the molecular methodologies available and our increasing understanding of the biochemical basis for crop improvement should allow even more rapid progress in the practical improvement of crop plants during the next ten years.

References

Barry, G. F. and Kishore, G. M. (1995). A method of decreasing the oil content of seeds by expression of ADP glucose pyrophosphorylase. *PCT Application 94870118*.

Bennett, J. (1993). Genes for crop improvement. In *Genetic Engineering*, Vol. 15, ed. J. K. Setlow, Plenum Press, New York, pp. 165–89.

Browse, J. A., Slack, C. R. (1983). The effect of temperature and oxygen on the rates of fatty acid synthesis and oleate desaturation in safflower (*Carthamus tinctorius*) seed. *Biochem Biophys. Acta* **753**, 145–52.

Browse, J. and Somerville, C. (1991). Glycerolipid synthesis: biochemistry and regulation. *Annu. Rev. Plant Physiol. Plant Mol. Bio.* **42**, 467–506.

Bryan, J. K. (1980). Synthesis of the aspartate family and branched chain amino acids. In *The Biochemistry of Plants*, Vol. 5, ed. B. J. Miflin, Academic Press, New York, pp. 403–52.

Burrell, M. M., Coates, S. A., Mooney, P. J. and ap Rees, T. (1993). Prospects for the modification of metabolism. In *Opportunities for Molecular Biology in Crop Production*, eds D. J. Beadle, D. H. L. Bishop, L. G. Copping, G. K. Dixon and D. W. Hollomom, British Crop Protection Council Monograph Series, No 55. Major Print Ltd, Nottingham. pp. 197–200.

Chaleff, R. S. and Ray, T. B. (1984). Herbicide-resistant mutants from tobacco cell cultures. *Science* **223**, 1148–51.

Falco, S. C., Guida, T., Locke, M., Mauvis, J., Sanders, C.,

Ward, R. T. and Webber, P. (1995). Transgenic canola and soybean seeds with increased lysine. *Biotechnology*, **13**, 577–82.

Hagai, K, Shaul, O. and Galili, G. (1994). Lysine synthesis and catabolism are coordinately regulated during tobacco seed development. *Proc. Natl. Acad. Sci. USA* **91**, 2577–81.

Haughn, G. W., Smith, J., Mazur, B. and Somerville, C. (1988). Transformation with a mutant Arabidopsis acetolactate synthase gene renders tobacco resistant to sulfonylurea herbicides. *Mol. Gen. Genet.* **211**, 266–71.

Hitz, W. D., Yadav, N., Reiter, N. S., Mauvis, C. J. and Kinney, A. J. (1994). *Reducing Polyunsaturation in Oils of Transgenic Canola and Soybean*, 11th International Plant Lipid Meeting, Paris.

Kacser, K. and Porteous, J. W. (1987). Control of metabolism: what do we have to measure? *Trends Biochem. Sci.* **12**, 7–14.

Kay, R., Chan, A., Daly, M. and McPherson, J. (1987). Duplication of CaMV 35S promoter sequences creates a strong enhancer for plant genes. *Science* **236**, 1299–303.

Kishore, G. M., Padgette, S. R. and Fraley, R. T. (1992). History of herbicide-tolerant crops, methods of development and current state of the art–Emphasis on glyphosate tolerance. *Weed Technology* **6**, 626–34.

Knutzon, D. S., Thompson, G. A., Radke, S. E., Johnson, W. B., Knauf, V. C. and Kridl, J. C. (1992). Modification of Brassica seed oil by antisense expression of a stearoyl-ACP desaturase gene. *Proc. Natl. Acad. Sci. USA* **89**, 2624–28.

Loader N. M., Woolner, E. M., Hellyer, A., Slabas, A. R. and Safford, R. (1993). Isolation and characterization of two Brassica napus embryo acyl-ACP thioesterase cDNA clones. *Plant Mol. Biol.* **23**, 769–88.

Martin, B. A., Wilson, R. F. and Rinne, R. W. (1986). Temperature effects upon the expression of a high oleic trait in soybean, *J. Am. Soc. Oil Chem.* **63**, 346–52.

Nawarth, C., Poirier, Y. and Somerville, C. (1994). Targeting of the polyhydroxybutyrate biosynthetic pathway to the plastids of *Arabidopsis thaliana* results in high levels of polymer accumulation. *Proc. Natl. Acad. Sci. USA* **91**, 12760–64.

Neuhaus H. E. and Stitt, M. (1990). Control analysis of photosynthetic partitioning: impact of reduced activity of ADPglucose pyrophosphorylase or plastid phophoglucomutase on the fluxes to starch and sucrose in *Arabidopsis*. *Planta* **182**, 445–54.

Page, R. A., Okada, S. and Harwood, J. L. (1994). Acetyl-CoA carboxylase exerts strong flux control over lipid synthesis in plants. *Biochim. Biophys. Acta* **1210**, 369–72.

Perlak, F. J., Fuchs, R. L., Dean, D. A., McPherson, S. L. and Fischhoff, D. A. (1991). Modification of the coding sequence enhances plant expression of insect control protein genes. *Proc. Natl. Acad. Sci.* USA **88**, 3324–28.

Perlak, J. F., Stone, T. B., Muskopf, Y. M., Peterson, L. J.,

Parker, G. B., McPherson, S. A., Wyman, J., Love, S., Reed, G., Biever, D. and Fischhoff, D. A. (1993). Genetically improved potatoes: protection from damage by Colorado potato beetles. *Plant Mol. Biol.* **22**, 313–21.

Poirier, Y., Dennis, D. E., Klomparens, K. E. and Somerville, C. (1992). Polyhydroxybutyrate, a biodegradable thermoplastic, produced in transgenic plants. *Science* **256**, 520–23.

Post-Biettenmiller, D., Jaworski, J. G. and Ohlrogge, J. B. (1991). *In vivo* pools of free and acylated acyl carrier proteins in spinach: evidence for sites of regulation of fatty acid biosynthesis. *J. Biol. Chem.* **266**, 1858–65.

Preiss, J. (1988). Biosynthesis of starch and its regulation In *The Biochemistry of Plants*, Vol. 14, ed. J. Preiss, Academic Press, New York, pp. 181–254.

Riazuddin, S. (1994). Plant genetic engeering and future agriculture. In *Genetic Engineering,* Vol. 16, ed. J. K. Setlow, Plenum Press, New York, pp. 93–113.

Saari, L. L. and Mauvais, C. J. (1995). Sulfonylurea resistant crops. In *Herbicide-Resistant Crops: Agricultural, Environmental, Economic, Regulatory and Technical Aspects*, ed. S. O. Duke, Lewis Publishers, Chelsea, MI.

Shaul, O. and Galili, G. (1993). Concerted regulation of lysine and threonine synthesis in tobacco plants expressing bacterial feedback-insensitive aspartate kinase and dihydrodipicolinate synthase. *Plant Mol. Biol.* **23**, 759–68.

Stalker, D. M., McBride, K. E. and Malyj, L. D. (1988). Herbicide resistance in plants expressing a bacterial detoxification gene. *Science* **243**, 419–22.

Stark, D. M., Timmerman, K. P., Barry, G. F., Preiss, J. and Kishore, G. M. (1992). Regulation of the amount of starch in plant tissues by ADP glucose pyrophosphorylase. *Science* **258**, 287–92.

Stitt, M. and Schulze, D. (1994). Does Rubisco control the rate of photosynthesis and plant growth? An exercise in molecular ecophysiology. *Plant Cell Environ.* **17**, 465–87.

Yadav, N., Wierzbicki, A., Knowlton, S., Pierce, J., Ripp, K., Hitz, W., Aegerter, M. and Browse, J. (1993). Genetic manipulation to alter fatty acid profiles of oilseed crops. In *Biochemistry and Molecular Biology of Membrane and Storage Lipids of Plants,* eds N. Murata and C. R. Somerville, The American Society of Plant Physiologists, Rockville, MD, pp. 60–66.

Index